GEO-DATA

The World Geographical Encyclopedia

Edited by
George Thomas Kurian

Gale Research Company
Book Tower
Detroit, Michigan 48226

Published in Cooperation with Geo-Data Publications
Box 519, Baldwin Place, NY 10505–0519

Geo-Data: World Geographical Encyclopedia

ISBN 0–914746–31–6

Distributed worldwide by Gale Research Company,
Book Tower, Detroit, Michigan 48226.

Library of Congress Cataloging-in-Publication Data

Kurian, George Thomas.
Geo-data : the world geographical encyclopedia / edited by George Thomas Kurian.
p. cm.
"Published in cooperation with Geo-Data Publications."
Includes bibliographical references (p.)
ISBN 0–914746–31–6 : $120.00
1. Gazetteers. I. Title.
G103.5.K87 1989
910'.3—dc20 89–23799
CIP

Printed in the United States of America.

PREFACE

Geo-Data: World Geographical Encyclopedia is designed as a descriptive encyclopedia of the countries, continents and oceans of the world. Geography has been described as the gateway to the study of nations. *Geo-Data* provides profiles of the physical environment of 204 nations and territories of the world in its first part. Each country chapter describes the terrain, landforms, topographical regions, mountains and rivers. The second part presents information on continents, climate and maritime features such as oceans, seas and gulfs. Also included are rankings of major geographical features, sources, and 204 country maps. Measurements are given in both metric and customary systems.

The editor wishes to acknowledge the help and encouragement received from Ms. Elizabeth Geiser of Gale.

TABLE OF CONTENTS

Part II

PART I

AFGHANISTAN

BASIC FACTS

Official Name: Democratic Republic of Afghanistan

Abbreviation: AFN

Area: 652,225 sq km (251,825 sq mi)

Area—World Rank: 40

Population: 14,480,863 (1988) 21,664,000 (2,000)

Population—World Rank: 53

Capital: Kabul

Boundaries: 5,770 km (3,585 mi); USSR 2,383 km (1,481 mi); China 71 km (44 mi); Pakistan 2,466 km (1,532 mi); Iran 850 km (528 mi)

Coastline: Nil

Longest Distances: 1,239 km (770 mi) NE–SW; 563 km (350 mi) SE–NW

Highest Point: Nowshak 7,485 m (24,557 ft)

Lowest Point: Amu Darya River Valley 255 m (837 ft)

Land Use: 12% arable land; 46% meadows and pastures; 39% other; Negligible % permanent crops

Afghanistan is a completely landlocked country in Central Asia surrounded by the Soviet Union to the north, Iran to the west and by Pakistan to the south and southeast. The nearest seaport is Karachi, Pakistan. Its extreme length from west to east includes a narrow corridor-shaped extension (Wakhan) in the east.

The main pitch of the land is from the northeast to the southwest, following the general trend of the Hindu Kush massif from the Pamir Knot on the border with Communist China to Iran. To the north, west and southwest there are no mountain barriers toward neighboring countries. The northern plains pass almost imperceptibly into the plains of Soviet Turkistan. In the west and southwest the plateaus and deserts merge into those of Iran.

Since much of the country is covered by deserts and receives little precipitation, water has been a dominant factor in determining the location and distribution of human settlement. Villages, oases and many of the historic towns are located near rivers and streams. Kabul, the country's capital and a meeting place of trade routes from east, west, north, and south lies on the well-watered plains of the river by the same name.

The boundaries that exist in 1989 were established in the late nineteenth century as the British and Russian empires approached each other in their expansionist periods. The greater part of the border with the Soviet Union and a small section of the border with Pakistan are marked by rivers; the remaining boundary lines are political rather than natural.

The frontier with the Soviet Union extends approximately 1,689 km (1,050 mi) southwestward from the point where the Pamir Knot touches Communist China to the desert and hilly region at the eastern border of Iran. Along the Amu Darya (classically, the Oxus) River, the boundary line has been established at the deepest continuous channel in the navigable portion and at the median line farther upstream.

The border with Iran runs generally southward almost 820 km (510 mi) from the Hari Rud River, across the swamp and desert regions bordering Iranian Khorasan and Sistan before reaching the northwestern tip of Pakistan. Its southern section crosses the Helmand River which occasionally changes its course in this area, causing boundary disputes between Iran and Afghanistan.

The border with Pakistan, totaling 1,810 km, (1,125 mi) runs eastward from Iran through the Chagai Hills and the southern end of the Registan Desert, then northward through mountainous country to a point near Chaman, the terminus of the railroad from Quetta in Pakistan. The boundary then follows an irregular course in a northeasterly direction some 281 km (175 mi) before reaching the Durand Line established in 1893 by agreement with British authorities. This line, which defines the border, continues on through mountainous regions to the Khyber Pass area. Beyond this point it rises to the crest of the Hindu Kush which it follows eastward to the Pamir Knot.

The last section of the border runs roughly parallel to the eastern end of the northern border and with it forms the Wakhan corridor, the narrowest portion of which is only 11 km (7 mi) wide. The extreme northeastern part of the Wakhan forms a 80 km (50-mi)-long common border with Sinkiang province of China.

The Durand Line divides Pashtun tribes between the two nations cutting off migration routes to important pasturelands. Its creation has caused much dissatisfaction among Afghans and has given rise to bitter border disputes between Afghanistan and Pakistan.

The country is made up of regions and provinces which have attained prominence in the historical past. The fertile northern stretch between the mountains and the Amu Darya, for example, was once part of Bactria, the seat of a rich and ancient civilization, and is still often referred to by that name. Other regions, notably Nuristan, eastward from Kabul, and the Hazarajat to the west were named after the people inhabiting the areas. In some instances, as in Badakhshan, in the northeast, the traditional names of natural or geographical regions, correspond with the names of current administrative units (provinces). Their geographical boundaries, however, do not necessarily correspond.

The dominant physical feature is the mountainous Central Highlands formed by the Hindu Kush and its subsidiary ranges. Traversing the country for 965 km (600 mi) from east to west, the towering peaks alternate with precipitous gorges and barren slopes. In the eastern portion some of the ridges have spectacular forest cover of cedar, pine, and larch. Farther west the forest cover yields sparse grass covering sought by nomadic shepherds when they ascend from the parched plains of the south.

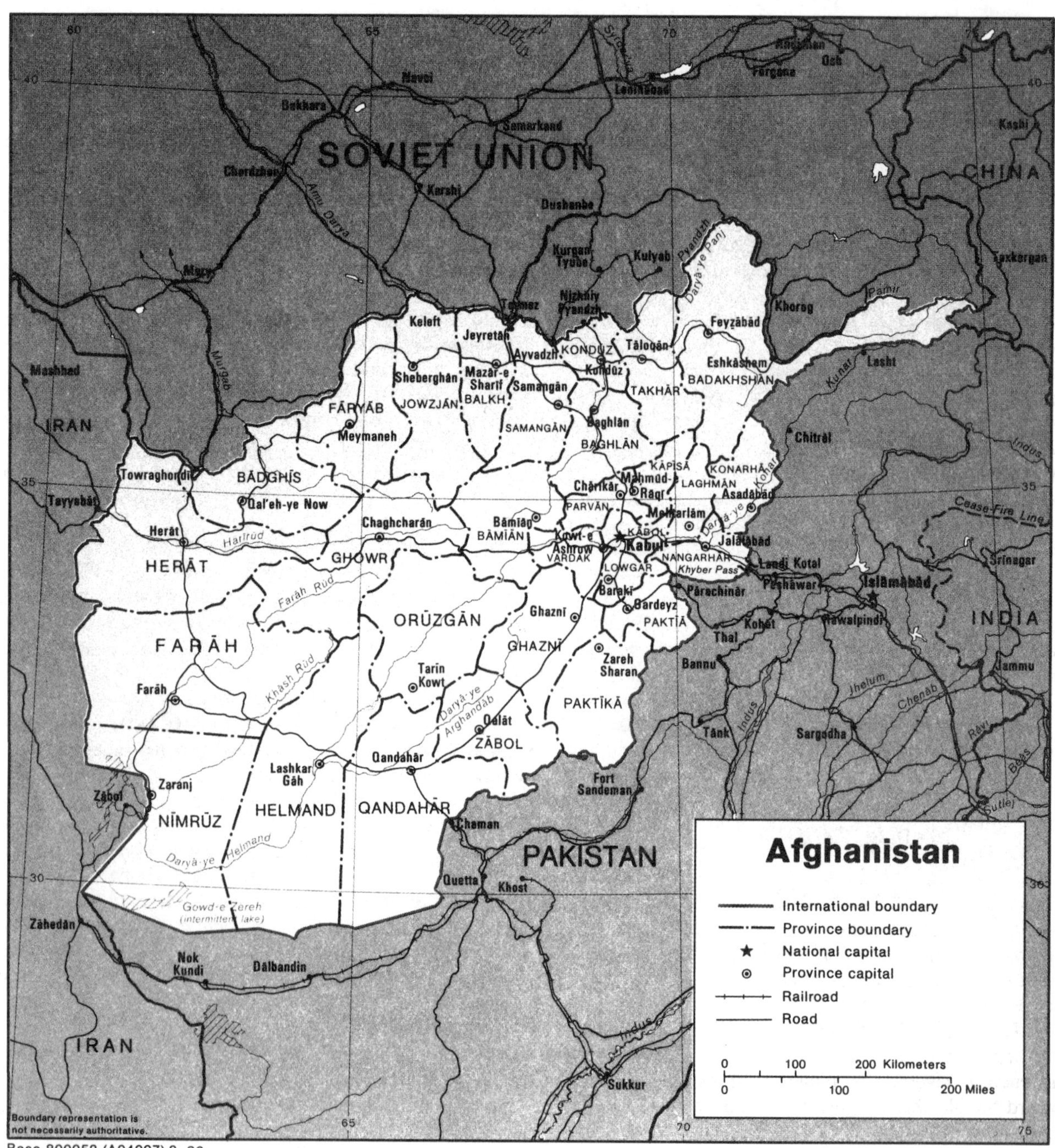

Base 800052 (A04007) 8-86

Small crops are cultivated in some of the narrow valleys, watered by swift-running snow-fed streams. Stretches of grassland affording good pasture are found in some of the higher valleys. The mountains are threaded by numerous passes. Some of these, including the Anjuman, Salang, and the Hajigak, have been used by conquering armies and by caravans in historical times.

North of the mountainous divide the land descends gradually into the plains of the Amu Darya. The area was once famous for the ancient city of Balkh a trading center and meeting ground of caravans on their way to China along the historic silk route. The land is fertile but requires water to make it cultivable and productive. The easternmost portion is mountainous, settled only in the valleys which are isolated among the tangled ranges. The Kunduz Valley, north of Salang Pass, however, has become an important center for growing sugar beets and cotton. To the west are rich pasturelands where large flocks of sheep are raised. Still farther west along the foothills of the Hindu Kush is the fertile Hari Rud Valley.

The southwestern plateaus comprise desolate, arid deserts and steppes, covered with sand dunes, dry streambeds and salt lakebeds. Around the northeastern rim of the Dasht-i-Margo and Registan Deserts in this region, high, serrated ridges rise precipitously from the desert floor. Northeast of the Registan, the area becomes more fertile with dense orchards and vineyards. Cultivable oases are found mainly around the city of Kandahar, from where agricultural commodities move to Pakistan and India.

Earthquakes are frequent, occurring as often as 50 times a year in some areas. They are most violent and most frequent in the northeast, where in 1955 the town of Faizabad, capital of Badakhshan province, was almost totally demolished. In 1956 violent tremors shook the area of the Central Hindu Kush, including the provinces of Bamian and Samangan.

MAJOR REGIONS

Central Highlands

The country's mountainous core, the Central Highlands, is part of the great Alpine-Himalayan mountain chain. An intricately interwoven pattern of ridges and valleys, the Central Highlands extend over an area of approximately 414,400 sq km (160,000 sq mi). The mountains fan out from the Pamir Knot in the east toward Iran in the west and enclose several arid plateaus. Transverse ridges point toward the Northern Plains; the southern ridge complex crosses over to Pakistan.

The east-west trending mountain axis is composed of three high ridges with the altitudes descending toward Iran. The main ridge begins in China and runs southwestward some 482 km (300 mi) as the Eastern Hindu Kush with peaks over 6,400 m (21,000 ft) high, and mountain passes at altitudes vary between 3,657 m and 4,572 m (12,000 and 15,000 ft). Most important of these is the Weran Pass (4,572 m, 15,000 ft) which connects the valley of the Kunar River on the south with the Kokcha River on the north. At the 3,962 m (13,000-ft) Anjuman Pass, crossed by a trail leading from Kabul northeastward to Faizabad, the Eastern Hindu Kush becomes the Central Hindu Kush. The former is a cold desert highland with snow-covered peaks and practically no vegetation. The climate of the Central Hindu Kush is less severe, and part of the mountains are forested. Flanked by huge, longitudinal valleys the highest crest (5,075 m, 16,650 ft) of the Central Hindu Kush is just west of the important Salang Pass (approximately 3,681 m, 12,000 ft) on the main road from Kabul to the Northern Plains.

The Koh-i-Baba Range with peaks at 4,572 m (15,000 ft) runs parallel to and south of the western end of the Central Hindu Kush, to which it is connected by two transverse ridges. Together with the Hindu Kush, the Koh-i-Baba extended westward by the Band-i-Bayan and the Kasa Murgh ranges south of the Hari Rud River constitute the country's main watershed. Other important ranges include the Koh-i-Hisar (extending northward from the upper reaches of the Murghab River), the Firoz Koh and the Paropamisus north of the broad Hari Rud Valley. The city of Herat is situated near the western end of this valley which runs into an intermontane plateau situated astride the boundary with Iran. The Zarmast Pass (elevation 2,499 m, 8,200 ft) on the road from Herat to Maimana is the major pass in the barren, deforested Paropamisus Range.

A similar series of ranges, at lower altitudes, runs parallel to the Paropamisus-Hindu Kush axis along the southern rim of the Northern Plains. They are broken by deep valleys of rivers which start in the Koh-i-Baba or in the western part of the Central Hindu Kush. Only the westernmost of these lower ranges, the Band-i-Turkistan, represents a real barrier to northward cross-country movement.

Several mountain chains fan out southwestward from the Koh-i-Baba, the Band-i-Bayan and the Kasa Murgh Ranges. These decrease in altitude as they approach the Southwestern Plateau Region where they yield to gently undulating deserts. Two of the ranges, the Koh-i-Mazar and the Koh-i-Khurd, are of special importance since they enclose the valley for the upper course of the Helmand River which collects the waters needed for irrigating the arid areas along its banks in the Southwestern Plateau Region. South of the Koh-i-Baba and the Band-i-Bayan lies a maze of ridges, plateaus and valleys, known as the Hazarajat.

Southeast of the Koh-i-Mazar Range, a series of lower ridges enclose long valleys which run parallel to the boundary with Pakistan. These mountains converge in the west-east trending Safed Koh along the Pakistan frontier southeast of Kabul and turn northeastward at the 1,036 m (3,400-ft) Khyber Pass. Bounded by this range system and the Hindu Kush is the intermontane valley region with Kabul, the capital city. Several tributary ranges of the Hindu Kush lend this area a rugged appearance in spite of the lush vegetation.

Northern Plains

North of the central mountain core are the Northern Plains, stretching from the Iranian border to the western foothills of the Pamir Knot. The area, a part of the Central Asia steppe, is demarcated on its eastern half from the Soviet Union by the Amu Darya River. Extending over an area of approximately 103,600 sq km, (40,000 sq mi) the Northern Plains are situated at an average elevation of 609 m (2,000 ft) except for the Amu Darya valley floor where it drops to as low as 183 m (600 ft). A considerable portion of the area consists of fertile, loess-covered plains. Intensely cultivated and densely settled, these plains are of outstanding agricultural importance and provide food for a considerable portion of the country's population. They also have rich natural gas resources. A flat strip of desert and steppe extends along the banks of the Amu Darya. Desert and desert-like steppe areas are likewise found west of Badakhshan along the foothills of the Central Hindu Kush and also west of Mazar-i-Sharif. The southernmost fringe of the area passes gradually into elevated plains which provide excellent pastureland.

The Southwestern Plateau

The Southwestern Plateau, situated southwest of the Central Highlands is a high, arid plateau country extending into Pakistan in the south and into Iran in the west. With an altitude of about 914 m (3,000 ft) it slopes gently to the southwest. Comprised of deserts and semideserts, it is crossed by a few large rivers among which the Helmand and its major tributary the Arghandab are the most important. The region covers approximately 129,500 sq km (50,000 sq mi) a fourth of which forms the Registan Desert. The latter is bounded by the Helmand and Arghandab Rivers, the Pakistan frontier and the easternmost portion of the Sistan depression. Sand ridges and dunes alternate with expansive desert plains, devoid of vegetation and covered with windblown sand changing here and there into barren gravel and clay areas. West of the Registan Desert lies the Dasht-i-Margo, a desolate steppe with salt flats. To the south the region is bounded by the slowly flowing Helmand River, its water depleted by evaporation and by the irrigation canals in the area of Girishk. The Khash Rud River forms the northern boundary.

On the northwestern rim of the Plateau, between the Khash Rud and Farah Rud Rivers, lies Khash Desert just south of the road from Kandahar to Farah. Both the Khash Desert and the Dasht-i-Margo steppe, slope southwestward almost imperceptibly into the Sistan depression the seat of an ancient civilization before Timur-i-Lang (commonly known as Tamerlane) destroyed the extensive irrigation system of the Helmand River. The lowest portion of this depression is covered by the salt flats of Gaud-i-Zirreh. When precipitation in the northern mountains is abundant, the Gaud-i-Zirreh becomes dotted with shallow ponds. In such years the overflow of Hamun-i-Saberi Lake, which straddles the Iranian border, runs into the Daryacheh-ye-Sistan Lake in Iran. The latter, in turn, overflows and discharges into the Gaud-i-Zirreh.

RIVERS

The drainage system is landlocked. Only a few rivers, all in the eastern part of the country and drained mainly by the Kabul, reach the ocean after emptying into Indus River in Pakistan. Most of the rivers and streams end in shallow desert lakes or oases inside or outside the country's boundaries.

About 10 percent of the surface area of eastern Afghanistan has no rivers. In the western part of the Northern Plains many rivers disappear in the soil before emptying into the Amu Darya. In the west the sandy deserts along the Iranian frontier have no watercourses.

Nearly half of the country's total area is drained by watercourses south of the Hindu Kush-Kasa Murgh ridge line, and half of this area is drained by the Helmand and its tributaries alone. The Amu Darya, the country's other major river, has the next largest drainage area.

In the south regions without drainage cover about 55,944 sq km (21,600 sq mi). Nearly half of this area is formed by the great sand desert of Registan, the rest by the Dasht-i-Margo and the Khash Deserts and by other smaller deserts in the Sistan depression.

Melting glaciers feed the rivers in the northeast. The water yield of rivers in the rest of the country depends on the seasonal pattern of rainfall or on sudden torrential downpours. Many of the smaller streams flow only for short periods following a rain. Because of their variable water levels the rain-fed rivers are unsuitable for waterborne communication. On the other hand, there is no need for bridges on the roads crossing them since the majority are fordable during most of the year. The exceptions are the roads crossing the Amu Darya and the lower course of the Kabul.

Rivers rise at the end of the winter. Their minimum yield is at the end of summer or the beginning of autumn. Some small rivers carry no water from April to May and November to December. The dry period is often prolonged by various forms of irrigation which causes many tributaries to disappear before reaching their principal river.

The flow of rivers which collect the runoff from the mountains along the northeastern border is affected by the summer rains sent by the monsoon of India. Their maximum yield, therefore, occurs twice a year, notably from July to September, and again from January to April.

Rivers which spring in the glaciers of the Pamir Knot or in Nuristan (the mountainous region along the Pakistan border northeast of Kabul) have a minimum yield in the winter during severe freezing. The maximum yield occurs in July and August, when the snows have melted.

Because of the great variation in the water yield during the different seasons and because of the absence of large lakes which may be used as reservoirs, the rivers are generally unsuitable for the generation of hydroelectric power. Only a few power stations have been built, mostly during the 1930's, including the plant on the Panjshir River, a small tributary of the Kabul, which provides electricity for the capital.

The 2,661 km (1,654-mi)-long Amu Darya originates in the glaciers of the Pamir Knot. Some 965 km (600 mi) of its upper course constitutes the Afghanistan-Soviet Union border. Nearly half of the surface area of area of its tributaries are in Afghanistan. Two of the larger tributaries, the Kokcha and the Kunduz, rise in the mountains of Badakhshan. Flowing in rapid torrents in its upper course, the Amu Darya becomes calmer below the mouth of the Kokcha, 96 km (60 mi) west of Faizabad. Along its banks there is often abundant vegetation of trees and bushes. During its flood period the upper course of the Amu Darya, swollen by snow and melting ice, carries along much gravel and large boulders. In the central and lower courses the flow diminishes because of evaporation during the summer months and because of the extensive use of the water for irrigation.

Two important rivers of the northern region are the Murghab and the Hari Rud. Their respective drainage areas are separated by the Paropamisus and Firoz Koh mountain chains. The Murghab flows about 321 km (200 mi) in a northwesterly direction from the eastern end of the Firoz Koh watershed before entering the Soviet Union. It drains the large, high valley between the Paropamisus chain and the Band-i-Turkistan, a parallel chain farther north. The Murghab has few fords and usually must be crossed by boat or ferry.

The Hari Rud rising at an altitude of 2,743 m (9,000 ft) on the western slopes of the Koh-i-Baba Range takes a steady westerly course, first in the alpine valley between the Paropamisus and the Firoz Koh chains on the north and then in the Kasa Murgh and the Band-i-Bayan chain in the south. The river continues across a broad plateau valley and flows south of Herat through a fertile oasis district which thrives under artificial irrigation from its waters. About 129 km (80 mi) west of Herat, the Hari Rud turns north and constitutes the border between Iran and Afghanistan for 104 km (65 mi) before it crosses the junction of the Afghan-Soviet boundaries.

In the southeastern region, the Kabul River, flowing eastward from the northern slopes of the Koh-i-Mazar chain, has the largest drainage area. Its major tributary on the south is the Logar. On the north are three additional important tributaries, the Panjshir, Alingar and the Kunar. They originate in the northeastern slopes of the Hindu Kush, in the region of Nuristan. The Kabul after traversing the capital city turns southeast, crosses the border into Pakistan, and becomes a major tributary of the Indus River east of Peshawar. The Kabul has a steady flow of water, and together with its north-bank tributaries, is intensively used for irrigation. The upper course of the river has many rapids, but downstream from the capital city it becomes navigable for rafts and flat-bottomed barges although water levels tend to be low during summer.

The Helmand is the principal river in the southwest; with its many tributaries the most important of which is the Arghandab, it drains more than 258,998 sq km (100,000 sq mi). Starting some 80 km (50 mi) west of Kabul in the Koh-i-Baba mountains, the Helmand is approximately 1,400 km (870 mi) long. In its upper and central course, the river cuts through deep alpine valleys. Before entering the Registan Desert about 80 km (50 mi) northwest of Kandahar, the waterflow is collected in a large natural valley reservoir. A similar reservoir exists on the Helmand's main west-bank tributary, the Arghandab, about 40 km (25 mi) northeast of Kandahar. The two dams and the irrigation canal systems were constructed by the Helmand Valley Authority, the country's only important hydrological project.

Below its confluence with the Arghandab, the Helmand becomes a slow, meandering stream without affluents. It crosses the desertlands of Registan and Dasht-i-Margo, continues its course to the desert lakes of the Sistan depression. Three major rivers form part of the Helmand drainage basin: the Kash Rud, Farah Rud and the Harut Rud. All originate in the Band-i-Bayan and Kasa Murgh Ranges and flow into the Sistan depression. The longest of these rivers, the Farah Rud, irrigates the oasis of Farah.

There are few lakes in the country. The largest are the Dary-acheh-i-Namakzar, the Hamun-i-Pusak and the Hamun-i-Saberi, all situated on the southwestern border. Most of the surface of the Daryacheh-i-Namakzar and of the Hamun-i-Saberi are in Iran. The swamps of the salt-steppe of Gaud-i-Zirreh rarely carry a sheet of open water. Lake Sar-i-Köl (Zorkul) is located in the Wakhan corridor near the border with the Soviet Union. Ab-i-Istada, situated on a plateau about 193 km (120 mi) northeast of Kandahar, is a salt lake.

PRINCIPAL TOWNS (estimated population at March 1982)

Kabul (capital)	1,036,407	Kunduz	57,112
Qandahar	191,345	Baghlan	41,240
Herat	150,497	Maymana	40,212
Mazar-i-Sharif	110,367	Pul-i-Khomri	32,695
Jalalabad	57,824	Ghazni	31,985

Area and population

Regions	area sq mi	area sq km	population 1984 estimate
Eastern	28,664	74,240	1,923,081
North-central	20,461	52,994	2,062,677
North-east	29,911	77,468	1,442,099
North-west	50,581	131,005	2,368,323
South-central	32,963	85,375	1,140,390
South-east	12,546	32,494	3,875,364
Western	76,699	198,649	1,554,500
TOTAL	251,825	652,225	14,366,434

ALBANIA

BASIC FACTS

Official Name: People's Socialist Republic of Albania

Abbreviation: ALB

Area: 28,748 sq km (11,000 sq mi)

Area—World Rank: 126

Population: 3,147,352 (1988) 4,040,000 (2000)

Population—World Rank: 109

Capital: Tirane

Boundaries: 1,204 km (748 mi); Yugoslavia 476 km (296 mi); Greece 256 km (159 mi)

Coastline: 472 km (293 mi)

Longest Distances: 340 km (211 mi) N-S; 148 km (92 mi) E-W

Highest Point: Korabit 2,751 m (9,026 ft)

Lowest Point: Sea level

Land Use: 21% arable land; 4% permanent crops; 15% meadows and pastures; 38% forest and woodland; 22% other

NATURAL REGIONS

Seventy percent of Albania is mountainous and often inaccessible. The remaining alluvial plain receives its precipitation seasonally, is poorly drained, is alternately arid or flooded, and much of it is devoid of fertility. Far from offering a relief from the difficult interior terrain, it is often as inhospitable to its inhabitants as are the mountains. Good soil and dependable precipitation occur, however, in river basins within the mountains, in the lake district on the eastern border, and in a narrow band of slightly elevated land between the coastal plains and the higher interior mountains.

North Albanian Alps

The mountains of the far north of Albania are an extension of the Dinaric Alpine chain and, more specifically, the Montenegrin limestone (karst) plateau. They are, however, more folded and rugged than the more typical portions of the plateau. The rivers have deep valleys with steep sides and do not furnish arable valley floors; most of the grazing and farming are done on the flatter mountaintops. The rivers provide little access into the area and are barriers to communication within it. Roads are few and poor. Lacking internal communications and external contacts, a tribal society flourished within this Alpine region for centuries. Only after World War II were serious efforts made to incorporate the people of the region into the remainder of the country.

Southern Mountains

The extent of the region occupied by the southern mountains is not settled to the satisfaction of all authorities. Some include all of the area in a large diamond shape roughly encompassing all the uplands of southern Albania beneath lines connecting Vlore, Elbasan, and Korce. Although this area has trend lines of the same type and orientation, it includes mountains that are associated more closely with the systems in the central part of the-country. Other authorities confine the area to the mountains that are east of Vlore and south of the Vijose River. These have features generally common to southern Albania and the adjacent Greek Epirus. This demarcation is considered preferable because it more nearly defined a traditional area that tends to lose some of the more purely national character of the lands north of it.

The southern ranges revert again to the northwest to southeast trend lines characteristic of the Dinaric Alps. They are, however, more gentle and accessible than the serpentine zone, the eastern highlands, or the North Albanian Alps. Transition to the lowlands is less abrupt, and arable valley floors are wider. Limestone is predominant, contributing to the cliffs and clear water along the Albanian Riviera. An intermixture of softer rocks has eroded and become the basis for the sedimentation that has resulted in wider valleys between the ridges than are common in the remainder of the country. This terrain encouraged the development of larger landholdings, thus influencing the social structure of the area.

PRINCIPAL TOWNS (population at mid-1983)

Town	Population
Tiranë (Tirana, the capital)	206,100
Durrës (Durazzo)	72,400
Shkodër (Scutari)	71,200
Elbasan	69,900
Vlorë (Vlonë or Valona)	61,100
Korcë (Koritsa)	57,100
Fier	37,000
Berat	36,600
Lushnjë	24,200
Kavajë	22,500
Gjirokastër	21,400

Source: *40 Years of Socialist Albania.*

Area and Population

		area		population
Provinces	**Capitals**	sq mi	sq km	1983 estimate
Berat	Berat	396	1,026	157,300
Dibër	Peshkopi	605	1,568	137,800
Durrës	Durrës	327	848	220,600
Elbasan	Elbasan	572	1,481	213,200
Fier	Fier	454	1,175	216,400
Gjirokastër	Gjirokastër	439	1,137	61,200
Gramsh	Gramsh	268	695	39,300
Kolonjë	Ersekë	311	805	22,500
Korçë	Korçë	842	2,181	201,300
Krujë	Krujë	234	607	94,600
Kukës	Kukës	514	1,331	88,400
Lezhë	Lezhë	185	479	54,200
Librazhd	Librazhd	391	1,013	64,100
Lushnjë	Lushnjë	275	712	117,800
Mat	Burrel	397	1,028	68,700
Mirditë	Rrëshen	335	867	45,800
Përmet	Përmet	359	930	37,100
Pogradec	Pogradec	280	725	62,700
Pukë	Pukë	399	1,033	46,100
Sarandë	Sarandë	424	1,097	78,200
Shkodër	Shkodër	976	2,528	210,200
Skrapar	Corovoda	299	775	42,500
Tepelenë	Tepelenë	315	817	46,100
Tiranë	Tiranë	478	1,238	316,100
Tropojë	Bajram	403	1,043	40,900
Vlorë	Vlorë	621	1,609	158,200
TOTAL		11,100	28,748	2,841,300

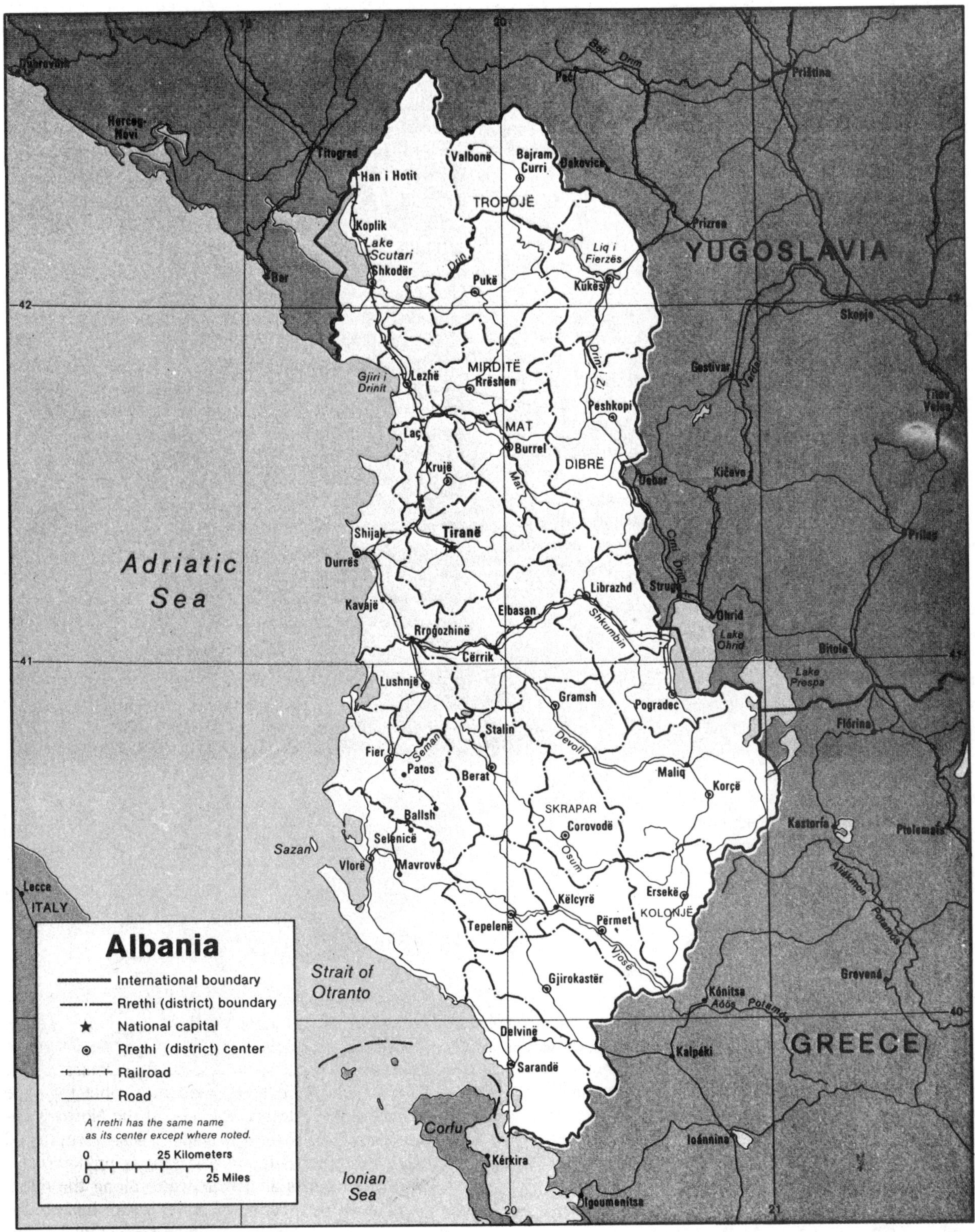

Base 800141 (A01792) 5-84

Lowlands

A low coastal belt extends from the northern boundary southward to about Vlore. It averages less than ten miles deep but widens to about thirty miles in the Elbasan area. In its natural state it is characterized by low scrub vegetation, varying from barren to dense. There are large areas of marshland and other areas of bare eroded badlands. Where elevations rise slightly and precipitation is regular—in the foothills of the central uplands, for example—the land is excellent. Marginal land is being reclaimed wherever irrigation is possible.

The land itself is of recent geological origin. It has been, and is being, created by sediments from the many torrents that erode the interior mountains. New alluvial deposits tend to be gravelly, without humus, and require many years before sufficient vegetation to make them fertile can be established. The sedimentation process, moreover, raises river channels above the level of the nearby terrain. Channels change frequently, devastating areas that have not been stabilized and creating marshes in others by blocking off the drainage. Roadbuilders are confronted with difficult and constantly changing conditions.

Rainfall is heavy during the winter and is infrequent to nonexistent during nearly half the year. Mosquitoes thrive in the hot, humid, and marshy land. Only since about 1930 have there been effective measures to control malaria. Before then no extensive working of areas near the marshes could be seriously considered. For these reasons the coastal zone, in addition to supporting few people, has until relatively recently acted as a barrier, hindering, rather than encouraging, contact with the interior.

Coastal hills descend abruptly to Ionian Sea beaches along the Albanian Riviera from Vlore Bay southward to about Sarande. The 152 and 304 m (500- and 1,000-ft) contour lines are within a mile or so of the water along nearly the entire distance. In the northern portion a 1,219 m (4,000-ft) ridge is frequently only two to three miles inland. South of Sarande is another small area of coastal lowlands fronting on the Ionian Sea and separated from the Greek island of Corfu (Kerkira) by a mile-wide channel. Climate and soil conditions permit the cultivation of citrus fruits in this southernmost area of Albania.

Central Uplands

The central uplands region extends south from the Drin River valley, which marks the southern boundary of the North Albanian Alpine area, to the southern mountains. It is an area of generally lower mountain terrain immediately east of the lowlands. In the north, from the Drin River to the vicinity of Elbasan, it constitutes an area about twenty miles wide. It narrows to practically nothing in the vicinity of Elbasan, then widens into a broader triangular shape with its base against the southern mountains. Earth shifting along the faultline that roughly defines the western edge of the central uplands causes frequent and occasionally severe earthquakes. Major damage occurred over wide areas in 1967 and 1969.

Softer rocks predominate in the uplands. The most extensive are flysch, a soft crumbly rock that is usually sandstone but frequently contains shales, sandy limestones, and marl. This type of formation erodes rapidly and is the basis of much of the poor alluvial lowland soil. The ridges of the uplands are extensions of the Dalmatian coastal range that enters Albania from Yugoslavia. Elevations are generally moderate, between 304 and 914 m (1,000 and 3,000 ft) with a few reaching above 1,524 m (5,000 ft).

Serpentine Zone

Although there are rugged terrain and high points in the central uplands, the first major mountain range inland from the Adriatic is an area of predominantly serpentine rock. The serpentine zone extends nearly the length of the country, from the North Albanian Alps to the Greek border south of Korce, an area 16 to 38 km (10 to 20 mi) wide and over 201 km (125 mi) in length lying generally between the central uplands and the eastern highlands. At Elbasan, however, it makes nearly direct contact with the coastal plain, and it reaches the eastern border for nearly 80 km (50 mi) in, and north of, the lake region. Within its zone there are many areas in which sharp limestone and sandstone outcroppings predominate over the serpentine, although the ranges as a whole are characterized by rounded mountain features.

The serpentine rock derives its name from its dull green color and often mottled or spotted appearance. It can occur in several states. Iron, nickel, or other metals can substitute in its chemical formula for the more prevalent magnesium and will cause color variations.

Eastern Highlands

The mountains east of the serpentine zone are the highest in the country and are the basis for part of the eastern boundary. They occupy a narrow strip south of Lakes Ohrid and Prespa, and a similar one, also running north and south, lies between the White Drin River and the Yugoslav city of Debar. A peak in the Korab range, on the border north of Debar, exceeds 2,743 m (9,000 ft). The ranges have north-south trend lines. Geologically young and composed largely of hard limestone rocks, the eastern highlands, together with the North Albanian Alps and the serpentine zone, are the most rugged and inaccessible of any terrain on the Balkan Peninsula.

Lake Region

The three lakes of easternmost Albania are part of the Macedonian lake district. The Yugoslav border passes through Lake Ohrid; all but a small tip of Little Lake Prespa is in Greece; and the point at which the boundaries of all three states meet is in Lake Prespa. The two larger lakes have areas of about 259 sq km (100 sq mi) each, and Little Lake Prespa is about one-fifth as large. These are total surface areas, including the portions on both sides of the national boundary lines. The surface elevation is about 696 m (2,285 ft) for Lake Ohrid and about 853 m (2,800 ft) for the other two. The lakes are remote and picturesque. Lake Ohrid is fed primarily from underground springs and is blue and very clear. At times its transparency can approach 21 m (70 ft). A good percentage of the terrain in the vicinity of the lakes is not overly steep, and it supports a larger population than any other inland portion of the country.

DRAINAGE

All but a very small portion of the precipitation drains through the rivers to the coastline without leaving the country. With the exception of a few insignificant trickles, only one small stream in the northern part of the country escapes Albania. As the divide is on the eastern side of the borders with Yugoslavia and Greece, however, a considerable amount of water from those countries drains through Albania. A quite extensive portion of the White Drin River basin is in the Kosovo area across the northeastern Yugoslav border. The three lakes shared with Yugoslavia and Greece, as well as all the streams that flow into them, drain into the Drin River. The watershed divide in the south also dips nearly forty miles into Greece at one point. Several tributaries of the Vijose River rise in that area.

With the exception of the Drin River, which flows northward and drains nearly the entire eastern border region before it turns westward to the sea, most of the rivers in the northern and central parts of the country flow much more directly westward to the sea. In the process they cut through the ridges rather than flowing around them. This apparent impossibility came about because the highlands were originally lifted without much folding. The streams came into existence at that time and antedate the ridges because the compression and folding of the plateau occurred later. The folding process was rapid enough in many instances to block the rivers temporarily, forming lakes that existed until the downstream channel was cut sufficiently to drain them. This sequence created the many interior basins that are typically a part of the landforms. During the lifetimes of the temporary lakes enough sediment was deposited in them to form the basis for fertile soils. Folding was only infrequently rapid enough to force the streams to radically different channels.

The precipitous fall from higher elevations and the highly irregular seasonal flow patterns that are characteristic of nearly all streams in the country reduce the immediate value of the streams. They erode the mountains and deposit the sediment that created, and continues to add to, the lowlands, but the rivers flood during the seasons when there is local rainfall. When the lands are parched and need irrigation, the rivers are usually dry. Their violence makes them difficult to control, and they are unnavigable. The Buene is an exception. It is dredged between Shkoder and the Adriatic and is navigable for small ships. In contrast to their histories of holding fast to their courses in the mountains, the rivers have constantly changed channels on the lower plains, making wastes of much of the land they have created.

The Drin River is the largest and most constant stream. Fed by melting snows from the northern and eastern mountains and by the more evenly distributed seasonal precipitation of that area, its flow does not have the extreme variations characteristic of nearly all other rivers in the country. Its normal flow varies seasonally by only about one-third. Along its length of about 282 km (175 mi) it drains nearly 5,956 sq km (2,300 sq mi) within Albania. As it also collects from the Adriatic portion of the Kosovo watershed and the three border lakes (Lake Prespa drains to Lake Ohrid via an underground stream), its total basin is around 15,539 sq km (6,000 sq mi).

The Semen and Vijose are the only other rivers that are more than 160 km (100 mi) in length and have basins larger than 2,590 sq km (1,000 sq mi). These rivers drain the southern regions and, reflecting the seasonal distribution of rainfall, are torrents in winter and nearly dry in the summer, in spite of their relatively long lengths. This is also the case with the many shorter streams. In the summer most of them carry less than a tenth of their winter averages, if they are not altogether dry.

The sediment carried by the mountain torrents continues to be deposited but, having created the lowlands, new deposits delay their exploitation. Stream channels rise as silt is deposited in them and eventually become higher than the surrounding terrain. Changing channels frustrate development in many areas. Old channels become barriers to proper drainage and create swamps or marshlands. It has been difficult to build roads or railroads across the lowlands or to use the land.

Irrigation has been accomplished ingeniously by Albanian peasants for many years, to the degree that they and their expertise have been sought after throughout Europe. Projects required to irrigate or to reclaim large areas of the lowlands, however, are on a scale that probably cannot be accomplished without financial assistance from outside the country.

Although water is available in quantitites adequate for irrigation and it has the amount of fall necessary for hydroelectric power production, terrain and seasonal factors are such that major capital investment would be required for both irrigation and power projects. Snow stabilizes drainage of the higher northern and eastern mountains but, unfortunately, the only major snow accumulations are in the Drin basin, influencing only the one river system.

ALGERIA

BASIC FACTS

Official Name: Democratic and Popular Republic of Algeria

Abbreviation: ALG

Area: 2,381,141 sq km (919,595 sq mi)

Area—World Rank: 10

Population: 24,194,777 (1988) 34,064,000 (2000)

Population—World Rank: 36

Capital: Algiers

Boundaries: 7,476 km (4,645 mi); Tunisia 958 km (595 mi); Libya 982 km (610 mi); Niger 956 km (594 mi); Mali 1,376 km (855 mi); Mauritania 463 km (288 mi); Morocco 1,637 km (1,017 mi)

Coastline: 1,104 km (686 mi)

Longest Distances: 2,414 km (1,500 mi) E-W 2,092 km (1,300 mi) N-S

Highest Point: Tahat 2,908 m (9,541 ft)

Lowest Point: Melrhir Shat 46 m (131 ft) below sea level

Land Use: 3% arable land; 13% meadows and pastures; 2% forest and woodland; 82% other; negligible % permanent crops

Algeria's Arabic name, Barr al Jazair (land of the islands), is believed to derive from the rocky islands along the Mediterranean coastline that posed a threat to ships. The northern portion, an area of mountains, valleys, and plateaus between the Mediterranean Sea and the Sahara Desert, forms an integral part of the section of North Africa known as the Maghrib, which includes the country's western and eastern neighbors—Morocco, Tunisia, and the northwestern portion of Libya, known historically as Tripolitania.

Geographic Regions

The Tell

The fertile Tell is the country's heartland, containing most of its cities and 90 percent of its inhabitants. Made up of hills and plains of the narrow littoral, the several Tell Atlas mountain ranges, and the intermediate valleys and basins, the Tell extends eastward from the Moroccan border to the mountains of the Grande Kabylie and the plain of Bejaïa on the east. Its eastern terminus is the Soummam River.

The best agricultural areas are the gentle hills extending 100 km (62 mi) westward from Algiers, the Mitidja Plain, which was a malarial swamp before its clearing by the French, and the Bejaïa Plain. The alluvial soils in these areas permitted the French to establish magnificent vineyards and citrus groves. By contrast, in the great valley of the Chelif River and other interior valleys and basins, aridity and excessive summer heat have limited the development of agriculture. The Grande Kabylie is a zone of impoverished small farm villages tucked into convoluted mountains.

The High Plateaus and the Saharan Atlas

Stretching more than 600 km (372 mi) eastward from the Moroccan border, the High Plateaus consist of undulating, steppe-like plains lying between the Tell and Saharan Atlas ranges. Averaging between 1,100 and 1,300 m (3,609 and 4,265 ft) in elevation in the west, the plateaus drop to 400 m (1,312 ft) in the east. So dry that they are sometimes thought of as part of the Sahara, they are covered by alluvial debris formed by erosion of the mountains, an occasional ridge projecting through the alluvial cover to interrupt the monotony of the landscape.

PRINCIPAL TOWNS (estimated population at 1 January 1983)

Town	Population	Town	Population
Algiers (El-Djezaïr, capital)	1,721,607	Tlemcen (Tilimsen)	146,089
Oran (Ouahran)	663,504	Skikda	141,159
Constantine (Qacentina)	448,578	Béjaia	124,122
Annaba	348,322	Batna	122,788
Blida (El-Boulaïda)	191,314	El-Asnam (Ech-Cheliff)	118,996
Sétif (Stif)	186,978	Boufarik	112,000*
Sidi-Bel-Abbès	146,653	Tizi-Ouzou	100,749
		Médéa (Lemdiyya)	84,292

*1977 figure.

April 1987 (census results, not including suburbs): Algiers 1,483,000; Constantine 438,000; Oran 590,000.

Population 1987 Census

Wilāyat	population	Wilāyat	population
Adrar	216,931	Médéa	650,623
Ain Defla	536,205	Mila	511,047
Ain Temouchent	271,454	Mostaganem	504,124
Alger	1,687,579	M'Sila	605,578
Annaba	453,951	Naâma	112,858
Batna	757,059	Oran	916,578
el-Bayadh	155,494	Ouargia	286,696
Béchar	183,896	el-Oued	379,512
Bejaïa	697,669	Oum el-Bouaghi	402,683
Biskra	429,217	Relizane	545,061
Blida	704,462	Saïda	235,240
Bordj Bou Arreridj	429,009	Sétif	997,482
Bouira	525,460	Sidi bel-Abbès	444,047
Boumerdes	646,870	Skikda	619,094
ech-Chief	679,717	Souk Ahras	298,236
Constantine	662,330	Tamanrasset	94,219
Djelfa	490,240	el-Tarf	276,836
Guelma	353,329	Tébessa	409,317
Ghardaïa	215,955	Tiaret	574,786
Illizi	19,698	Tindouf	16,339
Jijel	471,319	Tipaza	615,140
Khenchela	243,733	Tissemsilt	227,542
Laghouat	215,183	Tizi Ouzou	931,501
Mascara	562,806	Tiemcen	707,453
		TOTAL	22,971,558

Higher and more continuous than the Tell Atlas, the Saharan Atlas is formed of three massifs: the Ksour near the Moroccan border, the Amour, and the Ouled Nail south of Algiers. The mountains are better watered than the High Plateaus, and the highland topography includes some good grazing land.

Eastern Algeria

Eastern Algeria is made up of a compact massif area extensively dissected into mountains, plains, and basins. It differs from the western portion of the country in that its prominent topographic features do not parallel the coast. In its southern sector the steep cliffs and long ridges of the Aurès create an almost impenetrable refuge that has played an important part in the history of the Maghrib since Roman times. Near the northern coast the Petite Kabylie Mountains are separated from the Grande Kabylie range at the eastward limits of the Tell by the Soummam River. In the far east the coast is predominantly mountainous, but limited plains provide hinterlands for the port citiesof Bejaïa, Skikda, and Annaba. In the interior of the region the extensive high plains of Sétif and Constantine

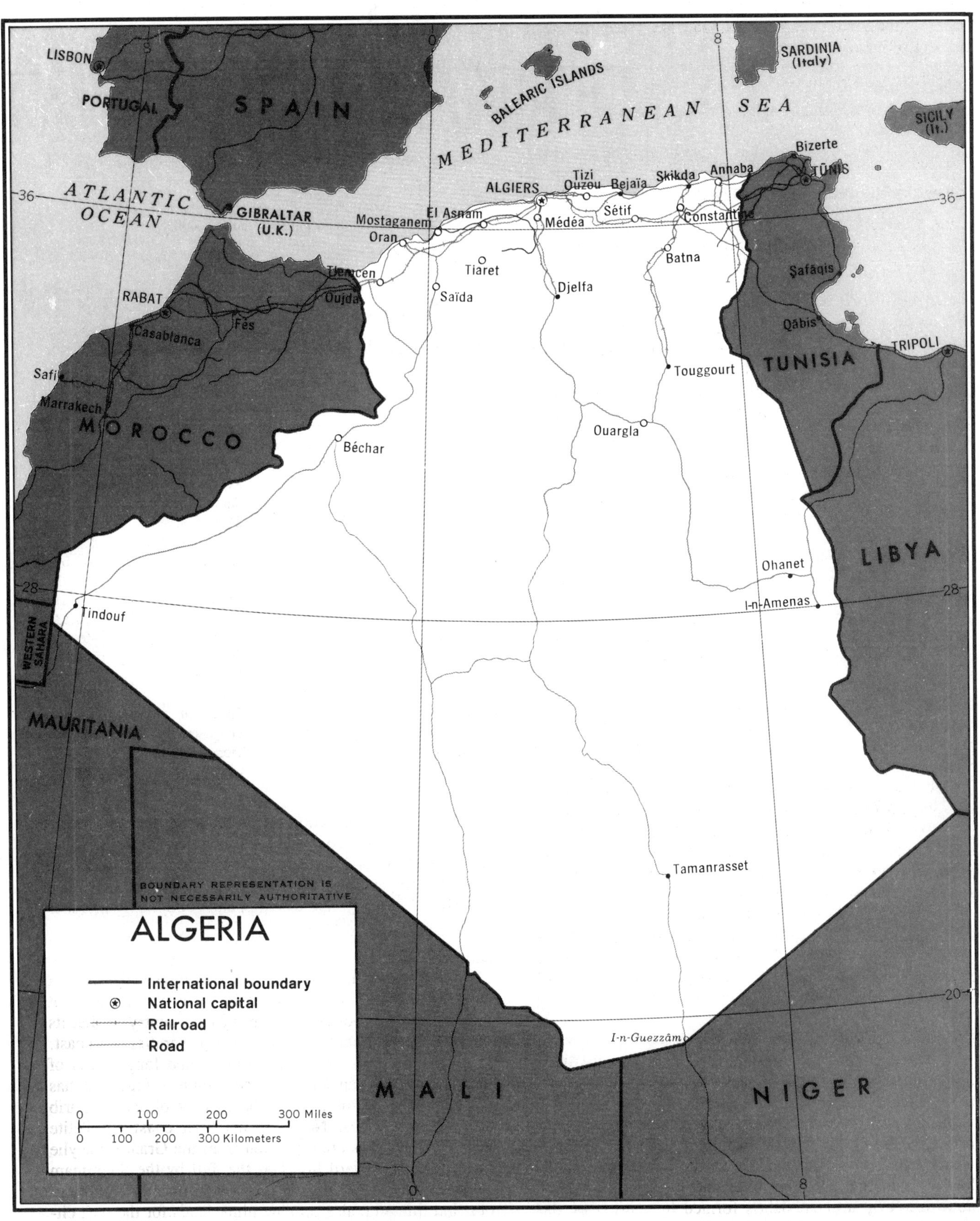

ALGERIA
International boundary
National capital
Railroad
Road
0 100 200 300 Miles
0 100 200 300 Kilometers
BOUNDARY REPRESENTATION IS NOT NECESSARILY AUTHORITATIVE
LISBON
PORTUGAL
SPAIN
BALEARIC ISLANDS
SARDINIA (Italy)
SICILY (It.)
MEDITERRANEAN SEA
ATLANTIC OCEAN
GIBRALTAR (U.K.)
ALGIERS
Tizi Ouzou
Bejaïa
Skikda
Annaba
Bizerte
TUNIS
Sétif
Constantine
Médéa
El Asnam
Mostaganem
Oran
Tiaret
Batna
Saïda
Djelfa
Şafāqis
Qābis
TRIPOLI
RABAT
Casablanca
Fès
Oujda
Safi
Marrakech
MOROCCO
TUNISIA
Touggourt
Ouargla
Béchar
LIBYA
Ohanet
I-n-Amenas
Tindouf
WESTERN SAHARA
MAURITANIA
Tamanrasset
I-n-Guezzâm
MALI
NIGER

were developed during the French colonial period as the principal centers of cereal cultivation. Near the city of Constantine, salt marshes offer seasonal grazing grounds to seminomadic sheepherders.

The Sahara

South of the Saharan Atlas the Algerian portion of the Sahara Desert extends southward for 1,500 km (931 mi) to the Niger and Mali frontiers. The desert is a land apart, scarcely considered a part of the country. Far from being covered wholly by sweeps of sand, however, it is a region of great diversity. Immense areas of sand dunes called *areg* (sing., *erg*) occupy only about one-fourth of the territory. The largest is the Grand Erg Oriental (Great Eastern Erg) where enormous dunes two to five meters high are spaced about forty meters apart. Much of the remainder is covered by rocky platforms called *humud* (sing., *hamada*) elevated above the sand dunes, and almost the entire southeastern quarter is taken up by the high, complex mass of the Ahaggar Mountains with irregular heights reaching above 2,000 m (6,561 ft). Surrounding the Ahaggar are sandstone plateaus, cut by deep gorges formed by ancient rivers, and to the west a lifeless desert of pebbles stretches to the Mali frontier.

The desert can be divided readily into northern and southern sectors, the northern one extending southward a little less than half the distance to the Niger and Mali frontiers. The north, less arid than the south, supports most of the few persons who live in the region; most oases are found in the northern sector. Sand dunes are the most prominent features of its topography, but between the desert areas of the Grand Erg Oriental and the Grand Erg Occidental (Great Western Erg) are plateaus, including a complex limestone structure called the M'zab where the M'zabite Berbers have settled. The southern zone of the Sahara is almost totally arid and is inhabited only by Tuareg nomads and, recently, by oil camp workers. Barren rock predominates, but in some parts of the Ahaggar range alluvial deposits permit a little gardening.

None of the country's rivers is navigable, and a large majority flow only seasonally or irregularly. The longest and best known is the Chelif, which wanders for 230 km (143 mi) across the Tell on a winding journey from its source in the Tell Atlas to the Mediterranean. Most of the Tell streams diminish to trickles or go dry in summer, but in the west many have been dammed for irrigation and to prevent erosion. In better watered Eastern Algeria numerous short, rapid streams have cut deeply into the mountains during the course of their descent to the Mediterranean.

In the High Plateaus the few streams have uncertain and irregular flows. The surface of the land, however, is dotted by salt lakes and salt marshes into which these seasonal streams from the two Atlas chains drain. Watercourses reaching southward from the Saharan Atlas into the desert are usually dry. Some carry water seasonally, however, and subterranean water persists the year around. In particular, the Saoura Wadi in the northwest of the Sahara has an occasional flow, and subterranean water follows the old course of the wadi for some 640 km (398 mi), forming a line of oases.

ANDORRA

BASIC FACTS

Official Name: Principality of Andorra

Abbreviation: AND

Area: 199 sq km (77 sq mi)

Area—World Rank: 177

Population: 49,422 (1988) 96,000 (2,000)

Population—World Rank: 187

Capital: Andorra la Vella

Boundaries: 125 km (77.7 mi); France 60 km (37.3 mi); Spain 65 km (40.4 mi)

Coastline: Nil

Longest Distances: 30.1 km (18.7 mi) E-W 25.4 km (15.8 mi) N-S

Highest Point: Coma Pedrosa Peak 2,946 m (9,665 ft)

Lowest Point: Valira River Valley 840 m (2,756 ft)

Land Use: 2% arable land 56% meadows and pastures; 22% forest and woodland; 20% other

Andorra is a landlocked country on the southern slopes of the Pyrenees Mountains between the French departments of Ariege and Pyrenees-Orientales to the north and the Spanish provinces of Gerona and Lerida to the south.

Most of Andorra's rugged terrain consists of gorges, narrow valleys and defiles surrounded by mountain peaks rising higher than 2,900 m (9,500 ft) above sea level. There is little level ground. All the valleys are at least 914 m (3,000 ft) high and the mean altitude is over 1,829 m (6,000 ft). Of the several lofty peaks, the highest is Coma Pedrosa.

The country is drained by a single basin whose main stream, Riu Valira, has two branches and six smaller open basins; hence the term valleys is used to describe the republic. The section of the river flowing through La Serrat by way of Ordino and La Massana is the Valira del Norte and that flowing through Canilo, Encamp, and Les Escaldes is the Valira del Orien.

Area and population

Parishes	Capitals	area sq mi	area sq km	population 1986 census
Andorra la Vella	Andorra la Vella	49	127	18,463
Canillo	Canillo }	74	191	1,153
Encamp	Encamp }			5,766
La Massana	La Massana	25	65	3,229
Les Escaldes-Engordany	—			11,734
Ordino	Ordino	33	85	1,096
Sant Julià de Lòria	Sant Julià de Lòria			5,535
TOTAL		181	468	46,976

Base 504784 (546491) 2-82

ANGOLA

BASIC FACTS

Official Name: People's Republic of Angola

Abbreviation: ANG

Area: 1,246,700 sq km (481,350 sq mi)

Area—World Rank: 21

Population: 8,236,461 (1988) 13,280,000 (2000)

Population—World Rank: 67

Capital: Luanda

Boundaries: 6,607 km (4,106 mi); Zaire 2,285 km (1,420 mi); Zambia 1,086 km (675 mi); Namibia 1,376 km (855 mi); Cabinda: Zaire 225 km (140 mi); Congo 201 km (125 mi)

Coastline: 1,434 km (891 mi)

Longest Distances: Angola Proper: 1,758 km (1,092 mi) SE-NW; 1,491 km (926 mi) NE-SW Cabinda: 166 km (103 mi) NNE-SSW; 62 km (39 mi) ESE-WNW

Highest Point: Mount Moco 2,620 m (8,596 ft)

Lowest Point: Sea level

Land Use: 2% arable land; 23% meadows and pastures; 43% forest and woodland; 32% other

Angola is located on the western coast of southern Africa, south of the equator.

Most of the country's territory is constituted by the westernmost extension of the great Central Plateau, partly edged in the west by mountains emerging out of the coastal lowland. These highland areas sometimes reach altitudes of more than 2,500 m (8,201 ft). In the north the plateau—the Portuguese term *planalto* is commonly used—reaches the coastal fringe in a one-step or gradual descent; elsewhere the descent is usually a two-step process, the last step sometimes quite precipitous.

The coastal lowland varies in width from perhaps 25 km (15 mi) near Benguela to more than 150 km (93 mi) in the Cuanza River valley just south of Angola's capital, Luanda, and differs markedly from Angola's highland mass. The presence of the cold, northward-flowing Benguela Current substantially reduces precipitation along the coast, making the region arid or nearly so south of Benguela (where it forms the northern extension of the Namib Desert—locally called the Moçâmedes Desert) and quite dry even in its northern reaches. Even where, as around Luanda, the average annual rainfall may be as much as fifty centimeters, it is not uncommon for the rains to fail. Given this pattern of precipitation, the far south is marked by sand dunes, which give way to dry scrub in the middle sections. Some parts of the northern coastal plain are marked by thick brush.

PRINCIPAL TOWNS (population at 1970 census)

Town	Population	Town	Population
Luanda (capital)	480,613*	Benguela	40,996
Huambo (Nova Lisboa)	61,885	Lubango (Sá de Bandeira)	31,674
Lobito	59,258	Malanje	31,559

*1982 estimate: 1,200,000.

Area and population

		area		population
Provinces	**Capitals**	sq mi	sq km	1987 estimate
Bengo	Caxito	14,173	36,708	147,000
Benguela	Benguela	15,115	39,151	699,000
Bié	Kuito	27,149	70,317	938,000
Cabinda	Cabinda	2,744	7,107	112,000
Huambo	Huambo	12,796	33,141	1,267,000
Huíla	Lubango	30,499	78,992	815,000
Kuando Kubango	Menongue	76,671	198,577	169,000
Kuanza Norte	N'Dalatando	7,717	19,988	463,000
Kuanza Sul	Sumbe	21,281	55,117	692,000
Kunene	N'Giva	29,327	75,956	245,000
Luanda	Luanda	570	1,477	1,134,000
Lunda Norte	Lucapa	39,685	102,784	303,000
Lunda Sul	Saurimo	29,860	77,336	145,000
Malanje	Malanje	33,686	87,247	829,000
Moxico	Lwena	77,870	201,683	227,000
Namibe	Namibe	22,043	57,090	77,000
Uige	Uige	23,728	61,455	589,000
Zaire	M'Banza Kongo	14,281	36,989	203,000
TOTAL		481,350	1,246,700	9,105,000

The average altitude of the *planalto* ranges from 1,000 to 1,800 m (3,280 to 5,905 ft) but parts of the Benguela-Bié Plateau in the center and the Humpata Highland area of the Huíla Plateau in the south reach heights of 2,500 m (8,201 ft) and more. The Malange Plateau to the north rarely extends beyond 1,000 m (3,281 ft). The Benguela-Bié highlands and the coastal area in the immediate environs of Benguela and Lobito, the Malange Plateau, and a small section of the Huíla Plateau near the town of Lubango have long been among the most densely settled areas in Angola. The climate and soils of the central *planalto* in particular proved attractive to those Europeans who lived outside the cities.

Vegetation in the Central Plateau varies with altitude: the predominant cover is savanna with isolated baobabs and acacias, but precipitation at the highest points on the western edge permitted the growth of deciduous forest, much depleted by the demand for timber and fuel. Tropical savanna—elephant grass and more frequent baobab—marks the Malange Plateau in the north. In a region of relatively sparse rainfall, the flat Huíla Plateau and the rocky Humpata Highland have no natural forest cover.

The southern desert-steppe is sandy and dry and has sparse vegetation except along the courses of major rivers. East of the Okavango River (in Angola it is the Cubango) the streams are more likely to be permanent, and elephant grass and scrubby forest cover the surface of the sandy clay floodplains; but the grass, good pasture in the rainy season, disappears in the dry season.

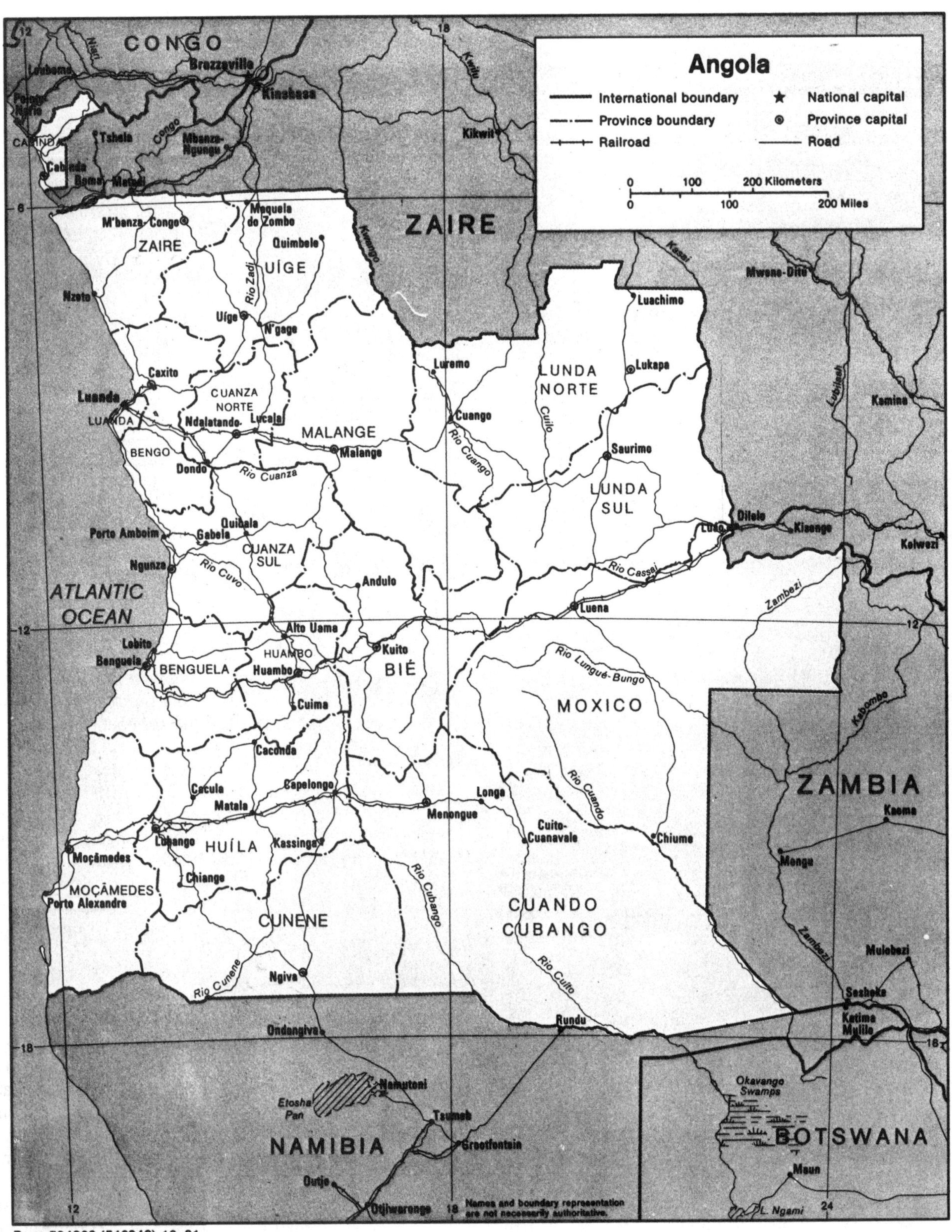

Base 504863 (546940) 10-81

Most of the eastern half of Angola is relatively flat and open plateau characterized by sandy soils. Running through its center into Zambia is the Lunda Divide, a set of low ridges marking the division between north-flowing and south- and east-flowing rivers. To the north the plateau slopes gently toward Zaïre; to the south, toward Namibia, Botswana, and Zambia. For the most part the cover is savanna, but galley forests occur along the courses of the major rivers of the north.

The Mayombe Hills in northeast Cabinda were under rain forest cover, but they have been heavily cut. The remainder of the enclave is an extension of the coastal plain, but it is warmer and wetter than much of the coastal fringe on Angola proper.

Drainage

Most of the country's many rivers originate in central Angola, but the pattern of flow is quite diverse and their ultimate outlets varied. A number of rivers flow in a more or less direct westerly course to the Atlantic Ocean, providing water for irrigation in the dry coastal strip and the potential for hydroelectric power, some of it realized. Two of Angola's most important rivers, the Cuanza and the Cunene, take a more indirect route to the Atlantic, the Cuanza flowing north and the Cunene south before turning west. The Cuanza is the only river other than the Congo (running for a short distance between Angola and Zaïre) that is navigable—for nearly 200 km (126 mi) from its mouth—by boats of some size.

North of the Lunda Divide a number of important tributaries of the Congo River flow north to join it, draining Angola's northeast quadrant. South of the divide some rivers flow to join the Zambezi and thence to the Indian Ocean, others to the Okavango and thence to the Okavango Swamp in Botswana. The tributaries of the Okavango and several of the southern rivers flowing to the Atlantic are seasonal, completely dry much of the year.

ANGUILLA

BASIC FACTS

Official Name: Anguilla

Abbreviation: ANA

Area: 91 sq km (35 sq mi)

Area—World Rank: 203

Population: 6,875 (1988)

Population—World Rank: 207

Capital: The Valley

Boundaries: Nil

Coastline: 61 km (38 mi)

Longest Distances: N/A

Highest Point: Crocus Hill 70 m (225 ft)

Lowest Point: Sea Level

Land Use: N/A

Anguilla is the most notherly of the Leeward Islands, lying 113 km (70 mi) north of St. Christopher, 8 km (5 mi) north of St. Maarten's and 270 km (168 mi) east of Puerto Rico. Also included in the territory are the island of Sombrero, 48 km (30 mi) north of Anguilla and several other uninhabited islands. Anguilla has a flat terrain, mostly rock and coral limestone, with sparse scrub oak, few trees, and some commercial salt ponds.

ANTIGUA AND BARBUDA

BASIC FACTS

Official Name: Antigua and Barbuda

Abbreviation: AAB

Area: 442 sq km (171 sq mi)

Area—World Rank: 179

Population: 70,925 (1988) 98,000 (2000)

Population—World Rank: 177

Capital: Saint John's

Boundaries: Nil

Coastline: 153 km (95 mi)

Longest Distances: N/A

Highest Point: Boggy Peak 402 m (1,319 ft)

Lowest Point: Sea Level

Land Use: 18% arable; 7% meadows and pastures; 16% forest and woodland; 59% other

Antigua and Barbuda consists of three islands along the outer edge of the Leeward Islands chain in the West Indies: Antigua (280 sq km; 108 sq mi); Barbuda (160 sq km; 62 sq mi); and the uninhabited rocky islet of Redonda (1.6 sq km; 0.6 sq mi). Barbuda, the most northerly island, lies 40 km (25 mi) north of Antigua while Redonda lies an equal distance to the southwest. The capital is St. John's on the northwestern coast of Antigua. The only other town is Codrington.

Capital: St. John's (36,000)

Area and population

Parishes	area sq mi	area sq km	population 1986 estimate
Saint George	10.2	26.4	80,000
Saint John's	26.2	67.9	
Saint Mary	25.1	65.0	
Saint Paul	17.7	45.8	
Saint Peter	12.8	33.2	
Saint Phillip	16.0	41.4	
Islands			
Barbuda	62.0	160.6	1,500
Redonda	0.5	1.3	
TOTAL	170.5	441.6	81,500

ARGENTINA

BASIC FACTS

Official Name: Argentine Republic

Abbreviation: ARG

Area: 2,780,092 sq km (1,073,399 sq mi)

Area—World Rank: 8

Population: 31,532,538 (1988) 37,197,000 (2000)

Population—World Rank: 29

Capital: Buenos Aires

Boundaries: 14,346 km (8,914 mi) Bolivia 742 km (461 mi); Paraguay 1,699 km (1,056 mi) Brazil 1,132 km (703 mi); Uruguay 495 km (308 mi) Chile 5,308 km (3,298 mi)

Coastline: 4,970 km (3,088 mi)

Longest Distances: 3,650 km (2,268 mi) N-S 1,430 km (889 mi) E-W

Highest Point: Cerro Aconcagua 6,959 m (22,831 ft)

Lowest Point: Salinas Chicas 42 m (138 ft) below sea level

Land Use: 9% arable; 4% permanent crops; 52% meadows and pastures; 22% forest and woodland; 13% other

Argentina, the eighth largest country in the world, is more than one-fourth the size of Europe. It is the second largest country in South America, both in area and in population, and its relief features range from the highest peak in the Western Hemisphere to a dry lake more than 30 m (100 ft) below sea level. Buenos Aires is South America's largest metropolis, but some parts of Argentina's southern plains have less than one inhabitant per sq mi. In different parts of the country, the continent's highest and lowest temperatures have been recorded. Vegetation includes desert scrub as well as lush jungle foliage, and the species of wildlife are correspondingly varied. The soils of the central plains are among the deepest and richest in the world, but there are also deserts and swamps in the northeastern reaches and, with the changing of the seasons, portions of the northern plains change from one to the other.

On the west the Andes mountains that separate the country from Chile extend the full length of both countries. Broad and lofty in the north, they are narrower and progressively lower in the south. East of the Andes and its foothills lies a great plain that covers two-thirds of the national territory. In the south its arid and windswept expanse is sparsely peopled and devoted principally to sheepherding. In the north its soils range from good to poor, its vegetation from jungle to scrub, and its population densities from scanty to dense. In the central lowlands the broad and fertile plains of the Pampa make up the country's economic and demographic heartland. It is in the Pampa that Buenos Aires and all of the other largest urban centers are located.

PRINCIPAL TOWNS (population at 1980 census)

Town	Population	Town	Population
Buenos Aires (capital)	2,922,829*	Mar del Plata	414,696
Córdoba	983,969	Santa Fé	291,966
Rosario	957,301	San Juan	291,707
Mendoza	605,623	Salta	260,744
La Plata	564,750	Bahia Blanca	223,818
San Miguel de Tucumán	498,579	Resistencia	220,104
		Corrientes	180,612
		Paraná	161,638

*The population of the metropolitan area was 9,967,826.

Area and population

Province	Capitals	area sq mi	area sq km	population 1986 estimate
Buenos Aires	La Plata	118,754	307,571	12,226,000
Catamarca	San Fernando del Valle de Catamarca	38,984	100,967	230,000
Chaco	Resistencia	38,469	99,633	791,000
Chubut	Rawson	86,752	224,686	316,000
Córdoba	Córdoba	65,161	168,766	2,629,000
Corrientes	Corrientes	34,054	88,199	724,000
Entre Rios	Paraná	30,418	78,781	968,000
Formosa	Formosa	27,825	72,066	338,000
Jujuy	San Salvador de Jujuy	20,548	53,219	487,000
La Pampa	Santa Rosa	55,382	143,440	231,000
La Rioja	La Rioja	34,626	89,680	183,000
Mendoza	Mendoza	57,462	148,827	1,344,000
Misiones	Posadas	11,506	29,801	690,000
Neuquén	Neuquén	36,324	94,078	315,000
Rio Negro	Viedma	78,384	203,013	477,000
Salta	Salta	59,759	154,775	768,000
San Juan	San Juan	34,614	89,651	520,000
San Luis	San Luis	29,633	76,748	234,000
Santa Cruz	Rio Gallegos	94,187	243,943	138,000
Santa Fe	Santa Fe	51,354	133,007	2,675,000
Santiago del Estero	Santiago del Estero	52,222	135,254	660,000
Tucumán	San Miguel de Tucumán	8,697	22,524	1,112,000
Other federal entities				
Distrito Federal	Buenos Aires	77	200	2,924,000
Tierra del Fuego	Ushuaia	8,210	21,263	50,000
TOTAL		1,073,399	2,780,092	31,030,000

Leeward Islands

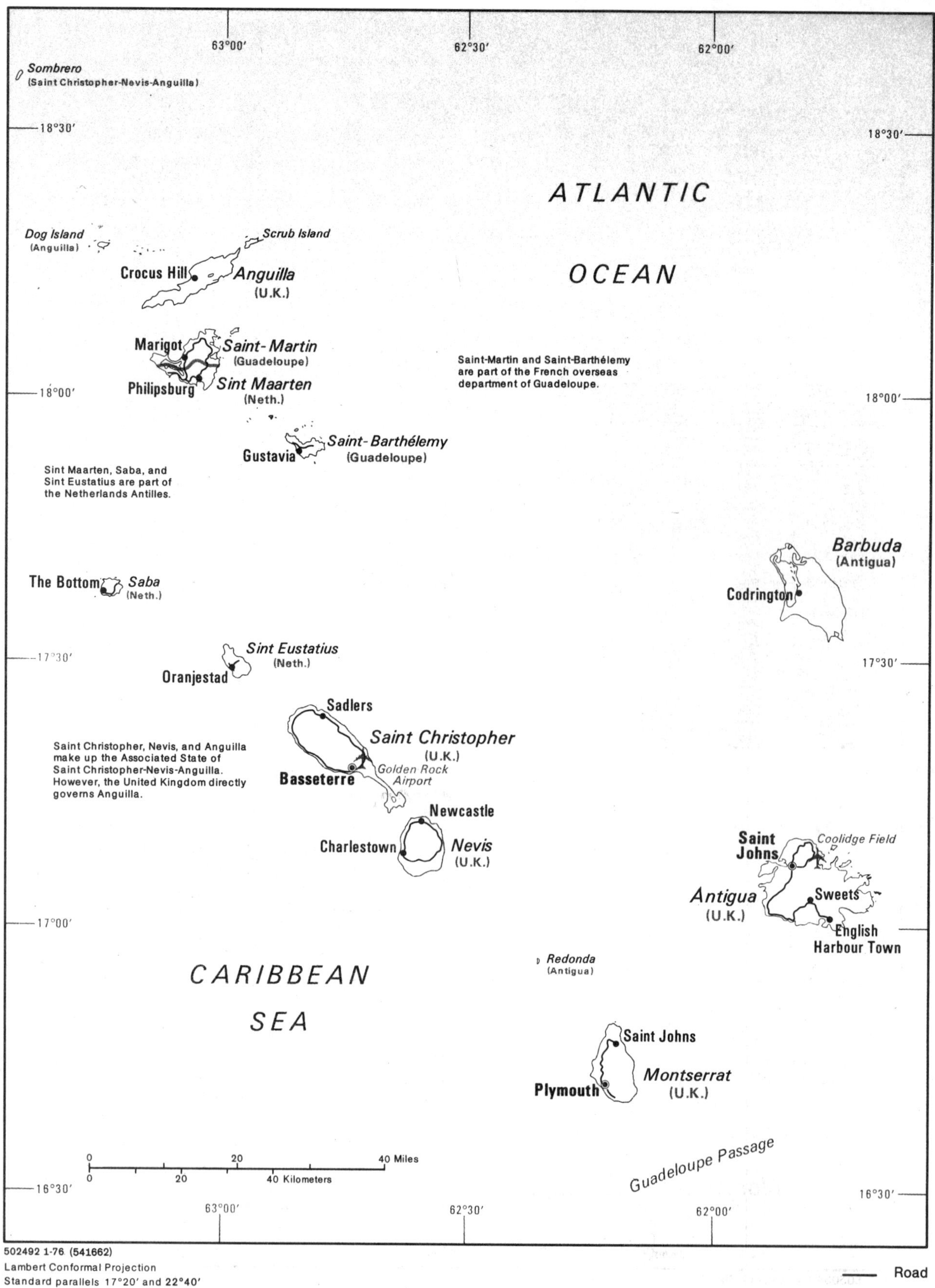

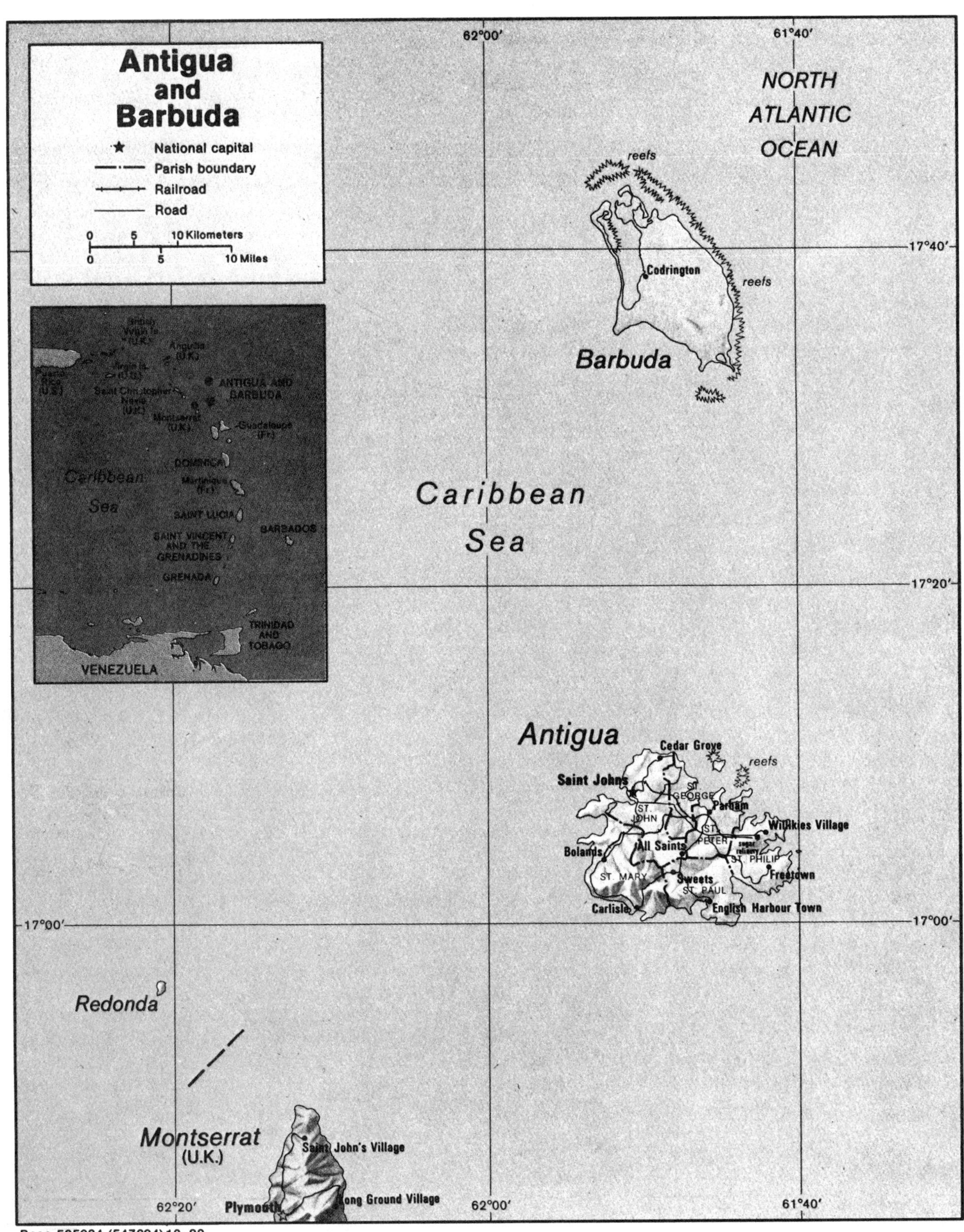

Base 505034 (547694) 12-82

Although several ways have been devised for dividing the country into geographic regions, the windswept southern region is universally recognized as Patagonia, and the fertile central plains are universally recognized as the Pampa. The remainder of the country is frequently divided into two additional regions by drawing an imaginary line southward from the border tripoint with Paraguay and Bolivia to the northwest corner of the Pampa and southwest to the limits of Patagonia. To the west of this imaginary line is the region of the Northwest Andes and Piedmont; to the east is the lowland region of the Northeast.

The Pampa

The Pampa, a great oval extending more than 804 km (500 mi) both north to south and east to west, is the heartland of the country. Consisting of only about one-fifth of the national territory, it includes well over half of the population, approximately 60 percent of the railroad network, 70 percent of the paved roads, and 80 percent of the industrial establishments. Nearly all the cereal crops, truck farm and dairy products, and cattle are produced in it.

Pampa is a Quechuan Indian word meaning level land or space, and the region consists almost entirely of an unbroken and very fertile plain. To the north its limits are determined by juncture with the scrub forests and savannas of the less fertile Chaco plain. Southward the line corresponds roughly to the Colorado River, which bisects the country on a slanting course from the Andes to the sea between latitudes 35° and 40° S. To the west it corresponds to the beginnings of the Andean piedmont and, along the northern part of the western edge, to a low mountain chain called the Sierra de Córdoba.

Lying beneath the soils of the region is an immense basin-like substructure of granites, gneisses, and quartzites, regarded as a southern fragment of the old Brazilian massif. The low profile of the Sierra del Tandil and Sierra de la Ventana, which range along the southern border of Buenos Aires Province, and the island of Martin García in the Rio de la Plata estuary are outcroppings of this crystalline substructure. The rich overlay of soils that fills the basin to varying and sometimes enormous depths is an accumulation of clays, loesses, and sands. Some are apparently of marine origin, but most are products of the leaching of soils from mountain and piedmont, or they have been deposited as dust from windstorms out of the southwest. There is a virtual absence of rock formations and stone over most of the plain. The surface is monotonously smooth, and there is a tilting toward the Atlantic coast that results in a pattern of drainage from the northwest.

There is a natural division into humid and dry subregions. The Humid Pampa fans out some 322 km (200 mi) from Buenos Aires, and the Dry Pampa occupies the outer reaches of the plain where rainfall becomes progressively scantier but where cooler weather reduces the rate of evaporation of surface water. Neither indigenous wildlife nor indigenous vegetation is distinctively different in the two subregions, and the distinction drawn between them dates from the extensive planting of the better watered Humid Pampa to food crops during the nineteenth century. When an Argentine refers to *la pampa,* he is referring to the gaucho and cattle country that lies largely within the Dry Pampa.

Three other regional names are frequently applied to areas lying all or in part within the Pampa. The Littoral (Litoral) is the area immediately to the north of Buenos Aires that lies along the west bank of the Paraná River, the site of the major cities of Rosario and Santa Fe. The Comahue is an area comprising the province of La Pampa, the southern Panhandle of Buenos Aires Province, and the Patagonian provinces of Neuquén and Río Negro. The third regional name is applied to an area vaguely referred to as the "camp." An anglicization of the Spanish word *campo* (country), the "camp" can be defined generally as rural Argentina. More exactly, however, it seems to refer principally or exclusively to the rural parts of the Pampa, Patagonia, and the Piedmont. It is less the name of any specific part of Argentina than of the parts of the countryside where pioneer British, Irish, Scottish, and Welsh immigrants settled. Native Argentines are familiar with the term, but it is encountered most frequently in English-language publications by writers who appear able to understand but not to define this term, which concerns as much a state of mind as it does geographical coordinates.

Patagonia

Occupying more than one-fourth of the country's continental territory, Patagonia extends southward some 1,931 km (1,200 mi) from the Colorado River to the Strait of Magellan and laterally from the Atlantic to the Chilean frontier. It contains all of the provinces of Neuquén, Río Negro, Chubut, and Santa Cruz. The Argentine portion of the Tierra del Fuego is also usually considered a part of the region.

Patagonia is an arid land of windy plateaus, crisscrossed at intervals by valleys carved out by floods from melting Andean ice fields. The coast is fringed by steep escarpments that truncate the interior plateau at elevations ranging from 61 m to 610 m (200 to nearly 2,000 ft) near the port of Colodoro Rivadavia.

Rising in tiers toward the west, the heavily eroded terrain reaches elevations of up to 1,524 m (5,000 ft). In the north, however, the Granbajo de Gualicho, a natural depression between the Colorado and Río Negro rivers descends to 32 m (105 ft) below sea level. The Patagonian soils, probably resting on a crystalline foundation similar to that of the Pampa, are considerably shallower and less productive. Made up of continental deposits accumulated over long periods, they consist principally of sandstones but include clays and marls in areas near tidewater.

The southern Andes, usually considered part of Patagonia, have natural features entirely different from those of the plateau but have little population or economic significance. Between the Río Negro and Colorado rivers they are dotted with lakes, on one of which the world famous resort town of San Carlos de Bariloche is located. In the extreme south there are glaciers, ice fields, and moraine-blocked lakes that drain into the Pacific.

Base 504807 (546795) 6-81

The Northeast

The Northeast region is made up of the lowlands lying north of the Pampa and east of the Andes mountains and their foothills. The Northeast provinces of Entre Ríos, Corrientes, and Misiones lie between the Paraná and Uruguay rivers and, because of their interriverine location, are frequently referred to as the Argentine Mesopotamia. The rolling grasslands of Entre Ríos Province, however, have characteristics similar to those of the Pampa, and some authorities include it in that region. The remainder of the Northeast, lying to the west of the Paraná River, is made up of the Argentine portion of the Gran Chaco Plain of South America that extends northward across Paraguay and into Bolivia. The Chaco subregion includes the provinces of Chaco and Formosa and portions of the provinces of Santiago del Estero and Santa Fe.

The fertile soils of Mesopotamia extend to depths of twenty feet or more. The contours are low and flat, and the slow course of the rivers results in seasonal floods that drain southwestward through the Parana-Uruguay river system. Topographical features are similar on the Chaco plain, but the shallower topsoil is made up of materials washed down from the Andean highlands. During the seasonal rains the rivers overflow their banks and flood extensive areas, which are converted into parched wastes during the dry months.

The Northwest Andes and Piedmont

The region formed by the central and northern Andes and their piedmont approaches is approximately the size of the Pampa. It extends along the western half of the country from Bolivia in the north to Patagonia in the south. On the west it is bordered by the Chilean frontier; and on the east, by the Chaco-Pampa sector of the great South American plain that extends from the tip of the continent northward the full length of the country and continues across Paraguay and into Boliva.

In the far northwest there is little to distinguish the terrain from the high plateau area of neighboring Bolivia. Along the frontier with Chile, peaks in Jujuy, Slata, Catamarca, and Tucumán provinces rise to 6,096 m (20,000 ft) and the highest plateaus have elevations of from 3,352 to 3,962 m (11,000 to 13,000 ft). The topography includes dry, sandy, and clay-filled salt basins, residual mountain systems, and volcanic cones and debris. The countryside is cut by deep river valleys as well as some broader valleys known as *quebradas,* historically important as colonial routes of penetration to the highest Andes.

To the south the Andes narrow as central Argentina is approached. In some areas the Chilean frontier follows the mountain crests; in others it veers eastward, and only piedmont areas remain in Argentina. South of Mendoza the elevations of the highest crests are lower but, in that province, on the Chilean border, is Aconcagua—the loftiest peak of the Western Hemisphere.

In the northern part of the region most of the population is located in isolated high valleys and on piedmont slopes where corn and wheat are grown and cattle are raised. Southward, in Tucumán Province, natural conditions have made possible extensive sugarcane culture. South of Tucumán rainfall becomes scanty, and the populations of the provinces of Catamarca, San Juan, and La Rioja are found along the principal streams on oasis-like patches of fertile land in the generally inhospitable countryside.

The old cities of San Juan, Mendoza, and San Luis were among the earliest centers of colonial settlement, and they, together with their hinterlands, became known as the Cuyo. The name survives in the National University of Cuyo, located in the city of Mendoza; and the provinces of San Juan, Mendoza, and San Luis are still considered by some authorities to be a separate Cuyo region. The Cuyo zone of the Piedmont is separated from the Pampa by the natural barrier of the Sierra de Córdoba, a low range of mountains extending along a north-to-south axis west of the city of Córdoba.

The Cuyo is an area of violent natural phenomena. There are volcanoes in its northern sector, and it lies entirely in an earthquake zone. The city of Mendoza was founded in the sixteenth century, but in its third centennial year it suffered a disastrous earthquake. To prevent a recurrence of the disaster occasioned by the tremor, the city was rebuilt with extraordinarily wide streets and solid buildings. In the early 1970s the city of San Juan had barely recovered from a major convulsion that had nearly destroyed it in 1944.

Hydrography

Except in the Northeast there are few large rivers, and many have only seasonal flows. Nearly all watercourses drain eastward toward the Atlantic, but a large number terminate in lakes and swamps or become lost in the thirsty soils of the Pampa and Patagonia. The four major riverine systems are those that feed into the Rio de la Plata estuary, those made up of the Andean streams, those of the central river system, and those of the southern system.

The system of the Rio de la Plata drainage basin is made up of the Paraná and Uruguay rivers and their tributaries. The basin is generally known as the Cuenca de la Plata (Basin of the Plata) and extends into Uruguay, Paraguay, Bolivia, and Brazil. These countries have joined with Argentina in a collective program for river development of the basin and are known as the River Plate Basin Group, the name deriving from an anglicization of Río de la Plata as River Plate rather than River of Silver—the literal translation.

The Uruguay River serves as the border with Uruguay and as the border between the Mesopotamian provinces and Brazil. The Paraná, together with its principal tributary streams, marks the entire border with Paraguay. Navigation of the upper Uruguay is blocked by falls near the town of Concordia in Entre Ríos Province, but the Paraná is navigable by vessels of varying depths across northern Argentina and—via the tributary Paraguay River—across Paraguay and into Brazil. During flood season, oceangoing vessels can proceed as far as Asunción, the capital of Paraguay. Shallow-draft vessels can continue up the Paraguay as far as Corumbá in Brazil, the terminus of a railroad that extends westward across the nearby Bolivian border and connects with the Bolivian capital of La Paz.

The economic importance of the Paraná-Paraguay system as a navigable stream was measurably increased by shipment of the first iron ore from the newly opened Mutún iron mines in Bolivia for reduction in Argentine furnaces.

The rivers of the Río de la Plata basin are of economic significance principally for drainage of the provinces of the Northeast and as means of cheap transportation. Their potential for production of hydroelectric power, however, was estimated at not less than 18 million kilowatts. About a dozen hydroelectric projects involving Argentine sections of the rivers were in various stages of planning, and others were planned for the upper stretches of the rivers as part of the program of the River Plate Basin Group. Some projects were to provide economic benefits far beyond the production of electric power. Plans for development of the Bermejo River—a major tributary of the Paraná—called for damming, flood and silt control, and canal construction measures that were to open up more than 1.6 million ha (4 million ac) of land in the Chaco for farming and to provide cheap transportation for ores from the mines of Salta and Jujuy provinces.

The Andean system of rivers is made up of rapid-flowing but short streams that originate high in the cordilleras and terminate in lakes and lagoons of interior basins. The largest are the Desaguadero, Tunuyan, San Juan, Diamante, and Ituel. For the most part, the rivers flow seasonally in the north and perennially in the south. Watercourses of this river system provide irrigation for the vineyards of Mendoza Province, and it is along them that the population clusters of the Piedmont are located. Since most of the Piedmont is semiarid and idle, the more than 50 billion cu yd of water estimated to lie underground in Mendoza and San Juan provinces are of potential importance.

The central system of rivers is made up of small streams that drain eastward toward the Atlantic but which, as a consequence of scanty rainfall and lack of gradient in the terrain, are in many instances lost in the porous soils of the plains. One of the largest, the Dulce, drains into the saline Mar Chiquita (Little Sea), Argentina's largest lake and a popular vacation spot. The other principal rivers of the system bear names that may lead to some confusion. The Salado River that separates the Chaco subregion from the Pampa is sometimes known as the Salado del Norte (Salado of the North) to distinguish it from the Salado del Sur (Salado of the South), which flows across the southern part of the Pampa. In all, however, there are no less than eighteen rivers of the country bearing the name Salado. In addition, five streams of the central system bearing the name Córdoba flow eastward from headwaters in the Sierra de Córdoba. They are denominated Primero (First) through Quinto (Fifth) Córdoba rivers. The largest, the Tercero (Third) and the Cuarto (Fourth), unite before joining the Paraná above Rosario. The remaining three are absorbed by the land. Drinkable surface water is in scanty supply in the lands of the central drainage system, and windmills are prominent features of the landscape.

The southern river system drains most of the Pampa and all of Patagonia. The principal streams that flow across the Pampa are the Salado del Sur, the Quequén Grande, and the Sauce Grande, and yet another Salado River discharges into the Atlantic immediately to the south of the Río de la Plata estuary.

The Colorado River is generally accepted as marking the imaginary boundary between the Pampa and Patagonia. Within the Patagonian region, the principal rivers are the Río Negro and the Chubut. At the southern extremity of the continent the Gallegos River flows from an Andean source near the Río Turbio coal mines to the Atlantic. Most of the Patagonian watercourses are small and flow parallel to one another eastward from the Andes. Several, the Río Negro and the Colorado in particular, rise close to the Chilean border, but the configuration of the Andean terrain causes them to follow tortuous courses of up to 160 km (100 mi) before escaping through the mountain walls. They have few tributaries and lose much of their water before reaching tidewater as they flow across the arid plains of Patagonia. Many of the numerous mountain lakes discharge toward the Pacific, but they are situated almost at the continental divide, and heavy rains sometimes cause them to overflow and send excess water eastward toward the Atlantic.

The Río Negro is easily the most important of the rivers in the southern system. It is the only watercourse outside of the Paraná-Uruguay system suitable for shallow-draft navigation. It is also a center for alfalfa and fruit culture and for colonization, and the great Chocón hydroelectric complex is located on one of its tributaries. The reservoir created by the Chocón dam is the country's second largest lake.

Except in the Northwest Andes and Piedmont, the drainage systems of Argentina tend to be poor. The terrain is flat or gently rolling throughout most of the Northeast, the Pampa, and Patagonia; gradients are minimal; and rivers and streams are relatively few. Although the rainfall is nowhere excessive, shallow lakes and seasonal swamps form in many parts of the country; and there are permanent swamps in much of the Mesopotamian provinces, in some coastal areas of Buenos Aires Province, to the north of the Mar Chiquita, in La Pampa Province, and in the drainage zone of the Bermejo River. The characteristic porosity of the soils makes possible the absorption of excess surface waters during normal times, but most of the country is susceptible to flooding during times of heavy rainfall, even in semiarid Patagonia. Floods during the winter of 1973, believed to be the worst in history, were reported to have inundated nearly 2 million ha (5 million ac) of land in western Buenos Aires Province and to have caused the equivalent of US $200 million in damage. In Patagonia the loss of up to 500,000 head of sheep was reported.

ARUBA

BASIC FACTS

Official Name: Aruba

Abbreviation: ARU

Area: 193 sq km (75 sq mi)

Area—World Rank: 196

Population: 62,322 (1988) 66,000 (2000)

Population—World Rank: 182

Capital: Oranjestad

Boundaries: Nil

Coastline: 72 km (45 mi)

Longest Distances: N/A

Highest Point: N/A

Lowest Point: N/A

Land Use: 0% arable, permanent crops, meadows and pastures, forest and woodland; 100% other

Aruba is an island lying in the southern Caribbean Sea, 25 km (15 mi) north of Venezuela and 68 km (42 mi) west of Curacao. It has a flat terrain with a few hills and scant vegetation.

Area and Population

		area		population
Island	**Capital**	sq mi	sq km	1981 census
Aruba	Oranjestad	75	193	60,312
TOTAL		75	193	60,312

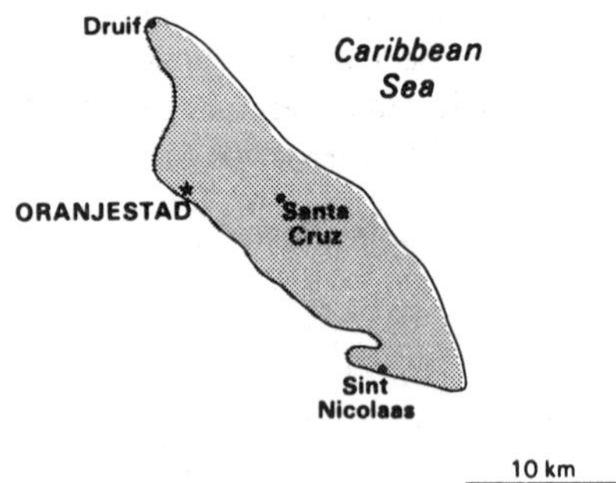

AUSTRALIA

BASIC FACTS

Official Name: Commonwealth of Australia

Abbreviation: AUS

Area: 7,682,300 sq km (2,966,200 sq mi)

Area—World Rank: 6

Population: 16,260,436 (1988) 19,078,000 (2000)

Population—World Rank: 48

Capital: Canberra, ACT

Boundaries: Nil

Coastline: 36,735 km (22,831 mi)

Longest Distances: 4,000 km (2,485 mi) E-W 3,837 km (2,384 mi) N-S

Highest Point: Mt. Kosciusko 2,228 m (7,310 ft)

Lowest Point: Lake Eyre (north) 16 m (52 ft) below sea level

Land Use: 6% arable; 58% meadows and pastures; 14% forest and woodland; 22% other

Australia is the world's smallest continent, lying southeast of Asia between the Pacific and Indian oceans. Its area is nearly as large as that of the United States excluding Hawaii and Alaska and half the size of Europe excluding the Soviet Union. The commonwealth also includes numerous external territories, such as Norfolk Island; the Territory of Ashmore and Cartier Islands; the Coral Islands Territory; the Territory of Cocos (Keeling) Islands; the Territory of Christmas Island; the Territory of Heard and McDonald Islands; and the Australian Antarctic Territory, administered by the Australian Department of Science. Australia is bounded on the north by the Timor and Arafura seas; on the northeast by the Coral Sea; on the east by the Pacific Ocean; on the southeast by the Tasman Sea; and on the south, west, and northwest by the Indian Ocean, with a total coastline of 36,735 km (22,831 mi). The continental shelf extends north to Papua New Guinea and south around Tasmania, with a width varying from 30 km (19 mi) to more than 240 km (149 mi). On the eastern coast the Great Barrier Reef extends for 2,000 km (1,243 mi) from Breaksea Spit, near Maryborough, Queensland, to the Gulf of Papua. The reef, which encloses an area of 207,000 sq km (79,902 sq mi) is an important marine ecosystem, a vast complex of islands and coral reefs containing many rare forms of life.

The country is customarily divided into three principal topographical regions: the large Western Plateau, underlain by the ancient rock shield, the Central Eastern

Lowlands, or Central Plains, underlain by horizontal sedimentary rock, and the Eastern Highlands, which are of considerably more complex geological origin. A smaller fourth region sometimes is known as the Southern Faultlands east of the Great Australian Bight. These regions are known popularly by different terms. The Outback applies generally to the interior and somewhat more specifically to the arid center of the Western Plateau and its semiarid northern plains. The name Red Center is applied to an area with characteristic red, brown and tan soils, tinged scarlet at sunrise and sunset, located in the heart of the continent.

Australia is one of the flattest continents with a mere 6% of its area lying over 610 m (2,000 feet) above sea level. Mount Lofty, ironically named, behind Adelaide reaches only 711 m (2,334 ft); only in the southeastern corner is there any considerable area over 1,524 m (5,000 ft) and here Mt Kosciusko attains 2,229 m (7,314 ft). But these misnamed Australian Alps have nothing alpine about them, and it is possible to drive up to Mt. Kosciusko's summit in an ordinary car. The mountains form part of an extensive area of high relief known as the Great Dividing Range which consists of fretted margins of plateaus and cones and plugs of long-extinct volcanoes. The wetter seaward flanks of the highlands are much more dissected than the Western slopes which fall away to the interior plains in long ridges or undulations. The coast, especially in New South Wales, often is backed by massive scarps cut by wild gorges. There is no volcanic activity in the country and crustal movements are minor.

The vast desert and semidesert region of the Western Plateau covers almost two-thirds of the continent. Averaging about 305 m (1,000 ft) above sea level, it is relieved by widely separated mountains. The highest of these elevations to the west occur in the Hamersley Range, whose tallest peak is over 1,219 m (4,000 ft). Toward the north-central rim of the plateau are the irregular ranges of the Kimberley Plateau—whose highest elevations reach 914 m (3,000 ft)—and the less rugged uplands and broad valleys of Arnhem Land. Toward the east of the plateau are three notable ranges—the Macdonnell, Musgrave and Petermann—with a general east-to-west orientation cut through by gorges. High points in both the Macdonnell and Musgrave ranges rise to over 1,493 m (4,900 ft). From the rocky ranges surrounding much of its west, the plateau stretches eastward as a continuous flatland with occasional stark outcroppings of granite or sandstone, the most impressive of which is Ayers Rock, a massive rounded monolith rising over 335 m (1,100 ft). Four major deserts—Gibson, Great Sandy, Great Victoria and Tanamy—are located on the plateau.

The plateau is rimmed by escarpments, one of the most unusual of which is the Nullarbor Plain, a flat, smooth, barren limestone lowland riddled by numerous underground caves stretching inland along the Great Australian Bight. The Darling Scarp in the far southwestern part of the continent separates the plateau from the coastal plain.

The plateau is broken to the southeast along a relatively small area near the coast near Adelaide, by heavy faulting and some folding. The faulting has produced a roughly north-to-south parallel series of highlands and gulfs consisting of, from east to west, the Mount Lofty Range, Gulf Saint Vincent, the Cape York Peninsula Hills, the Flinders Range, Spencer Gulf and the Eyre Peninsula.

The Central Eastern Lowlands extend from the Gulf of Carpentaria in the north to Western Victoria in the south. The lowland belt averages only about 152 m (500 ft) above sea level and falls to almost 12 m (40 ft) below sea level at Lake Eyre. The featureless plains are intersected by low rises that divide the region into three drainage basins: Lake Eyre, the Murray-Darling rivers and the Gulf of Carpentaria.

Beneath the Central Eastern Lowlands are several artesian basins, one of which, the Great Artesian Basin, underlies approximately one-fifth of the entire continent and is the largest in the world. Over 18,000 artesian bores have been drilled in this basin.

The Eastern Highlands are customarily but inaccurately described as the Great Dividing Range, but they do not form a true range and have an average altitude of under 914 m (3,000 ft). It consists of a complex belt of tablelands, ridges and coastal ranges from Cape York in northern Queensland to southern Victoria remerging in Tasmania on the southern side of the Bass Strait. Despite their generally low elevation, they are rugged and spectacular in parts. Between the highlands and the Pacific Ocean lies a coastal lowland strip varying in width from 48 to 402 km (30 to 250 miles).

PRINCIPAL TOWNS (estimated population at 30 June 1985)*

Canberra (national capital)	273,600†
Sydney (capital of NSW)	3,391,600
Melbourne (capital of Victoria)	2,916,600
Brisbane (capital of Queensland)	1,157,200
Perth (capital of W Australia)	1,001,000
Adelaide (capital of S Australia)	987,100
Newcastle	423,300
Wollongong	236,800
Gold Coast	208,100
Hobart (capital of Tasmania)	178,100
Geelong	147,100
Darwin (capital of Northern Territory)	68,500

*Figures refer to metropolitan areas, each of which normally comprises a municipality and contiguous urban areas.
†Includes Queanbeyan, in NSW.

Area and population

States	Capitals	area sq mi	area sq km	population 1987 estimate
New South Wales	Sydney	309,500	801,600	5,581,300
Queensland	Brisbane	666,900	1,727,200	2,616,300
South Australia	Adelaide	379,900	984,000	1,378,900
Tasmania	Hobart	26,200	67,800	448,600
Victoria	Melbourne	87,900	227,600	4,188,300
Western Australia	Perth	975,100	2,525,500	1,458,700
Territories				
Australian Capital Territory	Canberra	900	2,400	267,600
Northern Territory	Darwin	519,800	1,346,200	150,300
TOTAL		2,966,200	7,682,300	16,090,000

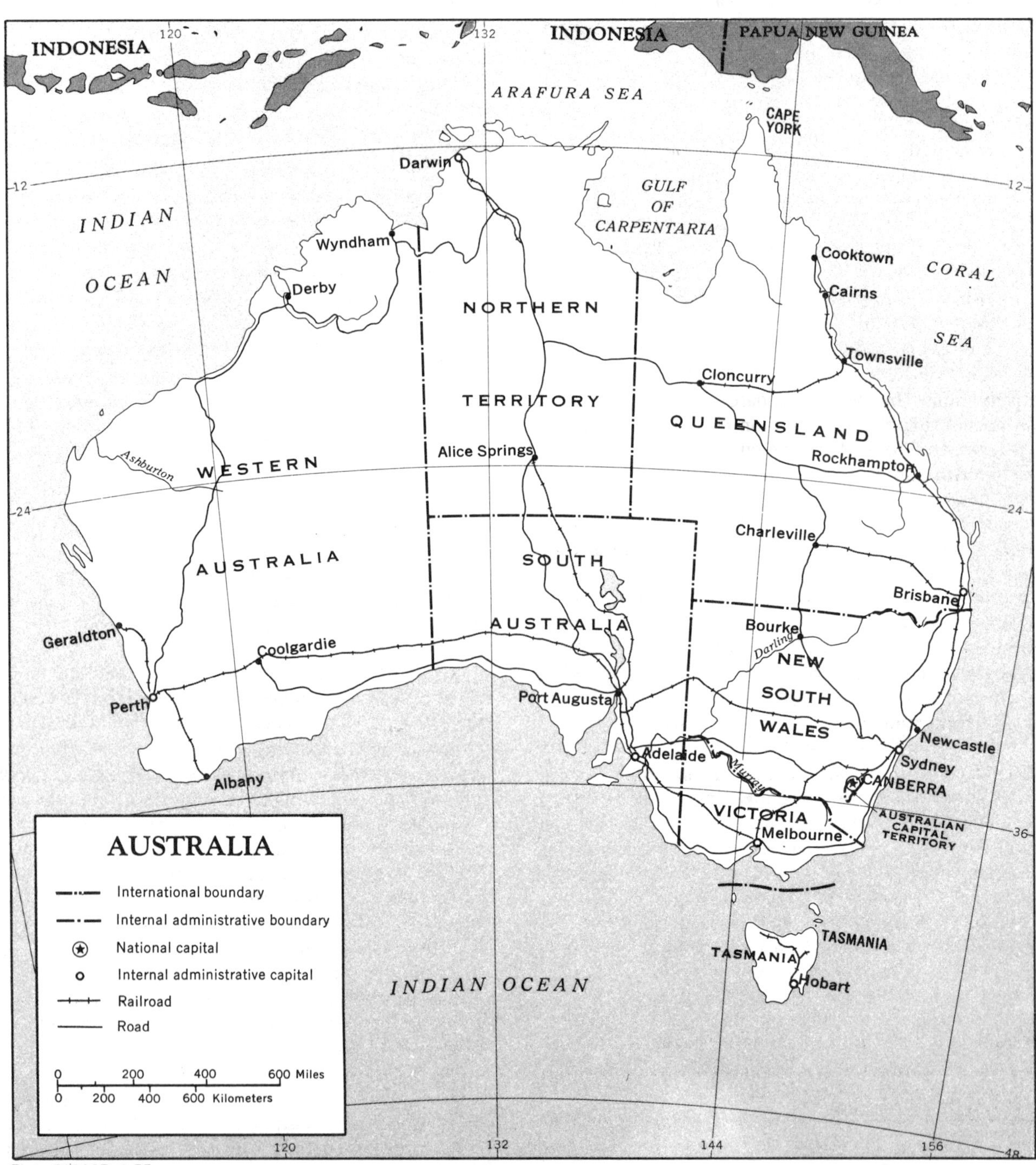

Base 503207 1-77

The highlands are low and broad in the north and rise as they progress southward. In the north the highest point is near Cairns in Queensland. The central portion is characterized by tablelands, such as the New England Plateau and the Blue Mountains. In the southern part, the highlands pass through the Australian Alps and the Snowy Mountains and across Victoria and end in the Grampians. The island of Tasmania, lying about 241 km (150 miles) southeast of the continent, is geologically part of the Eastern Highlands. The island has a rugged terrain, with a large central plateau and with some mountains rising to over 1,524 m (5,000 ft).

Australia is one of the world's driest continents with only a few permanent rivers and streams. More than 60% of the annual runoff is from the Tasmania, Gulf of Carpentaria and Timor Sea drainage basins. In roughly three-fourths of the country, stream flow is intermittent or dependent on seasonal rains and for months or for even years streams and rivers dry up.

The Murray-Darling river system which empties into Lake Alexandrina, a coastal lake east of Adelaide, is the most important drainage system in the country. Although relatively minor by world standards with an estimated mean annual runoff of 22.7 billion cu. m. (801.6 billion cu ft) (compared to 5.84 trillion cu. m. (206.2 trillion cu ft) for the Amazon) it has a large catchment area of 1.063 million sq km (410,318 sq mi) and is the main source of water for 80% of the country's irrigated land. The Darling is Australia's longest river and the Murray the second longest—2,739 km (1,702 mi) and 2,589 km (1,609 mi) respectively. The other tributaries of the Murray are the Lachlan, the Murrumbidgee and the Goulburn, which drain the western slopes of the southeastern highlands.

The rivers and streams that flow into the Gulf of Carpentaria from the northern part of the Central Eastern Lowlands form numerous small drainage systems of which only the Gregory River is permanent.

Lake Eyre and its surrounding inland lakes receive drainage from a series of seasonal, interconnected sluggish streams and rivers, principally Cooper's Creek or Barcoo River and the Warburton-Diamantina and Georgina rivers. These waterways make up a region known as the Channel Country which is dry for most of the year (except for permanent, scattered waterholes known as *billabongs*) and extensively flooded during summer rains.

AUSTRIA

BASIC FACTS

Official Name: Republic of Austria

Abbreviation: AUT

Area: 83,855 sq km (32,376 sq mi)

Area—World Rank: 109

Population: 7,577,072 (1988) 7,530,000 (2000)

Population—World Rank: 77

Capital: Vienna

Boundaries: 2,637 km (1,638 mi) West Germany 784 km (487 mi); Czechoslovakia 568 km (353 mi) Hungary 346 km (215 mi); Yugoslavia 311 km (193 mi) Italy 430 km (267 mi) Liechtenstein 36 km (22 mi) Switzerland 162 km (101 mi)

Coastline: Nil

Longest Distances: 579 km (360 mi) E-W 295 km (183 mi) N-S

Highest Point: Grossglockner 3,797 m (12,457 ft)

Lowest Point: Neusiedler See 115 m (377 ft)

Land Use: 17% arable land; 1% permanent crops; 24% meadows and pastures; 39% forest and woodland; 19% other

GEOGRAPHICAL FEATURES

Austria is a small landlocked Alpine country in south-central Europe.

Austria's geographical importance lies as the crossroads of Europe. Austrians place the country in the exact center rather than in the south-central part of Europe. The Hahneckkogel, a peak in the province of Salzburg is not only claimed to be the highest wooded mountain in Europe but is also said to mark the geographical center of the continent. Few physical frontiers follow natural physical barriers. The Danube flows along the Austrian-Czech border for only about 16 km (10 miles) and is an internal provincial boundary for only a slightly greater distance. In similar fashion, the high Alpine ridge across the country does not separate the provinces except at the Grossglockner mountain, where three of them meet.

Austria generally is divided topographically into the Eastern Alpine region, the North Alpine Forelands, the Bohemian Plateau, the Vienna Basin, and the eastern and southeastern lowlands. Although nearly three-quarters of the country are mountainous, 18.7% is arable, 26.6% is meadow and pasture, 38.6% is forest, a little more than 1.6% is vineyards, orchards and small garden plots, and about 14.5% consists of built-up areas and wasteland.

The Eastern Alps consist of a group of mountains that begin at the Swiss border and become three ranges that fan out as they cross the country. The central range is separated from the northern range by a line from the west to east from the Arlberg Pass, along the Inn River Valley in the Tirol and follows the upper valleys of the generally eastward-flowing Salzach, the Enns, the Mur and the Murz Rivers. The division ends to the east of the Semmering Pass in the province of Lower Austria. The Drau River separates the central and northern ranges. The central range is the largest with the highest elevations in Austria. The Eastern Alps rises sharply from the Bodensee in

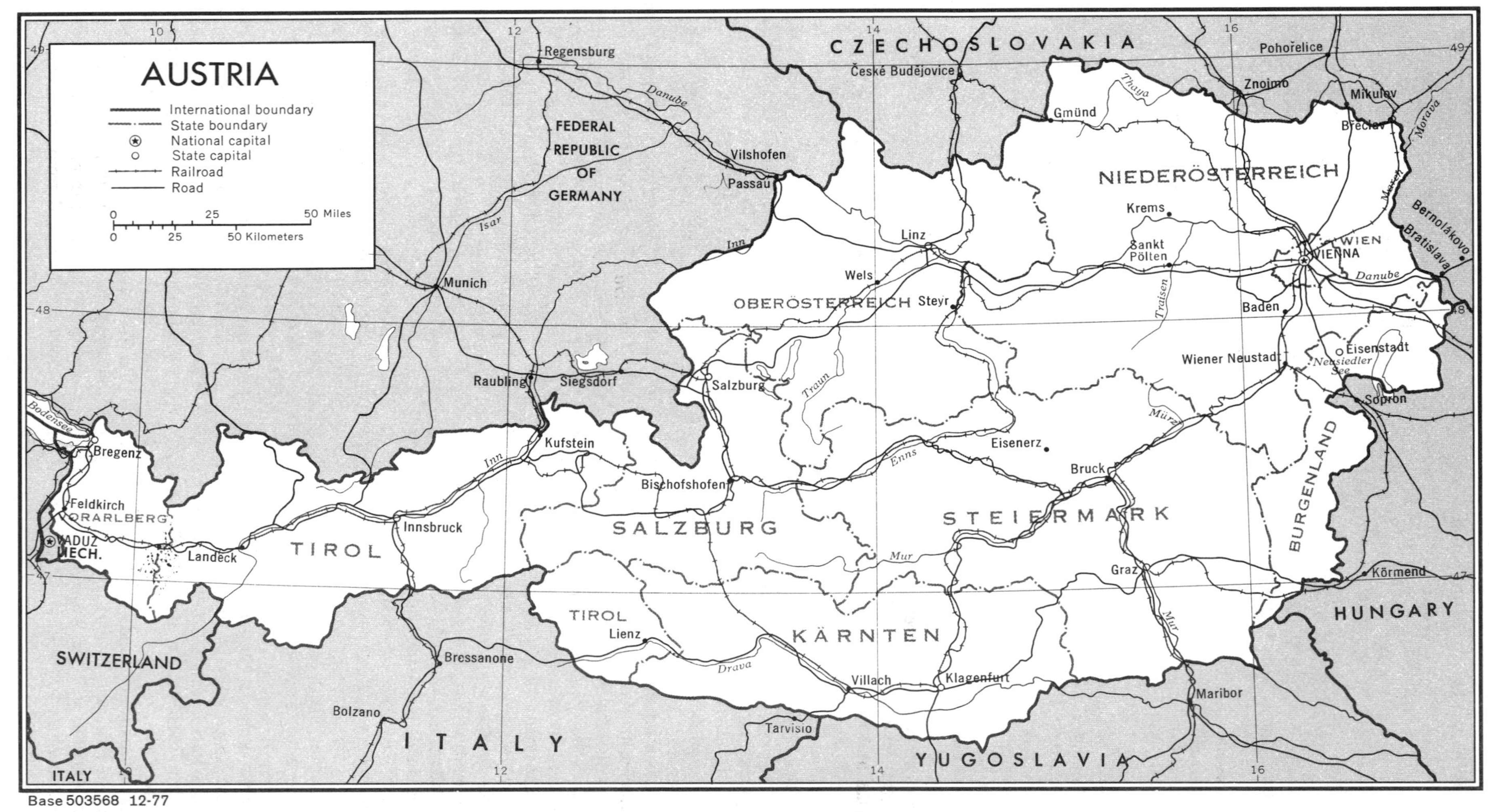

Base 503568 12-77

PRINCIPAL TOWNS (population at 1981 census)

Vienna (capital)	1,531,346	Klagenfurt	87,321
Graz	243,166	Villach	52,692
Linz	199,910	Wels	51,060
Salzburg	139,426	Sankt Pölten	50,419
Innsbruck	117,287	Steyr	38,942

Area and population

		area		population
States	**Capitals**	sq mi	sq km	1985 estimate
Burgenland	Eisenstadt	1,531	3,965	267,686
Kärnten	Klagenfurt	3,681	9,534	540,342
Niederösterreich	Sankt Pölten	7,402	19,172	1,423,741
Oberösterreich	Linz	4,626	11,980	1,285,955
Salzburg	Salzburg	2,762	7,154	456,502
Steiermark	Graz	6,327	16,387	1,183,383
Tirol	Innsbruck	4,883	12,647	601,618
Vorarlberg	Bregenz	1,004	2,601	309,287
Wien	—	160	415	1,489,153
TOTAL		32,376	83,855	7,557,667

Vorarlberg Province with a number of peaks exceeding 3,048 m (10,000 ft) in eastern Tirol. The range is lower at the Brenner Pass and then rises to its highest elevation, the Grossglockner, at 3,798 m (12,457 ft). The Hohe Tauern and Niedere Tauern are the major ranges of the central group which lowers generally to level land near the Hungarian border.

The northern and central ranges separate near Landeck where the Inn River takes up its generally eastward course. Much of the northern range consists of either limestone or dolomite, like the southern range.

The North Alpine Forelands are foothills of the Alps that extend between the mountains and the Danube River from the country's border north of the city of Salzburg to the Vienna Basin. Nearly all of the highlands northwest of Vienna and north of the Danube River are a part of the Bohemian massif rather than of the Alpine system. They are referred to as the Bohemian Plateau as they form a ring around the Bohemian portion of Czechoslovakia. The Alpine Forelands ultimately terminate in the foothills bordering the Vienna Basin. The basin itself is not completely flat, but the terrain is gentle. The basin extends into the Leitha River valley in a southeasterly direction toward the Semmering Pass, and is separated from the Nieusiedler See by the Leitha Mountains.

Because it contains a greater part of the Eastern Alps, precipitation drains in all directions, but all except a minute fraction of it eventually reaches the Danube River. A very small area in the far west drains westward from Vorarlberg to the Rhine River. A small area of the Austrian Tirol drains into Italian streams. An even smaller area northwest of Vienna drains northward, into streams that join the Elbe River.

The Danube is Austria's Grand River. The major Alpine tributaries of the Danube—including the Inn River which joins the Danube near the West German border—flow eastward through central Austria. To the east of the Inn are the Salzach and the Enns. The Danube bisects lower Austria and receives a large number of lesser streams. To the south, the Lietha flows northeast, draining the area from the Semmering Pass to the Hungarian border. From the other side of the Semmering Pass the Murz drains to the south, joining the Danube further downstream. Other southflowing streams in Carinthia and Styria, such as the Gail, the Gurk and the Mur, also flow into the Danube.

BAHAMAS

BASIC FACTS

Official Name: The Commonwealth of the Bahamas

Abbreviation: BAM

Area: 13,939 sq km (5,382 sq mi)

Area—World Rank: 140

Population: 242,983 (1988) 327,000 (2000)

Population—World Rank: 156

Capital: Nassau

Land Boundaries: Nil

Coastline: 2,543 km (1,580 mi)

Longest Distances: (The Archipelago) 950 km (590 mi) SE-NW 298 km (185 mi) NE-SW

Highest Point: Mount Alvernia 63 m (206 ft)

Lowest Point: Sea level

Land Use: 1% arable land; 32% forest and woodland; 67% other

The Bahama Islands is an archipelago of 700 islands (of which some 40 are inhabited) between SE Florida and N Hispaniola. From the north of the chain, which lies 96 km (60 mi) off the Florida coast, the archipelago extends 950 km (590 mi) SE to NW and 298 km (185 mi) NE to SW. The most populous island is New Providence (150 sq km, 58 sq mi). Of the other islands, known together as the Family Islands, the 15 largest are Grand Bahama, Biminis, Berry Islands, the Abacos group, Andros, Eleuthera, the Exumas, Cat Island, Rum Cay, San Salvador (also known as Watlings Islands), Long Island, Acklins Island, Mayaguana, the Ragged Island Range, and the Inaguas. In addition, there are thousands of small cays and rocks protruding from the shallow seas (the name Bahamas being derived from the Spanish bajamar, shallow sea).

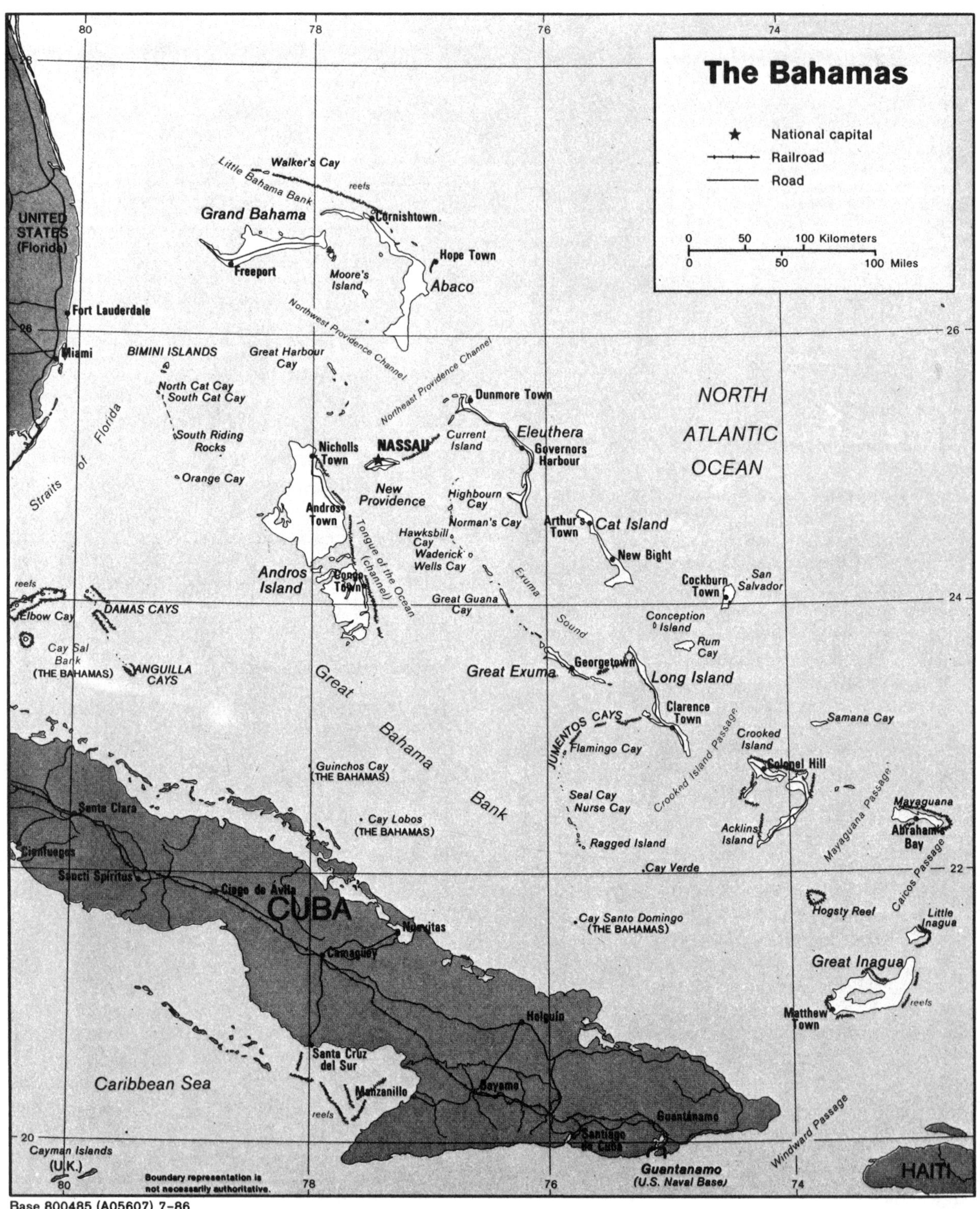

Base 800485 (A05607) 7-86

The islands are for the most part low and flat with the terrain only occasionally broken by small lakes and mangrove swamps. The shoreline is marked by coral reefs. There are no navigable rivers.

Area and population

Islands and Island Groups	Residence of Commissioner	area sq mi	area sq km	population 1980 census
Abaco, Great and Little, and Mores Island and cays	Marsh Harbour	649	1,681	7,324
Acklins Island	Pompey Bay	192	497	616
Andros Island	Kemps Bay	2,300	5,957	8,397
Berry Islands	Nicolls Town	12	31	509
Biminis, North and South, Cay Lobos, and Cay Sal	Alice Town	11	28	1,432
Cat Island	Arthur's Town	150	388	2,143
Crooked Island	Colonel Hill	84	218	517
Eleuthera, Harbour Island, and Spanish Wells	Rock Sound	200	518	10,600
Exuma, Great and Little, and cays	George Town	112	290	3,672
Grand Bahama	Freeport	530	1,373	33,102
Inagua, Great and Little	Matthew Town	599	1,551	939
Long Cay	. . .	9	23	33
Long Island	Clarence Town	230	596	3,358
Mayaguana	Abraham's Bay	110	285	476
New Providence	Nassau	80	207	135,437
Ragged Island and cays	Duncan Town	14	36	146
San Salvador and Rum Cay	Cockburn Town	90	233	804
TOTAL		5,382	13,939	209,505

BAHRAIN

BASIC FACTS

Official Name: State of Bahrain

Abbreviation: BAN

Area: 691 sq km (267 sq mi)

Area—World Rank: 169

Population: 480,383 (1988) 715, 074 (2000)

Population—World Rank: 142

Capital: Manama

Land Boundaries: Nil

Coastline: 126 km (78 mi)

Longest Distances: 48 km (30 mi) N-S 19 km (12 mi) E-W

Highest Point: Mount Al-Dukhan 134 m (440 ft)

Lowest Point: Sea level

Land Use: 2% arable; 2% permanent crops; 6% meadows and pastures; 0% forest and woodland; 90% other

Bahrain comprises an archipelago of thirty-three islands. The main island accounts for 85 percent of the total land area; it lies at the entrance of the Gulf of Bahrain, an inlet of the Persian Gulf between the coast of Saudi Arabia and the Qatar Peninsula. For centuries this position has given the island group regional importance as a trade and transportation center. Only Bahrain, Sitrah, Umm Nasan, and Al Muharraq are of significant size; the remainder are little more than exposed rock and sandbar.

Bahrain, from which the archipelago takes its name, is about thirty miles long and about ten miles wide at its broadest part. Most of the island is desert; low outcroppings of limestone form rolling hills, stubby cliffs, and shallow ravines. The interior contains an escarpment that rises about 137 m (450 ft) to form Jabal Dukhan, a steep-sided hill where oil was first found in the archipelago.

Manama is linked with the island of Al Muharraq and its major city of the same name by a causeway; the international airport is also on Al Muharraq. The island of Sitrah is linked to Bahrain by a bridge spanning a shallow channel. North of Sitrah is An Nabi Salih, where fresh-water springs irrigate numerous date groves. Northwest of Bahrain is the rocky islet of Jiddah, which serves as a prison settlement, and south of Jiddah the larger island of Umm Nasan, the personal property of the ruler and his private game preserve. About twelve miles southeast of the island of Bahrain and close to the coast of Qatar are the Hawar Islands, the subject of a territorial dispute between Bahrain and Qatar.

Principal Towns (population at 1981 census): Manama (capital) 121,986; Muharraq Town 61,853.

Area and population

Regions	area sq mi	area sq km	population 1987 estimate
Central	13.6	35.2	29,034
al-Ḥadd	2.0	5.2	8,509
Judd Ḥafṣ	8.4	21.6	46,741
al-Manāmah	9.8	25.5	146,994
al-Muḥarraq	5.9	15.2	75,579
Northern	14.2	36.8	34,364
Rifāʿ	112.6	291.6	45,530
Sitrah	11.0	28.6	35,188
Western	60.2	156.0	19,711
Towns with special status			
Ḥammād	5.1	13.1	. . .
Madīnat ʿĪsā	4.8	12.4	39,783
Islands			
Ḥawār and other	19.3	50.0	. . .
TOTAL	266.9	691.2	481,433

BANGLADESH

BASIC FACTS

Official Name: People's Republic of Bangladesh

Abbreviation: BGD

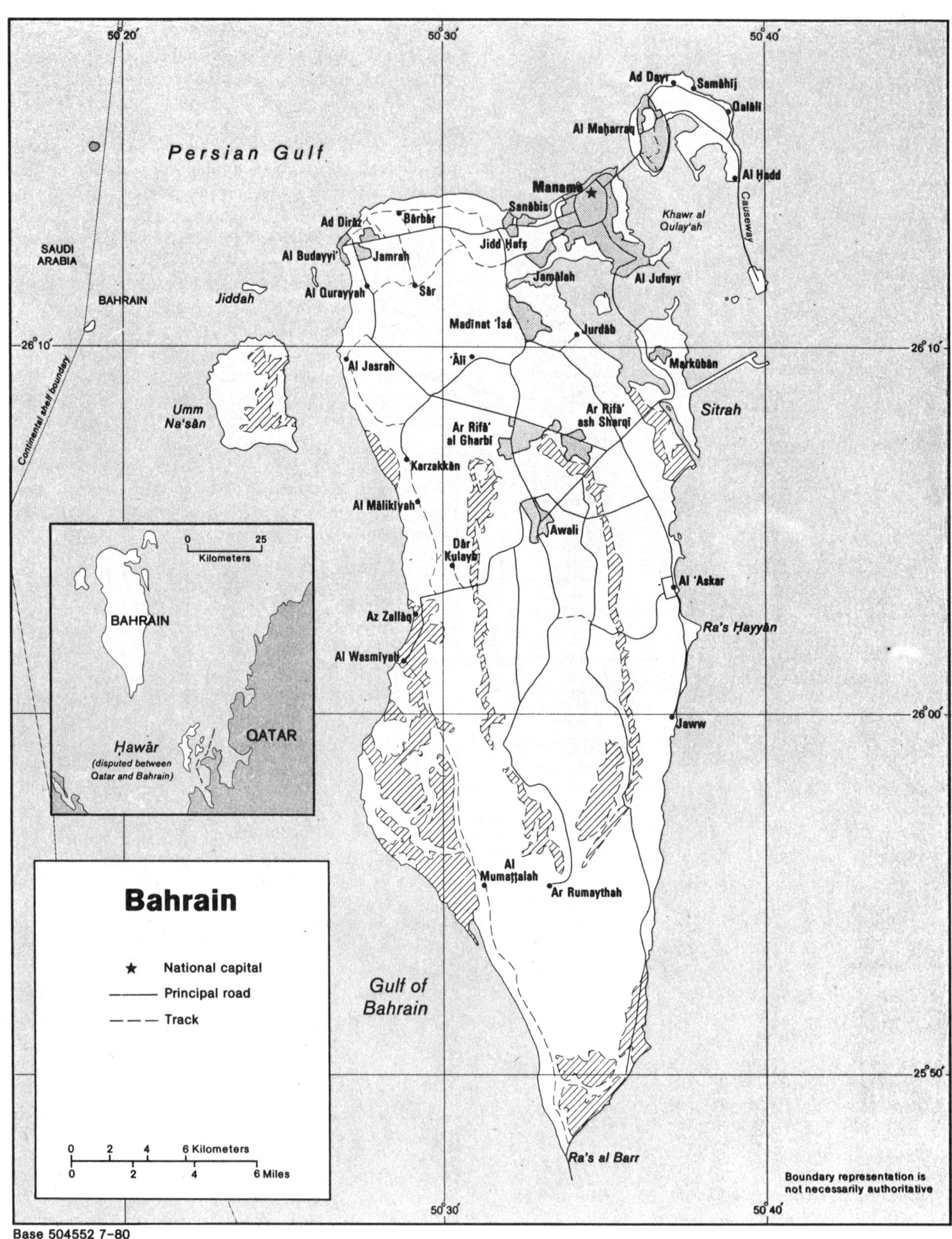

Base 504552 7-80

Area: 143,998 sq km (55,598 sq mi)

Area—World Rank: 90

Population: 109,963,551 (1988) 139,693,000 (2000)

Population—World Rank: 9

Capital: Dhaka

Boundaries: 3,390 km (2,107 mi); India 2583 km (1,605 mi); Myanmar 233 km (145 mi)

Coastline: 574 km (357 mi)

Longest Distances: 767 km (477 mi) SSE-NNW 429 km (267 mi) ENE-WSW

Highest Point: Reng Mountain 957 m (3,141 ft)

Lowest Point: Sea level

Land Use: 67% arable; 2% permanent crops; 4% meadows and pastures; 16% forest and woodland; 11% other

The distinguishing characteristic of both the geography and the population of Bangladesh is uniformity. Although the area of land—about the size of Wisconsin—is essentially fixed, the population, already ninth largest and believed to be densest of the countries of the world, was continuing to grow at the high rate of about 3 percent annually.

The country is bounded on the west, north, and east by a long land border with India, continued in the southeast by a short land and water border with Myanmar. On the south is a highly irregular deltaic coastline fissured by many rivers and streams flowing into the Bay of Bengal.

Except for the Chittagong Hill Tracts in the southeast and the low hills of Sylhet District in the northeast, the country is a flat alluvial plain. It is part of the Plain of Bengal at the eastern end of the great Indo-Gangetic Plain, running across the northern part of the South Asian subcontinent. Broadly speaking, the country is almost entirely a delta of deposited soils; the older alluvium is in the north. Most of these soils are highly fertile.

The dominant feature of the flat plain is the profusion of rivers, flowing generally north to south. Chief among these, and lying like a fan on the face of the land, are the Ganges-Padma, the Brahmaputra-Jamuna, the Meghna, and the river junction stem and estuary on the Bay of Bengal. These rivers and their innumerable tributaries and distributaries drain a vast area of the subcontinent and the Himalaya Mountains on its northern rim.

The physical geography of Bangladesh may be considered in terms of two principal divisions: the rivers and alluvial plain comprising most of the country and the much smaller area of Chittagong District and the Chittagong Hill Tracts on the extreme southeast. Countrywide, the most significant feature of the landscape is the extensive network of large and small rivers that are of primary importance in the economic and social life of the nation. This system, ever-changing in detail, provides drainage, determines the kind and extent of agricultural production, furnishes vast supplies of fish, forms an unparalleled grid of inland waterways for cheap and convenient transportation, and continually fertilizes much of the country by carrying away old soil and bringing in new.

This great river system, although it makes road, bridge, and airport construction difficult, is probably the country's principal physical resource; at the same time it is the greatest hazard. Seasonal flooding often brings widespread loss of life, crops, and property. Not only is Bangladesh a country of heavy rainfall, but its rivers also carry down to the Bay of Bengal runoff waters from the snows in the Himalaya Mountains far to the north. Cyclones from the bay occasionally intensify seasonal flooding to the point of major national disaster, as in November 1970. In 1974 winds and flooding contributed materially to the critical food shortage of that year.

The Bangladesh Plain

The alluvial plain of Bangladesh constitutes about 80 percent of the greater Plain of Bengal—also called the Lower Gangetic Plain, because it is the terminus of the 2,414-km-long (1,500-mi-long) Indo-Gangetic Plain. It lies between the Indian foothills of the Himalaya Mountains on the north and the Bay of Bengal on the south; this plain is often regarded as deltaic in its entirety, making it the largest delta in the world. By common usage this is an acceptable definition, but in a stricter geographical sense the true delta is smaller and harder to define. Generally, it lies south of 24°N latitude—that is, to include the junction of the Padma and Jamuna rivers, the Dhaka area between the Padma and the Meghna rivers, and the rest of the plain to the south. The country to the north is often called the paradelta.

The land characteristics of the Bangladesh Plain from north to south have sometimes been summed up by geographers as "old mud, new mud, and marsh." To this succinct, if oversimplified, description should be added the word flat. Most elevations are less than 9 m (30 ft) above sea level, although altitudes up to 106 m (350 ft) occur in the northern part of the plain and some even higher elevations are found along the eastern borders with the Indian states of Meghalaya and Tripura. General elevation decreases from north to south; below the Padma and Meghna rivers the terrain is essentially sea level.

The River System and Relief in the Plain

If the level of the plain were thirty feet lower, most of it would be under water. The essential importance of the river system, not simply for economic and communications reasons but for the very existence of the country, is illustrated by separate estimates that this same amount of water—thirty feet covering the plain—is actually the amount carried off annually by the rivers. Bangladesh is a country that exists by virtue of a rough balance between water input and output.

The river network of the plain moves generally from north to south but includes untold numbers of feeder and effluent streams, like capillaries, flowing east, west, southeast, and southwest in such profusion as to be virtually unique. The medium and smaller streams are by no means all tributary to the larger rivers. Great numbers of them especially in the true delta, are distributaries from the larger rivers.

Accurate, detailed description of this riverine labyrinth with its associated proliferation of lakes (*bhils* or *haors*) and marshes is made practically impossible not only by its magnitude but because many rivers, large and small, modify or change their channels from time to time. Some streams "die," and their courses dwindle to low capacity or none at all; others are "born," and the process is continuous. Description is further complicated by the fact that rivers often have different names along different segments, taking up new names after merger with other rivers or according to different usages from one area to another.

This highly complex river system is based upon three major rivers: the Ganges-Padma, the Brahmaputra-Jamuna, and the Meghna. The Meghna is deepest and swiftest; the other two, by contrast are broader and much longer. The outline of these rivers is in diagram roughly like a fan lying on the plain, the line of the Ganges-Padma running to the southeast, the Brahmaputra-Jamuna running south down the middle of the plain, and the Meghna coming from the northeast to the southeast. The flared handle of the fan is their junction stem and the estuary channels on the Bay of Bengal. The strategic and developmental significance of this river system is illustrated by the fact that Dhaka, the capital and principal city, lies centrally within it, a few miles north of the Padma-Meghna junction and inside the apex angle formed by it.

PRINCIPAL TOWNS (population at 1981 census)

Dhaka (capital)	3,430,312*	Barisal	172,905
Chittagong	1,391,877	Sylhet	168,371
Khulna	646,359	Rangpur	153,174
Rajshahi	253,740	Jessore	148,927
Comilla	184,132	Saidpur	126,608

*Including Narayanganj (population 270,680 in 1974).

The Ganges-Padma

The Ganges (also called the Ganga) River of India, flowing southeastward, comes to the boundary of Bangladesh in the northwest of the country. A distributary, the Bhagirathi-Hooghly River of India, then turns south. Formerly this was the main channel of the Ganges. By the late eighteenth century, however, if not earlier, the upper Bhagirathi channel had become heavily silted, and the Ganges took up its main channel to the southeast. Oskar H. K. Spate, a prominent geographer, has observed that "the cardinal factor in the later history of the Delta has been the eastward shift of the Ganga waters." For about 145 km (90 mi) the Ganges is the boundary between India and Bangladesh; it then continues to the southeast across the alluvial plain. In Bangladesh the Ganges—particularly in its lower course—is more commonly called the Padma.

East of the boundary with India and south of the Padma, elevations do not exceed twenty-five feet; most are lower. The hundreds of rivers and streams in this true delta segment of the plain are virtually all distributary from the Padma and flow south. Principal among them are the Madhumati River, which leaves the Padma west of its junction with the Jamuna, and the Arial Khan (also known as the Bhubanswar), which exists to the east of that point. This network dissects the lower delta in a multiplicity of channels entering the Bay of Bengal through a crumbled seacoast. These channels have names but are sometimes collectively referred to as "the many mouths of the Ganges." The delta land and islands along the coast from the Indian border east to the Padma-Meghna estuary and extending five to twenty-five miles inland are called the Sundarbans. This is a forested, tidal-flushed, salt-marsh region, much of it so shifting, low, and swampy as to preclude permanent habitation.

Area and population

Divisions	Administrative centres	area sq mi	area sq km	population 1985 estimate
Chittagong	Chittagong	17,535	45,415	26,062,000
Dhākā	Dhākā	11,881	30,772	29,043,000
Khulna	Khulna	12,963	33,574	19,792,000
Rājshāhi	Rājshāhi	13,219	34,237	24,383,000
TOTAL		55,598	143,998	100,468,000

The western part of the true delta, between the Indian border and the Madhumati River, is sometimes called the old delta, since it contains the beds of many dead or dying rivers. Here drainage projects have been undertaken in an effort to create more croplands. The new delta, east of the Madhumati, is somewhat lower than the old and is being slowly built up by silt and mud from the younger, swift-flowing streams and seasonal floods, but in forms subject to frequent change by deposit and erosion.

The Brahmaputra-Jamuna

Like the Indus River of Pakistan and the Ganges of India, the Brahmaputra River rises in the high Himalayas. It flows east across the southern Tibetan region of the People's Republic of China (PRC), turns abruptly south and then west, and rushes through the Assam Valley of India roughly opposite to its former direction across Tibet. Upon reaching the northern border of Bangladesh, the river turns due south and enters the country in multiple, narrowly separated channels. Within the next twenty-five miles the river receives two principal tributaries: first the Dharla River and then the larger Tista River, both coming from the northwest. Below the Tista junction the Brahmaputra becomes known as the Jamuna (not to be confused with a river of a similar name, a tributary of the Ganges, in India). The wide Jamuna, in often-shifting subchannels, flows south to its junction with the Padma about 72 km (45 mi) west of Dacca. Shortly before this junction the Jamuna receives from the northwest two large tributaries: the Karatoya River and then the Atrai River.

The northwest segment of the Bangladesh Plain, lying between the Padma and Jamuna rivers, is the internal governmental division of Rajshahi. This has been called by geographers the paradelta; it is an extensive plain falling from about 91 to 106 m (300 to 350 ft) general elevation in the north to about 30 m (100 ft) centrally and down to about 9 m (30 ft) in the south. It is cut by many old river courses as well as by newer, active rivers and, like

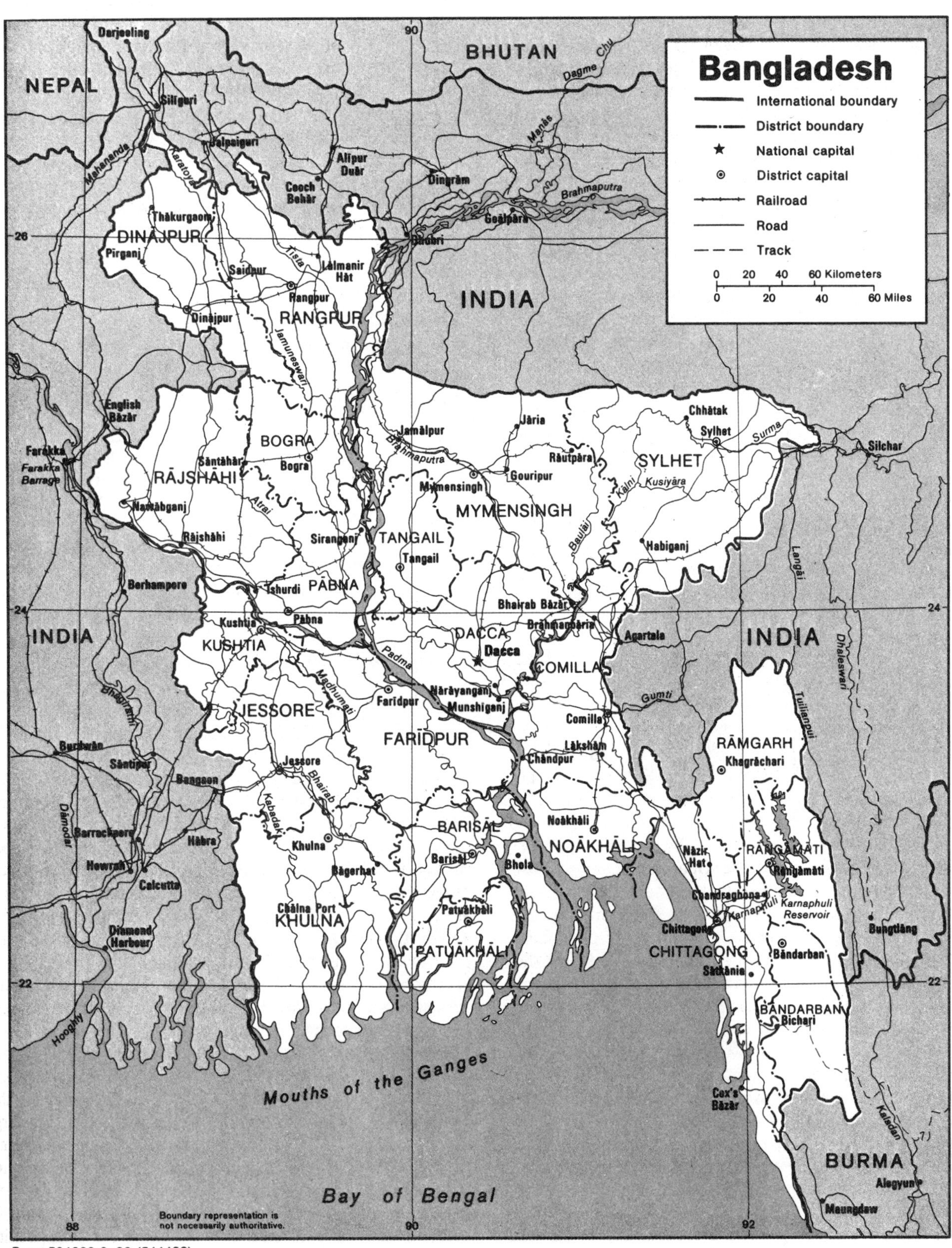

Base 504222 3-80 (544488)

the rest of the country, is subject to disastrous flooding. The central two-thirds of this area is called the Barind, a large land island of older, firm alluvial soils marked by ravines and low ridges. Earthquakes are not a major hazard in Bangladesh, but some have occurred in the Barind, and the detection of seismic vibrations is most common here. After their junction, the combined waters of the Jamuna and Padma continue southeast for about 96 km (60 mi) to the even wider junction with the third of the great rivers, the Meghna, flowing from the northeast.

The Meghna

In the northeast corner of Bangladesh, about 64 km (40 mi) east of the town of Sylhet, two branches of the Barak River enter the country from India. The larger, the Surma, moves west and then south; the second, the Kusiyara, turns southwest and then west. These rivers, with smaller tributaries, form the Kalni River about 56 km (35 mi) southwest of Sylhet. After a short run to the southwest, the Kalni becomes the upper Meghna and is soon reinforced by a major tributary from the north, the Baulai. From this point the Meghna continues southwest in a twisting, multichanneled course to the junction with the lower Padma about sixty-five air miles away. Until the early 1800s the Meghna's main channel did not join the Padma but ran separately east of the Padma down to the Bay of Bengal.

In its lower course, before joining the Padma, the Meghna receives two distributaries from the Jamuna: the Old Brahmaputra River and its branches and the Dhaleswari (sometimes called the Burhi Ganga). Like those of the Ganges-Padma and the Meghna, the main channel of the Brahmaputra-Jamuna changed in the late eighteenth or early nineteenth century. In northern Bangladesh, about 56 km (35 mi) down the Jamuna from the border, the Old Brahmaputra exits to the southeast, approximately along the old channel of the Brahmaputra. West of the town of Bajitpur the river divides into two branches. The first of these, the Lakhya, receives the Banar River from the northwest and then continues south to join the Dhaleswari at Narayanganj. The second and smaller branch moves southeast and is caught by the Meghna at the town of Kalipur.

Between the Banar River and the Jamuna is another slightly elevated tract—somewhat like the Barind—of firm soil, with elevations of 15 to 30 m (50 to 100 ft) above sea level. This area, south of Mymensingh and north of Dhaka, is called the Madhupur Jungle. Some geographers have suggested that it should be called the Madhupur Tract; its vegetation is not of true jungle character, and much of it has been cleared for cultivation. Below the Madhupur Tract, including Dhaka at an elevation of about 6 m (20 ft), the country along both banks of the lower Meghna is much like the new delta to the south of the Padma.

The upper drainage area of the Meghna lies to the east of the Old Brahmaputra. This is an extensive flatland with a few low hills north and east of Sylhet that exceed 91 m (300 ft) in elevation. South of the Kusiyara River six low hill ranges project northward into Sylhet District from the Tripura Hills of India. In these ranges the maximum elevation, near the Bangladesh-Indian border, is about 335 m (1,100 ft). Generally, in this northeast corner of Bangladesh, enclosed on the north and east by the hills of eastern India, plain elevations decline southward from about 150 ft to about thirty ft. Around Sylhet the country is known as the Sylhet Plain, or the Surma River Plain. Here, and throughout the upper Meghna-Surma drainage area, the most distinguishing feature is the profusion of large and small lakes, abounding in fish. The whole northeast quadrant of the Bangladesh Plain is even more vulnerable to flooding than the other parts of the country. Spate has noted that "the Meghna-Surma embayment is perhaps the most amphibious part of Bengal during the rains."

Far to the south of the Brahmaputra, the Dhaleswari River, the second principal distributary of the Jamuna, leaves the parent river above its junction with the Padma. It flows southeast, below Dhaka and roughly parallel to the lower Padma, receives the Lakhya at Narayanganj, and then joins the Meghna a few miles above its junction with the Padma. Thus the branches of the Jamuna also illustrate the general drainage of the plain. Those branches on the west bank are inputs; those on the east bank are effluents, eventually caught by the Meghna.

The Padma-Meghna Stem and Estuary

From the wide Padma-Meghna junction, the combined rivers move south in an S-shaped stem channel for about 64 km (40 mi) and then spread out into the Bay of Bengal through one of the largest estuaries in the world. This triangular estuary has a seaward base, from the shores of Khulna northeast across to Chittagong, of more than 160 km (100 mi) long. In this huge triangle are a number of permanent islands, including many that surface only at low tide and many that are temporary or that shift in outline from time to time. Many of these are *chars*, land forms built up by silting that may become permanent or erode. They are not limited to the Padma-Meghna estuary; *chars* occur in many places along the larger rivers and have frequently been the subject of dispute as to ownership.

In the estuary the largest of the permanent islands, from west to east, are Shahbazpur, North Hatia, South Hatia, and Sandwip. Separating them are the estuary channels of the Padma-Meghna. Some have several names. Principally, however, from west to east, these channels are the Tetulia, the Shahbazpur, the Hatia, and the Bamni-Sandwip.

The Meghna experiences tidal currents more strongly and further inland than any other South Asian river. The tides from the Bay of Bengal are effectively felt on the Meghna up to Kalipur (also called Bhairab Bazaar) and are noticeable for as much as 48 to 64 km (30 to 40 mi) farther upriver. In the stem of the Padma-Meghna and in the estuary, the regular rise of the tide is from ten to eighteen feet. Tidal waves called bores recur from time to time in the estuary and the stem rivers. Bores tend to form high waves with abrupt fronts when the incoming surge of water at flood tide encounters a resistance, such as a sandbar or a defile. The funnel shape of the Padma-Meghna estuary and the many channels between the islands are highly favorable to bore formation. A typical bore rushes in with thunderous noise as a wall of water that may be twenty feet or more in height. Velocity and height are magnified if the bore is backed by strong winds from the south. Regularly, bores impede or prevent navigation at the time of the equinoxes. When they are backed by winds of hurricane force, disastrous flooding results.

The three main rivers of the Bangladesh Plain drain a total area of about 1,553,993 sq km (600,000 sq mi); within Bangladesh, about 116,549 sq km (45,000 sq mi). The annual sediment load carried down to the estuary has been estimated at 2.4 billion tons; and the annual peak flood in the Padma-Meghna stem discharges water at about 5 million cu ft per second—twice the rate of the Mississippi River.

Chittagong and the Chittagong Hill Tracts

The districts of Chittagong, Ramgarh, Rangamati, and Bandarban, together constituting roughly one-sixth of the country, provide some geographical variation from the plain and the only significant hill system. Administratively the Chittagong Division also includes the districts of Noakhali, Comilla, and Sylhet, curving northward along the eastern boundary with India, but these three are part of the Bangladesh Plain and are mostly in the Meghna River and estuary stem drainage system. Chittagong District proper, along the Bay of Bengal and southeast of the estuary, is only in small part deltaic. East of it the Ramgarh, Rangamati, and Bandarban districts lie on the border with Myanmar and India.

On the western fringe of the north-south mountain ranges of Myanmar and eastern India that meet the Himalayan extremities far to the north, the Chittagong Hills are a series of narrow, roughly parallel, forested hill chains. These hills rise steeply to narrow ridge lines, generally no wider than 36 m (120 ft) and 609 to 914 m (2,000 to 3,000 ft) above sea level. Individual ridge lines, except at their northern and southern extremities, do not vary greatly in height within themselves; those closer to the 320 km-(200-mi) long eastern border with Burma and India are higher. The greatest elevation in Bangladesh is 1229 m (4,034 ft) at Keokradong, in the southeast extremity of the Chittagong Hill Tracts. Between the generally north-south hill lines lie fertile valleys devoted almost entirely to rice cultivation.

West of the Chittagong Hill Tracts, in Chittagong District, is a broad plain, cut by rivers draining to the Bay of Bengal, that rises to a final chain of low coastal hills with a maximum spot elevation of 351 m (1,152 ft) but mostly below 213 m (700 ft). West of these hills is a narrow, wet coastal plain, narrowest to the north of Chittagong city and to the south of Cox's Bazaar, where the low coastal hills in some stretches form precipitous sea cliffs. The coastal strip is widest and lowest at the ports of Chittagong and Cox's Bazaar and to the east of both of them.

Chittagong and the Chittagong Hill Tracts are high rainfall areas and have numerous rivers but not in the proliferation of the river system of the plain. None is of the size or importance of the three great rivers of the Bangladesh Plain or their major tributaries and distributaries. The Karnaphuli, the most important river of the area, originates in the east-central Chittagong Hill Tracts from inputs coming from both north and south and proceeds west and southwest to the Bay of Bengal at its inlet. On the Karnaphuli, some thirty air miles inland from Chittagong, the Karnaphuli Dam and the associated power plant are installations of national importance. The dam impounds the river's input waters in a vast reservoir lying up two of the longitudinal valleys to the north of the dam.

BARBADOS

BASIC FACTS

Official Name: Barbados

Abbreviation: BAR

Area: 430 sq km (166 sq mi)

Area—World Rank: 180

Population: 256,784 (1988) 263,000 (2000)

Population—World Rank: 155

Capital: Bridgetown

Land Boundaries: Nil

Coastline: 101 km (63 mi)

Longest Distances: 34 km (21 mi) N-S 23 km (14 mi) E-W

Highest Point: Mount Hillaby 340 m (1,115 ft)

Lowest Point: Sea level

Land Use: 77% arable; 9% meadows and pastures; 14% other

Barbados is the most easterly of the Caribbean Islands, lying east of the Windward Islands about 322 km (200 mi) NNE of Trinidad and 161 km (100 mi) ESE of St. Lucia. It is the second smallest country in the Western Hemisphere.

From the south and west the island presents a flat appearance rising in a series of ridges up to about 300 m (1,000 ft) and then falling steeply toward the sea. The highest point is Mount Hillaby (336 m, 1,105 ft) near the center of the island. The coast is encircled with coral reefs.

There are no rivers in the conventional sense but only gullies, water courses and underground channels. The best known of the underground channels is Cole's Cave in the middle of the island. Two rivulets known as Indian River and Joes River are of no use for either fishing or navigation.

Area and population	area		population
			1980
Parishes	sq mi	sq km	census
Christ Church	22	57	40,790
St. Andrew	14	36	6,731
St. George	17	44	17,361
St. James	12	31	17,255
St. John	13	34	10,330
St. Joseph	10	26	7,211
St. Lucy	14	36	9,264
St. Michael	15	39	99,953
St. Peter	13	34	10,717
St. Phillip	23	60	18,662
St. Thomas	13	34	10,709
TOTAL	166	430	248,983

BELGIUM

BASIC FACTS

Official Name: Kingdom of Belgium

Abbreviation: BEL

Area: 30,518 sq km (11,783 sq mi)

Area—World Rank: 124

Population: 9,880,522 (1988) 9,875,000 (2000)

Population—World Rank: 66

Capital: Brussels

Boundaries: 1,446 km (899 mi); Netherlands 450 km (280 mi); West Germany 162 km (101 mi); Luxembourg 148 km (92 mi); France 620 km (385 mi)

Coastline: 66 km (41 mi)

Longest Distances: 278 km (173 mi) SE-NW 181 km (112 mi) NE-SW

Highest Point: Botrange 694 m (2,277 ft)

Lowest Point: Sea level

Land Use: 24% arable; 1% permanent crops; 20% meadows and pastures; 21% forest and woodland; 34% other

Belgium is located in northwestern Europe. It has no natural frontiers. The Belgian Lorraine is a continuation of the French Lorraine and Luxembourg, the high plateau is a continuation of West Germany's Eifel uplands, and the Kempenland continues into the Netherlands, which also shares the coastal polders and deltas of the Schelde and Rhine rivers. The present-day borders are more or less those of the 19th-century Austrian Netherlands, the bishopric of Liège and the duchies of Brabant and Luxembourg. The border with the Netherlands dates from the Peace of Westphalia of 1648, the border with France from the Treaty of Utrecht of 1713, and the border with Luxembourg from that country's independence in 1857. The German-speaking cantons on the eastern border with West Germany were acquired in 1919 as war reparation included in the Treaty of Versailles.

Almost one-fifth of the country was reclaimed from the North Sea between the eighth and the 13th centuries. Salt marshes became rich plowland behind a lengendary barrier of dikes for which Dante himself expressed admiration in the *Inferno*. A coastal strip of a depth of 48 km (30 mi) was thus added to the country; at the same time rivers like the Schelde, which had spread out in broad, shallow deltas, were made navigable.

Belgium can be divided into three topographic zones: the northern lowlands, the central low plateaus and the southern hilly region. It is drained by two rivers, the Meuse and the Schelde, both of which rise in France and flow into the sea through the Netherlands. Most Belgian waterways drain toward the northeast, although a few streams in the eastern Ardennes flow to the Rhine, and the IJzer River cuts through the dunes in West Vlaanderen Province. All waterways are linked through a system of canals to the port of Antwerp.

The western part of the northern lowlands is divided into maritime Flanders (the coastal fringe of beaches, dunes and the belt of polders) and interior Flanders (the gently rising terrain of West Vlaanderen, Oost-Vlaanderen, northern Hainaut and northern Brabant provinces). The coastline is nearly straight, and the beach is white; practically free of pebbles; and stabilized by fences called groins, which reach from the higher beach into the water. Behind the beach lie the dunes and behind them the polders. Access of coastal cities to the sea has been affected by constant silting. Both Ghent and Brugge, once coastal cities, are now reached by canals.

The eastern part of the northern lowlands, called Kempenland, is bounded by the Meuse, Schelde, Demer, Rupel and Dijle rivers. It consists of sparsely populated, barren heathlands. Industry has developed near the Kempen coalfield because of its proximity to coal, cheap land, the Albert Canal and the port of Antwerp.

PRINCIPAL TOWNS (population at 31 December 1986)

Bruxelles (Brussel, Brussels)	973,499*
Antwerpen (Anvers, Antwerp)	479,748†
Gent (Gand, Ghent)	233,856
Charleroi	209,395
Liège (Luik)	200,891
Brugge (Bruges)	117,755
Namur (Namen)	102,670
Mons (Bergen)	89,693
Kortrijk (Courtrai)	76,216
Mechelen (Malines)	75,808
Oostende (Ostend)	68,318
Hasselt	65,563

*Including Schaerbeek, Anderlecht and other suburbs.
†Including Deurne and other suburbs.

The gently undulating central low plateaus include northern and southern areas. The northern section covers southern Brabant and Hainaut provinces; the Plateau of Hesbaye; the mixed region that follows the Demer River

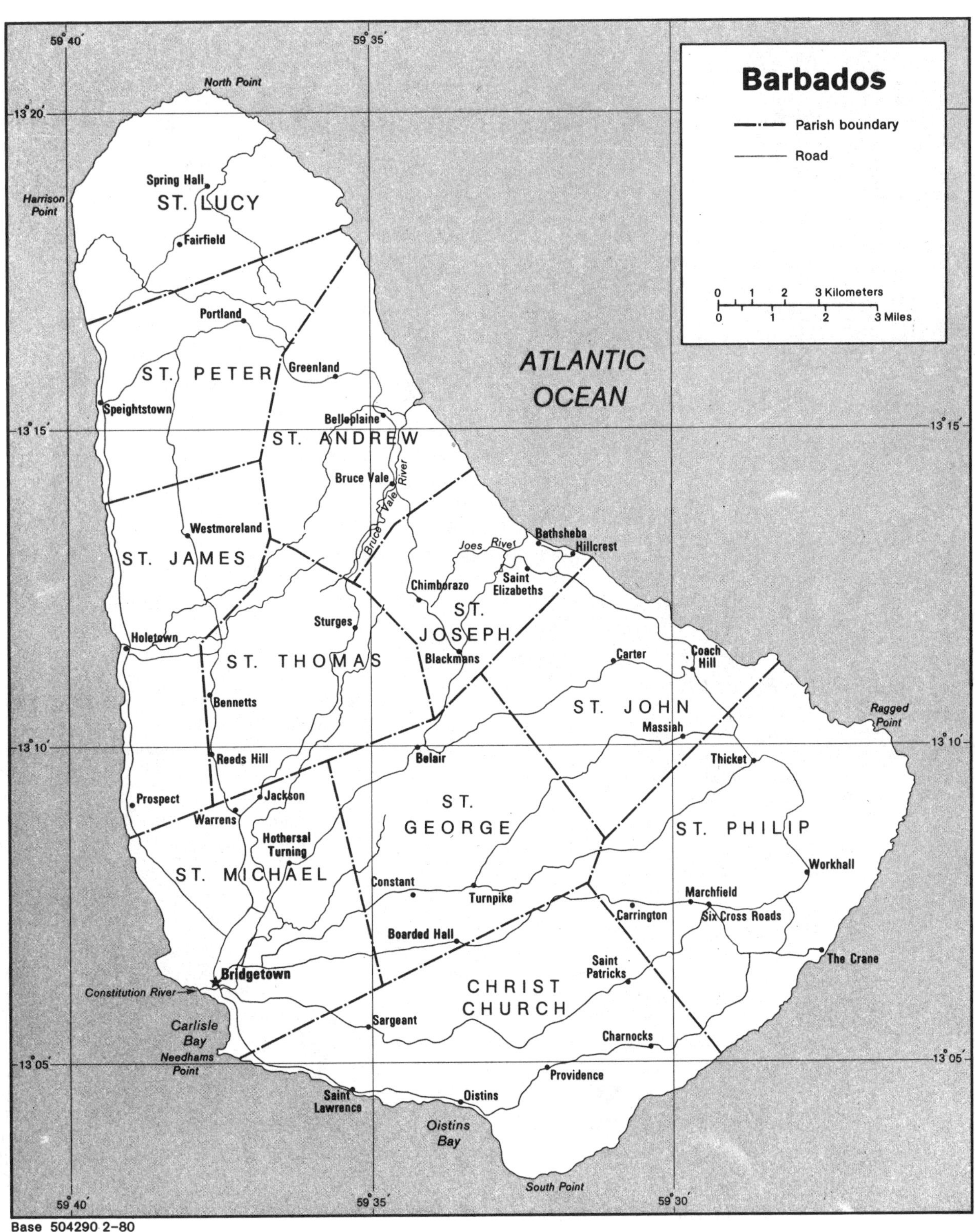

Base 504290 2-80

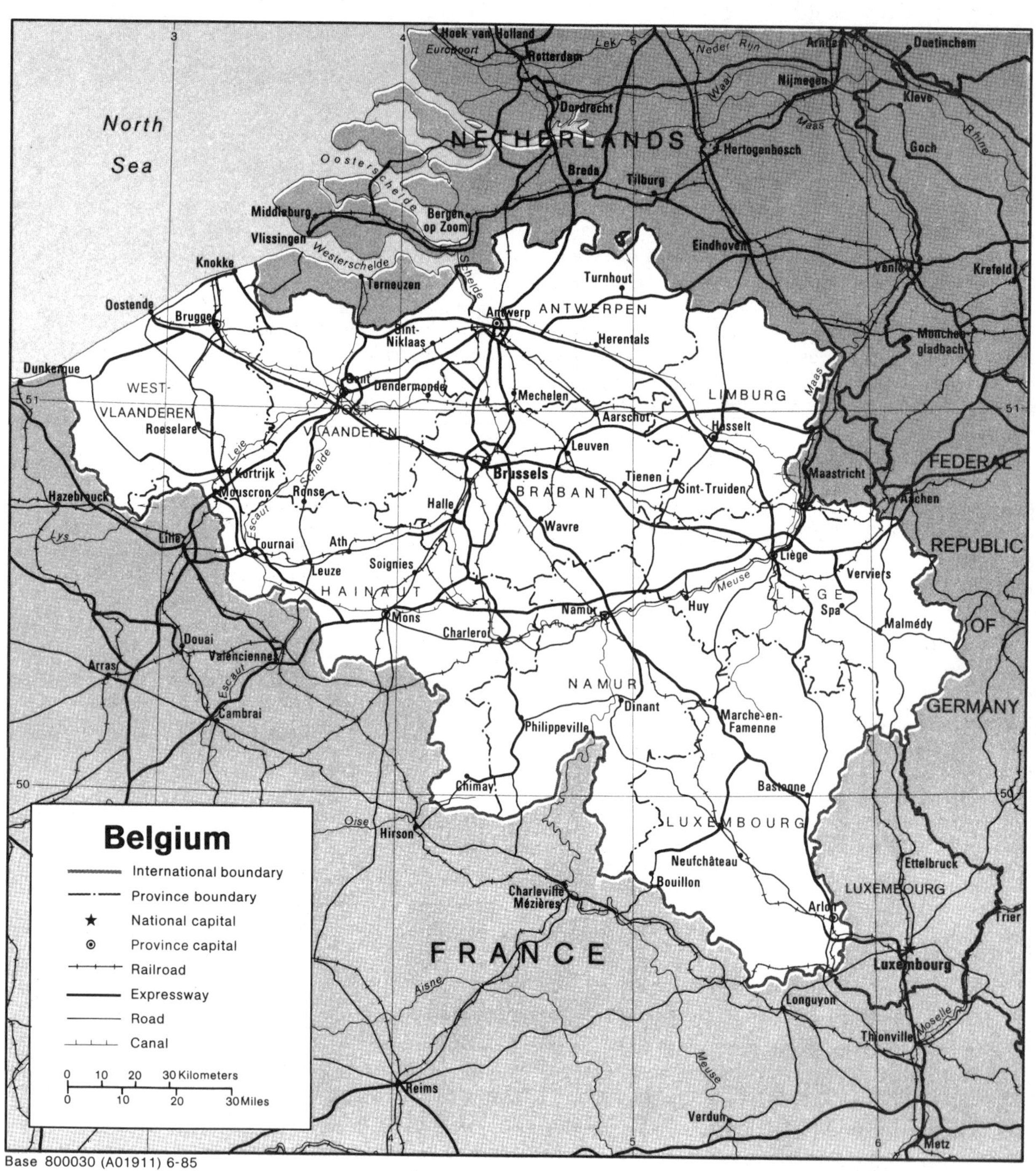

Base 800030 (A01911) 6-85

Valley; and the Pays de Herve, east of the city of Liège. The region includes Brussels. Midway through the central low plateaus runs the Sambre-Meuse Valley, which divides the northern from the southern section. The valley contains the industrial cities of Mons, Namur, Liège and Charleroi. The southern plateau, including the Entre-Sambre-et-Meuse region, is slightly higher than the northern one and has more forests and pastureland. To the east of this region lies the broken hill country of Condroz Plateau, which sometimes is considered part of the Ardennes. Its northern edge is defined by the Famenne Depression.

The thinly populated southern hilly region consists of the high plateau of the Ardennes and the Belgian Lorraine. The highest points of the region are in the Hautes Fagnes near the German border. Although an important dairying area, one-half of the Ardennes still is forested, as are the scarplands of the Belgian Lorraine at the southern end of the country.

As in other developed countries, water supply is being overburdened by excessive use. Belgium has an annual supply of some 800 million cu m (612 million cu ft) of groundwater, but three-quarters of the supply originates from the southern half of the country. About 80% of the water is being consumed, and the amount is steadily rising. Water pumped from underground sources accounts for three-quarters of total supplies; the remainder is surface water from rivers, canals, basins and barrages. The Meuse is by far the most important source of surface water, along with its tributaries the Semois, the Sambre and the Ourthe.

Area and population

Provinces	Capitals	area sq mi	area sq km	population 1986 estimate
Antwerp	Antwerp	1,107	2,867	1,582,786
Brabant	Brussels	1,297	3,358	2,218,349
East Flanders	Ghent	1,151	2,982	1,328,805
Hainaut	Mons	1,462	3,787	1,277,939
Liège	Liège	1,491	3,862	991,535
Limburg	Hasselt	935	2,422	731,875
Luxembourg	Arlon	1,715	4,441	224,988
Namur	Namur	1,415	3,665	412,231
West Flanders	Brugge	1,210	3,134	1,090,387
TOTAL		11,783	30,518	9,858,895

BELIZE

BASIC FACTS

Official Name: Belize

Abbreviation: BEZ

Area: 22,965 sq km (8,867 sq mi)

Area—World Rank: 133

Population: 171,735 (1988) 249,000 (2000)

Population—World Rank: 163

Capital: Belmopan

Boundaries: 995 km (618 mi); Mexico 251 km (156 mi); Guatemala 269 km (167 mi)

Coastline: 475 km (295 mi)

Longest Distances: 288 km (179 mi) NNE-SSW 109 km (68 mi) WNW-ESE

Highest Point: Victoria Peak 1,122 m (3,680 ft)

Lowest Point: Sea level

Land Use: 2% arable; 2% meadows and pastures; 44% forest and woodland; 52% other

Belize is on the eastern coast of Central America, just below the Yucatan Peninsula. The waters immediately offshore are shallow; they are sheltered by the second largest barrier reef in the world, dotted with a large number of islands known as cays.

The most notable topographic feature is the Maya Mountains, rising to a height of 1,000 to 1,100 m (3,400 to 3,700 ft) running northeast to southwest across the central and southern parts of the country. The country north of Belize City is mostly level, interrupted only by the Manatee Hills. The coast is flat and swampy and indented by many lagoons. The country is drained by 17 rivers.

Principal Towns (estimated population at mid-1985): Belmopan (capital) 4,500; Belize City (former capital) 47,000; Corozal 10,000; Orange Walk 9,600; Dangriga (formerly Stann Creek) 7,700.

Area and population

Districts	Capitals	area sq mi	area sq km	population 1985 estimate
Belize	Belize City	1,624	4,206	54,500
Cayo	San Ignacio	2,061	5,338	27,400
Corozal	Corozal	718	1,860	28,000
Orange Walk	Orange Walk	1,829	4,737	26,600
Stann Creek	Dangriga	840	2,176	16,500
Toledo	Punta Gorda	1,795	4,649	13,400
TOTAL		8,867	22,965	166,400

BENIN

BASIC FACTS

Official Name: People's Republic of Benin

Abbreviation: BEN

Base 504826 (546942) 9-81

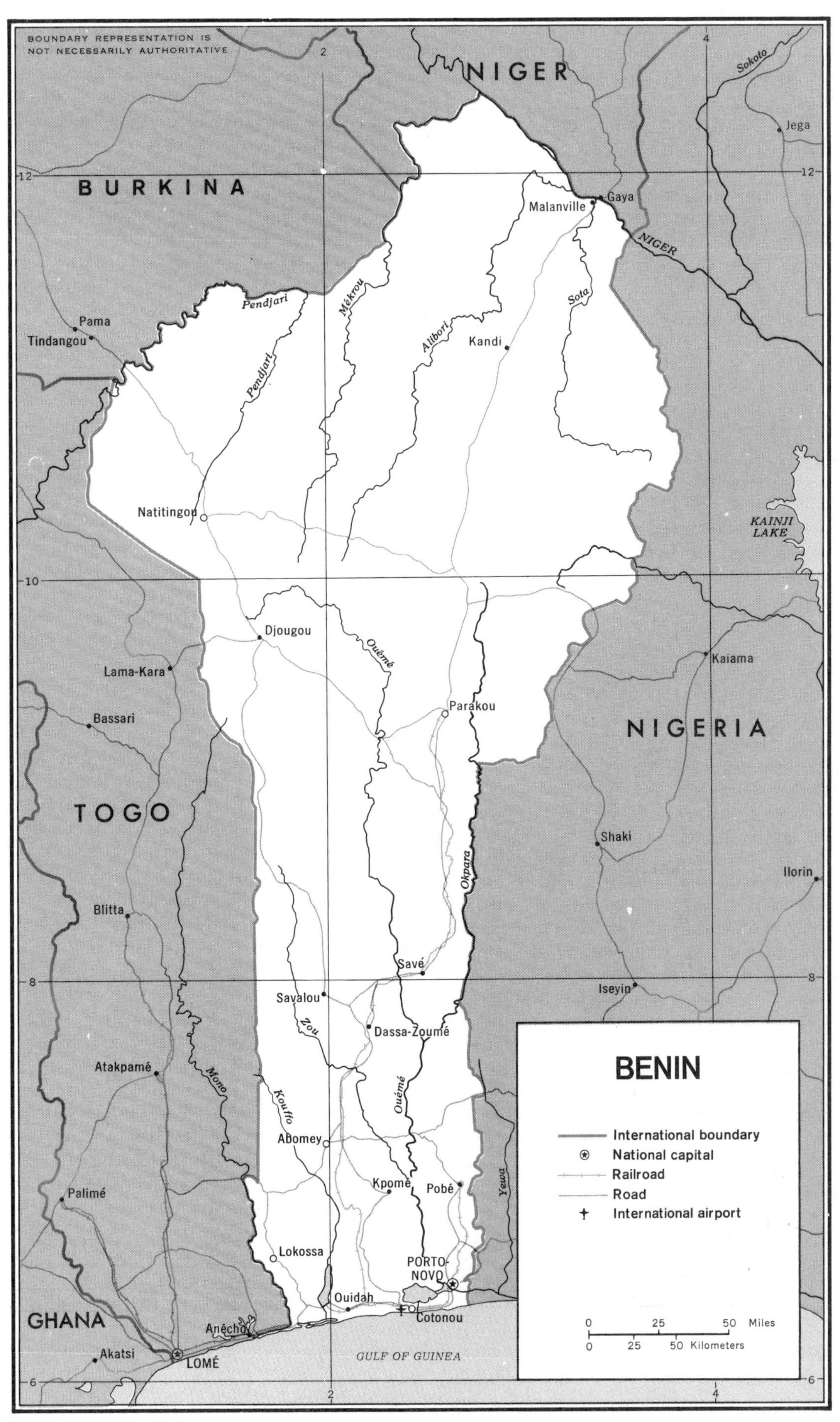
BOUNDARY REPRESENTATION IS NOT NECESSARILY AUTHORITATIVE
NIGER
BURKINA
NIGERIA
TOGO
GHANA
BENIN
International boundary
National capital
Railroad
Road
International airport
0 25 50 Miles
0 25 50 Kilometers
GULF OF GUINEA
KAINJI LAKE
Malanville
Gaya
Jega
Kandi
Pama
Tindangou
Natitingou
Djougou
Lama-Kara
Bassari
Parakou
Kaiama
Shaki
Ilorin
Blitta
Savé
Savalou
Iseyin
Dassa-Zoumé
Atakpamé
Abomey
Kpomé
Pobé
Palimé
Lokossa
PORTO-NOVO
Ouidah
Cotonou
Anécho
Akatsi
LOMÉ
Sokoto
Pendjari
Mékrou
Alibori
Sota
Ouémé
Okpara
Zou
Mono
Kouffo
Yewa
12
10
8
6
2
4

Area: 112,600 sq km (43,450 sq mi)

Area—World Rank: 96

Population: 4,497,150 (1988) 6,532,000 (2000)

Population—World Rank: 97

Capital: Porto Novo (official) Cotonou (de facto)

Boundaries: 1,955 km (1,215 mi); Niger 190 km (118 mi); Nigeria 750 km (466 mi); Togo 620 km (385 mi); Burkina Faso 270 km (168 mi)

Coastline: 125 km (78 mi)

Longest Distances: 665 km (413 mi) N-S 333 km (207 mi) E-W

Highest Point: 681 m (2,235 ft)

Lowest Point: Sea level

Land Use: 12% arable; 4% permanent crops; 4% meadows and pastures; 35% forest and woodland; 45% other

Benin is located in southern West Africa. Benin has four natural topographical regions. The first is the coastal belt. Behind this belt is a lagoon region with four lagoons (Cotonou or Nadoue, Ouidah, Grand Popo and Porto Novo) all of which join the sea at Grand Popo. Farther north is the terre de barre, a fertile clay plateau divided by a wide marshy depression, known as the Lama. The third region is composed of the Atakora Mountains in the northwest with an elevation of 654 m (2,146 ft). Lastly, there are the eastern plains, the Borgu and the plain of Kandi, sloping toward the Niger basin.

Benin's major rivers are the Niger, which forms the boundary with Niger, the Mono, which forms the border with Togo, the Couffo, which flows into Lake Aheme, and the Oueme, the longest of all. The Mono is navigable for 100 km (62 mi), the Niger for 89 km (55 mi) and the Oueme for 200 km (124 mi). Benin's northern rivers, the Mekrou, Alibory and Sota, and the Pandjari are subject to torrential floods.

PRINCIPAL TOWNS (estimated population at 1 July 1981) Cotonou 383,250; Porto-Novo (capital) 144,000.

Area and population

		area		population
Provinces	**Capitals**	sq mi	sq km	1985 estimate
Atacora	Natitingou	12,050	31,200	568,000
Atlantique	Cotonou	1,250	3,200	824,000
Borgou	Parakou	19,700	51,000	577,000
Mono	Lokossa	1,450	3,800	560,000
Quémé	Porto-Novo	1,800	4,700	738,000
Zou	Abomey	7,200	18,700	670,000
TOTAL		43,450	112,600	3,937,000

BERMUDA

BASIC FACTS

Official Name: Bermuda

Abbreviation: BER

Area: 54 sq km (21 sq mi)

Area—World Rank: 206

Population: 58,137 (1988) 64,000

Population—World Rank: 185

Capital: Hamilton

Land Boundaries: Nil

Coastline: 103 km (64 mi)

Longest Distances: 23 km (14 mi) long; 1.6 km (1 mi) average width

Highest Point: Town Hill 79 m (259 ft)

Lowest Point: Sea level

Land Use: 20% forest and woodland; 80% other

The Bermudas or Somers Islands are an isolated archipelago comprising about 150 islands, in the Atlantic Ocean, about 917 km (570 mi) off the coast of South Carolina. The islands are mostly flat and rocky, with luxuriant semitropical vegetation.

Area and population

	area		population
Municipalities	sq mi	sq km	1980 census
Hamilton	0.3	0.8	1,617
St. George	0.5	1.3	1,647
Parishes			
Devonshire	1.9	4.9	6,843
Hamilton	2.0	5.2	3,784
Paget	2.0	5.2	4,497
Pembroke	1.8	4.7	10,443
St. George's	1.7	4.4	2,940
Sandys	1.9	4.9	6,255
Smith's	1.9	4.9	4,463
Southampton	2.2	5.7	4,613
Warwick	2.2	5.7	6,948
TOTAL	21.0	54.0	54,050

BHUTAN

BASIC FACTS

Official Name: Kingdom of Bhutan

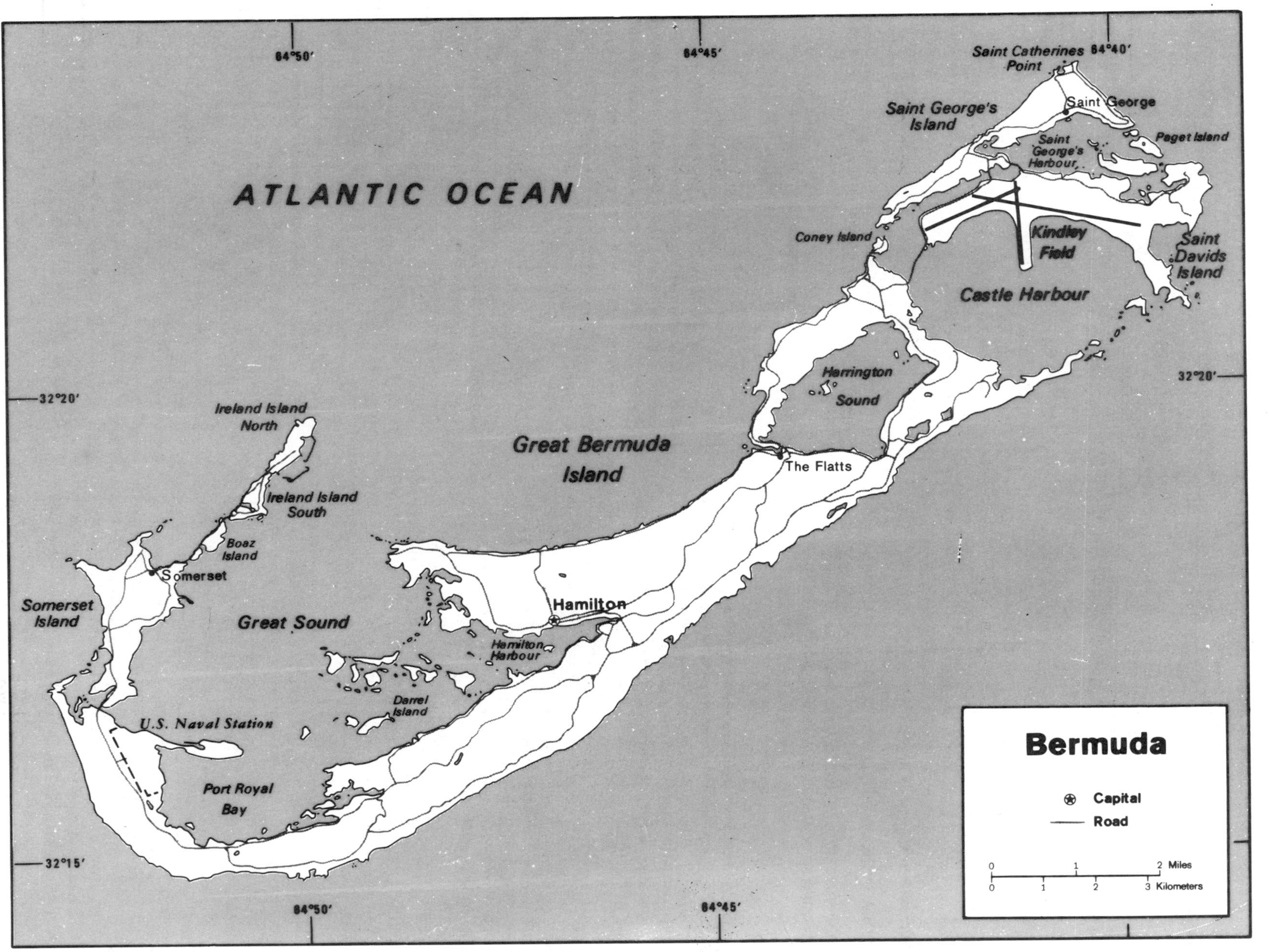
ATLANTIC OCEAN
Bermuda
Capital
Road
0 1 2 Miles
0 1 2 3 Kilometers
84°50′
84°45′
84°40′
32°20′
32°15′
Saint Catherines Point
Saint George
Saint George's Island
Saint George's Harbour
Paget Island
Coney Island
Kindley Field
Saint Davids Island
Castle Harbour
Harrington Sound
The Flatts
Great Bermuda Island
Hamilton
Hamilton Harbour
Ireland Island North
Ireland Island South
Boaz Island
Somerset
Somerset Island
Great Sound
Darrel Island
U.S. Naval Station
Port Royal Bay

Abbreviation: BHN

Area: 47,000 sq km (18,150 sq mi)

Area—World Rank: 118

Population: 1,503,180 (1988) 1,731,000 (2000)

Population—World Rank: 126

Capital: Thimphu (Paro Dzong is the administrative capital)

Land Boundaries: 1,019 km (633 mi); India 607 km (377 mi); China 412 km (256 mi)

Coastline: Nil

Longest Distances: 306 km (190 mi) E-W 145 km (90 mi) N-S

Highest Point: Kula Kangri 7,554 m (24,784 ft)

Lowest Point: Mans River Valley 97 m (318 ft)

Land Use: 2% arable; 5% meadows and pastures; 70% forest and woodland; 23% other

The boundaries of Bhutan cannot be identified by easily recognizable natural features. The indefinite common border with Tibet extends along the snow-capped and almost inaccessible crest of the main Himalayan Range, except for short distances in the northeast, northwest and west, where it lies south of the watershed.

The border with India was established by the British in the eighteenth and nineteenth centuries. The eastern section of this border, following spurs and valleys on the southern slopes of the main Himalayan Range, runs between Bhutan and the Indian territory formerly known as the North East Frontier Agency. On the south, the border with Assam and with West Bengal extends generally along the southern base of the abruptly rising Himalayan foothills. In some areas it includes the northern rim of the malarial Brahmaputra River lowlands, called the Duars Plain in this area.

Strategic Passes

The five passes through the Himalayas have determined the course of Bhutan's caravan routes with Tibet. Bhutanese chieftains formerly competed fiercely for control of these passes and the trade through them; the winners usually became rulers or powerful contenders for supreme authority. Since Bhutan stopped trading with Tibet in 1953 to impede the spread of Communist influence, the passes have lost their earlier significance. They now serve as escape routes for Tibetan refugees, and Bhutanese authorities regard them with concern as potential invasion routes for Chinese Communist forces.

With elevations ranging from approximately 4,572 m (15,000 ft) to more than 6,096 m (20,000 ft) the passes are negotiable only by pack animals or porters. The three most important appear to be those on routes leading from Paro Dzong in Bhutan across the northwestern frontier into the Chumbi Valley of Tibet. The southernmost of these leads to Yatung; the next, Tremo La pass, to Phari Dzong; and the northernmost, just beyond Lingshi Dzong, to Düna.

On the eastern section of the frontier, the pass north of Thunkar leads to Lhasa over a Tibetan motorable road reportedly completed by the Chinese Communists in 1965. The pass at Shingbe is on the shortest route from Tibet to Tashi Gang Dzong, the most important community in eastern Bhutan.

Other important passes include those within Bhutan which lead across the mountain spurs jutting southward from the main Himalayan Range. Tashi Gang Dzong in eastern Bhutan and Paro Dzong in the west are connected by the country's only lateral communication route, which must cross a series of valleys and ridges, averaging one for each 24 km (15 mi) of trail. The two most critical passes on this east-west axis route are within 40 km (25 mi) on either side of Tongsa Dzong, the most important community in central Bhutan; on the east is Rudong Pass, elevation 3,840 m (12,600 ft); on the west is Pele Pass, elevation 3,369 m (11,055 ft).

Area and population

Districts	Capitals	area sq mi	area sq km	population 1985 estimate
Bumthang	Jakar	1,150	2,990	23,900
Chirang	Damphu	310	800	108,800
Dagana	Dagana	540	1,400	28,400
Gasa	Gasa	2,000	5,180	16,900
Gaylegphug	Gaylegphug	1,020	2,640	111,300
Haa	Paro	830	2,140	16,700
Lhuntsi	Lhuntshi	1,120	2,910	39,600
Mongar	Mongar	710	1,830	73,200
Paro	Paro	580	1,500	45,600
Pema Gatsel	Pema Gatsel	150	380	37,100
Punakha	Punakha	330	860	16,700
Samchi	Samchi	830	2,140	172,100
Samdrup Jongkhar	Samdrup Jongkhar	900	2,340	73,100
Shemgang	Shemgang	980	2,540	44,500
Tashigang	Tashigang	1,640	4,260	177,700
Thimphu	Thimphu	630	1,620	58,700
Tongsa	Tongsa	570	1,470	26,000
Wangdi Phodrang	Wangdi Phodrang	1,160	3,000	47,200
TOTAL		18,150	47,000	1,285,300

Mountains

Bhutan is entirely mountainous, except for narrow strips of the Duars Plain which protrude across the southern borders into the Himalayan foothills at several places. Elevations vary from approximately 305 m (1,000 ft) in the south to almost 7,620 m (25,000 ft) in the north. The people are proud of their mountains, which are relatively unknown to the outside world, mainly because of the country's past policy of seclusion. Visitors are impressed with the scenic beauty of the snow-capped peaks and the variegated pattern of the rugged and deeply carved landscape.

Along the northern border there are four peaks with elevations above 6,096 m (20,000 ft). The highest is Gangri, elevation 7,540 m (24,740 ft), north of Tongsa Dzong; next in height is the country's most famous peak, pictur-

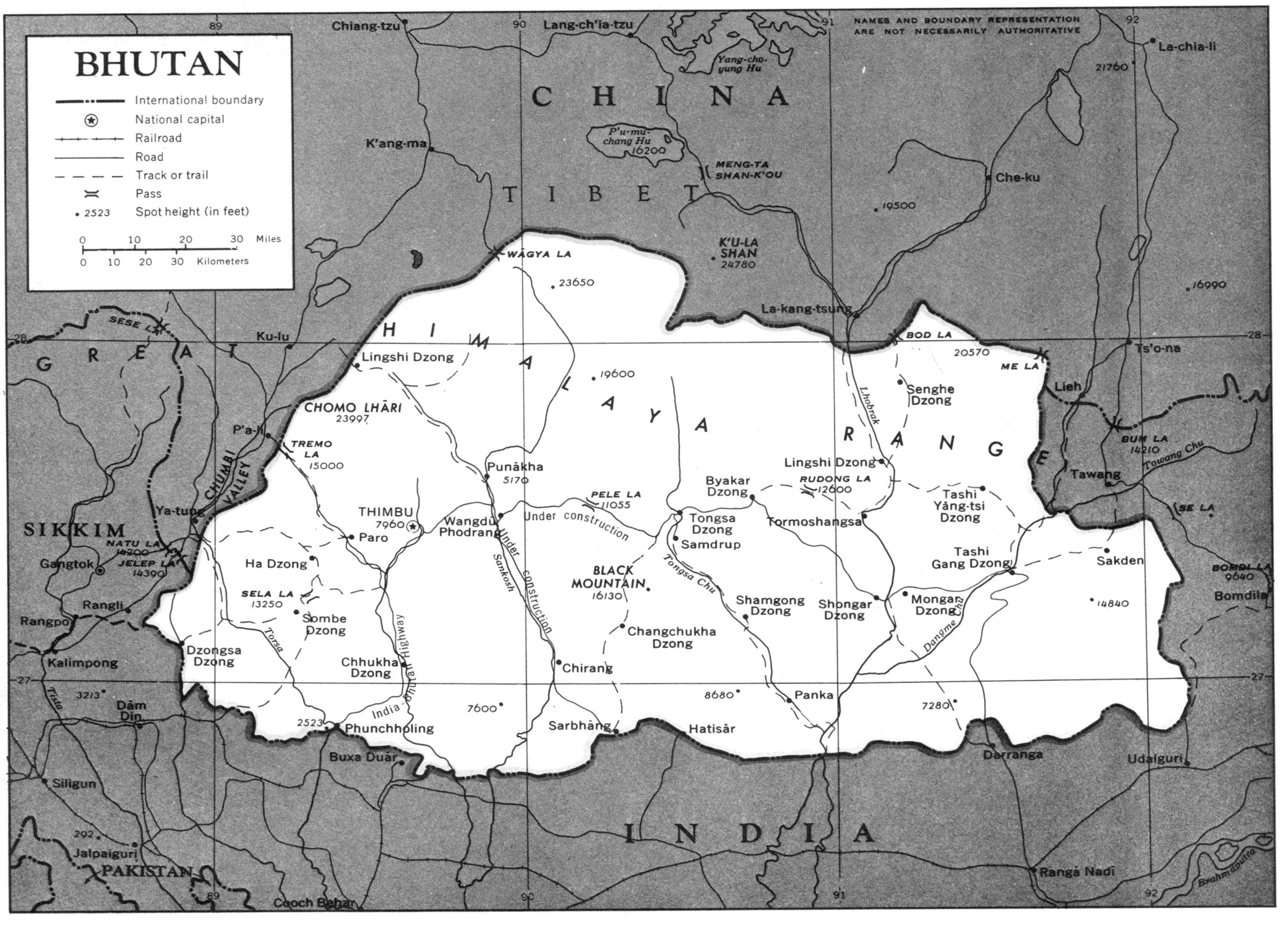

BHUTAN
International boundary
National capital
Railroad
Road
Track or trail
Pass
2523 Spot height (in feet)
0 10 20 30 Miles
0 10 20 30 Kilometers
NAMES AND BOUNDARY REPRESENTATION ARE NOT NECESSARILY AUTHORITATIVE
CHINA
TIBET
INDIA
SIKKIM
PAKISTAN
GREAT HIMALAYA RANGE
Chiang-tzu
Lang-ch'ia-tzu
Yang-cho-yung Hu
La-chia-li
21760
K'ang-ma
P'u-mu-chang Hu 16200
MENG-TA SHAN-K'OU
Che-ku
19500
K'U-LA SHAN 24780
16990
WĀGYA LA
23650
Lа-kang-tsung
SESE LA
Ku-lu
BOD LA
20570
ME LA
Ts'o-na
Lieh
BUM LA 14210
Towang Chu
Tawang
SE LA
BOMDI LA 9640
Bomdila
Lingshi Dzong
19600
Senghe Dzong
Lhobrak
CHOMO LHĀRI 23997
P'a-li
TREMO LA 15000
CHUMBI VALLEY
Punākha 5170
Lingshi Dzong
RUDONG LA 12600
Byakar Dzong
PELE LA 11055
Tashi Yāng-tsi Dzong
THIMBU 7960
Paro
Wangdu Phodrang
Under construction
Tongsa Dzong
Samdrup
Tormoshangsa
Tashi Gang Dzong
Sakden
Ya-tung
NATU LA 14200
JELEP LA 14390
Gangtok
Ha Dzong
SELA LA 13250
Sombe Dzong
Rangli
Rangpo
Sankosh
BLACK MOUNTAIN 16130
Tongsa Chu
Shamgong Dzong
Shongar Dzong
Mongar Dzong
14840
Dangme Chu
Kalimpong
Dzongsa Dzong
Torsa
Chhukha Dzong
Bhutan Highway
Changchukha Dzong
Chirang
3213
Dām Din
India
2523
Phunchholing
7600
Sarbhāng
8680
Hatisār
Panka
7280
Buxa Duār
Darranga
Udalguri
Siligun
Tista
292
Jalpaiguri
Cooch Behar
Rangā Nadī
Brahmaputra
89
90
91
92
28
27

52895 2-66

esque Chomo Lhari, elevation 7,314 m (23,997 ft), northwest of Punakha. The most prominent spur jutting southward into Bhutan is the Black Mountain Range, which separates the country into two almost equal parts, with the dividing line at Pele Pass.

Rivers

All Bhutan's numerous rivers flow generally southward through gorges and narrow valleys, eventually to drain into the Brahmaputra some 80 km (50 mi) south of the boundary with India. Except in the east and in the west, the headwaters of the streams are in the regions of permanent snow along the Tibetan border. None of the rivers is navigable, but many are potential sources of hydroelectric power.

The area east of the Black Mountain watershed is drained by the Tongsa Chu and its tributaries, the Bumtang and Dangme. West of the Black Mountain Range the drainage pattern changes to a series of parallel streams, beginning with the Sankosh River and its tributary, the Pho Chu. Farther west are the Paro Chu and Wong Chu which join to form the Raidak before it flows through the Sinchu La pass into India. Still farther west are the Torsa and Jaldhaka Rivers which rise in the Chumbi Valley and in Sikkim, respectively.

BOLIVIA

BASIC FACTS

Official Name: Republic of Bolivia

Abbreviation: BOL

Area: 1,098,581 sq km (424,164 sq mi)

Area—World Rank: 27

Population: 6,448,297 (1988); 9,837,000 (2000)

Population—World Rank: 81

Capital: La Paz

Land Boundaries: 6,532 km (4,029 mi); Brazil 3,125 km (1,942 mi); Paraguay 756 km (470 mi); Argentina 742 km (461 mi); Chile 861 km (535 mi); Peru 1,048 km (651 mi)

Coastline: Nil

Longest Distances: 1,529 km (950 mi) N-S 1,448 km (900 mi) E-W

Highest Point: Nevado Illimani 6,882 m (22,579 ft)

Lowest Point: Paraguay River Valley 100 m (325 ft)

Land Use: 3% arable; 25% meadows and pastures; 52% forests and woodland; 20% other

Bolivia is fifth in size among the countries of South America. Ringed by Peru, Brazil, Paraguay, Argentina, and Chile, it is entirely landlocked in the heart of the continent.

Structurally the country consists of a high plateau flanked on west and east by separate cordilleras of the Andes Mountains and by an extensive tropical lowland farther to the east. The ridges and valleys of the mineral-rich mountain systems separating the plateau from the lowlands form portions of the national territory so extensive that they constitute intermediate geographical regions.

Bolivia lies entirely within the tropics, but the extreme differences in elevation—as low as 91 m (300 ft) along the Brazilian border and more than 6,400 m (21,000 ft) on the highest mountain peaks—produce a great variety of climatic conditions. These, coupled with a wide diversity in soils, result in kinds of vegetation ranging from the sparse cover of scrub in the semiarid highlands to lush rain forest in the abundantly watered plains of the east.

Landform and Drainage

Rising to their greatest average elevations in Bolivia, the Andes Mountains consist of two chains separated by the lofty intermontane plateau that is the country's heartland.

On the west, the Cordillera Occidental (Western Cordillera) forms the border with Chile and has crests in excess of 3,962 m (13,000 ft) above sea level. Various perpetually snowcapped peaks rise to more than 4,876 m (16,000 ft) and the loftiest in the chain—Sajama—has a 6,519 m (21,391-ft) elevation. For the most part, they are either quiescent volcanoes or solfataras—volcanic vents producing sulfurous gases. The scanty number of streams originating in the chain drain both eastward to the interior plateau and westward to the Pacific Ocean, although the Loa River is the only watercourse to the Pacific that maintains a year-round flow. The Lauca River, which rises in the Chilean Andes, drains eastward to Lake Coipasa, the remnants of a much larger lake that once existed at the foot of the Cordillera Occidental. Chilean diversion of the waters of this river in the early 1960s has been the subject of a lively controversy between the two countries.

Eastward, on the other flank of the high plateau of the Altiplano, the Andes rise to even greater heights in the narrower Cordillera Real (Royal Cordillera), which extends southeastward to a line drawn approximately between the cities of Oruro and Cochabamba. Beyond this line, the Andes broaden into a complex mountain system that bends southward to the Argentine border. This southern extension of the Cordillera Real includes several ranges, the most prominent of which are the Cordillera Central (Central Cordillera) and Cordillera Oriental (Eastern Cordillera). Occasionally, the name Cordillera Real or Cordillera Oriental is given to identify the entire system of the eastern Andes.

The eastern mountains are of Palezoic origin and are considerably older than the volcanic mountain system to the west. The Cordillera Real forms an impressive array of peaks rising above the snow line, with an average elevation of over 5,486 m (18,000 ft) for most of its extent. The best known of the crests are Illampu, at 6,553 m (21,500 ft) and the triple crown of Illimani, which rises to 6,492 m (21,300 ft) behind the city of La Paz. In the southern extensions, the crests rarely rise above 4,876 m (16,000 ft) in the vicinity of Lake Poopó, but peaks again rise to 5,791 m (19,000 ft) or more still farther to the south. Various small streams drain westward onto the high plateau, but the larger watercourses drain eastward toward the Amazon Basin or southward toward the La Plata Basin.

PRINCIPAL TOWNS (estimated population at mid-1986)

La Paz (administrative capital)	1,033,288
Santa Cruz de la Sierra	457,619
Cochabamba	329,941
Oruro	184,101
Potosí	117,010
Sucre (legal capital)	88,774
Tarija	62,985

Area and population

		area		population
Departments	**Capitals**	sq mi	sq km	1985 estimate
Beni	Trinidad	82,458	213,564	240,000
Chuquisaca	Sucre	19,893	51,524	463,000
Cochabamba	Cochabamba	21,479	55,631	979,000
La Paz	La Paz	51,732	133,985	2,091,000
Oruro	Oruro	20,690	53,588	413,000
Pando	Cobija	24,644	63,827	47,000
Potosí	Potosí	45,644	118,218	878,000
Santa Cruz	Santa Cruz	143,098	370,621	1,048,000
Tarija	Tarija	14,526	37,623	270,000
TOTAL		424,164	1,098,581	6,429,000

Geographers are at odds concerning how best to divide Bolivia into regions. It is variously divided into two, three, or four regional components, and lines of demarcation vary. With the exception of a few scattered and tiny agricultural communities in the north, the Cordillera Occidental on the west has virtually no population and is ordinarily considered a boundary rather than part of a region. The high plateau of the interior, however, is universally identified as the Altiplano, and the lowlands to the east of the eastern Andes make up the Oriente (East). This great area is also known as the *llanos* (plains), and its extensive open areas are sometimes known as pampas. English-language writers occasionally refer to the region as a whole as the Eastern Plains.

There is nothing resembling a consensus with respect to the proper means of identifying the eastern Andes that separate the Altiplano from the Oriente. Often, this transitional territory is referred to generally as the intermediate region or zone. The eastern flank of the Cordillera Real, however, is specifically identifiable as a series of steep semitropical valleys called the Yungas, a word of Aymara origin roughly definable as "warm lands." South of Cochabamba, the physical characteristics of the intermediate region have characteristics very different from those of the Yungas, and it is often known as the region of the Valles (valleys). Although a majority of writers choose to treat the Yungas and the Valles as a single region, the topographical, climatic, and demographic characteristics are so different that the two may be considered as separate regional entities.

There is no consensus with respect to the size of the several regions. Most writers consider the Altiplano to constitute 14 to 16 percent of the national territory, but some choose to count the Yungas as part of the Altiplano. A majority regards the Yungas and the Valles as constituting about 15 percent of the territory, with about two-thirds of the total belonging to Valles. At the foot of the eastern Andes, however, lies a belt of sub-Andean terrain that makes up about 8 percent of the national territory. It is considered part of the Oriente but is sometimes assigned to the Yungas and Valles. Accordingly, the great eastern lowlands of the Oriente can be considered to constitute as much as 70 percent or as little as 62 percent of the total area of Bolivia.

Altiplano

The forbidding lunar landscape of the Altiplano extends southward from Peru to the Argentine frontier for a distance of 804 km (500 mi) and an average 50 km (80 mi) width at altitudes varying between 3,657 m and 4,267 m (12,000 and 14,000 ft). It tilts upward from the center toward both eastern and western cordillera systems and descends gradually from north to south. The plateau floor is made up of sedimentary debris washed down from the adjacent mountains and reaches enormous depths that are probably greater in the north where rainfall is heavier and the more numerous mountain streams have created larger alluvial fans.

The material frequently appears to consist of rock, but it is in fact made up of compressed sandy materials, clays, and gravels. It powders readily, is highly susceptible to erosion, and has so great a capacity to absorb moisture that the scanty rainfall drains away promptly to leave the topsoil permanently parched. There is so little organic material in the soil's content that fields are left fallow for periods of up to eight years.

The Altiplano is the largest basin of inland drainage in South America. Lake Titicaca, straddling the Peruvian border in the north, is at once South America's largest lake and the world's highest body of navigable water. The remainder of a much larger ancient body of water, Titicaca has an extent of 9,065 sq km (3,500 sq mi) and contains twenty-five islands, which played an important role in Inca mythology.

Its volume of contained water is sufficient to exercise influence on the climate in its vicinity, and the mediterranean conditions prevailing in the surrounding territory have made it the most heavily populated and the most agriculturally productive section of the Altiplano.

Lake Titicaca is drained to the south by the Desaguadero River. Flowing southward for 322 km (200 mi) to Lake Poopó, the river is the only major stream on the surface of the Altiplano. Titicaca has depths of up to 213 m (700 ft) and its icy waters are only slightly saline. In con-

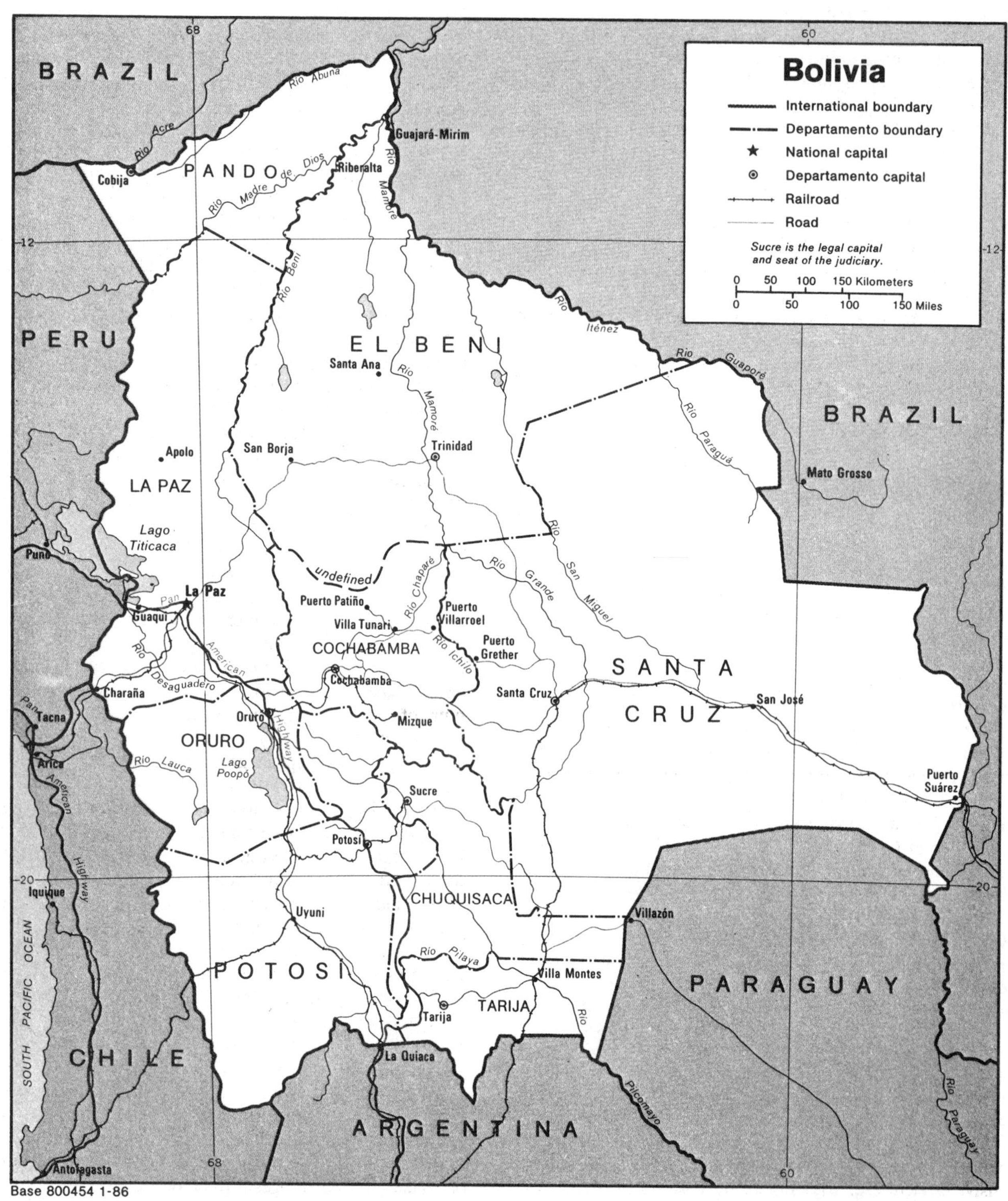

Base 800454 1-86

trast, Poopó is a shallow body of very brackish water with depths of ten feet or less. After heavy rains it occasionally overflows toward the Coipaso saltpan, immediately to the southwest, but it more customarily loses its water through evaporation and, as a consequence, is saline. Some geologists believe the Altiplano lakes and saltpans are remnants of a single ancient inland sea. Underground water of good quality but unknown extent is reported still to exist in the vicinity of the town of Patacamaya, about halfway between the cities of La Paz and Oruro.

In the southern part of the Altiplano the land becomes progressively more barren and the rainfall and vegetation scantier. The landscape is somber, and it is possible to travel for hours without seeing man or habitation. Much of the land is occupied by the great saltpans of Coipasa, Uyuni, and Chiguana. Elsewhere, stones adorn the surface in such profusion that, although many have been moved and formed into cairns and fences, the fields have the look of dried-up riverbeds. At the southern extremity of the plateau near the Argentine border, transverse hill systems span the gap between the eastern and western cordilleras of the Andes.

Yungas

The Yungas is a region made up of the sharply tilted mountain valleys that descend eastward from the crests of the Cordillera Real. The region is sometimes referred to by English-language writers as the *montaña* (mountain), the customary name of the corresponding eastern Andean slopes in Peru. The Yungas consists of three zones. The highest, sometimes called the *ceja* (eyebrow), has characteristics similar to those of the Altiplano; the lowest, or tropical zone, is related to the Oriente. Between, at levels of from 914 to 2,743 m (3,000 to 9,000 ft) lies the true semi-tropical Yungas—a series of narrow river valleys with rich soils, heavy rains, and lush vegetation.

All three elevation zones, including the heavily forested ridges that separate the valleys, are generally thought of as part of the Yungas region. It lies almost entirely in La Paz Department, but a transitional area in the south lies in Cochabamba Department. The Yungas lands are the most fertile in Bolivia, and their location is relatively close to the city of La Paz, but the region's steep slopes and deep and eroded gorges have challenged highway and railroad builders. In the early 1980s it remained a thinly populated region in an early stage of development.

Valles

South of the Yungas, the descent from the Andean heights to the lowlands is more gradual and occurs in a series of steps. This is the Valles region, in which the valleys and basins are broader and more extensive than in the Yungas. Their alluvial soils—notably those of Cochabamba, Sucre, and Tarija—support the country's most intensive agriculture. The high ridges and plateaus of the region, known as the *puna,* are unproductive and sparsely peopled. The removal of much of the vegetation has resulted in poor water retention leading to alternating droughts and floods, which have ruined many terraces and fields.

Although broader than the valleys of the Yungas, those of the Valles region are also deeply entrenched. The drainage pattern is divided between the system of the Pilcomayo River, which rises in the heart of the region and flows southward to join the Paraguay River near Asunción in Paraguay, and the system of the Chaparé, Ichilo, and Río Grande rivers, which join in the Oriente to form the Mamoré, whose waters eventually reach the Amazon.

Oriente

The Oriente includes all of the country located to the east of the Andes. The region slopes gradually from south to north, and from elevations of 610 to 762 m (2,000 to 2,500 ft) at the foot of the Andean piedmont in the west to as little as 91.4 m (300 ft) along parts of the Brazilian border in the east. Politically, it includes all of Pando Department, nearly all of the departments of El Beni and Santa Cruz, and parts of the departments of La Paz, Cochabamba, Tarija, and Chuquisaca. Containing about two-thirds of the national territory and less than one-sixth of the population, the Oriente is frontier Bolivia.

The flat northern part of the region—made up of the departments of Pando and El Beni—consists of tropical rain forest and the Plains of Moxos (Llanos de Moxos), which cover most of the basin of the Mamoré River. Because the topsoil is underlaid by a clay hardpan, drainage is poor, and much of the area is converted seasonally to swamp. In the plains that make up the northern part of Santa Cruz Department, the terrain becomes gently rolling and better drained. Savannas alternate with forests, and cultivated areas are more numerous. In general, however, it is estimated that no more than one-tenth of 1 percent of the Oriente is under cultivation. In the southern part of Santa Cruz, a low line of hills marks the beginning of the Bolivian part of the Chaco that continues southward into Paraguay.

The luxuriant tropical growth that covers much of the Oriente has given rise to a belief that the region is extraordinarily fertile. Alluvial soils of good quality exist along the courses of rivers and support high forest growth. Even these soils, however, are generally less fertile than those of the Yungas, and between the watercourses much of the land consists of *pampa negra* (black plain) that supports only sparse natural vegetation. About half of vast Santa Cruz Department consists of poor quality sandy soil known as *pampa blanca* (white plain).

The few streams of the Chaco drain southward into Paraguay. Elsewhere, the drainage of the entire region is part of the system of the Madeira River, one of the major tributaries of the Amazon. It is formed by the junction of the Beni and Mamoré rivers near the extreme northeast corner of the country. The rivers of the plains are wide and sluggish but of sufficient depth to permit navigation in certain sections by shallow-draft paddle wheel steamers or launch-towed barges. Interconnection between the greater watercourses of the Mamoré-Madeira system is prevented, however, by a series of granite outcrops that produce rapids and cascades, which effectively block them to navigation. The upper limits of power-driven navigation of any sort are reached at considerable distances short of the Andean foothills; as a consequence, water transport is of local importance to the river settlements but contributes little to the national distribution system.

BOTSWANA

BASIC FACTS

Official Name: Republic of Botswana

Abbreviation: BOT

Area: 581,730 sq km (224,607 sq mi)

Area—World Rank: 45

Population: 1,189,900 (1988) 1,817,400 (2000)

Population—World Rank: 131

Capital: Gaborone

Land Boundaries: 4,502 km (2,518 mi); Zimbabwe 813 km (505 mi); South Africa 1,778 km (1,105 mi); Namibia 1,461 km (908 mi)

Coastline: Nil

Longest Distances: 1,115 km (693 mi) NNE-SSW 951 km (591 mi) ESE-WNW

Highest Point: 1,515 km (4,969 ft)

Lowest Point: Confluence of the Shashi and Limpopo Rivers 513 m (1,684 ft)

Land Use: 2% arable; 75% meadows and pastures; 2% forest and woodland; 21% other

Botswana is a landlocked country located in southern Africa. It is a vast tableland at a mean altitude of 1,000 m (3,300 ft). A gently undulating plateau running from the South African border near Lobatse to a point west of Kanye and from there northward to Bulawayo on the Zimbabwean border forms the watershed between the two main natural divisions of Botswana. The fertile land to the south of this plateau is hilly bush country and grassland, or veld. To the west of the plateau, stretching over the border into Namibia, is the Kalahari (also known as Kgalagadi) Desert. The Kalahari is more accurately a semi-desert, or sandy tract, covered with thorn bush and grass. In the extreme northwest lies the area known as Ngamiland dominated by the Okavango Swamps, a great inland delta of some 16,835 sq km (6,500 sq mi), and the Makgadikgadi Salt Pans. Around the swamps and along the northeastern border from Kasane to Francistown there is forest and dense bush.

Most of Botswana is without surface drainage, and apart from Limpopo and Chobe Rivers the country's rivers never reach the sea. The major interior river system is the Okavango, which flows into Botswana from the Angolan Highlands in the northwest to form the Okavango Swamps, a delta covering about 3% of the total land area of the country. About half of this area is perennially flooded and the rest is seasonally flooded. From this marsh there is a seasonal flow of water into the ephemeral Lake Ngami and, along the Botletle River, to Lake Dow and the Makgadikgadi Salt Pans. Much of the water is, however, lost through evaporation. The Chobe River in the north flows into the Zambezi after marking the border of the Caprivi Strip for part of its course.

PRINCIPAL TOWNS (population at 1981 census)

Town	Population	Town	Population
Gaborone (capital)	59,657	Kanye	20,215
Francistown	31,065	Lobatse	19,034
Selebi-Phikwe	29,469	Mochudi	18,386
Serowe	23,661	Maun	14,925
Mahalapye	20,712	Ramotswa	13,009
Molepolole	20,565		

Area and population

		area		population
Districts	**Capitals**	sq mi	sq km	1984 estimate
Central	Serowe	57,039	147,730	355,000
Ghanzi	Ghanzi	45,525	117,910	21,000
Kgalagadi	Tsabong	41,290	106,940	26,000
Kgatieng	Mochudi	3,073	7,960	49,000
Kweneng	Molepolole	13,857	35,890	128,000
North East	Masunga	1,977	5,120	40,000
North West				
Chobe	Kasane	8,031	20,800	9,000
Ngamiland	Maun	42,135	109,130	75,000
Southern	Kanye	10,992	28,470	138,000
South East	Ramotswa	687	1,780	34,000
Towns				
Francistown	—	31	79	36,000
Gaborone	—	37	97	79,000
Lobatse	—	12	30	22,000
Orapa	—	4	10	5,800
Selebi-Pikwe	—	19	50	33,000
TOTAL		224,607	581,730	1,051,000

BRAZIL

BASIC FACTS

Official Name: Federative Republic of Brazil

Abbreviation: BRL

Area: 8,511,965 sq km (3,286,488 sq mi)

Area—World Rank: 5

Population: 150,685,145 (1988) 179,487,000 (2000)

Population—World Rank: 6

Capital: Brasilia

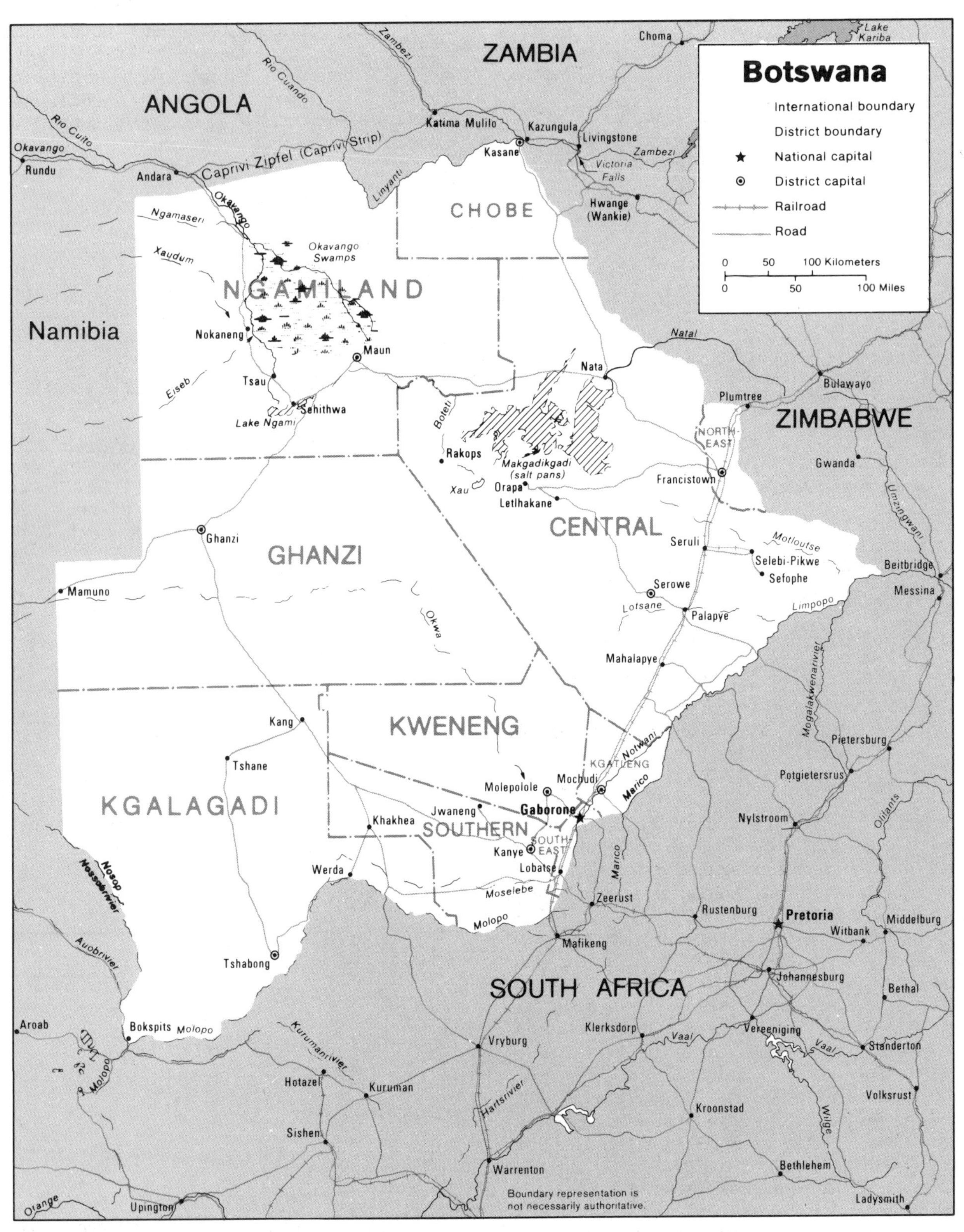
Botswana
International boundary
District boundary
National capital
District capital
Railroad
Road
0 50 100 Kilometers
0 50 100 Miles
ZAMBIA
ANGOLA
Namibia
ZIMBABWE
SOUTH AFRICA
CHOBE
NGAMILAND
CENTRAL
GHANZI
KWENENG
KGALAGADI
SOUTHERN
KGATLENG
SOUTH-EAST
NORTH-EAST
Gaborone
Maun
Kasane
Nata
Francistown
Serowe
Palapye
Mahalapye
Molepolole
Mochudi
Kanye
Lobatse
Jwaneng
Ghanzi
Mamuno
Kang
Tshane
Khakhea
Werda
Tshabong
Bokspits
Nokaneng
Tsau
Sehithwa
Lake Ngami
Okavango Swamps
Makgadikgadi (salt pans)
Rakops
Orapa
Letlhakane
Seruli
Selebi-Pikwe
Sefophe
Plumtree
Bulawayo
Livingstone
Victoria Falls
Hwange (Wankie)
Kazungula
Katima Mulilo
Choma
Lake Kariba
Caprivi Zipfel (Caprivi Strip)
Andara
Rundu
Pretoria
Johannesburg
Mafikeng
Zeerust
Rustenburg
Vryburg
Kuruman
Upington
Boundary representation is not necessarily authoritative.

Brazil
The islands of Fernando de Noronha, Rocas, Trindade, Martin Vaz, and Penedos de São Pedro e São Paulo are not shown.
Trindade and Martin Vaz are under the jurisdiction of Espírito Santo. Fernando de Noronha has territorio status.
ATLANTIC OCEAN
PACIFIC OCEAN
VENEZUELA
COLOMBIA
GUYANA
SURINAME
French Guiana (Fr.)
PERU
BOLIVIA
CHILE
ARGENTINA
PARAGUAY
URUGUAY
TRINIDAD AND TOBAGO
AMAZONAS
PARÁ
RORAIMA
AMAPÁ
ACRE
RONDÔNIA
MATO GROSSO
MATO GROSSO DO SUL
GOIÁS
DISTRITO FEDERAL
MARANHÃO
PIAUÍ
CEARÁ
RIO GRANDE DO NORTE
PARAÍBA
PERNAMBUCO
ALAGOAS
SERGIPE
BAHIA
MINAS GERAIS
ESPÍRITO SANTO
RIO DE JANEIRO
SÃO PAULO
PARANÁ
SANTA CATARINA
RIO GRANDE DO SUL
Brasília
Caracas
Bogotá
Georgetown
Paramaribo
Cayenne
La Paz
Asunción
Santiago
Buenos Aires
Montevideo
Santa Marta
Maracaibo
Barquisimeto
San Cristóbal
Ciudad Bolívar
Skeldon
Nieuw Nickerie
Santa Elena
Boa Vista
Diapoque
Macapá
Belém
São Luís
Manaus
Santarém
Careiro
Itaituba
Benjamin Constant
Jacare-Acanga
Humaitá
Pôrto Velho
Cachimbo
Rio Branco
Santa Inês
Teresina
Fortaleza
Pôrto Franco
Picos
Natal
João Pessoa
Recife
Juàzeiro
Maceió
Aracaju
Gurupi
Barreiras
Ibotirama
Salvador
Vilhena
Cuiabá
Goiânia
Vitória da Conquista
Pirapora
Montes Claros
Pôrto Seguro
Uberlândia
Belo Horizonte
Vitória
Campo Grande
Dourados
Volta Redonda
Niterói
Rio de Janeiro
São Paulo
Santos
Foz do Iguaçu
Curitiba
Barracão
Florianópolis
São Borja
Osório
Pôrto Alegre
Pelotas
Urubamba
Cusco
Nazca
Puno
Arequipa
Guaqui
Matarani
Santa Cruz
Sucre
Arica
Iquique
Uyuni
Antofagasta
San Miguel de Tucumán
La Serena
Córdoba
Santa Fe
Paysandú
Valparaíso
Mendoza
Rosario
Ibicuy
Colonia
Concepción
Zapala
Bahía Blanca
Mar del Plata
Rio Orinoco
Rio Negro
Amazon
Rio Tocantins
Rio Xingu
Rio Araguaia
Rio São Francisco
Rio Paranaíba
Rio Grande
Rio Paraná
Rio Ucayali
International boundary
Estado or território boundary
National capital
Estado or território capital
Railroad
Paved road
Unpaved road
Airport
0 500 Kilometers
0 500 Miles
Boundary representation is not necessarily authoritative.
5/85

Boundaries: 23,127 km (14,370 mi); Venezuela 1,495 km (929 mi); Guyana 1,606 km (998 mi); Suriname 593 km (368 mi); French Guiana 655 km (407 mi); Uruguay 1,003 km (623 mi); Argentina 1,263 km (785 mi); Paraguay 1,339 km (832 mi); Bolivia 3,126 km (1,942 mi); Peru 2,995 km (1,861 mi); Colombia 1,644 km (1,022 mi)

Coastline: 7,408 km (4,603 mi)

Longest Distances: 4,328 km (2,689 mi) N-S 4,320 km (2,684 mi) E-W

Highest Point: Neblina Peak 3,014 m (9,888 ft)

Lowest Point: Sea level

Land Use: 7% arable; 1% permanent crops; 19% meadows and pastures; 67% forest and woodland; 6% other

Brazil borders on all but two of the countries of South America, makes up about half of that continent's landmass, and is the home of half of its population. Its territory is larger than that of the forty-eight contiguous states of the United States.

The Amazon, the world's largest river in terms of flow of water, cuts laterally across the country's northern flank, and countless tributary streams drain a vast basin that encompasses three-fifths of the national territory. The entire basin, including fringes in neighboring countries, supports a tropical forest that provides natural replacement for 15 percent or more of the world's oxygen.

Although there are no high mountains—the highest elevations are at less than 3,048 m (10,000 ft)—most of the territory outside the Amazon Basin consists of a great highland block. The highlands drop precipitously to a narrow Atlantic coastal plain.

Brazil has fewer high elevations than any other country of South America except Uruguay and Paraguay. If slopes of up to 35° are considered suitable for farming, not more than one-fifth of the terrain is beyond the limits of agricultural usefulness. There are, however, many low mountain systems, rounded hills, and deep valleys. The drainage is generally good, but much of the landscape is highly vulnerable to erosion.

Highlands

Brazil officially defines its highlands as areas with elevations in excess of 200 m (656 ft). Some 59 percent of the national territory is in the highlands according to this definition, but only 0.5 percent is at more than 1,200 m (3,937 ft).

The principal highland zone, the Central Highlands, is an enormous block covering all of Brazil south of the Amazon Basin with the exception of a narrow coastal belt and a swampy section of western Mato Grosso State that is an extension of the Bolivian and Paraguayan Chaco plain. It is tilted almost imperceptibly westward and northward so that rivers rising near its eastern rim, almost within sight of the Atlantic, flow inland for hundreds of miles before veering northward or southward to join larger streams. Near the coastline clearly defined escarpments rise as high as 792 m (2,600 ft).

The highland block displays a variety of minor relief features. The northern and western half is made up of broad, rolling terrain punctuated irregularly by low, rounded hills. Frequently these hills are formed into systems that are given range names but are not high enough to be considered mountains. Southward from the Amazon Basin to about the middle of Goiás State, the terrain becomes extremely rough and formidable. Basic elevations are not loftier than those elsewhere in the highland block, but the crystalline rock and its cover of softer materials have been folded and eroded into a complex mass with ridges and ranges extending in all directions. Gradients are precipitous, and passage through them is tortuous and difficult. Roads frequently traverse many miles to reach destinations only short linear distances apart. The journey from Rio de Janeiro to Belo Horizonte in Minas Gerais, which can be completed in little more than an hour by plane, takes nearly a day by rail.

In this rugged complex, remnants of the crystalline rock have been weathered by nature and stand above the general level as mountains, but only in a few places do they extend above the timberline, 1,889 to 1,981 m (6,200 to 6,500 ft). The Serra do Mar parallels the coast for 1,609 km (1,000 mi) from Santa Catarina to Rio de Janeiro and continues northward into Espírito Santo as the Serra dos Orgãos. This extended range has a mean crest of about 1,524 m (5,000 ft) topped by peaks above 2,133 m (7,000 ft) including Pedra Acu, which rises to 2,318 m (7,605 ft) just west of Rio de Janeiro.

The Serra do Mar is so near the tidewater in many places that it rises almost directly from the shore. In others it recedes to leave a narrow littoral varying from twenty to forty miles in width. These are passes below 914 m (3,000 ft) only in two places where rivers have cut their way through the coastal escarpment short distances north of the city of Rio de Janeiro. The valleys of these streams, however, are blocked off from the interior plateau by a second range of mountains in Minas Gerais, the Serra da Mantiqueira. This range is the highest and most rugged of the Central Highlands; it includes the Pico dos Banderientes, which at 2,890 m (9,482 ft) is the highest elevation in the Central Highlands and is frequently but incorrectly believed to be the highest in Brazil.

A third significant range of mountains traverses a north-south axis from central Minas Gerais northward into Bahia State. Appropriately named the Serra do Espinhaço, which means Backbone Mountains, it forms a spine that determines the drainage divide between the São Francisco River to the west and short streams that tumble eastward to the Atlantic. It is important because of the great wealth of minerals that it contains. Sometimes the Serra do Espinhaço and the Serra da Mantiqueira together with that range's southward extending spurs are referred to collectively as the Serra Geral.

Much smaller uplands are the Guiana Highlands. Occupying about 2 percent of the national territory, the area is completely undeveloped with the exception of isolated mining settlements and is peopled only by a few scattered Indian tribes. Its location, immediately to the north of the equator in a zone of heavy rainfall, makes it the source of countless rivers and streams that descend in rushing falls and rapids to the Amazon. None of these watercourses is navigable very far upstream, and they contribute little to the development of the considerable mineral and forest wealth believed to exist near their headwaters.

This highland forms part of an immense plateau extending into Venezuela, Guyana, Suriname and French Guiana and is much higher than the Central Highlands. The crests of its ranges constitute the divide between drainage northward to the Orinoco River in Venezuela and southward to the Amazon and define the national borders. With an elevation of 3,013 m (9,888 ft) the Pico da Neblina in the Imeri range is Brazil's highest mountain.

Lowlands

The most extensive of the lowlands lie in the Amazon Basin, a vast drainage area that covers nearly three-fifths of Brazil. Not all of the Amazon Basin, however, is lowland. The Guiana Highlands and the northern flank of the Central Highlands are located there.

Most of the terrain of the lowlands of the Amazon is gently undulating. There are stretches of flat, swampy land called *varzea* located along the courses of major rivers and subject to frequent flooding. Swampland, however, is limited; most of the region is referred to as *terra firme* (high ground in a flooded area) and has elevations sufficiently varied to permit rainwater to drain off into streams.

PRINCIPAL TOWNS (estimated population at mid-1985)

Town	Population	Town	Population
Brasília (capital)	1,576,657	Natal	512,241
São Paulo	10,099,086	Maceió	484,094
Rio de Janeiro	5,615,149	Teresina	476,102
Belo Horizonte	2,122,073	Santos	461,096
Salvador	1,811,367	São João de Meriti	459,103
Fortaleza	1,588,709	Niterói	442,706
Nova Iguaçu	1,324,639	Jaboatão	411,341
Recife	1,289,627	João Pessoa	397,715
Curitiba	1,285,027	Campo Grande	386,520
Porto Alegre	1,275,483	Contagem	386,272
Belém	1,120,777	Ribeirão Preto	384,604
Goiânia	928,046	São José dos Campos	374,526
Campinas	845,057	Campos	367,134
Manaus	834,541	Aracaju	361,544
São Gonçalo	731,061	Feira de Santana	356,660
Guarulhos	717,723	Juíz de Fora	350,687
Duque de Caxias	666,128	Londrina	347,707
Santo André	637,010	Olinda	335,889
Osasco	594,249	Sorocaba	328,787
São Bernardo do Campo	565,620		
São Luís	564,434		

A government research and experiment organization calculates that 80 percent of the soils in the Amazon lowlands lose about one pound per acre of topsoil annually through erosion. This annual loss mounts alarmingly to fourteen tons when the forest cover is removed for agriculture. Moreover, a preliminary study released by the National Colonization and Agrarian Reform Institute early in 1974 concluded pessimistically that most of the land might be suitable only for grazing rather than for the mass colonization by subsistence farmers planned by the government. The better soils result from the decomposition of dibasic and basaltic rocks and are known as *terra roxa* (purple earth). Similar to the rich coffee-growing soil of São Paulo and Paraná states, *terra roxa* occurs principally in a stretch of land north of the Amazon between the Xingu and Tapajós rivers.

Area and population

States	Capitals	area sq mi	area sq km	population 1987 estimate
Acre	Rio Branco	58,915	152,589	374,000
Alagoas	Maceió	10,707	27,731	2,335,000
Amazonas	Manaus	604,035	1,564,445	1,833,000
Bahia	Salvador	216,613	561,026	11,170,000
Ceará	Fortaleza	58,159	150,630	6,100,000
Espírito Santo	Vitória	17,605	45,597	2,381,000
Goiás	Gioânia	247,913	642,092	4,659,000
Maranhão	São Luís	126,897	328,663	4,863,000
Mato Grosso	Cuiabá	340,156	881,001	1,599,000
Mato Grosso do Sul	Campo Grande	135,347	350,548	1,687,000
Minas Gerais	Belo Horizonte	226,708	587,172	15,021,000
Pará	Belém	482,906	1,250,722	4,476,000
Paraíba	João Pessoa	21,765	56,372	3,102,000
Paraná	Curitiba	77,048	199,554	8,228,000
Pernambuco	Recife	37,946	98,281	6,992,000
Piauí	Teresina	96,886	250,934	2,532,000
Rio Grande do Norte	Natal	20,469	53,015	2,204,000
Rio Grande do Sul	Pôrto Alegre	108,952	282,184	8,732,000
Rio de Janeiro	Rio de Janeiro	17,092	44,268	13,278,000
Rondônia	Pôrto Velho	93,840	243,044	818,000
Santa Catarina	Florianópolis	37,060	95,985	4,256,000
São Paulo	São Paulo	95,714	247,898	31,263,000
Sergipe	Aracaju	8,492	21,994	1,339,000
Other federal entities				
Distrito Federal	Brasīlia	2,245	5,814	1,720,000
Amapá	Macapá	54,161	140,276	227,000
Fernando de Noronha	Fernando de Noronha	10	26	1,000
Roraima	Boa Vista	88,844	230,104	112,000
TOTAL		3,286,487	8,511,965	141,302,000

This lowland includes rolling savanna areas, particularly in the northwestern corner, but the prevailing cover is tropical rain forest known as *selva*. It is generally thought of as one of the most remote and primitive regions of the world, but almost half of its scanty population lives in cities and towns. The story is told of a visitor who brought with him such equipment as snake boots and hammock but who found himself staying each night in an air-conditioned cabin, complete with private bathroom.

A second lowland area, the Pantanal, exists in western Mato Grosso. The floor of this lowland is largely swamp and marshland, with an average elevation of 152 m (500 ft) above sea level. Away from its many streams, which make up the headwaters of the Paraguay River, sedimentary deposits have left a soil suitable for varied agriculture.

The area is too wet to support forest, except for lightwoods on patches of higher ground, and it bears some resemblance to the flooded Amazon grasslands. A long dry season with lower temperatures, coupled with the presence of fertile alluvial soils, however, is conducive to the growth of good pasture. Cattle raising is, therefore, the principal economic activity.

The third and most highly developed lowland is the narrow strip of coastal plain that extends along the Atlantic seaboard from French Guiana to Uruguay. North of the state of Ceará it is fairly wide and is more or less an extension of the Amazon Basin, possessing most of the luxuriant growth of that region. South of Cape São Roque near Natal it remains tropical, merging into subtropical, and narrows to a mere ribbon at the foot of the highland escarpment. In some places, particularly between the city of Rio de Janeiro and Santos in São Paulo State, it disappears entirely. Near its southern extremity in Rio Grande do Sul, the plain widens into the rolling grasslands of the Rio Grande Plains that merge south and west with the pampas of Uruguay. At no other place does this thread of lowland offer any large level areas except at the deltas of the Dôce and Paraíba rivers in Espírito Santo and Rio de Janeiro, respectively.

Geographic Regions

The country covers so vast an area and has characteristics so varied that observers of the Brazilian scene customarily make their analyses in terms of specific regions. The system of internal regional division was established in 1970 by the Brazilian Institute of Geography and Statistics (Instituto Brasileiro de Geografia e Estatística—IBGE). According to this plan, the country is divided into five regions—the North, Northeast, Southeast, South, and Central-West. The system previously in effect differed from the current one in that the region generally corresponding to the Southeast was called the East and differed in composition. It included the states of Bahia and Sergipe, which are now assigned to the Northeast. The state of São Paulo, now in the Southeast, was a part of the South. Because of this change it is sometimes difficult to reconcile statistical and other regional information covering periods before and after 1970.

To facilitate statistical reporting, the IBGE shows the regions as made up of entire states and territories. As a consequence, they do not fully conform to natural features. In particular, each of the five includes territory in both the highland and the lowland areas.

The North

The largest of the five regions, the North, covers 42 percent of the country and is made up of the states of Amazonas, Acre, and Pará and the territories of Rondônia, Roraima, and Amapá. It lies primarily in the lowlands but includes the Guiana Highlands. On the south and east its perimeter corresponds roughly to the transition between the lowlands of the Amazon and the Central Highlands.

The three states and three territories that make up the North lie wholly within the Amazon Basin, but northern portions of the Central Highlands states of Pará, Goiás, and Mato Grosso are also located within that drainage system. The states and territories of the North, however, are actually oriented toward the Amazon River. The other areas of the basin are politically, socially, and economically oriented toward population centers to the east and south.

In this sense, the territorial limits of the North are rational ones. There is a tendency, however, to use the terms *North* and *Amazon Basin* indiscriminately and—further to confuse matters—the entire basin is sometimes referred to by the term *Legal Amazon* because of the benefits from certain federal tax incentives that are applicable throughout it. Moreover, when the term *Amazon Basin* is used, it is sometimes difficult to determine whether reference is intended to be to the Brazilian portion of the basin or to the entire drainage system including those portions located in neighboring countries. Finally, terms without specific meaning concerning geographic extent, such as *Greater Amazon, Amazonia, Amazon Valley,* or simply *Amazon,* are in common use.

The Northeast

The Northeast comprises 18 percent of the national territory and is made up of the states of Maranhão, Piauí, Ceará, Rio Grande do Norte, Paraíba, Pernambuco, Alagoas, Sergipe, and Bahia and the small island territory of Fernando de Noronha, located in the Atlantic Ocean 402 km (250 miles) northeast off the coast of Rio Grande do Norte. About two-thirds of the region is made up of the *sertão,* a dry interior zone. The remainder is divided into the *agreste,* a semihumid transitional zone, and the *zona de mata,* a humid coastal strip with a width of up to about 96 km (60 mi). Offshore, the waters of the continental shelf are extremely shallow, and the shoreline is rimmed by reefs and sandbars. The port of Belém is hemmed in by sandbars that prevent the entry of the largest vessels, and the ports of Salvador and Recife are flanked by reefs. The name of the latter city means reef.

The irregular landscape of the *sertão* consists of broken scrubland marked by rock outcroppings. Its soils are generally fertile and respond well to irrigation, but a dense rural population, depletion of the soil, and irregularity of rainfall have made it a land of poverty. Pressure for land causes relatively poor hillsides as well as flatlands to be worked and, after being planted for two years, the hillside plots must be allowed to lie fallow for as many as eight.

The transitional *agreste* zone is a cattle-raising area in which a few large estates are mixed with a great many tiny plots of land. Its soils are of fair fertility. The soils of the *zona de mata* are red and yellow and of low fertility. It is in this coastal belt, however, that many large cities and the heaviest concentration of rural population are located. In the region as a whole the best land, known as *coroa,* is located along the margins of rivers where periodic flooding replaces the topsoil with silt. In general, the better soils are in low-lying areas, and the soils of hillsides tend to be marginal.

The Northeast has been described as the cultural hearth of Brazil. It was the first focus of European settlement. During the twentieth century, however, it has continuously been a place of poverty and social problems. Its troubles, attributed to intermittent droughts, severe soil depletion, uneven land distribution, and population pressure, have led to a massive and sustained out-migration to other and more promising regions.

The Southeast

The densely populated Southeast is made up of the states of Minas Gerais, Espírito Santo, Rio de Janeiro, São Paulo, and Guanabara. The last consists principally of the city of Rio de Janeiro and was formed when the country's capital was transferred to the newly created Brasília in 1960. As a result, the city of Rio de Janeiro is now surrounded by, but not a part of, the state of Rio de Janeiro, a circumstance that may confuse some geography students.

Containing only 11 percent of the national territory, the Southeast is Brazil's economic, social, and political heartland. In 1988 the region was estimated to contribute 85 percent of the country's industrial production and 70 percent of its industrial workers, half of the voters in national elections, over 80 percent of the federal revenue, 50 percent of the large banks, and 80 percent of the bank deposits.

Topographically, the region is made up of a narrow coastal plain and an extensive interior plateau. In contrast to the tropical humidity of the coast, the upland interior is dry and invigorating. It has been suggested that the invigorating climate resulting from São Paulo's 762 m (2,500-ft) elevation has contributed substantially to that city's spectacular prosperity and growth. The state of São Paulo has large expanses of *terra roxa*. The state of Minas Gerais, which makes up over half of the region, has soils of relatively low fertility and is a center of cattle raising. The most heavily populated state after São Paulo, Minas Gerais suffers from rural crowding and, like the Northeast, is an area of heavy out-migration. During the 1980s, as a consequence, its population was the slowest growing the Brazilian states.

The South

Made up of the states of Paraná, Santa Catarina, and Rio Grande do Sul, the South contains seven percent of the national territory and is the smallest of the five regions. It is also the only one having a predominantly temperate climate. The Serra do Mar separates a narrow coastal strip of the region as far south as the midpoint of the southernmost state, Rio Grande do Sul. Inland lies the southern tip of the Central Highlands. South of Pôrto Alegre, however, the Serra do Mar mountain system almost disappears, the highlands dip downward, and the terrain of the southern half of Rio Grande do Sul is a cattle raising pampa related topographically more to contiguous Uruguay than to Brazil.

Generally a prosperous area, the South is, after the Southeast, the region most fully integrated into the market economy. It has little virgin land. The population, which is fairly evenly distributed, has a high proportion of persons of recent European origin.

Each of the three states has its own distinctive characteristics. The northern part of Paraná has Brazil's fertile and famous *terra roxa* soil that supports the coffee *fazenda* (large estates), and the plains of Rio Grande do Sul make it so much a cattle area that *gaúcho* is generally recognized as the nickname for any native of that state. Between the two, the diminutive state of Santa Catarina is a tumbled mass of highlands that has been described as a poor filling to a rich sandwich. Its convoluted terrain, however, includes most of Brazil's coal deposits.

The Central-West

This region is made up of the states of Mato Grosso and Goiás, plus the Federal District that was carved out of the latter state. Comprising 22 percent of the national territory, it is the country's second largest region.

With the exception of the Pantanal, the swampy lowland in western Mato Grosso, the entire region is located in the Central Highlands. Its northern portion lies in the Amazon Basin, but the transitional slope from highland to lowland is gradual, and the land is covered for the most part less by rain forest than by savanna and a combination of trees and shrubs known as *mata*. Much of the soil is the rich *terra roxa*. The southern portion of the region is a rolling prairie with rivers draining southward. Its soils are said to be sufficiently productive to support agriculture for ten years without the use of technical aids. The soils of the central portion are poorer, but in the mid-1970s the zone had already become an important cattle-raising area. Less isolated than the North, the Central-West as a whole was a fast-growing region attracting both spontaneous and planned settlement.

Hydrography

Brazil has no lakes of any consequence, but its river systems are among the world's most extensive, and the Amazon is the world's mightiest river. No other three rivers of the world combined equal the flow of 80 million gallons of water per second that the Amazon discharges into the Atlantic, and silt discoloration can be observed 321 km (200 mi) seaward from its mouth. The Amazon has suffered virtually no pollution, and it has a chemical purity superior to that of most tapwater in the United States.

The Amazon rises high in the Peruvian Andes but, during its 3,218 km (2000-mi) course eastward across northern Brazil, it drops only about 65 m (215 ft). It is navigable by oceangoing vessels as far as Iquitos in Peru, and Manaus, in the middle of the South American continent, is a seaport of importance. Smaller craft can reach Pôrto Velho, near the Bolivian frontier on the tributary Madeira River. During most of its course the river is slow flowing, but at Monte Alegre about 643 km (400 mi) from the mouth it is constricted by hills to a width of about one mile, and the flow reaches six miles per hour.

The Amazon has eighteen major tributaries, including ten larger than the Mississippi River. Altogether there are more than 200 rivers in the system, which drains about 59 percent of the Brazilian territory. Slow-flowing like the Amazon in their lower courses, the tributary streams meander intricately, and oxbow lakes are numerous.

Other major streams rise in the Central Highlands. The São Francisco River, the longest contained entirely in Brazil, rises near Belo Horizonte and is navigable for 1,609 km (1,000 mi). It flows northeastward along a line parallel to the coast before turning eastward toward the sea at the border between the states of Sergipe and Alagoas, where it drops 80 m (265 ft) at the spectacular Paulo Afonso Falls about 241 km (150 mi) from the coast.

Only two other rivers of the Central Highlands cut through the escarpments of the Atlantic coastal ranges. The Dôce River has carved an escape route through which a railroad has been built to connect the mines near Belo Horizonte with the port of Vitória. The Paraíba River, which has its source near São Paulo, flows northward parallel to the coast in a rift between two coastal mountain ranges before entering the sea at Cape São Tomé. Its valley forms the best line of communication between Rio de Janeiro and São Paulo and is the site of the great Volta Redonda steelworks.

Although many other rivers of the Central Highlands originate close to the sea, the coastal mountains and the westward inclination of the plateau cause them to flow westward to join major streams of the Río de la Plata drainage basin. Of the three major rivers forming part of the Río de la Plata basin, the Paraná is the largest and the one receiving the waters from most of the tributary streams. Rising in Goiás, it flows southward across the Central Highlands and forms parts of the frontiers with Paraguay and Argentina on its course to the Río de la Plata estuary. The important Sete Quedas hydroelectric power project is located on the Paraná.

The magnificent Iguaça Falls, which eclipse Niagara Falls in magnitude, are located on the Iguaça River close to the point at which it joins the Paraná. The complex is three miles wide and 82 m (270 ft) high and consists of some 300 cataracts.

The second of the Río de la Plata basin rivers, the Uruguay, is fed by streams originating in the Serra do Mar. It forms the border between the states of Santa Catarina and Rio Grande do Sul before turning southward to form a portion of the border with northeastern Argentina. The third river, the Paraguay, rises near Cuiabá in Mato Grosso and flows southward into Paraguay and Argentina.

BRITISH INDIAN OCEAN TERRITORY

BASIC FACTS

Official Name: British Indian Ocean Territory

Abbreviation: BIO

Area: 60 sq km (23 sq mi)

Area—World Rank: N/A

Population: Nil

Population—World Rank: N/A

Capital: None

Land Boundaries: Nil

Coastline: 120 km (74 mi)

Longest Distances: N/A

Highest Point: N/A

Lowest Point: Sea level

Land Use: 100% other

The British Indian Ocean Territory (BIOT) comprises the Chagos Archipelago of some 2,300 islands of which Diego Garcia is the largest and southernmost. The islands are flat and low with an elevation of less than 4 m (13 ft).

BRUNEI

BASIC FACTS

Official Name: Negara Brunei Darussalam

Abbreviation: BRU

Area: 5,765 sq km (2,226 sq mi)

Area—World Rank: 150

Population: 316,565 (1988) 388,000 (2000)

Population—World Rank: 158

Capital: Bandar Seri Begawan

Boundaries: Malaysia 381 km (237 mi)

Coastline: 160 km (100 mi)

Longest Distances: N/A

Highest Point: Mount Pagon 1,850 m (6,070 ft)

Lowest Point: Sea level

Land Use: 1% arable; 1% permanent crops; 1% meadows and pastures; 79% forest and woodland; 18% other

Brunei is located on the northwestern coast of the Sound of Borneo (Kalimantan). It is comprised of two enclaves separated by the Limbang River Valley, a salient of Sarawak. Apart from the heavily populated narrow coastal strip in the west, the land is primarily tropical rain forest.

Area and population

		area		population
Districts	Capitals	sq mi	sq km	1984 estimate
Belait	Kuala Belait	1,053	2,727	57,000
Brunei and Muara	Bandar Seri Begawan	220	570	129,400
Temburong	Bangar	503	1,303	6,800
Tutong	Tutong	450	1,165	22,700
TOTAL		2,226	5,765	215,900

Diego Garcia

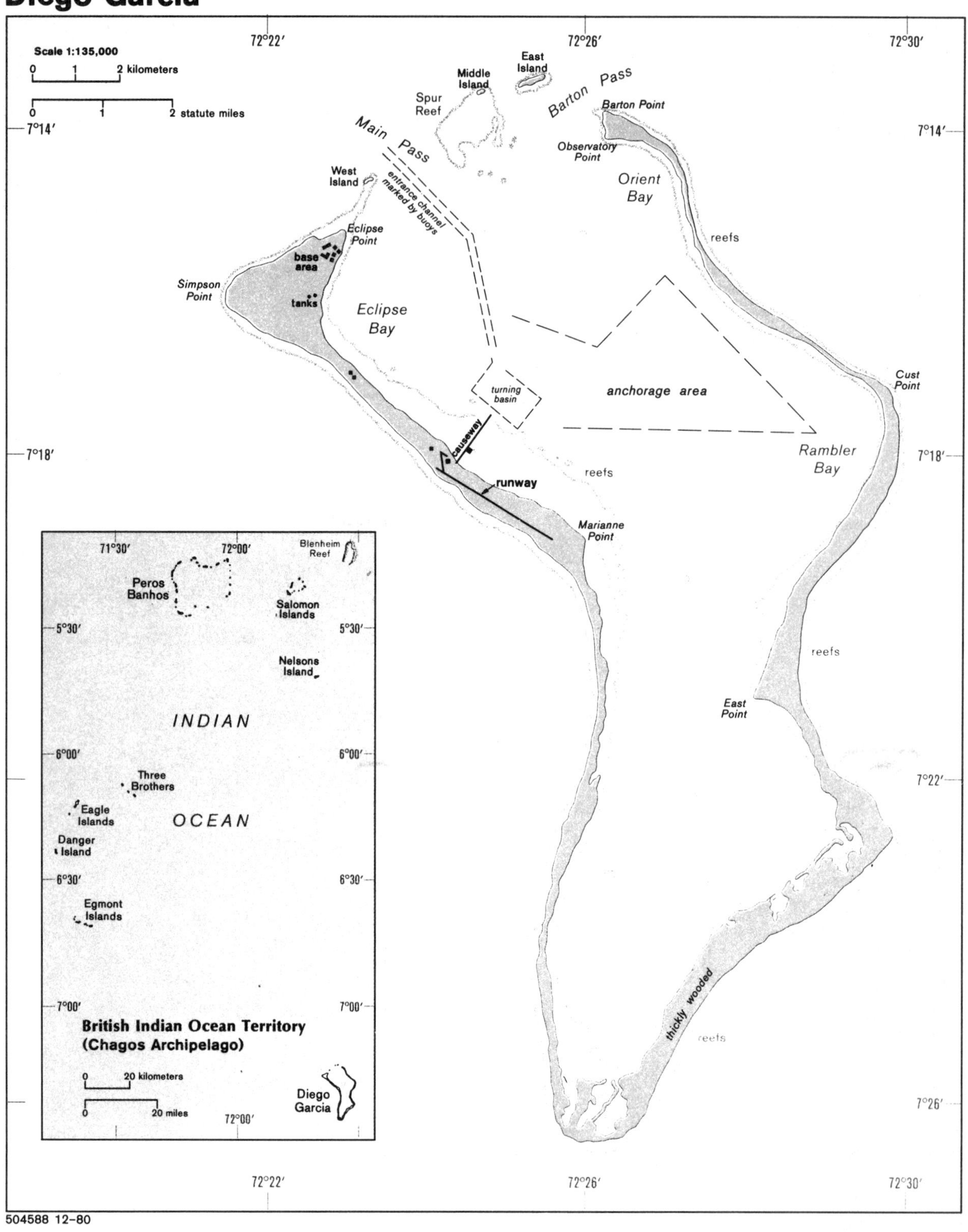

504588 12-80

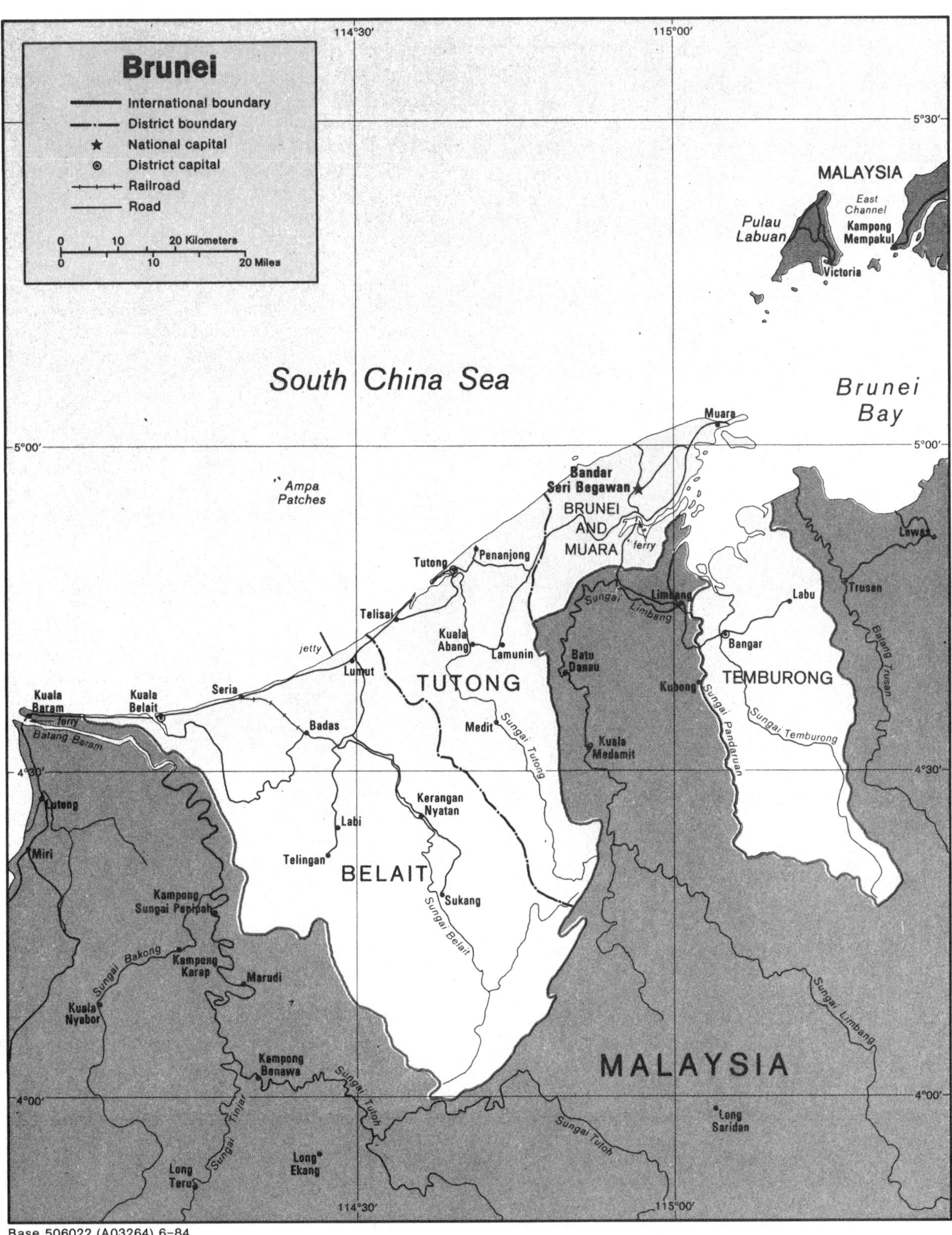

Base 506022 (A03264) 6-84

BULGARIA

BASIC FACTS

Official Name: People's Republic of Bulgaria

Abbreviation: BUL

Area: 110,912 sq km (42,823 sq mi)

Area—World Rank: 98

Population: 8,966,927 (1988) 9,276,000 (2000)

Population—World Rank: 68

Capital: Sofia

Boundaries: 2,145 km (1,332 mi); Romania 509 km (316 mi); Turkey 259 km (161 mi); Greece 493 km (306 mi); Yugoslavia 506 km (314 mi)

Coastline: 378 km (235 mi)

Longest Distances: 330 km (205 mi) N-S 520 km (323 mi) E-W

Highest Point: Musala 2,925 m (9,596 ft)

Lowest Point: Sea level

Land Use: 34% arable land; 3% permanent crops; 18% meadows and pastures; 35% forest and woodland; 10% other

Bulgaria is a land of unusual scenic beauty, having picturesque mountains, wooded hills, beautiful valleys, grain-producing plains, and a seacoast that has both rocky cliffs and long sandy beaches.

Topography

Alternating bands of high and low terrain extend generally east and west across the country. The four most prominent of these from north to south are the Danubian plateau, the Stara Planina (Old Mountain), or Balkan Mountains, the central Thracian Plain, and the Rodopi (or Rhodope Mountains). The western part of the country, however, consists almost entirely of higher land, and the individual mountain ranges in the east tend to taper into hills and gentle uplands as they approach the Black Sea.

The Danubian plateau, also called a plain or a tableland, extends from the Yugoslav border to the Black Sea. It encompasses the area between the Danube River, which forms most of the country's northern border, and the Stara Planina to the south. The plateau rises from cliffs along the river, which are typically 91 to 182 m (300 to 600 ft) high, and abuts against the mountains at elevations on the order of 365 to 457 m (1,200 to 1,500 ft). The region slopes gently but perceptibly from the river southward to the mountains. The western portion is lower and more dissected; in the east it becomes regular but somewhat higher, better resembling a plateau. Bulgarians name local areas within it, but they do not name the region as a whole. It is a fertile area with undulating hills and is the granary of the country.

The southern edge of the Danubian plateau blends into the foothills of the Stara Planina, the Bulgarian extension of the Carpathian Mountains. The Carpathians resemble a reversed *S* as they run eastward from Czechoslovakia across the northern portion of Romania, swinging southward to the middle of that country, where they run westward and cross Romania as the Transylvanian Alps. At a famous gorge of the Danube River known as the Iron Gate, which forms part of the Romania-Yugoslavia border, the Carpathians again sweep eastward, becoming Bulgaria's Stara Planina range.

Considered in its local context, the Stara Planina originates at the Timok Valley in Yugoslavia, continues southeastward as it becomes the northern boundary of the Sofia Basin, and then turns more directly eastward to terminate at Cape Emine on the Black Sea. It is some 595 km (370 mi) in length, and some 19 to 68 km (twelve to thirty mi) in width. It retains its height well into the central part of the country, where Botev Peak, its highest point, rises to about 2,377 m (7,800 ft). The range is still apparent until its rocky cliffs fall into the Black Sea. Over most of its length, its ridge is the divide between drainage to the Danube River and to the Aegean Sea. In the east small areas drain directly to the Black Sea.

Sometimes considered a part of the foothills of the Stara Planina, but separated from the main range by a long geological trench that contains the Valley of Roses, is the Sredna Gora (Middle Forest). The Sredna Gora is a ridge running almost precisely east to west, about 160 km (100 mi) long. Its elevations run to only a little more than 1,524 m (5,000 ft) but it is narrow and achieves an impression of greater height.

The southern slopes of the Stara Planina and the Sredna Gora give way to the Thracian Plain. The plain is roughly triangular in shape, originating at a point east of the mountains that ring the Sofia Basin and broadening as it proceeds eastward to the Black Sea. It encompasses the Maritsa River basin and the lowlands that extend from it to the Black Sea. As is the case with the Danubian plateau, a great deal of this area is not a plain in strict terms. Most of its terrain is moderate enough to allow cultivation, but there are variations greater than those of a typical plain.

The Rodopi occupies the area between the Thracian Plain and the Greek border. This range is commonly described as including the Rila mountain range south of Sophia and the Pirin range in the southwestern corner of the country. As such, the Rodopi is the most outstanding topographic feature, not only of the country, but also of the entire Balkan Peninsula. The Rila contains Mount Musala—called Mount Stalin for a few years—whose 2,895 m (9,500-ft) peak is the highest in the Balkans. About a dozen other peaks in the Rila are over 2,743 m (9,000 ft). They feature a few bare rocks and remote lakes above the tree line, but the lower peaks are covered with Alpine meadows, and the general aspect of the range is one of green beauty.

The Vitosha range is an outlier of the Rila. A symmetrical, 2,286 m (7,500-ft) high, isolated peak in the range is a landmark on the outskirts of Sofia. Snow covers its conical summit most of the year, and its steep sides are forested.

The Pirin is characterized by rocky peaks and stony slopes. An impression of the landscape is provided by a local legend, which says that when the earth was being created God was flying over the peninsula with a bag of huge boulders. The rocks were too heavy for the bag, and it broke over southwestern Bulgaria.

Some Bulgarian geographers refer to the western Rodopi and the Pirin as the Thracian-Macedonian massif. In this context, the Rodopi includes only the mountains south of the Maritsa River basin. There is some basis for such a division. The Rila is largely volcanic in origin. The Pirin was formed at a different time by fracturing of the earth's crust. The uplands east of the Maritsa River are not of the same stature as the major ranges.

Sizable areas in the western and central Stara Planina and smaller areas in the Pirin and in Dobrudzha have extensive layers of limestone. There are some 2,000 caves in these deposits. The public has become more interested in the caves during the past three or four decades, but only about 400 of them have been completely explored and charted.

To the east of the higher Rodopi and east of the Maritsa River are the Sakar and Strandzha mountains. They extend the length of the Rodopi along the Turkish border to the Black Sea but are themselves comparatively insignificant. At one point they have a spot elevation of about 853 m (2,800 ft), but they rarely exceed 457 m (1,500 ft) elsewhere.

Formation of the Balkan landmasses involved a number of earth crust foldings and volcanic actions that either dammed rivers or forced them into new courses. The flat basins that occur throughout the country were created when river waters receded from the temporary lakes that existed while the rivers were cutting their new channels. The largest of these is the Sofia Basin, which includes the city and the area about 24 km (15 mi) wide and 96 km (60 mi) long to its northwest and southeast. Other valleys between the Stara Planina and the Sredna Gora ranges contain a series of smaller basins, and similar ones occur at intervals in the valleys of a number of the larger rivers.

Drainage

From a drainage standpoint, the country is divided into two nearly equal parts. The slightly larger one drains into the Black Sea, the other to the Aegean. The northern watershed of the Stara Planina, all of the Danubian plateau, and the 48 to 80 km (30 to 50 mi) inland from the coastline drain to the Black Sea. The Thracian Plain and most of the higher lands of the south and southwest drain to the Aegean Sea. Although only the Danube is navigable, many of the other rivers and streams have a high potential for the production of hydroelectric power and are sources of irrigation water. Many are already being exploited.

Insignificant when compared with the watersheds that drain to the seas, about 323 sq km (125 sq mi) of the country drain into a few small salt lakes that have no outflowing water. The largest such lake has a surface area of 6 sq km (2.5 sq mi).

By far the greater part of the country that drains to the Black Sea does so through the Danube. Most of its major tributaries in the country (from west to east, the Ogosta, Iskur, Vit, Osum, Yantra, and Lom) carry more water than do the combination of the Provadiyska, Kamchiya, Fakiyska, and Veleka rivers, all of which flow directly into the Black Sea. Of the Danube's Bulgarian tributaries, all but the Iskur rise in the Stara Planina. The Iskur rises in the Rila and flows northward through a narrow basin. Territory not far from the river on both sides of it drains in the opposite direction, to the south. The Iskur passes through Sofia's eastern suburbs and cuts a valley through the Stara Planina on its way to join the Danube.

The Iskur and the other of the Danube's north-flowing tributaries have cut deep valleys through the Danubian plateau. The eastern banks tend to rise sharply from the rivers; the western parts of the valleys may have broad fields with alluvial soils. The peculiar, though consistent,

PRINCIPAL TOWNS

(estimated population at 31 December 1983)

Sofia (capital)	1,093,752	Pleven	140,440
Plovdiv	373,235	Shumen	104,089
Varna	295,218	Tolbukhin	102,292
Burgas (Bourgas)	183,477	Sliven	102,037
Ruse (Roussé)	181,185	Pernik	96,431
Stara Zagora	144,904		

Area and population

Provinces	Capitals	area sq mi	area sq km	population 1986 estimate
Blagoevgrad	Blagoevgrad	2,506	6,490	346,266
Burgas	Burgas	2,972	7,697	449,314
Gabrovo	Gabrovo	786	2,035	175,120
Khaskovo	Khaskovo	1,547	4,007	301,249
Kŭrdzhali	Kŭrdzhali	1,558	4,036	302,578
Kyustendil	Kyustendil	1,174	3,041	190,410
Lovech	Lovech	1,597	4,136	202,708
Mikhaylovgrad	Mikhaylovgrad	1,393	3,609	223,292
Pazardzhik	Pazardzhik	1,720	4,455	326,315
Pernik	Pernik	923	2,391	174,419
Pleven	Pleven	1,673	4,332	362,130
Plovdiv	Plovdiv	2,177	5,639	754,393
Razgrad	Razgrad	1,030	2,668	198,007
Ruse	Ruse	992	2,570	304,443
Shumen	Shumen	1,309	3,390	254,789
Silistra	Silistra	1,097	2,842	174,052
Sliven	Sliven	1,395	3,614	239,479
Smolyan	Smolyan	1,360	3,523	164,223
Sofiya	Sofia (Sofiya)	2,766	7,165	305,251
Stara Zagora	Stara Zagora	1,956	5,066	411,506
Tolbukhin	Tolbukhin	1,816	4,704	257,298
Tŭrgovishte	Tŭrgovishte	1,055	2,732	171,167
Varna	Varna	1,477	3,825	464,701
Veliko Tŭrnovo	Veliko Tŭrnovo	1,807	4,680	339,120
Vidin	Vidin	1,161	3,006	166,388
Vratsa	Vratsa	1,527	3,955	287,841
Yambol	Yambol	1,587	4,110	203,754
City Commune				
Sofiya	Sofia (Sofiya)	461	1,194	1,199,405
TOTAL		42,823	110,912	8,949,618

Base 505374 (A01105) 6-83

pattern is caused by forces resulting from the earth's rotation; these forces give the water a motion that tends to undercut the right banks of the streams. Some of these rivers are sizable streams, but the Danube gets only a little more than 4 percent of its total volume from its Bulgarian tributaries. As it flows along the northern border, the Danube averages one to 2.5 km (1.5 mi) in width. Its highest water levels are usually reached during June floods, and in normal seasons it is frozen over for about forty days.

Several major rivers flow directly to the Aegean Sea, although the Maritsa with its tributaries is by far the largest. The Maritsa drains all of the western Thracian Plain, all of the Sredna Gora, the southern slopes of the Stara Planina, and the northern slopes of the eastern Rodopi. Other than the Maritsa, the Struma in the west and the Mesta (which separates the Pirin from the main Rodopi ranges) are the two largest of the rivers that rise in Bulgaria and flow to the Aegean. Most of these streams fall swiftly from the mountains and have cut deep, scenic gorges. The Struma and Mesta reach the sea through Greece. The Maritsa forms most of the Greek-Turkish border after it leaves Bulgaria.

About 9,712 sq km (3,750 sq mi) of agricultural land have access to irrigation waters. Dams provide the water for about one-half of the acreage; diversions from rivers and streams serve about one-third; and water pumped from the ground and from streams accounts for the remainder.

Of the dams, ninety-two are termed large state dams. Their combined capacity is three times that of some 2,000 smaller dams. The sources of four large rivers—the Maritsa, Iskur, Mesta, and Rilska (a major tributary of the Struma)—are within a few miles of each other in the high Rila. Water from the upper courses of these and several other streams supplies the Sofia area with both water and electricity, and they have a potential for further development. There are major dams on the Tundzha, Iskur, Rositsa, and Struma rivers. The Danube is too massive a stream to harness, and damming the Maritsa along most of its course would flood too much valuable land. The rivers flowing north across the Danubian plateau also tend to be overly difficult to use in the areas where they are most needed.

The Vucha River, flowing from the Rodopi into the Maritsa River, is often used to illustrate how rivers have been effectively harnessed to provide a variety of benefits. Its cascade system of hydroelectric development employs six dams having the capacity to generate over 600,000 kw of electricity. The water they back up serves the municipal water systems in Plovdiv and a number of other towns in its vicinity, and the dams provide irrigation water for nearly 101,172 ha (250,000 ac) of cropland. The reservoirs themselves are being developed as recreational areas and mountain resorts.

Where a stream is difficult to dam or to divert, water is pumped from it. This has been feasible only since about 1950, when low-cost diesel engines and sufficient hydroelectric power became available from newly constructed dams on other streams. About eighty-five huge pumping stations have been set up along the Danube River, which furnished about three-quarters of the water acquired by this method; and in 1988 there were about 1,200 lesser stations operating on smaller streams, most of them on the Thracian Plain.

BURKINA

BASIC FACTS

Official Name: Burkina Faso

Abbreviation: BUF

Area: 274,200 sq km (105,869 sq mi)

Area—World Rank: 68

Population: 8,485,737 (1988) 11,719,000 (2000)

Population—World Rank: 73

Capital: Ougadougou

Land Boundaries: 3,301 km (2,051 mi); Niger 628 km (390 mi); Benin 270 km (168 mi); Togo 126 km (78 mi); Ghana 533 km (338 mi); Ivory Coast 531 km (330 mi); Mali 1,202 km (747 mi)

Coastline: Nil

Longest Distances: 873 km (542 mi) ENE-WSW 474 km (295 mi) SSE-NNW

Highest Point: Tena Kourou 747 m (2,451 ft)

Lowest Point: Volta Noire River Valley 200 m (650 ft)

Land Use: 10% arable; 37% meadows and pastures; 26% forest and woodland; 27% other

Burkina Faso is a landlocked nation located in West Africa. The country is one vast plateau tilted toward the south. The average altitude is 400 m (1,312 ft), reaching the highest point at Tenankourou 750 m (2,461 ft).

The plateau is carved by the valleys of the three Voltas, the Black Volta, the White Volta and the Red Volta, and their main tributaries, the Sourou and the Pendjari. The Volta, so named by the Portuguese because of its winding course, is a slow meandering river, so unhealthy that people who live near its banks are highly susceptible to a variety of diseases. The most important river of the Volta system is the Black Volta, which rises not far from Bobo-Dioulasso as two streams called the Plandi and the Dienkoa. Before it reaches the border with Ghana at Ouessa, it receives a number of tributaries such as the Kou, the Sourou, the Tui and the Bougouriba. The Pendjari, rising in Benin, forms the border between Benin and

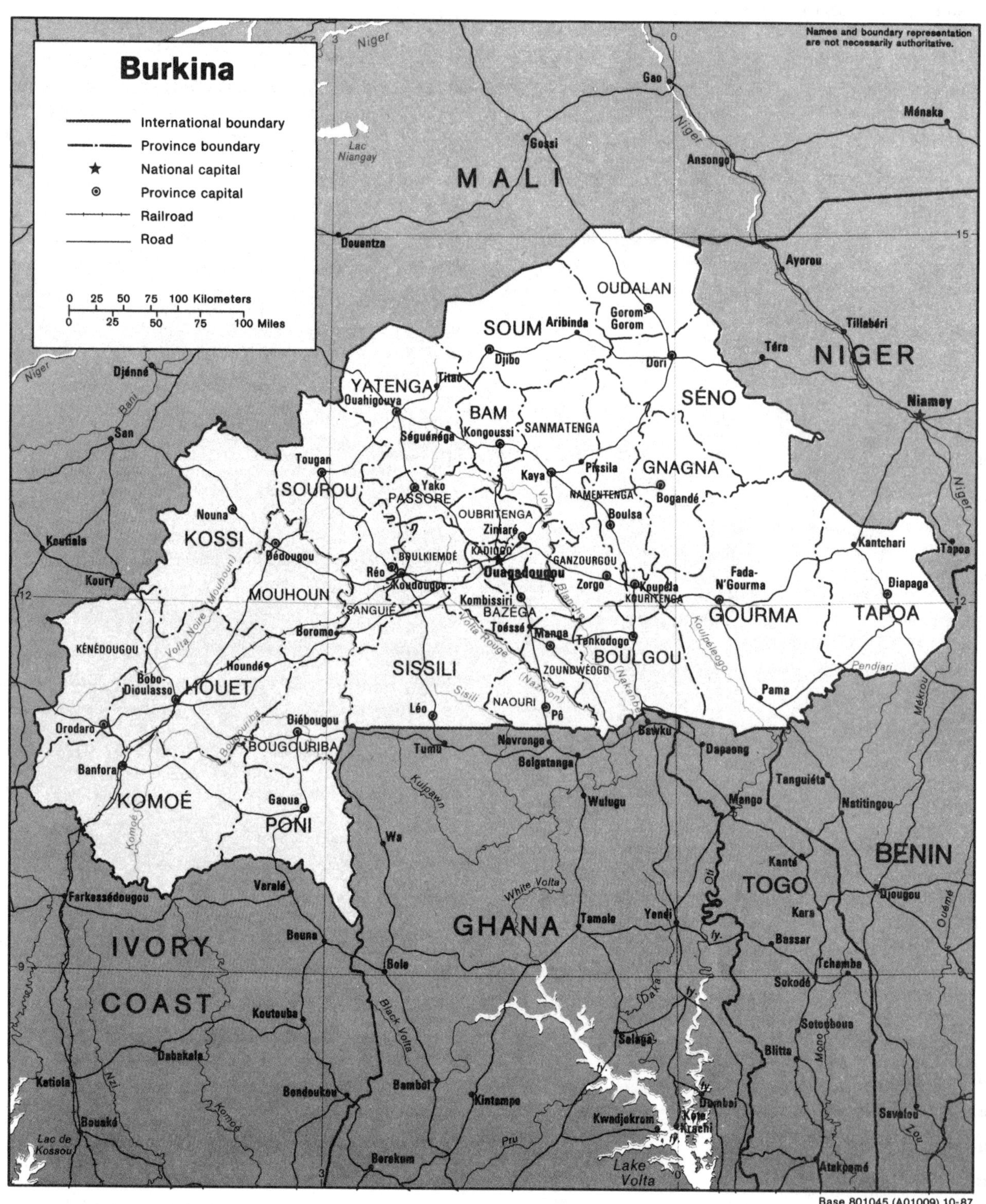

Burkina
International boundary
Province boundary
National capital
Province capital
Railroad
Road
0 25 50 75 100 Kilometers
0 25 50 75 100 Miles
Names and boundary representation are not necessarily authoritative.
MALI
NIGER
BENIN
TOGO
GHANA
IVORY COAST
OUDALAN
SOUM
YATENGA
BAM
SANMATENGA
SÉNO
GNAGNA
SOUROU
PASSORE
NAMENTENGA
OUBRITENGA
KOSSI
BOULKIEMDÉ
KADIOGO
GANZOURGOU
KOURITENGA
MOUHOUN
SANGUIÉ
BAZÈGA
GOURMA
TAPOA
KÉNÉDOUGOU
HOUET
SISSILI
ZOUNDWÉOGO
BOULGOU
NAOURI
BOUGOURIBA
KOMOÉ
PONI
Ouagadougou
Base 801045 (A01009) 10-87

Burkina for 170 km (110 mi). Burkina has one of the few permanent lakes in West Africa, the Bama, on the White Volta.

PRINCIPAL TOWNS (population at 1985 census)

Ouagadougou (capital)	442,223	Ouahigouya	38,604
Bobo-Dioulasso	231,162	Banfora	35,204
Koudougou	51,670	Kaya	25,779

Area and population

		area		population
Provinces	**Capitals**	sq mi	sq km	1985 census
Bam	Kongoussi	1,551	4,017	164,263
Bazéga	Kombissiri	2,051	5,313	306,976
Bougouriba	Diébougou	2,736	7,087	221,522
Boulgou	Tenkodogo	3,488	9,033	403,358
Boulkiemde	Koudougou	1,598	4,138	363,594
Comoé	Banfora	7,102	18,393	250,510
Ganzourgou	Zorgho	1,578	4,087	196,006
Gnagna	Bogandé	3,320	8,600	229,249
Gourma	FadaN'Gourma	10,275	26,613	294,123
Houet	Bobo-Dioulasso	6,360	16,472	585,031
Kadiogo	Ouagadougou	451	1,169	459,138
Kénédougou	Orodara	3,207	8,307	139,722
Kossi	Nouna	5,088	13,177	330,413
Kouritenga	Koupéla	628	1,627	197,027
Mouhoun	Dédougou	4,032	10,442	289,213
Nahouri	Pô	1,484	3,843	105,273
Namentenga	Boulsa	2,994	7,755	198,798
Oubritenga	Ziniaré	1,812	4,693	303,229
Oudalan	Gorom Gorom	3,879	10,046	105,715
Passoré	Yako	1,575	4,078	225,115
Poni	Gaoua	4,000	10,361	234,501
Sanguie	Réo	1,994	5,165	218,289
Sanmatenga	Kaya	3,557	9,213	368,365
Sèno	Dori	5,202	13,473	230,043
Sissili	Léo	5,303	13,736	246,844
Soum	Djibo	5,154	13,350	190,464
Sourou	Tougan	3,663	9,487	267,770
Tapoa	Diapaga	5,707	14,780	159,121
Yatenga	Ouahigouya	4,746	12,292	537,205
Zoundwéogo	Manga	1,333	3,453	155,142
TOTAL		105,869	274,200	7,967,019

BURUNDI

BASIC FACTS

Official Name: Republic of Burundi

Abbreviation: BUR

Area: 27,834 sq km (10,747 sq mi)

Area—World Rank: 128

Population: 5,155,665 (1988) 7,170,000 (2000)

Population—World Rank: 92

Capital: Bujumbura

Land Boundaries: 974 km (605 mi); Rwanda 290 km (180 mi); Tanzania 451 km (280 mi); Zaire 233 km (145 mi)

Coastline: Nil

Longest Distances: 263 km (163 mi) NNE-SSW 194 km (121 mi) ESE-WNW

Highest Point: 2,760 m (9,055 ft)

Lowest Point: Lake Tanganyika 772 m (2,534 ft)

Land Use: 43% arable; 8% permanent crops; 35% meadows and pastures; 2% forest and woodland; 12% other

Burundi is a landlocked nation located south of the equator in east-central Africa.

The geologic base of the country is an irregularly shaped area of the Great East African Plateau. Much of the countryside is covered by savanna grasslands and small farms extending over rolling hills, but there are also areas of both swamps and mountains. The divide between two of Africa's great watersheds, the Zaire and Nile Basins, extends from north to south through western Burundi at an average elevation of 2,438 m (8,000 ft). On the western slopes of this Zaire-Nile ridgeline, the land slopes abruptly into the Great East African Rift Valley, where the Rusizi Plain and Lake Tanganyika mark the western border of the country. The eastern slopes are more moderate, with rolling hills extending across the central uplands, at gradually reduced altitudes, to the dry plains and small plateaus of the eastern and southern border regions.

Except for the eastern and western border areas, Burundi lies at fairly high altitudes. The extremes of the tropical savanna climate are moderated because most of the land is at least 914 m (3,000 ft) above sea level; much of the central plateau has an average altitude of 1,524 to 1,981 m (5,000 to 6,500 ft) and the average for the entire country is about 1,615 m (5,300 ft). The heaviest concentrations of people are located in the central uplands, which are from 1,524 to 2,286 m (5,000 to 7,500 ft) in elevation. Trade winds from the Indian Ocean tend to hold temperatures down, providing these plateaus and rolling hills with a climate that is more comfortable and healthful for human beings than the higher altitudes of the Zaire-Nile Divide in the west or the lower altitudes of the Rift Valley or along the eastern and southern borders.

There are three natural regions: the Rift Valley area, consisting of the narrow plains along the Rusizi River and the shores of Lake Tanganyika (the Imbo area), together with the belt of foothills and slopeland on the western face of the Zaire-Nile Divide; the range of peaks which form this Divide; and the extensive central and eastern plateaus and savannas, separated by wide valleys and sloping into the warmer, drier plains of the eastern and southeastern borders.

The Rift Valley—Western Slope Area

West of the mountains of the Zaire-Nile Divide, a narrow plain extends southward along the Rusizi River from the Rwanda border through Bujumbura, at the north corner of Lake Tanganyika, then extends southward for another 48 km (30 mi) along the eastern shore of the lake. All of this plain (the Imbo) is below 1,066 m (3,500 ft) in elevation, and tropical temperatures prevail, with average annual temperatures of 70° to 73° F. Rainfall is 76 to 101 cm (30 to 40 in) per year and is often erratic. Until recently this expanse of acacia and palm trees, thorny plants, and grasses was primarily a homeland for elephants, antelopes, and various other savanna animals. Since 1950 resettlement programs have brought in farmers who produce corn, rice, cotton, peanuts, and vegetables. Much of this produce is sold in Bujumbura markets.

PRINCIPAL TOWNS

Bujumbura (capital), population 172,201 (census of August 1979); Gitega 15,943 (1978).

Area and population

Provinces	Capitals	area sq mi	area sq km	population 1987 estimate
Bubanza	Bubanza	422	1,093	200,420
Bujumbura	Bujumbura	515	1,334	584,812
Bururi	Bururi	971	2,515	374,660
Cankuzo	Cankuzo	749	1,940	129,275
Cibitoke	Cibitoke	633	1,639	235,279
Gitega	Gitega	768	1,989	561,950
Karuzi	Karuzi	563	1,459	258,811
Kayanza	Kayanza	475	1,229	446,219
Kirundo	Kirundo	661	1,711	359,485
Makamba	Makamba	761	1,972	155,676
Muramvya	Muramvya	591	1,530	437,846
Muyinga	Muyinga	705	1,825	315,008
Ngozi	Ngozi	567	1,468	476,408
Rutana	Rutana	733	1,898	179,302
Ruyigi	Ruyigi	913	2,365	206,933
TOTAL LAND AREA		10,026	25,967	4,922,084
INLAND WATER		721	1,867	
TOTAL AREA		10,747	27,834	

Above these flat plains, the belt of foothills and steep slopelands that forms the western face of the Divide is a mixture of farmlands and rough gullies and valleys. These rougher areas contain gallery forests, in which tree branches tend to form canopies over open aisles, or galleries. Temperatures decrease with elevation, and rainfall increases to as much as 127 cm (50 in) annually near the top of the watershed.

The Zaire-Nile Divide

Eastward from Lake Tanganyika and the Rusizi-Imbo Plains, which lie in the Great East African Rift Valley, the mountains associated with this Rift system rise steeply; an elongated series of ridges, generally less than 16 km (10 mi) wide, averages about 2,438 km (8,000 ft) in elevation, but none of the peaks within Burundi exceed 2,590 m (8,500 ft). Extending from south to north throughout the western part of the country, this Zaire-Nile crest is the dominant geographic feature. The "source of the Nile" is the headwaters of the Ruvubu River, on the eastern slope of this range in south-central Burundi. Annual rainfall is 127 to 152 cm (50 to 60 in) in the upper elevations, rising to about 178 cm (70 in) in the north, near the Rwanda border. Average annual temperatures are in the low sixties, with a high daily range, and nights are distinctly cool for much of the year. Farmers in these upper elevation produce principally peas, corn, and barley.

Plateaus and Savannas

The central and eastern plateaus, from the Zaire-Nile Crest to the eastern settlements of Kirunda, Muhinga, and Cankuza, are the areas of heaviest settlement. Almost half of the population lives on these uplands, between 1,524 and 1,981 m (5,000 and 6,500 ft) above sea level. The climate in the upper elevations is cool and rainy. In central Burundi average annual temperatures are approximately 65° to 67° F. and rainfall is 101 to 139 cm (40 to 55 in). As there is little trouble with tsetse fly above 1,524 m (5,000 ft) cattle herders as well as farmers have crowded these middle and higher altitude plateaus. All usable land is heavily farmed or grazed. Farmers grow wheat, barley, sorghum, corn, various beans and peas, bananas, manioc, and other tropical- and temperate-zone food crops; coffee and cotton are the most important commercial crops.

The savannas of the eastern border are less than 1,524 m (5,000 ft) above sea level, and the Mosso plains along the Muragarazi, Rumpungu, and Rugusi Rivers, which mark much of the southeastern border, average about 1,036 m (3,400 ft). Average temperatures are between 68° and 73° F., hotter than the central uplands. Rainfall ranges from 76 to 114 cm (30 to 45 in) per year, but is irregular, and these areas take on a semidesert appearance during the annual dry seasons. These plains are not as heavily populated as the higher plateaus, but crops include cotton, peanuts, cassava, coffee, and peppers. Before 1900 this was a stockraising area, but *nagana,* the bovine sleeping sickness carried by the tsetse fly, became more prevalent, and these hot savanna lands, many already overgrazed into semidesert, were left to the wild game until new settlers began to arrive during the 1950's. Outside the settled areas there are large areas of savanna land, swamplands, scattered forests, and bamboo groves which still provide a home for lions, leopards, warthogs, various antelopes, and other large and small animals and birds.

DRAINAGE

The country includes areas along both slopes of the Zaire-Nile Divide, the range of mountains associated with the western extension of the Rift Valley. Westward from this north-south ridgeline, short, swift rivers carry runoff waters down Burundi's narrow western watershed into the Rusizi River or Lake Tanganyika, which together form the boundary with Zaire. Since the Rusizi flows into Lake Tanganyika, which drains into the Zaire River systems, rainfall from this western slope eventually flows into the Atlantic Ocean. The short western-slope rivers tend to have shallow beds, and falls and rapids are common.

The areas east and southeast of the ridgeline, constituting four-fifths of the country, are part of the upper reaches of the Nile River system. From the small surviving forests near the mountain tops, elevations decrease eastward in a series of plateaus and rounded hills to about 1,524 m (5,000 ft) in the border areas of the northeast. Principal rivers in the central plateaus include the Ruvironza and

Burundi

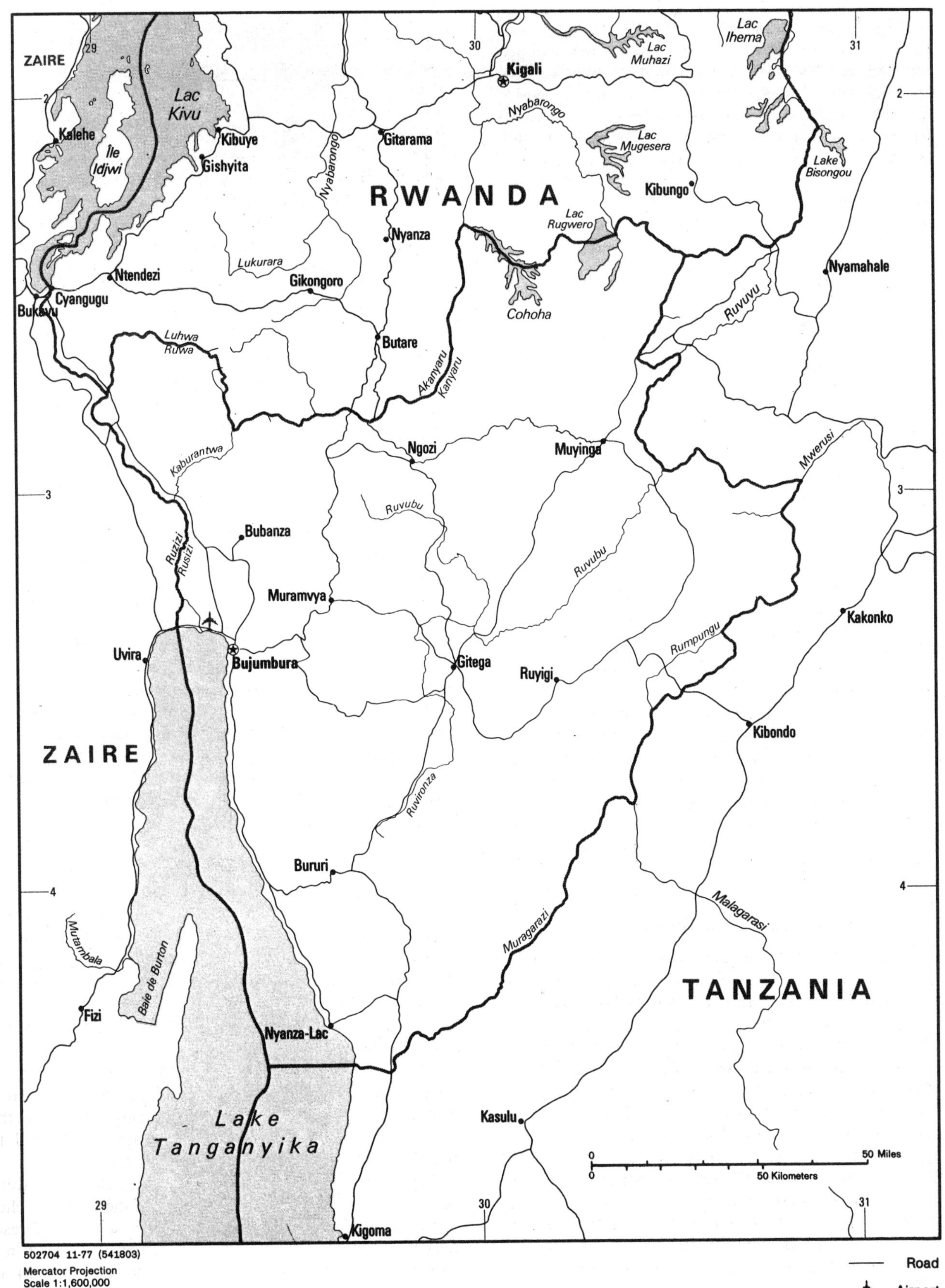

502704 11-77 (541803)
Mercator Projection
Scale 1:1,600,000

Ruvubu, the main channels of extremely complex networks. On the average, slopes are not as steep as those on the western face of the Divide, but cropland is subjected to serious erosion. Drainage channels in the southeastern areas carry surplus water south or east to the two principal rivers on the border with Tanzania, the Muragarazi and the Rumpungu. The Muragarazi flows northeastward, then southward and westward in Tanzania, eventually emptying into Lake Tanganyika. The Rumpungu extends northeastward into the Kagera River, which empties into Lake Victoria, a part of the Nile system.

Rivers vary greatly in different areas and seasons. Some segments are rushing torrents; other stretches of the same river, crossing a nearly level plateau, may be almost stagnant and lined with papyrus grasses. Flow rates depend upon wide local variations in rainfall and upon the vagaries of the two annual wet and dry seasons.

CAMBODIA

BASIC FACTS

Official Name: Cambodia

Abbreviation: CAM

Area: 181,035 sq km (69,898 sq mi)

Area—World Rank: 85

Population: 6,685,592 (1988) 9,772,000 (2000)

Population—World Rank: 74

Capital: Phnom Penh

Boundaries: 2,715 km (1,687 mi); Laos 541 km (336 mi); Vietnam 982 km (610 mi); Thailand 803 km (499 mi)

Coastline: 389 km (242 mi)

Longest Distances: 730 km (454 mi) NE-SW 512 km (318 mi) SE-NW

Highest Point: Mount Aoral 1,813 m (5,948 ft)

Lowest Point: Sea level

Land Use: 16% arable; 1% permanent crops; 3% meadows and pastures; 76% forest and woodland; 4% other

Cambodia in the southwestern part of the Indochinese peninsula lies completely within the tropics; its southernmost points are only slightly more than 10° above the equator. Roughly square in shape, the country is bounded on the north by Thailand and Laos, on the east and southeast by Vietnam and on the west by the Gulf of Thailand and by Thailand.

Much of the country's area is a rolling plain. Dominant features are the large, approximately centrally located Tonle Sap (Great Lake) and the Mekong River, which traverses the country from north to south.

Three-quarters of the area of the Khmer Republic consists of the Cambodian Basin. This basin's economic importance and population density and the general homogeneity of the people inhabiting it are such as to warrant considering the country essentially a single region. Rimming the basin on the southwest are the Cardamom and Elephant mountain ranges. On the north is the Dangrek Range, and to the northeast and east lies higher land that merges into the Central Highlands of Vietnam.

The basin's terrain consists chiefly of plains with elevations generally under 91 m (300 ft). Where the altitude increases slightly, the terrain becomes more rolling or dissected. The Cardamom Range, trending generally in a northwest-southeast direction, has elevations rising to over 1,524 m (5,000 ft); an eastern spur of this range reaches a height of 1,771 m (5,810 ft), the highest point in the country. The Elephant Range, running south and southeastward from the Cardamom, has elevations above 914 m (3,000 ft). These two ranges are bordered on the Gulf of Thailand side by a narrow coastal plain that was largely isolated until the opening of the port of Kampong Saom (Sihanoukville) in 1960 and the construction of a connecting road and rail line from the port of Phnom Penh.

The Dangrek Range at the northern rim of the basin consists of a steep escarpment with an average elevation of about 487 m (1,600 ft). The escarpment faces southward and constitutes the southern edge of the Khorat Plateau in Thailand. The watershed along the escarpment marks the boundary between Thailand and the Khmer Republic. In general the escarpment impedes easy communication between the two countries. Between the western part of the Dangrek and northern part of the Cardamom ranges, however, lies an extension of the Cambodian Basin that merges into lowlands in Thailand, allowing easy access to Bangkok.

The eastern end of the Dangrek Range and the northeastern highlands are separated by the valley of the Mekong, which offers a communication route between Cambodia and Laos. To the southeast the basin joins the Mekong delta, which, extending into Vietnam, provides both water and land communications between the two countries.

The drainage pattern of the Cambodian Basin is centered on the Mekong River and the Tonle Sap. With the exception of a small part of southeastern Cambodia, rivers in the basin drain into these two dominant features of the drainage system. In the southwest the Cardamom and Elephant ranges form a separate drainage divide. To the east of this divide the rivers flow into the Tonle Sap, whereas those to the west debouch into the Gulf of Thailand. Toward the southern end of the Elephant Range, however, some small rivers flow southward on the eastern side of the divide, emptying into the gulf because of the topography.

The Mekong in Cambodia flows southward from the Cambodia-Laos border to below the provincial capital of

Base 800460 (546756) 11-85

Kracheh, where it turns westward then southwestward to Phnom Penh. There are extensive rapids above Kracheh. From Kampong Cham the gradient becomes extremely low, and inundation of areas along the river occurs at flood stage—June through November—through breaks in the natural levees that have built up along its course. From Phnom Penh the river flows generally southeastward. It divides at this point into two principal channels, the new one being known as the Tonle Basak River, which flows independently from there on through the delta area into the South China Sea.

The river's flow is monsoonal—that is, its high and low level points are related to the wet and dry climatic periods that characterize its course through Southeast Asia. Its annual flood waters and the great expansion of the Tonle Sap, which reaches its maximum area in September or early October, are natural factors around which revolve the lives of a considerable part of the country's population. Tonle Sap was once an arm of the sea but was made an inland lake by the gradual silting up of the Mekong delta. The raising of the shore level prevented entry of tidal waters and eventually resulted in the conversion of the Tonle Sap into a freshwater body.

Area and Population

Provinces	Capitals	area sq mi	sq km	population 1981 census
Bătdâmbâng	Bătdâmbâng	7,407	19,184	719,000
Kâmpóng Cham	Kâmpóng Cham	3,783	9,799	1,070,000
Kâmpóng Chhnăng	Kâmpóng Chhnăng	2,132	5,521	221,000
Kâmpóng Saôm	Kâmpóng Saôm	26	68	53,000
Kâmpóng Spoe	Kâmpóng Spoe	2,709	7,017	340,000
Kâmpóng Thum	Kâmpóng Thum	10,657	27,602	379,000
Kâmpôt	Kâmpôt	2,320	6,008	354,000
Kândal	. . .	1,472	3,812	720,000
Kaôh Kŏng	Krŏng Kaôh Kŏng	4,309	11,161	25,000
Krâchéh	Krâchéh	4,283	11,094	157,000
Môndól Kiri	Senmonorom	5,517	14,288	16,000
Phnom Penh	Phnom Penh	18	46	329,000
Poŭthĭsăt	Poŭthĭsăt	4,900	12,692	175,000
Preăh Vihéar	Phnum Tbéng Meanchey			70,000
Prey Vêng	Prey Vêng	1,885	4,883	672,000
Rôtânokiri	Lumphăt	4,163	10,782	45,000
Siĕmréab	Siĕmréab	6,354	16,457	477,000
Stŏeng Trêng	Stŏeng Trêng	4,283	11,092	39,000
Svay Riĕng	Svay Riĕng	1,145	2,966	292,000
Takêv	Takêv	1,376	3,563	531,000
TOTAL LAND AREA		68,721	177,987	6,684,000
INLAND WATER		1,177	3,048	
TOTAL AREA		69,898	181,035	

Connected to the Mekong River by the Tonle Sab River, the great lake acts as a natural reservoir during the Mekong's flood period. The cresting waters of the Mekong are too great for discharge by its silted channels and those of its tributaries below the confluence with the Tonle Sab River. During this time the Mekong's waters back up, reversing the flow of the Tonle Sab, and enlarge the area of the Tonle Sap from a low of about 2,589 sq km (1,000 sq mi) in the dry season to 10,359 sq km (4,000 or more sq mi) at the height of flooding.

When the Mekong's level lowers, the Tonle Sab River again discharges the waters of the Tonle Sap. As the lake retreats, it leaves a new layer of sediment. Poor drainage immediately around the lake and the annual flooding have made a considerable area around it marshland during the dry season that is unusable for agriculture. The amount of sediment carried into the lake during the Mekong's flood stage appeared to be greater than that discharged later through the Tonle Sab River. Gradual silting of the lake would seem to be occurring—the Tonle Sap during low-water level is only about five feet deep, compared with between forty-five and fifty feet at flood stage.

CAMEROON

BASIC FACTS

Official Name: Republic of Cameroon

Abbreviation: CMN

Area: 465,468 sq km (179,714 sq mi)

Area—World Rank: 49

Population: 10,531,954 (1988) 15,801,000 (2000)

Population—World Rank: 59

Capital: Yaounde

Boundaries: 5,159 km (3,207 mi); Chad 1,047 km (651 mi); Central African Republic 822 km (511 mi); Congo 520 km (323 mi); Gabon 302 km (188 mi); Equatorial Guinea 183 km (114 mi); Nigeria 1,921 km (1,194 mi)

Coastline: 364 km (226 mi)

Longest Distances: 1,206 km (749 mi) N-S 717 km (446 mi) E-W

Highest Point: Mount Cameroon 4,100 m (13,451 ft)

Lowest Point: Sea level

Land Use: 13% arable; 2% permanent crops; 18% meadows and pastures; 54% forest and woodland; 13% other

Cameroon in west-central Africa, forms an irregular wedge extending northeastward from a coastline on the Gulf of Guinea, an arm of the Atlantic Ocean, to Lake Chad, 1,126 km (700 mi) inland. Behind the swamps and the lowlands generally referred to as the southwestern coastal zone, the land rises to mountains and plateaus extending more than 804 km (500 mi) inland before descending to a flat plain of moderate elevation in the far north.

Four loosely defined regions provide a useful descriptive framework: the northern plains, the central and southern plateaus, the western highlands and mountains, and the lowlands along the coast.

Northern Plains

The northernmost area of the country extends to Lake Chad, where the borders of Nigeria, Chad, and Cameroon intersect. The narrow neck of territory south of the lake is part of a shallow inland basin, several hundred miles wide, which extends in all directions from Lake Chad. Broad areas of low, rolling hills bear little vegetation on their thin soils; other areas are flat, marked here and there by scattered outcrops or hills of resistant rock rising above the general erosional level. West of Maroua the scattered rocky mounds and minor ridges are more numerous, rising westward to hills and elongated ridges.

Central and Southern Plateaus

The Adamaoua Plateau, lying between 7°N and 9°N latitude, extends from the eastern to the western border of Cameroon at elevations that are everywhere more than 914 m (3,000 ft) above sea level and average about 1,371 m (4,500 ft). Surface features in the central parts of this high plateau include small hills or mounds capped by erosion-resistant granite or gneiss. Along the western and, to a lesser degree, the eastern borders, old eruptions from fissures and volcanoes have covered thousands of square miles of the underlying granite with lava.

South of the Adamaoua Plateau begins a series of lower plateaus that extend throughout most of South Central and Eastern provinces at elevations averaging about 914 m (3,000 ft) but descending gradually southward to the border and westward toward a series of terraces leading downward to the coastal plain. The surfaces of these extensive southern plateaus are primarily mixtures of ancient granite and sedimentary rock. Soils are shallow in most areas and, for the most part, have been formed from the underlying granite.

PRINCIPAL TOWNS

1976 (population at census): Douala 458,426, Yaoundé (capital) 313,706, Nkongsamba 71,298, Maroua 67,187, Garoua 63,900, Bafoussam 62,239, Bamenda 48,111, Kumba 44,175, Limbe (formerly Victoria) 27,016.

Mid-1985 (estimated population): Douala 852,705, Yaoundé 583,470.

Western Highlands

The Cameroon Mountains, the highest range in the country, extend southeastward from the Cameroon-Nigeria border area at about 7°N latitude to Mount Cameroon on the coast. The major mountain range and the upland areas on its eastern and western slopes were built up by volcanic activity associated with a series of faults in the granite substructures underlying the African continent.

All of the ancient volcanoes in this complex had subsided before the dawn of recorded history except for Mount Cameroon, which has been active on four occasions during this century: 1909, 1922, 1954, and 1959. In 1922 and 1959 molten lava flowed several miles, destroying plantations on the lower slopes. The mountain is a complex of several connected fissures and cones, one of which reaches 4,069 m (13,350 ft) above sea level, more than half again as high as any other peak in the country. Elsewhere in the Cameroon Mountains, elevations range between 1,676 m and 2,438 m (5,500 and 8,000 ft).

Other ranges of lower elevation stand in the north near the western border of Northern Province. The most important of these are the Alantika Mountains, which mark the border for a short distance at about 8°30′ N latitude, and the Mandara Hills, which extend northward from the town of Garoua and the Bénoué River to about 11°N latitude.

Area and population

		area		population
Provinces	**Capitals**	sq mi	sq km	1984 estimate
Adamoua	Ngaoundéré	23,979	62,105	355,800
Centre	Yaoundé	26,655	69,035	1,764,400
Est	Bertoua	42,086	109,002	420,000
Extrême-Nord	Maroua	12,477	32,316	1,400,000
Littoral	Douala	7,810	20,229	1,829,900
Nord	Garoua	26,134	67,686	508,200
Nord-Ouest	Bamenda	6,722	17,409	1,009,100
Ouest	Bafoussam	5,360	13,883	1,197,700
Sud	Ebolowa	18,200	47,137	356,400
Sud-Ouest	Buea	9,540	24,709	700,900
LAND AREA		178,963	463,511	9,542,400
INLAND WATER		751	1,947	
TOTAL AREA		179,714	465,458	

Coastal Zone

Most of the coastal zone is a flat area of sedimentary soils that front on the Gulf of Guinea for about 257 km (160 mi). It is less than 32 km (20 mi) wide in most of the area northwest of Mount Cameroon, which divides the northwestern coastal plain from the broader lowlands in the central coastal area. Around Douala, the lowlands extend inland as much as 80 km (50 mi) narrowing to as little as five miles farther south. Along its seaward edges the central segment of the coastal zone is a series of many adjoining deltas.

Numerous rivers, fed by heavy rains during most of the year in the coastal zone and the adjacent plateaus and mountains, continue to expand the deltas with erosional debris. Near their mouths the major rivers, which are partially choked by this debris, divide into numerous sluggish channels. The various estuaries are tangled complexes of these channels and are also fed by small local streams.

Close to the coast the older deltas and flat swamplands are covered with mangrove trees and other swampland vegetation. From these coastal flats the plains rise very gradually to about 91 m (300 ft) above sea level. At approximately this level, a relatively abrupt increase in slope marks the first of several steps, or benches, leading upward to the inland plateaus.

Drainage Patterns

Most surface runoff from Cameroon eventually flows westward to the Atlantic Ocean—much of it by circuitous routes and by way of external river systems, such as the Niger and the Zaire. Three primary watershed ridgelines divide the country: the centrally located Adamaoua Plateau, a poorly defined north-south ridgeline in Eastern province, and the Cameroon Mountains.

CAMEROON
Road
Railroad
0 50 100 Miles
0 50 100 Kilometers
BOUNDARY REPRESENTATION IS NOT NECESSARILY AUTHORITATIVE
NIGER
NIGER
LAKE CHAD
N'DJAMENA
Kousseri
CHAD
Chari
Maiduguri
Bama
Yedseram
Maroua
Logone
Yagoua
Bongor
Kaélé
Fianga
Kabia
Pala
Gombe
Jos
NIGERIA
Benue
Jimeta
Garoua
Bénoué
Moundou
Makurdi
Donga
Takum
Ngaoundéré
Mbéré
Tibati
Wum
Bamenda
Mbam
Ikom
Cross
Mamfe
Baboua
Bouar
CENTRAL AFRICAN REPUBLIC
Bafoussam
Bafang
Noun
Lom
Nkongsamba
Bélabo
Kadei
Kumba
Bafia
Bertoua
Mambéré
Ntui
Nanga-Eboko
Wouri
Batouri
Berbérati
Buea
Douala
Sanaga
Victoria
Edéa
YAOUNDÉ
MALABO
Nyong
Mbalmayo
Isla de Bioko (Fernando Po)
Kribi
Ebolowa
Sangmélima
Dja
Sangha
EQUATORIAL GUINEA
Ntem
BIGHT OF BIAFRA
Bitam
Bata
GABON
CONGO
Ngoko
12
8
4
16

Rivers in Northern Province, where annual wet and dry seasons occur, exhibit major seasonal fluctuations in volume. Practically all rivers in the other six provinces carry a heavy flow for most of the year; most areas south of 5°N latitude have two rainfall and runoff maximums per year, but the variations in flow are within much narrower limits than those observed on northern streams.

The extreme north is a complex, relatively flat area of inland basins that have no outlet to the sea. The Logone and Chari river systems along the northeastern border annually inundate a broad area before emptying into Lake Chad, the major inland basin in this part of the continent. Most of these floodwaters are not collected locally but have been brought from high rainfall areas farther south in eastern Cameroon, in CAR, and in Chad. Most other rivers in the north flow only during the rainy half of the year and disappear in the sands and swamps of other shallow inland basins.

Some of the runoff originating on the Adamaoua Plateau flows northward into the upper tributaries of the Bénoué River, which winds across the western border and joins the Niger, the major river in Nigeria. By way of several tributaries, the Niger also receives a heavy flow from the Cameroon Mountains and from associated highlands in Northwestern and Southwestern provinces.

Various streams originating in the southern part of the Adamaoua Plateau feed into the Sanaga, the largest river in the southwestern part of the country. This river also collects a huge flow from heavy rainfall areas on the eastern slopes of the Cameroon Mountains and channels this heavy runoff into the Atlantic Ocean south of Douala. Three other major rivers—the Wouri, Dibamba, and Nyong—also feed into the tangled complex of deltas on the central Atlantic coast. Farther south, near the borders with Equatorial Guinea and Gabon, the Campo River watershed extends inland for about 321 km (200 mi).

Both the Sanaga and the Nyong rivers collect runoff from parts of Eastern province, but most of this very wet forest area is drained by various tributaries of the Sangha River, which for a short distance marks the border with the Congo and then flows southward into the Zaire River.

CANADA

BASIC FACTS

Official Name: Canada

Abbreviation: CAN

Area: 9,970,610 sq km (3,849,675 sq mi)

Area—World Rank: 2

Population: 26,087,536 (1988) 29,028,000 (2000)

Population—World Rank: 31

Capital: Ottawa

Boundaries: 37,636 km (23,386 mi); United States: lower 48 6,416 km (3,987 mi); Alaska 2,477 km (1,539 mi)

Coastline: Arctic Ocean 9,286 km (5,770 mi); Atlantic Ocean 9,833 km (6,110 mi); Pacific Ocean 2,543 km (1,580 mi); Hudson Bay and Strait 7,081 km (4,400 mi)

Longest Distances: 5,187 km (3,223 mi) E-W 4,627 km (2,875 mi) N-S

Highest Point: Mount Logan 5,951 m (19,524 ft)

Lowest Point: Sea level

Land Use: 5% arable; 3% meadows and pastures; 35% forest and woodland; 57% other

GEOGRAPHICAL FEATURES

Canada occupies all of the North American continent north of the United States except Alaska and the small French islands of St. Pierre and Miquelon. The most striking geographical characteristic of Canada is its immense size. It is the largest country in the Western Hemisphere and the second-largest in the world, next to the Soviet Union. Canada's size is about the same as that of the continent of Europe. Canada also encompasses the Canadian continental margin, including Hudson Bay, the Gulf of St. Lawrence, the Pacific Coast Straits and the channels of the Arctic Archipelago. Canada's longitudinal extent is such that it requires six time zones.

Topographically, Canada is divided into the Atlantic provinces, the Great Lakes-St. Lawrence Lowlands, the Canadian Shield, the Interior Plains, the Western Cordillera and the Northwest Territories.

The foundation of Canadian geology is the Canadian Shield (sometimes called the Precambrian Shield or the Laurentian Plateau), which takes up almost half of Canada's total area. It extends beyond the Canadian boundary into the United States in two limited areas: at the head of Lake Superior and in the Adirondack Mountains. Structurally, the shield may be thought of as a huge saucer, the center of which is occupied by Hudson and James bay, which have breached the northeastern rim to drain into the Atlantic Ocean through Hudson Strait. Most of the shield is relatively level and less than 612 m (2,000 ft.) above sea level. Only along the dissected rim of the saucer rim are there major hills and mountains: the Torngat Mountains in northeastern Labrador, the Laurentian Highlands, and along the northern shores of Lake Superior. Except for the plains, the rest of the shield is composed of undulating terrain with rocky, knoblike hills, the hollows between which are occupied by lakes interconnected by rapid streams.

The Canadian Shield is surrounded by a series of lowlands, the Atlantic region and the Great Lakes-St. Lawrence Lowlands to the east, the Interior Plains to the west and the Arctic Lowlands to the north. The Atlantic provinces have rugged, indented coasts. The Great Lakes-St. Lawrence Lowlands constitute the heartland of the country's population. This region has the largest area of level land easily accessible by water from the east.

In the Far West is the Western Cordillera, composed of relatively young, folded and faulted mountains and plateaus. It is only some 805 km. (500 mi) in Canada, much narrower than in the United States, with less extensive interior plateaus. Generally the mountains are much higher in Canada and contain some of the most beautiful alpine scenery in the world. The only other parts of Canada with comparable spectacular mountains are Baffin and Ellesmere islands in the northeastern Arctic.

Between the Western Cordillera and the Canadian Shield is the region broadly known as the West, including the Manitoba and Mackenzie lowlands. The Manitoba Lowland (leading to the Saskatchewan and Alberta plains) is the only part of Canada that is as flat as a tabletop. The boundary between the Manitoba Lowland and the Saskatchewan Plain is marked by the Manitoba Escarpment. The Saskatchewan and Alberta plains are divided in the south by the Missouri Couteau. The landscape of the two plains is similar to that of the U.S. Great Plains, with rolling plains; deeply incised rivers; water-filled depressions called sloughs; dry streambeds called coulees; and in the drier areas, mesas, buttes and badlands.

The Northwest Territories is a political rather than a geographical term. It covers the region east of the Western Cordillera and north of the Interior Plains and the Canadian Shield. Within this large area there are two distinct sub-regions: the subarctic Mackenzie River Valley to the west, and the arctic area of the islands and north-central mainland.

The highest point in Canada is Mount Logan (5,951 m; 19,525 ft); in the St. Elias Mountains of Yukon Territory. Most peaks in the Canadian Western Cordillera are over 4,500 m (14,765 ft) high. Rossland, B.C., is the highest city (1,056 m; 3,465 ft); Louisa, Alta, the highest hamlet (1,540 m; 5,053 ft), and Chilco Lake, B.C., with an area of 158 sq km (61 sq mi), the highest lake.

About 7.6% of Canada's total area is covered by lakes and rivers, making surface water the source of 90% of freshwater. The Central Canadian Shield is drained by the Nelson-Saskatchewan, Churchill, Severn and Albany rivers, flowing into Hudson Bay. The 4,290-km-long (2,635-mi-long) Mackenzie River, with its tributaries and three large lakes—Great Bear, Great Slave, and Athabasca—drains an area of more than 2.6 million sq km (1 million sq mi) into the Arctic Ocean. The Columbia, Fraser and Yukon rivers are the principal drainage systems of British Columbia and Yukon Territory. The Great Lakes drain into the broad St. Lawrence River, which flows into the Gulf of St. Lawrence.

Lakes are the natural regulators of river flow, smoothing out peak flows during flooding and sustaining the flow during dry seasons. Among the largest freshwater bodies in the world are the Great Lakes, of which 36% is in Canada. Groundwater is the principal source of water for streams during the frequent dry weather periods in the Prairies. In hot summer months, glaciers may contribute up to 25% of the flow of the Saskatchewan and Athabasca rivers. On an average annual basis, Canadian rivers discharge roughly 107 million sq m (1.152 billion sq ft) per second, or nearly 9% of the world's renewable water supply and equivalent to about 60% of Canada's mean annual precipitation.

PRINCIPAL TOWNS (census results, 3 June 1986)*

Ottawa (capital)	819,263†	St Catharine's-Niagara	343,258
Toronto	3,427,168	London	342,302
Montréal	2,921,357	Kitchener	311,195
Vancouver	1,380,729	Halifax	295,990
Edmonton	785,465	Victoria	255,547
Calgary	671,326	Windsor	253,988
Winnipeg	625,304	Oshawa	203,543
Québec	603,267	Saskatoon	200,665
Hamilton	557,029		

*Including Canadian residents temporarily in the USA, but excluding US residents temporarily in Canada.

†Including Hull.

Area and population

Provinces	Capitals	area sq mi	area sq km	population 1986 census
Alberta	Edmonton	248,800	644,390	2,375,278
British Columbia	Victoria	358,971	929,730	2,889,207
Manitoba	Winnipeg	211,723	548,360	1,071,232
New Brunswick	Fredericton	2,834	72,090	710,422
Newfoundland	Saint John's	145,510	371,690	568,349
Nova Scotia	Halifax	20,402	52,840	873,119
Ontario	Toronto	344,090	891,190	9,113,515
Prince Edward Island	Charlottetown	2,185	5,660	126,646
Quebec	Quebec	523,859	1,356,790	6,540,276
Saskatchewan	Regina	220,348	570,700	1,010,198
Territories				
Northwest Territories	Yellowknife	1,271,442	3,293,020	52,238
Yukon Territory	Whitehorse	184,931	478,970	23,504
TOTAL LAND AREA		3,558,096	9,215,430	25,354,064
INLAND WATER		291,579	755,180	
TOTAL AREA		3,849,675	9,970,610	

Canada's coastlines of nearly 244,000 km (151,647 mi) on the mainland and offshore islands are among the largest of any country in the world. On the Atlantic coast the submerged continental shelf has great width and diversity of relief. From the coast of Nova Scotia its width varies from 97 to 161 naut km (60 to 100 naut mi), from Newfoundland 161 to 451 naut km (100 to 280 naut mi) at the entrance of Hudson Strait, and northward it merges with the submerged shelf of the Arctic Ocean. The outer edge varies in depth from 183 to 366 m (620 to 10,201 ft). The overall gradient is slight, but the shelf is studded with shoals, ridges and banks. Hudson Bay is a shallow inland sea, 822,325 sq km (317,417 sq mi) in area, having an average depth of 128 m (422 ft). Hudson Strait separates Baffin Island from the continental coast and connects Hudson Bay with the Atlantic Ocean. It is 796 km (495 mi) long and from 69 to 222 km (43 to 138 mi) wide. The Pacific coast is strikingly different and is characterized by bold, abrupt relief—a repetition of the mountainous landscape. Numerous inlets penetrate the coast up to 140 km (89 mi) usually 1.6 naut km (1 naut mi) wide, with deep side canyons. From the islet-strewn coast the continental shelf extends from 80 to 161 naut km (50 to 100 naut mi) except on the western slopes of Vancouver Island and the Queen Charlotte Islands, where the seafloor drops rapidly.

Canada

800611 4-86

The Arctic Archipelago lies on a submerged plateau whose floor varies from flat to gently undulating. From the Alaskan border eastward to the mouth of the Mackenzie River the shelf is shallow and continuous, with its outer edge at a depth of 64 m (210 ft) about 69 naut km (40 naut mi) from the shore. Near the western edge of the Mackenzie River delta it is indented by the Mackenzie Trough (formerly known as the Herschel Sea Canyon), whose head comes within 24 naut km (15 naut mi) of the coast. The submerged portion of the Mackenzie Delta forms a pock-marked undersea plain, most of it less than 55 m (180 ft) deep and up to 121 naut km (75 naut mi) wide and 402 naut km (250 naut mi) wide. Most of the continental shoulder is over 549 m (1,801 ft) deep, sloping to the abyssal Canada Basin at 3,658 m (12,002 ft). The deeply submerged continental shelf runs along the entire western coast of the Arctic Archipelago from Banks Island to Greenland.

The largest islands are those in the Arctic Archipelago, extending from St. James Bay to Ellesmere. The largest on the western coast are Vancouver Island and the Queen Charlotte Islands. The largest on the eastern coast are Newfoundland; Prince Edward Island; Cape Breton Island; Grand Manan and Campobello Islands of New Brunswick; and Anticosti Island and the Îles de la Madeleine of Quebec. Notable islands of the inland waters include Manitoulin Island, in Lake Huron; the so-called Thirty Thousand Islands, in Georgian Bay; and the Thousand Islands, in the outlet from Lake Ontario into the St. Lawrence River.

CAPE VERDE

BASIC FACTS

Official Name: Republic of Cape Verde

Abbreviation: CAV

Area: 4,033 sq km (1,557 sq mi)

Area—World Rank: 152

Population: 353,885 (1988); 478,800 (2000)

Population—World Rank: 149

Capital: Praia

Land Boundaries: Nil

Coastline: 832 km (517 mi)

Longest Distances: 332 km (206 mi) SE-NW; 299 km (186 mi) NE-SW

Highest Point: Pico da Cano 2,829 m (9,281 ft)

Lowest Point: Sea level

Land Use: 9% arable; 6% meadows and pastures; 85% other

The Cape Verde Islands consist of an archipelago of 10 islands and five islets in the Atlantic Ocean.

The archipelago is divided into two districts: Barlavento, or Windward Islands, and Sotavento, or Leeward Islands, according to the direction of the prevailing northeasterly wind. Barlavento is composed of Santo Antao (754 sq km, 291 sq mi); Sao Vicente (228 sq km, 88 sq mi); Santa Luzia (34 sq km, 13 sq mi); Sao Nicolau (342 sq km, 132 sq mi); Boa Vista (622 sq km, 240 sq mi); and Sal (215 sq km, 83 sq mi). Sotavento is composed of Maio (267 sq km, 103 sq mi); Sao Tiago (992 sq km, 383 sq mi); Fogo (477 sq km, 184 sq mi); and Brava (65 sq km, 25 sq mi). Except for the low-lying islands of Sal, Boa Vista and Maio, the Cape Verde Islands are mountainous with rugged cliffs and deep ravines. The highest peaks are Pico da Cano, an active volcano reaching 2,828 m (9,281 ft), and two peaks reaching 1,935 m (6,350 ft) and 1,310 m (4,300 ft) on Santo Antao and Sao Tiago, respectively.

Area and Population

Islands / Counties	Capitals	area sq mi	area sq km	population 1980 census
Boa Vista		239	620	3,372
Boa Vista	Sal Rei			
Brava		26	67	6,985
Brava	Nova Sintra			
Fogo		184	476	30,978
Fogo	São Filipe			
Maio		104	269	4,098
Maio	Porto Inglês			
Sal		83	216	5,826
Sal	Santa Maria			
Santiago		383	991	145,957
Praia	Praia			57,748
Santa Catarina	Assomada			41,012
Santa Cruz	Pedra Badejo			22,995
Tarrafal	Tarrafal			24,202
Santo Antão		301	779	43,321
Paúl	Pombas			7,983
Porto Novo	Porto Novo			13,236
Ribeira Grande	Porto Sol			22,102
São Nicolau		150	388	13,572
São Nicolau	Riberia Brava			
São Vincente		88	227	41,594
São Vincente	Mindelo			
TOTAL		1,557	4,033	295,703

CAYMAN ISLANDS

BASIC FACTS

Official Name: Cayman Islands

Abbreviation: CAY

Area: 264 sq km (102 sq mi)

Area—World Rank: 191

Population: 23,037 (1988) 38,000 (2000)

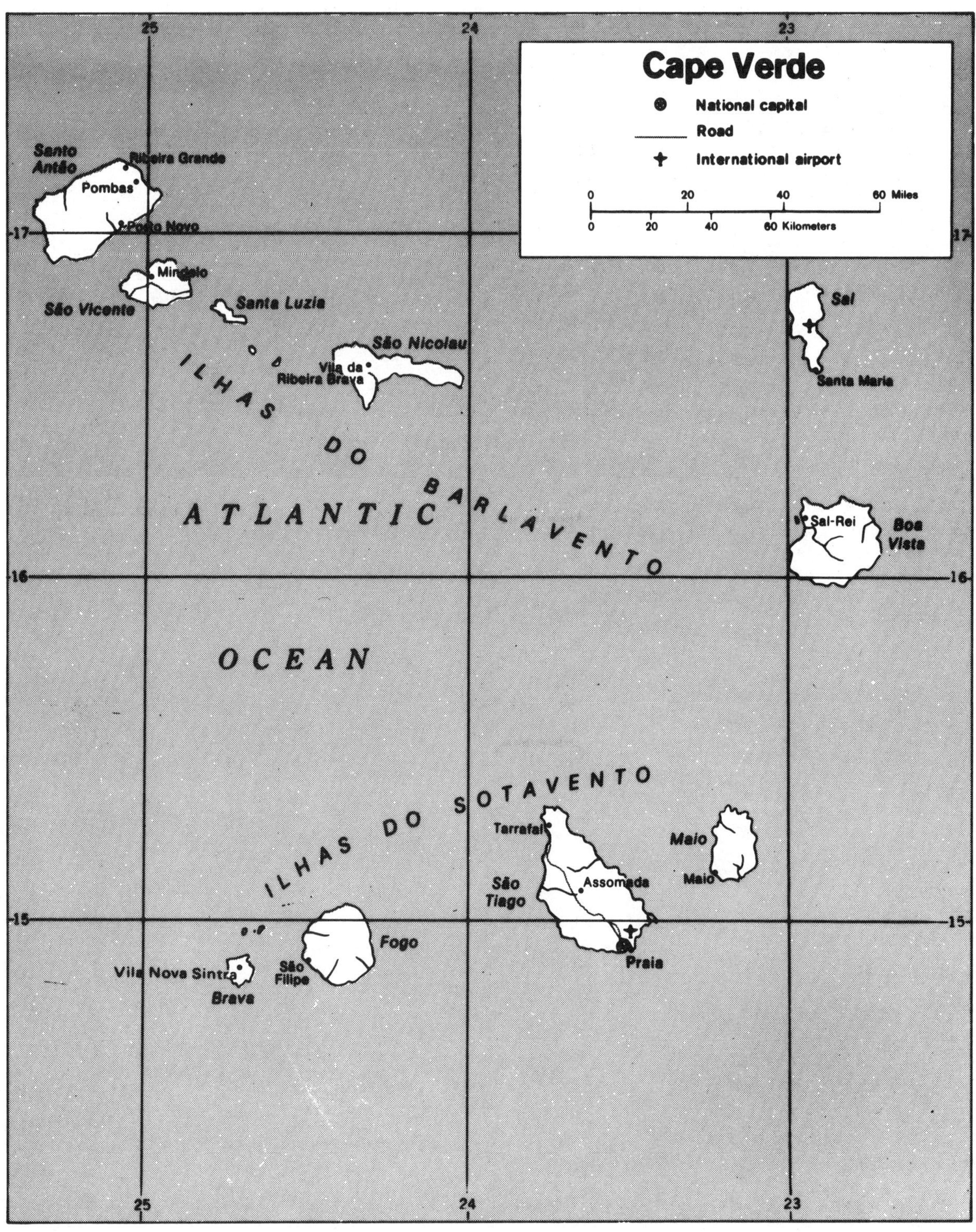

Cape Verde
National capital
Road
International airport
0 20 40 60 Miles
0 20 40 60 Kilometers
25
24
23
17
16
15
Santo Antão
Ribeira Grande
Pombas
Porto Novo
Mindelo
São Vicente
Santa Luzia
São Nicolau
Vila da Ribeira Brava
Sal
Santa Maria
Sal-Rei
Boa Vista
ILHAS DO BARLAVENTO
ATLANTIC
OCEAN
ILHAS DO SOTAVENTO
Tarrafal
Maio
Maio
Assomada
São Tiago
Praia
Fogo
São Filipe
Vila Nova Sintra
Brava

Cayman Islands

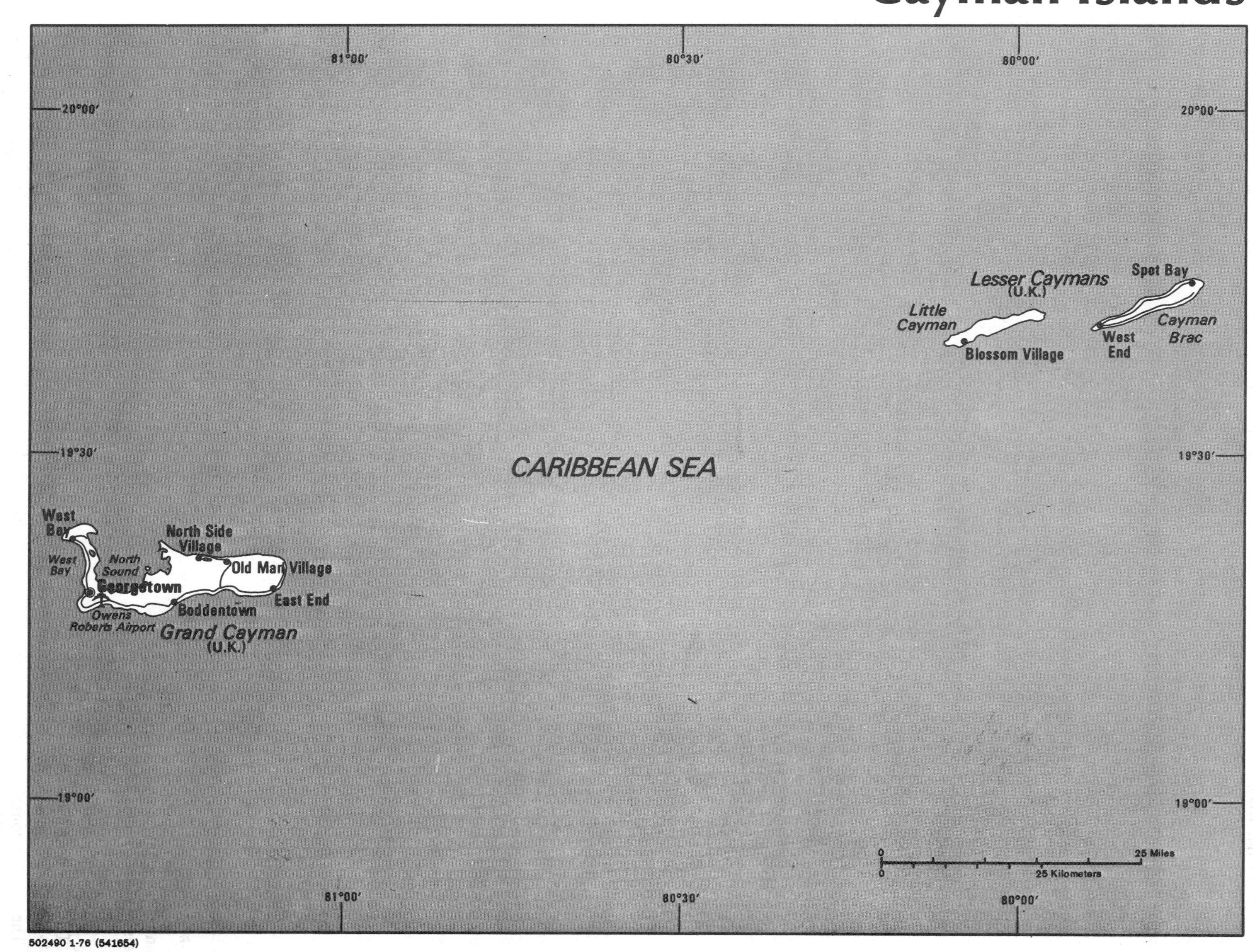

Population—World Rank: 195

Capital: George Town

Land Boundaries: Nil

Coastline: 160 km (100 mi)

Longest Distances: N/A

Highest Point: 42 m (138 ft)

Lowest Point: Sea level

Land Use: 8% meadows and pastures; 23% forest and woodland; 69% other

The Cayman Islands lie about 290 km (180 mi) northwest of Jamaica and consist of three main islands, Grand Cayman, Little Cayman and Cayman Brac. The islands have a low-lying limestone base and are surrounded by coral reefs.

CENTRAL AFRICAN REPUBLIC

BASIC FACTS

Official Name: Central African Republic

Abbreviation: CAR

Area: 622,436 sq km (240,324 sq mi)

Area—World Rank: 42

Population: 2,736,478 (1988) 3,823,000 (2000)

Population—World Rank: 112

Capital: Bangui

Land Boundaries: 5,232 km (3,251 mi); Chad 1,199 km (745 mi); Sudan 1,167 km (725 mi); Zaire 1,577 km (980 mi); Congo 467 km (290 mi); Cameroon 822 km (511 mi)

Coastline: Nil

Longest Distances: 1,437 km (893 mi) N-S 772 km (480 mi) E-W

Highest Point: Mont Ngaoui 1,410 m (4,626 ft)

Lowest Point: Ubangi River 335 m (1,100 ft)

Land Use: 3% arable; 5% meadows and pastures; 64% forest and woodland; 28% other

The Central African Republic is a landlocked country in the heart of Africa.

The country consists of a plateau with an average altitude of between 600 and 700 m (1,969 and 2,297 ft). The most prominent topographical features are the Bongo Massif (1,370 m, 4,495 ft) in the northeast, the Yade Massif (1,400 m, 4,593 ft) in the northwest and the Fertit Hills (1,280 m, 4,200 ft).

The Central African Republic is drained by two river systems: one flowing south, the other flowing north. Of those flowing south, the Chinko, Mbari, Kotto, Ouaka and Lobaye are tributaries of the Ubangi River and the Mambere and Kadei are tributaries of the Zaire. Two northern rivers, the Ouham and Bamingui, are tributaries of the Chari River.

PRINCIPAL TOWNS

Bangui (capital), population 473,817 (Dec. 1984 estimate); Berbérati 100,000, Bouar 55,000 (1982 estimates).

Area and population

Prefectures	Capitals	area sq mi	area sq km	population 1985 estimate
Bamingui-Bangoran	Ndélé	22,471	58,200	29,400
Bangui	Bangui	26	67	473,800
Basse-Kotto	Mobaye	6,797	17,604	187,200
Gribingui-Économique	Kaga-Bandoro	7,720	19,996	85,700
Haut-Mbomou	Obo	21,440	55,530	52,200
Haute-Kotto	Bria	33,456	86,650	233,100
Haute-Sangha	Berbérati	11,661	30,203	37,400
Kemo-Gribingui	Sibut	6,642	17,204	78,300
Lobaye	Mbaïki	7,427	19,235	160,700
Mbomou	Bangassou	23,610	61,150	132,900
Nana-Mambere	Bouar	10,270	26,600	197,600
Ombella-Mpoko	Bimbo	12,292	31,835	127,900
Ouaka	Bambari	19,266	49,900	216,200
Ouham	Bossangoa	19,402	50,250	269,300
Ouham-Pendé	Bozoum	12,394	32,100	242,100
Sangha-Économique	Nola	7,495	19,412	59,600
Vakaga	Birao	17,954	46,500	24,200
TOTAL		240,324	62,436	2,607,600

CHAD

BASIC FACTS

Official Name: Republic of Chad

Abbreviation: CHD

Area: 1,284,000 sq km (495,755 sq mi)

Area—World Rank: 20

Population: 4,777,963 (1988) 7,308,000 (2000)

Population—World Rank: 90

Capital: N'Djamena

Land Boundaries: 5,923 km (3,681 mi); Libya 1,054 km (655 mi); Sudan 1,360 km (845 mi); Central African Republic 1,199 km (745 mi); Cameroon 1,047 km (651 mi); Nigeria 88 km (55 mi); Niger 1,175 km (730 mi)

Coastline: Nil

Central African Republic

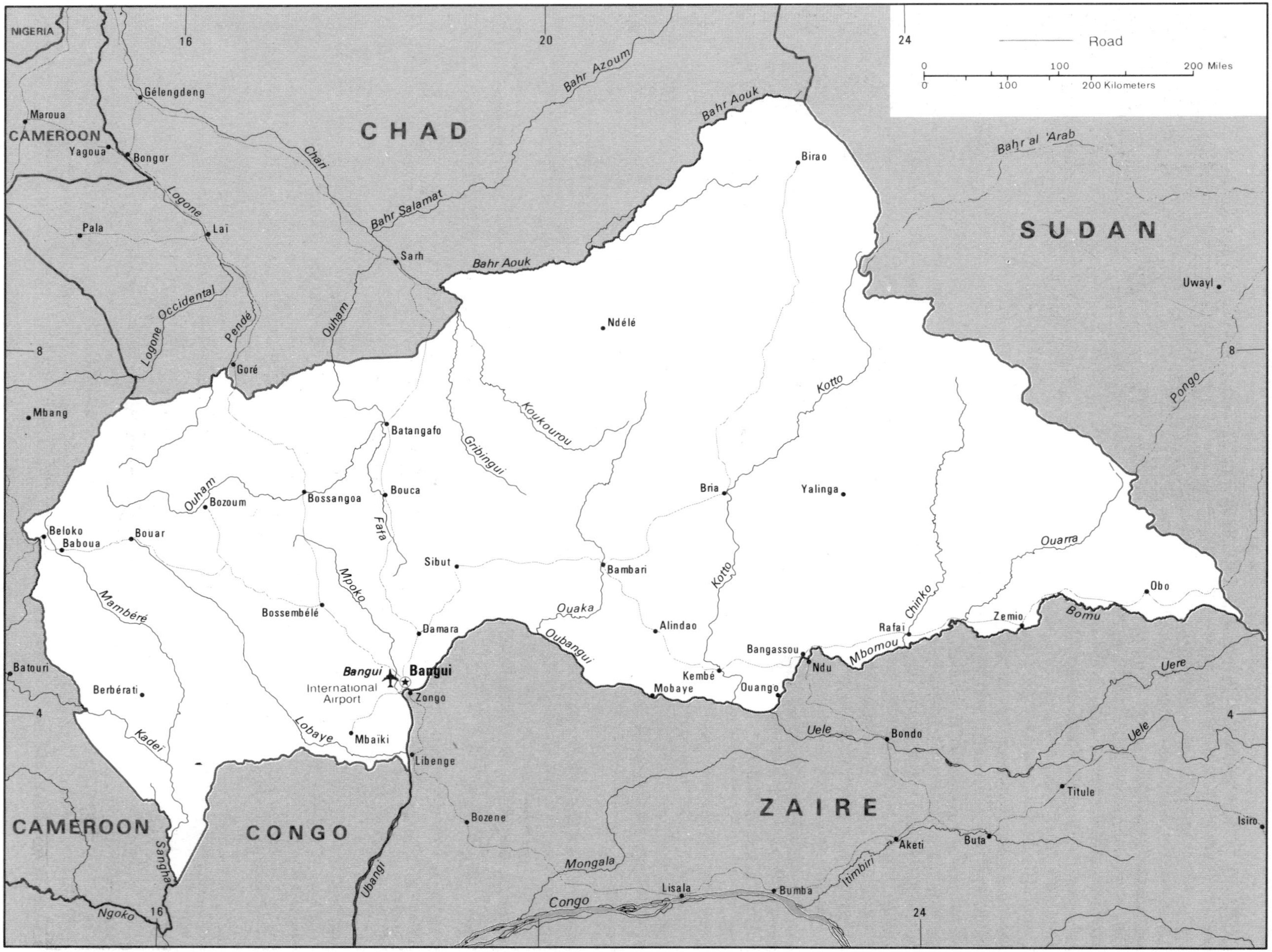

NIGERIA
CAMEROON
CHAD
SUDAN
ZAIRE
CONGO
CAMEROON
Road
0 100 200 Miles
0 100 200 Kilometers
24
16
20
8
4
Maroua
Yagoua
Bongor
Gélengdeng
Pala
Laï
Goré
Mbang
Sarh
Birao
Uwayl
Ndélé
Batangafo
Bouca
Bossangoa
Bozoum
Beloko
Baboua
Bouar
Sibut
Bria
Yalinga
Bambari
Alindao
Bossembélé
Damara
Bangui
Bangui International Airport
Zongo
Mbaïki
Libenge
Batouri
Berbérati
Bozene
Mobaye
Kembé
Ouango
Bangassou
Ndu
Rafai
Zemio
Obo
Bondo
Titule
Isiro
Buta
Aketi
Bumba
Lisala
Chari
Logone
Logone Occidental
Pendé
Ouham
Bahr Salamat
Bahr Azoum
Bahr Aouk
Bahr al 'Arab
Pongo
Koukourou
Gribingui
Kotto
Fafa
Mpoko
Mambéré
Kadéï
Lobaye
Ouaka
Oubangui
Ubangi
Mbomou
Chinko
Ouarra
Bomu
Uere
Uele
Itimbiri
Mongala
Congo
Sangha
Ngoko

Base 505125 (A00109) 5-82

Longest Distances: 1,765 km (1,097 mi) N-S 1,030 km (640 mi) E-W

Highest Point: Emi Koussi 3,415 m (11,204 ft)

Lowest Point: Bodele Depression 160 m (525 ft)

Land Use: 2% arable; 36% meadows and pastures; 11% forest and woodland; 51% other

A relatively large country in north-central Africa, Chad extends north-south for more than 1,609 km (1,000 mi) from the Tropic of Cancer, which crosses the heart of the Sahara Desert at 23.5° N., through broad transitional zones of subarid and humid savanna to the edge of the tropical rain forest about 7.5° N. The nation is landlocked, having no easy or direct access to the sea. N'Djamena, the nation's capital and the only major city, lies 1,126 km (700 mi) from the nearest seaport—Douala, a Cameroon port on the Atlantic Ocean.

The most important structural features are a broad, shallow central bowl and Lake Chad, together with the lake's major water source—the Chari-Logone river system. This drainage network collects considerable flow from the uplands along the southern border and adjacent areas in the Central African Republic and Cameroon. Part of this great volume of water is retained in swampy flood plains along the way, but much of the annual flow reaches the lake. The water is not highly mineralized, and the lake has continued to be an economically important reservoir of fresh water; the lake is an inland basin with no outlet to the sea and, in the subarid climate, there is a high rate of surface evaporation.

From the lake area and the central plains, the land rises very gradually to plateaus and ridges in the south and east and to arid plateaus and extinct volcanoes in the north. Central Chad is an area of mixed farming and grazing, a transition zone between the well-watered south and the barren north.

PRINCIPAL TOWNS (estimated population in 1979)

N'Djamena (capital)*	402,000	Bongor	69,000
Sarh*	124,000	Doba	64,000
Moundou	87,000	Laï	58,000
		Abêché	47,000

*Fort-Lamy was renamed N'Djamena in November 1973, and Fort-Archambault was renamed Sarh in July 1972.

LANDFORMS AND DRAINAGE PATTERNS

Much of Chad consists of the eastern half of a shallow, gently sloping basin several hundred miles wide. Lake Chad, on the western border of the country, lies in the lower reaches of this bowl. Northward from Lake Chad and the other depressions and swamps of the western border area, the basin extends for more than 804 km (500 mi) to the plateaus and mountain ranges associated with the Tibesti massif in northern Chad, a major landmark of the Sahara Desert. Eastward and southward from the lake, the relatively flat sedimentary basin extends for several hundred miles before rising gently to the rolling plateaus and scattered low mountains of the eastern and southern border areas.

Area and population

Préfectures	Capitals	area sq mi	area sq km	population 1984 estimate
Batha	Ati	34,285	88,800	410,000
Biltine	Biltine	18,090	46,850	200,000
Borkou-Ennedi-Tibesti	Faya	231,795	600,350	103,000
Chari-Beguirmi	N'Djamena	32,010	82,910	719,000
Guéra	Mongo	22,760	58,950	234,000
Kanem	Mao	44,215	114,520	234,000
Lac	Bol	8,620	22,320	158,000
Logone Occidental	Moundou	3,355	8,695	324,000
Logone Oriental	Doba	10,825	28,035	350,000
Mayo-Kebbi	Bongor	11,625	30,105	757,000
Moyen-Chari	Sarh	17,445	45,180	582,000
Ouaddai	Abéché	29,435	76,240	411,000
Salamat	Am Timan	24,325	63,000	121,000
Tandjilé	Laï	6,965	18,045	341,000
TOTAL		495,755	1,284,000	4,944,000

The high point of the Tibesti massif—the highest point in Chad—is a defunct volcano rising to an altitude of more than 3,413 m (11,200 ft), nearly 3,048 m (10,000 ft) above the surrounding plateau. Elsewhere the uplands and lesser mountains in northern, eastern, and southern areas of the country average from 305 to 914 m (1,000 to 3,000 ft) in elevation, sloping toward the central basin and the western border area. The Ennedi and Biltine highlands of the eastern border form the divide between this great inland basin and the Nile River drainage system in Sudan.

From the central bowl to southern Chad, the land slopes upward almost imperceptibly to rolling plateaus, which for the most part are less than 610 m (2,000 ft) above sea level. The plateaus are marked here and there by mountains, such as the Guéra massif near Mongo, which has at least one peak above 1,493 m (4,900 ft).

There are no permanent streams in northern or central Chad. Summer rainfall collected by the various shallow *wadis* (seasonally dry streambeds) flows toward inland basins; most of these streams disappear in the sands soon after the end of the brief rainy season.

The Chari and Logone rivers in southern Chad flow throughout the year, although they become shallow and sluggish toward the end of the dry season. Their headwaters are in the equatorial rainbelt in Cameroon and the Central African Republic, where rainfall averages more than fifty inches a year. A huge volume of water is carried from these upland areas into southern Chad, where the annual rainfall ranges between thirty and fifty inches. Tributaries, such as the Mayo-Kebbi in southwestern Chad, the Bahr Salamat in the southeast, and innumerable smaller streams, add to the flow as the six-month rainy season progresses. The result is an annual flood covering broad areas, especially in the lower reaches of the Chari and Logone rivers. These two major rivers join at N'Djamena and inundate the flat delta area for more than 80 km (50 mi) between the capital city and Lake Chad. Much of the land around N'Djamena and for a considerable distance upstream (southeastward) is also under water during the average autumn flood season. Major flood plains and swampy areas are also found farther upstream in the areas of Sarh and Moundou and elsewhere on various tributary streams.

Much of the great volume carried by this river system eventually reaches Lake Chad—an internal basin having no outlet to the sea—making it the seventh largest permanent lake in the world. It is, however, less than six feet deep during some dry seasons and less than twelve feet deep in most areas during the annual flood stage. The surface area varies seasonally between 12,950 and 25,900 sq km (5,000 and 10,000 sq mi).

The area covered by the shallow waters depends primarily upon the balance between the rate of evaporation—about eight inches annually from the surface of the lake—and the flow of the Chari-Logone river system. Losses into the shore areas, especially into the dunes on the northeastern shores, may account for no more than one foot of depth a year. Rivers other than the Chari and the Logone originate in semiarid areas and carry relatively small inputs; rainfall over Lake Chad adds about fifteen inches annually.

Other very shallow bowls, similar to Lake Chad in geologic origin, are scattered across the flat plains northeast, east, and southeast of the lake. Almost all of these sandy depressions are dry before the end of the annual dry (winter) season. One of the largest, the Bahr el Ghazal, receives some overflow from Lake Chad during its flood stage.

CHILE

BASIC FACTS

Official Name: Republic of Chile

Abbreviation: CHL

Area: 756,026 sq km (292,135 sq mi)

Area—World Rank: 37

Population: 12,638,046 (1988); 15,768,000 (2000)

Population—World Rank: 55

Capital: Santiago

Boundaries: 11,676 km (7,255 mi); Peru 169 km (105 mi); Bolivia 861 km (535 mi); Argentina 5,308 km (3,298 mi)

Coastline: 5,338 km (3,317 mi)

Longest Distances: 4,270 km (2,653 mi) N-S 356 km (221 mi) E-W

Highest Point: Mount Ojos del Salado 6,893 m (22,615 ft)

Lowest Point: Sea level

Land Use: 16% meadows and pastures; 21% forest and woodland; 7% arable; 56% other

Richly varied natural features exert a profound influence on the development of Chile. Mountains and deserts, rich in minerals, and a long coastline have made Chile a country of miners and seafarers. The Mediterranean climate, even terrain, and rich soils of central Chile, contrasting with the inhospitable climate and geography in the north and the south have attracted a heavy concentration of population in that region. This, in turn, has made possible the development of political, economic, and social cohesiveness. The evolution of a strong national identity and an ethnically homogeneous population have also been made possible by an isolation imposed by mountain, desert, and ocean. The origin of the country's name is uncertain, but one explanation, consistent with its physical setting, is that Chile is a corruption of the Aymará Indian word meaning "where the land ends."

The country is stretched along many degrees of longitude. Extending more than 4,184 km (2,600 mi) between northern and southern extremities, it averages not much more than 161 km (100 mi) in width. At the widest, its breadth is only about 354 km (220 mi), and at one point in the far south the Argentine frontier is within 22 km (14 mi) of the Pacific. Although the country forms a fringe along the western side of the South American continent, the capital city of Santiago lies almost due south of New York and, because of a peculiarity in the world's time chart, its time is 1 hour later than New York's. As a consequence of its great length, the hours of daylight and darkness vary substantially. The longest day includes 13 hours of daylight at the Peruvian frontier in the north and 17 hours on the Tierra del Fuego archipelago in the south.

There are three outstanding parallel natural features—the coastal range of mountains; the great cordillera of the Andes; and the Central Valley, which lies between the two mountain systems. The coastal range consists of a plateau-like series of rounded hills, broken occasionally by gorges and rivers. In the northern half of the country its peaks reach as high as 2,130 m (7,000 ft) but the system deteriorates south of Valparaiso and plunges into the sea in the far south, although its peaks reappear as the islands of the southern archipelagoes. There are few beaches and natural harbors along the narrow coastal lowlands, and in the north the coastal mountains rise close to the shoreline in steep escarpments.

The spine of the Andes extends almost the full length of the country and is a strong influential factor in the lives of the people. Almost nowhere is it possible to face eastward without seeing its peaks. There are more than 100 Andean volcanos, and the recentness of creation of the mountain system explains the frequency of damaging earthquakes.

The cordillera rises more abruptly on its Chilean side than on the easier eastern gradients in Argentina, and the crests are higher along the northern half of the range where all passes are at more than 3,048 m (10,000 ft). In this northern sector is Ojos del Salado, Chile's loftiest peak—more than 6,857 m (22,500 ft) high. It is only about 91 m (300 ft) lower than Aconcagua, the highest peak in the Western Hemisphere, which rises in Argentina but can be seen from Santiago on clear days. South from Santiago the peaks become progressively lower, and passes are as low as 1,524 m (5,000 ft). In the far south the Andes continue to deteriorate and become lost in the lowlands of Chilean Patagonia on both sides of the Strait of Magellan. The system makes a final appearance at Cape Horn, which is the crest of a submerged mountain.

Between the Andes and the coastal range, the long Central Valley is poorly defined in the far north, where it takes the form of almost rainless and barren plateau basins. Much of this region lies in the southern tropics, but the climate is modified by the cooling effect of Antarctic waters carried northward by the Humboldt Current. There is almost no rainfall and no natural vegetation other than occasional desert scrub, but it is sometimes called "the fertile desert," for in it are found the nitrate deposits which provide employment for most of the northern desert's scanty population. Immediately southward the Central Valley almost disappears in a series of transverse Andean spur ridges and intervening fertile valleys watered by intermittent rivers and streams fed by mountain snows. Farms and towns in these transverse valleys give the region a relatively denser population than that of the far north.

The Central Valley becomes well defined in the midland sector, which is the country's prosperous agricultural and industrial heartland. Well over half of the population live clustered in this area where fertile land and mild climate combine with easy communication routes. Here is the Chile of history. Southward from the heartland the Central Valley at the Bío-Bío River becomes a land of heavy rain and thick forest, which abruptly terminates at about 31° S., where the Gulf of Reloncaví submerges what was once a further extension of the valley. This was the country's old frontier, occupied by hostile Indians until late in the 19th century. Beyond the Gulf of Reloncaví there are only countless uninhabited islands, channels, fjords, heavily forested cordillera, and continuous wind and rain, until the foot of the continent is reached. At the foot there is a low grassy plain and, across the Strait of Magellan, the Tierra del Fuego archipelago.

Much of Chile's desert north is an arid region, roughly comparable to Baja California. The Mediterranean climate, vegetation, and heavy concentration of population of central Chile are very much like those of the state of California between the Mexican border and San Francisco. In particular, the general setting of California's San Joaquín Valley resembles that of the Central Valley near Santiago. The Pacific Northwest is reminiscent of the south-central Chilean land of forests and lakes; and the Alaskan panhandle suggests the archipelagoes of the extreme Chilean south.

GENERAL SETTING

Strung out along the western fringe of the Andes and cut off in the north by rainless desert, Chile is probably the most nearly isolated maritime nation in the world, for the sea provides a minimum of linkage with other countries. During the 19th century, Valparaíso was a vital port of call for ships of all nations moving between the Atlantic and the Pacific, but the opening of the Panama Canal in 1915 brought an end to that. The long coastline has few good harbors to facilitate coastal transport and, in the Central Valley, only between Santiago and Concepción are natural conditions conducive to easy movement.

These apparently disadvantageous natural conditions, however, have been advantages. While other countries have been divided by the Andes, Chile has been united by them. The Andean barrier, coupled with the barriers of desert and ocean, has permitted the country to develop its society with a minimum of outside influence and permitted its people to develop a racial homogeneity greater than in any other west coast country. Internally, the imperfection of natural communications lines in the north and south has been less significant than the easy communications in the central portion, which proved also to have the richest soils and the best climate. As a consequence, a population nucleus was able to develop into a coherent society in central Chile early in the country's history, leaving the development of peripheral areas for later dates. In addition, although the settlement pattern was one of large estates, the compactness of the nuclear area encouraged landowners actually to live on their estates more frequently than in other west coast countries and to take a part in the life of the rural communities.

Chile is a country of natural violence. Flash floods caused by sudden melting of Andean snowfields throw walls of silt and debris-laden water on subsistence farming villages, lining the course of streams to endanger crops, property, and sometimes life. Sailors and fishermen must accommodate themselves to inadequately protected harbors, storms, and uncertain offshore currents. Natural violence at its most serious, however, is seismic in origin. The geologically young Andes, in a line parallel to the ocean and close to the coastline, provide the environment for earthquakes in the narrow band of land lying between.

Well over 100 major earthquakes have been recorded since compilation of records began in 1575, many accompanied by fires and tidal waves. A majority of these disturbances have had epicenters north of Valparaíso, but the most serious have occurred to the south. Valparaíso itself was almost leveled by an earthquake and tidal wave early in the present century and was severely shaken in 1965. Concepción, in particular, has been leveled or damaged many times, and once its location was changed following its virtual destruction. It and other nearby population centers were affected by a 1939 earthquake which took some 20,000 lives, and a 1960 seismic disturbance and tidal wave damaged an area from Valdivia to Concepción, leaving 350,000 homeless.

In addition to earthquakes and tidal waves, the country often suffers from volcanic eruptions, floods, avalanches, landslides, and violent storms. Floods, avalanches, and landslides were reported to have affected 22 of the 25 provinces and to have destroyed the homes of 80,000 people in 1965 alone. A frequent occurrence during earlier years of the present century was the sudden destruction by flash floods of entire Santiago squatter settlements along the Mapocho River, which runs through the middle of the city.

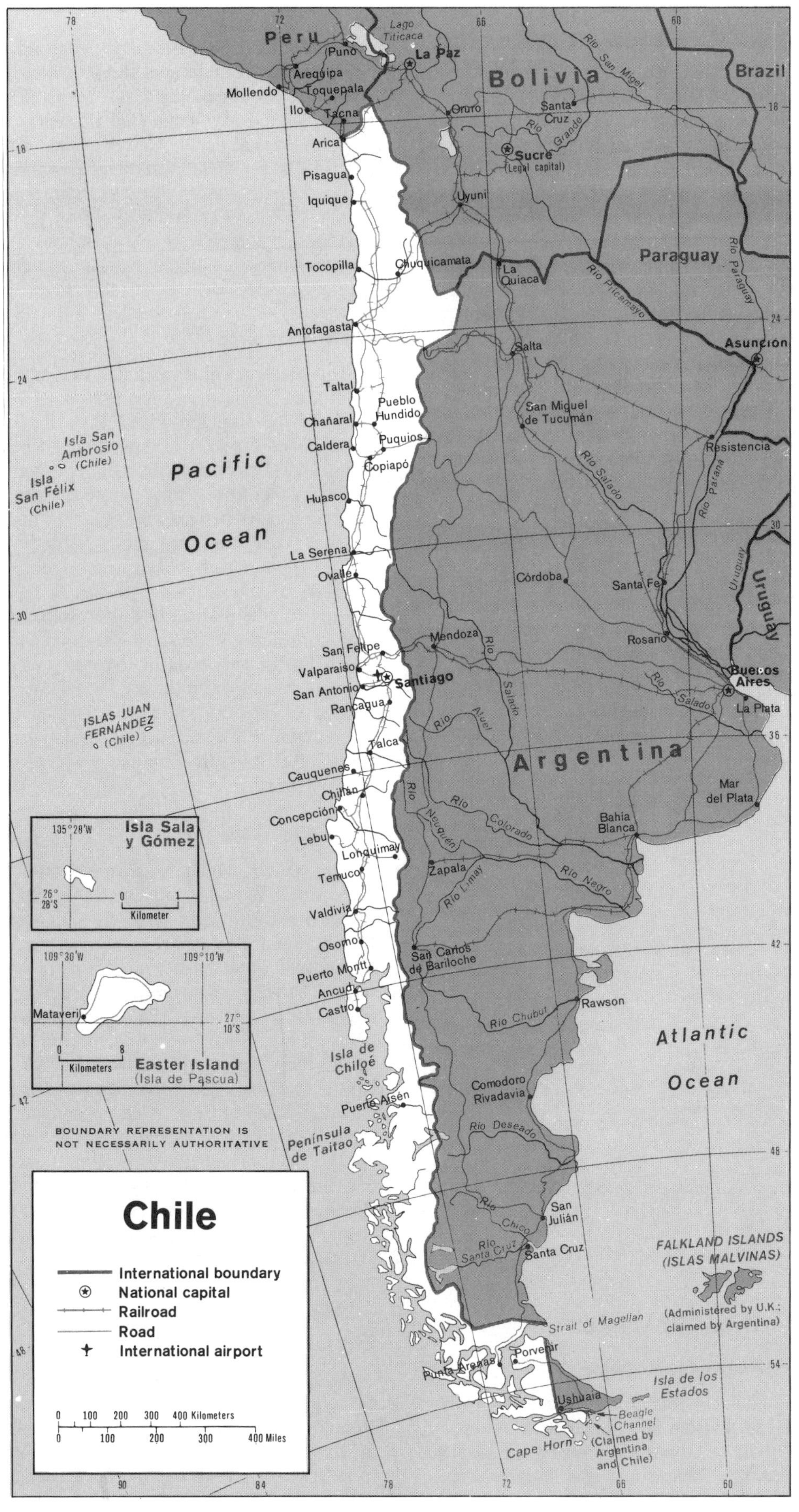
Chile
International boundary
National capital
Railroad
Road
International airport
Pacific Ocean
Atlantic Ocean
Peru
Bolivia
Brazil
Paraguay
Argentina
Uruguay
Santiago
La Paz
Sucre
(Legal capital)
Asunción
Buenos Aires
Isla Sala y Gómez
Easter Island
(Isla de Pascua)
Mataveri
Islas Juan Fernández
(Chile)
Isla San Ambrosio
(Chile)
Isla San Félix
(Chile)
Isla de Chiloé
Península de Taitao
Strait of Magellan
Beagle Channel
(Claimed by Argentina and Chile)
Cape Horn
Isla de los Estados
FALKLAND ISLANDS
(ISLAS MALVINAS)
(Administered by U.K.; claimed by Argentina)
BOUNDARY REPRESENTATION IS NOT NECESSARILY AUTHORITATIVE
Kilometers
Miles
Arica
Iquique
Antofagasta
Valparaíso
Concepción
Temuco
Valdivia
Puerto Montt
Punta Arenas
Ushuaia
Mendoza
Córdoba
Rosario

Drought is not one of the recurrent natural afflictions. Rainfall is sparse in the northern half, but agriculture is less dependent on rainfall than on irrigation water from streams carrying the melting snows of the Andes, which represent what is considered to be an almost inexhaustible natural reservoir. In 1968, however, the country experienced an unprecedented drought. Only a single rainfall had been recorded in Santiago during the year, and the snowcap was gone from the cordillera as far south as Linares Province in south-central Chile. Estimates of losses in crops and livestock ran as high as US$1 million, and the national electric company, heavily dependent on hydroelectric power, was in a state of emergency.

Popular reactions to these natural phenomena has tended to be fatalistic. The homeless and impoverished have been assisted, debris has been cleared away, and ruined buildings rebuilt, sometimes in what appears to be a safer locality or designed to be more resistant to the next catastrophe. In particular, since the disastrous 1939 earthquake, increasing attention has been given to raising stronger buildings, and quake-resistant construction has become a major concern of architects.

PRINCIPAL TOWNS (population at 30 June 1985)

Gran Santiago (capital)	4,318,305*	Temuco	171,831
		Rancagua	152,132
Viña del Mar	315,947	Talca	144,656
Valparaíso	267,025	Chillán	128,920
Talcahuano	220,910	Arica	127,925
Concepción	217,756	Iquique	120,732
Antofagasta	175,486	Valdivia	119,977

*Including suburbs.

Area and population

Regions	Capitals	area sq mi	area sq km	population 1986 estimate
Tarapacá	Iquique	22,697	58,786	314,800
Antofagasta	Antofagasta	48,360	125,253	366,300
Atacama	Copiapó	28,844	74,705	194,500
Coquimbo	La Serena	15,697	40,656	452,500
Valparaíso	Valparaíso	6,331	16,396	1,314,300
Libertador General Bernardo O'Higgins	Rancagua	6,354	16,456	621,700
Maule	Talca	11,839	30,662	792,900
Bío-Bío	Concepción	14,262	36,939	1,609,700
Araucanía	Temuco	12,334	31,946	745,200
Los Lagos	Puerto Montt	26,350	68,247	896,200
Aisén del General Carlos Ibáñez del Campo	Coihaique	42,084	108,997	73,000
Magallanes y de la Antártica Chilena	Punta Arenas	50,979	132,034	141,700
Región Metropolitana de Santiago	Santiago	6,003	15,549	4,804,200
TOTAL		292,135	756,626	12,327,000

NATURAL FEATURES

Geographic Regions

Because of the country's extreme length and narrowness, it lends itself to division into geographic regions by drawing imaginary lines crosswise between the Pacific and the Andean frontier. There is no general agreement as to exactly where these lines should be drawn, but the far northern desert is customarily referred to as the Great North (Norte Grande) and the transitional region immediately to its south as the Little North (Norte Chico). The fertile and heavily populated provinces of the midlands may be referred to as Central Chile, and the land of forests and lakes to their south as South-Central Chile. Finally, there are more remote archipelagoes, fjords, and channels of the Far South. Because the country has several insular possessions and maintains a claim to a large part of Antarctica, there is also a catchall region which may be referred to as Other Territories.

The Great North

This desert region includes the provinces of Antofagasta and Tarapacá and the portion of Atacama Province north of the Copiapó River. Here the Central Valley is made up of a series of high plateau basins with sands, clays, and salt deposits. In them is found the *caliche* (natural saltpeter) from which nitrate is extracted. The coastal range is usually between 609 and 914 m (2,000 and 3,000 ft) above sea level and drops abruptly to the ocean in cliffs, below which coastal plains are occupied by the northern coastal cities and towns. In the southern half of this region, the plateau basins are separated from the Andean cordillera by the Domeyko Mountain Range, beyond which lie the salt beds of the Atacama Puna. Numerous Andean streams drain toward the Pacific, but, with the exception of the 434 km (270 mi) long Loa River, which reaches the sea near Tocopilla, all are absorbed by the desert after providing limited amounts of water to oasis farm villages.

The Little North

This region, which is also sometimes known as the Andean fringe, is made up primarily of Coquimbo Province, but is usually thought of as including southern Atacama Province and all or part of Aconcagua Province. It is a transitional zone made up of short transverse valleys formed between spurs from the Andes which make longitudinal travel difficult. The most important of these fertile interruptions in an otherwise barren countryside is the Elqui River Valley, inland from La Serena. In the Little North the Andes are at their most majestic, with Chile's Ojos del Salado rising at its upper extremity and Argentina's Aconcagua at its lower. There are numerous rivers, but they feed primarily on thawing snows, and only three, the Huasco, Coquimbo, and Limari, flow the year round. Harbors are poor and the only port town of importance is Caldera, near Copiapo, which serves as the outlet for an extensive mining region.

Central Chile

At the lower extremity of Aconcagua Province lies the basin of the Aconcagua River, an extraordinarily fertile area known as the Vale of Chile, the southernmost point of advance of the Incas. From the abrupt Chacabuco Ridge, which marks the southern boundary of the Vale, the Central Valley stretches uninterruptedly for about 643 km (400 mi) to the Bío-Bío River. This is the country's heartland, consisting of a 64 to 72 km (40 to 45 mi) wide sloping plain made up of alluvial fans of soil of vast thickness built up by the region's principal rivers, the Mapocho, Maule, and Maipo. The soil is heavily mineralized and produces fruits and vegetables of great size and excellent quality grown under irrigation.

South of Santiago the Andes grow progressively lower and the snowline is at an average of about 3,657 m (12,000 ft). The coastal range also grows lower in the south after reaching peaks of up to 1,829 km (6,000 ft) between Santiago and Valparaíso. It rises sharply from the Central Valley and continues for 40 or 48 km (25 or 30 mi) to the west where it drops to the sea, except in narrow river valleys which pierce the range at irregular intervals.

South-Central Chile

South of the Bío-Bío the structural pattern of the countryside remains essentially the same for 644 km (400 mi) to the Gulf of Reloncaví. The Andes continue lower, however, and many volcanoes lie west of the main spine of the cordillera. The southern half of the zone is the Lake District. Twelve major lakes fill the glacially dammed Andean approaches. Lake Llanquihue, with about 777 sq km (300 sq mi) of surface, is the largest, but undoubtedly the most famed sight in this Alpine area is of the perfect snowcapped cone of 2,651 m (8,700 ft) Mount Osorno, rising behind lake Todos los Santos. The Central Valley, less continuous in this region, is crossed by numerous small rivers which serrate the residual coastal range. The coastline grows more irregular and more good harbors appear.

The Far South

Puerto Montt on the Gulf of Reloncaví is the terminus of the longitudinal railroad. It is also the terminus of South-Central Chile, for beyond it the Central Valley drops into the sea and the coastal range continues only as a great number of offshore islands in an area which is sometimes called Archipelagic Chile. The shoreline is deeply embayed, the terrain extremely broken, and glaciers lie below the mountain peaks. Except where the Taitao Peninsula juts seaward at about 46° S., it is possible to navigate at considerable hazard almost the full distance to the Strait of Magellan in a series of channels which run between the broken mainland coast and the islands of the archipelagoes. The Far South has very little arable land and almost no population except at the tip of the continent where both coasts of Tierra del Fuego belong to Chile. This lowland zone, sometimes called Patagonian or Atlantic Chile, is an undulating plain of glacial-morainic formation used largely for pasture.

Other Territories

Chile claims a wedge-shaped section of Antarctica, which is also claimed in part by Argentina and the United Kingdom, and owns Easter Island, the Juan Fernandez Islands, and several other small Pacific Islands. Easter Island, also known by its Polynesian name of Rapa Nui, lies 3,749 km (2,330 mi) due west from the port of Caldera. Inhabited by people of Polynesian origin, it is the Latin American possession most remote from the parent country. The Juan Fernandez Islands, about 579 km (360 mi) west of Valparaíso, are the site of a small fishing colony. They are famous for their lobsters and for the fact that Alexander Selkirk, marooned on them, was the inspiration for the novel *Robinson Crusoe*.

CHINA

BASIC FACTS

Official Name: People's Republic of China

Abbreviation: PRC

Area: 9,572,900 sq km (3,696,100 sq mi)

Area—World Rank: 3

Population: 1,088,169,192 (1988) 1,253,000,000 (2000)

Population—World Rank: 1

Capital: Beijing

Boundaries: 28,072 km (17,445 mi); Mongolia 4,673 km (2,904 mi); USSR 7,520 km (4,673 mi); North Korea 1,416 km (880 mi); Hong Kong 84 km (52 mi); Macao 1.6 km (1 mi); Vietnam 1,281 km (796 mi); Laos 425 km (264 mi); Myanmar 2,185 km (1,358 mi); India 1,194 km (742 mi); Bhutan 412 km (256 mi); Nepal 1,078 km (670 mi); Jammu and Kashmir 698 km (434 mi); Pakistan 523 km (325 mi); Afghanistan 71 km (44 mi)

Coastline: 6,511 km (4,046 mi)

Longest Distances: 4,845 km (3,011 mi) ENE-WSW 3,350 km (2,082 mi) SSE-NNW

Highest Point: Mount Everest 8,848 m (29,028 ft)

Lowest Point: Turfan Depression 154 m (505 ft) below sea level

Land Use: 10% arable; 31% meadows and pastures; 14% forest and woodland; 45% other

China is geographically diverse and includes vast areas of rugged inhospitable terrain. The third largest country in the world, it reaches across some 5,000 km (3,106 mi) of the East Asian landmass, an erratically changing configuration of broad plains, expansive deserts, and lofty mountain ranges. Eastern China, its seacoast fringed with offshore islands, is a region of fertile lowlands, foothills and mountains, desert and steppe. Western China is a realm of sunken basins, rolling plateaus, and towering massifs, including a portion of the highest tableland on the globe.

PRINCIPAL TOWNS (Wade-Giles or other spelling in brackets)

Population at 31 December 1985 (official estimates in '000)*

Town	Population	Town	Population
Shanghai (Shang-hai)	6,980	Yancheng	1,250
Beijing (Pei-ching or Peking, the capital)	5,860	Qingdao (Ch'ing-tao or Tsingtao	1,250
Tianjin (T'ien-chin or Tientsin)	5,380	Hangzhou (Hang-chou or Hangchow)	1,250
Shenyang (Shen-yang or Mukden)	4,200	Fushun (F'u-shun)	1,240
Wuhan (Wu-han or Hankow)	3,400	Yulin	1,230
Guangzhou (Kuang-chou or Canton)	3,290	Chaozhou	1,210
Chongqing (Ch'ung-ch'ing or Chungking)	2,780	Dongguan	1,210
Harbin (Ha-erh-pin)	2,630	Xiaogan	1,200
Chengdu (Ch'eng-tu)	2,580	Fuzhou (Fu-chou or Foochow)	1,190
Xian (Hsi-an or Sian)	2,330	Suining	1,170
Zibo (Tzu-po or Tzepo)	2,300	Xintai	1,160
Nanjing (Nan-ching or Nanking)	2,250	Changsha (Chang-sha)	1,160
Liupanshui	2,220	Shijiazhuang (Shih-chia-chuang or Shihkiachwang)	1,160
Taiyuan (T'ai-yüan)	1,880	Jilin (Chi-lin or Kirin)	1,140
Changchun (Ch'ang-ch'un)	1,860	Nanchang (Nan-ch'ang)	1,120
Dalian (Ta-lien or Dairen)	1,630	Baotau (Pao-t'ou or Paotow)	1,100
Zaozhuang	1,590	Puyang	1,090
Zhengzhou (Cheng-chou or Chengchow)	1,590	Huainan (Huai-nan or Hwainan)	1,070
Kunming (K'un-ming)	1,490	Zhongshan	1,060
Jinan (Chi-nan or Tsinan)	1,430	Luoyang (Lo-yang)	1,050
Tangshan (T'ang-shan)	1,390	Weifang	1,040
Guiyang (Kuei-yang or Kweiyang)	1,380	Laiwu	1,040
Linyi	1,370	Leshan	1,030
Lanzhou (Lan-chou or Lanchow)	1,350	Jingmen	1,020
Taian	1,330	Ningbo	1,020
Pinxiang	1,290	Urumqi (Urumchi)	1,000
Suzhou (Su-chou or Soochow)	1,280	Heze	1,000
Anshan (An-shan)	1,280	Datong (Ta-t'ung or Tatung)	1,000
Qiqihar (Chi'-ch'i-ha-erh or Tsitsihar)	1,260		

*Data refer to municipalities, which may include large rural areas as well as an urban centre.

Terrain and Drainage

The country exhibits great variation in terrain and vegetation. Mountains cover more than two-thirds of the nation's territory, impeding communciation and leaving only limited areas of level land for agriculture. Most ranges, including all the major ones, trend east-west. In the Southwest region the Himalayas and the Kunlun Mountains enclose the Plateau of Tibet, the most extensive plateau in the world, where elevations average more than 4,000 m (13,123 ft) above sea level and the loftiest summits rise to over 7,200 m (23,622 ft).

From the Plateau of Tibet other less elevated highlands, rugged east-west trending mountains and plateaus interrupted by deep depressions, fan out to the north and east. A continental scarp marks the eastern margin of this territory extending from the Greater Khingan Range in Manchuria, through the Taihang Shan (a range of mountains overlooking the North China Plain), to the eastern edge of the Yunnan-Guizhou Plateau in the South. Virtually all of the low-lying areas of China, the regions of dense population concentration and intensive cultivation, are found east of this scarp line.

East-west ranges include some of Asia's greatest mountains. In addition to the Himalayas and the Kunlun Mountains, there are the Kailas Range and the Tian Shan (the latter standing between two great basins), the massive Tarim Basin on the south, and the Dzungarian Basin to the north. Rich deposits of coal, oil, and metallic ores lie in the area. The largest inland basin in China, the Tarim measures 1,500 km (932 mi) from east to west and 600 km (373 mi) from north to south at its widest parts.

The Himalayas form a natural boundary on the southwest as the Altai Mountains do on the northwest. Lesser ranges branch out, some at sharp angles to the main system, from the major ranges. The mountains form the chief watersheds of all the principal rivers.

The spine of the Kunlun mountain range separates into several branches as it runs eastward from the Pamirs. The northernmost of the branches, the Altun Shan and the Qilian Shan, rim the Plateau of Tibet in west central China and overlook the Tsaidam Basin, a sandy and swampy basin containing many salt lakes. A southern branch of the Kunlun forms the watershed between the Huang He and the Yangtze River. The Gansu Corridor, west of the great bend in the Huang He, has traditionally been an important communications link with Central Asia.

Most of the Great Wall along the country's northern flank, the actual length of which is more than 3,300 km (2,050 mi) was built about 220 B.C., under the Qin Dynasty. North of the Great Wall, between Gansu Province on the west and the Greater Khingan Range on the east, lies the Plateau of Inner Mongolia, at an average elevation of 1,000 m (3,280 ft) above sea level. The Yin Shan, a system of mountains with average elevations of 1,364 m (4,475 ft), extends east-west through the center of this vast desert and steppe peneplain. To the south is the largest loess plateau in the world, covering 600,000 sq km (308,881 sq mi) in Shaanxi Province and parts of Gansu and Shanxi provinces and of Ningxia-Hui Autonomous Region. The plateau is veneered by a layer of loess—a yellowish soil blown in from the deserts of Inner Mongolia. The loose, loamy deposit travels easily on the wind,

Base 800028 (545114) 1-84

and over the centuries the Huang He has become choked with silt.

Area and population

Provinces	Capitals	area sq mi	area sq km	population 1986 estimate
Anhwei (Anhui)	Ho-fei (Hefei)	54,000	139,900	51,560,000
Chekiang (Zhejiang)	Hangchow (Hangzhou)	39,300	101,800	40,300,000
Fukien (Fujian)	Foochow (Fuzhou)	47,500	123,100	27,130,000
Heilungkiang (Heilongjiang)	Harbin (Harbin)	179,000	463,600	33,110,000
Honan (Henan)	Cheng-chou (Zhengzhou)	64,500	167,000	77,130,000
Hopeh (Hebei)	Shih-chia-chuang (Shijiazhuang)	78,200	202,700	55,480,000
Hunan (Hunan)	Ch'ang-sha (Changsha)	81,300	210,500	56,220,000
Hupeh (Hubei)	Wu-han (Wuhan)	72,400	187,500	49,310,000
Kansu (Gansu)	Lan-chou (Lanzhou)	141,500	366,500	20,410,000
Kiangsi (Jiangxi)	Nan ch'ang (Nanchang)	63,600	164,800	34,600,000
Kiangsu (Jiangsu)	Nanking (Nanjing)	39,600	102,600	62,130,000
Kirin (Jilin)	Ch'ang-ch'un (Changchun)	72,200	187,000	22,980,000
Kwangtung (Guangdong)	Canton (Guangzhou)	89,300	231,400	62,530,000
Kweichow (Guizhou)	Kuei-yang (Guiyang)	67,200	174,000	29,680,000
Liaoning (Liaoning)	Shen-yang (Shenyang)	58,300	151,000	36,860,000
Shansi (Shanxi)	T'ai-yüan (Taiyuan)	60,700	157,100	26,270,000
Shantung (Shandong)	Tsinan (Jinan)	59,200	153,300	76,950,000
Shensi (Shaanxi)	Sian (Xi'an)	75,600	195,800	30,020,000
Szechwan (Sichuan)	Ch'eng-tu (Chengdu)	219,700	569,000	101,880,000
Tsinghai (Qinghai)	Hsi-ning (Xining)	278,400	721,000	4,070,000
Yunnan (Yunnan)	K'un ming (Kunming)	168,400	436,200	34,060,000
Autonomous regions				
Inner Mongolia (Nei Monggol)	Hu-ho-hao-t'e (Hohhot)	454,600	1,177,500	20,070,000
Kwangsi Chuang (Guangxi Zhuang)	Nan-ning (Nanning)	85,100	220,400	38,730,000
Ningsia Hui (Ningxia Hui)	Yin-ch'uan (Yinchuan)	25,600	66,400	4,150,000
Sinkiang Uighur (Xinjiang Uygur)	Urumchi (Urumqi)	635,900	1,646,900	13,610,000
Tibet (Xizang)	Lhasa (Lhasa)	471,700	1,221,600	1,990,000
Municipalities				
Peking (Beijing)	—	6,500	16,800	9,600,000
Shanghai (Shanghai)	—	2,400	6,200	12,170,000
Tientsin (Tianjin)	—	4,400	11,300	8,080,000
TOTAL		3,696,100	9,572,900	1,045,320,000

Because the level of the river drops precipitously as it reaches the North China Plain, where it continues a sluggish course across the delta, it brings down a heavy load of sand and mud from the upper reaches, much of which is deposited on the flat plain. The flow is channeled mainly by a constant repair of man-made embankments; the river now actually flows on a raised ridge, the riverbed having risen to 50 m (164 ft) or more above the plain. As a result waterlogging, floods, and course changes have been recurrent in past centuries. Emperors were judged by their concern or indifference to preservation of the embankments. In the modern era the new leadership has evidenced a deep commitment to dealing with the problem and has undertaken extensive flood control and conservation measures.

Flowing from its source in the Tibetan highlands, the Huang He courses toward the sea through the North China Plain, the historic center of Chinese expansion and influence. Han people have farmed the rich alluvial soils of the plain since ancient times, constructing the Grand Canal for north-south transport. The plain itself is actually a continuation of the central Manchurian Plain to the northeast, but separated from it by the Bo Hai, an extension of the Yellow Sea.

Like other densely populated areas of China, the plain is subject not only to floods but to earthquakes. For example, the mining and industrial center of Tangshan, about 165 km (102 mi) east of Beijing, was leveled by an earthquake in July 1976. The Chinese reported 242,000 persons killed and 164,000 injured.

The Qin Ling, a continuation of the Kunlun Mountains, divides the North China Plain from the Yangtze delta and accounts for the major geographic boundary between the two great parts of China Proper. It is in a sense a cultural boundary as well, influencing the distribution of custom and language. South of the Qin Ling divide, lie the densely populated and highly developed areas of the lower and middle plains of the Yangtze River and, on its upper reaches, the Szechwan Basin, an area encircled by a high barrier of mountain ranges.

The New Spelling of Chinese Place Names

Municipalities directly under the central authorities:
- Beijing (Peking)
- Shanghai (Shanghai)
- Tianjin (Tientsin)

Provinces, autonomous regions for minority nationalities and some well-known cities and other places:

- Anhui (Anhwei) Province
  - Hefei (Hofei)
  - Bengbu (Pengpu)
- Fujian (Fukien) Province
  - Fuzhou (Foochow)
  - Xiamen (Amoy)
- Gansu (Kansu) Province
  - Lanzhou (Lanchow)
- Guangdong (Kwangtung) Province
  - Guangzhou (Kwangchow)
  - Shantou (Swatow)
- Guangxi Zhuang (Kwangsi Chuang) Autonomous Region
  - Nanning (Nanning)
  - Guilin (Kweilin)
- Guizhou (Kweichow) Province
  - Guiyang (Kweiyang)
  - Zunyi (Tsunyi)
- Hebei (Hopei) Province
  - Shijiazhuang (Shihchiachuang)
  - Tangshan (Tangshan)
- Heilongjiang (Heilungkiang) Province
  - Harbin (Harbin)
  - Daqing Oilfield (Taching Oilfield)
  - Qiqihar (Chichihar)

Henan (Honan) Province
Zhengzhou (Chengchow)
Louyang (Loyang)
Kaifeng (Kaifeng)
Hubei (Hupeh) Province
Wuhan (Wuhan)
Hunan (Hunan) Province
Changsha (Changsha)
Jiangsu (Kiangsu) Province
Nanjing (Nanking)
Suzhou (Soochow)
Wuxi (Wuhsi)
Jiangxi (Kiangsi) Province
Nanchang (Nanchang)
Jiujiang (Chiuchiang)
Jilin (Kirin) Province
Changchun (Changchun)
Liaoning (Liaoning) Province
Shenyang (Shenyang)
Anshan (Anshan)
Luda (Luta)
Nei Monggol (Inner Mongolia) Autonomous Region
Hohhot (Huhehot)
Baotou (Paotou)
Ningxia Hui (Ningsia Hui) Autonomous Region
Yinchuan (Yinchuan)
Qinghai (Chinghai) Province
Xining (Sining)
Shaanxi (Shensi) Province
Xian (Sian)
Yanan (Yenan)
Shandong (Shantung) Province
Jinan (Tsinan)
Qingdao (Tsingtao)
Yantai (Yentai)
Shanxi (Shansi) Province
Taiyuan (Taiyuan)
Dazhai (Tachai)
Sichuan (Szechuan) Province
Chengdu (Chengtu)
Chongqing (Chungking)
Taiwan (Taiwan) Province
Taibei (Taipei)
Xinjiang Uygur (Sinkiang Uighur) Autonomous Region
Urumqi (Urumchi)
Xizang (Tibet) Autonomous Region
Lhasa (Lhasa)
Yunnan (Yunnan) Province
Kunming (Kunming)
Dali (Tali)
Zhejiang (Chekiang) Province
Hangzhou (Hangchow)

The country's longest and most important waterway, the Yangtze is navigable over much of its length and has a vast hydroelectric potential. Rising on the Plateau of Tibet, it traverses 6,300 km (3,914 mi) through the heart of the country, draining an area of 1.8 million sq km (0.694 million sq mi) before emptying into the East China Sea. The roughly 300 million persons who live along its middle and lower reaches cultivate a great rice and wheat producing area. The Szechwan Basin—favored by a mild, humid climate and a long growing season—produces a rich variety of crops; it is also a leading silk producing area and an important industrial region with substantial mineral resources.

Second only to the Qin Ling as an internal boundary is the Nan Ling, the southernmost of the east-west mountain ranges. The Nan Ling defines that part of China where a tropical climate permits two crops of rice to be grown each year. Southeast of the mountains lies a coastal, hilly region of small deltas and narrow valley plains; the drainage area of the Zhu Jiang and its associated network of rivers occupies much of the region to the south. West of the Nan Ling stands the Yunnan-Guizhou Plateau, more or less rising in two steps, averaging respectively 1,200 and 1,800 m (3,739 and 5,905 ft) in elevation toward the precipitous mountain regions of the eastern Plateau of Tibet.

The Hai He, like the Zhu Jiang and other major waterways, flows from west to east. Its upper course consists of five rivers which converge near Tianjin, then flow 70 km (43 mi) before emptying into the Bo Hai. Another major river, the Huai He, rises in Henan Province and flows through several lakes before joining the Yangtze.

Inland drainage involving a number of upland basins in the North and Northeast accounts for less than 40 percent of the country's total drainage area. Many such rivers and streams flow into lakes or diminish in the desert. Some are useful for irrigation.

China's extensive territorial waters are principally marginal seas of the western Pacific, washing a long and much indented coastline and having many islands. Taiwan and Hainan are the largest and second largest, respectively, of the 5,000 islands. The Yellow, East China, and South China seas are marginal seas of the Pacific Ocean. More than half the coastline (predominantly in the south) is rocky; most of the remainder is sandy. The Bay of Hangzhou roughly divides the two types of shoreline.

CHRISTMAS ISLAND

BASIC FACTS

Official Name: Territory of Christmas Island

Abbreviation: CHD

Area: 135 sq km (52 sq mi)

Area—World Rank: 200

Population: 2,278 (1988) 3,000 (2000)

Population—World Rank: 212

Capital: The Settlement

Land Boundaries: Nil

Coastline: 54 km (33 mi)

Longest Distances: N/A

Highest Point: Murray Hill 361 m (1,184 ft)

Lowest Point: Sea Level

Land Use: 100% other

Christmas Island is situated in the Indian Ocean, directly south of the Western tip of Java, about 2,330 km (1,450 mi) northwest of Perth. Almost completely surrounded by a reef, the island has steep cliffs along the coast that rise abruptly to a central plateau.

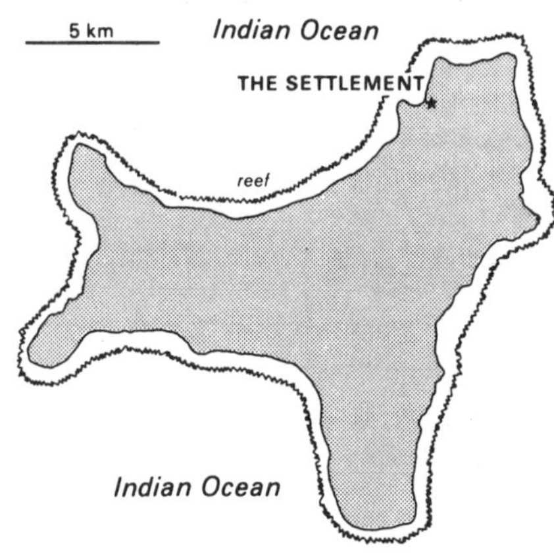

COCOS (KEELING) ISLANDS

BASIC FACTS

Official Name: Territory of Cocos (Keeling) Islands

Abbreviation: CKI

Area: 14.4 sq km (5.6 sq mi)

Area—World Rank: 211

Population: 600 (1988), 1,000 (2000)

Population—World Rank: 214

Capital: West Island

Land Boundaries: Nil

Coastline: Undetermined

Longest Distances: N/A

Highest Point: N/A

Lowest Point: N/A

Land Use: 100% other

Cocos (Keeling) Islands is a group of coral atolls consisting of 27 islands in the Indian Ocean about 2,770 km (1,720 mi) northwest of Perth. Two coral atolls are thickly covered with coconut palms and other vegetation.

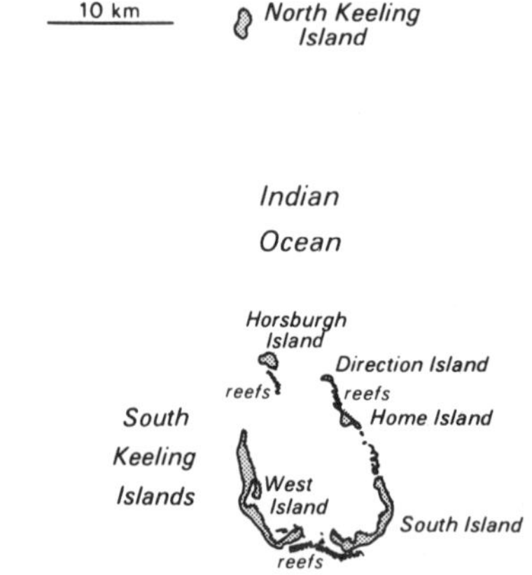

COLOMBIA

BASIC FACTS

Official Name: Republic of Colombia

Abbreviation: COL

Area: 1,141,748 sq km (440,831 sq mi)

Area—World Rank: 26

Population: 31,298,803 (1988) 35,436,000 (2000)

Population—World Rank: 30

Capital: Bogota

Boundaries: 9,242km (5,742 mi); Venezuela 2,219 km (1,379 mi); Brazil 1,645 km (1,022 mi); Peru 1,626 km (1,010 mi); Ecuador 586 km (364 mi); Panama 266 km (165 mi)

Coastline: Caribbean Sea 1,600 km (994 mi); Pacific Ocean 1,300 m (808 mi)

Longest Distances: 1,700 km (1,056 mi) NNW-SSE 1,210 km (752 mi) NNE-SSW

Highest Point: Cristobal Colon Peak 5,800 m (19,029 ft)

Lowest Point: Sea Level

Land Use: 2% permanent crops; 29% meadows and pastures; 4% arable; 49% forest and woodland; 16% other

Located in the northwest corner of the South American continent, Colombia is the only country of South America with both Atlantic and Pacific coastlines. It is the fifth largest in size of the Latin American countries.

The country is dominated by three ranges of the Andes Mountains that cross the country on a northwesterly axis in the form of a trident. The greatest concentration of population is in plateaus and basins scattered among these ranges and in the valleys of the two great rivers that separate them. West of the Andes are two coastal lowlands. The Atlantic Lowlands consist largely of open land, much of it swampy; there is extensive cattle raising and plantation agriculture there, as well as heavy concentrations of population around the several port cities. The narrower Pacific Lowlands are swampy, heavily forested, and sparsely populated. East of the Andes lies a great interior plain crossed from east to west by many large rivers. Consisting principally of open land in the north and tropical jungle in the south, it has few settlements and remains largely undeveloped.

Some 45 percent of the country is forested, about half of it in exploitable timber. Precipitous slopes and convoluted terrain in the highlands and the prevalence of swamplands and jungle at lower elevations limit the utilization of land; less than 3 percent of the national territory consists of farmland and, of this proportion, 13 percent is in permanent and seasonal crops, 53 percent is in pasture, 23 percent is fallow, and the remainder is forested or unused.

Steeply sloping terrain, the composition of the soil, and heavy precipitation have made over 30 percent of the country highly susceptible to erosion; in an area equivalent in size to about one-fourth of all farmland, erosion is far advanced. A government official estimated in 1976 that 5 percent of the territory in the department of Cundinamarca had become so eroded as to be useless.

The prevailing pattern of land use is one in which the more fertile flatlands tend to be devoted to commercial farming and cattle raising and smaller farm plots reach far up steep mountain sides. Land has been farmed successfully on plots with slopes of 50 percent or more. Soils vary radically in characteristics and fertility, and the complexity of the physical environment makes the matching of soils to climatic conditions difficult. The most productive soils are in the river valleys and the basins of the highland interior. Soils of the Atlantic Lowlands are generally superior to those of the Pacific coast, and the predominantly acid soils of the eastern plains are relatively infertile.

PRINCIPAL TOWNS (population at 15 October 1985)

Town	Population	Town	Population
Bogotá, DE (capital)	3,982,941	Cúcuta	379,478
Medellín	1,468,089	Bucaramanga	352,326
Cali	1,350,565	Manizales	299,352
Barranquilla	899,781	Ibagué	292,965
Cartagena	531,426	Pereira	287,999

Area and population

		area		population
Commissariats	**Capitals**	sq mi	sq km	1985 census
Amazonas	Leticia	42,342	109,665	30,327
Guainía	San Felipe (Obando)	27,891	72,238	9,214
Guaviare	Guaviare	16,342	42,327	35,305
Vaupés	Mitú	25,200	65,268	18,935
Vichada	Puerto Carreño	38,703	100,242	13,770
Departments				
Antioquia	Medellín	24,561	63,612	3,888,067
Atlántico	Barranquilla	1,308	3,388	1,428,601
Bolívar	Cartagena	10,030	25,978	1,197,623
Boyacá	Tunja	8,953	23,189	1,097,618
Caldas	Manizales	3,046	7,888	838,094
Caquetá	Florencia	34,349	88,965	214,473
Cauca	Popayán	11,316	29,308	795,838
Cesar	Valledupar	8,844	22,905	584,631
Chocó	Quibdó	17,965	46,530	242,768
Córdoba	Montería	9,660	25,020	913,636
Cundinamarca	Bogotá	8,735	22,623	1,382,360
Huila	Neiva	7,680	19,890	647,756
La Guajira	Riohacha	8,049	20,848	255,310
Magdalena	Santa Marta	8,953	23,188	769,141
Meta	Villavicencio	33,064	85,635	412,312
Nariño	Pasto	12,845	33,268	1,019,098
Norte de Santander	Cúcuta	8,362	21,658	883,884
Quindío	Armenia	712	1,845	377,860
Risaralda	Pereira	1,598	4,140	625,451
Santander	Bucaramanga	11,790	30,537	1,438,226
Sucre	Sincelejo	4,215	10,917	529,059
Tolima	Ibagué	9,097	23,562	1,051,852
Valle	Cali	8,548	22,140	2,847,087
Intendancies				
Arauca	Arauca	9,196	23,818	70,085
Casanare	Yopal	17,236	44,640	110,253
Putumayo	Mocoa	9,608	24,885	119,815
San Andrés y Providencia	San Andrés	17	44	35,936
Special District				
Bogotá		613	1,587	3,982,941
TOTAL		440,831	1,141,748	27,867,326

Geographic Regions

Geographers have devised different ways to divide Colombia into regions. The simplest is to see the country as consisting of three parts—a highland core, coastal lowlands to the west, and plains to the east. The plan is consistent with the pattern of the terrain, but the coastal lowlands of the Atlantic have characteristics very different from those of the Pacific. It is therefore appropriate to divide the country into four geographic regions: the Central Highlands, consisting of the three Andean ranges and intervening valley lowlands; the Atlantic Lowlands and the Pacific Lowlands coastal regions, separated by swamps at the base of the Isthmus of Panama; and eastern Colombia, the great plain that lies to the east of the Andes.

Central Highlands

Near the Ecuadorian frontier the Andes Mountains divide into three distinct chains, called cordilleras, that extend northwestward almost to the Caribbean Sea. Altitudes reach more than 5,791 m (19,000 ft) and mountain peaks are permanently covered with snow. The elevated basins and plateaus of these ranges have a moderate climate that provides pleasant living conditions and enables farmers in many places to harvest twice a year. Torrential rivers on the slopes of the mountains produce a large hydroelectric power potential and add their volume to the navigable rivers in the valleys. The great majority of the population lives in these mountainous regions where, before the appearance of the white man, Indians had developed a culture almost as complex and elaborate as that of the Incas to the south and the Aztecs to the north.

The three parallel chains of the Andes—the Cordillera Occidental in the west, the Cordillera Central, and the Cordillera Oriental in the east—present differing characteristics. In the Cordillera Oriental at elevations between 2,438 and 2,743 m (8,000 and 9,000 ft) three large fertile basins and a number of small ones provide suitable areas for settlement and intensive economic production. In the basin of Cundinamarca, where the Spanish found the Chibchas—settled tribes of Indians practicing agriculture—the white invaders founded the town of Santa Fé (later Bogotá) at an elevation of 2,639 m (8,660 ft) above sea level.

To the north of Bogotá, on the densely populated plateaus of Chiquinquirá and Boyacá, are fertile fields, rich mines, and large industrial establishments that produce a great part of the national wealth. Still farther north, where the Cordillera Oriental makes an abrupt turn to the northwest near the Venezuela border, the highest point of this range, the Sierra Nevada de Cocuy, rises to 5,581 m (18,310 ft) above sea level. In the department of Santander the valleys on the western slopes are more spacious, and agriculture is intensive in the area around Bucaramanga. The northernmost region of the range, around Cúcuta and Ocaña, is so rugged that historically it has been found easier to maintain communication and transportation toward Venezuela from this area than toward the adjacent parts of Colombia.

The Cordillera Central, also called the Cordillera del Quindío, is the loftiest of the mountain systems. Its crystalline rocks form a 805 km (500 mi) long towering wall dotted with snow-covered volcanoes. There are no plateaus in this range and no passes under 3,352 m (11,000 ft). The highest peak, the Nevado del Huila, reaches 5,750 m (18,865 ft) above sea level. Toward its northern end this cordillera separates into several branches that descend toward the Caribbean coast.

The Cordillera Occidental is separated from the Cordillera Central by the deep rift of the Cauca River valley; this range is the lowest and the least populated of the three and supports little economic activity. A pass about 1,524 m (5,000 ft) above sea level provides the major city of Cali with an outlet to the Pacific Ocean. The relatively low elevation of the cordillera permits dense vegetation, which on the western slopes is truly tropical.

The two rivers that separate the lines of the Andean trident have formed fertile floodplains in valleys that reach deep into the highlands. The Magdalena River rises near a point some 177 km (110 mi) north of Ecuador where the Cordillera Oriental and the Cordillera Central diverge. Its spacious drainage area is fed by numerous mountain torrents originating high in the snowfields, where for millennia glaciers have planed the surface of folded and stratified rocks. The Magdalena is navigable from the Caribbean Sea as far as the town of Neiva, deep in the interior, but is interrupted midway by rapids at the town of Honda. The valley floor is very deep; nearly 805 km (500 mi) from the river's mouth the elevation is no more than about 305 m (1,000 ft).

Running parallel to the Magdalena and separated from it by the Cordillera Central, the Cauca River has headwaters not far from those of the Magdalena, which it eventually joins in swamplands of the Atlantic region. The area known customarily as the Cauca Valley does not include all of the lands flanking the river. This tropical valley, a fertile sugar zone that includes the best farmland in the country, follows the course of the river for about 241 km (150 mi) southward from a narrow gorge at about its midpoint near the town of Cartago. The cities of Cali and Palmira are situated on low terraces above the floodplain of the Cauca Valley.

Atlantic Lowlands

The Atlantic lowlands consist of all Colombia north of an imaginary line extending northeastward from the Gulf of Urabá to the Venezuelan frontier at the northern extremity of the Cordillera Oriental. The region corresponds generally to one often referred to by foreign writers as the Caribbean lowland or coastal plain; in Colombia, however, it is consistently identified as Atlantic rather than Caribbean. The semiarid Guajira Peninsula, in the extreme north, bears little resemblance to the rest of the region. In the southern part rises the Sierra Nevada de Santa Marta, an isolated mountain system with peaks reaching a height of over 5,791 m (19,000 ft) and slopes generally too steep for cultivation.

The Atlantic Lowlands region is in roughly the shape of a triangle, the longest side of which is the coastline. Most of the country's commerce moves through Barranquilla, Cartagena, Santa Marta, and the other ports located along this important coast. Beyond these cities are swamps, hidden streams, and shallow lakes that support banana and cotton plantations, countless small farms and, in higher places, cattle ranches.

The coastal cities were the first places settled by the Spanish. The Atlantic region merges into and is connected with the Central Highlands through the two great river valleys. It is the second most important region, after the Central Highlands, in population and economic activity.

Pacific Lowlands

The Pacific Lowlands are a thinly populated region of jungle and swamp with considerable but little-exploited potential in minerals and other resources. Buenaventura is the only port of any size on the more than 1,287 km (800 mi) of coastline. On the east the Pacific Lowlands are bounded by the Cordillera Occidental, from which run numerous streams. A majority of these streams flow westward to the Pacific, but the largest, the navigable Atrato River, flows northward to the Gulf of Urabá, a circumstance that makes the river settlements accessible to the major Atlantic ports and commercially related primarily to the Atlantic Lowlands hinterland. To the west of the Atrato River rises the Serranía de Baudó, an isolated chain that occupies a large part of the coastal plain. Its highest elevation is less than 1,829 m (6,000 ft), and its vegetation resembles that of the surrounding tropical forest.

Caribbean
Sea
Certain islands of the Archipiélago de San Andres y Providencia (13°00'N, 81°30'W) and the Isla de Malpelo 3°58'N, 81°35'W) belonging to Colombia are not shown on this map.
NETHERLANDS ANTILLES
(Netherlands)
Willemstad
North
Pacific
Ocean
PANAMA
Panama
VENEZUELA
Maracaibo
Lago de Maracaibo
La Guaira
Caracas
Valencia
ECUADOR
Quito
PERU
Iquitos
BRAZIL
LA GUAJIRA
Riohacha
Santa Marta
Barranquilla
ATLÁNTICO
Ciénaga
Cartagena
Valledupar
MAGDALENA
CESAR
Tolú
Sincelejo
SUCRE
Montería
CÓRDOBA
Turbo
BOLÍVAR
NORTE DE SANTANDER
Cúcuta
Pamplona
Bucaramanga
SANTANDER
ANTIOQUIA
Medellín
ARAUCA
Arauca
Barbosa
Paz de Río
BOYACÁ
CASANARE
Tunja
Yopal
Quibdó
CHOCÓ
CALDAS
Manizales
RISARALDA
Pereira
CUNDINAMARCA
Armenia
Ibagué
Bogotá
Girardot
QUINDÍO
Villavicencio
Puerto López
VICHADA
Puerto Carreño
Buenaventura
VALLE DEL CAUCA
TOLIMA
DISTRITO ESPECIAL
Cali
META
Neiva
HUILA
Puerto Inírida
GUAINÍA
CAUCA
Popayán
San José del Guaviare
GUAVIARE
Tumaco
NARIÑO
Florencia
Pasto
Mocoa
Ipiales
PUTAMAYO
CAQUETÁ
VAUPÉS
Mitú
AMAZONAS
Leticia
Río Magdalena
Río Cauca
Río Atrato
Río Arauca
Río Meta
Río Orinoco
Río Guaviare
Río Guainía
Río Vaupés
Río Caquetá
Río Putumayo
Río Napo
Río Japurá
Río Içá
Río Negro
Amazon
Río Marañón
Río Ucayali
Río Javari
Highway
12
8
4
0
78
74
70
Colombia
International boundary
Internal administrative boundary
National capital
Internal administrative capital
Railroad
Road
0 50 100 150 Kilometers
0 50 100 150 Miles
Boundary representation is not necessarily authoritative.

The Atrato swamp—in Chocó Department adjoining the Panama frontier—is a bottomless muck forty miles in width that for years has challenged engineers seeking to complete the Pan-American Highway. Except where the highway is interrupted by this stretch, known as the Tapón del Chocó (Chocón Plug), and an adjoining swampland in southern Panama known as the Darien Gap, it reaches without interruption from Alaska to the Straits of Magellan. In 1986 it was anticipated that this final segment of the highway would be completed in perhaps another six years by rock filling and bridge construction. A second major transportation project under consideration involving Chocó Department, for which various schemes had intermittently been proposed, was the construction of a second interoceanic canal by dredging the Atrato River and other streams and digging short access canals. Completion of either or both of these projects would do much to transform this somnolent region.

Eastern Colombia

The area east of the Andes includes about 699,297 sq km (270,000 sq mi), three-fifths of the country's total area, but Colombians refer to it almost as if it were an alien land. The entire area is known as the Eastern plains. The Spanish term for plains (*llanos*) can be applied only to the open plains in the northern part where cattle raising is practiced, particularly in piedmont areas near the Cordillera Oriental.

The subregion is unbroken by highlands except in Meta Department, where the Macarena Sierra, an outlier of the Andes, is of interest to scientists because its vegetation and wildlife are believed to be reminiscent of those that once existed throughout the Andes. The numerous large rivers of eastern Colombia include many that are navigable. The Guaviare River and the streams to its north flow eastward and drain into the basin of the Orinoco, the largest river in Venezuela. Those south of the Guaviare flow into the basin of the Amazon. The Guaviare serves as a border for five political subdivisions, and it divides eastern Colombia into the Eastern Plains subregion in the north and the Amazonas subregion in the south. In the south the plains give way to largely unexplored tropical jungle, where the scanty population becomes scantier still.

Islands

Colombia possesses a few islands in the Caribbean and some in the Pacific Ocean, the combined areas of which do not exceed twenty-five square miles. Off Nicaragua about 644 km (400 mi) northwest of the Colombian coast, an archipelago of some thirteen small cays grouped around two larger islands forms the San Andrés y Providencia Intendency. Other islands in the same area—the sovereignty of which has been in dispute—are the small islands, cays, or banks of Santa Catalina, Roncador, Quita Sueño, Serrana, and Serranilla. Off the coast south of Cartegena are several small islands, among them the islands of Rosario, San Bernardo, and Fuerte.

The island of Malpelo lies in the Pacific Ocean about 434 km (270 mi) west of Buenaventura. Nearer the coast a prison colony is located on Gorgona Island. Gorgonilla Cay is off its southern shore.

COMOROS

BASIC FACTS

Official Name: Federal Islamic Republic of the Comoros

Abbreviation: COM

Area: 1,862 sq km (719 sq mi)

Area—World Rank: 158

Population: 429,479 (1988) 650,000 (2000)

Population—World Rank: 145

Capital: Moroni

Land Boundaries: Nil

Coastline: 340 km (211 mi)

Longest Distances: N/A

Highest Point: Kartala 2,561 m (8,402 ft)

Lowest Point: Sea level

Land Use: 35% arable; 8% permanent crops; 7% meadows and pastures; 16% forest and woodland; 34% other

Comoros is located in the northern entrance of the Mozambique Channel about halfway between the northern tip of Madagascar and the mainland coast of Mozambique. It includes three main islands—Grande Comore, Anjouan and Moheli—and several islets with a total land area of 1,795 sq km (693 sq mi). Grande Comore, the largest of the islands comprises 1,148 sq km (443 sq mi); Moheli, 290 sq km (112 sq mi); and Anjouan, 357 sq km (138 sq mi).

PRINCIPAL TOWNS (population at 1980 census): Moroni (capital) 17,267; Mutsamudu 13,000; Fomboni 5,400.

Area and population

Governorates/Islands	Capitals	area sq mi	area sq km	population 1987 estimate
Moili (Mohéli)	Fomboni	112	290	21,803
Ngazidja (Grande Comore)	Moroni	443	1,148	226,874
Ndzouani (Anjouan)	Mutsamudu	164	424	173,494
TOTAL		719	1,862	422,171

The Comoros are volcanic in origin, and Mount Kartala on Grande Comore (2,561 meters, 8,402 ft) is an active volcano. The center of Grande Comore is a desert lava field. The black basalt relief rises 1,200 to 1,600 m (3,950 to 5,250 ft) on Anjouan and 500 to 800 m (1,650 to 2,600 ft) on Moheli. Mayotte has a lagoon fringed with coral reefs.

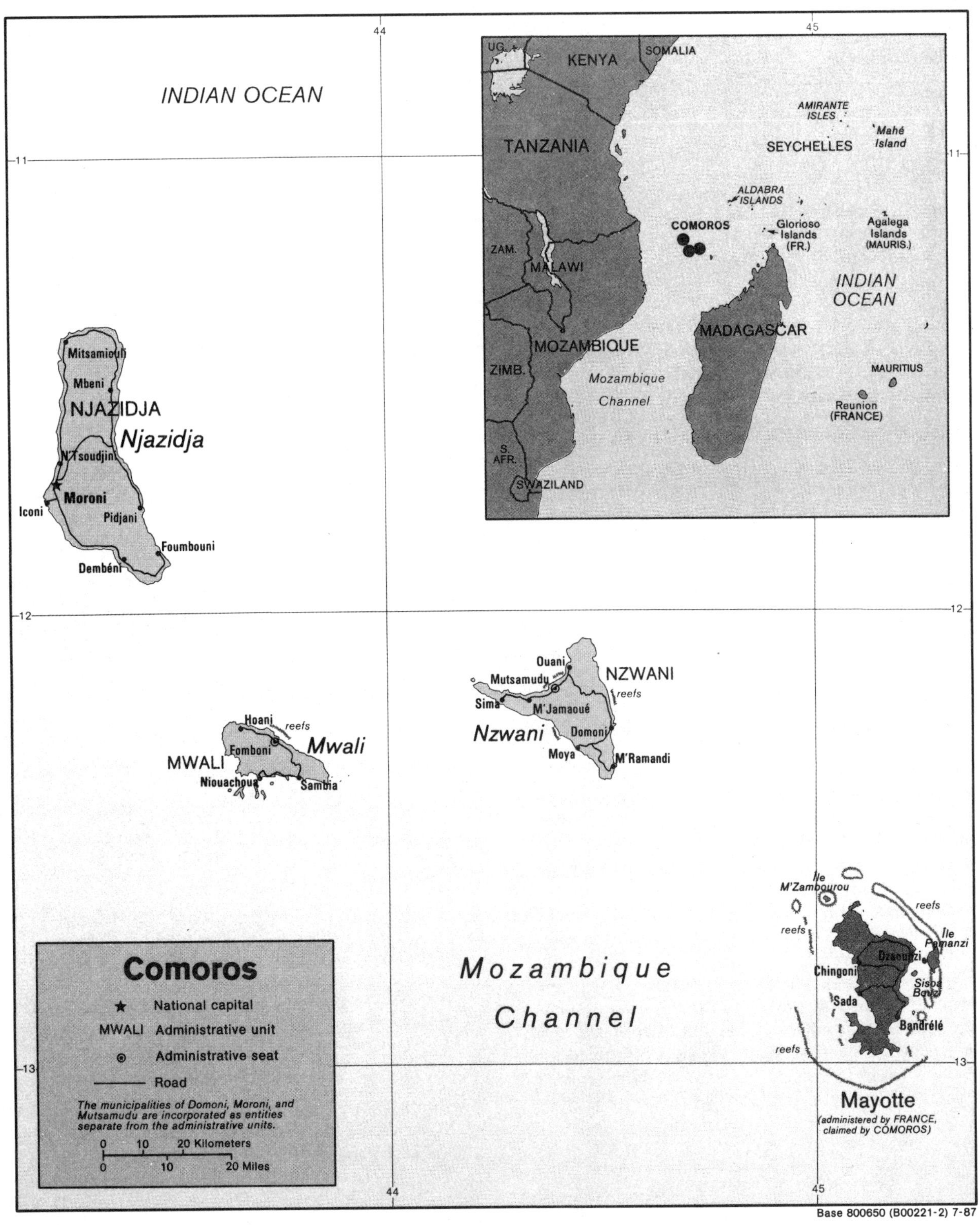
INDIAN OCEAN
NJAZIDJA
Njazidja
Mitsamiouli
Mbeni
N'Tsoudjini
Moroni
Iconi
Pidjani
Foumbouni
Dembéni
MWALI
Mwali
Hoani
reefs
Fomboni
Niouachoua
Sambia
NZWANI
Nzwani
Ouani
Mutsamudu
Sima
M'Jamaoué
Domoni
Moya
M'Ramandi
Mozambique
Channel
Ile M'Zambourou
reefs
Chingoni
Sada
Bandrélé
Ile Pamanzi
Sisoa Bouzi
Mayotte
(administered by FRANCE, claimed by COMOROS)
KENYA
SOMALIA
UG.
TANZANIA
AMIRANTE ISLES
Mahé Island
SEYCHELLES
ALDABRA ISLANDS
COMOROS
Glorioso Islands (FR.)
Agalega Islands (MAURIS.)
ZAM.
MALAWI
INDIAN OCEAN
MOZAMBIQUE
MADAGASCAR
ZIMB.
Mozambique Channel
MAURITIUS
Reunion (FRANCE)
S. AFR.
SWAZILAND
Comoros
National capital
MWALI Administrative unit
Administrative seat
Road
The municipalities of Domoni, Moroni, and Mutsamudu are incorporated as entities separate from the administrative units.
0 10 20 Kilometers
0 10 20 Miles
Base 800650 (B00221-2) 7-87

CONGO

BASIC FACTS

Official Name: People's Republic of the Congo

Abbreviation: CON

Area: 342,000 sq km (132,047 sq mi)

Area—World Rank: 57

Population: 2,153,685 (1988) 3,600,000 (2000)

Population—World Rank: 120

Capital: Brazzaville

Boundaries: 4,625 km (2.874 mi); Cameroon 520 km (323 mi); Central African Republic 467 km (290 mi); Zaire 1,625 km (1,010 mi); Angola 201 km (125 mi); Gabon 1,656 km (1,029 mi)

Coastline: 156 km (97 mi)

Longest Distances: 1,006 km (625 mi) NNE-SSW; 402 km (250 mi) ESE-WNW

Highest Point: 903 m (2,963 ft)

Lowest Point: Sea level

Land Use: 2% arable; 29% meadows and pastures; 62% forest and woodland; 7% other

Congo is an irregularly shaped equatorial country located on the west coast of Africa. It is the third largest and the least generously endowed by nature of the four former French territories—Moyen-Congo (Middle Congo), Gabon, Chad, and Oubangui-Chari (now the Central African Republic)—that formed the Federation of French Equatorial Africa (Afrique Equatoriale Française—AEF) from 1910 to 1958.

This elongated country has a maximum north-south distance of nearly 1,287 km (800 mi) but extends only 483 km (300 mi) from east to west at its broadest point. Its perimeter measures slightly more than 4,667 km (2,900 mi) of which only 160 km (100 mi) are coastline along the south Atlantic Ocean, between Gabon and the Angolan exclave of Cabinda. Approximately 60 percent of the country is lowlands covered by forest, and the remainder consists chiefly of plateaus, valleys, and savanna table lands.

PRINCIPAL TOWNS (population at 1974 census)

Town	Population	Town	Population
Brazzaville (capital)	302,459	Loubomo (Dolisie)	28,577
Pointe-Noire	140,367	Ngamaba-Mfilou	18,944
Nkayi (Jacob)	28,957	Loandjili	15,738
		Mossendjo	11,913

1980 estimates: Brazzaville 422,400; Pointe-Noire 185,110; Nkayi 32,520; Loubomo 30,830.

Area and population

Regions	Capitals	area sq mi	area sq km	population 1984 census
Bouenza	Madingou	4,734	12,260	150,603
Cuvette	Owando	28,900	74,850	135,744
Kouilou	Pointe-Noire	5,274	13,660	74,870
Lékoumou	Sibiti	8,089	20,950	68,287
Likouala	Impfondo	25,500	66,044	49,505
Niari	Loubomo	10,011	25,930	110,003
Plateaux	Djambala	14,826	38,400	109,663
Pool	Kinkala	13,124	33,990	184,263
Sangha	Ouesso	21,544	55,800	34,213
Communes				
Brazzaville	—	25	65	585,812
Loubomo	—	5	12	49,134
Mossendjo	—	—	—	14,469
Nkayi	—	2	5	36,540
Ouesso	—	—	—	11,939
Pointe-Noire	—	13	34	294,203
TOTAL		132,047	342,000	1,909,248

Coastal Plain

The coastal area, lying to the southwest between Gabon and the exclave of Cabinda, is an undulating plain fringed with sandy shores. It stretches for about 160 km (100 mi) along the south Atlantic coast and reaches inland approximately forty miles to the Mayombé escarpment to the northwest and to the foothills of the Crystal Mountains, which extend southeastward into the Congo from Gabon. The area is bisected by the Kouilou River, which drains the area between these two mountain ranges from Makabana to the sea.

The effect of the Benguela (Antarctic) Current flowing from the south is to enhance the formation of sandspits along the coastal plain, which is virtually treeless except in scattered areas. In addition to the mangrove-fringed lagoons, the area is marked by lakes and rivers with accompanying marshland and heavy vegetation in low-lying areas. Extensive swampy areas exist both to the northwest and southeast of the mouth of the Kouilou River.

Inland from the seacoast the land rises somewhat abruptly to a series of eroded hills and plateaus, which run parallel to the coastline. From the lower reaches of the Crystal Mountains on the Gabon border, this area rises southeasterly in a succession of sharp ridges of the Mayombé range that reach elevations of 487 to 610 m (1,600 to 2,000 ft). Deep gorges have been cut in these ridges by the swift Kouilou River or its tributaries.

Niari Valley

Lying between the Chaillu Mountains to the north and the Mayombé Mountains to the south, the Niari Valley extends in a generally east-west direction for almost 322 km (200 mi). Originally covered with tall grasses and savanna, it has been extensively cleared to permit a great variety of agricultural pursuits and diversified industrial activity. Topographical conditions and the cooling effect of the Benguela Current account for a particularly favorable climate, which permits the harvesting of two crops per year.

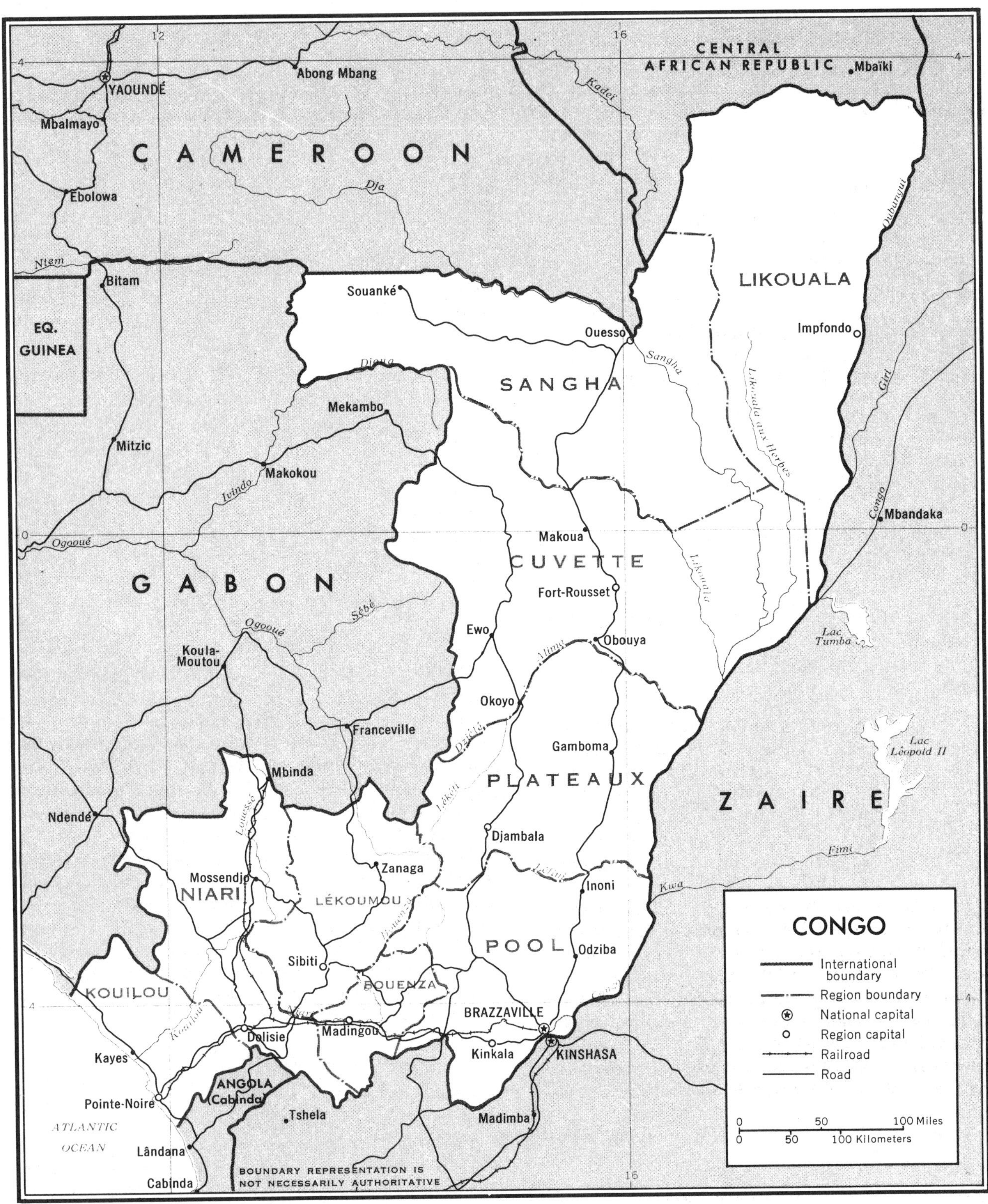

Base 500030 11-71

While the northern slope of the valley rises somewhat abruptly to the Chaillu Mountains, the southern and eastern slopes increase more gently in elevation toward the treeless plains that form the Plateau of the Cataracts south of Brazzaville. Small streams and rivers afford good drainage for the valley floor, except in the northwest, where the joining of the Niari River with the Kouilou and the Louessé rivers permanently inundates much of the low-lying land south of Makabana. The valley contains the route of both the Congo-Ocean Railway and one of the main roads that connects Brazzaville with the port of Pointe-Noire, thus facilitating the transport of its products.

Central Highlands

The Central Highlands encompass the area generally known as the Batéké Plateau and extend for approximately 129,500 sq km (50,000 sq mi) over the south-central portion of the country. This region is characterized predominantly by rounded, low hills of less than 305 m (1,000 ft) elevation and scattered rolling plains. In the northern part of this sector, however, toward the lower Gabon border, the hills are more peaked, and crests rise as high as 823 m (2,700 ft) above sea level. For the most part the lower hills are untimbered and the plateaus grass covered.

The Central Highlands are not heavily settled and contain few roads and little industrial activity. The climate is tropical and includes distinct wet and dry seasons. A considerable portion of the area north of the Niari River and extending as far west as the vicinity of Zanaga is covered with dense tropical forest. Subsistence agriculture is practiced, and only a limited amount of fertile cropland is available.

To the northwest along the Gabon border and running almost to the equator, a region of hills and plateaus forms the western rim of the Zaire Basin. The plateaus are separated from each other by deep valleys that carry the eastward-flowing tributaries of the Zaire River. Savanna and tropical rain forest are interspersed throughout the entire area.

Zaire Basin

The northeast section of the country, covering an area of approximately 155,400 sq km (60,000 sq mi), lies within the Zaire Basin. It consists of flat, swampy valleys and low divides descending east and southeast from the western hills to the Zaire River. The region is covered with dense equatorial rain forest, and large portions lying northeast and southwest of the Sangha River are permanently inundated. Flooding occurs seasonally almost everywhere, and in areas south of Mossaka, along the Zaire River, extensive marshland covered with swamp vegetation exists.

The rivers of the basin had been the primary communication routes between the interior and the population centers to the south, and the Oubangui and Zaire rivers formed the principal trade artery for the entire region, as well as for countries immediately to the north. Aside from the production of palm oil, the principal economic activities of the region were limited to wood and wood products. Considerable portions of the Zaire Basin were virtually uninhabited, and the sparse settlements that existed engaged in hunting, fishing, trading, and subsistence agriculture.

COOK ISLANDS

BASIC FACTS

Official Name: Cook Islands

Abbreviation: CKS

Area: 236 sq km (91 sq mi)

Area—World Rank: 194

Population: 17,995 (1988) 18,000 (2000)

Population—World Rank: 198

Capital: Avarua

Land Boundaries: Nil

Coastline: 120 km (74 mi)

Longest Distances: N/A

Highest Point: Te Manga 653 m (2,142 ft)

Lowest Point: Sea level

Land Use: 4% arable; 22% permanent crops; 74% other

Cook Islands consist of 15 islands lying more than 3,220 km (2,000 mi) northeast of New Zealand. The southern islands are coral atolls and the northern islands are volcanic. The southern group consists of eight islands of which the largest are Rarotonga and Mangaia. The northern group comprises seven islands of which the largest is Penrhyn and the smallest Nassau. Except for Rarotonga, the islands lack streams and wells.

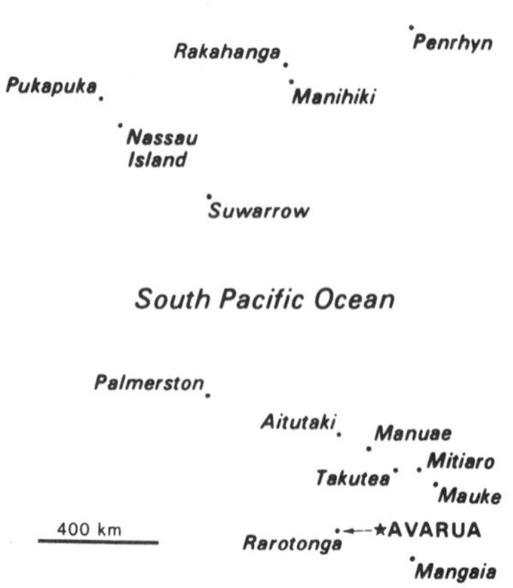

COSTA RICA

BASIC FACTS

Official Name: Republic of Costa Rica

Abbreviation: CSR

Area: 51,100 sq km (19,730 sq mi)

Area—World Rank: 116

Population: 2,888,277 (1988) 3,587,000 (2000)

Population—World Rank: 115

Capital: San José

Boundaries: 1,891 km (1,175 mi); Nicaragua 300 km (186 mi); Panama 363 km (226 mi)

Coastline: Caribbean Sea 212 km (132 mi); Pacific Ocean 1,016 km (631 mi)

Longest Distances: 464 km (288 mi) N-S 274 km (170 mi) E-W

Highest Point: Mount Chirripo 3,819 m (12,530 ft)

Lowest Point: Sea level

Land Use: 6% arable; 7% permanent crops; 45% meadows and pastures; 34% forests and woodland; 8% other

Costa Rica, bordered by Nicaragua, Panamá, the Pacific Ocean, and the Caribbean Sea, is one of the smallest and least populated of all Latin American countries, but its climate and topography are diverse, ranging from hot coastal lowlands to cold mountain peaks. Most of the population, however, is concentrated in a temperate highland valley called the Meseta Central (Central Basin), which contains most of the country's large cities as well as densely populated areas.

Costa Rica has an area slightly smaller than the state of West Virginia. With coasts on both the Atlantic and Pacific oceans, the country occupies part of the Central American isthmus. It is bordered on the north by Nicaragua and on the southeast by Panamá. The flat, open Caribbean coast, 209 km (130 mi) long, contrasts sharply with the irregular Pacific coast, 1,014 km (630 mi) of hilly or mountainous peninsulas, coastal lowlands, and deep gulfs and bays. A series of three mountain ranges flanked by lower hilly sections bisects the country from northwest to southeast and is partly responsible for the different climatic conditions of the two coasts. The mountains block the rain-bearing northeast trade winds, which cause the heavy and continual rainfall of the Caribbean coastal area. The Pacific coast receives its rain from May through October when the southwest winds blow on shore. Because of these rainfall patterns, rivers to the Caribbean flow constantly throughout the year, while those to the Pacific either dry up or decrease significantly in volume during the dry season. However, because of the relatively complicated topography of the Pacific coast, there are dramatic local exceptions to the general pattern, particularly in the northern lowlands of Guanacaste, which lie in the lee of the coastal mountains of the Nicoya peninsula. This area sometimes suffers severe drought even in the rainy season. On the other hand, the mountains north of the port of Golfito block the southwest winds, and a luxuriant rain forest and a poorly defined dry season characterize the area.

Climatic conditions are also dependent to a great degree upon altitude, in addition to proximity to one or the other of the coasts. On the Pacific side the hot country (*tierra caliente*) extends from sea level to about 457 m (1,500 ft) above and experiences daytime temperatures of 85° to 90°F. The temperate country (*tierra templada*) extends from 457 to 1,524 m (1,500 to 5,000 ft) above, with average daytime temperatures of 75° to 80°F. It is in this temperate zone that the densest population concentrations are to be found. The Meseta Central, an upland basin which has been the focus of settlement since the time of the Spanish Conquest, lies in this temperate zone. Four of the provincial capitals are located here, within a radius of 48 km (30 mi). The cold country (*tierra fría*) composes the land above 1,524 m (5,000 ft) and experiences average daytime temperatures of 75° to 80°F., but nighttime temperatures of 50° to 55°F. The cold country of Costa Rica is limited to a few mountain and volcanic peaks. Because of the greater rainfall on the Caribbean side and the warmer waters of the Caribbean, which affect the coastal air temperatures, the hot country and the temperate country climatic zones extend to higher altitudes on the Caribbean side than on the Pacific side.

Upland Basins: The Meseta Central and the General Valley

The most important area of Costa Rica is the Meseta Central, actually two upland basins separated by low volcanic hills. About 9,065 sq km (3,500 sq mi) in area and located in the temperate country between 914 and 1,524 m (3,000 and 5,000 ft) above sea level, it lies between the Cordillera Central to the north and low mountains and hills to the south. It was originally built up by volcanic ejecta and overlaid with basaltic lavas which weathered and contributed naturally rich soils to the basins. Four volcanoes overlook the Meseta from the north: Poás 2,760 m (9,055 ft above sea level), Barba 2,929 m (9,612 ft), Irazú 3,480 m (11,417 ft), and Turrialba 3,420 m (11,220 ft). Before the end of the 18th century the vegetation of the Meseta was dominated by subtropical forest, but a lucrative coffee market developed for Costa Rica around 1850, and since then much of the area has been planted to that crop. Cultivation has weathered and leached the soils, resulting in soils of medium to low natural fertility. The land surface of the Meseta is generally level or rolling and thus amenable to agriculture, except near the headwaters of rivers, where it is hilly and occasionally too steep for agriculture.

The slightly higher and smaller eastern basin, called the Cartago basin, about 1,524 m (5,000 ft) above sea level, was the economic and cultural center of the colonial

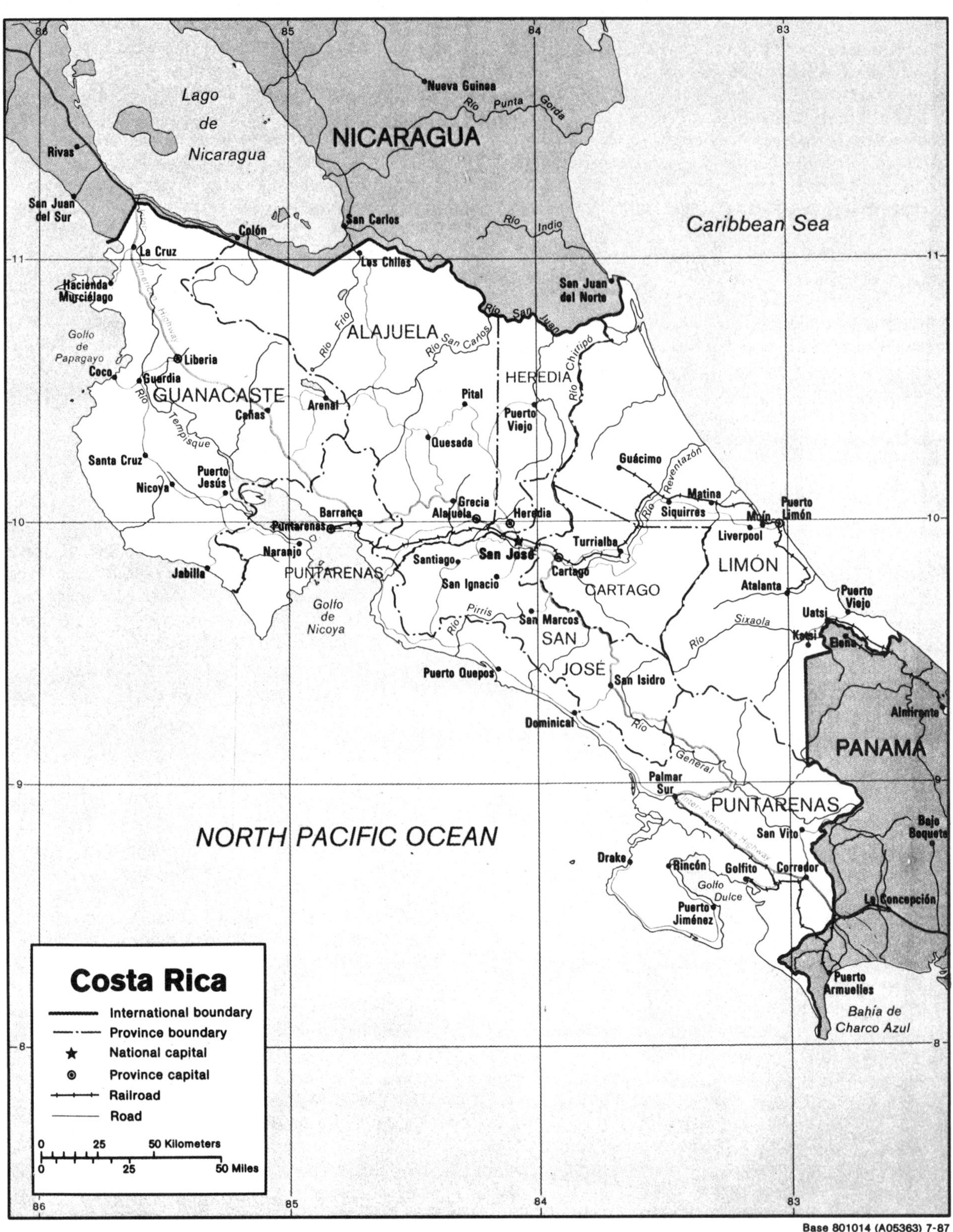
Costa Rica
International boundary
Province boundary
National capital
Province capital
Railroad
Road
0 25 50 Kilometers
0 25 50 Miles
NICARAGUA
PANAMA
Caribbean Sea
NORTH PACIFIC OCEAN
Lago de Nicaragua
Golfo de Papagayo
Golfo de Nicoya
Golfo Dulce
Bahía de Charco Azul
GUANACASTE
ALAJUELA
HEREDIA
LIMÓN
CARTAGO
SAN JOSÉ
PUNTARENAS
San José
Liberia
Alajuela
Heredia
Cartago
Puntarenas
Puerto Limón
Nueva Guinea
Rivas
San Juan del Sur
San Carlos
Colón
La Cruz
Los Chiles
San Juan del Norte
Hacienda Murciélago
Coco
Guardia
Cañas
Arenal
Pital
Puerto Viejo
Quesada
Santa Cruz
Nicoya
Puerto Jesús
Barranca
Grecia
Guácimo
Matina
Siquirres
Moín
Liverpool
Turrialba
Naranjo
Jabilla
Santiago
San Ignacio
Atalanta
San Marcos
Uatsi
Katsi
Elena
Almirante
Puerto Quepos
San Isidro
Dominical
Palmar Sur
San Vito
Bajo Boquete
Drake
Rincón
Golfito
Corredor
Puerto Jiménez
La Concepción
Puerto Armuelles
Río Punta Gorda
Río Indio
Río San Juan
Río Frío
Río San Carlos
Río Chirripó
Río Tempisque
Río Reventazón
Río Pirrís
Río Sixaola
Río General
Inter-American Highway
86
85
84
83
11
10
9
8
Base 801014 (A05363) 7-87

period. It is drained by the headwaters of the Reventazón River, which flows through its deeply gorged valley to the Caribbean. The climate of the Cartago basin has greater rainfall and is less comfortable than that of the lower San José basin to the west because it is not protected from Caribbean influences by the volcanoes.

A well-developed dairy industry exists in the basin and on the southern slopes of the Irazú and Turrialba volcanoes. Much pastureland, however, was destroyed by the 1963–65 eruptions of Irazú volcano. Volcanic activity was limited to the emission of ash, but this ash dammed a small river that rises on Irazú. A torrential rain accompanied by high winds struck the area, the dam gave way, and the muddy waters flooded the city of Cartago, tearing down houses, destroying roads, and resulting in some loss of life.

The eruptions of Irazú also took a great toll in the San José basin to the west. Fallen ash greatly damaged the area's valuable coffee crop, creating conditions that aided the spread of dangerous crop plagues. It destroyed pastureland and consequently starved the livestock. Ash-carrying flood waters plugged the water pumping station in San José, and water had to be carried to the city by truck. An intermittent rain of high-acid-content ash fell on San José for two and a half years, corroding metal, damaging machinery, and causing an increase in respiratory diseases.

The San José basin is the site of the cities of San José, Heredia, and Alajuela. The nation's main coffee belt surrounds San José, where climate and soil conditions are nearly ideal for its growth with relatively rich soils, mild temperatures, an average annual rainfall of 178 cm (70 in) and a January through April dry season in which to harvest and process the beans. West of the San José area coffee gives way to maize, rice, and sugar at lower altitudes of the Meseta.

The San José basin is drained by the Grande de Tárcoles River that empties into the Pacific just south of Puntarenas, the port and provincial capital. This river valley has become the nation's major rice growing area.

The only other upland valley of major importance is the General Valley, drained by the General River, which becomes the Grande de Térraba River and empties into the Bay of Coronado. It lies between the granitic Cordillera de Talamanca to the north and the coastal mountains of the southwest. The Cordillera extends for almost 322 km (200 mi) from just southeast of the Meseta Central to the Panamanian border. Ten of its peaks rise to more than 2,987 m (9,800 ft) above sea level, the highest of them being the Chirripó Grande 3,810 m (12,500 ft), the highest point in the country. It was to these mountains that forest Indians of South American origin fled when the Spanish arrived, and remnants of them are still found there.

The General Valley, almost as large as the Meseta Central, is a relatively isolated structural depression that ranges from 183 to 1,066 m (600 to 3,500 ft) above sea level. River flood plains, terraces, rolling hills, and savanna dominate the landscape. Settlement of the valley began on a significant scale about 50 years ago. The first arrivals were subsistence cultivators who destroyed much of the original forest to make way for their small fields. Erosion subsequently stripped many hills of the relatively infertile red topsoils.

Since the completion of the Inter-American Highway, which connected the General Valley with the Meseta Central, there has been a massive influx of settlers who find the fertile river flood plains and alluvial terraces suitable for farming and pasturing livestock. Rainfall is abundant, averaging between 203 to 256 cm (80 and 140 in) annually, with a short dry season from December through April. It appears that, with the newly accessible markets and the growing population, the forests of the Cordillera de Talamanca will be overexploited. Many stands of oak trees along the Inter-American Highway south of Cartago have already been burned for charcoal, despite laws against such indiscriminate exploitation.

The Pacific Coastal Region: Puntarenas and Guanacaste

Across the coastal mountains from the General Valley lies the Palmar lowland complex. This area is unique on the Pacific coast; its climate is similar to that of the Caribbean coast, but structurally it has more in common with the Guanacaste lowland to the northwest. Although there is a short 3-month dry season, the coastal mountains act as a watershed for the on-shore winds and annual rainfall is between 343 to 498 cm (135 and 200 in) enough to support tropical rain forest vegetation of commercial value. When the Caribbean banana plantations were overcome with disease in the 1930's much banana cultivation was shifted to this part of the Pacific where there are rich alluvial soils. The three-month dry season necessitates irrigation in some places, however, and violent tropical storms sometimes take their toll in crops.

The Pacific port of Golfito was built during the 1930's to accommodate the need for new banana-shipping facilities, as was the port of Quepos to the northwest. The movement of the banana industry from the Caribbean to these areas on the Pacific stimulated the immigration of people from the Meseta Central, Guanacaste Province, Nicaragua, and the nearby Chiriquí Province in Panamá.

PRINCIPAL TOWNS (1984 population)

Alajuela	(33,929)
Cartago	(23,884)
Liberia	(14,093)
Heredia	(20,867)
Limón	(43,158)
Puntarenas	(47,851)
San José	(245,370)

The three Pacific peninsulas of Burica, Osa, and Nicoya are mountainous and sparsely populated. A narrow, alluvial coastal plain extends from the Osa peninsula to the important port of Puntarenas, although it is squeezed out in some places by the relatively low coastal mountains. The plain becomes wider north of Puntarenas and eventually merges with the Guanacaste plain, which extends inland from the head of Nicoya gulf and along the eastern side of the Nicoya peninsula. The northern reaches of this plain are separated from the Pacific Ocean and Nicaragua only by low hills.

Parallel to the Guanacaste plain, the Cordillera de Guanacaste, volcanic in origin, stretches for 112 km (70 mi) from the western border with Nicaragua to the Cordillera Central from which it is separated by low mountains. The highest peak in the Guanacaste chain is the Miravalles volcano 2,024 m (6,640 ft, above sea level). The eastern slope of this chain blocks the northeast trade winds, protecting the plain from heavy, constant Caribbean rains. The 914 m (3,000-ft) coastal mountains of the Nicoya peninsula also act as a watershed for the rainbearing southwest winds that blow from May through October. Thus, the lowlands of Guanacaste, lying in the lee of both mountain ranges, often suffers from drought even during the rainy season. The valley of the Tempisque River, which drains the area, receives almost no rain during the dry season.

Area and population

		area		population
Provinces	**Capitals**	sq mi	sq km	1984 census
Alajuela	Alajuela	3,766	9,753	427,962
Cartago	Cartago	1,206	3,125	271,671
Guanacaste	Liberia	3,915	10,141	195,208
Heredia	Heredia	1,026	2,656	197,575
Limón	Limón	3,548	9,188	168,076
Puntarenas	Puntarenas	4,354	11,277	265,883
San José	San José	1,915	4,960	890,434
TOTAL		19,730	51,100	2,416,809

The vegetation of the plain was once predominantly open deciduous forest, but most of this has been transformed into tropical savanna grasslands by slash-and-burn cultivators. Large-scale cattle raising has been characteristic of the Guanacaste plain since the 16th century, although recently many landholders of the east coast of the Nicoya peninsula have turned to the cultivation of cereals and other crops. In early August 1968, the Arenal volcano in the Cordillera de Guanacaste erupted and hot volcanic ash fell on rich cattle-grazing lands. Few animals died as a direct result of the ash, but about 80,000 had to be slaughtered for immediate sale and another 100,000 had to be moved to new pastures.

Most soils in the Guanacaste area are fertile. Those of the northern plains and the areas east of the Tempisque River valley are of recent volcanic origin, while those of the Tempisque River valley itself are alluvial. Soils on most of the Nicoya peninsula, however, are intensely weathered soils of the lateritic (iron and aluminum bearing) groups.

The Guanacaste region has remained relatively thinly settled because of its inaccessibility and because of certain unfavorable climatic and edaphic features and the low labor requirement for cattle raising. Nevertheless, small subsistence farmers have been moving to the Nicoya peninsula since World War II.

The Caribbean

The extensive Caribbean lowland is a continuation of the vast Nicaraguan lowland. Widest in the north along the border with Nicaragua, it is gradually pinched out by coastal mountains just south of Limón. Still farther south, it opens up again where sediment from the Sixaola River has built a small delta.

Much of the land is covered with tropical rain forests and is sparsely populated and relatively inaccessible, except in and around Limón, the port and provincial capital. Nevertheless, in recent years there has been a rapid influx of settlers into the San Carlos district, just east of the Guanacaste Range. Most of the newcomers are dairy farmers from the Meseta Central or subsistence cultivators.

Annual rainfall is heavy on the entire plain, averaging between 150 and 200 in on the San Juan River delta, and between 120 and 150 in farther inland and on the coast farther south. Much of the plain is poorly drained, except in the gently sloping piedmont zone just east of the volcanic highlands.

The northern part of the lowland is drained by the San Juan River and its three main tributaries, which rise in the volcanic highlands. An extensive delta has built up around the mouth of the San Juan, which is in flood from September through November. Although the San Juan River lies within Nicaraguan territory, Costa Rica has, by treaty, full rights of navigation. The lower reaches of the river are shallow, although it is navigable all the way from the Caribbean to Lake Nicaragua. The remaining rivers that drain the highlands and the Caribbean lowland south of the San Juan are relatively short. They drop precipitously from the highlands, but are not long enough to have built up extensive flood plains in their lower courses.

The Sixaola River delta and the San Juan delta southeast to Limón have rich, alluvial soils. Those of the San Carlos district and the coastal mountains just south of Limón, however, are intensely weathered and subject to continuous leaching because of heavy rains. The natural fertility of soils in this area is not being renewed because the prevailing northeast trade winds blow the ash from erupting volcanoes toward the Pacific.

For many centuries the Caribbean coast attracted few settlers because of impenetrable vegetation, excessive heat and humidity, and disease-carrying insects. Around 1880, however, banana planters moved into the area, believing that the heavy rains, fertile soils, and infrequency of destructive windstorms made the area ideal for cultivation. The banana plantations moved to the Pacific coast, but the majority of the descendants of the Jamaican Negroes who had been brought in to work the first plantations remained and planted such crops as cacao and abacá, a sort of hemp.

CUBA

BASIC FACTS

Official Name: Republic of Cuba

Abbreviation: CUB

Area: 110,861 sq km (42,804 sq mi)

Area—World Rank: 99

Population: 10,353,932 (1988) 11,727,000 (2000)

Population—World Rank: 63

Capital: Havana

Land Boundaries: Nil

Coastline: 3,235 km (2,010 mi)

Longest Distances: 1,223 km (760 mi) E-W 89 km (55 mi) N-S

Highest Point: Turquino Peak 1,974 m (6,476 ft)

Lowest Point: Sea level

Land Use: 23% arable; 6% permanent crops; 23% meadows and pastures; 17% forest and woodland; 31% other

Cuba, the largest island of the Greater Antilles, lies some 145 km (90 mi) south of the Florida keys and a slightly greater distance to the east of the Yucatán Peninsula. It is flanked by Jamaica on the south, Hispaniola on the southeast, and the Bahamas on the northeast. The principal trade routes to the Gulf of Mexico skirt its northern and southern coasts, and in the sixteenth century the island received from the Spanish monarchy the designation of "key to the Gulf of Mexico." This strategic location is memorialized at the top of the national coat of arms by a key that hangs suspended between the two headlands, Florida and the Yucatán Peninsula.

The most extensive mountainous zone of the long, narrow island lies near its eastern extremity. Smaller mountain zones with lower elevations occur near the midsection and in the far west, but well over half of the terrain consists of flat or rolling plains with a great deal of rich soil well suited to the cultivation of sugarcane, the dominant crop. Numerous rivers running northward or southward from an interior watershed are for the most part short and rapid; they provide good drainage but are not generally suitable for navigation.

Satellite cays and islets are strung along both the northern and the southern coasts of Cuba, and except near its western tip a wealth of excellent harbors indent the shoreline. The submerged shelf on which the island rests, however, is capped by extensive coral reef development that poses a serious hazard to navigation.

The long, narrow island of Cuba extends some 1,200 km (746 mi) from Cape Maisi on the east to Cape San Antonio on the west, about the distance from New York to Chicago. The largest of the West Indian islands, Cuba has a territorial extent about matching that of all the other islands combined. In addition to the main island, the Cuban archipelago includes the Isle of Pines near the south coast in the Gulf of Batabano and some 1,600 coastal cays and islets. The main island occupies 94.7 percent of the national territory, and the Isle of Pines and the other cays and islets occupy respectively 2.0 percent and 3.3 percent of the total.

Cuba was raised from the seafloor by geological action occurring about 20 million years ago and at one time was connected with other Antillean islands. The mountains of southeastern Cuba are related to those of southern Mexico, Jamaica, and Hispaniola; and the limestone formations that make up much of the island resemble those of Florida, Jamaica, and the Yucatán Peninsula.

Cuba's topography has resulted from the interaction of constructive forces that determined the basic structure and alignment of landforms and destructive forces of wind and water that sculpted the structure into its present configurations. Soil erosion, however, has been less severe than on most other Antillean islands. The island is still subject to some crustal instability, and its history has been marked by earthquakes of varying intensity. The zone of maximum instability occurs in the southeastern mountains. Light quakes are recorded frequently in the southeast, and a severe tremor suffered by Santiago de Cuba in 1578 was followed by another exactly one century later. Havana recorded several strong disturbances during the nineteenth century, and a severe tremor struck Pinar del Río Province in 1880. None, however, has occurred in the twentieth century.

The least mountainous of the Greater Antilles, Cuba has a median elevation of no more than 91 m (300 ft) above sea level, and its three principal mountainous zones are isolated and separated by plains. The most extensive highland zone occupies much of the island's eastern extremity, the second rises near the center of the island, and the third rises in the extreme west.

The loftiest mountain system is the Sierra Maestra, which skirts the southeastern coastline west of Guantánamo Bay except where it is broken by the small lowland depression on which Santiago de Cuba is located. It is the most heavily dissected and steepest of the Cuban ranges, and its peaks include Pico Turquino 1,993 m (6,540 ft), which is the country's highest elevation.

On the east, the Sierra Maestra terminates in a low area around the United States Naval Base at Guantánamo Bay. The lowlands around Guantánamo mark the termination of the Central Valley, which is some 96 km (60 mi) in length and merges with plains to the west. Most of the island east of a line from north to south between Nipe Bay and Santiago de Cuba is mountainous, however, and includes such ranges as the Sierra de Nipe, the Sierra de Nicaro, the Sierra del Cristal, and the Cuchillas de Toa. The port of Baracoa on the northeast coast is the most isolated urban center on the island.

The mountains of central Cuba, less extensive and of lower elevation than those of the east, occupy the southern portion of Las Villas Province. The two principal systems, known collectively as the Escambray mountains and separated by the Agabama River, are the Sierra de Trinidad in the west and the Sierra de Sancti Spíritus in the east. The principal ranges of the western highlands are the Sierra del Rosario, which commences near the town of Guanajay west of Havana and extends southwestward along the spine of the island for about 96 km (60 mi), and the Sierra de los Organos, which continues in the same direction almost to the tip of the island. These western highlands are limestone formations weathered into strange shapes. Ranks of tall erosion-resistant limestone columns resembling organ pipes gave the Sierra de los Organos its name. The numerous shapes, sinkholes, and underground caverns and streams are limestone developments known as

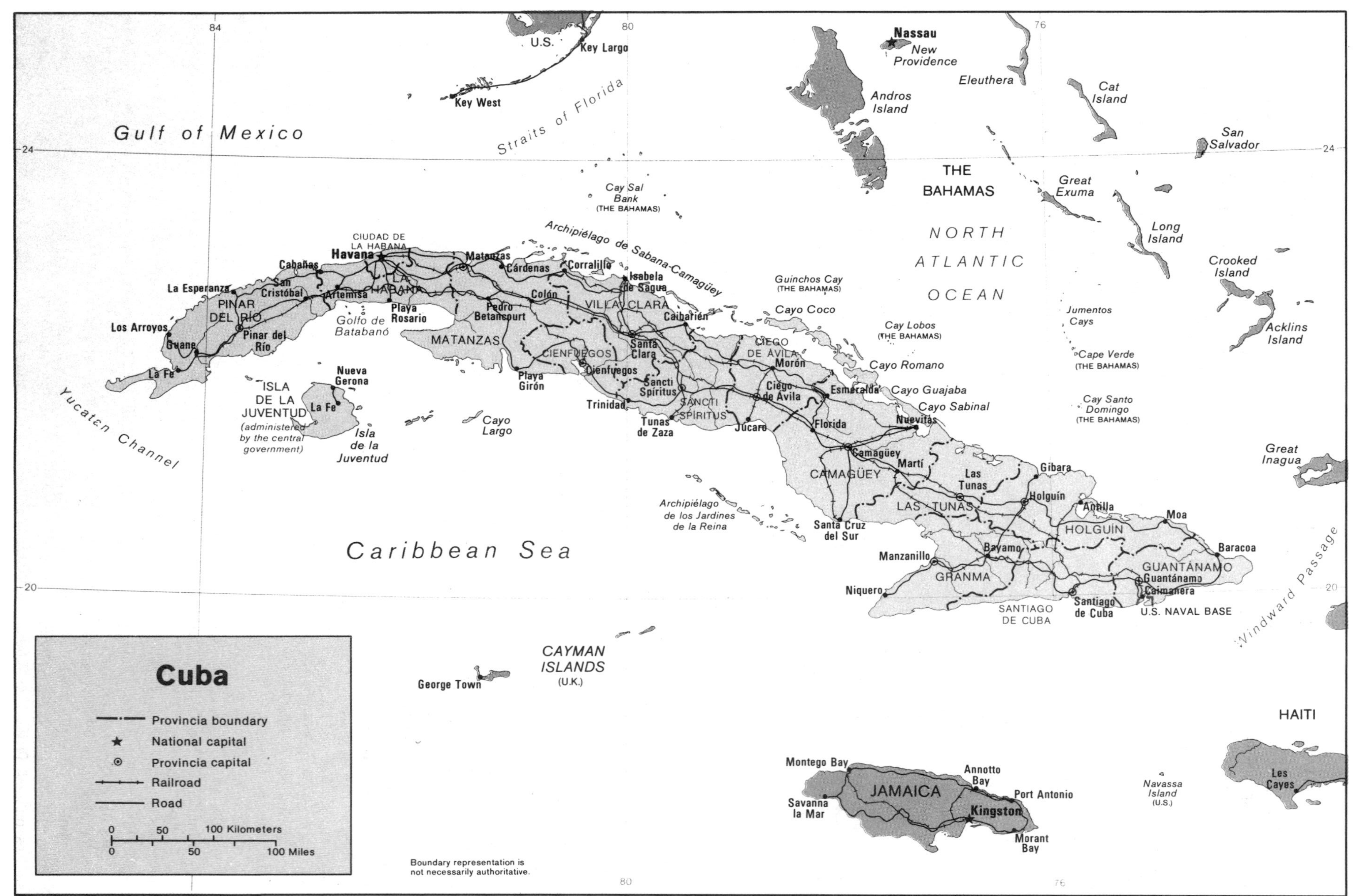

Base 505031 (547623) 3-82

karst. Karst landscape is most characteristic of the western highlands but is widely distributed about the island.

Most of the country's more than 200 rivers originate in the interior near the island's watershed and flow northward or southward to the sea. Smaller streams and arroyos that remain dry during most of the year are also numerous. River levels rise significantly during the rainy season, when 80 percent of their flow occurs, and seasonal flooding is common. The watercourses for the most part are not navigable, and their potential as sources of hydroelectric power has yet to be realized. The longest and heaviest flowing is the Cauto River of Oriente Province, which rises in the Sierra Maestra near Santiago de Cuba and flows westward to the Gulf of Guacanayabo. Rivers are most numerous in Oriente Province and are fewest in La Habana Province and in Camagüey Province west of the city of Camagüey.

Seven subterranean river basins constitute the sources of many surface rivers, and there are extensive reservoirs of fresh and brackish groundwater. Sulfide mineral springs are located in Pinar del Río and Matanzas provinces, and there are radioactive thermal springs in Las Villas Province and on the Isle of Pines. There are no large lakes, but coastal swamplands are numerous and extensive; the largest covers more than 4,403 sq km (1,700 sq mi) on the Zapata Peninsula. Others occur on both the northern and the southern coasts and on the Isle of Pines.

No overall inventory of the country's surface water has been undertaken, and the only hydrometeorological station is the one established in 1955 at the headwaters of the Toa River in Oriente Province. The revolutionary government is said to have installed more than 160 such stations during its first ten years in power. It has shown a keen interest in the country's river network as an important asset to industry and agriculture and has engaged extensively in the construction of dams, weirs, reservoirs, and systems of diversion and irrigation channels. The amount of land under irrigation is said to have been tripled during the first ten years of the Castro government.

Almost two-thirds of the Cuban landscape consists of flatlands and rolling plains. With hills and the lower and gentler slopes of the mountains, they make up as much as three-fourths of the national territory. The generally easy gradients minimize the hazards of land erosion and facilitate both development of the transportation network and land tillage, including the use of mechanized equipment.

Limestone, usually with a high clay content, is the basic ingredient of most Cuban soils. In particular a red limestone earth known as Matanzas clay extends in a wide and continuous belt from a point west of Havana to near Cienfuegos on the southern coast and reappears in extensive patches in western Camagüey Province. The dark red material, which undergoes little chemical or physical alteration to depths of as much as twenty feet, permits a good downward percolation of moisture that minimizes runoff and has so little stickiness that land can be plowed within a few hours after a heavy rain. The rich Matanzas clay would be suitable for a wide variety of crops, but the traditional Cuban practice, somewhat modified since 1959, has been to plant sugarcane on the better soils and to plant other crops on less productive lands. A second zone of fertile soil occurs on the plains north of Cienfuegos between the Sierra de Sancti Spíritus and the Caribbean coast. Scattered alluvial soils also have a high degree of fertility but occur only in the narrow floodplains of watercourses. Certain soils of Camagüey Province and those of the Guantánamo Basin are suitable for cane production but are of lower fertility.

Extensive areas of sandy soil in Pinar del Río, western Las Villas, and portions of Camagüey are characterized by an inability to hold moisture. This soil can support only palm trees, xerophytic shrubs, and grasses suitable for pasture. Soils suitable for coffee cultivation occur in all three major mountain zones; tobacco grows well in eastern Las Villas and in portions of Pinar del Río and Oriente. The mangrove-dotted coastal swamps and cays have soils of limited fertility made up of organic silt, organic clay, and peat, each underlain with clay of high plasticity.

The main island of Cuba rests on a subsurface shelf from which the numerous cays, coral islets, and reefs rise. Submerged about 91 to 182 m (300 to 600 ft) below sea level, the shelf varies in breadth off the north coast, is almost nonexistent off the southeast coast, and attains its maximum breadth off the remainder of the southern coastline, where it extends to the limits of the gulfs of Batabano, Ana María, and Guacanayabo. Its outer rim is flanked on the southeast by the deep Bartlett Trough, which separates Cuba and Jamaica, and on the southwest by the Cayman or Yucatán depression. The two troughs are separated by the shallows of the submerged Cayman ridge, a continuation of the Sierra Maestra range that reemerges as the Cayman Islands. Off the northern coast of Camagüey Province the sea-lane of the Old Bahama Channel at some points is only ten miles wide as it passes between the Cuban shelf and the shallows of the Great Bahama Bank.

PRINCIPAL TOWNS (estimated population at 31 December 1986)

Town	Population	Town	Population
La Habana (Havana, the capital)	2,036,799	Cienfuegos	112,225
Santiago de Cuba	364,554	Bayamo	108,716
Camagüey	265,588	Pinar del Río	108,109
Holguín	199,861	Matanzas	106,954
Santa Clara	182,349	Las Tunas	96,846
Guantánamo	179,091	Ciego de Avila	82,942
		Sancti Spíritus	78,154

Except where the precipitous cliffs of the Sierra Maestra plunge into the sea, most of the Cuban shoreline is fringed with coral reefs and archipelagoes of cays. In the north an almost unbroken chain of cays extends from Cárdenas to Nuevitas. In the south the Isle of Pines is the largest member of the Canarreos Archipelago, and the Jardines de la Reina chain flanks the Gulf of Ana María. Coral reefs are interlaced with many of the cays, clog the gulfs of Ana María and Guacanayabo, and form a chain of unborn islets off the western extremity of the island.

Cuba's approximately 3,540 km (2,200-mi) coastline is indented by some of the world's finest natural harbors. There are about 200 in all, and many are of the pouch or

Area and population

Provinces	Capitals	area sq mi	area sq km	population 1986 estimate
Camagüey	Camagüey	6,174	15,990	706,675
Ciego de Avila	Ciego de Avila	2,668	6,910	343,290
Cienfuegos	Cienfuegos	1,613	4,178	344,642
Ciudad de la Habana	—	281	727	2,013,746
Granma	Bayamo	3,232	8,372	763,975
Guantánamo	Guantánamo	2,388	6,186	476,858
Holguín	Holguín	3,591	9,301	953,457
La Habana	Havana	2,213	5,731	614,533
Las Tunas	Las Tunas	2,544	6,589	464,327
Matanzas	Matanzas	4,625	11,978	584,256
Pinar del Río	Pinar del Río	4,218	10,925	666,815
Sancti Spíritus	Sancti Spíritus	2,604	6,744	414,244
Santiago de Cuba	Santiago de Cuba	2,382	6,170	951,084
Villa Clara	Santa Clara	3,345	8,662	787,461
Special municipality				
Isla de la Juventud	Nueva Gerona	926	2,398	67,522
TOTAL		42,804	110,861	10,152,885

bottleneck variety with narrow entrances that broaden into spacious deepwater anchorages. Among the ports on the north coast that have harbors of this kind are Mariel, Havana, Nuevitas, Manati, Puerto Padre, Gibara, and Antilla. South coast bottleneck ports include Guantánamo, Santiago de Cuba, and Cienfuegos. The principal open bay ports, Cárdenas and Matanzas, are located close to one another on the north coast of Matanzas Province.

Most of these ports were developed primarily for the export of sugar, and on the eve of the Revolution there were twenty-three sugar-loading facilities. Nuevitas ranked first and Havana second in volume of sugar handled, but all of the twenty-three handled commercially important quantities. There were no important harbors west of Cienfuegos on the south coast or Mariel on the north. Shallow waters, coral formations, and a lack of good natural harbors are characteristic of the coasts of western Cuba, but since sugarcane is not grown in the west the need for ports is reduced accordingly. Elsewhere along the coastline good ports are lacking only along the 241 km (150 mi) between Caibarién and Nuevitas, where cays are numerous and there is extensive coral development.

Many geographers have found it convenient to divide Cuba into four geographical regions. Different names have been used to identify these regions, and there is no exact agreement as to where the regional boundaries should be set. In general the western region is made up of the provinces of Pinar del Río, La Habana, Matanzas, and a portion of Las Villas; the central region includes most of Las Villas and the western part of Camagüey; the east-central region takes up most of Camagüey and the extreme west of Oriente; and the eastern region covers the mountainous eastern end of the island. These regions, however, do not represent significantly different patterns of natural features, land use, or population distribution, and their value as a study aid is diminished accordingly. Moreover, under the revolutionary government the term *region* has been reserved for the administrative subdivisions of a province.

CYPRUS

BASIC FACTS

Official Name: Republic of Cyprus

Abbreviation: CYP

Area: 9,251 sq km (3,572 sq mi)

Area—World Rank: 147

Population: 691,966 (1988) 926,000 (2000)

Population—World Rank: 138

Capital: Nicosia

Land Boundaries: Nil

Coastline: 538 km (334 mi)

Longest Distances: 227 km (141 mi) ENE-WSW 97 km (60 mi) SSE-NNW

Highest Point: Olympus 1,951 m (6,401 ft)

Lowest Point: Sea level

Land Use: 40% arable; 7% permanent crops; 10% meadows and pastures; 18% forest and woodland; 25% other

The largest Mediterranean island after Sicily and Sardinia, Cyprus is in the extreme northeastern corner of the Mediterranean 71 km (44 mi) south of Turkey, 105 km (65 mi) west of Syria and 370 km (230 mi) north of Egypt. It includes the small island outposts of Cape Andreas known as the Klidhes. The average width is between 56 km and 72 km (35 mi and 45 mi). The narrow peninsula known as the Karpas extending 74 km (46 mi) northeastward to Cape Andreas is nowhere more than 16 km (10 mi) wide. The length of the coastline is 538 km (334 mi). The northern part of Cyprus, north of the so-called Attila Line or Green Line, is under Turkish occupation. The physical setting is dominated by two mountain masses and the central plain they encompass, the Mesaoria. The Troodos Mountains cover most of the southern and western portions of the country, accounting for roughly half the total area, including the southwestern Nicosia District, all of the Paphos and Limassol districts except their coastal plains, and the western Larnaca District. The narrow Kyrenia Range, extending along the northern coastline, occupies substantially less area with lower elevations. The two mountain systems generally run parallel to the Taurus Mountains on the Turkish mainland, whose silhouette is visible from northern Cyprus. Coastal lowlands, varying in width, surround the island.

The rugged Troodos Mountains, whose principal range stretches from Pomos Point on the northwest almost to Larnaca Bay on the east, are the single most conspicuous

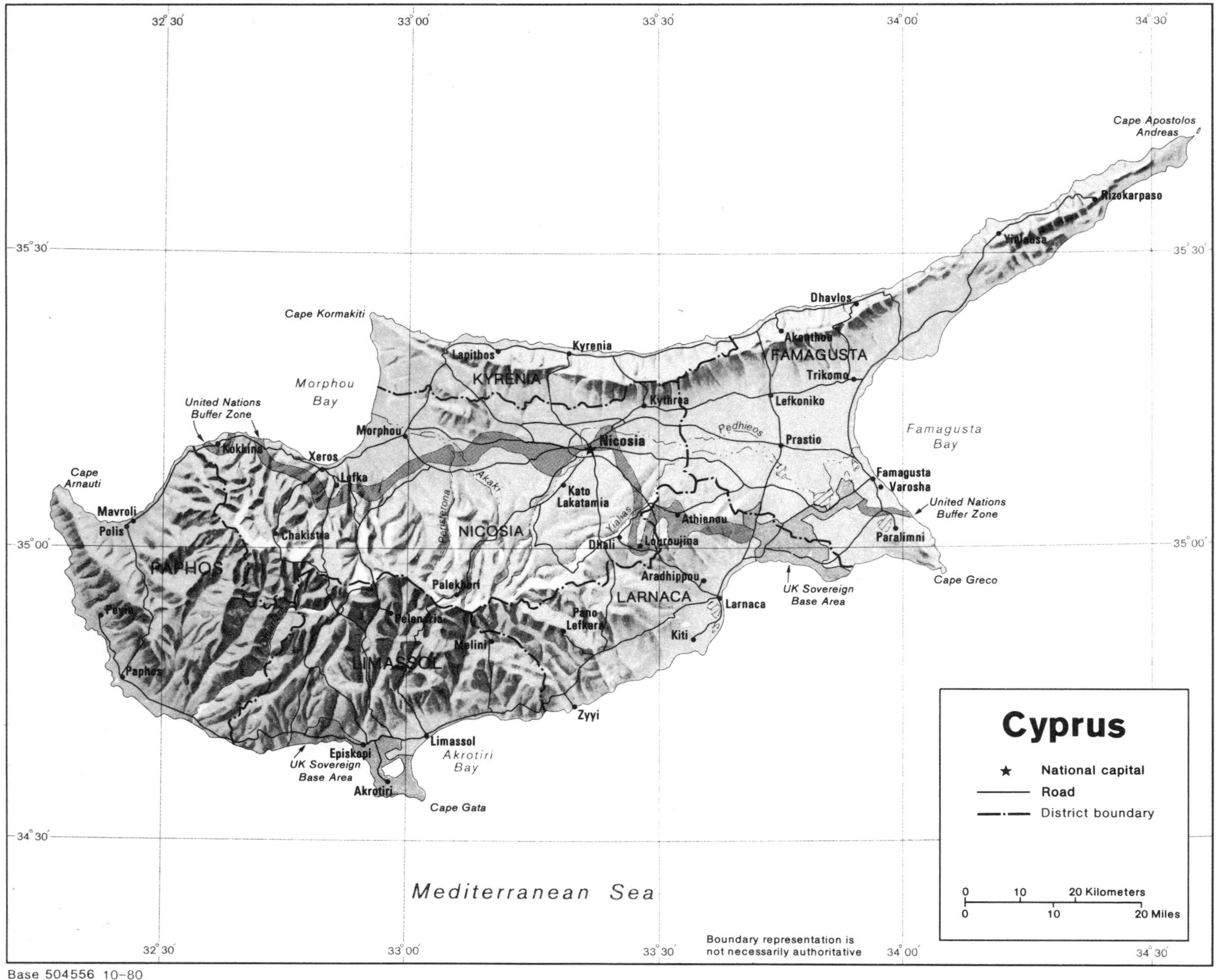

Base 504556 10-80

features of the landscape. Intensive uplifting and folding in the formative period left the area highly fragmented so that subordinate ranges and spurs veer off at many angles, their slopes incised by steep-sided valleys. To the southwest, the mountains descend in a series of stepped foothills to the coastal plain.

PRINCIPAL TOWNS (population at 1 October 1982): Nicosia (capital) 149,100 (excl. Turkish-occupied portion); Limassol 107,200; Larnaca 48,300; Famagusta (Gazi Mağusa) 39,500 (mid-1974); Paphos 20,800.

Greek Sector

Area and population (under government control; includes UN Buffer Zone and U.K. Sovereign Base Areas)

		area		population
Districts	**Capitals**	sq mi	sq km	1982 census
Famagusta	Famagusta	...	...	24,187
Larnaca	Larnaca	433	1,121	83,151
Limassol	Limassol	538	1,393	143,847
Nicosia	Nicosia	...	...	207,290
Paphos	Paphos	539	1,396	45,023
TOTAL		2,276	5,896	503,498

Whereas the Troodos Mountains are an extensive massif formed of molten igneous rock, the Kyrenia Range is a narrow, limestone ridge that rises abruptly from the plains on either side. Its easternmost extension becomes a series of foothills on the Karpasian Peninsula. That peninsula points toward Asia Minor, to which Cyprus belongs geologically.

Turkish Sector

Area and population

		area		population
Provinces	**Administration centres**	sq mi	sq km	1978 estimate
Lefkoşe (Nicosia)	Lefkoşe	...	...	68,286
Gazimagosa (Famagusta)	Gazimagosa	...	...	55,647
Girne (Kyrenia)	Girne	247	640	22,807
TOTAL		1,295	3,355	146,740

Even the highest peaks of the Kyrenia Range, including St. Hilarion and Buffavento, are hardly more than half the height of the great dome of the Troodos Massif, Mount Olympus 1,951 m (6,401 ft), but their jagged slopes make them considerably more spectacular.

The Mesaoria Plain is the agricultural heartland of the country, and in its middle lies Nicosia, the divided capital of the island. Nicosia is unique among island capitals in not being situated on the water. In general, settlements are found concentrated inland, away from the water's edge. Cypriots share few of the maritime interests characteristic of most island people.

A network of winter rivers rises in the Troodos Mountains and flows radially in all directions. Even the Yialias and the Pedhieos, the two rivers that drain eastward across the Mesaoria to empty into Famagusta Bay, become dry courses in summer. The Seraghis flows northwest through the Morphou Plain.

Roughly 37% of the island is under Turkish occupation. Since the de facto partition of the island in 1974, U.N. forces have manned the buffer zone between the cease-fire line. Movement across the line is permitted only to government officials or outside observers, who enter and leave at the single official transit point in the old town of Nicosia.

CZECHOSLOVAKIA

BASIC FACTS

Official Name: Czechoslovak Socialist Republic

Abbreviation: CZK

Area: 127,905 sq km (49,384 sq mi)

Area—World Rank: 92

Population: 15,620,722 (1988); 16,194,000 (2000)

Population—World Rank: 49

Capital: Prague

Land Boundaries: 3,472 km (2,157 mi); Poland 1,310 km (814 mi); USSR 98 km (61 mi); Hungary 679 km (422 mi); Austria 570 km (354 mi); West Germany 356 km (221 mi); East Germany 459 km (285 mi)

Coastline: Nil

Longest Distances: 767 km (477 mi) E-W 468 km (291 mi) N-S

Highest Point: Gerlachovka 2,655 m (8,711 ft)

Lowest Point: Bodrog River 94 m (308 ft)

Land Use: 40% arable; 1% permanent crops; 13% meadows and pastures; 37% forest and woodland; 9% other

Czechoslovakia divides topographically as well as historically into three major areas: Bohemia, Moravia, and Slovakia. Bohemia consists of the five western political divisions (known as regions or *krajs)* with names ending in český; Moravia, of the two central political divisions with names ending in *moravský;* and Slovakia, of the three eastern political divisions with names ending in *slovenský.* The area known as Slovakia is coterminous with the Slovak Socialist Republic; the other seven divisions constitute the Czech Socialist Republic.

The areas of western Bohemia and eastern Slovakia are portions of different mountain and drainage systems. All but a minute fraction of the Bohemian region drains to the North Sea by way of the Elbe River. The hills and low mountains that enclose the area are part of the north-central European uplands that are north of the Danube Basin and extend from southern Belgium, through the central German lands, and into the Moravian regions of Czechoslovakia. These uplands, which are distinct from the Alps to the south and the Carpathians to the east, are known geologically as the Hercynian Massif. Most of the Slovak area drains into the Danube River, and its mountains are part of the Carpathian range that continues eastward and southward into Romania.

The uplands of Moravia are a transition between the Hercynian Massif and the Carpathians and contrast with them in having more nearly north-south ridge lines. Most of Moravia drains southward to the Danube, but the Oder River rises in its northeast area and drains a sizable portion of the northern region.

Bohemia

Bohemia has a distinct individuality that derives in large measure from its topography. It is ringed with low mountains or high hills that, although less confined to the immediate border area along the southern and southeastern sides, are sufficient to serve as a watershed along almost the entire periphery. Streams flow from all directions toward Prague. These features have fostered local solidarity and a common set of economic interests.

PRINCIPAL TOWNS (estimated population at 1 January 1986)

Praha (Prague, capital)	1,193,513	Hradec Králové	99,571
Bratislava	417,103	Pardubice	94,451
Brno	385,684	České Budějovice	94,206
Ostrava	327,791	Havířov	91,873
Košice	222,175	Ústí nad Labem	91,703
Plzeň (Pilsen)	175,244	Žilina	91,444
Olomouc	106,086	Gottwaldov (Zlín)	86,210
Liberec	100,917	Nitra	85,276

In the northwest the Krušné Hory (Ore Mountains) border on the German Democratic Republic (East Germany) and are known to the Germans as the Erzgebirge; the Sudeten Mountains in the northeast border on Poland in the area that was a part of Germany before World War II. Germans outnumbered Czechs in both of these localities before they were relocated after 1945. The large German population was accounted for, in part, by the fact that the hills—particularly those of the Krušné Hory—were more gentle in the north, promoting German movement into the area, whereas the rugged escarpment in the south inhibited Czech movement outward.

The Český Les (Bohemian Forest), bordering on the Federal Republic of Germany (West Germany), and the Šumava, bordering on West Germany and Austria, are mountain ranges that form the western and southwestern parts of the ring around the Bohemian Basin, and both are about as high as the Krušné Hory in the northwest. Bohemia's mountainous areas differ greatly in population. The northern regions are densely populated, whereas the less hospitable Bohemian Forest and Šumava are among the most sparsely populated areas in the entire country.

The middle lands of the basin are lower, but their features vary widely. There are small lake areas in the middle south region and in the Vltava Basin north of Prague. Some of the western grainlands are gently rolling. By contrast, other places have streams, such as the Vltava River, that have cut deep gorges. This has given a large area southwest of Prague a broken relief pattern that is typical of several other districts.

Area and population

Republics / Regions	Capitals	area: sq mi	area: sq km	population: 1986 estimate
Czech Socialist Republic	Prague			
Jihočeský	České Budějovice	4,380	11,345	695,066
Jihomoravský	Brno	5,802	15,028	2,058,020
Severočeský	Ústí nad Labem	3,019	7,819	1,183,145
Severomoravský	Ostrava	4,273	11,067	1,958,877
Středočeský	Prague	4,245	10,994	1,137,086
Východočeský	Hradec Králové	4,340	11,240	1,244,452
Západočeský	Plzeň	4,199	10,875	873,239
Slovak Socialist Republic	Bratislava			
Středoslovenský	Banská Bystrica	6,944	17,986	1,581,144
Východoslovenský	Košice	6,253	16,196	1,463,333
Západoslovenský	Bratislava	5,595	14,492	1,715,861
Capital Cities				
Prague	—	192	496	1,193,513
Bratislava	—	142	367	417,103
TOTAL		49,384	127,905	15,520,839

Slovakia

Slovakia's landforms do not make it as distinctive a geographic unit as Bohemia. Its mountains generally run east-west across the land; their various ranges tend to segregate groups of people. Population clusters are most dense in the river valleys; the highest elevations are rugged, have the most severe weather, and are the most sparsely settled. Some of the flatlands in southern Slovakia are poorly drained and support only a few people.

The main mountain ranges are the High Tatras (Vysoké Tatry) and Slovenské Rudohorie, both of which are part of the Carpathians. The High Tatras extend in a narrow ridge along the Polish border and are attractive as a summer resort area. The highest peak in the country, Gerlachovka (also known as Stalinov Štít), which has an elevation of about 2,655 m (8,711 ft) is in this ridge. Snow persists at the higher elevations well into the summer months and all year long in some sheltered pockets. The tree line is at about 1,500 m (4,921 ft). An ice cap extended into this area during glacial times, leaving pockets that became mountain lakes.

The Slovakian lowlands in the south and southeast that border on Hungary and contain substantial Magyar populations are part of the greater Danube Basin. From a point a few kilometers south of the Slovakian capital of Bratislava, the main channel of the Danube River demarcates the border between Czechoslovakia and Hungary for about 175 km (108 mi). As it leaves Bratislava, the Danube divides into two channels, the main channel being the Danube proper and the northern channel, the Little Danube (Malý Dunaj). The Little Danube flows eastward to join the Váh River just north of Komárno, where the Váh converges with the main Danube. The land between the Little Danube and the Danube is known as the Great Rye Island (Velký Žitný Ostrov), a marshland maintained

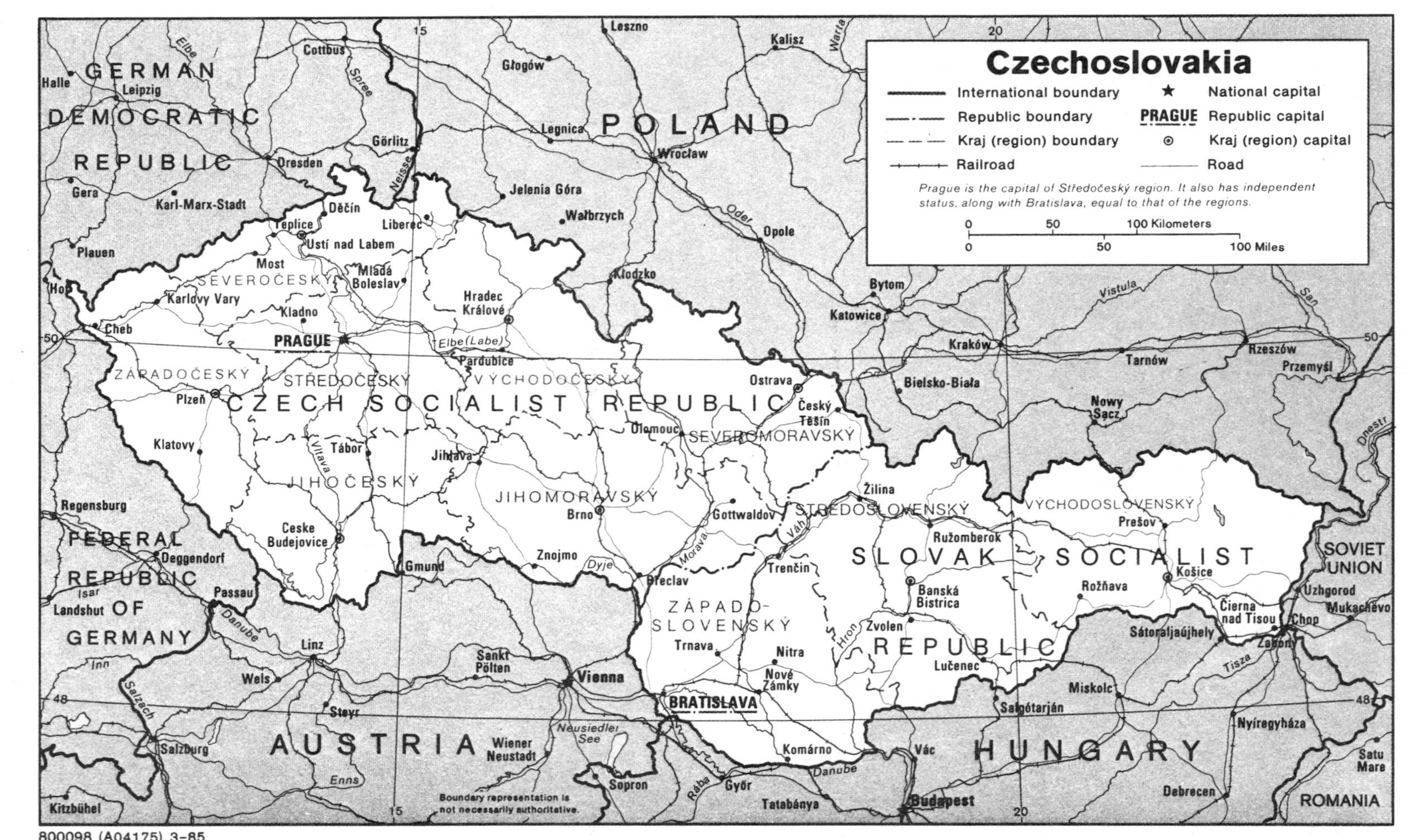

800098 (A04175) 3-85

for centuries as a hunting preserve for the nobility. Dikes and artificial drainage have made the land cultivable for grain production, but it is still sparsely settled.

Moravia

Moravia is a topographic borderland situated between Bohemia and Slovakia. Its southwest-to-northeast ridge lines and lower elevations made it useful as a communication and commercial route from Vienna to the north and northeast during the periods when Austria was dominant in Central Europe.

Central and southern Moravian lowlands are part of the Danube Basin and are similar to the lowlands they join in southern Slovakia. The upland areas are smaller and more broken than those of Bohemia and Slovakia. The northwest hills are soft sandstone and are cut by deep gorges. South of them, but north of Brno, is a karst limestone area with underground streams and caves. These and the other uplands west of the Morava River are associated with the Hercynian Massif. To the east of the river the land is called Carpathian Moravia.

DENMARK

BASIC FACTS

Official Name: Kingdom of Denmark

Abbreviation: DEN

Area: 43,092 sq km (16,638 sq mi)

Area—World Rank: 119

Population: 5,125,676 (1988) 5,165,000 (2000)

Population—World Rank: 91

Capital: Copenhagen

Boundaries: 7,471 km (4,642 mi); West Germany 68 km (42 mi)

Coastline: 7,403 km (4,600 mi)

Longest Distances: 402 km (250 mi) N-S 354 km (220 mi) E-W

Highest Point: Yding Skovhoj 173 m (568 ft)

Lowest Point: Lammefjord 7 m (23 ft) below sea level

Land Use: 61% arable land; 6% meadows and pastures; 12% forest and woodland; 21% other

Located in southern Scandinavia, the Kingdom of Denmark consists of Denmark proper, the Faroe Islands and Greenland. Denmark itself comprises the peninsula of Jutland (Jylland) and some 406 islands. The country is surrounded by water on the other three sides: the Skaggerak on the north; the Kattegat, the Øresund and the Baltic Sea on the east; and the North Sea on the west.

PRINCIPAL TOWNS (population at 1 January 1985)

Town	Population	Town	Population
København (Copenhagen, the capital)	1,358,540*	Horsens	46,586
Århus (Aarhus)	194,348	Vejle	43,867
Odense	136,803	Helsingør (Elsinore)	43,581
Ålborg (Aalborg)	113,865	Kolding	41,770
Esbjerg	70,975	Roskilde	39,663
Randers	55,780	Næstved	38,177

*Copenhagen metropolitan area.

The precise size of Denmark proper is subject to constant variation owing to marine erosion and deposit and reclamation work. Moreover, when the tides are active, the coastline shifts twice daily, up to 16 km. (10 mi.) on the western coast of South Jutland. Not included in the land area are inlets or fjords directly connected with the sea, among them the lagoon of Ringkøbing, which is linked with the North Sea by the sluices of Hvide Sande. The 406 islands (of which only 97 are inhabited) account for over one-third of the land area. The largest are Zealand (Sjaelland, 7,015 sq km; 2,709 sq mi), Fünen (Fyn, 2,984 sq km; 1,152 sq mi), Lolland (1,234 sq km; 480 sq mi), Bornholm (588 sq km; 227 sq mi) and Falster (514 sq km; 198 sq mi). Geographically, Denmark may be described as a virtual archipelago, the only one in northern Europe.

Area and population

		area		population
Counties	**Capitals**	sq mi	sq km	1987 estimate
Århus	Århus	1,761	4,561	589,108
Bornholm	Rønne	227	588	46,839
Frederiksborg	Hillerød	520	1,347	339,627
Fyn	Odense	1,346	3,486	456,483
København	—	203	526	606,870
Nordjylland	Ålborg	2,383	6,173	483,381
Ribe	Ribe	1,209	3,131	216,967
Ringkøbing	Ringkøbing	1,874	4,853	266,088
Roskilde	Roskilde	344	891	213,476
Sønderjylland	Åbenrå	1,520	3,938	249,805
Storstrøm	Nykøbing	1,312	3,398	257,880
Vejle	Vejle	1,157	2,997	328,849
Vestjælland	Sorø	1,152	2,984	282,397
Viborg	Viborg	1,592	4,122	230,760
Cities				
Copenhagen (København)	—	34	88	469,706
Frederiksberg	—	3	9	86,558
TOTAL		16,638	43,092	5,124,794

Denmark is a low-lying country, with its highest point, Yding Skovhoj in East Jutland, only 173 m (568 ft) above sea level. The surface relief is characterized by glacial moraine deposits, which form undulating plains with gently rolling hills interspersed with lakes. The largest lake is Arreso, (40.6 sq km; 15.7 sq mi). The moraines consist of a mixture of clay, sand, gravel and boulders, carried by glaciers from the mountains of Scandinavia and raised from the bed of the Baltic, with an admixture of limestone and other rocks. During the last glaciation, only the northern and eastern parts of the country lay under the icecap. The ice limit followed a line running from Viborg

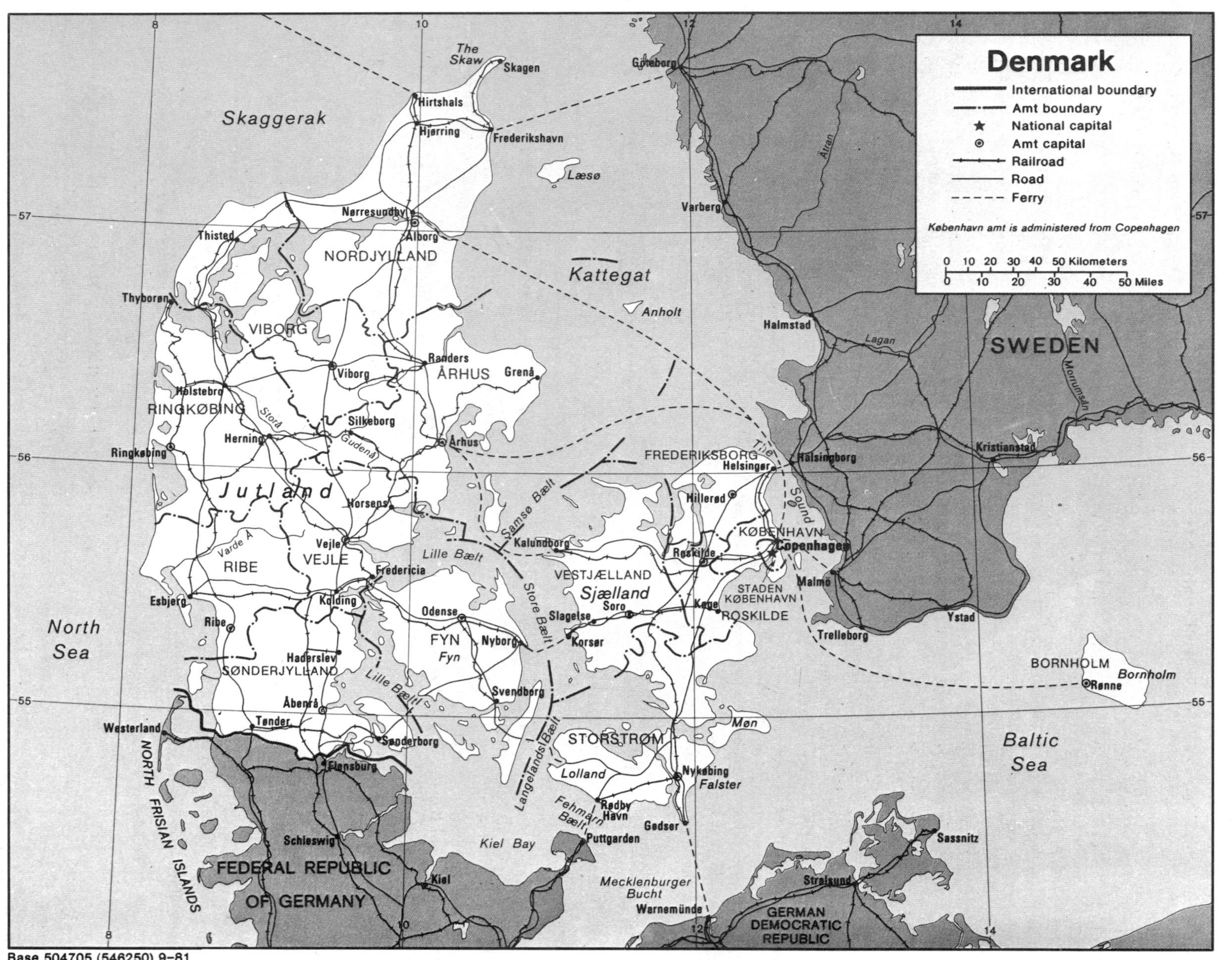
Denmark
International boundary
Amt boundary
National capital
Amt capital
Railroad
Road
Ferry
København amt is administered from Copenhagen
0 10 20 30 40 50 Kilometers
0 10 20 30 40 50 Miles
Skaggerak
Kattegat
North Sea
Baltic Sea
SWEDEN
FEDERAL REPUBLIC OF GERMANY
GERMAN DEMOCRATIC REPUBLIC
NORTH FRISIAN ISLANDS
The Skaw
Skagen
Hirtshals
Hjørring
Frederikshavn
Læsø
Anholt
Nørresundby
Ålborg
Thisted
Thyborøn
NORDJYLLAND
VIBORG
Viborg
Randers
ÅRHUS
Grenå
Holstebro
RINGKØBING
Storå
Silkeborg
Gudenå
Herning
Ringkøbing
Århus
Jutland
Horsens
Vejle
VEJLE
Varde Å
RIBE
Esbjerg
Ribe
Fredericia
Kolding
Haderslev
SØNDERJYLLAND
Åbenrå
Tønder
Westerland
Sønderborg
Flensburg
Schleswig
Kiel
Kiel Bay
Lille Bælt
Odense
FYN
Fyn
Nyborg
Svendborg
Samsø Bælt
Store Bælt
Langelands Bælt
Kalundborg
VESTJÆLLAND
Sjælland
Slagelse
Sorø
Korsør
FREDERIKSBORG
Helsingør
Hillerød
Roskilde
KØBENHAVN
Copenhagen
STADEN KØBENHAVN
Køge
ROSKILDE
The Sound
STORSTRØM
Møn
Lolland
Nykøbing
Falster
Rødby Havn
Gedser
Fehmarn Bælt
Puttgarden
Mecklenburger Bucht
Warnemünde
Stralsund
Sassnitz
Göteborg
Varberg
Halmstad
Ätran
Lagan
Morrumsån
Hälsingborg
Kristianstad
Malmö
Trelleborg
Ystad
BORNHOLM
Bornholm
Rønne
8
10
12
14
55
56
57
Base 504705 (546250) 9-81

to Bovbjerg on the western coast and southward to the national boundary near Tinglev. The country west and south of this line formed a polar landscape during the final glaciation. Between the hills are extensive level outwash plains of the meltwater formed from stratified sand and gravel outside the ice limit. These heathland plains are the site of the country's densest settlements. The boundary line between the sandy West Jutland and the loam plains of East and North Denmark is the most important geographical dividing line in the country. West of the line is a region of scattered farms; to the east, villages with high population density.

Valleys, both tunnel and regular, furrow the moraine landscape. The East Jutland inlets were created by the intruding sea in the lowest part of the tunnel valleys, to which glacial erosion also contributed. The inlets form natural harbors, making maritime activities easy means of livelihood. The Gudenå River, the longest river in Denmark (158 km; 98 mi), follows the intersecting valley systems.

Flat sand and gravel tracts make up one-tenth of the total land area. They are particularly numerous in the northern part of the country, such as in the Limfjorden area. Where draining is rendered difficult in these flat and low-lying regions, the land often is swampy. Along the coast of South Jutland, where there is a strong tidal range, there are salt marshes formed by clay deposited by tidal waters. Dune landscapes form an almost unbroken belt along the entire coast of Jutland.

The substratum on which Danish soils have developed are chiefly moraine and meltwater sand. There are two basic types: loam and podsol. Loam is dark-colored, porous and highly organic and thus is rich in plant food. It is common in eastern Denmark. Podsol occurs chiefly in open, sandy tracts, where the earth is subject to drying out, as in West Jutland. Tillage and regulation of the water table have considerably improved the Danish soil, which is not intrinsically well suited to agriculture. With deep plowing and the use of fertilizers, more than three-fourths of the land surface can be efficiently cultivated.

DJIBOUTI

BASIC FACTS

Official Name: Republic of Djibouti

Abbreviation: DJB

Area: 23,200 sq km (8,950 sq mi)

Area—World Rank: 132

Population: 320,444 (1988) 690,000 (2000)

Population—World Rank: 143

Capital: Djibouti

Boundaries: 517 km (321 mi); Ethiopia 459 km (285 mi); Somalia 58 km (36 mi)

Coastline: 314 km (195 mi)

Longest Distances: N/A

Highest Point: Moussa Ali 2,063 m (6,768 ft)

Lowest Point: Lake Assal 155 m (509 ft) below sea level

Land Use: 9% meadows and pastures; 91% other

Djibouti (formerly known as French Somaliland and later as the French Territory of the Afars and the Issas) is located in northeast Africa.

The entire country is mostly a sand and stone desert broken in places by lava streams and salt lakes. The three principal geographic regions are the coastal plain, less than 200 m (650 ft) in elevation, the mountains backing the plains with lofty peaks, such as the Moussa Ali 2,010 m (6,694 ft), and the plateau behind the mountains rising from 300 to 1,500 m (1,000 to 5,000 ft). The coastline is deeply indented by the Gulf of Tadjoura which is 45 km (28 mi) across at its entrance and penetrates 58 km (36 mi) inland.

PRINCIPAL TOWNS (1981): Djibouti (capital), population 200,000; Dikhil; Ali-Sabieh; Tadjourah; Obock.

Area and population		area		population
Districts	**Capitals**	sq mi	sq km	1982 estimate
ʿAlī Sabīḥ (Ali-Sabieh)	ʿAlī Sabīḥ	925	2,400	15,000
Dikhil	Dikhil	2,775	7,200	30,000
Djibouti	Djibouti	225	600	200,000
Obock	Obock	2,200	5,700	15,000
Tadjouri (Tadjourah)	Tadjoura	2,825	7,300	30,000
TOTAL		8,950	23,200	335,000

Few streams flow above the surface except following rains. The drainage is partly eastward to the coast and partly inland to the Lake Assal and Lake Abbe.

DOMINICA

BASIC FACTS

Official Name: Commonwealth of Dominica

Abbreviation: DOM

Area: 750 sq km (290 sq mi)

Area—World Rank: 166

Population: 97,763 (1988) 114,000 (2000)

Population—World Rank: 176

Capital: Roseau

Djibouti

502715 9-77 (542150)
Mercator Projection
Scale 1:1,300,000

Railroad

Road

Airport

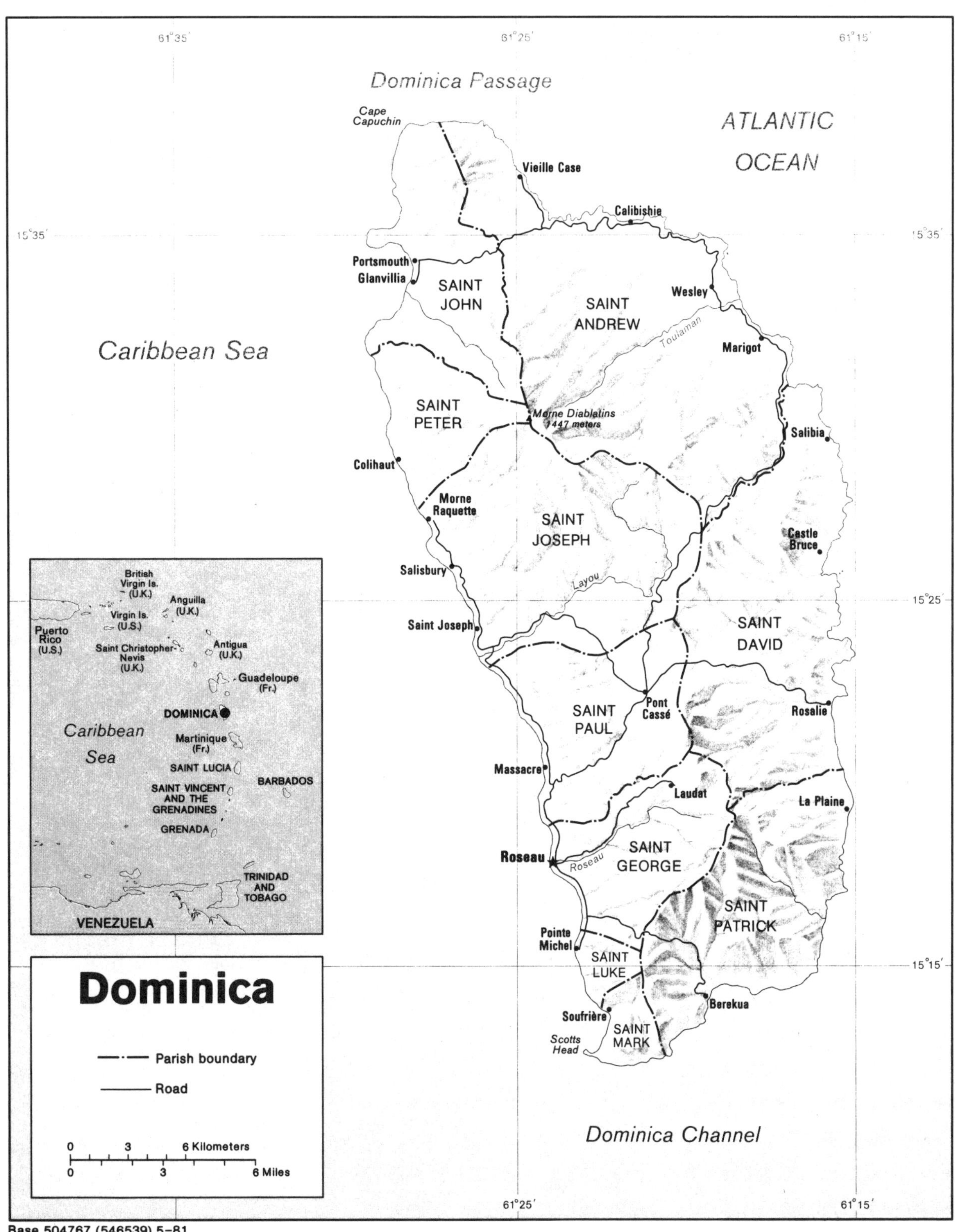

Base 504767 (546539) 5-81

Land Boundaries: Nil

Coastline: 148 km (91 mi)

Longest Distances: N/A

Highest Point: Morne Diablotin 1,447 m (4,747 ft)

Lowest Point: Sea level

Land Use: 9% arable; 13% permanent crops; 3% meadows and pastures; 41% forest and woodland; 34% other

The largest of the Windward Islands, Dominica is located at the northern end of the Windward chain of the Lesser Antilles between Guadeloupe and Martinique. It is roughly rectangular in shape. The island is dominated by a high mountain range running like a spine with lateral spurs on either side. The highest peak is Morne Diablotin 1,446 m (4,747 ft) and there are a number of other peaks, such as Morne au Diable, Morne Brule, Morne Couronne, Morne Anglais and Morne Plat Pays. None of the many rivers is navigable but they give limited access to the interior. The main ones are Indian, Espangnol, Layou, Roseau and Queens running west to the Caribbean Sea and Hodges, Tweed, Clyde, Maclaralin, Grand Bay, Rosalie, and Wanerie running east to the Atlantic.

Principal Towns (population at 1981 census): Roseau (capital) 8,279; Portsmouth 2,200.

Area and population	area		population
Parishes	sq mi	sq km	1981 census
St. Andrew	69	179	12,748
St. David	49	127	7,337
St. George	21	54	20,501
St. John	23	60	5,412
St. Joseph	46	119	6,606
St. Luke	4	10	1,503
St. Mark	4	10	1,921
St. Patrick	32	83	9,780
St. Paul	26	67	6,386
St. Peter	11	29	1,601
TOTAL	290	750	73,795

DOMINICAN REPUBLIC

BASIC FACTS

Official Name: Dominican Republic

Abbreviation: DNR

Area: 48,443 sq km (18,704 sq mi)

Area—World Rank: 117

Population: 7,136,748 (1988) 9,247,000 (2000)

Population—World Rank: 83

Capital: Santo Domingo

Boundaries: 892 km (554 mi); Haiti 290 km (180 mi)

Coastline: 602 km (374 mi)

Longest Distances: 386 km (240 mi) E-W; 261 km (162 mi) N-S

Highest Point: Duarte Peak 3,175 m (10,417 ft)

Lowest Point: Lake Enriquillo 40 m (131 ft) below sea level

Land Use: 23% arable; 7% permanent crops; 43% meadows and pastures; 13% forest and woodland; 14% other

Dominican Republic occupies the eastern two-thirds of Hispaniola, the second-largest island in the Caribbean; the western one-third is occupied by the Republic of Haiti. Lying about 966 km (600 mi) southeast of Florida, the island is separated from Puerto Rico on the east by the Mona Passage, and from Cuba on the west by the Windward Passage. Because these two seaways are the principal water routes linking North America and Europe with Central and South America, the Dominican Republic and Haiti have repeatedly been subjected to external influences.

The Dominican Republic has a rugged and mountainous terrain in which the dominant relief features are four parallel ranges that extend in a northwesterly direction in the western part of the country and a single range of low mountains that extends from east to west in the eastern part of the island. Satellite ranges, extensive valleys between the larger ranges, narrow and broken coastal lowlands, and a lowland plain that covers most of the eastern extremity make up the remainder of the Dominican landmass.

Landform and Drainage

The border between the Dominican Republic and Haiti follows an irregular line extending from north to south, but the relief features of Hispaniola follow an east-west axis. As a consequence, the principal mountain ranges and intervening valleys are shared by the two countries. The laterally extending mountains have in the past made internal communication difficult. The arrangement of the intervening valleys, however, has served to furnish easy movement from one country to the other and has led to border incidents and illegal crossings of the frontier that have played important parts in the histories of the two nations.

The geography of the Dominican Republic does not lend itself readily to the division of the country into a few regions with similar characteristics of topography, climate, vegetation, and land use. A geomorphic study prepared by the Organization of American States divides the country into no less than twenty regions, too many to be used in a general study. The Dominican Republic can be studied generally, however, in terms of its highland and its lowland areas.

Highlands

The principal mountain systems are four parallel ranges extending in a northwesterly direction in the western part of the country and a single minor chain—the

Cordillera Oriental—in the east. The core of the system is the Cordillera Central, which rises in the east near Santo Domingo and veers northwestward into Haiti, where it becomes the Massif du Nord. It is flanked near the northern coast by the Cordillera Septentrional, and on the south by the Sierra de Neiba. Still farther south, the Sierra de Baoruco forms an extension of the southern mountain ranges of Haiti.

The dissected and complex Cordillera Central divides the country into two parts. Its convoluted ridges crest between 1,524 and 2,438 m (5,000 and 8,000 ft), but there are individual peaks with considerably greater heights. Pico Duarte, which has an elevation of 3,174 m (10,414 ft), is the highest in the West Indies.

The rugged slopes, precipitous and sometimes faceted, have gradients of as high as forty degrees and constitute the principal watershed of the country. The Cordillera Central is composed of a mixture of volcanic, metamorphic, and sedimentary rock. Together with the Massif du Nord in Haiti, including the various intermont valleys, the range makes up about one-third of the landmass of Hispaniola. Within it, innumerable streams have carved a jigsaw puzzle of yawning canyons and rocky gulches that in many places makes transit almost impossible.

The Cordillera Septentrional also has precipitous slopes and deeply etched valleys but is a less imposing range in which elevations do not exceed 1,219 m (4,000 ft). Rising in the west near the city of Monte Cristi, it extends parallel to the Atlantic coastline across the northern part of the country and is separated from a related series of hills on the Samaná Peninsula (Peninsula de Samaná) by swamps that surround the mouth of the Yuna River (Río Yuna).

The two ranges that lie to the south of the Cordillera Central, the Sierra de Neiba and the Sierra de Baoruco, begin as escarpments flanking Neiba Bay (Bahía de Neiba) in the southwest and continue northwestward to join corresponding ranges in Haiti. Both crest generally at elevations of between 914 and 1,219 m (3,000 and 4,000 ft) but have peaks as high as 1,828 m (6,000 ft). The eastern part of the Sierra de Neiba is separated from the remainder of the range by the Yaque del Sur River (Río Yaque del Sur) and is known as the Sierra de Martín García.

The Cordillera Oriental is less a range of mountains than a narrow band of hills representing an eastward terminal spur of the Cordillera Central. It extends westward some 137 km (85 mi) from the Atlantic coast along the southern shore of Samaná Bay (Bahía de Samaná) to the foothills of the Cordillera Central about 48 km (30 mi) north of Santo Domingo. The western third of the range is rolling rather than craggy and permits fairly easy access from the capital city to the interior lowlands. The remainder is craggy and dissected by small streams that flow from its northern slopes to Samaná Bay and larger ones that flow southward to the Caribbean. Elevations are generally under 305 m (1,000 ft) except in the extreme east where a few isolated promontories rise to over 610 m (2,000 ft).

PRINCIPAL TOWNS (Population at 12 December 1981)

Santo Domingo, DN (capital)	1,313,172
Santiago de los Caballeros	278,638
La Romana	91,571
San Pedro de Macorís	78,562
San Francisco de Macorís	64,906
Concepción de la Vega	52,432
San Juan	49,764
Barahona	49,334
San Felipe de Puerto Plata	45,348

Lowlands

The country's lowlands consist for the most part of long valleys that, like the mountains that define them, extend in a northwesterly direction from origins close to the Caribbean Sea to corresponding lowlands in Haiti. In these, the fertile alluvial soils of their flood plains and terraces are suitable for intensive agriculture, and shallower soils provide good pasture.

The most extensive of the valleys, the Cibao Valley is the breadbasket of the country. Covering nearly 5,180 sq km (2,000 sq mi) or 10 percent of the national territory, it has a longitudinal extent of about 225 km(140 mi) from Samaná Bay on the east to the Haitian frontier; there, as the Northern Plain of Haiti, it continues into the neighboring republic. This heartland area lies between the Cordillera Central on the south and the Cordillera Septentrional on the north and has a valley floor width that varies between nine and seventeen miles. It breaks apart at the city of Santiago de los Caballeros. The portion to the east is known as the Vega Real, or the Royal Plain. This portion of the Cibao includes the country's richest and deepest agricultural land. The section on the western flank of Santiago is also fertile, but most of its land is arid and is becoming productive only as a result of the program for increasing irrigation.

Directly to the south of the Cibao, and separated from it by the crests of the Cordillera Central, is the San Juan Valley (Valle de San Juan), which extends across the frontier into Haiti as the Central Plateau. The valley, which is defined on its south by the Sierra de Neiba, has an extent of about 1,813 sq km (700 sq mi) and a length of 112 km (70 mi). It has a width of nearly 48 km (30 mi) miles at the approaches to the Haitian border, but it narrows progressively to no more than five miles at its funnel extremity near the Caribbean Sea.

The third of the parallel valleys lies between the Sierra de Neiba and the Sierra de Baoruco and is known both as the Neiba Valley (Valle de Neiba) and as the Enriquillo Basin (Hoya de Enriquillo). It is about 1,942 sq km (750 sq mi) in extent, and most of the valley floor is below sea level. It extends about 96 km (60 mi) from the Neiba Bay on the southeast to the frontier where it merges with the Cul-de-Sac of Haiti. The area is semiarid to arid, and the soils are generally of limited productivity. At one time this region was covered by a portion of the Caribbean Sea that separated the Sierra de Baoruco and the corresponding Haitian range from the mainland. At a later period, this strait was closed off at both ends by geological action, and gradual evaporation of the trapped waters has left brackish lakes and a series of ridge-like terraces on

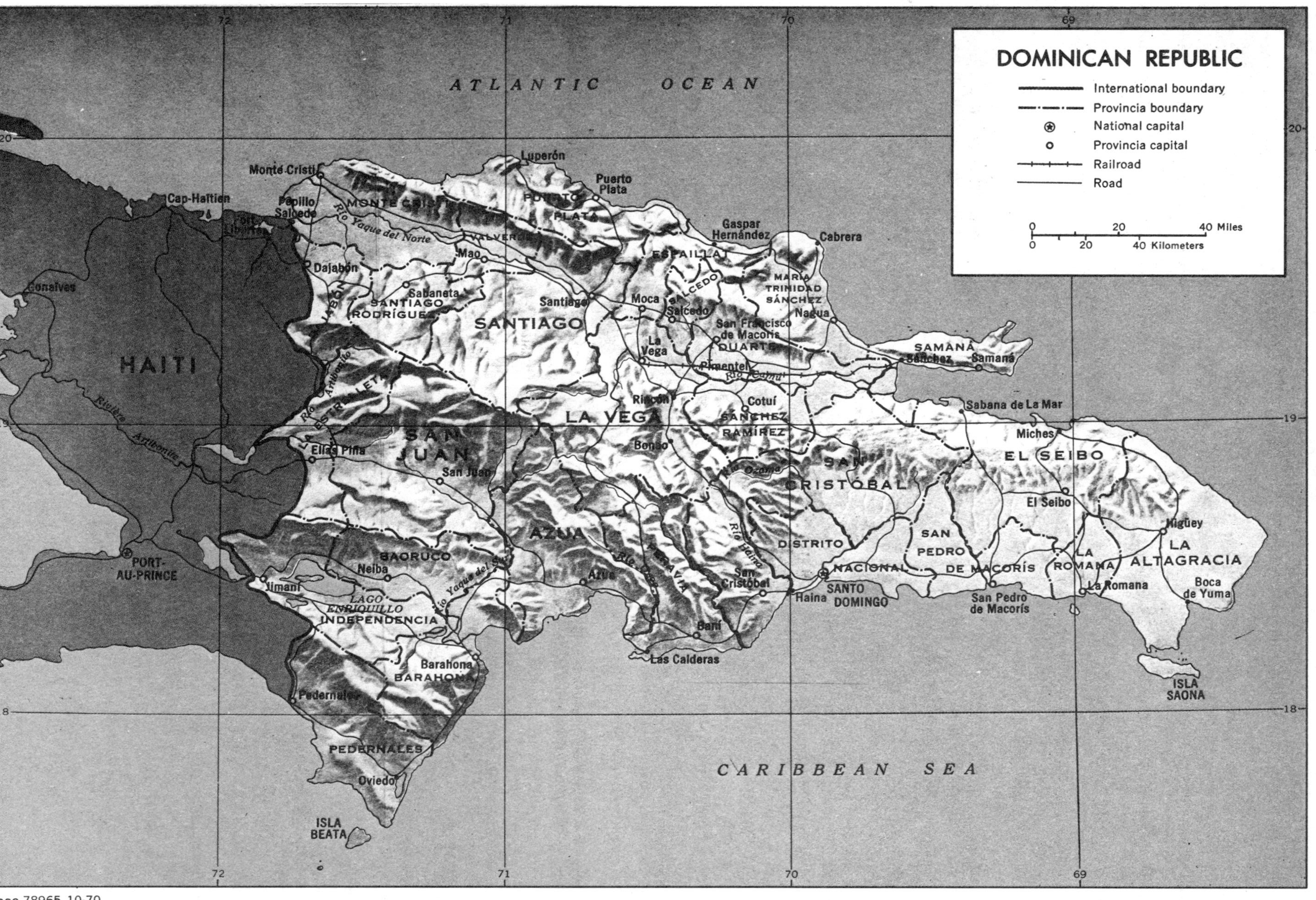

Base 78965 10-70

both flanks of the valley. The valley floor is made up of alluvial deposits with outcroppings of sedimentary and other rocks.

The largest of the other lowland regions is the Caribbean Coastal Plain (Llanura Costera del Caribe) that together with the foothills of the Cordillera Oriental mark its northern boundary. The plain covers more than 2,849 sq km (1,100 sq mi). It is composed principally of a limestone platform formed by corals and alluvial deposition. Inland, there are calcerous soils of high fertility, and to the west of Santo Domingo there are infertile soils derived from acid clays. The region is the center of the country's cattle-raising and sugar industries.

In addition to the major lowland regions, there are numerous small valleys and basins, particularly in the Cordillera Central. Along the northern coastline, a discontinous coastal plain lies between the Atlantic Ocean and the Cordillera Septentrional. The soils are of good quality, but the lack of good passes through the mountains leaves the north coast relatively isolated from the rest of the country. In the far south, an arid coastal plain covers the lower part of the Pedernales Peninsula, and coastal plains of only moderate fertility surround the towns of Baní and Azua. In the northeast a narrow coastal plain separates the Cordillera Oriental and Samaná Bay. To its east, a low-lying area known as Los Haitises makes up most of San Cristóbal Province. It is composed of limestone and some alluvial soil and has altitudes reaching a maximum of about 244 m (800 ft).

Area and population

Provinces	Capitals	area: sq mi	area: sq km	population: 1987 estimate
Azua	Azua	938	2,430	178,877
Bahoruco (Baoruco)	Neiba	531	1,376	85,356
Barahona	Barahona	976	2,528	148,881
Dajabón	Dajabón	344	890	62,640
Duarte	San Francisco de Macorís	499	1,292	255,672
El Seibo	El Seibo	641	1,659	95,333
Espaillat	Moca	386	1,000	178,033
Hato Mayor	Hato Mayor	514	1,330	76,023
Independencia	Jimaní	719	1,861	42,081
La Altagracia	Higüey	1,191	3,084	108,667
La Estrelleta	Elías Piña	690	1,788	70,971
La Romana	La Romana	209	541	149,652
La Vega	La Vega	916	2,373	296,039
María Trinidad Sánchez	Nagua	506	1,310	122,253
Monseñor Nouel	Bonao	388	1,004	121,906
Monte Cristi	Monte Cristi	768	1,989	90,534
Monte Plata	Monte Plata	841	2,179	170,758
Pedernales	Pedernales	373	967	18,459
Peravia	Baní	626	1,622	182,489
Puerto Plata	Puerto Plata	726	1,881	224,425
Salcedo	Salcedo	206	533	107,667
Samaná	Samaná	382	989	71,313
Sánchez Ramirez	Cotuí	453	1,174	137,382
San Cristóbal	San Cristóbal	604	1,564	313,497
San Juan	San Juan	1,375	3,561	260,462
San Pedro de Marcorís	San Pdero de Marcorís	450	1,166	184,078
Santiago	Santiago de los Caballeros	1,205	3,122	657,729
Santiago Rodríguez	Sabaneta	394	1,020	60,146
Santo Domingo	—	570	1,477	2,127,496
Valverde	Mao	220	570	108,891
TOTAL		18,704	48,443	6,707,710

Hydrography

The rivers of the Dominican Republic for the most part are shallow, subject to wide seasonal change in flow and consequently of little use for transportation. Flowing out of the several highlands in varying directions, they form a variety of drainage systems. The Cibao Valley has two systems. On its western flank it is drained into the Atlantic near Monte Cristi by the Yaque del Norte, the country's longest river; east of Santiago de los Caballeros in the Vega Real, the Yuna River drains eastward into Samaná Bay.

South of the Cordillera Central, the San Juan Valley is also divided between two hydrographic systems with opposite watersheds. Near the frontier it is drained by a tributary of the Artibonito River (Río Artibonito), a stream that continues westward across the border as the principal watercourse of Haiti; to the southeast, it is drained by the Yaque del Sur River, which flows into the Caribbean at Neiba Bay. The principal tributary of the Yaque del Sur, the San Juan, has a flow that increases fourfold between the driest and the wettest months.

Surface water in the arid Neiba Valley still farther to the south is lost through evaporation rather than runoff, although the Yaque del Sur River provides drainage for its southeastern portion as well as for the Neiba Valley. At the eastern end of the island, numerous rivers flowing southward from the Cordillera Oriental drain the Caribbean Coastal Plain. Among these, the Ozama River (Río Ozama), on which Santo Domingo is located, and the Macorís River (Río Macorís) are navigable for several miles from their mouths and are used for the transportation of sugar.

In 1972 dam construction was completed on the Yaque del Norte River at Tavera, located in the Cibao south of Santiago de los Caballeros, and was in progress on the Nizao River (Río Nizao) at Valdesia. The dams created large artificial lakes; the first was to provide irrigation for an extensive acreage in the central Cibao, and the second was to supplement power and water supplies in Santo Domingo as well as to irrigate agricultural land along the Caribbean coast.

The largest of the country's natural lakes is Lake Enriquillo (Lago Enriquillo) in the Neiba Valley. A remnant of the strait that once occupied the area, its waters are 43 m (140 ft) below sea level. Although it is fed by many streams from the surrounding mountains and has no outlet, the high rate of evaporation in the valley is causing its waters gradually to recede. The only other sizable body of inland water in the country is a smaller lake, located in the same valley near Neiba Bay. The most extensive marshland extends inland from the delta of the Yuna River on Samaná Bay. At the opposite end of the Cibao, there are salt marshes south of Monte Cristi Bay.

Coasts and Coastal Waters

Hispaniola, both in the Dominican Republic and in Haiti, is girded by an offshore rocky platform. On the Atlantic coast of the Dominican Republic, the platform is highly developed in the shallow waters of Samaná Bay, and continues in a westerly direction to the frontier, and along the Haitian coastline. Along the Atlantic coast it extends seaward from a few hundred yards to more than 48 km (30 mi) at a maximum depth of 61 m (200 ft).

At irregular intervals the shelf rises to form tiny islands and jagged coral reefs that lie close to the surface, and represent hazards to navigation in waters to the east of Monte Cristi. The northern coast is marked by many sandy beaches and occasional rocky escarpments. Inland, the low but craggy ridges of the Cordillera Septentrional make access to the interior difficult. There are a few sheltered roadsteads, but the combination of dangerous offshore waters and the absence of ready access from the coast to the interior hinterland has discouraged port development, although most of the produce of the Cibao moves through these ports.

The Caribbean coast is better suited to port development. The submarine shelf is generally lower than along the northern coast, reefs and islets are relatively few, and access from ports to the interior is easier. A majority of the ports scattered along the 1,600 km (1,000-mi) coastline of the country are found on the Caribbean side. The best of the natural harbors are located on the broad estuaries of rivers that meet the Caribbean at Santo Domingo, San Pedro de Macorís, and La Romana.

Among the numerous islands scattered off the Dominican coastline, only three are permanently inhabited, and none is of significant economic importance. The largest, Saona Island (Isla Saona), has maximum dimensions of 24 by 6 km (fifteen by four mi) and is located at the southeastern tip of Hispaniola. The 52 sq km (twenty square mi) of Beata Island (Isla Beata) lie off the Pedernales Peninsula in the extreme west, and tiny Catalina Island (Isla Catalina) lies a few miles west of the sugar port of La Romana. In addition, off the coasts lie some 7,770 sq km (3,000 sq mi) of fishing banks with depths of less than 183 m (600 ft).

ECUADOR

BASIC FACTS

Official Name: Republic of Ecuador

Abbreviation: ECU

Area: 269,176 sq km (103,130 sq mi)

Area—World Rank: 69

Population: 10,231,630 (1988) 13,939,000 (2000)

Population—World Rank: 65

Capital: Quito

Boundaries: 2,702 km (1,679 mi); Colombia 538 km (334 mi); Peru 1,316 km (818 mi)

Coastline: 848 km (527 mi)

Longest Distances: 714 km (444 mi) N-S; 658 km (409 mi) E-W

Highest Point: Chimbarazo 6,310 m (20,702 ft)

Lowest Point: Sea level

Land Use: 6% arable; 3% permanent crops; 17% meadows and pastures; 51% forest and woodland; 23% other

Ecuador is located in the northwestern part of the South American continent. It is bordered on the north by Colombia, on the east and south by Peru, and on the west by the Pacific Ocean. The equator, which gives the country its name, passes a few miles north of the capital city of Quito, where the sun rises at 6:00 A.M. and sets at 6:00 P.M. throughout the year.

The dominant topographical features are two parallel ranges of the lofty Andes mountains that separate a fertile coastal lowland on the west and a more extensive lowland of the Amazon Basin on the east. Ecuador has been described as a jungle interrupted by a double line of formidable mountain peaks. Between the two Andean ranges transverse hill systems form a series of high basins, in which both pre-Columbian settlement and the early European settlement were concentrated. Streams that rise in the Andes flow westward to the Pacific or eastward toward the Amazon to form the drainage systems.

Geographical Regions

The country's mainland divides naturally into a coastal lowland, a central mass made up of the Andean highlands, and an interior lowland that forms part of the Amazon Basin. These three regions are customarily known as the Costa, the Sierra, and the Oriente. A fourth region is made up of the offshore Galápagos Islands. The borders of the continental regions are not exactly defined. Most geographers consider the two lowland regions to extend up the approaches of the Andes to an elevation of about 487 m (1,600 ft). On the basis of this definition, the Oriente would include about half of the territory, and the remainder would be divided equally between the other two. The National Planning Board (Junta Nacional de Planificación), formerly Ecuadorian National Board of Planning and Economic Coordination, however, considers the Costa to contain 16.5 percent of the national territory, the Sierra to contain 24.3 percent, the Oriente to contain 57.4 percent, and the remaining 1.8 percent to be made up of the landmass of the Galápagos Islands. This arrangement places each province entirely in a single region. The Costa is made up of Esmeraldas, Manabí, Los Ríos, Guayas, and El Oro; the Oriente is made up of Napo, Pastaza, Morona-Santiago, and Zamora-Chinchipe; and the remaining ten provinces are located in the Sierra. This regional arrangement is customarily used in census reports and other public statistical compilations.

The Costa

The Costa is sometimes identified in English as the Coastal Lowlands and in Spanish as the Litoral (Littoral). It includes Ecuador's largest and fastest growing city, Guayaquil, and the basin made up of the Guayas River drainage system is the country's richest agricultural zone. The coastal lowlands are the fastest growing and most prosperous of the country's regions and enjoy the prospect of substantial further development.

The Costa is widest in a central belt between Cape Pasado and the Santa Elena Peninsula. Near both the northern and southern extremities of the region, the Sierra highlands intrude close to tidewater. At intervals, subtropical river valleys that are physical extensions of the Costa penetrate far into the Sierra.

Pacific tides, ranging from six to nine feet from ebb to flood, are more moderate than the average along the Pacific coast. Like the rest of the Pacific coast, that of Ecuador has few good natural harbors, but the great roadstead of Guayaquil—53 km (thirty-three mi) up the Guayas River from the Gulf of Guayaquil—is the best and largest on South America's western flank. Esmeraldas, near the Colombian frontier, is the country's second seaport.

The diversity of natural features of the Costa is so great that it can be considered to be a single geographic region only because the terrain rises abruptly from it to the Andean Sierra. Multiple climatic conditions, soils, forms of vegetation, and settlement patterns set it apart from the more homogeneous Sierra and Oriente.

The rivers and streams that form the region's drainage pattern spill precipitously out of the Sierra, but their currents slow as they meander to the Pacific across the lowlands. It is at this point that many of the watercourses, particularly in the Guayas Basin, have formed interior alluvial fans composed of loose topsoil washed away from the floors of the Andean basins.

The principal drainage systems are those of the Guayas River in the south and the Esmeraldas River in the north. The Santiago River drains the rain forest of the far north. Both of the northern rivers are navigable by light craft in their lower courses and have been panned for gold since the Pre-Colombian era. The Guayas is the largest and most important of the region's watercourses. From its mouth to the city of Guayaquil, it is less a natural river than a commercially developed waterway. Above Guayaquil, it divides into several navigable streams, and a multitude of its tributaries enrich the Guayas Basin with their annual quotas of alluvium carried down from the Sierra.

The Sierra

The principal features of the region are two parallel spines of the Andes. On the west, the Cordillera Occidental is a compact high range extending roughly north to south the full length of the country. To its east, the Cordillera Central is less a true mountain range than a series of lofty peaks. Both ranges are of volcanic origin. Between them, an original crystalline formation has collapsed to form a trench with elevations from 2,133 to 3,048 m (7,000 to 10,000 ft), and with high rims that are from 40 to 64 km (twenty-five to forty mi) apart.

The trench, called the Avenue of the Volcanoes by the nineteenth-century naturalist Baron Alexander von Humboldt and popularly referred to as the Interandean Lane (Callejón Interandino), makes up about three-eighths of the Sierra. It is broken by transverse hill systems to form a series of *hoyos* (intermont basins) in which most of the region's population lives. In all, there are about a dozen, descending in altitude from north to south. In the southern quarter of the Sierra the terrain is increasingly broken, the soil is poorer, and the *hoyos* are valleys spilling into the Costa or Oriente rather than true basins.

East of the crests of the Cordillera Central, the downward slope to the floor of the Amazon Basin is interrupted by lower mountains. They are broken at the midpoint by the wide valley of the Pastaza River, but aerial mapping undertaken during the 1960s indicates that they form a third parallel range with elevations of as high as 3,962 m (13,000 ft). The system is identified by Ecuadorian geographers as the Cordillera Oriental or, simply, as the Third Cordillera. This transitional Andean area is sometimes regarded as a separate region lying between the Sierra and the Oriente.

The highest of the Andean peaks is Chimborazo at 6,271 m (20,576 ft) and, in all, there are twenty-two peaks with elevations in excess of 4,267 m (14,000 ft). Many are active or dormant volcanoes, and in many places the glow of molten lava can often be seen reflected in the night skies. Cotopaxi, at 5,896 m (19,344 ft) is the loftiest active volcano in the world, and the twin peaks of dormant Pichincha overlook Quito. South of Azuay Province the volcanoes disappear, the mountain chains are lower and less symmetrical, and the path of the Interandean Lane becomes obscured by a more complex mountain pattern.

The irregular pattern of Cotopaxi's eruptions is typical of Ecuador's active volcanoes. Its first recorded activity occurred in 1534. It was then quiescent for two centuries, but its renewed violence in 1742 and 1768 caused severe damage. After another century of rest, the most recent series of eruptions occurred between 1878 and 1880. For the most part, the very loftiness of the volcanoes limits the hazard of their activity, but an 1886 eruption of Tungurahura laid waste the town of Baños, and lava flows that do not themselves reach arable land often cause mud avalanches.

All major rivers rise in the Sierra. They have carved deep trenches that interfere with transportation and limit the amount of land suitable for cultivation. The original soils of the region were derived from weathered lava and volcanic ash, which reached depths of eight or more feet. The depositing of ash is a continuing process that makes frequent sweeping of some Andean streets necessary. The soil is fertile, but its extremely loose and porous composition is nonresistant, and primitive farming methods practiced over centuries have in many instances led to erosion reaching a point of complete elimination of the topsoil.

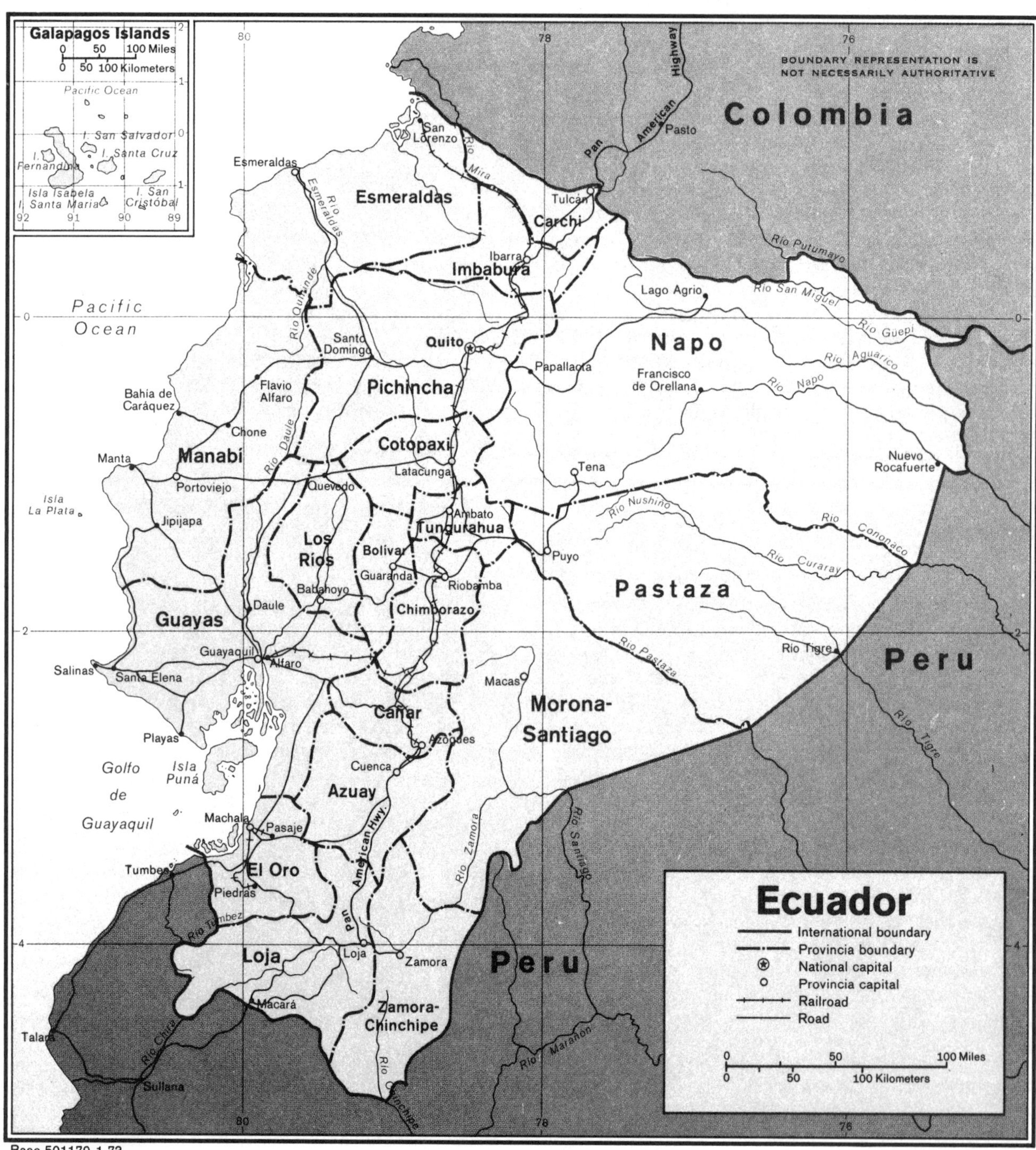

Base 501179 1-73

In localities where the topsoil has been completely eroded, a hard and erosion-resistant clay subsoil has been exposed. In some of the long-established farm communities this material has been brought into marginal production, but much of it remains wasteland, which is enlarged by the runoff from each rain. Frequently, picks and digging sticks are used laboriously to break up the clay subsoil, and night soil and compost are mixed with it to make some production possible.

The Oriente

There is no consensus with respect to the exact elevation on the convoluted eastern slopes of the Andes at which the Oriente region begins. It encompasses 50 percent or more of Ecuador and consists principally of an alternately flat and gently undulating expanse of tropical rain forest. The scanty population consists principally of tribal people, but the discovery of oil near its northern frontier during the 1960s, coupled with the building of roads through the jungle that resulted from this discovery, stimulated what in 1972 gave promise of becoming an important colonization movement.

The tangled and luxuriant jungle growth of the region is supported by moisture and heat rather than by intrinsic fertility of earth that has been leached of its original mineral content. The rich jungle growth is a natural one replenished by the decay of its own organic material. In the few places where slash-and-burn agriculture has been practiced, the demand on the land, coupled with the denial of enrichment provided by natural decay, has ruined the earth in a few planting seasons.

The region is watered by a multitude of rivers and streams, but the low gradient of the terrain after they pour into the Amazon Basin from their sources in the Andes results in generally poor drainage. The heavy forest cover coupled with the intricate topography of the Andean slopes leaves few good access routes to the lowlands. The widest and best is provided by the valley of the Pastaza River, which rises west of the town of Baños in the Sierra. The second major access route follows the valley of the Paute River, which is located farther to the south. The largest of the rivers is the Napo. It originates at the foot of the Andes with the confluence of two smaller streams whose valleys do not provide significant access routes. All Oriente waters eventually find their way to the Atlantic through the Amazon River.

The Galápagos Islands

The fourth and smallest of the geographic regions lies some 1,048 km (650 mi) westward from the mainland. It consists of the Galápagos Islands that make up the Archipiélago de Colón (Columbus Archipelago). The island group covers a span of about 402 km (250 mi) from east to west, and its collective landmass of about 7,769 sq km (3,000 sq mi) is made up of a dozen islands of some size and several hundred islets and reefs. Only five of the islands have permanent populations, and over half of the people live on San Cristóbal (or Chatham) Island, where the town of Puerto Baquerizo is located. The highest elevation on the island is a 1,368 m (4,490-ft) volcanic peak.

The equator passes across the northern flank of the archipelago, but the cool waters of the Peru, or Humboldt, Current modify its temperatures. The same current lessens rainfall, but there is enough precipitation above the 305 m (1,000-ft) level to support tree life and some crops. Fishing is excellent, and the isolation of the island group has permitted the development of unique flora and fauna, which were studied by Charles Darwin as a basis for his theory of natural selection.

PRINCIPAL TOWNS (estimated population at mid-1986)

Town	Population	Town	Population
Guayaquil	1,509,108	Portoviejo	134,393
Quito (capital)	1,093,278	Manta	129,578
Cuenca	193,012	Ambato	122,139
Machala	137,321	Esmeraldas	115,138

Area and population

Regions / Provinces	Capitals	area sq mi	area sq km	population 1986 estimate
Coastal				
El Oro	Machala	2,281	5,908	406,800
Esmeraldas	Esmeraldas	5,854	15,162	297,400
Guayas	Guayaquil	8,256	21,382	2,485,800
Los Rios	Babahoyo	2,459	6,370	533,700
Manabi	Portoviejo	6,990	18,105	1,039,400
Eastern				
Morona-Santiago	Macas	10,200	26,418	85,600
Napo	Tena	20,200	52,318	151,800
Pastaza	Puyo	11,687	30,269	38,500
Zamora-Chinchipe	Zamora	7,102	18,394	59,100
Sierra				
Azuay	Cuenca	3,124	8,092	513,300
Bolívar	Guaranda	1,599	4,142	164,700
Cañar	Azogues	1,344	3,481	198,300
Carchi	Tulcan	1,446	3,744	143,300
Chimborazo	Riobamba	2,338	6,056	369,200
Cotopaxi	Latacunga	2,007	5,198	312,700
Imbabura	Ibarra	1,921	4,976	281,000
Loja	Loja	4,429	11,472	404,000
Pichincha	Quito	6,404	16,587	1,710,300
Tungurahua	Ambato	1,201	3,110	374,300
Island territory				
Galápagos Islands	Puerto Baquerizo Moreno	3,086	7,994	8,000
TOTAL		103,930	269,178	9,647,100

EGYPT

BASIC FACTS

Official Name: Arab Republic of Egypt

Abbreviation: EGT

Area: 997,739 sq km (385,229 sq mi)

Area—World Rank: 29

Population: 53,347,679 (1988) 63,941,000 (2000)

Population—World Rank: 21

Capital: Cairo

Boundaries: 2,689 (1,670 mi); Gaza Strip 11 km (7 mi); Israel 255 km (158 mi); Sudan 1,273 km (790 mi); Libya 1,150 km (714 mi)

Coastline: Mediterranean Sea 957 km (595 mi) Red Sea 1,368 km (850 mi)

Longest Distances: 1,572 km (997 mi) SE-NW 1,196 km (743 mi) NE-SW

Highest Point: Mount Katrina 2,642 m (8,668 ft)

Lowest Point: Qattara Depression 133 m (436 ft) below sea level

Land Use: 3% arable; 2% permanent crops; 95% other

Most of the terrain of Egypt is hot, dry desert, covering almost 96 percent of the country. Over 96 percent of the population, however, find shelter and food in the remaining territory—the long, narrow, fertile Nile Valley and its delta. Two dominant characteristics of life in Egypt—overpopulation and the preeminence of the Nile River (Nahr an Nil)—overshadow all others.

The entire country lies within the wide band of desert that stretches from the Atlantic coast across North Africa to the Red Sea. The topographic channel through which the Nile flows across the Sahara causes an interruption in the desert, and the contrast between the Nile Valley and the rest of the country is abrupt and dramatic.

Natural Regions

The total land area utilized by the people is about 38,850 sq km (15,000 sq mi) (about the size of Switzerland); the remaining unoccupied land is twenty-five times greater in area. Unlike most of North Africa and Arabia, the country has had an extremely disturbed geological history that has produced four major subregions, or distinct physical divisions: the Nile Valley and Nile Delta; the Western Desert; the Eastern Desert and Red Sea Highlands; and the Sinai Peninsula. The region that includes the Nile valley and delta is the most important, since it supports over 96 percent of the people on the country's only cultivable land.

Nile Valley and Delta

The Nile valley and delta (encompassing twenty-one of the twenty-five governorates of Egypt) is the world's most extensive oasis, a result of the second longest river in the world and its seemingly inexhaustible sources. The topographic channel permitting the Nile to flow across the Sahara makes possible the existence of Egypt as something other than a wasteland and desert.

The Nile River extends about 1,609 km (1,000 mi) across Egypt and is a combination of the White Nile, the Blue Nile, and the Atbara streams. The White Nile, furnishing an average 28 percent of the Nile's waters in Egypt, begins in Uganda at Lake Victoria as the Victoria Nile and becomes the White Nile as it enters Sudan. After a drop of almost 610 m (2,000 ft) from Lake Victoria to Juba in southern Sudan, the river passes through a wide, flat area in south-central Sudan where the rate of flow slows almost to stagnation; it descends only 76 m (250 ft) over the 1,609 km (1,000 mi) to Khartoum, the capital of Sudan.

The Blue Nile, stemming from Lake Tana in Ethiopia, has a steeper gradient and forms a confluence with the White Nile at Khartoum, furnishing an average 58 percent of the Egyptian Nile's waters. For some miles downstream from Khartoum, the silt-laden Blue Nile water on the east bank is distinguishable from the clearer water of the White Nile on the west.

The Atbara feeds into the main Nile between the sixth and fifth cataracts below Khartoum. Although it shrinks to a number of pools during the low-water season (January to June), the Atbara provides 22 percent of the river flow in late summer, when the torrential summer rains of the Ethiopian plateau result in floodwaters. The Blue Nile has a similar pattern; it contributes 17 percent in the low-water season and 68 percent during the flood season.

In contrast the continuous flow of the White Nile contributes only 10 percent during the flood season but over 80 percent throughout the low-water period. Thus before the Aswan High Dam was completed, the rains of equatorial Africa flowing through the White Nile watered the Egyptian stretch of the river throughout the year, whereas the seasonal rains on the Ethiopian Highlands through which the Blue Nile flows caused the river to overflow its banks and deposit a layer of fertile mud over adjacent fields. The great flood of the main Nile occurred in Egypt during the months of August, September, and October, although it sometimes began as early as June at Aswan and often did not completely wane until January. There were rare years when the flood hardly occurred.

The Nile enters Egypt near Wadi Halfa, a Sudanese town that was completely rebuilt on high ground after the original site was submerged in the reservoir created by the Awsan High Dam. In fact the river now begins its flow into Egypt and through the valley as Lake Nasser, which extends south from the dam some 322 km (200 mi) to the border and an additional 159 km (99 mi) into Sudan. Lake Nasser's waters fill the area through Lower Nubia (Upper Egypt and northern Sudan), within the narrow groove between the cliffs of sandstone and granite created by the flow of the river over many centuries. Below Aswan the cultivated floodplain strip widens to an average of ten to twelve mi; below Isna (161 km [100 mi] north of Aswan) the plateau on both sides of the valley rises as high as 548 m (1,800 ft) above sea level; at Qena (96 km [60 mi] north of Isna) the 304 m (1,000-ft) limestone cliffs force the river to change course to the southwest for about 64 km (40 mi) before turning northwest for about 161 km (100 mi) to Asyut. From Asyut north the escarpments on both sides diminish, and the valley widens to 22 km (14 mi) at one point. Finally the Nile reaches the delta at Cairo.

From the Sudan border to the sea there is an average gradient of one to 3,962 m (13,000 ft) and an average flow of two to four mi per hour. The river's tendency to hug the east bank has produced a wider cultivable area on

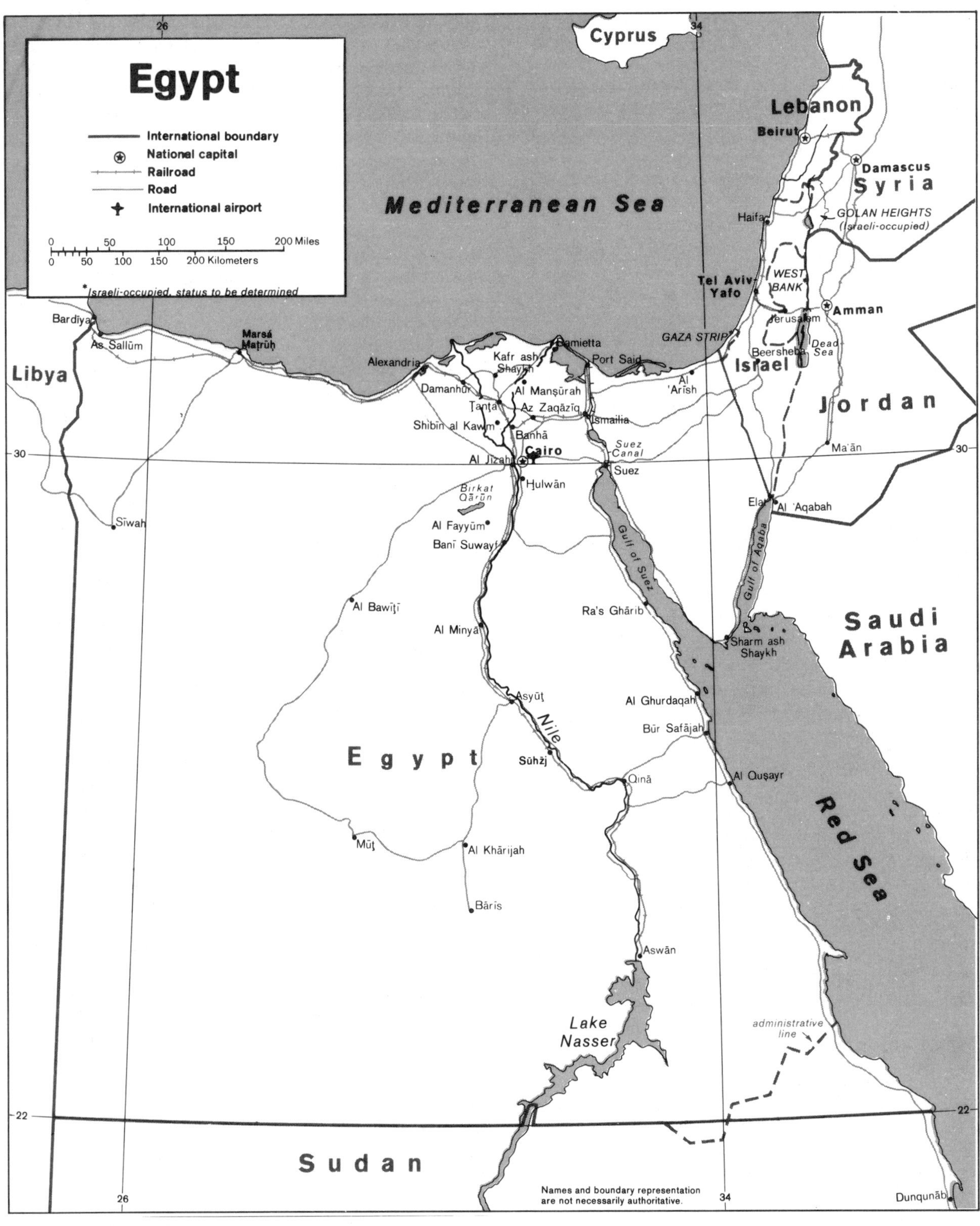

Egypt
International boundary
National capital
Railroad
Road
International airport
0 50 100 150 200 Miles
0 50 100 150 200 Kilometers
*Israeli-occupied, status to be determined
Cyprus
Mediterranean Sea
Lebanon
Beirut
Damascus
Syria
Haifa
GOLAN HEIGHTS (Israeli-occupied)
Tel Aviv-Yafo
WEST BANK
Amman
Jerusalem
GAZA STRIP
Dead Sea
Beersheba
Israel
Jordan
Ma'ān
Libya
Bardīyah
As Sallūm
Marsá Maṭrūḥ
Alexandria
Damietta
Kafr ash Shaykh
Port Said
Damanhūr
Al Manṣūrah
Al 'Arīsh
Ṭanṭā
Az Zaqāzīq
Ismailia
Shibīn al Kawm
Banhā
Suez Canal
Al Jīzah
Cairo
Suez
Ḥulwān
Birkat Qārūn
Elat
Al 'Aqabah
Sīwah
Al Fayyūm
Banī Suwayf
Gulf of Suez
Gulf of Aqaba
Al Bawīṭī
Ra's Ghārib
Saudi Arabia
Al Minyā
Sharm ash Shaykh
Asyūṭ
Al Ghurdaqah
Nile
Būr Safājah
Egypt
Sūhāj
Qinā
Al Quṣayr
Red Sea
Mūṭ
Al Khārijah
Bārīs
Aswān
Lake Nasser
administrative line
Sudan
Names and boundary representation are not necessarily authoritative.
Dunqunāb
26
34
30
22

the west bank. Throughout its length in Egypt no tributary streams enter the Nile.

PRINCIPAL TOWNS (population at census of November 1976, excluding nationals abroad)

El-Qahira (Cairo, the capital)	5,074,016	Asyut	213,751
		Zagazig	202,575
El-Iskandariyah (Alexandria)	2,317,705	Es-Suweis (Suez)	193,965
		Damanhur	170,633
El-Giza	1,230,446	El-Fayoum	166,910
Shoubra el-Kheima	394,223	El-Minya (Menia)	146,366
El-Mahalla el-Koubra	292,114	Kafr ed-Dawar	146,248
Tanta	283,240	Ismailia	145,930
Bur Saʿid (Port Said)	262,760	Aswan	144,654
El-Mansoura	259,387	Beni Suef	117,910

Area and population

Regions **Governorates**	**Capitals**	area sq mi	area sq km	population 1986 estimate
Desert				
al-Baḥr al-Aḥmar	al-Ghurdaqah	78,643	203,685	74,000
Maṭrūḥ	Marsā Maṭrūḥ	81,897	212,112	180,000
Sinā' al-Janūbiyah	aṭ-Ṭūr	12,796	33,140	24,000
Sinā' ash-Shamālīyah	al-ʿArish	10,646	27,574	156,000
al-Wādi al-Jadīd	al-Kharijah	145,369	376,505	118,000
Lower Egypt				
al-Buḥayrah	Damanhūr	3,911	10,130	3,271,000
ad-Daqahliyah	al-Manṣūrah	1,340	3,471	3,540,000
Dumyāṭ	Dumyāṭ	227	589	747,000
al-Gharbiyah	Ṭanṭā	750	1,942	2,904,000
Kafr ash-Shaykh	Kafr ash-Shaykh	1,327	3,437	1,839,000
al-Minūfiyah	Shibin al-Kawm	592	1,532	2,194,000
al-Qalyūbīyah	Banhā	387	1,001	2,262,000
ash-Sharqiyah	az-Zaqāziq	1,614	4,180	3,394,000
Upper Egypt				
Aswān	Aswān	262	679	796,000
Asyūṭ	Asyūṭ	600	1,553	2,222,000
Bani Suwayf	Bani Suwayf	510	1,322	1,452,000
al-Fayyūm	al-Fayyūm	705	1,827	1,527,000
al-Jizah	al-Jizah	32,878	85,153	3,279,000
al-Minyā	al-Minyā	873	2,262	2,746,000
Qinā	Qinā	715	1,851	2,194,000
Sawhāj	Sawhāj	597	1,547	2,490,000
Urban				
Būr Saʿid (Port Said)	—	28	72	382,000
al-Iskandariyah (Alexandria)	—	1,034	2,679	2,893,000
al-Ismāʿīlīyah (Ismailia)	—	557	1,442	484,000
al-Qāhirah (Cairo)	—	83	214	6,325,000
as-Suways (Suez)	—	6,888	17,840	265,000
TOTAL		385,229	997,739	47,758,000

At Cairo the Nile spreads out over what was once a broad estuary that has been filled by riverine deposits to form a fertile delta 250 km (155 mi) wide at the seaward base and about 160 km (100 mi) from north to south. According to reports written in the first century A.D., seven branches of the Nile ran through the delta. Twelfth-century records mention six. Since then nature and man have closed all but two main outlets: the east branch, Damietta (177 km (110 mi) long), and the west branch, Rosetta (111 km (69 mi) long)—both named after the ports at their mouths. These are supplemented by a network of drainage and irrigation canals. In the north near the coast the delta embraces a series of salt marshes and lakes; most notable among them are lakes Maryut, Idku, Burullus, and Manzala.

The value of the Nile in relation to the fertility and the productivity of the adjacent land is largely attributable to the silt deposits laid by the flooding waters. Archaeological research indicates that men once lived at a much higher elevation along the river than they do in the modern period, suggesting that the river was higher or the floods more severe in ancient eras. The precise timing and the amount of annual flow have always been unpredictable. A year's flow has been measured at as low as 42 billion cu ft and as high as 150 billion cu ft. For centuries Egyptians have attempted to predict these factors and to moderate the severity of the floods while taking maximum advantage of their benefits.

Until the recent erection of the Aswan High Dam, the fertility of the Nile Valley was dependent not only on the water that was brought to the arable land but also on the materials brought by the water. Researchers estimate that the deposit of beneficial silt in the valley began about 10,000 years ago. The average deposit of arable soil through the course of the river valley is about twenty-seven ft. The upper part of the valley has less, the section between Qena and Aswan averaging about twenty-two feet, and some places have thirty-seven ft. Analysis of the flow revealed that 10.7 million tons of solid matter passed Cairo each year. Most sediment, but not all, is now obstructed by the Aswan High Dam and retained in Lake Nasser.

The Western Desert

The Western Desert accounts for almost three-fourths of the land area. To the west of the Nile this immense desert spans the area from the Mediterranean south to the Sudanese border. The Gilf Kebir, near the southwest boundary with Libya, has an altitude of over 914 m (3,000 ft), an exception to the uninterrupted territory of basement rocks covered by slight layers of horizontally bedded sediments forming a massive plain or low plateau. A major feature of this plain is the Great Sand Sea, extending from the Siwa (Siwah) Oasis to Gilf Kebir. Scarps (ridges) and deep depressions (basins) are found intermittently in the Western Desert, and there is an absence of drainage lines. The importance of the depressions for occupance makes the Western Desert the foremost region after the Nile valley and delta.

The Egyptian government considers this area a frontier region, dividing it into two frontier governorates at about the twenty-sixth parallel: Marsa Matruh to the north and New Valley to the south. There are seven important depressions in the Western Desert, and all are considered oases except the largest, Qattara, which contains only salt water. Halfway between the Nile and the Libyan border and 80 km (50 mi) from the Mediterranean coast, the Qattara Depression is approximately 18,129 sq km (7,000 sq mi) (about the size of New Jersey) and drops at times to almost 137 m (450 ft) below sea level. Uninhabited, it contains badlands, salt marshes, and lakes, offering little hope for future development.

Limited agriculture, some natural resources, and permanent habitation characterize the remaining six depressions. As oases these depressions have fresh water in sufficient quantities, provided either by the Nile waters or from local groundwater sources. The Siwa Oasis, close to the Libyan border and west of Qattara, is isolated from the rest of the country but has sustained life since ancient times; Herodotus and Alexander the Great both visited the Temple of Amon located there.

The other major oases form a topographic chain extending from the Al Fayyum Oasis, sometimes called the Fayyum Depression—64 km (40 mi) southwest of Cairo—south to the Bahariya (Bahriyah), Farafra (Farafirah), and Dakhla (Dakhilah) oases before reaching the country's largest oasis, Kharga. Around 3,600 years ago a canal was constructed from the Nile to the Al Fayyum Oasis, probably to divert excessive floodwaters there. Over time this has produced an irrigated area of over 1,813 sq km (700 sq mi), enhanced because the basin gently slopes toward a shallow, brackish lake, Birkat Qarun, at its northern reaches.

On the floors of the remaining depressions, water is available to support limited populations. The Bahariya Oasis, primarily important because of iron ore deposits, lies 338 km (210 mi) southwest of Cairo, and the Farafra Oasis, larger but sparsely populated, lies directly south. The Dakhla and Kharga oases complete the chain to the south. These two were the first to be affected by the ambitious New Valley project, which proposes the watering of the entire oasis belt of depressions and vast land reclamation.

The Eastern Desert and Red Sea Highlands

The topographic features of the region east of the Nile are quite dissimilar to those of the Western Desert. The Eastern Desert rises abruptly from the Nile. The upward-sloping plateau of sand gives way within 80 to 128 km (50 to 80 mi) to arid, defoliated, rocky hills running north and south between the Sudan border and the delta and reaching elevations of 2,133 m (7,000 ft). This region's most prominent feature is the easterly chain of rugged mountains, the Red Sea Highlands, extending from the Nile Valley eastward to the Gulf of Suez and the Red Sea. This elevated region has a natural drainage pattern that, because of insufficient rainfall, rarely functions; a complex of irregular, sharply cut wadis, or dry streambeds, extend westward toward the Nile.

The region is considerably isolated from the rest of the country and, because of the difficulty in sustaining any form of agriculture, no initiative has been taken toward oasis cultivation. Without permanent settlements, except for a few villages on the Red Sea coast, the importance of this region to Egypt lies in its natural resources. The entire Eastern Desert is administered under a single governorate, Red Sea, with its capital at Hurghada (Al Ghurdaqah) on the Red Sea.

Sinai Peninsula

Another frontier governorate, Sinai Peninsula, is a triangular-shaped area of 61,124 sq km (23,000 sq mi), situated east of the Gulf of Suez and northeast of the Eastern Desert. Closely akin to the desert, it contains mountains in its southern sector that are a geological extension of the Red Sea Hills, the low range along the Red Sea coast that includes the highest point in the country, Mount Catherine (Jabel Katrinah) at 2,642 m (8,668 ft), adjacent to Mount Sinai. These mountains, red in color, give the Red Sea its name.

The southern side of the peninsula has a sharp escarpment that subsides after a narrow coastal shelf into the Red Sea and the Gulf of Aqaba. The land tilts downward from this area to the north toward the coast of the Mediterranean. This sloping limestone plateau begins at an elevation of over 914 m (3,000 ft), occupies about two-thirds of the peninsula, and ends in a flat, sandy coastal plain. This flat area bends north with the eastern shore of the Mediterranean from the Suez Canal area without interruption into the lowlands of western Israel.

EL SALVADOR

BASIC FACTS

Official Name: Republic of El Salvador

Abbreviation: ESR

Area: 21,041 sq km (8,124 sq mi)

Area—World Rank: 134

Population: 5,388,644 (1988) 6,717,000 (2000)

Population—World Rank: 93

Capital: San Salvador

Boundaries: 842 km (523 mi); Honduras 335 km (208 mi); Guatemala 203 km (126 mi)

Coastline: 304 km (189 mi)

Longest Distances: 269 km (167 mi) WNW-ESE; 106 km (66 mi) NNE-SSW

Highest Point: Mount El Pital 2,730 m (8,957 ft)

Lowest Point: Sea level

Land Use: 27% arable; 8% permanent crops; 29% meadows and pastures; 6% forest and woodland; 30% other

El Salvador is located on the southern slopes of the Central American cordillera in the isthmus between Mexico and Panama. The smallest of the American republics, it is also the smallest mainland nation of the Western Hemisphere. It is the only country in the isthmus without a coastline on the Caribbean Sea. Its varied terrain ranges from tropical lowlands on the Pacific coastal plain to arid semidesert in the mountainous region of the north.

Ninety percent of the land is of volcanic origin, but there are no very high peaks. Santa Ana volcano, in the mountains of the northwest, is the loftiest, rising to a height of 2,377 m (7,800 ft) and there are four active volcanoes as well as a number of extinct ones.

Physical Zones

The mountain ranges run east and west across the country and roughly divide it into three distinct physical zones, usually characterized by their general climatic conditions, based primarily on elevation. The southern coastal plain is called the torrid land (*tierra caliente*). It consists of the narrow, relatively flat coastal belt extending the length of the country from the Guatemalan border on the west to the Gulf of Fonseca on the east. It is an area of extensive agriculture, with some developing industry and fishing, particularly in the vicinity of Acajutla in the west, one of the country's most rapidly growing ports. There are two other important harbors, La Libertad, which serves as the main port for the capital, and La Unión on the Bay of La Unión off the Gulf of Fonseca.

The southern mountains, called the Coastal Range, form the northern demarcation of the torrid zone, which is bounded by the Pacific Ocean on the south. As the mountains extend toward the sea, they leave a narrow strip of fertile soil, which is edged mostly by sandy beaches and estuaries but in some areas ends in steep cliffs and promontories set off by rugged mountains and volcanoes. The coastal strip averages about ten miles in width, with a maximum of about twenty miles around the Gulf of Fonseca. The country's principal watercourse, the Lempa River, which rises in the mountains of Guatemala to the northwest, empties into the Pacific near the central point of the coast after flowing over 128 km (80 mi) eastward before turning south to traverse the entire country.

The northernmost latitudinal zone is known as the cold land (*tierra fria*). It comprises the northern lowlands formed by the wide valley of the Lempa River and the northern east-west range, the Sierra Madre, which extends to the Honduran border. The designation of cold land is not very appropriate, as the area is predominantly warm along the river valley and has a pleasant, temperate climate on the upper mountain slopes. It was once heavily covered with forest, but overexploitation and damaging agricultural practices have eroded its potential, and it has become arid and semibarren. As a result, it is the country's most sparsely populated zone, and there is little farming or other development of the area.

Between the two mountain-demarcated zones lie the central highlands called the temperate land (*tierra templada*), which includes most of the national territory. It is also known as the central valley but is actually a plateau that runs the length of the country between the two mountain ranges to the north and south. The plateau averages some 609 m (2,000 ft) in elevation, and its rolling country is interspersed with mountains and volcanoes. The capital, San Salvador, and most cities and sizable towns are located in its lush valleys or rich volcanic soil, giving it a concentration of the country's agricultural and industrial activities, as well as its population.

Administrative Zone

In addition to the division of the country into physical zones dictated by geographic considerations, it is also commonly divided into three administrative zones, which reflect economic and political factors. These are designated the Western, Central, and Eastern Regions and are made up by grouping together a varying number of departments, as the country's provinces are called. The regional boundaries run generally north and south, and natural features are used where possible. The Lempa River forms the border between the Western and Central Regions in the northwest and constitutes the entire border between the Central and Eastern Regions.

The Western Region is the smallest of the three and is made up of only three departments, constituting 22 percent of the national area. It includes the department of Santa Ana, which occupies an extensive fertile valley that is the richest coffee-growing area in the country. The region also has the nation's most active port, Acajutla. There are two major volcanoes in the region, one of them being the country's highest peak, the Santa Ana, and the other, Izalco, the largest active volcano.

The Central Region is the largest, most densely populated, and most economically active. Making up 42 percent of the land area, it comprises seven departments, including the department of San Salvador, in which the national capital is situated. Almost three-fourths of the nation's industrial output and approximately one-third of its coffee are produced in the region. As throughout the country, there are a number of volcanoes in the area.

The Eastern Region, with four departments, makes up 36 percent of the national territory. Divided into a tropical coastal strip and an inland plateau, the region is almost entirely agricultural. It produces 12 percent of the coffee but less than 16 percent of the nation's industrial output. It is the largest cotton-producing area.

All three regions share the mountain ranges, coastal strip, and central plateau that traverse them from west to east with minor, but dramatic, changes in structure and contour. The overall impression of the country is one of great natural beauty, with volcanoes, lakes, and fields of rich brown soil. Nowhere in the country are the ubiquitous volcanoes lost to view and, along with sloping hills green with coffee bushes and giant shade trees, they offer a varied panorama of striking natural contrasts.

As its entire territory lies within the Central American volcanic axis, El Salvador is called a land of volcanoes. Southeast from Guatemala the volcanic range decreases in elevation, and Salvadorean peaks are generally somewhat smaller than those in neighboring countries. Between the composite volcanic cones that are widely dispersed throughout the area are low, flat alluvial basins and rolling hills eroded from ash deposits that constitute a highly fertile volcanic soil. In addition to the four active volcanoes,

El Salvador
International boundary
Department boundary
National capital
Department capital
Railroad
Road
0 10 20 30 Kilometers
0 10 20 30 Miles
GUATEMALA
HONDURAS
NICARAGUA
Pacific Ocean
Golfo de Fonseca
AHUACHAPÁN
SANTA ANA
SONSONATE
CHALATENANGO
LA LIBERTAD
SAN SALVADOR
CUSCATLÁN
CABAÑAS
SAN VICENTE
LA PAZ
USULUTÁN
SAN MIGUEL
MORAZÁN
LA UNIÓN
Ahuachapán
La Hachadura
Santa Cruz
Chalchuapa
Lago de Coatepeque
Santa Ana
Texistepeque
Metapán
Citalá
Lago de Güija
Río Lempa
Río Paz
Sonsonate
Acajutla
Armenia
Nueva San Salvador
La Libertad
Quezaltepeque
Aguilares
Apopa
San Salvador
Suchitoto
Tejutla
Embalse Cerrón Grande
Chalatenango
Río Sumpul
Victoria
Sensuntepeque
Ilobasco
Cojutepeque
Lago de Ilopango
Río Jiboa
Comalapa
Zacatecoluca
La Herradura
San Vicente
El Triunfo
Usulután
La Canoa
Puerto El Triunfo
Río Grande de San Miguel
Laguna de Olomega
Intipucá
El Cuco
Sesori
San Francisco
Río Torola
San Miguel
El Carmen
La Unión
Lislique
Santa Rosa de Lima
Río Goascorán
El Progreso
Chiquimula
Jalapa
Ipala
Esquipulas
Monjas
Río Ostúa
Asunción Mita
Jutiapa
Nueva Ocotepeque
Cucuyagua
Santa Rosa de Copán
Río Mejocote
Gracias
San Sebastián
San Juan
Río Guarajambala
Marcala
Lepaterique
Nacaome
San Lorenzo
90
89
88
14
13
Boundary representation is not necessarily authoritative.
Base 504510 8-80 (545437)

there are many fumarole and geyser fields marked by latent volcanism, with steam jets and hot springs issuing in numerous places.

Principal Towns 1985

San Salvador	459,902
Santa Ana	137,879

The land is frequently shaken by earthquakes, some of which cause extensive damage. The capital city was completely destroyed by an earthquake in 1854, and in 1919 it was struck by another, which did extensive damage but mostly to surrounding farmlands. Since that time, earthquake activity has been relatively minor, although a fairly serious tremor in 1965 again caused limited destruction in San Salvador. The Izalco volcano has been relatively inactive since 1957, but it still flares periodically in picturesque activity. For 150 years it maintained a regularity of eruption that earned it the name of "lighthouse of the Pacific." Its brilliant flares were clearly visible at sea, and at night its flowing lava turned it into a brilliant incandescent cone.

Drainage

The country's single river of any significance, the Lempa, connects with an irregular network of some 150 streams and minor watercourses that are of limited value for power, fishing, or local transportation. All the streams empty into the Pacific Ocean but are navigable only for short distances. Many streams flow down the southern slopes of the Coastal Range directly into the Pacific, whereas some course down the northern slopes to join others draining the cordillera and flow eventually into the Lempa.

Area and population

Departments	Capitals	area sq mi	area sq km	population 1985 estimate
Ahuachapán	Ahuachapán	479	1,240	271,990
Cabañas	Sensuntepeque	426	1,104	199,229
Chalatenango	Chalatenango	779	2,017	256,688
Cuscatlán	Cojutepeque	292	756	222,389
La Libertad	Nueva San Salvador	638	1,653	440,030
La Paz	Zacatecoluca	473	1,224	278,719
La Unión	La Unión	801	2,074	346,087
Morazán	San Francisco (Gotera)	559	1,447	235,632
San Miguel	San Miguel	802	2,077	480,486
San Salvador	San Salvador	342	886	1,094,249
Santa Ana	Santa Ana	781	2,023	490,367
San Vicente	San Vicente	457	1,184	220,630
Sonsonate	Sonsonate	473	1,226	364,075
Usulután	Usulután	822	2,130	437,325
TOTAL		8,124	21,041	5,337,896

The 257 km (160-mi) Lempa is the largest river in Central America. It rises in Guatemala, crosses a corner of Honduras, and cuts across the northern mountains before turning eastward for over 129 km (80 mi). It then turns south and flows for 105 km (65 mi), crossing the southern range, to empty into the Pacific. It drains a total of approximately 49,210 sq km (19,000 sq mi), of which almost 28,490 (11,000) are in El Salvador. The river is navigable for only short, disconnected stretches, the longest up to 40 km (25 mi) for shallow-draft vessels, but it serves as the nation's major source of electric power.

Two other less important river systems drain small portions of the west and east. The Paz River on the Guatemalan border drains the coastal plain in its western extremities, and the San Miguel River does the same in the east. There are several large lakes and lagoons, the principal ones being Lake Ilopango near the capital and Lake Guija on the border with Guatemala. Many other small lakes, lagoons, and sulfur springs dot the countryside. A few have been made into popular resort areas, and there is some sport fishing, but most commercial fishing is done in offshore waters.

EQUATORIAL GUINEA

BASIC FACTS

Official Name: Republic of Equatorial Guinea

Abbreviation: EQG

Area: 28,051 sq km (10,831 sq mi)

Area—World Rank: 127

Population: 346,839 (1988) 445,000 (2000)

Population—World Rank: 153

Capital: Malabo

Boundaries: 910 km (566 mi); Cameroon 183 km (114 mi); Gabon 386 km (240 mi)

Coastline: Rio Muni 167 km (104 mi); Bioko 174 km (108 mi)

Longest Distances: Rio Muni 248 km (154 mi) ENE-WSW; 167 km (104 mi) SSE-NNW; Bioko 74 km (46 mi) NE-SW; 37 km (23 mi) SE-NW

Highest Point: Santa Isabel Peak 3,808 m (9,868 ft)

Lowest Point: Sea level

Land Use: 5% arable; 4% permanent crops; 4% meadows and pastures; 61% forest and woodland; 26% other

Equatorial Guinea is located on the west coast of West-Central Africa and consists of a mainland province and five islands: Bioko (formerly Macias Nguema Biyogo), Fernando Po, Pigalu (formerly, Annabon), Elobey Grande, Elobey Chico and Corisco. Bioko is located 32 km (20 mi) from the coast of Cameroon, and Pigalu is located 350 km (220 mi) from mainland Gabon. Corisco and the Elobey Islands are off the SW coast of Bioko close to the Gabonese coast.

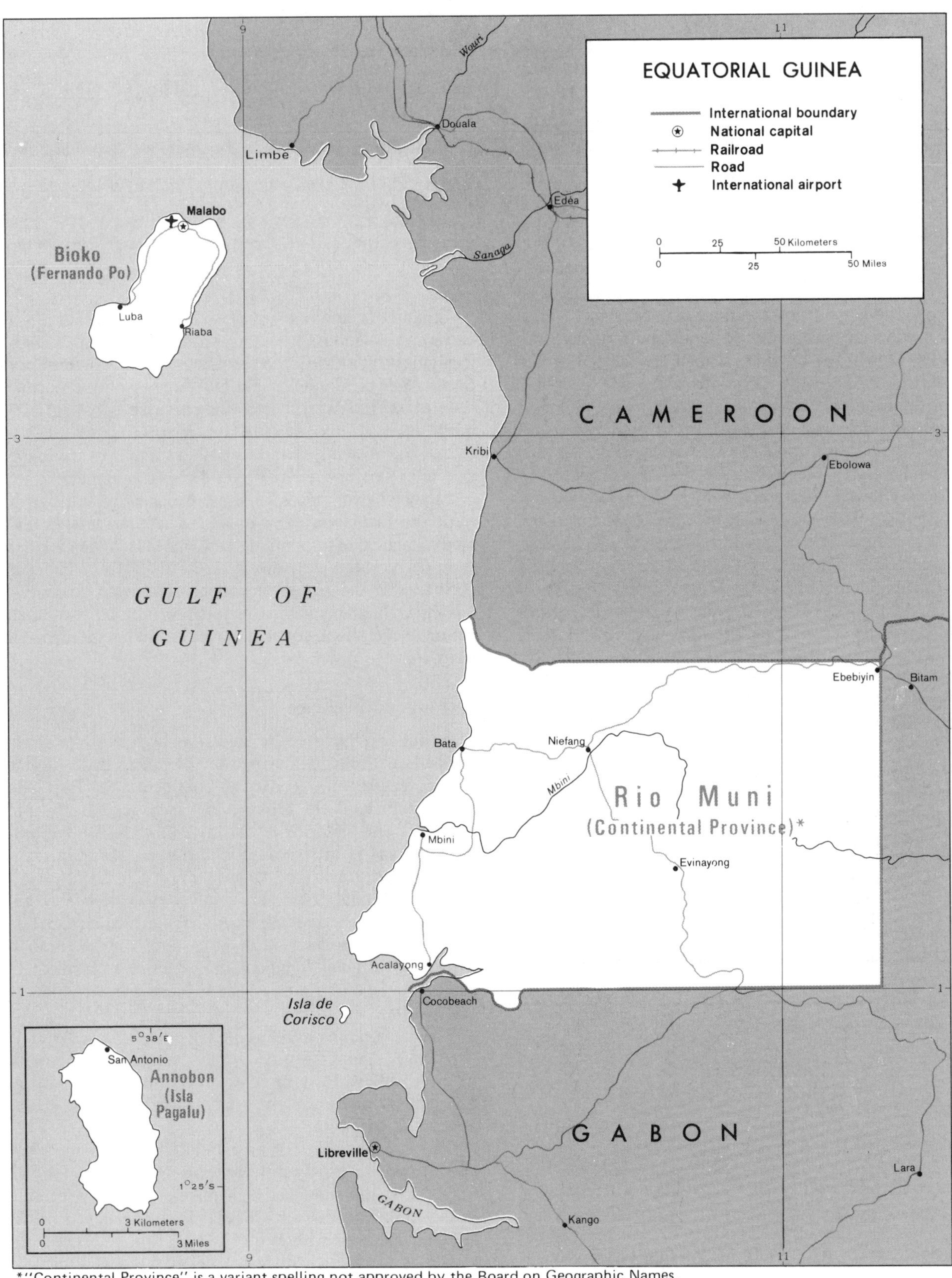

*"Continental Province" is a variant spelling not approved by the Board on Geographic Names.

Principal towns (population at 1983 census): Malabo (capital) 15,253, Bata 24,100.

Area and population	area		population
Regions / **Provinces**	sq mi	sq km	1983 census
Insular	785	2,034	59,196
Annobón	7	17	2,006
Bioko Norte	299	776	46,221
Bioko Sur	479	1,241	10,969
Continental	10,045	26,017	240,804
Centro-Sur	3,834	9,931	52,393
Kie-Ntem	1,523	3,943	70,202
Litoral	2,573	6,665	56,370
Wele-Nzas	2,115	5,478	51,839
TOTAL	10,830	28,051	300,000

Bioko is the largest island in the Gulf of Guinea. It has two large volcanic formations separated by a valley that bisects the island. The coastline is high and rugged in the south but lower and more accessible in the north. In the north of the island is Mount Malabo (3,007 m, 9,865 ft). In the center is the Pico de Moka with an alpine environment. In the south is Gran Caldera. Rio Muni on the African mainland is a jungle enclave with a coastal plain rising steeply toward the Gabon border. In the interior the plain gives way to a succession of valleys separated by low hills and spurs of the Crystal Mountains. The highest peaks are Monte Chocolate (1,100 m, 3,609 ft), the Piedra de Mzas, Monte Mitra and Monte Chime, rising to 1,200 m (3,937 ft). Corisco covering 15 sq km (6 sq mi) and the Great and Little Elobeys, each about 2.5 sq km (1 sq mi), are volcanic islands.

The main rivers are the Mbini (formerly Rio Benito), Rio Campo and Rio Muni. The Mbini, which divides mainland Rio Muni into two, is not navigable except for a 20-km (12-mi) stretch. The Kye, a tributary of the Rio Campo, forms the eastern border with Gabon. The Rio Muni is not properly a river at all but an estuary of several rivers among which the Utamboni is the most notable. The islands have only storm arroyos and small cascading rivers.

ETHIOPIA

BASIC FACTS

Official Name: Ethiopia

Abbreviation: ETH

Area: 1,223,500 sq km (472,400 sq mi)

Area—World Rank: 23

Population: 48,264,570 (1988) 66,509,000 (2000)

Population—World Rank: 22

Capital: Addis Ababa

Boundaries: 6,133 km (3,811 mi); Djibouti 426 km (265 mi); Somalia 1,645 km (1,022 mi); Kenya 785 km (488 mi); Sudan 2,266 km (1,408 mi)

Coastline: 1,011 km (628 mi)

Longest Distances: 1,639 km (1,018 mi) E-W; 1,577 km (980 mi) N-S

Highest Point: Ras Dashen Mountain 4,620 m (15,158 ft)

Lowest Point: Lake Asale 116 m (381 ft) below sea level

Land Use: 12% arable; 1% permanent crops; 41% meadows and pastures; 24% forest and woodland; 22% other

Ethiopia occupies the major portion of the Horn—the easternmost landmass of the African continent. Roughly triangular in shape, the country shares frontiers with Sudan, Kenya, Somalia, and Djibouti. From the northern apex at 18° north latitude to the southern border at 3°30′ north latitude, the landmass measures roughly 932 mi (1,500 kilometers), a distance about equal to its longest east-west axis between 33° and 48° east longitude.

Major physiographic features, which offer some of the most spectacular scenery in Africa, are a massive highland complex of mountains and plateaus divided by the deep Great Rift Valley and a series of lowlands along the periphery of the higher elevations. The wide diversity of terrain is fundamental to regional variations in climate, natural vegetation, soil composition, and settlement patterns.

Terrain and Drainage

About two-thirds of the land rises high over the coastal lowlands to form part of one of the most clearly defined terrain features of the African continent—the East African Rift Plateau. In Ethiopia it has a general elevation ranging from 1,500 to 3,000 m (932 to 1,864 ft) above sea level and is interspersed with higher mountain ranges and cratered cones, the highest of which is the 462 m (1,515 ft) Mount Rasdajan in the northeastern corner of Gonder administrative region. The northwestern part of the dissected high plateau is the historic core of Ethiopia, where the original kingdom of Aksum was formed; the national capital of Addis Ababa is located in its southeastern section.

Millennia of erosion have produced steep-sided valleys, in places 1.6 km (1 mi) deep and several kilometers wide. In these valleys flow rapid waters unsuitable for navigation but adequate as potential sources of hydroelectric power and irrigation.

The highland that comprises much of the country consists of two essentially different physiographic regions: the rugged and mountainous high plateau—often referred to as the Ethiopian Plateau—bisected by the Great Rift Valley and, merging with it in the east, a more level region known as the Somali Plateau.

Northward from Addis Ababa the high plateau includes six of the country's fourteen administrative regions and part of a seventh—Eritrea. It inclines slightly toward the

Ethiopia
International boundary
Province boundary
National capital
Province capital
Railroad
Road
International airport
0 75 150 Kilometers
0 75 150 Miles
SAUDI ARABIA
Red Sea
Port Sudan
Atbarah
Nahr Atbarah
Nile
Khartoum
Kassala
Blue Nile
White Nile
Sannar
SUDAN
Ar Ruşayriş
ERITREA
Āk'ordat
Mits'iwa
Āsmera
Teseney
Tekezē
Adwa
Mek'elē
TIGRAY
GONDER
Gonder
T'ana Hāyk
WELO
Desē
GOJAM
Debre Mark'os
Blue Nile
WELEGA
Nek'emtē
SHEWA
Addis Ababa
Aswah Wenz
Dirē Dawa
Hārer
Jijiga
HĀRERGĒ
Nazret
ĀRSĪ
Āsela
Baro Wenz
Metu
Akobo
ĪLUBABOR
Jima
Bonga
KEFA
Omo
Awasa
Goba
BALĒ
OGĀDĒN REGION
Wābē Shebelē Wenz
Administrative line
Arba Minch'
GAMO GOFA
Kibre Mengist
SĪDAMO
Dawa
Dolo
Luuq
Beled Weyne
Lake Rudolf
UGANDA
Marsabit
KENYA
Webi Ganaane
Webi Shabeelle
Mogadishu
INDIAN OCEAN
YEMEN (SANAA)
Ta'izz
YEMEN (ADEN)
Aden
Aseb
Bab el Mandeb
DJIBOUTI
Djibouti
Gulf of Aden
Berbera
Hargeysa
SOMALIA
20
12
4
36
44
Boundary representation is not necessarily authoritative

519668 6-80

west and northwest, then abruptly descends near the boundary with Sudan. Its towering mountains and deep chasms provide a wide variety of physiography, climate, and natural vegetation. The plateau is marked by such mountain ranges as the Chercher, Aranna, and Chelalo. Given the rugged nature of these massifs and the surrounding tableland, foreigners receive a false impression of the country's topography when Ethiopians refer to the landform as a plateau. Few of these peaks' surfaces are flat except for a scattering of level-topped mountains known to Ethiopians as *ambas.*

South of Addis Ababa the plateau is also rugged, but its elevation is slightly lower than in its northwestern section. The eastern segment beyond the Great Rift Valley exhibits characteristics almost identical to its western counterpart.

Toward the southeast, beyond the continuation of the high plateau and its Ahmar and Mendebo mountain ranges, lies the Somali Plateau. This flat, arid, and rocky semidesert is sparsely populated.

Some geographers—particularly those in Ethiopia—consider the Great Rift Valley a third physiographic region. This most extensive fault on the earth's surface extends from the Middle East's Jordan Valley to the Shire tributary of the Zambezi River in Mozambique. The vast segment that runs through the center of Ethiopia is marked in the north by the Denakil Depression and the coastal lowlands or Afar Plain, as they are sometimes known. To the south, at approximately 9° north latitude, the rift becomes a deep trench slicing through the high plateau from north to south, its varying width averaging 48 km (30 mi). The southern half of the Ethiopian segment of the valley is dotted by a chain of relatively large lakes. Some are fresh water, fed by small streams from the east; others contain various salts and minerals.

Tropical lowlands that form the periphery of the high plateau, particularly in the far north and along the western frontier, provide marked contrast to the upland terrain. In the north the Great Rift Valley broadens into funnel-shaped saline plain. The Denakil Depression, a large triangular-shaped basin that in places is 115 m (377 ft) below sea level, is said to be one of the hottest places on earth. North of the depression the maritime hills, also called the Denakil Alps, border a hot, arid, and treeless coastal strip of land 16 to 80 km (10 to 50 mi) wide. These coastal hills drain inland into the saline lakes from which commercial salt is extracted. Along the coast of the Red Sea, mainly opposite the Buri Peninsula, are the small and sparsely inhabited Dahlac Islands.

In contrast to the steep scarps of the plateau along the Great Rift Valley and in the north, the western and southwestern slopes descend somewhat less abruptly and are broken more often by the river exits. Between the high plateau and the Sudanese border in the west lies a narrow strip of sparsely populated, tropical lowland that belongs politically to Ethiopia but whose people are related to those in Sudan.

The existence of small volcanoes, hot springs, and many deep gorges indicates that large segments of the landmass are still geologically unstable. A number of volcanoes occur in the Denakil area, and hot springs and steaming fissures are found in other northern areas of the Great Rift Valley. A line of seismic belts extends along the length of Eritrea and the Denakil Depression, but serious earthquakes have not been recorded in the area during the twentieth century.

All of the country's rivers originate in the highlands and flow outward in many directions through deep gorges. Because of the general slope of the highlands, a number of the larger rivers are tributaries of the Nile system. Most notable of these is the Blue Nile (Abay), the country's largest river; it and its tributaries account for two-thirds of the Nile River flow below Khartoum in Sudan. The system drains an extensive area of the central portion of the high plateau. The Blue Nile, the Tekeze, and the Baro account for about half of the outflow of water from the country. The Blue Nile has its source in Laka Tana, the country's largest lake, which lies in a depression in the west-central section of the plateau.

In the northern half of the Great Rift Valley between steep cliffs descending several thousand feet flows the Awash River, originating some 80 km (50 mi) west of Addis Ababa. Coursing northward, it is joined by several tributaries until it becomes a river of major importance, only to disappear in the saline lakes of the Denakil Depression.

In the southeast portions of the Somali Plateau, seasonally torrential rivers provide drainage toward the southeast. Chief of these is the Wabi Shebele, which forms the regional border between Bale and Harerge administrative regions and flows into Somalia. In the southern part of the area, many small streams combine to form the Wabi Gestro, which also crosses into Somalia, where it becomes the Juba.

PRINCIPAL TOWNS (population at 1984 census)

Town	Population	Town	Population
Addis Ababa (capital)	1,412,577	Dessie	68,848
Asmara	275,385	Harar	62,160
Dire Dawa	98,104	Mekele	61,583
Gondar (incl. Azeso)	80,886	Jimma	60,992
Nazret	76,284	Bahir Dar	54,800
		Akaki	54,146
		Debre Zeit	51,143

Area and Population

Regions	Capitals	area: sq mi	area: sq km	population: 1987 estimate
Arsi	Asela	9,500	24,600	1,808,512
Bale	Goba	49,500	128,300	1,095,129
Eritrea	Asmera	45,300	117,400	2,951,080
Gemu Gofa	Arba Minch	15,400	40,100	1,356,687
Gojam	Debre Markos	24,900	64,400	3,530,540
Gonder	Gonder	28,300	73,400	3,178,692
Hararge	Harer	98,400	254,800	4,527,423
Ilubabor	Metu	19,600	50,800	1,048,293
Kefa	Jima	20,500	53,000	2,664,557
Shewa	Addis Ababa	33,000	85,500	10,394,448
Sidamo	Awasa	45,100	116,700	4,123,352
Tigray	Mekele	25,400	65,700	2,624,362
Welega	Nekemte	27,000	69,800	2,693,623
Welo	Dese	30,500	79,000	3,962,018
TOTAL		472,400	1,223,500	45,958,716

FALKLAND ISLANDS

BASIC FACTS

Official Name: Colony of the Falkland Islands

Abbreviation: FLK

Area: 12,173 sq km (4,700 sq mi)

Area—World Rank: 142

Population: 1,821 (1988) 2,000 (2000)

Population—World Rank: 210

Capital: Stanley

Land Boundaries: Nil

Coastline: 1,288 km (800 mi)

Longest Distances: N/A

Highest Point: Mount Usborne 705 m (2,312 ft)

Lowest Point: Sea level

Land Use: 99% meadows and pastures; 1% other

The Falkland Islands comprise two large islands—East Falkland and West Falkland—and about 200 smaller ones in the southwestern Atlantic Ocean, about 770 km (480 mi) northeast of Cape Horn. The terrain is rocky with some boggy undulating plains and hilly moorlands. Some smaller islands are volcanically active. The coastline is deeply indented.

FAROE ISLANDS

BASIC FACTS

Official Name: Faroe Islands

Abbreviation: FAR

Area: 1,399 sq km (540 sq mi)

Area—World Rank: 160

Population: 46,853 (1988) 53,000 (2000)

Population—World Rank: 188

Capital: Torshavn

Land Boundaries: Nil

Coastline: 764 km (417 mi)

Longest Distances: 120 km (74 mi) N-S; 79 km (49 mi) NE-SW

Highest Point: Slættaratindur Eysturoy 882 m (2,894 ft)

Lowest Point: Sea level

Land Use: 2% arable; 98% other

The Faroe Islands are in the Atlantic, six degrees to the northwest of Denmark. Of the Faroes' 19 islands, 18 are inhabited. Among the larger islands are Strømø (374 sq km; 174 sq mi), Østerø (266 sq km; 110 sq mi), Vågø (178 sq km; 69 sq mi), Syderø (153 sq km, 59 sq mi) and

Sandø (114 sq km; 44 sq mi). The maximum length of the islands is 120 km (70 mi) north to south and 79 km (49 mi) northeast to southwest.

The Faroe landscape is characterized by a stratified series of basalt sheets, with intervening thinner layers of solidified volcanic ash (tufa). Glacial action has carved the valleys into trough-shaped hollows and formed steep peaks, the highest being Slaettaretindur (882 m.; 2,894 ft) on Østerø. Millions of seabirds nest on the rocky coastal ledges.

FIJI

BASIC FACTS

Official Name: Fiji

Abbreviation: FIJ

Area: 18,274 sq km (7,056 sq mi)

Area—World Rank: 137

Population: 740,761 (1988) 936,000 (2000)

Population—World Rank: 137

Capital: Suva

Land Boundaries: Nil

Coastline: 1,129 km (701 mi)

Longest Distances: (Excluding Rotuma) 595 km (370 mi) SE-NW; 454 km (282 mi) NE-SW

Highest Point: Tomanivi 1,323 m (4,341 ft)

Lowest Point: Sea level

Land Use: 8% arable; 5% permanent crops; 3% meadows and pastures; 65% forest and woodland; 19% other

Fiji is located in the South Pacific about 2,735 km (1,700 mi) NE of Sydney, 1,769 km (1,100 mi) N of Auckland, and 4,466 km (2,776 mi) SW of Honolulu. The country consists of over 822 islands stretched over a total area of 647,497 sq km (250,000 sq mi).

Only 105 of the islands are inhabited; over 500 are islets and some are mere rocks a few meters in circumference. The largest islands are Vanua Levu, with 5,535 sq km (2,137 sq mi), and Viti Levu, with 10,386 sq km (4,010 sq mi).

Fiji's larger islands are mountainous and of volcanic origin, often rising precipitously from the shore. On the southeastern windward sides the islands are covered with dense tropical forests. The highest elevation is Mt. Victoria (1,323 meters, 4,341 ft); there are 28 other peaks over 900 meters (3,000 ft). Most islands are surrounded by coral reefs.

The major river is the Rewa on Viti Levu, which is navigable by small boats for 113 km (70 mi).

Area and population		area		population
Divisions Provinces	Capitals	sq mi	sq km	1986 census
Central	Suva			
Naitasiri		643	1,666	100,227
Namosi		220	570	4,836
Rewa		105	272	97,442
Serua		320	830	13,356
Tailevu		369	955	44,249
Eastern	Levuka			
Kandavu		185	478	9,805
Lau		188	487	14,203
Lomaiviti		159	411	16,066
Rotuma		18	46	2,688
Northern	Labasa			
Mathuata		774	2,004	74,735
Mbua		532	1,379	13,986
Thakaundrove		1,087	2,816	40,433
Western	Lautoka			
Mba		1,017	2,634	197,633
Nandronga-Navosa		921	2,385	54,431
Ra		518	1,341	31,285
TOTAL		7,056	18,274	715,375

FINLAND

BASIC FACTS

Official Name: Republic of Finland

Abbreviation: FIN

Area: 338,145 sq km (130,559 sq mi)

Area—World Rank: 58

Population: 4,949,716 (1988) 5,255,000 (2000)

Population—World Rank: 94

Capital: Helsinki

Boundaries: 3,671 km (2,282 mi); USSR 1,269 km (789 mi); Sweden 586 km (364 mi); Norway 716 km (445 mi)

Coastline: 1,100 km (684 mi)

Longest Distances: 1,165 km (724 mi) N-S; 542 km (337 mi) E-W

Highest Point: Haltia Mountain 1,328 m (4,357 ft)

Lowest Point: Sea level

Land Use: 8% arable; 76% forest and woodland; 16% other

Finland is a far northern country on the European continent. One-third of its land area lies above the Arctic Circle.

The Finnish landscape is characterized by a rather asymmetric distribution of hills and plains, with the highest elevation (1,324 m; 4,344 ft) in the extreme Northwest. Most of the higher landforms, referred to as

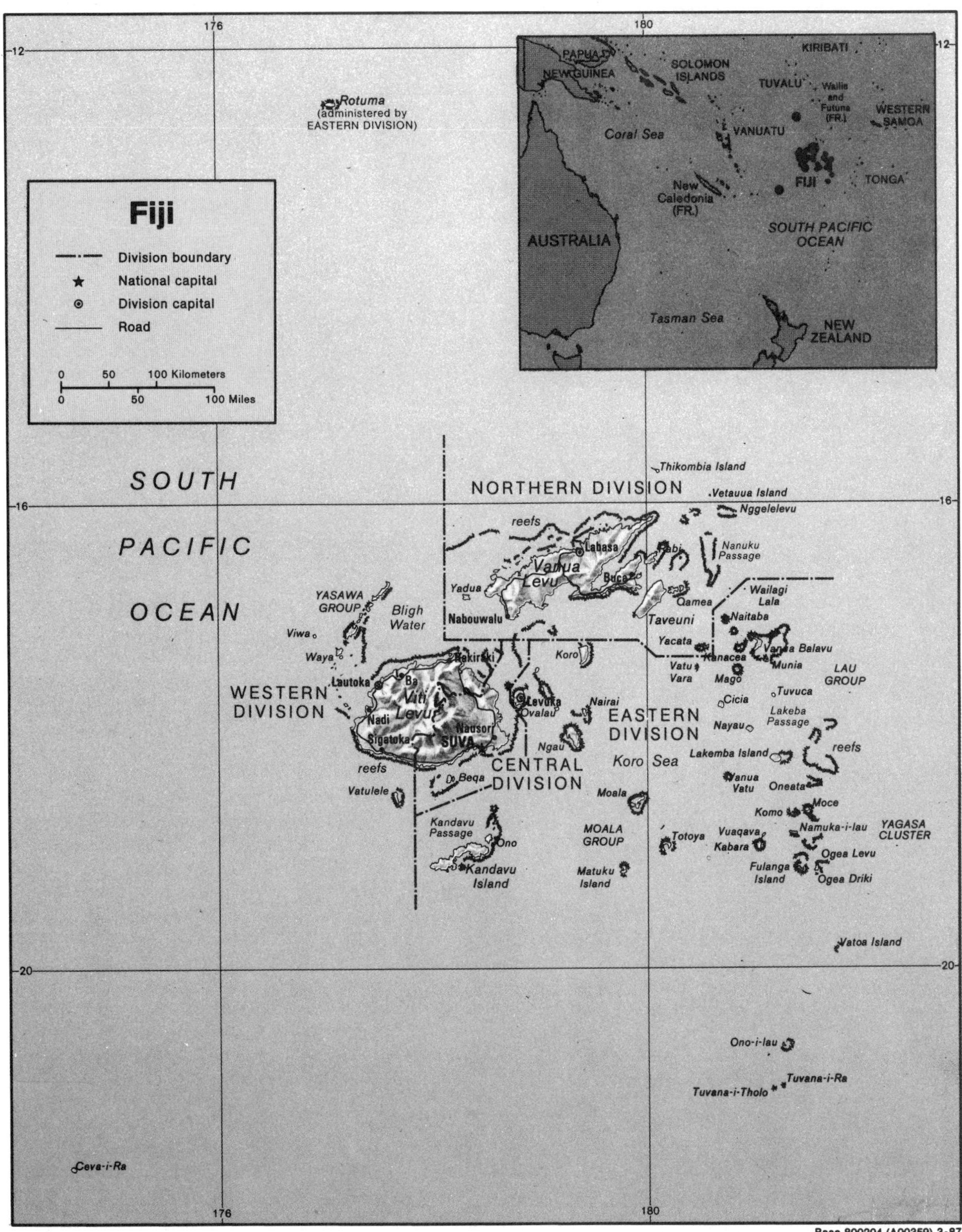
Fiji
Division boundary
National capital
Division capital
Road
0 50 100 Kilometers
0 50 100 Miles
Rotuma
(administered by
EASTERN DIVISION)
PAPUA
NEW GUINEA
SOLOMON
ISLANDS
KIRIBATI
TUVALU
Wallis
and
Futuna
(FR.)
WESTERN
SAMOA
Coral Sea
VANUATU
New
Caledonia
(FR.)
FIJI
TONGA
AUSTRALIA
SOUTH PACIFIC
OCEAN
Tasman Sea
NEW
ZEALAND
SOUTH
PACIFIC
OCEAN
NORTHERN DIVISION
Thikombia Island
Vetauua Island
Nggelelevu
reefs
Labasa
Nanuku
Passage
Vanua
Levu
Buca
Yadua
Wailagi
Lala
YASAWA
GROUP
Bligh
Water
Qamea
Nabouwalu
Taveuni
Naitaba
Viwa
Yacata
Vanua Balavu
Waya
Koro
Rakiraki
Kanacea
Munia
Vatu
Vara
Mago
LAU
GROUP
Lautoka
Ba
Viti
Levu
Levuka
Ovalau
Nairai
Tuvuca
WESTERN
DIVISION
Nadi
Cicia
Lakeba
Passage
EASTERN
DIVISION
Nayau
Nausori
Sigatoka
SUVA
Ngau
reefs
Lakemba Island
Koro Sea
reefs
CENTRAL
DIVISION
Beqa
Vanua
Vatu
Oneata
Vatulele
Moala
Moce
Komo
Kandavu
Passage
MOALA
GROUP
Totoya
Vuaqava
Namuka-i-lau
YAGASA
CLUSTER
Ono
Kabara
Ogea Levu
Kandavu
Island
Matuku
Island
Fulanga
Island
Ogea Driki
Vatoa Island
Ono-i-lau
Tuvana-i-Ra
Tuvana-i-Tholo
Ceva-i-Ra
12
16
20
176
180
Base 800204 (A00359) 3-87

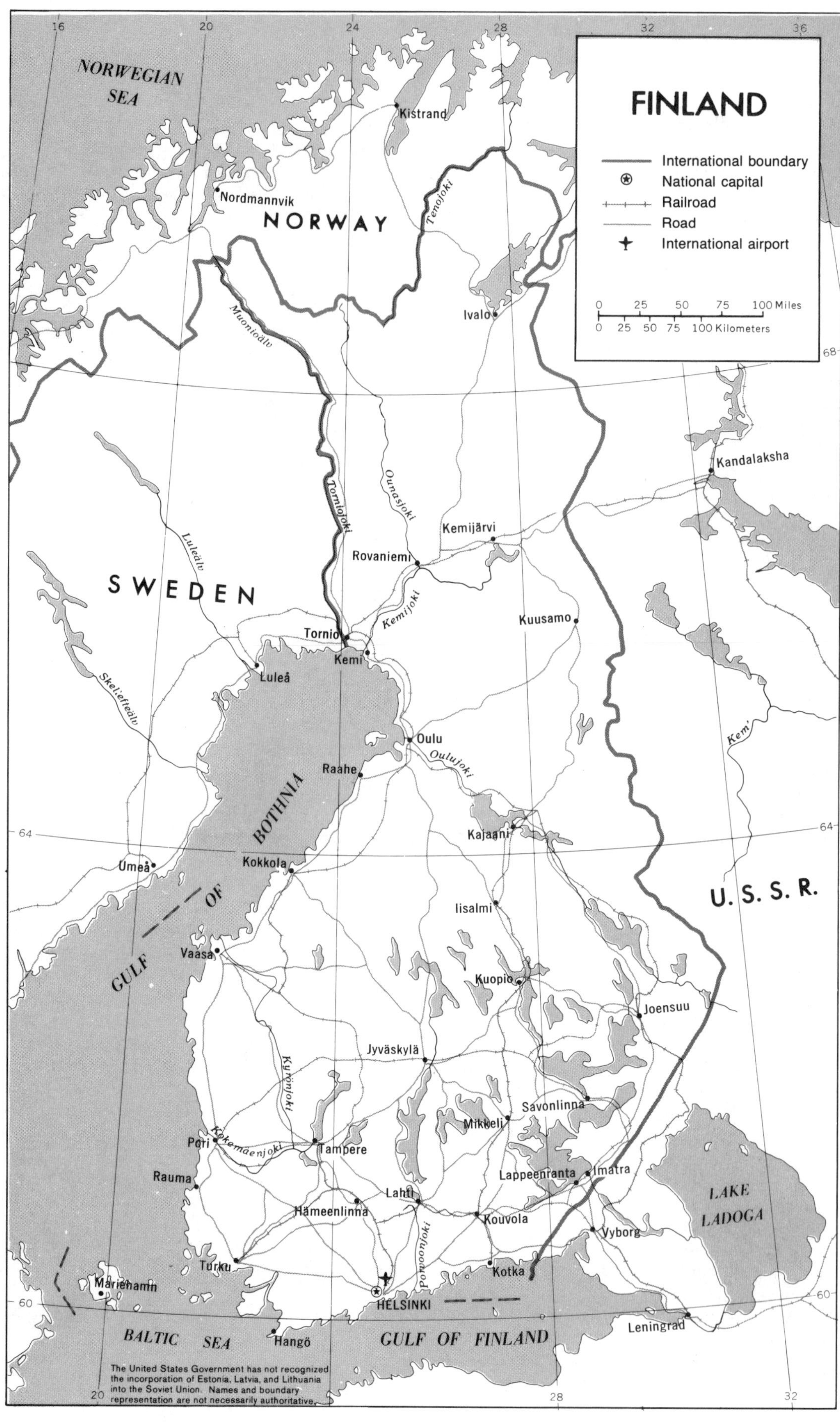

FINLAND
International boundary
National capital
Railroad
Road
International airport
0 25 50 75 100 Miles
0 25 50 75 100 Kilometers
NORWEGIAN SEA
NORWAY
SWEDEN
U. S. S. R.
Kistrand
Nordmannvik
Tenojoki
Ivalo
Muonioälv
Tornionjoki
Ounasjoki
Kandalaksha
Kemijärvi
Rovaniemi
Luleälv
Kemijoki
Kuusamo
Tornio
Kemi
Luleå
Skellefteälv
Oulu
Oulujoki
Kem'
Raahe
GULF OF BOTHNIA
Kajaani
Umeå
Kokkola
Iisalmi
Vaasa
Kuopio
Joensuu
Jyväskylä
Kyrönjoki
Savonlinna
Mikkeli
Pori
Kokemäenjoki
Tampere
Lappeenranta
Imatra
Rauma
Lahti
Hämeenlinna
Kouvola
LAKE LADOGA
Vyborg
Porvoonjoki
Turku
Kotka
Mariehamn
HELSINKI
BALTIC SEA
Hangö
GULF OF FINLAND
Leningrad
The United States Government has not recognized the incorporation of Estonia, Latvia, and Lithuania into the Soviet Union. Names and boundary representation are not necessarily authoritative.

mountains, have rounded ridgetops averaging between 457 and 762 m (1,500 and 2,500 ft) above sea level, but there is a major interruption around Lake Inari, which occupies a plain at elevations of 91 to 183 m (300 to 600 ft). More than half of eastern Finland is hilly, with the land gently sloping toward the southwest. Paralleling the coast of the Gulf of Bothnia is a belt of plains about 97 km (60 mi) wide, with elevations from 5 to 18 m (15 to 60 ft). The separation between these plains and the hills is rather sharp in the North compared to the Southwest.

The surface of the land has been scoured and gouged in recent geological times by glaciers that have left thin deposits of gravel, sand and clay. The relief of Finland has been considerably affected by the continental glacier, which on retreating left the bedrock littered almost everywhere with morainic deposits. The resulting formations can be seen most clearly in the shape of complex features such as the Salpausselka ridges and of numerous eskers running north to west to south to east. Another reminder of the Ice Age is the fact that Finland still is emerging from the sea, so that its area grows by 7 sq km (2.7 sq mi) annually. In Ostrobothnia the land rises by 90 cm (3 ft) and in the Helsinki area by 30 cm (1 ft) every 100 years.

The entire Finnish coast is paralleled by an island zone. The zone reaches its greatest breadth and complexity in the Southwest. Finland's offshore islands are numbered by the tens of thousands—the Åland Archipelago alone has nearly seven thousand. The Åland and Turku archipelagoes are rich in flora and fauna and abound in fish. They also were the first inhabited parts of the country. However, in more modern times some of the less accessible islands have become deserted while others have become the haunts of sportsmen and tourists. In the inner skerries, summer cottages have proliferated, used by those commuting between the coastal cities and their tributary islands. A network of ferries also brings a summer surge of visitors. In the outer skerries, deserted farmsteads have been restored by vacationers.

Finland's coastal zone is known appropriately as the "golden horseshoe." It is dominated by the two cities of Helsinki and Turku (Åbo), the former captial of the country, situated on the mouth of the Aurajoki. Turku was the site of the first university and Finland's first place of religious pilgrimage. The developed coastal zone extends northward from Turku through the so-called *vakka suomi* and on to the Kokemäki River, which drains the lakes of Häme to the port of Pori. Eastward the coastal plain extends to the Russian border. Here successive little rivers open up fertile valleys behind old, established settlements.

Ostrobothnia (Pohjanmaa to the Finns) also has its coastal zone. It is a land of little relief but of many rivers. Its southern coastal plains, the broadest in Finland, are traversed by a series of parallel flowing rivers that originate in Suomenselka, the highest point. Between the Oulu Valley and the Swedish border, the character of the rivers changes. Oulujoki, its formerly impressive falls now harnessed for power, is Finland's most impressive river. Beyond its broad and rapid-strewn estuary are those of the Ijoki, Simojoki, Kemijoki and Tornionjoki. Ostrobothnia's coastline is actively changing as the waves of a tideless sea break upon its coast.

Lakes cover the greater part of southern Finland. In relation to its size, Finland has more lakes than any other country. There are 55,000 lakes that are at least 200 m (656 ft) in breadth and 19 large lakes, including the artificial reservoirs of Lokka and Portipahtta, that are more than 200 sq km (77 sq mi) in area. The largest, Lake Saimaa (4,440 sq km, 1,698 sq mi) is the fifth-largest lake in Europe. Most of these lakes are quite shallow, the average depth being only 7 m (23 ft) and the greatest depth just over 100 m (328 ft). The lakes are dominated by long, sinuous esker ridges, rising scores of feet above the lake surface and generally clad with lofty pines and flanked by sandy beaches. Such ridges as Punkaharju, Pyynikki and Pulkkila are nationally renowned. The lake districts have more forests than the coasts do.

PRINCIPAL TOWNS (estimated population at 31 December 1986)

Helsinki (Helsingfors) (capital)	487,521
Tampere (Tammerfors)	169,994
Turku (Åbo)	161,188
Espoo (Esbo)	160,406
Vantaa (Vanda)	146,425
Oulu (Uleåborg)	97,869
Lahti	94,205
Kuopio	78,529
Pori (Björneborg)	77,805
Jyväskylä	65,442
Kotka	58,367
Vaasa (Vasa)	54,253
Lappeenranta (Villmanstrand)	53,917
Joensuu	47,017
Hämeenlinna (Tavastehus)	42,326

The easternmost part of Finland is Karelia, part of which was ceded to the Soviet Union by the Armistice of 1944 and the Peace Treaty of 1947. It is dominated by the Saimaa Canal, one of the most impressive structures in Finland.

Nearly half of Finland's land area is described as North Country or Nordkalotten (Pohjoiskalotti in Finnish), including the most elevated parts of the country. It is the land of the Lapps. It also is one of the coldest zones in Europe, with the timberline passing through it. Below and above the treeline, the North Country has extensive swamps, and about a third of the area is covered with bogland. The vast expanses of swamp are the least attractive elements in the northern landscape.

The North Country also is intersected with some of the country's longest and most impressive rivers, such as the Kemi and the Tornionjoki, the latter shared with Sweden. Many of these rivers empty into the freshwater Bothnian Gulf, but some, such as the Paatsjoki and the Tenojoki, drain into the Arctic, and others have carved dramatic gorges through to Russian Karelia. These torrents are among the most unspoiled in the country.

Drainage patterns are directly related to the surface features. The North is drained by long rivers, such as the Muonio, the Tornio and the Kemi. In the central part of the country the streams become shorter, except for the Oulu. They also are more sluggish and flow across land that must be ditched before it can be used for cultivation. In the lake district in the Southeast, rivers are long and narrow and dammed by the great east to west double ridge called the Salpausselka, which runs parallel to the

Gulf of Finland coast eastward from Helsinki. The area south of the lake district and westward along the coast is drained mostly by a series of short streams..

Area and population

Provinces	Capitals	land area sq mi	land area sq km	population 1987 estimate
Åland (Ahvenanmaa)	Mariehamn (Maarianhamina)	590	1,527	23,627
Häme	Hämeenlinna	6,568	17,010	680,445
Keski-Suomi	Jyväskylä	6,266	16,230	247,905
Kuopio	Kuopio	6,375	16,511	256,157
Kymi	Kouvola	4,163	10,783	338,537
Lappi	Rovaniemi	35,930	93,057	200,275
Mikkeli	Mikkeli	6,310	16,342	208,416
Oulu	Oulu	21,956	56,866	433,116
Pohjois-Karjala	Joensuu	6,866	17,782	177,199
Turku ja Pori	Turku	8,559	22,170	714,134
Uusimaa	Helsinki	3,822	9,898	1,204,510
Vaasa	Vaasa	10,211	26,447	444,466
TOTAL LAND AREA		117,616	304,623	4,928,787
INLAND WATER		12,943	33,522	
TOTAL AREA		130,559	338,145	

FRANCE

BASIC FACTS

Official Name: French Republic

Abbreviation: FRN

Area: 543,965 sq km (210,026 sq mi)

Area—World Rank: 46

Population: 55,798,282 (1988) 58,707,000 (2000)

Population—World Rank: 17

Capital: Paris

Boundaries: 7,660 km (4,759 mi); Belgium 630 km (391 mi); Luxembourg 75 km (47 mi); West Germany 435 km (270 mi); Switzerland 550 km (324 mi); Italy 500 km (311 mi); Andorra 60 km (37 mi); Spain 690 km (492 mi)

Coastline: Mediterranean Sea 1,700 km (1,056 mi); Bay of Biscay and Atlantic Ocean 1,800 km (1,118 mi); North Sea and English Channel 1,220 km (758 mi)

Longest Distances: 962 km (598 mi) N-S; 950 km (590 mi) E-W

Highest Point: Mont Blanc 4,807 m (15,771 ft)

Lowest Point: Lac de Cazaux et de Sanguinet 3 m (10 ft) below sea level

Land Use: 32% arable; 2% permanent crops; 23% meadows and pastures; 27% forest and woodland; 16% other

France is the largest country in Western Europe. Shaped in the form of a hexagon, it is distinguished topographically by clearcut features and divisions. The present landscape is dominated by four Hercynian massifs, variously composed of granite, sandstone or shale. Between these massifs lie undulating floors of sedimentary formation, linked to each other by a series of lowland corridors. Beyond these, on the southeast and southwest rise the high walls of mountain ranges—the Jura, the Alps and the Pyrenees—that form the frontiers of France.

Of the Hercynian massifs, the first, the Ardennes Plateau occupying 1,554 sq km (500 sq mi) is the western tip of a block that is part of the Middle Rhine uplands of Germany with an elevation of less than 457 m (1,500 ft). The Vosges, on the other hand, rises in the south to rounded granite summits over 1,219 m (4,000 ft) in elevation. The Ardennes and the Vosges enclose the Paris Basin on its eastern side, separating it from the Plain of Alsace where the Rhine flows. Much greater in extent is the Armorican massif which protects the Paris Basin in the west. It covers 64,750 sq km (25,000 sq mi) thrusting out into the Atlantic and the English Channel in two rocky promontories, Brittany and the Cottentin Peninsula. But its hills, trending east to west in a series of ridges, seldom exceed 365 m (1,200 ft) in height. Finally there is the Massif Central which covers 77,700 sq km (30,000 sq mi) and whose summit rises to 1,524 m (5,000 ft) or more. This granitic mass separates northern from southern France.

Between these four pillars lie the major lowlands. The largest of these, the Paris Basin, occupies no less than one-third of French territory—155,400 sq km (70,000 sq mi). It lies open on the northwest to the English Channel and on the northeast it merges with the western extremity of the Great North European Plain. Formed in a series of concentric beds, it comprises parts of the drainage basins of the Loire, the Meuse, and the Moselle, as well as the Seine and the smaller rivers, as the Somme, which drain to the Channel coast. West of the Paris Basin, the Poiteau Gate, a col in the basement rocks between Armorica and the Massif Central links the Paris Basin with the Basin of Aquitaine. It is bordered on the east by the Massif Central and on the south by the Pyrenees and it drains to the Atlantic by way of the Garonne and smaller rivers, as the Charente and the Adour.

The Pyrenees, whose uplift occurred before that of the Alps, form a barrier which rises above 3,048 m (10,000 ft) and more or less hermetically seals off the border with Spain. From their foothills there stretches eastward to the Alps the southern fringe of France—the coastal plain between the Massif Central and the sea. Here and there, especially in Provence, the plains are broken by chains of low hills, such as the Massif des Maures and the Massif de l'Esterel which reappears from beneath the Mediterranean in the island of Corsica.

The French Alps represent only a small part of the whole chain, but even so they occupy 38,849 sq km (15,000 sq mi) of French territory and include the highest peak in Europe, Mont Blanc 9,810 m (15,782 ft). The French section of the Alps represents the broad outer slope of the great chain at its western extremity. The Jura Mountains rise to 1,524 m (5,000 ft) along the border completing the great line of natural fortification. It diverges westward from the main curve of the Alpine system and the mountains cover an area of some 12,950 sq km (5,000 sq mi) with serried ranks of hills in the south and high plateaux in the north. West of the mountains and curving with them is a great downfold drained by the Rhone and the Saone. In the west, it reaches up to the Plateau of Langres where it meets the eastern rim of the Paris Basin and in the north to the Lorraine Gate. Eastward, it extends as far as the Belfort Gap—through which the Rhine once flowed—and there links up with the Plain of Alsace and the present valley of the Rhine.

The drainage system of France is based on five major rivers. Across the northern plains, with their regular rainfall and gentle slopes, flows the Seine, the most gentle and regular of French rivers, and the one best adapted to navigation. The ratio of its winter maximum and summer minimum is 3:1.

The Loire is the longest of the French rivers and its basin occupies the central part of the country. About one-fifth of the basin is mountainous, however, and as a result, the Loire is the least predictable and least useful of the rivers, apart from its tidal stretch. Carrying a heavy load of debris and meandering between islands and sandbanks, with sudden floods generated in its own upper course and that of its tributary, the Allier, the Loire represents a constant threat to the basinlands, to neutralize which levees with a maximum height of 6 m (20 ft) in Touraine and Anjou, have been constructed along its banks.

The Garonne is the shortest river. Unnavigable above Langon, it is essentially a mountain torrent, despite the length of its lowland course. Draining the northern slopes of the Pyrenees with their high rainfall and rapid runoff, it is reinforced by tributaries similar in character, flowing off the Massif Central. It has a seasonal maximum in spring and is capable of flooding catastrophically, with sudden rises in level over 9 m (30 ft).

The Rhone is the largest and most complex of French rivers with very rapid currents and a volume three times that of the Seine. Fed principally by Alpine tributaries, it also receives waters from the Saone in winter and the Doubs, flowing from the Jura, in spring. The net result is that the Rhone has a constant flow although floods sometimes occur. However, because of the breadth of the Rhone Valley the floods tend to spread across it rather than build up to dangerous heights.

Lastly, there is the Rhine which is considered more a European river than a specifically French one. It flows along the borders of Alsace at an early stage in its career when it is still largely Alpine. The Rhine is connected to the lands to the west by the depression known as Col de Saverne.

France shows a rare combination of national unity and regional variety. Ten regions have been identified based on geographical and cultural factors. They are from north to south: the Nord, the Paris Basin, the East, Burgundy and the Upper Rhine, the Alps, Mediterranean France, Aquitaine and the Pyrenees, the Massif Central, the Loire Valley and Atlantic France, and Armorica.

The Nord is the terminal point of the lowlands bordering the English Channel, including the coastal Flanders and the Walloon areas of Cambresis and Hainaut. Dunkirk (Dunkerque) is the region's principal outlet to the sea and Lille its principal conurbation. South of Lille lies the "black country" of the Scarpe and Escaut Valleys and still further south Cambrai. The northern coastlands end where the hills of Artois mark the northern edge of the Paris Basin.

The Paris Basin is the cradle of France, occupying one-third of the national territory. Over the greater part of the Basin the surface is low plateau cut by a few broad valleys. Although parts of the Basin are infertile and dry, the whole center and north are covered by a rich alluvial deposit. Two fifths of the Basin are drained by the Seine while the Moselle and the Meuse cut across the plateaux on the eastern rim and the Loire cuts across the southern edge. At the center of the basin lies Paris which throughout history has been a virtual synonym for France and an epitome of all things French. The Paris District stretches for 64 km (40 mi) although its tentacles reach up to 160 km (100 mi). It includes the Ile-de-France, parts of the Orleans regions, Upper Normandy and Picardy. Upper Normandy lies across the lower Seine and stretches to the coast. East of the river it embraces the Vexin Plateau and on the west the Valley of the Eure. Between Vexin and the Channel coast lies the Pays de Caux. On one of the lower loops of the Seine stands the city of Rouen which serves as a port for Paris. The region's other ports are Dieppe and Le Havre, both known as scenes of fierce fighting in the Allied invasion of Normandy in 1944. Picardy adjoins the Ile-de-France on the Middle Oise. The Champagne lies across three of the geological belts that encircle the Seine Basin and bestrides the Argonne and the Ardennes. It terminates on the west at the foot of the edge of the innermost saucer of the Paris Basin. The Marne is the river of Champagne rising on the Plateau of Langres and providing the principal channel between the Rhine and the Seine. Another river is the Meuse joined on its right bank by its tributary, the Chiers.

The East comprises Lorraine and the Vosges and Alsace regions. Lorraine is the basin of the Moselle which is the French Rhineland, the far eastern zone of France. The Vosges is an upland mass separating Lorraine and Alsace. The Vosges Mountains, long worn by erosion into squalid, rather squat, rounded domes (the Ballons) rise to form the highest points in the northern half of France. Further east, between the Rhine and the almost parallel course of the Ill (a river which gave its name to Alsace) the plains are intersected by the valleys of Rhine tributaries.

Burgundy and the Upper Rhine encompasses the corridor between Alsace and Central France. The unifying element in this eastern region is supplied by the curving trough of the Saone Plain, a line carried on, south of Lyons, by the narrowing valley of the Rhone. On the west side this lowland is bounded successively by the hills of the Cote d'Or, Maconnais, Beaujolais, Lyonnais, and the fringe of the Cevennes. The Saone divides this lowland into two provinces: on the right bank, Burgundy proper and on the left bank, Franche Comte which became French territory only in the 18th century. Doubs is the river of Franche-Comte, emerging from the northern Jura and after virtually encircling the mountains, it reverses its direction in a sharp elbow bend near Montbeliard. In its

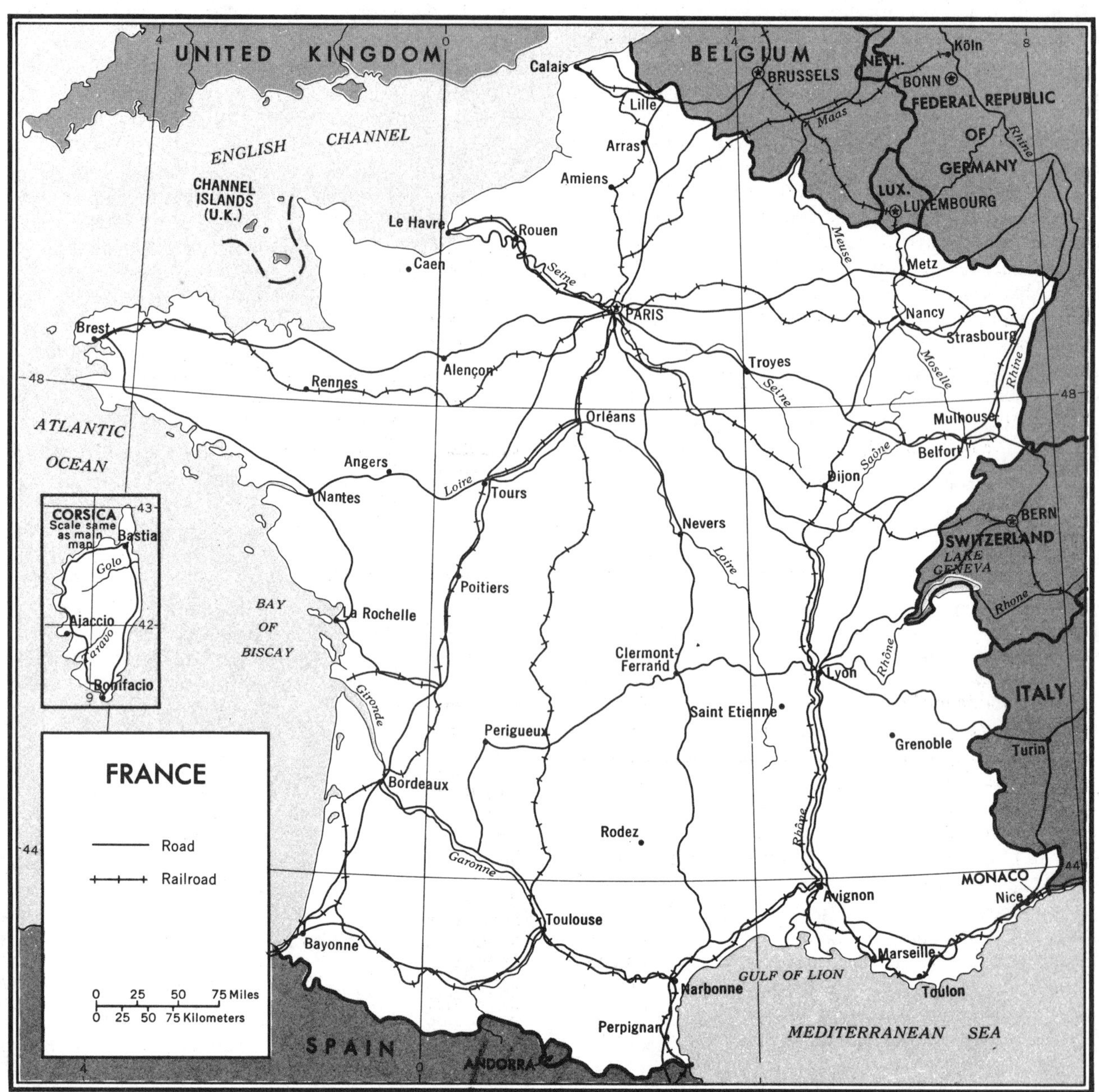

Base 57554 1-68

upper course Doubs forms the border with Switzerland for part of its length.

Where the rivers Saone and Doubs join the lowlands of Burgundy are some 80 km (50 mi) wide. Below Dijon, the hillsides that slope down to the Saone form a long, sunny rim to the plain. Here are found the great vineyards of Burgundy and Beaujolais. Parallel with the Saone a series of ridges project north and northeast from the main body of the Massif Central increasing in elevation toward the south and reaching over 914 m (3,000 ft) in the star-shaped cluster of the mountains of Beaujolais. Closeby is the region of Charolais, its close-set pastures sloping down toward the Loire. Like its companion region to the north, Bazois, it is noted for its large white cattle. Downstream, the Loire enters Nivernais, between which and Burgundy rises the great mass of Moravan, a tilted granite block that forms a detached bastion of the Massif Central. South of Dijon, is the Rhone-Alps region with its Mediterranean influences. This region is enclosed on the east by the Alps which shelters the three ancient provinces of Savoy, Dauphine and Provence. The Alps are represented here by the massifs of Belledonne and Pelvoux, the latter rising to over 3,657 m (12,000 ft) and the Prealpes, between which lie the valleys of the Drac and the Romanche, the tributaries of the Isere. In southern Dauphine, the Massifs of Champsaur and Devoluy as well as the slopes of the Durance Basin present a desolate appearance.

Languedoc and Provence, situated between the borders of Italy and Spain, are Mediterranean regions. Here Marseilles is the gateway of France on the Mediterranean. Eastern Provence is a country of low mountains and its coastline is steep and indented. Beyond Toulon to the east the effect of the cold Mistral rarely is felt and the sun holds undisputed sway. This is the Cote d'Azur, the Mecca of holidaymakers, including St. Tropez, Cannes, Nice and Monaco. Although separated by over 160 km (100 mi) of sea from Provence, Corsica exhibits structurally the same features as Provence. The island rises to over 1,676 m (5,500 ft) and possesses a coastal plain only on its eastern side. Languedoc on the west bank of the Rhone is totally different from Provence. Where one is mountainous, the other is flat and where one has a rocky coast of headlands and bays, the other has a shoreline of sandbars and lagoons. Whereas Provence beaches are jammed with tourists, those of Languedoc are practically deserted. The plains of Languedoc are a sea of vines although the quality of the wines is much inferior to those from vineyards further west and north. South and west of the River Aude lies Roussillon, a plain enclosed on three sides by hills: on the north Corbieres and the Little Pyrenees, on the west Canigo and on the south Alberes Chain, an extension of the Pyrenees.

Aquitaine covers the basins of the Garonne and the Adour. The center is occupied by the wide Plain of the Garonne and the west by the Plain of Landes which borders the Bay of Biscay. Southward the abrupt rise of the Pyrenees is modified by the huge alluvial cone that forms the Plateau of Lannemezan. On the east Aquitaine ends where the granite Cevennes rises above the Plains of Albigeois and Lauraguais and the narrow Gate of Carcassonne leads into the Mediterranean. Aquitaine was the granary of France during the Middle Ages but in the past century its vitality has been draining out. The upper basin of the Garonne Basin is known historically as Upper Languedoc but popularly as Midi-Pyrenees. It constitutes the economic hinterland of Toulouse. Bordeaux stands at the neck of the Gironde estuary into which the Garonne flows. The rectilinear coastline of Aquitaine is formed by the long dune ridge of the Gascon Landes. In the foothills zone of the Pyrenees there are three small distinctive *pays,* separated from each other by the Adour and its two tributaries, Gave d'Oloron and Gave de-Pau. The first is the Pays-Basque, the second Bearn and the third Bigorre.

The French side of the Pyrenees is much steeper than the Spanish, but there are variations between east and west in the distribution of valleys and lowlands. The western side has high valleys which drop straight down the slope. But toward the east the mountains are penetrated by broad valleys and interrupted by high grassy plateaux.

The name Massif Central is given to a triangular area from which many French rivers flow. Although called central, it is situated almost entirely in the southern half of the country and its southern edge is less than 80 km (50 mi) from the Mediterranean. Further, it is by no means a uniformly mountainous area but rather a series of uplands. The Massif lies open to the north and west penetrated by the long valleys of the Garonne and the Loire. By contrast the southern side rises abruptly above the plains of the Mediterranean coast. Although generally infertile and dry, the region is noted for its scenery. The Massif proper is divided into three subregions: Limousin, Upper Languedoc and Auvergne. In addition, the Bourbonnais and the Cevennes are also regarded as part of the Massif.

PRINCIPAL TOWNS (population at 1982 census)

Town	Population	Town	Population
Paris (capital)	2,188,918	Le Mans	150,331
Marseille (Marseilles)	878,689	Dijon	145,569
Lyon (Lyons)	418,476	Limoges	144,082
Toulouse	354,289	Angers	141,143
Nice	338,486	Tours	136,483
Strasbourg	252,264	Amiens	136,358
Nantes	247,227	Nîmes	129,924
Bordeaux	211,197	Aix-en-Provence	124,550
Saint-Etienne	206,688	Besançon	119,687
Montpellier	201,067	Metz	118,502
Le Havre	200,411	Villeurbanne	118,330
Rennes	200,390	Caen	117,119
Reims (Rheims)	181,985	Mulhouse	113,794
Toulon	181,405	Perpignan	113,646
Lille	174,039	Orléans	105,589
Brest	160,355	Rouen	105,083
Grenoble	159,503	Boulogne-Billancourt	102,595
Clermont-Ferrand	151,092	Roubaix	101,886

While the Massif Central divides north and south France, the Loire Valley and the Atlantic region unites them. Apart from the upland areas of Berry, the Loire Basin is made up of four parts: Boischot in the south, Brenne in the west, the Champagne of Berry in the center, and Sologne in the north. South of the Loire are two main regions: Poitou and Charente. Together, they make up the central section of the Atlantic seaboard between the Loire

Area and population

Regions / Departments	Capitals	area sq mi	area sq km	population 1986 estimate
Alsace				
Bas-Rhin	Strasbourg	1,836	4,755	938,000
Haut-Rhin	Colmar	1,361	3,525	661,700
Aquitaine				
Dordogne	Périgueux	3,498	9,060	380,100
Gironde	Bordeaux	3,861	10,000	1,166,400
Landes	Mont-de-Marsan	3,569	9,243	302,900
Lot-et-Garonne	Agen	2,070	5,361	302,300
Pyrénées-Atlantiques	Pau	2,952	7,645	566,500
Auvergne				
Allier	Moulins	2,834	7,340	365,000
Cantal	Aurillac	2,211	5,726	160,400
Haute-Loire	Le Puy	1,922	4,977	207,100
Puy-de-Dôme	Clermont-Ferrand	3,077	7,970	601,900
Basse Normandie				
Calvados	Caen	2,142	5,548	604,500
Manche	Saint-Lô	2,293	5,938	473,500
Orne	Alen	2,356	6,103	295,400
Bretagne				
Côtes-du-Nord	Saint-Brieuc	2,656	6,878	544,600
Finistère	Quimper	2,600	6,733	840,600
Ille-et-Vilaine	Rennes	2,616	6,775	774,300
Morbihan	Vannes	2,634	6,823	604,700
Bourgogne				
Côte-d'Or	Dijon	3,383	8,763	482,700
Nièvre	Nevers	2,632	6,817	236,200
Saône-et-Loire	Mâcon	3,311	8,575	571,100
Yonne	Auxerre	2,868	7,427	317,300
Centre				
Cher	Bourges	2,793	7,235	322,500
Eure-et-Loire	Chartres	2,270	5,880	378,800
Indre	Châteauroux	2,622	6,791	238,800
Indre-et-Loire	Tours	2,366	6,127	520,900
Loiret	Orléans	2,616	6,775	561,600
Loir-et-Cher	Blois	2,449	6,343	301,800
Champagne-Ardenne				
Ardennes	Charleville-Mézières	2,019	5,229	299,200
Aube	Troyes	2,318	6,004	292,100
Haute-Marne	Chaumont	2,398	6,211	210,400
Marne	Châlons-sur-Marne	3,151	8,162	550,700
Corse				
Corse-du-Sud	Ajaccio	1,550	4,014	113,300
Haute-Corse	Bastia	1,802	4,666	135,400
Franche-Comté				
Doubs	Besançon	2,021	5,234	468,900
Haute-Saône	Vesoul	2,070	5,360	237,700
Jura	Lons-le-Saunier	1,930	4,999	245,500
Territoire de Belfort	Belfort	235	609	133,800
Haute-Normandie				
Eure	Évreux	2,332	6,040	486,700
Seine-Maritime	Rouen	2,424	6,278	1,206,000
Île-de-France				
Essonne	Évry	696	1,804	1,027,300
Hauts-de-Seine	Nanterre	68	176	1,362,700
Paris	Paris	40	105	2,127,900
Seine-et-Marne	Melun	2,284	5,915	976,300
Seine-Saint-Denis	Bobigny	91	236	1,332,400
Val-de-Marne	Créteil	95	245	1,182,600
Val-d'Oise	Pontoise	481	1,246	973,800
Yvelines	Versailles	882	2,284	1,267,800
Languedoc-Roussillon				
Aude	Carcassonne	2,370	6,139	286,000
Gard	Nîmes	2,260	5,853	558,100
Hérault	Montpellier	2,356	6,101	745,200
Lozère	Mende	1,995	5,167	73,500
Pyrénées-Orientales	Perpignan	1,589	4,116	349,100
Limousin				
Corrèze	Tulle	2,261	5,857	242,000
Creuse	Guéret	2,149	5,565	136,800
Haute-Vienne	Limoges	2,131	5,520	357,000
Lorraine				
Meurthe-et-Moselle	Nancy	2,024	5,241	711,700
Meuse	Bar-le-Duc	2,400	6,216	198,300
Moselle	Metz	2,400	6,216	1,009,400
Vosges	Épinal	2,268	5,874	393,800
Midi-Pyrénées				
Ariège	Foix	1,888	4,890	134,700
Aveyron	Rodez	3,373	8,736	277,900
Gers	Auch	2,416	6,257	173,000
Haute-Garonne	Toulouse	2,436	6,309	851,500
Haute-Pyrénées	Tarbes	1,724	4,464	227,100
Lot	Cahors	2,014	5,217	156,700
Tarn	Albi	2,223	5,758	339,700
Tarn-et-Garonne	Montauban	1,435	3,718	194,500
Nord-Pas-de-Calais				
Nord	Lille	2,217	5,742	2,501,300
Pas-de-Calais	Arras	2,576	6,671	1,421,900
Pays de la Loire				
Loire-Atlantique	Nantes	2,631	6,815	1,029,700
Maine-et-Loire	Angers	2,767	7,166	700,100
Mayenne	Laval	1,998	5,175	276,700
Sarthe	Le Mans	2,396	6,206	511,500
Vendée	La Roche-sur-Yon	2,595	6,720	499,700
Picardie				
Aisne	Laon	2,845	7,369	535,500
Oise	Beauvais	2,263	5,860	689,000
Somme	Amiens	2,382	6,170	549,500
Poitou-Charentes				
Charente	Angoulême	2,300	5,956	341,600
Charente-Maritime	La Rochelle	2,650	6,864	519,500
Deux-Sèvres	Niort	2,316	5,999	344,600
Vienne	Poitiers	2,699	6,990	377,900
Provence-Côte d'Azur				
Alpes-Maritimes	Nice	1,660	4,299	894,800
Alpes-de-Haute-Provence	Digne	2,674	6,925	122,400
Bouches-du-Rhône	Marseille	1,964	5,087	1,740,900
Hautes-Alpes	Gap	2,142	5,549	107,000
Var	Toulon	2,306	5,973	754,000
Vaucluse	Avignon	1,377	3,567	439,700
Rhône-Alpes				
Ain	Bourg-en-Bresse	2,225	5,762	443,100
Ardèche	Privas	2,135	5,529	271,600
Drôme	Valence	2,521	6,530	404,000
Haute-Savoie	Annecy	1,694	4,388	522,000
Isère	Grenoble	2,869	7,431	980,600
Loire	Saint-Étienne	1,846	4,781	738,200
Rhône	Lyon	1,254	3,249	1,460,900
Savoie	Chambéry	2,327	6,028	333,200
TOTAL		210,026	543,965	55,279,100

and the Gironde. Toward the coast, Poitou drops away to fertile alluvial plains and, west of Niort, these in turn give way to marshes. The northern gatepost of the Poitou Gap is provided by the hills of Gatine that represent the southern extremity of the Armorican Massif. Also part of Poitou is Vendee, a region of hills, plains and marshes.

Charente lies to the south of Poitou, an open land with little forest and less relief. On the coasts of Charente, old saltmarshes have been converted into oyster beds.

Armorica comprises Brittany and Lower Normandy. Eastern lower Normandy is geologically part of the Paris Basin, but southwest of Caen, the region is hilly and is therefore called the "Norman Switzerland." Extending far out into the English Channel toward the coast of Britain is the Peninsula of Cottentin at the tip of which lies Cherbourg with its splendid harbor. To the south of Normandy lies the Celtic world of Brittany. Upper Brittany is dominated by the city of Rennes. Lower Brittany, beyond the River Vilaine, is the stronghold of the Bretons. Bare rocky ridges run in a double longitudinal line along the backbone of the peninsula. Brittany is the sole province of France, with the exception of Vendee, where the rural areas are still overpopulated. This is true of Argoat, the interior plateau, and even more so, of the Armor, the coastal strip.

FRENCH GUIANA

BASIC FACTS

Official Name: Department of Guiana

Abbreviation: FGN

Area: 86,504 sq km (33,399 sq mi)

Area—World Rank: 108

Population: 91,641 (1988) 142,000 (2000)

Population—World Rank: 175

Capital: Cayenne

Boundaries: 1,183 km (734 mi); Brazil 673 km (418 mi); Suriname 510 km (317 mi)

Coastline: 378 km (235 mi)

Longest Distances: 400 km (249 mi) N-S; 300 km (186 mi) E-W

Highest Point: 830 m (2,723 ft)

Lowest Point: Sea level

Land Use: 82% forest and woodland; 18% other

French Guiana lies on the north coast of South America separated from Brazil by the Oyapock River in the east and the Tumuc-Humuc Mountains in the south and from Suriname by the Moroni River on the west. The territory includes several islands, such as Iles du Salut (Devil's Island, Royale and Saint-Joseph), the Pere and Mere Islands, the Malingre and Remire Islands, and the two Connetables.

French Guiana consists of a small swampy coast called "terres basses," varying from 16 to 48 km (10 to 30 mi) in width, and a vast interior, the "terres hautes," including grassy plateaus, equatorial forests and mountains.

Area and population		area		population
Arrondissements	**Capitals**	sq mi	sq km	1982 census
Cayenne	Cayenne	17,590	45,559	61,587
Saint-Laurent-du-Maroni	Saint-Laurent-du-Maroni	15,809	40,945	11,435
TOTAL		33,399	86,504	73,022

FRENCH POLYNESIA

BASIC FACTS

Official Name: Territory of French Polynesia

Abbreviation: FPN

Area: 3,521 sq km (1,359 sq mi)

Area—World Rank: 153

Population: 190,939 (1988) 256,000 (2000)

Population—World Rank: 161

Capital: Papeete

Land Boundaries: Nil

Coastline: 2,525 km (1,568 mi)

Longest Distances: N/A

Highest Point: Mount Orohena 2,241 m (7,352 ft)

Lowest Point: Sea level

Land Use: 1% arable; 19% permanent crops; 5% meadows and pastures; 31% forest and woodland; 44% other

French Polynesia comprises 120 islands in the South Pacific Ocean. It includes five groups:

1 Society Islands including Tahiti, Moorea and Raiatea.

2 Marquesas Islands about 1,200 km (750 mi) NE of Tahiti.

3 Tuamotu Islands about 480 km (300 mi) south and southwest of Marquesas.

4 Gambier Islands including Mangareva, Taravai and Akamaru.

5 Tubuai or Austral Islands south of Society Islands.

Some of the islands have a rugged terrain while others are low islands with reefs.

French Guiana

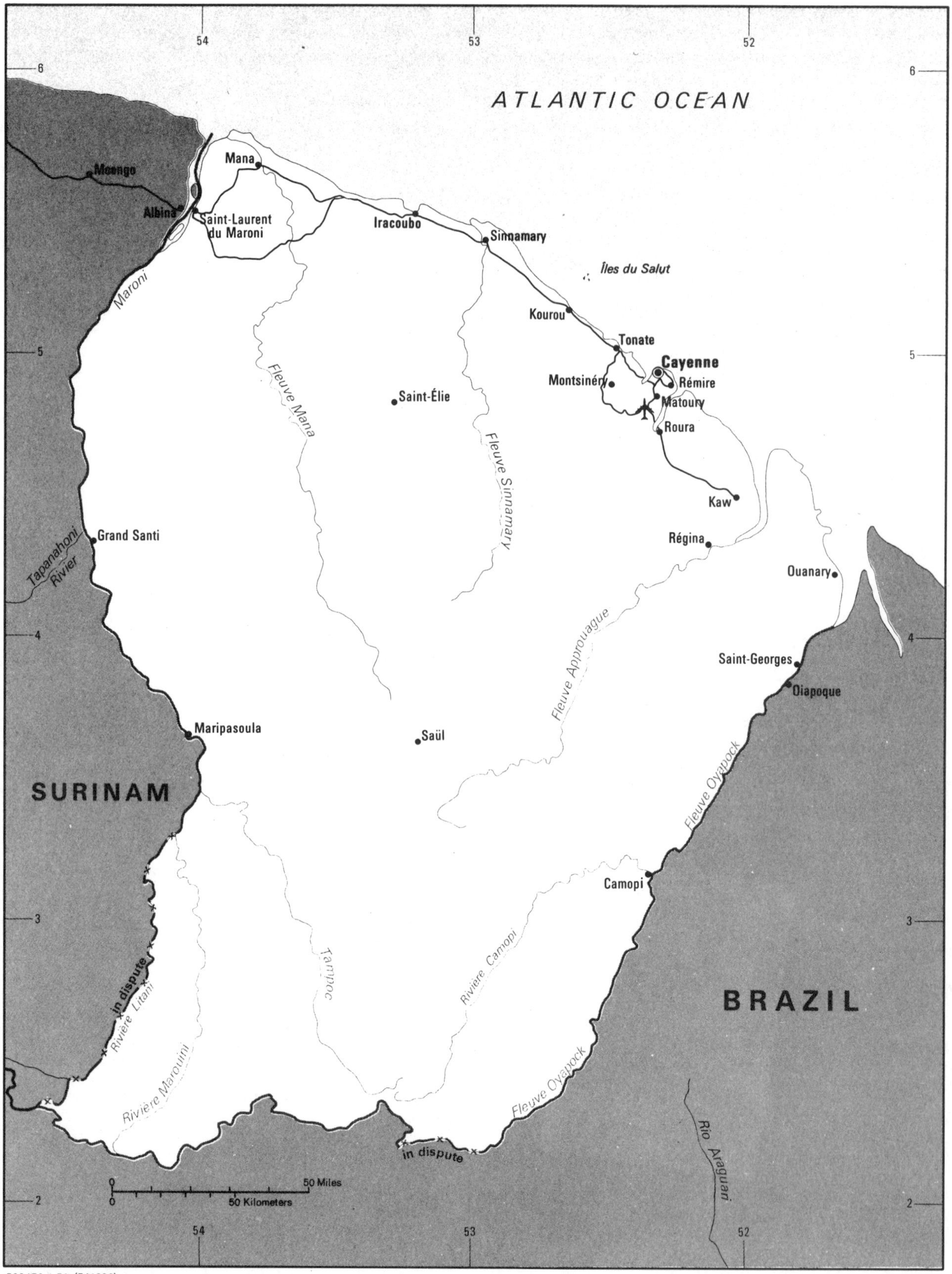

ATLANTIC OCEAN
Mana
Moengo
Albina
Saint-Laurent du Maroni
Iracoubo
Sinnamary
Îles du Salut
Kourou
Tonate
Cayenne
Rémire
Montsinéry
Matoury
Roura
Kaw
Régina
Ouanary
Saint-Georges
Oiapoque
Saint-Élie
Saül
Grand Santi
Maripasoula
Camopi
Maroni
Tapanahoni Rivier
Fleuve Mana
Fleuve Sinnamary
Fleuve Approuague
Fleuve Oyapock
Rivière Camopi
Tampoc
Rivière Litani
Rivière Marouini
Rio Araguari
in dispute
in dispute
SURINAM
BRAZIL
54
53
52
6
5
4
3
2
0 50 Miles
0 50 Kilometers

502470 1-76 (541396)
Mercator Projection
Scale 1:2,300,000
Boundary representation is not necessarily authoritative

—— Road

 Airport

Area and population		area		population
Circumscrip-tions	Capitals	sq mi	sq km	1983 census
Îles Australes	Mataura	57	148	6,283
Îles Marquises	Taiohae	405	1,049	6,548
Îles sous le Vent	Uturoa	156	404	19,060
Îles Tuamotu et Gambier	Papeete	280	726	11,793
Îles du Vent	Papeete	461	1,194	123,069
TOTAL		1,550	4,000	166,753

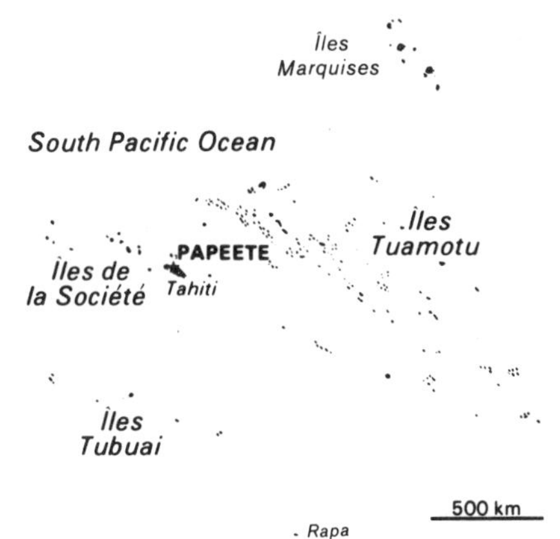

GABON

BASIC FACTS

Official Name: Gabonese Republic

Abbreviation: GAB

Area: 267,667 sq km (103, 347 sq mi)

Area—World Rank: 70

Population: 1,051,937 (1988) 1,603,000 (2000)

Population—World Rank: 130

Capital: Libreville

Boundaries: 3,083 km (1,916 mi); Cameroon 302 km (188 mi); Congo 1,656 km (1,029 mi); Equatorial Guinea 386 km (240 mi)

Coastline: 739 km (459 mi)

Longest Distances: 717 km (446 mi) NNE-SSW; 644 km (400 mi) ESE-WNW

Highest Point: 1,575 m (5,167 ft)

Lowest Point: Sea level

Land Use: 1% arable; 1% permanent crops; 18% meadows and pastures; 78% forest and woodland; 2% other

Gabon is located on the west coast of Africa, straddling the equator.

The low-lying coastal plain is narrow in the north (29 km, 18 mi) and south but broader in the estuary regions of Ogooue and Gabon. South of the Ogooue are numerous lagoons, such as N'Dogo, M'Goze and M'Komi, along the coastline. The interior relief is more complex, though nowhere dramatic. In the north the Crystal Mountains enclose the valleys of the Woleu and N'Tem Rivers and the Ivindo Basin. In southern Gabon the coastal plain is dominated by granitic hills. Between N'Gounie and the Ogooue the Chaillus Massif rises to 915 m (3,000 ft). The highest point in Gabon is Mount Iboundji (1,575 m, 5,167 ft).

Virtually the entire country is contained in the basin of the Ogooue River with its two major tributaries, the N'Gounie and Ivindo. All these rivers are perennially navigable.

PRINCIPAL TOWNS (population in 1975)

Libreville (capital)	251,400	Lambaréné	22,682
Port-Gentil	77,611		

Area and population		area		population
Provinces	Capitals	sq mi	sq km	1978 estimate
Estuaire	Libreville	8,008	20,740	359,000
Haut-Ogooué	Franceville	14,111	36,547	213,000
Moyen-Ogooué	Lambaréné	7,156	18,535	49,000
Ngounié	Mouila	14,575	37,750	118,000
Nyanga	Tchibanga	8,218	21,285	98,000
Ogooué-Ivindo	Makokou	17,790	46,075	53,000
Ogooué-Lolo	Koulamoutou	9,799	25,380	49,000
Ogooué-Maritime	Port-Gentil	8,838	22,890	194,000
Woleu-Ntem	Oyem	14,851	38,465	166,000
TOTAL		103,347	267,667	1,300,000

THE GAMBIA

BASIC FACTS

Official Name: Republic of The Gambia

Abbreviation: GAM

Area: 10,689 sq km (4,127 sq mi)

Area—World Rank: 145

Population: 779,488 (1988) 1,156,000 (2000)

Population—World Rank: 136

Capital: Banjul

Gabon

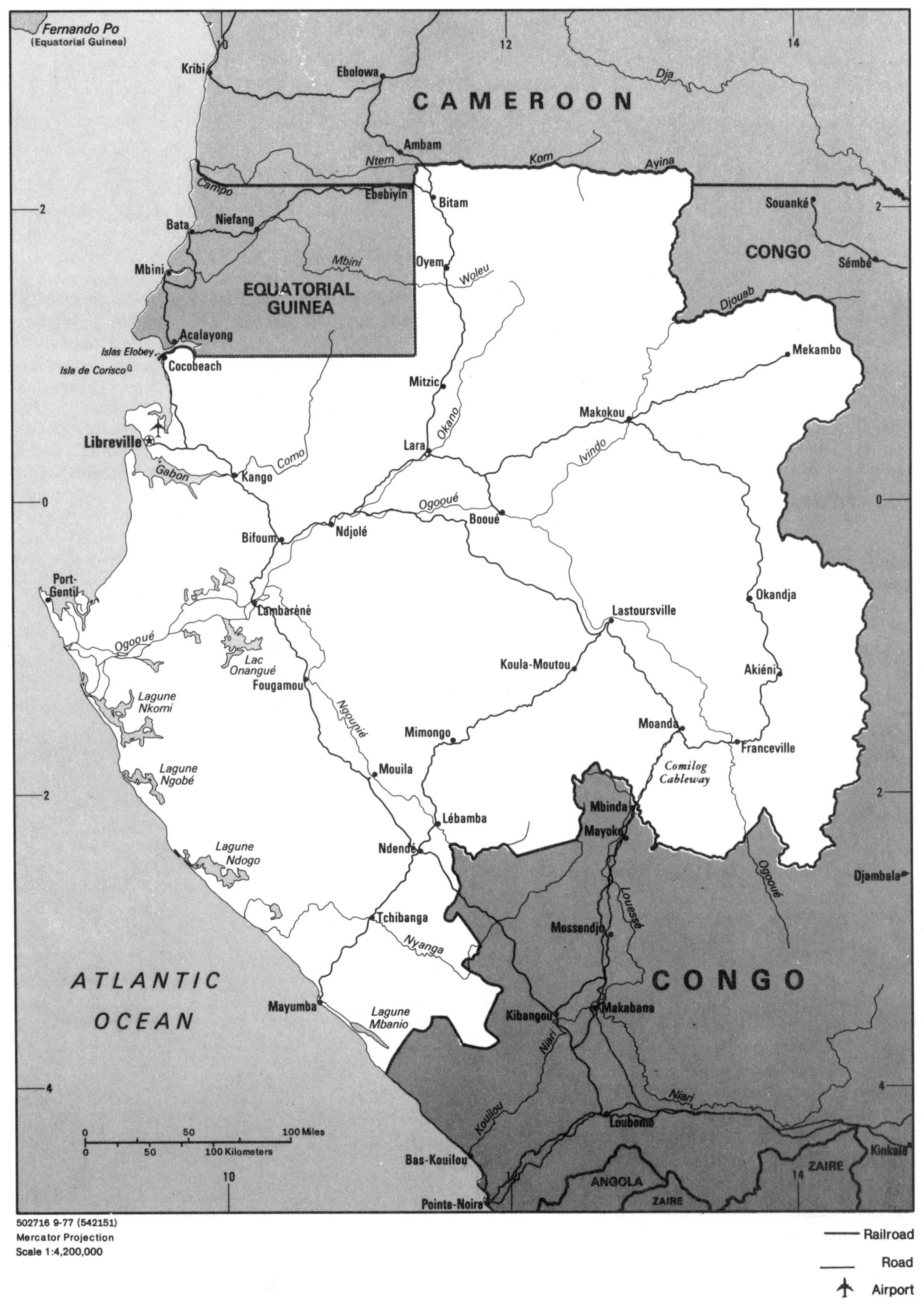

502716 9-77 (542151)
Mercator Projection
Scale 1:4,200,000

Railroad
Road
Airport

The Gambia

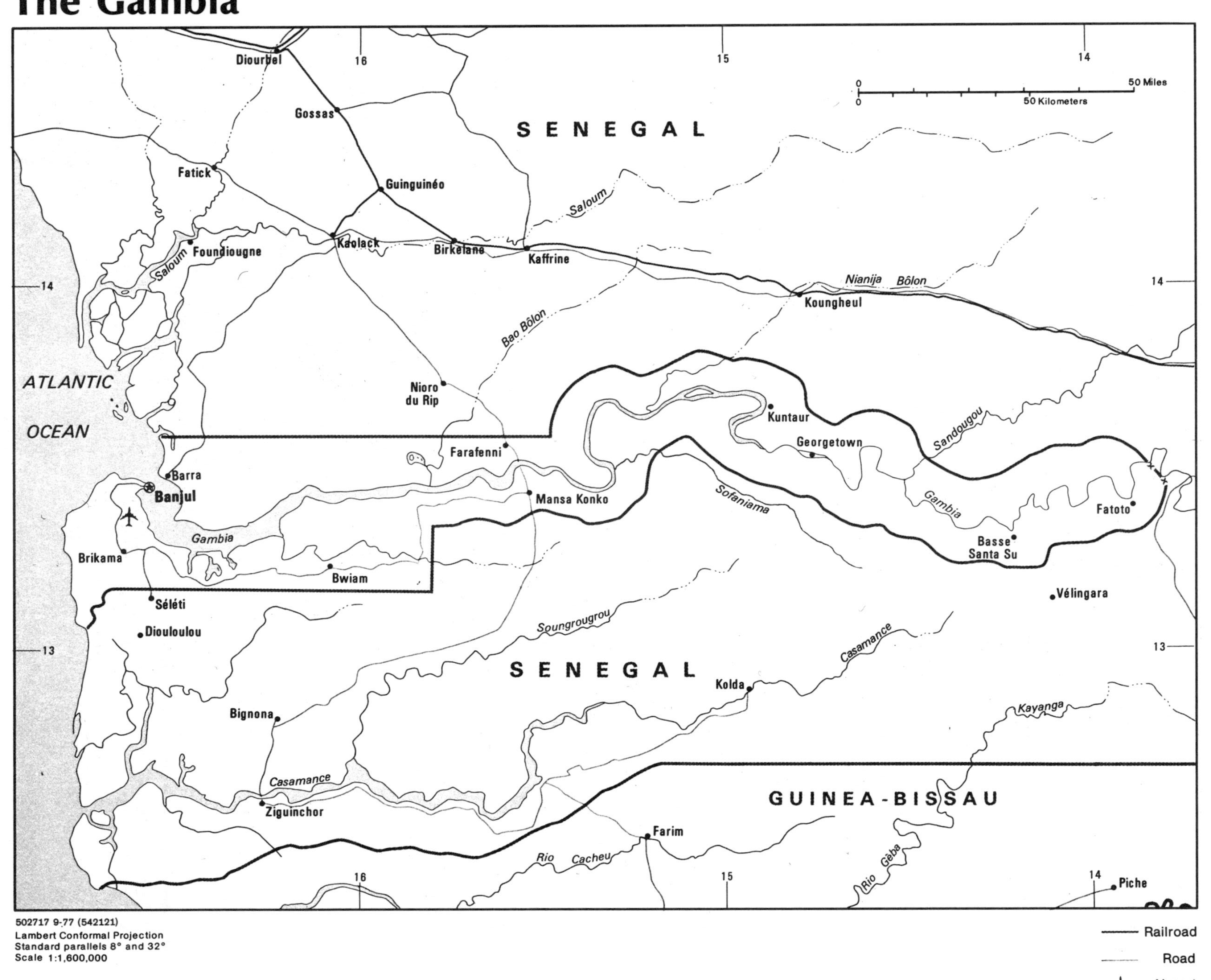

502717 9-77 (542121)
Lambert Conformal Projection
Standard parallels 8° and 32°
Scale 1:1,600,000

Boundaries: 827 km (514 mi); Senegal 756 km (470 mi)

Coastline: 71 km (44 mi)

Longest Distances: 338 km (210 mi) E-W; 47 km (29 mi)

Highest Point: 50 m (150 ft)

Lowest Point: Sea level

Land Use: 16% arable; 9% meadows and pastures; 20% forest and woodland; 55% other

The Gambia is located on the West coast of West Africa along both banks of the River Gambia.

Most of the Gambia is low-lying but it is generally divided into three regions on the basis of topographical features: the valley floor built up of alluvium with areas known as Bango Faros, a dissected plateau edge, consisting of sandy and often precipitous hills alternating with broad valleys, and a sandstone plateau which extends, in parts, across the border into Senegal.

Principal Towns (1983 census): Serrekunda 68,433, Banjul (capital) 44,188, Brikama 19,584, Bakau 19,309, Farafenni 10,168, Sukuta 7,227, Gunjur 7,115.

The Gambia River is one of the finest waterways in West Africa and is navigable as far as Kuntaur, 240 km (150 mi) upstream, by seagoing vessels and as for as Koina by vessels of shallow draft. Thick mangrove swamps border the lower reaches of the river and behind these mangroves are the "flats", which are submerged completely during the wet season. Near Banjul the river has a width of 4.8 km (3 mi).

Area and population

		area		population
Divisions	**Capitals**	sq mi	sq km	1983 census
Kombo Saint Mary	Kanifing	29	76	101,504
Lower River	Mansakonko	625	1,618	55,263
MacCarthy Island	Kuntaur/ Georgetown	1,117	2,894	126,004
North Bank	Kerewan	871	2,256	112,225
Upper River	Basse	799	2,069	111,388
Western	Brikama	681	1,764	137,245
City				
Banjul	—	5	12	44,188
TOTAL		4,127	10,689	687,817

EAST GERMANY

BASIC FACTS

Official Name: German Democratic Republic

Abbreviation: GDR

Area: 108,333 sq km (41,627 sq mi)

Area—World Rank: 101

Population: 16,596,875 (1988) 16,303,000 (2000)

Population—World Rank: 44

Capital: East Berlin

Boundaries: 2,607 km (1,619 mi); Poland 456 km (283 mi); Czechoslovakia 430 km (267 mi); West Germany 1,381 km (858 mi)

Coastline: 340 km (211 mi)

Longest Distances: 533 km (331 mi) NNE-SSW; 324 km (201 mi) ESE-WNW

Highest Point: Fichtelberg 1,214 m (3,983 ft)

Lowest Point: Sea level

Land Use: 45% arable; 3% permanent crops; 12% meadows and pastures; 28% forest and woodland; 12% other

East Germany lies in the heart of the northern European plain. The terrain is gentle, and the landscape is marked by few sharp contrasts. Landform areas merge into one another; no significant natural boundaries bar communications or distinguish one section of the country from another. The country, however, can be roughly divided into two geographic regions. The northern plain covers most of the country and contains the coastal area in the far north and the lowlands in the center. The uplands consist of mountains and rolling hills that cover the southern section.

The district of Rostock stretches along the entire length of the Baltic coast. The coastline is uneven but generally flat and sandy. The continuous action of wind and waves has created sand dunes and ridges along the coast, and sandbars have formed that connect the mainland with some of its offshore islands. The northern sections of Schwerin and Neubrandenburg districts, which are also categorized as coastal, are dotted with marshes and numerous lakes. Much of East Germany contains soils of poor quality. The coastal section is no exception; soils are sandy, porous, and low in nutrients. Nonetheless, with the exception of the Börderland in the south, the coastal region contains some of the most intensively cultivated agricultural land in the country. About nine-tenths of the area is under cultivation; it produces mainly rye and potatoes. The region enjoys a maritime climate that is moderate and marked by few extremes in temperature. Average annual rainfall is between 24 and 25 in (sixty-one and sixty-four cm) close to the national average.

Most of the country lies in an area of the northern plains known as the central lowlands. This includes the districts of Frankfurt, Potsdam, and Cottbus as well as portions of Schwerin, Neubrandenburg, Magdeburg, Halle, Leipzig, and Dresden. This region (together with the coastal area) covers about 80 to 85 percent of the land and was formed by glaciation during the Quaternary period. The lowlands are dominated by rolling hills and low ridges that rarely reach elevations in excess of 91 m (298 ft) above sea level. Numerous lakes, varying in size,

Base 504830 (546840) 2-82

shape, and depth, cover the landscape, particularly in western Neubrandenburg and around Berlin. In general these lakes are of little commercial value because of their shallow depth. Broad valleys, carved as glaciers receded, crosscut the plains, providing natural transportation routes. Soils of gravel and coarse sand predominate, and as a result, much of the area, especially around Berlin, is forest and pastureland. The most fertile soils of clay and sand loam are found in the Elbe basin and along the rivers bordering Poland, but only slightly more than half the region is under cultivation. The climate exhibits greater extremes of temperature as the maritime climate of the coast gives way to a continental climate where the rivers freeze in winter. In general, however, the weather is moderate. Rainfall approximates the national average of sixty-four centimeters annually.

The Börderland, a fertile belt of rolling countryside, forms a transition zone from the central lowlands to the uplands in the south. The country's most valuable agricultural land is found here. Loess, a fine silt, provides a thick soil cover that is favorable for intensive cultivation of crops such as wheat, barley, and sugar beets. The Börde area forms an arc extending from Magdeburg and Halle southeast through parts of Leipzig and Dresden. Its broadest section lies along the Elbe and Saale rivers. Much of the country's mineral wealth, including sizable reserves of brown coal and potash, is found in this area of the country. The climate is continental but moderate, and the growing season is relatively long.

The uplands cover about 20 percent of the southern section. The landscape consists of hills and high ridges. Included in this region are portions of the districts of Magdeburg, Halle, Leipzig, and Dresden as well as Erfurt, Suhl, Gera, and Karl-Marx-Stadt. The Harz forms the northwest section of the uplands. It highest peak, Brocken, reaches a height of 1,141 m (3,743 ft). In the southwest, extending some 104 km (66 ft) is the Thüringer Wald, a narrow ridge of thick woodland. To the southeast, forming the border with Czechoslovakia, are the Erzgebirge. Elevations in this range reach 1,213 m (3,980 ft). Many major industrial centers are situated along the base of the Erzgebirge. Traditional passages into the region lie between the Harz and the Thüringer Wald and between the Thüringer Wald and the Erzgebirge. Good agricultural land is found at the base of the Thüringer Wald surrounding Erfurt, but soils in the southernmost districts are poor and are not favorable for cultivation. Temperatures depend on elevation and exposure and will dip in the higher mountain areas. Rainfall varies; in the Harz, for example, rainfall averages as high as 58 in (147 cm) per year while at the base of the Thüringer Wald where the uplands merge with the Börderland, rainfall averages about fifty-one centimeters.

PRINCIPAL TOWNS (estimated population at 31 December 1985)

Town	Population	Town	Population
East Berlin (capital)	1,215,600	Erfurt	216,000
Leipzig	553,700	Potsdam	135,500
Dresden	519,800	Gera	131,800
Karl-Marx-Stadt (Chemnitz)	315,500	Schwerin	127,500
Magdeburg	285,000	Cottbus	124,800
Rostock	244,400	Zwickau	120,200
Halle an der Saale	235,200	Jena	107,400
		Dessau	103,600

Area and population

Districts	Capitals	area sq mi	area sq km	population 1986 estimate
Berlin, capital city	—	156	403	1,215,586
Cottbus	Cottbus	3,190	8,262	883,308
Dresden	Dresden	2,602	6,738	1,775,574
Erfurt	Erfurt	2,837	7,349	1,235,546
Frankfurt	Frankfurt	2,774	7,186	707,100
Gera	Gera	1,546	4,004	741,320
Halle	Halle	3,386	8,771	1,790,835
Karl-Marx-Stadt	Karl-Marx-Stadt	2,320	6,009	1,875,918
Leipzig	Leipzig	1,917	4,966	1,378,456
Magdeburg	Magdeburg	4,450	11,526	1,252,143
Neubrandenburg	Neubrandenburg	4,227	10,948	619,623
Potsdam	Potsdam	4,853	12,568	1,121,099
Rostock	Rostock	2,732	7,075	901,722
Schwerin	Schwerin	3,348	8,672	592,231
Suhl	Suhl	1,489	3,856	549,598
TOTAL		41,827	108,333	16,640,059

WEST GERMANY

BASIC FACTS

Official Name: Federal Republic of Germany

Abbreviation: FRG

Area: 248,708 sq km (96,026 sq mi)

Area—World Rank: 74

Population: 60,980,202 (1988) 59,107,000 (2000)

Population—World Rank: 13

Capital: Bonn

Boundaries: 4,816 km (2,993 mi); Denmark 67 km (42 mi); East Germany 1,381 km (858 mi); Czechoslovakia 356 km (221 mi); Austria 801 km (498 mi); Switzerland 334 km (208 mi); France 450 km (280 mi); Luxembourg 129 km (80 mi); Belgium 152 km (94 mi); Netherlands 574 km (357 mi)

Coastline: 572 km (355 mi)

Longest Distances: 853 km (530 mi) N-S; 453 km (281 mi) E-W

Highest Point: Zugspitze 2,963 m (9,721 ft)

Lowest Point: Freepsum Lake 2 m (7 ft) below sea level

Land Use: 30% arable; 1% permanent crops; 19% meadows and pastures; 30% forest and woodland; 20% other

West Germany is located in West-Central Europe. It is divided topographically into five major regions: the northern lowlands, the central uplands, the Alpine foothills, the Bavarian Alps, and the South German Scraplands. The northern lowlands encompass the territory of three lander: Schleswig-Holstein, Hamburg, and Bremen, and most of a fourth, Lower Saxony. The lowlands are part of a great plain that extends across north-central Europe, broadening from Belgium and the Netherlands until it reaches the Ural Mountains. Elevation in this region rarely exceeds 150 m (492 ft) and that in the central and western part 100 m (328 ft). The land slopes imperceptibly toward the sea. The North Sea portion of the coastline is devoid of cliffs and has wide expanses of sand, marsh and mud flats (*Watten*). In the western area the former line of inshore sand dunes became the East Frisian Islands when the shoreline sank during the 13th century. The mud flats between the islands and the shore are exposed at very low tides and are crossed by innumerable channels. The mud and sand are constantly shifting making navigation treacherous.

The offshore islands have maximum elevations of less than 35 m (115 ft) and have been subject to eroding forces that have washed away whole sections during storms. In 1854, for example, the only village on Wangerooge, the easternmost of the main East Frisian Islands, was washed away. While the East Frisian Islands are strung along the Coast in a nearly straight line, having long axes roughly parallel to the coast, those in the North Frisian Islands are irregularly shaped and haphazardly positioned.

The Schleswig-Holstein coast on the Baltic Sea differs markedly from that on the North Sea side. The former is indented by a number of small fjords with steep banks. The deep water and shelter of the fjords provide safe sailing conditions and fishing villages are common on the coast.

The Central Uplands, or the *Mittelgebirge,* forms the second largest of the five divisions. Its varied topography contrasts broad, tilted blocks of sedimentary rocks with deep, trough-like valleys and lowlands. In this way, the Black Forest, Odenwald, and Taunus Hills directly overlook the flat expanse of the Upper Rhine Valley while the Harz Mountains, the Eifel, the Hunsruck Hills and the Sauerland Plateau stand out respectively over the Lower Saxony Flatland, the Cologne embayment, the Nahe depression, and the plain of Westphalia. In a few parts the mountain summits rise well over 1,000 m (3,280 ft). The highest peak is in the Black Forest where the Feldberg reaches an elevation of 1,496 m (4,908 ft). Almost as high are the Grosser Arber 1,456 m (4,771 ft) and the Rachel 1,453 m (4,767 ft) but elsewhere the highest points are much lower. The Taunus Mountains are low and their highest point is only 875 m (2,870 ft) above sea level. The Westerwald, to the north and northwest of the Taunus, more nearly resembles the Eifel area across the Rhine. The terrain is lower as it blends into the northern lowlands and rarely reaches 600 m (1,986 ft). The most prominent peak, the Drachenfels, reaches a height of only 303 m (994 ft).

The commonest types of valleys are those that have been produced by the superimposition of the river systems flowing northward to the North Sea or the Baltic Sea. The most outstanding is that of the Rhine which, after leaving the Swiss frontier, first cuts through the southernmost parts of the central uplands by way of the Rift Valley between Lorrach and Bingen and then wends its course through meandering gorges. The main tributaries of the Middle and Upper Rhine, the Mosel, the Sieg, the Lahn, the Main, and the Neckar, also have incised themselves into the plateaus. The 300-km (186 mi) Rhine Rift Valley, also is known as terraced country in its upper section. This valley ends abruptly in the vicinity of Frankfurt, Wiesbaden and Mainz, where the river turns sharply to the west. Another important lowland in the *Mittelgebirge* area is the Saar basin which has changed hands between France and Germany on several occasions. It is noted for its rich coalfield about 40 km (25 mi) long and 13 km (8 mi) wide. To the east and north of the coalfield the terrain increases in elevation and to the north it blends into the Hunsruck mountain range. Across the Moselle River is the considerably steeper and more rugged Eifel region which extends to the lowland region north of Bonn and Aachen.

PRINCIPAL TOWNS (estimated population at 30 June 1986)

Town	Population	Town	Population
West Berlin	1,868,700	Gelsenkirchen	284,400
Hamburg	1,575,700	Münster	268,900
München (Munich)	1,269,400	Karlsruhe	267,600
Köln (Cologne)	914,000	Wiesbaden	266,700
Essen	617,700	Mönchengladbach	254,700
Frankfurt am Main	593,400	Braunschweig	
Dortmund	569,800	(Brunswick)	247,300
Stuttgart	564,500	Augsburg	245,600
Düsseldorf	561,200	Kiel	244,700
Bremen	524,700	Aachen (Aix-la-	
Duisburg	516,600	Chapelle)	238,600
Hannover		Oberhausen	222,100
(Hanover)	506,400	Krefeld	216,700
Nürnberg		Lübeck	209,800
(Nuremberg)	466,500	Hagen	206,100
Bochum	381,000	Mainz	188,500
Wuppertal	375,300	Saarbrücken	185,100
Bielefeld	299,200	Freiburg im	
Mannheim	295,500	Breisgau	184,800
Bonn (capital)	290,800	Kassel	184,200

The third physical region is the South German scarplands, a succession of escarpments and intervening valleys stretching across the country from southern Baden-Wurttemberg to the north-eastern corner of Bavaria, and presenting great environmental diversity. The most impressive of these scarplands begin in southern Baden where they are separated from the Swiss Jura by the valley of the High Rhine. From here the outcrop begins to broaden out to form the bleak plateau of the Swabian Alps. At their most spacious point, these hills are about 40 km (25 mi) in breadth and in several places exceed altitudes of 900 m (2,953 ft). Further northeast, the Swabian Alps merge into the Franconian Jura which run in a more northerly direction from the Altmuhl River to the valley of the upper Main. The southern part of the Franconian Jura is known as the Franconian Alps while the stretch running northeast of Nuremberg is known as the Franconian Switzerland. In both areas the summit surface is remarkably uniform between 550 and 650 m (1,804 to 2,132 ft) yet the deep incision of the valleys and their spectacular landscapes make these uplands the most distinctive region in southern Germany.

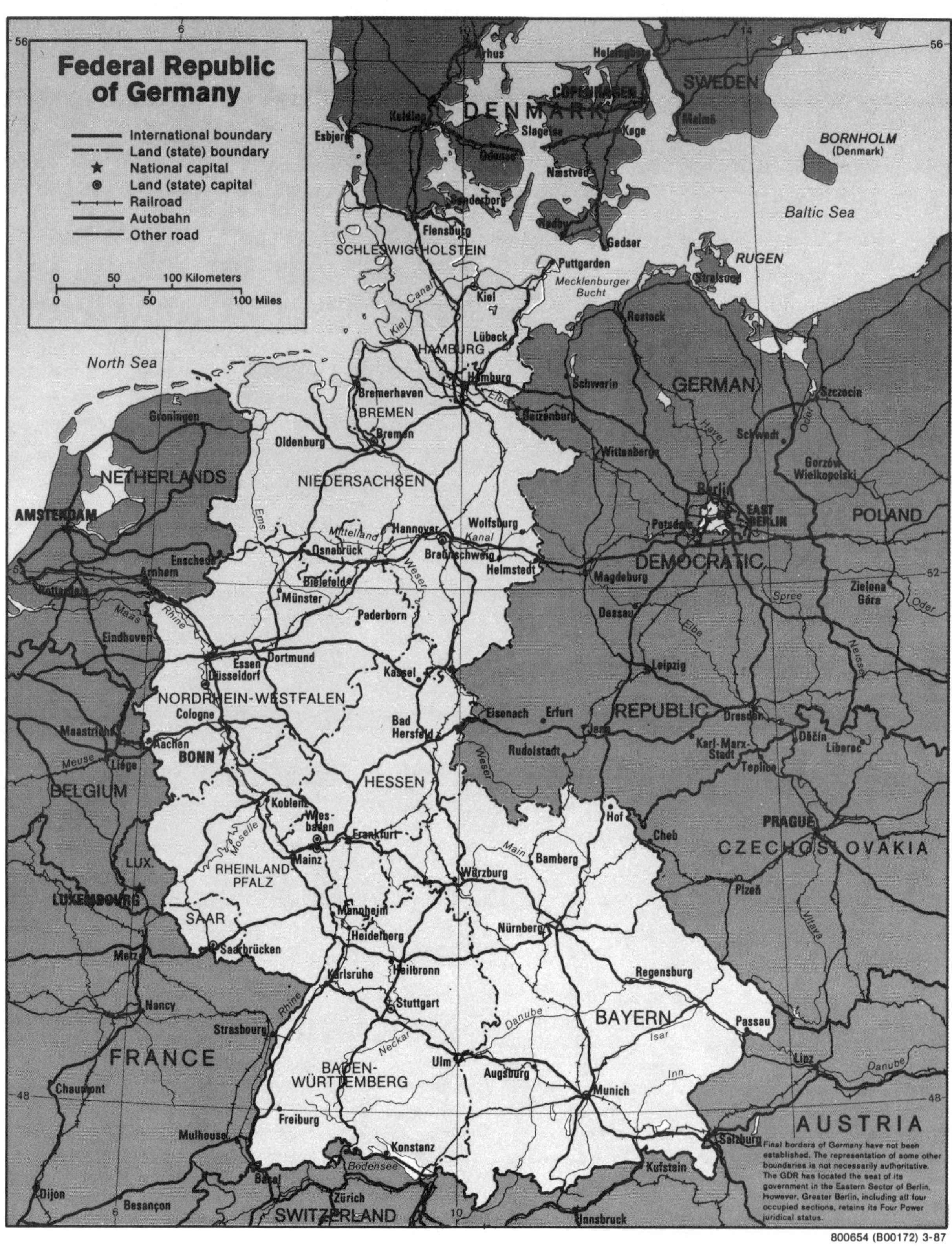

800654 (B00172) 3-87

The fourth physical region is the Alpine Foothills comprising all of Bavaria and the eastern portion of Baden-Wurttemberg. Relatively little of the area is forest and a high proportion is extremely productive cropland and pastureland. Most of the region is in the upper Danube River basin. Because the Danube itself has been pushed well to the north of its original course by the extension of alpine glaciers, each of its main Alpine tributaries—the Iller, the Lech, the Isar and the Inn—cross nearly the full breadth of the region. The Foothills also is speckled with many lakes with clear, clean water and steep, wooded banks.

The fifth region is the Bavarian Alps, the small fringe of high mountains that extend in a narrow strip along the country's southern boundary. They range eastward from Bodensee on the Swiss border to just west of Salzburg on the Austrian border. The highest point is the Zugspitze 2,962 m (9,718 ft) but there are several others with altitudes of more than 2,500 m (8,202 ft) as for example: Watzmann 2,713 m (8,901 ft); Hochfrottspitze 2,649 m (8,691 ft), and Madelgabel 2,645 m (8,678 ft).

The greater part of the country drains into the North Sea via the Rhine, the Ems, the Weser and the Elbe Rivers. A small area north and northeast of Hamburg drains into the Baltic Sea and a considerable area in the southeast lies in the Danube River basin. The divide between the watersheds of the Danube and Rhine basins winds round Baden Wurttemberg and Bavaria, most of which drains into the former. A small area north of the Bodensee, however, drains into the Rhine.

Area and population

States	Capitals	area sq mi	area sq km	population 1987 estimate
Baden-Württemberg	Stuttgart	13,804	35,751	9,321,200
Bayern	Munich	27,241	70,553	11,023,000
Bremen	Bremen	156	404	654,600
Hamburg	Hamburg	292	755	1,571,900
Hessen	Wiesbaden	8,152	21,114	5,542,900
Niedersachsen	Hannover	18,320	47,450	7,197,200
Nordrhein-Westfalen	Düsseldorf	13,153	34,068	16,678,300
Rheinland-Pfalz	Mainz	7,663	19,847	3,612,500
Saarland	Saarbrücken	992	2,568	1,042,700
Schleswig-Holstein	Kiel	6,072	15,727	2,613,500
Berlin (West)	Berlin (West)	185	480	1,878,700
TOTAL		96,030	248,717	61,136,500

Rivers, lakes, mountains, islands

Rhine (below Konstanz)	865 km	537 mi
Donau (as far as Passau)	647 km	402 mi
Elbe (below Schnackenburg)	227 km	141 mi
Dortmund-Ems Canal	269 km	167 mi
Mittelland	259 km	161 mi
Edersee	538 sq km	208 sq mi
Bodensee (Lake Constance) (total area) (German part)	305 sq km	118 sq mi
Zugspitze (Bavarian Alps)	2,962 m	9,718 ft
Watzmann (Bavarian Alps)	2,713 m	8,901 ft
Feldberg (Black Forest)	1,493 m	4,898 ft
Grosser Arber (Bavarian Forest)	1,456 m	4,777 ft
Fehmarn Island	185 sq km	71.4 sq mi
Sylt Island	99 sq km	38 sq mi

GHANA

BASIC FACTS

Official Name: Republic of Ghana

Abbreviation: GHN

Area: 238,533 sq km (92,098 sq mi)

Area—World Rank: 78

Population: 14,360,121 (1988) 18,730,000 (2000)

Population—World Rank: 54

Capital: Accra

Boundaries: 2,617 km (1,626 mi); Togo 877 km (545 mi); Ivory Coast 668 km (415 mi); Burkina Faso 544 km (338 mi)

Coastline: 528 km (328 mi)

Longest Distances: 458 km (285 mi) NNE-SSW; 297 km (1,985 mi) ESE-WNW

Highest Point: Afadjato 885 km (2,905 ft)

Lowest Point: Sea level

Land Use: 5% arable; 7% permanent crops; 15% meadows and pastures; 37% forest and woodland; 36% other

Ghana is one of the southern tier of countries that face on the Gulf of Guinea in the great bulge of West Africa. It is situated between Togo on the east, Ivory Coast on the west, and Burkina Faso on the north and northwest. The country lies, entirely in the tropics, just north of the point at which the Greenwich prime meridian intersects the equator. The coast is characterized by strong surfs, which make landing difficult except at artificially constructed harbors. Average elevation is relatively low, mostly between sea level and about 305 m (1,000 ft). The highest elevation is under 914 m (3,000 feet) and is located in the Akwapim-Togo Ranges along the eastern border.

Five major geographical regions can be distinguished. In the southern part of the country are the low plains, part of the belt that extends along the entire coastal area of the Gulf of Guinea. To the north of these plains lie three distinct regions—the Ashanti Uplands, the Volta Basin, and the Akwapim-Togo Ranges. The fifth region, the high plains, occupies the northern and northwestern parts of the country. These plains also are part of a belt stretching generally eastward and westward through West Africa.

The Low Plains

The low plains are divided into the coastal savannas, the Volta Delta, and the Akan Lowlands. The coastal savannas form a narrow strip of land along most of the southern coast from near Takoradi to the Togo border. They range in width from about five miles at their western extreme to over 80 km (50 mi) in the eastern Accra Plains section. It is generally undulating country covered with grass and scrub. There is a pronounced differentiation, however, between the lagoon-fringed coastal part and the more northeastern part, the Accra Plains, which lie at the foot of the Akwapim-Togo Ranges.

The Accra Plains are mostly flat and almost featureless, with the land descending gradually to the gulf from a height of about 152 m (500 ft) at the base of the Akwapim-Togo Ranges. The monotony of the scenery is broken, however, by occasional, isolated, steep-sided hill clusters that rise to heights of between 274 and 475 m (900 and 1,500 ft). Near the coast the intermittent drainage from these plains empties into the gulf through a series of river valleys. The valleys are often swampy during the rainy seasons, and their outlets are periodically blocked by sandbars to form lagoons. The plains, particularly a section nearer the coast that receives less rainfall, are generally free of the tsetse fly and are suitable for livestock breeding. The most favorable areas for agriculture are at the foot of the Akwapim-Togo Ranges.

To the west of Accra the coastal plain varies from about five to ten miles in width. The land is more undulating than in the Accra Plains with wide valleys and rounded, low hills; rocky headlands also occur. A number of larger commercial centers are found in this area, including Winneba, Saltpond, Cape Coast, and Sekondi. Fishing is important, and many fishing villages line the coast. Agriculture is practiced by farmers in settlements away from the coast; the area, however, is infested by the tsetse fly and is unsafe for cattle.

The Volta Delta, which forms a distinct subregion of the low plains, projects out into the Gulf of Guinea in the extreme southeast. As this delta grew outward over the centuries, sandbars developed across the mouth of the Volta River and also of some smaller rivers that empty into the gulf in the same area, resulting in the formation of numerous lagoons, some of large size. These lagoons and swamps long made road communication difficult with the rather considerable population living in the delta. This situation was ameliorated during the 1960s through improvement of important roads in the area.

The land is flat and generally covered with grass and scattered fan palms. Along the coastal part dense groves of coconut palms also are found, and at places inland in the drier, older section of the delta, oil palms grow in profusion. Soils are easily worked, and staple crops, such as cassava and corn, are grown, as are a variety of vegetables. In the vicinity of Keta, the intensive, commercial cultivation of shallots is carried on. The main occupation in the delta, however, is fishing, and this industry supplies dried and salted fish to other sections of the country.

The Akan Lowlands, which make up the greater part of the low plains, have a general elevation between sea level and 152 m (500 ft). A number of hill ranges occur, mainly oriented in a northeast-southwest direction, which have a few peaks exceeding 609 m (2,000 ft), although most high points rarely rise above 304 m (1,000 ft). The lowlands contain the basins of the Densu, Pra, Ankobra, and Tano rivers, all of which have played important roles in the development of the country.

The Densu River basin in the eastern section of the lowlands has an undulating topography. Many of the hills found there have craggy summits, which give a striking appearance to the landscape. The important commercial centers of Koforidua and Nsawam are located in this basin. The Pra River basin, to the west of the Densu, has a relatively flat relief in its upper section; the lower part resembles the topography of the Densu basin. It is a rich cocoa- and food-producing region. The valley of the Birim River, a main tributary of the Pra, is the country's most important diamond-producing area.

The combined Ankobra River basin and the middle and lower basins of the Tano River in the west of the lowlands form the largest subdivision. The general relief is much the same as in the other principal river basins; however, in the northwest the land has a plateau appearance, and the average elevation is above 152 m (500 ft).

Akwapim-Togo Ranges

The Akwapim-Togo Ranges in the eastern part of the country consist of a generally rugged complex of folded strata, with many prominent heights composed of volcanic rocks. The ranges begin west of Accra and continue in a northeasterly direction through the Volta Region and finally cross the frontier in the upper part of that region completely into the Republic of Togo.

In their southeastern part the ranges are bisected by a deep, narrow gorge that has been cut by the Volta River. The head of this gorge is the site of the Akosombo Dam, which impounds the water of the river to form Lake Volta. The ranges south of the gorge form the Akwapim section of the mountains. The average elevation in this section is about 475 m (1,500 ft) and the valleys are generally deep and relatively narrow. North of the gorge for about 80 km (50 mi), the Togo section has broader valleys and generally low ridges. Beyond this point, the folding becomes more complex, and heights increase greatly, with several peaks rising above 762 m (2,500 ft). The country's highest point, Mount Afadjato 885 m (2,905 ft), is located in this area.

The ranges are generally covered with deciduous forests, and their higher elevation gives them a relatively cooler and more pleasant climate. Small-scale subsistence farming is typical and includes cultivation of the usual staples. In parts of the Togo section coffee plantations also exist.

Ashanti Uplands

The Ashanti Uplands lie just to the north of the Akan Lowlands area and extend from the Ivory Coast border, through western and part of northern Brong-Ahafo Region and the Ashanti Region (excluding its eastern section), to the eastern end of the Kwahu Plateau. With the exception of the Kwahu Plateau, the uplands slope gently toward the south, gradually decreasing in elevation from about 304 m to 152 m (1,000 to 500 ft). Erosion of the crystalline rocks that underlie this area has left a number of hills and ranges, trending generally southwest to northeast, which

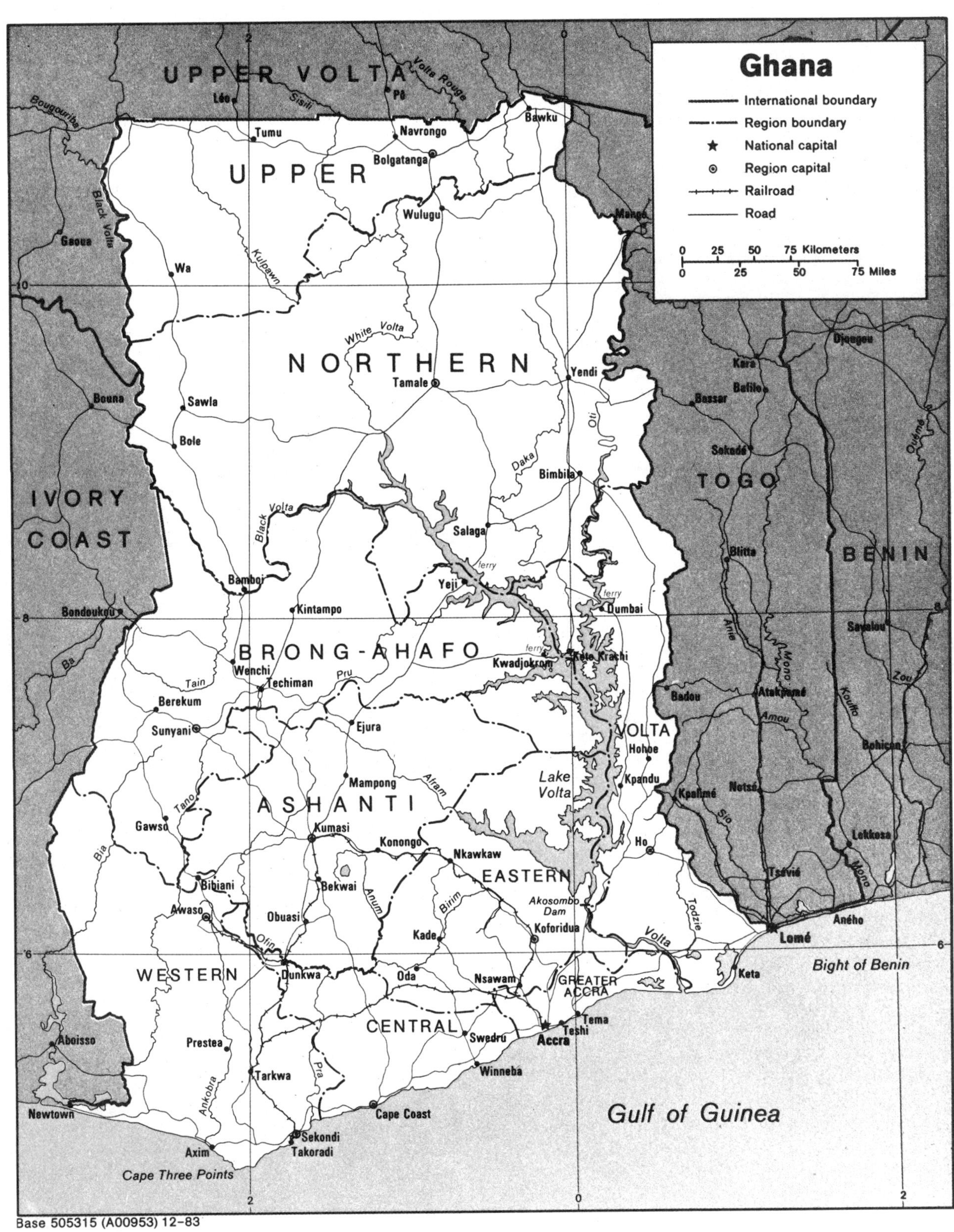

Base 505315 (A00953) 12-83

in places reach heights between 457 and 762 m (1,500 and 2,500 ft). In the southernmost part, their valleys become more open, and the region merges imperceptibly into the Akan Lowlands.

The Kwahu Plateau, forming the northeastern and eastern part of the uplands, has a quite different geologic structure and consists largely of relatively horizontal sandstones. Elevation averages 457 m (1,500 ft) and high points rise to over 762 m (2,500 ft). The greater height of the plateau gives it a comparatively cooler climate.

The uplands lie across the rain-bearing winds, and the entire region receives substantial amounts of rainfall. The uplands were originally covered by deciduous forests; however, many areas have been cleared for cocoa farms. It also has important mineral deposits; for instance, Obuasi, in the southern part, has long been considered the richest gold-mining town in the country.

Kumasi, the country's second largest city and formerly the capital of the Ashanti Confederation, is located in this region. The principal roads between the coast and the northern part of the country converge on the city, which is also the meeting point for rail lines reaching it from Accra and Sekondi-Takoradi. Almost all products from the north pass through Kumasi on their way to the more populous south.

Volta Basin

The Volta Basin region occupies the central part of the country and covers about 45 percent of the country's total area. Much of the southern and southwestern part of this basin is under 152 m (500 ft) in elevation, whereas in the northern section, lying above the upper part of Lake Volta and the Black Volta, elevations are from about 152 to 228 m (500 to 750 ft). The edges of the basin are characterized by high scarps. The Kwahu Plateau marks the southern end of the basin, although it forms a natural part of the Ashanti Uplands. The Konkori Scarp, on the western edge of the basin, and the Gambaga Scarp, in the north, have elevations from about 304 to 457 m (1,000 to 1,500 ft).

Much of the basin is characterized by poor soil conditions, and the area generally experiences a long, hot dry season, with rainfall decreasing northward to less than forty-five inches annually. The population, principally made up of farmers, has a quite low density, especially in the central and northwestern areas of the basin. Archaeological finds indicate, however, that the region once was more heavily populated. Extensive areas appear to have undergone periodic burning for perhaps more than a millennium, exposing the soil to excessive drying and erosion and making the area less attractive to cultivators.

A rather striking divergence from the rest of the basin is seen in the Afram Plains, which make up its southeastern corner. The terrain is low, averaging 61 to 152 m (200 to 500 ft) in elevation, and the rainfall is between forty-five and fifty-five inches (114 and 139 cm) a year. Stands of high forest occur on islands of higher ground. The land is very flat in the vicinity of the Afram River; this area is swampy or flooded during the rainy season, and it was largely submerged in the formation of Lake Volta. The entire area is thinly populated, and communications were poor until the mid-1960s, when new, all-weather roads were constructed in connection with the resettlement in the plains of communities and people displaced by the lake's rising waters.

PRINCIPAL TOWNS (population at 1984 census)

Town	Population
Accra (capital)	964,879
Kumasi	348,880
Tamale	136,828
Tema	99,608
Takoradi	61,527
Cape Coast	57,700
Sekondi	32,355

The High Plains

The general terrain in the northern and northwestern part of the country outside the Volta Basin region consists of a dissected plateau area, which averages between 152 and 304 m (500 and 1,000 ft) in elevation and in some places is even higher. The rainfall is between forty and forty-five inches (102 and 114 cm) annually, although in the northwest it is closer to fifty inches (127 cm). Soils in the high plains have generally greater fertility than in the Volta Basin, and the population density is considerably higher. Grains are a major crop, but farming is chiefly for local consumption. The tsetse fly is virtually absent, and livestock raising is a major occupation. The region is the largest producer and exporter of cattle to the rest of the country.

Area and population

		area		population
Regions	**Capitals**	sq mi	sq km	1984 census
Ashanti	Kumasi	9,417	24,389	2,089,683
Brong-Ahafo	Sunyani	15,273	39,557	1,179,407
Central	Cape Coast	3,794	9,826	1,145,520
Eastern	Koforidua	7,461	19,323	1,679,483
Greater Accra	Accra	1,253	3,245	1,420,066
Northern	Tamale	27,175	70,384	1,162,645
Upper East	Bolgatanga	3,414	8,842	771,584
Upper West	Wa	7,134	18,476	439,161
Volta	Ho	7,942	20,570	1,201,095
Western	Sekondi-Takoradi	9,236	23,921	1,116,930
TOTAL		92,098	238,533	12,205,574

Drainage

The entire country is interlaced by a net of streams and rivers. The stream pattern is closest in the moister south and southwest. North of the Kwahu Plateau, in the eastern part of the Ashanti Region, and in the western part of the Northern Region, the pattern is much more open and makes access to water more difficult. Stream flow is not regular throughout the year, and during the dry seasons the smaller streams and rivers dry up or have greatly reduced flow, even in the wetter areas of the country.

A major drainage divide runs from the southwestern part of the Akwapim-Togo Ranges northwestward through the Kwahu Plateau and then irregularly westward to the Ivory Coast border. Almost all streams and rivers north and east of this divide are part of the vast Volta drainage system, which covers some 157,989 sq km (61,000 sq mi) or more than two-thirds of the country. To the south and southwest of the plateau several smaller independent river systems flow directly into the Gulf of Guinea. The most important are the Pra, the Ankobra, and the Tano. Only the Volta, Ankobra, and Tano rivers are navigable by launches or lighters, and this is possible only in their lower sections.

The largest river, the Volta, has three branches, all of which originate in Burkina Faso. The Black Volta forms the northwest border to just below the ninth parallel, then flows southeastward into Ghana to the east. The White Volta and the Red Volta both enter the country in the northeast. About twenty-five miles inside the border, the Red Volta joins the westward-flowing White Volta, which eventually turns and flows southward through approximately the central part of the country.

Until the latter half of the 1960s the Black Volta and the White Volta came together near the middle of the country to form the Volta River, which from this confluence flowed first southeastward, then south, for about 499 km (310 mi) to the Gulf of Guinea. In 1964 the closing of a dam across the Volta at Akosombo, roughly some 80 km (50 mi) upstream from its mouth, created a vast lake along the entire former course of the river above this point. Arms of the lake extend into the lower courses of the Black Volta and the White Volta, which now flow separately into it. The Oti and Daka rivers, the principal tributaries of the Volta in the eastern part of the country, and the Pru, Sene, and Afram rivers, major tributaries located north of the Kwahu Plateau, also now empty into long extensions of the lake into their river valleys.

The Pra is the easternmost and most extensive of the three principal rivers that drain the area south of the Volta system divide. It rises south of the Kwahu Plateau and, flowing generally southward, enters the Gulf of Guinea near Shama. In the early part of the twentieth century it was used extensively to float timber to the coast for export, but this trade has been taken over by road and rail transport.

The Ankobra, to the west of the Pra, has a relatively small drainage basin. It rises in a hilly region south of Bibiani and flows in a generally southerly direction to enter the gulf just west of Axim. Small craft can navigate approximately 80 km (50 mi) inland from its mouth. At one time it served for the transport of machinery to the gold-mining areas in the vicinity of Tarkwa. The Tano, which is the westernmost of the three rivers, rises near Techiman in the Brong-Ahafo Region. It flows almost directly south, emptying into the broad Aby Lagoon in the southeast corner of the Ivory Coast. Navigation by steam launch is possible on the Tano as far inland as Tanosu.

A number of short, small rivers are also found in the southern part of the country. Two of these, the Densu and Ayensu, are important as sources of water for Accra and Winneba, respectively. The country has one large natural lake, Lake Bosumtwi, located about twenty miles southeast of Kumasi. It occupies the steep-sided caldera of a former volcano and has an area of about 46 sq km (18 sq mi). Several small streams flow into the lake; there is no drainage out of it, however, and its level is gradually rising.

GIBRALTAR

BASIC FACTS

Official Name: Gibraltar

Abbreviation: GIB

Area: 5.8 sq km (2.3 sq mi)

Area—World Rank: 213

Population: 29,141 (1988) 29,000 (2000)

Population—World Rank: 193

Capital: Gibraltar

Land Boundaries: 1.2 km (0.7 mi) with Spain

Coastline: 12 km (7 mi)

Longest Distances: N/A

Highest Point: 426 m (1,398 ft)

Lowest Point: Sea level

Land Use: 100% other

Gibraltar is a rocky promontory (hence its original name, Jebel el Tariq, of which Gibraltar is an elliptical form), actually a narrow peninsula running southward from the southwestern coast of Spain, to which it is connected by an isthmus. About 8 km (5 mi) across the bay to the west lies the Spanish port of Algeciras, and 32 km (20 mi) to the south, across the Strait of Gibraltar, lies Morocco. The Mediterranean Sea is to the east and the Atlantic Ocean to the west.

From a low, sandy plain in the North, the land rises sharply to the 427 m (1,400 ft) Rock of Gibraltar, a shrub-covered mass of limestone with huge caves.

GREECE

BASIC FACTS

Official Name: Hellenic Republic

Abbreviation: GRE

Area: 131,957 sq km (50,949 sq mi)

Area—World Rank: 91

Population: 10,015,041 (1988) 10,608,000 (2000)

Population—World Rank: 64

Capital: Athens

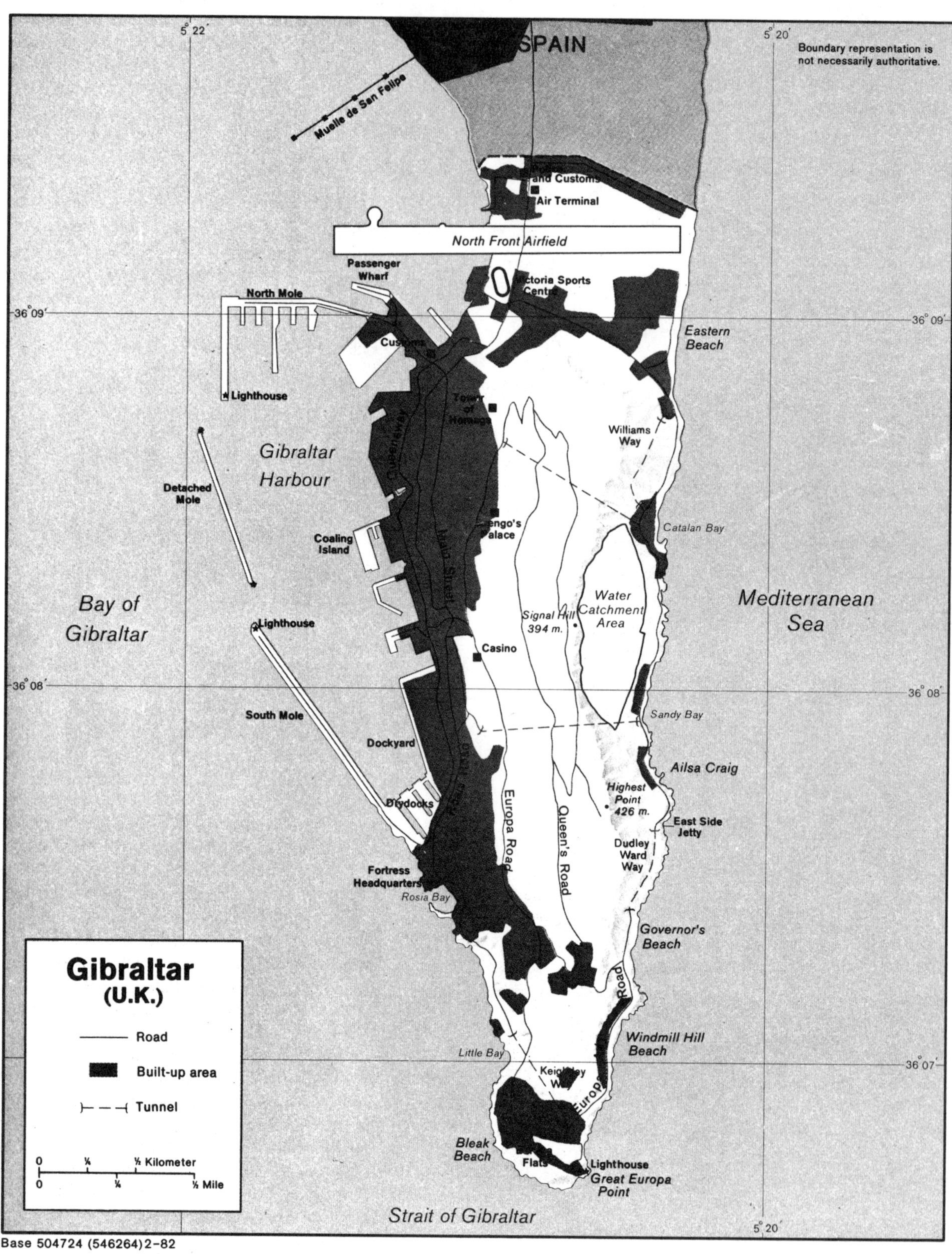

Base 504724 (546264) 2-82

Boundaries: 16,191 km (10,060 mi); Yugoslavia 246 km (153 mi); Bulgaria 475 km (295 mi); Turkey 203 km (126 mi); Albania 247 km (153 mi)

Coastline: 15,020 km (9,333 mi); Continental Greece 4,078 km (2,534 mi); Ionian Islands 1,004 km (624 mi); Aegean Islands 8,892 km (5,525 mi); Cretan Islands 1,046 km (650 mi)

Longest Distances: 940 km (584 mi) N-S; 772 km (480 mi) E-W

Highest Point: Mount Olympus 2,917 m (9,570 ft)

Lowest Point: Sea level

Land Use: 23% arable; 8% permanent crops; 40% meadows and pastures; 20% forest and woodland; 9% other

Greece is the southernmost country in the Balkan Peninsula. Of its total land area about a fifth is composed of several hundred islands in the lonian and Aegean seas. Because islands make up such a large percentage of its national territory, Greece appears smaller on the map than it actually is, while in fact it is larger than Czechoslovakia, Hungary or Bulgaria. On the basis of geographical coordinates, the village of Ormenio in the prefecture of Evros is the northernmost point; Gavdos, south of Crete, is the most southerly point; the island of Strongyll, east of Kastellorizo, is the most easterly point; and the island of Othoni, northwest of Corfu, the most westerly point.

Except in the North, the country is bounded on three sides by seas—Ionian, Mediterranean, Sea of Crete and Aegean—which occupy deep indentations all along the coasts so that there is no point on the mainland that is very far from the sea. The total coastline is comprised of continental Greece, 4,078 km (2,554 mi); the Ionian Islands, 1,004 km (624 mi); the Aegean Islands, 8,892 km (5,525 mi); and the Cretan Islands, 1,046 km (650 mi).

Greece is a series of peninsulas and a galaxy of islands, the latter numbering more than 1,400 and ranging from small barren rocks to Crete, the fifth-largest island in the Mediterranean. About 170 of the islands are inhabited. Mountains dominate the landscape everywhere and also have determined the course of Greek history. The topography of the land can best be described as rugged, and so is the appearance of the Greek peasant.

Greece is divided into nine geographical regions—six mainland and three insular. Although these are not strictly administrative units, official statistics are conventionally reported by regions.

The northern tier of regions—Thrace, Macedonia and Epirus—border on the country's northern neighbors. The remaining mainland regions are Thessaly, central Greece and Peloponnesus. The insular regions are the Ionian Islands, hugging the western coast from Albania to the Peloponnesus; Crete, the position of which makes it a divider between the Aegean and the Mediterranean seas; and the Aegean Islands, scattered about the sea of the same name.

The region of Thrace, in the northeastern corner of Greece, often is referred to as Western Thrace or Greek Thrace, to distinguish it from Bulgarian and Turkish Thrace. The region includes the island of Samothrace, about 40 km (25 mi) southeast of Alexandroupolis, Thrace's major port. The Maritsa River, which rises in western Bulgaria, flows generally south from the Turkish city of Edirne (formerly Adrianople) to the Aegean, forming the border between Turkey and Greece. Known to the Greeks as Evros (and to the Turks as Meric), its valley is one of the country's most arable lowlands. The northern border of the region with Bulgaria runs through the Rhodope Mountains, which, to the west of the Evros Valley, becomes a wide plateau reaching south almost to Alexandroupolis, thus effectively dividing Thessaly. Lake Vistonis, actually a lagoon, further divides the region to the west. These features are the basis of the division of the region into three *nomos:* Alexandroupolis, Rhodope and Xanthi. Other than the Evros and Nestos rivers on the region's borders, there are no major rivers in Greek Thrace. The Thracian coastline is generally smooth and uniform.

The island of Samothrace is administered as part of the *nomos* of Evros. The island measures 179 sq km (69 sq mi) but has only a small population because of its poor soils and lack of good harbors.

Macedonia, which bears the name of what was historically a much larger area, is the largest geographical region, with 13 *nomoi* and the small, autonomous district of Mount Athos. Part of Macedonia is in Yugoslavia. The Greek region extends from the Nestos River to the Albanian border in the West. Its northern boundary runs through the mountain ranges that Greece shares with Yugoslavia and Bulgaria, while the southern boundary is the Aegean coastline, the length of which is greatly extended by the irregular configuration of the Chalcidice Peninsula, from which three narrow peninsulas jut into the Aegean. The easternmost peninsula is the location of the famed Monastery of Mount Athos, where celibate monks have lived for over 1,000 years.

Like the rest of the country, Macedonia is very mountainous, but the mountains are interrupted by valleys and coastal plains. The Plain of Drama, Strimon Valley and Axios Valley are important agricultural areas. Drama contains some of the best tobacco-growing areas in the country. The port city of Salonika is at the head of the Gulf of Salonika on a natural harbor. Second only to Athens both as a port and as a city, Salonika has become in effect the capital of the North, where the minister for northern Greece is stationed.

Thasos, a 378 sq km (146 sq mi) island about 6.4 km (4 mi) out from the delta of the Nestos River, is administered by the *nomos* of Kavala. Once one of the wealthiest islands in the Mediterranean, with gold and marble quarries, it is now inhabited mostly by poor subsistence farmers.

West of Macedonia and south of Albania lies the region of Epirus, part of a greater Balkan territory carved up after the breakup of the Ottoman Empire.

The southern limit of Epirus is the same as it always has been—Ambracian Gulf, site of Octavian's naval victory over Antony and Cleopatra in 31 B.C. The present-day northern limit is the border of Albania, although historically the region stretched into Albania. Lowlands are rare in Epirus, and the high Pindus Mountains to the

GREECE
International boundary
National capital
Railroad
Road
International airport
0 25 50 75 Miles
0 25 50 75 Kilometers
YUGOSLAVIA
BULGARIA
BLACK SEA
ALBANIA
TIRANË
Lake Ohrid
Lake Prespa
Crni Drim
Vardar
Nestós Potamós
Struma
Maritsa
Dráma
Xánthi
Komotiní
Sérrai
Kilkís
Kavála
Alexandroúpolis
İpsala
İstanbul
SEA OF MARMARA
Édhessa
Flórina
Kastoría
Véroia
Thessaloníki
THÁSOS
Políyiros
Kozáni
Kateríni
Aliákmon Potamós
MOUNT ATHOS
LEMNOS
Piniós Potamós
Kérkira
Ioánnina
Lárisa
Igoumenítsa
Tríkala
Kardhítsa
Vólos
AEGEAN
SEA
TURKEY
Árta
Prevéza
Levkás
Karpenísion
Lamía
EUBOEA
LESBOS
Mitilíni
Ámfissa
Khalkís
Mesolóngion
Levádhia
Khíos
İzmir
Argostólion
Pátrai
Corinth
ATHENS
Piraiévs
Corinth Canal
Zákinthos
Pírgos
Lávrion
SÁMOS
Tripolis
Návplion
Alfiós Potamós
Ermoúpolis
IONIAN SEA
CYCLADES
DODECANESE
Sparta
Kalámai
Ródhos
RHODES
SEA OF CRETE
KÁRPATHOS
Kastéllion
Khaniá
Réthimnon
Iráklion
CRETE
Áyios Nikólaos
20
24
28
40
36

East have hindered communications, with the result that the region is the most isolated mainland region. It also is one of the most backward regions, with a population constantly declining because of emigration to other regions.

Across the Pindus watershed from Epirus and south of Macedonia is the region of Thessaly, which stretches to the Aegean Sea. The plains of Thessaly are the most extensive lowlands and the most fertile agricultural lands in the country. Larissa, Volos, Trikala and Kardhista are the capitals of Thessaly's four *nomoi*. Volos is at the head of the almost landlocked Gulf of Pagasai, from which Jason and the Argonauts are said to have sailed in search of the Golden Fleece. Mount Olympus, the legendary home of the gods, is at the northeast corner of Thessaly, almost on the Macedonian boundary. Several islands of the northern Sporades archiepelago, including Skiathos, Skopelos, Ikiodhromia and Pelagos, are administered by the *nomos* of Magnisia. The island of Skiros, farther to the southeast, also is one of the northern Sporades but is included in the *nomos* of Euboea.

PRINCIPAL TOWNS (population at 1981 census)

Athinai (Athens, the capital)	885,737	Larissa	102,426
		Iraklion	102,398
Thessaloniki (Salonika)	406,413	Volos	71,378
		Kavala	56,705
Piraeus	196,389	Canea	47,451
Patras	142,163	Serres	46,317

The Ionian Islands are grouped as a region more for administrative convenience than for any geographic unity. Constituting about 1.8% of the total land area, they are spread out along the coast from Albania to the Peloponnesus. Most of them are uninhabited rocks, but each of the four largest and the surrounding satellites are grouped into a *nomos*. The northernmost and the second-largest of the islands is Corfu, part of which lies off the southern coast of Albania. The Strait of Corfu, which separates the island from the mainland, varies in width from about 3.2 km (2 mi) to about 24 km (15 mi). Most of the population of the *nomos* live on the island of Corfu. Corfu has a mountain range stretching across its northern and widest area that almost separates the northern coastal plain from the remainder of the island. Included in the range is Pantakroto—at 914 m (3,000 ft) above sea level, the highest peak on the island. From north to south the terrain changes from mountainous to hilly to lowland, and much of it is fertile and arable. Several miles to the south, off the coast of central Greece, is the island of Levkas. About 303 sq km (117 sq mi) in area, Levkas is the smallest of the four major islands in the region and the smallest *nomos* in the country. The center of the island is mountainous, restricting agriculture to the coastal lowlands, but there are upland basins where the soils are fertile and the mountainsides have been terraced for extensive cultivation of grape vines. Cephalonia, at 782 sq km (302 sq mi) the largest of the Ionian Islands, is directly west of the Gulf of Patras. The capital of the *nomos* is Argostolion. Cephalonia is very mountainous, and arable land there is scarce; nevertheless, with a favorable climate the island produces important crops. The island of Ithaca, to the northeast, is included in the *nomos* of Cephalonia. Ithaca also is very mountainous and has very little arable land. The southernmost *nomos* in the region is Zante, about 12.9 km (8 mi) off Cephalonia and 14.5 km (9 mi) west of the Peloponnesus. Zante has a wide, fertile plain flanked on the west by low mountains that drop precipitously to the sea. With more than 25% of its land under cultivation, Zante is the most agriculturally developed of the Ionian Islands.

Area and population	area		population
Regions	sq mi	sq km	1981 census
Aegean Islands	3,522	9,122	428,533
Central Greece and Évvoia	9,417	24,391	1,099,841
Crete	3,219	8,336	502,165
Greater Athens	165	427	3,027,331
Ionian Islands	891	2,307	182,651
Ípiros	3,553	9,203	324,541
Macedonia	13,066	33,841	2,120,481
Pelopónnisos	8,254	21,379	1,012,528
Thessalía	5,420	14,037	695,654
Thráki	3,312	8,578	345,220
Autonomous administration			
Ayion Oros (Mt. Athos)	130	336	1,472
TOTAL	50,949	131,957	9,740,417

South of Epirus and Thessaly, stretching from the Ionian Sea to the Aegean Sea and including the elongated island of Euboea, is central Greece. On most maps, Euboea appears almost to be connected to the mainland, but it is actually separated by a strait that in some places is only 61 m (200 ft) long. Central Greece is made up of seven *nomoi*, all of which still bear their pre-Christian names: Attica, Aetolia, Euboea, Boeotia, Evritania, Phocis and Phthiotis. Attica, which includes Athens and Piraeus, is mountainous in the North but levels off to plains that extend from Athens to Cape Sounion at the end of the Attica Peninsula. The plains, though less fertile than those of Thessaly, are arable enough for the production of grains, olives, grapes, figs and cotton, while the foothills make good pastures. The *nomos* of Attica also administers the islands of Aegina, Hydra and Kethira, the last named lying far to the south, between Peloponnesus and Crete.

The Peloponnesus, a peninsula of 21,446 sq km (8,278 sq mi), is connected to Attica by an isthmus that is only 6.4 km (4 mi) across at its narrowest point. Although mountainous, it has a narrow coastal plain around its entire periphery. The coastal plain in the Northwest is considerably wider and also is the most fertile area of the region. The highest point, Mount Killini (2,376 m; 7,795 ft), is in the Corinth *nomos*. The Corinth Canal cuts through the isthmus to connect the Gulf of Corinth with the Saronic Gulf, thus shortening the voyage from the Ionian Sea to Piraeus by about 322 km (200 mi). The canal is 6.4 km (4 mi) long and 21 m (69 ft) wide at water level. There are four major cities in the region: Patras, the largest and one of the most important ports, and Kalamata, both on the coast: and Tripolis and Sparta in the interior. The alluvial plains of the peninsula—Argolis to the east and the lowlands of Lakonia and Messinia to the south—are fertile.

Crete, the site of the first European civilization, is the largest of Greek islands and the fifth-largest Mediterranean island, with an area of 8,308 sq km (3,207 sq mi). Crete is long and narrow and, like all of Greece, mountainous. In width the island ranges from about 11.2 km (7 mi) to 58 km (36 mi) and is about 245 km (152 mi) long. The mountains that cover the island are separated into four distinct masses by depressions called isthmuses. The northern coast is heavily indented but has some fine natural harbors. Along the southern coast the mountains often drop directly to the sea as sheer cliffs, but the coast is very regular. Mount Ida, at 2,498 m (8,195 ft) above sea level, is the highest point on the island; the other three mountain blocks decrease in size from west to east.

The Aegean Sea is an arm of the Mediterranean demarcated by Crete to the south and connected to the Black Sea to the northwest through the Dardanelles (the ancient Hellespont), the Sea of Marmara and the Bosporus. Many of the islands are remnants of a land bridge that once connected Greece and Asia Minor. The five *nomoi* of the region are the Dodecanese, Cyclades, Samos, Chios and Lesbos. Cyclades is a group of 29 islands a few miles southeast of Attica. At the center of the group is Delos, considered as a sacred island in ancient times and revered as the birthplace of Apollo and his sister Artemis. Many of the other islands are dry, rocky and infertile, but a few, such as Naxos, the largest, and Siros, the capital of the *nomos,* grow some crops. On one of the islands, Milos, one of the world's great works of art, the Venus de Milo, was discovered in 1820.

East of the Cyclades and just off the southwestern coast of Turkey is the Dodecanese archipelago of 18 inhabited islands covering a land area of 2,705 sq km (1,044 sq mi). The largest (1,399 sq km; 540 sq mi) and the most famous of these islands is Rhodes. This island also is mountainous, with a long central ridge running from one end to the other. However, the foothills and the coastal plains are generally well watered and fertile. Other important islands are Kos, Kalimnos and Patmos, the last named noted as the place where St. John wrote the Book of Revelation as an exile.

The *nomos* just north of the Dodecanese is Samos, consisting of the islands of Samos, Ikaria and the Fournoi group, with its capital at Vathi. Samos, the home of the mathematician Pythagoras, has better agricultural land than most Aegean islands and is blessed with trees such as firs and cypresses. Ikaria is less fertile, and the Fournoi group is almost barren.

The next *nomos* to the north is Chios and a few nearby islets. Chios claims to be birthplace of Homer, an honor claimed by several other places. Much of the island is mountainous, but the open plains to the south and east are fertile, and the climate is favorable to the production of citrus fruits, figs and grapes. The northernmost *nomos* is Lesbos, consisting of the islands of Lesbos, Lemnos and Eustratios. With an area of 1,630 sq km (629 sq mi), Lesbos is the third-largest Greek island. Two large gulfs indent the coasts of Lesbos. The terrain is rugged but the coastal lowlands and lower foothills are fertile. However, the island lacks water and is subject to searing winds.

There are at least 20 mountains of more than 2,000 m (6,562 ft). Most of them are only a few miles from the sea. The highest are:

Mountain	M.	Ft.
Olympus	2,917	9,571
Smolikas	2,637	8,652
Voras	2,524	8,281
Grammos	2,520	8,268
Giona	2,510	8,235
Tyrphi	2,497	8,193
Tzoumerka	2,469	8,101
Parnassos	2,457	8,061
Psiloritis	2,456	8,058
White Mountains	2,453	8,048

There are few large rivers because of the mountainous terrain and the low rainfall. The major ones are:

River	Km	Mi.
Aliakmonas	297	185
Acheloos	220	137
Pinelos	205	127
Evros	204	127
Nestos	130	81
Strymonas	118	73
Thiamis	115	71
Alpheios	110	68
Arachthos	110	68
Evrotas	82	51

The ten largest lakes in order of size are as follows:

Lake	Sq. Km	Sq. Mi.
Thrichonis	96.5	37.2
Volvi	75.6	29.2
Vegoritis	72.5	28.0
Koronela	47.9	18.5
Vistonis	45.6	17.6
Small Prespa	43.1	16.6
Large Prespa	38.3	14.8
Yliki	28.7	11.1
Lake Kastoria	28.6	11.0
Pamvotis	19.2	7.4

Greece is an area of frequent earthquakes and earth tremors. The famous statue known as the Colossus of Rhodes was toppled by an earthquake in 227 B.C. In the ninth century Corinth was destroyed by an earthquake that reportedly killed 45,000. In the 1950s the country was again ravaged by earthquakes, particularly along the Ionian Islands, along the western coast and on some of the Aegean Islands.

GREENLAND

BASIC FACTS

Official Name: Greenland

Abbreviation: GLD

Area: 2,175,600 sq km (840,000 sq mi)

Greenland

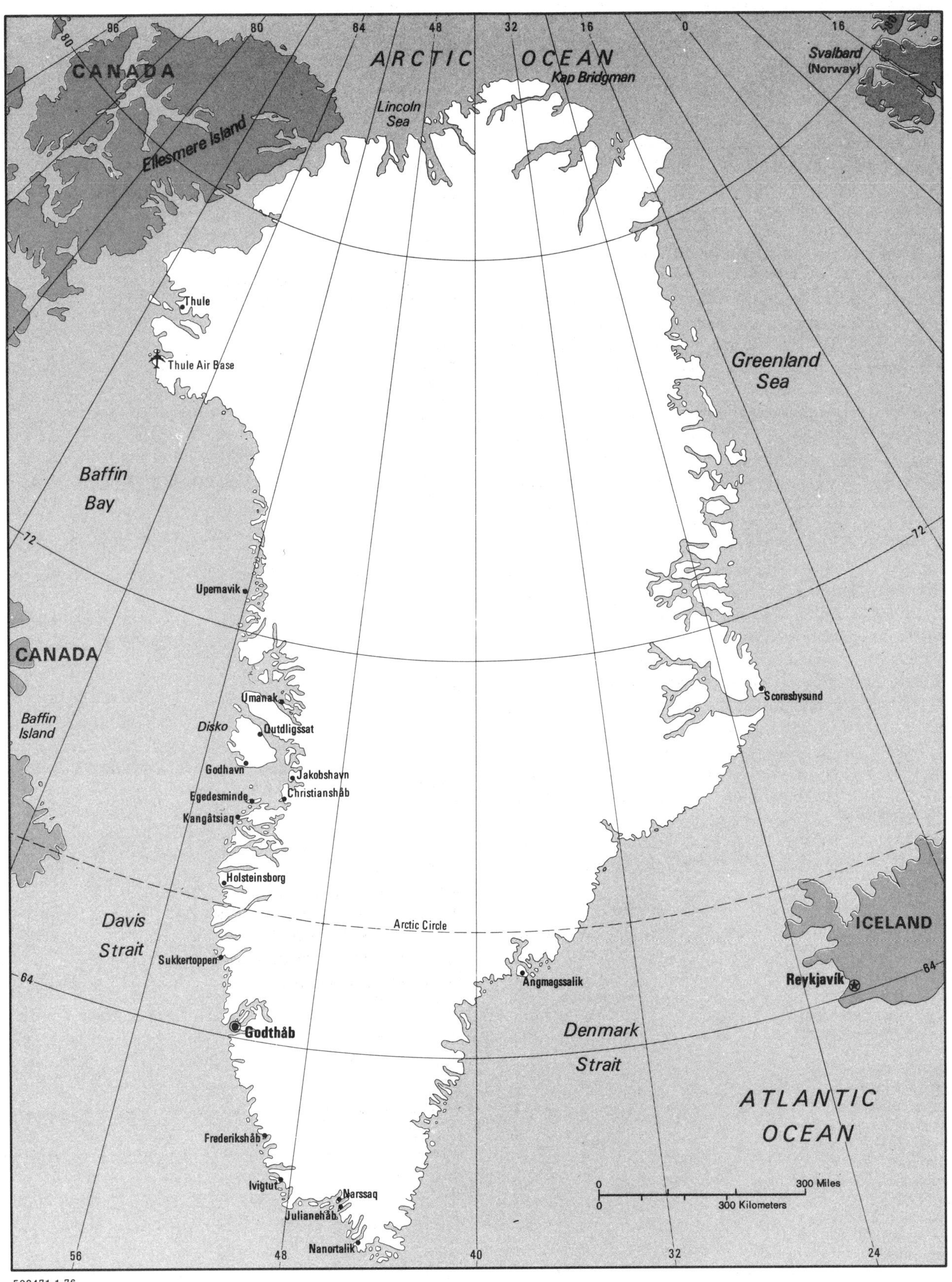

502471 1-76
Scale 1:13,000,000

Airport

Area—World Rank: 13

Population: 54,790 (1988) 61,000 (2000)

Population—World Rank: 186

Capital: Godthab (nuuk)

Land Boundaries: Nil

Coastline: 44,087 km (27,378 mi)

Longest Distances: 2,670 km (1,660 mi) N-S; 1,290 km (800 mi) E-W

Highest Point: Gunnbjorn Mountain 3,700 m (12,139 ft)

Lowest Point: Sea level

Land Use: 1% meadows and pastures; 99% other

Greenland is the largest island in the world. Of its total land area 84% lies under its icecap. Of the ice-free area, some 150,000 sq km (38,000 sq mi) are inhabited. Greenland is bounded on the north by the Arctic Ocean, on the east by the Greenland Sea, on the southeast by Denmark Strait, on the south by the Atlantic Ocean and on the west by Baffin Bay and Davis Strait.

Greenland is a mountainous country with lofty fringes, the highest point of which is Gunnbjorns Fjaeld, 7,700 m (12,140 ft). The average thickness of the ice field is 1,515 m (4,971 ft).

Area and population	area		population
Counties	sq mi	sq km	1987 estimate
Communes			
Avanersuaq (Nordgrønland)	41,200	106,700	
Qaanaaq (Thule)	...	...	794
Kitaa (Vestgrønland)	46,000	119,100	
Aasiaat (Egedesminde)	...	...	3,524
Ilulissat (Jakobshavn)	...	...	4,522
Ivittuut (Ivigtut)	...	...	29
Kangaatsiaq (Kangâtsiaq)	...	...	1,263
Maniitsoq (Sukkertoppen)	...	...	3,992
Nanortalik	...	...	2,653
Narsaq (Narssaq)	...	...	2,131
Nuuk (Godthåb)	...	...	11,649
Paamiut (Frederikshåb)	...	...	2,611
Qaqortoq (Julianehåb)	...	...	3,436
Qasigiannguit (Christianshåb)	...	...	1,778
Qeqertarsuaq (Godhavn)	...	...	1,076
Sisimiut (Holsteinsborg)	...	...	4,948
Upernavik	...	...	2,229
Uummannaq (Umanaq)	...	...	2,583
Tunu (Østgrønland)	44,800	115,900	
Illoqqortoormiut (Scoresby-sund)	...	...	549
Tasiilaq (Angmagssalik)	...	...	2,817
TOTAL (ICE-FREE)	131,900	341,700	53,733
Permanent ice	708,100	1,833,300	
TOTAL	840,000	2,175,000	

GRENADA

BASIC FACTS

Official Name: Grenada

Abbreviation: GRD

Area: 345 sq km (133 sq mi)

Area—World Rank: 186

Population: 84,455 (1988) 134,000 (2000)

Population—World Rank: 172

Capital: Saint George's

Land Boundaries: Nil

Coastline: 121 km (75 mi)

Longest Distances: 34 km (21 mi) NE-SW; 19 km (12 mi) SE-NW

Highest Point: Mount St. Catherine 840 m (2,756 ft)

Lowest Point: Sea level

Land Use: 25% arable; 26% permanent crops; 3% meadows and pastures; 9% forest and woodland; 47% other

Grenada is located about 241 km (150 mi) SW of Barbados and 145 km (90 mi) NW of Trinidad. The country consists of the island of Grenada, the most southerly of the Windward Islands, the islands of Carriacou and Petit Martinique, and a number of smaller islets of the Grenadines, extending in an arc from Grenada to North St. Vincent. The small islets include Diamond, Ronde, Green, Sandy, Caille, Les Tantes, Frigate, Large and Saline Islands.

The island is almost wholly volcanic. The mountain mass in the center of the main island consists of a number of ridges, some of which contain crater basins and one a large crater lake, Grand Etang. Close to the NE coast there are two other crater lakes, Lake Antoine and Levera Pond. The highest peak is Mount St. Catherine, 840 m (2,756 ft) above sea level. The coastline is indented with beautiful beaches and bays.

Area and population		area		population
Parishes	**Capitals**	sq mi	sq km	1981 census
Carriacou	—	13	34	4,671
St. Andrew	—	35	91	22,425
St. David	—	18	47	10,195
St. George's	—	26	67	29,369
St. John	—	15	39	8,328
St. Mark	—	9	23	3,968
St. Patrick	—	17	44	10,132
TOTAL		133	345	89,088

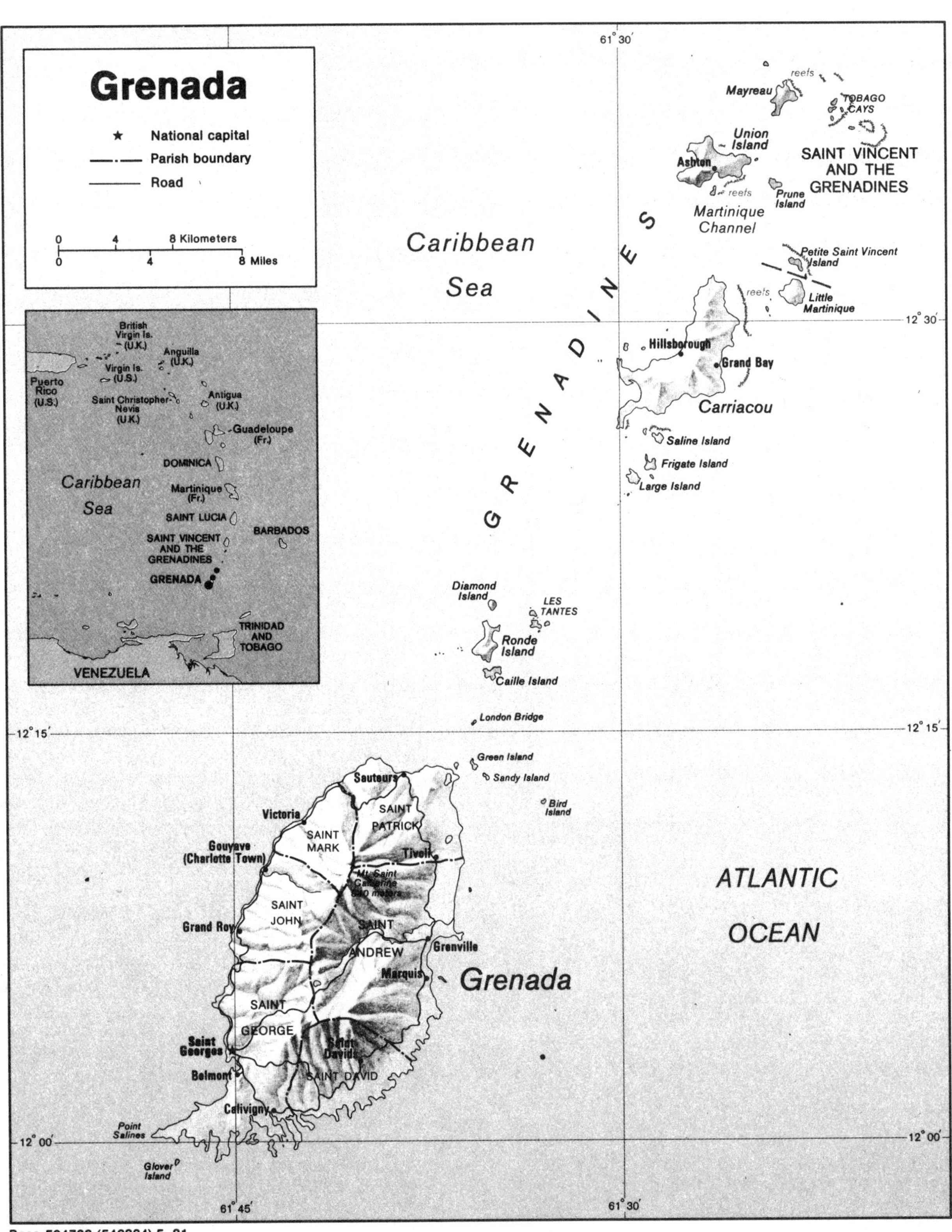

Base 504728 (546324) 5-81

GUADELOUPE

BASIC FACTS

Official Name: Department of Guadeloupe

Abbreviation: GDP

Area: 1,780 sq km (687 sq mi)

Area—World Rank: 159

Population: 338,730 (1988) 353,000 (2000)

Population—World Rank: 151

Capital: Basse-Terre

Land Boundaries: Nil

Coastline: 656 km (408 mi)

Longest Distances: 67 km (42 mi) E-W; 60 km (37 mi) N-S

Highest Point: Soufriere 1,467 m (4,813 ft)

Lowest Point: Sea level

Land Use: 18% arable; 5% permanent crops; 13% meadows and pastures; 40% forest and woodland; 24% other

Guadeloupe is the most northern of the Windward Islands group in the West Indies. It is formed by two large islands, Grande-Terre and Basse-Terre, separated by a narrow sea channel with a smaller island, Marie Galante, to the southeast and another, La Desirade, to the east. There are also a number of small dependencies, mainly Saint-Barthelemy and the northern half of Saint-Martin. Basse-Terre is volcanic; its highest peak, La Soufriere, erupted in the 18th and 19th centuries, and is still active. Grande-Terre is low, consisting of limestone plains encircled by a coral reef. Les Saintes and Saint-Barthelemy are of volcanic formation while the other islands are composed of limestone. Grande-Terre has no regular streams, but Basse-Terre's mountains receive much rainfall and feed numerous rivers.

Principal Towns: Basse-Terre (capital) 13,656, Pointe-á-Pitre 25,310 (1981 estimates).

Area and population

Arrondissements	Capitals	area sq mi	area sq km	population 1982 census
Basse-Terre	Basse-Terre	332	861	138,242
Pointe-á-Pitre	Pointe-á-Pitre	297	769	179,027
Saint-Martin–Saint Barthélemy	Marigot	29	75	11,131
TOTAL		687	1,780	328,400

GUATEMALA

BASIC FACTS

Official Name: Republic of Guatemala

Abbreviation: GTM

Area: 108,889 sq km (42,042 sq mi)

Area—World Rank: 100

Population: 8,831,148 (1988) 12,222,000 (2000)

Population—World Rank: 70

Capital: Guatemala

Boundaries: 1,931 km (1,200 mi); Belize 261 km (162 mi); Mexico 925 km (575 mi); Honduras 248 km (154 mi); El Salvador 167 km (104 mi)

Coastline: Caribbean Sea 85 km (53 mi); Pacific Ocean 245 km (152 mi)

Longest Distances: 457 km (284 mi) NNW-SSE; 428 km (266 mi) ENE-WSW

Highest Point: Tajumulco Volcano 4,220 m (13,845 ft)

Lowest Point: Sea level

Land Use: 12% arable; 4% permanent crops; 12% meadows and pastures; 40% forest and woodland; 32% other

The area of Guatemala, is slightly larger than the state of Tennessee. It is bordered on the north and west by Mexico, on the south by the Pacific Ocean, on the northeast by Belize, on the southeast by El Salvador, and on the east by Honduras and the Gulf of Honduras.

The country has great variety in climate and landforms. The climate ranges from hot and humid in parts of the lowlands to very cold in the highlands, where frosts are common in some months and where snow falls occasionally. The climatic variation makes possible the cultivation of any crop grown in the Western Hemisphere. The altitude varies from sea level to over 3,962 m (13,000 ft) in the volcanic highlands.

Much of the country is comprised of highlands, and this is where the great majority of people continue to live, as they did when the Spanish arrived. The mountain systems are more related to those of the West Indies than to those of North and South America, trending west-east rather than north-south. They are generally highest in the west-central Departments and gradually slope eastward to the coast and to the lower mountains along the border with El Salvador and Honduras and northeastward to the lowlands of El Petén. The slopes to the Pacific incline from the volcanic axis, the backbone of the country, and are more abrupt.

Drainage to the Caribbean predominates, although there are many short and unnavigable rivers which flow to the Pacific from the southern highlands.

Dominica, Guadeloupe, and Martinique

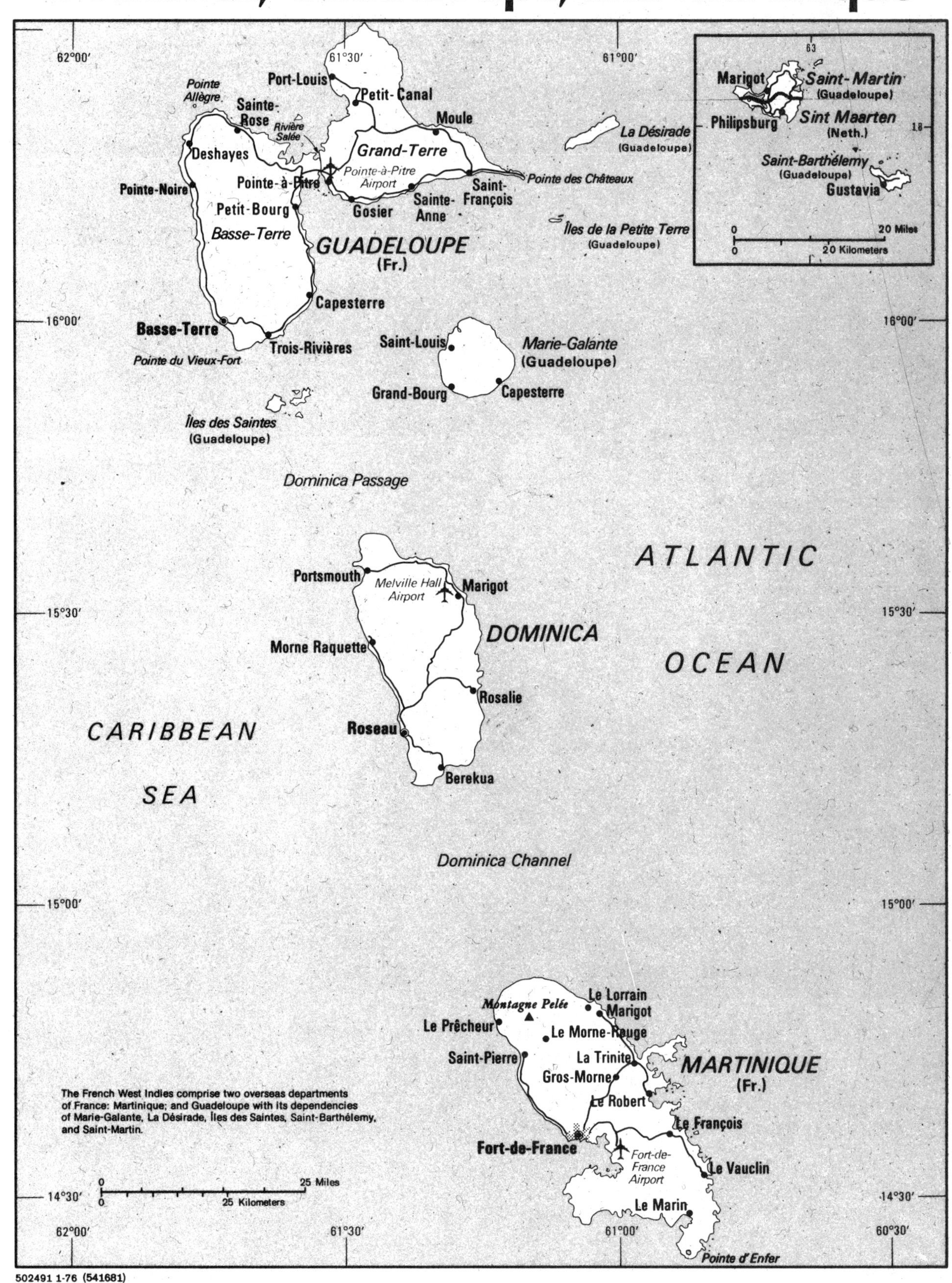

502491 1-76 (541681)
Lambert Conformal Projection
Standard parallels 14°40' and 16°10'
Scale 1:1,100,000

—— Road
✈ Airport

Base 504916 (547180) 2-82

Pacific Coastal Lands

The Pacific coast is straight and open, with no natural harbors and relatively shallow offshore waters. Long stretches of black sand line the coast.

Wet lagoons filled with mangrove lie inland from the sandy shore. The Chiquimulilla Canal, which runs 112 km (70 mi) from the port of San José to the Salvadorean border, is part of this coastal lagoon, but has been dredged to allow river traffic.

The coastal plain is predominantly savanna, interspersed with semideciduous forests which line the streams originating in the highlands. Most of the savanna is given over to cattle ranching, but there are independently owned banana holdings along the Nahualate River and in the Retalhuleu area.

Farther inland the plain begins a somewhat steeper, more dissected ascent to the highlands through the upper piedmont, 609 to 1,524 m (2,000 to 5,000 ft) above sea level. Tropical broadleaf forests once covered these upper slopes.

The Highlands

The highlands above 1,524 m (5,000 ft) are covered by the remnants of a once extensive pine and oak forest which was cleared for the highland subsistence agriculture now prevailing. The forest cover disappears above 3,048 m (10,000 ft) and high-altitude bunch grass, called *páramo,* predominates.

Sierra Madre

In a system of mountains and high plateaus extending from Mexico to El Salvador and Honduras, more than 30 volcanoes, some still active, dot the southern escarpment of the Sierra Madre. The two highest are the Tajumulco 4,208 m (13,809 ft) above sea level and the Tacaná 4,053 m (13,300 ft) volcanoes. The second capital city now called Ciudad Vieja, is located between two others, Fuego 3,839 m (12,579 ft) and Agua 3,751 m (12,307 ft).

Earthquakes, related both to volcanism and to the deeply seated fault zones which lie off the southern coast, are frequent and sometimes destructive in this area. In 1773 earthquakes destroyed the old capital of Antigua. The present capital, Guatemala City, is just under 1,524 m (5,000 ft) above sea level, and is located in the Valle de la Ermita (Valley of the Hermit); but it has not proved to be safe from severe quakes. In early 1918 a series of earthquakes did great damage to the city. Buildings constructed since have enough flexibility to resist all but the most severe quakes. Much of Quezaltenango, which is almost 2,438 m (8,000 ft) above sea level, and is the nation's second largest city, was destroyed by an earthquake in 1902.

The lava plateaus and ash-filled basins, frequently as high as 2,438 m (8,000 ft) above sea level in the western section of the Sierra Madre, are often separated by deep ravines difficult to cross even on foot. Rivers falling abruptly from the mountains have cut these canyons out of the soft volcanic soil. Pockets of dense population are often isolated from one another by these ravines. Guatemala City is located in a highland valley farther east, where the monthly averages range from 61° F. in December to 67° F. in April. Average rainfall is between 101 and 152 cm (40 and 60 in) annually, with a distinct 6-month dry season.

Eighteen principal, though relatively short, rivers flow from the mountains to the Pacific Ocean. They are navigable only for short distances in small boats, but they have great potential for the production of hydroelectric power and, in fact, serve to supply the major portion of electric power available in the country.

There are two important lakes of volcanic origin in the Sierra Madre highlands. Lake Atitlán in the Department of Sololá is said to be one of the most beautiful lakes in the world. The volcanoes Atitlán, San Pedro, and Tolimán line its shores, as do numerous Indian villages. The inhabitants of these villages use the lake for fishing and transport between villages. The lake, over 304 m (1,000 ft) deep in places, receives a number of rivers, but its drainage is underground. Lake Amatitlán, just south of Guatemala City, is smaller and less spectacular. Steam rises from this warm-water lake, and medicinal sulfur springs are found along the banks. The nearby volcano Pacaya, which erupted in 1964, produces these effects. The lake has its outlet in the Michatoya River.

The Sierra de Chuacús branches due east from the Sierra Madre in the southern part of the Department of El Quiché. East-north-east of these mountains lie the Sierra de las Minas and the Montañas del Mico. These two chains serve as a natural barrier to communication between the Motagua River valley and the Verapaz Departments.

Sierra de los Cuchumatanes

The other mountain chain enters Guatemala from Mexico in the Department of Huehuetenango. This is the Sierra de los Cuchumatanes, a great limestone massif. The height of the Cuchumatanes plateau ranges between 2,743 and 3,352 m (9,000 and 11,000 ft).

The mountains slope away in northern and western Huehuetenango to Mexico. The topography is very rough, restricting the area available to agriculture, although there are some flood terraces which catch the alluvial soil and provide small fertile patches for corn and even some sugar, bananas, and other crops.

To the east, but separated from the Cuchumatanes by the valley of the Salinas River, lies the Sierra de Chamá. Some coffee is grown in the Cobán district of Alta Verapaz on the slopes of the Sierra, but the area is relatively isolated from major transportation routes, and good soils occur only in small wet hollows and valleys, although rainfall is heavier than in the Cuchumatanes. Still farther east and extending nearly to Livingston on the Bay of Amatique lies the Sierra de Santa Cruz, just north of the Polochic River-Lake Izabal lowland.

Caribbean Coast and River Valleys

The coast along the Gulf of Honduras is flat and open to Caribbean storms. The Bay of Amatique, however, which is 16 km (10 mi) wide and 40 km (25 mi) long, is sheltered, and the country's major port, Puerto Barrios, is located on its shores, along with the ports of Matías de Gálvez and Livingston.

Three valley corridors extend inland from the Carribbean coast. They serve to link various parts of the interior, particularly the highlands, with the Caribbean coast, but they are separated from one another by mountain ranges.

The Motagua River rises near Chichicastenango in the Department of El Quiché and flows for about 402 km (250 mi) until it empties into the Gulf of Honduras. On the last few miles of its course it serves as the boundary between Guatemala and Honduras. It is navigable for the last 193 km (120 mi) of its length. It receives a number of rivers, one being the Hondo River which serves to supply the city of Zacapa with electricity. The Motagua River valley approaches Guatemala City from the Caribbean coast but branches west in the Department of El Progreso.

The swampy Polochic River-Lake Izabal lowland lies north of the Sierra de las Minas and the Montanãs del Mico and is separated from the Motagua River valley by them. The Polochic River rises in Alta Verapaz and flows west, emptying into Lake Izabal, the largest lake in the country. The lake empties into the Dulce River, which in turn empties into the Bay of Amatique at the port of Livingston. The lake, which is 43 km (27 mi) long and 19 km (12 mi) wide, and the Dulce River are navigable throughout their entire lengths. This is the main corridor linking the Caribbean coast with the Verapaz Departments.

The Sarstún River rises in Alta Verapaz and flows east, emptying into the Bay of Amatique. It serves, in the latter part of its course, as the boundary between Belize and Guatemala and links El Petén with the coast. The terrain and climate surrounding it are much like those of El Petén and the northern parts of the Departments of El Quiché and Alta Verapaz.

PRINCIPAL TOWNS (population at 1981 census)

Town	Population	Town	Population
Guatemala City (capital)	754,243	Puerto Barrios	46,882
		Retalhuleu	46,652
Escuintla	75,442	Chiquimula	42,571
Quezaltenango	72,922	Mazatenango	38,181

Source: CELADE.

El Petén

The vast area of El Petén, comprising about one-third of the national territory, extends as a distinct appendage into the Yucatán Peninsula. It is a rolling limestone plateau, between 152 and 213 m (500 and 700 ft) above sea level, covered with dense tropical rainforest, occasionally interspersed with wide savannas. The soils are relatively poor for agriculture. The annual rainfall is heavy, averaging 203 cm (80 in) in the north and 441 cm (150 in) in the south.

Because of the porosity of the soil, much of the drainage is underground, though there are many lake basins which overflow and flood the land when the rains are particularly heavy. Most of the rivers flow either through Mexico, emptying into the Gulf of Mexico, or through Belize emptying into the Gulf of Honduras.

Area and population

Departments	Capitals	area sq mi	area sq km	population 1986 estimate
Alta Verapaz	Cobán	3,354	8,686	506,800
Baja Verapaz	Salamá	1,206	3,124	161,500
Chimaltenango	Chimaltenango	764	1,979	298,100
Chiquimula	Chiquimula	917	2,376	227,800
El Progreso	Progreso	742	1,922	98,200
Escuintla	Escintla	1,693	4,384	467,300
Guatemala	Guatemala City	821	2,126	1,747,500
Huehuetenango	Huehuetenango	2,857	7,400	609,600
Izabal	Puerto Barrios	3,490	9,038	278,600
Jalapa	Jalapa	797	2,063	168,600
Jutiapa	Jutiapa	1,243	3,219	318,800
Petén	Ciudad Flores	13,843	35,854	192,800
Quezaltenango	Quezaltenango	753	1,951	485,700
Quiché	Santa Cruz	3,235	8,378	491,700
Retalhuleu	Retalhuleu	717	1,856	206,100
Sacatepéquez	Antigua Guatemala	180	465	155,400
San Marcos	San Marcos	1,464	3,791	609,600
Santa Rosa	Cuilapa	1,141	2,955	242,000
Sololá	Sololá	410	1,061	207,400
Suchitepéquez	Mazatenango	969	2,510	316,100
Totonicapán	Totonicapán	410	1,061	258,100
Zacapa	Zacapa	1,039	2,690	147,400
TOTAL		42,042	108,889	8,195,100

The Salinas River rises in Huehuetenango and flows north to contribute to the Usumacinta River, which empties into the Gulf of Mexico. The two rivers form part of the border between Mexico and El Petén. The Pasión River, which rises in northern Alta Verapaz and flows north and west in El Petén, serves as a link between Cobán, the capital of Alta Verapaz, and El Petén. It also contributes to the Usumacinta, as does the San Pedro River, which rises north of Flores, capital of El Petén. The Belize River and the Azul River both rise in El Petén and empty into the Caribbean. Flores is located on an island in Lake Petén Itzá, which is 15 mi long, 2 mi wide, and about 50 m (165 ft) deep. The lake has no visible outlet because its drainage is underground.

GUINEA

BASIC FACTS

Official Name: Republic of Guinea

Abbreviation: GUN

Area: 245,857 sq km (94,926 sq mi)

Area—World Rank: 75

Population: 6,909,298 (1988) 8,879,000 (2000)

Population—World Rank: 86

Capital: Conakry

Boundaries: 3,820 km (2,374 mi); Senegal 330 km (205 mi); Mali 932 km (579 mi); Ivory Coast 605 km (376 mi); Liberia 563 km (350 mi); Sierra Leone 652 km (405 mi); Guinea-Bissau 386 km (240 mi)

Coastline: 352 km (219 mi)

Longest Distances: 831 km (516 mi) SE-NW; 493 km (306 mi) NE-SW

Highest Point: Mount Nimba 1,752 m (5,748 mi)

Lowest Point: Sea level

Land Use: 6% arable; 12% meadows and pastures; 42% forest and woodland; 40% other

Guinea is situated on the southwestern edge of the great bulge of West Africa, between roughly 7° and 12.5° north of the equator. From its westernmost limit bordering on the Atlantic Ocean, the country curves inland in a great southeasterly-bearing arc that averages some 241 to 321 km (150 to 200 mi) in width. On the arc's inner perimeter lie Sierra Leone and northern Liberia, and on its outer edge are the states of Guinea-Bissau, Senegal, Mali, and Ivory Coast.

The country has a varied terrain that ranges from wide coastal marshes and an inner lying plain along the Atlantic Ocean to high central plateaus, a region of broad savannas in the east, and a combination of mountains and plains in the southeast.

The government distinguishes four geographic regions each of which is characterized by different morphological features and a somewhat different climate. Additionally, in each of three regions a different major ethnic group predominates, and in the fourth region are groups having cultural and organizational similarities. The regions include Lower Guinea (also known as Maritime Guinea); Middle Guinea, consisting principally of the central highland area (the Fouta Djallon); Upper Guinea, a region of savannas; and the Forest Region, which includes the Guinea Highlands and rain forests of southeastern Guinea.

Lower Guinea

Lower Guinea stretches inland from the Atlantic Ocean to the main mass of the Fouta Djallon. The coast is of the submerged kind lined by broad marshes through which drowned rivers (estuaries) open onto the sea. This coastal strip is broken at only two points where spurs of resistant rock formations jut into the ocean; one is found at Cape Varga in the north, and the other is the Camayenne (or Kaloum) Peninsula on which Conakry is situated. Tides are high along the entire coast, reaching fifteen or more feet, which results in brackish water in estuaries many miles inland.

Behind the coastal swamps lies an alluvial plain averaging about thirty miles wide but considerably narrower in its central section. Soils tend to be soggy during periods of heavy rain, but the continuous, generally humid heat favors agriculture. Crops consist of rice, fonio (a variety of millet), and maize (corn); kola trees and oil palms are also widely grown, the latter being a characteristic feature of the landscape. Banana plantations have been developed in the southern part, and pineapples are also cultivated there.

To the east of this plain, the region rises in a series of foothills that merge into the Fouta Djallon. In the south, in particular, these foothills occur in steep steps having escarpments from several hundred to well over 304 m (1,000 ft) high. The foothills area was included in the maritime region primarily because of the greater ethnic and economic relationships it has with this region than with the Fouta Djallon.

Middle Guinea

The Fouta Djallon highland mass constituting most of Middle Guinea consists of a complex of elevated, relatively level plateaus. About 12,950 sq km (5,000 sq mi) of this area are over 914 m (3,000 ft) above sea level. The plateaus are deeply cut in many places by narrow valleys, many of which run at roughly right angles, giving the region a checkerboard appearance. A number of major valleys extend for long distances, providing important lines of communication; the railroad from Conakry to Kankan runs in part through one of these valleys.

Much of the plateau area is inhabited by the Peul, who raise large numbers of cattle there. Agriculture is frequently difficult because of the hard lateritic soil crust, and the main crops are grown in the valleys. In certain places, as near Labé, soils derived from igneous rocks are rich and permit cultivation of coffee, bitter oranges, and jasmine (used in perfumes). In some areas pineapples are grown, and banana plantations exist in wetter valleys able to provide adequate moisture for growth throughout the dry season.

Upper Guinea

The principal feature of Upper Guinea, which lies to the east of the Fouta Djallon, is the extensive lightly wooded, tall grass savannas. This savanna area is interrupted, however, by a long rocky spur extending eastward along the Mali border, from the Fouta Djallon, for over 160 km (100 mi). Shorter spurs are also found east of Dabola, and in the area west of Siguiri rounded granite domes rise above the plain. The southern limit of the region is generally marked by the northwest-southeast trending Guinea Highlands.

The region has an average altitude of about 304 m (1,000 ft). Hard lateritic crust underlies much of the savanna, and agriculture is practiced mainly in the river valleys, which in the case of the principal tributaries of the Niger River extend for hundreds of miles. The main crops include wet rice, fonio, groundnuts (peanuts), and sweet potatoes. Cattle raising is an important industry on the savanna but comparatively less so than in the Fouta Djallon.

Forest Region

The Forest Region encompasses the southeastern corner of Guinea. Its major feature is the Guinea Highlands, which have general elevations ranging from about 457 m

Guinea
International boundary
Region boundary
National capital
Region capital
Railroad
Road
Regions have the same names as their capitals.
0 100 Kilometers
0 100 Miles
SENEGAL
MALI
GUINEA-BISSAU
THE GAMBIA
SIERRA LEONE
LIBERIA
COTE D'IVOIRE
North Atlantic Ocean
Conakry
Freetown
Monrovia
Bamako
Tambacounda
Georgetown
Dialakoto
Velingara
Kolda
Farim
Mansôa
Bafatá
Gabú
Buba
Catió
Koundara
Kédougou
Kali
Mali
Gaoual
Koubia
Tougué
Labé
Lélouma
Pita
Sangarédi
Boké
Télimélé
Kamsar
Dalaba
Fria
Boffa
Mamou
Kindia
Dubréka
Coyah
Forécariah
Dinguiraye
Bissikrima
Dabola
Sisséla
Kouroussa
Siguiri
Niandakoro
Mandiana
Kankan
Faranah
Falaba
Kabala
Kissidougou
Kérouané
Beyla
Guéckédou
Macenta
Voinjama
Nzérékoré
Lola
Yomou
Yekepa
Gahnpa
Gbarnga
Mange
Makeni
Lunsar
Sefadu
Pepel
Waterloo
Bo
Kenema
Momaigi
Sulima
Robertsport
Tubmanburg
Harbel
Buchanan
Pyne Town
Guiglo
Touba
Odienne
Bougouni
Koulikoro
Kita
Niagassola
Bafoulabé
Dialafara
Niger
Gambie
Bafing
Bakoye
Baoulé
Tinkisso
Sankarani
Niantan
Milo
Diani
Mongo
Little Scarcies
Bagbe
Sewa
Jong
Mano
Gbeya
Saint Paul
Saint John
Dube
Koliba
Falémé
Sandougou
Rio Geba
Kogon
Tominé
Konkouré
ferry
15
12
9
6

(1,500 ft) above sea level in the west to over 914 m (3,000 ft) in the east; peaks at several points attain 1,219 m (4,000 ft) and above. The topography, consisting mainly of rounded and dome-shaped surfaces, contrasts sharply with the plateau terrain of the Fouta Djallon. The difference is emphasized further by the dense rain forest—now largely secondary growth—that is the usual cover in areas below 609 m (2,000 ft). Higher areas are more lightly forested, and some detached hills have crests of bare rocks.

PRINCIPAL TOWNS (population at December 1972)
Conakry (capital) 525,671 (later admitted to be overstated); Kankan 60,000.

Area and population

Regions	Capitals	area sq mi	area sq km	population 1983 census
Beyla	Beyla	6,738	17,452	161,347
Boffa	Boffa	1,932	5,003	141,719
Boké	Boké	3,881	10,053	225,207
Conakry	Conakry	119	308	705,280
Coyah (Dubréka)	Coyah	2,153	5,576	134,190
Dabola	Dabola	2,317	6,000	97,986
Dalaba	Dalaba	1,313	3,400	132,802
Dinguiraye	Dinguiraye	4,247	11,000	133,502
Faranah	Faranah	4,788	12,400	142,923
Forécariah	Forécariah	1,647	4,265	116,464
Fria	Fria	840	2,175	70,413
Gaoual	Gaoual	4,440	11,500	135,657
Guéckédou	Guéckédou	1,605	4,157	204,757
Kankan	Kankan	7,104	18,400	229,861
Kérouané	Kérouané	3,070	7,950	106,872
Kindia	Kindia	3,409	8,828	216,052
Kissidougou	Kissidougou	3,425	8,872	183,236
Koubia	Koubia	571	1,480	98,053
Koundara	Koundara	2,124	5,500	94,216
Kouroussa	Kouroussa	4,647	12,035	136,926
Labé	Labé	973	2,520	253,214
Lélouma	Lélouma	830	2,150	138,467
Lola	Lola	1,629	4,219	106,654
Macenta	Macenta	3,363	8,710	193,109
Mali	Mali	3,398	8,800	210,889
Mamou	Mamou	2,378	6,160	190,525
Mandiana	Mandiana	5,000	12,950	136,317
Nzérékoré	Nzérékoré	1,460	3,781	216,355
Pita	Pita	1,544	4,000	227,912
Siguiri	Siguiri	7,626	19,750	209,164
Télimélé	Télimélé	3,119	8,080	243,256
Tougué	Tougué	2,394	6,200	113,272
Yomou	Yomou	843	2,183	74,417
TOTAL		94,926	245,857	5,781,014

The areas around Beyla and Nzérékoré consists of rolling plains. At one time probably covered by rain forest, the plains' present vegetation is mainly derived savanna. Southeast of Nzérékoré are the Nimba Mountains on the Liberian and Ivory Coast frontiers. Guinea's highest point, Mount Nimba 1,752 m (5,748 ft) is in this range.

Agriculture in the Guinea Highlands includes the cultivation of rice, maize, cassava, kola and oil palm trees, bananas, and coffee. Tobacco is also grown in the plains areas, and cattle are also raised. The arbitrary nature of the colonial political division of West Africa is well illustrated in the Forest Region, where the easiest line of communication to the coast are through neighboring Liberia and Sierra Leone rather than through the rest of Guinea. The borders with those two countries also artificially divide peoples of the same principal ethnic groups in the area.

DRAINAGE

Guinea is the source of over one-half of West Africa's principal rivers and many of their tributaries, which rise either in the Fouta Djallon or the Guinea Highlands of the Forest Region. The two highlands, moreover, form the drainage divide between the upper Niger River basin and the rivers that flow westward through Guinea, Sierra Leone, and Liberia to the Atlantic Ocean. In the north the Fouta Djallon also separates the watersheds of the Niger, Gambie (in The Gambia, known as the Gambia), and Sénégal rivers. The Gambie actually rises in the Fouta Djallon, and a major tributary of the Sénégal, the Bafing (in reality the upper course of the Sénégal), also has its origin there.

The fan-shaped system of the Niger River, which originates in the Guinea Highlands, drains over one-third of the country's total area including most of Upper Guinea and the Forest Region. In the west Lower Guinea is crossed by many usually short rivers, which originate either in the Fouta Djallon or in its foothills. Among the more important for navigation purposes are the Rio Nunez, which debouches through the Rio Nunez estuary; the Fatala, emptying into the Rio Pongo estuary; and the Mélikhouré, near the Sierra Leone border. The Konkouré River, situated north of Conakry, has little navigation value but has important potential for hydroelectric power development.

GUINEA-BISSAU

BASIC FACTS

Official Name: Republic of Guinea-Bissau

Abbreviation: GBS

Area: 36,125 sq km (13,948 sq mi)

Area—World Rank: 122

Population: 950,742 (1988) 1,200,000 (2000)

Population—World Rank: 133

Capital: Bissau

Boundaries: 1,122 km (697 mi); Senegal 338 km (210 mi); Guinea 386 km (240 mi)

Coastline: 398 km (247 mi)

Longest Distances: 336 km (209 mi) N-S; 203 km (126 mi) E-W

Highest Point: 310 m (1,017 ft)

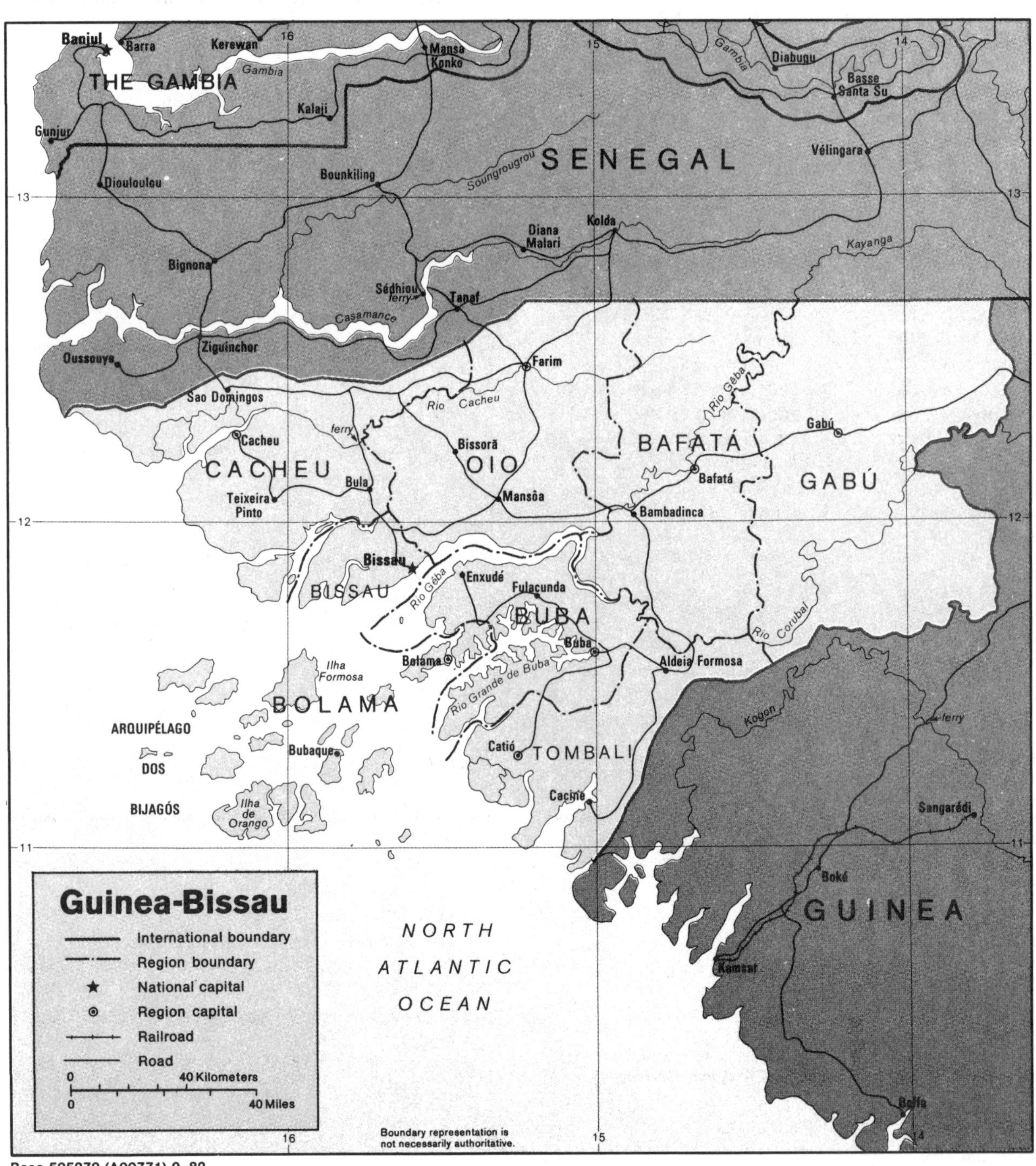

Base 505270 (A00771) 9-82

Lowest Point: Sea level

Land Use: 9% arable; 1% permanent crops; 46% meadows and pastures; 38% forest and woodland; 6% other

Guinea-Bissau is located on the west coast of West Africa. It consists of a mainland, the Bijagos Archipelago, and various coastal islands, such as Jeta, Bolama, Melo, Pecixe, Bissau, Areicas and Como. The Bijagos Archipelago consists of over 18 islands, among them Caravela, Caraxe, Formosa, Uno, Orango, Orangozinho, Bubaque and Roxa. The mainland relief consists of a coastal plain and a transition plateau forming the planalto de Bafata in the center and the planalto de Gabu abutting on the Fouta Djallon. The highest elevation is 244 m (800 ft) in the southeast.

The country is drained by a number of meandering rivers flowing into the Atlantic through wide estuaries. The main rivers are Cacheu, also known as Farim for part of its course, the Mansoa, the Geba, the Corubal, the Rio Grande and, on the southern border with Guinea, the Cacine. These rivers provide the principal means of transportation. Ocean-going vessels of shallow draught can reach most of the main towns, and flat-bottomed tugs and barges can reach smaller settlements except those in the northeast.

Principal Towns: Bissau (capital) 109,214, Bafatá 13,429, Gabú 7,803, Mansôa 5,390, Catió 5,170, Cantchungo 4,965, Farim 4,468 (census of April 1979).

Area and population

		area		population
Regions	**Capitals**	sq mi	sq km	1979 census
Bafatá	Bafatá	2,309	5,981	115,656
Bissau	Bissau	324	840	51,796
Bolama	Bolama	1,013	2,624	25,449
Cacheu	Cacheu	1,998	5,175	127,514
Gabú	Gabú	3,533	9,150	103,683
Oio	Farim	2,086	5,403	131,271
Quinara	Fulacunda	1,212	3,138	35,567
Tombali	Catió	1,443	3,736	55,088
Autonomous Sector				
Bissau	—	30	78	107,281
TOTAL		13,948	36,125	753,305

GUYANA

BASIC FACTS

Official Name: Cooperative Republic of Guyana

Abbreviation: GUY

Area: 215,000 sq km (83,000 mi)

Area—World Rank: 81

Population: 765,796 (1988) 888,000 (2000)

Population—World Rank: 135

Capital: Georgetown

Boundaries: 3,014 km (1,873 mi); Suriname 726 km (451 mi); Brazil 1,208 km (751 mi); Venezuela 650 km (404 mi)

Coastline: 430 km (267 mi)

Longest Distances: 807 km (501 mi) N-S; 436 km (271 mi) E-W

Highest Point: Mount Roraima 2,772 m (9,094 ft)

Lowest Point: Sea level

Land Use: 3% arable; 6% meadows and pastures; 83% forest and woodland; 8% other

The Guiana region (including Guyana, Suriname, French Guiana and parts of Brazil and Venezuela) is an area bounded by the Orinoco River on the West and Northwest, by the Amazon and Rio Negro on the South, and by the Atlantic Ocean on the East and Northeast. Guyana occupies approximately 10 percent of this area. Guyana, in one of the native Amerindian dialects, means "land of waters." The country has an area equal in size to the state of Idaho, and extends between 1° and 9° North latitude and from 56° to 62° West longitude.

Although Guyana is more than 322 km (200 miles) from the Caribbean, it can be viewed as a Caribbean "sugar island" perched on the northeast shoulder of South America but separated from the "mainland" by swamps, a few miles inland.

Guyana is divided into three main geographical zones, within which there are several additional geological features. On the coast there is a narrow belt of alluvial soils most of which lies below high-tide level and is protected by a system of sea defenses and canals. This coastal plain occupies approximately five percent of the total land area. South of this zone, Guyana is covered by a lush equatorial forest extending to the borders of Brazil and Venezuela over 181,299 sq km (70,000 sq mi) territory or 84 percent of the total area. The third main geographical area is the savannah grassland which lies behind the coastal belt in the northeast and beyond the forests in the southwest. These savannahs occupy the remaining 11 percent of Guyana and support only low grasses useful for land-extensive cattle grazing.

THE COASTAL PLAIN

The coastal plain of northeastern Guyana was first settled in the early 18th century when the Dutch, having exhausted the riverside soils along the near interior rivers, were forced to begin reclamation of the plain from the sea tides and inland river swamps. The plain became heavily settled in the 19th century after the British gained control. They introduced large numbers of East Indians to work with the sugar plantations after slavery was abolished in 1833. Due to the extensive system of sea defenses, the coastal plain, if it were not for the many palms and other tropical vegetation, looks much like the coast of the Netherlands. The sea defenses include 225 km (140 mi) of seawall, and canals for irrigation and drainage.

The plain ranges from 4 to 6 km (3 to 4 mi) wide and extends from the Corentyne River on the border with Suriname to the Venezuelan border in the northwest, but it is not settled beyond the Pomeroon River on the "near" northwest coast where the plain narrows and the soils become less fertile. The plain, much of it below sea level is rarely cultivated beyond 16 km (10 mi) inland, and the first mile behind the seawall is generally used for either pasturage or rice, which can survive the higher water table and higher salinity of the soil.

The coastal plain is made up largely of alluvial muds from the Amazon River which have been carried and deposited along the northeast shoulder of South America by the south equatorial current as it splits on the horn of Brazil and travels along the coast past the Guianas. This mud, a rich clay of high fertility, overlays the white sands and clays formed from the erosion of the interior bedrock and carried seaward by the rivers of Guyana.

The northeastern coastal plain is divided into four distinct sections, and travel is made difficult by the three major Guyanese rivers. They are, from east to west the Berbice, the Demerara and the Essequibo. The plain is cut off from the interior of forest zone by a barrier of swamps which has formed between the white sandy hills of the interior and the "back-dams" of the coastal plain drainage and irrigation works. These swamps, prevented from intruding into the croplands by the "back-dams," serve a useful purpose as water conservancies or natural reservoirs during periods of drought.

THE FOREST ZONE

By far the largest of Guyana's regions, this 181,299 sq km (70,000 sq mi) equatorial forest is also the most geologically complex. It includes nearly all of Guyana's known mineral resources. Its vast hardwood lumber reserves are largely inaccessible but still represent a great natural resource.

Geologically, from the coastal plain or about 64 km (40 mi) from the coast, the forest zone consists of a peneplain (an eroded plateau), which, as a whole, presents a relief of softly undulating, forest-covered hills running toward the northeast and ranging from 122 m (400 ft) in elevation in the interior to 15 m (50 ft) near the coastal plain. This area, also known in part as the white sands (or *zanderij*) belt is from 128 to 160 km (80 to 100 mi) in width and extends from the Pomeroon River into Suriname. These low sandy hills are laced with metamorphic and igneous intrusions of basic rocks and the relief is accentuated by ranges of rockier hills running out from the Kaieteurian Plateau and the Pakaraima Mountains. These hills, not yet worn away by erosion, are marked by rapids and falls as the rivers descend through them.

Although very infertile, being almost pure quartz, the white sands support a dense hardwood forest. If these trees are removed, erosion is rapid and severe. Guyana's valuable bauxite reserves are also concentrated in the white sands belt. They lie in strata, 24 km (15 mi) wide and 9 m (30 ft) thick from the Pomeroon into Suriname. Lying under nearly 30 m (100 ft) of sand, the deposits may be associated with an ancient coastline. The white sands occupy an area of over 15,540 sq km (6,000 sq mi) in the central near interior of the forest zone. They thin out in the northwest and merge into other loose sedimentary soils in the higher elevations of the zone.

The Kaieteurian Plateau and the Pakaraima Mountains dominate west-central Guyana. The plateau is generally about 610 m (2,000 ft) in elevation, and mountains rise as high as 2,743 m (9,000 ft) near Venezuela and Brazil. This ancient crystalline plateau was once below sea level in geologic times and is overlain with sandstones and shales of sedimentary origin. The mountains and hills above the plateau are the result of intrusions of basic eruptive rocks. This mountain range separates Guyana from the Orinoco River watershed, and is the source of several of the main tributaries of the Essequibo, especially the Mazaruni, which may prove to have considerable hydroelectric potential.

In the northwest of Guyana this heavily forested plateau of weathered bedrock extends, although much lower in elevation, almost to the coast itself, leaving a very narrow coastal plain near the river mouths.

THE SAVANNAHS

The savannahs of Guyana occupy, when mountain areas encircled by them are included, nearly 25,890 sq km (10,000 sq mi). The largest savannah, the Rupununi, divided into a northern and southern section by the Kanuku Mountains, covers approximately 15,540 sq km (6,000 sq mi) and lies in the extreme southwestern part of Guyana. The Kanuku Mountains reach nearly 914 m (3,000 ft) and cover 4,531 sq km (1,750 sq mi) while the savannah itself is approximately 76 m (250 ft) above sea level on either side of the Rupununi River which floods and covers the savannah during the rainy season. A second area of savannah, the "intermediate savannah," lies about 96 km (60 mi) inland from the Berbice coast and, enclosed by the forests of the white sands belt, covers nearly 5,180 sq km (2,000 sq mi).

Both savannahs support only sparse grasses for pasturage on a land extensive basis—from 16 to 28 ha (40 to 70 ac) per head. There is also some shifting cultivation by the Amerindians along the Rupununi's banks and in the foothills of the Kanuku Mountains. The Rupununi is very isolated; Lethem, the administrative headquarters, near the Brazilian border in the heart of the savannah country, is over 322 km (200 mi) from the nearest paved road and air freight is the only practical way to get the locally produced beef to market in Georgetown. Yet, because it is flat and not covered with equatorial forest, many Guyanese, while still living on the coastal plain, have high hopes for development in the Rupununi, about which more is known than perhaps any other area of the interior.

Guyana

International boundary
National capital
Railroad
Road
International airport

Boundary representation is not necessarily authoritative

DRAINAGE

Drainage in Guyana is generally from the south and southwest toward the Atlantic in the north and northeast. The four principal rivers, from east to west, the Corentyne (which forms the border with but lies within Suriname), the Berbice, the Demerara and the Essequibo, form the main coastal plain of 193 km (120 mi). The remainder of the coastline 241 km (150 mi) in the northwest is also cut by several smaller rivers. Between the four major rivers there are several rivers (the Canje, Mahaicomy, Abary and Mahaica), of local importance, that originate in the "intermediate savannah." The main rivers and their tributaries originate well into the interior. The Essequibo, its headwaters in Brazil, is over 965 km (600 mi) long, drains more than half of Guyana and has an estuary 22 km (14 mi) wide. The four main tributaries of the Essequibo, the Mazaruni, the Cuyuni, the Potaro and the Rupununi all flow into the river from the west, and with the exception of the Rupununi, all drain the Kaieteurian Plateau.

Area and population		area		population
Administrative Regions		sq mi	sq km	1980 census
Region 1	(Barima/Waini)	. . .	. . .	18,297
Region 2	(Pomeroon/Supenaam)	. . .	. . .	42,268
Region 3	(Essequibo Islands/ West Demerara)	. . .	. . .	104,747
Region 4	(Demerara/Mahaica)	. . .	. . .	318,952
Region 5	(Mahaica/Berbice)	. . .	. . .	53,862
Region 6	(East Berbice/Corentyne)	. . .	. . .	152,517
Region 7	(Cuyuni/Mazaruni)	. . .	. . .	14,142
Region 8	(Potaro/Siparuni)	. . .	. . .	4,265
Region 9	(Upper Takutu/ Upper Essequibo)	. . .	. . .	13,051
Region 10	(Upper Demerara/ Berbice)	. . .	. . .	36,518
TOTAL		83,000	215,000	758,619

Guyana, as a whole, is part of the Amazon-Orinoco watershed and the headwaters of the Essequibo and its tributaries are often confused with those of the Amazon. Overall drainage is rather poor as the average gradient of the main rivers is only one foot per mile. Even in the mountains and savannahs of Guyana there are extensive swamps and flooding, and all new land projects require extensive drainage control before they are suitable for agricultural use. The rivers of Guyana are also limited to access routes to the interior as falls or low water stop navigation, for all but the smallest craft, at from 64 km (40 mi) upstream on the Essequibo, to 96 km (60 mi) on the Demerara and Corentyne, and a little more than 160 km (100 mi) on the Berbice. Their estuaries hinder movement along the coastal plain and there is periodic debate over bridging the Demerara—at present a venture of high cost.

Stream flow, but not stream level, varies widely with the seasons, June and July being the peak "flood" months, March and April being the months of least flow. Tides have considerable effect on coastal drainage problems. Used irrigation water is exhausted by gravity or pump at low tide along most of the coastal plain. The tides, during the lunar months, range from 1 m (3.5 ft) during the first and third quarters to 2.2 m (7.75 ft) during the second and fourth. This tidal flow carries upriver from 64 to 80 km (40 to 50 mi) on the four main rivers. On the smaller rivers tidal flow is negligible because the sand-bars off the coast prevent the tide from entering their small estuaries, which due to excessive silting have cut no channels to the sea.

Guyana has no beaches in the usual sense. Mud from the Amazon and from her own rivers keeps the water off Guyana a brown churning mass of sandbars, semi-liquid muds, and clouded water. At Georgetown there is a sandbar two miles at sea which limits vessels to 14 m (9 ft) of draft at low tide and 32 (20) at high. New Amsterdam's bar at the mouth of the Berbice is 5 km (3 mi) at sea and limits vessels to 2 and 5 m (7 and 16 ft) at low and high tide respectively. These bars, always shifting, force the "half-loading" of ships at Georgetown and New Amsterdam. Mud flats continue up to 24 km (15 mi) off-shore before navigation is considered free.

The drainage and irrigation system of Guyana's coastal plain is extensive and costly—so costly as to make small-scale operations not worth the effort in some areas. Both drainage and irrigation systems grew haphazardly throughout the 18th and 19th centuries. Many areas have had to be abandoned and new areas have proven very costly to open due to the poor drainage of the plain and the needs for irrigation water for flood-fallowing to control salinity, for the rice and sugar crops during "dry" years and for transport of goods. Poor drainage has also made mechanization difficult.

HAITI

BASIC FACTS

Official Name: Republic of Haiti

Abbreviation: HAT

Area: 27,400 sq km (10,579 sq mi)

Area—World Rank: 130

Population: 6,295,570 (1988) 7,118,000 (2000)

Population—World Rank: 89

Capital: Port-au-Prince

Boundaries: 1,400 km (870 mi); Dominican Republic 360 km (224 mi)

Coastline: 1,040 km (646 mi)

Longest Distances: 485 km (301 mi) ENE-WSW; 386 km (240 mi) SSE-NNW

Highest Point: La Selle Peak 2,674 m (8,773 ft)

Lowest Point: Sea level

Land Use: 20% arable; 13% permanent crops; 18% meadows and pastures; 4% forest and woodland; 45% other

Haiti occupies the western one-third of Hispaniola, the second-largest island in the Caribbean; the eastern two-thirds is occupied by the Dominican Republic. Lying about 965 km (600 mi) southeast of Florida, the island is separated from Puerto Rico on the east by the Mona Passage, and from Cuba on the west by the Windward Passage. Because these two seaways are the principal water routes linking North American and Europe with Central and South America, the histories of Haiti and the Dominican Republic have been affected by external influences with unusual frequency.

In the aboriginal language, the word *haiti* means high land. The name is appropriate, for although the highest crests do not reach elevations as great as those of neighboring Dominican Republic, intricately convoluted mountains and hills cover most of the countryside. Less than 20 percent of the land lies at elevations below 183 m (600 ft), and about 40 percent is at elevations in excess of 457 m (1,500 ft). The mountain ranges which follow a roughly east-west axis, make internal communication difficult and have contributed to the development of regionalism.

Landform and Drainage

The border between Haiti and the Dominican Republic follows an irregular line extending from north to south, but the relief features of Hispaniola follow an east-west axis. As a consequence, the principal mountain ranges and intermontane valleys are shared by the two countries. The laterally extending mountains have made internal communications difficult and have given rise to regionalism in both countries. The arrangement of intervening valleys, however, has also served to furnish easy access from one country to the other and has led to border incidents and illegal crossings of the troubled frontier that have played an important part in the histories of the two nations.

Highlands

The intricate highland pattern that covers more than three-fourths of Haiti is characterized by narrow-crested east-west ranges and spurs extending in random directions. Although there are at least five major systems and numerous spurs, the ranges meet one another to form a highland conglomerate that is discontinuous only in the south where the Cul-de-Sac lowland extends eastward from the Gulf of Gonâve (Golfe de la Gonâve) at Port-au-Prince to the Dominican frontier. The slopes of the mountains are often precipitous, but the demand for agricultural land has been so great that the steepest of mountainside plots have been tilled, and jocular but vivid tales are told of farmers falling to their deaths off their cornfields. Intensive utilization of these slopes has in many localities resulted in complete removal of the original forest cover, and erosion of the landscape is so extensive that only remnants of the natural topsoil remain.

In the north, the most extensive of the mountain systems is the Massif du Nord, which slants southeastward from the Atlantic Ocean near Port-de-Paix across the Dominican border to become the Cordillera Central. This range forms part of the Caribbean Antillean system that extends from Puerto Rico and the Virgin Islands westward across Hispaniola to Cuba. Nowhere in Haiti does it reach 1,219 m (4,000 ft) in elevation, but it is rugged and intricately dissected. Its complex geology includes sedimentary, magmatic, and plutonic rock, and limestone cliffs scar its slopes. To its west at the extremity of the island, satellite ranges extend to Môle-St.-Nicolas. To the southwest, the range called the Montagnes Noires has altitudes up to 610 m (2,000 ft) and extends laterally across the country to a point where its approaches are separated by the Artibonite River (Rivière de l'Artibonite) from the Chaîne de Mateaux, a range with a southwesterly axis that extends from the Gulf of Gonâve to the frontier and into the Dominican Republic as the Sierra de Neiba.

The Chaîne de Mateaux is separated by the Cul-de-Sac from the mountain system in the far south that extends the full length of the long southern peninsula of Haiti to the frontier and into the Dominican Republic as the Sierra de Bahoruco. In the west it is the Massif de la Hotte, and in the east it is the Massif de la Selle. The latter range has several peaks with elevations of over 2,133 m (7,000 ft), and the Morne de la Selle at 2,680 m (8,793 ft) is the country's highest peak. Extensive pine forests on the higher slopes of this range constitute the country's principal remaining timber reserve.

Lowlands

The most important of the lowland regions of the country are the Northern Plain (Plaine du Nord), the Central Plateau, Artibonite Plain (Plaine de l'Artibonite), and the Cul-de-Sac. There are also scattered stretches of narrow coastal plain and small coastal basins, as well as pockets of level land tucked into the mountains where small groups of people practice subsistence agriculture in virtually complete isolation.

The Northern Plain, which has an area of about 362 sq km (150 sq mi) located between the Atlantic Ocean and the Massif du Nord, extends eastward from near Cap-Haïtien to the Dominican border. Its rich soils are formed in part by abrasion and in part by alluvial deposition. The heartland of the plantation economy of the French colonial era, the plain is a geographical extension of the Cibao Valley in the Dominican Republic.

Southward from the Massif du Nord, the Central Plateau extends eastward from the Montagnes Noires to the Dominican frontier, where it joins the San Juan Valley. Its more than 1,351 km (840 mi) of rolling terrain make it the largest of the country's flatlands. Slightly dissected and composed of consolidated and unconsolidated sediments, the plateau has an average elevation of about 305 m (1,000 ft) and its relatively thin soils are useful principally for pasturage.

Separated from the Central Plateau by the Montagnes Noires and located to the north of the Chaîne de Mateaux, the funnel-shaped Artibonite Plain has an area of about

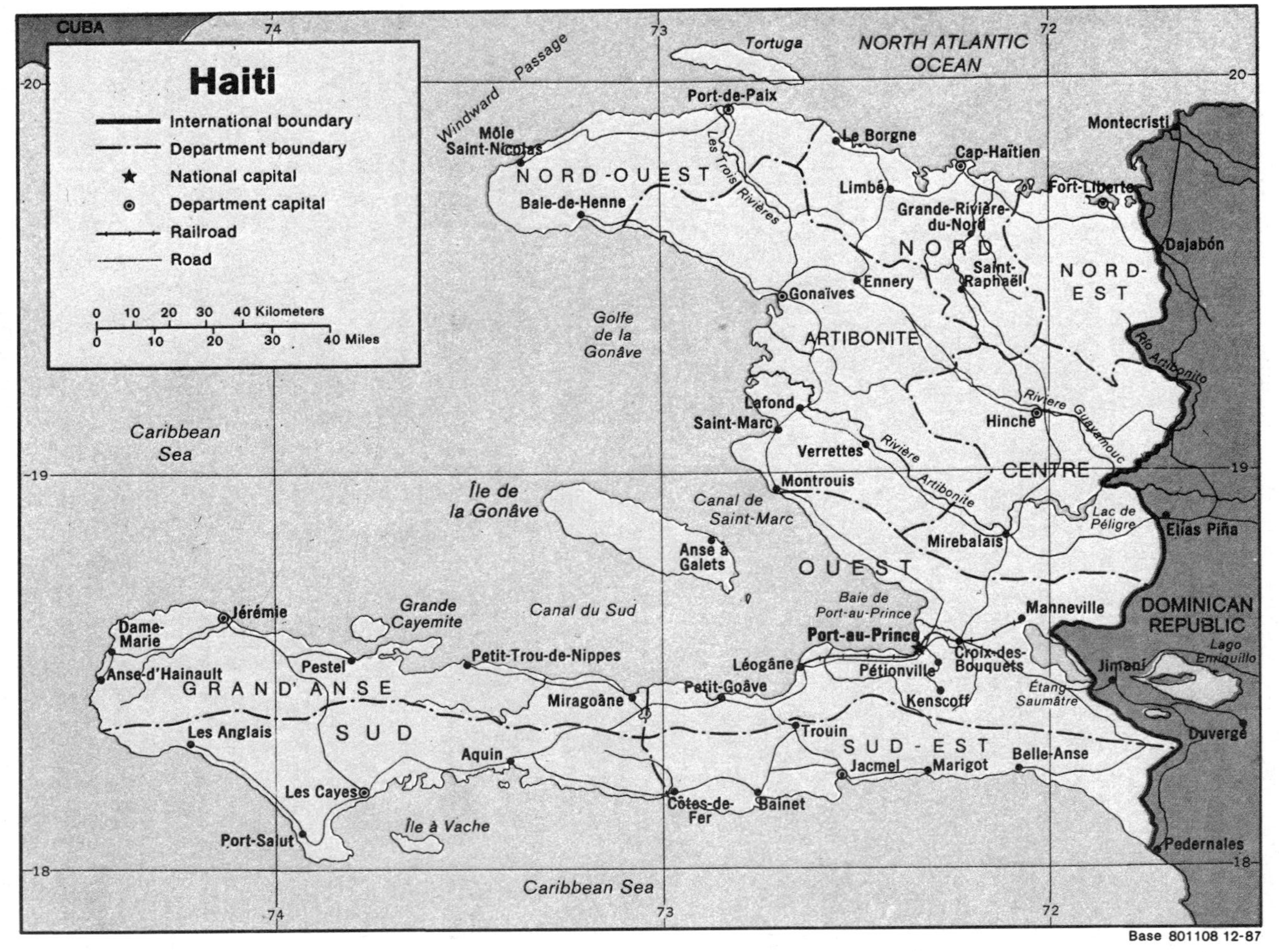
Haiti
International boundary
Department boundary
National capital
Department capital
Railroad
Road
0 10 20 30 40 Kilometers
0 10 20 30 40 Miles
CUBA
Windward Passage
Tortuga
NORTH ATLANTIC OCEAN
Caribbean Sea
Golfe de la Gonâve
Île de la Gonâve
Canal de Saint-Marc
Canal du Sud
Baie de Port-au-Prince
Grande Cayemite
Île à Vache
Caribbean Sea
NORD-OUEST
NORD
NORD-EST
ARTIBONITE
CENTRE
OUEST
GRAND' ANSE
SUD
SUD-EST
DOMINICAN REPUBLIC
Port-de-Paix
Môle Saint-Nicolas
Baie-de-Henne
Les Trois Rivières
Le Borgne
Cap-Haïtien
Limbé
Fort-Liberté
Montecristi
Grande-Rivière-du-Nord
Dajabón
Saint-Raphaël
Ennery
Gonaïves
Rio Artibonito
Lafond
Saint-Marc
Verrettes
Rivière Artibonite
Hinche
Rivière Guayamouc
Montrouis
Lac de Péligre
Elías Piña
Anse à Galets
Mirebalais
Manneville
Port-au-Prince
Croix-des-Bouquets
Pétionville
Léogâne
Lago Enriquillo
Jimaní
Étang Saumâtre
Kenscoff
Petit-Goâve
Miragoâne
Petit-Trou-de-Nippes
Jérémie
Dame-Marie
Anse-d'Hainault
Pestel
Les Anglais
Aquin
Les Cayes
Port-Salut
Côtes-de-Fer
Bainet
Trouin
Jacmel
Marigot
Belle-Anse
Duvergé
Pedernales
74
73
72
20
19
18
Base 801108 12-87

777 sq km (300 sq mi). Drained by the Artibonite River that crosses the central part of the country after rising in the Dominican Republic, it is broadest along the coast of the Gulf of Gonâve and narrows progressively to the east as the adjacent mountains encroach progressively on the river valley. The region is generally fertile, but near the coast its soils are too alkaline for intensive agriculture.

In the far south, the 388 sq km (150 sq mi) that make up the Cul-de-Sac lie between the Chaîne de Mateaux and the Massif de la Selle. Extending eastward from Port-au-Prince to the frontier, the Cul-de-Sac becomes the Neiba Valley in the Dominican Republic. It is a down-faulted depression once filled by the waters of an ocean channel that separated the mountain ridges to the south from the mainland. Later, alluvial fans formed gradually by rivers at both ends blocked off the waters of the channel, causing them to evaporate and to leave a series of sedimentary terraces and brackish lakes.

According to the Haitian Statistical Institute, there are sixteen other plains, valleys, and basins, ranging in extent from 44 to 297 sq km (seventeen to 115 sq mi). These, together with other smaller lowland areas including those on the adjacent islands, have a total area of a little more than 777 sq km (300 sq mi). In all, the lowlands cover about 22 percent of the country's territory.

Principal City (1987)
Port-au-Prince (472,895)

Area and population

Departements	Capitals	area sq mi	area sq km	population 1987 estimate
Artibonite	Gonaïves	1,750	4,532	789,019
Centre	Hinche	1,429	3,700	393,217
Grande Anse	Jérémie	1,268	3,284	514,962
Nord	Cap-Haïtien	790	2,045	602,336
Nord-Est	Fort-Liberté	676	1,752	197,669
Nord-Ouest	Port-de-Paix	899	2,330	320,632
Ouest	Port-au-Prince	1,795	4,649	1,808,274
Sud	Les Cayes	1,117	2,894	526,420
Sud-Est	Jacmel	855	2,215	379,273
TOTAL		10,579	27,400	5,531,802

Hydrography

More than 100 rivers and streams form an intricate tracery as they flow from their mountain headwaters into the Atlantic, into the Gulf of Gonâve which is formed between the extended arms of the northern and southern peninsulas, and into the Caribbean Sea. None of any size flows eastward into the Dominican Republic. In the highlands the flow is rapid and permanent, but the movement tends to slow and to meander as the watercourses reach the lowlands. The flow becomes subject to considerable seasonal change, and in many instances it is dissipated by evaporation before reaching tidewater. None of the rivers is navigable, but they are important for crop irrigation and for their hydroelectric power potential.

Much the largest of streams is the Artibonite River. It is shallow, as are the other Haitian watercourses; but it is the longest, and its flow averages ten times that of any of the others. Second in length is the Trois Rivières, which spills into the Atlantic at the town of Port-de-Paix at the gap between the Massif du Nord and the smaller ranges that cover the tip of the northern peninsula. Next is the Grande Anse, which reaches tidewater near the town of Jérémie on the southern peninsula. The Massacre River (Rivière du Massacre, better known as the Rio Dajabon) and the Pedernales River (Rivière Pedernales), both of which rise in the Dominican Republic, form portions of the Haiti-Dominican Republic border before they flow into the Atlantic Ocean and the Caribbean Sea, respectively.

The largest of the lakes, 181 sq km (70 sq mi), is the brackish Lake Saumâtre (Etang Saumâtre), which is located in the Cul-de-Sac close to the frontier and is the habitat of many exotic species of tropical wildlife. There are also several smaller natural lakes and a reservoir known as Lake Péligre (Lac de Péligre), formed by the damming of the upper Artibonite River at the point of convergence between the Montagnes Noires and the Chaîne de Mateaux. Initiated in the 1930s as a flood control project, the project also involves irrigation and hydroelectric schemes that have progressed slowly. Completion of the dam in 1956 resulted in the creation of a massive artificial lake and made possible some control over the flow of the Artibonite River, which had previously changed seasonally from a raging torrent to an uncertain trickle.

Coastal Waters and Islands

Much of the Haitian coastline is rimmed by an underwater sedimentary platform that extends around the island of Hispaniola. There are many protected anchorages, but waters close to the shoreline tend to be shallow. These depths range from about 1.2 m (four ft) at Port-de-Paix on the Atlantic coast to 3 m (ten ft) or more at Les Cayes on the Caribbean coast and Gonaïves on the Gulf of Gonâve. The platform is widest at Port-au-Prince—the country's principal port—where it spreads across most of the adjacent bay as far as Gonâve Island (Ile de la Gonâve). The platform extends continuously along the Atlantic coast where, off the port towns of Cap-Haïtien and Port-de-Paix, there are also coral reefs. A reef adjacent to Cap-Haïtien is believed to hold the remains of the flagship of Columbus, the *Santa Maria*.

The Haitian government classifies six places as maritime ports, four placed as secondary maritime ports, and an additional sixteen as ports for coastal traffic. About half are located on the Gulf of Gonâve, and the remainder are distributed equally between the Atlantic and Caribbean coasts. They tend to be shallow, however, and port improvements, including dredging, might be of considerable importance to the transportation system in a country where internal transportation is notably deficient.

The largest of the islands is Gonâve, located in a gulf of the same name off Port-au-Prince. Its area of approximately 207 sq km (eighty sq mi) is made up of rugged terrain, and its highest point, Morne la Pierre, rises to more than 762 m (2,500 ft). Second in size is Tortue Island (Ile de la Tortue), better known by its Spanish name of Tortuga. Having an area of 181 sq km (seventy sq mi), it lies in the Atlantic Ocean off Port-au-Paix. It was a major pirate stronghold during the colonial era.

Among the remaining islands, the largest are Vache Island (Ile á Vache) located off the town of Les Cayes in the Caribbean, and the Cayemites (Les Cayemites) in the Gulf of Gonâve west of the town of Jérémie. Both are surrounded by dangerous coral reefs.

HONDURAS

BASIC FACTS

Official Name: Republic of Honduras

Abbreviation: HON

Area: 112,088 sq km (43,277 sq mi)

Area—World Rank: 97

Population: 4,972,287 (1988) 6,978,000 (2000)

Population—World Rank: 95

Capital: Tegucigalpa

Boundaries: 2,170 km (1,348 mi); Nicaragua 922 km (573 mi); El Salvador 335 km (208 mi); Guatemala 248 km (154 mi)

Coastline: Caribbean Sea 591 km (367 mi); Gulf of Fonseca 74 km (46 mi)

Longest Distances: 663 km (412 mi) ENE-WSW; 317 km (197 mi) NNW-SSE

Highest Point: Mount Las Minas 2,849 m (9,347 ft)

Lowest Point: Sea level

Land Use: 14% arable; 2% permanent crops; 30% meadows and pastures; 34% forest and woodland; 20% other

Honduras is the second largest Central American republic, stretching latitudinally from the Guatemalan border on the west to the Segovia River (also known as the Coco or the Wanks River) on the east, which separates it from Nicaragua. It is mountainous but has lowland areas along both coasts and within the eastern department of Gracias a Dios.

Particularly attractive to settlers has been the deep, flat basin that winds between the mountains from the plains around San Pedro Sula to the Gulf of Fonseca. This relatively flat trail is also the site of the country's major interoceanic road system.

Over 80 percent of the land is mountainous, thereby limiting the area suitable for cultivation and pastures. Much of the small amount of cultivated area is located in the flatlands, lofty plateaus, and river valleys that are between, and parallel to, the mountains. These temperate valleys and flatlands are also the primary areas of settlement except for the north coast banana district, which was retrieved from tropical forests in the twentieth century.

Both the Atlantic and Pacific coasts have tropical climates, and the coastlands are called hot lands *(tierra caliente)* after their climate. The distance to which these tropical lands extend into the interior is less on the Pacific side, where the mountains are closer to the shore. Inland, the mountains also serve to block the penetration of moisture-laden winds from the coasts into the interior.

Topography

Two distinct series of mountain ranges divide the country roughly into two halves, the north and the south. In the north, mountain ranges extend from the Guatemala border on the west to the Plátano River on the east. These northern ranges are all extensions of the Central American Cordillera, a mountain chain that traverses Central America from Mexico to Nicaragua.

The southern half of the country is elevated by a series of mountain ranges called the Volcanic Highlands, which extend from the border with El Salvador in the southwest and across the southern part of the country to the border with Nicaragua in the east.

Mountain chains of the Central American Cordillera trend east-northeast and west-southwest. They run largely parallel to the coast and to each other. Offshore, northeast of La Ceiba on the north coast, one extension of these ranges forms the island department of the country, the Bay Islands (Islas de la Bahía).

The northern mountain ranges were formed by changes in the earth's surface several million years ago. Underneath the surface cover of limestone and sandstone, the mountains are composed of granite, mica, slate, and other materials. Some limestone and sandstone fragments have eroded from the mountain slopes to form the soil materials of the northern valleys that run between the ranges.

The Volcanic Highlands in the southern half of the country have no recognizable trend. Unlike the mountains of the north, these southern ranges are newer, consisting of lava formed by volcanic eruption some 12 million years ago. Volcanic material has both eroded and been ejected from these highlands and forms the fertile soil on which the agricultural industries of the south in Choluteca Department thrive. The last volcanic eruption occurred in 1854.

The southern Volcanic Highlands are, overall, higher than the northern Central American Cordillera chains. Of the two highest peaks in the country, one is found in the mountain chain bordering the western side of Lake Yojoa and is 2,834 m (9,300 ft) above sea level. The other is southwest of Gracias, the capital of Lempira Department, and is 2,865 m (9,400 ft) high.

In the areas between one mountain range and the other, in both the Central American Cordillera ranges and the Volcanic Highlands, are plateaus, river valleys, and savannas. These constitute the only arable level land in the country except for the north coast areas, which are planted in bananas.

These intermontane flatlands average two to seven miles in width and are flanked by mountains usually 914 to 2,133 m (3,000 to 7,000 ft) in height. In the northern half of the country they are found interspersed between

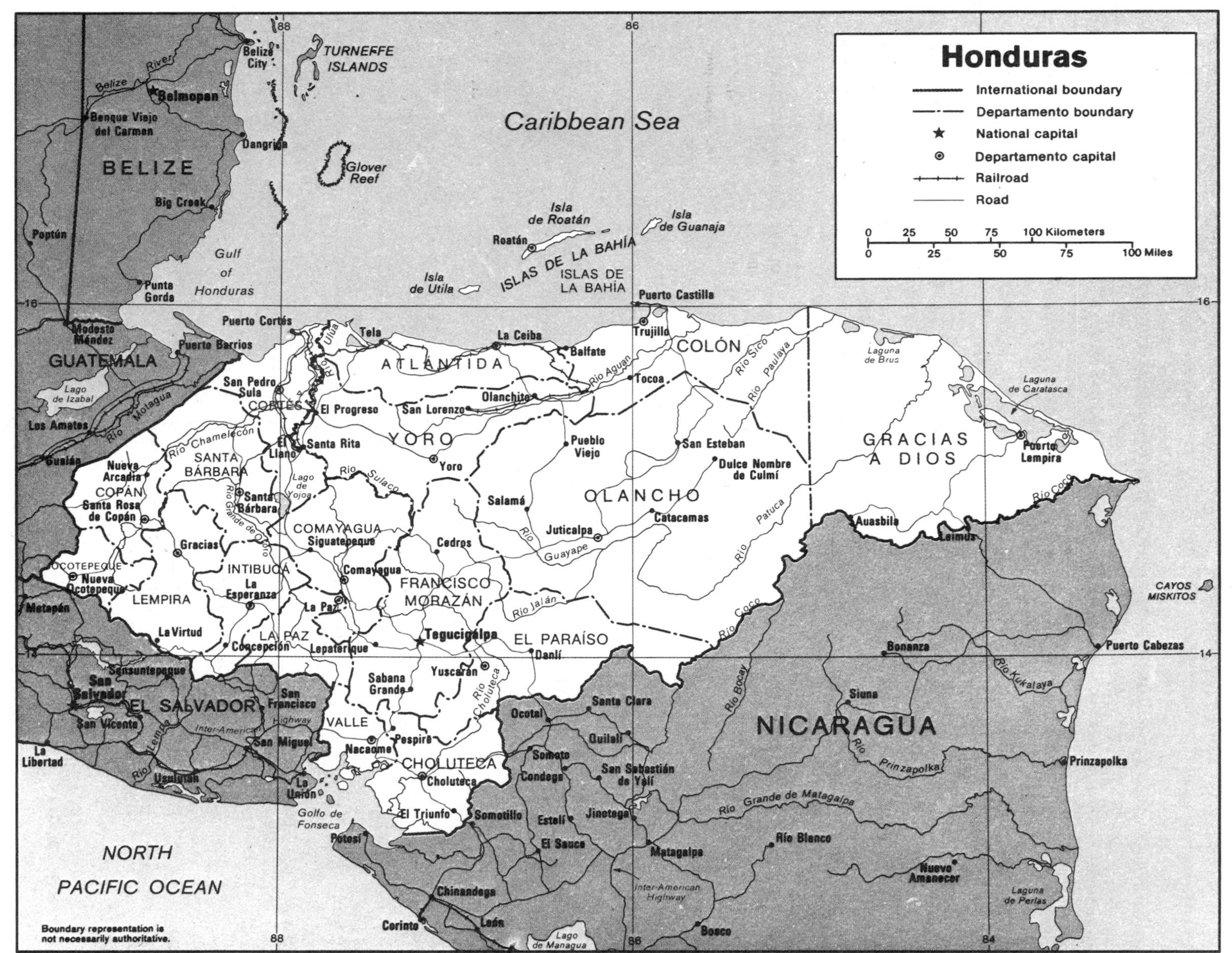

Base 800321 (A00162) 3-85

mountain ranges from the western border with Guatemala to the Plátano River, where the northern mountains terminate. In the south the flatlands are interspersed between the volcanic ranges from El Salvador in the west to the Segovia River in the east.

Historically, these level lands have been the most highly populated regions. Two examples are the river valley of Ulúa, where San Pedro Sula is located, and the Aguán river valley around Trujillo. A series of valleys at various elevations, drained by the Ulúa and its tributary, the Humuya, in the north and the Nacaome and Choluteca in the south, extends the entire north-south distance of the country. It was by traveling along these linked valleys that Spanish settlers were able to found San Pedro Sula, Comayagua, Tegucigalpa, and other urban centers in the sixteenth century.

PRINCIPAL TOWNS (Preliminary mid-1986 population estimate, excluding suburbs)

Town	Population	Town	Population
Tegucigalpa	604,600	Tela	27,200
San Pedro Sula	399,700	Siguatepeque	25,200
La Ceiba	63,800	Santa Rosa de Copán	20,000
Choluteca	60,700	Danlí	18,800
El Progreso	58,300	Juticalpa	13,900
Puerto Cortés	40,900	Olanchito	13,000
Comayagua	30,100		

Area and population

Departments	Administrative centres	area sq mi	area sq km	population 1983 estimate
Atlántida	La Ceiba	1,641	4,251	242,200
Choluteca	Choluteca	1,626	4,211	289,600
Colón	Trujillo	3,427	8,875	128,400
Comayagua	Comayagua	2,006	5,196	211,500
Copán	Santa Rosa de Copán	1,237	3,203	217,300
Cortés	San Pedro Sula	1,527	3,954	624,100
El Paraíso	Yuscarán	2,787	7,218	206,600
Francisco Morazán	Tegucigalpa	3,068	7,946	736,300
Gracias a Dios	Puerto Lempira	6,421	16,630	35,500
Intibucá	La Esperanza	1,186	3,072	111,400
Islas de la Bahía	Roatán	100	261	18,700
La Paz	La Paz	900	2,331	86,600
Lempira	Gracias	1,656	4,290	174,900
Ocotepeque	Nueva Ocotepeque	649	1,680	64,100
Olancho	Juticalpa	9,402	24,351	228,100
Santa Bárbara	Santa Bárbara	1,975	5,115	286,800
Valle	Nacaome	604	1,565	125,600
Yoro	Yoro	3,065	7,939	304,300
TOTAL		43,277	112,088	4,092,200

Tropical lowland areas are found on both coasts and in the far eastern Garcias a Dios Department. Between Guatemala and Nicaragua, the north coast extends over 724 km (450 mi), and the lowlands run some 121 km (75 mi) inland to the downward slopes of the northern mountains. The south coast is much shorter, about 145 km (90 mi) in all, and the lowlands extend about 40 km (25 mi) inland.

Unlike the north coast mountains that gradually merge into the coastline, the Pacific Volcanic Highlands drop quickly to the sea, pinching out the south coast and bringing cooler temperatures to lower elevations than those in the north. The farther away from the mountains, however, the more tropical foliage and inclement weather have made the coastal plains unattractive to settlers.

Eastern Gracias a Dios Department is a wide extension of the northern coastline with similar topographical features. Nearest the coast it is swampy and overgrown with mangrove forest. Only on the largest lagoon, Caratasca, has any important commercial port developed—Puerto Lempira. Farther inland, great stands of Caribbean pines cover large portions of land.

Except for the banana plantations, more settlement has taken place in Gracias a Dios Department along the-Segovia River than on the coasts. Much of the settlement devolved from the dispersion of indigenous tribes that had traditionally lived in the area and from the demarcation of the border with Nicaragua in the 1960s. At that time those local inhabitants who chose Honduran nationality moved to that country's side of the riverbanks.

There are many large river systems that drain the country and whose alluvial deposits have contributed to the fertility of the soil. In the north, from west to east, are the Chamelecón, the Ulúa, the Aguan, the Sico, the Paulaya, the Plátano, the Sicre, the Patuca, and the Segovia rivers. All the rivers in the north flow into the Caribbean Sea. The Ulúa and its tributaries drain one-third of the country.

In the south, from west to east, are the Lempa and Sumpul rivers, which run nearly the entire length of El Salvador's northern border with Honduras; the Goascorán, which marks El Salvador's eastern border with the country; the Nacaome, which arises north of Nacaome (capital of Valle Department) and runs south through Nacaome into the Gulf of Fonseca; and the Choluteca, which drains the area around Tegucigalpa and proceeds irregularly south through Choluteca (capital of Choluteca Department), and runs into the Gulf of Fonseca. All the rivers that arise in the south flow toward the Pacific Ocean.

HONG KONG

BASIC FACTS

Official Name: Hong Kong

Abbreviation: HKG

Area: 1,037 sq km (400 sq mi)

Area—World Rank: 162

Population: 5,651,193 (1988) 6,665,000 (2000)

Population—World Rank: 88

Capital: Victoria

Land Boundaries: China 30 km (18.6 mi)

Coastline: 733 km (455 mi)

Longest Distances: N/A

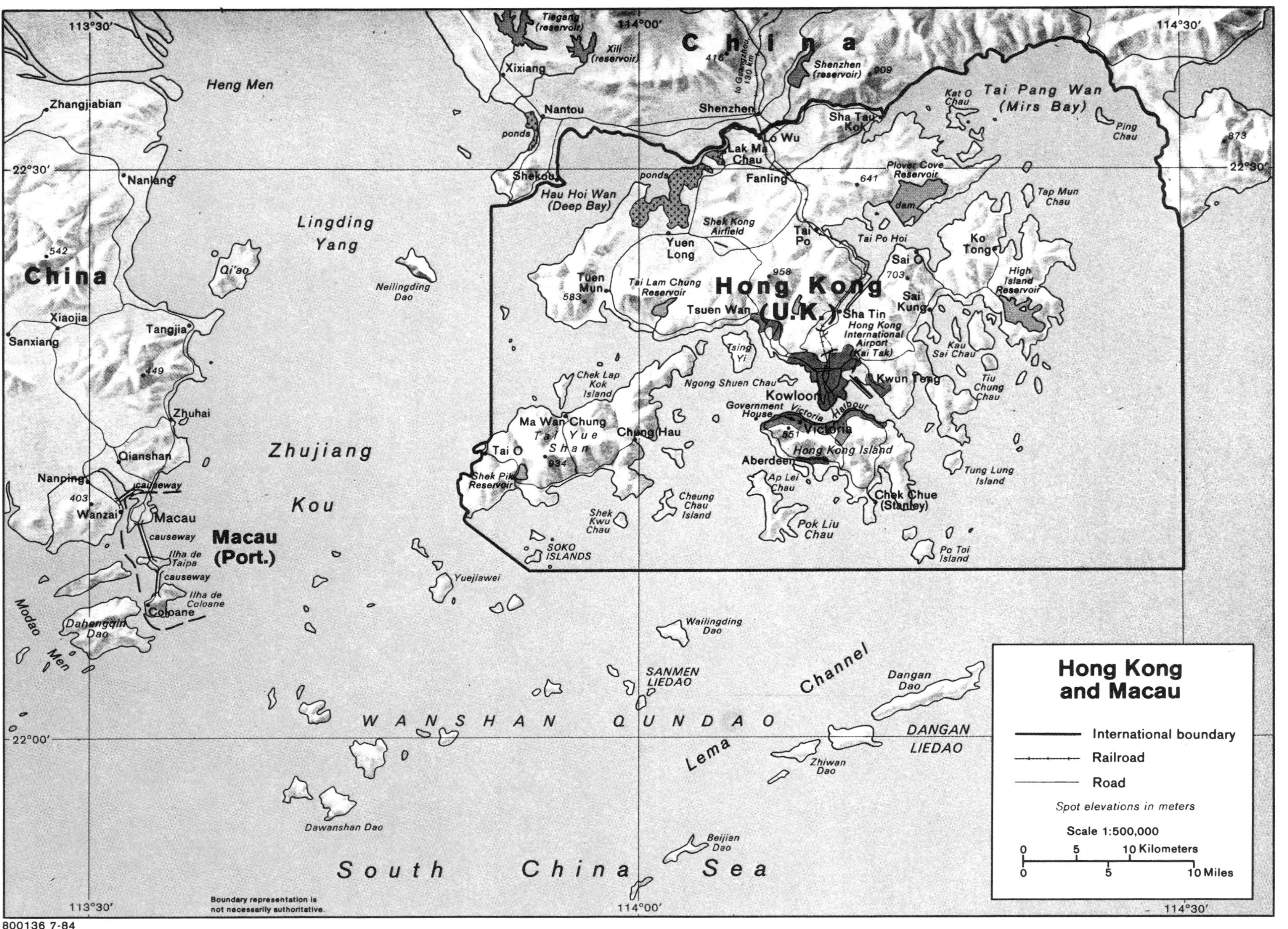
Hong Kong
and Macau
International boundary
Railroad
Road
Spot elevations in meters
Scale 1:500,000
0 5 10 Kilometers
0 5 10 Miles
China
Hong Kong
(U.K.)
Macau
(Port.)
Tai Pang Wan
(Mirs Bay)
Lingding
Yang
Zhujiang
Kou
Heng Men
Modao Men
South China Sea
WANSHAN QUNDAO
Lema Channel
DANGAN
LIEDAO
SANMEN
LIEDAO
Shenzhen
Lo Wu
Lak Ma Chau
Fanling
Sha Tau Kok
Yuen Long
Tuen Mun
Tsuen Wan
Tai Po
Sha Tin
Sai Kung
Kowloon
Victoria
Aberdeen
Chek Chue
(Stanley)
Hong Kong Island
Victoria Harbour
Kwun Tong
Hong Kong International Airport (Kai Tak)
Government House
Shekou
Nantou
Xixiang
Hau Hoi Wan
(Deep Bay)
Zhuhai
Tangjia
Qianshan
Wanzai
Nanping
Nanlang
Xiaojia
Sanxiang
Zhangjiabian
Ilha de Taipa
Ilha de Coloane
Coloane
causeway
Dahengqin Dao
Qi'ao
Neilingding Dao
Dawanshan Dao
Wailingding Dao
Beijian Dao
Zhiwan Dao
Dangan Dao
Yuejiawei
Lantau: Tai Yue Shan
Chek Lap Kok Island
Ma Wan Chung
Chung Hau
Tai O
Shek Pik Reservoir
SOKO ISLANDS
Shek Kwu Chau
Cheung Chau Island
Pok Liu Chau
Ap Lei Chau
Po Toi Island
Tung Lung Island
Tiu Chung Chau
Kau Sai Chau
High Island Reservoir
Ko Tong
Sai O
Tap Mun Chau
Ping Chau
Kat O Chau
Plover Cove Reservoir
dam
Tai Po Hoi
Tsing Yi
Ngong Shuen Chau
Tai Lam Chung Reservoir
Shek Kong Airfield
Shenzhen (reservoir)
Xili (reservoir)
Tiegang (reservoir)
ponds
to Guangzhou 130 km
113°30′
114°00′
114°30′
22°30′
22°00′
Boundary representation is not necessarily authoritative.
800136 7-84

Highest Point: Tai Mo Shan 957 m (3,140 ft)

Lowest Point: Sea level

Land Use: 7% arable; 1% permanent crops; 1% meadows and pastures; 12% forest and woodland; 79% other

The Territory of Hong Kong lies in east Asia, off the south coast of mainland China, and consists of the island of Hong Kong, Stonecutters Island, the Kowloon Peninsula, and the New Territories, which are partly on the mainland. Most of Hong Kong is rocky, hilly, and deeply eroded. Some 46 sq km (18 sq mi) have been developed for industrial, commercial and residential uses, some 50 sq km (19 sq mi) are cultivated, and the remaining 318 sq km (123 sq mi) are mainly hillside and swamp.

Area and population	area		population
Districts	sq mi	sq km	1986 census
Hong Kong Island	30.4	78.7	1,175,800
Kowloon	16.3	42.2	2,301,700
New Territories	356.6	923.7	1,918,500
TOTAL	403.3	1,044.6	5,396,000

HUNGARY

BASIC FACTS

Official Name: Hungarian People's Republic

Abbreviation: HUN

Area: 93,036 sq km (35,921 sq mi)

Area—World Rank: 105

Population: 10,588,271 (1988) 10,369,000 (2000)

Population—World Rank: 60

Capital: Budapest

Land Boundaries: 2,242 km (1,393 mi); Czechoslovakia 608 km (378 mi); USSR 215 km (134 mi); Romania 432 km (268 mi); Yugoslavia 631 km (392 mi); Austria 356 km (221 mi)

Coastline: Nil

Longest Distances: 268 km (167 mi) N-S; 528 km (328 mi) E-W

Highest Point: Kekes 1,015 m (3,330 ft)

Lowest Point: Tisza River Valley 259 m (79 ft)

Land Use: 54% arable, 3% permanent crops; 14% meadows and pastures; 18% forest and woodland; 11% other

Hungary is a landlocked country of east-central Europe that has access to the Black Sea via the Danube River. Its central location places the capital city, Budapest, within 2,414 km (1,500 mi) of all other European and North African capitals.

About two-thirds of the land has an elevation of less than 198 m (650 ft) and the highest mountain point is only a little more than 1,005 m (3,300 ft). Of the lower land, plains country predominates, but there is much low, gently undulating hilly ground. The southeastern section of the country and the extreme northwest are generally flat; low hills prevail over the major part of the land west of the Danube River, which roughly bisects the country from north to south. Low mountains extend in a roughly straight line northeast from the Austrian border. East of the Danube the mountains intercept the Czechoslovak border and follow it to the northernmost points in the country. Lower elevations have great expanses of fertile soils.

Although the area is small and the whole of it is within the middle-Danube, or Carpathian, basin, the country is usually divided into four major topographic regions. The Great Hungarian Plain (Nagy Magyar Alfold), more commonly called the Great Plain, accounts for about one-half of the total area and comprises the lowlands east of the Danube River. Transdanubia (Dunantul), the hillier region west of the Danube, is approximately two-thirds the size of the Great Plain. The remainder of the country is accounted for in the Little Plain (Kisalfold) in the extreme northwest and the low Northern Mountains that range along the Czechoslovak border east of the Danube and north of the Great Plain.

The Great Plain has a mean elevation of a little more than 91 m (300 ft). Except for its river valleys, its landscape is relieved largely by hillocks and sand dunes. The largest deviations from the average elevation occur on a plateau between the Danube and Tisza rivers and in an area in the northeast along the Romanian border. The plateau is from 30 to 45 m (100 to 150 ft) higher than the floodplains of the rivers, and the gentle northeastern hills reach approximately 182 m (600 ft).

Dunantul, the Hungarian name for Transdanubia, means the land beyond the Danube. It consists of approximately 32,375 sq km (12,500 sq mi) of rolling country. Its flatter lands are less monotonous than those of the Great Plain and occupy less of the region than do its hilly and low mountainous sections. The uplands in the west are foothills of the Alps.

Lake Balaton, the last remnant of the ancient Pannonian Sea, is roughly in the center of the region. To its east, and extending to the Danube River, is the Mezofold, a lowland similar in many respects to the Great Plain. The higher lands immediately south of the lake are the Transdanubian Hills, or the Somogy Hills. Farther south are the coal-bearing and uranium-rich Mecsek and Villany mountains.

Most elevations in the alpine foothills to the west of the lake are below 304 m (1,000 ft) but a few isolated spots on the Austrian border rise to nearly 914 m (3,000 ft). The Transdanubian Central Mountains extend along the northern side of the lake and continue northeastward into the residential areas of Budapest and into the Danube Nook (Dunazug). The chain consists of several minor ranges, many of them having much relatively unspoiled beauty. The scenery is not of the rugged mountain type; most of the hillcrests are rounded, and elevations range from about 213 to less than 762 m (700 to less than 2,500 ft).

The Little Plain, and area of about 5,956 sq km (2,300 sq mi), is so small that it is sometimes considered a part of Transdanubia. Its mean elevation is approximately 30 m (100 ft) higher than that of the Great Plain. Most of this region consists of rich agricultural land, but swampland is also prevalent.

The Northern Mountains constitute the remaining 4,988 sq km (3,100 sq mi) of the country's area. They are the lower volcanic fringe of the Carpathian Mountains, the only uplands in the country that are part of the Carpathian system.

The individual ranges in the group extend northeastward from the gorge of the Danube River near Esztergom for about 225 km (140 mi). Although the highest point in the mountains—Mount Kekes in the Matra range—is only about 1,015 m (3,330 ft) above sea level, many of the slopes in the area are steep and give a false impression of their height. Points at the upper elevations are sunny, have many springs and small streams, and are popular resort areas.

Drainage

The entire country is located in the middle-Danube basin. The drainage pattern follows the Danube River, which eventually flows into the Black Sea. Local streams in northern Transdanubia and the Little Plain flow to the Danube within Hungary, but those in larger portions of the country, including southern Transdanubia and most of the Great Plain, drain to its tributaries—the Drava and Tisza—and join the Danube in Yugoslavia.

The middle course of the Danube stretches for 965 km (600 mi) between the Deveny Gate (just above Bratislava, Czechoslovakia) and the Iron Gate on the boundary between Yugoslavia and Romania. About 386 km (240 mi) of the middle river is in, or borders, Hungary. Although it is a great stream as it flows through Budapest and continues southward through Hungary, its size within this portion is only about 35 percent of that which it attains by the time it enters the Black Sea. Its flow averages 3,050 cubic yards per second at Budapest, as compared with 8,420 cubic yards per second at its exit into the Black Sea.

As it flows through Hungary, the Danube falls very little. It is about 134 m (440 ft) above sea level as it reaches Hungary and about 85 m (280 ft) above sea level as it leaves the country, but by far the greater portion of the fall is along the Czechoslovak border and in the gorge of the river north of Budapest. In the Great Plain between the capital and the Yugoslav border it falls only about fifty feet, or less than six inches per mile.

The slight fall and irregular waterflow account for the devastation that the river has periodically inflicted upon the river cities and the adjacent plains. Floodwaters are expected twice during each normal year. The first, its white flood, occurs in April or May when snow melts at lower elevations up the river and its ice breaks up. The water released in the white flood is usually not its most dangerous factor. Ice plugs can develop, and the backed-up water and ice then can be much more destructive than the high water alone. An ice plug during the white flood of 1838, for example, destroyed the cities of Obuda, Pest, and much of Buda.

The second, or green flood, usually occurs in June. Most of the up-river tributaries are fed by heavy rains during the late spring, and much of the snow at higher elevations melts at about the same time; the result is usually higher water levels than are experienced during the early flood. At Budapest the river may rise twenty-five feet or more above its normal level during the green flood. Devastation from the ice-free water is usually less, however, because the flood stages are more predictable and flood controls are more effective.

The Tisza and Drava are also major rivers, although they are not in the same class with the Danube. The Tisza is the Danube's second largest tributary. Similar to the Danube, during the seasons when its flow is low or moderate, it meanders slowly across the Great Plain but, in contrast to the Danube, its streambed is flat, and it has little or no valley. Its flow is highly irregular, and during early and late spring floods it may carry fifty times as much water as it does during the summer. Like the Drava it may also have a lesser flood in October, but in most respects it is more similar to the Danube, and its highest waters usually occur during the second, or June, flood. The earlier one may also be its more dangerous but for different reasons. The Tisza's ice melts first in the south and does not accumulate in increasing quantities downstream; consequently, it has had no serious ice plugs in

PRINCIPAL TOWNS (population at 1 January 1987)

Town	Population	Town	Population
Budapest (capital)	2,093,487	Nyíregyháza	118,179
Debrecen	214,836	Székesfehérvár	112,703
Miskolc	211,156	Kecskemét	103,944
Szeged	185,559	Szombathely	86,682
Pécs	179,051	Szolnok	80,921
Győr	130,194	Tatabánya	76,463

Area and population

Counties	Capitals	area sq mi	area sq km	population 1986 estimate
Baranya	Pécs	1,732	4,487	432,000
Bács-Kiskun	Kecskemét	3,229	8,362	558,000
Békés	Bekéscsaba	2,175	5,632	422,000
Borsod-Abaúj-Zemplén	Miskolc	2,798	7,248	791,000
Csongrád	Szeged	1,646	4,263	457,000
Fejér	Székesfehévár	,689	4,374	426,000
Györ-Sopron	Györ	1,549	4,012	428,000
Hajdú-Bihar	Debrecen	2,398	6,212	551,000
Heves	Eger	1,404	3,637	342,000
Komárom	Tatabánya	869	2,250	321,000
Nógrád	Salgótarján	982	2,544	233,000
Pest	Budapest	2,469	6,394	985,000
Somogy	Kaposvár	2,331	6,036	353,000
Szabolcs-Szatmár	Nyiregyháza	2,293	5,938	578,000
Szolnok	Szolnok	2,165	5,608	436,000
Tolna	Szekszárd	1,430	3,704	266,000
Vas	Szombathely	1,288	3,337	280,000
Veszprém	Veszprém	1,810	4,689	388,000
Zala	Zalaegerszeg	1,461	3,784	313,000
Capital City				
Budapest		203	525	2,080,000
TOTAL		35,921	93,036	10,640,000

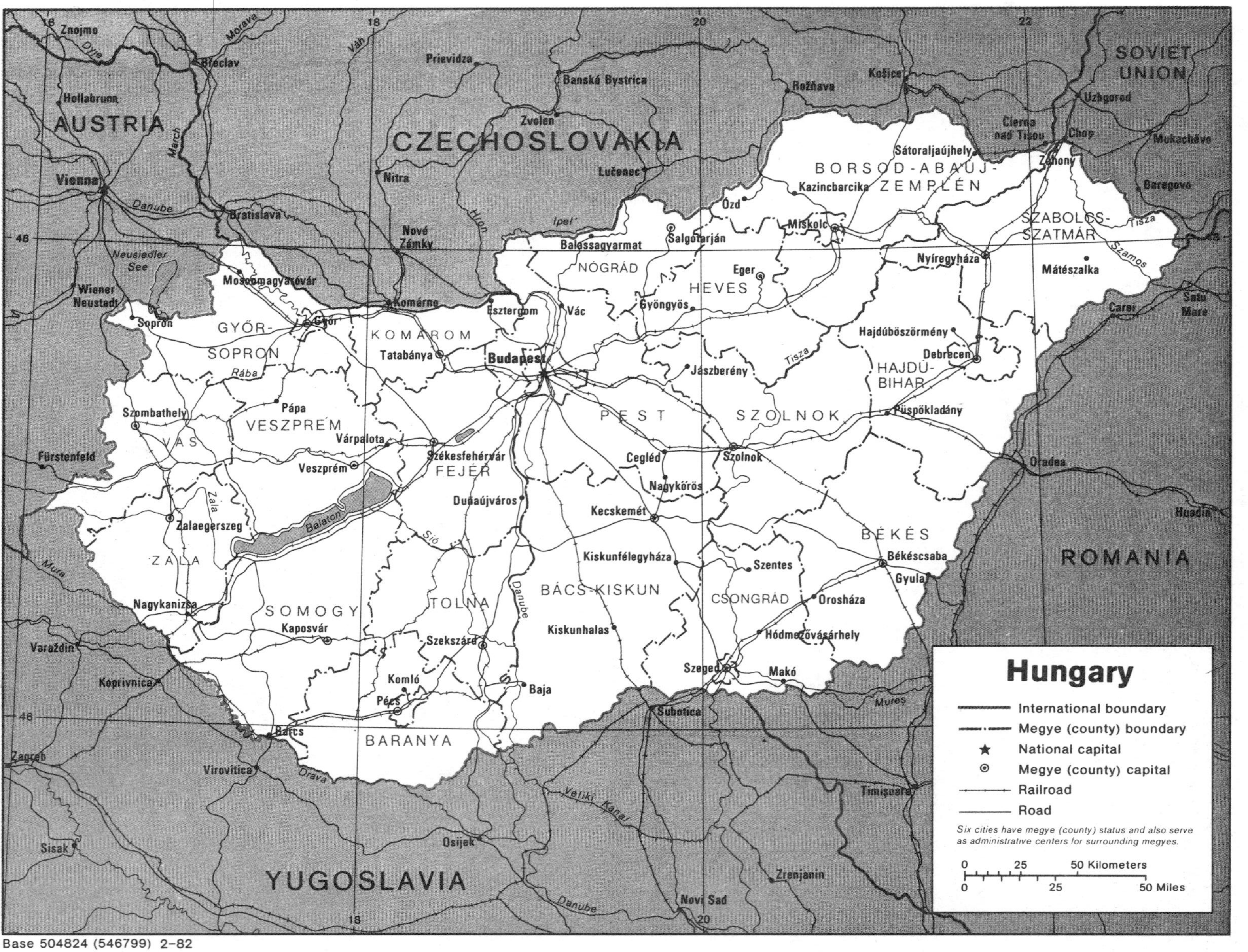

Base 504824 (546799) 2-82

recent history. Its high water, however, may concur with that of the Danube. When this happens, the swollen Danube cannot accept the Tisza's waters, and they back up for the river's entire length in Yugoslavia and into vast areas of Hungary's Great Plain. It was such a flood in 1879 that devastated the city of Szeged, which is located about ten miles north of the Hungarian-Yugoslav border.

The Tisza's early summer and greater flood usually does not concur with that of the Danube. Caused almost exclusively by heavy seasonal precipitation, the greater part of which falls in the Carpathians, this flood is most severe in northern areas and diminishes in the south. The southern tributaries, fed from less extreme amounts of rainfall, are relatively less swollen.

The Drava is smaller than the Tisza by only a slight margin. It accumulates most of its volume in Austria, flows across the northern tip of Yugoslavia, forms a part of the Hungarian-Yugoslav border for about 129 km (80 mi), and turns back into Yugoslavia again for about 64 km (40 mi) before joining the Danube. It may have three annual floods—the third in early autumn—but it flows adjacent to higher terrain on its northern shore and is a significant feature in the life of far fewer Hungarians than are the Danube and Tisza.

Both the Danube and Tisza rivers have been controlled since the mid-nineteenth century. The Tisza had been an especially capricious stream. It frequently had inundated enough territory to make its valley resemble an inland sea, and it had changed channels often and unpredictably. Control measures are not completely effective, but they have permitted the populating of several millions of acres of hitherto worthless land. In addition to maintaining river channels, irrigation water has been made available, swamps have been drained, and all-season roads have been built where transport was formerly paralyzed for much of the year.

Lake Balaton, 120 km (75 mi) southwest of Budapest, is about 72 km (45 mi) long; its width varies, never reaching more than eight miles. It averages a little more than ten feet in depth, and its deepest point is only about thirty-five feet. The deeper northern end drops off steeply from the shore, providing excellent fishing but few good beaches for bathing. In the south there are beaches with fine sand and water depths that increase gradually. In some places bathers may wade on soft sand for half a mile or more from shore.

Lake Ferto (Neusiedler See in German) on the northwestern border is shared with Austria, but Hungary's portion is only about one-fourth of the total. Although its depth varies, the lake averages only approximately three feet. The shallow lake frequently freezes entirely and is unsuitable for recreational purposes and fish-breeding. Reeds grown in and around it have some commercial value, however, and it provides a haven for many species and great numbers of waterfowl.

Lake Velence, between Lake Balaton and Budapest, is worked artificially to maintain water depths of from three to six feet and to keep it suitable for holiday bathing, even though it does not have good natural beaches. It is excellent for fishing and is stocked largely with sporting varieties for anglers. Marshes on the southwestern end of the lake have been drained, and the reeds have been cleared from about half of its surface.

ICELAND

BASIC FACTS

Official Name: Republic of Iceland

Abbreviation: ICD

Area: 103,000 sq km (39,769 sq mi)

Area—World Rank: 102

Population: 246,526 (1988) 277,000 (2000)

Population—World Rank: 157

Capital: Reykjavik

Land Boundaries: Nil

Coastline: 4,970 km (3,088 mi)

Longest Distances: 490 km (304 mi) E-W; 312 km (194 mi) N-S

Highest Point: Hvannadalshnukur 2,119 m (6,952 ft)

Lowest Point: Sea Level

Land Use: 23% meadows and pastures; 1% forest and woodland; 76% other

The westernmost country of Europe, Iceland is an island in the North Atlantic Ocean just below the Arctic Circle, 322 km (200 mi) east of Greenland, 1,038 km (645 mi) west of Norway and 837 km (520 mi) northwest of Scotland. The mainland comprises of 102,950 sq km (39,748 sq mi) (including 408 sq km (157 sq mi) of lakes). The islands and skerries comprise 150 sq km (58 sq mi). Of the many islands, the most notable are the Westman Islands off the southern coast.

Iceland consists mainly of a central volcanic plateau with elevations ranging from 700 to 800 m (2,297 to 2,625 ft) ringed by mountains, the highest of which is Hvannadalshnukur (2,119 m; 6,952 ft) in the Oraefajokull glacier. Lava fields cover about one-ninth of the country; glaciers, about one-eighth. Geologically, the country is still very young and bears signs of still being in the making. It appears on the whole roughly hewn, abrupt and jagged, without the softness of outline that characterizes more mature landscapes. The average height is 500 m (1,640 ft) above sea level, and one-quarter of the country lies below the 200 m (656 ft) contour line. The largest lowland areas include Arnessysla, Rangarvallasysla, and Vestur-Skaftafellssysla in the South and Mýrar in the West. In the plateaus, land is broken into more or less tilted blocks, with most leaning toward the interior of the country. Glacial erosion has played an important role in giving the valleys their present shape. In some areas, such as between Eyjafjorour and Skagafjorour, the landscape possesses alpine characteristics. There are numerous and striking gaping fissures within the post glacially active volcanic belts.

Icelandic coasts can be divided into two main types. In regions not drained by the debris-laden glacial rivers, the coasts jut irregularly, incised with numerous fjords and smaller inlets. They offer many good natural harbors where the fjords have been deepened by glacial erosion. The other type of coast is sand, with smooth outlines featuring extensive offshore bars with lagoons behind them. The beaches from Djupivogur in the Southeast to Olfusa in the Southwest belong in this category.

Glaciers cover an area of 11,200 sq km (4,323 sq mi) or 11% of the total land area. Nearly all types of glaciers, from small cirque glaciers to extensive plateau icecaps are represented. The biggest of these icecaps, Vatnajökull, with an area of 8,300 sq km (3,204 sq mi) and a maximum thickness of 1,000 m (3,281 ft) is larger than all the glaciers in continental Europe put together. One of its southern outlets, Breidamerkurjökull, reaches more than 120 m (394 ft) below sea level. Other large icecaps are Langjökull (1,025 sq km; 396 sq mi) and Hofsjökull (953 sq km; 368 sq mi) in the Central Highlands, Mýrdalsjökull (700 sq km; 270 sq mi) in the South and Drangajökull (160 sq km; 62 sq mi) in the Northwest. The altitude of the glaciation limit is lowest, about 600 m (1,961 ft) in the Northwest and highest, over 1,500 m (4,922 ft) in the highlands north of Vatnajökull. Since about 1890, the glaciers have greatly thinned and retreated, and some of the smaller ones have almost disappeared. During the 1960s the retreat began to slow down, but some of the glaciers are now advancing again.

Because of the heavy rainfall, Icelandic rivers are numerous and relatively large. Pjorsá, the longest river, has a length of 237 km (147 mi) and Jökulsá á Fjöllum, the second longest, a length of 206 km (128 mi). Other major rivers include Skjálfandafljót, Jökulsá á Brú, Lagarfljót, Skeidara and Kuoafljót. These rivers belong to three types. The debris-laden glacial rivers usually divide into a great number of interlinked tributaries that constantly change course and flow through the outwash plains that lie below the glaciers. Skeidara is a typical example. The maximum discharge in the glacial rivers usually occurs in July or early August. Direct-runoff rivers drain the old basalt areas in summer and autumn, while spring-fed rivers drain the regions covered by postglacial lava fields. Swift currents make Icelandic rivers unnavigable, for the most part.

The wealth of waterfalls is typical of the young landscape. The largest, Dettifoss (44 m; 144 ft) is in Jökulsá á Fjöllum. The others are Gullfoss (32 m; 105 ft) in Hivita, Godafoss in Skjálfandafljót, Skogafoss and Fjallfoss.

Iceland possesses numerous lakes, mostly of tectonic origin. Others resulted from the deepening of valleys by glacial erosion or damming by lava flows, glacial deposits and rock slides. Small crater lakes are common in explosion craters, especially in the Landmannalaugar-Veidivotn area, where the Lake Oskjuvatn caldera has an area of 11 sq km (4.2 sq mi) and a depth of 217 m (712 ft). On the sandy shores lagoon lakes are common, the biggest being Hop 30 sq km (12 sq mi).

The soils may be roughly grouped into mineral soils and organic soils, with a number of intermediate types. The mineral soils are loessial, formed by materials supplied by explosive volcanic eruptions and glacier erosion. Because of the cool climate, the chemical and biological processes in the soil are slow, and their properties strongly reflect those of volcanic rocks. Most soils are suitable for agriculture but require heavy fertilizer applications.

Earthquakes are frequent in the country. The largest ones occur within a fracture zone that runs through the Southern Lowlands, where disastrous earthquakes were recorded in 1784 and 1896. Earthquakes also are frequent on the Reykjanes Peninsula and in the districts around Skjálfanda, Eyjafjordur, and Skagafjordur in the North.

Area and Population

Regions Counties	Administrative centres	area sq mi	 sq km	population 1986 estimate
Austurland		8,683	22,490	13,131
Austur-Skaftafellssýsla	Höfn	2,347	6,080	2,216
Nordhur-Múlasýsla	Seydhisfjördhur	4,799	12,430	3,253
Sudhur-Múlasýsla	Eskifjördhur	1,537	3,980	7,662
Nordhurland eystra		8,370	21,680	25,764
Eyjafjardharsysla	Akureyri	1,602	4,150	18,853
Nordhur-Thingeyjarsýsla	Húsavík	2,077	5,380	1,648
Sudhur-Thingeyjarsýsla	Húsavík	4,691	12,150	5,263
Nordhurland vestra		4,973	12,880	10,676
Austur-Húnavatnssýsla	Blönduós	1,900	4,920	2,604
Skagafjard-harsýsla	Saudhárkrókur	2,077	5,380	6,504
Vestur-Húnavatnssýsla	Blönduós	996	2,580	1,568
Rekjavíkursvaedhi og Reykjanessvaedhi		741	1,920	148,883
Gullbringusýsla	Keflavik	405	1,050	33,977
Kjósarsýsla	Hafnarfjördhur	336	870	114,906
Sudhurland		9,649	24,990	20,065
Árnessýsla	Selfoss	3,401	8,810	10,491
Rangárvallasýsla	Hvolsvöllur	3,197	8,280	8,250
Vestur-Skaftafellssýsla	Vík	3,050	7,900	1,324
Vestfirdhir		3,676	9,520	10,193
Austur-Bardhastran-darsýsla	Patreksfjördhur	444	1,150	387
Nordhur-Ísafjardharsýsla	Isafjördhur	1,181	3,060	5,113
Strandsýsla	Hólmavík	1,015	2,630	1,146
Vestur-Bardhastran darsýsla	Patreksfjördhur	598	1,550	1,979
Vestur-Ísafjardharsýsla	Isafjördhur	436	1,130	1,568
Vesturland		3,676	9,520	14,940
Borgarfjard-harsýsla	Borgarnes	753	1,950	6,768
Dalasýsla	Budhardalur	815	2,110	1,044
Mýrasýsla	Borgarnes	1,262	3,270	2,569
Snaefellsnessýsla	Stykkishólmur	846	2,190	4,559
TOTAL		39,768	103,000	243,698

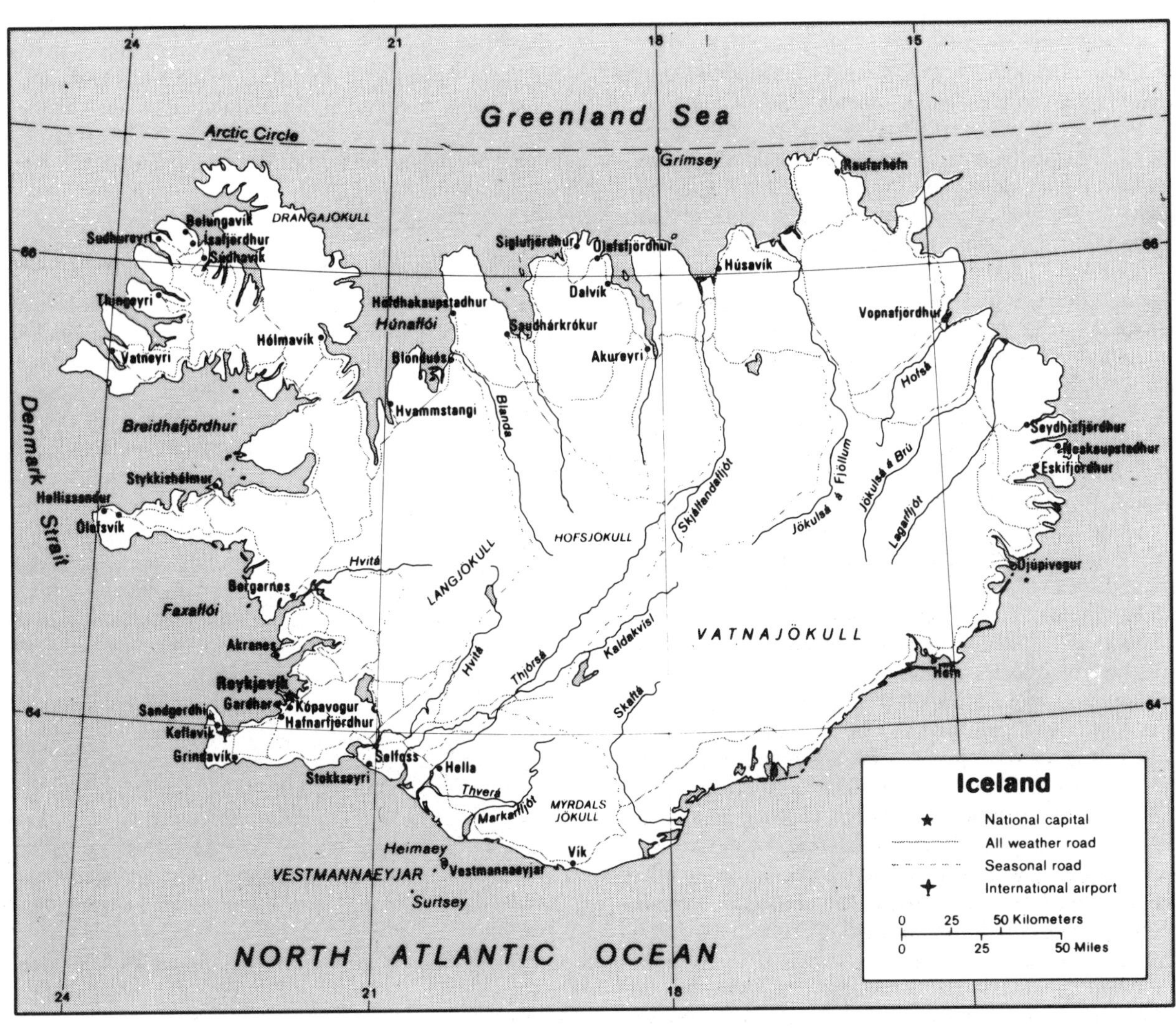
Greenland Sea
Arctic Circle
Grimsey
Raufarhöfn
DRANGAJÖKULL
Bolungavik
Sudhureyri
Ísafjördhur
Súdhavik
Siglufjördhur
Ólafsfjördhur
Húsavík
Dalvík
Thingeyri
Höfdhakaupstadhur
Húnaflói
Saudhárkrókur
Vopnafjördhur
Hólmavík
Akureyri
Vatneyri
Blönduós
Hvammstangi
Blanda
Hofsá
Denmark Strait
Breidhafjördhur
Jökulsá á Fjöllum
Jökulsá á Brú
Seydhisfjördhur
Neskaupstadhur
Eskifjördhur
Skjálfandafljót
Stykkishólmur
Hellissandur
Ólafsvík
HOFSJÖKULL
Lagarfljót
Hvítá
LANGJÖKULL
Djúpivogur
Borgarnes
Faxaflói
Kaldakvisl
VATNAJÖKULL
Akranes
Hvítá
Thjórsá
Reykjavík
Gardhar
Kópavogur
Hafnarfjördhur
Sandgerdhi
Keflavík
Skaftá
Höfn
Grindavík
Selfoss
Hella
Stokkseyri
Thverá
Markarfljót
MYRDALS JÖKULL
Heimaey
Vík
VESTMANNAEYJAR
Vestmannaeyjar
Surtsey
NORTH ATLANTIC OCEAN
Iceland
National capital
All weather road
Seasonal road
International airport
0 25 50 Kilometers
0 25 50 Miles
24
21
18
15
66
64

Iceland is very rich in natural heat, as the regional heat flow within the neovolcanic areas averages two or three times the global average. Two main types of thermal areas can be distinguished on the basis of the maximum subsurface temperature of thermal water. The low-temperature areas where the maximum temperature is less than 150°C (302°F) have few springs. The hot springs are common all over the country except in the East and Southeast and number over 300. The largest is Deildartunguhver in Borgarfjordur.

The English word "geyser" comes from the most famous Icelandic hot spring, at Geysir in Haukadalur in South Iceland. At times Geysir is quiescent, while at others it spouts water to a height of about 60 m (197 ft). A nearby famous spouter is called Strokkur. Beautiful silica sinter terraces are displayed around the hot springs at Hveravellir in the Central Highlands. The total natural heat discharge of the high-temperature areas reaches several thousand megawatts.

Nearly every type of volcano is found in Iceland. Lava-producing fissures forming so-called center rows are the most common. It was one of these, the Lakagigar, that in 1793 poured out the most extensive lava flow in the world in historical times. It covers 565 sq km (218 sq mi) and has a volume of about 12 sq km (4 sq mi). The crater rows follow the same direction as the Pleistocene ridges. Shield volcanoes such as the Skjáldbreidur also have produced a great amount of lava but have not been active for the past 1,000 years. Iceland also has active volcanoes of the central type fed by magma chambers higher up in the earth's crust. Many of them are blanketed by perpetual ice, including two that have erupted most frequently in historical times, Grimsvotn and Katla. The latter, which has Iceland's largest caldera 80 sq km (31 sq mi), has erupted about 20 times in the past 1,000 years. Each eruption of these volcanoes is accompanied by a water flood (*jokulhlaup*). These floods occur every five to 10 years, sometimes without volcanic eruptions. The eruption of Oraefajökull in 1362 devastated the settlement at the foot of the volcano. Another large caldera, at Askja in the Dyngjufjoll massif, erupted in 1875, causing great damage in East Iceland. The most famous of Icelandic volcanoes is Hekla, which was renowned during the Middle Ages as the abode of the damned. Since its first recorded eruption in 1104, Hekla has erupted 14 times, the last in 1970.

Explosion pits are found throughout the country. Submarine eruptions often occur off the coasts, especially in the Reykjanes ridge. The last one, which began in 1963 and lasted until 1967, built up the island of Surtsey, which now covers an area of 2.8 sq km (1.1 sq mi). On January 23, 1973, the Vestmannaeyjar crater erupted, burying one-third of the town of the same name.

INDIA

BASIC FACTS

Official Name: Republic of India

Abbreviation: IND

Area: 3,166,414 sq km (1,222,559 sq mi)

Area—World Rank: 7

Population: 816,828,360 (1988) 941,008,000 (2000)

Population—World Rank: 2

Capital: New Delhi

Boundaries: 15,098 km (9,381 mi); China 1,893 km (1,176 mi); Nepal 1,508 km (937 mi); Bhutan 573 km (356 mi); Myanmar 1,403 km (872 mi); Bangladesh 2,583 km (1,605 mi); Pakistan 2,028 km (1,260 mi)

Coastline: 5,110 km (3,175 mi)

Longest Distances: 3,214 km (1,997 mi) N-S; 2,933 km (1,822 mi) E-W

Highest Point: Kanchenjunga 8,598 m (28,208 ft)

Lowest Point: Sea level

Land Use: 55% arable; 1% permanent crops; 4% meadows and pastures; 23% forest and woodland; 17% other

India is the seventh largest country in area in the world. It resembles, roughly, two triangles with a common base lying across the middle of the country. The northeast side of the northern triangle is the massive Himalayan mountain wall. Southwest of the Himalayan foothills lies the Indo-Gangetic Plain, a broad alluvial lowland fed by the Ganges River and by tributaries of the Indus River of Pakistan.

The southern triangle, or peninsula of India, projects south into the Indian Ocean; it is bounded on the southeast by the Bay of Bengal and on the southwest by the Arabian Sea. The common base of the two great triangles is a general west-east chain of hills and associated rivers across central India, with the central peninsular Deccan (Deccan Plateau) to the south.

Isolated geographically from the rest of Asia by the mountain wall, deserts, and seas, the Indian subcontinent is a distinct entity. In the northwest and west, however, are a number of passes. India's early history reflects the absorption of migrant peoples who entered from the northwest and moved down into the Indo-Gangetic Plain. The wider coastal plains on the southeastern side of the peninsula resulted historically in the major European lodgments' being made from the Bay of Bengal rather than from the Arabian Sea frontier.

The Indo-Pakistani subcontinent has three main geographic regions: the northern mountain area, or Himalayas; the Indo-Gangetic Plain; and the southern tableland. The Indo-Gangetic Plain and the portions of the Himalayas that fall within India's political borders are collectively known as northern India. The tableland constitutes what is known as southern India or, more simply, the peninsula. Across the central part of the country lies a complex line of hills and mountains running in a generally east-west direction. Historically, these ranges were of great significance because they hindered—though they did not completely prevent—movement of peoples and communication between the two areas and thus contributed to certain historical and cultural distinctions between the peoples of northern and southern India that still persist.

The Himalayas

The name Himalaya, which means "abode of snow" in Sanskrit, is given to the tremendous system of mountain ranges, the loftiest in the world, that extends along the northern frontiers of Pakistan, India, Nepal, Sikkim, and Bhutan. Jutting southward from the main body of the Himalayas at each extremity and bordering the subcontinent on both east and west are lesser ranges. In the west, extending through Pakistan, run the Sulaiman and Kirthar ranges; in the east lie the Patkai and Naga hills in India and the Arakan and Pegu ranges in Myanmar. They are considered by some geographers part of the Himalayas because they originated in the same geological upheaval and by others as separate systems because they strike off from the main body of the Himalayas.

The Himalayan system, about 2,414 km (1,500 mi) in length and varying in width from 241 to 321 km (150 to 200 mi), is made up of three more or less parallel ranges. The average height of the northernmost range, the Greater Himalayas, is 6,096 m (20,000 ft); the range has the three highest altitudes on earth. These famous peaks are Mount Everest (8,847 m; 29,028 ft) in Nepal; K2, also called Mt. Godwin-Austen or Dapsang (8,611 m; 28,250 ft) in Pakistan-controlled northern Kashmir; and Kanchenjunga (8,597 m; 28,208 ft) on the border between Nepal and Sikkim. The peaks of the Lesser Himalayas are mostly between 1,524 and 3,657 m (5,000 and 12,000 ft) in height; some exceed 4,572 m (15,000 ft). The Outer, or southern, Himalayas form a system of low foothills, averaging between 914 and 1,219 m (3,000 and 4,000 ft) in height, between the Lesser Himalayas and the Indo-Gangetic Plain.

In each of the main Himalayan ranges the southward slopes are too steep either to accumulate snow or to support more than sparse tree growth. The northward slopes, which are much gentler, are generally forest clad below the snow line. Between some of the ranges are extensive high plateau areas; elsewhere are found some of the deepest gorges on earth, and in still other parts are fertile and beautiful valleys, such as those of Kashmir and Kulu.

Separating India from the Asiatic mainland, the Himalayas bar entry to the freezing winds from Tibet; serve as a screen within which the monsoons operate; and provide a steady source for the three great river systems—the Indus, the Ganges, and the Brahmaputra—that water the alluvial plains below. As a result of erosion, the rivers also carry a vast quantity of silt, which constantly enriches the plains. Because these mountain ranges are geologically young, earthquakes are not infrequent. Periodically they cause extensive damage in more densely settled sub-Himalayan areas of India, Pakistan, and Nepal.

Overland travel through the Himalayan mountain shield has always been difficult and dangerous and continued to be in the 1990s, although to a lesser degree than in earlier periods. Transport of goods is ordinarily on the back of man or beast (mules, yaks, and sheep) through a number of passes that have been known from ancient times. Of these, the Karakoram Pass is one of the best known and most important. It is situated at 5,575 m (18,290 ft) at the point where the India-Pakistan Cease-Fire Line meets the PRC frontier and about 322 km (200 mi) northeast of Srinagar.

Other important passes include the Jelep La and Natu La, northeast of Darjeeling on the north-south trade route through the Chumbi Valley, and Shipki La, about thirty miles northeast of Kalpa in Himachal Pradesh, near the PRC line. Only in recent years has extensive sustained overland penetration been attempted through the Himalayas. In 1956 and 1957, as part of its supply route to Tibet, the PRC built the Aksai Chin road across the northeastern tip of Kashmir in territory under dispute between PRC and India. This road was the first trans-Himalayan route to be motorable throughout the year.

Culturally the Himalayas have been of great significance to India. This land of difficulty, mystery, and towering grandeur has in all ages impressed the Indians. Around Mount Kailas and the sources of the Ganges, the tales of Siva and Parvati were woven, and here also are located some of the most sacred shrines of Hinduism. The eternal snow has been seen as symbolic of an ideal serenity. To the great poets and religious sages of the past and to Indians today, the mountains represent strength, stability, and security.

The Indo-Gangetic Plain

At the foot of the Himalayan mountain barrier, extending from Assam and the Bay of Bengal on the east to the Afghan border and the Arabian Sea on the west, lies one of the most densely populated areas of the world: the Indo-Gangetic Plain, sometimes called the plain of northern India. Covering some 776,996 sq km (300,000 sq mi), it extends roughly 2,414 km (1,500 mi) from east to west, varying in width from about 161 km (100 mi) in the east to about 482 km (300 mi) in the west. The Indo-Gangetic plain, once a gulf between the peninsula and the Himalayas, is the product of the continual deposits of alluvium borne by the great river systems. The thickness of the alluvial deposit has never been conclusively ascertained. It may be as deep as 3,048 m (10,000 ft) in places and is thought to be deepest close to the mountains. The trough has been so filled over the ages that it looks to the eye like a level plain. Between Delhi and the Bay of Bengal, about 1,528 km (950 mi) to the east, there is a drop of only about 201 m (660 ft) in elevation.

The Indo-Gangetic Plain is sometimes described by geographers under four approximate subdivisions: the Indus Valley (entirely in Pakistan), the Punjab, the middle Ganges, and the lower Ganges. The plains region in India, nevertheless, may be regarded topographically as a unit, since there are no strong physical bases for division. There are, however, regional variations caused primarily by the availability of water. The five plains states of Punjab, Haryana, Uttar Pradesh, Bihar, and West Bengal have a reasonably good and dependable supply of water either through rainfall, which increases from west to east, or through irrigation. These comparatively rich states, which include only about 10 percent of the country's area, contain about 40 percent of its population.

The plain of the Ganges River, extending from Delhi on the banks of the Yamuna (Jumna) River in the west to Bangladesh, is known as Hindustan. The land presents little variation in relief. The gradations in the change from the dry, wheat-growing Punjab to the wet, rice- and jute-growing delta area of Bengal in the east are gradual, but the contrasts between the two ends of the plain are marked. Between the Gangetic Plain and the lower foothills of the Himalayas lies a level or gently undulating strip of jungle land only slightly higher than the plain itself. Called the Tarai, it is a swampy, malarial region, which is being drained and brought under cultivation. This long, narrow strip along the southern base of the mountains adds to the barrier effect of the Himalayas; especially before the mid-twentieth century, the Tarai was noted for its wild animal life, virulent malaria, and paucity of population as compared to the plain immediately to the south.

The plain of the Brahmaputra, or Assam Valley in the east is much narrower than that of the Ganges. Like the Ganges delta area, it is a region of heavy rainfall. At the western extremity of the Indo-Gangetic Plain, the plain of the Indus River and its tributaries lies almost entirely in Pakistan, though some of the upper Indus tributaries drain part of the Indian state of Punjab (the name Punjab is derived from *panch ab,* meaning five waters, or rivers). It is a region of very low or irregular rainfall, and agriculture in the area depends largely on irrigation from the rivers.

Below the present state of Punjab and extending southwest along the Pakistani border is the sparsely populated Thar Desert, which covers the larger part of the state of Rajasthan. No rivers of any significance flow through these wastes to compensate for the scarcity of rain. The Thar contains a large supply of white marble that has been used in many of India's notable buildings. There also are sizable deposits of limestone, gypsum, lignite, fuller's earth, and salt. Most of the soils are not alluvial; some are suitable for cultivation if water can be brought in. The Indian government's ambitious project of carrying water from the Punjab rivers to Rajasthan for irrigation is opening large areas for cultivation and halting the advance of the desert to surrounding areas. To the southwest of the Thar Desert, in the state of Gujarat, lies the desolate Rann of Kutch, scene of the Indo-Pakistani conflict in the spring of 1965. It is a region of hard salt flats for half the year, and it is under water for the other half.

The Peninsula

South of the Indo-Gangetic Plain lies the third of the three principal subregions of India: the peninsula that includes about two-fifths of the country. Projecting southward like a great triangle, its base stretching across India from the state of Gujarat in the west to West Bengal in the east, the peninsula is bounded on the east by the Bay of Bengal and on the west by the Arabian Sea. Just off the apex of the triangle to the southeast lies the island country of Sri Lanka.

The peninsula is an area of great complexity and diversity. It may be described generally as a plateau region roughly bounded by ranges of low mountains and hills that geographically block the peninsula off not only from the plain to the north but also from the deserts and semi-arid regions of Rajasthan to the northwest and the coastal flats of Gujarat to the west. The jumbled hills, small, scarped plateaus, and broken valleys stretching across the base of the peninsular triangle have historically acted as a physical impediment and, more important, as a cultural line of division between northern and southern India. The largest of the identifiable plateaus are the central Malwa Plateau, the Chota Nagpur in the northeast of the peninsula, and the main plateau, or Deccan, in the south-central part. The peninsula is flanked on both sides by coastal plains inward from the sea to the coastal hill lines. This plain is narrow on the west but distinctly broader along the eastern side.

Geologically the peninsula is a region of great stability, relatively immune from earthquakes of any consequence. Composed chiefly of highly metamorphosed rocks from the earliest periods of the earth's history, the peninsula is traversed by many fault depressions, some of which are filled with ancient sediments that today form India's major coal deposits. Because of a slight downward tilting of the peninsula to the east, believed to have occurred at the geological time of the Himalayan uplift, all but two of the peninsular rivers flow to the east into the Bay of Bengal.

The Hills of Central India

The complex of hills and low mountain ranges, varying in elevation from about 457 to 1,219 m (1,500 to 4,000 ft) that occupies the northern part of the peninsula constitutes what is sometimes called central India. The main ranges in the area are the Aravallis, the Vindhyas, and the Satpuras in the west; the Mahadeo, Maikal, and Bhanrer hills in the center; and the Kaimur, Hazaribagh, and jungle hills of the Chota Nagpur to the east.

On the west the gaunt, worn ridges of the Aravallis jut northward across Rajasthan, acting as a barrier to the encroaching desert, and thrust into the Indo-Gangetic Plain. Their last visible spur is the Ridge of Delhi. Pronounced east-west structure lines are evident in the hill ranges of the Vindhyas, Satpuras, and Kaimurs and in the faulted troughs of the Narbada and Tapti rivers, which flow into the Gulf of Cambay. These two rivers are the only major peninsular rivers that flow to the west.

Two plateaus are important in this region. The Malwa plateau in the west, between the Aravallis and the Vindhyas, has historically been the route to southern India for rulers based in Delhi. In the east the Chota Nagpur Plateau of southern Bihar contains the most important mineral concentrations in India and is the site of extensive mining and industrial development centered in Jamshedpur. The broken, jungle-like country of this region has tended to make it a "shatter zone," restricting, but not preventing, communication between the peoples of the north and south.

The Peninsular Interior

The peninsular interior is a large area of many varying physical components. The name Deccan, which means "south," is often applied loosely to all the elevated land of southern India. More properly, however, it refers to the western portions of the irregular central plateau. The northeastern section is composed of the heavily forested Orissa Hills, which have been inhabited mainly by tribal people. With the completion of the giant Hirakud multipurpose dam project on the Mahanadi River, the pattern of land use and settlement throughout this underdeveloped area is undergoing fundamental change.

The Deccan is actually not a single plateau but a series of plateaus topped by rolling hills and intersected by many rivers. It is a geologically stable region; seismic disturbances are rare. In general, it is based on geologically old crystalline rocks covered with thin soil, except where the rivers have deposited alluvial silt along their banks. In the northwestern portions, however, are beds of fertile black soil known as the Deccan lavas. The area of the Deccan lavas corresponds roughly with the Marathi-speaking region. The plateau averages about 762 m (2,500 ft) in elevation in the west and about 305 m (1,000 ft) in the eastern parts. Its major rivers, the Godavari, the Krishna (sometimes called the Kistna), and the Cauvery, and their main tributaries all rise in the hills along the western side of the peninsular triangle, in some cases within 80 km (50 mi) of the Arabian Sea, and flow eastward into the Bay of Bengal.

The Coastal Plains

The narrow alluvial plains bordering India's coastline of some 5,632 km (3,500 mi) (less inlets) were formed by the erosion of the edges of the ancient peninsular tableland. The plains are bordered and separated from the body of the peninsula by an irregular fringe of mountains on each side of the triangle: the Western and Eastern Ghats. The shore is characteristically flat and sandy with few ports or inlets in relation to its length.

The western and eastern coastal plains vary considerably. The western coast falls into three main sections: the Gujarat coast, circling the Gulf of Cambay and extending south to the city of Bombay; the Konkan-Kanara coast, from Bombay to Mangalore; and the Malabar coast, from Mangalore to Cape Comorin at the southern tip of the peninsula. Ranging from five to fifty miles in width, the western coastal plain is widest at Gujarat, which was formerly the major outlet of overland trade through Malwa from northern India and now contains the important industrial and textile center of Ahmadabad. Most of the coastal plain is composed of fertile alluvial soil that supports flourishing agriculture. The Malabar, or Kerala, section is one of the most densely populated regions in all India.

The east coast stands in distinct contrast to the western littoral. By and large, it is composed of alternating coastal plains, which often approach 161 km (100 mi) in width; delta plains, which stretch inland for 80 km (fifty mi) or more; and sporadic, isolated remnant hills. Along its 1,931 km (1,200 mi) from Orissa to the southern tip of the peninsula, the climatic variation of the east coast is distinct and is reflected in varying patterns of agriculture and population density. The middle and southern parts of- the east coast, that is, the coastlines of the states of Andhra Pradesh and Tamil Nadu, are collectively referred to as the Coromandel coastal plain.

The Plateau Ranges

Two ranges, quite different in character, parallel the coasts of the peninsula and separate the interior plateau from the littoral plains. The Bluff mountains called the Western Ghats have an average elevation of 1,066 m (3,500 ft) although rising in some places to 2,348 m (8,000 ft), and face the ocean like a wall. At the tip of the peninsula the range converges with the Eastern Ghats. The Eastern Ghats are disconnected and low, averaging only about 610 m (2,000 ft) in elevation. Whereas the Western Ghats acted as a real barrier to European penetration of the interior from the sea, the eastern lowlands, especially those of Madras, were more open.

The hills to the south of the Deccan plateau proper include the Nilgiris, whose highest point is 2,670 m (8,760 ft) above sea level; the Anaimalais, dominated by the highest peak in the peninsula, which is 2,694 m (8,841 ft) high; and the Cardamom Hills, which extend south to within a few miles of Cape Comorin. The Cardamom Hills are sometimes classed as a southward extension of the Western Ghats. In these groups of hills live most of the distinctively primitive or near-primitive peoples of South India.

The Offshore Islands

Two groups of islands belonging to India—each group classed and governed as union territory—lie one on either side of the southern tip of the country. The area and population of these islands are so small, however, in comparison to the mainland as to preclude their classification as a fourth major region of the country.

The eastern group, the Andaman and Nicobar Islands, are in the Bay of Bengal. The Andamans form a generally north-south chain some 1,287 km (800 mi) east of the city of Madras. The Nicobar chain of fewer and smaller islands stretches south-southeast from the Andamans and is separated from them by the Ten Degree Channel. The overall length of the Andaman-Nicobar group is about 804 km (500 mi). The total land area is roughly 8,287 sq km (3,200 sq mi). Approximately 90 percent of these low-lying, lightly populated islands is forest, and timber production is the main economic activity.

In the Arabian Sea off the Malabar coast is the second group, composed of the Laccadive, Minicoy, and Amindivi Islands, collectively named the union territory of Lakshadweep on November 1, 1973. The distance from Mangalore southwest to the center of the island group is about 402 km (250 mi). This irregular scattering of small islands stretches about 160 km (100 mi) northwest from the center point and the same distance to the southeast. The total area is only about 50 sq km (12 1/2 sq mi). Most, although not all, of these low-lying small islands are occupied. Population density is high on the inhabited islands. Fishing and coconut farming are the main occupations.

RIVERS

Rivers are regarded with love and reverence and figure prominently in the epic and folk literature. Several, espe-

Base 504177 6-79

cially the Ganges, the Yamuna, and the Godavari, are regarded as sacred by Hindus and are visited and bathed in by millions of pilgrims annually. All the rivers provide water for irrigation and, when harnessed, provide power for homes and factories. The Ganges and the Brahmaputra are the main carriers of freight and travelers.

The river systems may be classed under four headings as Himalayan; plains; Deccan, or peninsular; and coastal. As a group, the coastal rivers are comparatively short and shallow; most are absorbed in salt lakes, basins, or sand areas before reaching the sea. The major rivers are those of the Himalayas and the plain in northern India and, to the south, those of the Deccan region and the hills of the peninsula.

Principal Rivers of Northern India

The Indus River, rising in the Tibetan Himalayas, flows generally northwest for about 804 km (500 mi), then turns generally southwest for about 1,609 km (1,000 mi) and empties through a delta area into the Arabian Sea. The long southwest course of the Indus lies entirely in Pakistan, and about half of its northwest course is in Tibet; it flows through India only in crossing the state of Jammu and Kashmir. The Indus, in addition to other branches, has five principal tributaries, also of Himalayan origin, that are of importance to India: the Sutlej, Beas, Ravi, Chenab, and Jhelum. These rivers are sometimes referred to as the five rivers of the Punjab, but in this older usage "the Punjab" means the general region so known in pre-1947 India rather than the smaller Indian state of Punjab. Only the relatively short Beas River is wholly within India. The political boundary cuts across the other four and the Indus. In this area, where irrigation is of major importance on both sides of the boundary, disputes over the distribution of these waters have been one of the main sources of friction between India and Pakistan.

South of the Punjab and east of the semiarid and desert region of western India, the main feature of the Indo-Gangetic Plain, as its name implies, is the most revered and mightiest of India's rivers—the Ganges. This famous river, about 2,510 km (1,560 mi) in length, carries more water than the six main peninsular rivers combined. The Ganges has two headstreams in the Indian Himalayas, the Alakananda and Bhagirathi rivers. The latter is usually recognized as the main headstream, and the origin of the Ganges is identified in an ice cave about 48 km (30 mi) north of Nanda Devi, almost on the line of the India-PRC frontier and 3,139 m (10,300 ft) above sea level.

The Ganges cuts south through the Himalayas and enters the plain at the city of Hardwar. Proceeding south and then eastward, it winds through Uttar Pradesh and is joined by its major tributary, the Yamuna, at Allahabad. The Yamuna, companion to the Ganges, also rises in the Himalayas and through much of its length flows only 80 to 120 km (50 to 75 mi) to the west, and later to the south, of the Ganges. The confluence of these two rivers is considered sacred and is visited by hundreds of thousands of pilgrims every year. The Ganges then continues in a southeasterly direction and has numerous tributaries, and finally, in Bangladesh, it is joined by the Brahmaputra. Thereafter the waters of the Ganges-Brahmaputra are distributed into many channels to form the vast Bengal delta before emptying into the Bay of Bengal.

Area and population

States	Capitals	area sq mi	area sq km	population 1981 census
Andhra Pradesh	Hyderābād	106,204	275,068	53,549,673
Arunāchal Pradesh	Itanagar	32,333	83,743	631,839
Assam	Prāgjyotiṣapura	30,285	78,438	19,896,843
Bihār	Patna	67,134	173,877	69,914,734
Goa	Panaji	1,430	3,702	1,007,749
Gujarāt	Gāndhingar	75,685	196,024	34,085,799
Haryāna	Chandigarh	17,070	44,212	12,922,618
Himāchal Pradesh	Simla	21,495	55,673	4,280,818
Jammu and Kashmir	Srinagar	39,145	101,387	5,987,389
Karnātaka	Bangalore	74,051	191,791	37,135,714
Kerala	Trivandrum	15,005	38,863	25,453,680
Madhya Pradesh	Bhopāl	171,215	443,446	52,178,844
Mahārāshtra	Bombay	118,800	307,690	62,784,171
Manipur	Imphāl	8,621	22,327	1,420,953
Meghālaya	Shillong	8,660	22,429	1,335,819
Mizorām	Aizawl	8,140	21,081	493,757
Nāgāland	Kohima	6,401	16,579	774,930
Orissa	Bhubaneswar	60,119	155,707	26,370,271
Punjab	Chandigarh	19,445	50,362	16,788,915
Rājasthān	Jaipur	132,140	342,239	34,261,862
Sikkim	Gangtok	2,740	7,096	316,385
Tamil Nādu	Madras	50,216	130,058	48,408,077
Tripura	Agartala	4,049	10,486	2,053,058
Uttar Pradesh	Lucknow	113,673	294,411	110,862,013
West Bengal	Calcutta	34,267	88,752	54,580,647
Union Territories				
Andaman and Nicobar Islands	Port Blair	3,185	8,249	188,741
Chandigarh	Chandigarh	44	114	451,610
Dādra and Nagar Haveli	Silvassa	190	491	103,676
Daman and Diu		43	112	78,981
Delhi	Delhi	572	1,483	6,220,406
Lakshadweep	Kavaratti	12	32	40,249
Pondicherry	Pondicherry	190	492	604,471
TOTAL		1,222,559	3,166,414	685,184,692

The 2,897 km (1,800 mi) long Brahmaputra rises in southwestern Tibet, not far from the separate sources of both the Indus and the Ganges, and flows 1,126 km (700 mi) eastward through southern Tibet, receiving many tributaries during its course through the Himalayas. It turns briefly to the northeast and rushes through several steep gorges before turning to the south and southwest and entering the Assam Valley. At the boundary with Bangladesh, the Brahmaputra turns due south to meet the Ganges near Dhaka.

The Peninsular Rivers

Four of the six major rivers of the peninsula, the Mahanadi, the Godavari, the Krishna, and the Cauvery, flow into the Bay of Bengal. All have reached their base level of erosion and thus have broad shallow valleys that tend to flood during the rainy season. Several of the rivers have waterfalls and cascades in their upper courses; the Cauvery Falls at Sivasamudram and the Paikari Falls in the Nilgiris have both been harnessed for electric power. All of these rivers rise in the Western Ghats, flow eastward, and deposit silt in fertile deltas at their mouths. Most of the rivers are fast flowing during the rainy season but are reduced to a fraction of their size or may even be dry during the hot dry season.

The Mahanadi, which rises in Madhya Pradesh, is an important river of Orissa and is the supply source for the huge multipurpose Hirakud Dam project begun in 1949. The Godavari has its source north of Bombay and follows a general southeasterly course for 1,488 km (900 mi) to its mouth on the coast of Andhra Pradesh. Its river basin area is second in size only to the Ganges. The fertile Godavari delta is one of India's principal rice-growing areas.

The Cauvery, sometimes called the Ganges of the south, rises in southern Karnataka and flows irregularly southeastward across that state and the state of Tamil Nadu to the east. Harnessed for irrigation since ancient times, an estimated 95 percent of the water of the Cauvery is put to use before the river empties into the Bay of Bengal. It flows through a major delta in the highly fertile Thanjavur district.

PRINCIPAL TOWNS (population at 1981 census*)

Town	Population	Town	Population
Greater Bombay	8,243,405	Gwalior	539,015
Delhi	4,884,234	Hubli-Dharwar	527,108
Calcutta	3,305,006	Sholapur	514,860
Madras	3,276,622	Jodhpur	506,345
Bangalore	2,218,593	Trivandrum	499,531
Hyderabad	2,187,262	Ranchi	489,626
Ahmedabad	2,159,127	Mysore	479,081
Kanpur		Vijaywada	
(Cawnpore)	1,486,522	(Vijayavada)	461,772
Nagpur	1,219,461	Jamshedpur	457,061
Pune (Poona)	1,203,351	Rajkot	445,076
Jaipur (Jeypore)	997,165	Meerut	417,395
Lucknow	916,954	Jalandhar	408,196
Indore	829,327	Bareilly	394,938
Madurai	820,891	Kozhikode (Calicut)	394,447
Patna	813,963	Chandigarh	379,660
Surat	776,876	Ajmer	375,593
Howrah	744,429	Guntur	367,699
Vadodara (Baroda)	734,473	Tiruchirapalli	362,045
Varanasi (Banaras)	720,755	Salem	361,394
Coimbatore	704,514	Kota	358,241
Agra	694,191	Kolhapur	340,625
Bhopal	671,018	Raipur	338,245
Jabalpur		Warangal	335,150
(Jubbulpore)	649,085	Faridabad	330,864
Allahabad	619,628	Moradabad	330,051
Ludhiana	607,052	Aligarh	320,861
Amritsar	594,844	Durgapur	311,798
Srinagar	594,775	Thane	309,897
Visakhapatnam	584,166	Bhavnagar	308,642
Cochin	551,567		

*Figures refer to the city proper in each case. For urban agglomerations, the following populations were recorded: Calcutta 9,194,018; Delhi 5,729,283; Madras 4,289,347; Bangalore 2,921,751; Ahmedabad 2,548,057; Hyderabad 2,545,836; Pune (Poona) 1,686,109; Kanpur 1,639,064; Nagpur 1,302,066; Jaipur 1,015,160; Lucknow 1,007,604; Coimbatore 920,355; Patna 918,903; Surat 913,806; Madurai 907,732; Varanasi (Banaras) 797,162; Jabalpur 757,303; Agra 747,318; Vadodara (Baroda) 744,881; Cochin 685,836; Dhanbad 678,069; Jamshedpur 669,580; Allahabad 650,070; Ulhasnagar 648,671; Tiruchirapalli 609,548; Srinagar 606,022; Visakhapatnam 603,630; Gwalior 555,862; Kozhikode (Calicut) 546,058; Vijaywada 543,008; Meerut 536,615; Trivandrum 520,125; Salem 518,615; Sholapur 514,860; Ranchi 502,771.

Capital: New Delhi, population 273,036 in 1981.

Only two major rivers of the peninsula flow into the Arabian Sea—the Narbada and the Tapti. The Narbada rises in eastern Madhya Pradesh and traverses the state in enclosed valley between the Vindhyas on the north and the spurs of the Satpuras to the south. As it enters the state of Gujarat, the river widens to about one mile and flows calmly through fertile fields. It forms a thirteen-mile-wide estuary and flows into the Gulf of Cambay. The shorter Tapti follows a companion course, 80 to 160 km (50 to 100 mi) south of the Narbada, and is a tidal river for the last 51 km (32 mi) of its 724 km (450 mi) length.

INDONESIA

BASIC FACTS

Official Name: Republic of Indonesia

Abbreviation: IDO

Area: 1,919,443 sq km (471,101 sq mi)

Area—World Rank: 15

Population: 184,015,906 (1988) 222,753,000 (2000)

Population—World Rank: 5

Capital: Jakarta

Boundaries: 38,889 km (24,165 mi); Malaysia 1,496 km; Papua New Guinea 777 km (483 mi)

Coastline: 36,616 km (22,752 mi)

Longest Distances: 5,271 km (3,275 mi) E-W; 2,210 km (1,373 mi) N-S

Highest Point: Jaya Peak 5,030 m (16,503 ft)

Lowest Point: Sea Level

Land Use: 8% arable; 3% permanent crops; 7% meadows and pastures; 67% forest and woodland; 15% other

Indonesia consists of more than 3,000 islands scattered for about 5,149 km (3,200 mi) along the equator—the largest archipelago in the world encompassed by a single nation and the third largest country in Asia, after the People's Republic of China (PRC) and India. Superimposed on a map of the United States, the island chain would overlap New York and San Francisco. Along the length of the island chain are found huge volcanic mountains, some of which are still active; dense jungle; graceful palm trees; long ribbons of sand beach; mountain lakes; rushing rivers and waterfalls; rice paddies nourished by rich volcanic soil; and plantations of rubber, oil palms, coffee, tea, sugar, tobacco, and spices.

Indonesia's geographic location at one of the world's major crossroads makes it of strategic importance to all of the great powers and particularly to those who rely on sea and air to maintain their national interests in Asia. The country forms a natural barrier between the Indian Ocean and the South China Sea-Pacific Ocean. It shares with Malaysia command of the Strait of Malacca and the Indian Ocean to Mainland China, Japan, and North America. Between its islands pass the alternate routes to the north, including routes from Australia to the Orient and North America. The islands lie on both sides of the equator. The northernmost portion of the archipelago is only 160 km (100 mi) from the Philippine island of Mindanao; the Sulu island chain, extending southwestward from Mindanao, is much closer. To the south Indonesia is separated from Australia by the narrow Timor and Arafura seas.

The islands are part of the Malay archipelago, which also includes the Philippines. The main islands of Java, Bali, Sumatra (Sumatera), and Kalimantan, the lesser islands between them, and the adjacent Asian mainland rise from the submerged Sunda shelf (no more than 91 to 182 m (300 to 600 ft) below sea level), considered a continuation of the Asian continent. The island of New Guinea, of which Irian Djaja is a part, and a few associated island groups are exposed parts of another submerged platform that is part of the Australian continental shelf. Between these two shelves is a sea area of great depth, where the islands are in waters as deep as 1,829 to 6,400 m (6,000 to 21,000 ft).

Between what are known as the Asian and the Australian regions of the archipelago is the Wallace line, the boundary between the deep and the shallow water running through the channel between Bali and Lombok and to the west of Sulawesi (Celebes). Animals, plants, and even ethnic groups show differences between the two regions that support the theory of geographic and geologic origin on two separate continents.

NATURAL FEATURES

The archipelago includes 13,667 islands, many of them only a few acres in size. Not all of these islands have been officially named, and only about 1,000 are inhabited. Five of the islands include nine-tenths of the nation's land area—Java, the center of the country's population; Sumatra, the top economic producer; Sulawesi, the second ranking producer; Kalimantan, the two-thirds of the island of Borneo that is within Indonesia; and Irian Djaja, a permanent possession confirmed through an "act of free choice" conducted in 1969 and known before 1973 as Irian Barat (West Irian), or west New Guinea.

Formerly the islands of the Netherlands Indies were nominally grouped as the Greater Sunda Islands (a term no longer in official use), including Java, Sumatra, Sulawesi, and Kalimantan; the Lesser Sunda Islands (Lesser Sundas), stretching to the east from Bali to Timor, a group now known officially as Nusa Tenggara (Southeastern Islands); and Maluku (formerly Moluccas, or Spice Islands), which includes Halmahera, Buru, Ceram (Seram), and Ambon.

A series of agreements was concluded in the early 1970s relating to territorial waters and continental shelf and seabed boundaries between Indonesia and Malaysia, Thailand, and Australia, and discussions were held with authorities in the Philippines and Vietnam concerning boundaries in the Celebes and South China seas. An Indonesian continental shelf law was promulgated in January 1973.

Of considerably greater international interest was Indonesian policy regarding the Strait of Malacca, one of the busiest sea routes in the world. In late 1971, the Indonesian and Malaysian governments declared that, in implementing their respective twelve-mile territorial water limits, which overlap in the narrowest part of the strait, they did not regard the strait as an international waterway. The declaration evoked a strong reaction from Japan and the Soviet Union.

The landscape is highly varied, and mountains stand out in sharp relief on the larger islands. A majority of Indonesians live in sight of volcanic cones or craters. Tidal swamps extend far into the interior of parts of Java, Sumatra, and Kalimantan; the principal dry flatlands are located on Java and in North Sumatra (Sumatera Utara). The dominant physical characteristics of the other islands are either heavily forested rolling hills and mountains or humid morasses hidden from the sun by dense tropical vegetation. The immense trees of the jungles are laced with lianas, and orchids and ferns hang from crevices in their bark.

Rivers and rivulets are found in every part of the islands. Although most rivers are short, they are often important for irrigation and occasionally destructive in periods of exceptionally heavy rainfall. The largest river, the Mamberamo, is in Irian Djaja. Other major rivers on Sumatra and Kalimantan—some navigable for part of their length—were associated with the development of old trading centers that served as river ports and later as provincial capitals, such as Djambi, Palembang, and Bandjarmasin. Of the lakes scattered through the islands, by far the largest is Lake Toba in North Sumatra, which covers over 1,295 sq km (500 sq mi) between towering cliffs that once were the rim of a volcanic crater. This area has become one of the country's major tourist attractions and a center of developing recreational activity.

The Mountains and Plains

In earlier geological periods the eastern and southeastern regions of Asia went through several stages of folding, which can be discerned in two long mountain systems that intersect each other in Sulawesi and Halmahera. The first, consisting of two parallel ridges that are a continuation of the western Burmese chain, runs through Sumatra, Java, Bali, and Timor and curves sharply in a great hook through the southeastern islands to Ceram and Buru. The second runs southwest through the Philippines into eastern Indonesia; hence the complexity of the mountain structure and the peculiar shape of Sulawesi and Halmahera, which have ranges running north-south and east-west.

The hills and mountains of Sumatra follow the west coast and are intersected by short but rapid streams that water the narrow coastal plains; to the east the mountains level out to broad expanses of lowland cut by sluggish

PRINCIPAL TOWNS (population)

	1980 Census	1983*
Jakarta (capital)	6,503,449	7,347,800
Surabaya	2,027,913	2,223,600
Bandung	1,462,637	1,566,700
Medan	1,378,955	1,805,500
Semarang	1,026,671	1,205,800
Palembang	787,187	873,900
Ujung Pandang (Makassar)	709,038	840,500
Malang	511,780	547,100
Padang	480,922	656,800
Surakarta	469,888	490,900
Yogyakarta	398,727	420,700
Banjarmasin	381,286	423,600
Pontianak	304,778	342,700

*Revised official estimates for 31 December.

streams that run through the interminable marshes to the Strait of Malacca and the Java Sea. In Java, too, the mountains lie close to the shoreline of the Indian Ocean-but, being less continuous than those of Sumatra, they allow frequent access from north to south, which has facilitated the development of a network of the roads and railroads.

The mountains of Kalimantan run mainly along the border with Sabah and Sarawak; south of the foothills Kalimantan consists of vast lowlands and swamps. Communication with the interior is maintained almost entirely by riverboat. Sulawesi, on the other hand, is extremely mountainous, its peaks rising in places to well over 2,438 m (8,000 ft). The few coastal plains are very narrow, and many of their inhabitants, notably the Makasarese and Buginese, seek their livelihood from the sea.

To the east Nusa Tenggara and Maluku are spread over a sea area of more than 2,589,988 sq km (1 million sq mi), although the islands themselves are only 155,399 sq km (60,000 sq mi) in land area. Many of the islands of Nusa Tenggara and Maluku are mountainous; on Bali, Lombok, and Ceram there are peaks of over 3,048 m (10,000 ft) but, along the hook-shaped bend formed by the island chain, the mountains are discontinuous. The 7,700 sq km (3,000 sq mi) expanse of the Aru archipelago is in no place more than 91 m (300 ft) above sea level.

Irian Djaja is covered with dense forests, vast swamps, and towering nonvolcanic mountains, the highest in the nation. A great mountain system, extending almost the entire length of the province in a general northwest to southeast direction, constitutes a massive barrier between the northern and southern parts of Irian Djaja. Some peaks are covered with snow throughout the year, including the 5,029 m (16,500 ft) Puntjak Djaja, the country's loftiest peak.

Indonesia is the most highly volcanic region in the world; over 100 peaks either are active or were active until recently. The volcanic ash enriches the soil, and the greatest population density is to be found in these regions. Thus Java, with the most volcanoes, is by far the most densely populated of the islands; in Kalimantan and in Irian Djaja, where volcanic activity is rare or absent, the population is sparse although not only for this reason. The greatly superior productivity of a volcanic region apparently justifies the risk of living there; at times large crop areas are burned out or badly scorched, and villages

Area and Population

Metropolitan district	Capitals	area sq mi	area sq km	population 1986 estimate
Jakarta Raya	Jakarta	228	590	8,164,400
Provinces				
Bali	Denpasar	2,147	5,561	2,709,200
Bengkulu	Bengkulu	8,173	21,168	985,600
Irian Jaya	Jayapura	2,928	421,981	1,363,500
Jambi	Jambi	7,345	44,924	1,822,200
Jawa Barat	Bandung	7,877	46,300	31,876,400
Jawa Tengah	Semarang	3,207	34,206	27,755,900
Jawa Timur	Surabaya	8,503	47,922	31,639,300
Kalimantan Barat	Pontianak	6,664	146,760	2,827,000
Kalimantan Selatan	Banjarmasin	4,541	37,660	2,328,000
Kalimantan Tengah	Palangkaraya	8,919	152,600	1,159,000
Kalimantan Timur	Samarinda	8,162	202,440	1,690,500
Lampung	Tanjung Karang	12,860	33,307	6,422,100
Maluku	Ambon	28,767	74,505	1,659,100
Nusa Tenggara Barat	Mataram	7,790	20,177	3,107,700
Nusa Tenggara Timur	Kupang	18,485	47,876	3,048,900
Riau	Pakanbaru	36,511	94,562	2,583,900
Sulawesi Selatan	Ujung Pandang	28,101	72,781	6,669,500
Sulawesi Tengah	Palu	26,921	69,726	1,604,800
Sulawesi Tenggara	Kendari	10,690	27,686	1,122,400
Sulawesi Utara	Menado	7,345	19,023	2,406,400
Sumatera Barat	Padang	19,219	49,778	3,851,500
Sumatera Selatan	Palembang	40,034	103,688	5,586,900
Sumatera Utara	Medan	27,331	70,787	9,667,500
Timor Timur	Dili	5,743	14,874	618,500
Special autonomous districts				
Aceh	Banda Aceh	21,387	55,392	3,078,400
Yogyakarta	Yogyakarta	1,224	3,169	2,913,400
TOTAL		741,101	1,919,443	168,662,000

threatened by the lava streams are deserted. The villagers, however, always return. In 1883 the Krakatau (Krakatoa) volcano in the Sunda Strait (Selat Sunda) erupted, killing 30,000 people, the greatest volcanic disaster in history. It then lay dormant, and renewed growth appeared on its slopes until 1927, when it threatened once more. In 1928 a new cone arose, forming a crater island. The most recent eruption was in 1952. There is no human habitation on Krakatau, and even navigation within a certain distance is usually forbidden. In 1963 Gunung Agung (the Great Mountain) erupted on Bali, killing thousands and rendering others homeless. In addition to volcanic disasters, Indonesia has suffered from earthquakes.

The Indonesian Seas

Located too near the equator to feel the force of the typhoons, the seas of Indonesia are generally calm. Consequently, navigation in the Java, Banda, Molucca, and Celebes seas offers few dangers. The fearsome typhoons of the South China Sea spend themselves before reaching Indonesian waters, and the gales that blow from time to time through the Torres Strait, between Australia and New Guinea, seldom move farther than the extreme southeast

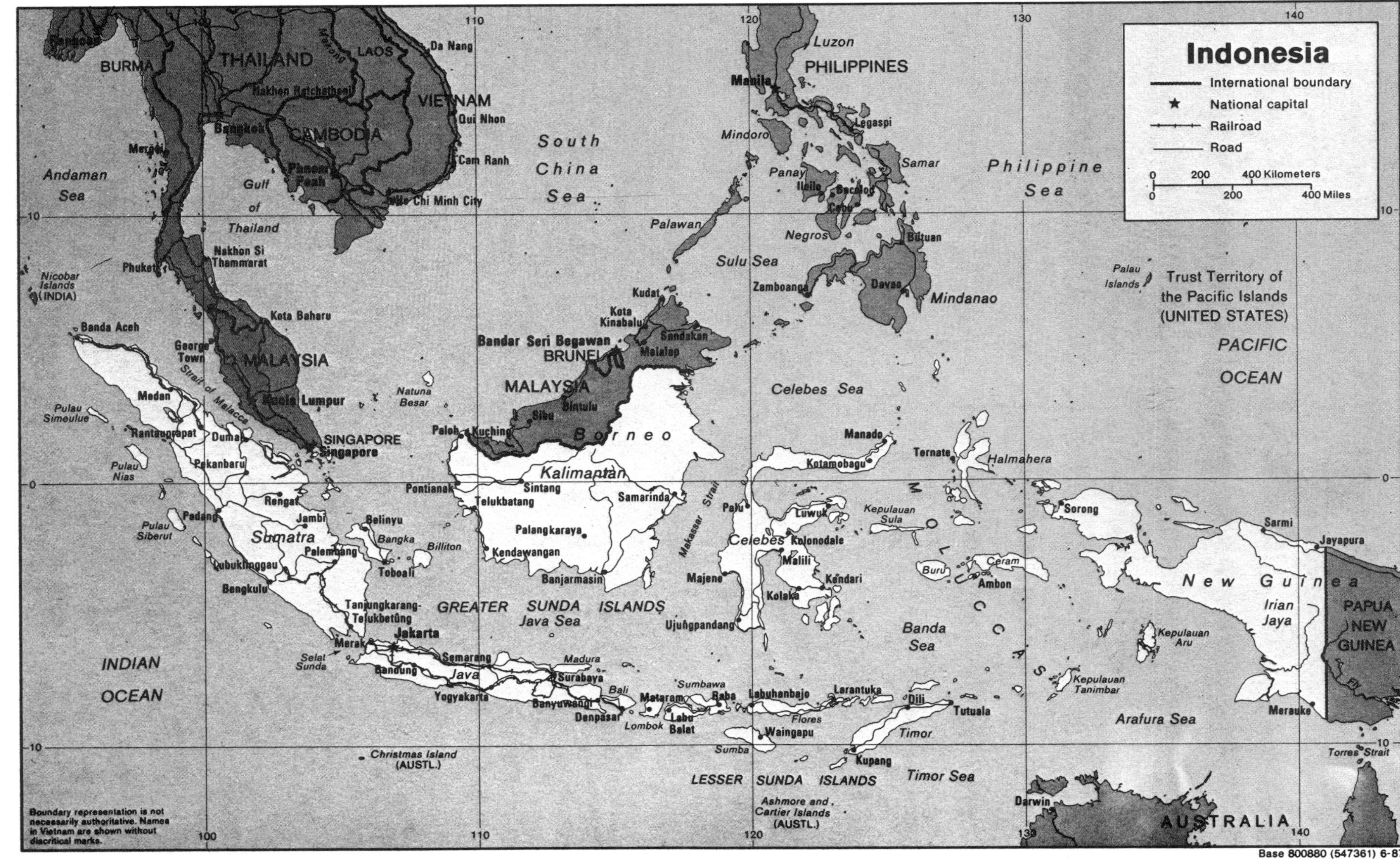

Indonesia
International boundary
National capital
Railroad
Road
0 200 400 Kilometers
0 200 400 Miles
BURMA
THAILAND
LAOS
CAMBODIA
VIETNAM
Da Nang
Qui Nhon
Cam Ranh
Bangkok
Andaman Sea
Gulf of Thailand
South China Sea
PHILIPPINES
Luzon
Manila
Mindoro
Legaspi
Samar
Panay
Negros
Palawan
Sulu Sea
Zamboanga
Davao
Butuan
Mindanao
Philippine Sea
Palau Islands
Trust Territory of the Pacific Islands (UNITED STATES)
PACIFIC OCEAN
Nicobar Islands (INDIA)
Phuket
Nakhon Si Thammarat
Kota Baharu
MALAYSIA
Kuala Lumpur
SINGAPORE
Singapore
Strait of Malacca
Banda Aceh
Medan
Pulau Simeulue
Rantauprapat
Dumai
Pekanbaru
Pulau Nias
Padang
Pulau Siberut
Rengat
Jambi
Sumatra
Palembang
Lubuklinggau
Bengkulu
Natuna Besar
Belinyu
Bangka
Billiton
Toboali
Tanjungkarang-Telukbetung
Merak
Selat Sunda
Jakarta
Bandung
Semarang
Java
Yogyakarta
Madura
Surabaya
Banyuwangi
Bali
Denpasar
INDIAN OCEAN
GREATER SUNDA ISLANDS
Java Sea
Christmas Island (AUSTL.)
Kudat
Kota Kinabalu
Sandakan
Bandar Seri Begawan
BRUNEI
MALAYSIA
Bintulu
Sibu
Kuching
Paloh
Borneo
Kalimantan
Pontianak
Sintang
Telukbatang
Palangkaraya
Kendawangan
Banjarmasin
Samarinda
Makassar Strait
Celebes Sea
Manado
Kotamobagu
Palu
Celebes
Luwuk
Kolonodale
Malili
Kendari
Majene
Kolaka
Ujungpandang
Ternate
Halmahera
Kepulauan Sula
MOLUCCAS
Buru
Ceram
Ambon
Banda Sea
Sorong
Sarmi
Jayapura
New Guinea
Irian Jaya
PAPUA NEW GUINEA
Kepulauan Aru
Kepulauan Tanimbar
Arafura Sea
Merauke
Torres Strait
Sumbawa
Mataram
Lombok
Raba
Labuhanbajo
Flores
Larantuka
Dili
Tutuala
Timor
Waingapu
Sumba
Kupang
Timor Sea
LESSER SUNDA ISLANDS
Ashmore and Cartier Islands (AUSTL.)
Darwin
AUSTRALIA
Boundary representation is not necessarily authoritative. Names in Vietnam are shown without diacritical marks.
Base 800880 (547361) 6-87

ern islands of the archipelago. Other areas of infrequent disturbance are to the north of Sulawesi and in the Strait of Malacca, where the usually moderate winds occasionally reach gale force.

Visibility is good, and tides are moderate, and in northwestern Java the difference between high and low water is only eighteen inches. Currents run slowly and with little force, except in some of the narrower straits. Waters are shallow, rarely reaching a depth of more than sixty fathoms (110 m; 360 ft), except in the channel between the Asian and Australian land shelves. Generally, weather is less a danger to navigation than shallow seas and coral reefs.

The islands have many good harbors and anchorages for light craft; good natural harbors capable of handling heavy shipping also exist, but they are not located in areas considered desirable for development. Many harbors, especially those associated with ports at the mouths of rivers, are affected by heavy silting; through neglect, many became inaccessible to heavy shipping in the post-World War II period, a condition gradually being remedied under the government's economic development program.

IRAN

BASIC FACTS

Official Name: Islamic Republic of Iran

Abbreviation: IRN

Area: 1,648,196 sq km (636,372 sq mi)

Area—World Rank: 17

Population: 51,923,689 (1988) 64,822,000 (2000)

Population—World Rank: 20

Capital: Tehran

Boundaries: 7,680 km (4,772 mi); USSR 1,740km (1,081 mi); Afghanistan 850 km (528 mi); Pakistan 830 km (516 mi); Iraq 1,280 km (795 mi); Turkey 470 km (292 mi)

Coastline: Caspian Sea 630 km (392 mi); Gulf of Oman and Persian Gulf 1,880 km (1,168 mi)

Longest Distances: 2,250 km (1,398 mi) SE-NW 1,400 km (870 mi) NE-SW

Highest Point: Mount Demavend 5,604 m (18,386 ft)

Lowest Point: Caspian Sea 28 m (92 ft) below sea level

Land Use: 8% arable; 27% meadows and pastures; 11% forest and woodland; 54% other

Iran is located on a high triangular plateau. In general the plateau is part of a larger plateau that includes parts of Afghanistan and Pakistan. Geologists explain that the region was formed and shaped—and continues to be influenced—by the uplifting and folding effect of three giant blocs or plates pressing against each other; the Arabian plate, the Eurasian plate, and the Indian plate. The squeezing and pressing resulted in considerable folding at the edges and some folding in the interior, thus forming the mountain ranges. Subterranean shifts produced numerous faults in the earth's crust, and it is along or near these faults that the country's frequent and devastating earthquakes occur. Although the figures are rough estimates, Iranian government sources in 1976 stated that more than 60,000 people had been killed in earthquakes in the preceding fifteen years.

For centuries the level of the Caspian Sea, the world's largest landlocked body of water, declined irregularly, and the coastal plain represents former sea floor plus new top soil. Official elevation of the sea is 28 m (92 ft) below sea level. The fall of the water level has been halted by the construction of a canal system connecting the Caspian and Black seas. Generally the coast is smooth and shallow, but two lagoons, requiring constant dredging, make possible the well-sheltered ports of Bandar Nowshahr, on the eastern end of the shoreline, and Bandar Anzali, near the western end.

About half of the country is within the sparsely settled Central Plateau. Summer temperatures in the basin are extremely high, sometimes rising to more than 125°F, but winters can be bitterly cold. The deserts are among the driest and most barren in the world, and black, muddy salt marshes present dangers to the traveler. Settlement is confined to oases and to the flanks of small mountain chains; pastoralism is almost impossible.

The region, having no outward drainage, occupies a series of closed basins with elevations of 610 to 916 m (2,000 to 3,000 ft), and is almost completely surrounded by mountains. Rain falls only in winter and never measures more than a few inches a year. The basins have central drainage areas that may remain dry for months or years. The lowest parts may have lakes a few inches deep that dry up and leave salt crests known as *kavirs.* As evaporation proceeds, the thick plates of crystallized salt increase in size, press against each other, break up, and give the appearance of glaciers; underneath the crust lie marshes of black, slimy mud. Because of the danger of

PRINCIPAL TOWNS (estimated population 1982)

Town	Population	Town	Population
Tehran (Teheran)	6,022,029*	Abadan	294,068‡
Mashad (Meshed)	1,500,000*	Orumiyeh	262,588
Isfahan	1,000,000*	Rasht	259,638
Tabriz	852,296	Qazvin	244,265
Shiraz	800,416	Kerman	238,777
Bakhtaran		Hamadan	234,473
(Kermanshah)	531,350	Ardebil	221,970
Karaj	526,272†	Arak	209,970
Ahwaz	470,927	Yazd	193,282
Qom	424,048	Khorramshahr	140,490‡

*Population at 8 October 1986 census, including suburbs.
†Including suburbs.
‡Population at November 1976 census.

Base 800763 (544499) 8-86

crossing the marshes, many are unexplored. The firm, hard gravel plains that may surround the salt flats are called *dashts*. The extensive Dasht-i-Kavir and Dasht-i-Lut deserts are located in the Central Plateau.

The broken and irregular ranges of the eastern mountains, extending from the Soviet Union in the north to Pakistan in the south, are barren, but the valleys that intersperse the area are fertile. Violent winds blow dust all summer and bring severe blizzards in the winter. The temperature runs to extremes, both hot and cold, and there is very little rain.

In the north, where the mountains reach 2,133 to 2,743 m (7,000 to 9,000 ft), cattle grazing and settlements can be found above 1,219 m (4,000 ft). Wheat, barley, vegetables, and fruit are common crops on irrigated lands. A few farmers grow dates and cereal crops, and farther south some people earn a living from fishing in Lake Helmand along the Afghanistan border. This lake region of the Sistan basin in the southeast usually has sufficient water for crops, but farming is handicapped by occasional drought years.

Throughout most of the country, water has been such a problem that all water resources were nationalized in October 1967. Less than 14 percent of the land receives over 52 percent of the precipitation. In some areas precipitation is lacking for long periods of time. Sudden storms with heavy rains a few times a year may provide the entire annual rainfall. Beyond the local damage that these storms bring about, the rapid runoff precludes the use of precipitation for agricultural purposes.

Area and population

Provinces	Capitals	area sq mi	area sq km	population 1984 estimate
Azārbāijān-e Gharbi	Orūmiyeh	15,000	38,850	1,915,000
Azārbāijān-e Sharqi	Tabriz	25,909	67,103-	4,097,000
Bakhtarān	Bakhtarān	9,138	23,667	1,177,000
Boyer Aḥmad-e Kohkilūyeh	Yāsūj	5,506	14,261	390,000
Būshehr	Būshehr	10,677	27,653	519,000
Chahār Mahāl-e Bakhtiāri	Shahr Kord	5,741	14,870	601,000
Eṣfahān	Eçfahān	40,405	104,650	3,012,000
Fārs	Shirāz	51,467	133,298	2,806,000
Gilān	Rasht	5,679	14,709	2,069,000
Hamadān	Hamadān	7,639	19,784	1,407,000
Hormozgān	Bandar-ʿAbbās	25,819	66,871	694,000
Ilām	Ilām	7,353	19,044	308,000
Kermān	Kermān	69,466	179,916	1,535,000
Khorāsān	Mashhad	120,980	313,337	4,441,000
Khūzestān	Ahvāz	25,978	67,282	2,284,000
Kordestān	Sanandaj	9,652	24,998	906,000
Lorestān	Khorramābād	11,121	28,803	1,306,000
Markazi	Arāk	15,403	39,895	1,430,000
Māzandarān	Sari	18,292	47,375	2,880,000
Semnān	Semnān	34,764	90,039	370,000
Sistān-e Balūchestān	Zāhedān	70,108	181,578	997,000
Tehrān	Tehrān	7,381	19,118	7,243,000
Yazd	Yazd	27,031	70,011	569,000
Zanjān	Zanjān	14,053	36,398	1,488,000
TOTAL		634,562	1,643,510	44,444,000

Where rivers and precipitation were insufficient, the solution was irrigation or storage; if these measures failed or were lacking, inhabitants moved away. In food production, water is regarded as more important than land itself. Where rainfall is inadequate, three sources of supplementary water for crops are available—wells, rivers, and either springs, reservoirs, or *qanats*. A *qanat* consists of a line of shafts sunk to intercept a man-made underground conduit originating at a water collection point, usually the base of a mountain. The shafts are used as wells at intermediate habitation points. In more arid areas, underground stone structures built with domed masonry roofs, known as *birkehs,* serve as storage pools; they are not wholly satisfactory, however, since their vulnerability to pollution is high.

Evaporation compounds the problem. Specialists estimate that 50 to 60 percent of the water evaporates in the intense summer heat. Ground absorption varies according to soil composition and structure, but it is considerable. Experts estimate that the rivers and streams carry only one-fourth of the precipitation the country receives. Absorption into the ground, however, gives promise of long-range advantages; government studies have been made to locate subsurface accumulation areas, and several thousand producing wells have been sunk.

The major drainage basins are the Persian Gulf and Gulf of Oman in the south, the Central Plateau, Lake Orumiyeh in the northwest, and the Caspian Sea, into which the Atrak and Gorgan rivers flow, northeast of Tehran. Because of the mountain drainage structure in the marginal regions, more than half of the country's surface loses its water to the interior. Scattered subbasins or sumps, which are low points in the larger inland basins, can be used to collect water, naturally or artificially, in sufficient amounts to support habitation. There is evidence that a more extensive network of rivers existed in earlier eras, but in modern times the drainage system by rivers and streams is abbreviated and, in some places, intermittent. The absence of any river with a substantial volume of flow emphasizes the extremely arid, segmented conditions of the terrain. There are only six rivers that drain the interior slopes, and these dwindle to brooks or dry up completely during the hot months.

IRAQ

BASIC FACTS

Official Name: Republic of Iraq

Abbreviation: IRQ

Area: 438,317 sq km (169,235 sq mi)

Area—World Rank: 53

Population: 17,583,467 (1988) 25,151,000 (2000)

Population—World Rank: 46

Capital: Baghdad

Boundaries: 3,738 km (2,323 mi) Turkey 305 km (190 mi); Iran 1,515 km (941 mi); Kuwait 254 km (158 mi); Saudi Arabia 895 km (556 mi); Jordan 147 km (91 mi); Syria 603 km (375 mi)

Coastline: 19 km (12 mi)

Longest Distances: 984 km (611 mi) SSE-NNW; 730 km (454 mi) ENE-WSW

Highest Point: 3,607 m (11,835 ft)

Lowest Point: Sea Level

Land Use: 12% arable; 1% permanent crops; 9% meadows and pastures; 3% forest and woodland; 75% other

Iraq is divided by most geographers, including those of Iraq's government, into four main zones or regions: the desert in the west and southwest; the rolling upland between the upper Tigris and Euphrates rivers; the highlands in the north and northeast; and the alluvial plain through which the Tigris and Euphrates flow. The desert zone, an area lying west and southwest of the Euphrates River, is a part of the Syrian Desert, which covers sections of Syria, Jordan, and Saudi Arabia. The region, sparsely inhabited by pastoral nomads, consists of a wide, stony plain interspersed with rare sandy stretches. A widely ramified pattern of wadis—watercourses that are dry most of the year—runs from the border to the Euphrates. Some are over 400 km (248 mi) long and carry brief but torrential floods during the winter rains.

The uplands region between the Tigris north of Samarra and the Euphrates north of Hit is known as Al Jazirah (the island) and is a part of a larger area that extends westward into Syria between the two rivers and into Turkey. Water in the area flows in deeply cut valleys, and irrigation is much more difficult than it is in the lower plain. Much of this zone may be classified as desert.

The northeastern highlands begin just southwest of a line drawn from Mosul to Kirkuk and extend to the borders with Turkey and Iran. High ground, separated by broad, undulating steppes, gives way to mountains ranging from 1,000 to nearly 4,000 m (3,280 to 13,123 ft) near the Iranian and Turkish borders. Except for a few valleys, the mountain area proper is suitable only for grazing. In the foothills and steppes, however, adequate soil and rainfall make cultivation possible.

The alluvial plain begins north of Baghdad and extends to the Persian Gulf. Here the Tigris and Euphrates rivers lie above the level of the plain in many places, and the whole area is a delta interlaced by the channels of the two rivers and by irrigation canals. Intermittent lakes, fed by the rivers in flood, also characterize southeastern Iraq. A fairly large area 15,600 sq km (6,023 sq mi) just above the confluence of the two rivers at Al Qurnah and extending east of the Tigris beyond the Iranian border is marshland, the result of centuries of flooding and inadequate drainage. Much of it is a permanent marsh, but some parts dry out in the early winter, and other parts become marshland only in years of great flood.

Because the waters of the Tigris and Euphrates above their confluence are heavily silt laden, irrigation and fairly frequent flooding deposit large quantities of silty loam in much of the delta area. Windborne silt contributes to the total deposit of sediments. It has been estimated that the delta plains are built up at the rate of nearly twenty centimeters a century. In some areas major floods lead to the deposit, in temporary lakes, of as much as thirty centimeters of mud.

PRINCIPAL TOWNS (population at 1977 census)

Baghdad (capital)	3,236,000*	Mosul	1,220,000
Basrah (Basra)	1,540,000	Kirkuk	535,000

*The population of Baghdad at the 17 October 1987 census was 3,844,608.

Area and Population		area		population
Governorates	**Capitals**	sq mi	sq km	1985 estimate
al-Anbār	ar-Ramādi	53,175	137,723	582,058
Bābil	al-Ḥillah	2,030	5,258	739,031
Baghdād	Baghdād	1,992	5,159	4,648,609
al-Baṣrah	Basra	7,363	19,070	1,304,153
Dhi Qār	an-Nāṣiriyah	5,261	13,626	725,913
Diyālā	Baʿqūbah	7,449	19,292	691,350
Karbalāʾ	Karbalāʾ	1,944	5,034	329,234
Maysān	al-ʿAmārah	5,445	14,103	411,843
al-Muthannā	as-Samāwah	19,702	51,029	253,8G5
an-Najaf	an-Najaf	10,751	27,844	472,103
Ninawā	Mosul	14,555	37,698	1,358,082
al-Qādisiyah	ad-Diwāniyah	3,285	8,507	511,799
Ṣalāh ad-Din	Sāmarrāʾ	11,198	29,004	442,782
at-Taʾmim	Kirkūk	4,012	10,391	650,965
Wasiṭ	al-Kūt	6,683	17,308	483,716
Kurdish Autonomous Region				
Dahūk	Dahūk	2,368	6,120	330,356
Irbīl	Irbīl	5,587	14,471	742,682
as-Sulaymānīyah	as-Sulaymānīyah	6,083	15,756	906,495
LAND AREA		168,878	437,393	15,584,987
INLAND WATER		357	924	
TOTAL AREA		169,235	438,317	

The Tigris and Euphrates also carry large quantities of salts. These, too, are spread on the land by sometimes excessive irrigation and flooding. A high water table and poor surface and subsurface drainage tend to concentrate the salts near the surface of the soil. In general the salinity of the soil increases from Baghdad south to the Persian Gulf and severely limits productivity in the region south of Ar Amarah.

The Euphrates originates in Turkey, is augmented by the Nahr (river) al Khabur in Syria, and enters Iraq in the northwest. Here it is fed only by the wadis of the western desert during the winter rains. It then winds through a gorge, which varies from two to sixteen kilometers in width, until it flows out on the plain at Ar Ramadi. Beyond there the Euphrates continues to the Hindiyah Barrage, which was constructed in 1914 to divert the river into the Hindiyah Channel; the present-day Shatt al Hillah had been the main channel of the Euphrates before 1914. Below Al Kifl the river follows two channels to

Base 504065 3-79 (544444)

As Samawah, where it reappears as a single channel to join the Tigris at Al Qurnah.

The Tigris also rises in Turkey but is significantly augmented by several rivers in Iraq, the most important of which are the Khabur, the Great Zab, the little Zab, and the Uzaym, all of which join the Tigris above Baghdad, and the Diyala, which joins it about 36 km (22 mi) below the city. At the Kut Barrage much of the water is diverted into the Shatt al Gharraf, which was once the main channel of the Tigris. Water from the Tigris thus enters the Euphrates through the Shatt al Gharraf well above the confluence of the two main channels at Al Qurnah.

Both the Tigris and the Euphrates break into a number of channels in the marshland area, and the flow of the rivers is substantially reduced by the time they come together at Al Qurnah. Moreover the swamps act as silt traps, and the Shatt al Arab is relatively silt free as it flows south. Below Basra, however, the Karun River enters the Shatt al Arab from Iran, carrying large quantities of silt that present a continuous dredging problem in maintaining a channel for oceangoing vessels to reach the port at Basra.

The waters of the Tigris and Euphrates are essential to the life of the country, but they may also threaten it. The rivers are at their lowest level in September and October and at flood in March, April, and May they may carry forty times as much water as at low mark. Moreover one season's flood may be ten or more times as great as that in another year. In 1954, for example, Baghdad was seriously threatened, and dikes protecting it were nearly topped by the flooding Tigris.

IRELAND

BASIC FACTS

Official Name: Ireland

Abbreviation: IRE

Area: 70,285 sq km (27,137 sq mi)

Area—World Rank: 113

Population: 3,531,502 (1988) 3,817,000 (2000)

Population—World Rank: 103

Capital: Dublin

Boundaries: 3,603 km (2,239 mi); Northern Ireland 434 km (270 mi)

Coastline: 3,169 km (1,969 mi)

Longest Distances: 486 km (302 mi) N-S; 275 km (171 mi) E-W

Highest Point: Carrauntoohil 1,038 m (3,406 ft)

Lowest Point: Sea Level

Land Use: 14% arable; 71% meadows and pastures; 5% forest and woodland; 10% other

Ireland is located on an island in the eastern part of the North Atlantic on the continental shelf of Europe. It is bounded on the northeast by Northern Ireland, on the east and southeast by the Irish Sea and St. George's Channel and on the north and west by the Atlantic Ocean.

Ireland has been compared to a saucer in which a limestone plateau is rimmed by coastal highlands of varying geological origins. Maritime influences penetrate the land, which consists of a framework of south-to-west Caledonian hills and east-to-west Armorian chains around which lowlands spread in unbroken continuity. The central plain area is characterized by many lakes, bogs and scattered low ridges averaging about 91 m (300 ft) above sea level. The principal mountain ranges are the Mourne Mountains, the Wicklow Mountains in the East and the Macgillicuddy's Reeks in the Southwest. The highest peaks are Carrauntuohill (1,041 m; 3,414 ft), and Mount Brandon (953 m; 3,127 ft) near Killarney, and Lugnaquilla (926 m; 3,039 ft), 64 km (40 mi) south of Dublin.

Variety is the main feature of the lowlands, which may roughly be delimited by a line drawn from Dundalk to Boyle and thence in two concave curves to Galway and from Galway to Dublin. The Central Lowland is the heart of Ireland. Easy passageways along valley and lowland corridors lead from it to every Irish shore. South of the Dublin-Galway line, for example, tongues of lowland spread around the various uplands and lead to the fertile valleys of Munster. Around the lowland there are several types of landscape, each owing its distinctive character to its structure and history. The stretch of country between Galway Bay and Killarney has little economic value or aesthetic attraction and lies off the main commercial and tourist routes. To the west the limestones give way to a variety of sedimentary and metamorphic rocks, and the green fields of the lowland are replaced by bogs or towering mountains in West Connacht. In Clew Bay, however, a stretch of the Central Lowland reaches the coast to divide West Connacht into two parts. In the lower Moy Basin the limestone lowland is separated from the center of the country by the Ox Mountains. The finest plateau lies between Curlew Hills and Donegal Bay, where the lake-strewn Erne Valley opens a natural avenue from Donegal to the Ulster basin around Lough Neagh.

The rivers of Ireland have courses of considerable variety and are among the most attractive features of the scenery. They cross the Central Lowlands as slow-moving streams, frequently surrounded by bogs and marshes and in many cases reach their estuaries through valleys. Shannon, the most important, rises in the plateau near Sligo Bay, flows sluggishly over the western part of the lowland, receiving tributaries of similar lethargic habits, and then fills Lough Derg before beginning its final spurt through rapids to its estuary. The Boyne and the Barrow also are rivers of the lowland. The Slaney cuts through the Leinster chain and flows through a steep-walled granite gorge between the Wicklow Mountains and the Blackstairs

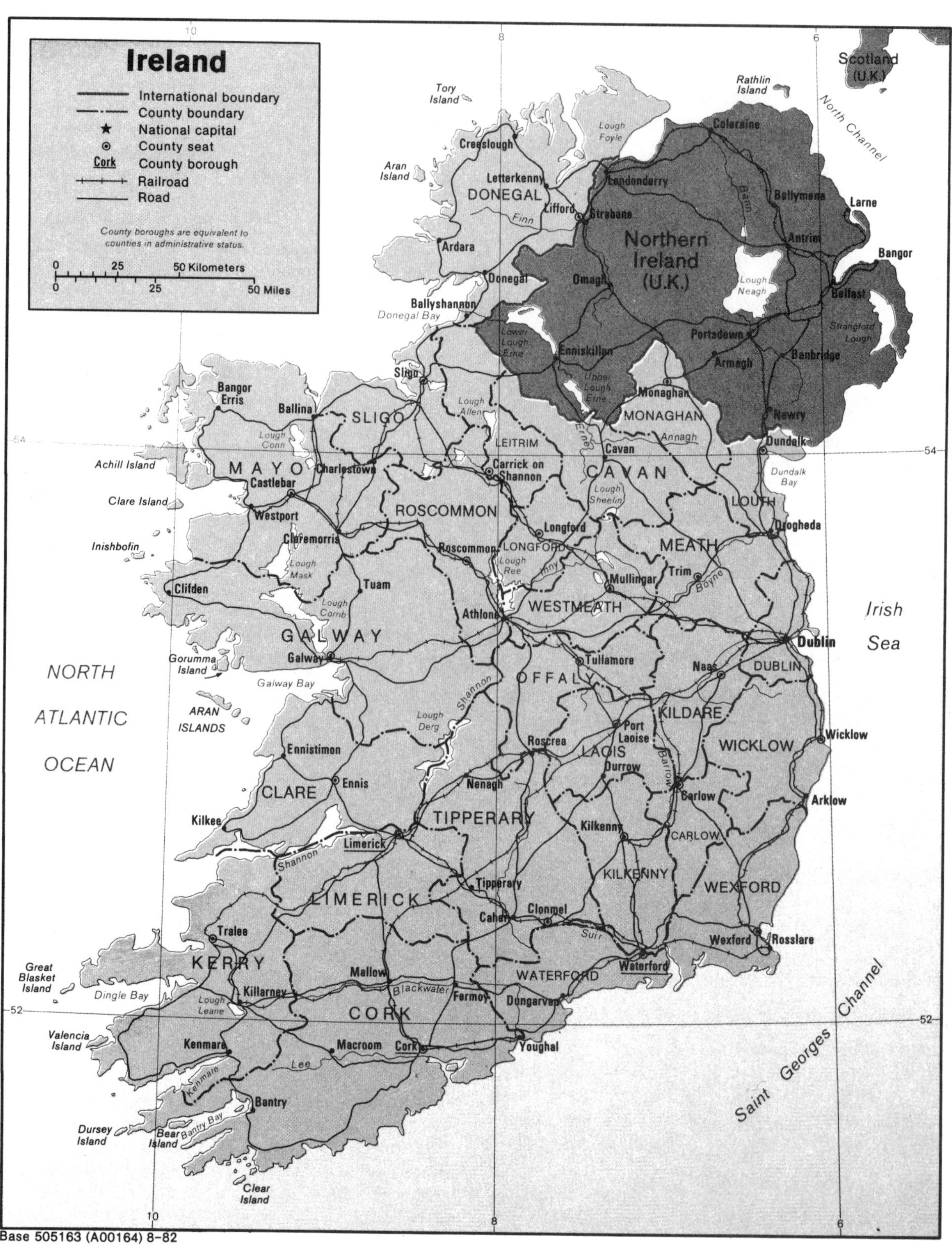

Base 505163 (A00164) 8-82

Mountains. It rises in Lugnaquilla and crosses the Wexford Lowlands, receiving a number of left-bank tributaries. The Bann rises in the Mourne Mountains, and fed by tributaries from the hilly country, flows northward. The Lagan rises in the Slieve Croob, runs toward Lough Neagh, then bends sharply northeast and flows to Lough Belfast through a depression. The Foyle, with its major tributary, the Strule, rises in the low hills to the northeast of Enniskillen, receives numerous tributaries from the Sperrin Mountains and Donegal and passes through Derry as a broad river to the lough. The Erne is a river of the drumlin country from its headwater streams in the Central Lowland and the hilly area of Monaghan to the sea at Donegal Bay after a sharp fall at Ballyshannon. The Moy runs parallel to the Ox Mountains and then swings northward, reinforced by several tributaries from the bog-covered lowlands of Mayo, into Killala Bay. The Corrib, on the lowland east of Connemara, has a completely dry channel between Lough Mask and Lough Corrib and receives many subterranean waters before it reaches the sea at Galway. Its tributaries include the Clare. Other Irish rivers include the Suir, the Liffey, the Suck, the Blackwater, the Lee and the Nore.

The Irish coast, with its striking cliffs, are among the most impressive in Europe. In the South and West, the coast is heavily indented where the ranges of Donegal, Mayo and Munster end in bold headlands and rocky islands, forming long, narrow, fjordlike inlets or wide-mouthed bays. The coastal precipices of the West are almost 610 m (2,000 ft) high in Slieve League and higher in Achill Island, where it is possible to look down 183 m (600 ft) into the water. From Cape Clear to Waterford Harbor there are long stretches of cliff over 61 m (200 ft) high, broken only by the inlets of small streams. For nearly 80 km (50 mi) south of Dundalk the coastline is a smooth stretch of low drift cliffs or sand dunes, with a wide and gently shelving sandy shore. The coast between Dundalk Bay and Belfast Lough consists mainly of low cliffs. In the North, the coasts are of great magnificence. The columnar basalts of the Giant's Causeway are visited by thousands yearly.

As Ireland was completely covered by ice sheets during the Ice Age, all extant flora and fauna are migrant species from other parts of Europe. However, with the submergence of the land bridge between Ireland and Britain in about 6000 B.C., this migration ceased. As a result, Ireland has a much lower variety of flora and fauna than Britain. The flora is similar to that of Britain but numerically poorer. It is estimated that there are only 1,300 species in Ireland, as against 2,300 in Britain. At the same time, Ireland has an interesting admixture of Alpine, Mediterranean and American flora. The Mediterranean species, however, are the most conspicuous in all parts of Ireland.

PRINCIPAL TOWNS

(population, including suburbs or environs, at 1986 census)

Dublin (capital)*	920,956	Galway	47,104
Cork	173,694	Waterford	41,054
Limerick	76,557		

*Greater Dublin area, including Dún Laoghaire (population 54,715 in 1986).

Bogs remain the most significant feature of Irish landscape; they occur on all the mountains and cover large areas of the lowlands. The remaining types of vegetation include a small but significant area of woodland, patches of primeval forest, shelter belts, demesne woodlands and modern plantations. The natural forest cover is sessile oak mixed with ash, wych, elm, birch and yew. Pine is dominant on poorer soils.

Area and Population	area		population
Provinces Counties	sq mi	sq km	1986 census
Connacht	6,611	17,122	430,726
Galway	2,293	5,940	178,180
Leitrim	581	1,525	27,000
Mayo	2,084	5,398	115,016
Roscommon	951	2,463	54,551
Silgo	693	1,796	55,979
Leinster	7,580	19,633	1,851,134
Carlow	346	896	40,958
Dublin	356	922	1,020,796
Kildare	654	1,694	116,015
Kilkenny	796	2,062	73,094
Laoighis	664	1,719	53,270
Longford	403	1,044	31,491
Louth	318	823	91,698
Meath	902	2,336	103,762
Offaly	771	1,998	59,806
Westmeath	681	1,763	63,306
Wexford	908	2,351	102,456
Wicklow	782	2,025	94,482
Munster	9,315	24,127	1,019,694
Clare	1,231	3,188	91,343
Cork	2,880	7,460	412,623
Kerry	1,815	4,701	123,922
Limerick	1,037	2,686	164,204
Tipperary North Riding	771	1,996	59,453
Tipperary South Riding	872	2,258	77,051
Waterford	710	1,838	91,098
Ulster	3,093	8,012	235,641
Cavan	730	1,891	53,881
Donegal	1,865	4,830	129,428
Monaghan	498	1,291	52,332
TOTAL LAND AREA	26,600	68,895	3,537,195
INLAND WATER	537	1,390	
TOTAL AREA	27,137	70,285	

ISRAEL

BASIC FACTS

Official Name: State of Israel

Abbreviation: ISR

Area: 20,700 sq km (7,992 sq mi) Excluding Gaza Strip and West Bank

Area—World Rank: 135

Population: 4,297,379 (1988) 5,475,000 (2000)

Population—World Rank: 96

Capital: Jerusalem (Not recognized by other countries)

Boundaries: 1,006 km (625 mi) Excluding Gaza Strip and West Bank); Egypt 255 km (159 mi); Jordan 238 km (148 mi); Lebanon 79 km (49 mi); Syria 76 km (47 mi); West Bank 307 km (191 mi); Gaza Strip 51 km (32 mi)

Coastline: 273 km (169 mi)

Longest Distances: 322 km (200 mi) N-S; 111 km (69 mi) E-W

Highest Point: Mount Meron 1,208 m (3,963 ft)

Lowest Point: Dead Sea 400 m (1,312 ft) below sea level

Land Use: 17% arable; 5% permanent crops; 40% meadows and pastures; 6% forest and woodland; 32% other

Israel is in southwestern Asia along the eastern end of the Mediterranean Sea. At its narrowest point, north of Tel Aviv, it is only 19 km (12 mi) across. Israel also administers areas taken from Syria and Jordan in the Six-Day war of 1967, including the Golan Heights (1,710 sq km; 660 sq mi) and the West Bank or Judea (5,878 sq km; 2,270 sq mi).

Except for the border with Egypt, which was settled at Camp David, all the other boundaries are disputed and have not been acknowledged in peace treaties. Israel has no de jure rights to occupied areas in the West Bank, the Gaza Strip, and the Golan Heights.

Topographically, Israel comprises four distinct divisions: coastal plains, mountains, valleys and deserts. Nearly two-thirds of the population is concentrated along a narrow strip of Mediterranean coastal plain, including the three urban zones of Acre-Haifa, Tel Aviv-Jaffa and Ashkelon-Ashdod as well as the occupied Gaza Strip.

The largest topographical division comprises mountains, softstone or dolomite in the North, pierced by natural caves and intersected by wadis (mostly dry riverbeds) and lofty granite peaks in the South. Mount Hermon in the North, with several peaks rising up to over 2,743 m (9,000 ft), towers over the basalt plateau of the Golan Heights. It has a true mountainous climate, with a snow-cap and ski slopes. The mountains of Upper Galilee rise to 1,208 m (3,962 ft) at Meron, while Lower Galilee is about half as high. Mount Carmel, rising to 546 m (1,790 ft), consists of two ranges with a rich lowland between. To the south lie the fabled hills of Samaria and Judea, with the coastal plain to their west and the Jordan Valley and the Dead Sea basin to the east. Nestled in these hills lies Jerusalem, Nablus and Hebron. Parts of Samaria are 960 m (3,149 ft) above sea level. The Judean Hills are a compact range 80 km (50 mi) long and 14 to 19 km (9 to 12 mi) across, with an average altitude of 750 m (2,460 ft).

Descending eastward to the Dead Sea, the Judean Hills turn into the Judean Desert, a wilderness of savage scenic splendor. To the south is Masada, where Jews made a heroic stand against Romans in the first century. At the junction of the desert and the Dead Sea is the retreat of the Qumran sect, where the Dead Sea Scrolls were discovered in the late 1940s. Jericho, the only town in the region, also is the world's oldest, dating back 9,000 years.

PRINCIPAL TOWNS (population at 4 June 1983)

Town	Population	Town	Population
Jerusalem (capital)	428,668*	Petach-Tikva	123,868
Tel-Aviv—Jaffa	327,625	Ramat Gan	117,072
Haifa	235,775	Beersheba	110,813
Holon	133,460	Bene Beraq	96,150

*Including East Jerusalem, annexed in June 1967.

Where the Judean Hills end in the South, the Negev Desert begins. It makes up two-thirds of the land area but contains only 6% of the population. The northern Negev is a region of low sandstone hills, steppes and fertile plains abounding in canyons and wadis. It contains three towns: Beersheba, city of the Patriarchs, whose population grew from 5,000 in 1948 to over 100,000 in 1984, Dimona, and Arad. In the central Negev, to the south, the mountains are higher and the climate drier. It is a tract of bare, rocky peaks and craters. Mitzpe Ramon, the region's only town, is perched opposite the crest of Mount Ramon (1,035 m; 3,395 ft). The Arava, an extremely parched stretch of desert between the Dead Sea and the Red Sea, has an average annual rainfall of less than 25 mm (1 in) and its summer temperatures are high. The Eliat Mountains form the southern tip of the Negev triangle, with sharp pinnacles of gray and red granite broken by dry gorges and sheer cliffs. Eliat is the region's major city and port. The scenic attractions of Eliat, where Solomon's galleys once anchored, its warm winters; and the teeming fish and corals of its subtropical waters have made it a major vacation resort and scuba diving center.

Fringing these mountains are four valleys: Hula, Capernaum, Jordan and Jezreel. The Hula, the northernmost of the valleys, is between the mountains of Upper Galilee and the Golan Heights. The narrow valley of Capernaum by Lake Kinneret (the Sea of Galilee) is famous for its Christian shrines. The Jordan Valley, running from north to south, includes the Beisan Valley. A huge depression or geological fault 400 m (1,312 ft) below sea level, it is part of the great Syrian-African rift that split the earth's crust millions of years ago. The Jezreel Valley, or the Plain of Esdraelon, separating Galilee from Samaria, is the country's most extensive valley, covering 365 sq km (141 sq mi).

Israel's only permanent rivers are the Jordan, the Yarkon, the Na'aman, the Kishon, the Taninim, the Alexander and the Ga'aton. All, except the Jordan, flow into the Mediterranean. The Jordan is the largest and best known and is the country's main source of water supply. It derives its name—Yawrden or Down-Rusher—from its

Area and population

		area		population
Districts	**Capitals**	sq mi	sq km	1985 estimate
Central (Ha Merkaz)	Ramla	479	1,242	889,100
Haifa (Hefa)	Haifa	330	854	592,700
Jerusalem (Yerushalayim)	Jerusalem	215	557	506,200
Northern (Ha Zafon)	Tiberias	1,347	3,490	706,700
Southern (Ha Darom)	Beersheba	5,555	14,387	510,100
Tel Aviv	Tel Aviv—Yafo	66	170	1,015,300
TOTAL		7,992	20,700	4,220,100

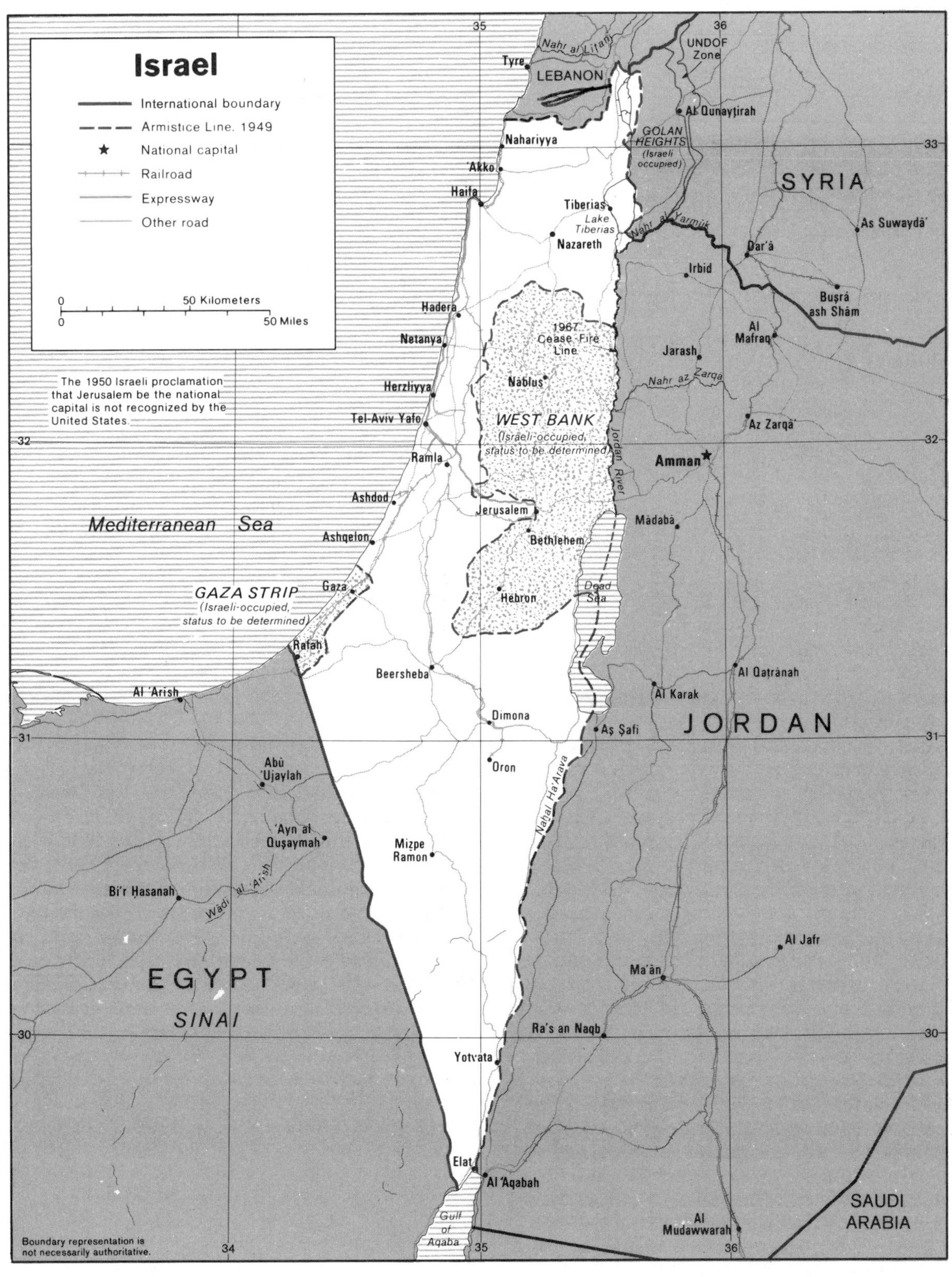

Israel
International boundary
Armistice Line, 1949
National capital
Railroad
Expressway
Other road
0 50 Kilometers
0 50 Miles
The 1950 Israeli proclamation that Jerusalem be the national capital is not recognized by the United States.
Mediterranean Sea
LEBANON
Nahr al Līţānī
UNDOF Zone
GOLAN HEIGHTS (Israeli occupied)
SYRIA
Tyre
Nahariyya
'Akko
Haifa
Tiberias
Lake Tiberias
Nazareth
Nahr al Yarmūk
Al Qunayţirah
As Suwaydā'
Dar'ā
Irbid
Buşrá ash Shām
Al Mafraq
Jarash
Nahr az Zarqā
Az Zarqā'
Hadera
Netanya
Herzliyya
Tel-Aviv Yafo
Ramla
Ashdod
Ashqelon
Gaza
Rafah
1967 Cease-Fire Line
Nāblus
WEST BANK (Israeli-occupied, status to be determined)
Jordan River
Amman
Jerusalem
Bethlehem
Hebron
Dead Sea
Mādabā
GAZA STRIP (Israeli-occupied, status to be determined)
Beersheba
Al Qaţrānah
Al Karak
Al 'Arīsh
Dimona
Aş Şāfī
JORDAN
Oron
Abū 'Ujaylah
'Ayn al Quşaymah
Mizpe Ramon
Nahal Ha'Arava
Bi'r Ḩasanah
Wādī al 'Arīsh
Al Jafr
EGYPT
SINAI
Ma'ān
Ra's an Naqb
Yotvata
Elat
Al 'Aqabah
Gulf of Aqaba
Al Mudawwarah
SAUDI ARABIA
Boundary representation is not necessarily authoritative.
35
36
34
33
32
31
30

headlong drop from source to mouth. Three of its sources—the Banyas, the Dan and the Hasban—rise on Mount Hermon. Along its 254 km (158 mi) course, the Jordan descends over 701 m (2,300 ft) to the lowest point on earth. While it swells during the rainy season, it is for most of the year a small, muddy stream that can easily be forded at several points. For the last section of its course it forms Israel's border.

One inland lake, Lake Kinneret (the Sea of Galilee) is wholly within Israel's borders; another, the Dead Sea, is only partly so. Lake Hula has been drained and survives in part as a nature reserve.

ITALY

BASIC FACTS

Official Name: Italian Republic

Abbreviation: ITL

Area: 301,278 sq km (116,324 sq mi)

Area—World Rank: 65

Population: 57,455,362 (1988) 57,388,000 (2000)

Population—World Rank: 15

Capital: Rome

Boundaries: 9,392 km (5,835 mi);Switzerland 744 km (462 mi); Austria 430 km (267 mi); Yugoslavia 209 km (130 mi); France 514 km (319 mi); San Marino 34 km (21 mi); Vatican City 3.2 km (2 mi)

Coastline: 7,458 km (4,634 mi)

Longest Distance: 1,185 km (736 mi) SE-NW;281 km (237 mi) NE-SW

Highest Point: Mont Blanc 4,807 m (15,771 ft)

Lowest Point: Sea Level

Land Use: 32% arable; 10% permanent crops; 17% meadows and pastures; 22% forest and woodland; 19% other

Situated in southern Europe, Italy is the smallest of the great peninsulas forming southern Europe. Italy stands out because of her boot-shaped configuration and its central place in the Mediterranean Sea. With its slender and elongated form, it has a maximum length of 1,185 km (736 mi) from Mont Blanc to the Salentine Peninsula in the extreme Southeast, while its greatest width is 630 km (391 mi) from Rocca Chardonnet to Monte Nevoso. The peninsula is surrounded by four seas: the Ligurian Sea, the Tyrrhenian Sea, the Ionian Sea and the Adriatic Sea. Besides Sardinia and Sicily, the national territory includes many islands and archipelagoes of which the largest are Elba, Asinara, Ischia, San Pietro, San Antioco, Salina, Lipari, Panetlleria, Giglio and Vulcano.

Italy is predominantly mountainous, and plains comprise less than one-third of its area. The two principal mountain ranges are the Alps and the Apennines. The Alps are made up of a series of massifs and chains running almost parallel to each other. The average elevations are higher toward the west, and the distances between them increase toward the east, assuming the shape of a fan. The Alps are commonly divided into three ranges. The Western Alps begin a short distance west of Genoa at Cadibona Pass near Savona and sweep in a great arc to Lake Maggiore. They include over 50 peaks over 3,048 m (10,000 ft) in height, including the Mont Blanc (9,810 m; 15,782 ft) on the French border and the highest peak entirely within Italy, and Gran Paradiso (4,061 m; 13,323 ft). The Western Alps resemble cliffs rather than slopes, and in some places less than 24 m (15 mi) separate the plains from the peaks. The Central Alps, extending from Lake Maggiore to the Adige River, also have over 50 peaks of more than 3,048 m (10,000 ft), but in contrast with the Western Alps, have valleys between the mountain ranges, making communications easier. The Central Alps cover a larger area than the Western Alps, collect more precipitation and have large glaciers. The Eastern Alps run from the Adige River to the Tarvis Pass on the Yugoslavian border. Their valleys tend to be wider and the terrain less steep. Also called the Venetian Alps, the Eastern Alps are divided into the Dolomites, the Carnic Alps and the Julian Alps. The Dolomites have 18 peaks over 3,048 m (10,000 ft), but to their east, all of their peaks are less than 3,048 m (10,000 ft).

The Apennine system is not formed of consecutive chains, as are the Alps, but by obliquely placed sections joined by traverses. Formed of softer rock, they are gentler in form and less elevated than the Alps, their highest elevation at Monte Carno in the Gran Sasso Range being only 2,895 m (9,500 ft). They are divided into the Northern (or Ligurian) Apennines and the Tuscan-Emilian Apennines, both of which have rounded and flattened mountains. On the other hand, the Apuan Hills, formed of crystalline rock, have notched and craggy shapes. The middle ranges, the Central Apennines, have a shorter and steeper Adriatic side with valleys generally running lengthwise. The Southern Apennines, formed by large isolated massifs, are more varied, lower in altitude and more barren. The Apennine foothills, with only small peaks, descend to the Tyrrhenian Sea.

Sicily's Peloritani, Nebrodi and Madonie ranges are geologically similar to the Apennines. Except for Mount Etna, they are not very high. The relief diminishes toward the east, forming a series of hill ranges and individual hills, while the island's barren central plateau slopes toward the south. Sardinia consists mostly of plateaus interspersed with mountain ranges. On the eastern coast, mountains rise steeply from the sea, particularly the Limbara Mountains, the Marghine Range and the Gennargentu Mountains.

The development of volcanic phenomena bears witness to Italy's geologic youth. Vesuvius near Naples, Etna in Sicily, and Stromboli and Vulcano in the Aeolian Islands

Italy
Railroad
Road
International airport
0 25 50 75 Miles
0 25 50 75 Kilometers
Fed. Rep. of Germany
Austria
France
Switzerland
Liech.
Yugoslavia
Algeria
Tunisia
Monaco
San Marino
Corsica
Sardinia
Sicily
Ligurian Sea
Adriatic Sea
Tyrrhenian Sea
Ionian Sea
Mediterranean Sea
Bern
Zürich
Innsbruck
Villach
Klagenfurt
Ljubljana
Zagreb
Postojna
Rijeka
Pula
Geneva
Lake Geneva
Bodensee
Bellinzona
Bolzano
Tarvisio
Udine
Gorizia
Trieste
Como
Bergamo
Brescia
Vicenza
Treviso
Venice
Milan
Verona
Padova
Torino
Piacenza
Parma
Genoa
Savona
Bologna
Ravenna
La Spezia
Rimini
Nice
Pisa
Livorno
Florence
Siena
Ancona
Perugia
Terni
Pescara
Civitavecchia
Rome
Foggia
Barletta
Bari
Naples
Salerno
Brindisi
Taranto
Otranto
Olbia
Porto Torres
Sassari
Oristano
Cagliari
Messina
Reggio di Calabria
Palermo
Trapani
Catania
Agrigento
Siracusa
Annaba
Tūnis
Rhine
Rhône
Inn
Drava
Po
Adige
Arno
Tiber
Ofanto
Tirso
Salso
8
12
16
46
42
38
Boundary representation is not necessarily authoritative

PRINCIPAL TOWNS (population at 31 December 1986)

Town	Population	Town	Population
Roma (Rome, the capital)	2,815,457	Perugia	146,713
Milano (Milan)	1,495,260	Ferrara	143,950
Napoli (Naples)	1,204,211	Ravenna	136,016
Torino (Turin)	1,035,565	Pescara	131,027
Genova (Genoa)	727,427	Rimini	130,698
Palermo	723,732	Reggio nell'Emilia	130,086
Bologna	432,406	Siracusa (Syracuse)	122,857
Firenze (Florence)	425,835	Monza	122,064
Catania	372,486	Sassari	120,152
Bari	362,524	Bergamo	118,959
Venezia (Venice)	331,454	Terni	111,157
Messina	268,896	Forli	110,482
Verona	259,151	Vicenza	110,449
Taranto	244,997	La Spezia	108,937
Trieste	239,031	Cosenza	106,026
Pavoda (Padua)	225,769	Piacenza	105,626
Cagliari	222,574	Torre del Greco	105,066
Brescia	199,286	Ancona	104,409
Reggio di Calabria	178,842	Pisa	104,384
Modena	176,880	Novara	102,742
Parma	175,842	Catanzaro	102,558
Livorno (Leghorn)	174,065	Bolzano (Bozen)	101,515
Prato	164,595	Lecce	100,981
Foggia	159,051	Udine	100,211
Salerno	154,848	Trento (Trent, Trient)	100,202

are active volcanoes. There are numerous extinct volcanoes dating from the Quaternary Period, such as Monte Amiata in southern Tuscany; the Volsini, Cimini and Sabatini mountains; and the Alban Hills of Latium. In Central Italy, lakes have formed in the craters of several extinct volcanoes, such as Lake Bolsena, Lake Vico, Lake Bracciano and Lake Albano. In addition, secondary volcanic activity is present in the Phleghrian Fields, with their many sulphurous springs. Seismic activity is widespread and strong throughout the entire Apennine region and in southern Italy. Periodically Italy has suffered from major earthquakes, such as the Calabrian-Sicilian earthquake which leveled the cities of Reggio Calabria and Messina, claiming 100,000 lives in 1908.

Area and Population

Regions / Provinces	Capitals	area sq mi	area sq km	population 1986 estimate
Abruzzi	L'Aquila	4,168	10,794	1,250,057
Chieti	Chieti	999	2,587	381,521
L'Aquila	L'Aquila	1,944	5,034	297,791
Pescara	Pescara	473	1,225	292,709
Teramo	Teramo	752	1,948	278,036
Basilicata	Potenza	3,858	9,992	618,647
Matera	Matera	1,331	3,447	207,188
Potenza	Potenza	2,527	6,545	411,459
Calabria	Catanzaro	5,823	15,080	2,131,412
Catanzar	Catanzaro	2,026	5,247	769,461
Cosenza	Consenza	2,568	6,650	772,620
Reggio di Calabria	Reggio di Calabria	1,229	3,183	589,331
Campania	Naples	5,249	13,595	5,651,200
Avellino	Avellino	1,078	2,792	445,670
Benevento	Benevento	800	2,071	297,277
Caserta	Caserta	1,019	2,639	796,381
Napoli	Naples	452	1,171	3,064,607
Salerno	Salerno	1,900	4,922	1,047,265
Emilia-Romagna	Bologna	8,542	22,123	3,939,289
Bologna	Bologna	1,429	3,702	919,591
Ferrara	Ferrara	1,016	2,632	374,341

Area and Population

Regions / Provinces	Capitals	area sq mi	area sq km	population 1986 estimate
Forlì	Forlì	1,123	2,910	607,297
Modena	Modena	1,039	2,690	596,437
Parma	Parma	1,332	3,449	397,827
Piacenza	Piacenza	1,000	2,589	274,726
Ravenna	Ravenna	718	1,859	354,600
Reggio nell'Emilia	Reggio nell'Emilia	885	2,292	414,470
Friuli-Venezia Giulia	Trieste	3,030	7,847	1,219,556
Gorizia	Gorizia	180	467	142,232
Pordenone	Pordenone	878	2,273	276,354
Trieste	Trieste	82	212	272,327
Udine	Udine	1,890	4,895	528,643
Lazio	Rome	6,642	17,203	5,101,641
Frosinone	Frosinone	1,251	3,239	476,611
Latina	Latina	869	2,251	457,978
Rieti	Rieti	1,061	2,749	145,088
Roma	Rome	2,066	5,352	3,747,335
Viterbo	Viterbo	1,395	3,612	274,629
Liguria	Genoa	708	5,416	1,771,319
Genova	Genoa	2,091	1,834	1,015,540
Imperia	Imperia	446	1,155	222,925
La Spezia	La Spezia	341	882	237,868
Savona	Savona	596	1,545	294,986
Lombardia	Milan	9,211	23,857	8,881,683
Bergamo	Bergamo	1,066	2,760	910,009
Brescia	Brescia	1,846	4,782	1,028,999
Como	Como	798	2,067	783,007
Cremona	Cremona	684	1,771	329,616
Mantova	Mantova	903	2,339	373,815
Milano	Milan	1,066	2,762	3,984,538
Pavia	Pavia	1,145	2,965	504,171
Sondrio	Sondrio	1,240	3,212	175,898
Varese	Varese	463	1,199	791,630
Marche	Ancona	3,743	9,694	1,425,734
Ancona	Ancona	749	1,940	437,881
Ascoli Piceno	Ascoli Piceno	806	2,087	357,971
Macerata	Macerata	1,071	2,774	294,476
Pesaro e Urbino	Pesero	1,117	2,893	335,406
Molise	Campobasso	1,713	4,438	333,502
Campobasso	Campobasso	1,123	2,909	240,345
Isernia	Isernia	590	1,529	93,157
Piemonte	Turin	9,807	25,399	4,394,312
Alessandria	Alessandria	1,375	3,560	455,263
Asti	Asti	583	1,511	211,825
Cuneo	Cuneo	2,665	6,903	547,694
Novara	Novara	1,388	3,594	503,079
Torino	Turin	2,637	6,830	2,289,054
Vercelli	Vercelli	1,159	3,001	387,397
Puglia	Bari	7,470	19,348	4,005,226
Bari	Bari	1,980	5,129	1,507,476
Brindisi	Brindisi	710	1,838	404,688
Foggia	Foggia	2,774	7,185	697,531
Lecce	Lecce	1,065	2,759	803,342
Taranto	Taranto	941	2,437	592,189
Sardegna	Cagliari	9,301	24,090	1,638,172
Cagliari	Cagliari	2,662	6,895	755,771
Nuoro	Nuoro	2,720	7,044	276,982
Oristano	Oristano	1,016	2,631	159,250
Sassari	Sassari	2,903	7,520	446,169
Sicilia (Sicily)	Palermo	9,926	25,708	5,084,311
Agrigento	Agrigento	1,175	3,042	487,311
Caltanissetta	Caltanissetta	822	2,128	294,098
Catania	Catania	1,371	3,552	1,051,380
Enna	Enna	989	2,562	197,301
Messina	Messina	1,254	3,247	684,703
Palermo	Palermo	1,927	4,992	1,241,357
Ragusa	Ragusa	623	1,614	286,224
Siracusa	Siracusa	814	2,109	406,574

Area and Population		area		population
Regions				1986
Provinces	Capitals	sq mi	sq km	estimate
Trapani	Trapani	951	2,462	435,363
Toscana	Florence	8,877	22,992	3,576,508
Arezzo	Arezzo	1,248	3,232	313,631
Firenze	Florence	1,498	3,879	1,198,400
Grosseto	Grosseto	1,739	4,504	220,255
Livorno	Livorno	468	1,213	346,257
Lucca	Lucca	684	1,773	383,588
Massa-Carrara	Massa-Carrara	447	1,157	205,716
Pisa	Pisa	945	2,448	389,048
Pistoia	Pistoia	373	965	265,637
Siena	Siena	1,475	3,821	253,976
Trentin-Alto Adige	Bolzano	5,259	13,620	878,590
Bolzano-Bozen	Bolzano	2,857	7,400	434,361
Trento	Trento	2,402	6,220	444,229
Umbria	Perugia	3,265	8,456	816,939
Perugia	Perugia	2,446	6,334	589,954
Terni	Terni	819	2,122	226,985
Valle d'Aosta	Aosta	1,259	3,262	113,714
Veneto	Venice	7,090	18,363	4,370,533
Belluno	Belluno	1,420	3,678	217,418
Padova	Padova	827	2,142	815,586
Rovigo	Rovigo	691	1,789	251,544
Treviso	Treviso	956	2,477	730,708
Venzia	Venice	950	2,460	838,000
Verona	Verona	1,195	3,096	781,850
Vicenza	Vicenza	1,051	2,721	735,427
TOTAL		116,324	301,277	57,202,345

Except for the Po Plain, the plains of Italy are small in area and few in number. The Po Plain, flanked by the northern Apennines and the Alps, is slightly inclined toward its central axis and toward the Adriatic Sea. The other plains, however, are important demographically if not geographically. They include the Lower Valdano Plain, corresponding to the Lower Arno River; Versilia, between the Apuan Hills and the Tyrrhenian Sea; the Tuscan Maremma, in the lower valley of the Ombrone River near Grosseto and formed partly by marshland and partly by reclaimed land, the Agro Romano, intersected by the Tiber and Aniene rivers; the Pontine Marshes, once a malarial swamp; the Campanian Plain, adjacent to the Gulf of Gaeta; the Gulf of Naples; the Sele Plain at the mouth of the Sele River, the Apulian Tavoliere, the plains of Metaponto and Sibari, overlooking the Gulf of Taranto; the Plain of Catalia and the Conca d'Oro in Sicily; and the Campidano in Sardinia.

Of the many rivers, the Po is preeminent not only in length (642 km; 399 mi), but also in the size of its basin. Due to sedimentation, the Po's riverbed has risen to a higher level than the surrounding plain. Thus floods are a chronic problem. The Po is navigable for almost its entire length. Its tributaries from the west—the Dora Riparia, the Dora Baltea, the Ticino, the Adda, the Oglio and the Mincio—are shallow in the winter but crest in the spring and summer due to spring rains and melting of glaciers. Its tributaries from the east—the Tanaro, the Scrivia, the Trebbia, the Taro, the Sacchia and the Panaro—are almost dry during summer and full in the fall and spring. This explains the Po's constant water level and its spring floods. The most important rivers of Venetia are the Adige (410 km; 255 mi), the Brenta, the Piave, the Tagliamento and the Isonzo. The Reno is the most important river flowing into the Adriatic. Others are the Tronto, the Aterno-Pescara and the Ofanto, all of which are short, rushing streams with low levels in summer and high levels in spring. The principal river draining the Ionian slope is the Bradano. Three Ligurian rivers flowing into the Tyrrhenian Sea are the Rota, Lavagna and Magra. The Arno and its tributaries flow through Tuscany, and south of it flows the Ombrone-Grossetano, feeding the marshes of the Tuscan Maremma and Marta.

The Tiber, the Liri-Garigliano, the Volturno and the Sele flow through Latium and Campania with fairly constant levels. The Simeto and the Salso are Sicily's most important rivers, and the Tirso is Sardinia's principal one.

Italian lakes are of three types: Alpine valley lakes, crater lakes in peninsular regions, and coastal lakes. Lakes Garda, Maggiore, Como and Iseo are examples of the first; Lakes Bolsena, Bracciano, Vico, Albano and Nemi of the second; and Lakes Orbetello, Massaciuccoli, Fondi, Lesina and Varano of the third.

IVORY COAST

BASIC FACTS

Official Name: Republic of the Ivory Coast (Côte d'Ivoire)

Abbreviation: IVR

Area: 320,763 sq km (123,847 sq mi)

Area—World Rank: 63

Population: 11,184,847 (1988) 16,194,000 (2000)

Population—World Rank: 57

Capital: Abidjan

Boundaries: 3,542 km (2,201 mi); Mali 515 km (320 mi); Burkina Faso 531 km (330 mi); Ghana 668 km (415 mi); Liberia 716 km (445 mi); Guinea 605 km (376 mi)

Coastline: 507 km (315 mi)

Longest Distances: 808 km (502 mi) SE-NW; 780 km (485 mi) NE-SW

Highest Point: Mount Nimba 1,752 m (5,748 ft)

Lowest Point: Sea Level

Land Use: 9% arable; 4% permanent crops; 9% meadows and pastures; 26% forest and woodland; 52% other

Ivory Coast lies almost wholly between 5° and 10° north of the equator and is bisected by the fifth meridian west of Greenwich. Westernmost of the African countries that border the Gulf of Guinea, its outline is a rough square, oriented in the cardinal directions. The country is bounded for the entire length of its eastern border by Ghana. The northern boundary is shared for approximately equal distances by Burkina Faso on the east and Mali on the west. Again for approximately equal distances, the western border, from north to south, abuts Guinea and Liberia. The shore of the Gulf of Guinea forms the southern boundary.

From the sea, the ground slopes gently and with very little bold relief to elevations of about 426 m (1,400 ft) along the northern border. The only mountain masses of any consequence are along the western border and in the northwest where some of the higher peaks exceed 914 m (3,000 ft) in elevation. The four parallel drainage basins formed by the four main rivers of the country run generally north to south, but except for the westernmost, the divides between them are not sharply defined.

Dense forest characterizes the southern third of the country, but farther inland the woodlands become more and more sparse and grassy, with the heaviest growth bordering the water courses or dispersed in isolated pockets. Toward the north, there is no forest in a strict sense. Scattered trees and shrubs dot the grasslands, their size and frequency diminishing progressively from south to north.

Viewed as a whole, almost all of the country is little more than a wide plateau, sloping gradually southward to the sea. There are no large rivers, mountain barriers or marked climatic differences dividing the land into distinctive geographic regions. More than by any other physical feature, the land is differentiated by zones of natural vegetation, extending roughly east and west across the entire country, parallel to the coastline. Three main regions, corresponding to these zones, are commonly recognized.

The Lagoon Region

The Lagoon region (*zone lagunaire*) is a narrow coastal belt extending along the Gulf of Guinea from the Ghana border to the vicinity of Fresco, near the mouth of the Sassandra River. For its entire length, the coast of this region is fringed by a strip of low, sandy islands or sandbars, known as the *cordon littoral.* Built by the combined action of the heavy surf and the ocean current which sets eastward, the sand barrier has closed all but a few of the river mouths and formed a series of lagoons between itself and the true continental shore.

The fringing bars and sandy islands vary in width from a few hundred feet to 5 or 6 km (3 or 4 mi) and seldom rise more than 30.4 m (100 ft) above sea level. On the seaward side, their smooth, steep beaches are pounded by surf, heavy at all seasons, but particularly so in July and November. Behind the beaches, the sandy soil supports a luxuriant growth of coconut palm and salt-resistant coastal shrubs.

Most of the lagoons are narrow, salty and shallow and are parallel to the coastline, linked to one another by small watercourses or canals, built by the French. But occasionally, where the larger rivers empty, they become broad estuaries which may extend 16 or 24 km (10 or 15 mi) inland. Mudbanks, sandbars, and wooded islands dot the sheltered surface of the lagoons; their landward shores are indented with little forested bays and steep rocky headlands. The only permanent natural exits to the open sea are at Assinié, Grand-Bassam, Grand-Lahou, and Fresco, where the flow of sizable rivers has prevented the formation of barrier islands. However, the exits are so shallow and so impeded by shifting bars and strong currents that they are not navigable by vessels of deep draft and are often dangerous or unusable even by small craft.

Inland, there are no sharp variations in relief, and the terrain, broken mainly by swampy depressions, is generally flat. Watercourses are sluggish, except in time of flood. Viewed from the seaward side, the occasional hills, which may rise to several hundred feet a few miles from the shore, have an exaggerated prominence because of contrast with the low monotony of the skyline.

As a result of concentrated development under the French, the dense rain forest which once came down to the water's edge along the continental side of the lagoons has been largely supplanted by clearings for plantations and farms and by second-growth woodlands or areas of brush and grass, extending from 8 to 24 km (5 to 15 mi) inland. Some of the less favorable areas have been left undisturbed, and, in particular, the low shores along the estuaries and the edges of marshy inlets are frequently choked by dense mangrove thickets.

The Dense Forest Region

Variously referred to in the Ivory Coast and by French writers as the *zone de la forêt* or the *forêt dense,* the dense forest region forms a broad belt that covers roughly a third of the country north of the lagoon region and extends from Ghana on the east to Liberia on the west. West of Fresco, it reaches all the way to the sea. Its northern boundary, although well-defined, is very irregular, descending in the form of a wide V from points on the Ghana and Guinea borders some 322 km (200 airline mi) inland, to within about 121 km (75 mi) of the sea north of Grand-Lahou.

The region gains its identity from the heavy tropical forest that flourishes throughout, except where it has been disturbed by man. Its northern limit is marked by a distinct but irregular transition to open, grassy woodlands. This division is only in small part caused by climatic differences. The limits are primarily the result of persistent cutting and burning by man encroaching on the forest from the north. No such limits exist on the east and west where the forest continues into the adjacent countries.

For over 160 km (100 mi) between Fresco and the mouth of the Cavally River, the forest reaches the Gulf of Guinea. Unlike the lagoon region farther east, this western coastline is relatively bold. There are no sheltered bays or anchorages suitable for ocean-going vessels, and surf makes most of the beaches dangerous. The foreshore is a succession of small sandy beaches, separated by rocky points and backed by a broken line of hills several hundred feet high. Except where it has been cleared, the forest extends to the water's edge.

Inland, from the Ghana border west to beyond the Sassandra River, the gently rolling relief of the region is only rarely broken by small hill masses or isolated tors, none of which rises as high as 304 m (1,000 ft). The main drainage is toward the south. The only bold relief is west of the Sassandra, along the Liberian border, where a series

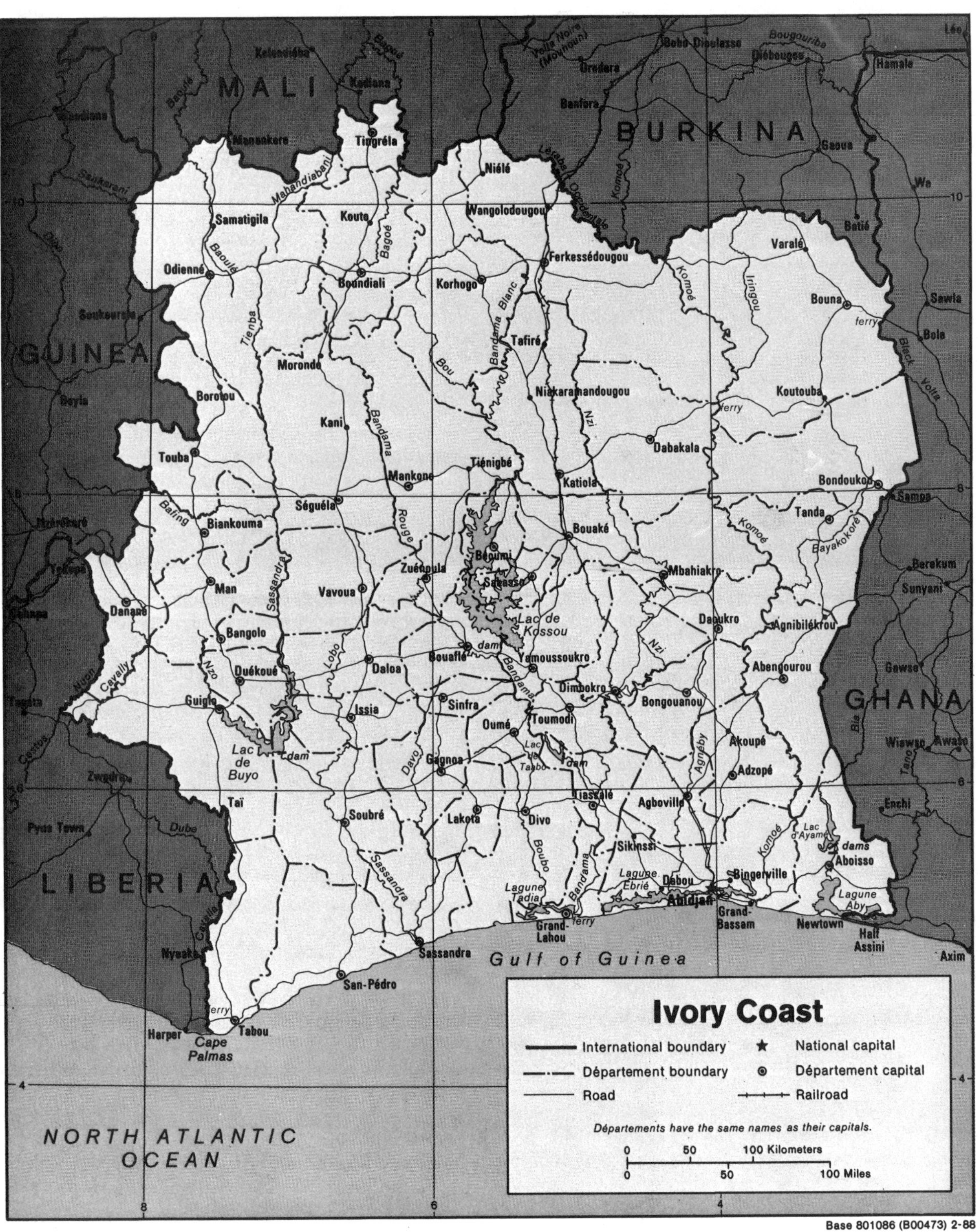

Ivory Coast
International boundary
Département boundary
Road
National capital
Département capital
Railroad
Départements have the same names as their capitals.
0 50 100 Kilometers
0 50 100 Miles
MALI
BURKINA
GUINEA
GHANA
LIBERIA
NORTH ATLANTIC OCEAN
Gulf of Guinea
Tingréla
Niélé
Samatigila
Kouto
Wangolodougou
Ferkessédougou
Odienné
Boundiali
Korhogo
Varalé
Bouna
Tafiré
Morondo
Niakaramandougou
Koutouba
Borotou
Kani
Dabakala
Touba
Mankono
Tiénigbé
Katiola
Bondoukou
Séguéla
Tanda
Biankouma
Bouaké
Béoumi
Man
Vavoua
Zuénoula
Sakassou
Mbahiakro
Danané
Lac de Kossou
Daoukro
Agnibilékrou
Bangolo
Bouaflé
Yamoussoukro
Abengourou
Duékoué
Daloa
Dimbokro
Bongouanou
Guiglo
Issia
Sinfra
Oumé
Toumodi
Akoupé
Lac de Buyo
Gagnoa
Adzopé
Taï
Tiassalé
Agboville
Soubré
Lakota
Divo
Sikensi
Bingerville
Aboisso
Dabou
Abidjan
Grand-Bassam
Grand-Lahou
Newtown
Half Assini
Sassandra
San-Pédro
Tabou
Harper
Cape Palmas
Abidjan
Lac de Kossou
Lagune Ebrié
Lagune Aby
Lagune Tadio
Bandama
Sassandra
Komoé
Nzi
Cavally
Base 801086 (B00473) 2-88

of low mountains extends east and south from the Nimba Mountains, west of Man, to within about 80 km (50 mi) of the sea, north of Tabou. At the northern limits, some of the higher mountains exceed 914 m (3,000 ft) in elevation. At the southern end, Mt. Niénokoué rises prominently to 762 m (2,500 ft)—over 304m (1,000 ft) above the surrounding country.

PRINCIPAL TOWNS (population at 15 June 1979)
Abidjan 1,423,323; Bouaké 272,640

Area and population

		area		population
Departments	**Capitals**	sq mi	sq km	1975 census
Abengourou	Abengourou	2,664	6,900	177,692
Abidjan	Abidjan	5,483	14,200	1,389,141
Aboisso	Aboisso	2,413	6,250	148,823
Adzopé	Adzopé	2,019	5,230	162,837
Agboville	Agboville	1,486	3,850	141,970
Biankouma	Biankouma	1,911	4,950	75,711
Bondoukou	Boundoukou	6,382	16,530	296,551
Bongouanou	Bongouanou	2,151	5,570	216,907
Bouaflé	Bouaflé	2,189	5,670	164,817
Bouaké	Bouaké	9,189	23,800	808,048
Bouna	Bouna	8,290	21,470	84,290
Boundiali	Boundiali	3,048	7,895	96,449
Dabakala	Dabakala	3,734	9,670	56,230
Daloa	Daloa	4,483	11,610	265,529
Danané	Danané	1,776	4,600	170,249
Dimbokro	Dimbokro	3,293	8,530	258,116
Divo	Divo	3,058	7,920	202,511
Ferkessedougou	Ferkessde-dougou	6,845	17,728	90,423
Gagnoa	Gagnoa	1,737	4,500	174,018
Guiglo	Guiglo	5,463	14,150	137,672
Issia	Issia	1,386	3,590	104,081
Katiola	Katiola	3,637	9,420	77,875
Korhogo	Korhogo	4,826	12,500	276,816
Lakota	Lakota	1,054	2,730	76,105
Man	Man	2,722	7,050	278,659
Mankono	Mankono	4,116	10,660	82,358
Odienné	Odienné	7,954	20,600	124,010
Oumé	Oumé	927	2,400	85,486
Sassandra	Sassandra	6,768	17,530	116,644
Séguéla	Séguéla	4,340	11,240	75,181
Soubré	Soubré	3,193	8,270	75,350
Tingréla	Tingréla	849	2,200	35,829
Touba	Touba	3,367	8,720	77,786
Zuénoula	Zuénoula	1,093	2,830	98,792
TOTAL		123,847	320,763	6,702,866

The Savanna Woodland Region

This region comprises all of the Ivory Coast lying north of the closed forest. It is characterized by a scattering of single trees or clumps of trees over continuous stretches of grass and low shrubbery and by narrow strips of heavier timber bordering the watercourses or lines of drainage. The size, types, and density of tree growth diminish progressively from south to north, but so gradually that changes are apparent only over distances of many miles, and important differences are evident only along the southern and northern extremities of the region.

As a consequence, some authorities, particularly the French, divide the region into two approximately equal belts, roughly at 8°30'N. Those who recognize the division differ as to nomenclature, referring to the southern belt as the forest-savanna mosaic, the *zone de transition* or the savanna. Similarly the northern belt is sometimes known as the woodlands and savanna (relatively moist type), the *zone soudanienne* or the *zone soudanienne meridionale*. The distinction, however, is only of interest to specialists in vegetation. For practical purposes, it suffices to bear in mind the transitional nature of the region.

Except in the northwest, the relief is a continuation of the generally featureless, gradually rising terrain of the dense forest region. The only mountainous area is along the Guinea border, reaching north from the vicinity of Man in continuation of the eastern slopes of the Guinea highlands. In this area, Mt. Tonkui, near Man, and a few other peaks rise above 914 m (3,000 ft). The highlands, in less broken form, extend east past Odienné nearly to Korhogo and then smooth into a flat divide, northeast into Mali, forming a watershed between the Niger River basin on the north and the rivers that flow south through the Ivory Coast. Along this divide, elevations vary between 305 and 609 m (1,200 and 2,000 ft) with only a few of the highest hill masses in the western part approaching 914 m (3,000 ft).

Elsewhere, the gently rolling plains are broken only by occasional, isolated granite domes or small hill masses. Among the latter, the Komonos Hills, midway between Firkéssédougou and the Ghana border, are the most extensive. The ground east of the Comoé River forms part of a major divide which extends from Burkina Faso across the northeastern corner of the Ivory Coast and separates the main drainage system of that country from the Volta River basin. Near Bondoukou, where the divide crosses the Ghana border, a mass of broken hills emerges to reach elevations of about 609 m (2,000 ft).

RIVERS

Four main rivers cut across the geographic regions from north to south to the Gulf of Guinea—the Comoé, Bandama, Sassandra and Cavally Rivers. Although they are permanent streams with a good volume of water, they are commercially navigable only for short distances inland from the sea coast because of rocky ledges and shifting shoals which form rapids and prevent passage even of small canoes. Uncontrolled and given to seasonal flooding, they are not only of little use as communications lines but are also obstacles to east-west travel. In fact, development of a good lateral road system has been much impeded by the cost and difficulty of bridging the main rivers or installing ferries.

The drainage system of the country is not complex, and all four of the main rivers and their principal affluents share the same characteristics. Overall gradients are gentle, averaging no more than 1.2 m (4 ft) per mile. Fall is irregular, however, and sometimes abrupt at rapids. Even near their sources, the rivers follow meandering courses. Because of the pronounced seasonal rainfall in the north, most of the upper tributaries, except those in the far west, are intermittent and change from dry stream beds to swollen torrents when heavy rains set in. As a consequence, the rivers are subject to wide variations in flow and to flooding over their entire length. In the lower reaches, where the rivers are used to float timber to the coast, this condition requires careful calculation to take advantage of high water and at the same time avoid loss of logs in areas of overflow during exceptional crests.

Easternmost of the main rivers, the Comoé has its sources in the Sikasso Plateau of Burkina Faso. Its basin is narrow, hemmed in between the broad basin in the Bandama River system on the west and the basins of the Black Volta and Bia Rivers on the east. Its principal tributaries are the Léraba River from the west and the Iringo and Kongo Rivers from the east. The Comoé River empties into the Ébrié Lagoon near Grand-Bassam, where it reaches the sea through a shallow gap between the barrier islands. Inland from the lagoon, it is navigable for vessels of light draft for about 48 km (30 mi) to Alépé.

The Bandama River system, whose basin drains about half the entire country, covers the flat central part of the Ivory Coast. The system consists of three main streams which follow very winding courses southward from the Niger River divide in the extreme north. The Bandama River proper is the axis of the system and the longest river in the country. Its companion on the west is the Marahoué or Bandama Rouge River, and on the east, the Nzi River. All three ultimately converge in the south and form a single river for the last 96 km (60 mi) to the Tagba Lagoon, opposite Grand-Lahou, where it enters the sea through a shallow pass between the barrier islands. Depending on the season, the head of navigation for small craft is 48 to 64 km (30 to 40 mi) inland from Grand-Lahou.

The Sassandra River basin is west of the Bandama system. The main source is the Tiémba River which rises in the high ground between Odienné and Boundiali and flows directly south. East of Touba, it unites with Féréodougouba River from the Guinea highlands to become the Sassandra. The river reaches the coast and the open sea through a narrow estuary which extends some 16 km (10 mi) inland. The exit is shallow and much obstructed by shifting sand bars, current and surf. Navigation by small boats is generally possible inland for about 80 km (50 mi).

The Cavally River has its headwaters in the Nimba Mountains of Guinea and forms the border with Liberia for over half its length. Its upper basin is narrow and hemmed in by confused masses of forested hills. It has no large tributaries. In its southern reaches, it winds through comparatively flat country and crosses the last of its many rapids about 48 km (30 mi) from the sea. Although it is navigable by small boats as far inland as this point, its exit to the sea near Cape Palmas is narrow and is obstructed by one of the most dangerous combinations of rock and sand bars on the entire coast.

Of the lesser streams, the Bia River, which flows from Ghana across the southeastern corner of the Ivory Coast, furnishes a useful shallow-draft waterway from the agricultural center of Aboisso southward for 16 km (10 mi) to the head of the Aby Lagoon.

JAMAICA

BASIC FACTS

Official Name: Jamaica

Abbreviation: JAM

Area: 10,991 sq km (4,244 sq mi)

Area—World Rank: 144

Population: 2,458,102 (1988) 2,882,000 (2000)

Population—World Rank: 116

Capital: Kingston

Land Boundaries: Nil

Coastline: 518 km (322 mi)

Longest Distances: 235 km (146 mi) N-S; 82 km (51 mi) E-W

Highest Point: Blue Mountain Peak 2,256 m (7,402 ft)

Lowest Point: Sea level

Land Use: 19% arable; 6% permanent crops; 18% meadows and pastures; 28% forest and woodland; 29% other

Jamaica is the third largest island of the Greater Antilles—after Cuba and Hispaniola—and is the largest of the West Indian islands of the Commonwealth Caribbean.

Lying some 137 km (85 mi) south of Cuba, Jamaica is situated in the middle of the Caribbean Sea on direct trade routes between North and South America and between Europe and Panama. It was used as a way station by merchant ships and freebooters during the early colonial period, and throughout its history the island's location at a crossroads of sea communication routes has significantly influenced its political, economic, and social development.

The maximum elevation is more than 2,133 m (7,000 ft) and there are numerous ranges of mountains and hills. Coastal plains and valleys fringe an interior plateau of limestone composition that covers most of the island. The plateau surface is broken by sharply crested ridges, twisting valleys, and broad basins.

Landform and Drainage

Jamaica does not lend itself readily to separation into geographic regions. It can, however, be considered generally as divided into coastal lowlands and valleys and a great interior plateau of white limestone. The uneven plateau surface is broken by many interior valleys, by ranges of limestone hills and mountains, and by two mountain ranges of different composition and appearance.

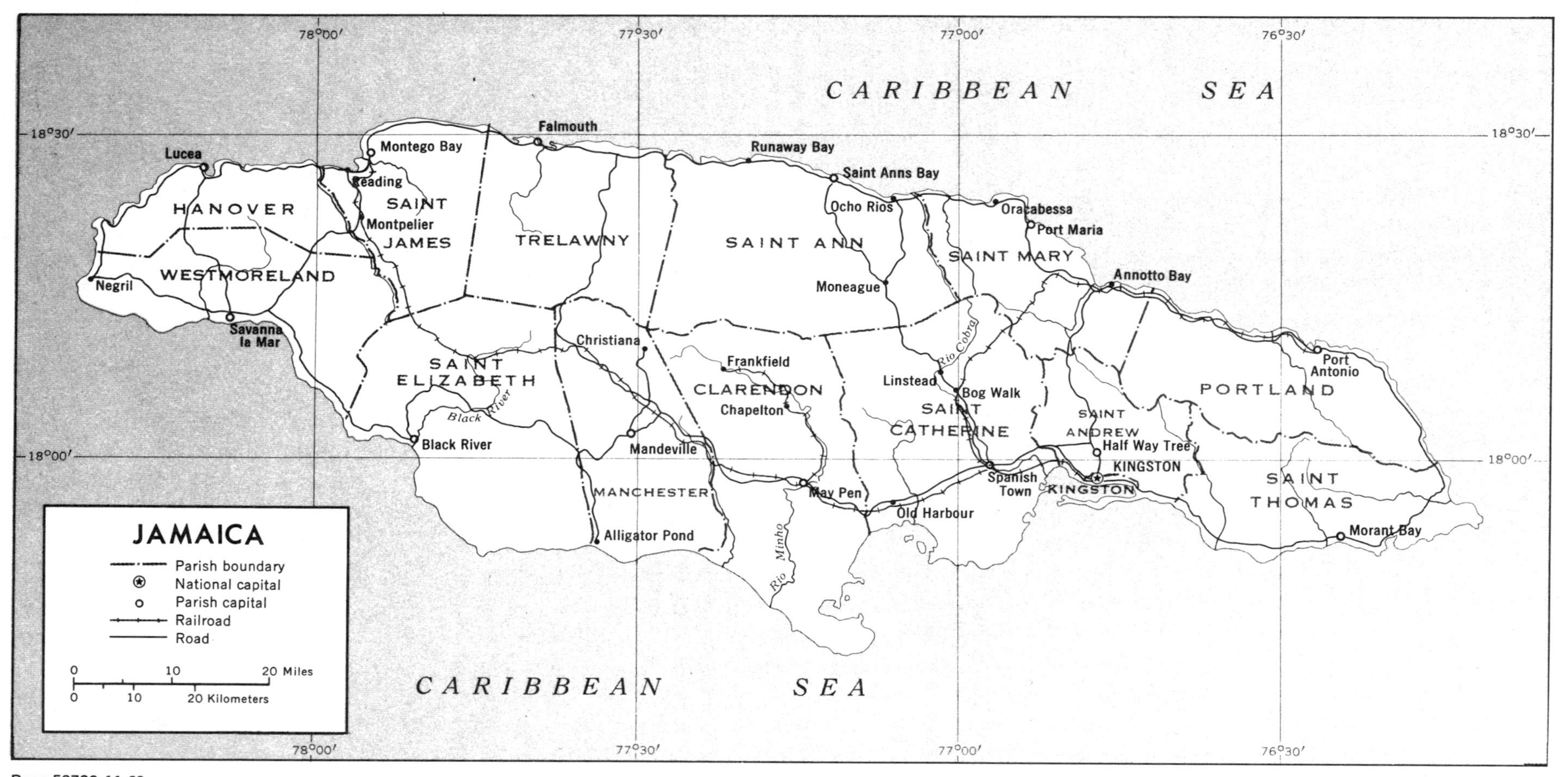

Base 58782 11-68

Elevations on the plateau range from near sea level to a maximum of about 914 m (3,000 ft) in the crests of the limestone uplands. Along much of the coastline the plateau extends almost to tidewater, and in parts of the parishes of Saint Ann and Saint Elizabeth it rises in steep coastal cliffs that reach as high as 609 m (2,000 ft). The plateau is ruggedly irregular, and its most characteristic landscape is known to geologists as karst—an irregular, limestone terrain with sinks, underground caverns and streams, steep hills, and caves. The sinkholes, which are often of considerable depth and appear frequently in remarkably straight rows, were caused by the collapse of the roofs of the caverns beneath. The karst features of the landscape are most distinctive in an area of about 518 sq km (200 sq mi) located largely in Trelawney Parish and known as the Cockpit Country because of the evident suitability of the countless sinkholes as arenas for cockfights. It is a forbidding land of strange plant forms and often bizarrely shaped limestone cones or haystack hills interspersed with caves. It was once called the ''Land of Look Behind'' because its scanty population was made up of hostile runaway slaves, and Spanish horsemen venturing into the area were said to have ridden two to a mount, one rider facing to the rear to keep a precautionary watch.

The karst landscape is broken by numerous interior valleys and basins with flat floors and steep limestone slopes and by numerous limestone hill and mountain features. The loftiest and largest of the limestone ranges are the John Crow Mountains, which rise in the extreme northeast of the island between the Rio Grande and the sea; their rugged terrain remains only partially explored. In addition, a series of limestone mountains and hills extend across the central and western portions of the interior, low limestone ranges flank portions of both northern and southern coasts, and other limestone ranges and isolated hills are scattered about the plateau.

The crests of the Blue Mountains, the country's principal mountain system, form part of the boundary between the parishes of Portland and Saint Thomas near the eastern end of the island, where they follow a southeast to northwest axis for a distance of about 80 km (fifty miles). These mountains were formed by an uplifting and warping of the limestone plateau that has exposed much of the older underlying strata. The relief is also quite different from that of the limestone ranges because of extensive gully development by streams in the older strata; in the limestone uplands all of the drainage is subterranean.

The system is composed of two ranges, the first of which is the more northerly and extensive of the two, and in it Blue Mountain Peak rises to more than 2,255 m (7,400 ft), the country's highest elevation. From this central peak, narrow ravines and sharp ridges descend like spokes of a wheel. The second range, also known as the Port Royal Mountains, extends southeastward from the principal range. It rises from the Liguanea Plain north of Kingston and reaches elevations of up to about 1,219 m (4,000 ft).

The northern coastal plain is narrow but extends almost continuously from the vicinity of Annotto Bay in the east to Montego Bay in the west. It is broadest to the south of Falmouth, where the Queen of Spain's Valley juts into the interior. Almost the entire coastline east of Falmouth is thickly fringed with coconut palms. White sand beaches are more numerous here than elsewhere, and it is this part of Jamaica that attracts three-fourths of the heavy flow of tourists from abroad. Montego Bay, Ocho Rios, and Port Antonio are the principal tourist centers.

On the southern coast the plains are discontinuous but much more extensive than on the north. They are relatively dry, and savanna landscape is characteristic. The city of Kingston lies on the broad Liguanea Plain, an expanse of 337 sq km (130 sq mi) that extends inland and westward and includes also the old city of Spanish Town. The plain is the economic heartland of the country. It was formed as a delta of numerous streams flowing southward from the Blue Mountains, and the Kingston area is subject to intermittent flooding.

West of the Liguanea Plain a second extensive coastal lowland stretches inland from the coast in Saint Catherine and Clarendon parishes. It is a sugarcane-producing area, and the small city of May Pen is located in it. Still farther to the west in Manchester and much of Saint Elizabeth parishes the limestone plateau drops directly to the sea in steep cliffs. The main highway around the island, which runs close to the coast elsewhere, turns inland and does not again swing toward the coast until it reaches the town of Black River, in Saint Elizabeth Parish, where an extensive swampy plain surrounds the river. The plateau again approaches the coast in the western part of Saint Elizabeth and the eastern part of Westmoreland, but the Westmoreland Plain occupies much of the western extremity of the island.

Vestiges of volcanic activity occur in Jamaica in the form of lava cones and hot springs, and there have been occasional serious earthquakes. A violent tremor in 1692 caused most of the buccaneer town of Port Royal to slide into the sea, and a 1907 earthquake followed by a tidal wave wrecked much of the Kingston area and took 800 lives. Although the principal epicenter of seismic activity

Capital: Kingston (population 104,000 at 1980 census).
Other towns (1970 census): Montego Bay (42,800); Spanish Town. (41,600).

Area and population

Parishes	Capitals	area sq mi	area sq km	population 1985 estimate
Clarendon	May Pen	462	1,196	212,100
Hanover	Lucea	174	450	64,000
Kingston		8	22	
Manchester	Mandeville	321	830	153,800
Portland	Port Antonio	314	814	76,200
Saint Andrew		166	431	625,800
Saint Ann	Saint Ann's Bay	468	1,213	144,600
Saint Catherine	Spanish Town	460	1,192	388,000
Saint Elizabeth	Black River	468	1,212	142,400
Saint James	Montego Bay	230	595	145,300
Saint Mary	Port Maria	236	611	109,900
Saint Thomas	Morant Bay	287	743	83,800
Trelawny	Falmouth	338	875	72,200
Westmorland	Savanna-la-Mar	312	807	125,600
TOTAL		4,244	10,991	2,343,700

is probably located somewhere off the north coast, the Kingston metropolitan area lies on unconsolidated alluvium and is thus particularly vulnerable to tremors.

The prevailing drainage pattern is one in which watercourses flow either northward or southward to the sea from springs in the interior highlands so numerous that Jamaica is sometimes called the Isle of Springs. In the north the watercourses are rapid, short, and blue-green in color as they flow through narrow valleys. In the south the valleys are broader, and the rivers are slower and muddier in color and follow looping courses after emerging onto the coastal plains. Many are dry during half of the year.

There are no rivers on the plateau; the porous sandstone permits the water to descend to the caverns beneath, and a network of streams lies beneath the plateau surface. Only the interior basins have watercourses; water emerges from one side of the floor as springs and disappears at the opposite side in sinkholes.

The Black River is navigable for about 40 km (25 mi) from its mouth. No other river is navigable for a significant distance, but the Rio Cobre, the White River, and several other of the more rapid streams are important sources of hydroelectric power. There are several radio active hot springs, and one—the Milk River Bath—is said to have the highest level of radioactivity in the world. Swamps, partially drained, are located along the lower course of the Black River and in the vicinity of Morant Point at the eastern tip and South Negril Point at the western tip of the island.

The relatively low fertility and excessive porosity of the limestone soils make most of the plateau unsuitable for crop production. It is good grazing land, however, and extensive areas are used as pasture. The fertile soils of the interior valleys are known as *tierra rosa* (reddish earth) and are similar to some of the best Cuban soils. In addition, where accessibility and configuration of terrain permit, the soils of the Central Range and the Blue Mountains are sufficiently fertile to support numerous small farms.

Coastal valleys and plains are made up in large part of alluvium washed down from the interior highlands. Much of the transported material has been washed out to sea, however, and valleys jutting into the plateau—the Yallahs River valley east of Kingston in particular—have been seriously eroded by heavy agricultural use. The most common coastal soil composition is one of alluvial sands, gravels, and loams, which tend to be somewhat more fertile in the south than in the north.

At the eastern end of the island and along the northern coast the ocean plunges to great depths not far from the shoreline. Near the resort town of Port Antonio the water drops to a depth of 182 m (600 ft) no more than one-half mile offshore, and the Bartlett Trough that separates Jamaica from Cuba reaches a depth of 7,010 m (23,000 ft). West of Kingston, however, relatively shallow water off the southern coast covers a sunken highland that extends southwestward to the coast of Central America.

The 885 km (550 mi) shoreline is indented by numerous harbors, of which six are rated primary and thirteen are rated secondary, and the harbor at Kingston is the world's seventh largest. It is admirably sheltered on its southern flank by the Palisadoes Peninsula; this eight-mile-long sandspit connecting several coral islands is the site of Port Royal and of the Norman W. Manley International Airport.

Near the southeast coast there are extensive coral reefs recently formed on a coastal limestone platform on which the Palisadoes Peninsula, the Portland Bight cays, and numerous small sand and coral islands have been built. Elsewhere a few scattered coral formations occur, particularly on the eastern extremity of the island. Jamaica's only offshore territories are the Morant Cays, 64 km (forty statute mi) southeast of Morant Point, and the more extensive Pedro Cays, about 96 km (sixty mi) south of the southwestern coast. The Pedro Cays are located in the extensive shallows of the Pedro Banks and serve as a base for fishermen.

JAPAN

BASIC FACTS

Official Name: Japan

Abbreviation: JAP

Area: 377,801 sq km (145,870 sq mi)

Area—World Rank: 56

Population: 122,626,038 (1988) 132,589,000 (2000)

Population—World Rank: 7

Capital: Tokyo

Land Boundaries: Nil

Coastline: 9,387 km (5,833 mi)

Longest Distances: 3,008 km (1,869 mi) NE-SW; 1,645 km (1,022 mi) SE-NW

Highest Point: Mount Fuji 3,776 m (12,388 ft)

Lowest Point: Hachiro-gata 4 m (13 ft) below sea level

Land Use: 11% arable; 2% permanent crops; 2% meadows and pastures; 68% forest and woodland; 17% other

The Japanese archipelago forms a convex crescent off the eastern coast of Asia, bounded on the north by the Sea of Okhotsk, on the east and south by the Pacific Ocean, on the southwest by the East China Sea, and on the west by the Sea of Japan. No point in Japan is more than 150 km (93 mi) from the sea. The distance between Japan and the Asian continent, of which the nearest point is the Korean Peninsula, is about 200 km (124 mi).

The country consists of four principal islands—Hokkaido, Honshu, Shikoku, and Kyushu; over 3,000 small adjacent islands and islets including Oshima in the Nanpo chain and over 200 other smaller islands including those of the Amami, Okinawa and Sakishima chains of the Ryukyu Island archipelago. The national territory also includes the small Bonin (Ogasawara) Islands, Iwo Jima, and the Volcano Islands located in the Pacific Ocean some 1,100 km (683 mi) south of central Honshu. A territorial dispute with the Soviet Union concerns the two southernmost of the Kuril Islands, Etorofu and Kunashiri and the smaller Shikotan and Habomai Island Group, northeast of Hokkaido. The four major islands are separated only by narrow straits and form a natural entity.

The Japanese islands essentially are the summits of mountain ridges that have been uplifted near the outer edge of the Asian continental shelf. Consequently, the country is extremely mountainous and the plains and intermontane basins scattered throughout make up only 25% of the national territory. A long range of mountain ranges run down the middle of the archipelago dividing it into two halves—the "face" fronting the Pacific Ocean and the "back" facing the Sea of Japan. Although the mountains are precipitous, most of them are only a few hundred meters high and present a somewhat monotonous profile. Central Japan, however, is marked by the convergence of three mountain chains that form the Japanese Alps (Hida Mountains) that extend from north to south over a distance of 96 to 112 km (60 to 70 mi) with about 15 peaks that exceed 3,048 m (10,000 ft). The highest point in the country is Mount Fuji, a dormant volcano that rises to 3,776 m (12,890 ft). Snow lingers here late into spring and the mountainsides are streaked by fallen rocks, but there are no true glaciers.

Japanese plains, unlike those of Europe, are not basins of sedimentary deposits but depressed zones in which great masses of alluvium have accumulated. They have a fairly level profile, sloping rather steeply to the sea, where they end in a line of dunes. Most of the plains are located along the coast, including the largest, Kanto, where Tokyo is located, Nobi surrounding Nagoya, Kinki plain in the Osaka-Kyoto area, the Sendai plain in northeastern Honshu, and Ishikarai and Tokachi plains on Hokkaido. Wide river beds lined by high banks cut across these lowlands. Ascending terraces are a familiar feature of the interfluvial areas. On the approaches to the mountains, the slopes steepen and are laced by numerous watercourses, sometimes isolating a group of hills. Many plains are found in the interior, particularly in the mountainous region of central Honshu, known as Tosan, where enclosed basins (such as Kofu, Nagano, Lake Suwa and Matsumoto) are joined by valleys.

Japan has 1.6 km (1 mi) of coastline for every five sq km (1.9 sq mi) of land area, a ratio equal to that of Great Britain. The shores are of tectonic origin which explains why the pattern is different on the Sea of Japan and the Pacific coasts. Along the latter, the fracture cuts obliquely across the shore line, forming large indentations, such as the Bozo, Izu, and Kii peninsulas, and the bays of Sendai, Tokyo and Ise (Nagoya). The deep trough that contains the Inland Sea and the two openings at either end have also been bent perpendicularly which accounts for the violin shape of the island of Shikoku and the bulge at the center of the body of water. In contrast the shore is less indented on the Sea of Japan side. The entire southwest coast of Kyushu from Kagoshima Bay to the Straits of Shimonoseki is deeply fragmented and fractured. The northern half of the coastline is uplifted from the sea while the southern half has subsided and been invaded by the sea. Flat shores terminating in dunes are common along the Sea of Japan. On the Pacific side, flat shores are found at the head of the principal bays where the great metropolises are situated. North of Tokyo Bay is a type of landscape called *suigo* or "land of water." Here the plain is exactly at mean sea level, protected by levees and locks and by a system of pumps, as in Holland.

Topographically Japan is divided into eight major regions of which the islands of Hokkaido, Shikoku and Kyushu each form a region and the island of Honshu forms five. Hokkaido constitutes more than one-fifth of Japan. It was long looked upon as a remote frontier area because of its forests and rugged climate. Hokkaido is divided along a line extending from Cape Soya to Cape Erimo. The eastern half includes the Daisetsu Mountain range at the foot of which lie the plains of Tokachi and Konsen. This is the most inhospitable section of the whole country. The western half is milder and less hilly.

The Tohoku (literally, northeast) occupies the northeastern part of Honshu above approximately the 27th parallel. It is a dry bright region with flat, well drained alluvial plains. The region is still considered a rural back country where traditional dialects and customs persist.

The central zone, corresponding to the widest parts of the archipelago, consists of Kanto, Chubu, and Chugoku regions. Kanto is the most industrialized region in Japan and includes the Tokyo-Yokohama industrial complex. The Chubu region, lying west of the Kanto, is characterized by the greater height and ruggedness of its mountains. It comprises three distinct districts: Hokuriku, a coastal strip on the Sea of Japan, Tosan, the central Highlands and Tokai, a narrow corridor lying along the Pacific coast. Hokuriku is the "snow country" so dear to Japanese romanticism, where the winters are swept by violent squalls and the summers are hot and humid. It rains the year round. The Tosan, often called the Roof of Japan, includes the Japanese Alps. The population is chiefly concentrated in six elevated basins connected by narrow valleys. The western part of the Tokai district includes the Nobi plain. The Kinki region lies to the west and consists of a comparatively narrow area of Honshu, stretching from the Sea of Japan on the north to the Pacific Ocean on the south. It includes Japan's second largest commercial-industrial complex centered on Osaka and Kobe, and the two former imperial cities of Nara and Kyoto. The other important lowland area of Kinki, the Osaka Plain, forms the Harshin commercial industrial complex. The Chugoku region occupies the western end of Honshu and is divided into two distinct districts by mountains running through the central part. The northern, somewhat narrower, part is known as San'in or the shady side and the southern part as San'yo or sunny side.

The Shikoku region also is divided by mountains into a narrow northern subregion that fronts the Inland Sea and a wider southern part that faces the Pacific Ocean. Most of the population lives in the northern zone. The southern part is mostly mountainous and sparsely populated.

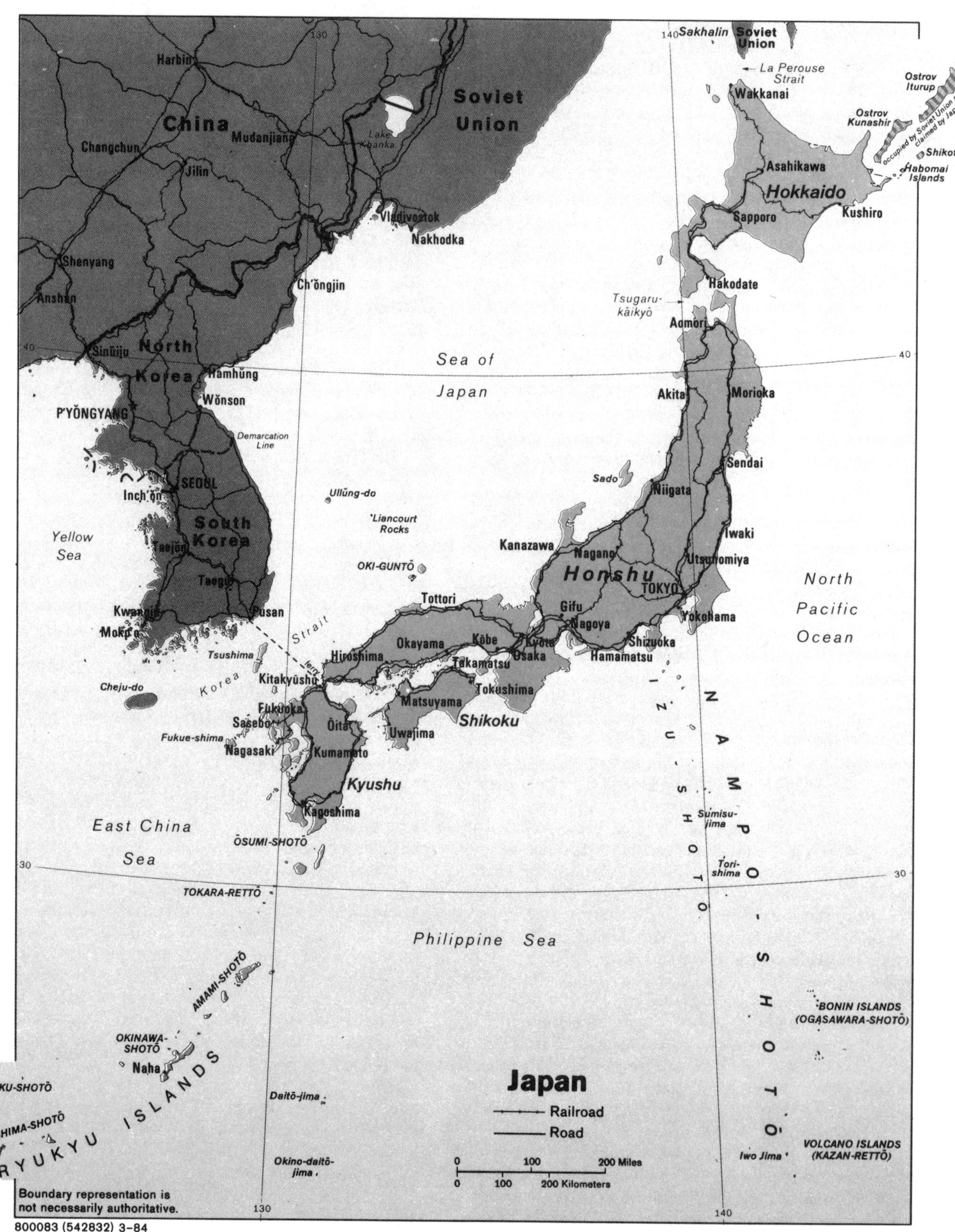

800083 (542832) 3-84

Kyushu, the southernmost of the main islands, is divided into northern and southern parts by the Kyushu Mountains which run diagonally across the middle of the island. The northern part is one of Japan's most industrialized regions and includes the Kitakyushu industrial region.

The Ryukyu Islands include well over 200 islands and islets of which fewer than half are populated. They extend in a chain generally southwestward from the Tokara Strait to within 193 km (120 mi) of Taiwan.

A tenth of the world's volcanoes are found in Japan. Of the 265 known volcanoes, 20 have been active since the beginning of the century. They are particularly numerous in Hokkaido, the Fossa Magna and Kyushu. The mountainous area of Kyushu resembles a lunar landscape with wide craters and cones of every form ranging from the ash cone of Mount Fuji to the volcanic dome of Daisetsu. Volcanic eruptions cause few deaths because they can be anticipated. Among the great eruptions in modern times were those of Mount Bandai in 1888 and Mount Aso in 1953 and 1958.

Japan also is subject to periodic earthquakes, with over 1,500 recorded annually. Minor tremors occur almost daily in one part of the country or another causing a slight shaking of buildings. Major earthquakes, although rarer, result in thousands of deaths, as the Kanto earthquake in 1923 in which 130,000 were killed. There are three particularly sensitive seismic zones: the Bozo Peninsula, less than 56 km (35 mi) from Tokyo, the Sea of Japan, some 105 km (65 mi) north of Osaka, and the Pacific coast in northern Tohoku and Hokkaido. Because of the danger they pose, Japan has become a world leader in research on the causes and prediction of earthquakes and the construction of earthquake-proof buildings. Extensive civil defense efforts focus on training in protection against earthquakes, in particular against accompanying fires which represent the greatest danger. Undersea earthquakes also expose the Japanese coastline to danger from tidal waves. An equally great hazard are movements of the earth that shake loose entire mountainsides. These landslides, generally composed of clay, may be from 6 to 23 m (20 to 75 ft) deep, several hundred feet wide, and up to 4 km (2.5 mi) long. The speed of the flow varies with the season and the water content of the soil. Landslides are especially numerous in Hokuriku on the Sea of Japan side where more than 10,000 have been counted. The largest sometimes carry with them entire paddy fields, tree groves and even houses. This makes it necessary for the communities affected to redistribute the land periodically as the property at the tip of the flow steadily shrinks in size.

At the head of most of the bays where the great cities are located, the land is sinking slowly causing buildings to subside up to 45 cm (1.5 in) annually. Roads and water-mains also are affected. Since 1935 the port area of Osaka has subsided almost 3 m (10 ft).

Japan is perhaps the only country in the world to suffer from both typhoons and snow. Typhoons cause an average of 1,500 deaths and destroy 20,000 dwellings a year. Southern Shikoku is a particularly vulnerable area. In regions bordering the Sea of Japan, the winter monsoon, laden with snow can be as destructive as typhoons. Snowfall is heavy along the western coast where it covers the ground for almost four months depositing a very thick blanket on the streets and rooftops. The cost of keeping railway tracks and roads open is a heavy burden on the municipalities.

Although the country is exceptionally well watered, the absence of large plains has precluded the formation of a good river system. The longest river, the Tone, is only 322 km (200 mi) long. Rivers tend to be steep and swift and hence unsuitable for navigation. Moreover, the mountainous terrain and the absence of glaciers make the flow highly irregular. The early summer ''plum rains'' account for a large part of the annual precipitation and turn slow streams into raging torrents. In winter, these riverbeds are transformed into wide stretches of gravel furrowed by thin trickles of water. The peak rainy season is from May to October with regional variations. Rivers are used mostly for hydroelectric production and for irrigation.

The landscape is speckled with lakes of every description and size. The biggest is Biwa, 673 sq km (260 sq mi) in area, which fills a large fault basin east of Kyoto. Celebrated in song and legend, Biwa is threatened today by pollution as well as the demands for fresh water from Osaka (connected to Biwa by the Yodo River, its outlet) and Kyoto connected by a tunnel.

PRINCIPAL CITIES* (population at 1 October 1986)

City	Population	City	Population
Tokyo (capital)†	8,354,615	Oita	390,096
Yokohama	2,992,926	Takatsuki	384,784
Osaka	2,636,249	Hirakata	382,253
Nagoya	2,116,381	Urawa	377,235
Sapporo	1,542,979	Omiya	373,022
Kyoto	1,479,218	Asahikawa	363,631
Kobe	1,410,834	Fukuyama	360,261
Fukuoka	1,160,440	Iwaki	350,569
Kawasaki	1,088,624	Suita	348,948
Kitakyushu	1,056,402	Nagano	336,973
Hiroshima	1,044,118	Fujisawa	328,387
Sakai	818,271	Nara	327,702
Chiba	788,930	Takamatsu	326,994
Sendai	700,254	Toyohashi	322,142
Okayama	572,479	Machida	321,188
Kumamoto	555,719	Hakodate	319,194
Kagoshima	530,502	Toyama	314,111
Higashiosaka	522,805	Kochi	312,241
Hamamatsu	514,118	Toyoda	308,111
Amagasaki	509,115	Naha	303,674
Funabashi	506,966	Koriyama	301,673
Sagamihara	482,778	Akita	296,400
Niigata	475,630	Aomori	294,045
Shizuoka	468,362	Kawagoe	285,437
Himeji	452,917	Okazaki	284,966
Nagasaki	449,382	Miyazaki	279,114
Kanazawa	430,481	Maebashi	277,319
Matsudo	427,473	Yao	276,394
Yokosuka	427,116	Tokorozawa	275,168
Matsuyama	426,658	Kashiwa	273,128
Hachioji	426,654	Fukushima	270,762
Nishinomiya	421,267	Shimonoseki	269,169
Kurashiki	413,632	Akashi	263,363
Toyonaka	413,213	Yokkaichi	263,001
Gifu	411,743	Neyagawa	258,228
Utsunomiya	405,375	Tokushima	257,884
Kawaguchi	403,015	Ichinomiya	257,388
Wakayama	401,352	Kasugai	256,990
Ichikawa	397,822	Koshigaya	253,479

*Except for Tokyo, the data for each city refer to an urban county (*shi*), an administrative division which may include some scattered or rural populations as well as an urban centre.

†The figure refers to the 23 wards (*ku*) of Tokyo. The population of Tokyo-to (Tokyo Prefecture) was 11,829,363.

Area and population		area		population
Regions Prefectures	Capitals	sq mi	sq km	1986 estimate
Chūbu				
Aichi	Nagoya	1,984	5,138	6,507,000
Fukui	Fukui	1,618	4,191	820,000
Gifu	Gifu	4,091	10,596	2,036,000
Ishikawa	Kanazawa	1,620	4,197	1,155,000
Nagano	Nagano	5,245	13,585	2,144,000
Niigata	Niigata	4,857	12,579	2,479,000
Shizuoka	Shizuoka	3,001	7,773	3,596,000
Toyama	Toyama	1,642	4,252	1,119,000
Yamanashi	Kōfu	1,723	4,463	838,000
Chūgoku				
Hiroshima	Hiroshim	3,269	8,466	2,831,000
Okayama	Okayama	2,737	7,090	1,923,000
Shimane	Matsue	2,559	6,628	794,000
Tottori	Tottori	1,349	3,493	617,000
Yamaguchi	Yamaguchi	2,358	6,106	1,599,000
Hokkaidō				
Hokkaidō (Territory)	Sapporo	32,247	83,519	5,678,000
Kantō				
Chiba	Chiba	1,988	5,150	5,216,000
Gumma	Maebashi	2,454	6,356	1,930,000
Ibaraki	Mito	2,353	6,094	2,746,000
Kanagawa	Yokohama	927	2,402	7,542,000
Saitama	Urawa	1,467	3,799	5,950,000
Tochigi	Utsunomiya	2,476	6,414	1,879,000
Kinki				
Hyōgo	Kōbe	3,235	8,378	5,302,000
Mie	Tsu	2,231	5,778	1,757,000
Nara	Nara	1,425	3,692	1,319,000
Shiga	Ōtsu	1,551	4,016	1,167,000
Wakayama	Wakayama	1,824	4,725	1,085,000
Kyūshū				
Fukuoka	Fukuoka	1,915	4,960	4,740,000
Kagoshima	Kogoshima	3,539	9,165	1,817,000
Kumamoto	Kumamoto	2,860	7,408	1,842,000
Miyazaki	Miyazaki	2,986	7,735	1,175,000
Nagasaki	Nagasaki	1,588	4,112	1,591,000
Ōita	Ōita	2,447	6,337	1,250,000
Saga	Saga	939	2,433	880,000
Ryukyu				
Okinawa	Naha	870	2,254	1,190,000
Shikoku				
Ehime	Matsuyama	2,190	5,672	1,529,000
Kagawa	Takamatsu	727	1,882	1,024,000
Kōchi	Kōchi	2,744	7,107	838,000
Tokushima	Tokushima	1,600	4,145	835,000
Tohoku				
Akita	Akita	4,483	11,612	1,249,000
Aomori	Aomori	3,713	9,617	1,520,000
Fukushima	Fukushima	5,322	13,784	2,085,000
Iwate	Morioka	5,899	15,279	1,431,000
Miyagi	Sendai	2,815	7,292	2,190,000
Yamagata	Yamagata	3,601	9,327	1,262,000
Metropolis				
Tōkyō	Tōkyō	835	2,162	11,893,000
Urban prefectures				
Kyōto	Kyōto	1,781	4,613	2,596,000
Ōsaka	Ōsaka	721	1,868	8,706,000
TOTAL		145,870	377,801	121,672,000

Floods are common, especially in the Pacific coastal areas where the subsidence of land makes it necessary to raise large embankments and dikes against rivers that flow at a level well above the surrounding plains. During periods of heavy rains angry waters bearing great masses of alluvium break through the embankments inundating the adjacent fields and covering them with a thick carpet of gravel and sand. Sometimes, typhoons bringing fresh torrents of water to the rivers convert whole plains into vast lakes and sweep away roads and railroads.

JORDAN

BASIC FACTS

Official Name: Hashemite Kingdom of Jordan

Abbreviation: JOR

Area: 89,206 sq km (34,443 sq mi)

Area—World Rank: 107

Population: 2,850,482 (1988) 4,705,000 (2000)

Population—World Rank: 111

Capital: Amman

Boundaries: 1,586 km (985 mi); Iraq 134 km (83 mi); Israel 238 km (148 mi); Saudi Arabia 742 km (461 mi); Syria 375 km (233 mi); West Bank 97 km (60 mi)

Coastline: 26 km (16 mi)

Longest Distances: 562 km (349 mi) NE-SW; 349 km (217 mi) SE-NW

Highest Point: Mount Ramm 1,754 m (5,755 ft)

Lowest Point: Dead Sea 400 m (1,312 ft) below sea level

Land Use: 4% arable; 0.5% permanent crops; 1% meadows and pastures; 0.5% forest and woodland; 94% other

Jordan is a landlocked country except at the southern extremity where nearly twenty-six kilometers (16 mi) of shoreline along the Gulf of Aqaba provide access to the Red Sea. A great north-south geological rift, forming the depression of Lake Tiberias (Sea of Galilee), the Jordan Valley, and the Dead Sea, is the dominant topographical feature.

Topography

The country consists mainly of a fairly high plateau—divided into ridges by valleys and gorges—and a few mountainous areas. Fractures of the earth's surface are evident in the great geological rift that extends from the Jordan Valley through the Red Sea southward, gradually disappearing south of the lake country of East Africa. Although formerly an earthquake-prone region, no severe shocks have been recorded for several centuries.

By far the greatest part of the East Bank is desert, displaying the land forms and other features associated with great aridity. Most of the land is in fact part of the great Syrian (or North Arabian) Desert. There are broad expanses of sand and dunes, particularly in the south and

PRINCIPAL TOWNS (including suburbs)
Population in December 1986; Amman (capital) 972,000; Zarqa 392,220; Irbid 271,000; Salt 134,100.

Area and population		area		population
Governorates	**Capitals**	sq mi	sq km	1986 estimate
ʿAmmān	Amman	...	...	1,160,000
al-Balqāʾ	aṣ-Ṣalt	...	...	193,800
Irbid	Irbid	...	...	680,200
al-Karak	al-Karak	...	...	120,100
Maʿān	Maʿān	...	...	97,500
al-Mafraq	al-Mafraq	...	...	98,600
aṭ-Ṭafilah	aṭ-Ṭafilah	...	...	41,400
az-Zarqāʾ	az-Zarqāʾ	...	...	404,500
TOTAL		34,443	89,206	2,796,100

southeast, together with salt flats. Occasional jumbles of sandstone hills or low mountains support only meager and stunted vegetation that comes to life for a short period after the scanty winter rains.

The drainage network is coarse and incised. In many areas the relief is such as to provide no eventual outlet to the sea, so that sedimentary deposits accumulate in basins where moisture evaporates or is absorbed in the ground. Toward the depression in the western part of the East Bank, the desert rises and gradually becomes the Jordanian Highlands—a steppe country of high, deeply cut limestone plateaus with an average elevation of about 900 m (2,952 ft); occasional summits reach 1,200 m (3,937 ft) in the north and over 1,500 m (4,921 ft) in the south.

The western edge of this plateau country forms an escarpment along the eastern side of the Jordan River-Dead Sea depression and its continuation south of the Dead Sea. Most of the wadis that provide drainage from the plateau country into the depression carry water only during the short season of winter rains. But they are sharply incised with deep, canyonlike walls, and wet or dry the wadis can be formidable obstacles to travel.

The Jordan River is short, but from its mountain headwaters, some 160 km (99 mi) in an air line north of the river's mouth at the Dead Sea, the stream bed drops from an elevation of about 3,000 m (9,842 ft) above sea level to nearly 435 m (1,427 ft) below sea level. Before reaching Jordanian territory the river forms Lake Tiberias, the surface of which is nearly 190 m (623 ft) below sea level. The river's principal tributary is the Yarmuk River, which near its junction forms the boundary between Israel on the northwest, Syria on the northeast, and Jordan on the south. The Az Zarqa River, the second main tributary of the Jordan, rises and empties entirely within the East Bank.

The Dead Sea occupies the deepest depression on the land surface of the earth. The depth of the depression is accentuated by the mountains and highlands paralleling it that rise to elevations of 800 to 1,200 m (2,625 to 3,937 ft) above sea level. The sea's greatest depth is about 430 m (1,411 ft) and it thus reaches a point nearly 900 m (2,953 ft) below sea level. Toward the southern end, a narrow peninsula juts from the east to divide the sea into a larger and much deeper northern basin and a shallow southern basin scarcely three m (10 ft) deep. The entire body of water is some eighty km (49 mi) long and has a maximum width of slightly over ten km (6 mi).

South of the Dead Sea a continuation of the depression, known popularly as the Wadi al Arabah but officially called the Wadi al Jayb, rises gradually through a barren desert until it reaches sea level about halfway to the Gulf of Aqaba. The wadi continues to rise to an elevation of about 300 m (984 ft) above sea level some sixty-five km (40 mi) from the gulf. At the summit there is a divide between drainage lines running north to the Dead Sea and south to the gulf.

KENYA

BASIC FACTS

Official Name: Republic of Kenya

Abbreviation: KEN

Area: 582,646 sq km (224,961 sq mi)

Area—World Rank: 44

Population: 23,341,638 (1988) 36,950,000 (2000)

Population—World Rank: 38

Capital: Nairobi

Boundaries: 3,969 km (2,467 mi); Sudan 306 km (190 mi); Ethiopia 779 km (484 mi); Somalia 682 km (424 mi); Tanzania 769 km (478 mi); Uganda 772 km (480 mi)

Coastline: Indian Ocean 523 km (325 mi); Lake Victoria 138 km (86 mi)

Longest Distances: 1,131 km (703 mi) SSE-NNW; 1,025 km (637 mi) ENE-WSW

Highest Point: Mount Kenya (Kirinyaga) 5,199 m (17,058 ft)

Lowest Point: Sea level

Land Use: 3% arable; 1% permanent crops; 7% meadows and pastures; 4% forest and woodland; 85% other

Kenya lies astride the equator on the east coast of Africa with its northernmost and southernmost points about equidistant—somewhat over 4° north and south—from that imaginary line. The total area includes almost 13,468 sq km (5,200 sq mi) of water, contained mainly in Lake Rudolf, known in Kenya as Lake Turkana 6,405 sq km (2,473 sq mi), and the country's portion of Lake Victoria, known in Kenya as Victoria Nyanza 3,784 sq km (1,461 sq mi).

The country has a great diversity of terrain, ranging from barrier reefs off the Indian Ocean coast to sandy desert, forested uplands, and perpetually snow-covered Mount Kenya. A particularly prominent feature is the section of the great Rift Valley of East Africa that runs

Jordan
International boundary
1949 Armistice line
1967 Cease-Fire line
Governorate boundary
National capital
Governorate capital
Railroad
Road
0 25 50 Kilometers
0 25 50 Miles
Boundary representation is not necessarily authoritative.
SYRIA
IRAQ
SAUDI ARABIA
SAUDI ARABIA
ISRAEL
ISRAEL
LEBANON
Mediterranean Sea
Damascus
UNDOF Zone
GOLAN HEIGHTS (Israeli occupied)
WEST BANK (Israeli occupied-status to be determined)
1967 Cease-Fire Line
1949 Armistice Line
Lake Tiberias
Dead Sea
Gulf of Aqaba
SINAI
EGYPT
Nahr al Yarmūk
Jordan River
IRBID
AL MAFRAQ
AL BALQĀ'
AZ ZARQĀ'
AMMAN
AL KARAK
AṬ ṬAFĪLAH
MA'ĀN
Amman
Irbid
Al Mafraq
Jarash
As Salt
Az Zarqā'
Mādabā
Al Mazra'ah
Al Qaṭrānah
Al Karak
Aṣ Ṣāfī
Aṭ Ṭafīlah
Ash Shawbak
Al Jafr
Ma'ān
Ra's an Naqb
Al 'Aqabah
Al Mudawwarah
Bā'ir
Azraq ash Shīshān
Maḥaṭṭat al Ḥafīf
Maḥaṭṭat al Jufūr
Ṭurayf
Kāf
An Nabk Abū Qaṣr
Al Bi'r
Ḥaql
Elat
Yotvata
Mizpe Ramon
Oron
Dimona
Beersheba
Hebron
Jerusalem
Ramla
Tel Aviv-Yafo
Netanya
Nāblus
Janīn
Nazareth
Haifa
Tiberias
Nahariyya
Tyre
Al Qunayṭirah
Dar'ā
As Suwaydā'
Buṣrá ash Shām
36
38
32
30

through Kenya. The most striking physiographic distinction, however, is between the large area of higher land encompassing roughly the southwestern one-third of the country and the remaining two-thirds consisting of low plateaus and plains. Conditions governing the climate have made this distinction even more pronounced, as the highlands include the only large area—other than the coastal belt—that can count on generally reliable rainfall; they also contain almost all of the good soils, and the great majority of the settled population lives there. In contrast, the outer arc of lower land is peopled almost entirely by nomadic pastoralists living in a terrain covered mostly by semiarid vegetation and receiving highly variable rainfall that in many places rarely exceeds ten inches a year.

Kenya is characterized by a highly diverse terrain and climatic conditions that range from moist to arid. Climatically three principal subdivisions, based on the reliability of rainfall, can be distinguished. They include a narrow coastal belt along the Indian Ocean that has a rainfall generally adequate for agriculture; a central highlands region and to the west the adjacent lower plateaus that border Lake Victoria, which also have adequate precipitation for crops; and the remaining over 70 percent of the country that usually receives less than twenty inches of rain annually and is usable chiefly as rangeland.

Geographically the country may be conveniently divided into seven major regions. These include a coastal belt, a region of plains behind the coastal strip, a low eastern plateau region, a region of arid plainlands covering approximately the northern one-fifth of the country, the fertile Kenya Highlands—bisected by the north-south Rift Valley Region—and an area of western plateaus that forms part of the Lake Victoria basin.

Coastal Region

The Coastal Region, extending some 402 km (250 mi) from the Tanzanian border in the south to the Somalia border in the north, exhibits somewhat different features in its southern and northern parts. The shoreline in the larger southern part—below the Tana River delta—is formed largely of coral rock and sand and is broken frequently by bays, inlets, and branched creeks. Mangrove swamps line these indentations, but along the ocean are many stretches of coral sand that form attractive beaches. A rise in the ocean level relative to the land in this area brought the drowning of deep cuts, made at an earlier period by creeks, in the now slightly elevated coral platforms along the coast. At certain places islands were formed, the most notable of which is Mombasa. The sheltered nature and deep water of the creeks around Mombasa led centuries ago to its use as a port.

Not far off the shoreline a barrier reef, broken only at a few points, parallels the coast. Immediately inland from the coast a narrow plain is succeeded by a low plateau area that reaches an elevation of about 152 m (500 ft) and after a few miles terminates in a line of discontinuous ridges. Rainfall throughout this subregion is adequate for agriculture and, together with the fertile soils of dried-up lagoons and certain parts of the low plateau, has resulted in the growth of a rather dense farming population. The principal staples are maize (corn), cassava, and bananas. Coconuts are extensively cultivated, as are citrus fruits and mangoes; cashew nuts and kapok are also important. Other crops include sisal and sugarcane, and there is some truck farming.

The principal physiographic feature of the smaller northern section of the region is the Lamu Archipelago, formed also by the drowning of the coastline as a result of a rise in the ocean level. This area was historically a major center of Arab trade and continues to have a considerable Arab population. Overall, however, the northern part is less heavily populated than the south, and farming is mainly of a subsistence nature. This is in part because of the smaller amount of rainfall. In considerable part, however, it is related to historical factors, including periodic early raids on the inhabitants by ethnic groups from farther north and dissension among Arab traders and local Arab rulers that brought a decline of trade. British antislavery activities in the nineteenth century also had a major impact by destroying the slave trade and the use of slaves on plantations, both then primary sustaining elements of the area's economy.

Coastal Hinterland and Tana Plains Region

The coastal hinterland, forming the southern part of this region, is a relatively featureless erosional plain broken only in a few places by small, somewhat higher hill groups. The plain rises very gradually westward from an elevation of about 152 m (500 ft) at the so-called coastal ranges on its eastern edge to about 304 m (1,000 ft) where it meets the Eastern Plateau Region. Rainfall is usually below twenty inches a year, and the area is sparsely populated, mostly by nomadic pastoralists. Part of this hinterland falls within the *nyika* (Swahili for wilderness), an area of bushland and thicket inhabited largely by wild animals.

The Tana Plains section of the region is mainly a depositional plain—equally featureless and deficient in rainfall—extending northward from the upper part of the Coastal Region to the northern plainlands. The plain's eastern edge is the border of Somalia, into which country it actually extends, and its western limits are marked by the higher elevation of the Eastern Plateau Region. The vegetation is mainly bush and scattered trees. The population of under ten people per square mile consists of nomadic pastoralists, most of them Cushitic-speaking Galla and Somali.

The perennial Tana River flows across the plain on its course from the Kenya Highlands to the Indian Ocean. Eventually considerable areas of land along the river in this ordinarily low rainfall region are expected to come under irrigation; some land is already irrigated. A major feature of the plain is the great Lorian Swamp into which the perennial flow of another river, the Ewaso Ngiro, empties.

Eastern Plateau Region

The Eastern Plateau Region consists of a belt of plains extending north and south to the east of the Kenya Highlands. Elevations run mainly between 304 and 916 m (1,000 and 3,000 ft); notable exceptions are the Chyulu Range and the Taita Hills, which rise to over 2,134m (7,000 ft). The region has a generally monotonous appearance except for the numerous scattered inselbergs (isolated hills and pinnacles, some craggy, others domed and smooth) that were left in many places during the erosional development of the plains. The southern part of the region includes the Amboseli, Serengeti, and Aruba plains. This portion is well known as the site of the Amboseli and Tsavo national parks, which hold some of the largest concentrations of big game in East Africa.

Rainfall averages between ten and twenty inches a year, the more northerly sections of the region having the lesser amounts. Both the Chyulu Range and the Taita Hills have greater precipitation, although rainfall amounts received in the Chyulu Range are highly unreliable. At certain times of the year the seasonal Namanga River pours a large amount of water onto the extremely flat Amboseli Plain to form Lake Amboseli, which at its fullest has an area of about 114 sq km (44 sq mi). During the dry season the lake completely disappears, and the area becomes a dusty plain. Much of the vegetation in the region is bushed grassland and thicket that gradates in the north into semidesert bush and grass. The higher elevations of the Taita Hills, however, have woodland growth, and woodland and some forest remnants are found in the Chyulu Range.

Northern Plainlands Region

The vast Northern Plainlands Region stretches from the Uganda border on the west to Somalia in the east. It consists of a series of arid plains of differing origins—chiefly erosional or formed by great outpourings of lava—and includes within its limits Lake Rudolf and the Chalbi Desert. Lake Rudolf, approximately 250 km (155 mi) long and having a maximum width of some 56 km (35 mi), has no present outlet. An earlier connection with the Nile River system has been postulated, however, based in part on the contemporary presence of the giant Nile perch in its waters.

The area west of the lake is quite arid, and rainfall is under ten inches; in some years it is almost negligible. Rivers and streams flowing through the area—including the large Turkwel and Kerio rivers, which originate in the Kenya Highlands and in the rainy season empty into Lake Rudolf—dry up at certain times of the year. Water holes remain, however, and at various other points water lies only a short distance below riverbeds. The area is inhabited by the nomadic Turkana, who rear camels, goats, and sheep, grazing them on the scattered bush and grass that form the main vegetation.

Significant features east of Lake Rudolf include the Chalbi Desert, which has Kenya's only environment classifiable as true desert. This extensive area was once covered by a lake that resulted from the damming of natural drainage by volcanic activity in the Mount Marsabit area. The plains themselves around Mount Marsabit consist of a vast lava plateau. Those farther eastward have developed on the continental basement rock; erosion here, as in the Eastern Plateau Region, has dotted the landscape with inselbergs of varying shapes and sizes. The plains closest to the Somalia border are underlain by sedimentary deposits.

The entire area east of the Chalbi Desert is equally arid and ordinarily supports vegetation only of the semidesert type. Certain spots are more favored climatically, however, including Mount Marsabit, which at higher elevations may receive thirty or more inches of rain annually and has an upper forest cover. Foothills of the southern Ethiopian highlands extending into Kenya also have more rainfall, and several perennial rivers flow south-southeastward from these hills onto the plains. The sparse nomadic population of Cushitic language speakers, who raise cattle on the plains east of Lake Rudolf, move their herds to these areas of relatively good water supply during the dry season.

Kenya Highlands Region

The Kenya Highlands Region, a large section of which was known as the White Highlands during the colonial period because of the concentration of the European population there, comprises the complex of high land found in the country's west-central part. The region actually consists of two major divisions lying respectively east and west of the great north-south Rift Valley. Each of these major sections has a number of subdivisons differing in origin, but the whole area is tied together by the common denominator of markedly higher altitude, cooler temperatures, and generally greater precipitation than found in other regions.

Tectonic activity played a major part in the formation of the highlands. This included the early upwarping that resulted in what is known in geologic terminology as the Kenya dome and the faulting and displacement both major and minor across this dome—much apparently in relatively recent geological time—that produced the great Rift Valley and many of the region's numerous escarpments. Great outpourings of lava, much also of comparatively recent origin, have added thousands of feet to the elevation over broad areas.

A striking feature on the eastern edge of the highlands is Mount Kenya, the country's highest point, which rises to 5,199 m (17,058 ft). An extinct volcano, the mountain was apparently much higher at one time since all traces of its former crater have eroded away. Among the more important subdivisons of the eastern highlands is the area east of the Aberdare Range, which is populated by the Kikuyu, the country's largest ethnic group. Much of the original forest in this area has been cut down, and the land is cropped intensively both for subsistence and for cash crops. The principal staples include bananas, maize, potatoes, and pulses. Cash crops include coffee, pyrethrum (the oil is used in insecticides), and tea. Livestock, including dairy cattle, is also part of the economy.

To the west of the Rift Valley many upper elevations in the southern part of the highlands remain covered by forest. Small farms dot the area at somewhat lower levels, and tea is grown both on plantations and by individual farmers. Forest also still covers large areas of the northern part of the western highlands. The local population grows wheat and maize, and cattle are grazed on the area's rich grassland. On the northwestern edge of the highlands lies

Kenya

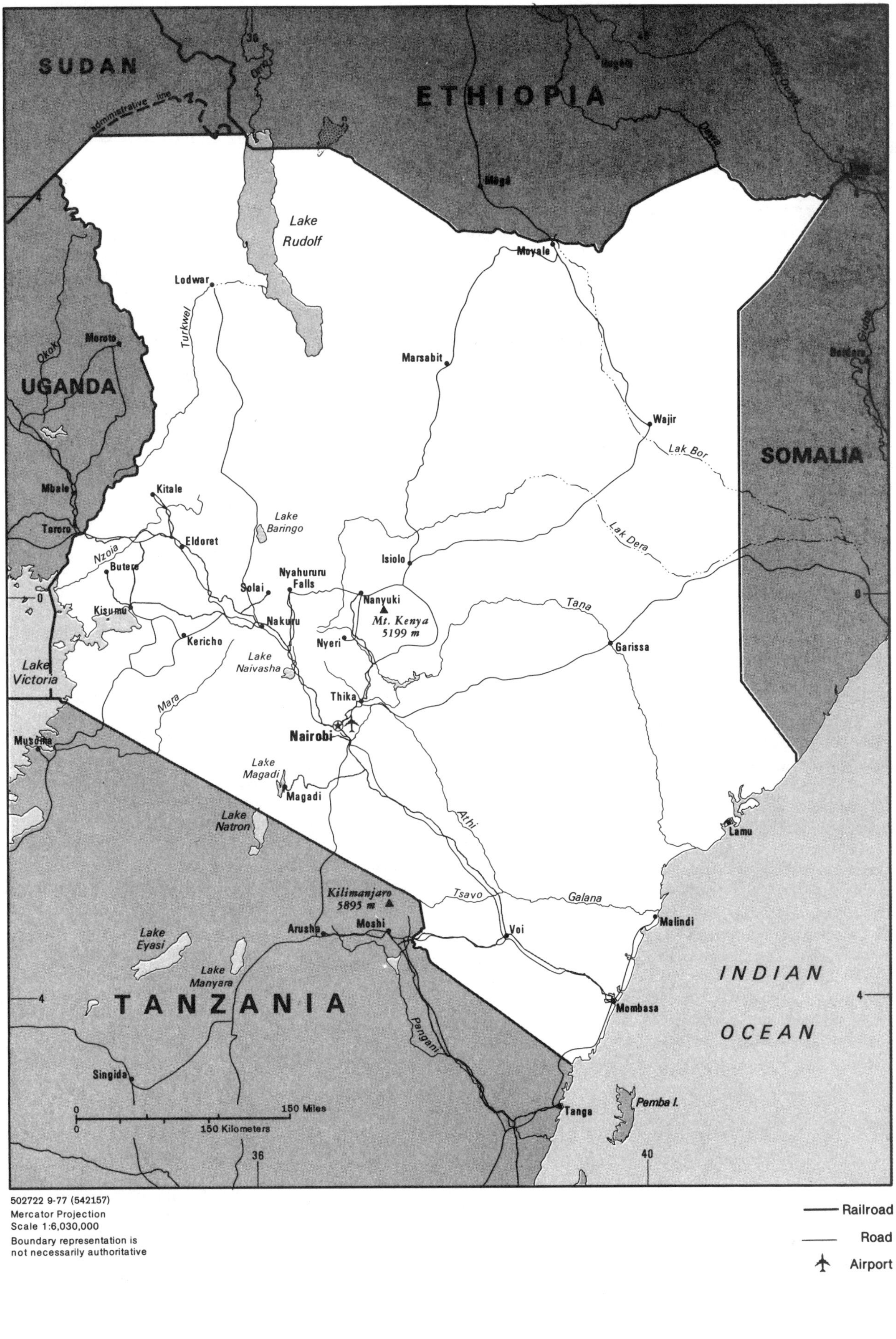

SUDAN
administrative line
ETHIOPIA
SOMALIA
UGANDA
TANZANIA
INDIAN OCEAN
Lake Rudolf
Lodwar
Turkwel
Moroto
Okok
Mbale
Tororo
Kitale
Eldoret
Nzoia
Butere
Kisumu
Kericho
Lake Victoria
Lake Baringo
Solai
Nyahururu Falls
Nakuru
Lake Naivasha
Mara
Nanyuki
Mt. Kenya 5199 m
Nyeri
Thika
Nairobi
Lake Magadi
Magadi
Lake Natron
Musoma
Kilimanjaro 5895 m
Arusha
Moshi
Lake Eyasi
Lake Manyara
Pangani
Singida
Tanga
Pemba I.
Moyale
Marsabit
Wajir
Lak Bor
Lak Dera
Isiolo
Tana
Garissa
Athi
Tsavo
Galana
Voi
Malindi
Mombasa
Lamu
Dawa
Mēga
36
40
0
4
150 Miles
150 Kilometers

502722 9-77 (542157)
Mercator Projection
Scale 1:6,030,000
Boundary representation is not necessarily authoritative

Railroad
Road

Airport

Mount Elgon, an extinct volcano rising to over 4,267 m (14,000 ft) above sea level.

Rift Valley Region

The great Rift Valley of eastern Africa, formed by a long series of faulting and differential rock movements, extends in Kenya from the Lake Rudolf area in the north generally southward through the Kenya Highlands and into Tanzania. In the vicinity of Lake Rudolf the valley floor is under 457 m (1,500 ft) above sea level, but southward it rises steadily until in its central section in the area of Lake Naivasha the elevation is close to 1,889 m (6,200 ft). From that point southward it drops off to about 610 m (2,000 ft) at the Kenya-Tanzania border.

The central section of the valley, about 64 km (40 mi) in width, is rimmed by high escarpments. On the east is the Kinangop Plateau, some 610 m (2,000 ft) above the valley floor, and east of that plateau lies the Aberdare Range, which has elevations above 3,962 m (13,000 ft). On the valley's western side are the Mau Escarpment, rising to nearly 3,048 m (10,000 ft) and farther north the Elgeyo Escarpment and the Cherangani Hills, the latter having elevations over 3,352 m (11,000 ft). The valley floor has been subjected to extensive volcanic activity, and several cones rise high above it; the area remains one of latent volcanism, hot springs and steam emerging at numerous spots. Volcanoes and lava heaps divide the central section into compartments in which lie a series of lakes that are remnants of an earlier larger lake or lakes. These present-day lakes have no outlets. Their content ranges from the alkaline but relatively fresh water of lakes Baringo and Naivasha, both of which support fish populations to the higher soda content of lakes Elmenteita, Bogoria (formerly Lake Hannington), and Nakuru and the almost solid and commercially exploited soda ash of Lake Magadi.

The northern and southern parts of the valley receive a yearly rainfall averaging between ten and twenty inches. They have semidesert vegetation consisting of grass, bush, and scattered trees and are inhabited by nomadic pastoralists. The most important of these nomads are the Masai cattle raisers in the southern valley area. The more elevated central section around Nakuru has higher precipitation, and the vegetation includes wooded grassland. The heavier rainfall in this part, especially close to the escarpments, allows the cultivation of grain crops. Beef and wool production and dairy farming also are carried on. The railway from Mombasa and Nairobi to western Kenya and Uganda crosses the valley at this point.

Western Plateaus Region

The Western Plateaus Region forms part of the extensive downwarped basin in which Lake Victoria lies. In Kenya the region consists mainly of faulted plateaus marked by escarpments that descend in a gentle slope from the Kenya Highlands Region to the shore of the lake. The region is divided by the secondary Kano Rift Valley into northern and southern components having somewhat different features. This faulted valley lies at a right angle to the main rift running through the highlands and is separated from that valley by a great lava mass. The lake intrudes into the Kano rift for about 80 km (50 mi) to form Winam Bay (formerly Kavirondo Gulf), at the eastern end of which is Kisumu, the country's fourth largest town and a major lake port. East of this arm of the lake is the low-lying Kano Plain, a heavily populated agricultural area that suffers periodically from drought and floods.

Various factors, including good soils and generally adequate rainfall for crops, have led to a very high concentration of population in this region. The area is farmed extensively, the principal staples being cassava, maize, various millets, and pulses. Coffee and cotton are also grown. The region is generally free of the tsetse fly and has the country's largest concentration of cattle relative to land area.

PRINCIPAL TOWNS (population at census of August 1969)

Town	Population	Town	Population
Nairobi (capital)	509,286	Nanyuki	11,624
Mombasa	247,073	Kitale	11,573
Nakuru	47,151	Malindi	10,757
Kisumu	32,431	Kericho	10,144
Thika	18,387	Nyeri	10,004
Eldoret	18,196		

1979 census: Nairobi 827,775; Mombasa 341,148; Kisumu 152,643.
Mid-1984 estimates: Nairobi 1,103,554; Mombasa 425,634.

DRAINAGE

The principal drainage pattern centers in the Kenya Highlands Region, from which streams and rivers radiate in a generally eastward direction toward the Indian Ocean, westward to Lake Victoria, and northward either to Lake Rudolf or to the arid terrain of northern Kenya, where they disappear. A secondary drainage system is formed by rivers in the southern highlands of Ethiopia, which extend into Kenya along the eastern section of their mutual boundary. These rivers are all seasonal, and those that receive sufficient water at flood times to reach the sea do so through Somalia. Minor internal systems are associated with the lakes in the Rift Valley.

The two largest perennial rivers, and the only navigable ones, are the Tana and the Galana (known locally in its course from west to east as the Athi, Galana, and Sabaki), both of which empty into the Indian Ocean. The Tana River, approximately 724 km (450 mi) long, rises in the southeastern part of the Kenya Highlands. From there it flows in a great arc northeastward along the highlands,then eastward across arid low plateaus and plains, and finally south

Area and population

Provinces	Provincial headquarters	area sq mi	area sq km	population 1984 estimate
Central	Nyeri	5,087	13,176	2,926,200
Coast	Mombasa	32,279	83,603	1,688,000
Eastern	Embu	61,734	159,891	3,423,500
Nairobi	Nairobi	264	684	1,103,600
North Eastern	Garissa	48,997	126,902	484,700
Nyanza	Kisumu	6,240	16,162	3,508,500
Rift Valley	Nakuru	67,131	173,868	4,132,400
Western	Kakamega	3,228	8,360	2,269,400
TOTAL LAND AREA		220,625	571,416	19,536,300
INLAND WATER		4,336	11,230	
TOTAL AREA		224,961	582,646	

to enter the sea at Kipini. About 322 km (200 mi) of the lower length are navigable by shallow-draft launches.

The Tana basin covers about 62,160 sq km (24,000 sq mi) and includes much of the flow from the Aberdare Range and Mount Kenya. In the Tana River's lower reaches the gradient is extremely gentle, banks are low, and flooding occurs during high water. As the river nears the coast, it develops many backwaters and at times may change course.

The Galana River rises in the southeastern part of the Kenya Highlands and with its tributaries (including the Tsavo, which rises on the eastern slope of Kilimanjaro in Tanzania) flows into the Indian Ocean north of Malindi. It is navigable by canoe for approximately 161 km (100 miles) inland, where further travel is effectively impeded by the rapids of Lugards Falls.

Several smaller rivers originate in the foothills of the eastern Kenya Highlands area within the Tana River basin, including the Lagh Thua and Mkondo Wa Kokani. Their flow usually disappears in the semiarid region east of the highlands but at times of heavy flooding manages to cross the area and empty into the Tana River. South of the Galana River the Goshi River (in its upper course called the Voi) has a length of about 209 km (130 mi), but only about 81 km (fifty mi) of the lower course is perennial.

The western slopes of the Kenya Highlands are drained by a number of generally parallel rivers that empty into Lake Victoria. The largest rivers include the Nzoia, about 257 km (160 mi) long, which drains the Cherangani Hills and the eastern slope of Mount Elgon, and the Yala, which has a length of about 177 km (110 mi) and eventually reaches the lake through Lake Kanyaboli and the Yala Swamp. Yala Falls—and Selby Falls on a tributary of the Nzoia—have considerable hydroelectric power potential. The Mara River, having its source in the Mau Escarpment in the southwest part of the highlands, flows southward for about 161 km (100 mi); it then enters Tanzania and turns westward to flow for almost another 161 km (100 mi) to Lake Victoria.

The northern part of the Kenya Highlands east of the Rift Valley is drained by small rivers that disappear in the arid land to the north and by the larger, eastward-flowing system of the Ewaso Ngiro, which has a drainage basin of about 56,980 sq km (22,000 sq mi). The Ewaso Ngiro usually terminates in the great Lorian Swamp in the Tana Plain, about 290 km (180 mi) from its source in the highlands; but at times a heavy runoff floods the swamp, and the waters flow eastward as the Lagh Dera into Somalia.

KIRIBATI

BASIC FACTS

Official Name: Republic of Kiribati

Abbreviation: KIR

Area: 849 sq km (328 sq mi)

Area—World Rank: 164

Population: 67,638 (1988) 77,000 (2000)

Population—World Rank: 180

Capital: Tarawa

Boundaries: Nil

Coastline: 1,143 km (710 mi)

Longest Distances: N/A

Highest Point: 81 m (265 ft)

Lowest Point: Sea level

Land Use: 51% permanent crops; 3% forest and woodland; 46% other

Area and population		area		population
Island Groups / **Islands**	**Capitals**	sq mi	sq km	1985 census
Gilberts Group	Bairiki Islet	110	285	61,226
Abaiang	Tuarabu	7	17	4,386
Abemama	Kariatebike	11	27	2,966
Aranuka	Takaeang	4	12	984
Arorae	Roreti	4	9	1,470
Banaba	Anteeren	2	6	46
Beru	Taubukinberu	7	18	2,702
Butaritari	Butaritari	5	13	3,622
Kuria	Tabontebike	6	15	1,052
Maiana	Tebangetua	6	17	2,141
Makin	Makin	3	8	1,777
Marakei	Rawannawi	5	14	2,693
Nikunau	Rungata	7	19	2,061
Nonouti	Teuabu	8	20	2,930
Onotoa	Buariki	6	16	1,927
Tabiteuea North	Utiroa	10	26	3,171
Tabiteuea South	Buariki	5	12	1,322
Tamana	Bakaka	2	5	1,378
Tarawa North	Abaokoro	6	15	3,205
Tarawa South	Bairiki	6	16	21,393
Line Group	Kiritimati	207	535	2,633
Northern		167	432	2,633
Kiritimati (Christmas)	London	150	388	1,737
Tabuaeran (Fanning)	Paelau	13	34	445
Teraina (Washington)	Washington	4	10	451
Southern (Caroline, Flint, Malden, Starbuck, Vostok)		40	103	
Phoenix Group (Birnie, Enderbury, Kanton, [Canton], McKean, Manra [Sydney], Nikumaroro [Gardner], Orona [Hull], Rawaki [Phoenix])	Kanton	11	29	24
TOTAL		328	849	63,883

Kiribati comprises three island groups of 33 low atolls and Ocean Island or Banaba, a raised atoll in the west. The three island groups are dispersed over an area of

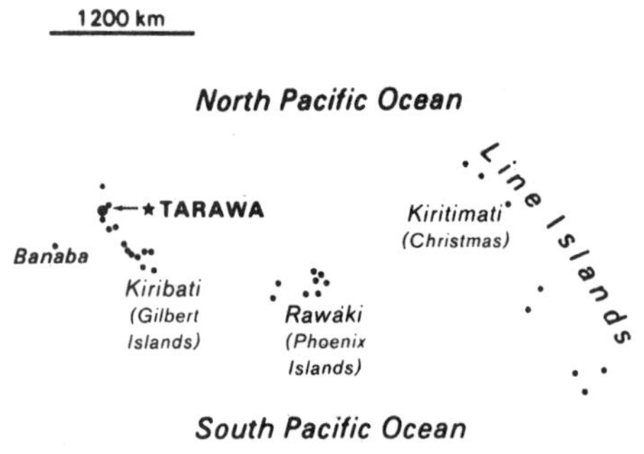

1200 km

North Pacific Ocean

★TARAWA
Banaba
Kiribati (Gilbert Islands)
Rawaki (Phoenix Islands)
Kiritimati (Christmas)
Line Islands

South Pacific Ocean

3 million sq km (1.1 million sq mi) in mid Pacific: the Gilbert Islands on the equator; the Phoenix Islands to the east and the Line Islands to the north of the equator.

The Gilbert group consists of Abaiang, Abemama, Aranuka, Arorae, Beru, Butaritari, Kuria, Maiana, Makin, Marakei, Nicunau, Nonouti, Onotoa, Tabiteuca, Tamana, and Tarawa. The Phoenix group comprises Birnie, Canton, Enderbury, Gardner, Hull, McKean, Phoenix, and Sydney. The Line Group encompasses Christmas, Fanning, Malden, Starbuck, Vostock, Washington, Caroline and Flint, the last two leased to commercial interests on Tahiti. Only some of the islands are inhabited.

NORTH KOREA

BASIC FACTS

Official Name: Democratic People's Republic of Korea

Abbreviation: KRN

Area: 122,370 sq km (47,250 sq mi)

Area—World Rank: 94

Population: 21,983,795 (1988) 28,166,000 (2000)

Population—World Rank: 39

Capital: Pyongyang

Boundaries: 2,702 km (1,679 mi); China 416 km (880 mi); USSR 18 km (11 mi); South Korea 240 km (149 mi)

Coastline: 1,028 km (639 mi)

Longest Distances: 719 km (447 mi) NNE-SSW; 371 km (231 mi) ESE-WNW

Highest Point: Mount Paektu 2,744 m (9,003 ft)

Lowest Point: Sea level

Land Use: 18% arable; 1% permanent crops; 74% forest and woodland; 7% other

The Korean peninsula projects to within 193 km (120 mi) of the principal Japanese islands of Honshu and Kyushu and separates the Yellow Sea on the west from the Sea of Japan on the east. The northern frontier shares a 1,030 km (640 mi) border with the People's Republic of China (PRC). The border with the Soviet Union in the extreme northeastern corner, about 121 km (seventy-five mi) southwest of Vladivostok, is only 17 km (10.4 mi) long. Most of the frontier is marked by two rivers, the Yalu and the Tumen (Amnok-kang and Tuman-gang, respectively, in Korean). The Yalu River originates on the slopes of an extinct volcano called Mount Paektu, Korea's highest peak at 2,743 m (9,000 ft) and flows southwest to the Yellow Sea. The Tumen River originates in the same area and flows first northeast and then southeast into the Sea of Japan.

The peninsula and its associated continental landmass below these rivers occupy the same latitude as the United States from New England to South Carolina. The area is broadest at its northern border and narrowest about 217 km, (135 mi) at its center, where the Military Demarcation Line, most of which winds slightly north of the thirty-eighth parallel, divides North Korea (Democratic People's Republic of Korea) from South Korea (Republic of Korea). As a result of the armistice agreement concluding the Korean War, about 55 percent of the area or some 121,729 sq km (47,000 sq mi) falls above the demarcation line and is occupied by North Korea. The demarcation line divides the 4,000 m wide (or about 2.5 mi wide) Demilitarized Zone (DMZ), which is uninhabited save for two small villages near P'anmunjŏm and some cultivated land near bordering areas.

North Korea is largely mountainous, although the mountains are generally not high. Sixty-five percent of the total area is composed of mountains and hills that are less than 1,000 m (3,280 ft) in height; only 15 percent rise above 3,280 ft; and the remaining 20 percent is made up of plains and lowlands. The major mountain ranges are located in the north-central and northeastern sections and along the east coast. On the east coast the hills drop sharply to a narrow coastal plain, whereas toward the west the slope is more gradual. The peninsula lies in a stable geologic zone; there are no active volcanoes, and only a few minor earth tremors occur yearly.

Terrain

Mountains and uplands cover 80 percent of the territory; the proportion is as high as 90 percent in Chagang Province and Yanggang Province. The major mountain ranges form a crisscross pattern extending from northwest to southeast and northeast to southwest. The Mach'ŏllyŏng Range extends from the vicinity of Mount Paektu in

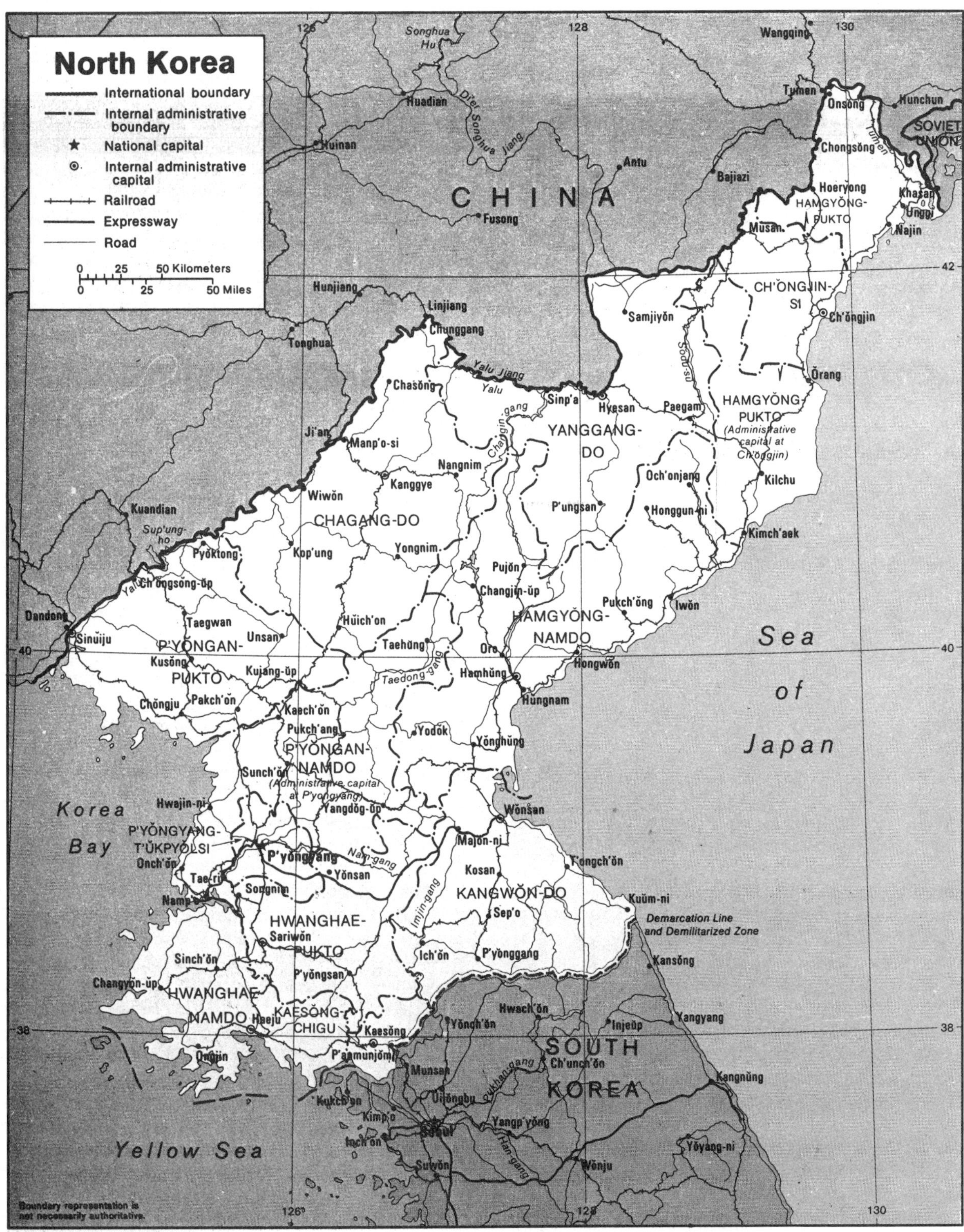

Base 504883 (546979) 3-82

a southeasterly direction toward the east coast. This range has peaks over 1,981 m (6,500 ft) in altitude. Running northeasterly from the center of the Mach'ŏl-lyŏng Range toward the Tumen River valley is the Hamgyŏng Range, which also has a number of peaks over 6,500 ft, including Mount Kwanmo 2,540 m, (8,334 ft). The southwest extension of the Hamgyŏng Range is known as the Pujŏl-lyŏng Range. To the west of the Hamgyŏng and Pujŏl-lyŏng ranges lies the relatively low averaging 999 m, (3,280 ft) Kaema Plateau, a heavily forested basaltic tableland. West of the Kaema Plateau is the Nangnim Range, averaging 1,499 m (4,920 ft) and extending southward. To the west of the Nangnim Range are two less prominent ranges, the Chŏgyu-ryŏng and the Myŏhyang, both of which are from 500 to 1,000 m (1,640 to 3,280 ft) in height. The Ch'ŏngch'ŏn River flows in the valley between them.

Korea's other major mountain chain, the T'aebaek Range, rises south of Wŏnsan and extends down the eastern side of the peninsula; it is often called the "backbone of Korea." Only a short portion of its length is in North Korea, but this section includes the scenic Kŭmgang Mountains (Diamond Mountains), known throughout Korea for their unusually shaped hills and striking vistas.

The terrain east of the Hamgyŏng, Pujŏl-lyŏng, and T'aebaek ranges consists of short, parallel ridges that extend from these mountains to the Sea of Japan, creating in effect a series of isolated valleys accessible only by rail lines branching off from the main coastal track. West of the T'aebaek Range, the terrain of central Korea is characterized by a series of lesser ranges and hills that gradually level off into plains along the west coast.

The plains regions are important to the nation's economy, although they constitute only one-fifth of the total area. Most of the plains are alluvial, built up from silt deposited on their banks by rivers in their middle and lower courses. Other plains, such as the P'yŏngyang peneplain, were formed by aeons of erosion from surrounding hills. A number of plains areas exist on the west coast, including the P'yŏngyang peneplain and the Unjon, Anju, Chaeryŏng, and Yŏnbaek plains. Of these, the Chaeryŏng and the P'yŏngyang are the most extensive, each covering an area of about 618 sq km (200 sq mi). They are followed in size by the Yŏnbaek Plain, which is about 315 sq km (120 sq mi); the rest are about 207 sq km (80 sq mi) each. The mountains along the east coast drop abruptly to the sea and, as a result, few plains are found. The most important are the Hamhŭng, Yŏngch'on, Kilchu, Yŏnghŭng, and Susŏng plains, of which the 311 sq km (120 sq mi) Hamhŭng Plain is the largest. The plains support most of the country's farmlands, and their small size indicates the severe physical limitations placed on agriculture.

Rivers

The mountainous areas are located mainly in the northern and eastern parts of the country, whereas the western region consists of lower hills and plains; thus, the major rivers flow westward into the Yellow Sea. These include the Yalu, Ch'ŏngch'ŏn, Taedong, Yesŏng, and Imjin rivers. On the east coast there are many short, swift-flowing rivers but only two of major proportions, the Tumen and the Sŏngch'ŏn. The rivers and streams flow strongly during summer, fed by seasonal rainfall and melting snow in the mountains, but the volume drops considerably during the dry winters. The rivers serve a threefold economic function: they provide a source of water for irrigation during the dry months from October through April; they are utilized as an auxiliary means of transportation to ease the strain on the railroads; and they are used to generate electricity. In order to bring power to certain regions on the east coast, the course of several westward-flowing streams has been reversed by means of mountain tunnels.

The major river is the Yalu, which flows from Mount Paektu to the Yellow Sea, a distance of almost 805 km (500 mi). Because its course cuts through rocky gorges for much of its length, its alluvial plains are less extensive than its size would suggest. Oceangoing vessels can dock at Sinŭiju, and small craft can travel upstream as far as Hyesan. Although it is important for transportation and irrigation, the Yalu's main value lies in its hydroelectric power potential. Dams have been built on the Yalu and four of its tributaries, the Changjin, Hŏch'ŏn, Pujŏn, and Tongno rivers; in some instances, the electric power generated is shared with the PRC.

The Taedong River is the most important waterway in the west-central region. It waters the farmlands of the extensive P'yŏngyang peneplain and serves as a major transportation artery for internal commerce. Three principal cities, P'yŏngyang, Namp'o, and Songnim, have developed along its banks. The Taedong is susceptible to flooding. The Tumen is of negligible value as an inland waterway since it is too narrow. The value of the Imjin River to the country is limited, as it flows through the DMZ and part of South Korea before reaching the sea south of Kaesŏng.

Coastlines and Ports

Of the Korean peninsula's 8,690 km (5,400 mi) of coastline, excluding over 3,000 offshore islands, North Korea accounts for about one-fifth of the total, or roughly 1,046 km (650 mi) along the west coast and 805 km (500 mi) along the east coast. The west coast is highly indented and irregular, and it is studded with a multitude of small offshore islands. The Korea Bay is shallow and has an unusually great tidal range—from 6 to 12 m (twenty to forty ft). A considerable portion of the tidelands have potential value as agricultural land, reed fields, and salt-evaporation facilities. Some reclamation has been undertaken in North and South P'yŏngan provinces, and in 1974 the government announced ambitious plans to reclaim up to 101,272 ha (250,000 ac) of tidelands for agricultural use. The main port on the west coast is Namp'o, which is located at the mouth of the Taedong River and is a center for both international and domestic trade.

The east coast is relatively straight with few islands; and the coastal waters of the Sea of Japan are very deep, averaging about 1,676 m (5,500 ft). The coast is washed by both warm and cold currents, contributing to a wide variety of marine life; it is also a region of frequent fog.

The principal eastern ports are Unggi, Wŏnsan, Ch'ŏngjin, Kimch'aek, and Najin.

PRINCIPAL CITIES (estimated population, 1976)

Pyongyang (capital)	1,500,000	Hungnam	260,000
Chongjin	300,000	Kaesong	240,000

Area and population

		area		population
Provinces	**Capitals**	sq mi	sq km	1968 estimate
Changang-do	Kanggye	6,300	16,200	780,000
Hamgyŏng-namdo	Hamhŭng	7,400	19,200	1,315,000
Hamgyŏng-pukto	Ch'ŏngjin	6,100	15,900	1,110,000
Hwanghae-namdo	Haeju	2,900	7,600	1,340,000
Hwanghae-pukto	Sariwŏn	3,300	8,600	1,060,000
Kangwŏn-do	Wŏnsan	4,100	10,700	1,030,000
P'yŏngan-namdo	P'yŏngsan	4,700	12,300	2,250,000
P'yŏngan-pukto	Sinŭiju	4,600	12,000	1,760,000
Yanggang-do	Hyesan	5,400	14,100	435,000
Special cities				
Ch'ŏngjin-si	—	700	1,900	385,000
Hamhŭng-si	—	300	800	530,000
P'yŏngyang-si	P'yŏngyang	700	1,800	1,275,000
Special district				
Kaesŏng-chigu	Kaesŏng	500	1,200	289,000
TOTAL		47,300	122,400	13,559,000

SOUTH KOREA

BASIC FACTS

Offical Name: Republic of Korea

Abbreviation: ROK

Area: 99,143 sq km (38,279 sq mi)

Area—World Rank: 103

Population: 42,772,956 (1988) 49,914,000 (2000)

Population—World Rank: 23

Capital: Seoul

Boundaries: 1,558 km (968 mi); North Korea 240 km (149 mi)

Coastline: 1,318 km (819 mi)

Longest Distances: 624 km (399 mi) NNE-SSW; 436 km (271 mi) ESE-WNW

Highest Point: Mount Halla 1,950 m (6,398 ft)

Lowest Point: Sea level

Land Use: 21% arable; 1% permanent crops; 1% meadows and pastures; 67% forest and woodland; 10% other

The Republic of Korea (South Korea) is part of a divided country, a fact that has had a profound effect on the course of its recent history and economic development. The peninsula as a whole has been used in the past as a bridge by larger and more powerful neighbors. Its rugged terrain—though diverse and often scenically beautiful—prevented until modern times easy transportation and communication between remote regions. These conditions make the homogeneity and national identity of the people all the more remarkable.

The Korean peninsula projects to within 193 km (120 mi) of the principal Japanese islands of Honshū and Kyūshū on the southeast. The north of the peninsula borders a section of the northeastern provinces of the People's Republic of China (PRC) over a distance of some 1,368 km (850 mi) and borders the Soviet Union for about 18 km (eleven mi). Elongated and irregular in shape, the peninsula separates the Sea of Japan from the Yellow Sea; the seas are known in Korea as the Eastern Sea and the Western Sea, respectively.

The peninsula has a north-south extent of about 965 km (600 mi) and occupies the same latitudes as the United States from New England to South Carolina. It is broadest at its northern border and narrowest (about 217 km, 135 mi) in its center, where the Military Demarcation Line, most of which winds slightly north of the thirty-eighth parallel, divides South Korea from the Democratic People's Republic of Korea (North Korea). As a result of the armistice agreement concluding the Korean War, about 45 percent of the Korean peninsula, or 98,873 sq km (38,175 sq mi) falls below the demarcation line. The demarcation line divides the 4,000 m (about 2.5 mi) wide Demilitarized Zone (DMZ), which is uninhabited save for two small villages near P'anmunjŏm and some cultivated land near bordering areas. Excluding the southern half of the DMZ, the territory under effective administration by South Korea amounts to 98,425 sq km (38,002 sq mi). A strip of territory south of the DMZ is under South Korean military rather than civilian operational control.

Topography

The peninsula is very rugged and mountainous, although only one peak, Paektu Mountain in the extreme north, exceeds 2,743 m (9,000 ft) in elevation. Only about 15 percent of the land may be considered plains, and these plains are mostly coastal, small in area, and isolated from one another; the southern half of the peninsula contains much more level land than the northern half. The rugged mountain ranges had traditionally acted as a barrier to man's movement and cut Korea off from mainland Asia except for the maritime margins of the peninsula. North Korea and South Korea are structurally separated by a depression extending across the narrowest part of the peninsula from Wŏnsan almost directly south to Seoul. The demarcation line crosses this depression about 48 km (30 mi) north of Seoul.

Base 504885 (546980) 3-82

From the watershed divide close to the east coast the land slopes sharply and abruptly to the narrow and discontinuous lowland of the Eastern Littoral. This coast is fairly regular, with only two major bays and few islands, although local landforms are varied. The slope toward the west is much more gradual than that on the east, and the rivers are relatively long and meandering. These rivers have built up more extensive plains near the western and southern coast than were produced by the short streams east of the watershed. The west coast is very irregular, characterized by numerous small bays and peninsulas and studded with hundreds of islands. The seas adjacent to the east and west coasts are also very dissimilar. The very deep Sea of Japan has a small tidal range. The shallow Yellow Sea off the west coast, with the second greatest tidal range in the world, exposes vast silted mud flats at low tide; these conditions and the strong currents are a handicap to navigation.

The mountainous character of Korea lends special importance to the few lowlands and plains, which have become the primary areas of human habitation and economic activity. Passes through the mountains are critical focal points of transportation routes that constitute the avenues along which cultural interchange has historically taken place. For the peninsula as a whole, the major transport routes lie along the east and west coasts and the central depression, the latter providing the best transpeninsular route. In South Korea the most important lines of movement extend through a series of small basins and hill areas along the Southern Littoral and through the central part of the Sobaek Mountains.

Dominating the country's topography are the T'aebaek Mountains, running some 257 km (160 mi) parallel and close to the east coast and forming the backbone of the country. Other mountain groups branch off from this range like ribs. Although not high, the mountain terrain is very rugged and often spectacular, with ridge following ridge, and the overall topography is highly dissected.

Fourteen distinct physiographic regions have been identified in South Korea. These can be grouped into five major regions: the large Central Region, the Eastern Littoral, the Southern Mountain and Valley Region, the Naktong River Basin, and the Southern Littoral. A sixth region in the northwest, the Imjin River Basin, which is part of the transpeninsular depression, lies mostly in North Korea. Cheju Island, a volcanic island, is distinctive in its natural and cultural features.

The Central Region is an upraised area that slopes westward from the T'aebaek Mountains. The eastern part of this region is rugged, with steep slopes and swiftly flowing streams. Closer to the west coast the topography is more subdued, less mountainous and more hilly, and the rivers have built fairly extensive alluvial plains. Among the larger of these plains are those of the Han River near Seoul and of the Kŭm River. Smaller river plains make up the rest of a coastal lowland extending south of Seoul and from 16 to 80 km (ten to fifty mi) inland from the Yellow Sea. The lowland is interspersed with hilly tracts extending from the interior mountains to the sea. The coast itself is extremely indented, flooded lower courses of streams alternating with rocky headlands degenerating into offshore islands. The shoreline is indefinite; a tidal range of up to thirty feet alternately covers and exposes mud flats, shoals, and low-lying islands. Many shallow arms of the sea have been diked off, and the land has been reclaimed for agricultural use. The lower courses of the rivers are suitable for limited navigation, and there are many small coastal ports despite the dangers and navigational difficulties posed by the rocky coast and very high tidal range. Some of the more extensive agricultural areas lie in this western coastal lowland.

The Eastern Littoral is an elongated strip of steep foothills about 32 to 40 km (twenty to twenty-five mi) wide along the eastern flank of the T'aebaek Mountains. Short streams flowing in narrow, steep valleys have formed a succession of tiny plains separated from one another by extensions of the hills to the coast. The entire coast is rather isolated and has a relatively low population density, restricted level land available for agriculture, and a number of small fishing ports. In contrast to the western coast, the water is deep immediately offshore, and the tidal range is low.

The Southern Mountain and Valley Region is dominated by the Sobaek Mountains, which extend southwestward from the southern end of the T'aebaek Mountains. The Sobaek Mountains separate into a series of parallel ridges and valleys that extend to a complex of coastal indentations and offshore islands at the southwestern tip of the country. The mountains form an interior divide, separating the northwest and Seoul from the southeast and Pusan; they also divide the Chŏlla provinces from the Kyŏngsang provinces. The mountains include the highest peak on mainland South Korea.

The Naktong River Basin in the southeast is a complex of structural basins and river floodplains separated from one another by low hills. The Naktong River forms an extensive delta where it reaches the sea a few miles west of Pusan, South Korea's major port. The coastline of the Naktong River Basin may be divided at the river mouth near Pusan. To the north of this point the coast is relatively smooth and, like that of the Eastern Littoral, consists of alternating headlands and bays; the latter have small lowlands at their heads, but they are not as isolated from the interior as their counterparts farther north. To the west of the Naktong River mouth the coast has the complexity of the Southern Littoral.

In the Southern Littoral, where the various arms of the Sobaek Mountains reach the sea, a number of small structural basins are to be found. Offshore the basins contain deep water and create an extremely intricate coastline of extensive, highly irregular peninsulas flanked by abruptly rising islands. At times the peninsulas almost enclose equally irregular bays deeply penetrating the land. Inland, alluvium replaces the water, and the plains are fertile and agriculturally productive. The hills rise abruptly from the plains, many of which are extremely small.

Of a comparatively large number of rivers and streams, four are of major importance: the Han River, the Kŭm River, the Naktong River, and the Sŏmjin River. In addition, the Yŏngsan and Tongjin rivers water South Korea's main ricegrowing areas. Because of their very low gradients, the rivers to the west of the T'aebaek Mountains watershed were used for transportation; river navigation declined in importance in modern times with the introduction of new means of transport, the diversion of water for irrigation, and the construction of dams. River flow is highly seasonal, the heaviest flows occurring in the summer months. Floods are common in the basins associated with the major river systems, particularly in estuary areas along the west coast. During much of the year, however, the rivers are shallow, exposing very wide, gravelly riverbeds.

PRINCIPAL TOWNS (population at 1985 census)*

Sŏul, (Seoul, the capital)	9,639,110	Sŏngnam	447,692
Pusan	3,514,798	Suwŏn	430,752
Taegu	2,029,853	Chŏnju	426,473
Inchon	1,386,911	Chŏngju	350,256
Kwangju		Mokpo	236,085
(Gwangju)	905,896	Chinju (Jinju)	227,309
Taejon	866,148	Cheju	202,911
Ulsan	551,014	Kunsan (Gunsan)	185,649
Masan	448,746	Yŏsu	171,933
		Chunchŏn	162,988

*Preliminary data.

Area and population

Provinces	Capitals	area sq mi	area sq km	population 1985 census
Cheju-do	Cheju	705	1,825	489,458
Chŏlla-namdo	Kwangju	4,729	12,249	3,748,442
Chŏlla-pukto	Chŏngju	3,108	8,050	2,202,218
Ch'ungch'ŏng-namdo	Taejŏn	3,411	8,835	3,001,538
Ch'ungch'ŏng-pukto	Ch'ŏngju	2,870	7,433	1,391,084
Kangwŏn-do	Ch'unch'ŏn'	6,523	16,894	1,726,029
Kyŏnggi-do	Inch'ŏn	4,193	10,859	4,794,240
Kyŏngsang-namdo	Masan	4,577	11,855	3,013,276
Kyŏngsang-pukto	Taegu	7,506	19,441	3,519,121
Special cities				
Inch'ŏn-si	Inch'ŏn	80	207	1,387,475
Pusan-si	Pusan	168	435	3,516,768
Sŏul-t'ŭkpyŏlsi	Seoul	234	605	9,645,824
Taegu-si	Taegu	176	455	2,030,649
TOTAL		38,279	99,143	40,466,577

KUWAIT

BASIC FACTS

Official Name: State of Kuwait

Abbreviation: KUW

Area: 17,818 sq km (6,880 sq mi)

Area—World Rank: 138

Population: 1,938,075 (1988) 3,007,000 (2000)

Population—World Rank: 122

Capital: Kuwait

Boundaries: 632 km (393 mi) Saudi Arabia 163 km (101 mi); Iraq 257 km (160 mi)

Coastline: 212 km (132 mi)

Longest Distances: 205 km (127 mi) SE-NW 176 km (109 mi) NE-SW

Highest Point: 290 m (951 ft)

Lowest Point: Sea level

Land Use: 8% meadows and pastures; 92% other

Located at the northwestern corner of the Persian Gulf, Kuwait is bounded on the east by the Gulf, on the north and west by Iraq, and on the southwest by Saudi Arabia. To the south, also bordering on the Gulf, is the 5,180 sq km (2,000 sq mi) Neutral Zone, owned jointly by Kuwait and Saudi Arabia. The area of the state itself is less than the size of New Jersey or Wales. It is roughly rectangular in shape. Included in this territory are a number of large offshore islands, the largest of which is Bubiyan, separated from the northwest mainland by a narrow creek. Of these, only the island of Failaka at the mouth of Kuwait Bay is inhabited. This island is believed to have been a center of civilization in antiquity and is the site of an ancient Greek temple built by the forces of Alexander the Great. The only prominent geographic feature is Kuwait Bay, which indents the shoreline for about 40 km (25 mi), providing natural protection for the Port of Kuwait and accounting for nearly half of the state's some 193 km (120 mi) of Gulf shoreline.

Most of Kuwait and the Neutral Zone consists of waterless desert. There is one small oasis at Jaharah at the western end of Kuwait Bay and a few wells in the coastal villages. There are no permanent streams, but a few *wadis* (watercourses) are filled by winter rain. Notable among these is the Wadi al-Batin, the broad shallow valley forming the western boundary of the country. At Raudhatain in the north, several shallow *wadis* converge and provide a temporary repository for winter flood water, which quickly evaporates or sinks into the porous subsoil. The land's level character is explained in part by the absence of direct drainage to the sea, but fundamentally the monotony is caused by the area's geographic structure. Unfolded sedimentary rocks of Miocene and Pleistocene Age lie almost horizontally on gently-folded Cretaceous and Jurassic Rocks. There is a continual redistribution of the land surface by flood runoff and erosion.

From Umm Qasr in Iraq south along the coastline to beyond the indentation of Kuwait Bay there is a narrow coastal lowland. To the northeast, overlooking the Bay, the land rises to 61 to 91 m (200 to 300 ft) above sea level

PRINCIPAL TOWNS (population at 1985 census)

Kuwait City (capital)	44,335	Hawalli	145,126
Salmiya	153,369	Faranawiya	68,701
		Abraq Kheetan	45,120

KUWAIT
International boundary
National capital
Railroad
Road
0 25 Miles
0 25 Kilometers
47
48
30
29
Al Başrah
Khorramshahr
Abadan
IRAQ
IRAN
Shaţţ al Arab
Şafwān
Khawr Zubayr
Umm Qaşr
Al Fāw
WARBAH
Khawr 'Abd Allah
BŪBIYĀN
IRAQ
Kuwait Bay
MASKĀN
FAYLAKAH
KUWAIT
'AWHAH
Al Jahrah
Ḩawallī
Ash Shamiyah
Wādī al Bāţin
Persian Gulf
Al Maqwa'
N.Z
Al Fuḩayḩīl
Al Aḩmadī
Ash Shu'aybah
Al Ad'amī
Khawr al Mufattah
SAUDI ARABIA
Names and boundary representation
are not necessarily authoritative

502915 3-76

and slopes gently northwards and eastwards. This low plateau, known as the Zor, terminates at the escarpment of Az in the north and at the oasis of Jaharah in the east. The central part of the country is low and flat and is separated from the Gulf Shore by the 122 m (400 ft) high Ahmadi Ridge which runs north to south.

The Kuwaiti desert is undulating and gravelly, with few low hills or ridges. The only trees apart from those found in the date gardens at Jaharah are tamarisks found in Kuwait City and a few villages. The most common desert shrub is the *arfaj,* which grows to a maximum height of 0.7 m (2.5 ft) and is used for firewood by nomadic tribesmen in the vicinity of the Wadi al-Batin. For a brief period in the spring, if winter rains have been adequate, the *wadi* beds are covered with grass and flowering desert shrub. The spring flush, however, is very brief, for the May sun dries the grass and withers the shrubs and the June winds soon envelop everything in driving sand.

Area and population

		area		population
Governorates	**Capitals**	sq mi	sq km	1987 estimate
al-Aḥmadī	al-Aḥmadī	1,984	5,138	345,783
al-Jahrā'	al-Jahrā'	4,372	11,324	329,588
Capital	Kuwait City	38	98	160,860
Ḥawalli	Ḥawalli	138	358	1,036,337
Islands	—	348	900	. . .
TOTAL		6,880	17,818	1,872,568

LAOS

BASIC FACTS

Official Name: Lao People's Democratic Republic

Abbreviation: LAO

Area: 236,800 sq km (91,400 sq mi)

Area—World Rank: 80

Population: 3,849,752 (1988) 4,906,000 (2000)

Population—World Rank: 102

Capital: Vientiane

Land Boundaries: 4,513 km (2,804 mi); Vietnam 1,555 km (966 mi); China 425 km (264 mi); Cambodia 541 km (336 mi); Thailand 1,754 km (1,090 mi); Myanmar 238 km (148 mi)

Coastline: Nil

Longest Distances: 1,162 km (722 mi) SSE-NNW; 478 km (297 mi) ENE-WSW

Highest Point: Mount Bia 2,820 m (9,252 ft)

Lowest Point: 70 m (230 ft)

Land Use: 4% arable; 3% meadows and pastures; 58% forest and woodland; 35% other

Laos is a landlocked, largely mountainous country lying in the heart of the Indochinese peninsula. It is located entirely in the tropics—its southern tip being about fourteen degrees north of the equator and its most northerly point being just slightly below the Tropic of Cancer.

Less than three-fifths of the national territory is contained in the northern section of the country and over two-fifths in the country's southern panhandle. The highly mountainous nature and elevations in the northern part, as well as in much of the Annamite Chain extending the length of the panhandle, long acted as barriers to communication with the countries lying to the north, northwest, and east. The Mekong River on the west, however, has served as a link with related peoples of Thailand and with Cambodia.

Topography

The northern section of the country lying above the Laotian panhandle is characterized by rugged mountain terrain. This terrain is a continuation of the folded mountains that sweep generally southward into Southeast Asia from eastern Tibet. The main ranges trend chiefly northeast to southwest. Their mountains are sharp crested and steep sloped, and the slopes are greatly dissected. Valleys are V-shaped, and there are many narrow gorges; adjacent crests rise 809 to 1,219 m (2,000 to 4,000 ft) above valley floors. Several ranges are around 1,524 m (5,000 ft) in height, and many peaks are well over 1,829 m (6,000 ft). The country's highest mountain, Phou Bia, rising to 2,819 m (9,249 ft) above sea level, is situated in Xieng Khouang Province in this northern section. This section also contains a number of scattered relatively large hill areas that are moderately to highly dissected; the hills usually have rounded tops.

A prominent feature of the northern part of the country is the comparatively large plain, known as the Plain of Jars, located on the Tran Ninh Plateau in Xieng Khouang Province. The plateau, lying mostly between about 1,015 and 1,219 m (3,330 and 4,000 ft) above sea level, has relatively infertile soils. The Plain of Jars, a term applying correctly to only a part of the plateau and derived from prehistoric large stone jars found in the area, however, has become well known since the early 1960s as the site of periodic fighting between the forces of the Royal Lao Government and of the Lao People's Liberation Army.

The chief topographic feature of the Laotian panhandle is the Annamite Chain, which runs along the entire eastern side of this section of the country. The chain extends northwest to southeast, paralleling the direction of flow of the Mekong River. In its upper portion, mountains resemble those in northern Laos, having rugged peaks and deep valleys; peaks are from about 1,524 and 2,438 m (5,000 to 8,000 ft) in elevation. In general, this portion of the chain presents a formidable barrier to movement between Laos and Vietnam.

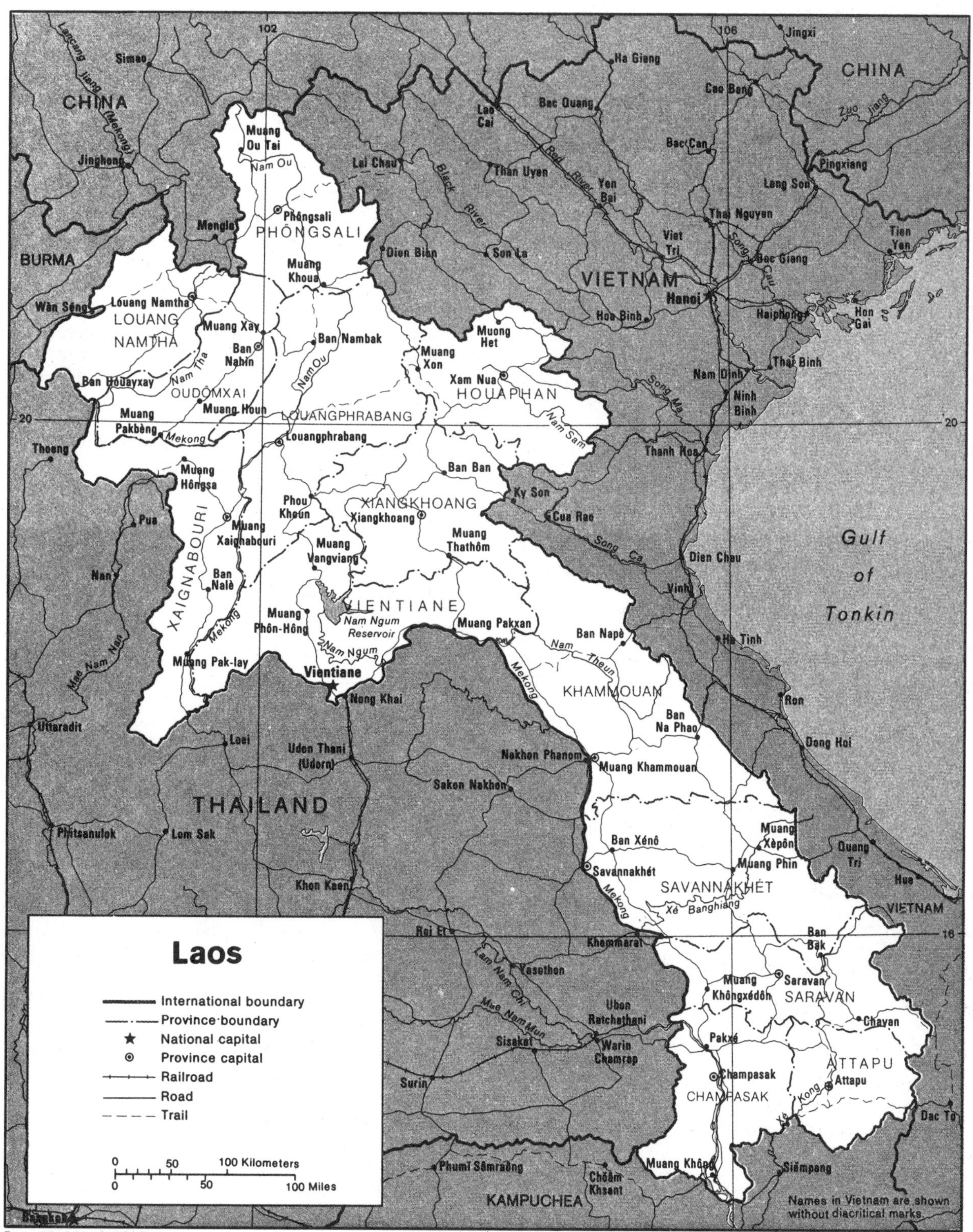

Base 504833 (546896) 3-82

In central-east Khammouane Province elevations are somewhat lower, and passes allow easier crossing. Farther south, at about the latitude of the city of Khammouane, the chain enters a limestone region characterized by steep ridges and peaks, sinkholes, and disappearing streams. Then a comparative flat area occurs. From this point to the southern end of Laos the chain again becomes very rugged, and its elevations rise to 1,981 m (6,500 ft); the high point is over 2,286 m (7,500 ft).

At the neck of the panhandle section the Annamite Chain extends to the Mekong River. Below this the mountains are buttressed to the west by several plateaus, the best known being the Cammon Plateau in Khammouane Province. The terrain is a generally rolling plain. From the plateaus the land slopes more gently westward to the alluvial plains along the Mekong. Prominent in the southern part of the country is the fertile Bolovens Plateau. Almost encircled by a high escarpment, the plateau has an elevation of about 1,067 m (3,500 ft). Its terrain is also generally rolling, and there are large patches of grassland.

PRINCIPAL TOWNS (population in 1973)

Vientiane (capital)*	176,637	Luang Prabang	44,244
Savannakhet	50,690	Saya Bury	13,775
Pakse	44,860	Khammouane	12,676

*The population of the Vientiane municipality at the 1985 census was 377,409.

Area and population		area		population
Provinces	**Capitals**	sq mi	sq km	1985 census
Attapu	Attapu	. . .	. . .	69,631
Bokeo	Houayxay	. . .	. . .	54,925
Bolikhamxay	Pakxan	. . .	. . .	122,300
Champasak	Pakxé	. . .	. . .	403,041
Houaphan	Xam Nua	. . .	. . .	209,921
Khammouan	Thakhek	. . .	. . .	213,462
Louang Namtha	Louang Namtha	. . .	. . .	97,028
Louangphrabang	Louangphrabang	. . .	. . .	295,475
Oudomxay	Xay	. . .	. . .	187,115
Phôngsali	Phôngsali	. . .	. . .	122,984
Saravan	Saravan	. . .	. . .	187,515
Savannakhét	Savannakhét	. . .	. . .	543,611
Vientiane	Vientiane	. . .	. . .	264,277
Xaignabouri	Xaignabouri	. . .	. . .	223,611
Xékong	Thong	. . .	. . .	50,909
Xiangkhoang	Phônsavan	. . .	. . .	161,589
Municipalities				
Vientiane	—	. . .	. . .	377,409
TOTAL		91,400	236,800	3,584,803

Rivers

The Mekong River and its tributaries drain all of Laos except parts of eastern Houa Phan Province and northern Xieng Khouang Province that lie east of the main mountain divide; these drain into the Gulf of Tonkin. The Mekong is the center of the country's economic life. Its flood plains provide the major wet-rice lands; its waters furnish fish, the main protein element in the diet; and the system carries the pirogues, sampans, and light barges that transport the country's freight. All of the larger towns are on or near its banks.

The Mekong borders on, or flows through, Laos for more than 1,609 km (1,000 mi). In the northern part of the country it is characterized by narrow valleys and deep gorges. Downstream from Vientiane the valley widens, and alluvial plains of varying width are found on each side. The river is navigable from Luang Prabang to Savannakhet. Downstream from Savannakhet, formidable rapids extend for more than 129 km (80 mi) but are passable at high water. In the extreme south the double falls of Khone are impassable at all times.

The Mekong's tributaries in the north include the Nam Tha and the Nam Ou, which flow in deep, narrow valleys. Both, however, offer some possibilities for water travel by small craft, and both contain alluvial pockets where wet rice can be cultivated. A major tributary draining part of northern Laos is the Nam Ngum, which enters the Mekong east of Vientiane through an alluvial plain.

Below the bend of the Mekong east of Vientiane where the river turns southward, the tributaries generally empty through relatively level alluvial plains for the last 64 or 80 km (40 or 50 mi) of their courses. The main tributaries in this part of the country are the Nam Kading, Se Bang Fai, Se Bang Hieng, and Se Done. Another large tributary, the Se Kong, which drains southeast Laos, joins the Mekong inside Cambodia.

LEBANON

BASIC FACTS

Official Name: Republic of Lebanon

Abbreviation: LEB

Area: 10,230 sq km (3,950 sq mi)

Area—World Rank: 146

Population: 2,674,385 (1988) 3,617,000 (2000)

Population—World Rank: 113

Capital: Beirut

Boundaries: 656 km (407 mi); Syria 359 km (223 mi); Israel 102 km (63 mi)

Coastline: 195 km (121 mi)

Longest Distances: 217 km (135 mi) NE-SW; 56 km (35 mi) SE-NW

Highest Point: Mount Sauda 3,083 m (10,115 ft)

Lowest Point: Sea level

Land Use: 21% arable; 9% permanent crops; 1% meadows and pastures; 8% forest and woodland; 61% other

Lebanon is located on the eastern edge of the Mediterranean Sea. The winding 322-km (200-mi) boundary with

Syria follows a rift in the north and a mountain crest in the east; it is recognized internationally. On the south and southeast is a cease-fire line about 72 km (45 mi) long between Lebanon and Israel. This frontier, although it follows more or less the old boundary with Palestine, does not follow the natural limit of the coastal range, which is the Plain of Esdraelon in Palestine; it is unstable and is not recognized internationally.

The country is roughly rectangular in shape, becoming narrower toward the south. At no point is it wider than 88 km (55 mi), nor narrower than 32 (20); the average width is about 56 km (35 mi).

Geographical Regions

There are four geographical regions: the coastal plain, the coastal mountain range, the central plateau or valley, and the interior mountain range. Despite the small size of the country, these regions are well demarcated and are characterized by differences in climate, soil and minerals, water supply, fauna and flora, settlement density and, to a certain extent, religious communities and ethnicity.

The coastal plain is narrow and sloping. It is widest (about 13 km; 8 mi) in the north along the Syrian border. In places it disappears where the mountains come directly to the sea. Because of its proximity to the sea and international communications, it is the site of the principal cities and was the home of the ancient Phoenicians. It has few good natural harbors but has, instead, many shallow, curved bays.

The coastal mountain range, called the Lebanon Mountains, rises rather abruptly from the coastal plain, with few transitional foothills. It is separated from the Nusayriyah Range along the Syrian coast by a geological rift, composed of the Akkar Plain near Tripoli and the Buqay'ah Plain near Homs in Syria, which forms the northern boundary and the bed of the Nahr al Kabir. The range runs the entire length of the Mediterranean Coast, diminishing in height from north to south, merging into the Hills of Galilee in Palestine. The highest peaks are around Basharri, southeast of Tripoli; of this group, the highest is Qurnat as Sawda (3,083 m, 10,115 ft). Another block of peaks rises east of Beirut; of these, Jabal Sannin (2,608 m, 8,557 ft) is the highest. The Lebanon Mountains and the coastal plain together are formed by an upfold, complicated by numerous small faults, which is bounded on the east by a sharp fault which forms the western boundary of the Biqa Valley.

East of the Lebanon Mountains is the narrow central valley or plateau, whose ancient name was Coelosyria, or Hollow Syria. In modern times it is called the Biqa Valley and, from its watershed near Baalbak (1,100 m, 3,609 ft), it slopes to the north and south. In the south it is separated from the Jordan Valley by a range of hills and by a southwestward extension of the interior range, whereas in the north it opens into the Syrian plain at Homs. Formed by a sharp fault on the west and by steeply inclined strata on the east, it is filled by geologically recent sediments. It is about 120 km (75 mi) long and varies from 8 to 12 km (5 to 8 mi) in width.

To the east of the Biqa Valley rise the interior mountain range, called the Anti-Lebanon Mountains (Jabal ash Sharqi), and its southern extension, Mount Hermon (Jabal ash Shaikh), both forming the eastern boundary of Lebanon with Syria. Mount Hermon rises to 2,760 m (9,055 ft) and, in the northern sector, Tal'at Musa rises to 2,658 m (8,721 ft); the average height of the range is about 2,286 m (7,500 ft). The range is formed, like the coastal range, of a simple upfold; on the north it merges into the Syrian plain, and on the south it is terminated by the Jordan Valley rift.

The entire region is subject to earthquakes produced by movement along local fault lines. Shock waves from the Asia Minor earthquake zone are also felt in Lebanon. In A.D. 555 Roman Beirut was devastated by an earthquake, and in 1758 the temples at Baalbak were demolished. As recently as 1956, a severe earthquake wrought much destruction in southern Lebanon. The Government maintains seismological stations at several locations in order to study the problem.

Drainage

Although the country is well watered and there are many rivers and streams, there are no navigable rivers, nor is any one river the sole source of irrigational water for its agriculture. Drainage patterns have been determined by geological features and climate; although rainfall is seasonal, most streams are perennial.

Most rivers in Lebanon originate in the springs, often quite large, which emerge from the permeable limestone strata cropping out at the 916 to 1524 m (3,000 to 5,000 ft) level in the Labanon Mountains. Few emerge in the Anti-Lebanon Mountains in this manner. Other springs emerge from alluvial soil and join to form rivers. Whatever their source, the rivers are fast-moving and straight, most of them cascading down narrow mountain canyons to the sea.

From north to south, the principal rivers along the Mediterranean coast are: the Nahr al Kabir, which forms the northern boundary with Syria; the Nahr al Barid; the Nahr Abu Ali, which flows through Tripoli; the Nahr al Jawzah; the Nahr Ibrahim, the ancient Adonis River which still flows red in the springtime; the Nahr al Kalb; the Nahr Bayrut; the Nahr ad Damur; the Nahr al Awwali; the Nahr az Zahrani; and the Nahr al Litani.

The Biqa Valley is watered by two rivers which rise in the watershed near Baalbak: the Orontes flowing north (in

PRINCIPAL TOWNS (estimated population in 1975)

Beirut (capital) 1,500,000; Tarabulus (Tripoli) 160,000; Zahleh 45,000; Saida (Sidon) 38,000; Sur (Tyre) 14,000.

Area and population

Governorates	Capitals	area sq mi	area sq km	population 1970 estimate
Bayrūt	Beirut(Bayrūt)	7	18	474,870
al-Biqā[c]	Zaḥiah	1,653	4,280	203,520
Jabal Lubnān	B[c]abdā	753	1,950	833,055
al-Janūb	Sidon (Ṣaydā)	364	943	249,945
an-Nabaṭiyah	an-Nabaṭiyah	408	1,058	. . .
ash-Shamāl	Tripoli (Ṭarābulus)	765	1,981	364,935
TOTAL		3,950	10,230	2,126,325

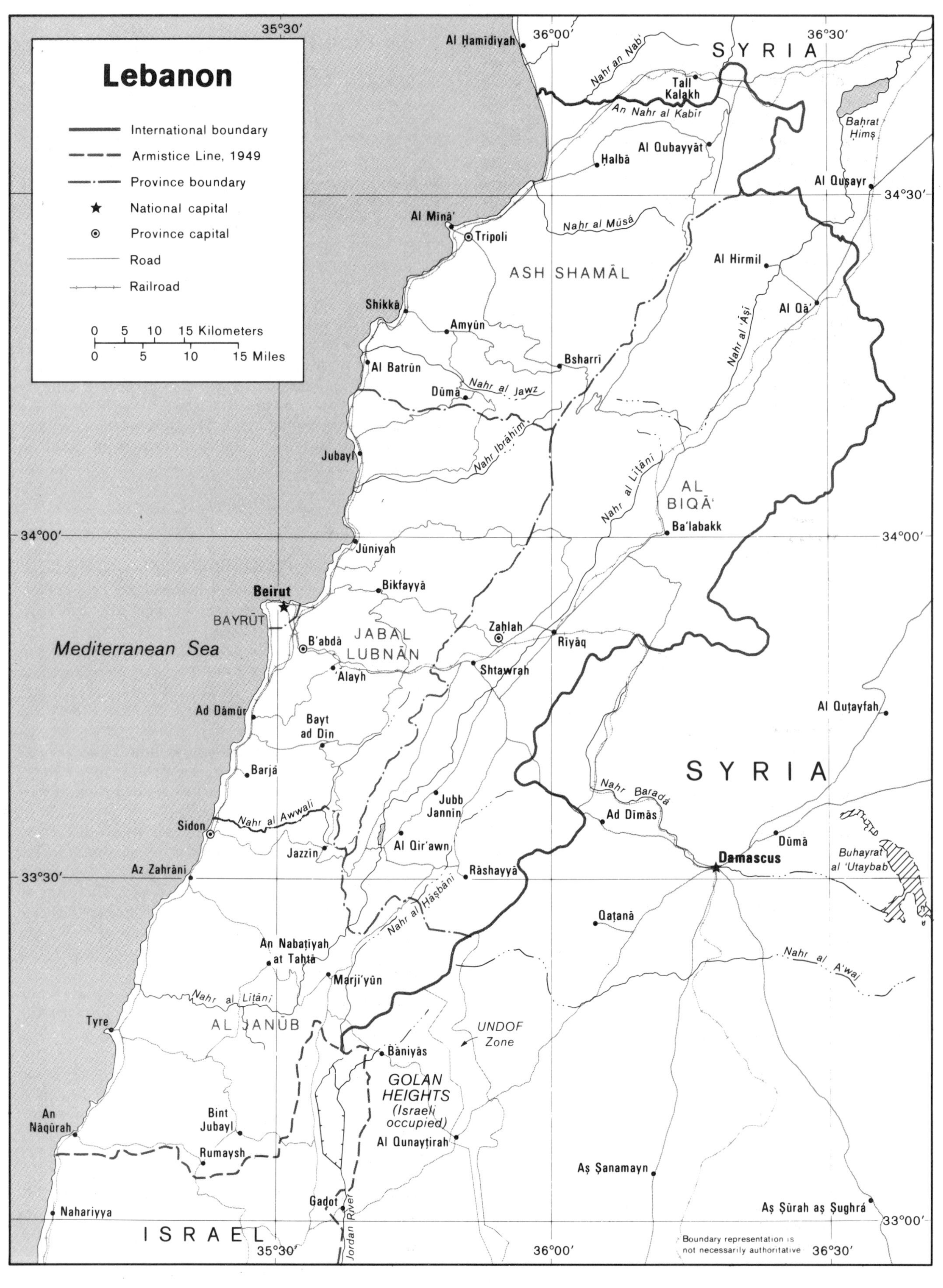
Lebanon
International boundary
Armistice Line, 1949
Province boundary
National capital
Province capital
Road
Railroad
0 5 10 15 Kilometers
0 5 10 15 Miles
35°30′
36°00′
36°30′
34°30′
34°00′
33°30′
33°00′
SYRIA
ISRAEL
Mediterranean Sea
Al Ḥamīdiyah
Nahr an Nab'
Tall Kalakh
An Nahr al Kabīr
Baḥrat Ḥimṣ
Al Qubayyāt
Ḥalbā
Al Quṣayr
Al Mīnā'
Tripoli
Nahr al Mūsā
ASH SHAMĀL
Al Hirmil
Shikkā
Al Qā'
Amyūn
Nahr al 'Āṣī
Bsharri
Al Batrūn
Dūmā
Nahr al Jawz
Nahr Ibrāhīm
Jubayl
Nahr al Līṭānī
AL BIQĀ'
Ba'labakk
Jūniyah
Bikfayyā
Beirut
BAYRŪT
B'abdā
JABAL LUBNĀN
Zaḥlah
Riyāq
'Alayh
Shtawrah
Ad Dāmūr
Bayt ad Dīn
Al Quṭayfah
Barjā
Nahr Baradā
Jubb Jannīn
Sidon
Nahr al Awwalī
Ad Dīmās
Al Qir'awn
Jazzīn
Dūmā
Damascus
Buhayrat al 'Utaybah
Az Zahrānī
Rāshayyā
Nahr al Ḥāṣbānī
Qaṭanā
An Nabaṭīyah at Taḥtā
Nahr al A'waj
Marji'yūn
Nahr al Līṭānī
Tyre
AL JANŪB
UNDOF Zone
Bāniyās
GOLAN HEIGHTS (Israeli occupied)
An Nāqūrah
Bint Jubayl
Al Qunayṭirah
Rumaysh
Aṣ Ṣanamayn
Gadot
Jordan River
Aṣ Ṣūrah aṣ Ṣughrá
Nahariyya
Boundary representation is not necessarily authoritative

Arabic it is called Nahr al 'Asi, the Rebel River, because this direction is unusual), and the Litani flowing south into the hill region of the southern Biqa Valley, where it makes an abrupt turn to the west and is thereafter called Qasimiyah. The Orontes continues to flow north into Syria and eventually reaches the Mediterranean in Turkey. Its waters, for much of its course, flow through a channel considerably lower than the surface of the ground. The Nahr Barada, which waters Damascus, has as source a spring in the Anti-Lebanon Mountains.

Smaller springs and streams serve as tributaries to the principal rivers. Since the rivers and streams have such steep gradients and are so fast moving, they are erosive instead of depositary in nature, a process which is aided by the soft character of the limestone which composes much of the mountains, the steep slopes of the mountains, and the heavy rainstorms. There is one seasonal lake, fed by springs, on the eastern slopes of the Lebanon Mountains near Yammunah, about 40 km (25 mi) southeast of Tripoli.

LESOTHO

BASIC FACTS

Official Name: Kingdom of Lesotho

Abbreviation: LES

Area: 30,355 sq km (11,720 sq mi)

Area—World Rank: 125

Population: 1,666,012 (1988) 2,282,000 (2000)

Population—World Rank: 125

Capital: Maseru

Land Boundaries: 909 km (565 mi) South Africa

Coastline: Nil

Longest Distances: 248 km (154 mi) NNE-SSW; 181 km (112 mi) ESE-WNW

Highest Point: Thabana Ntlenyana 3,482 m (11,425 ft)

Lowest Point: Sea level

Land Use: 10% arable; 66% meadows and pastures; 24% other

Lesotho is an enclave located within the east-central part of South Africa. It is the only country in the world entirely surrounded by another.

Area and population		area		population
Districts	**Capitals**	sq mi	sq km	1986 census
Berea	Teyateyaneng	858	2,222	194,631
Butha-Buthe	Butha-Buthe	682	1,767	100,644
Leribe	Leribe	1,092	2,828	257,988
Mafeteng	Mafeteng	818	2,119	195,591
Maseru	Maseru	1,652	4,279	311,159
Mohale's Hoek	Mohale's Hoek	1,363	3,530	164,392
Mokhotlong	Mokhotlong	1,573	4,075	74,676
Qacha's Nek	Qacha's Nek	907	2,349	63,984
Quthing	Quthing	1,126	2,916	110,376
Thaba-Tseka	Thaba-Tseka	1,649	4,270	104,095
TOTAL		11,720	30,355	1,577,536

Lesotho has three distinct geographical regions extending longitudinally across the country. The Western Lowlands cover approximately a quarter of the country's land area between the Caledon River and the Cave Sandstone foothills. They consist of undulating basins and plains ranging in width from 10 km (6 mi) to 64 km (40 mi) with altitudes averaging between 1,520 and 1,820 m (5,000 to 6,000 ft). The Cave Sandstone Terrace is an intermediate region between the highlands and the lowlands with an average altitude of more than 1,820 m (6,000 ft). The Maluti Mountains, spurs of the main Drakensberg, with some peaks over 3,000 m (10,000 ft) high, mark the western edge of the eastern highlands, which are South Africa's main watershed. Located in this high plateau is Lesotho's highest point, Thabana Ntlenyana, 3,482 m (11,425 ft).

Lesotho is drained by tributaries of the Orange and Caledon Rivers and the Tugela River.

LIBERIA

BASIC FACTS

Official Name: Republic of Liberia

Abbreviation: LBR

Area: 99,067 sq km (38,250 sq mi)

Area—World Rank: 104

Population: 2,463,190 (1988) 3,642,000 (2000)

Population—World Rank: 117

Capital: Monrovia

Boundaries: 2,123 km (1,319 mi) Guinea 563 km (350 mi); Ivory Coast 716 km (445 mi) Sierra Leone 306 km (190 mi)

Coastline: 538 km (334 mi)

Longest Distances: 548 km (341 mi) ESE-WNW; 274 km (170 mi) NNE-SSW

Highest Point: Mount Wuteve 1,380 m (4,528 ft)

Lesotho

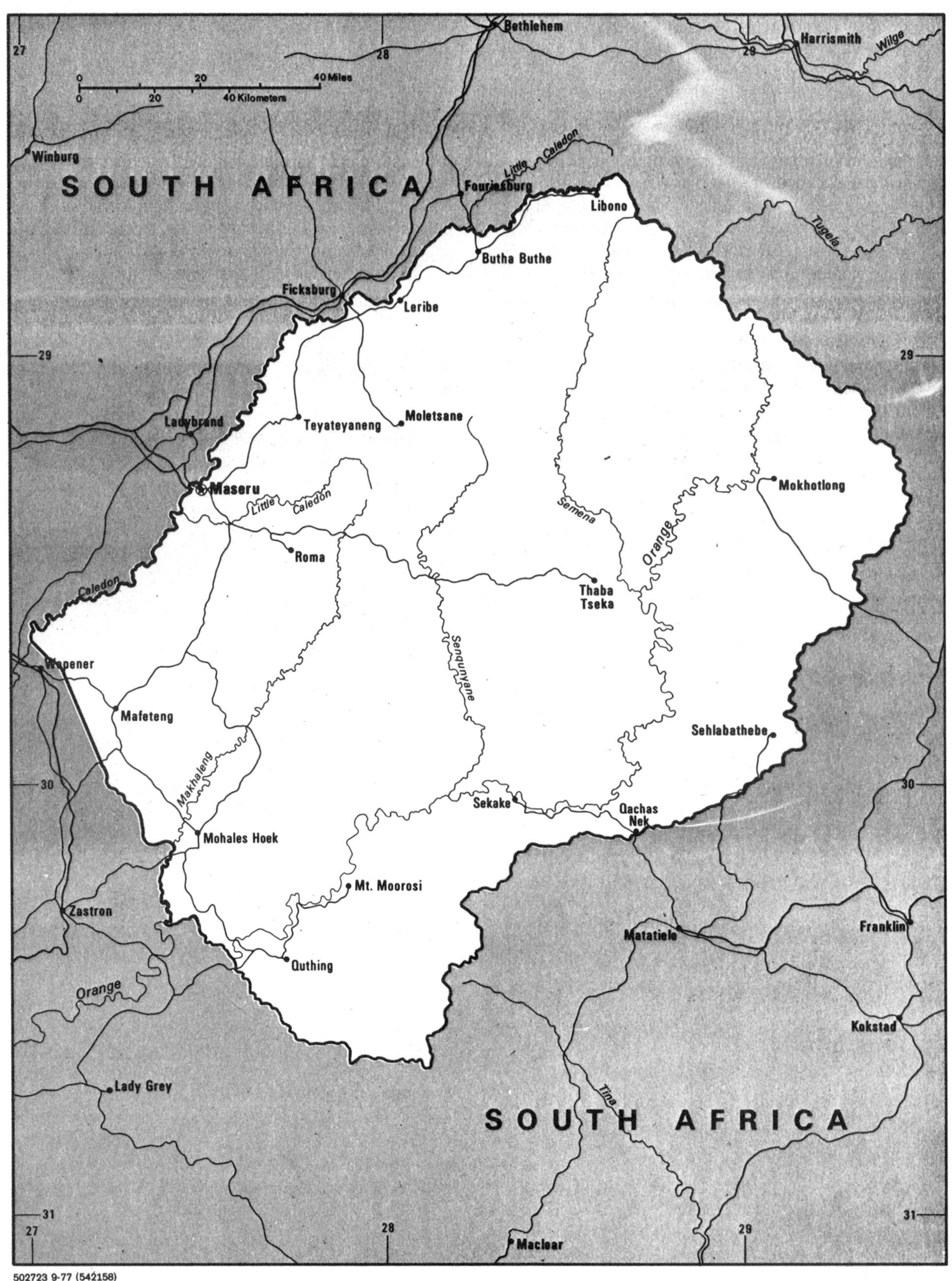

502723 9-77 (542158)
Lambert Conformal Projection
Standard parallels 6° and 30°
Scale 1:1,500,000

Railroad
Road

Lowest Point: Sea level

Land Use: 1% arable; 3% permanent crops; 2% meadows and pastures; 39% forest and woodland; 55% other

Liberia lies a few degrees north of the equator on the great western bulge of Africa. It extends along the coast between Sierra Leone on the northwest and the Ivory Coast on the southeast near Cape Palmas. Here at the country's southern extremity, the shoreline of West Africa turns eastward and faces the Gulf of Guinea. Inland the country ascends the seaward slopes of the Guinea Highlands to a very irregular border with Guinea on the north. Its width from the sea varies from about 161 to 322 km (100 to 200 mi). The coastal region is sometimes referred to as the Grain Coast, so called from the "Grains of Paradise," or malagueta peppers, that attracted European traders in the early days.

Behind a low coastal plain most of the country is a rolling plateau, much broken by disjointed outliers of the mountain ranges of the Guinea Highlands, which lie parallel to and close beyond the inland borders, where a few elevations exceed 1,219 m (4,000 ft). The land is well watered, and a number of narrow, roughly parallel river basins drain the country southwest toward the sea. Although almost all of Liberia was once a region of tropical rain forest and there are extensive areas of virgin growth, much of the forest along the coast and inland across the center of the country has been modified by cutting and burning and is either fairly open or covered by secondary woodlands. Some of the country has never been thorougly explored, and most of it is inadequately mapped. Contrasts that exist between various parts of the country are more a function of relief than of any other physical characteristics. There is less rainfall inland than along the coast, but the yearly pattern is the same, and the decrease in quantity is not enough to result in marked differences in vegetation or in water resources. Temperatures are fairly uniform throughout, and there are no great regional differences in the soil. Taken as a whole, the interior of the country has such a complex mixture of topographic features that it cannot be divided into unbroken geographic areas along clear-cut lines.

Coastal Region

The coastal region, a belt of gently rolling low plains extending 32 or 48 km (20 or 30 mi) inland, is broken along the shore by river estuaries, tidal creeks and swamps, and a few prominent, rocky capes and promontories that appear as landmarks from the sea. The rather straight, sandy shoreline is only slightly indented by the mouths of rivers that are so obstructed by shifting bars, submerged rocks and sandpits trending northwest as to provide no natural harbors. The surf is normally heavy all along the coast but is worst at the height of the rainy season. The tidal range is moderate.

In the northwest, not far from the border with Sierra Leone, Cape Mount rises steeply from the sea to an elevation of over 305 m (1,000 ft) and overlooks Fishermans Lake, a broad tidal lagoon, on the landward side. Cape Mesurado, site of Monrovia, and the lagoon that lies behind it, are similar features on a smaller scale. Farther to the southeast, several other fairly prominent headlands break the monotony of the low shoreline. Before they enter the sea, many of the rivers run parallel to the coast only a few miles behind the beach, and their minor tributaries form swampy reaches or tidal creeks bordered by extensive mangrove, pandanus and palm thickets.

The coastal plain is the region of greatest development, most altered by man, where the settlers from America established their communities and made their farms. Nearly all of the land, except in some parts of the southeast where the high forest still stands near the sea, has been cut over and is now a mosaic of farm plots or larger plantations, grassy open spaces and patches of second-growth woodland or thick brush. Bottom lands along the streams are frequently flooded by the rains, and low ground forms many more-or-less permanent swamps.

Interior Hills and Plateaus

Behind the coastal region the terrain becomes more broken as it rises toward the high ground along the country's inland border. Relief is characterized by many discontinuous ranges of hills or low mountains, broad, rather shallow valleys and occasional escarpments or more gradual steps in elevation that break the ground into poorly defined plateaus at irregular distances inland. In the western and central parts of this region, the ridge lines between major drainage basins are not prominently defined, but run from the northeast to the southwest, generally perpendicular to the coast. In the southeast, on the other hand, a major divide crosses the region almost parallel to the coast before turning south to Cape Palmas. Average elevations run roughly from 152 to 304 m (500 to 1,000 ft) with the hills and mountains reaching 457 to 762 m (1,500 to 2,500 ft).

Although most of the region is covered by tropical rain forest, a strip that extends across the center from the direction of Monrovia to the northeast has been denuded of woodlands by continuous clearing and cultivation. A good deal of the forest is of second growth, but large stands of virgin timber spread across the west of the region to the Sierra Leone border, and there are even more extensive areas of untouched forest in the southeast. With respect to the coastal region, this is the hinterland, a region of aboriginal hamlets or larger villages with their surrounding farm plots, only beginning to be opened up by roads and the spread of economic development. Some of the heavily forested areas are very sparsely inhabited and have yet to be systematically explored.

The Guinea Highlands

The far northern corner of the country and a salient reaching northward farther to the east extend into the seaward flanks of the Guinea Highlands, a broken mass of high ground that stretches across Guinea and forms part of a major watershed between streams that flow across Liberia to the Atlantic Ocean and the great Niger River basin on the northeast. Spurs from the central axis of the

Guinea Highlands extend generally southwest across the border into Liberia and drop off quite abruptly in some places. Outlying mountains or ranges of mountains rise above the rolling uplands to elevations that occasionally exceed 1,219 m (4,000 ft).

The Walo and Wangisi Mountains in the far north, and the Nimba Mountains, rich in iron ore and lying farther east where the borders of Guinea, the Ivory Coast and Liberia meet, are among the most prominent features of the relief. Parts of the region have very little tree cover, and there are many areas of open grassland with scattered trees or thickets of brush. Dense woodlands are found mainly in the valleys. Like the region of interior hills and plateaus, this region is a hinterland, little touched by modern development, but it supports a population of higher than average density.

RIVERS

The features of the river system form a remarkably uniform pattern from one end of the country to the other. Major drainage, with only one exception—the middle reaches of the Cavalla River—is toward the southwest, perpendicular to the coast. The river basins are narrow, and the major streams are spaced at regular intervals across the country. There are no well developed valleys or flood plains. Except close to the coast, gradients are fairly steep and irregular, and the flow is over bedrock.

Although all of the main rivers carry a good volume of water the year round, there are great variations in flow because of the rainfall pattern and rapid runoff in the watersheds. Most streams overflow their banks after the heavy downpours that mark the rainy season. Even at their highest, however, the rivers are not navigable for any important distance inland because of rapids. In their middle and upper reaches, tortuous channels, small islands and rock-strewn rapids prevent much use of the rivers for any great distance, even by canoes, and make it impracticable to float logs to the coastal areas. On the whole, the country's rivers are obstacles to, rather than lines of, travel.

The Mano River, which rises in the Guinea Highlands and flows southwest through the Gola Forest, is the most western of the country's main rivers. Downstream from a point about 80 km (50 airline mi) from the coast, it forms the border with Sierra Leone. Its estuary, running northwest behind a narrow spit of land for about 8 km (5 mi), is of little use for navigation. The Loffa River roughly parallels the Mano River at distances varying from about 16 to 32 km (10 to 20 mi) to the southeast. The lower reaches are much braided and obstructed by rapids to within some 32 km (20 mi) of the sea.

The Saint Paul River, which reaches the sea about 40 km (25 mi) southeast of the Loffa estuary, was an important axis of penetration inland by the American settlers at Monrovia in the early days. Its headwaters are in Guinea, in the mountains east of Macenta. After forming part of the border between Guinea and Liberia, it enters Liberia and follows a southwesterly course. Until the river crosses wholly into Liberia, it is known as the Diani or Nianda River. Its exit to the ocean about 11 km (7 mi) north of Monrovia is badly obstructed, but upstream from there the river is navigable by small craft for about 25 km (18 mi) to the first rapids at White Plains.

The Farmington River, a comparatively short stream that rises in the center of the country, empties into the sea about 58 km (35 mi) southeast of the mouth of the Saint Paul River. The estuary is formed by the confluence of the Farmington River, the Junk River, which joins from the west on a course parallel to and a few miles inland from the shore, and the Gbage (Little Bassa) River, which comes in from the east on a similar course. The Farmington River is the only inland waterway in the country that has much commercial importance. From Harbel, about 24 km (15 mi) upstream, the Firestone Plantations Company ships rubber by shallow-draft diesel lighters to the sea and thence along the coast to Monrovia.

The Saint John River drains a somewhat wider basin than the country's other rivers. Rising in Guinea northwest of the Nimba Mountains, it flows southwest and reaches the sea about 40 km (25 mi) southeast of the Farmington estuary. Its upper reach is known as the Mani River and forms part of the border with Guinea.

The Cess (Cestos) River, about 80 km (50 mi) southeast of the Saint John River, is a counterpart of the latter but drains the eastern side of the Nimba Mountains. Its upper course, known as the Nuon River, forms part of the border with the Ivory Coast. Between the Cess River and Cape Palmas, some 241 km (150 mi) farther to the southeast, are a number of narrow basins that drain to the sea, but they are of not great length because of an irregular divide that heads them about 80 km (50 mi) inland and runs more-or-less parallel to the coast. From west to east in this area, the main streams are the Sangwin, Sinoe, Dugbe, Dubo and Grand Cess Rivers.

Along the eastern side of the country, the right bank of the Cavalla (Cavally) River forms well over half the border with the Ivory Coast. Headwaters are in Guinea and

PRINCIPAL TOWN

Monrovia (capital), population (including Congotown) 171,580 in 1974; 208,629 in 1978.

Area and population		area		population
Counties	**Capitals**	sq mi	sq km	1984 census
Bong	Gbarnga	3,127	8,099	255,813
Grand Bassa	Buchanan	3,382	8,759	159,648
Grand Cape Mount	Robertsport	2,250	5,827	79,322
Grand Gedeh	Zwedru	6,575	17,029	102,810
Grand Kru	Barclayville			
Lofa	Voinjama	7,475	19,360	247,641
Margibi	Kakata	1,260	3,263	97,992
Maryland	Harper	2,066	5,351	132,058
Montserrado	Bensonville	1,058	2,740	544,878
Nimba	Saniquillie	4,650	12,043	313,050
Sinoe	Greenville	3,959	10,254	64,147
Territories				
Bomi	Tubmanburg	755	1,955	66,420
Rivercess	Rivercess City	1,693	4,385	37,849
TOTAL		38,250	99,067	2,101,628

Liberia
International boundary
County boundary*
National capital
County seat
Railroad
Road
*Internal boundaries shown do not reflect the addition of several counties since 1984.
0 25 50 Kilometers
0 25 50 Miles
SIERRA LEONE
GUINEA
IVORY COAST
NORTH ATLANTIC OCEAN
LOFA
BONG
NIMBA
GRAND CAPE MOUNT
MONTSERRADO
GRAND BASSA
GRAND JIDE
SINO
MARYLAND
Monrovia
Voinjama
Zorzor
Noway Camp
Bomi Hills
Tubmanburg
Bong Town
Robertsport
Klay
Gbarnga
Palala
Totota
Kakata
Harbel
Gardnersville
Paynesville
Buchanan
River Cess
Greenville
Grand Cess
Harper
Cape Palmas
Plibo
Nyaaka
Pyne Town
Zwedru
Tapeta
Ganta
Sanniquellie
Yekepa
Diecke
Yomou
Nzérékoré
Lola
Macenta
Guéckédou
Beyla
Kailahun
Pendembu
Sefadu
Panguma
Kenema
Joru
Zimmi
Sulima
Borotou
Touba
Biankouma
Man
Danané
Toulepleu
Guiglo
Duékoué
Taï
Lake Piso
Lofa
Gbeya
Saint Paul
Saint John
Cestos
Dube
Cavally
Nuon
Mani
Nianda
Via
Diani
Makona
Mano
Moa
Sewa
Meli
Milo
Dion
Gouan
Bafing
Nzo
Tienba
Pampana
10
8
6
4
Base 800638 (A02839) 5-87

the Ivory Coast northeast of the Nimba Mountains. It reaches the Gulf of Guinea about 24 km (15 mi) east of Cape Palmas. Navigable to its first rapids about 80 km (50 mi) upstream, it is used to some extent by the Firestone Cavalla Plantation and the people along the Liberian side.

LIBYA

BASIC FACTS

Official Name: Socialist People's Libyan Arab Jamahiriya

Abbreviation: LBY

Area: 1,775,500 sq km (685,524 sq mi)

Area—World Rank: 16

Population: 3,956,211 (1988) 7,292,000 (2000)

Population—World Rank: 99

Capital: Tripoli

Boundaries: 6,032 km (3,748 mi) Egypt 1,115 km (693 mi); Sudan 383 km (238 mi) Chad 1,054 km (655 mi); Niger 354 km (220 mi) Algeria 982 km (610 mi); Tunisia 459 km (285 mi)

Coastline: 1,685 km (1,047 mi)

Longest Distances: 1,989 km (1,236 mi) SE-NW; 1,502 km (933 mi) NE-SW

Highest Point: Bette 2,267 km (7,438 ft)

Lowest Point: Sabkhat Ghuzzayil 47 m (154 ft) below sea level

Land Use: 1% arable; 8% meadows and pastures; 91% other

The Mediterranean coast and the Sahara Desert are Libya's most prominent natural features. There are several highlands but no true mountain ranges except in the empty southern desert near the Chad border, where the Tibesti Massif rises to over 3,000 m (9,842 ft). A relatively narrow coastal strip and highland steppes immediately south of it are the most productive agricultural regions. Still farther southward a pastoral zone of sparse grassland gives way to the vast Sahara Desert, a barren wasteland of rocky plateaus and sand. It supports minimal human habitation, and agriculture is possible only in a few scattered oases.

Between the productive lowland agricultural zones lies the Gulf of Sidra, where along the coast for a stretch of 500 km (310 mi) of hopeless wasteland the desert extends northward to the sea. This barren zone, known as the Sirtica, has great historical significance. To its west the area known as Tripolitania has characteristics and a history similar to those of nearby Tunisia, Algeria, and Morocco. It is frequently considered with these states to constitute a supranational region called the Maghrib. To the east the area known historically as Cyrenaica has been closely associated with the Arab states of the Middle East. In this sense the barren Sirtica marks the dividing point between the North African Maghrib and the Middle Eastern Mashriq.

Along the shore of Tripolitania for more than 300 km (186 mi) coastal oases alternate with sandy areas and lagoons. Inland from these lies the Jifarah Plain, a triangular area of some 15,000 sq km (5,791 sq mi). About 120 km (74 mi) inland the plain terminates in an escarpment that rises to form the Jabal Nafusah, a plateau with elevations of up to 1,000 m (3,280 ft).

In Cyrenaica there are fewer coastal oases, and the Marj Plain—the lowland area corresponding to the Jifarah Plain of Tripolitania—covers a much smaller area. It forms a crescent about 210 km (130 mi) long between Benghazi and Darnah and extends inland a maximum of less than 50 km (31 mi). Elsewhere along the Cyrenaican coast the precipice of an arid plateau reaches to the sea. Behind the Marj Plain the terrain rises abruptly to form Jabal al Akhdar (Green Mountain), so called because of its leafy

PRINCIPAL TOWNS (population at 1973 census)

Town	Population	Town	Population
Tripoli (capital)*	481,295	Darna	30,241
Benghazi	219,317	Sebha	28,714
Misurata	42,815	Tubruq (Tobruk)	28,061
Az-Zawia (Azzawiya)	39,382	Al-Marj	25,166
Al-Beida	31,796	Zeleiten (Zliten)	21,340
Agedabia	31,047		

*In January 1987 Colonel Qaddafi designated Hun, a town 650 km to the south-east of Tripoli, as the administrative capital of the country.

Area and population

Baladiyāt	Capitals	area: sq mi	area: sq km	population: 1984 census
Ajdābiyā	Ajdābiyā	. . .	. . .	100,547
Awbāri	Awbāri	. . .	. . .	48,701
al-ʿAziziyah	al-ʿAziziyah	. . .	. . .	85,068
Banghāzi	Banghāzi	. . .	. . .	485,386
Darnah	Darnah	. . .	. . .	105,031
al-Fataḥ	al-Marj	. . .	. . .	102,763
Ghadamis	Ghadamis	. . .	. . .	52,247
Gharyān	Gharyān	. . .	. . .	117,073
al-Jabal al-Akhḍar	al-Baydā'	. . .	. . .	120,662
al-Khums	al-Khums	. . .	. . .	149,642
al-Kufrah	al-Kufrah	. . .	. . .	25,139
Marzuq	Marzuq	. . .	. . .	42,294
Miṣrātah	Miṣrātah	. . .	. . .	178,295
Nigāt al-Khums	Zuwārah	. . .	. . .	181,584
Sabhā	Sabhā	. . .	. . .	76,171
Sawfajjīn	Banī Walīd	. . .	. . .	45,195
ash-Shāti	Birāk	. . .	. . .	46,749
Surt	Surt	. . .	. . .	110,996
Ṭarābulus	Tripoli (Ṭarābulus)	. . .	. . .	990,697
Tarhunah	Tarhunah	. . .	. . .	84,640
Ṭubruq	Ṭubruq	. . .	. . .	94,006
Yafran	Yafran	. . .	. . .	73,420
az-Zāwiyah	az-Zāwiyah	. . .	. . .	220,075
Zlīṭan	Zlīṭan	. . .	. . .	101,107
TOTAL		685,524	1,775,500	3,637,488

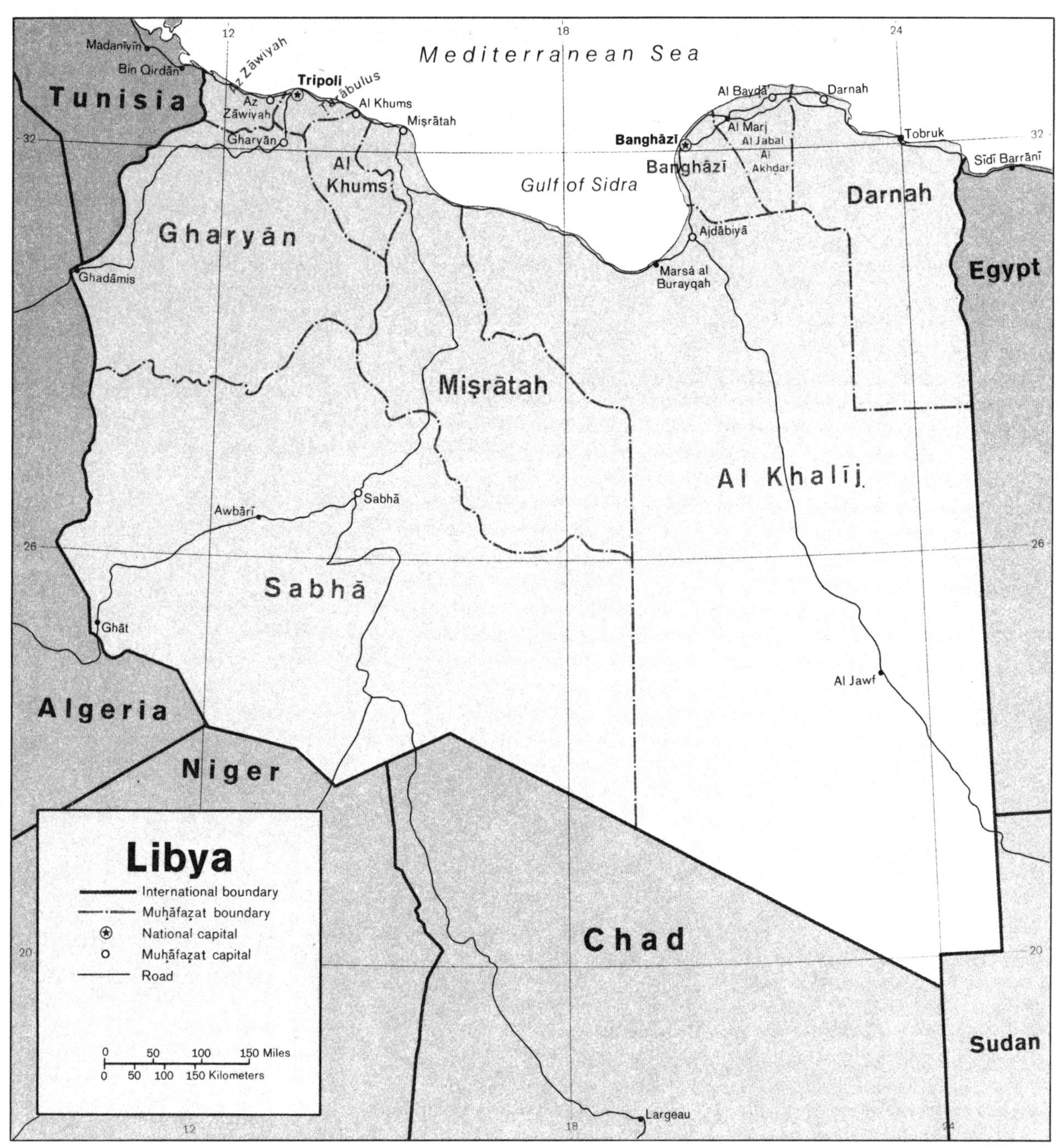

Base 501566 4-73

cover of pine, juniper, cypress, and wild olive. It is a limestone plateau with maximum altitudes of about 900 m (2,952 ft).

From Jabal al Akhdar, Cyrenaica extends southward across a barren grazing belt that gives way to the Sahara Desert, which extends still farther southwest across the Chad frontier. Unlike Cyrenaica, Tripolitania does not extend southward into the desert. The southwestern desert, known as Fezzan, was administered separately during both the Italian regime and the federal period of the Libyan monarchy. In 1969 the revolutionary government officially changed the regional designation of Tripolitania to Western Libya, of Cyrenaica to Eastern Libya, and of Fezzan to Southern Libya; but the old names were intimately associated with the history of the area, and during the 1970s they continued to be used frequently. Cyrenaica comprises 51 percent, Fezzan 33 percent, and Tripolitania 16 percent of the country's area.

LIECHTENSTEIN

BASIC FACTS

Official Name: Principality of Liechtenstein

Abbreviation: LIE

Area: 160 sq km (62 sq mi)

Area—World Rank: 198

Population: 27,825

Population—World Rank: 194

Capital: Vaduz

Land Boundaries: 76 km (47.2 mi) Austria 34.9 km (21.7 mi); Switzerland 41.1 km (25.5 mi)

Coastline: Nil

Longest Distances: 28 km (17.4 mi) N-S; 11.3 km (7 mi) E-W

Highest Point: Vorder Grauspitz 2,599 m (8,527 ft)

Lowest Point: Ruggeller Riet 430 m (1,411 ft)

Land Use: 25% arable; 38% meadows and pastures; 19% forest and woodland; 18% other

The fourth smallest country in Europe, Liechtenstein is a landlocked country, roughly triangular in shape, situated in the Rhine River Valley.

One third of the country lies in the upper Rhine valley, occupying a narrow slice of land. The remainder of the country is part of the Alps mountain range that runs east to west through Switzerland. The greatest elevation is Grauspitz 2,392 m, (7,848 ft) in a spur of the Rhaetian Alps.

Principal Towns (estimated population in 1986): Vaduz (capital) 4,920; Schaan 4,757; Balzers 3,477; Triesen 3,180; Eschen 2,844; Mauren 2,713; Triesenberg 2,277.

Area and population

Communes	area sq mi	area sq km	population 1986 estimate
Balzers	7.6	19.6	3,477
Eschen	4.0	10.3	2,844
Gamprin	2.4	6.1	907
Mauren	2.9	7.5	2,713
Planken	2.0	5.3	290
Ruggell	2.9	7.4	1,362
Schaan	10.4	26.8	4,757
Schellenberg	1.4	3.5	672
Triesen	10.2	26.4	3,180
Triesenberg	11.5	29.8	2,277
Vaduz	6.7	17.3	4,920
TOTAL	61.8	160.0	27,399

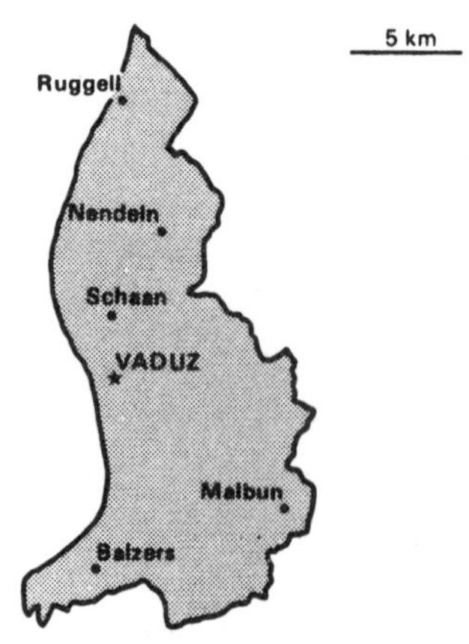

LUXEMBOURG

BASIC FACTS

Official Name: Grand Duchy of Luxembourg

Abbreviation: LUX

Area: 2,586 sq km (999 sq mi)

Area—World Rank: 155

Population: 366,232 (1988) 372,000 (2000)

Population—World Rank: 148

Base 504738 (546430) 12-81

Capital: Luxembourg

Land Boundaries: 356 km (221 mi); West Germany 135 km (84 mi); France 73 km (45 mi); Belgium 148 km (92 mi)

Coastline: Nil

Longest Distances: 82 km (51 mi) N-S; 57 km (35 mi) E-W

Highest Point: Buurgplaatz 559 m (1,834 ft)

Lowest Point: 427 m (130 ft)

Land Use: 24% arable; 1% permanent crops; 20% meadows and pastures; 21% forest and woodland; 34% other

Luxembourg is a landlocked country in Western Europe.

Luxembourg consists of two distinct geographical regions: the rugged uplands of the Ardennes in the North, with elevations ranging from 396 to 549 m (1,300 to 1,800 ft) above sea level; and the fertile lowlands of the South called Bon Pays, with a mean altitude of 229 m (750 ft) above sea level. Crisscrossing the area are deep valleys. Most rivers drain into the Sauer, which, in turn, flows into the Moselle on the eastern border. The northern region, comprising one-third of the country, is heavily forested. Along the Moselle is a very fertile wine-growing region. The soil is generally sandy, with sandstone and limestone prevalent.

Principal Towns (population at census of 31 March 1981): Luxembourg-Ville (capital) 78,900; Esch-sur-Alzette 25,100; Differdange 16,700; Dudelange 14,100; Petange 12,100.

Area and population	area		population
Districts / **Cantons**	sq mi	sq km	1986 estimate
Diekirch	447	1,157	54,420
Clervaux	128	332	9,710
Diekirch	92	239	22,390
Redange	103	267	10,500
Vianden	21	54	2,790
Wiltz	102	265	9,030
Grevenmacher	203	525	40,030
Echternach	72	186	10,990
Grevenmacher	82	211	16,910
Remich	49	128	12,130
Luxembourg	349	904	272,250
Capellen	77	199	28,790
Esch	94	243	112,250
Luxembourg (Ville et Campagne)	92	238	113,570
Mersch	86	224	17,640
TOTAL	999	2,586	366,700

MACAU

BASIC FACTS

Official Name: Macau

Abbreviation: MAC

Area: 16.9 sq km (6.5 sq mi)

Area—World Rank: 210

Population: 432,232 (1988) 837,000 (2000)

Population—World Rank: 144

Capital: Macau

Land Boundaries: China 0.34 km (0.2 mi)

Coastline: 41 km (25 mi)

Longest Distances: 5 km (3 mi) E-W; 1.6 km (1 mi) N-S

Highest Point: Coloane Alto 174 m (571 ft)

Lowest Point: Sea level

Land Use: 100% other

Macau is situated on the south coast of China, on the west side of the Canton River, almost directly opposite Hong Kong which is 56 km (35 mi) away. It consists of a peninsula and two small islands, Taipa and Coloane.

Area and population		area		population
Districts / **Parishes**	**Capital**	sq mi	sq km	1986 estimate
Marine Area	—	—	—	. . .
Islands		4.2	10.9	10,200
São Francisco Xavier (Coloane)	—	2.7	7.1	3,700
Nossa Senhora Carmo (Taipa)	—	1.5	3.8	6,500
Macau	Macau	2.4	6.1	416,200
Santo António	—	. . .	. . .	. . .
São Lázaro	—	. . .	. . .	. . .
São Lourenço	—	. . .	. . .	. . .
Sé	—	. . .	. . .	. . .
Nossa Senhora Fátima	—	. . .	. . .	. . .
TOTAL		6.5	16.9	426,400

MADAGASCAR

BASIC FACTS

Official Name: Democratic Republic of Madagascar

Abbreviation: MAD

Area: 587,041 sq km (226,658 sq mi)

Area—World Rank: 43

Population: 11,073,361 (1988) 15,550,000 (2000)

Population—World Rank: 61

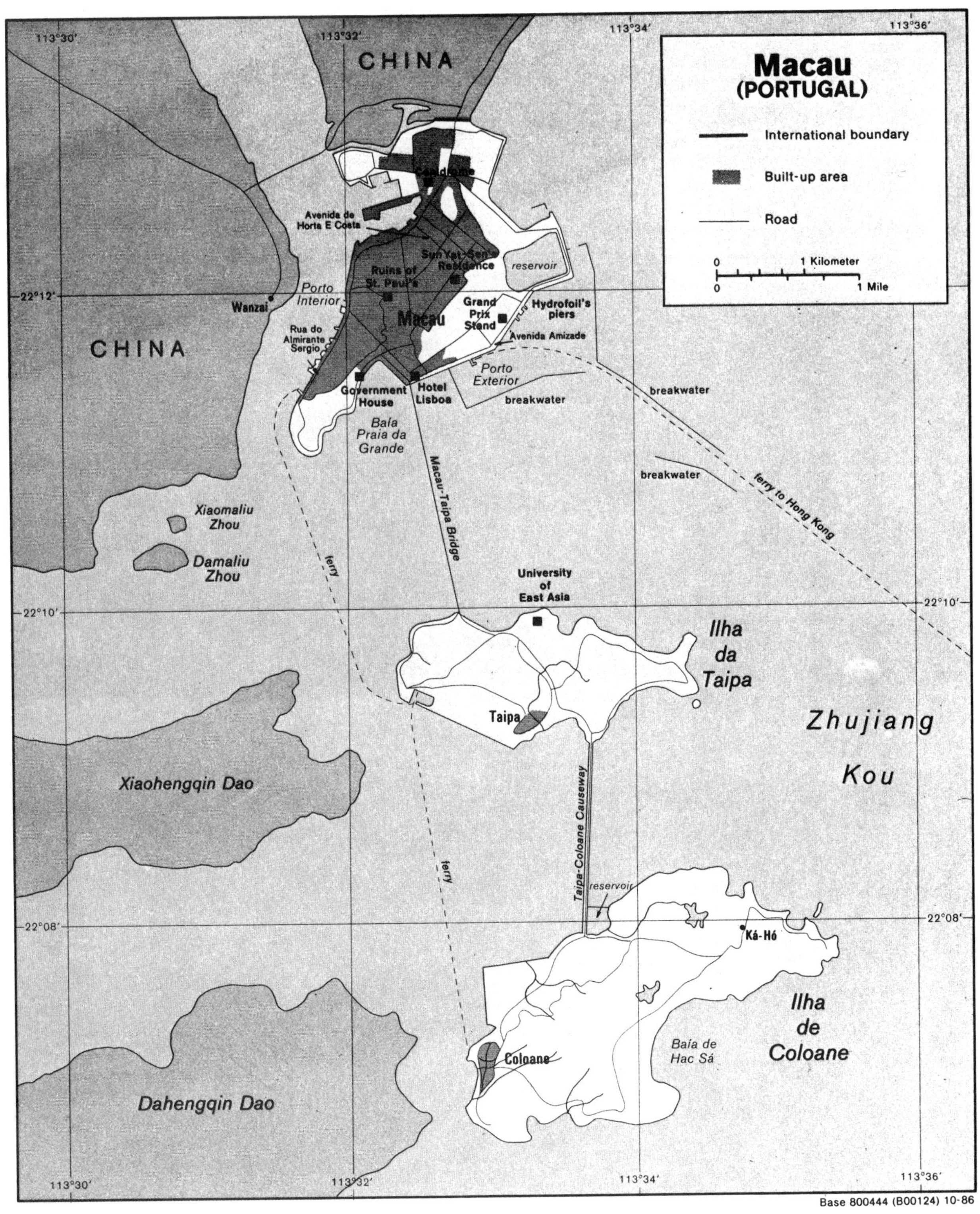
Macau
(PORTUGAL)
International boundary
Built-up area
Road
0
1 Kilometer
0
1 Mile
CHINA
CHINA
Avenida de Horta E Costa
Sun Yat-Sen's Residence
reservoir
Ruins of St. Paul's
Porto Interior
Wanzai
Macau
Grand Prix Stand
Hydrofoil's piers
Avenida Amizade
Rua do Almirante Sergio
Porto Exterior
Government House
Hotel Lisboa
breakwater
breakwater
breakwater
Baía Praia da Grande
Macau-Taipa Bridge
ferry to Hong Kong
Xiaomaliu Zhou
Damaliu Zhou
ferry
University of East Asia
Ilha da Taipa
Taipa
Zhujiang Kou
Xiaohengqin Dao
Taipa-Coloane Causeway
ferry
reservoir
Ká-Hó
Ilha de Coloane
Baía de Hac Sá
Coloane
Dahengqin Dao
113°30′
113°32′
113°34′
113°36′
22°12′
22°10′
22°08′
Base 800444 (B00124) 10-86

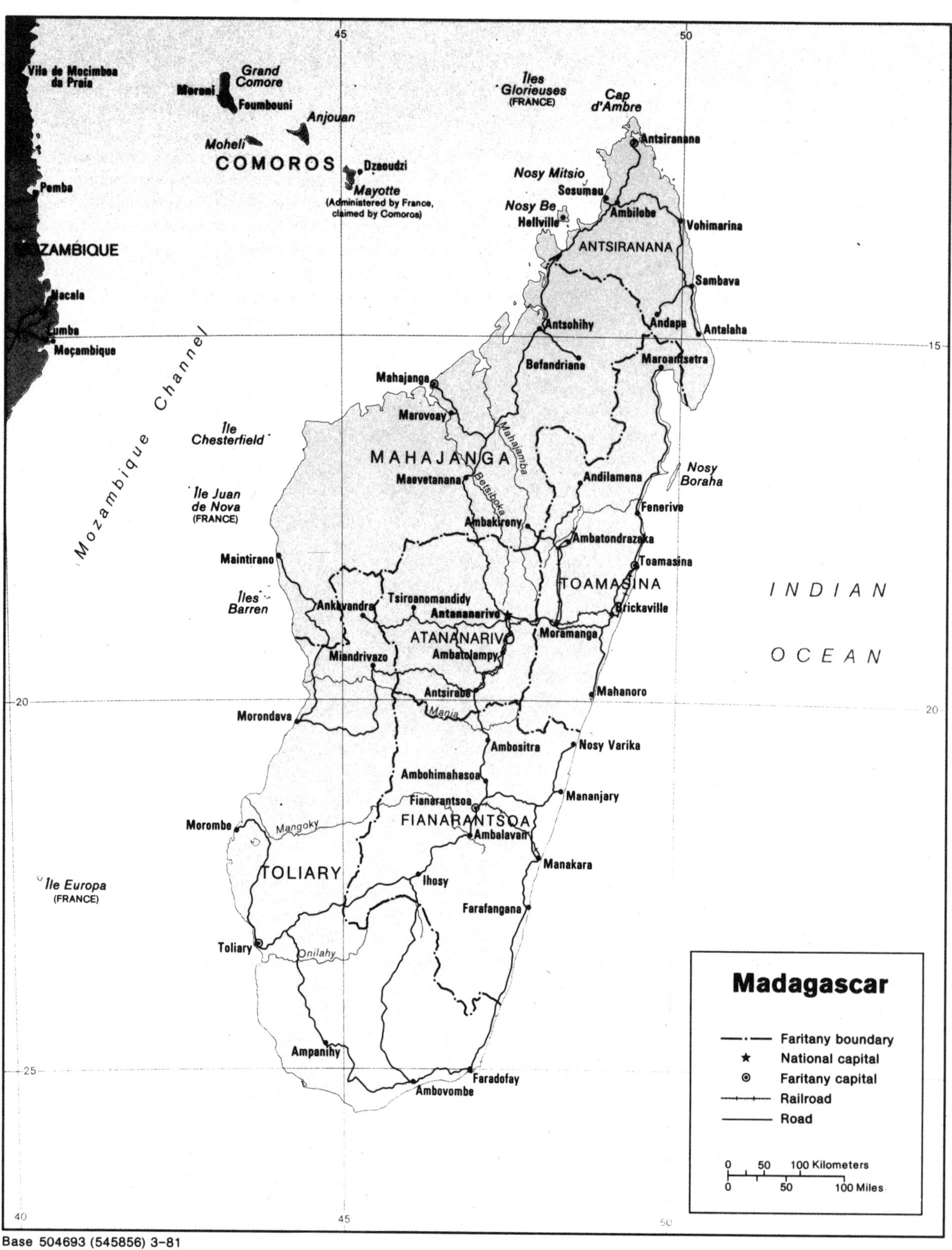

Base 504693 (545856) 3-81

Capital: Antananarivo

Land Boundaries: Nil

Coastline: 3,991 km (2,480 mi)

Longest Distances: 1,570 km (976 mi) NNE-SSW; 569 km (354 mi) ESE-WNW

Highest Point: Maromakotro 2,876 m (9,436 ft)

Lowest Point: Sea level

Land Use: 4% arable; 1% permanent crops; 58% meadows and pastures; 26% forest and woodland; 11% other

Madagascar is the fourth largest island in the world. Located in the Indian Ocean astride the forty-seventh meridian some 402 km (250 mi) east of southern Africa, the main island is about 1,609 km (1,000 mi) long and averages 563 km (350 mi) in width.

Landforms

Inside Madagascar's natural ocean boundaries, a broad, sparsely populated western coastal plain rises gradually to rolling foothills and low plateaus; the eastern plain is narrow and supports a comparatively dense population. Between these dissimilar plains stands a complex of mountain ranges, rounded hills, and open valleys, extending from southwest to northeast throughout most of the length of the island. Most of the higher elevations are located east of the southwest-northeast central axis, but they are generally referred to as the Central Highlands or, less accurately, as the high plateaus.

PRINCIPAL TOWNS (population at 1975 census)

Town	Population	Town	Population
Antananarivo (capital)	406,366	Mahajanga (Majunga)	65,864
Antsirabé	78,941	Toliary (Tuléar)	45,676
Toamasina (Tamatave)	77,395	Antsiranana (Diégo-Suarez)	40,443
Fianarantsoa	68,054		

The population of Antananarivo was estimated to be 662,585 in 1985.

The eastern slopes of these highlands rise abruptly above the narrow coastal plain—no more than ten miles wide in some areas—along the Indian Ocean. Mountain ranges above 1,219 m (4,000-ft) elevation stand within 96 km (sixty mi) of this shore, rising from the coastal plains in a series of escarpments separated by narrow, hilly plateaus or benches.

Many of the agricultural regions and population centers in the Central Highlands are at elevations between 1,219 and 1,524 m (4,000 and 5,000 ft) in a broad area of rolling grassy slopes, valleys, and numerous barren, rounded domes, overlooked by mountains rising in some cases above 2,438 m (8,000-ft) altitude. Monotonous landscapes of rounded and sometimes baretopped hills of approximately equal height cover wide areas, but there is nevertheless considerable diversity in the landforms, elevations, and geologic origins of the highlands. Westward, this elevated central area descends gradually into rolling foothills and dissected plateaus, then into swamps and estuaries along the coast and the Mozambique Channel.

Area and population

Provinces	Capitals	area sq mi	area sq km	population 1985 estimate
Antananarivo	Antananarivo	22,503	58,283	3,195,800
Antsiranana	Antsiranana	16,620	43,046	689,800
Fianarantsoa	Fianarantsoa	39,526	102,373	2,209,700
Mahajanga	Mahajanga	57,924	150,023	1,075,300
Toamasina	Toamasina	27,765	71,911	1,444,700
Toliara	Toliara	62,319	161,405	1,396,700
TOTAL		226,658	587,041	10,012,000

Much of the surface layer in the middle elevations and many prominent topographic features developed from the ancient granite structure underlying the island, which is similar to the substructure of the African continent. Most of the mountain ranges were apparently thrust upward by long-term lateral shifting in the foundation rock, but some of the highest mountains are of volcanic origin. Lava flows occurring in geologically recent times covered the older granite landforms in a few highland areas, providing the basis for some of the country's best soils. The Tsaratanana massif in northern Madagascar, surmounted at 2,880 m (9,450 ft) by the highest mountain on the island, is primarily granite but is partially covered by lava from eruptions in the immediate area. Rock and soil from other volcanic activity are associated with the Ankaratra massif in the center of the country.

Another structure apparently created by ancient shifting of the granite substructure is a rift valley lying between mountain ranges northeast of the city of Antananarivo. It is similar to the larger rift valleys of East Africa. Lake Alaotra, the largest of the lakes and swamps in this rift, lies below rock walls rising as much as 701 m (2,300 ft) above the lake surface.

On the western slopes lateral shifting has also lifted, tilted, and fractured old layers of sedimentary and alluvial rock and soil. Long-term erosion has worn down the resulting ridges, moving great quantities of decomposed rock and soil to the coastal plain.

Drainage

Rivers in the east are torrents rushing down steep slopes. Despite heavy forest growths, rainwater drains quickly to the main river channels and back to the ocean. There is some flow of water in these rivers even during the least rainy months (September to November), but the volume drops to about one-fifth that of the summer wet season. In delta areas alluvial deposits have been shifted by littoral currents, distorting river mouths.

On the western slopes of the uplands, typical river gradients are steep in some stretches and shallow in others. Some sections may be empty during the height of the western dry season in July or August. Swamps in flood plains or near stagnant sections of the river are green oases in a red-brown landscape as the dry season advances; they support spots or strips of mosses, reeds, grasses, and trees. Western rivers are sluggish on the broad coastal plain, and alluvial soil material carried down from the highlands is deposited behind the coast or in the deltas along the Mozambique Channel.

Streams are ephemeral in the arid south and southwest, flowing briefly after sporadic rains. Problems of low rainfall are compounded in some areas by the porous nature of limestone soils and substructure, which allow the rainfall to seep away in underground channels.

There are about nineteen lakes, of which only a few are of significant size. The highland area west of Tamatave is marked by a series of former lake basins, of which the largest is Lake Alaotra. These various basins, which support a relatively dense population, were formed by lava flows that partially filled and divided an existing rift valley.

MALAWI

BASIC FACTS

Official Name: Republic of Malawi

Abbreviation: MLW

Area: 118,484 sq km (45,747 sq mi)

Area—World Rank: 95

Population: 7,679,368 (1988) 11,631,000 (2000)

Population—World Rank: 78

Capital: Lilongwe

Land Boundaries: 2,768 km (1,720 mi); Tanzania 451 km (280 mi); Mozamobique 1,497 km (930 mi); Zambia 820 km (510 mi)

Coastline: Nil

Longest Distances: 853 km (530 mi) N-S; 257 km (160 mi) E-W

Highest Point: Sapitwa 3,002 m (9,849 ft)

Lowest Point: 37 m (120 ft)

Land Use: 25% arable; 20% meadows and pastures; 50% forest and woodland; 5% other

Malawi, an inland nation in southeastern Africa, is well within the southern tropics. Its territory extends north-south for 901 km (560 mi) at an average width of less than 161 km (100 mi) in a southern segment of the East African Rift Valley. Within its borders are 24,410 sq km (9,425 sq mi) of water area, mostly in Lake Malawi and two smaller lakes. Lake Malawi is the catch basin for runoff from most of the land in two of the country's three major administrative divisions, the Northern and Central regions, which lie on its western flank. The lake drains southward into the Shire River, which also collects surface drainage from most of the Southern Region. Thus, almost the entire land area and Lake Malawi are parts of one complex watershed, from which runoff waters flow across the southern border, via the Shire, to the Zambezi River.

A complex geologic history has contributed to the formation of a landscape of great diversity in elevations and relief features. Floodplains, marshes, hills, plateaus, escarpments, and mountains range from a few hundred feet above sea level in the lower valley of the Shire River to rugged peaks more than 2,590 m (8,500 ft) in elevation in several widely separated sections of the country.

Plateaus and Mountains

In terms of human habitation, Malawi's most important geographic features are the plateaus, which form three-fourths of the land area. The surface of most of these elevated plains has been eroded into low rolling hills and open, shallow valleys. Here and there such surfaces are interrupted by ranges of rugged hills or isolated peaks of resistant rock.

The Shire Plateau in the Southern Region, best known to European settlers as the Shire Highlands, covers about 7,251 sq km (2,800 sq mi). It is heavily cultivated, producing both commercial and subsistence crops. Areas under cultivation are interrupted by low escarpments and ranges of rocky hills, which are remnants of ancient erosion cycles. Altitudes vary from 762 to nearly 1,219 m (2,500 to 4,000 ft). Blantyre, Malawi's largest town, and Zomba stand in the western sections of this plateau, which slopes gently eastward toward Lake Chilwa.

A much broader plateau in the Central Region, covering about 23,309 sq km (9,000 sq mi) is commonly known as the Lilongwe Plain. It has numerous broad valleys and dambos (areas of moist soils on impermeable subsurface layers) separated by low rounded hills. Elevations vary between 762 and 1219 m (2,500 and 4,000 ft).

The Nyika Plateau in the Northern Region is the highest formation of this kind in Malawi. It covers some 23,309 sq km (9,000 sq mi) at elevations between 2,133 and 2,438 m (7,000 and 8,000 ft). It is sparsely populated and is less productive than the plateaus farther south.

The country has about six other plateaus, some of which are known locally as hills or plains. Whatever their names or differences in local landscapes and soils, most of them are extensive flat or rolling surfaces, between 762 and 1,371 m (2,500 and 4,500 ft) above sea level, close to the median altitude for all of Malawi.

A few groups or ranges of mountains rise above the level of the highest plateaus. In the North, several peaks on the Nyika Plateau reach 2,590 m (8,500 ft). None of the mountains in the Central Region exceeds Dedza Mountain, which is just over 2,255 m (7,400 ft) above sea level. In the Southern Region, Zomba Mountain, standing north of the town of Zomba, has peaks above 2,072 m (6,800 ft). The Mulanje Mountains (also called the Mulanje Plateau, or the Mulanje Massif) near the southeastern border is the highest mountain complex in the country. Various sections of several sq mi each are above 2,743 m (9,000 ft) and the highest pinnacle rises to 3,000 m (9,840 ft).

Lake Malawi and Its Shoreline

Lake Malawi, one of the largest and deepest lakes in the world, extends from north to south for more than 563 km (350 mi) occupying the floor of a major southern segment of the East African Rift Valley system. Its surface is

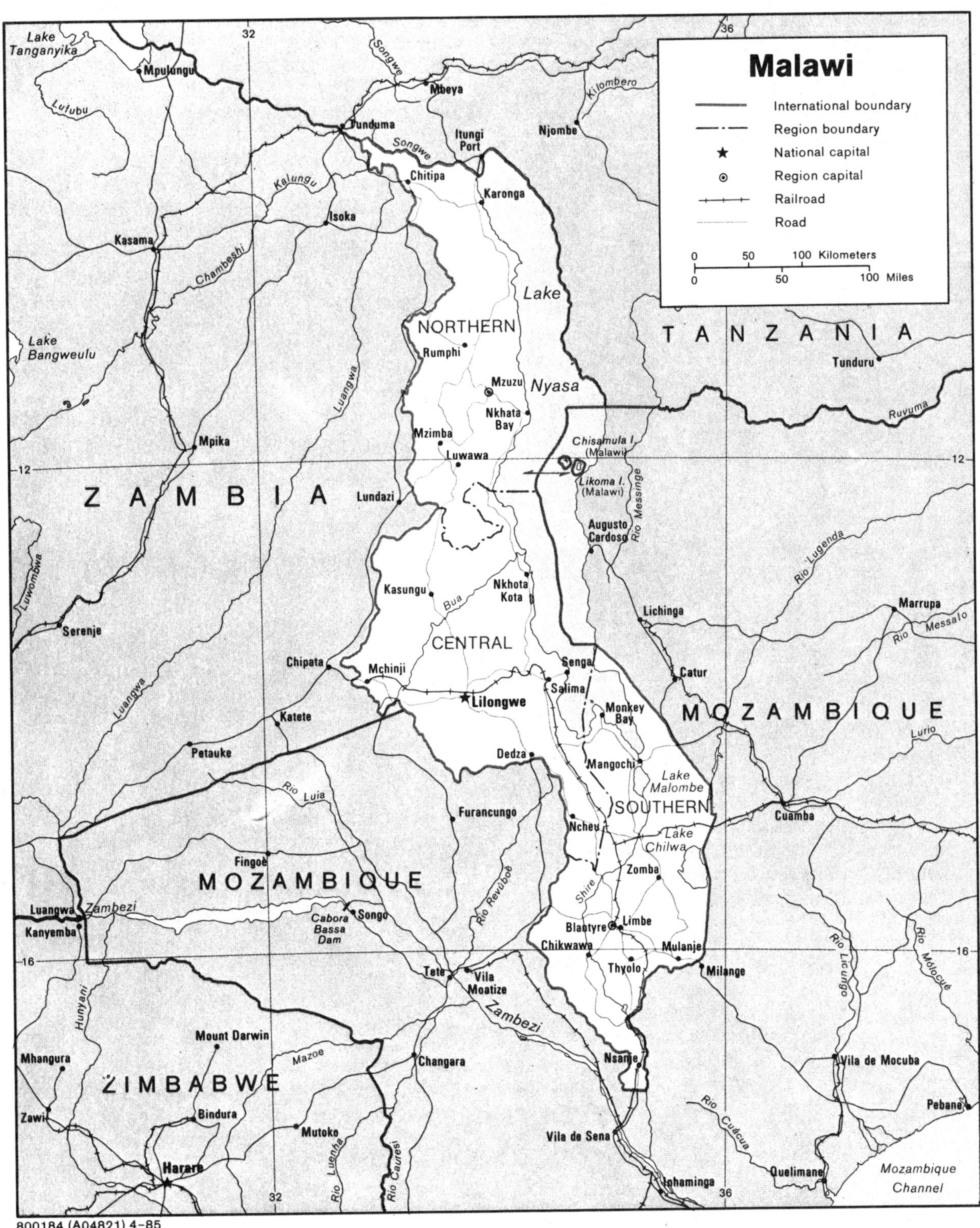

800184 (A04821) 4-85

Capital: Lilongwe, estimated population 75,000 (including suburbs) in 1976.
Other Principal Town: Blantyre, population 219,011 (incl. Limbe) at 1977 census.

Area and population

Regions Districts	Capitals	area sq mi	area sq km	population 1986 estimate
Central	Lilongwe	13,742	35,592	2,938,300
Dedza	Dedza	1,399	3,624	409,300
Dowa	Dowa	1,174	3,041	339,100
Kasungu	Kasungu	3,042	7,878	266,800
Lilongwe	Lilongwe	2,378	6,159	964,900
Mchinji	Mchinji	1,296	3,356	217,400
Nkhotakota	Nkhotakota	1,644	4,259	129,300
Ntcheu	Ntcheu	1,322	3,424	310,300
Ntchisi	Ntchisi	639	1,655	120,200
Salima	Salima	848	2,196	181,000
Northern	Mzuzu	10,398	26,931	815,000
Chitipa	Chitipa	1,353	3,504	90,500
Karonga	Karonga	1,141	2,955	134,500
Mzimba	Mzimba	4,027	10,430	378,300
Nkhata Bay	Nkhata Bay	1,579	4,090	133,000
Rumphi	Rumphi	2,298	5,952	78,700
Southern	Blantyre	12,260	31,753	3,525,600
Blantyre	Blantyre	777	2,012	522,500
Chikwawa	Chikwawa	1,836	4,755	248,500
Chiradzulu	Chiradzulu	296	767	225,600
Machinga	Machinga	2,303	5,964	437,200
Mangochi	Mangochi	2,422	6,272	386,800
Mulanje	Mulanje	1,332	3,450	611,300
Mwanza	Mwanza	886	2,295	91,700
Nsanje	Nsanje	750	1,942	139,300
Thyolo	Thyolo	662	1,715	411,800
Zomba	Zomba	996	2,580	450,900
TOTAL LAND AREA		36,400	94,276	
INLAND WATER		9,347	24,208	
TOTAL		45,747	118,484	7,278,900

about 472 m (1,550 ft) above sea level. As parts of the lake bottom lie more than 213 m (700 ft) below sea level, the average depth of the lake is about 685 m (2,250 ft); near the town of Nkhata Bay on the western shore, depths of 704 m (2,310 ft) have been recorded. In some coastal areas nearly perpendicular rock walls rise from the lake bottom to a height of 609 m (2,000 ft) above the surface—sheer cliffs rising more than a mile from the lake bed, of which less than one-half is visible above the average surface level.

Lesser cliffs and gentler slopes also are found at various places along the lake, interspersed with segments of flat lakeshore plains. These flat littoral plains vary from narrow coastal ribbons to widths of 24 km (15 mi) and are marked by swamps and dambos lying on impermeable clay and rock substructures. In some areas along the western shore, the land rises in a series of terraces to foothills. The hills lead upward to the plateaus and mountains of the western watershed, which drains into the lake and which also constitutes most of the land territory of the Northern and Central regions.

The Shire Valley

The huge trough or rift occupied by Lake Malawi continues southward from the present shoreline of the lake to the southern border of Malawi, where it becomes so shallow that it merges into the plains of Mozambique. This section of the rift, some 386 km (240 mi) long, is known as the Shire Valley.

In an earlier geologic era Lake Malawi extended nearly 129 km (80 mi) southward beyond its present limits. The land area exposed by the northward retreat of the lake comprises the upper Shire Valley, which slopes gently southward. Through this broad upper segment of the valley, the Shire River drains the overflow from Lake Malawi, passing through Lake Malombe, which occupies a shallow bowl, 29 km (18 mi) in length, in the valley floor. Various swamps extend east and west for considerable distances from the river bottom. Beyond them, flat-topped alluvial terraces—old lake shore remnants—rise in shallow tiers to foothills and escarpments.

The average gradient of the valley floor increases sharply south of the boundary of the ancient lake, descending more than 365 m (1,200 ft) during the next 96 km (sixty mi). This middle segment of the valley is a rough transition zone of rocky hills and ledges, having little in common with either the upper or lower segments of the valley. In some places it is no more than one mile wide. As soils are thin and stony, this part of the valley supports only a sparse population. In this area the waters of the Shire River race through the Murchison Rapids, the collective name for a series of rocky cataracts extending intermittently over most of this central section of the river.

The lower Shire Valley extends from the lowest of these rapids to the southern border, a distance of about 177 km (110 mi). This very flat segment of the valley is the lowest land area in the country. It slopes gently southward, descending only about 70 m (230 ft)—from a 107 m (350-ft) elevation at the foot of the last rapids to 36 m (120 ft) near the southern border crossing. With such a minimal gradient, the current in this segment is very slow.

Floodplains and riverine swamps cover extensive areas, the largest of which is the Elephant Marsh, which is about 64 km (forty mi) long and as much as 14 km (nine mi) wide. Beyond the broad floodplains, most of the immediate boundaries of the valley are little more than low hills or eroded escarpment walls of limited height. Two other major valleys join the lower Shire Valley: the Mwanza, from the west; and the Ruo, which collects runoff waters from much of the Shire Highlands, from the east. Topsoil carried down from these valleys and from most of the Southern Region settles out of the sluggish floodwaters of the lower Shire, enriching floodplains and marshes. Some of these have been converted to farming and have supported a relatively dense population for centuries.

Drainage

Malawi is a part of the Zambezi River basin, which includes the southern segment of the series of large-scale geologic faults known as the East African Rift System. In this segment, the rift is a broad, elongated bowl, and Lake Malawi occupies a long section of its lower depths. The part of the rift just south of the present-day shores of Lake Malawi was also formerly a part of the lake, and Lake Malombe is a shallow remnant of this larger ancient lake bed. The Shire River drains the overrun from Lake Malawi, flowing southward across the old lake bed, through Lake Malombe, and then continuing southward toward the Zambezi River. West of Lake Malawi, the western border of the country follows the top of the watershed ridgeline at the western edge of this segment of the rift system. Approximately eight sizable rivers and hundreds of smaller streams carry surface runoff from the plateaus and hills of the Northern and Central regions toward Lake Malawi. Most of them empty into the lake, but a few sink into the sands of the flat lakeshore plains. Some streams are intermittent, becoming dry during the latter months of the annual dry season. The Lilongwe River, for example, is dry for nearly one month each year, and its average flow throughout the dry season is less than 5 percent of its average rainy season volume.

As runoff from most of the Southern Region feeds into tributaries of the Shire, adding their volume to that of the runoff from Lake Malawi and its catch basin, surface runoff from almost the entire country is carried away by this single river. It flows slowly in the flat plains just south of Lake Malawi. West of Zomba the current increases greatly for about 96 km (60 mi), as the river descends a series of escarpments of a much lower altitude. Numerous rapids and cataracts in this segment necessitated a long portage by nineteenth century explorers and settlers and have continued to restrict the use of the river for transportation during more recent times.

South of this rough central section, the lower segments of the river are sluggish. The Shire joins the Zambezi some 64 km (40 mi) south of the Malawi border, and its waters are carried eastward across Mozambique to the Indian Ocean.

An internal drainage basin near the southeastern border is the only area of the country from which surface runoff does not flow into streams that eventually lead to the Shire River and therefore to the Zambezi. This area of a few hundred sq mi drains into Lake Chilwa, a complex of lakes and marshes that has no outlet to the sea.

MALAYSIA

BASIC FACTS

Official Name: Malaysia

Abbreviation: MLY

Area: 330,434 sq km (127,581 sq mi)

Area—World Rank: 61

Population: 16,398,306 (1988) 23,271,000 (2000)

Population—World Rank: 45

Capital: Kuala Lumpur

Boundaries: 2,685 km (1,667 mi); Thailand: 506 km (314 mi); Singapore 16 km (10 mi); Indonesia 1,782 km (1,106 mi); Brunei 381 km (236 mi)

Coastline: Peninsular Malaysia 1,931 km (1,200 mi); Sarawak 888 km (552 mi); Sabah 1,558 km (968 mi)

Longest Distances: Peninsular Malaysia: 748 km (465 mi) SSE-NNW and 322 km (200 mi) ENE-WSW; Sarawak 679 km (422 mi) NNE-SSW and 254 km (158 mi) ESE-WNW Sabah: 412 km (256 mi) E-W and 328 km (204 mi) N-S

Highest Point: Mount Kinabalu 4,101 m (13,455 ft)

Lowest Point: Sea level

Land Use: 3% arable; 10% permanent crops; 63% forest and woodland; 24% other

Malaysia is composed of two major segments divided by the southernmost portion of the South China Sea.

Peninsular Malaysia

The Malay Peninsula is a long, narrow continuation of a range of mountains that extend southward from Thailand and Myanmar and separate the South China Sea from the Andaman Sea. Peninsular Malaysia occupies the southern portion of the peninsula and is connected to Singapore at the southern tip by a man-made causeway. It is separated from the Indonesian island of Sumatra by the Strait of Malacca.

Peninsular Malaysia measures approximately 722 km (480 mi) in its longest north-south dimension and some 322 km (200 mi) at its widest east-west axis and has a total area of 131,313 sq km (50,700 sq mi). Penang, a short distance off the northwestern mainland, is an economically important island whose harbor is a busy trading port. Other islands, often of great scenic beauty, are bases for fishing craft.

The mountain chain that forms the backbone of the Malay Peninsula and contains the Cameron Highlands continues southward from Thailand for about 482 km (300 mi). Various peaks exceed 1,828 m (6,000 ft) and the highest mountain in the range, Gunong Korbu, rises to a height of 2,182 m (7,160 ft) in Perak. The range, about 48 to 64 km (30 to 40 mi) wide, divides Peninsular Malaysia into two unequal parts. The larger is to the east of the range, and the smaller but more populated area is to the west. A long spur extending from the main chain at a point below the border with Thailand, the Bintang Range, divides the valley of the Perak River from the basin of the Muda River. Numerous rugged limestone hills are found on both sides of the mountain range, and many, especially in the north, contain caves.

East of the main chain, and separated from it by valleys, is a secondary mountain highland that intersects the states of Kelantan, Terengganu, and Pahang. Generally lower than the main range, it nevertheless contains the highest elevation in Peninsular Malaysia, Gunong Tahan 2,184 m (7,168 ft).

To the west, south, and east of the mountains are the coastal lowlands. The most significant of these, from the viewpoints of continuity, ease of communications, and number of inhabitants, is on the west coast. The width varies from roughly 16 to 80 km (ten to fifty mi) and the land is generally level to rolling. Few hills attain an elevation of more than 152 m (500 ft) but the characteristics of the slopes change abruptly near the mountains. The lowland on the east coast is more irregular, less densely populated, and less important in productivity and commerce. In the north this plain is some 64 km (forty mi) in width and tapers to no more than five mi as the range along the east coast approaches the sea; farther south it widens, but swamps and lagoons extending as far as 32 km (20 mi) inland reduce the importance of the area for settlement and economic usefulness. A plain with low hills at the southern end of the main range embraces western Pahang and interior areas of Negeri Sembilan and Johor (Johore).

The Pahang River, which extends for about 458 km (285 mi) is Peninsular Malaysia's longest river and is known by various names in different parts of its course. The Perak River, the second longest, extends for about 322 km (200 mi). Other rivers of significance are the Kelantan, Terengganu, and Endau, all flowing into the South China Sea and the Muar and the Muda, both flowing into the Strait of Malacca.

Most of the western rivers have relatively short courses. In their upper reaches many drop more than 1,219 m (4,000 ft) in less than 24 km (15 mi) before reaching the flat coastal plain. The extensive silting of the riverbeds near the sea limits navigation, but the light Malay outrigger sailing canoe can enter some estuaries for a few mi; above this, only canoes and rafts can be used. The silting also causes flooding of the poorly drained coastal areas and the formation of large swampy areas on the plains. The rivers flowing eastward are longer and have a more moderate gradient in their upper reaches. In most of the country drainage has an interlacing pattern, and hills are cut by small streams of gullies.

The western coastline is practically an unbroken succession of mangrove swamps and mudflats, but harbors have been developed in the sheltered areas. The Strait of Malacca in some ways resembles an inland sea. Port Kelang (formerly Port Swettenham), the chief port of Malaysia and near Kuala Lumpur, is the only port that does not require the use of lighters. Penang and Malacca are also important centers of navigation. Except for an occasional outcrop the east coast along the South China Sea is an almost continuous stretch of sand and surf. There is, however, a large swampy coastal stretch on the littoral of Pahang and northeast Johor. Since the east coast is exposed to heavy seas and high winds, coastal navigation is limited; sandbars block rivers during much of the year, and there are no important harbors.

Sarawak and Sabah

Sarawak and Sabah occupy the northern part of Borneo. The two states are contiguous and are about 1,078 km (670 mi) long, have a maximum width of 257 km (160 mi) and a total area of approximately 202,019 sq km (78,000 sq mi). Brunei, which chose not to become part of Malaysia in 1963, occupies two enclaves in northern Sarawak.

PRINCIPAL TOWNS (population, excluding underenumeration, at 1980 census)

Town	Population	Town	Population
Kuala Lumpur (capital)	919,610	Taiping	146,002
Ipoh	293,849	Seremban	132,911
George Town (Penang)	248,241	Kuantan	131,547
Johore Bahru	246,395	Kota Kinabalu	108,725
Petaling Jaya	207,805	Melaka (Malacca)	87,494
Kelang	192,080	Sibu	85,231
Kuala Trengganu	180,296	Butterworth	77,982
Kota Bahru	167,872	Kuching	72,555
		Sandakan	70,420

Area and population

Regions / States	Capitals	area: sq mi	area: sq km	population: 1985 estimate
East Malaysia				
Sabah	Kota Kinabalu	28,460	73,711	1,222,718
Sarawak	Kuching	48,050	124,449	1,477,428
West Malaysia				
Johor	Johor Baharu	7,330	18,985	1,867,333
Kedah	Alor Setar	3,639	9,425	1,263,155
Kelantan	Kota Baharu	5,765	14,931	1,048,420
Melaka	Melaka	640	1,658	524,028
Negeri Sembilan	Seremban	2,565	6,646	647,159
Pahang	Kuantan	13,884	35,960	921,360
Pinang	Pinang	398	1,031	1,049,282
Perak	Ipoh	8,110	21,005	2,020,135
Perlis	Kangar	307	795	166,948
Selangor	Shah Alam	3,072	7,956	1,731,090
Terengganu	Kuala Terengganu	5,002	12,955	638,830
Federal Territory				
Kuala Lumpur	—	94	243	1,103,228
TOTAL LAND AREA		127,317	329,750	15,681,114
INLAND WATER		264	684	
TOTAL AREA		127,581	330,434	

The coastline of some 2,253 km (1,400 mi) is rather regular in Sarawak but becomes broken and deeply indented in Sabah. The land border with Kalimantan (Indonesian Borneo), which occupies the rest of the island, is some 1,448 km (900 mi) long. Because of the wild, rugged, and unexplored country through which it passes, the border has never been surveyed; and because the area is constantly under cloud cover, aerial photography has not succeeded in determining a definite line. Indigenous peoples of the area are unaware of any boundary and move freely from one region to another.

The topography of Sarawak shows a flat coastal plain followed by a narrow belt of hills, then a sharp rise to the mountainous mass that borders Kalimantan. The mountainous interior is an area of irregular masses of dissected highlands; of almost unexplored ravines, gorges, and plateaus; and of complex, unconnected ranges with a mean elevation of about 1,524 m (5,000 ft) and with occasional peaks over 2,134 m (7,000 ft). Gunong Murud, near the junction of Sarawak, Sabah, and Kalimantan, is the state's loftiest mountain at 2,423 m (7,950 ft).

The hills that extend the full length of Sarawak are usually less than 305 m (1,000 ft) above sea level but are broken by a few mountain groups of about 762 m (2,500 ft); some of the mountains extend to the coast and terminate as sea-formed cliffs. Notable of this kind of formation are the Santubong Mountains that jut into the sea just

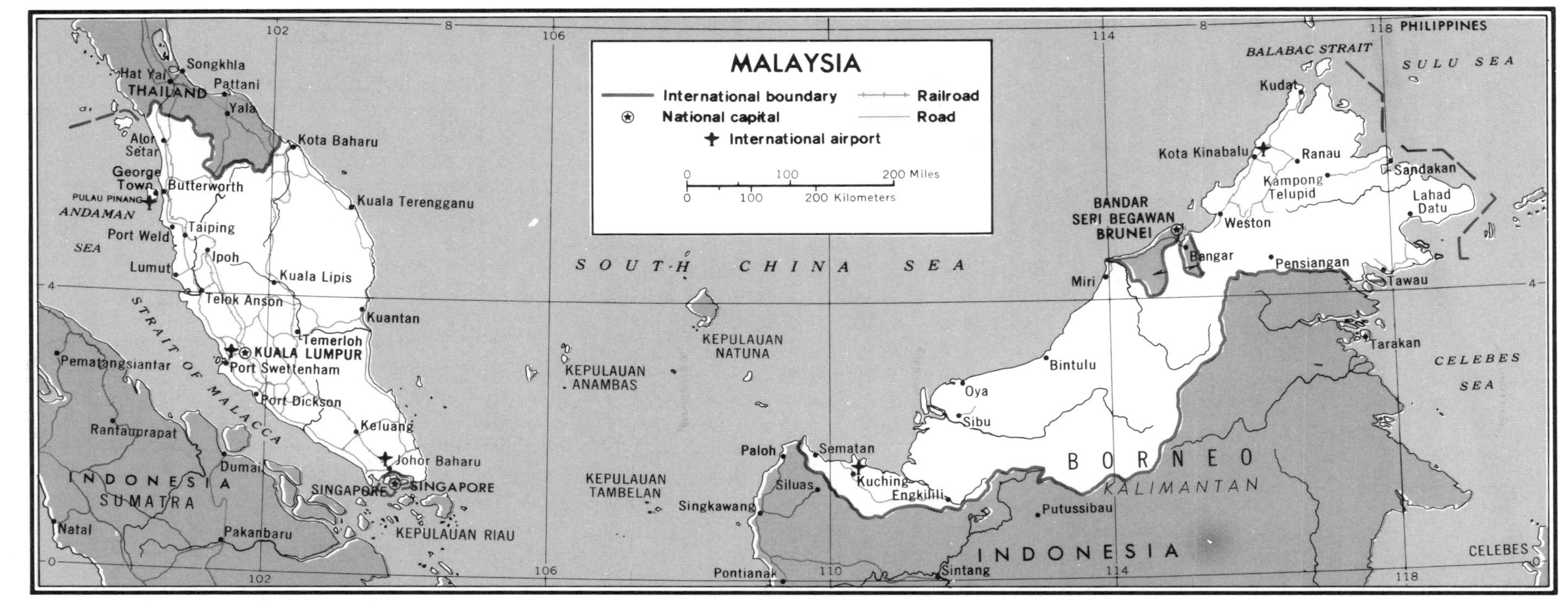
MALAYSIA
International boundary
National capital
International airport
Railroad
Road
0 100 200 Miles
0 100 200 Kilometers
SOUTH CHINA SEA
STRAIT OF MALACCA
ANDAMAN SEA
SULU SEA
CELEBES SEA
BALABAC STRAIT
PHILIPPINES
THAILAND
Hat Yai
Songkhla
Pattani
Yala
Kota Baharu
Alor Setar
George Town
PULAU PINANG
Butterworth
Kuala Terengganu
Port Weld
Taiping
Ipoh
Lumut
Kuala Lipis
Telok Anson
Kuantan
Temerloh
KUALA LUMPUR
Port Swettenham
Port Dickson
Keluang
Johor Baharu
SINGAPORE
SINGAPORE
KEPULAUAN RIAU
Pematangsiantar
Rantauprapat
Dumai
INDONESIA
SUMATRA
Natal
Pakanbaru
KEPULAUAN ANAMBAS
KEPULAUAN NATUNA
KEPULAUAN TAMBELAN
Paloh
Sematan
Siluas
Kuching
Engkilili
Singkawang
Pontianak
Sintang
Oya
Sibu
Bintulu
Miri
BORNEO
KALIMANTAN
Putussibau
INDONESIA
BANDAR SERI BEGAWAN
BRUNEI
Bangar
Kota Kinabalu
Weston
Kudat
Ranau
Kampong Telupid
Pensiangan
Sandakan
Lahad Datu
Tawau
Tarakan
CELEBES
102
106
110
114
118
8
4
0

north of the capital city of Kuching. These short ranges lend a rugged character to the terrain, with steep hills and narrow valleys or gorges where the rivers' rapids have cut through.

The coastal region is a flat alluvial plain several feet above sea level. The plain is often swampland, and its width varies from less than one mile at Miri in the North to over 160 km (100 mi) at some points farther south but is generally 32 to 64 km (20 to 40 mi) wide. The alluvial deposits are mud or a deep and extensive mantle of peat, and the soil is unsuitable for agriculture because of poor drainage and because about one-third of the area is subject to regular saltwater tidal or freshwater river flooding. Some moderately productive paddy land can be found on the banks of river deltas.

The Sarawak coastal waters are very shallow and, coupled with the regular coastline, afford no significant harbors. Moreover the siltladen rivers deposit large amounts of alluvium, which forms great bars that effectively obstruct the passage of large vessels upstream. The higher elevations, where the hills extend to the sea, can support some farming and permanent settlements, but the muddy beaches fringed by thick mangroves and nipa palm swamps discourage economic activity elsewhere along the coast.

The rivers of Sarawak are numerous and of considerable volume because of the abundance of rainfall in the area. They all rise in the interior and descend through gorges until they reach the flat coastal plain, where they change character completely and meander to the sea. Despite the rapid drop only one significant waterfall has been discovered, however, and there has been little opportunity for the development of hydroelectric power.

Major rivers of Sarawak include the Rajang and the Lupar, both in the south. The Rajang, which is 583 km (350 mi) long, is the country's largest and probably most important river. It is navigable by small ocean vessels for 96 km (sixty mi) and by shallow-draft riverboats for 241 km (150 mi) from its mouth. The Lupar extends inland for 228 km (142 mi). Farther north the Baram River, 402 km (250 mi) long, drains the north-central area and empties into the sea just north of Miri; the Limbang River flows for 195 km (122 mi) between and beyond the two enclaves of Brunei. These and other rivers are the only effective avenues for inland travel. Jungle paths connect their headwaters, and a few trails pass over the watershed into Kalimantan.

The terrain in Sabah is superficially similar to that in Sarawak, but several important differences can be noted. The coastline, for example, becomes progressively more irregular from west to east until, in the east, facing the Sulu Sea, it is boldly and deeply indented. In the Brunei Bay area the Sunda Shelf, which accounts for the shallowness of the sea off Sarawak, reaches its northern limit so that offshore water deepens considerably, and eastern Sabah offers many excellent deepwater ports and anchorages. On the South China Sea, Labuan Island, 56 km (thirty-five mi) from the mainland entrance at Brunei Bay, has a sheltered harbor that is the major terminus for seaborne traffic destined for Sarawak and Brunei as well as Sabah.

The mountains of Sabah are also different from those of Sarawak. The interior mass, particularly where it borders Kalimantan, is the same complex arrangement of highly dissected ranges and occasional peaks of over 2,133 m (7,000 ft) but the ranges lie much closer to the sea and tend to be less complex. The coastal plain in the west along the South China Sea reaches inland only ten to twenty mi and then gives way abruptly to the only continuous range in Malaysian Borneo. This backbone, known as the Crocker Range, is really the southwestern extension of the chain that created the islands of the Philippines. It averages somewhat higher than elevations in Sarawak, particularly in its northern reaches, where spectacular Gunong Kinabalu, 4,101 m (13,455 feet,) towers over everything else in the country; no other peak in Sabah reaches 2,438 m (8,000 ft).

East of the Crocker Range a series of parallel but lower ranges extend from the elevated interior core to the Sulu Sea. Wide valleys separate the ranges, and the sea has formed long, deep bays along the east coast. The ridges form a series of peninsulas, including Kudat and Benkoka, which enclose Maruda Bay in the extreme north, and Sandakan, Dent, and Simporna, which define the bays of Labuk, Sandakan, Darvel, and Cowie in the east.

Sabah is drained by many rivers. Those in western Sabah, with one exception, empty into the South China Sea and are relatively short because the mountains are so near the coast. The one exception is the Padas River, which has cut a deep gorge through the Crocker Range and drains a large section of interior lowland. All western rivers carry silt that is deposited on the narrow coastal plain where it forms good agricultural land. The other rivers of Sabah drain into the Sulu or Celebes seas. The 563 km (350-mile) long Kinabatangan is the largest and the most significant river of Sabah. Rising in the southern part of the Crocker Range, it flows generally east-west through the middle of Sabah and empties into Sandakan Bay. Launches can travel for about 193 km (120 mi) from the coast to reach the rich plantations and forest valleys.

MALDIVES

BASIC FACTS

Official Name: Republic of Maldives

Abbreviation: MLD

Area: 298 sq km (115 sq mi)

Area—World Rank: 188

Population: 203,187 (1988) 283,000 (2000)

Population—World Rank: 159

Capital: Male

Land Boundaries: Nil

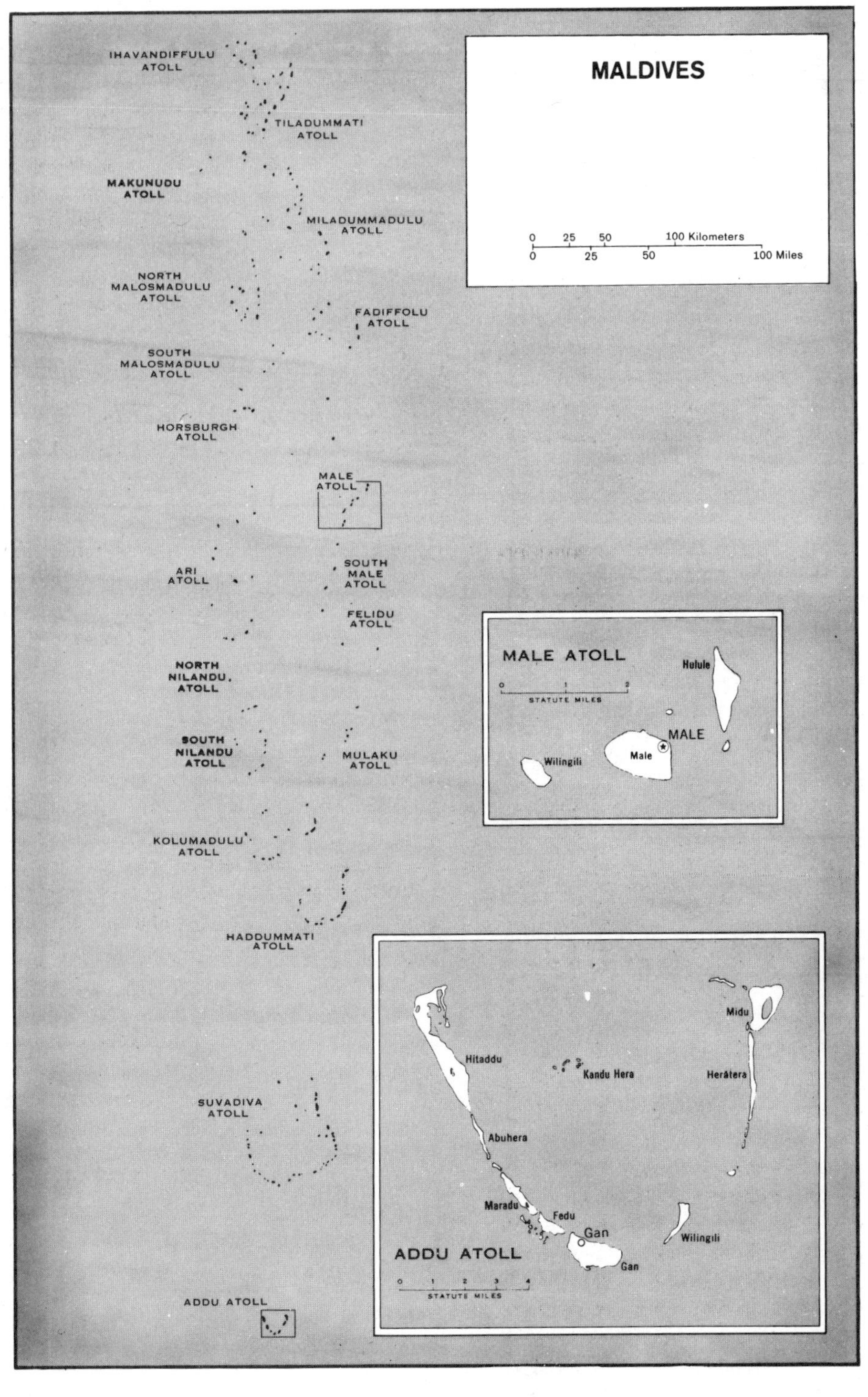
MALDIVES
0 25 50 100 Kilometers
0 25 50 100 Miles
IHAVANDIFFULU ATOLL
TILADUMMATI ATOLL
MAKUNUDU ATOLL
MILADUMMADULU ATOLL
NORTH MALOSMADULU ATOLL
FADIFFOLU ATOLL
SOUTH MALOSMADULU ATOLL
HORSBURGH ATOLL
MALE ATOLL
ARI ATOLL
SOUTH MALE ATOLL
FELIDU ATOLL
NORTH NILANDU ATOLL
SOUTH NILANDU ATOLL
MULAKU ATOLL
KOLUMADULU ATOLL
HADDUMMATI ATOLL
SUVADIVA ATOLL
ADDU ATOLL
MALE ATOLL
STATUTE MILES
Hulule
MALE
Male
Wilingili
ADDU ATOLL
STATUTE MILES
Midu
Hitaddu
Kandu Hera
Herätera
Abuhera
Maradu
Fedu
Gan
Gan
Wilingili

Coastline: 2,393 km (1,487 mi)

Longest Distances: 795 km (494 mi) N-S; 159 km (99 mi) E-W

Highest Point: 24 m (80 ft)

Lowest Point: Sea level

Land Use: 10% arable; 3% meadows and pastures; 3% forest and woodland; 84% other

Maldives is an archipelago of far flung tropical atolls in the northcentral Indian Ocean. Spanning an arc more than 805 km (500 mi) long, the islands rise only a few feet above the level of the sea. Isolated from India and Sri Lanka by more than 644 km (400 mi) of water and remotely situated from major trade routes, and Maldives survive outside the mainstream of world commerce and communication. Movement even within the archipelago is limited, and only Male, the centrally located capital, has a concentrated population.

The Maldives are a string of 19 coral atolls extending over 885 km (550 mi) in a north-south direction from 7°06′N to 0°42′S latitude and between 72°33′E and 73°44′E longitude. They are located about 644 km (400 mi) southwest of Sri Lanka and roughly the same distance from the southern tip of India. The atolls comprise approximately 2,000 islands but only 215 of them are inhabited. The islands are small, the average size is just over ½ sq km (¼ sq mile) and none is larger than 13 sq km (5 sq mi): the total land area of the Maldives constitutes a little over 259 sq km (100 sq mi).

Area and population		area		population
Administrative atolls	**Capitals**	sq mi	sq km	1985 census
Haa-Alifu	Dhidhdhoo	. . .	. . .	9,891
Haa-Dhaalu	Nolhivaranfaru	. . .	. . .	10,848
Shaviyani	Farukolhu Funadhoo	. . .	. . .	7,529
Noonu	Manadhoo	. . .	. . .	6,874
Raa	Ugoofaaru	. . .	. . .	9,516
Baa	Eydhafushi	. . .	. . .	6,945
Lhaviyani	Naifaru	. . .	. . .	6,402
Kaafu	Male	. . .	. . .	54,908
Alifu	Mahibadhoo	. . .	. . .	7,695
Vaavu	Felidhoo	. . .	. . .	1,423
Meemu	Muli	. . .	. . .	3,493
Faafu	Magoodhoo	. . .	. . .	2,148
Dhaalu	Kudahuvadhoo	. . .	. . .	3,576
Thaa	Veymandhoo	. . .	. . .	6,942
Laamu	Hithadhoo	. . .	. . .	7,158
Gaafu-Alifu	Viligili	. . .	. . .	6,081
Gaafu-Dhaalu	Thinadhoo	. . .	. . .	8,870
Gnyaviyani	Foah Mulah	. . .	. . .	6,189
Seenu	Hithadhoo	. . .	. . .	14,965
TOTAL		115	298	181,453

General Characteristics

Fourteen of the 19 atolls have lagoons that afford anchorages for boats and for ships of medium draft. Although the surrounding reefs are formed of thickly grown coral, there is no evidence of coral at the bottoms of the lagoons. East-west passage through the island chain is provided by a number of clear navigable channels, but most ship traffic passes through Eight Degree Channel between the northern extremity of the Maldives and Minicoy Island in the southern Laccadives.

The islands are low with elevations averaging 1.5 to 1.8 m (5 to 6 ft) above sea level. The higher elevations tend to be located toward the seaward sides of the islands; inner shores are frequently marshy and are probably the breeding areas for the anopheles mosquito which has spread malaria widely through the atolls.

The most important islands are Male on the southern end of North Male Atoll in the central part of the chain and Gan at the southwestern extremity of Addu, the southernmost of the Maldivian atolls. Of lesser importance are the islands in the atolls of Suvadiva, Tiladummati, Miladummadulu, and Malosmadulu.

MALI

BASIC FACTS

Official Name: Republic of Mali

Abbreviation: MAL

Area: 1,240,192 sq km (478,841 sq mi)

Area—World Rank: 22

Population: 8,665,548 (1988) 9,541,000 (2000)

Population—World Rank: 76

Capital: Bamako

Land Boundaries: 7,501 km (4,661 mi); Algeria 1,376 km (855 mi); Niger 821 km (510 mi); Burkina Faso 1,202 km (747 mi); Ivory Coast 515 km (320 mi); Guinea 932 km (579 mi); Senegal 418 km (260 mi); Mauritania 2,237 km (1,390 mi)

Coastline: Nil

Longest Distances: 1,852 km (1,151 mi) ENE-WSW; 1,258 km (782 mi) NNW-SSE

Highest Point: Hombori Tondo 1,115 m (3,789 ft)

Lowest Point: 22 m (72 ft)

Land Use: 2% arable; 25% meadows and pastures; 7% forest and woodland; 66% other

Mali is a landlocked country located in West Africa. It is generally flat, except in the south and the east. In the south the Futa Djallon Highlands and the Manding Mountains provide a barrier separating Mali from Guinea. The eastern region contains two spectacular mountain ranges: the Bandiagara Plateau and the Hombori Mountains, the

Mali

502728 9-77 (542164)
Lambert Conformal Projection
Standard parallels 8° and 32°
Scale 1:11,000,000
Boundary representation is not necessarily authoritative

Railroad
Road
Airport

highest point of which is the Hand of Fatima and Hombori Tondo 1,149 m (3,772 ft). The Adrar des Iforas is an eroded sandstone plateau in northeastern Mali that forms part of the Hoggar mountain system. The central part of Mali is filled by the flood plains of the Niger Delta, covering a surface area of some 103,599 sq km (40,000 sq mi). Northern Mali lies within the Sahara. In the extreme north are vast plains known as the Tanezrouft and Taoudenni covered in many areas by shifting sand dunes, known as ergs.

Mali is traversed by the Senegal and the Niger (known in Mali as Djoliba) Rivers and their tributaries. The Senegal is formed at the small town of Bafoulabe through the confluence of the Bafing and Bakoye Rivers. The Niger traverses Mali for 1,625 km (1,010 mi), nearly one-third of its total length. Beyond the town of Segou, the Niger forms a vast inland delta and then receives its main tributary, the Bani, at Mopti. Beyond Mopti it breaks up into two channels, the Bara Issa and the Issa Ber, that spread out to form a number of shallow seasonal lakes—Debo, Fati, Teli, Korientze, Tanda, Niangaye, Do, Garou, Aougoundou and others. Just above Dire, the two main branches join again, changing to an eastern direction beyond Kabara and making the great bend toward the southeast at Bourem. The Niger is navigable in Mali for large craft from Koulikoro and Garo during the high-water period (August to January).

PRINCIPAL TOWNS (population at 1976 census)

Bamako (capital)	404,000	Sikasso	47,000
Ségou	65,000	Kayes	45,000
Mopti	54,000		

Area and population

Regions	Capitals	area sq mi	area sq km	population 1987 census
Gao	Gao	124,323	321,996	383,734
Kayes	Kayes	76,356	197,760	1,058,575
Koulikoro	Koulikoro	34,685	89,833	1,180,260
Mopti	Mopti	34,257	88,752	1,261,383
Ségou	Ségou	21,671	56,127	1,328,250
Sikasso	Sikasso	29,529	76,480	1,308,828
Tombouctou	Tombouctou	157,907	408,977	453,032
Districts				
Bamako	Bamako	103	267	646,163
TOTAL		478,841	1,240,192	7,620,225

MALTA

BASIC FACTS

Official Name: Republic of Malta

Abbreviation: MAT

Area: 316 sq km (122 sq mi)

Area—World Rank: 187

Population: 369,240 (1988) 370,000 (2000)

Population—World Rank: 150

Capital: Valetta

Land Boundaries: Nil

Coastline: 179 km (111 mi)

Longest Distances: 45 km (28 mi) SE-NW; 13 km (8 mi) NE-SW

Highest Point: 253 m (829 ft)

Lowest Point: Sea level

Land Use: 38% arable; 3% permanent crops; 59% other

Malta is an archipelago of five islands in the central Mediterranean. Three of the islands (Malta, Gozo and Comino) are inhabited and two, Cominotto and Filfla, are uninhabited. Malta, the main island, lies 93 km (58 mi) south of Sicily and 290 km (180 mi) north of the Libyan coast extending 45 km (28 mi) southeast to northwest and 13 km (8 mi) northeast to southwest. Of the total area Malta accounts for 245.8 sq km (94.9 sq mi), Gozo 67.1 sq km (25.9 sq mi) and Comino 2.8 sq km (about 1 sq mi).

The islands have neither mountains nor rivers. The terrain consists of low hills running east to northwest to a height of 239 m (786 ft) with clefts that form deep harbors, bays, creeks and rocky coves.

Principal Towns (estimated population at 31 December 1984): Sliema 20,071, Birkirkara 18,041, Qormi 17,130, Valletta (capital) 14,013.

Area and population

Census regions	area sq mi	area sq km	population 1987 estimate
Gozo and Comino	27	70	25,112
Inner Harbour	6	15	97,765
Northern	30	78	31,738
Outer Harbour	12	32	101,631
South Eastern	20	53	42,653
Western	27	69	44,435
TOTAL	122	316	343,334

MARTINIQUE

BASIC FACTS

Official Name: Department of Martinique

Abbreviation: MRQ

Area: 1,091 sq km (421 sq mi)

Area—World Rank: 161

Population: 351,105

Population—World Rank: 152

Capital: Fort-de-France

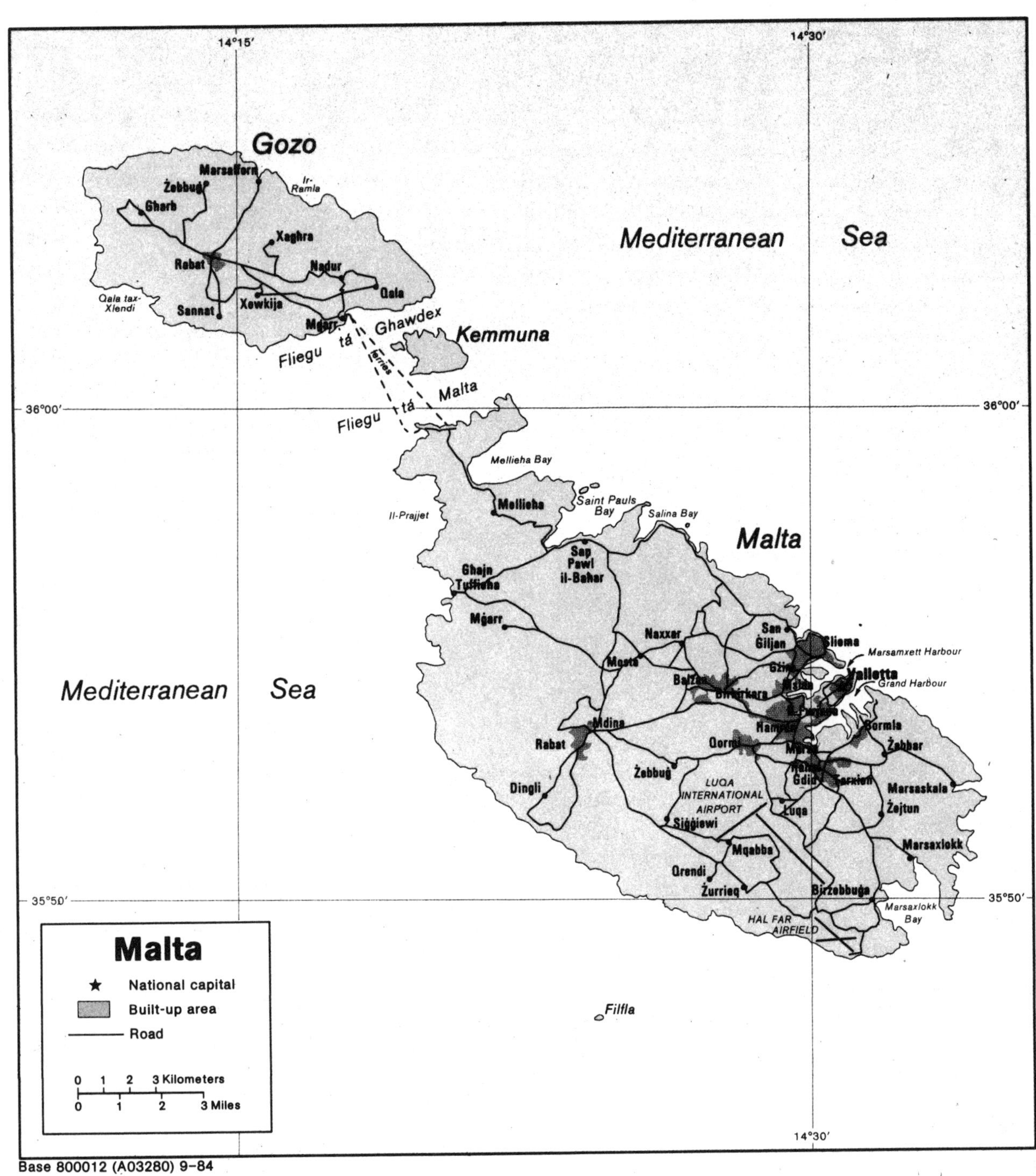

Base 800012 (A03280) 9-84

Dominica, Guadeloupe, and Martinique

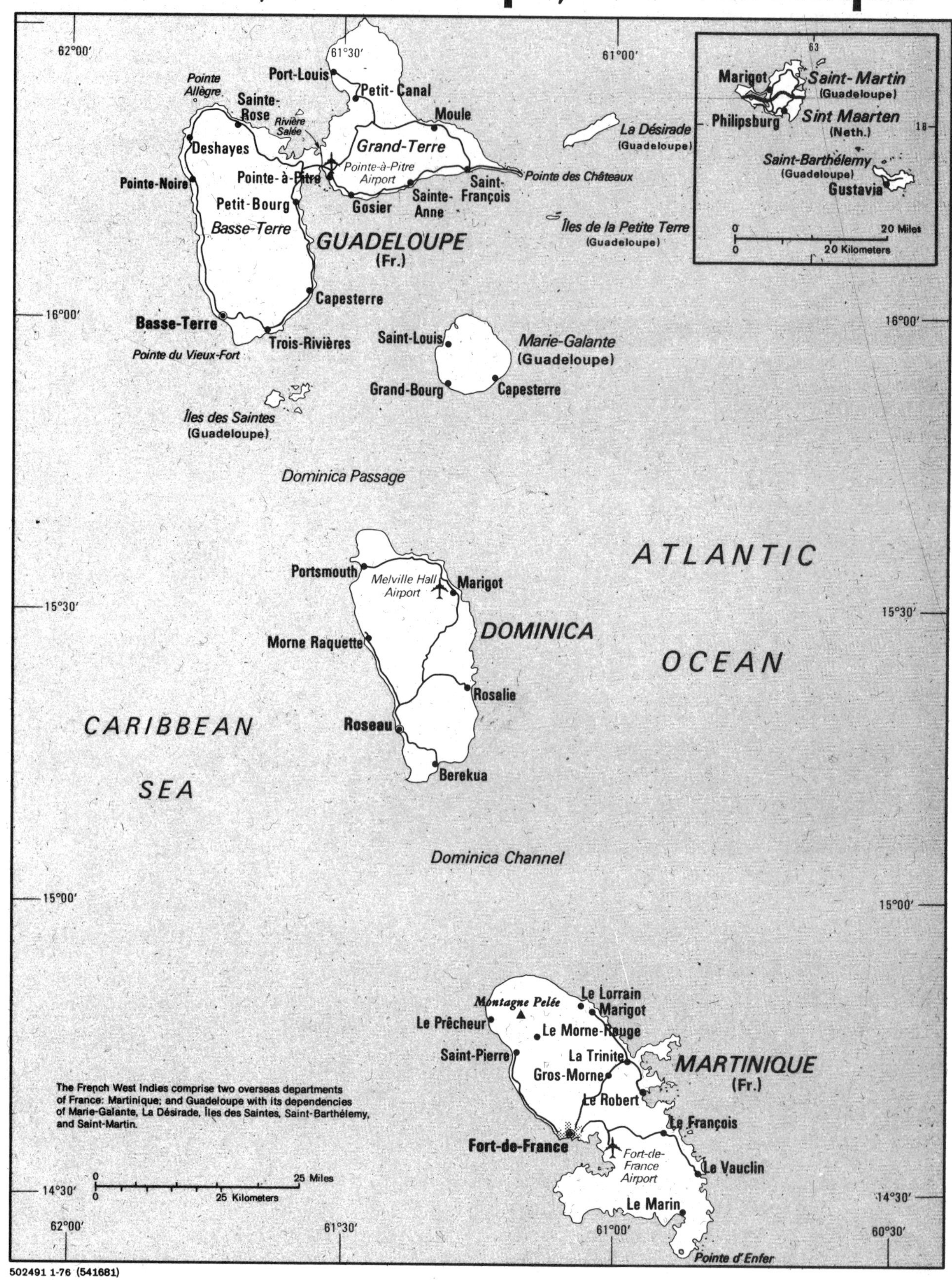

502491 1-76 (541681)
Lambert Conformal Projection
Standard parallels 14°40′ and 16°10′
Scale 1:1,100,000

Road
Airport

Land Boundaries: Nil

Coastline: 290 km (180 mi)

Longest Distances: 75 km (47 mi) SE-NW; 30 km (19 mi) NE-SW

Highest Point: Mount Pelee 1,397 m (4,583 ft)

Lowest Point: Sea level

Land Use: 10% arable; 8% permanent crops; 30% meadows and pastures; 26% forest and woodland; 26% other

Martinique is one of the Windward Islands in the West Indies with Dominica to the north and Saint Lucia to the south. Most of the island is mountainous. The two highest peaks, Pelee 1,350 m (4,428 ft) and Carbet 1,189 m (3,900 ft) are volcanoes. Mount Pelee erupted and completely destroyed the city of Saint-Pierre on May 8, 1902. The coastline is heavily indented.

Area and population

Arrondissements	Capitals	area sq mi	area sq km	population 1982 census
Fort-de-France	Fort-de-France	141	365	176,749
Le Marin	Le Marin	154	399	78,329
La Trinité	La Trinité	126	327	73,488
TOTAL		421	1,091	328,566

MAURITANIA

BASIC FACTS

Official Name: Islamic Republic of Mauritania

Abbreviation: MRT

Area: 1,030,700 sq km (398,000 sq mi)

Area—World Rank: 28

Population: 1,919,106 (1988) 2,673,000 (2000)

Population—World Rank: 124

Capital: Nouakchott

Boundaries: 5,392 km (3,351 mi); Algeria 463 km (288 mi); Mali 2,237 km (1,390 mi); Senegal 813 km (505 mi); Morocco 1,213 km (754 mi)

Coastline: 666 km (414 mi)

Longest Distances: 1,515 km (941 mi) NE-SW; 1,314 km (816 mi) SE-NW

Highest Point: Mount Jill 915 m (3,002 ft)

Lowest Point: Sebkha de Ndrhamcha 3 m (10 ft) below sea level

Land Use: 38% meadows and pastures; 15% forest and woodland; 47% other

Mauritania is a generally flat plain with occasional ridges and clifflike outcroppings. This plain is longitudinally bisected in the center by a series of plateaus, the western faces of which present formidable obstacles to east-west communications and transport. These ranges gradually fall eastward to the barren El Djouf or empty quarter, a vast region of large sandy dunes. To the west, between the ocean and the central range, are alternating areas of clayey plains, *regs,* and sandy dunes, *ergs,* some of which are mobile. Generally speaking these dunes increase in size and mobility toward the north.

The Maures traditionally divide the country into four regions: the Sahel, corresponding to the western one-third of the country and including the coast; the Trub el Hajra, the country of stone, corresponding to the Adrar and Tagant plateaus; El Djouf or vast eastern emptiness (called Cherg by the Maures); and the Guebla or south. Modern geographers, although differing to some extent, generally divide the country into zones corresponding to amounts of precipitation, vegetation belts, and temperature variations.

It is also common to distinguish the coastal area due to a different set of prevailing winds. This division also results in four zones: the Saharan Zone, the Sahelian Zone, the Senegal River Valley Zone, and the Coastal Zone. Although there are marked differences from one zone to another, there are no natural boundaries that clearly delineate where one region begins and the other ends. Thus boundaries can be drawn only in a general way.

Saharan Zone

The Saharan Zone comprises the northern two-thirds of the country with the boundary generally corresponding to the isohyet (a line on the earth's surface along which the rainfall is the same) that represents annual precipitation of 150 mm (5.8 in). Varying from 5.7 in at Tidjikdja in the Southern Saharan Zone to less than 1 in in the far north, rain usually falls during the *hivernage,* or rainy season, lasting from July to September and is the result of fortuitous storms that dump large amounts of water in short periods of time. Annual variations are also an important factor, Atar having recently gone five years without any rain and in 1927 receiving almost 10 in.

Physically, the Saharan region is dry and barren. Tiris-Zemmour in the north, and Adrar and northern Hodh-Occidental in the east are vast empty stretches of *ergs,* series of barren sandy dunes, alternating with granite outcroppings, which may, after a rain or in the presence of a well, support short-lived vegetation. In the plateaus of Adrar and Tagant, where population is relatively dense, springs and wells provide good pasturage and permit the raising of some crops. Some 809 ha (2,000 ac) are under cultivation on the Adrar plateaus, and larger seasonal ponds sustain extensive date palm groves. In the western portion of the Saharan Zone, extending toward Nouakchott, sandy dunes are aligned from northeast to southwest in ridges that are from 3 to 32 km (2 to 20 mi) wide. Between these ridges are depressions floored by limestone and clayey sand capable of producing vegetation after a rain. The dunes become more mobile from south to north.

Mauritania
International boundary
National capital
Railroad
Road
International Airport
Boundary representation is not necessarily authoritative
MOROCCO
ALGERIA
WESTERN SAHARA
(Claimed by Morocco)
MALI
SENEGAL
THE GAMBIA
ADRAR
TRARZA
Tan-tan
Oued Drâa
Tarfaya
Tindouf
Daora
El Aaiún
Saguia el Hamra
Uad el Jat
Bir Mogreïn
Guelta Zemmur
Dakhla
F'dérik
Tazadit
Uad Atui
Nouadhibou
Cap Blanc
Atar
Akjoujt
Île Tidra
Cap Timiris
Tidjikdja
Nouakchott
Aleg
Rosso
Bogué
Dagana
Senegal
Kaédi
Kiffa
'Ayoûn el 'Atroûs
Néma
Timbédra
Kankossa
Saint-Louis
Louga
Vallée du Ferlo
Linguère
Matam
Sélibaby
Nioro du Sahel
Thiès
Touba
Dakar
Diourbel
Saloum
Kaolack
Kayes
Bakoy
Baoulé
Tambacounda
Banjul
Gambia
Niger
Ségou
0 100 200 Kilometers
0 100 200 Miles
12
8
28
24
20
16

Sahelian Zone

South of isohyet 150 to within 32 km (20 mi) of the Senegal River is the Sahelian Zone. This zone of the Mauritanian steppes and savanna grasslands consist of an east-west belt with its axis running through Boutilimit, Aleg, Kiffa, 'Ayoun-el-'Atrous, and Néma. Great herds of cattle, sheep, and goats move across this zone from north to south following the season.

Precipitation varies from the minimum of 6 inches in the north to a maximum of 44 cm (17.8 in) at Kankossa in the south. The *hivernage* begins earlier in the Sahel than in the Sahara, about June, and lasts until October. Again, annual variations are of significance: Kiffa had 54 cm (7.2 in) in 1951, but 181 cm (21.5 in) in 1936. Because so many animals depend on annual rains, a delay of one month in the beginning of the rainy season can cause large losses and lead to mass migrations from Hodh-Oriental and Hodh-Occidental to Mali.

Although temperature extremes are somewhat less severe than in the Sahara, diurnal variations do range from 60° F. to 70° F. The *harmattan* is the prevailing wind, but here it has a slightly higher moisture content than in the Sahara and thus slows rapid evaporation of surface water.

Senegal River Valley

Variously known as the Chemama or the Pre-Sahel, this zone consists of a narrow belt of land extending 16 to 32 km (10 to 20 mi) north of the Senegal River (it is wider in Guidimaka) and is completely dominated by the seasonal cycle of the river. The river valley is the millet basket of the country, supplying over 80 percent of the country's agricultural production. Almost all of the economically active population is engaged in sedentary agriculture or fishing along the side channels of the river.

Rainfall in the Senegal River valley is generally the highest in the country, Ranging from 30 to 66 cm (12 to 26 in) per year, precipitation begins in May and lasts until September. This rainfall, combined with annual flooding of the river, provides the basis for extensive agriculture.

Temperatures are cooler and are subject to less annual and diurnal variation. The mean yearly maximum and minimum temperatures for Kaédi are 94° F. and 74° F., respectively. Humidity is much higher along the river and the air is often clouded by dense mists. During the rainy season tornadoes are a frequent occurrence. In this semi-tropical climate, unlike the rest of the country, malaria flourishes.

The Senegal is the only permanent river between southern Morocco and central Senegal. Rising in Guinea, it flows north and west 4,023 km (2,500 mi) to the sea at Saint Louis. For its entire length along the Mauritania-Senegal border there are no falls, and the river is navigable as far as Kayes in Mali during the rainy season and Podor in Senegal during the rest of the year. Heavy rains, beginning in April in Guinea and May and June in Senegal and Mali, bring the annual floods. The river crests at 13 m (45 ft) at Bakel by mid-September and 3.6 m (12 ft) at Rosso by mid-October. This flood covers the entire valley up to a width of 24 to 32 km (15 to 20 mi), filling numerous lakes and sloughs *(marigots)* that empty back into the river during the dry season. As soon as the flood recedes from the bottom lands, planting is begun. When the long dry season begins, the Chemama becomes a ribbon of green with the dry and brown Sahel on either side. By the end of the dry season, ocean tides may come up as far as Rosso.

PRINCIPAL TOWNS (population at census of January 1977)

Town	Population	Town	Population
Nouakchott (capital)	134,986*	Zouérate	17,474
Nouadhibou (Port-Etienne)	21,961	Rosso	16,466
Kaédi	20,848	Atar	16,326

*Estimated at 350,000 in 1984

Area and population

Regions	Capitals	area sq mi	area sq km	population 1987 estimate
el-ʾAçâba	Kiffa	13,900	36,000	160,000
Adrar	Atar	83,100	215,300	70,000
Brakna	Aleg	14,000	37,100	169,000
Dakhlet Nouadhibou	Nouadhibou	11,600	30,000	33,000
Gorgol	Kaédi	5,400	14,000	188,000
Guidimaka	Sélibaby	4,000	10,000	115,000
Hodh ech-Chargui	Néma	64,000	166,000	267,000
Hodh el-Gharbi	ʿAyoûn el-ʾAtroûs	22,000	57,000	140,000
Inchiri	Akjoujt	19,000	49,000	26,000
Tagant	Tidjikdja	36,000	93,000	80,000
Tiris Zemmour	Fdérik	98,600	255,300	37,000
Trarza	Rosso	26,000	67,000	249,000
District				
Nouakchott	Nouakchott	400	1,000	285,000
TOTAL		398,000	1,030,700	1,819,000

The Coastal Zone

The Coastal or Sub-Canarian Zone extends the length of the approximately 644-km (400-mi) long Atlantic coast. Prevailing oceanic trade winds from the Canary Islands tend to modify the influence of the hot dry winds of the Sahara with the result that the climate is humid but temperate. Rainfall is quite minimal; in Nouadhibou it averages less than 25 mm (1 in) annually and occurs during the Saharan *hivernage* in July–September. Temperatures are moderate, varying from mean maximums of 83°F. and 89°F. for Nouadhibou and Nouakchott, respectively, to mean minimums of 61°F. and 67°F.

Battering surf and shifting sand banks mark the entire length of the shoreline that sweeps in a great smooth arc from Cape Blanc to the mouth of the Senegal River. The Cape Blanc peninsula, jutting southward and forming Lévrier Bay to the east, is 48 km (30 mi) long and 13 km (8 mi) wide. The peninsula is divided between Western Sahara and Mauritania with the Mauritanian city, port, and railhead of Nouadhibou located on the eastern shore. One of the largest natural harbors on the West Coast of Africa, Lévrier Bay, measures 43 km (27 mi) long by 32 km (20 mi) wide at its broadest point. Sixty km (37 mi) south of Cape Blanc is the low-lying sandy island of Arguin (site of the first Portuguese installation south of Cape Bojador in 1455). Farther south is the coastline's only significant promontory, 7-m (23-foot) high Cape Timiris, forming, to the north, St. John Bay. From Cape Timiris to the marshy area around the mouth of the Senegal, the coast is smooth and marked only by an occasional high dune.

MAURITIUS

BASIC FACTS

Official Name: Mauritius

Abbreviation: MAS

Area: 2,040 sq km (788 sq mi)

Area—World Rank: 157

Population: 1,099,983 (1988) 1,202,000 (2000)

Population—World Rank: 132

Capital: Port Louis

Land Boundaries: Nil

Coastline: 217 km (135 mi)

Longest Distances: 61 km (38 mi) N-S; 47 km (29 mi) E-W

Highest Point: Petite Riviere Noire Peak 828 m (2,717 ft)

Lowest Point: Sea level

Land Use: 54% arable; 4% permanent crops; 4% meadows and pastures; 31% forest and woodland; 7% other

Mauritius lies 804 km (500 mi) off the east coast of Madagascar, situated strategically on the trade route between South Africa, India, and the Far East.

Mauritius is a picturesque island with rugged volcanic features and a large fertile plain. It is a compact island, 61 km (thirty-eight mi) long and 63 km (twenty-nine mi) broad, the worn and eroded base of an old volcanic formation. The surface of the island consists of a broad plateau that slopes toward a northern coastal plain from elevations of approximately 670 m (2,200 ft) near the southern coastline. Several low mountain groups and isolated peaks rise above the level of the plateau to give the appearance of a more rugged landscape. A coral reef nearly encircles the island.

PRINCIPAL TOWNS (estimated population at mid-1985)

Port Louis (capital)	136,323	Curepipe	63,181
Beau Bassin/Rose Hill	91,786	Vacoas-Phoenix	54,430
Quatre Bornes	64,506		

Area and population

Islands / Districts	area sq mi	area sq km	population 1986 estimate
Mauritius	720	1,865	998,471
Black River	100	259	38,336
Flacq	115	298	111,551
Grand Port	101	262	96,146
Moka	89	230	63,028
Pamplemousses	69	179	93,647
Plaines Wilhems	78	202	313,025
Port Louis	17	44	138,272
Rivière du Rampart	57	148	83,856
Savanne	94	243	60,610
Rodrigues	40	104	35,284
Agelega / Saint Brandon	27	71	500
TOTAL	788	2,040	1,034,255

MAYOTTE

BASIC FACTS

Official Name: Territorial Collectivity of Mayotte

Abbreviation: MAY

Area: 373 sq km (144 sq mi)

Area—World Rank: 183

Population: 66,882

Population—World Rank: 179

Capital: Dzaoudzi

Land Boundaries: Nil

Coastline: 165 km (102 mi)

Longest Distances: N/A

Highest Point: N/A

Lowest Point: Sea Level

Land Use: N/A

PRINCIPAL TOWNS Dzaoudzi (capital) 5,865, Mamoudzou 12,026, Pamanzi-Labattoir 4,106.

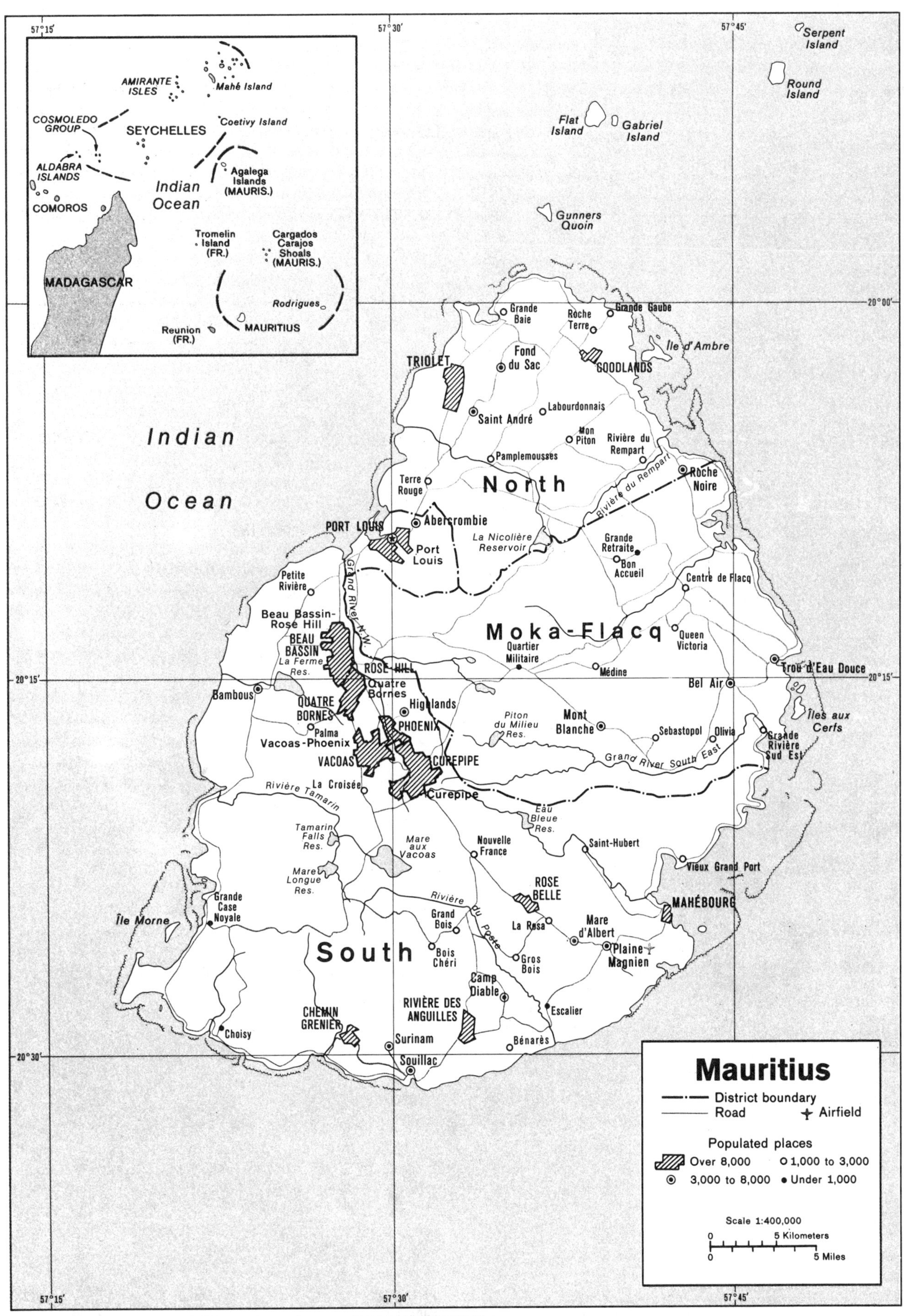

Base 503989 12-78 Map data is of 1972 except for country names which were updated in 1988.

Area and Population

Islands / Communes	Capitals	area sq mi	area sq km	population 1985 census
Grand Terre				
Acoua	Acoua	4.9	12.6	2,708
Bandraboua	Bandraboua	12.5	32.4	3,533
Bandrele	Bandrele	14.1	36.5	2,974
Boueni	Boueni	5.4	14.1	3,004
Chiconi	Chiconi	3.2	8.3	4,025
Chirongui	Chirongui	10.9	28.3	3,387
Dembeni	Dembeni	15.0	38.8	2,322
Kani-Keli	Kani-Keli	7.9	20.5	2,792
Koungou	Koungou	11.0	28.4	3,479
Mamoudzou	Mamoudzou	16.2	41.9	12,086
Mtsamboro	Mtsamboro	5.3	13.7	3,918
M'tsangamouji	M'tsangamouji	8.4	21.8	3,249
Ouangani	Ouangani	7.3	19.0	2,575
Sada	Sada	4.3	11.2	4,137
Tsingoni	Tsingoni	13.4	34.8	3,007
Petite Terre				
Dzaoudzi-Labattoir	Dzaoudzi	2.6	6.7	5,865
Pamandzi	Pamandzi	1.7	4.3	4,106
TOTAL		144.1	373.2	67,167

Mayotte is an island in the Comoro Archipelago in the Mozambique Channel between Madagascar and mainland Africa. It is surrounded by a coral reef which encloses the islets of M'Zambourou and Pamanzi. Beyond the island's coastal plain, a plateau reaches heights of 660 m (2,165 ft).

MEXICO

BASIC FACTS

Official Name: United Mexican States

Abbreviation: MEX

Area: 1,958,201 sq km (756,066 sq mi)

Area—World Rank: 14

Population: 83,527,567 (1988) 99,604,000 (2000)

Population—World Rank: 11

Capital: Mexico City

Boundaries: 14,243 km (8,850 mi); United States 3,125 km (1,942 mi); Belize 259 km (161 mi); Guatemala 962 km (598 mi)

Coastline: Gulf of Mexico and Caribbean Sea 2,749 km (1,708 mi); Pacific Ocean 7,148 km (4,441 mi)

Longest Distances: 3,220 km (2,001 mi) SSE-NNW; 1,060 km (659 mi) ENE-WSW

Highest Point: Orizaba Peak (Citlaltepetl Volcano) 5,610 m (18,406 ft)

Lowest Point: Lake Salada 8 m (26 ft) below sea level

Land Use: 12% arable; 1% permanent crops; 39% meadows and pastures; 24% forest and woodland; 24% other

Mexico is third in size and second in population among the countries of Latin America. Most of its mountain ranges, plateaus, and lowlands are continuations of landforms of the southwestern United States. Its demographic characteristics and patterns of settlement, however, are those of Middle America.

Hernán Cortés, the Spanish conqueror, is reputed to have described the topography of the country by crumbling a piece of paper, throwing it down, and saying, "This is the map of Mexico." Two-thirds of the country is mountainous, and Mexicans tend to think of directions as up and down rather than in terms of the four quarters of the compass. The mountains have tended to perpetuate regionalism, and access to many areas is so difficult they have remained economically undeveloped and culturally isolated.

Extending southeastward from its border with the United States, Mexico forms a generally narrowing cone, broken in the northwest by the long, narrow peninsula of Baja California and in the extreme southeast by the blunt peninsula of Yucatán. A fairly broad coastal plain borders the Gulf of Mexico, and a narrower and broken strip of lowland borders the Pacific Ocean. Great mountain ranges extend roughly parallel to the two coastal lowlands, and between them lies a great interior plateau region. The plateau narrows to the south and terminates in a transverse mountain range consisting of a series of volcanic cones, some of them still active. This zone of volcanoes is a part of the Circum-Pacific Volcanic Ring that circles the Pacific Ocean in a great arc from Japan to the Andes.

This highland region of volcanoes is the heartland of Mexico. Most of the large cities and the densest rural population are located in its basins and valleys. Southward, the range of volcanoes drops away to the basin of the Balsas River flowing into the Pacific. The south, however, is mountainous except in a few river valleys, along the coasts, on the peninsula of Yucatán, and on the low-lying Isthmus of Tehuantepec. Southern Mexico is a relatively undeveloped tropical region in which a predominantly Indian population lives in mountain pockets and along river valleys.

Highlands and Lowlands

The predominant structural features of Mexico are the southward continuations of the more northern parts of the North American continent. On the east, the Sierra Madre Oriental range is an extension of the Rocky Mountain range and, on the west, the Sierra Madre Occidental range is an extension of the Sierra Nevada range of the United States. Between the two Sierra Madres the lofty Northern Plateau, made up of highland and intermontane basins, extends northward from near Mexico City to become the western tablelands of the United States. Eastward from the Sierra Madre Oriental, the coastal lowlands of the Gulf of Mexico are a continuation of the coastal plain of Texas, and the structure of the Yucatán Peninsula resembles that of Florida. The depression known as the Salton Trench in the United States continues into Mexico and becomes the body of water known internationally as the Gulf of California and by Mexicans as the Mar de Cortés. On the western coast of the mainland the outliers of the massive Sierra Madre Occidental—the country's most extensive mountain system—descend almost directly to tidewater, leaving only a narrow and discontinuous coastal plain and a few river deltas. West of the Gulf of California the barren peninsula of Baja California is a continuation of the Coast Ranges of the United States.

Between the parallels of 18° and 20° north latitude lie a series of volcanic peaks that terminate the southward extension of North American physiographic features. Some geographers regard this volcanic fracture zone, a region of frequent and sometimes serious earthquakes, as the termination of the North American continent. The loftiest peaks lie on the southern flank of this zone and form a kind of mountain range that extends laterally from the Pacific almost to the Gulf of Mexico. It is known variously as the Cordillera Neovolcanica, the Sierra Volcanica Transversal, and the Cordillera de Anahuác.

The range is anchored on the east, not far from the Gulf of Mexico, by the perfect volcanic cone of Orizaba. With an elevation of nearly 5,791 m (19,000 ft), it is the country's highest mountain. The second and third in height, Ixtacihuatl and Popocatapetl, respectively, are visible from Mexico City. Most of the country's highest peaks are in this chain. Nearly all of the volcanoes of the fracture zone are extinct, but some volcanism continues. Ixtacihuatl last erupted in the eighteenth century, but its crest is still sometimes topped by a plume of smoke. The new volcano of Paracutín first erupted in 1943 in a cornfield 402 km (250 mi) southwest of Mexico City. Before ceasing activity in the early 1950s, it had reached an elevation of 518 m (1,700 ft).

South of the fracture zone, the land descends to a broad depression formed by the Balsas River and its tributaries, and between this depression and the Pacific coast lies the Sierra Madre del Sur range with altitudes of up to 3,048 m (10,000 ft). This range is not related to the mountain systems of northern Mexico but resembles the submerged chain of mountains whose peaks appear as the islands of the Greater Antilles. East of the Balsas River depression, the interior is occupied by a complex system of ranges known as the Oaxaca Highlands; on the Gulf of Mexico the coastal lowlands are broken by an isolated upland called the Tuxtlas mountains, the highest Atlantic coastal elevations south of Labrador.

At about 95° east longitude, Mexico narrows to form the lowlands of the Isthmus of Tehuantepec in which altitudes do not exceed 152 m (500 ft). Beyond this isthmus, the Sierra Madre de Chiapas extends to the Guatemalan border, separated from the Pacific by a fairly broad coastal plain. The remaining major mountain system, the Chiapas Highlands, occupies most of the interior east of the Isthmus of Tehuantepec and south of the peninsula of Yucatán. Along the northern flank of both the Chiapas and Oaxaca highlands lies a narrow band of folded ranges that resembles the Sierra Madre Oriental and may thus be considered a far southern extremity of the Rocky Mountains.

Geographic Regions

Geographers see Mexico as divided in different ways into geographical regions. For some statistical and administrative purposes, however, the Mexican government has devised a system of five regions consisting of the North Pacific, the North, Central Mexico, the Gulf Coast, and the South Pacific.

The arrangement is imperfect in the sense that the regions bear only a limited relationship to natural features, and the small South Pacific state of Colima is left as a coastal enclave flanked by two states of Central Mexico. Climate, land utilization, and population distribution patterns do generally conform to the arrangement, however, and it is adaptable for use in the preparation of statistical tables.

The North Pacific

Second largest but least populated of the regions, the North Pacific is made up of the three coastal states lying west of the Sierra Madre Occidental plus the peninsula of Baja California. It is located principally between the parallels of 20°N and 30°N and, like all west coast regions in this latitudinal range in both northern and southern hemispheres, it consists largely of desert.

Nearly half of the region is occupied by the large state of Sonora. Most of it is the Sonoran Desert, which continues southward into Sinaloa. Desert conditions terminate

PRINCIPAL TOWNS (estimated population at 30 June 1979)

Cuidad de México (Mexico City—capital)	9,191,925
Netzahualcóyotl	2,331,351
Guadalajara	1,906,145
Monterrey	1,064,629
Heróica Puebla de Zaragoza (Puebla)	710,833
Ciudad Juárez	625,040
León	624,816
Tijuana	566,344
Acapulco de Juárez (Acapulco)	462,144
Chihuahua	385,953
Mexicali	348,528
San Luis Potosí	327,333
Hermosillo	324,292
Veracruz Llave (Veracruz)	319,257
Culiacán Rosales (Culiacán)	306,843
Torreón	274,717
Mérida	269,582
Saltillo	258,492
Aguascalientes	257,179
Morelia	251,011
Tampico	248,369
Toluca de Lerdo (Toluca)	241,920
Cuernavaca	241,337
Reynosa	231,082
Victoria de Durango (Durango)	228,686
Nuevo Laredo	223,606
Jalapa Enríquez (Jalapa)	201,473
Poza Rica de Hidalgo (Poza Rica)	198,003
Matamoros	193,305
Mazatlán	186,290
Querétaro	185,821
Ciudad Obregón	181,733
Villahermosa	175,845
Irapuato	161,047

in Sinaloa with an increase in rainfall, but the physiography of the desert zone is repeated in southern Sinaloa and in Nayarit. It consists of ancient foothills of the Sierra Madre Occidental that have been buried by waste washed down from that range. A narrow coastal plain extends inland for a distance of only ten to fifteen miles. This plain, however, is one of the country's most productive commercial farming areas, particularly on the alluvial fans created by the Sonora, Yaqui, Fuerte, Sinaloa, and Culiacán rivers and several other rivers and streams.

The Morelos Dam on the Colorado River at the head of the Gulf of California has converted the desert land of the Mexicali Valley into an important agricultural area devoted primarily to cotton farming. In the far west the narrow finger of the Baja California Peninsula is a desert on which block mountains with elevations of up to 2,743 m (9,000 ft) drop abruptly into the Gulf of California and descend to a narrow plain only on the southern part of the Pacific coast. The upper half of the peninsula is the state of Baja California, and the lower half is the state of Baja California Sur.

The North

Extending eastward from the crests of the Sierra Madre Occidental to the Gulf of Mexico, the North has more than 40 percent of the country's territory but less than 19 percent of its population. The Sierra Madre Oriental lies almost entirely within the region. The coastal state of Tamaulipas is situated to the east of this range and is physically similar to the states that make up the Gulf Coast region. Its general aridity, however, is unlike that of the Gulf Coast but similar to the other states of the North.

The other six states of the region comprise the great Northern Plateau of Mexico. The terrain includes extensive flat areas, but it is broken by numerous hill ranges, most of them longitudinal. It narrows somewhat from north to south, and elevations increase as the plateau of Central Mexico is neared. At approximately its midpoint, the plateau is interrupted by a series of cross ranges between the two Sierra Madre systems.

South of these ranges, in an area occupied principally by the states of Zacatecas and San Luís Potosí, the land is more arid than that of Central Mexico, and the population is less dense. The pattern of settlement, however, is similar to it in the sense that it is made up of population clusters in intermontane basins. This portion of the Northern Plateau as a whole is sometimes referred to as the Central Meseta, or central tableland.

To the north of the cross ranges, the land becomes progressively more arid and the population scantier. The states of Chihuahua and Coahuila are at once the country's largest and least densely populated. It is a dry pastoral countryside of great desert basins known as bolsons. These depressions drain internally to salty lakes or salt flats bordered by gradually sloping shallow alluvial fans and rimmed by steeply sloped hill ranges. The largest of these, the Mayarán Bolson, has an elevation of about 1,097 m (3,600 ft); a chain of interconnected basins known collectively as the Mapimí Bolson straddles the borders of the states of Durango, Coahuila, and Chihuahua at elevations of a little more than 914 m (3,000 ft).

The poor quality of the soils, resulting in particular from the absence of humus, and the scanty rainfall limit farming to a few oases. The area does, however, produce herbaceous and xerophyte vegetation that permits extensive animal husbandry. Much of the land reclamation through irrigation projects has taken place in the North, particularly in areas sufficiently close to the United States border to use the waters of the Rio Grande. One of the largest irrigation projects fans out from Matamoros in Tamaulipas State and is located near the mouth of the river. Close to the city of Torreón the Coahuila Bolson includes an old lakebed that has become the extensive La Laguna irrigation district. It is the North's principal irrigation area away from the Rio Grande, but the irrigation complex was not carefully planned and neither its extent nor its productivity has met expectations.

Central Mexico

Third in size among the regions, Central Mexico contains almost half of the country's population. It is physiographically the southern end of the Northern Plateau and consists mainly of a volcanic zone that is sometimes called the Neovolcanic Plateau.

Central Mexico is the heartland of the country and borders on all other regions. Its southern perimeter is fairly well defined by the decline from the plateau to the Balsas River depression, but its other flanks merge imperceptibly into other regions. On the west, the coastal areas of the states of Jalisco and Michoacán are physiographically part of the South Pacific region and, on the east, the plateau slopes almost to the Gulf of Mexico in the Gulf Coast region. Northward, the plateau merges into the North region without perceptible change other than the gradual disappearance of the old volcanoes and the increasing aridity in the area.

Central Mexico was an ancient center of civilization and is one of the world's loftiest areas of concentrated population after those of Tibet and the Andean countries. The terrain consists principally of rolling hills and the dissected cones of thousands of old volcanoes interspersed by broad basins and valleys with floors at elevations of 1,524 to 2,438 m (5,000 to 8,000 ft). The seven largest and most important of these basins and valleys extend westward in a transverse chain from the Basin of Puebla on the east through the Valley of México; the basins of Toluca, Guanajuato, and Jalisco; and the valleys of Morelos and Aguascalientes. The distinction between the terms *basin* and *valley* is one based on common usage rather than on distinctive difference in configuration of the areas.

The Valley of México is drained internally, and the Basin of Puebla drains to the Pacific Ocean through streams of the Balsas River system. The remaining five valleys are drained to the Pacific through the Lerma River system. The basins west of Toloca are interconnecting and form a subregion that terminates in mountains west of the city of Guadalajara. This area of relatively level land is known as the Bajío and consists of rich volcanic ash and the lacustrine soils of old lakebeds. These soils and a relatively long growing season make the Bajío the most productive agricultural zone of Central Mexico.

Mexico
International boundary
Estado boundary
National capital
Estado capital
Railroad
Road
0 100 200 300 Kilometers
0 100 200 300 Miles
Key to estados in central Mexico
1 AGUASCALIENTES
2 GUANAJUATO
3 QUERÉTARO DE ARTEAGA
4 HIDALGO
5 MÉXICO
6 DISTRITO FEDERAL
7 MORELOS
8 TLAXCALA
Boundary representation is not necessarily authoritative.
UNITED STATES
PACIFIC OCEAN
Gulf of Mexico
Gulf of California
Bahía de Campeche
Golfo de Tehuantepec
GUATEMALA
BELIZE
HONDURAS
EL SALVADOR

Area and Population

States	Capitals	area sq mi	area sq km	population 1986 estimate
Aguascalientes	Aguascalientes	2,112	5,471	647,700
Baja California Norte	Mexicali	26,997	69,921	1,348,500
Baja California Sur	La Paz	28,369	73,475	291,000
Campeche	Campeche	19,619	50,812	553,000
Chiapas	Tuxtla Gutiérrez	28,653	74,211	2,435,300
Chihuahua	Chihuahua	94,571	244,938	2,206,000
Coahuila	Saltillo	57,908	149,982	1,840,900
Colima	Colima	2,004	5,191	405,500
Durango	Durango	47,560	123,181	1,347,100
Guanajuato	Guanajuato	11,773	30,491	3,404,400
Guerrero	Chilpancingo	24,819	64,281	2,469,500
Hidalgo	Pachuca	8,036	20,813	1,771,300
Jalisco	Guadalajara	31,211	80,836	5,049,700
México	Toluca	8,245	21,355	10,650,300
Michoacán	Morelia	23,138	59,928	3,281,900
Morelos	Cuernavaca	1,911	4,950	1,194,200
Nayarit	Tepic	10,417	26,979	824,200
Nuevo León	Monterrey	25,067	64,924	3,032,400
Oaxaca	Oaxaca	36,275	93,952	2,609,500
Puebla	Puebla	13,090	33,902	3,923,300
Querétaro	Querétaro	4,420	11,449	905,900
Quintana Roo	Chetumal	19,387	50,212	351,600
San Luis Potosí	San Luis Potosí	24,351	63,068	1,951,100
Sinaloa	Culiacán	22,521	58,328	2,254,500
Sonora	Hermosillo	70,291	182,052	1,743,700
Tabasco	Villahermosa	9,756	25,267	1,252,600
Tamaulipas	Ciudad Victoria	30,650	79,384	2,207,800
Tlaxcala	Tlaxcala	1,551	4,016	643,900
Veracruz	Jalapa	27,683	71,699	6,389,700
Yucatán	Mérida	14,827	38,402	1,254,000
Zacatecas	Zacatecas	28,283	73,252	1,235,200
Federal District				
Distrito Federal	—	571	1,479	10,051,500
TOTAL		756,066	1,958,201	79,579,900

There are many other smaller basins and valleys and pockets of arable land at higher elevations. Among the largest is the high and rather arid basin of Tlaxcala that lies directly to the east of Mexico City. The small but famous basin in which the picturesque tourist town of Taxco is situated lies between mountains not far from Mexico City in the South Pacific State of Guerrero.

The Lerma River rises in highlands northwest of Mexico City and flows westward to Lake Chapala on the outskirts of Guadalajara. It is the country's largest lake and a popular vacation and retirement center. Lake Chapala empties into the Santiago River, which flows northwestward to the Pacific. Although the Lerma-Santiago river system is only seventh in the country in terms of runoff, it is of particular importance because of its location in the country's heartland.

The Gulf Coast

The Gulf Coast is made up of the states of Veracruz and Tabasco plus the political subdivisions of the Yucatán Peninsula, which is sometimes considered a separate region. The interior of Veracruz extends to the watershed of the Sierra Madre Oriental, and its coastal slopes are patterned by a string of plantations and commercial farms. The fairly broad Veracruz coastal plain is interrupted by outliers of the range that extend almost to tidewater a little north of the city of Veracruz. These lowlands are fertile but hot and marshy and once so insect-infested that Veracruz was known as the "City of Death." Drainage of swamps and other health and sanitation programs, however, have made the coastal plain of both Veracruz and Tabasco an area of rapid population growth. Tabasco is for the most part a lowland state in which the countless streams that form the delta of the Grijalva and Usumacinta rivers have created Mexico's most extensive coastal swamplands.

The Yucatán Peninsula is a low platform of barren limestone soils on which elevations seldom exceed 152 m (500 ft) of undulating lowland. Physiographically it is similar to Florida and unlike anything else in Mexico. The northeastern tip is a desert, but rainfall increases to the south, and a dense tropical rain forest covers Quintana Roo Territory and the base of the peninsula.

There is virtually no surface drainage, but the northern part of the peninsula is pitted by countless small depressions called cenotes formed by the collapse of the ceilings of the enormous subterranean limestone reservoirs. Water is drawn to the surface by windmills. The cenotes become fewer to the south, and the population obtains water from rainwater ponds and numerous small lakes. Beyond a wave-built barrier reef off the northern and western coasts, the bank of Campeche is the most extensive part of the Mexican continental shelf.

The South Pacific

The South Pacific region is made up of the states of Guerrero, Oaxaca, and Chiapas plus the state of Colima, which is situated as a coastal enclave in the region of Central Mexico. It is the most rural and slowest growing of the regions and the one in which Indian life most nearly continues unaffected by the twentieth century. Cut off from the rest of the country by the plateau of Central Mexico, its natural features resemble those of Central America.

The South Pacific region is extremely mountainous; but the Grijalva and Usumacinta rivers that rise in Guatemala and flow across the region to the Gulf of Mexico, the Balsas river that crosses its northern flank, and many other rivers and streams give it a greater flow of water than the other regions of Mexico. The basin of the Papaloapán River in the state of Oaxaca is the site of the large Miguel Alemán Reservoir and an extensive farm colonization area.

MONACO

BASIC FACTS

Official Name Principality of Monaco

Abbreviation: MNC

Area 1.9 sq km (0.7 sq mi)

Area—World Rank: 215

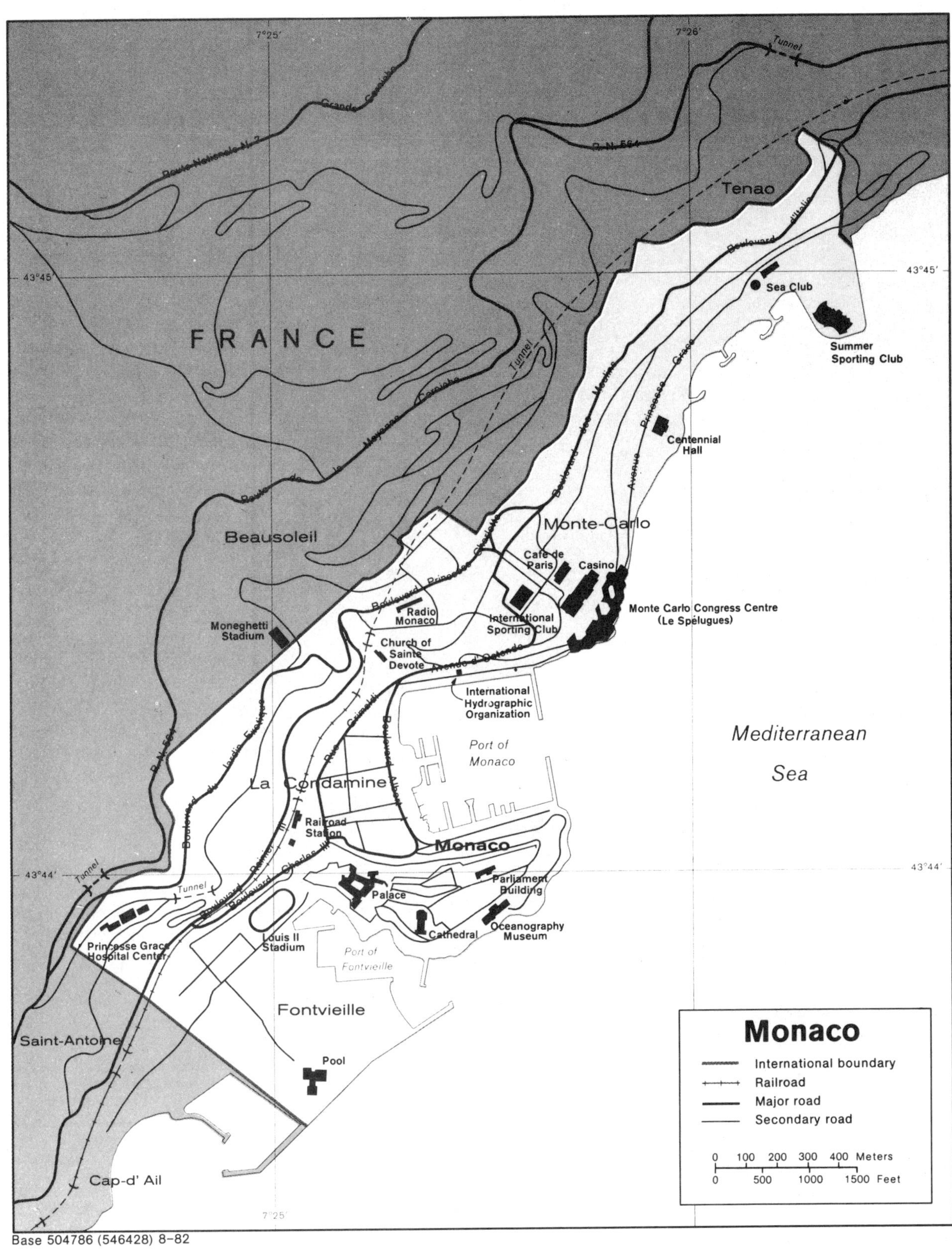

Base 504786 (546428) 8-82

Population: 28,917 (1988) 34,000 (2000)

Population—World Rank: 192

Capital: Monaco

Boundaries: 8.5 km (5.2 mi) France 4.4 km (2.7 mi)

Coastline: 4.1 km (2.5 mi)

Longest Distances: 3.18 km (1.98 mi) E-W; 1.1 km (0.68 mi) N-S

Highest Point: 140 m (459 ft)

Lowest Point: Sea Level

Land Use: 100% other

The second-smallest country in Europe and the world, Monaco is situated in the southeastern part of the French Department of Alpes-Maritimes.

The principality has three distinct divisions: la Condamine (the business district around the port) the casino, and Monaco-Ville on a rock promontory about 61 m (200 ft) above sea level.

MONGOLIA

BASIC FACTS

Official Name: Mongolian People's Republic

Abbreviation: MPR

Area: 1,565,000 sq km (604,000 sq mi)

Area—World Rank: 18

Population: 2,067,624 (1988) 2,764,000 (2000)

Population—World Rank: 121

Capital: Ulan Bator

Land Boundaries: 7,678 km (4,771 mi), USSR 3,005 km (1,867 mi); China 4,673 km (2,904 mi)

Coastline: Nil

Longest Distances: 2,368 km (1,471 mi) E-W; 1,260 km (783 mi) N-S

Highest Point: 4,374 m (14,350 ft)

Lowest Point: Hoh Lake 553 m (1,814 ft)

Land Use: 1% arable; 79% meadows and pastures; 10% forest and woodland; 10% other

From the Altai to the foothills of the Khingan range, Mongolia extends almost 2,414 km (1,500 mi) from west to east (87° 47′–119° 54′E). From north to south, it stretches almost 1,287 km (800 mi), the northernmost point (52° 16′N) being almost as far north as Moscow or Edmonton, and the southernmost point (41° 32′N) nearly as far south as New York City. Ulan Bator's latitude is that of Duluth.

Passageways across the intermontane Mongolian Plateau barrier vary from 1,931 to 3,218 km (1,200 to 2,000 mi) in length. Since the late 17th century, the main trade route across Mongolia proceeded from Russia to China, via Urga (Ulan Bator) and the Gobi—from Kyakhta to Kalgan and through a pass in the Great Khingan range to Peking. A branch of the Kalgan gateway follows the Kwang Ho and the Great Wall, to join the Sian gate from the south. Three branches of the Kalgan route diverge west (toward Balkhash), northwest (across the Altai into Siberia), and north (to Baikal). From Karakorum, in northern Mongolia, Genghis Khan used the Kalgan gateway to invade and conquer China in the 13th century. Subsequently, before the opening of north Chinese ports to Occidental trade, European goods were moved from the Baltic across Siberia to China, via the Kalgan passage. Direct Russian trade with and influence in Mongolia developed after the mid-19th century, but never attained the scale or importance of Chinese activity (Manchu officials, merchants, moneylenders, farmers, artisans, coolies) until the overthrow of the Manchu dynasty in 1911.

Until very recent times, Outer Mongolia has suffered from its underdevelopment, geographical remoteness, military weakness, dispersed and small population. Internally, the region as a whole has not been well suited to human use. Apart from the danger of storms, floods, droughts, and catastrophic earthquakes, the Mongolian nomad and herder had to cope with human disease (such as tuberculosis and venereal ills), cattle disease (epizootic

PRINCIPAL TOWNS (estimated population at January 1986)

Ulan Bator	515,000†	Erdenet	45,400
Darhan	74,000	Baga nuur	25,000*

*1984 figure.
†1987 figure.

Area and population

		area		population
Provinces	**Capitals**	sq mi	sq km	1984 estimate
Arhangay	Tsetserleg	21,000	55,000	83,800
Bayanhongor	Bayanhongor	45,000	116,000	69,500
Bayan-Ölgiy	Ölgiy	18,000	46,000	84,000
Bulgan	Bulgan	19,000	49,000	46,000
Dornod	Choybalsan	47,000	122,000	67,800
Dornogovi	Saynshand	43,000	111,000	47,400
Dundgovi	Mandalgov	30,000	78,000	44,100
Dzavhan	Uliastay	32,000	82,000	88,100
Govi-Altay	Altay	55,000	142,000	62,500
Hentiy	Öndörhaan	32,000	82,00	60,500
Hovd	Hovd	29,000	76,000	71,600
Hövsgöl	Mörön	39,000	101,000	96,600
Ömnögovi	Dalandzadgad	64,000	165,000	35,500
Övörhangay	Arvayheer	24,000	63,000	93,200
Selenge	Sühbaatar	16,000	42,000	76,400
Sühbaatar	Baruun-urt	32,000	82,000	47,200
Töv	Dzuunmod	31,000	81,000	90,200
Uvs	Ulaangom	27,000	69,000	81,400
Autonomous municipalities				
Darhan	—	100	200	63,600
Erdenet	—	300	800	40,500
Ulaanbaatar	—	800	2,000	470,500
Total	—	604,000	1,565,000	1,820,400

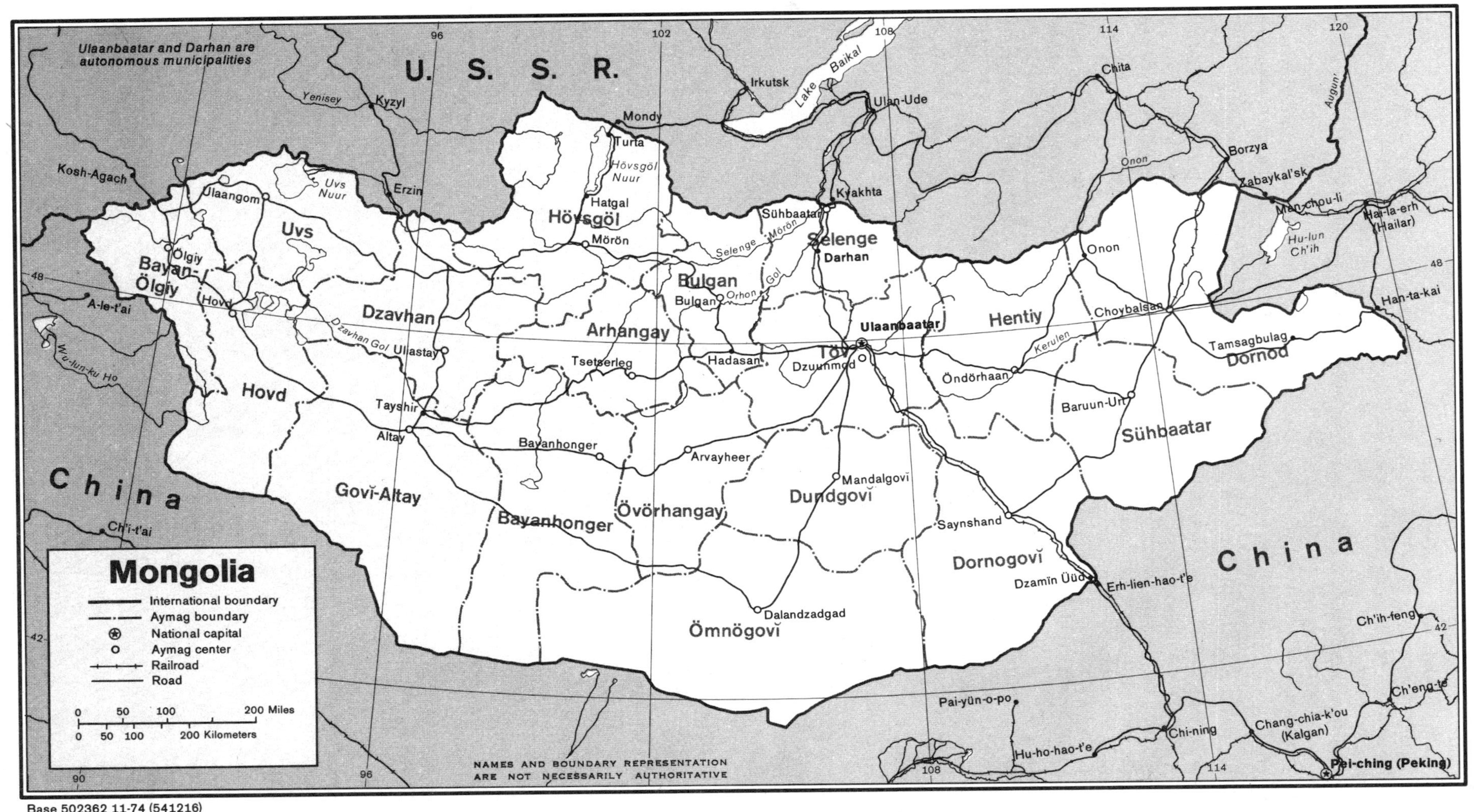

Base 502362 11-74 (541216)

epidemics, cyclical murain), bad sanitation, and a capriciously inhospitable climate which results in unstable grass crops for livestock herding and even worse conditions for cereal production.

The Steppe regions historically occupied by nomad herders are so close to the border between semi-arid and arid climatic conditions that even small variations in annual rainfalls can have a considerable effect on the growth of grass and hence the raising of livestock. They are furthermore subject to the swing of climatic cycles of considerable duration. A series of "bad years" could bring the nomadic peoples close to disaster. Numbers of geographers and historians have accepted the theory that cyclic dry periods have had much to do with the numerous eruptions of migration and conquest out of Central Asia.

Land Forms and Drainage

Most of north, west, southwest, and central Mongolia consists of mountains, plateaus, and depressions. Mongols speak of a Khangai zone and a Gobi zone, but a Russian geographer has depicted five natural regions; the mountainous Altai (the largest mountain system), the Great Lakes depression (lakes, some sand, mainly fertile plains), the mountainous Khangai-Khentei (medium-altitude, old mountains, gentle slopes and valleys), the uplifted eastern plains (smooth and rolling, best pastures and hayfields, forests and rivers), and the nearly treeless Gobi (one third of the country, hilly in west, saline lakes and marshes in flat lowlands, semi-arid or arid, monotonous yellow-green landscape of coarse, stunted bunchgrass, some sand and stony desert).

Nowhere lower than 457 m (1,500 ft) above sea level, Mongolia has an average elevation varying between 914 and 1,524 m (3,000 and 5,000 ft). In the extreme west, in the Mongolian Altai range, Tabun Bogdo peak soars to 4,358 m (14,298 ft). There is some glaciation. Other ranges include the Khentei, along the Russian frontier; and the Khangai, in the west-central sector. Elevations decline from northwest to southeast, from alpine snow peaks to rolling contours to flat tops to ridges and low hills to flat plains.

In the north, a considerable river system exists, belonging to the Arctic basin. The Selanga arises in the Khangai uplands and flows into Baikal; about 595 km (370 mi) of the river lie within Mongolia. Among the numerous tributaries are the Orkhon 1,126 km (700 mi) and the Tula 703 km (437 mi). The Selanga's watershed area is the greatest of the river basins—approximately 28,231 sq km (109,000 sq mi) inside Mongolia. Originating in this region are the great Yenisei and a minor affluent of the Irtysh. The hydrographic system is less developed in the east, the area of Pacific drainage, where the longest river is the Kerulen 1,086 km (675 mi). In the depressions of the Great Lakes, and in the Central Asian basin, flow the Dzabkhan 804 km (500 mi), Tes 563 km (350 mi), and Khobdo 499 km (310 mi). The river system in the Gobi district is negligible; the few small rivers in the northern portion of the desert zone rise in the Khangai range but vanish in salt lakes. Only two sizeable rivers are of exotic origin, the Khalkhyn, in the extreme east, and the Khobdo, in the extreme west.

Generally speaking, the water supply of Mongolia's mountainous rivers is not plentiful. In a land of low precipitation, rain must account for 80 to 85 percent of the river water. During some 6 months in winter, the rivers are covered with very thick ice; smaller waterways freeze to the bottom. Melt water provides 15 to 20 percent of the yearly runoff. The mountain districts of the north contain the greatest potential for hydroelectric power generation. Most waterways (with the exception of the Selenga) are not suitable for navigation, and are useful mainly for floating timber. A number of rivers may be used for the watering of cattle and for irrigation.

The many hundreds of lakes in Mongolia are ordinarily located high above sea level; those with outlets usually have fresh water. Most of the 16 biggest lakes are in the northwest. Fed by 200 rivers and rivulets, bitter Ubs Nur 3,366 sq km (1,300 sq mi) is the largest saline lake without an outlet. Forty-six rivers empty into alpine Khubsugul 2,590 sq km (1,000 sq mi), the deepest (maximum 518 m; 1,700 ft) and largest fresh-water lake, sometimes called the Mongol Sea. In the Great Lakes depression there are 300 lakes, as well as high waterfalls and springs. In the east, fresh-water Buir Nor 668 sq km (235 sq mi) is the largest lake. The country as a whole has over 200 developed sites of natural mineral-water springs, hot and cold.

Only 64 percent of the country has a supply of water. The rigors of winter cut off even this supply from surface waters and the many wells. Snow and ice are the sources of water for drinking and work purposes. The relatively best water situation is in the north—the region of major rivers and heavier precipitation. Where surface streams are lacking (as in the south), subsurface water becomes important. In the Gobi, for example, it is nowhere difficult to find water. Recent Hungarian geological surveys have revealed sizeable natural reservoirs of subterranean water—cold, clear, and sweet—often 2 or 3 yards below the surface of the desert, 90 percent less than 152 m (500 ft) down.

Approximately 200 wells have already been drilled. Artesian wells as deep as 183 m (600 ft) have also been located in underlying chalk deposits. New pumps will speed the process and cut down on manual requirements. About 60 percent of all Gobi pasturage is now watered. By 1990, it is planned to have over 2,500 wells in the Gobi; settled life would then at last be feasible in traditionally arid areas. No longer will the old Mongol adage be true, that "water is more difficult to get than a piece of fat."

MPR, Soviet, and East European geologists and surveyors have conducted detailed investigations of the water economy and hydro-geological prospecting. Water-supply and irrigation construction is underway in almost all *aimaks,* under the Water Economy Ministry, on the basis of new schematic pasture-capability maps compiled by 1961. Greater capital-investment budgets have been allocated under the succeeding 5-year plans. Between 1961 and 1966, the ministry asserted that 560 improved spring-fed pools and 994 dug wells were built. Another 8,500 wells and water sites were constructed by 1970, 6,000 old wells repaired, and 600 springs dammed. Thus the MPR expresses the intention of providing readily available sources of water for more than 300 million acres of pastureland. Additionally, 700 driven wells were opened between 1966 and 1970, to supply water to rural districts, industrial installations, and *aimak* centers.

MONTSERRAT

BASIC FACTS

Official Name: Montserrat

Abbreviation: MNT

Area: 102 sq km (40 sq mi)

Area—World Rank: 202

Population: 12,078

Population—World Rank: 202

Capital: Plymouth

Land Boundaries: Nil

Coastline: 40 km (25 mi)

Longest Distances: N/A

Highest Point: Chances Peak 914 m (3,000 ft)

Lowest Point: Sea level

Land Use: 20% arable; 10% meadows and pastures; 40% forest and woodland; 30% other

Montserrat is one of the Leeward Islands in the West Indies. It lies about 55 km (35 mi) north of Basse Terre, Guadeloupe, and about 43 km (27 mi) southwest of Antigua. The island is mostly volcanic and mountainous, with a small coastal lowland.

MOROCCO

BASIC FACTS

Official Name: Kingdom of Morocco

Abbreviation: MOR

Area: 458,730 sq km (177,117 sq mi)

Area—World Rank: 51

Population: 24,976,168 (1988) 33,018,000 (2000)

Population—World Rank: 35

Capital: Rabat

Boundaries: 5,007 km (3,112 mi); Algeria 1,617 km (1,005 mi); Mauritania 1,213 km (754 mi)

Coastline: 2,177 km (1,353 mi)

Longest Distances: 1,809 km (1,124 mi) NE-SW; 525 km (326 mi) SE-NW

Highest Point: Mount Toubkal 4,165 m (13,665 ft)

Lowest Point: Sebkha Tah 55 m (180 ft) below sea level

Land Use: 18% arable; 1% permanent crops; 28% meadows and pastures; 12% forest and woodland; 41% other

Located at the northwest corner of Africa, Morocco has coasts on both the Atlantic Ocean and the Mediterranean Sea. The country has been subjected to numerous political and cultural invasions that have contributed to its status as an important trading area and have given it an international flavor. Morocco and the neighboring states of Algeria and Tunisia constitute an area known as the Maghrib (''the time or place of sunset,'' ''the west''). Libya is also sometimes included in this grouping of Arab states with similar physical features and closely related histories and racial origins.

The topography is dominated by rugged chains of mountains. Among the several ways that have been devised for the regional subdivision of the country, the simplest is one consisting of three geographic regions: the mountainous interior, including plateaus and fertile valleys; the Atlantic coastal lowlands; and the semiarid and arid area of eastern and southern Morocco where the mountains descend gradually into the Sahara Desert. On the country's northern flank the Rif runs parallel to the Mediterranean coast. The peaks seldom exceed 2,100 m (7,000 ft), but erosion has carved deep ravines, particularly on the seaward slopes, that make penetration of these highlands difficult.

South of the Rif a series of three Atlas ranges somewhat overlap one another as they slant across the country on a generally northeast-southwest axis. The most northerly of the three, the Middle Atlas range, is separated from the Rif by only a narrow corridor. the city of Fès is located in this historically important Taza Gap through which successive waves of Arab invaders poured westward onto the Atlantic plains beginning in the seventh century A.D.

Immediately to the south of the Middle Atlas and parallel to it is the lofty High Atlas range, which is a whole interior region of elevations nowhere less than 1,200 m (4,000 ft). Some 720 km (450 mi) in length and up to sixty-four km (40 mi) in breadth, the High Atlas divides the country into two climatic zones—one that receives the westerly winds from the Atlantic and one that is influenced by the proximity of the Sahara. West of the city of Marrakech the High Atlas is a solid granite wall with peaks ranging from 3,000 m (10,000 ft) to over 3,900 m (13,000 ft). Eastward beyond the Tichka Pass the mountains drop to 2,400 m (8,000 ft) or lower, although one peak, Boulemane, exceeds 3,900 m (13,000 ft). The mountains are less symmetrical than those in the west, and their slopes and declivities are sharper.

Leeward Islands

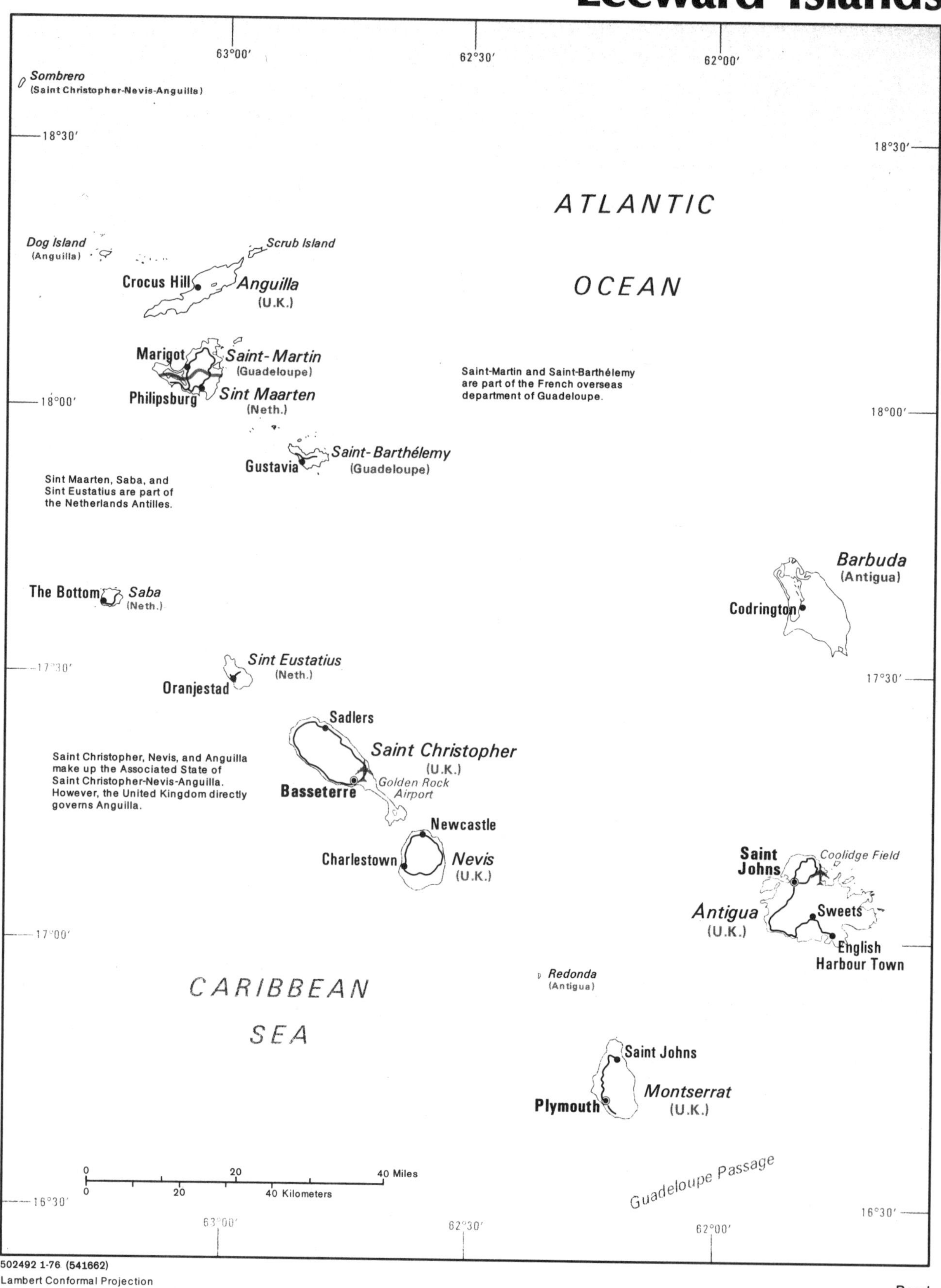

502492 1-76 (541662)
Lambert Conformal Projection
Standard parallels 17°20' and 22°40'
Scale 1:1,200,000

Road
Airport

Base 504118 7-79

Farther south and to the west lies the third of the Atlas systems, the Anti-Atlas. Backing on the High Atlas and separated from it in the west by the fertile valley of the Sous River, the Anti-Atlas has no peaks as high as those of the other Atlas ranges, and in those portions west of the Drâa River it has the appearance of a high, denuded plateau. East of the Drâa, it rises as a massif of old rocks with elevations of 2,400 m (8,000 ft) or more, and farther still to the east it drops gradually to rocky pre-Saharan uplands.

The coastal lowlands of western Morocco stretch from Tangier south to Essaouira. Small enclosed coastal plains occur south of Agadir and around the mouth of the Drâa River. The coast is flat and is bordered with sand dunes or marshes; the plain north of Essaouira is uninterrupted, although the depth varies widely. This plain is the fertile heart of Morocco, and because of the richness of the soil it has been the target of countless invasions and has formed a base for the rulers of Morocco throughout history.

Inland a large open but irregular plateau with elevations of between 540 and 900 m (1,800 and 3,000 ft) covers tens of thousands of sq mi. The soil is poor, but the area around Khouribga (known as the phosphates plateau) is the center of rich phosphate deposits that make Morocco one of the world's largest producers of this mineral.

Two major inland plains of agricultural importance lie between the plateau and the Atlas ranges. One, the Tadla Plain, is centered on the Oum er Rbia River, which has covered the plain with a rich deposit of silt. The other, the Haouz, is the basin of the Tensift River near Marrakech.

The entire area of the Atlantic plains and plateaus is relatively open and easily accessible, and transportation is well developed. Though low and regular, the coast offers few natural harbors, none of which is of significant size. Until the construction of man-made harbors at Casablanca—the principal port—and several smaller port facilities, landings were made by small boats through heavy surf.

Eastern Morocco is the lowland area between the Middle Atlas and the Algerian frontier, an extremely arid and sparsely populated area of monotonous tableland, valleys, and depressions that are the remnants of former lakes. Elevations reach 900 m (3,000 ft). In the north a narrow valley occupied by the Taza Gap extends westward between the Rif and Middle Atlas to the city of Fès.

The country has the most extensive system of rivers in North Africa. The streams generally flow northwest to the Atlantic or southeast to the Sahara; the Moulouya is an exception that flows 560 km (350 mi) northeast from the Atlas systems to the Mediterranean. The principal rivers with outlets to the Atlantic are the Sebou, Oum er Rbia, Bou Regreg, Tensift, and Sous. The Ziz and Rheris are the main rivers flowing southward into the Sahara. The largest of the rivers is the Sebou; together with its tributaries it represents nearly half of the country's surface water resources. None of the rivers is navigable, but in a country where seasonal droughts are common their value for irrigation is enormous.

PRINCIPAL TOWNS (estimated population, including suburbs, at mid–1981)

Casablanca	2,408,600	Kénitra	449,700
Rabat (capital)*	841,800	Tétouan	371,700
Fès (Fez)	562,000	Tanger (Tangier)	304,000
Marrakech (Marrakesh)	548,700	Safi	255,700
		Agadir	245,800
Meknès	486,600	Khouribga	229,600
Oujda	470,500	Béni-Mellal	204,800

*Including Salé.

Area and population:

Provinces	Capitals	area sq mi	area sq km	population 1985 estimate
Agadir	Agadir	2,282	5,910	647,000
Azilil	Azilil	3,880	10,050	399,000
Béni Mellal	Béni Mellal	2,732	7,075	749,000
Ben Slimane	Ben Slimane	1,066	2,760	185,000
Boulemane	Boulemane	5,558	14,395	140,000
Chaouen (Chefchaouen)	Chaouen (Chefchaouen)	1,680	4,350	328,000
Essaouira	Essaouira	2,446	6,335	409,000
Fès	Fès	2,085	5,400	877,000
Figuig	Figuig	21,618	55,990	105,000
Guelmim	Guelmim	11,100	28,750	141,000
al-Hoceima	al-Hoceima	1,371	3,550	331,000
Ifrane	Ifrane	1,278	3,310	106,000
el-Jadida	el-Jadida	2,317	6,000	818,000
el-Kelaa des Srarhna	el-Kelaa Srarhna	3,888	10,070	614,000
Kénitra	Kénitra	1,832	4,745	781,000
Khémisset	Khémisset	3,207	8,305	430,000
Khénifra	Khénifra	4,757	12,320	390,000
Khouribga	Khouribga	1,641	4,250	473,000
Marrakech	Marrakech	5,697	14,755	1,355,000
Meknès	Meknès	1,542	3,995	670,000
Nador	Nador	2,367	6,130	655,000
Ouarzazate	Ouarzazate	16,043	41,550	572,000
Oujda	Oujda	7,992	20,700	843,000
er-Rachidia	er-Rachidia	23,006	59,585	450,000
Safi	Safi	2,813	7,285	754,400
Settat	Settat	3,764	9,750	728,000
Sidi Kacem	Sidi Kacem	1,568	4,060	545,000
Tangier	Tangier	461	1,195	466,000
Tan-Tan	Tan-Tan	6,678	17,295	51,000
Taounate	Taounate	2,156	5,585	559,000
Taroudannt	Taroudannt	6,355	16,460	593,000
Tata	Tata	10,010	25,925	103,000
Taza	Taza	5,799	15,020	649,000
Tètouan	Tètouan	2,326	6,025	757,000
Tiznit	Tiznit	2,687	6,960	336,000
Perfectures				
Ain Chok-Hay Hassani	—			339,000
Ain Sebaa-Hay Mohammadi	—			471,000
Ben Msik-Sidi Othmane	—	623	1,615	737,000
Casablanca-Anfa	—			976,000
Mohammadia-Znata	—			173,000
Rabat-Saté	—	492	1,275	1,169,000
TOTAL		177,117	458,730	21,885,000

MOZAMBIQUE

BASIC FACTS

Official Name: People's Republic of Mozambique

Abbreviation: MOZ

Area: 799,380 sq km (308,542 sq mi)

Area—World Rank: 34

Population: 14,947,554 (1988) 20,463,000 (2000)

Population—World Rank: 52

Capital: Maputo

Boundaries: 7,003 km (4,351 mi); Tanzania 756 km (470 mi); South Africa 491 km (305 mi); Swaziland 108 km (67 mi); Zimbabwe 1,223 km (760 mi); Zambia 424 km (263 mi); Malawi 1,497 km (930 mi)

Coastline: 2,504 km (1,556 mi)

Longest Distances: 2,016 km (1,253 mi) NNE-SSW; 772 km (480 mi) ESE-WNW

Highest Point: Mount Binga 2,436 m (7,992 ft)

Lowest Point: Sea level

Land Use: 4% arable; 56% meadows and pastures; 20% forest and woodland; 20% other

When viewed in a regional perspective, the geographic significance of Mozambique comes into sharp relief: even a cursory glance at a topographical map of southern Africa reveals that the only convenient land routes connecting the fertile plains and highlands of south-central Africa with the Indian Ocean and the outside world run down the river valleys of Mozambique. Since prehistoric times these valleys—particularly the Rovuma, Save, Limpopo and, above all, the Zambezi—have served as highways for major population movements among the southern Bantu peoples.

Mozambique covers an area about twice that of California. Its 2,795 km (1,737 mi) of coastline run from the northern extremity of the country, at the mouth of the Rovuma River, to the southernmost point at Ponta Ouro, a distance of 1,963 km (1,220 mi) as the crow flies.

Topographically and in other ways the Zambezi Valley divides the country into northern and southern halves. North of the Zambezi Valley and to the east of the Malawi border and the Shire River a narrow littoral, in most places less than 32 km (twenty mi) wide, gives way to hills and low plateaus to the west, eventually rising to extensions of the great Zimbabwean Highlands, as does the westernmost part of all Mozambique. High points generally take the form of isolated peaks, or inselbergs, situated on plateaus, or of sharply rising systems of hills, escarpments, and plateaus. North of the Zambezi Valley the highest and most rugged features are found in three unconnected highland regions: the Livingstone-Nyasa Highland of Niassa Province; the Shire or Namuli Highlands in the western parts of Nampula and Zambezia provinces; and the Angonia Highlands of northeastern Tete Province. The Maconde Plateau, situated along most of the border between Tanzania and Cabo Delgado Province, is of lower elevation. At the head of the Zambezi Valley in Tete Province, north of the Zambezi and west of the Luia River, are the Tete Highlands.

South of the Zambezi Valley the littoral extends farther and farther inland until it takes up almost the entire width of the country—almost 322 km (200 mi) in Gaza and Inhambane provinces—with the exceptions of the small and isolated Gorongosa Highlands in west Sofala Province, the extension of the Mashonaland Plateau into westernmost Manica Province—which includes Mount Binga, the highest elevation in Mozambique at almost 2,438 m (8,000 ft)—and the thin strip of the Lebombo Mountains in Maputo Province (formerly Lourenco Marques). All told 44 percent of Mozambique is littoral lowlands and marshes, mostly south of the Zambezi and especially south of the Save; 17 percent is low plateaus and hills roughly between 183 and 610 m (600 and 2,000 ft) high; 26 percent is high plateaus and hills between 610 m and 1,005 m (2,000 and 3,300 ft) high; and 13 percent is mountains over 1,005 m (3,300 ft). The heaviest concentrations of population are along the littoral and its immediate hinterland, around the mouth of the Zambezi and near its confluence with the Shire, and in the Angonia Highlands.

Drainage

All the waters of Mozambique ultimately flow into the Indian Ocean. Five major basins and several smaller ones drain the country, although several have their major catchment areas as far away as eastern Angola. Flow tends to fluctuate, owing to the alternation of rainy and dry seasons. The greatest flow takes place between January and March, the least between June and August.

The largest and most important basin is that of the Zambezi River, historically the principal means of communication between inland central Africa and the coast; consequently the basin has been the home of a wide variety of both African and non-African populations. From the town of Tete downstream the valley is low lying and of a very gentle slope, Tete having an elevation of less than 152 m (500 ft). Farther upstream the river enters a narrow gorge, which prompted the construction of the Cabora Bassa dam—one of the largest in the world in power-

Area and population

		area		population
Provinces	**Capitals**	sq mi	sq km	1987 estimate
Cabo Delgado	Pemba	31,902	82,625	1,109,900
Gaza	Xai-Xai	29,231	75,709	1,128,700
Inhambane	Inhambane	26,492	68,615	1,167,000
Manica	Chimoio	23,807	61,661	756,900
Maputo	Maputo	9,944	25,756	544,700
Nampula	Nampula	31,508	81,606	2,837,900
Niassa	Lichinga	49,829	129,056	607,700
Sofala	Beira	26,262	68,018	1,257,700
Tete	Tete	38,890	100,724	981,300
Zambézia	Quelimane	40,544	105,008	2,952,200
City				
Maputo	—	232	602	1,006,800
TOTAL LAND AREA		303,623	786,380	14,360,800
inland water		5,019	13,000	
TOTAL AREA		308,642	799,380	

Mozambique
International boundary
Province boundary
National capital
Province capital
Railroad
Road
0 50 100 150 Kilometers
0 50 100 150 Miles
TANZANIA
ZAIRE
ZAMBIA
MALAWI
ZIMBABWE
BOTSWANA
SOUTH AFRICA
SWAZILAND
INDIAN OCEAN
Mozambique Channel
CABO DELGADO
NIASSA
NAMPULA
TETE
ZAMBEZIA
MANICA
SOFALA
INHAMBANE
GAZA
MAPUTO
Lake Nyasa
Lake Bangweulu
Bangweulu Swamps
Lake Kariba
Kariba Dam
Cabora Bassa Dam
Lago Chirua
Bassas da India (FRANCE)
Île Europa (FRANCE)
Kasama
Mansa
Mpika
Lubumbashi
Kitwe
Ndola
Serenje
Kapiri Mposhi
Kabwe
Lusaka
Kafue
Luangwa
Petauke
Chipata
Mchinji
Lilongwe
Mzuzu
Nkhata Bay
Nkhotakota
Senga
Fingoe
Furancungo
Songo
Tete
Vila Moatize
Zomba
Blantyre
Milange
Songea
Tunduru
Masasi
Mtwara
Mueda
Augusto Cardoso
Lichinga
Catur
Marrupa
Montepuez
Pemba
Cuamba
Nacala
Nampula
Lumbo
Moçambique
Alto Molócue
Angoche
Vila de Mocuba
Pebane
Quelimane
Vila de Sena
Inhaminga
Gorongosa
Chimoio
Catandica
Changara
Mutare
Vila do Dondo
Beira
Dombe
Espungabera
Nova Mambone
Inhassoro
Vilanculos
Chicualacuala
Massingir
Inhambane
Inharrime
Chokwe
Xai-Xai
Manhiça
Maputo
Zitundo
Chirundu
Kariba
Mount Darwin
Zawe
Shamva
Mtoko
Harare
Binga
Gokwe
Lupane
Eastnor
Gweru
Mvuma
Bulawayo
Masvingo
Gwanda
Rutenga
Beitbridge
Messina
Francistown
Selebi-Pikwe
Palapye
Pietersburg
Potgietersrus
Nylstroom
Nelspruit
Pretoria
Johannesburg
Vereeniging
Bethal
Mbabane
Vaal
Vaaldam
Luapula
Chambeshi
Kafue
Luangwa
Lunsemfwa
Zambezi
Sanyati
Mazoe
Shangani
Gwai
Lundi
Nuanetsi
Limpopo
Olifants
Komati
Rio Save
Save
Rio Púngoè
Rio Revuè
Búzi
Rio Changane
Rio Licungo
Rio Ligonha
Rio Lúrio
Rio Messalo
Rio Lugenda
Ruvuma
28
32
36
40
12
16
20
24

generating potential—near Songo. The dam has made the river navigable to that point, a distance of roughly 483 km (300 mi) and is expected to regulate the flow, thus making the river navigable year round. When flooding has been completed, the lake created by the dam will extend almost to the Zambian border, nearly 241 km (150 mi) away; the lake is expected to cover about 1,554 sq km (600 sq mi) and, because it fills in a very deep gorge, will reach an enormous capacity. The dam is expected to provide a considerable return in cheap hydroelectric power and in irrigation facilities, but there were indications that severe ecological problems would present themselves in the future, including difficulties in controlling choking water vegetation and the spreading of schistosomiasis, much as has taken place with respect to the Aswan High Dam in Egypt.

Other major basins from north to south include the Rovuma basin, fed by the Messinge, Luchulingo, Chiulezi, and Lugenda rivers; the Rovuma River is navigable from its confluence with the Lugenda to a point near its mouth. The Lurio basin comprises the Lurio River and its tributaries—the Maracoleta, Luleio, Macequesse, Nirongene, Nualo, Malema, and Lalana rivers. The Save River drains a generally low and marshy area. The Limpopo River, fed principally by the Changane River, drains the Limpopo basin. The Limpopo is a sluggish river other than at flood tide and was given the euphonious epithet of "the great gray-green, greasy Limpopo River, all set about with fever-trees" by Rudyard Kipling.

Three lakes in northern Mozambique, Lake Nyasa, Lake Chiuta, and Lake Shirwa, form part of the frontier with Malawi but have remained largely unexploited. Many low areas are marshy, particularly along the coast. Much of the area around the mouth of the Zambezi and a belt running along the lower reaches of the Púngoé River and its tributary the Mucombeze north to the Zambezi is marshy, hindering north-south communications and promoting the spread of disease.

MYANMAR

BASIC FACTS

Official Name: Socialist Republic of the Union of Myanmar

Abbreviation: MYA

Area: 676,577 sq km (261,228 sq mi)

Area—World Rank: 39

Population: 39,632,183 (1988) 48,553,000 (2000)

Population—World Rank: 24

Capital: Yangon

Boundaries: 8,134 km (5,055 mi); China 2,185 km (1,358 mi); Laos 238 km (148 mi); Thailand 1,799 km (1,118 mi); Bangladesh 233 km (145 mi); India 1,403 km (872 mi)

Coastline: 2,276 km (1,414 mi)

Longest Distances: 1,931 km (1,200 mi) N-S; 925 km (575 mi) E-W

Highest Point: Hkakabo Mountain 5,881 m (19,296 ft)

Lowest Point: Sea level

Land Use: 15% arable land; 1% permanent crops; 1% meadows and pastures; 49% forest and woodland; 34% other

In outline, Myanmar can be compared to a diamond-shaped kite with a long tail. At points of maximum extent the kite's body stretches some 1,287 km (800 mi) from north to south and 804 km (500 mi) from east to west. The tail, protruding from the southeastern flank of the kite, shares the Malay Peninsula with Thailand for another 804 km (500 mi) southward to the Isthmus of Kra.

The land frontiers, with the exception of a few water features, consist of a ring of tortuous uplands that become ranges, defining most of the borders. These mountains make overland transportation between Myanmar and its neighbors very difficult and, as a consequence, most foreign trade must be seaborne. The long coastline provides for this purpose several good harbors associated with river mouths. Internally, the principal means of communication is the Irrawaddy River system which drains the greater part of the country in a generally north-to-south pattern.

Topographic features divide the country into four north-to-south belts. Reading westward from the borders with Communist China and Thailand, they are the Shan Plateau, the Central Belt, the Western Mountain Belt, and the Arakan Coastal Strip.

The deeply dissected Shan Plateau rises to an average elevation of about 914 m (3,000 ft) above sea level. Its western edge is clearly marked off from the Central Belt by a north-south cliff, or fault scarp, which often rises 610 m (2,000 ft) in a single step. Much of the surface of this plateau is of a steeply rolling, hilly nature. In other portions mountain masses rise abruptly to heights of 1,829 m (6,000 ft) or more.

Several of the shorter streams in this plateau flow sluggishly through broad valleys, but the largest river, the Salween, is deeply entrenched. It flows in a series of rapids and waterfalls through steep, narrow valleys, and little or no land is available for cultivation in the valley bottoms. It has, however, considerable, but little developed, potential as a source of hydroelectric power. The Shan Plateau area available for cultivation is accordingly restricted to rolling plateaus and a few small river valleys.

To the south toward the Isthmus of Kra, the ranges of the Malay Peninsula are repeated northward to merge with the plateau. This area, roughly corresponding to the tail of the kite, is sometimes treated as a separate region. It is, however, topographically associated with the Shan Plateau.

The major part of the Central Belt is composed of ancient river valleys that have been covered by deep, alluvial deposits through which the Irrawaddy, its tributary the Chindwin, and the Sittang rivers flow. The lower valleys of the Irrawaddy and Sittang rivers form a vast, low-lying delta area of about 25,900 sq km (10,000 sq mi). The delta continues to move seaward at a rate of 4.8 km (3 mi) per century because of the heavy silting brought by the rivers. It is the most intensively populated and farmed area in the country and is the heartland of the nation's rice economy. The northern portion of the Central Belt above the 25th parallel is a relief stemming from a mountain mass where the ridges of the Himalaya Mountains curve southward and become the mountain system of Myanmar's eastern frontier. These mountains are very high and rugged and contain Hkakabo Razi on the northern frontier, which rises to almost 6,096 m (20,000 ft) and is the highest peak in the nation. Mount Saramati on the Indian border, at 3,810 m (12,500 ft) is the second highest.

The Western Mountain Belt is composed of ranges that originate in the northern mountain arc and continue southward to the extreme southwestern corner of the country. Here they disappear under the sea only to reappear some 321 km (200 mi) offshore as India's Andaman Islands. These ranges are known by several names along the Myanmar-Assam border, but in the southern portion of the belt where they lie entirely within Myanmar they are known as the Arakan Mountains. As in the Shan Plateau, the landscape is dominated by a series of parallel ridges separated by streams flowing in restricted valleys. Here, however, the slopes are very steep, and the mountains are far more rugged than any in the Shan Plateau. Mount Victoria, for example, rising to about 3,048 m (10,000 ft) is the highest peak in the Arakan Mountains and the third highest in the nation. Passage across the Western Mountain Belt is possible only by precipitous trails through a few traverse gaps. The region is hard to reach and has a very small proportion of land level enough for more than the most rudimentary kind of agriculture.

The Arakan Coastal Strip is a narrow, predominantly alluvial belt lying between the Arakan Mountains and the Bay of Bengal. In its northern portion there is a broad area of level land formed by the flood plains of the several short streams that come down from the mountains. In the south the coastal strip narrows and is displaced in many places by hill spurs that reach the bay. Offshore there are many large islands and hundreds of smaller ones, a number of which are low lying and level enough to permit intensive rice cultivation. The region as a whole has a high percentage of rich farmland, but it suffers from a serious lack of communications with the rest of Myanmar. Surface travel to Bangladesh, for example, is far easier than travel to other parts of the mother country.

River Systems

Myanmar's primary drainage system consists of the Irrawaddy River complex. This river and its tributaries and other streams that make up its basin drain some two-thirds of the country. It rises near the northernmost tip of Myanmar and flows the entire length of the country to its enormous delta where nine mouths empty into the Andaman Sea. The Irrawaddy's most important tributary, the Chindwin River, drains the northwest and is fed by tributary streams from the mountains of the Indian frontier. The Irrawaddy is navigable for a distance of over 1,287 km (800 mi) inland; and the Chindwin, for another 160 km (100 mi).

Area and population

Divisions	Capitals	area sq mi	area sq km	population 1983 census
Irrawaddy	Bassein	13,567	35,138	4,994,061
Magwe	Magwe	17,305	44,820	3,243,166
Mandalay	Mandalay	14,295	37,024	4,577,762
Pegu	Pegu	15,214	39,404	3,799,791
Yangon	Yangon	3,927	10,171	3,965,916
Sagaing	Sagaing	36,535	94,625	3,862,172
Tenasserim	Tavoy	16,735	43,343	917,247
States				
Chin	Falam	13,907	36,019	368,949
Kachin	Myitkyinā	34,379	89,041	904,794
Karen	Pa-an	11,731	30,383	1,055,359
Kayah	Loi-kaw	4,530	11,733	168,429
Mon	Moulmein	4,748	12,297	1,680,157
Rakhine (Arakan)	Sittwe (Akyab)	14,200	36,778	2,045,559
Shan	Taunggyi	60,155	155,801	3,716,841
TOTAL		261,228	676,577	35,307,913

PRINCIPAL TOWNS (population estimates, 1983)

Town	Population	Town	Population
Yangon	2,458,712	Moulmein	219,991
Mandalay	532,895	Akyab	143,000
Bassein	335,000	Taunggyi	80,678*

*Population at 1973 census

Also regarded as a portion of the Irrawaddy basin is the Sittang River, which rises just south of Mandalay and parallels the Irrawaddy on its eastern flank. The Sittang has suffered from excessive silting as a consequence of cultivation and forest clearing. Between 1910 and 1958 its depth at Toungoo, located about the midpoint of its 483 km (300-mi) length, dropped from an average of eighteen ft to three ft, but its width doubled. The Sittang can be used by small boats for short distances between its shoals, gorges, and rapids.

The other large river, the Salween, rises in Communist China and parallels the Irrawaddy as it courses through the eastern part of Myanmar. Until it reached the Shan Plateau it has no tributaries, but in that area it is fed by numerous streams, some of which are several hundreds of mi in length. The Salween has a narrow, ribbon-like drainage basin, and its course is so deeply incised and broken and suffers such enormous changes in level that it is perhaps the least useful of the major rivers for navigation purposes. In this river, too, there has been excessive silting, which has interfered with harborage around its mouth at Moulmein.

In Arakan Division there are four major rivers that flow from north to south and are separated by abrupt watersheds related to the Arakan Mountains fold structure. All empty into the Bay of Bengal. The courses of these rivers become trellis patterns near the coast, there is much silting, and all have deltaic extensions. Except for the Kaladan River, which is navigable upstream for about 88 km (55mi), they are of little use for navigation but are important because of population clusters at their estuaries.

The remaining river system consists of short streams that run westward to the Andaman Sea in Tenasserim.

Base 503538 12-77 (543580)

They flow as torrents separated by ranges that are repeated seaward as a string of offshore islands. All of these rivers carry heavy silt loads, none is navigable, and they are important primarily as settlement points.

There are few lakes. The largest is Lake Inle, which covers about 259 sq km (100 sq mi) in a basin area of the Shan Plateau. It is the residue of a much larger body of water that is still shrinking. Drained by a tributary of the Salween River, it abounds in fish and is surrounded by very fertile paddies and a cluster of farm villages. It is also a much-favored recreation spot. Other lakes and ponds are for the most part either closed bodies in the courses of former rivers of Upper Myanmar or are formed by reclaiming marshes of the deltas.

NAMIBIA

BASIC FACTS

Official Name: Namibia (South-West Africa)

Abbreviation: NMB

Area: 823,144 sq km (317,818 sq mi)

Area—World Rank: 33

Population: 1,301,598 (1988) 1,639,000 (2000)

Population—World Rank: 129

Capital: Windhoek

Boundaries: 3,935 km (2,443 mi); Angola 1,376 km (854 mi); Botswana 1,360 km (844 mi); South Africa 966 km (600 mi); Zambia 233 km (145 mi)

Coastline: 1,489 km (925 mi)

Longest Distances: 1,498 km (931 mi) SSE-NNW; 880 km (547 mi) ENE-WSW

Highest Point: Brandberg 2,579 m (8,461 ft)

Lowest Point: Sea Level

Land Use: 1% arable; 64% meadows and pastures; 22% forest and woodland; 13% other

For physical features, See South Africa

Area and population

Magisterial Districts	Capitals	area sq mi	area sq km	population 1987 estimate
Bethanien	Bethanien	6,951	18,004	3,000
Boesmanland	Tsumkwe	7,131	18,468	3,000
Caprivi Oos	Katima Mulilo	4,453	11,533	44,000
Damaraland	Khorixas	17,977	46,560	28,000
Gobabis	Gobabis	16,003	41,447	25,000
Grootfontein	Grootfontein	10,239	26,520	25,000
Hereroland-Oos	Otjinene	20,058	51,949	22,000
Hereroland-Wes	Okakarara	6,371	16,500	18,000
Kaokoland	Opuwo	22,467	58,190	20,000
Karasburg	Karasburg	14,717	38,116	11,000
Karibib	Karibib	5,108	13,230	10,000
Kavango	Rundu	19,674	50,955	122,000
Keetmanshoop	Keetmanshoop	14,788	38,302	20,000
Lüderitz	Lüderitz	20,488	53,063	16,000
Maltahöhe	Maltahöhe	9,874	25,573	6,000
Mariental	Mariental	18,413	47,689	24,000
Namaland	Giberon	8,154	21,120	15,000
Okahandja	Okahandja	6,811	17,640	15,000
Omaruru	Omaruru	3,253	8,425	6,000
Otjiwarongo	Otjiwarongo	7,934	20,550	19,000
Outjo	Outjo	14,951	38,722	10,000
Owambo	Ondangwa	20,000	51,800	520,000
Rehoboth	Rehoboth	5,476	11,182	33,000
Swakopmund	Swakopmund	17,258	44,697	18,000
Tsumeb	Tsumeb	6,340	16,420	22,000
Windhoek	Windhoek	12,920	33,489	129,000
TOTAL		317,818	823,144	1,184,400

NAURU

BASIC FACTS

Official Name: Republic of Nauru

Abbreviation: NAR

Area: 21.2 sq km (8.2 sq mi)

Area—World Rank: 209

Population: 8,902

Population—World Rank: 205

Capital: None

Land Boundaries: Nil

Coastline: 19 km (12 mi)

Longest Distances: 5.6 km (3.5 mi) NNE-SSW 4 km (2.5 mi) ESE-WNW

Highest Point: 64 m (210 ft)

Lowest Point: Sea level

Land Use: 100% other

Nauru is an oval-shaped island in the West-Central Pacific 53 km (33 mi) from the Equator, 3,539 km (2,200 mi) NE of Sydney and 3,934 km (2,445 mi) SW of Honolulu. It is the smallest nation in Asia.

There is no formal capital; there are no urban centers. Government offices are in the Domanaeb building in the Uaboe district.

The island is encircled by a sandy beach, which rises gradually forming a fertile section no wider than 275 m (300 y). A coral cliff rises from this belt to a central plateau about 60 m (200 ft) high. A brackish lagoon, known

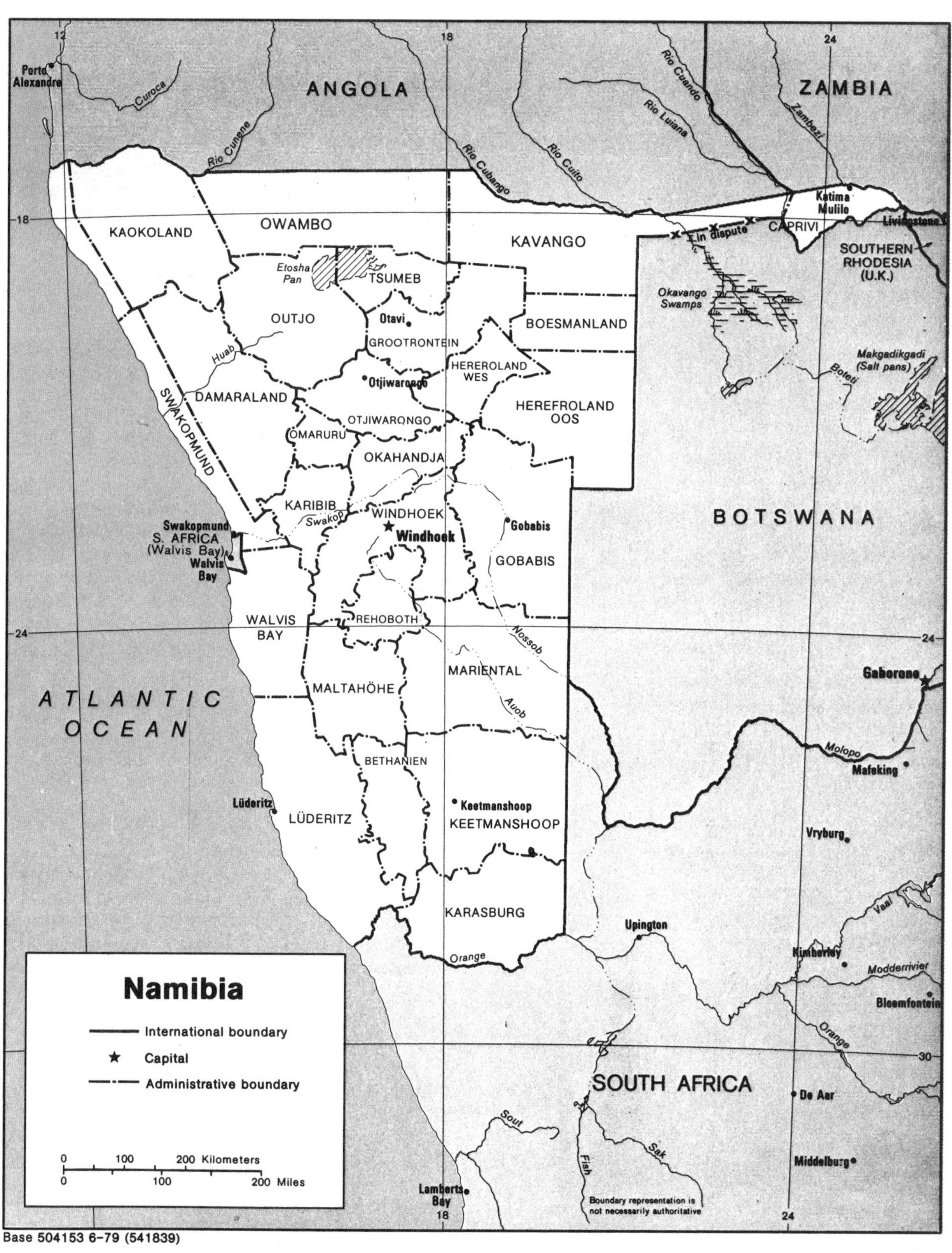

Base 504153 6-79 (541839)

as Buada, covers some 121 ha (300 ac) at the southeastern end of the plateau.

Area and population	area		population
Districts	sq mi	sq km	1983 census
Alwo	0.4	1.1	812
Anabar	0.6	1.5	226
Anetan	0.4	1.0	265
Anibare	1.2	3.1	87
Baitsi	0.5	1.2	363
Boe	0.2	0.5	578
Buada	1.0	2.6	467
Denigomodu	0.3	0.9	2,600
Ewa	0.5	1.2	269
Ojuw	0.4	1.1	132
Meneng	1.2	3.1	1,024
Nibok	0.6	1.6	338
Uaboe	0.3	0.8	272
Yaren	0.6	1.5	559
TOTAL	8.2	21.2	8,043

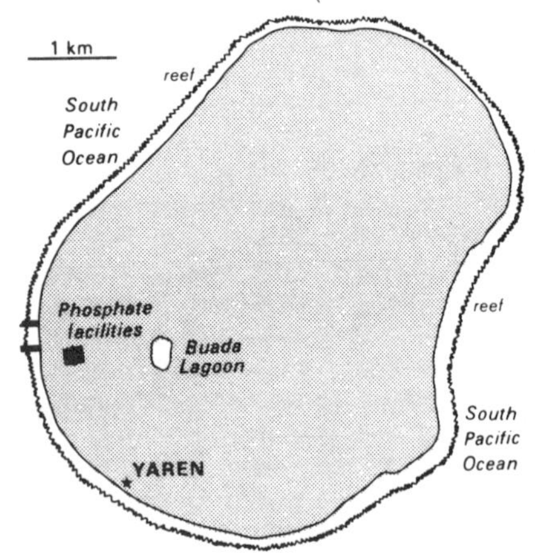

NEPAL

BASIC FACTS

Official Name: Kingdom of Nepal

Abbreviation: NEP

Area: 147,181 sq km (56,827 sq mi)

Area—World Rank: 89

Population: 18,252,001 (1988) 23,176,000 (2000)

Population—World Rank: 43

Capital: Kathmandu

Land Boundaries: 2,671 km (1,660 mi); China 1,078 km (670 mi); India 1,593 km (990 mi)

Coastline: Nil

Longest Distances: 885 km (550 mi) SE-NW; 201 km (125 mi) NE-SW

Highest Point: Mount Everest 8,848 m (29,028 ft)

Lowest Point: 60 m (197 ft)

Land Use: 17% arable; 13% meadows and pastures; 33% forest and woodland; 37% other

Nepal, a predominantly mountainous rectangle, is a landlocked country bounded by Tibet on the north, Sikkim on the east and India on the south and west. The nearest seaport is Calcutta, 644 km (400 mi) to the southeast at the head of the Bay of Bengal.

Nepal presents a wider range of physical diversity than probably any other country of comparable size. The complex mountain mass in the north contains some of the world's highest peaks—six are more than 7,924 m (26,000 ft) above sea level. To the south, no more than 161 km (100 mi) from these barren, icy heights, the cultivated fields and steaming jungles of the northern rim of the Gangetic Plain are less than 183 m (600 ft) above sea level.

Numerous streams and rivers flow generally southward out of the mountains, meander across the Tarai plain and finally join the Ganges in northern India. The presence of fertile alluvial soil at stream junctions and at other places in the valley bottoms is a major determinant in the settlement pattern, most of the largest population concentrations being along the rivers and their principal tributaries. Virtually the entire country is drained by three large river systems: in the east, the Kosi, with its seven large tributaries, three of which rise in Tibet, north of the main Himalayan range; in the center, the Narayani (called the Gandak in India); and in the west, the Karnali (called the Girwa in the Western Tarai and the Gogra in India).

The country can be divided into three main geographic regions: the Mountain Region, which constitutes almost three-fourths of the total area; the Katmandu Valley (sometimes called the Nepal Valley), a relatively small, disc-shaped area enclosed in the east-central part of the Mountain Region; and the Tarai Region, a narow belt which extends along the boundary with India in the northern part of the Ganges River plain. Characteristics of terrain peculiar to each region are associated with sharp contrasts in soil, vegetation, climate, and even in the economic and social patterns of the people.

The Mountain Region

The Mountain Region is part of the Himalayan range. Its major heights in northeastern Nepal generally define the boundary with Tibet, while in the northwest they lie just to the south of the boundary. The gigantic peaks and deep gorges of the region provide much of the subject matter of the myths and folklore of the various local ethnic groups. Even for Nepalese living in the Tarai flatlands, the mountains are a near presence, too large and spectacular ever to be entirely out of mind.

The whole Mountain Region is marked by a series of parallel north-south ridges flanking deep, narrow, southward-sloping valleys. Extending east-west across the southern edge of the Region are the subsidiary Mahabharat Lekh and the Siwalik Ranges, both much lower than the main Himalayan range. The rivers in the principal valleys rise some 80 to 160 km (50 to 100 mi) inside Tibet on the high plateau north of the boundary. These streams are older than the mountain mass through which they flow, having created their valleys by erosion as the mountain barrier lifted around them. Thus, the actual watershed is not generally the line of high peaks in the region itself, but the Tibetan plateau farther north. Drainage north of the main Himalayan range is into the Brahmaputra River in Tibet (where it is known as the Tsangpo); in the south, into the Ganges. The waters of both rivers virtually join in the delta region northeast of Calcutta before emptying into the Bay of Bengal.

The valleys, hills and slopes of the Mountain Region are densely populated wherever tillable soil can be found. Most communication routes are restricted to treacherous tracks feasible only for travel by foot. Precipitation in most places is sufficient to support dense forests at elevations up to about 3,962 m (13,000 ft). In settled areas, clearings have been made by cutting for timber or by burning to open up croplands. Large tracts of untouched timber remain in the more inaccessible areas.

Based mainly on differences in physical features and climate, the region may be subdivided into three general areas by two lines, one running generally northward from Kathmandu and the other, about 241 km (150 mi) to the west, extending northward from the foothills near the boundary with India. From east to west, these subdivisions are designated the Eastern Mountains, the Western Mountains and the Far Western Mountains.

Eastern Mountains

The Eastern Mountain area 26,195 sq km (10,114 sq mi) has four of the six highest peaks in the world: Mount Everest 8,847 m (29,028 ft), Mount Lhotse 8,500 m (27,890 ft), Mount Makalu 8,480 m (27,824 ft) and Mount Cho Oyu 8,189 m (26,867 ft). All are situated on a 77 km (30-mi) segment of the boundary with Tibet. Forming a massive, white, saw-toothed range, they are visible on clear days from points throughout eastern Nepal. Trails leading up the principal valleys cross the Tibetan frontier over four difficult but well-known passes: Khangla Deorali and Rakha La to the east of the peaks, and Nangpa La and Kodari (Kuti) to the west.

Western Mountains

The 28,686 sq km (11,076 sq mi) of the Western Mountain area present a jumble of ridges and deep valleys projecting at various angles from the main Himalayan range. Relatively heavy preciptation supports some of the lushest vegetation in the country. Two mountains dominate the area: Dhaulagiri 8.172 km (26,813 ft) and Annapurna 8,077 m (26,502 ft). Both are within 80 km (50 mi) of the town of Pokhara, which is less than 1,524 m (5,000 ft) above sea level. The principal pass from this part of the region into Tibet is at Rasua Garhi, commonly called the Girange Dzong (Kyirong) Pass after the nearby Tibetan town about 80 km (50 mi) north of Kathmandu. Other passes to the west include the Gya La at Larkya, the Kore La at Mustang and the Yansang Bhanjyang above Tingjegaon.

Far Western Mountains

The Far Western Mountain area is the driest and most sparsely inhabited section of the Mountain Region. Its 48,896 sq km (18,879 sq mi), comprising almost 35 percent of the country's total area, contain less than 20 percent of the country's population. The scattered settlements of subsistence farmers, animal herders or mountain porters are generally confined to the river valleys. The southward drainage pattern is interrupted in many places by east-west ranges around which the streams zigzag on their way to the Ganges. Three passes in this area lead into Tibet: the Namja La (above Mugu), the Takhu La and the Nara Lagna (on the Karnali River, west of Munchu), all at elevations of approximately 4,876 m (16,000 ft).

The Katmandu Valley

The Katmandu Valley, just south of the junction between the Eastern and Western Mountains, is a circular basin of only 565 sq km (218 sq mi), said to be a dried-up lakebed. In it are the kingdom's three largest towns, including the capital, and it is generally regarded by Napalese as the heart of the country. The valley floor, which is between 1,219 and 1,524 m (4,000 and 5,000 ft) above sea level, is protected from icy Tibetan winter winds by the Himalayan heights to the north. On the south it is shielded from the extreme effects of the summer monsoons by the encircling Mahabharat Lekh range and it is drained by the area's principal river, the Baghmati Nadi. With ample rainfall and virtually a year-round growing season, the intensively cultivated soil provides food for this densely populated area.

Area and population		area		population
Development regions **Geographic regions**	**Capitals**	sq mi	sq km	1981 census
Eastern	Dhankūtā	10,987	28,456	3,708,923
Mountain				338,439
Hill				1,257,042
Tarai				2,113,442
Central	Kāthmāndu	10,583	27,410	4,909,357
Mountain				413,143
Hill				2,108,433
Tarai				2,387,781
Western	Pokharā	11,351	29,398	3,128,859
Mountain				19,951
Hill				2,150,939
Tarai				957,969
Mid-western	Surkhet	16,362	42,378	1,955,611
Mountain				242,486
Hill				1,042,365
Tarai				670,760
Far-western	Dipāyal	7,544	19,539	1,320,089
Mountain				288,877
Hill				604,336
Tarai				426,876
TOTAL		56,827	147,181	15,022,839

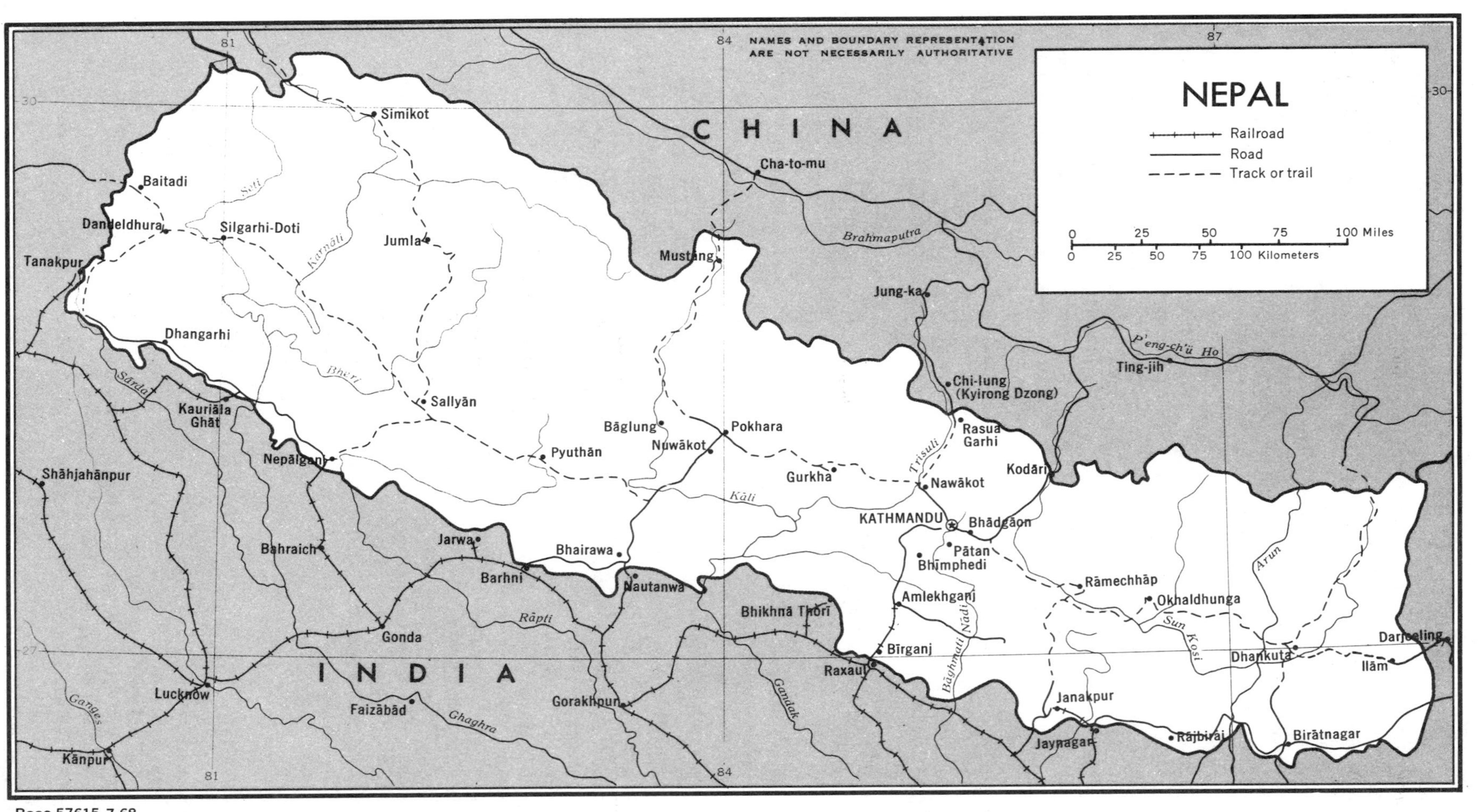

Base 57615 7-68

The Tarai Region

The Tarai Region, with a total area of 23,220 sq km (8,969 sq mi), consists mainly of a narrow belt of flat, alluvial land on the boundary with India. A northern extension of the Gangetic Plain, it varies between 46 and 183 m (150 and 600 ft) in altitudes and between 8 and 88 km (5 and 55 mi) in width. On the northern edge of this fertile strip is the Siwalik Range, sometimes called the Churia Hills or Churia Range, which rises to heights of almost 1,524 m (5,000 ft). This range is paralleled some 32 km (20 mi) to the north by the narrow Mahabharat Lekh, with elevations up to 3,048 m (10,000 ft). The Tarai is crossed by numerous streams which, particularly in the east, during the annual monsoon floods carry down tons of silt, sand, gravel and huge boulders from the mountains to the north.

Precipitation varies widely from east to west. In the east, heavy rainfall permits intensive cultivation of crops throughout the year, and uncultivated areas are covered with jungle vegetation or high grasses. In the west, relatively light and uncertain rainfall limits cultivation to small plots cleared from jungles, which line the streams. Malaria is endemic in the entire region, and the jungles, particularly those in the far western section are the habitat of a wide variety of tropical wild life, including several species of large game animals and poisonous snakes.

Six subregions of the Tarai can be distinguished on the basis of topography and climate: the Eastern Tarai, the Eastern Inner Tarai, the Center Inner Tarai, the Western Tarai, the Midwestern Tarai (sometimes called the West Inner Tarai) and the Far Western Tarai.

Eastern Tarai

The Eastern Tarai ranks next to the Katmandu Valley in the favorable conditions it offers for human habitation. It is generally level and well drained; the soil is fertile; and the rainfall is ample and dependable. Moreover, it benefits commercially from road and rail connections with nearby population centers in India.

Eastern Inner Tarai

The Eastern Inner Tarai consists mainly of the narrow eastern section of the Mahabharat Lekh range. Steep slopes limit cultivation to scattered patches of grain on the hillsides. The region is well covered with dense but often inaccessible forests. The valleys are deep, narrow, wet and infested with malaria-carrying mosquitoes. The sparse population is virtually isolated from the rest of the country.

Center Inner Tarai

The Center Inner Tarai is a transition area between the Eastern Tarai and the Katmandu Valley to the north. The terrain consists mainly of severely eroded hill slopes and forested mountains. Alluvial soil is found in the river valleys. Despite oppressive humidity and a high incidence of malaria, some of the country's major agricultural development and settlement expansion projects are underway in this subregion.

Western Tarai

The Western Tarai is a narrow strip of flat, fertile land devoted to the production of vegetables and grains. The landscape is dotted with villages, many of which are connected by cart roads. Population density ranks after that of the Katmandu Valley and the Eastern Tarai.

Midwestern Tarai

The Midwestern Tarai is the only region where the southern ridges of the Siwalik Range extend across the boundary with India. The characteristic terrain is rounded hills covered with open forests. Monsoon rains have severely eroded most of the slopes, and the valley bottoms are generally strewn with boulders, driftwood and other debris carried down by the runoff.

Far Western Tarai

The Far Western Tarai is the driest region in the country, but it still has ample rainfall for vegetation growth. Some monsoon seasons are marked by torrential rains and deep mud, leaving water standing in vast swampy areas during the periods between rains. Forests have been denuded in some accessible places by cutting to meet the demand of nearby Indian markets for timber or by burning to clear land for farming. Population density is almost as low as that of the sparsely settled mountainous area to the north.

MAJOR RIVERS

Three separate river systems, each having its headwaters on the Tibetan plateau, drain almost all of Nepal. The Kosi River drains the Eastern Mountains; the Narayani, the Western Mountains; and the Karnali, the Far Western Mountains. After plunging through deep gorges, the waters of these streams drop their heavy sediment and debris on the plains. Most rivers in the Tarai overflow their banks onto wide floodplains during the rainy season. Many shift their courses during this period, and the receding waters leave vast stagnant pools in former sections of the streambeds.

Besides providing fertile alluvial soil for cultivation, the heavy flow of water through a widespread network of narrow river channels presents great possibilities for hydroelectric development. Meanwhile, the deep gorges are formidable obstacles to communications and contribute to the virtual isolation of many upstream settlements.

The Kosi

The Kosi River has seven major affluents. The principal one, the Arun, rises almost 160 km (100 mi) inside the Tibetan plateau. Two other major tributaries, the Sun Kosi and the Tamur, flow generally eastward and westward, respectively, and join the Arun just north of the eastern section of the Mahabharat Lekh range, to form the southward-flowing Sapt Kosi. Its floodwaters move slowly across the Tarai within no defined banks from June to September each year and leave very heavy deposits of alluvium.

The Narayani

The Narayani cuts through the Western Mountains, and the gorges of some of its tributaries are the deepest in the world. Northwest of Pokhara, the Kali Gandaki tributary, with its streambed at the elevation of 1,106 m (3,630 ft) flows between the region's highest peaks, Dhaulagiri and Annapurna, which are only 35 km (22 mi) apart. Occasional shallow basins at high altitudes with intervening rapid descents present great potentialities for hydroelectric exploitation. The lower Narayani is navigable for small steamers and timber barges in the winter season or when the river is not in flood stages.

The Karnali

The Karnali River is noted for its deep gorges, which are generally too wide to be crossed by locally built suspension bridges. Moreover, its current is too rapid in most places to be negotiated safely by the dugout canoes of the area. Thus, the river tends to isolate rather than link the settlements along its course and to hinder rather than facilitate travel to other parts of the country.

NETHERLANDS

BASIC FACTS

Official Name: Kingdom of The Netherlands

Abbreviation: NET

Area: 41,785 sq km (16,133 sq mi)

Area—World Rank: 120

Population: 14,716,100 (1988) 15,245,000 (2000)

Population—World Rank: 51

Capital: The Hague (official) Amsterdam (de facto)

Boundaries: 1,605 km (997 mi); West Germany 556 km (345 mi); Belgium 407 km (253 mi)

Coastline: 642 km (399 mi)

Longest Distances: 312 km (194 mi) N-S; 264 km (164 mi) E-W

Highest Point: Vaalserberg 321 m (1,053 ft)

Lowest Point: Prins Alexander Polder 6.7 m (22 ft) below sea level

Land Use: 25% arable land; 1% permanent crops; 34% meadows and pastures; 9% forest and woodland; 31% other

The Netherlands forms part of the Northwest European Plain and is bounded in the east by West Germany, on the south by Belgium and on the north and west by the North Sea. It is crossed almost at its central point by latitude 52° North and longitude 5° East. The actual area is slightly larger than Belgium and slightly smaller than Denmark or Switzerland. The official name is Netherlands (Low Lands), but the country is commonly referred to as Holland, which, strictly speaking, applies only to the western coastal provinces of North and South Holland.

The highest point is in the extreme Southeast, 321 m (1,053 ft) above sea level. The areas in the North and West that lie below sea level account for nearly half of the total land area. The lowest area is a reclaimed polder northeast of Rotterdam and is 6.7 m (22.0 ft) below sea level. By the end of the present century, when the reclamation of the Zuyder Zee will be completed, an area roughly equivalent to the area relinquished to the sea in past centuries will have been reclaimed.

Despite its small size, the Netherlands has a varied topography as a result of its complex geological history. The country is divided into two main regions, one comprising areas below sea level (Low Netherlands) and the other those above sea level (High Netherlands). Although primarily based on elevation, this classification coincides with the broad division of the country according to its geological formation. The High Netherlands was formed mainly in the Pleistocene Age (which began about 2 million years ago and ended about 10,000 years ago) and is composed chiefly of sand and gravel. On the other hand, the Low Netherlands is relatively younger, having been formed in the Holocene Age (less than 10,000 years ago) and consists mainly of clay and peat. There are other differences. The High Netherlands is undulating and even hilly in places, with farms alternating with woodland and heath. The Low Netherlands is predominantly flat, and is intersected by natural and man-made waterways.

The Netherlands has nine distinct topographical regions:

1. The South Limburg Plateau is the only part of the country not classed as lowland. The hills, which rise to over 300 m (984 ft), are the foothills of the Central European Plateau. This is also virtually the only area of the country where rocks can be found at or near surface levels.
2. The ground moraine region of Drenthe and East Friesland covers the northern part of the country approximately from Haarlem to Nymegen.
3. The terminal moraine region covers the central part of the High Netherlands. There are parallel ranges of hills up to 100 m (328 ft) high in the provinces of Utrecht, Gelderland and Overyssel, dissected by the valley of the Yssel River.
4. The sandy region of North Brabant and Limburg.
5. The raised bog region, a transitional region between the High and Low Netherlands where marshy conditions were conducive to peat formation.
6. The peat regions of Holland and Utrecht, Friesland and Overyssel, an area historically subject to erosion by

the sea. The polders are more than 4 m (13 ft) below sea level.

7. The young marine clay regions in the southwestern and northern coastal districts, including areas reclaimed from the Zuyder Zee.

8. The alluvial clay regions.

9. The dunes created by the action of wind and water. The new dunes are at least 30 m (98 ft) high in places and several kilometers wide.

Much of the Low Netherlands has been wrested from the sea over the course of some eight centuries. Before the rise in the sea level after the last great ice age, large parts of the North Sea were dry, and Great Britain was joined to the Continent. Rivers such as the Rhine, the Thames and the Elbe flowed on well to the north of their present courses and did not empty into the sea until what is now the Dogger Bank. After the sea level had reached approximately the present coastline, the rise became slower and more irregular. At times of rapid rise, extensive coastal areas were swallowed by the sea, and only islands remained in the southwestern and northern Netherlands. The former Zuyder Zee reached its greatest extent about A.D. 1250. The struggle waged by the inhabitants against the encroachments of the sea was purely defensive. They first built homes or villages on artificial mounds known as "terps," which later were linked by dikes. In the 17th century some of the lakes in North Holland were drained with the use of windmills. The 180,000 ha (44,479 ac) Haarlem-mermeer, southwest of Amsterdam, and the area north of Rotterdam were drained in this fashion. In the archipelago of the Southwest and in the coastal areas of the North, reclamation took place in a different manner. The sea flowed in twice a day at high tide and left sand and silt behind as it retreated. When this process continued long enough, these areas came to be above sea level and were then surrounded by a dike. In the areas north of Groningen and in Friesland, low dams were built out into the sea and behind which sand and silt could quietly settle.

After 1900, land reclamation was undertaken on a larger scale. An ambitious plan was drawn up for reclaiming part of the Zuyder Zee. The first polder, Weiringermeer, was drained in 1930. In 1932, the Zuyder Zee—now called Usselmeer or Lake Ussel—was sealed off from the Wadden Zee by a 30-km (19 mi) barrier dam. Since then, other areas reclaimed from the sea include Lauwers Zee on the northern coast between Groningen and Friesland, and the Maasvlakte, south of the entrance to the new waterway. Environmental objections have delayed the reclamation of the Wadden Zee.

PRINCIPAL TOWNS (estimated population at 1 January 1987)

Amsterdam (capital)*	682,702	Nijmegen	146,639
Rotterdam	572,642	Apeldoorn	145,696
's-Gravenhage (The Hague)*	445,127	Enschede	144,227
		Zaanstad	128,388
Utrecht	229,326	Arnhem	127,671
Eindhoven	190,962	Breda	119,427
Groningen	168,019	Maastricht	115,272
Tilburg	153,625	Dordrecht	106,987
Haarlem	149,099	Leiden	106,808

*Amsterdam is the capital, while The Hague is the seat of government.

Area and population		area		population
Provinces	**Capitals**	sq mi	sq km	1986 estimate
Drenthe	Assen	1,025	2,654	431,997
Flavoland	Lelystad	548	1,420	177,334
Friesland	Leeuwarden	1,295	3,357	598,068
Gelderland	Arnhem	1,935	5,011	1,761,492
Groningen	Groningen	905	2,344	560,029
Limburg	Maastricht	838	2,170	1,088,331
Noord-Brabant	's-Hertogenbosch	1,910	4,946	2,124,656
Noord-Holland	Haarlem	1,031	2,672	2,322,708
Overijssel	Zwolle	1,289	3,339	998,751
Utrecht	Utrecht	514	1,331	944,372
Zeeland	Middelburg	691	1,790	355,781
Zuid-Holland	The Hague	1,123	2,908	3,164,652
TOTAL LAND AREA		13,105	33,943	14,529,430
INLAND WATER		3,028	7,842	
TOTAL AREA		16,133	41,785	

Dunes and dikes protect the Low Netherlands against flooding. Almost all the area to the west and north consists of polders where the water level is mechanically controlled at about 1 m (3 ft) below ground level, thus permitting cultivation. However, the more marshy soils of the older polders, reclaimed before 1850, can be used only for grazing. Polders do not necessarily lie below sea level, although this is the case with the Usselmeer polders, which are 3.5 m (11.5 ft) below sea level and with polders created by draining lakes, which lie often below 6.7 m (22.0 ft). In areas of young marine clay and along the rivers, many polders lie above the average sea level, which means that it is not always necessary to pump the water out. The total number of polders is about 5,000; the older ones small.

As early as the Middle Ages, the inhabitants of the polders were faced with the task of safeguarding their area by joint effort once it had been wrested from the sea. At that time, the first *waterschappen* (water control boards) were created. Numbering 200, they are public bodies under the provincial authorities but elected by the *ingelanden* (landowners). The executives of these bodies are appointed by the crown. The boards have authority over water quality, environmental quality and recreation in the polder areas. Their duties are funded by a water control tax supplemented by central government funds.

In January 1986 the Northeast Polder and the two Flavoland polders were constituted as the Province of Flavoland. Nearly half of the province will be set apart for agriculture, while Lelystad in eastern Flavoland and Almere in southern Flavoland are expected to have populations of 100,000 and 250,000, respectively, by the end of this century. The reclamation of the final Zuyder Zee polder, the Markerwaard, has been delayed because of political and environmental reasons.

The southwestern part of the Netherlands consists of an area of islands and peninsulas among which the Rhine, Maas, and Scheldt rivers and their tributaries find their way to the sea. Much of the area was reclaimed as a result of accretion and embanking, and small islands grew to become larger areas. Sometimes these gains were lost through fresh flooding. In 1953 a great storm inundated 64,800 ha (160,000 ac) and caused 1,800 deaths in the low-lying polders. Following this disaster the Delta

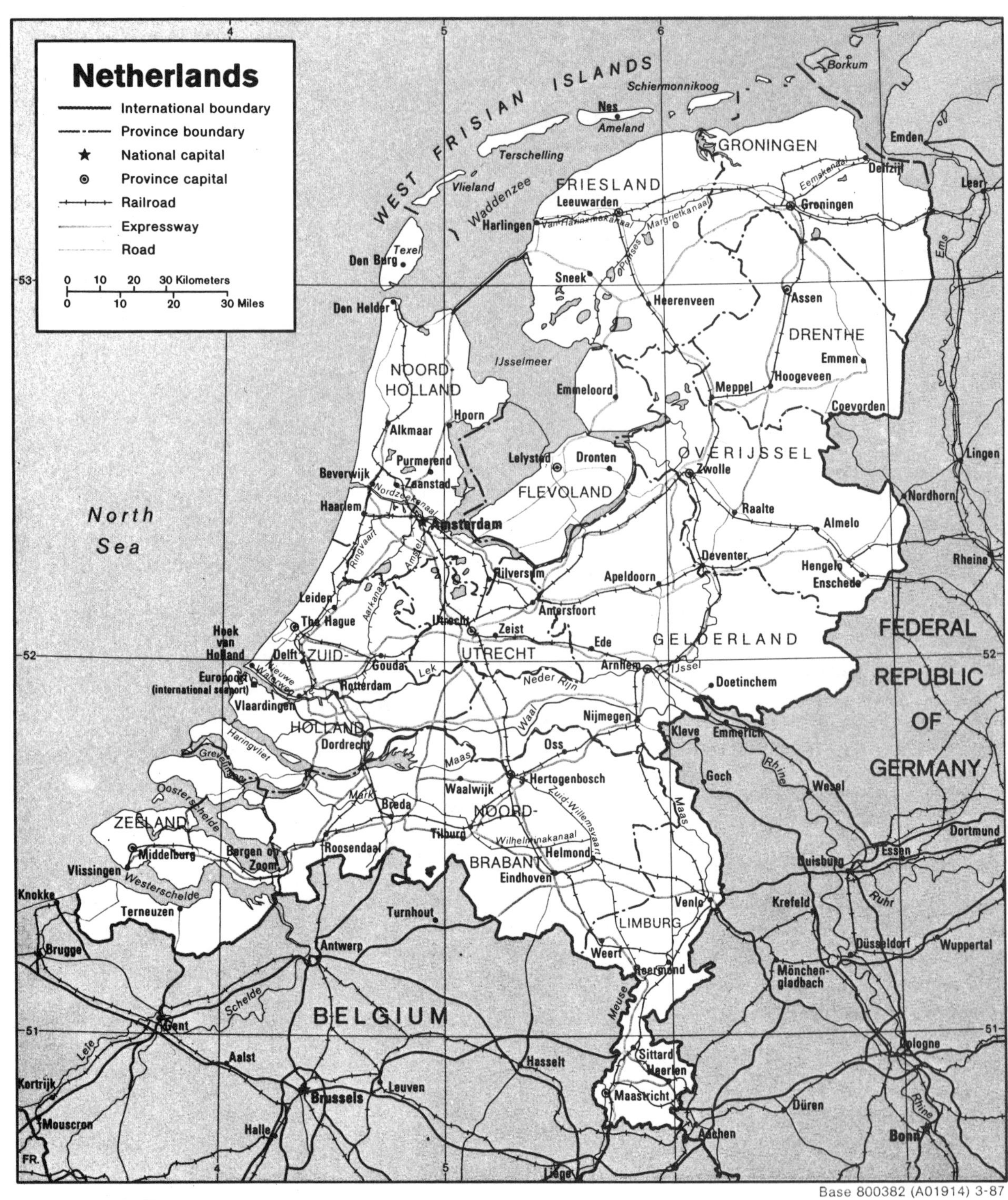

Base 800382 (A01914) 3-87

Project was launched to reduce danger of floods through construction of a number of dams. A movable water surge barrier was built in Hollandse Ussel east of Rotterdam. Thirteen other dams, sluices, bridges and canals were built before the project was completed in 1987. Strong tidal currents necessitated development of new techniques for sealing the estuaries. The project has not only eliminated the danger of floods in this region but also has opened up the previously isolated archipelago by roads built over the dams. The project has had serious drawbacks, the most important being damage to fisheries.

The Randstad, the western and most urbanized section of the Netherlands, is formed by the provinces of North Holland, South Holland and Utrecht and more specifically by the urban regions of Amsterdam, Rotterdam and Utrecht, forming a nearly complete ring of towns. The term Randstad is used to designate this area, although this term has no official status. Literally, the term is translated as "Rim City." The term "Greenheart City" also is used. However, the English term conurbation expresses the idea more clearly. Official reports refer to it as an urban zone.

The principal agglomerations of the Randstad cover 1,712 sq km (661 sq mi), roughly comparable to Greater Los Angeles or Greater London. But unlike them, the Randstad remains a congeries of cities, each with its own function; for example, The Hague is the seat of government; Amsterdam, the capital; Rotterdam, the chief port; and Utrecht, the transportation hub.

The Randstad is commonly subdivided into 21 functional regions:

1. Amsterdam, the capital region, at the confluence of the Amstel and the Ij rivers.

2. and 3. Het Gooi and South Kennemerland. Het Gooi is on the gently undulating land of the low hills and includes Hilversum and Bussum. The South Kennemerland area west of Amsterdam is centered in Haarlem.

4. The Zaan, an industrial region north of Amsterdam, and centered on the town of Zaandam.

5. Ijmond, a heavy industrial region located together with a fishing port complex at the seaward end of the North Sea Canal.

6. and 7. North Kennemerland and South Kennemerland/Rijnland districts in the coastal dune belt and, in the North, the polders behind them.

8. and 9. Bollenstreek and Rijnsburg Bollenstreek is the chief bulb-growing area, and Rijnsburg, at the mouth of the Old Rhine, west of Leiden, is a vegetable-growing area.

10–13. The Hague Agglomeration, Scheveningen, Wassenaar and Delft, cities on the western end of the southern belt of the Randstad.

14–17. Westland, Berkel, Boskoop and Aalsmeer, horticultural or market gardening areas.

18. Rotterdam, a continuous industrial and residential zone along both banks of the New Waterway and upstream along the Rhine tributaries of the Maas and Waal to Dordrecht and Gorinchem. It also includes the Hook of Holland and Europoort.

19. Utrecht. An old ecclesiastical and university town and transportation center.

20. Utrecht Glacial Ridge, a mainly residential area interspersed with wooded heathlands.

21. The Central Region, the rural "greenheart" of the Randstad.

NETHERLANDS ANTILLES

BASIC FACTS

Official Name: Netherlands Antilles

Abbreviation: NLA

Area: 800 sq km (308 sq mi)

Area—World Rank: 165

Population: 182,676 (1988) 186,000 (2000)

Population—World Rank: 162

Capital: Willemsted

Land Boundaries: Nil

Coastline: 364 km (226 mi)

Longest Distances: N/A

Highest Point: Saba 862 m (2,828 ft)

Lowest Point: Sea level

Land Use: 8% arable; 92% other

The Netherlands Antilles consists of two groups of islands in the Caribbean Sea, about 800 km (500 mi) apart. The main group, off the coast of Venezuela, consists of Bonaire and Curacao which (together with Aruba) are known as the Leeward Islands; to the north are the small volcanic islands of St. Eustatius, Saba and St. Maarten (the northern half of which is a dependency of the French overseas department of Guadeloupe) known as the Windward Islands (though actually in the Leeward group of the Lesser Antilles).

Area and population

Island councils	Capitals	area sq mi	area sq km	population 1981 census
Leeward Islands				
Bonaire	Kralendijk	111	288	8,753
Curaçao	Willemstad	171	444	147,388
Windward Islands				
Saba	The Bottom	5	13	965
Sint Eustatius or Statia	Oranjestad	8	21	1,358
Sint Maarten (Dutch part only)	Philipsburg	13	34	13,156
TOTAL		308	800	171,620

Netherlands Antilles

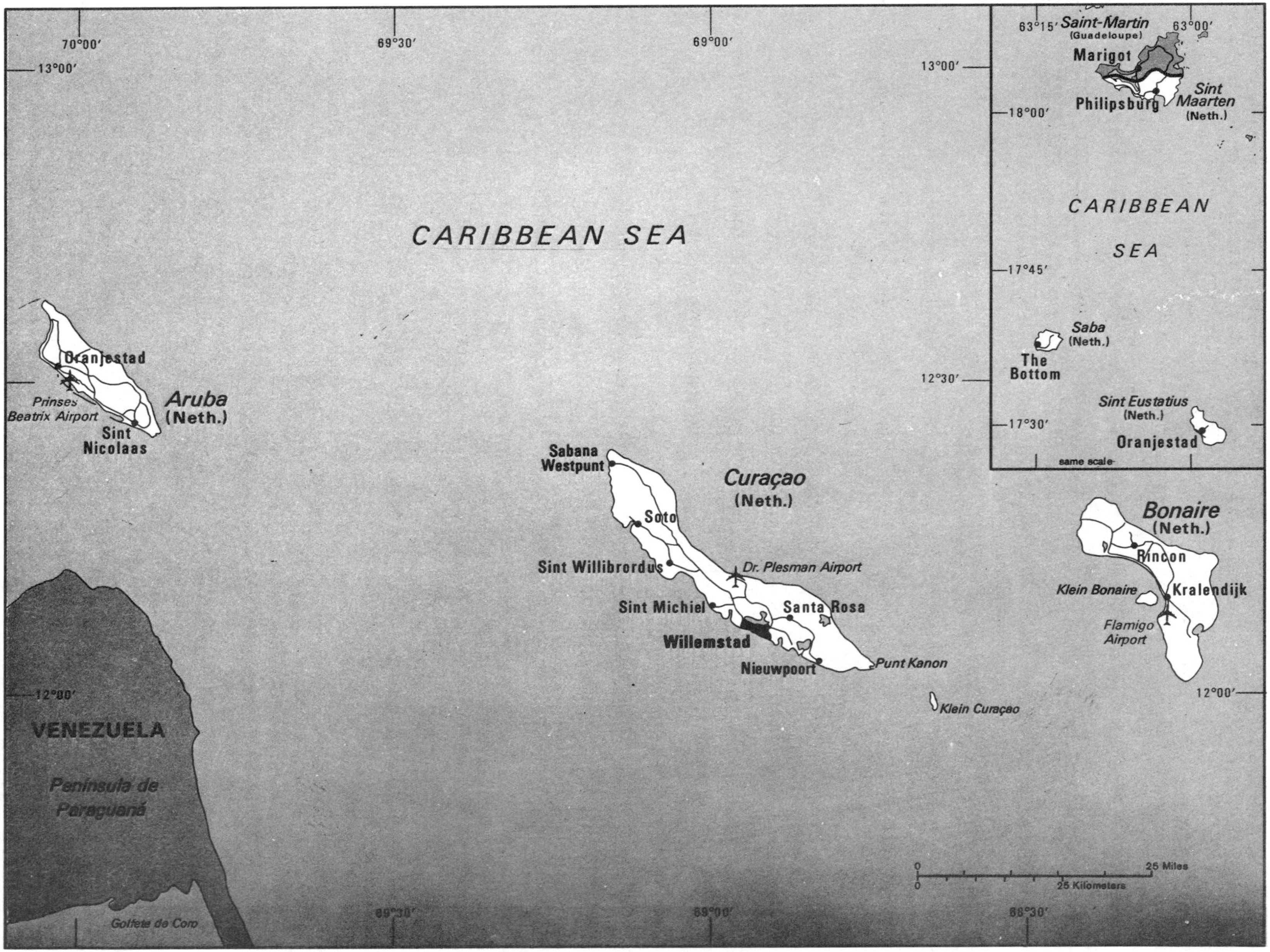

CARIBBEAN SEA
Aruba (Neth.)
Oranjestad
Prinses Beatrix Airport
Sint Nicolaas
Curaçao (Neth.)
Sabana Westpunt
Soto
Sint Willibrordus
Dr. Plesman Airport
Sint Michiel
Santa Rosa
Willemstad
Nieuwpoort
Punt Kanon
Klein Curaçao
Bonaire (Neth.)
Rincon
Kralendijk
Klein Bonaire
Flamigo Airport
VENEZUELA
Península de Paraguaná
Golfete de Coro
Saint-Martin (Guadeloupe)
Marigot
Philipsburg
Sint Maarten (Neth.)
CARIBBEAN SEA
Saba (Neth.)
The Bottom
Sint Eustatius (Neth.)
Oranjestad
same scale
70°00′
69°30′
69°00′
68°30′
13°00′
12°30′
12°00′
63°15′
63°00′
18°00′
17°45′
17°30′
0 25 Miles
0 25 Kilometers

502493 1-76 (541677)
Lambert Conformal Projection
Standard parallels 11°30′ and 13°30′
(inset 17°15′ and 18°15′)
Scale 1:1,000,000

Road
Airport

All the Windward Islands have volcanic bases, partly covered with coral reefs. They are semiarid and flat with little vegetation. The Leeward Islands are more mountainous. Saba, the most fertile, is an extinct volcano with luxuriant vegetation in its crater and on its sides.

NEW CALEDONIA

BASIC FACTS

Official Name: Territory of New Caledonia and Dependencies

Abbreviation: NCD

Area: 18,734 sq km (7,233 sq mi)

Area—World Rank: 136

Population: 150,981 (1988) 175,000 (2000)

Population—World Rank: 165

Capital: Noumea

Land Boundaries: Nil

Coastline: 2,254 km (1,400 mi)

Longest Distances: N/A

Highest Point: Mount Panie 1,628 m (5,341 ft)

Lowest Point: Sea level

Land Use: 14% meadows and pastures; 51% forest and woodland; 35% other

New Caledonia comprises one large island and several small ones in the South Pacific Ocean about 1,500 km (930 mi) east of Queensland, Australia. The main island,

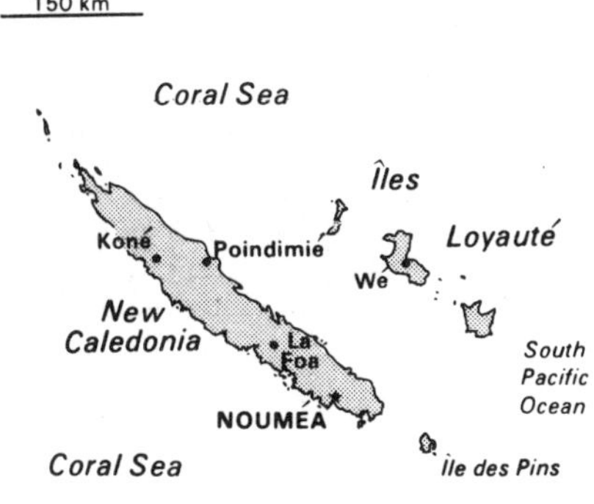

Islands of Huon and Chesterfield are not shown.

New Caledonia, is long and narrow. Rugged mountains divide the west of the island from the east and there is little flat land. The nearby Loyalty Islands and the uninhabited Chesterfield Islands lie about 400 km (248 mi) northwest of the main island.

Area and population

Regions	Capitals	area sq mi	area sq km	population 1983 census
Loyauté	. . .	765	1,981	15,510
Nord	. . .	2,837	7,348	21,512
Nouméa	Nouméa	637	1,650	85,098
Sud	. . .	2,995	7,757	23,248
TOTAL		7,233	18,734	145,368

NEW ZEALAND

BASIC FACTS

Official Name: New Zealand

Abbreviation: NZD

Area: 267,515 sq km (103,288 sq mi)

Area—World Rank: 71

Population: 3,343,339 (1988) 3,712,000 (2000)

Population—World Rank: 106

Capital: Wellington

Land Boundaries: Nil

Coastline: 9,173 km (5,700 mi)

Longest Distances: 1,600 km (994 mi) NNE-SSW; 450 km (280 mi) ESE-WNW

Highest Point: Mount Cook 3,764 m (12,349 ft)

Lowest Point: Sea level

Land Use: 2% arable; 53% meadows and pastures; 38% forest and woodland; 7% other

New Zealand lies in the southwestern Pacific Ocean and consists of two main and a number of smaller islands roughly the same size as Japan or the British Isles. The main North and South Islands are separated by the Cook Strait which is 26 to 145 m (16 to 90 mi) wide. They lie on an axis running from northeast to southwest, except for the lowlying Northland Peninsula.

The principal islands are North Island (114,669 sq km, 44,262 sq mi); Nearby Islands (69 sq km, 27 sq mi); South Island (149,883 sq km, 57,855 sq mi); Nearby Islands (4 sq km, 1.5 sq mi); Stewart Island (1,746 sq km, 674 sq mi); Chatham Islands (963 sq km, 372 sq mi); and

Outlying Islands, (including Raoul Island in the Kermadec Group and Campbell Island) 778 sq km (300 sq mi). New Zealand also has jurisdiction over Tokelau and the Ross Dependency.

New Zealand is very mountainous with less than a quarter of the land below 200 m (656 ft). In the North Island the main ranges run generally southwest, parallel to the coast from East Strait to Cook Strait with further ranges and four volcanic peaks to the northwest. The South Island is much more mountainous. A massive mountain chain, the Southern Alps, runs across the length of the island. Outlying ranges of the Southern Alps extend to the north and the southwest of the South Island. There are at least 223 named peaks higher than 2,300 m (7,546 ft). The 15 highest are

	m	ft
Cook	3,764	12,350
Tasman	3,497	11,474
Dampier	3,440	11,287
Silberhorn	3,279	10,785
Lendenfeldt	3,201	10,502
Mt Hicks	3,183	10,443
Torres	3,163	10,378
Teichelmann	3,160	10,368
Sefton	3,157	10,358
Malte Brun	3,155	10,352
Haast	3,138	10,296
Elie de Beaumont	3,117	10,227
Douglas Peak	3,085	10,122
La Perouse	3,079	10,102
Heidinger	3,069	10,069

In contrast the highest peak on North Island, Ruapehu, is only 2,797 m (9,177 ft) high. There also are 350 glaciers in the Southern Alps of which the largest are the Tasman (29 km, 18 mi); Murchison (17 km, 11 mi); Mueller (13 km, 8.1 mi); Fox (15 km, 9.3 mi); Franz Josef (13 km, 8.1 mi); Godley (13 km, 8.1 mi) and the Hooker (11 km, 6.8 mi).

Much of the lowland is broken and hilly. In neither island do plains figure prominently.

The natural divisions of the South Island are the Canterbury Plains, the Mountain Highland and the western coast. The Canterbury Plains are a huge accumulation of shingle waste derived from the mountains of the eastern flank. They rise steadily but imperceptibly inland until they abut abruptly against the foothills at a height of from 335 to 427 m (1,110 to 1,400 ft). Their smooth surface is without relief except for a gentle seaward slope and the deep, mile-wide trenches cut across them by the rivers covered with tussock. The plains are ideal sheep country. The eastern side of South Island is sheltered from the moist air masses of the Tasman Sea by the alpine barrier. Inland basins such as Mackenzie Country and central Otago are shut off not only to the west but to the east and the south as well. The south-central part of South Island is New Zealand's continental interior.

The mountain highland falls into three distinct parts, distinguished from each other by their rocks, their landforms and their geological history. The southernmost part of Fiordland, appropriately named for its deep, vertically walled canyon like valleys is occupied to the west by outward-radiating salt water fiords and occupied inland by the freshwater arms of the cold lakes—Te Anatu, Manapouri, and Monowai. This mountain region is the wettest in New Zealand. East of Fiordland is the mountain interior of Otago where the mountains are separated by terraced, gravel-floored basins, some of them occupied by lakes. This is the driest part of New Zealand, and the basin floors are arid and desert like. To the north of the frosty, windswept saucer of the Mackenzie Basin and from there to Cook Strait extends the so-called hill country, a succession of lofty alpine ranges crowned by the cloud-piercing summit of Mount Cook. The ranges, built of a blue-gray sandstone called greywacke, are separated by U-shaped valleys cut by glaciers.

Except for Fiordland, South Island's mountain interior is occupied by sheepmen right up to the snow line. Much of the former grasslands have deteriorated and in central Otago and parts of Mackenzie Country the parched floors of gravel basins look like deserts.

The main divide of the Southern Alps lies well to the west. From it, the land falls away rapidly to the Tasman Sea. Although some speak of a Westland plain, there is no plain as such but only a mountain slope. Only in the central waist of the island is the mountain rampart set back a little from the coast and here the coastal lowland is made up of broken hill country. In some places one or two valleys have cut back into the highland.

North Island is characterized by hill country. The mountain highland here is narrow and lies to the east. Its V-shaped valleys are not glaciated, and the narrow ridges are clothed to their summits in vegetation. On the eastern flank of the mountain axis lies a plaster of younger and softer rocks. From Hawke Bay southward these are separated from the Rimutakas and the Ruahines by the elongated Wairarapa lowland, but farther north the deeply corrugated hill country lies immediately on the flank of the mountain ranges to the west.

West of the ranges extends the wide sweep of the Volcanic Plateau. Above it to the southwest rise the masses of the volcanic piles of Ruapehu, Ngauruhoe and Tongariro. Deep-cut hill country with short but steep slopes occupies most of the rim of the Volcanic Plateau but especially between the Manawatu Plain and the northern edge of the King Country and the middle Waikato Valley. Its elevation decreases and its slopes become gentler toward the coast in the west. There is little coastal lowland and where it is widest—in Taranaki—the Egmont peak rises well over 2,438 m (8,000 ft).

The narrow northern peninsular section of the North Island is rarely elevated though it often is broken and irregular in surface. At the Tamaki Isthmus, the Tasman Sea and the Pacific Ocean are separated by no more than a mile or two of land. The characteristic landforms of this region are small plains, mangrove flats, swamp lowlands and peaty basins interspersed with low hill country and higher hills.

The rivers are swift-flowing and shallow, and only a few are navigable. They are, however, suitable for hydroelectric power generation, with their high rate of flow and reliable volume of ice-free water. The longest rivers are the Waikato (425 km, 264 mi), the Wanganui (290 km, 180 mi), and the Rangitikei (241 km, 150 mi) flowing into

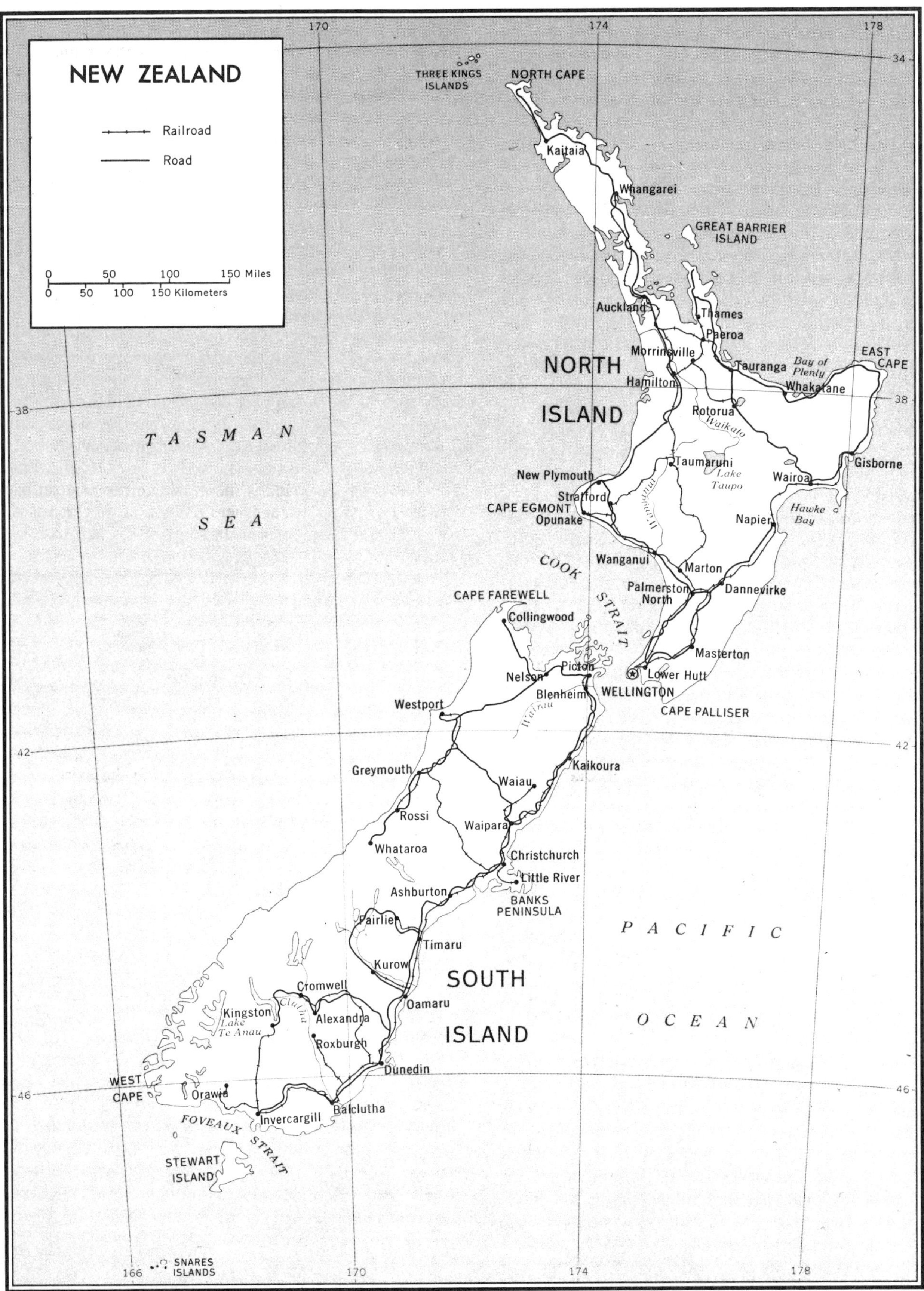

Base 59524 4-70

the Tasman Sea on North Island; Clutha (322 km, 200 mi), the Taìeri (288 km, 177 mi) and the Clarence (209 km, 130 mi) and the Waitaki (209 km, 130 mi) flowing into the Pacific Ocean on North Island and the Mataura (240 km, 149 mi) Waiau (217 km, 135 mi), and Oreti (203 km, 126 mi) flowing into the Foveaux Strait on South Island. New Zealand has many lakes, those in the south being particularly noted for their magnificent scenery. The largest are Taupo (606 sq km, 234 sq mi) on North Island and Te Anau (344 sq km, 133 sq mi) and Wakatipu (293 sq km, 113 sq mi) on South Island.

Volcanic activity of the past few million years has played an important part in shaping the landscape of which Banks Peninsula, a twin volcanic dome in Canterbury, is a notable specimen. The largest volcanic outpourings of late geological times has been in the region between Tongariro National Park and the Bay of Plenty coast. The major volcanoes, such as Ruapehu, Tongariro and Ngauruhoe were built up during the Pleistocene Epoch as also was the Volcanic Plateau, one of the largest and youngest accumulations of acid volcanic rocks in the world. Mount Taranaki, and the remnants of three other volcanic cones date from the Pleistocene Epoch. Pirongia, in the Waikato, is a cone of some 900 m (2,953 ft) high. Late Tertiary and Quaternary Period basaltic eruptions in North Auckland have built lava plateaus and many young cones.

Compared with some other parts of the almost continuous belt of earthquake activity around the rim of the Pacific, such as Japan, Chile and the Philippines, the level of seismic activity in New Zealand is moderate, although earthquakes are common. It may be compared roughly with that prevailing in California. A shock of Richter magnitude 6 or above occurs on an average about once a year, a shock of magnitude 7 or above once in 10 years, and a shock of magnitude 8 and above perhaps once a century. In historic times, only one earthquake (that in southwestern Wairarapa in 1855) approached that high magnitude. Other major earthquakes since 1840 have been the Hawke Bay earthquake of 1931 in which 256 people died and Butler earthquake of 1929, in which 17 people died. In regions where the majority of the earthquakes are very shallow, as in Calfornia, there is a tendency for earthquakes to cluster near geological fault traces, but in regions of deeper activity, as in New Zealand, this is not so. There is very little activity near the Alpine Fault which stretches for some 500 km (311 mi) from Milford Sound to Lake Rotoiti, considered one of the world's largest faults.

Within New Zealand, at least two separate systems of seismic activity can be distinguished. The main seismic region covers the whole of the North Island except Northland Peninsula and that part of the South Island north of a line passing roughly between Banks Peninsula and Cape Foulwind. The southern or Fiordland seismic region includes southern Westland, western Southland, and Western Otago. Less clearly defined activity covers the remainder of the two main islands and extends eastward from Banks Peninsula to include the Chatham Islands. In historic times, the main and Fiordland seismic regions have been significantly more active than the rest of New Zealand.

PRINCIPAL CENTRES OF POPULATION

(at census of 4 March 1986)

Wellington (capital)	352,035	Hamilton	167,711
Auckland	889,225	Dunedin	113,592
Christchurch	333,191		

Area and population

Statistical areas	area sq mi	area sq km	population 1986 census
North Island			
Central Auckland	2,154	5,578	889,225
East Coast	4,212	10,908	48,364
Hawke's Bay	4,356	11,283	150,744
Northland	4,883	12,646	127,558
South Auckland-Bay of Plenty	14,240	36,881	518,721
Taranaki	3,754	9,724	108,979
Wellington	10,715	27,751	598,024
South Island			
Canterbury	16,691	43,230	431,421
Marlborough	4,243	10,989	38,087
Nelson	6,768	17,530	81,160
Otago	14,209	36,801	186,142
Southland	11,160	28,905	104,817
Westland	5,903	15,289	23,842
TOTAL	103,288	267,515	3,307,084

NICARAGUA

BASIC FACTS

Official Name: Republic of Nicaragua

Abbreviation: NIC

Area: 127,849 sq km (49,363 sq mi)

Area—World Rank: 93

Population: 3,407,183 (1988) 5,261,000 (2000)

Population—World Rank: 104

Capital: Managua

Boundaries: 2,046 km (1,271 mi); Honduras 922 km (573 mi); Costa Rica 300 km (186 mi)

Coastline: Caribbean Sea 478 km (297 mi); Pacific Ocean 346 km (215 mi)

Longest Distances: 580 km (360 mi) NE-SW; 494 km (307 mi) NW-SE

Highest Point: Mount Mogoton 2,107 m (6,913 ft)

Lowest Point: Sea level

Land Use: 9% arable; 1% permanent crops; 43% meadows and pastures; 35% forest and woodland; 12% other

Nicaragua is part of Middle America, transitional between South America and Anglo-America and more specifically one of the five Central American states.

The country can be divided internally into four regions: The West—the heavily populated Pacific Highlands and the lowlands and lakes of the Great Rift depression; the Highland Frontier—the growing western and northern pioneer fringe of the Central Highlands; the Empty Lands—the essentially uninhabited forests of the eastern part of the highlands and the bulk of the Caribbean Coastal Plains; and the East—scattered settlements at the mouths of the rivers on the Mosquito Coast.

Nicaragua's Central Highlands (Cordilleras) are the southern extension of the Central American Highlands which begin in the north in the Mexican State of Chiapas and are separated from the new volcanics of the western coastal ranges in Nicaragua by the dominant physical features of the country, the Great Rift valley. The western ranges are both physiographically and geologically a northern extension of the Costa Rican Highlands. The lowlands of the rift merge with the coastal plain north of León and lead into the low shores of the Gulf of Fonseca in Honduras. The coastal plains and low hills of the swampy Mosquito Coast on the Caribbean side lie between the foothills of the Central Highlands and the sea. These lowlands are far more extensive than those of the Pacific coastal region, but are much less important in the life of the country. The bulk of the Great Rift, however, drains to the Caribbean via the Rio San Juan. The rainy eastern slopes of the Central Highlands generate several large, permanent rivers. The seasonally dry and narrow western watersheds produce no really significant streams. The two large lakes, Managua and Nicaragua, in the Great Rift are, however, the largest bodies of fresh water in Middle America. They drain into the Caribbean via the Rio San Juan.

NATURAL FEATURES

Nicaragua can be divided on grounds of geologic origins, landforms, soils, climate and natural vegetation and wildlife, into three fundamental regions: the wet and dry recent volcanic peaks and ash beds and alluvial lowlands of the Pacific Coast and the Great Rift; the older (Tertiary) volcanics and cordilleran ranges of the wetter and higher Central Highlands; and the wet tropical alluvial plains of the Caribbean coast. The relatively few occurrences of economically valuable minerals, insofar as this subject is presently understood, are found only in the Central Highlands.

Landforms

The landforms of Nicaragua break into four main groups. In the west there is a complex mixture of new volcanic peaks, lava and ash beds, and alluvium, comprising the coastal highlands and the Great Rift (valley). The Great Rift is a depression formed by subsidence of the land between two major fault lines. The faults here are part of that great set of faults rimming the Pacific Basin. The two sets of faults themselves extend into Honduras and San Salvador, across the Gulf of Fonseca on the northwest and into Costa Rica to the southeast, but the depression is primarily a Nicaraguan feature.

The Coastal Highlands

The only significant highlands in the Coastal Block is the Diriamba Highlands, a narrow, uplifted block west of the Great Rift. Structurally, it is a continuation of the Costa Rican highlands, however, in Nicaragua, it is only partly elevated. The Diriamba Highlands, locally called the *"Sierras,"* are between Managua, Granada, and the Pacific Ocean. Many of the larger and older coffee plantations are found here. To the north and west of Lake Managua, in the vicinity of the old capital of León, these uplands trail off into low hills and finally into an ash-blanketed coastal plain in the vicinity of the new deepwater ports of Corinto and Puerto Somoza and the old colonial harbor of Realejo. This lowland merges inland with the central lowlands to the south, and via a low saddle leads to the lowlying shores of the Gulf of Fonseca to the north. The central range of volcanic peaks reemerges after the lowland gap in a hilly area dominated by the volcanic peak of Coisigüina, capping the northern end of the fault block in Nicaragua, and forming the southern cape of the Gulf of Fonseca. It is in this fertile area of lowlands and low hills that the Nicaraguan cotton production is focussed. To the south of the Diriamba Highlands the hills again trail off into the low country of the Rivas Isthmus, lying between Lake Nicaragua and the sea, and containing the port of San Juan del Sur. The uplands then recommence to the south in Costa Rica.

The Great Rift (Central Lowland)

The floor of the Great Rift is partially taken up by the two largest fresh water lakes of Middle America, Lake Managua and Lake Nicaragua. The first is the smaller of the two, 61 km (38 mi) long and from 16 to 26 km (10 to 16 mi) wide. It drains via the 16 km (10-mi) long Rio Tipitapa into the big lake, which is approximately 160 km (100 mi) long and 72 km (45 mi) wide. An active chain of recent volcanos intrude the center of the depression and their peaks form several islands in Lake Nicaragua. The most important of these is the Isla de Ometepe with the twin cones of Madera and Concepcion. The cones and calderas of this chain are responsible for much of the fertile weathered ash which floors the rift. But they also have frequently caused damage to crops, livestock, people, and structures.

To the south and east of Lake Nicaragua the faults have sheared through the older (Tertiary) volcanics, some of which are exposed as low hills cut through by the San Juan river, which here forms the boundary between Costa Rica and Nicaragua. The old Caribbean port town of San Juan del Norte (Greytown) is at the mouth of the river.

The Central Highlands

To the north and east of the Great Rift are the volcanic ridges and plateaux of the Central Highlands. These older (Tertiary) volcanics (lava and ash) have almost entirely covered the underlying folded and faulted granitic and metamorphic rocks of the Antillean portions of the Cordilleran System. As such they comprise the Nicaraguan portion of that structural system which begins in the Brooks Range of Alaska, continues through the Rocky Mountains of Canada and the United States, the Mexican and Central American Cordilleras, the islands of the West Indies (Antilles), and the Andes of South America to Tierra del Fuego and Cape Horn. In Nicaragua as in Honduras to the north the structural trend is generally east-west rather than north-south since the system here is swinging toward Jamaica and Santo Domingo. On the western edge of the highlands, however, the topographic trend is rather from northwest to southeast, since it is a scarp determined by the fault line of the Great Rift. It is this apparent north-south axis which led the Spanish explorers and settlers to believe that the structures led southward into Costa Rica and thence via Panama to Colombia. This natural error is recorded in the name which is still often applied to the western part of the highlands, the Cordillera de los Andes. In point of fact, however, the Pacific uplands of Honduras, Nicaragua, Costa Rica and Panama are not Cordilleran (Andean) at all, but rather are part of the recent Coast Range system, itself part of the still active, Pacific "ring-of-fire" in both North and Middle America. This system begins in the Aleutians and includes the Alaskan Range, the mountains of Western British Colombia, the United States Pacific Coast, Baja California and the Southwestern Coast of Mexico, as well as the volcanic peaks of the Nicaraguan Great Rift and the Diriamba Highlands.

From the eroded western scarp and associated continental drainage divide, the trend of the several individual cordilleras is east-west, and they are separated by deep river valleys draining to the Caribbean. Particularly in the north—for instance, the Rio Coco (Segovia) system—the trends of these valleys are controlled by the underlying (Antillean) geologic structures. South of the Rio Coco valley is the S-shaped ridge of the Cordillera Isabelia. On the other side of the Rio Tuma valley from that ridge is the Cordillera Dariense, and south of there, across the upper Rio Grande (to which the Tuma is a tributary) are the Montanas de Huapi. Still further south, and the southernmost of these spurs of the cordillera, is the Cordillera de Yolaina. This ridge terminates at Punta del Mono (Monkey Point), south of the Siquia-Escondido drainage. The block then trails off to the south in the hills north of the San Juan valley. The shorter streams of the Punta Gordo and the Indio drain the eastern slopes of these hills.

The highest peaks of the Central Highlands are only some 2,134 m (7,000 ft) in elevation, and most of the ridges range from 1,981 m (6,500 ft) down to 1,066 m (3,500 ft). Nonetheless, they constitute an effective barrier to the moist winds from the Caribbean, and thus the western scarp is in a "rain shadow." The drier western and northern parts of the Central Highlands are now fairly well settled with both coffee and beef production and subsistence farming, but the rainforested eastern parts are still essentially empty.

Area and population

Zones Departments	Capitals	area sq mi	 sq km	population 1985 estimate
Atlantic				
Rio San Juan	San Carlos	2,876	7,448	34,330
Zelaya	Bluefields	22,816	59,094	325,454
North Central				
Boaco	Boaco	1,924	4,982	97,432
Chontales	Juigalpa	1,910	4,947	111,786
Estelí	Estelí	849	2,199	115,333
Jinotega	Jinotega	3,697	9,576	143,264
Madriz	Somoto	756	1,958	80,268
Matagalpa	Matagalpa	2,623	6,794	263,649
Nueva Segovia	Ocotal	1,290	3,341	139,116
Pacific				
Carazo	Jinotepe	398	1,032	97,106
Chinandega	Chinandega	1,800	4,662	385,506
Granada	Granada	372	964	136,068
León	León	2,021	5,234	257,815
Managua	Managua	1,389	3,597	903,998
Masaya	Masaya	224	581	179,114
Rivas	Rivas	830	2,149	101,825
National District				
Distrito Nacional				
TOTAL LAND AREA		45,775	118,558	3,272,064
INLAND WATER		3,588	9,291	
TOTAL AREA		49,363	127,849	

The Caribbean Lowlands

The ridges and plateaus of the Central Highlands merge to the east into the lowlands bordering the Caribbean Sea. This area, comprising nearly one-third of the country but including only about one percent of the population is also known as La Mosquitia or the Mosquito (Miskito) Coast. Originally, the Mosquito Coast referred to only that narrow strip of land bounded in the north by the Huahua (Wawa) River, on the south by the Escondido River and on the west by foothills of the Central Highlands—225 km (140 mi) long and 64 km (40 mi) wide. This was the area of the Mosquito Indians. Now the term has been extended to include the Caribbean Coast of Nicaragua, and parts of Costa Rica and Honduras as well. It is composed of alluvial plains and valleys separated by low and gentle watersheds of weathered and leached and recemented (laterized) volcanics. There are numerous shallow bays and lagoons and associated salt marshes along the low-lying coast.

The San Juan Valley provides a nearly sea-level pass from the Caribbean coast to Lake Nicaragua, and thus would provide the first part of the route for the Nicaraguan canal that has been proposed for almost a century.

Hydrology

The country divides generally into two hydrologic regimes, the relatively continuous runoff from the wet eastern slopes of the Central Highlands into the Caribbean, and the more seasonal runoff of the wet and dry west. The one notable exception to this is the San Juan drainage, which drains both walls of the Great Rift and the valley itself except for the two small portions to the north of Lake Managua which drain north into the Gulf of Fonseca, or west into the Pacific. The San Juan, however, since it drains lakes of Managua and Nicaragua which regulate its water supply, is more even in flow than would be

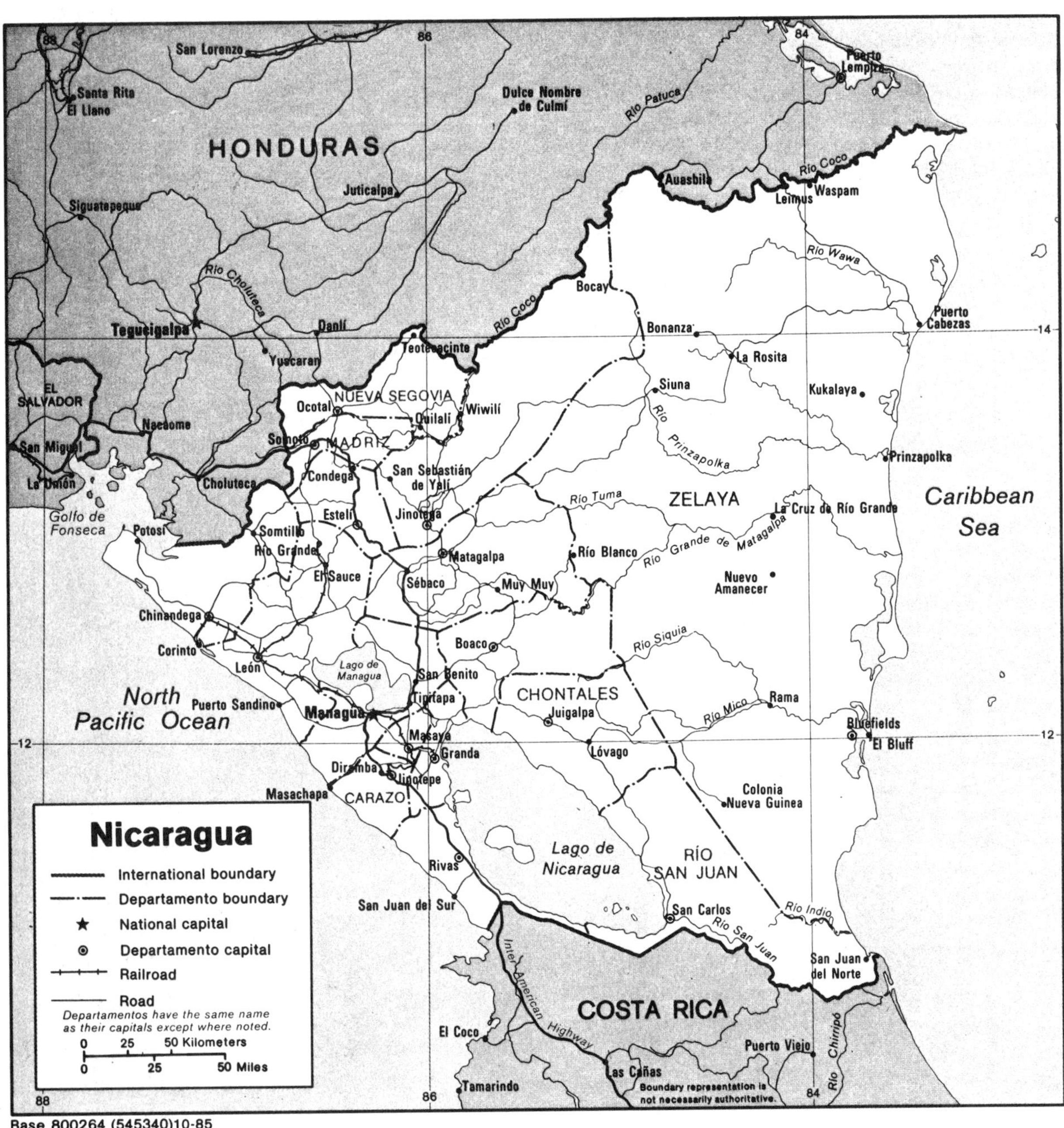

Base 800264 (545340)10-85

expected considering the seasonality of the runoff into the structural basin which it drains.

The other principal east-flowing streams are the Escondido, the Grande, the Prinzapolco and the Coco (Segovia). These, along with the lower San Juan and the estuaries of some of the many lesser streams, are navigable. Nicaragua's first major hydroelectric installation and artificial lake is near completion. It is being constructed on the upper Rio Tuma, a tributary of the Rio Grande. When completed it will be an additional major source of electricity for the country. The water table, in the eastern alluvial plains is always high, which is of little positive economic value in this wet climate, but does create problems for construction, and often for cultivation.

Local navigation on the lakes has been historically of great importance, but is expected to continue to decline. Only if an interoceanic canal were developed here would navigation of these waters assume major significance.

There are almost no important streams flowing west into the Pacific. The two largest intermittent streams which flow west out of the Central Highlands are the Rio Negro and the Viejo. In the north the Rio Negro crosses into Honduras and empties into the Gulf of Fonseca; the Viejo drains the other side of the divide from the Tuma headwaters and feeds Lake Managua. Except in the extreme north, the drainage divide is much closer to the Pacific than to the Caribbean. Therefore, when this is coupled with the rain shadow effect of the Central Highlands, there are only steep, short, and greatly fluctuating streams along the eastern face of the Cordillera.

The streams which drain the western slopes of Diriamba Highlands, the northern new volcanics, are not quite as steep but are all short and small, and nearly all are intermittent.

NIGER

BASIC FACTS

Official Name: Republic of Niger

Abbreviation: NGR

Area: 1,186,408 sq km (458,074 sq mi)

Area—World Rank: 25

Population: 7,213,945 (1988) 10,832,000 (2000)

Population—World Rank: 80

Capital: Niamey

Land Boundaries: 5,621 km (3,493 mi); Libya 354 km (220 mi); Chad 1,175 km (730 mi); Nigeria 1,497 km (930 mi); Benin 190 km (118 mi) Burkina Faso 628 km (390 mi); Mali 821 km (510 mi); Algeria 956 km (594 mi)

Coastline: Nil

Longest Distances: 1,845 km (1,146 mi) ENE-WSW; 1,025 km (637 mi) NNW-SSE

Highest Point: Indoukal-n-Taghes 2,022 m (6,634 ft)

Lowest Point: 200 m (650 ft)

Land Use: 3% arable; 7% meadows and pastures; 2% forest and woodland; 88% other

Landlocked Niger is the largest country in West Africa.

Four-fifths of Niger is an arid desert; the remaining fifth is savanna. Vast areas in the north are characterized by the same relief: sandy basins, low plateaus, isolated hills and peaks and limestone or sandstone bluffs. The Tamgak Mountains in the northwest rise 1,800 m (5,900 ft) above the Iferouane Valley. In the north-central region is the volcanic Air Massif pierced by deep valleys, called koris, where there is a dense vegetation of acacias and doum palms. Farther east is the Tenere, a sandy and arid desert.

The lifeline of the country is the Niger River, which flows through the south for 300 km (186 mi). Niger also shares Lake Chad with Nigeria and Chad.

PRINCIPAL TOWNS (population in 1977)

Niamey (capital)	225,314	Tahoua	31,265
Zinder	58,436	Agadez	20,475
Maradi	45,852	Birni N'Konni	15,227

1981 (estimates): Niamey 360,000; Zinder 75,000.

Area and population

Departments	Capitals	area sq mi	area sq km	population 1987 estimate
Agadez	Agadez	244,869	634,209	174,000
Diffa	Diffa	54,138	140,216	203,000
Dosso	Dosso	11,970	31,002	876,000
Maradi	Maradi	14,896	38,581	1,243,000
Niamey	Niamey	34,862	90,293	1,585,000
Tahoua	Tahoua	41,188	106,677	1,234,000
Zinder	Zinder	56,151	145,430	1,298,000
TOTAL		458,075	1,186,408	6,613,000

NIGERIA

BASIC FACTS

Official Name: Federal Republic of Nigeria

Abbreviation: NGA

Area: 923,768 sq km (356,669 sq mi)

Area—World Rank: 31

Population: 111,903,502 (1988) 139,230,000 (2000)

Population—World Rank: 10

Capital: Lagos

Boundaries: 4,821 km (2,996 mi); Chad 88 km (55 mi); Cameroon 1,690 km (1,050 mi); Benin 772 km (480 mi); Niger 1,497 km (930 mi)

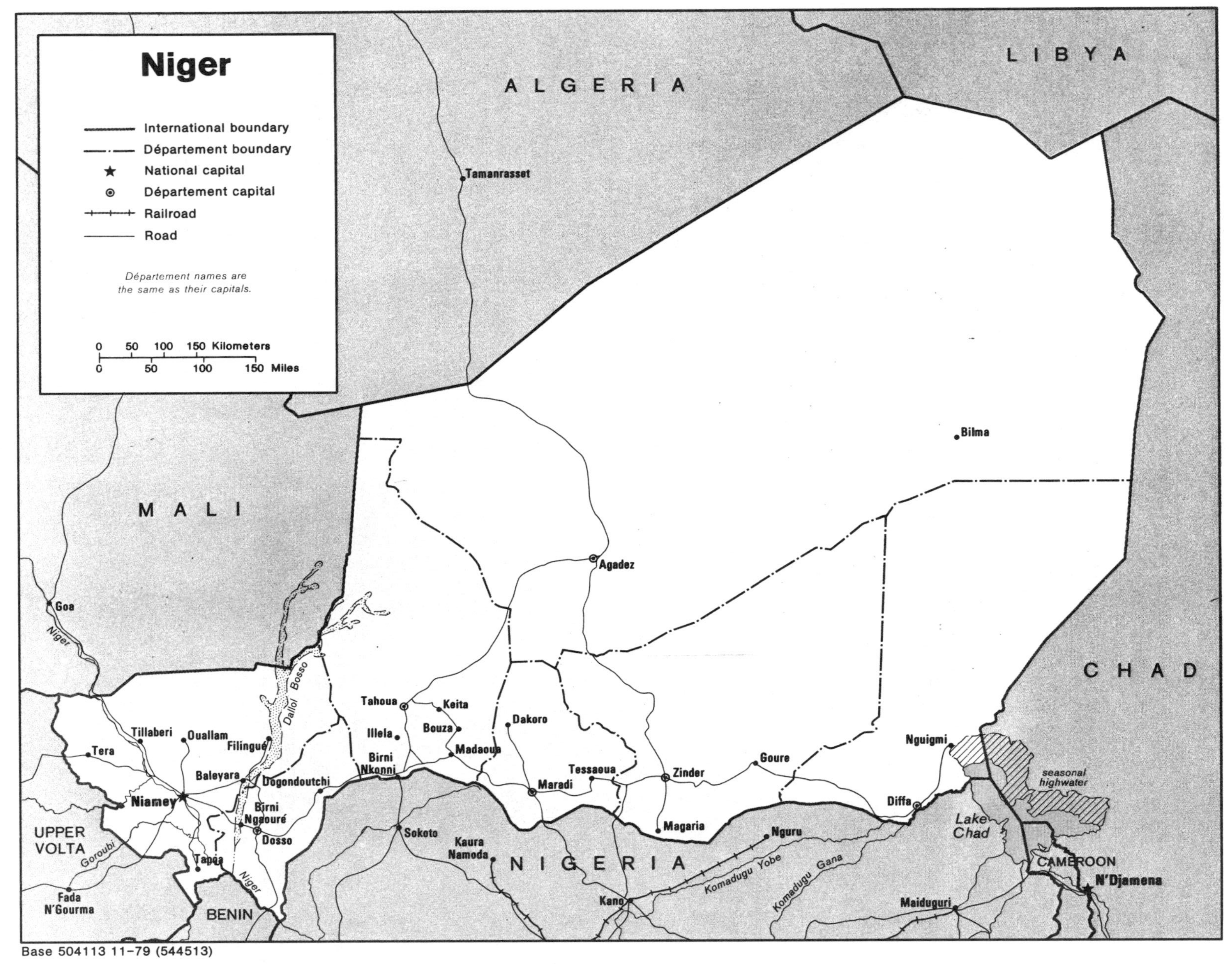

Base 504113 11-79 (544513)

Coastline: 774 km (481 mi)

Longest Distances: 1,127 km (700 mi) E-W; 1,046 km (650 mi) N-S

Highest Point: Mount Waddi 2,419 m (7,936 ft)

Lowest Point: Sea level

Land Use: 31% arable; 3% permanent crops; 23% meadows and pastures; 15% forest and woodland; 28% other

Nigeria is the easternmost of the countries that face on the Gulf of Guinea in the West African bulge. It lies entirely in the tropics, its southern edge being only a few degrees above the equator and its northern border well below the Tropic of Cancer.

The outstanding geographic feature is the basin of the Niger and Benue rivers, running east and west through the center of the country. South of the basin the elevation is generally under 304 m (1,000 ft) except for some plateau surfaces. To the north of it is a broad plateau region that occupies the country to its northern border and has elevations of 304 m to 1,219 m (1,000 to over 4,000 ft). On the east the country is bordered by mountainous regions, in which the highest point—of 2,042 m (6,700 ft)—is located.

GEOGRAPHIC REGIONS

A number of major geographic divisions are readily distinguishable, stretching in generally east-west zones across the country. Along the lower part is a low coastal zone, representing a general continuation of the belt of low plains that extends along the entire Gulf of Guinea coast of West Africa. North of this zone lies an area of hills and low plateaus. Through the middle of the country, bisecting it from east to west, extends the great valley of the Niger and Benue rivers. North of this valley the area as far as the country's northern border is occupied by a broad plateau. Along the eastern border is found a distinct zone of mountainous country.

Within these general divisions variations occur, forming separate geographic regions. At least twelve such regions have been recognized and in the southern part of the country consist of the Low Plains, the Lower Niger Valley, the Niger Delta, the Southeastern Scarplands, the Southeastern Lowlands, and the Western High Plains. In the central part is the Niger-Benue River Valley region, and in the north are the Northern High Plains, Jos Plateau, Sokoto Plains, and Chad Basin regions. The twelfth region comprises the Eastern Highlands.

Low Plains

The Low Plains lie mainly in Western and Mid-Western States. They are bordered on the Gulf of Guinea by a coastal stretch of low-lying, often swampy land interspersed with lagoons and creeks and varying up to 32 km (20 mi) or more in width. The outer edge of the coastal area consists of sand spits in its western part; however, it changes to mud as the coast nears the Niger Delta. Behind the outer spits and lagoons creeks of varying size generally parallel the coast and form a continuous waterway from the border with Benin on the west to the tributaries of the Niger Delta in the east.

Much of the land in this coastal area is below an elevation of 15 m (50 ft); however, it rises to above 45 m (150 ft) inland, and the average elevation above sea level is about 30 m (100 ft). Drainage on the mainland side of the lagoons is poor, and many smaller rivers lose themselves in the freshwater swamps that line the northern edges of the lagoons. The only major opening to the gulf from this area is at Lagos, the country's capital, located on a group of islands at the western end of Lagos Lagoon.

The plains in Western State rise gently from the lagoon and creek area northward toward the Western High Plains. In their southern part they consist of a dissected plain underlain by clay, limestone, and sandstone formations. Erosion has produced low, relatively flat-topped hills. This area is heavily forested, and its physical features are often masked by the vegetation. Elevations in this southern section are mostly below 183 m (600 ft). The northern part is underlain by crystalline rocks, and the land has an undulating appearance. Elevations range generally from about 121 to 304 m (400 ft to 1,000 ft) above sea level. Numerous isolated, steep-sided rock domes occur in this area, either singly or in groups. Some of these domes, left standing by erosion of the surrounding terrain, are quite high; a group east of Ondo includes points rising to 914 m (3,000 ft). Soil and rainfall conditions make this area highly suitable for cocoa cultivation, and it constitutes the country's major cocoa-growing region.

In Mid-Western State the land in the west and south consists of a low sandy coastal plain with gentle slopes and elevations mostly under 122 m (400 ft). The central and western sections contain extensive and luxuriant forest areas in protected reserves. To the northeast the terrain rises to elevations of from 228 to 304 m (750 to 1,000 ft) to form the comparatively flat Ishan Plateau. Many of the rivers in this section of the plains originate in the Ishan Plateau, where they have cut deeply incised valleys. In their lower courses the valleys widen greatly and are subject to extensive flooding during the rainy season.

Lower Niger Valley

The Lower Niger Valley region lies roughly east of the Low Plains and consists of a comparatively narrow valley extending from the confluence of the Niger and Benue rivers at Lokoja southward for about 297 km (185 mi) to the apex of the Niger Delta. In its upper section it comprises a number of quite narrow gorges between which lie somewhat broader stretches bordered by hills with elevations between 91 and 213 m (300 and 700 ft).

In its central section the valley widens. On the left bank there is a broad alluvial plain, which is intensively farmed during low-water periods. The section has numerous sandbanks and islands that are also farmed at certain times of the year. During the rainy season, however, the water level rises as much as 9 m (30 ft) and completely inundates these features. Near Onitsha, the largest and most important port on the Niger, the river again narrows. Below this point it flows through an alluvial flood plain that stretches to the delta. Much of this area is also flooded at high water.

Niger Delta

One of the major features of the West African coastline is the Niger Delta, which projects into the Gulf of Guinea from the southern coast of Nigeria between the Bight of Benin and the Bight of Biafra. This great bulge of sedimentary material, deposited by the Niger River, stretches some 120 to 128 km (75 to 80 mi) from its apex below the town of Aboh to the sea; it covers an area of about 25,900 sq km (10,000 sq mi). The water of the Niger flows through this delta in a series of radial tributaries. For navigational purposes, the two most important are the Forcados and Nun rivers.

The outer edge of the delta is fringed by sand spits and ridges, varying in width from a fraction of a mile to 10 or more mi. Behind these ridges are mangrove swamps covering about 10,360 sq km (4,000 sq mi), and farther inland an extensive area of freshwater swamps is found. Islands of solid ground within the delta are occupied by populated settlements. The delta is the site of large natural gas and oil deposits.

Southeastern Scarplands

Directly to the east of the Lower Niger Valley is a highly eroded plateau area characterized by prominent scarps. The region's chief geographic features are the Udi and Igala plateaus and the Akwa-Orlu Uplands. The Udi Plateau runs from north to south; its northern end merges with the Igala Plateau, which extends from east to west. The general elevation of these plateaus is about 304 m (1,000 ft) above sea level, with escarpments rising considerably higher. The scarps on the northern and eastern edges drop abruptly at various points in their 152 m (500-foot) descent in elevation. The plateau surface is dotted with flat-topped and domed hills that have produced one of the country's most highly scenic landscapes. Important coal-bearing formations are exposed in these scarps.

The scarp faces of both plateaus have been deeply gullied by stream erosion; many of these gullies are nearly 91 m (300 ft) deep. In the Awka-Orlu Uplands, which lie to the west of the plateaus and have nearly the same elevation, even more extensive erosion has occurred. Gullies in this area are hundreds of ft deep, and some cover more than a sq mile.

The western part of the region embraces the upper valley of the Anambra River and the valley of its major tributary, the Mamu River. Much of this area is a rather featureless plain about 61 to 121 m (200 to 400 ft) above sea level, which in places is heavily forested. Alluvial materials have been deposited along the rivers, as well as in the valleys of smaller tributary streams.

Southeastern Lowlands

The Southeastern Lowlands occupy the area between the Lower Niger Valley and the Niger Delta on the west, the Eastern Highlands on the east, and the Benue River valley on the north. The coastline of this region resembles that found in the Low Plains area; it is similarly fringed with sandy spits and mangrove swamps. Unlike the more westerly coastal areas, longitudinal lagoons are completely absent in this region.

The Southeastern Lowlands are divided into two subregions—an oil palm belt and the Cross River basin. The oil palm belt lies generally east and northeast of the Niger Delta. The area is underlain by the coastal plains sand formation, and its porous, leached soils are relatively infertile. The area formerly supported a high forest growth, but this was largely cut down and replaced with secondary vegetation, of which the oil palm is the dominant species.

The Cross River basin lies generally east of the scarplands and has a width in the Enugu area of about 80 km (50 mi). Extending from north to south for nearly 322 km (200 mi), the upper limit of its drainage system approaches the Benue River trough. Elevations of the low region vary from about 61 m (200 ft) in the south to 152 m (500 ft) in the north. Large areas are relatively flat, and land along streams and rivers becomes swampy in the rainy season.

Western High Plains

The Western High Plains, or Plateau of Yorubaland, are part of the belt of high plains that extends through West Africa. They lie between the low western coastal plains and the Niger-Benue River Valley region and are broadest in the west near the border with Benin, where the land has a general elevation of more than 305 m (1,000 ft). It displays a mature erosion surface surmounted by ridges of more resistant rock and numerous dome-shaped hills that project several hundred ft above the surrounding terrain. The area is covered largely with savanna parkland and grass.

The eastern section of the Western High Plains has two distinct plateau surfaces—one has an elevation of about 228 m (750 ft), and the other, of about 372 m (1,200 ft). The landscape is characterized by dome-shaped hills, some of which attain a height of 609 m (2,000 ft) above sea level. Drainage throughout the Western High Plains is generally good; however, many streams disappear in the harmattan (dry) period, and holes must be dug in streambeds to obtain water.

Niger-Benue River Valley

The combined valleys of the Niger and Benue rivers form a great east-west arc approximately across the middle of the country. The Niger River valley extends from the border with Benin on the west, and the Benue River valley extends from the eastern border with Cameroon. Near Lokoja in the center of the country the Benue River joins the Niger River, which changes its course at this point to flow southward to the Gulf of Guinea.

After entering the country from Benin the Niger River flows in a relatively open valley underlain by sedimentary rocks. Between Yelwa and Jebba the valley is characterized by open flood plains, separated by comparatively narrow, rocky stretches. In the area of Jebba the river traverses a rather narrow gorge. The gradient between

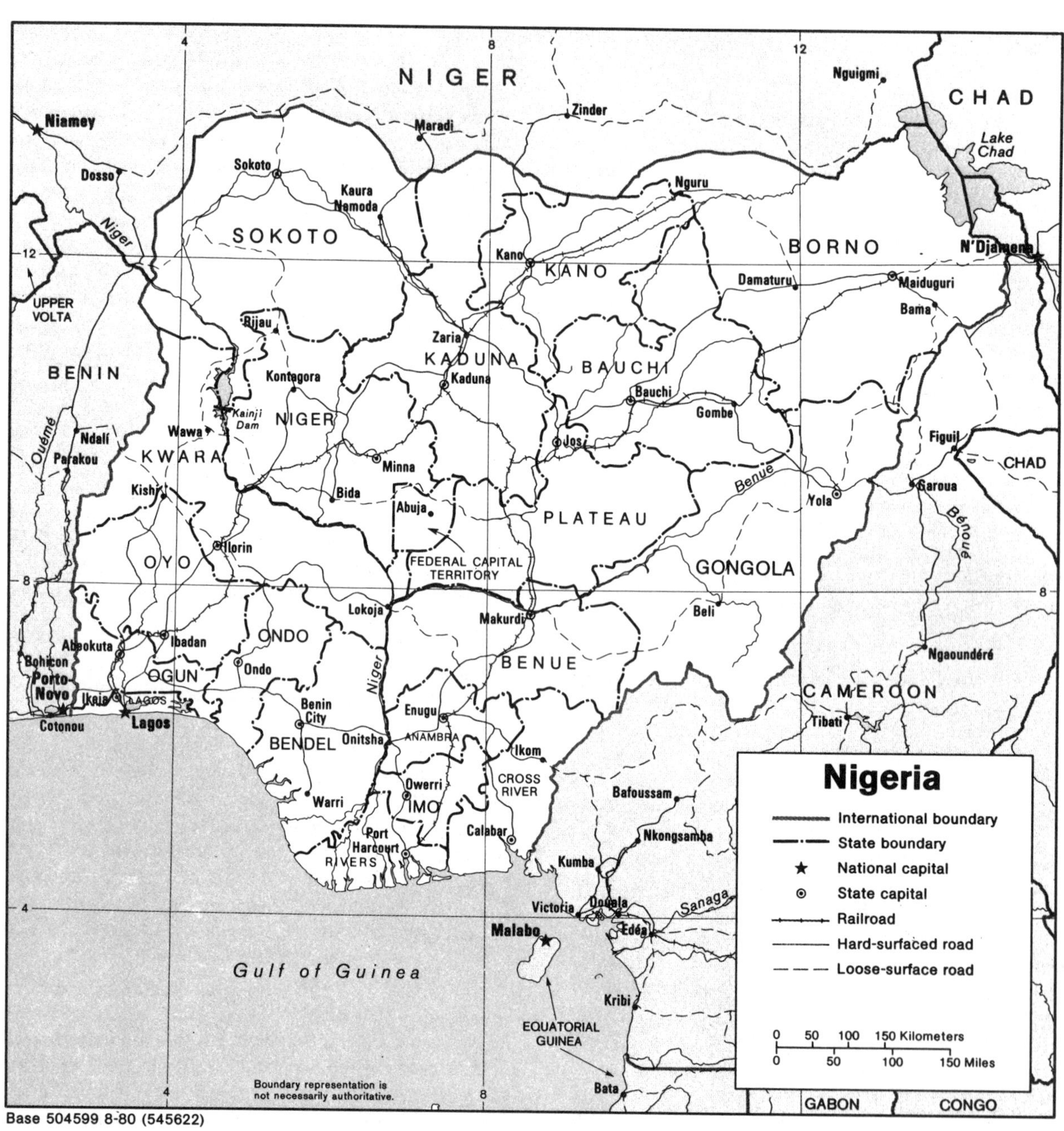

Base 504599 8-80 (545622)

Yelwa and Jebba and the harder underlying rock have produced a series of rapids and falls that limit use of the river for navigation. In the late 1960s this section of the valley was altered by construction of a dam at Kainji above Jebba. Flooding of an 137-km (eighty-five-mile) stretch of the valley produced Kainji Lake.

Eastward from Jebba to the confluence of the two rivers at Lokoja the valley broadens; in some places it is more than 80 km (50 mi) wide. At many points along this part of the river, extensive swampy plains are found. Sections of the undulating valley have elevations of about 122 m (400 ft) above sea level; in other places the land consists of eroded hill country.

The Benue River valley is underlain by sedimentary formations throughout its entire length from the Cameroon frontier to the confluence at Lokoja. For most of this distance the valley is broad and in some places attains a width of about 160 km (100 mi). Terrain features vary considerably in the erosional low plains that parallel the river's course. In some sections they have an undulating, but otherwise featureless, appearance. In other areas flat-topped hills are found, and near the confluence of the two rivers they become a dissected sandstone plateau.

The Benue River's course has not been affected by harder basement rocks as in the case of the Niger River. Rapids and falls are completely absent, and navigation is affected only by shifting sandbars. During flood periods the Benue River is navigable by flat-bottomed boats to Garoua in Cameroon. Most of the river's tributaries are comparatively shallow and are characterized by sandbars and islands. Entry to the Benue River is at a low gradient, and the tributaries may have more than one channel.

Northern High Plains

The Northern High Plains comprise a broad plateau area. The large central section of the plateau extends for about 482 km (300 mi) from east to west. Sometimes designated the High Plains of Hausaland, the region consists of a series of stepped plains that range from about 183 m (600 ft) above sea level at the outer edge to roughly 914 m (3,000 ft) in the area surrounding the Jos Plateau. Some plains' levels are separated by prominent escarpments.

The central section consists largely of undulating surfaces with occasional smooth, low ranges. At various points, however, steepsided, domed hills project more than 457 m (1,500 ft) above the surrounding countryside. Rivers and streams in these plains flow in broad, shallow valleys. Because of the low gradients, they generally contain numerous sandbars.

The Northern High Plains in the Gongola River basin east of Bauchi are highly dissected, and the rivers and streams in the basin flow through relatively deep, narrow valleys, paralleled by flat-topped hills. Much of the area is characterized by sandstone ridges.

East of the Gongola River basin lies the Biu Plateau, an area of about 5,180 sq km (2,000 sq mi). The upper level of the plateau, from 609 to 914 m (2,000 to 3,000 ft) is separated from the Northern High Plains by a pronounced scarp. Inactive volcanic cones are found in the northern part of the area.

PRINCIPAL TOWNS (estimated population at 1 July 1975)

Town	Population	Town	Population
Lagos (federal capital)	1,060,848	Ado-Ekiti	213,000
Ibadan	847,000	Kaduna	202,000
Ogbomosho	432,000	Mushin	197,000
Kano	399,000	Maiduguri	189,000
Oshogbo	282,000	Enugu	187,000
Ilorin	282,000	Ede	182,000
Abeokuta	253,000	Aba	177,000
Port Harcourt	242,000	Ife	176,000
Zaria	224,000	Ila	155,000
Ilesha	224,000	Oyo	152,000
Onitsha	220,000	Ikere-Ekiti	145,000
Iwo	214,000	Benin City	136,000

Area and population

States	Capitals	area sq mi	area sq km	population 1984 estimate
Anambra	Enugu	6,824	17,675	6,029,500
Bauchi	Bauchi	24,944	64,605	4,075,800
Bendel	Benin City	13,707	35,500	4,125,500
Benue	Makurdi	17,442	45,174	4,068,600
Borno	Maiduguri	44,942	116,400	5,025,000
Cross River	Calabar	10,516	27,237	5,830,800
Gongola	Yola	35,286	91,390	4,367,600
Imo	Owerri	4,575	11,850	6,157,000
Kaduna	Kaduna	27,122	70,245	6,868,800
Kano	Kano	16,712	43,285	9,681,800
Kwara	Ilorin	25,818	66,869	2,884,400
Lagos	Ikeja	1,292	3,345	2,825,200
Niger	Minna	25,111	65,037	1,961,800
Ogun	Abeokuta	6,472	16,762	2,596,000
Ondo	Akure	8,092	20,959	4,617,200
Oyo	Ibadan	14,558	37,705	8,732,300
Plateau	Jos	22,405	58,030	3,397,500
Rivers	Port-Harcourt	8,436	21,850	2,883,300
Sokoto	Sokoto	39,589	102,535	7,608,900
Federal Capital Territory		2,824	7,315	. . .
TOTAL		356,669	923,768	93,736,200

Jos Plateau

The Jos Plateau, surrounded by the south-central section of the Northern High Plains, includes distinctive features that set it aside as a separate geographic region. Covering an area of about 7,770 sq km (3,000 sq mi), it is separated from the surrounding plains area by pronounced escarpments. The area's general elevation is over 1,219 m (4,000 ft) above sea level, and hills in its eastern part attain heights of over 1,767 m (5,800 ft).

The plateau's cooler temperatures and difference in vegetation also set it off from the Northern High Plains. The region is the site of tin and other metals that have made the region economically important.

Sokoto Plains

The Sokoto Plains occupy the extreme northwestern part of the country and are generally monotonous in appearance. In the western part erosion has left occasional tablelike hills that stand about 30 m (100 ft) above the surrounding plain. Trenchlike stream and river valleys characterize this part. They have broad flat floors 2 or more mi in width and steep clifflike sides 30 to 60 m (100 to 200 ft) high. Water in the rivers is largely seasonal and is limited chiefly to the short rainy season.

Retarded waterflow and deposits have gradually raised the height of some riverbeds to levels above the surrounding valley floor.

Chad Basin

The Chad Basin consists of a broad plains area in the northeastern part of the country. In this area the land slopes gently eastward from the Northern High Plains to Lake Chad. The region has a general elevation of about 304 m (1,000 ft) above sea level and is largely featureless except for fixed dune formations in the northern section, many of which are covered with trees and grass.

Extremely low gradients affect riverflow in the region. During much of the year the flood plains are swampy areas, and the rivers disappear into them before reaching Lake Chad. During the rainy season the flat river valleys are flooded over broad areas, but water supply becomes a problem in the dry period. In this season shallow wells to underground water reservoirs, as well as artesian sources that underlie the basin, are used.

Eastern Highlands

The eastern boundary with Cameroon is characterized for about 804 km (500 mi) by mountainous country. The northern part of the highlands consists of several hill groups, with high points around 1,097 m (3,600 ft). To the south of these are the Mandara Mountains, which extend along the border to the Benue River valley. These mountains comprise a dissected plateau with a general elevation of about 1,219 m (4,000 ft). Encompassing an area of some 482 km (300 mi) in length with an average width of about 32 km (20 mi), they give the impression of being a collection of giant granite blocks.

The central part of the region consists of the Adamawa Highlands, a discontinuous series of mountain ranges and high plateau surfaces situated between the Benue River valley and the Donga River valley. They include the Alantika Mountains along the border and, separated in the west by a lower plains area, the Shebshi Hills. The Shebshi Hills, generally at an elevation of 1,066 m (3,500 ft), are a dissected plateau with highly eroded lower slopes. The highest surveyed point in the country, Vogel Peak, with an elevation of 2,042 m (6,700 ft) above sea level, is located in these hills.

To the southwest of the Adamawa Highlands and connected with it by a narrow belt of land ranging about 1,219 m (4,000 ft) in elevation is the Nigerian section of the Bamenda Highlands, most of which lie in Cameroon. Westward extensions of these highlands are known as the Obudu Uplands and the Oban Hills. High points in the uplands reach close to 1,828 m (6,000 ft). Transportation is difficult, and both of these areas are sparsely populated. They are heavily forested except for the uppermost levels of the uplands, which have grassland cover.

Drainage

The dominant feature of the drainage system is the Niger River and its principal tributary, the Benue River. The Niger enters the country in the northwest after traveling some 2,896 km (1,800 mi) through other areas of West Africa. It flows roughly southward and then eastward to the middle of the country, where it is joined by the Benue. Turning south at this point, it empties into the Gulf of Guinea, after a total traverse within the country of approximately 1,175 km (730 mi) through a network of tributaries in the great delta formed at its mouth. The Benue River, which rises in Cameroon, flows about 796 km (495 mi) inside Nigeria, in a generally westward direction, to its confluence with the Niger River.

The Niger drains all of western Nigeria north of the Western High Plains, and the Benue drains the east-central part of the country. The most important river outside the Niger-Benue system is the Cross River in the southeast. The Cross originates in southern Cameroon and enters the country through the Eastern Highlands. Until the early part of the twentieth century this river was the principal link between the Cross River basin area and the outside world. During the period of the slave trade it was a major route for that trade.

The country's overall drainage pattern is somewhat complex. In the southwest a drainage divide runs from the Benin border eastward through the Western High Plains to the Lower Niger Valley. South of this divide the rivers flow directly into the Gulf of Guinea or its fringe lagoons; those rivers to the north are tributaries of the Niger. East of the Lower Niger Valley another divide passes through the Igala and Udi plateaus and separates the Lower Niger Valley from the Cross River basin to the east. A northern extension of this divide to the Eastern Highlands marks the divide between the basin and the Benue River system.

In the north the watershed pattern centers on the Jos Plateau. From the ground this center appears quite insignificant; the area is rather featureless and consists of an undulating, swampy plain. From it, however, radiate the major tributaries of both the Niger and Benue rivers, as well as rivers that flow in the direction of Lake Chad. Thus, to the west and southwest from headwaters on the plateau flow the Sokoto and Kaduna rivers, each for well over 482 km (300 mi) before reaching the Niger. To the east, then south for a total distance of some 531 km (330 mi), the Gongola River flows to the Benue.

To the east and northeast above the Gongola River basin, the Komadugu Yobe River and its tributaries flow toward Lake Chad. This river passes through the Sahel savanna zone and loses large amounts of water to evaporation and seepage; only a trickle finally reaches the lake. In the dry season it usually contains no running water in this area but only pools of varying sizes.

Two large bodies of fresh water are found in the country. In the northeast a portion of Lake Chad lies within the country's borders; in the far western part is Kainji Lake, formed during the latter 1960s by the damming of the Niger River.

Lake Chad is subject to great seasonal variations in area and water level; these conditions result in its alternately advancing and receding over considerable distances in the flat plains area on the Nigerian side. The lake at its height, usually between December and January, may cover up to 25,900 sq km (10,000 sq mi) in central Africa. During the ensuing months it may diminish to less than 12,950 sq km (5,000 sq mi). Depths are between 3 to 5 m (10 to 16 ft) but in places may be only 1.2 m (4 ft) or less. Little water is supplied to the lake from rivers in Nigeria; its principal source is the Chari River in the Republic of Chad.

Kainji Lake was developed as a combined hydroelectric power and river navigation project. The lake itself extends for about 137 km (85 mi) in a section of the Niger River valley from Kainji to a point beyond Yelwa and has a width of from 14 to 24 km (9 to 15 mi). Filling of the lake was accomplished between August and October 1968. At maximum level it covers an area of about 1,243 sq km (480 sq mi).

NIUE

BASIC FACTS

Official Name: Niue

Abbreviation: NIU

Area: 258 sq km (100 sq mi)

Area—World Rank: 192

Population: 2,520 (1988) 2,000 (2000)

Population—World Rank: 209

Capital: Alofi

Land Boundaries: Nil

Coastline: 64 km (40 mi)

Longest Distances: N/A

Highest Point: N/A

Lowest Point: Sea level

Land Use: 61% arable; 4% permanent crops; 4% meadows and pastures; 19% forest and woodland; 12% other

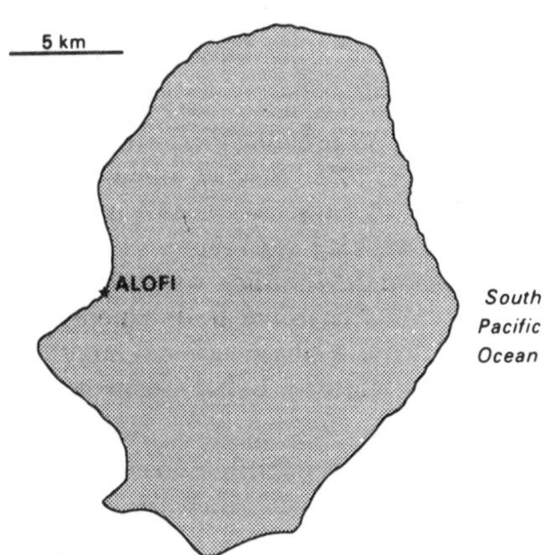

Niue is one of the world's largest coral islands and lies in the Pacific Ocean about 480 km (300 mi) east of Tonga and 930 km (580 mi) west of the southern Cook Islands. It has steep limestone cliffs along the coast around a central plateau.

NORFOLK ISLAND

BASIC FACTS

Official Name: Territory of Norfolk Island

Abbreviation: NOR

Area: 35 sq km (14 sq mi)

Area—World Rank: 207

Population: 2,448

Population—World Rank: 211

Capital: Kingston

Land Boundaries: Nil

Coastline: 32 km (20 mi)

Longest Distances: N/A

Highest Point: Mount Bates 319 m (1,047 ft)

Lowest Point: Sea level

Land Use: 25% meadows and pastures; 75% other

Norfolk Island lies off the eastern coast of Australia, about 1,400 km (869 mi) east of Brisbane. The island is mostly volcanic with some rolling plains.

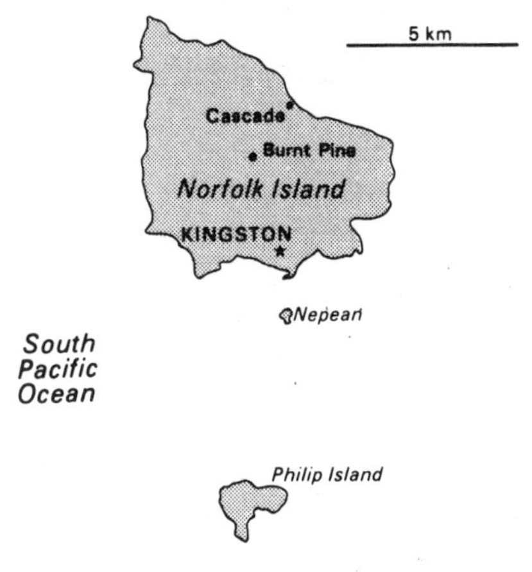

NORWAY

BASIC FACTS

Official Name: Kingdom of Norway

Abbreviation: NRW

Area: 323,878 sq km (125,050 sq mi)

Area—World Rank: 62

Population: 4,190,758 (1988) 4,356,000 (2000)

Population—World Rank: 98

Capital: Oslo

Boundaries: 4,526 km (2,813 mi); Sweden 1,619 km (1,006 mi); Finland 716 km (445 mi); USSR 196 km (122 mi)

Coastline: 1,995 km (1,240 mi)

Longest Distances: 1,752 km (1,089 mi) NNE-SSW; 430 km (267 mi) ESE-WNW

Highest Point: Glittertinden 2,472 m (8,110 ft)

Lowest Point: Sea level

Land Use: 3% arable; 27% forest and woodland; 70% other

Norway is in the western part of the Scandinavian peninsula, with almost one-third of the country north of the Arctic Circle. Svalbard, with 62,700 sq km (24,208 sq mi) constitutes 16.20% of the land area; Jan Mayen, with 380 sq km (147 sq mi), has 0.10% of the land area. Norway has the greatest length of any European country outside the Soviet Union. Norway's greatest width is 430 km (267 mi) ESE to WNW, but in places it is less than 8 km (4.9 mi). Norway is bounded on the north by the Arctic Ocean, on the northeast by Finland and the Soviet Union, on the west by the Norwegian Sea of the Atlantic Ocean, on the east by Sweden, on the south by the Skagerrak and on the southwest by the North Sea. Traveling northward from Skagerrak, the sea boundaries consist of the North Sea as far as the Stadt; the Atlantic, called the Norwegian Sea, as far as the North Cape; and then the Arctic Ocean. The total coastline is 1,995 km (1,238 mi) when measured as an unbroken line but more than seven times that length with the inclusion of the fjords and the greater islands.

The Scandinavian peninsula slopes abruptly toward the Atlantic coast, giving a rugged appearance to the Norwegian side of the border with Sweden. This ruggedness takes on a grander scale to the south of the Trondheim depression, which forms an effective boundary zone between the northern and southern halves of the country. In general, Norway is a mountainous country, with average elevations exceeding 457 m (1,500 ft). Its upland character is most clearly seen in the high fells. The fell area is commonly divided into six areas: Jostedalsbreen, the Jotunheim and the western *vidder* and Dovrefjell, Trollheimen and the eastern *vidder.* Fully 3,000 sq km (1,158 sq mi) of the fells are under icefields, the most extensive of which is the Jostedalsbreen. The ice surfaces range from 914 to 1,828 m (3,000 to 6,000 ft) in height. Mountains clutter the topographical map of Norway, and an array of suffixes indicate mountainous features: *egg* (crest); *hammer* (precipice); *hord* (a broad summit); *kamp* (a broad top); *knat* (a crag); *koll* (a rounded top).

Deep troughs are incised into the plateau surfaces. The troughs take the form of narrow gorges in the interior. Seaward, they broaden into fjords. The fjord heads commonly are characterized by deltaic flats; their central stretches are of great depth, but their mouths are relatively shallow. The fjord zone reaches its greatest breadth in Sogne and Hardanger. Arctic fjords such as Tana, Lakse and Varanger are broader and shorter. The fjords are complemented by deeply entrenched valleys on the landward side. Glaciated valleys such as Setesdal, Numdal, Hallingdal and Valdres, with their extended ribbon lakes, frequently give the illusion of inland fjords. The pattern of alternating valleys and plateaus suggests that Norway is a collection of fjords, as many geographers have described it.

Much of the landscape is dominated by steep gradients. Verticality also has resulted in deep scouring of the land and its continual denudation. Thus water moves in cataracts, scarring the weathered precipices. It is said that in western Norway the sound of the erosion can be heard in the form of a roar or a rumble.

Except in the Southwest and the Far North, the coast to which the plateaus fall has a girdle of islands. The island zone reaches its broadest width of over 60 km (37 mi) at the southern approaches to the Trondheim fjord. The outer islands, protruding from relatively shallow waters, rarely exceed 30 m (100 ft) in height, while the inner islands may rise to 305 m (1,000 ft). These islands are characterized by a series of rock terraces known as strandflats.

In the extreme Southwest, the Jaeren Peninsula has been described as a bit of Denmark clinging to Norway. Viewing the coast in its entirety, there is an interesting contrast between the longer fjords of the South and the shorter fjords of the North, and the larger islands of the North and the smaller islands of the South.

Complementary to the islands are numerous lakes, which are more diverse in form than those of Finland. Because of altitudinal variations, Norway's lakes are at many different levels. In the Southeast, Lake Mjosa is the largest, while Lake Rosvann is the highest in North Norway.

Between the southwestern massif and the southeastern fells of the Swedish borderlands is a rich valley complex focusing upon Gudbrandsdal and Østerdal. These two great dales open up routes to the Trondheim lowland and also drain into the Oslo fjord through flights of broad and fertile terraces. Here is the dividing line between Østfold and Vestfold, characterized by continuous levels of level lowland. Østland is the most Swedish part of Norway in its landscapes. Trondelag also bears resemblance to Østland in its lowlands. Extending from Stavanger fjord to Skien on the margins of Østland is Sorlandet. North Norway is described by geographers as the fifth major unit of

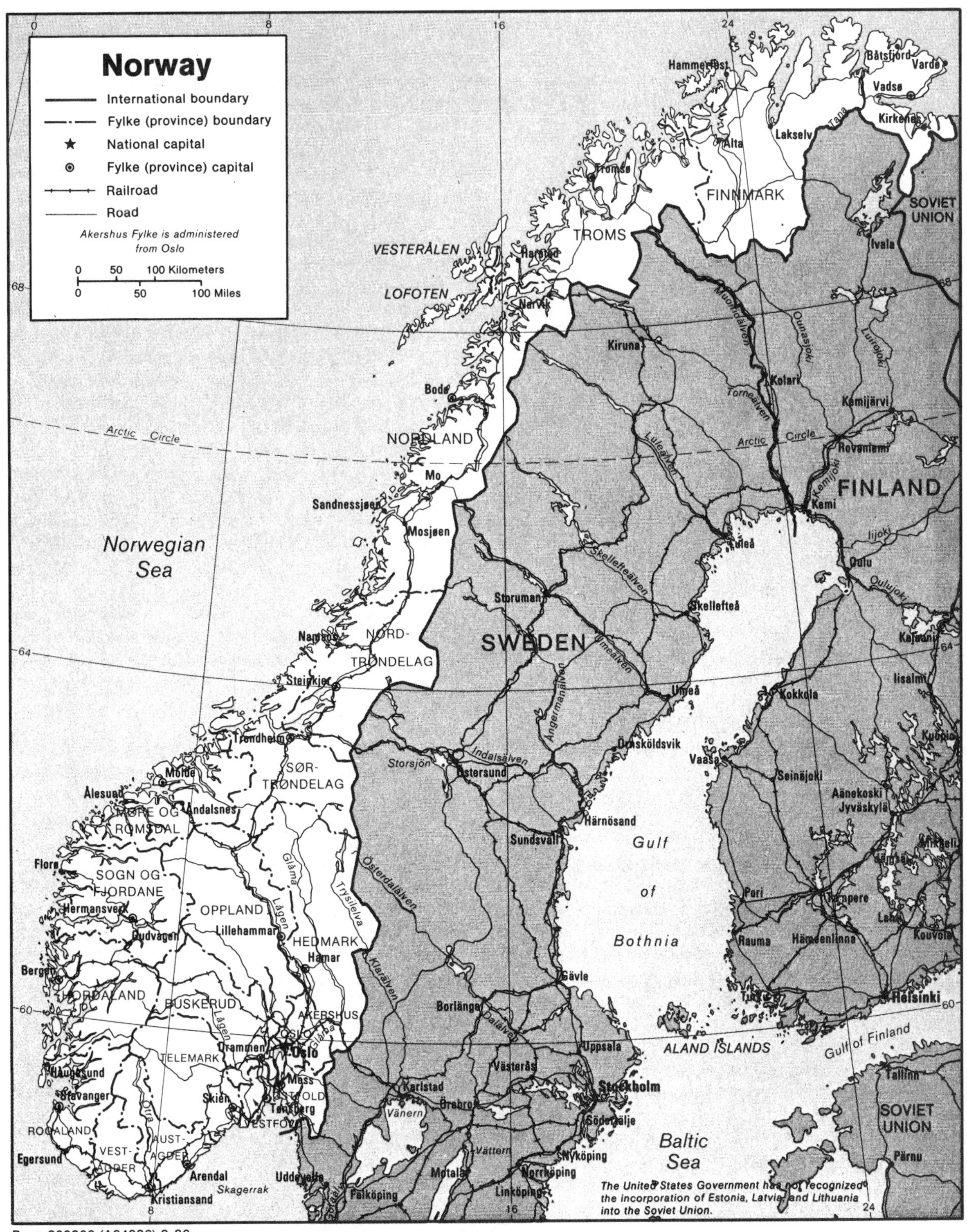

Base 800366 (A04936) 2-86

PRINCIPAL TOWNS (population at 1 January 1987)

Oslo (capital)	451,099	Stavanger	95,437
Bergen	208,809	Kristiansand	63,293
Trondheim	134,496	Drammen	51,324

Area and population

		area		population
Counties	**Capitals**	sq mi	sq km	1987 estimate
Akershus	—	1,898	4,917	399,797
Aust-Agder	Arendal	3,557	9,212	95,475
Buskerud	Drammen	5,763	14,927	221,384
Finnmark	Vardø	18,779	48,637	74,690
Hedmark	Hamar	10,575	27,388	186,305
Hordaland	Bergen	6,036	15,634	402,343
Møre og Romsdal	Molde	5,832	15,104	237,489
Nordland	Bodø	14,798	38,327	241,048
Nord-Trøndelag	Steinkjer	8,673	22,463	126,648
Oppland	Lillehammer	9,753	25,260	181,620
Oslo	Oslo	175	454	449,220
Østfold	Moss	1,615	4,183	235,813
Rogaland	Stavanger	3,529	9,141	326,611
Sogn og Fjordane	Leikanger	7,195	18,634	105,966
Sør-Trøndelag	Trondheim	7,271	18,831	247,354
Telemark	Skien	5,913	15,315	162,595
Troms	Tromsø	10,021	25,954	146,595
Vest-Agder	Kristiansand	2,811	7,281	141,284
Vestfold	Tønsberg	856	2,216	192,934
TOTAL		125,050	323,878	4,175,171

the country, and it is as dissected as any other part of Norway. Some of the largest islands in Scandinavia—Hinnøy, Senja, Kvaløy, and Sorey—are in this region. The two most extensive and detached island groups are those of Lofoten and Vesterålen. Much of the land is bare of vegetation.

Norwegian mountain ranges are roughly divided into three groups. The most northerly, the Kjölen, is by far the greatest and forms a natural barrier between Norway and Sweden, receding with decreasing height northward to the Finnish border. The Dovrefjell range marks the division between North and South Norway. The Langfjell, consisting of several ranges, contains the highest peaks in the peninsula. Galdhøpiggen, in this group is the highest peak in the country, at 2,560 m (8,399 ft). At an altitude of 1,830 m (6,004 ft) lies the Hardanger vidda, a desert plateau of some 6,474 sq km (2,500 sq mi), with steep sides scarred and grooved by waterfalls and valleys.

Throughout Norway, the sound of falling water is the natural accompaniment to the landscape. The melting snow from the great fields of the Hardanger and Josteldalsbreen mountains releases vast amounts of what the Norwegians call "white coal." Among the most magnificent falls are the 253 m (830 ft) Vettisfos, east of Sogne; and the 162 m (530 ft) Vøringfos, which falls from the heights of Fossli to the Maabodal Valley.

Most of the northern end of the Norwegian Plateau is covered by icecaps. They are, on the whole, uninterrupted by peaks rising above them and almost without crevasses. Jostedalsbreen, 1,502 sq km (580 sq mi) in area and possibly 457 m (1,500 ft) thick, is the largest glacier in Europe. The next two largest are the Svartisen, about 1,036 sq km (400 sq mi) in area; and Folgefonn, about 280 sq km (108 sq mi) in area, the top of which is over 1,524 m (5,000 ft) above sea level. Other large snowfields include Hallinskarvet in Hardanger, Snohetta in Dovrefjell, Store Borgefjell overlooking the Namsen Valley, Seiland near Hammerfest and Oksfjordjokel near Kvanangen. Most of the glaciers are clean, with little of the loose dirt that spoils the appearance of the Alpine glaciers.

The rivers of Norway are swift and turbulent, rushing through steep valleys and rocky gorges. Generally the rivers are not navigable, but they are valuable as flumes for the timber coming down from forest districts. The only navigable rivers are the Glomma and the Dramselv. The Glomma, the largest river in Scandinavia, is 563 km (350 mi) long and rises more than 610 m (2,000 ft) above sea level at Aursunden Lake, north of Røros. Many lakes widen the stream, and the river is famous for its fine waterfalls. Dramselv rises in Valdres and enters Oslofjord at Drammen. The other major rivers are the Numedalslagen, the Nidelv, the Rauma, the Driva, the Sand, the Bjoreia and the Evanger.

Lakes abound in Norway. Nearly one-twelfth of the country is under fresh water, sometimes so deep that the waterbed is far below the level of the sea. By far the greatest of these lakes is the Møjsa, 363 sq km (140 sq mi) in area and 452 m (1,982 ft) deep. Hornindaisvann, although only 32 km (20 mi) in area, has the same depth as Møjsa. Among the more beautiful lakes are Loen, Olden, Bygdin, Tyin, Femundsjon, Rosvann, Randsfjord, Tyrifjord, Snaasenvatn, Tansjon and Altevatn. The levels of the lakes differ greatly, by as much as 316 m (1,300 ft) in altitude. Most of the larger lakes are 122 m (400 ft) above the sea and were perhaps heads of fjords that have since disappeared.

In 1920 the Spitsbergen archipelago was placed under Norwegian sovereignty and given the name Svalbard. Bear Island (Björnoya, 178 sq km; 69 sq mi) usually is included in the Svalbard group. Norwegian sovereignty over Jan Mayen Island (372 sq km; 164 sq mi) was recognized in 1930. Compared with the mainland, the islands have a complex geology, with a much greater range of rocks. It shares some of the basalt features common to Iceland and the Faeroes. Icefields cover extended areas, and the surface is covered by permafrost. Svalbard is the most northerly human settlement in Europe.

OMAN

BASIC FACTS

Official Name: Sultanate of Oman

Abbreviation: OMN

Area: 300,000 sq km (100,000 sq mi)

Area—World Rank: 66

Population: 1,265,382 (1988) 1,973,000 (2000)

Population—World Rank: 127

Capital: Muscat

Boundaries: 3,234 km (2,010 mi); South Yemen 288 km (179 mi); Saudi Arabia 676 km (420 mi); United Arab Emirates 410 km (255 mi)

Coastline: 1,860 km (1,156 mi)

Longest Distances: 972 km (604 mi) NE-SW; 513 km (319 mi) SE-NW

Highest Point: Mount Sham 3,035 m (9,957 ft)

Lowest Point: Sea level

Land Use: 5% meadows and pastures; 95% other

Oman lies at the southeastern corner of the Arabian Peninsula. Its northernmost tip touches the Strait of Hormuz. Its coastline runs south along the Gulf of Oman and then west along the Arabian Sea. On the north it is bordered by United Arab Emirates; on the northwest and west by the shifting sands of the Rub al-Khali of Saudi Arabia; and on the southwest by Southern Yemen.

Muscat is the name of the capital city and refers also to the coastal strip stretching from the Strait of Hormuz to the tip of Arabia at Ras al-Hadd. Oman is a broader term and is used to describe the entire region forming the horn of southeastern Arabia. The Sultan claims sovereignty over all of Oman except that portion bordering the Persian Gulf and the belt of territory intersecting his lands on the Gulf of Oman. This region is known as Trucial Oman, comprising the territory of the United Arab Emirates. Also within the Sultan's domains are the offshore island of Masirah and the Province of Dhufar, which lies between Oman and Southern Yemen.

The territory claimed by the Sultan extends in some places to about 322 km (200 mi) inland from the coast. Estimates are not precise because the inland boundary with Saudi Arabia is undemarcated.

Geography has made the Sultanate a virtual island, bordered mostly by the sea and the wastes of the Rub al-Khali. Historically, the country's contacts with the rest of the world have been largely by way of the sea, which not only provided access to foreign countries, but also linked the coastal towns of Oman. The Rub al-Khali, difficult to cross until the era of modern desert transport, has served as a barrier between the Sultanate and the Arabian interior. Insularity has been reinforced by the formidable Hajar mountain chain, which forms a belt between the coast and the desert from the Musandam Promontory to the city of Sur. The desert and the mountains have combined to keep Muscat and Oman remarkably free of foreign encroachments—either military or cultural—from the interior.

Geographic Regions

The Sultanate is divided by natural features into several distinct districts: the tip of the Musandam Peninsula; the Batinah Plain; the Muscat-Matrah coastal area; inner Oman comprising the Jabal Akhdar (Green Mountain), its western foothills, and the desert fringes; the Province of Dhufar in the south; and the offshore island of Masirah. Until 1958 the Sultan also ruled Gwadur, a coastal enclave on the mainland of West Pakistan. This Persian Gulf port, a remnant of the Sultanate's far flung maritime empire of the early 19th century, was ceded to Pakistan in exchange for $8.4 million on September 4, 1958.

Area and population	area		population
Region **Area**	sq mi	sq km	1987 estimate
Dhofar	40,000	100,000	. . .
Southern	. . .	. . .	. . .
Musandam (R'ūs al-Jibāl)	800	2,000	13,000
Musandam	. . .	. . .	. . .
Other	79,200	198,000	. . .
al-Baṭīnah	. . .	. . .	. . .
al-Jaww and al-Buraymī	. . .	. . .	. . .
Dhahirah (aẓ-Ẓāhirah)	. . .	. . .	. . .
Capital	. . .	. . .	. . .
Eastern al-Ḥajar	. . .	. . .	. . .
Ja'lān and Sur (Ja'lān)	. . .	. . .	. . .
Sharqīyah	. . .	. . .	. . .
'Uman Interior	. . .	. . .	. . .
Western al-Ḥajar	. . .	. . .	. . .
TOTAL	120,000	300,000	1,331,000

The northernmost province, Ras al-Jabal, extends from Ras Musandam to Dibah. It touches the Strait of Hormuz, which links the Persian Gulf with the Gulf of Oman, and is separated from the rest of the Sultanate by a belt of territory belonging to the United Arab Emirates. The province consists entirely of low mountains forming the northernmost extremity of the Western Hajar. Two inlets—the Elphinstone and the Malcolm—cleave the coastline about one-third of the distance from the Strait of Hormuz and, at one point, are separated by only a few yards of land. The coastline is extremely rugged and the Elphinstone inlet, 16 km (10 mi) long and surrounded on all sides by cliffs 914 to 1,219 m (3,000 to 4,000 ft) in height, has frequently been compared to a Norwegian fjord.

The intervening belt of UAE territory separating Ras al-Jabal from the Sultan's other lands extends almost as far south as the coastal town of Shias. From the point at which the Sultanate is entered again to the town of Sib, 241 km (150 mi) to the southeast, runs a narrow well-populated coastal plain, known as Batinah. Across this plain a number of valleys descend from the Western Hajar to the south, which are heavily populated in their upper courses. A narrow ribbon of oases, watered by wells and underground channels, extends the length of the plain, averaging about 1 mile in from the coast. Dates, limes, mangoes, and other fruits are grown in irrigated coastal gardens and, along their edge, small quantities of cereal grains are produced. On the landward side of the Batinah Plain there are acacia trees, beyond which a barren, pebbly plain slopes gradually to the foothills of the Western Hajar about 16 km (10 mi) away. The coastal gardens abound in bird life. The grey partridge is common and, in the colder season, flocks of sand grouse appear on the adjoining plains and ducks are seen in pools and creeks where the *wadis* (watercourses) run to the sea.

South of Sib the coast changes character. For a distance of about 177 km (110 mi) from Sib to Ras al-Hadd, it is

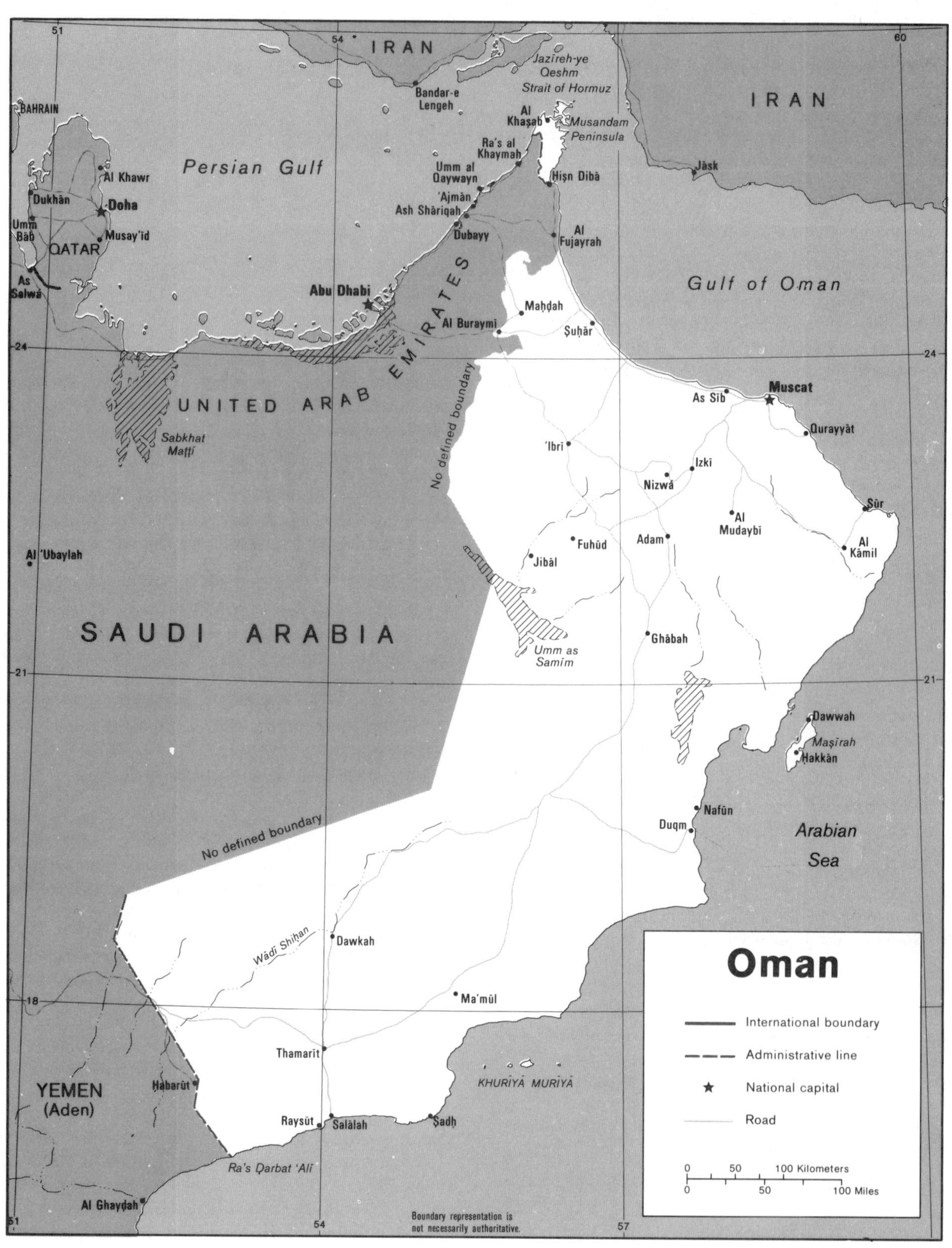

Oman
IRAN
IRAN
Persian Gulf
Gulf of Oman
Arabian Sea
SAUDI ARABIA
UNITED ARAB EMIRATES
QATAR
BAHRAIN
YEMEN (Aden)
Jazīreh-ye Qeshm
Strait of Hormuz
Musandam Peninsula
Bandar-e Lengeh
Al Khaşab
Ra's al Khaymah
Umm al Qaywayn
'Ajmān
Ash Shāriqah
Dubayy
Ḩişn Dibā
Al Fujayrah
Jāsk
Al Khawr
Dukhān
Doha
Umm Bāb
Musay'id
As Salwá
Abu Dhabi
Al Buraymī
Maḩḑah
Şuḩār
As Sīb
Muscat
Qurayyāt
'Ibrī
Izkī
Nizwá
Şūr
Al Mudaybī
Al Kāmil
Adam
Fuhūd
Jibāl
Ghābah
Sabkhat Maţţī
Umm as Samīm
No defined boundary
No defined boundary
Al 'Ubaylah
Dawwah
Maşīrah
Ḩakkān
Nafūn
Duqm
Wādī Shiḩan
Dawkah
Ma'mūl
Thamarīt
Ḩabarūt
KHURĪYĀ MURĪYĀ
Raysūt
Şalālah
Şadḩ
Ra's Ḑarbat 'Alī
Al Ghayḑah
Boundary representation is not necessarily authoritative.
International boundary
Administrative line
National capital
Road
0 50 100 Kilometers
0 50 100 Miles
51
54
57
60
24
21
18

barren and bounded almost throughout its length by cliffs. There is no cultivation and very little habitation along this stretch. Although the deep water off this coast renders navigation relatively easy, there are few harbors or safe anchorages. The two best are at Muscat and Matrah where natural harbors gave rise to the growth of cities centuries ago.

West of the coastal areas lies the high tableland of central Oman. The Hajar mountains form two ranges: the Hajar al-Gharbi or Western Hajar and the Hajar al-Sharqi or Eastern Hajar. They are divided by the Wadi Samail, a valley which forms the traditional route between Muscat and the interior. The general elevation is about 1,219 m (4,000 ft) but the peaks of a high ridge known as the Jabal Akhdar—which is considered as a separate district but is actually part of the Western Hajar geographically—rise to 3,048 m (10,000 ft) above sea level in some places.

Behind the Western Hajar are two inland districts, Dhahirah and inner Oman, which are separated from each other by the lateral range of Jabal Kour. Both recede into stony desert before meeting the wastes of the Rub al-Khali. Inland of the Eastern Hajar are the sandy districts of Sharqiyah and Jalan, bordering the desert. They are not separated from each other or from the plateau of inner Oman by any prominent natural feature.

The desolate coastal tract from Jalan to Ras Nus has no general name. Low hills and inhospitable wastelands meet the sea for miles on end. Halfway along this length of coast and separated from it by about 16 km (10 mi) is the barren island of Masirah. It is 64 km (40 mi) long, virtually uninhabited.

The dependency of Dhufar extends from Ras Sharbatat on the coast to the border of Southern Yemen. Its exact northern limit has never been defined but the territory claimed by the Sultan includes the Wadi Mughsin, about 241 km (150 mi) inland. The southwestern portion of Dhufar's coastal plain is regarded as one of the most beautiful spots in Arabia and its capital, Salala, is the permanent residence of the Sultan. The southwestern coastal plain contains splendid vegetation and birdlife, but only 16 km (10 mi) inland the low rugged foothills of the Qara mountain chain begin to rise. The highest peaks are between 914 and 1,219 m (3,000 and 4,000 ft) in elevation. They gradually slope downward to a narrow pebbly desert adjoining the Rub al-Khali to the north.

PAKISTAN

BASIC FACTS

Official Name: Islamic Republic of Pakistan

Abbreviation: PAK

Area: 796,095 sq km (307,374 sq mi)

Area—World Rank: 35

Population: 107,467,457 (1988) 137,651,000 (2000)

Population—World Rank: 8

Capital: Islamabad

Boundaries: 6,661 km (4,139 mi); China 523 km (325 mi); India 2,028 km (1,260 mi); Iran 830 km (516 mi); Afghanistan 2,466 km (1,532 mi)

Coastline: 814 km (506 mi)

Longest Distances: 1,875 km (1,165 mi) NE-SW; 1,006 km (625 mi) SE-NW

Highest Point: K2 8,611m (28,250 ft)

Lowest Point: Sea level

Land Use: 26% arable; 6% meadows and pastures; 4% forest and woodland; 64% other

Pakistan, a wedge-shaped country, is bounded on the south by the Arabian Sea; on the west, northwest, and north by Iran and Afghanistan; on the northeast by the People's Republic of China (PRC); and on the east by India. About one-third of the Pakistan-India frontier is the cease-fire line in the Jammu and Kashmir region, disputed between the two countries since their independence.

Pakistan can be divided into three major geographic areas: the northern highlands; the Indus River plain, with its two major subdivisions corresponding roughly to the provinces of Punjab and Sind; and the Baluchistan Plateau. Some geographers designate additional major regions; for example, the mountain ranges along the western border with Afghanistan are sometimes described separately from the Baluchistan Plateau, and on the eastern border with India, south of the Sutlej River, the Thar Desert may be treated separately from the Indus Plain. Nevertheless, the country may conveniently be visualized in general terms as divided in thirds by an imaginary line drawn eastward from the Khyber Pass and another drawn southwest from Islamabad down the middle of the country. Roughly, then, the northern highlands are north of the imaginary east-west line, the Baluchistan Plateau is to the west of the imaginary southwest line, and the Indus Plain lies to the east of that line.

The northern highlands are a region of some of the most rugged, formidable mountains in the world. The Himalayan Mountains stretch in a great convex arc from the northeast corner of India some 2,414 km (1,500 mi) across the top of that country and, at its northwest corner, merge into the Karakoram and the Pamir mountain ranges of Pakistan's northern agencies of Baltistan and Gilgit. South of the Wakhan Corridor and west of the Pamirs are the massive heights and steep valleys of the mountains called the Hindu Kush.

In the northern highlands, virtually all elevations are higher than 2,438 m (8,000 ft) above sea level; more than one-half are above 4,522 m (15,000 ft), and more than fifty peaks are above 6,705 m (22,000 ft). Travel through the area is difficult and dangerous. Because of the rugged topography and rigors of the climate, the northern mountains and the Himalayan chain to the east have throughout history been formidable barriers to movement into Pakistan.

South of the Khyber Pass in the NWFP is the east-west Safed Koh Range, generally lower than the northern highlands but nevertheless reaching 4,761 m (15,620 ft) in its western extension to the Afghanistan border. South of the Safed Koh and clustered near the border are the mountains of Waziristan. Beyond them, the Toba Kakar Range of about 2,743 m (9,000 ft) average crest elevation extends in an arc of about 482 km (300 mi) from northern Baluchistan to the Khojak Pass. The Ras Koh Range, west of Quetta, and the Chagai Hills, extending further west to the Pakistan-Iran-Afghanistan tripoint, complete the western highlands lying generally against the Afghan border.

Quetta, situated at 1,676 m (5,500 ft) elevation between the Khojak Pass and the interior Bolan Pass to the southeast, is the focal point for the mountain, road, and rail systems of Baluchistan. The Quetta region, as a geological focal point, has a complicated and unstable structure that is seismically important. A severe earthquake in 1931 was followed by one of major proportions on May 31, 1935, in which the city was virtually ruined and at least 20,000 lives were lost. Lesser tremors have occurred from time to time. The Central Brahui Range continues for 282 km (175 mi) south from Quetta and then divides into the Kirthar Range, extending southeast, and the Makran Range to the southwest and west as far as the Nihing River on the Iranian border. Southeast from Quetta, the Bugti Hills merge into the Sulaiman Range, which curves to the north along the line between the provinces of Baluchistan and the Punjab with summits of 1,828 to 2,133 m (6,000 to 7,000 ft).

The Baluchistan Plateau, generally at 914 to 1,219 m (3,000 to 4,000 ft) is thus defined and enclosed by the western ranges along the Afghan border and by those emanating southwest and southeast from Quetta. The plateau itself is an arid tableland of approximately 350,945 sq km (135,000 sq mi) with interior drainage. A large salt marshland called the Hamuni-Lora is at the eastern end of the Chagai Hills on the Afghan border. An even larger salt lake and swampy area, the Hamun-i-Mashkel, is south of the Chagai Hills on the border with Iran.

The principal river of Baluchistan is the Zhob, running along the southern slopes of the Toba Kakar Range and north into the Gumal (Gomal) River, a tributary of the Indus in the southern part of the NWFP. In southern Baluchistan several minor rivers flow into the Arabian Sea; from west to east these include the Dasht, the Mashkai, the Nal, and the Porali. Along the Arabian Sea below the line of the southern Makran Range is a coastal plain, and a lowland salient of less than 152 m (500 ft) elevation projects northward between the Central Brahui and Sulaiman mountains. Otherwise, the province of Baluchistan is determined by the mountain ranges and the plateau.

The Indus River, rising in the Tibetan Himalayas, flows generally northwest for about 804 km (500 mi). After crossing the Indian-held portion of Jammu and Kashmir, it enters Pakistan's Baltistan Agency, is joined by a tributary southeast of the town of Gilgit, and flows southwest for 1,609 km (1,000 mi) to the Arabian Sea. The vast drainage area of the Indus, one of the principal rivers of Asia, corresponds roughly in Pakistan to the provinces of the Punjab and Sind.

At Attock on the border between the Punjab and the NWFP, the Indus receives the waters of the Kabul River from the west. Continuing south and forming part of the boundary between the Punjab and the NWFP, the Indus passes west of the Potwar Plateau, which lies between Islamabad and the Salt Range. This roughly east-west range has some peaks of nearly 1,524 m (5,000 ft), but most do not exceed 762 m (2,500 ft). After the confluence of the Gumal and the Indus rivers a few mi south of Dera Ismail Khan, the Indus continues south for about 321 km (200 mi) to Mithanhot, where it is joined by its major tributary, the Panjnad.

The short Panjnad River, about 121 km (75 mi) long, is actually the combined input of the "five rivers of the Punjab,"—the word *Punjab* being derived from *panch* (five) and *ab* (water). The five rivers from north to south are the Jhelum, the Chenab, the Ravi, the Beas, and the Sutlej. The Beas is in fact a tributary of the Sutlej, entering the latter a few mi east of the point where the Sutlej enters Pakistan. The boundary between Pakistan and India, laid down by the Radcliffe Award of 1947, cuts across the four rivers. Because irrigation is of key importance on both sides of the boundary, disputes over the distribution of these waters formed one of the main sources of friction between the two countries. Assisted by mediation of the World Bank, Pakistan and India resolved this dispute by the Indus Waters Treaty of 1960, allocating water rights to the Indus, the Jhelum, and the Chenab to Pakistan and assigning rights to the Ravi, Sutlej, and Beas to India.

The upper Indus Plain in the Punjab varies from about 152 to 304 m (500 to 1,000 ft) in elevation and consists of fertile alluvium deposited by the rivers. The lower Indus Plain, corresponding generally to the province of Sind, is lower in altitude, declining to sea level at the coast. The lower plain differs from the upper in that it was formed by the deposits of one river rather than a network and the alluvium is more recent. The lower plain east of the city of Hyderabad is one of the most productive agricultural regions of the country.

PRINCIPAL CITIES (population at 1981 census)

City	Population	City	Population
Karachi	5,180,562	Peshawar	566,248
Lahore	2,952,689	Sialkot	302,009
Faisalabad (Lyallpur)	1,104,209	Sargodha	291,362
Rawalpindi	794,843	Quetta	285,719
Hyderabad	751,529	Islamabad (capital)	204,364
Multan	722,070	Jhang	195,558
Gujranwala	658,753	Bahawalpur	180,263

Area and population

Provinces	Capitals	area: sq mi	area: sq km	population: 1983 estimate
Baluchistān	Quetta	134,050	347,188	4,611,000
North-West Frontier	Peshāwar	28,773	74,522	11,658,000
Punjab	Lahore	79,284	205,345	50,460,000
Sind	Karāchi	54,407	140,913	20,312,000
Federally Administered Tribal Areas	...	10,510	27,221	2,329,000
Federal Capital Area				
Islāmābād	...	350	906	359,000
TOTAL		307,374	796,095	89,729,000

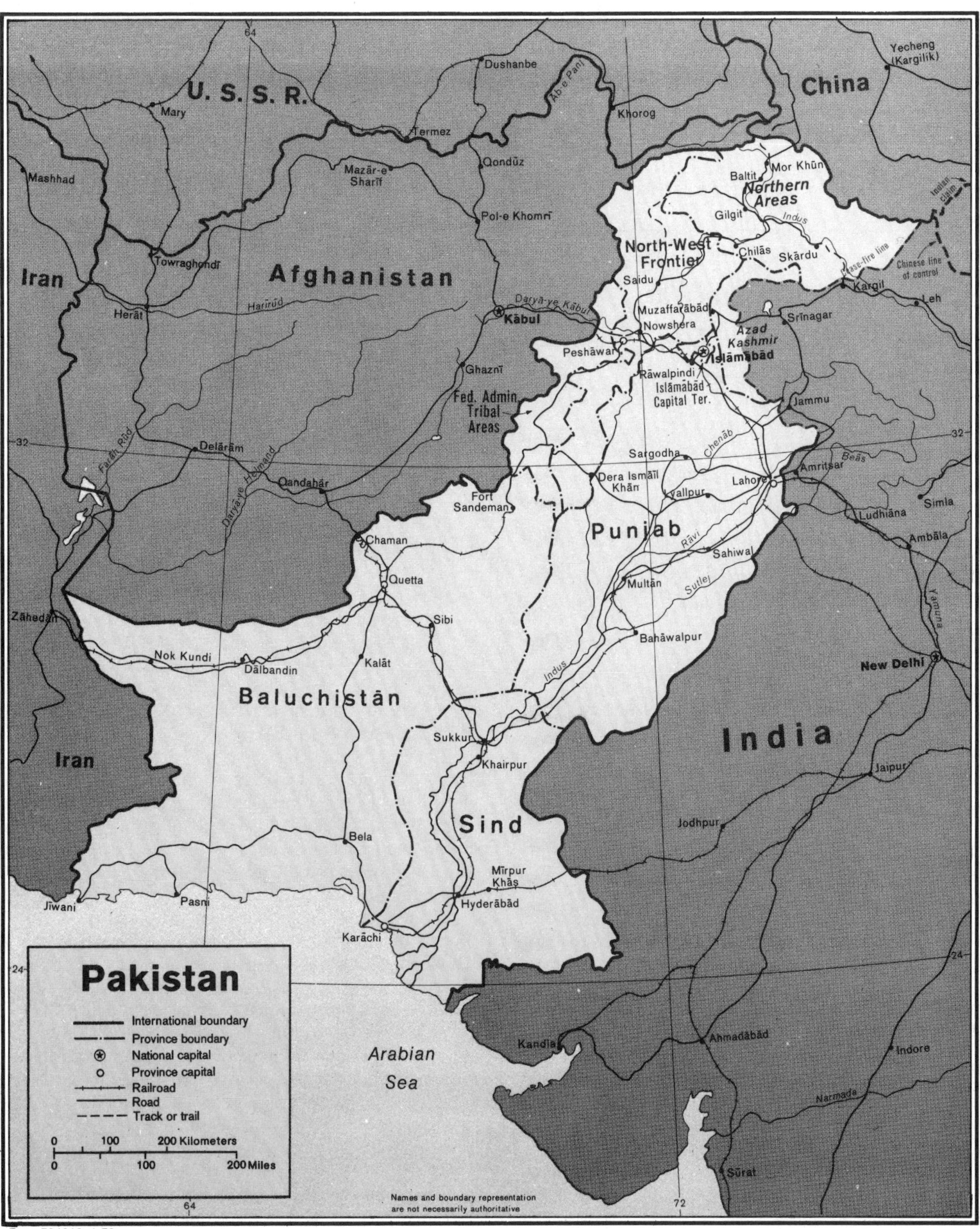

Base 504018 1-79

Despite the extensive development of irrigation in the Indus Plain, some desert areas remain unusable. South of the Salt Range and between the Indus and Jhelum rivers is the Thal Desert, extending about 160 km (100 mi) in length. Of far greater size is the Thar Desert, which lies south of the Sutlej along the Pakistan-India border all the way to the Rann of Kutch. These areas are slowly being reclaimed by canals, especially in the Thal Desert.

PANAMA

BASIC FACTS

Official Name: Republic of Panama

Abbreviation: PAN

Area: 77,082 sq km (29,762 sq mi)

Area—World Rank: 111

Population: 2,323,622 (1988) 2,893,000 (2000)

Population—World Rank: 119

Capital: Panama City

Boundaries: 2,417 km (1,501 mi); Colombia 266 km (165 mi); Costa Rica 364 km (226 mi)

Coastline: Caribbean Sea 624 km (388 mi); Pacific Ocean 1,188 km (738 mi)

Longest Distances: 398 km (247 mi) ESE-WNW and 272 km (169 mi) NNE-SSW; west of Canal Zone; 337 km (209 mi) ESE-WNW and 159 km (99 mi) NNE-SSW

Highest Point: Baru Volcano 3,475 m (11,401 ft)

Lowest Point: Sea level

Land Use: 6% arable; 2% permanent crops; 15% meadows and pastures; 54% forest and woodland; 23% other

Panama is located on the narrowest and lowest part of the Isthmus of Panama that links North and South America. That part of the isthmus within the country varies in width from about forty-eight km (30 mi) to over 185 km (115 mi). Because of the location and contour of the country, directions expressed in terms of the compass are often surprising. For example, a transit of the Panama Canal from the Pacific to the Atlantic involves travel not to the east but to the northwest, and in Panama City the sunrise is to the east over the Pacific. In practice compass points are not in general use as a means of describing locations. The two coastlines are referred to as the Atlantic and Pacific rather than north and south (Atlantic, rather than Caribbean, is the terminology more often used.)

The dominant feature of the country's landform is the central spine of mountains and hills that forms the continental divide. The divide does not form part of the great mountain chains of North America, and only near the Colombian border are there highlands related to the Andean system of South America. The spine that forms the divide is the highly eroded arch of an uplift from the sea bottom in which peaks were formed by volcanic intrusions.

The mountain range of the divide is called the Cordillera de Talamanca near the Costa Rican border. It becomes the Serrianía de Tabasara, and the portion of it closer to the lower saddle of the isthmus where the canal is located is often called the Sierra de Veraguas. As a whole, the range between Costa Rica and the canal is generally referred to by Panamanian geographers as the Cordillera Central.

The highest point in the country is the Volcán Barú (formerly known as the Volcán de Chiriquí) that rises to approximately 3,454 m (11,332 ft). The apex of a highland that includes the nation's richest soils, the Volcán Barú is still referred to as a volcano but has been inactive for millennia.

The more than 300 rivers emptying into the Pacific are longer and slower running than those of the Atlantic slopes, and their basins are more extensive. The largest of these is the Río Tuira, which flows into the Golfo de San Miguel and is the nation's only navigable river. The second largest, the Río Chepo in Panama Province, is the site of a major hydroelectric project.

The 1,246-km (773 mi) stretch of Atlantic coastline is marked by several good natural harbors, but there are few towns on the Atlantic, and the only important port facilities are those of Cristobal, located on Limon Bay at the Atlantic terminus of the canal. The numerous islands of the Bocas del Toro Archipelago near the Costa Rican border provide an extensive natural roadstead and shield the banana port of Almirante. The 366 San Blas Islands are strung out for more than 160 km (100 mi) along the sheltered coastline of San Blas.

The Pacific coastline is about 1,634 km (1,014 mi) in length. The principal islands are those of the Archipiélago de las Perlas in the middle of the Gulf of Panama, the prison island of Coiba in the Golfo de Chirquí, and the decorative island of Taboga, a tourist attraction that can be seen from Panama City. In all, there are some 1,000 islands off the Pacific coast.

The Pacific coastal waters are extraordinarily shallow. Depths of 180 m (590 ft) are reached only outside the perimeters of both the Gulf of Panama and the Golfo de Chiriquí, and wide mud flats extend up to 70 km (43 mi) seaward from the coastlines. As a consequence the tidal range is extreme. A range of about 70 cm (27 in) between high and low water on the Atlantic contrasts sharply with over 7 m (23 ft) on the Pacific, and some 130 km (80 mi) up the Río Tuira the range is over 5 m (16 ft).

Geographic Regions

There is no generally recognized group of regions, and no single set of names is in common use. One system often used by Panamanian geographers, however, portrays the country as divided into five regions that reflect population concentration and economic development as well as geography.

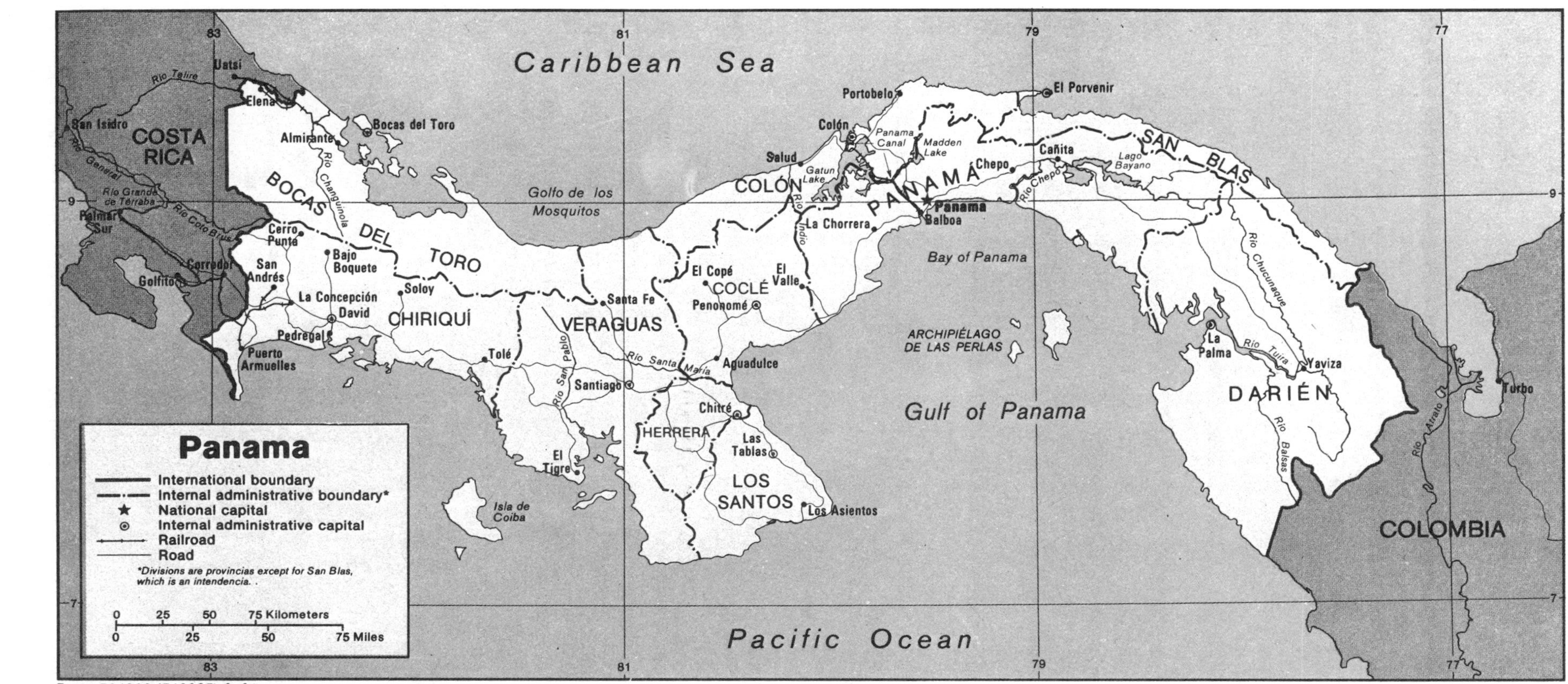

Base 504908 (540285) 8-81

Darién

The largest and most sparsely populated of the regions, Darién extends from the hinterlands of Panama City and Colón to the Colombian border and comprises more than one-third of the national territory. In addition to the province of Darién, it includes San Blas Territory and the eastern part of Panama Province. Darién—a name that was once applied to the entire isthmus—is a land of rain forest and swamp; roads are almost nonexistent.

Central Isthmus

The Central Isthmus does not have precisely definable boundaries. Geographically it is the low saddle of land that bisects the isthmus at the canal. It extends on the Pacific side as far as the town of La Chorrera. On the Atlantic, it includes small villages and clustered farms around Gatun Lake. East of the canal it terminates gradually as the population grows sparse, and the jungles and swamps of the Darién region begin. More a concept than a region, the Central Isthmus, with a width of about 100 m (62 ft), is the densely populated historic transportation route between Atlantic and Pacific.

Central Panama

This part lies to the southwest of the canal and is made up of all or most of the provinces of Veraguas, Coclé, Herrera, and Los Santos. It is located between the continental divide and the Pacific and is sometimes referred to as the Central Provinces. The sparsely populated Santa Fe District of Veraguas Province is located across the continental divide on the Atlantic side, however, and a frontier part of Coclé is also on the Atlantic side of the divide.

The hills and lowlands of Central Panama are dotted with farms and ranches and include most of the country's rural population. This is the traditional Panama that has undergone the least change during the years of the republic. Its heartland is a heavily populated rural arc that frames the Bahia de Parita and includes most of the country's largest market towns, including the provincial capitals of Penonomé, Santiago, Chitré, and Las Tablas. This agriculturally productive area has a relatively long dry season and is known as the dry zone of Panama.

PRINCIPAL TOWNS (population at 1980 census): Panama City (capital) 389,172; Colón 59,840; David 50,016.

Area and population		area		population
Provinces	**Capitals**	sq mi	sq km	1987 estimate
Bocas del Toro	Bocas del Toro	3,443	8,917	77,500
Chiriquí	David	3,381	8,758	360,300
Coclé	Penonomé	1,944	5,035	164,500
Colón	Colón	1,915	4,961	163,100
Darién	La Palma	6,488	16,803	38,400
Herrera	Chitré	937	2,427	101,300
Los Santos	Las Tablas	1,493	3,867	81,300
Panamá	Panama City	4,642	12,022	1,037,400
Veraguas	Santiago	4,280	11,086	210,000
Special territory				
Comarca de San Blas	El Porvenir	1,238	3,206	40,600
TOTAL AREA		29,762	77,082	2,274,400

Chiriquí

The remaining part on the Pacific side of the divide is taken up by Chiriquí Province. Some geographers regard it and Central Panama as a single region, but the lowlands of the two are separated by the hills of the Las Palmas Peninsula, and the big province of Chiriquí has enough individuality to warrant consideration as a separate region. The second largest and second most populous of the nine provinces, it is to some extent a frontier area as well as one of considerable economic importance. It is only in Chiriquí that the frontiers of settlement have pushed up well into the interior highlands, and the population has a particular sense of regional identity. A native of Chiriquí can be expected to identify himself, above all, as a Chiricano.

Atlantic Panama

This region includes all of Bocas del Toro Province, the Atlantic portions of Veraguas and Coclé, and the western districts of Colón. It is home to a scant 5 percent of the population, and its only important population concentrations are near the Costa Rican border where banana plantations are located.

PAPUA NEW GUINEA

BASIC FACTS

Official Name: Papua New Guinea

Abbreviation: PNG

Area: 462,840 sq km (178,704 sq mi)

Area—World Rank: 50

Population: 3,649,503 (1988) 4,673,000 (2000)

Population—World Rank: 105

Capital: Port Moresby

Boundaries: 4,572 km (2,841 mi); Indonesia 777 km (483 mi)

Coastline: 3,795 km (2,358 mi)

Longest Distances: 2,082 km (1,294 mi) NNE-SSW; 1,156 km (718 mi) ESE-WNW

Highest Point: Mount Wilhelm 4,509 m (14,793 ft)

Lowest Point: Sea level

Land Use: 1% permanent crops; 71% forest and woodland; 28% other

Papua New Guinea lies in the SW Pacific about 160 km (100 mi) NE of Australia. It includes the former Trust Territory of New Guinea, comprising the Western Bismarck Archipelago (Mussau, New Britain,

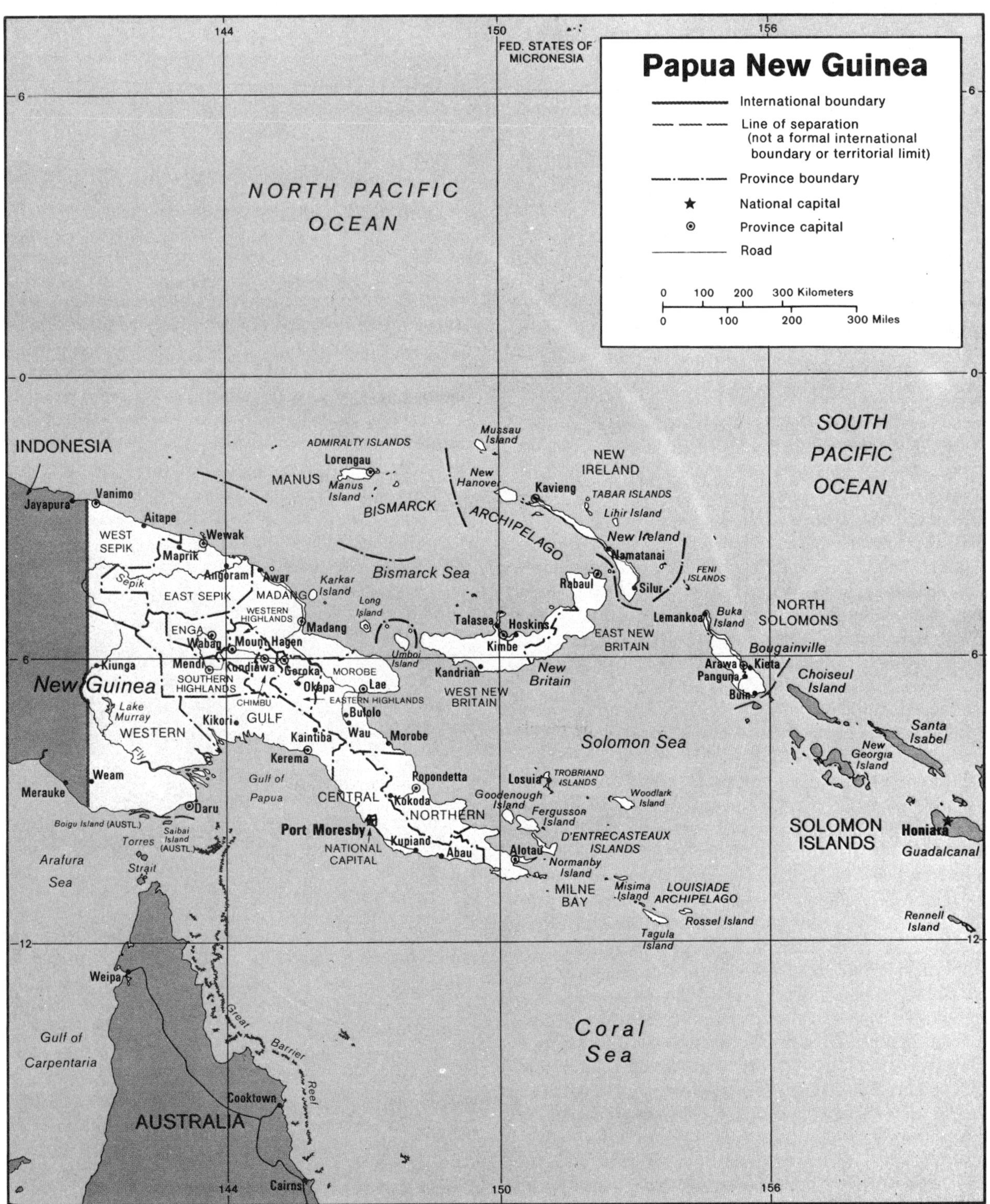

Base 800494 (A05601) 3-87

Area and population

Provinces	Administrative centres	area sq mi	area sq km	population 1986 estimate
Central	Port Moresby	11,400	29,500	132,800
Chimbu	Kundiawa	2,350	6,100	186,400
Eastern Highlands	Goroka	4,300	11,200	306,300
East New Britain	Rabaul	6,000	15,500	154,200
East Sepik	Wewak	16,550	42,800	254,900
Enga	Wabag	4,950	12,800	178,600
Gulf	Kerema	13,300	34,500	71,600
Madang	Madang	11,200	29,000	245,700
Manus	Lorengau	800	2,100	29,900
Milne Bay	Alotau	5,400	14,000	150,600
Morobe	Lae	13,300	34,500	357,100
National Capital District	Port Moresby	100	240	141,500
New Ireland	Kavieng	3,700	9,600	77,200
Northern	Popondetta	8,800	22,800	90,300
North Solomons	Kieta	3,600	9,300	154,500
Southern Highlands	Mendi	9,200	23,800	259,500
Western	Daru	38,350	99,300	91,700
Western Highlands	Mount Hagen	3,300	8,500	300,200
West New Britain	Kimbe	8,100	21,000	107,700
West Sepik	Vanimo	14,000	36,300	128,300
TOTAL		178,703	462,840	3,441,200

New Hanover and New Ireland), Karkar, Long, Northeast New Guinea, the Northern Solomon Islands (Bougainville and Buka), Umboi and the former Territory of Papua, comprising the D'Entrecasteaux Group (Dobu, Ferguson, Good enough, Normanby and Sanaroa), the Louisiade Group (Missina, Rossel and Tagula), Southeast New Guinea, the Trobriand Group (Kaileuna, Kiriwina, Kitava and Vakuta) and the Woodlark Group (Madau and Murua).

The main island, comprising about 85% of the land area, has a central mountain core that is not a single chain but a complex of ranges rising to 4,500 m (14,700 ft) interspersed by broad upland valleys at elevations of over 1,524 m (5,000 ft). The main mountain range contains the nation's highest peak, Mt. Wilhelm 4,509 m (14,793 ft). A second mountain chain fringes the north coast. Active volcanoes dot the landscape. Between the northern and central range of mountains is the Central Depression, including the valleys of the Sepik, Ramu and Markham Rivers. Swamps cover large areas of the country, and on the SW littoral the great delta plain of the Daru coast forms one of the most extensive swamps in the world.

The other islands are characterized by the same pattern of mountain ranges fringed by coastal plains and swamps or extensively developed barrier coral reefs. Manus in the Admiralty Group is 80 km (50 mi) long and 32 km (20 mi) wide; New Britain is 595 km (370 mi) long and 80 km (50 mi) wide; New Ireland is 354 km (200 mi) long and only 11 km (7 mi) wide; Bougainville in the Solomon Group is 204 km (127 mi) long and 80 km (50 mi) wide; and Buka is 56 km (35 mi) long and 14 km (9 mi) wide. Many of the 22 islands comprising the Trobriand Group are low coral types. Bougainville and New Britain are among the most active volcanic regions in Melanesia.

The largest rivers are the Fly, the Purari and the Kikori, which flow southward into the Gulf of Papua, and the Sepik and the Ramu, flowing northward into the Pacific. The Fly, over 1,126 km (700 mi) long, is navigable for 805 km (500 mi) by shallow-draft vessels.

PARAGUAY

BASIC FACTS

Official Name: Republic of the Paraguay

Abbreviation: PAR

Area: 406,752 sq km (157,048 sq mi)

Area—World Rank: 54

Population: 4,386,024 (1988) 5,405,000 (2000)

Population—World Rank: 100

Capital: Asuncion

Land Boundaries: 3,794 km (2,358 mi); Brazil 1,339 km (832 mi); Argentina 1,699 km (1,056 mi); Bolivia 470 km (756 mi)

Coastline: Nil

Longest Distances: 992 km (616 mi) SSE-NNW; 491 km (305 mi) ENE-WSW

Highest Point: 800 m (2,625 ft)

Lowest Point: 46 m (151 ft)

Land Use: 20% arable; 1% permanent crops; 39% meadows and pastures; 35% forest and woodland; 5% other

Located in the south-central interior of the South American continent and bisected laterally by the Tropic of Capricorn, Paraguay is separated on the west from Argentina by the Pilcomayo and Paraguay rivers and on the south by the Alto Paraná River. On the east it is separated from Argentina and Brazil by the higher reaches of the Alto Paraná, and on the north and northwest its border with Bolivia is marked by smaller streams and by surveyed boundary lines.

Paraguay is seventh in size among the South American states. Flowing almost directly southward out of Brazil, the Paraguay River splits the country into two distinct regions. To the west is the Chaco, a hot and semiarid plain with a scanty population. East of the river is the region of Eastern Paraguay with a rolling landscape, a wealth of fertile land, and an unevenly distributed population.

Geographical Regions

The river from which Paraguay takes its name divides the country into two regions with strikingly different characteristics. The three-fifths of the land north and west

PRINCIPAL TOWNS (population at 1982 census)

Asunción (capital)	455,517	Encarnación	27,632
San Lorenzo	74,359	Concepción	22,866
Fernando de la Mora	66,810	Coronel Oviedo	21,782
Lambaré	65,145	Villarrica	21,203
Puerto Presidente Stroessner	48,528	Caaguazú	19,027
Pedro Juan Caballero	37,331	Pilar	13,135

Area and population		area		population
Regions **Departments**	**Capitals**	sq mi	sq km	1985 estimate
Occidental		95,338	246,925	50,400
Alto Paraguay	Fuerte Olimpio	17,754	45,982	10,100
Boquerón	Dr. Pedro P. Peña	18,034	46,708	12,000
Chaco	Mayor Pablo Lagerenza	14,041	36,367	300
Nueva Asunción	General Eugenio A. Garay	17,359	44,961	200
Presidente Hayes	Pozo Colorado	28,150	72,907	27,800
Oriental		61,710	159,827	3,228,600
Alto Paraná	Puerto Presidente Stroessner	5,751	14,895	255,000
Amambay	Pedro Juan Caballero	4,994	12,933	69,400
Asunción	Asunción	45	117	477,100
Caaguazú	Coronel Oviedo	4,430	11,474	333,000
Caazapá	Caazapá	3,666	9,496	111,400
Canendiyú	Salto del Guairá	5,663	14,667	77,100
Central	Asunción	952	2,465	572,500
Concepción	Concepción	6,970	18,051	143,000
Cordillera	Caacupé	1,910	4,948	194,000
Guairá	Villarrica	1,485	3,846	149,600
Itapúa	Encarnación	6,380	16,525	284,500
Misiones	San Juan Bautista	3,690	9,556	80,100
Ñeembucú	Pilar	4,690	12,147	69,500
Paraguarí	Paraguarí	3,361	8,705	201,900
San Pedro	San Pedro	7,723	20,002	210,500
TOTAL		157,048	406,752	3,279,000

of the river is the Chaco. The two-fifths to the south and east is Eastern Paraguay. The home of some 96 percent of the population, it is sometimes referred to as Paraguay Proper. A 1945 decree law defines the two regions officially as Occidental (Western) and Oriental (Eastern) Paraguay.

The Chaco

Geologically, the Chaco region of Paraguay is a part of the South American Gran Chaco, which extends from Argentina in the south to the fringes of Bolivia and Brazil in the north. Because it is located in the northern part of the Gran Chaco, its complete name, seldom used, is Chaco Boreal. Bounded on the east by the Paraguay River and on the southwest by the Pilcomayo River, the expanse is thought to have been the bed of an ancient sea.

Except for low hills in the northeast, the featureless landscape is virtually flat, broken by a few constant and numerous intermittent rivers and streams and by extensive swamps in the south. From an altitude of over 305 m (1,000 ft) in parts of the northeast, the plain slopes southward imperceptibly, with a grade of less than 1 percent, to an altitude of some 91 m (300 ft) at the confluence of the Pilcomayo and the Paraguay rivers.

Eastern Paraguay

Eastern Paraguay is a region of spacious plains, broad valleys, and extensive low plateaus. Surfaces are flat to gently rolling; local relief features seldom exceed 30 m (100 ft); and the highest peak rises to some 701 m (2,300 ft) above sea level. About 80 percent of the region is below 304 m (1,000 ft), and the lowest elevation is 54 m (180 ft) in the extreme south at the point where the Paraguay and Alto Paraná rivers join. These two rivers, which bound the region on all sides except along the northern border with Brazil, are fed by countless tributary streams that make an intricate tracery across the landscape as they flow eastward to the Alto Paraná and westward to the Paraguay.

The region as a whole divides naturally into five physiographic subregions. The heavily wooded Paraná Plateau on the east occupies one-third of the region and extends its full length from north to south and up to 145 km (90 mi) westward from the Brazilian and Argentine frontiers. Its western terminus is defined by an abrupt escarpment that descends from an altitude of about 457 m (1,500 ft) at Pedro Juan Caballero in the north to about 183 m (600 ft) at its southern extremity. The plateau slopes moderately to east and south, and its surface is remarkably uniform. Wide interfluves are separated by narrow valleys that carry the waters of the eastward-flowing tributaries of the Alto Paraná.

The lower terrain lying between the escarpment and the Paraguay River is sometimes referred to as the Central Basin. Eroded extensions of the Paraná Plateau that reach almost to the Paraguay, however, further divide the area into subregions. The first of these extensions—the Northern Upland—occupies the portion northward from the Aquidabán River to the Apa River on the Brazilian frontier. For the most part it consists of a rolling plateau about 183 m (600 ft) above sea level and 76 to 91 m (250 to 300 ft) above the plain farther to the south. The second extension of the Paraná Plateau occurs in the vicinity of Asunción and is known as the Central Hill Belt. Although nearly flat surfaces are not lacking in this subregion, the rolling terrain is extremely uneven. Small, isolated peaks are numerous, and it is here that the only lakes of any size are found. Topographical irregularities begin in the city of Asunción, where several hill crests stand some 91 m (300 ft) above the river, and near Villarrica a narrow crest reaches above 610 m (2,000 ft).

Between these two upland subregions is the Central Lowland, an area of low elevation and relief, sloping gently upward toward the Paraná Plateau. Valleys of its westward-flowing rivers are broad and shallow, and their courses are subject to flooding that creates seasonal swamps. The most conspicuous features of this subregion are flat-topped hills covered with thick forest that project twenty to thirty ft from the grassy plain to cover areas ranging from a few acres to several sq mi. Apparently weathered remnants of rock related to geology farther to the east, they are called *islas de monte* (mountain islands), and their margins are known as *costas* (coasts).

Base 59904 10-68

The remaining subregion, in the southwest corner of the country, is the Neembucú Plain. This alluvial flatland has a slightly west and southwest incline obscured by gentle undulations marking its surface. A major tributary of the Paraguay River, the Tebicuary bisects the swampy lowland, which is broken in its central portion by rounded swells of land up to 30 m (100 ft) in height. It was on these that many of the early Jesuit mission stations were established.

Hydrography

The word *Paraguay* is derived from a Guaraní term for which various definitions have been proposed. The word *water* or *river* figures in all of them, however, and the full name of the riverside country is not Republic of Paraguay (República de Paraguay) but Republic of the Paraguay (República del Paraguay); the word *river* had been elided from the full name. The Paraguay and Alto Paraná rivers and their tributaries define most of the country's frontiers, provide all of its drainage, and serve as indispensable transportation routes. Most of the larger towns of the interior, as well as Asunción, are river ports.

Rising in the Mato Grosso of Brazil, the Paraguay River borders or passes through the country in a southward course of about 1,128 km (700 mi). It flows sluggishly but is shallow and sometimes overflows its low banks to form temporary swamps, to damage crops, or to destroy villages. River islands, meander scars, and oxbow lakes bear evidence to frequent changes in course. The average gradient is less than five inches per mile, the stream velocity is seldom over 4 km (3 mi) per hour; and the mean annual change in level at Asunción is about five ft. The lack of a more marked seasonal change in the volume of flow is attributed to the periodic flooding of the upper course over the vast Pantanal swamp in Brazil, which serves as a kind of reservoir that only gradually releases its runoff into the river's course southward.

The Brazilian town of Corumbá is usually considered the head of navigation, and during most years vessels with seven-foot draft can reach Concepción without difficulty. Medium-sized ocean vessels can sometimes reach Asunción, but a tortuous course and shifting sandbars make this transit difficult.

The Paraná River rises in Brazil. It flows some 804 km (500 mi) from the Brazilian frontier at the Guairá Falls, where it becomes known as the Alto Paraná River, to its juncture with the Paraguay. The Paraná River then continues southward to the Río de la Plata Estuary. At the Guairá Falls, the Alto Paraná tumbles 91 m (300 ft) in a series of steps into a narrow gorge. From this point it flows southward at a rate of more than 16 km (10 mi) per hour between gorge walls diminishing from 91 m (300 ft) in the north to about 30 m (100 ft) at Encarnación, where the river curves westward. The flow then slackens to 10 km (6 mi) per hour, the gorge disappears, and the stream widens to a maximum of 29 km (18 mi) as the course becomes braided and large river islands appear. In summer months the depth is usually sufficient to permit vessels with drafts of up to 3 m (10 ft) to reach the Guairá Falls, but the river's navigational value is severely limited by seasonal and occasional conditions. On the upper course, sudden floods may raise the water level by as much as 45 m (150 ft) in twenty-four hours; west of Encarnación, the rocks of the riverbed sometimes come within 1 m (3 ft) of the surface during winter and effectively sever communication between the upper river and Buenos Aires.

The third river in size, the Pilcomayo, is a tributary of the Paraguay and enters the parent river near Asunción after following the entire length of the frontier between the Chaco and Argentina. During most of its course the river is sluggish and marshy, although small craft can navigate the lower reaches. The Verde and Monte Lindo rivers also enter the Paraguay from the Chaco. Other Chaco streams overflow their badly defined channels in summer but are reduced to trickles or vanish altogether in winter to form salt marshes.

Major tributaries of the Paraguay River entering it from Eastern Paraguay are miniatures of the parent stream. They descend rapidly from their source in the Paraná Plateau to the lower lands; here they broaden and become sluggish where their gradients decrease as they meander westward. At low water their banks rise 1 to 7 m (5 to 25 ft) above the water level, but after heavy rains these miniature gorges fill and sometimes overflow to inundate nearby lowlands. The rivers flowing eastward across Eastern Paraguay as tributaries of the Alto Paraná have very different characteristics. Shorter, faster flowing, and narrower, they follow sinuous courses in gorges carved in the basalt of the Paraná Plateau and drop to the parent river in step-like falls. Some sixteen of these rivers and numerous smaller streams enter the Alto Paraná above Encarnación.

Drainage of the Chaco is generally poor because of the flatness of the land and the small number of important streams. In many parts of the region the water table is only a few ft beneath the surface of the ground, and there are numerous small ponds and seasonal marshes. As a consequence of the poor drainage, most of the water is too salty for drinking or irrigation.

The drainage pattern of Eastern Paraguay also reflects the essential flatness of much of the terrain. The eastward runoff is prompt through the countless streams of the Paraná Plateau, but most of the lowland portions of the region are inadequately drained because of the seasonal overflow of the numerous westward-flowing streams. Particularly in the Neembucú Plain, an almost impervious clay subsurface prevents the absorption of ground water. About 30 percent of Eastern Paraguay is flooded from time to time, and maps show extensive areas of marshlands. Most are seasonal, however, and only around the largest geographic depressions, such as those around the shallow Lake Ypacaraí in the Central Hill Belt and Lake Ypoá in the Neembucú Plain, are permanent bogs encountered.

PERU

BASIC FACTS

Official Name: Republic of Peru

Abbreviation: PER

Area: 1,285,216 sq km (496,225 sq mi)

Area—World Rank: 19

Population: 21,269,074 (1988) 27,952,000 (2000)

Population—World Rank: 40

Capital: Lima

Boundaries: 10,153 km (6,309 mi); Ecuador 1,529 km (950 mi); Colombia 1,506 km (936 mi); Brazil 2,822 km (1,754 mi); Bolivia 1,047 km (651 mi); Chile 169 km (105 mi)

Coastline: 3,079 km (1,913 mi)

Longest Distances: 1,287 km (800 mi) SE-NW; 563 km (350 mi) NE-SW

Highest Point: Mount Huascaran 6,746 m (22,133 ft)

Lowest Point: Sea level

Land Use: 3% arable; 21% meadows and pastures; 55% forest and woodland; 21% other

Peru is fourth in size of the Latin American states; only Brazil, Argentina, and Mexico are larger. The narrow, arid coastal region is the most densely populated part of the country. The extensive valleys and high plains of the Andes that were pre-Colombian population centers continue to be the home of about one-third of the people. The great lowlands to the east comprise about three-fifths of the national territory, but they remain sparsely settled and little developed.

Geographic Regions

Three generally recognized geographic regions are identified and their limits prescribed by legislation enacted in 1960. The Costa is defined as being the littoral and foothills of the Andes below the level of 2,000 m (6,561 ft), the Selva is made up of the tropical rain forest and mountain slopes east of the Andes below the 2,000 m (6,561 ft) level. Between these two regions lies the Sierra. For some purposes the Costa and Sierra are divided into northern, central, and southern regions. When this system of regional classification is employed, the Selva is considered to be the fourth region.

The Costa

The coastal region consists of the dry plains and sand dunes of the littoral below an altitude of about 260 m (853 ft) and at higher levels of dry Andean foothills intricately carved by erosion. The former is known as the Low Costa and the latter as the High Costa. In all, the region includes about 11 percent of the national territory. Along its coastline there are some twenty-five ports, but few of these have good harbors. There are no large islands, but off the central part of the coast are some forty small islands from which guano fertilizer, composed of the droppings of seabirds, is harvested.

The Sierra

The Andes form the backbone of South America, extending south from the Caribbean Sea along the entire Pacific Coast to Cape Horn. In Peru the ranges are only about 100 km (328 mi) wide at the Ecuadorean border; they widen gradually to the south until they are almost 320 km (199 mi) wide near Lago (Lake) Titicaca. The Sierra consists of a high plateau which gradually rises as it extends southward; upon it are imposed three mountain chains interconnected by a large number of random ridges that rise from 1,000 to 2,000 m (3,280 to 6,561 ft) above the plateau level. The Sierra region includes about 26 percent of the national territory. The pattern of the mountain system is highly mixed—tremendous canyons are cut below the high surfaces. The highest section, called the Cordillera Blanca, has a number of snowcapped peaks of which the tallest, Mount Huascarán, is 6,729 m (22,077 ft).

The Sierra was the center of the Inca Empire, and today the Indians still farm these highlands much as their ancestors did before the conquest, using some of the ancient irrigation systems and terraced fields. Settlements, found wherever there is arable land, are most abundant in the high mountain valleys and along the low streams before they drop into deep gorges. Because the level lands above 4,200 m (13,779 ft) in elevation can be used only as pasturage, the arable portion of the Sierra is small. It is, however, quite densely populated.

The largest habitable areas are found around Cajamarca, in the valley of the Río Santa between the Cordillera Negra and the Cordillera Blanca, in the broad basin of the Río Mantaro valley between Jauja and Huancayo (Junín Department), in the vicinties of Ayacucho and Cusco, in the high basin of the Lago Titicaca, and in the vicinity of Arequipa. The areas of relatively dense population vary in altitude from 2,000 to 3,400 m (6,561 to 11,154 ft) above sea level except near Cusco and around Lago Titicaca where they extend to about 3,900 m (12,795 ft.)

The Selva

Like the Costa, the Selva is made up of two zones. In all, it includes nearly 63 percent of the national territory. Below heights of about 1,000 m (3,280 ft) is found the tropical rainforest of the Amazon that makes up the Low Selva. Above these levels the mountain slopes and valleys are referred to as the High Selva or brow of the mountain (*ceja de montaña*). In practice the entire Selva is frequently referred to as la Montaña.

The High Selva is composed of a pattern of alternating hills and valleys which links the Andean landscape and the rain forest of the Low Selva. Ranging from eighty to 160 km (100 mi) in width, it is a zone of great variation in vegetation, depending on rainfall, relief, and soils. Rains and intensive river erosion have cut deeply into the steep slopes and created greater obstacles to ground com-

Base 76870 3-70

PRINCIPAL TOWNS (estimated population at mid-1985)

Lima (capital)	5,008,400	Piura	256,150
Arequipa	531,829	Chimbote	253,289
Callao	515,200	Cuzco	225,683
Trujillo	438,709	Iquitos	215,275
Chiclayo	347,702	Huancayo	186,724

Area and population

		area		population
Departments	**Capitals**	sq mi	sq km	1987 estimate
Amazonas	Chachapoyas	15,945	41,297	311,800
Ancash	Huaraz	14,158	36,669	936,600
Apurímac	Abancay	7,934	20,550	361,400
Arequipa	Arequipa	24,528	63,528	884,200
Ayacucho	Ayacucho	17,058	44,181	553,000
Cajamarca	Cajamarca	13,486	34,930	1,200,000
Cuzco	Cuzco	29,471	76,329	980,600
Huancavelica	Huancavelica	8,139	21,079	371,400
Huánuco	Huánuco	13,088	33,897	571,600
Ica	Ica	8,205	21,251	508,200
Junín	Huancayo	15,944	41,296	1,037,500
La Libertad	Trujillo	8,973	23,241	1,150,900
Lambayeque	Chiclayo	5,304	13,737	854,600
Lima	Lima	13,058	33,821	6,116,700
Loreto	Iquitos	146,342	379,025	605,900
Madre de Dios	Puerto Maldonado	30,271	78,403	44,500
Moquegua	Moquegua	6,065	15,709	123,400
Pasco	Cerro de Pasco	9,356	24,233	264,800
Piura	Piura	14,055	36,403	1,374,200
Puno	Puno	27,947	72,382	984,500
San Martín	Moyobamba	20,197	52,309	414,500
Tacna	Tacna	5,881	15,232	188,300
Tumbes	Tumbes	1,827	4,732	131,700
Ucayali	Pucallpa	38,931	100,831	211,700
Constitutional Province				
Callao	Callao	57	148	545,100
TOTAL		496,225	1,285,216	20,727,100

munication between the Costa and the Amazon than does the Andes. Although much of the soil is fertile and the rainfall is ample for agriculture, only a small percentage of the lower valleys have slopes suitable for cultivation. The High Selva is an area of transitional land use in which shifting agriculture and permanent fields exist side by side.

The Low Selva is very humid, and most of it supports dense vegetation consisting predominantly of rain forest. There are patches of open land, but movement is generally restricted to the extensive river network.

Hydrography

The more than 3,800-m-high (12,467 ft) Lago Titicaca is the country's most conspicuous water feature; its basin was a population center before Inca times. The lake is actually an inland sea 8,320 sq km (3,212 sq mi) in area, having a maximum depth of about 360 m (1,181 ft), it is about 220 km (136 mi) in length and ranges up to 60 km (37 mi) in width. It moderates the climate of the land areas in its basin, serves as the main avenue of trade with Bolivia, which shares title to its waters, and supports a sizable fishing community. There are no other Peruvian lakes of great size, but there are several smaller lakes in the Sierra, including Lago Lauricocha located slightly to the north of Cerro de Pasco. Because the Río Marañón, a major tributary of the Amazon, originates from it, this lake is often cited as the source of the Amazon.

Over sixty rivers flow from the Andes to the Pacific through the Costa. All flow generally westward except the largest, the Río Santa, which runs about 160 km (100 mi) from north to south through the picturesque Callejón de Huaylas valley before spilling seaward through a deep canyon to the Pacific. Also of particular significance if only for its name, is the Río Rímac that divides the capital city; the name of Lima is a corruption of the word Rímac. The majority of these rivers are 360 km (223 mi) or less in length, and most are seasonal streams that become swollen and turbulent during the summer months. The Costa agricutural land is irrigated by water from these streams.

A more complex and vastly more extensive system consists of the rivers that originate in the Sierra close to the continental divide but flow eastward into the Amazon Basin. The superior cutting power of these streams has captured most of the drainage of the region. The large rivers in the Selva are navigable for great distances, and the river ports of Iquitos, Pucallpa, Yurimaguas, Borja, and Puerto Maldonado are major settlements.

PHILIPPINES

BASIC FACTS

Official Name: Republic of the Philippines

Abbreviation: PHP

Area: 300,000 sq km (115,800 sq mi)

Area—World Rank: 67

Population: 63,199,307 (1988) 74,057,000 (2000)

Population—World Rank: 14

Capital: Manila

Land Boundaries: Nil

Coastline: 26,289 km (16,325 mi)

Longest Distances: 1,851 km (1,150 mi) SSE-NNW; 1,062 km (660 mi) ENE-WSW

Highest Point: Mount Apo 2,954 m (9,692 ft)

Lowest Point: Sea Level

Land Use: 26% arable; 11% permanent crops; 4% meadows and pastures; 40% forest and woodland; 19% other

The Philippine archipelago contains about 7,100 islands and extends over 1,609 km (1,000 mi) from north to south. Only 154 of these islands, however, exceed 13 sq km (5 sq mi) in area. The two largest islands, Luzon in the north and Mindanao in the south, comprise about 65 percent of the total land area of the archipelago. The very complex and volcanic origin of most of the islands is visible in their varied and rugged terrain. Mountain ranges divide the island surfaces into narrow coastal strips and shallow interior plains or valleys; no point on land is more than 40 km (25 mi) from the coast or from a mountain range. The sea and mountain barriers have been constant factors shaping the historical, cultural, and demographic development of the nation.

The sea has been both a barrier and a bridge and correspondingly both a source of isolation and a means of communication. It was the sea that isolated the archipelago from the mainstream of Asian history. As a result the islands never experienced the cycles of conquest and collapse of suzerain empires known in the rest of Asia. The sea also, however brought the migrants who became the ancestors of the majority of the nation's population, important contacts with Chinese and Arab traders, and eventually the Spanish. The proximity of the islands and the calmness of the interisland seas, straits, and other waterways, however, did not encourage development of the skills required for long, open sea voyages, and the Philippines has not become an important seafaring nation. Early migrants settled in narrow communities strung along the coast and established a settlement pattern still evident today. Among some groups arrival in the same boat or at the same time became a basis for identification and social organization. Fishing and interisland transport have remained important occupations; fish continues to provide a major portion of protein in the Filipino diet.

Most of the Philippine Islands are situated in a region of considerable geological instability and extreme climatic phenomena. The unusually complex origin of the archipelago—still not fully understood—continues to be expressed in minor earthquakes and occasionally in a major eruption. A major fault line extends along the eastern part of the archipelago roughly paralleling the very deep Philippine Trench and—like Japan to the far north and Indonesia to the southwest—most of the Philippines is considered part of the volcanically active regions of the western Pacific. In addition the northern islands are especially affected by typhoons that, although mainly seasonal, bring destructive winds and flooding rains.

This geological complexity and climatic activity nevertheless formed a relatively rich resource pattern. Ancient uptilting has exposed many sites rich in minerals; volcanic activity has enriched arable soils; seasonal rainfall and local rainfall variations support a diversified agriculture.

Most of the larger islands have exceptionally narrow coastal shelves. Population clusters have generally been restricted to the more irregular, wider, western coastal regions. The north-south mountain barriers have generally hindered the development of east-west passage. The road system follows the same north-south pattern, and neither Luzon nor Mindanao has a completely circumferential highway.

The territory of the Philippines can be conveniently divided into several geographic regions, three of which—Luzon, the Visayan Islands, and Mindanao—are based on geographic proximity of islands rather than any physical, social, or economic homogeneity. Separate from these regions are Mindoro Island and the islands of the Palawan group, which together form an island bridge stretching southwest from Luzon to Sabah (on the island of Borneo and part of the Federation of Malaysia), and the Sulu Archipelago, which extends similarly from western Mindanao to near Sabah.

The Luzon Region

The Luzon region consists of Luzon Island, many much smaller adjacent islands, and the small island groups of Batan and Babuyan lying to the north. The island of Luzon has an area of 104,687 sq km (40,420 sq mi), or more than one-third of the country's total area. In shape it resembles an upright rectangle with an irregular southeastern peninsular handle. The main part of the island is roughly 402 km (250 mi) in length and has a width generally between 120 and 160 km (75 and 100 mi), although the widest point is 222 km (138 mi).

The portion of Luzon that lies north of Lingayen Gulf is very mountainous; its highest peak, Mount Pulog, rises to 2,934 m (9,626 ft). The island has three mountain ranges that run roughly parallel in a north-south direction. A range in the east, the Sierra Madre, runs so close to the island's eastern shore that there is barely any coastal lowland. The valley of the north-flowing Cagayan River separates this eastern range from a large mountain complex to the west, the Cordillera Central; the Cagayan Valley, about 225 km (140 mi) long and averaging about 64 km (40 mi) in width, is an area of rich soil and is an important agricultural region. The valley is separated from the Central Luzon Plain by a cross flow of volcanic origin between the Sierra Madre and the southern extension of the Cordillera Central.

The rugged central mountain complex of the Cordillera Central covers most of the northern part of the island but affords a productive narrow coastal strip to the west. Below it lies the Central Luzon Plain, which extends south to Manila Bay. This plain, one of the country's most important agricultural areas, is approximately 160 km (100 mi) long and roughly 80 km (50 mi) wide. Its altitude is not much above sea level, and the plain includes a number of extensive swampy stretches.

The Central Luzon Plain is bordered on both sides by rugged coastal ranges. On the west the third range, the Zambales Mountains, extends from the Lingayen Gulf southward through the Bataan Peninsula and terminates at Manila Bay. The range crowds the shore and leaves only a narrow coastal lowland but is economically important because of its extensive mineral deposits. The mountains to the east, an extension of the Sierra Madre, drop off sharply to the Pacific Ocean and leave little accessible level terrain and no protective harbors against the heavy Pacific surf.

At the southern end of the central plain a large, natural, ample harbor is provided by Manila Bay. The city of Manila is located on the bay's eastern shore. Lying to the southeast of the bay and joined to it by the short Pasig River is the largest freshwater lake in the Philippines, Laguna de Bay, which at one time was probably an arm or extension of Manila Bay. It has a water surface of 922 sq km (356 sq mi). A few miles to the southwest of Laguna

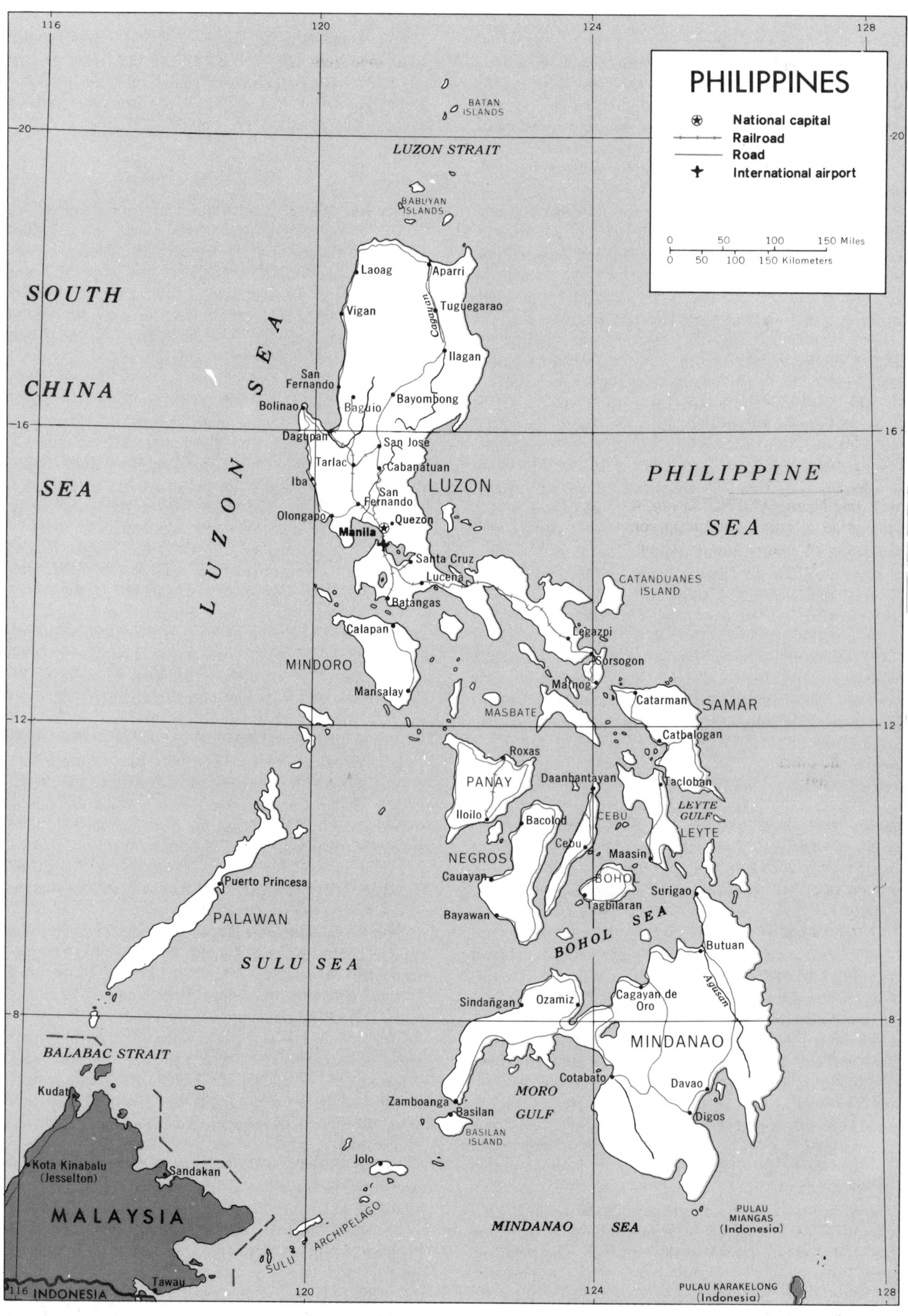
PHILIPPINES
National capital
Railroad
Road
International airport
0 50 100 150 Miles
0 50 100 150 Kilometers
BATAN ISLANDS
LUZON STRAIT
BABUYAN ISLANDS
SOUTH CHINA SEA
LUZON SEA
PHILIPPINE SEA
Laoag
Aparri
Vigan
Tuguegarao
Cagayan
Ilagan
San Fernando
Bolinao
Baguio
Bayombong
Dagupan
San Jose
Tarlac
Cabanatuan
Iba
San Fernando
LUZON
Olongapo
Manila
Quezon
Santa Cruz
Lucena
Batangas
CATANDUANES ISLAND
Calapan
Legazpi
MINDORO
Sorsogon
Mansalay
Matnog
MASBATE
Catarman
SAMAR
Catbalogan
Roxas
PANAY
Daanbantayan
Tacloban
LEYTE GULF
CEBU
Iloilo
Bacolod
LEYTE
Cebu
NEGROS
Maasin
Cauayan
BOHOL
Puerto Princesa
Surigao
Tagbilaran
Bayawan
PALAWAN
BOHOL SEA
Butuan
SULU SEA
Agusan
Sindañgan
Ozamiz
Cagayan de Oro
MINDANAO
BALABAC STRAIT
Cotabato
Davao
Kudat
MORO GULF
Zamboanga
Basilan
Digos
BASILAN ISLAND
Kota Kinabalu (Jesselton)
Jolo
Sandakan
MALAYSIA
SULU ARCHIPELAGO
MINDANAO SEA
PULAU MIANGAS (Indonesia)
Tawau
INDONESIA
PULAU KARAKELONG (Indonesia)
116
120
124
128
20
16
12
8

de Bay lies Lake Taal. This lake has an active volcano in its center that erupted in 1965, causing several deaths.

Southeastern Luzon consists of a large convoluted peninsula that accounts for one-third of the island's overall length. A mountainous volcanic area, it contains the periodically active 2,420 m (7,941 ft) Mount Mayon volcano, one of the world's most symmetrical large volcanic cones. Although the peninsula's terrain is extremely rugged, the fertility of the volcanic soil and heavy rainfall make over one-half of the area agriculturally productive. In addition there are many good, sheltered harbors available for interisland shipping.

The Visayan Islands

The Visayan Islands are grouped in a roughly circular pattern around the Visayan Sea. They include seven large, populated islands that range in size from Masbate, which is 3,268 sq km (1,262 sq mi) in area, to Samar, 13,079 sq km (5,050 sq mi). The others are Bohol (3,865 sq km; 1,492 sq mi), Cebu (4,421 sq km; 1,707 sq mi), Leyte (7,213 sq km; 2,785 sq mi), Panay (11,515 sq km; 4,446 sq mi), and Negros (12,703 sq km; 4,905 sq mi). Including a myriad of islets, the group comprises over half the total number of islands that make up the country.

The seven larger islands have a combined land surface of over 55,944 sq km (21,600 sq mi), representing about 19 percent of the national total. Most of the Visayan Islands have mountainous interiors, and in general lowlands are few and small, usually confined to the coastal strips of the larger islands. There are, however, three sizable lowland areas: the Leyte Valley, the Iloilo Plain on Panay, and the plains of western and northern Negros.

Samar and Leyte, the easternmost islands, act as a buffer for the other islands against the full force of storms and occasional typhoons originating from the Pacific Ocean. Samar's interior has rugged mountains, but elevations are generally below 304 m (1,000 ft): these mountains are covered with commercially exploitable forests.

Leyte is separated from Samar by a narrow strait and is actually a detached peninsula of Samar. A central mountain range running north and south divides the northern part of the island. To the east of the range is the fertile, alluvium-filled Leyte Valley, and to the west is a comparatively flat highland. Both are used for agriculture; the principal crops are rice, corn, abaca, and coconuts.

Cebu, Negros, and Panay are important agriculturally and also form the commercial center of the region; their ports contribute significantly to both interisland and overseas shipping. The long narrow island of Cebu is one of the most distinctive of the Visayan Islands. It is the center of corn cultivation, accounting for a quarter of all the country's acreage devoted to the crop. The island is the site of the country's largest copper mine and also produces low-grade coal and limestone for cement. Cebu has long been one of the most prosperous of the islands.

West of Cebu, Negros and Panay are roughly similar in area and population. Both islands have sizable lowland plains that permit intensive agriculture. Negros is an important producer of both sugarcane and corn; cultivation of rice is dominant on Panay.

The two smallest islands—Bohol and Masbate—are both under 3,884 sq km (1,500 sq mi) in area. Masbate is extremely productive, yielding rice, corn, and hemp. Bohol is covered mostly with secondary forest but has extensive rice cultivation on the coastal lowlands.

The Mindanao Region

The Mindanao region consists of the island of Mindanao and numerous small offshore islands, including a mineral-rich group off the northeastern coast. Mindanao, the second largest of the Philippine Islands, has an area of 94,630 sq km (36,537 sq mi). Its very irregular shape is characterized by a number of sizable gulfs and bays and several large peninsulas that give it an extremely long coastline. It has five major mountain systems, some formed by volcanic action, and a varied and complex topography that includes numerous rivers and a number of lakes.

The eastern edge of Mindanao is highly mountainous. This mountainous area, which is a southward continuation of the mountain range in Leyte, is heavily wooded. It shelters the interior from most of the storms of the Pacific but leaves almost no coastal lowlands. The northern part, known as the Diuata Mountains, contains several elevations above 1,828 m (6,000 ft), and the southern section rises to over 2,133 m (7,000 ft), reaching a high point of 2,804 m (9,200 ft).

West of this eastern range is an extensive longitudinal lowland plain area known as the Davao-Agusan Trough, through which the Agusan River flows south to north for almost its entire extent. The valley, over 160 km (100 mi) long and 19 to 59 km (12 to 37 mi) wide, has an excellent agricultural potential; however, its development has been slowed because of the scarcity of roads and the distance of markets. A large-scale influx of homesteaders into this frontier region occurred in the 1960s. The port of Davao, the country's third largest city, is at the southern end of the trough and is the island's principal city.

On the west the trough is bordered by a broad mass of rugged mountain ranges, one of which bisects the island from north to south. This range contains 2,954 m (9,692 ft) Mount Apo, the highest peak in the country, which overlooks Davao Gulf. Between these highlands and a southwestern coastal range lies another broad lowland area, the Cotabato Lowland, which covers over 6,475 sq km (2,500 sq mi). The most extensive part of the lowland is the Cotabato Valley, which is drained by the Mindanao River. The valley has large swampy stretches but comprises a region of great agricultural potential.

The central mountain complex extends into the northwest corner of the main body of the island, terminating in the Bukidnon-Lanao Highlands. An undulating expanse of plateau formations at approximately 609 m (2,000 ft) in elevation, the highlands are interspersed with a number of extinct volcanic peaks—some rising to heights of 2,895 m (9,500 ft). Roughly centered in these highlands is Lake Sultan Alonto (formerly Lake Lanao), 347 sq km (134 sq mi) in area and the second largest lake in the country. This area and the northern coastal province of Misamis Oriental have been extensively colonized by immigrants from the Visayan Islands. The plateau is the site of a number of large cattle ranches, and pineapples are grown there commercially.

Just west of the lake the island narrows to an isthmus ten miles wide, from which the long Zamboanga Peninsula protrudes to the southwest for some 273 km (170 mi). The peninsula is covered largely with mountains and possesses limited coastal lowlands. Soils in some of the lowlands and floodplains of the small rivers, however, are fertile. The peninsula has a number of rubber plantations, coconuts and abaca are grown commercially, and there are forests of commercial importance. The city of Zamboanga, the second largest city on Mindanao, is located on the southern tip of the peninsula, between the Moro Gulf and the Sulu Sea.

Mindoro, Palawan, and the Sulu Archipelago

Just south of Luzon lies Mindoro, the country's seventh largest island, having an area of 9,733 sq km (3,758 sq mi). The island is largely mountainous and has high peaks rising above 2,438 m (8,000 ft). Much of the island is covered with secondary forest, apparently the result of earlier destruction of the original forest by slash-and-burn cultivation. A moderately wide coastal lowland lies to the east and northeast of the mountain zone, and there is some agricultural exploitation of this area. The lowland contains Lake Naujan, one of the country's larger lakes.

Extending southwestward from Mindoro is Palawan, fifth in size among the Philippine Islands. It has a length of over 442 km (275 mi), a width varying from five to thirty miles, and an area of 11,655 sq km (4,500 sq mi); it is surrounded by well over 1,100 smaller islands and islets. A ridge of rugged mountains that run its entire length is bordered by narrow coastal strips. A good part of the island is covered with forests, which include some important commercial species. Agricultural development has been limited. Both Mindoro and Palawan are considered essentially as pioneer areas.

Southwest of the Zamboanga Peninsula of Mindanao is the Sulu Archipelago, a string of smaller islands of volcanic and coral origin protruding from a submarine ridge that joins Mindanao to Sabah in Malaysia. A chain some 322 km (200 mi) long, it has over 800 islands, including some 500 unnamed ones; its total area is about 4,144 sq km (1,600 sq mi). The three principal islands are Basilan, directly offshore from Zamboanga; Jolo, containing the capital city of the same name; and Tawitawi, near Sabah, which has one of the best fleet anchorages in the world—used during World War II by a large part of the Japanese fleet. Fishing is a major industry. Although most of the islands are heavily forested, local agriculture generally supports the population.

PRINCIPAL TOWNS (population at 1980 census)

Town	Population	Town	Population
Manila (capital)*	1,630,485	Cagayan de Oro City	227,312
Quezon City*	1,165,865	Angeles City	188,834
Davao City	610,375	Butuan City	172,489
Cebu City	490,281	Iligan City	167,358
Caloocan City*	467,816	Olongapo City	156,430
Zamboanga City	343,722	Batangas City	143,570
Pasay City*	287,770	Cabanatuan City	138,298
Bacolod City	262,415	San Pablo City	131,655
Iloilo City	244,827	Cadiz City	129,632

*Part of Metropolitan Manila.

Area and population

Regions	area: sq mi	area: sq km	population: 1985 estimate
Bicol	6,808	17,633	4,104,000
Cagayan Valley	14,055	36,403	2,648,000
Central Luzon	7,039	18,231	5,726,000
Central Mindanao	8,994	23,293	2,733,000
Central Visayas	5,773	14,951	4,362,000
Eastern Visayas	8,275	21,432	3,185,000
Ilocos	8,328	21,568	4,056,000
National Capital Region	246	636	7,354,000
Northern Mindanao	10,937	28,328	3,350,000
Southern Mindanao	12,237	31,693	4,032,000
Southern Tagalog	18,117	46,924	7,490,000
Western Mindanao	7,214	18,685	2,994,000
Western Visayas	7,808	20,223	5,323,000
TOTAL	115,800	300,000	57,357,000

Drainage

For the most part the country's mountainous terrain causes drainage systems characterized by short violent streams. The larger rivers are not navigable except for short distances. Most main streams and their tributaries are subject to extensive damaging floods during the heavy rainfall of the monsoon and typhoon seasons.

The largest river in Luzon is the Cagayan River, which drains the Cagayan Valley. Over 322 km (200 mi) in length, it flows northward and empties into the sea at Aparri. The low-lying Central Luzon Plain is interlaced by a network of rivers and streams. Two of the plain's more important rivers are the Agno, which flows northward into Lingayen Gulf, and the Pampanga, which empties into Manila Bay. The Pasig River flows through Manila and, although relatively short, is one of the country's most important rivers commercially.

Two large rivers, both over 322 km (200 mi) long, are found on Mindanao. The Agusan River flows northward through the Agusan Valley into the Mindanao Sea. The Mindanao River and its tributaries drain the Cotabato Lowland. This river empties into Moro Gulf northwest of the city of Cotabato.

PITCAIRN ISLANDS

BASIC FACTS

Official Name: Pitcairn, Henderson, Ducie and Oeno Islands

Abbreviation: PTN

Area: 4.5 sq km (1.8 sq mi)

Area—World Rank: 214

Population: 55 (1988)

Population—World Rank: 215

Capital: Adamstown

Land Boundaries: Nil

Coastline: 51 km (31.6 mi)

Longest Distances: N/A

Highest Point: N/A

Lowest Point: Sea Level

Land Use: N/A

Pitcairn Islands comprise, Pitcairn, Henederson, Ducie and Oeno Islands (of which the last three are uninhabited) in the south Pacific Ocean about halfway between Peru and New Zealand. The islands are volcanic with rocky cliffs along the coast.

100 km

Sandy
Oeno
Henderson
Ducie
ADAMSTOWN
Pitcairn

South Pacific Ocean

POLAND

BASIC FACTS

Official Name: Polish People's Republic

Abbreviation: PPR

Area: 312,683 sq km (120,727 sq mi)

Area—World Rank: 64

Population: 37,958,420 (1988) 42,094,000 (2000)

Population—World Rank: 26

Capital: Warsaw

Boundaries: 3,708 km (2,304 mi) USSR 1,224 km (773 mi) Czechoslovakia 1,310 km (814 mi); East Germany 460 km (286 mi)

Coastline: 694 km (431 mi)

Longest Distances: 689 km (428 mi) E-W 649 km (403 mi) N-S

Highest Point: Rysy 2,499 m (8,199 ft)

Lowest Point: Raczki Elblaskie 1.8 m (5.9 ft) below sea level

Land Use: 48% arable; 1% permanent crops; 13% meadows and pastures; 29% forest and woodlands; 9% other

A relief map of Europe depicts Poland as an unbroken plain extending from the Baltic shore to the Carpathian Mountains in the south. Close study, however, reveals greater variety and complexity. Differences in climate and terrain occur, with some local deviations, in bands that extend east to west across the country, accounting for the wide variations in land utilization and population density. The coastal area lacks natural harbors except those at Gdansk-Gdynia and Szczecin. The coast and the adjoining lake district have fewer natural resources, fertile soils, and people than areas to the south. The vast plains south of the lake district have more fertile soils, a longer growing season, and a denser population than the northern regions. The southern foothills and mountains contain most of the country's mineral wealth and much of the most fertile soils and have attracted the greatest concentration of industry and people.

Most of Poland lies in the North European plain that extends from the North Sea coast of the Netherlands to the Ural Mountains in the Soviet Union. In the far south and southwest are small highland areas in the Carpathian and Sedeten (Sudety) mountains, shared with Czechoslovakia.

Topography

Poland's average elevation is 173 m (567 ft); more than 90 percent of its area lies below 300 m (984 ft), and only 3 percent, chiefly in the south, rises above 500 m (1,640 ft). Rysy, the highest peak at 2,499 m (8,205 ft) is on the Czechoslovak border about 95 km (59 mi) south of Krakow in the Tatry range of the Carpathians. Six other peaks on the Polish side of the Tatry Mountains reach 1,900 m (6,233 ft) or more. The Sudety Mountains are lower, only one peak exceeding 1,600 m (5,249 ft). In both ranges a total of about 300 sq km (115 sq mi) rises about 1,000 m (3,280 ft). The lower land is found just south of the Gulf of Danzig, where approximately 60 sq km (23 sq mi) lie below sea level.

Geographers usually divide the country into five topographic zones, each extending from west to east. The largest, accounting for three-fourths of Poland's territory, is the great central lowlands area. It is narrow in the west but expands to both the north and the south as it extends eastward. At the eastern border it includes everything from near the northeastern tip of the country to about 200 km (124 mi) from the southeastern corner.

To the south of the central lowlands, extending across the country parallel to the southern border in a belt roughly 90 to 120 km (55 to 74 mi) wide, is an area of foothills of the two mountain ranges. The foothills blend into the mountains in the extreme south and in the southwestern corner of the country. Neither range is rugged enough over large areas to limit habitation, and in only a few isolated places is the population density below the country average. Most of the more rugged slopes are in the Tatry Mountains; many slopes in the Sudety range are gentle and can be cultivated or used as meadows and pastures on dairy farms.

North of the central lowlands are the hills, forests, and lakes created by the recession of the most recent glacier millennia ago. The area is usually referred to as the lake district. The effects of glaciation are the most prominent features of the terrain for 200 km (124 mi) or more inland from the Baltic Sea in the western part of the country but for a much shorter distance in the east. The earth and stone carried by glaciers embossed what would have been a nearly flat area. Glacial action led to the formation of many lakes and low hills. Much of the land is forested. The sea is the primary influence over a narrow band of the coastline that distinguishes it from interior regions.

Drainage

By far the greatest portion of the country drains northwestward to the Baltic Sea by way of the Wisla (Vistula) and Odra (Oder) rivers. Most other rivers join the Vistula and Oder systems, but a few streams in the northeastern region reach the sea through Soviet territory.

The Vistula and its tributaries drain the country's largest basin, an area almost double the Polish portion of the Oder basin. The Vistula basin includes practically all of the southeastern and east-central regions and much of the northeast. The Vistula rises in the Tatry range of the Carpathians and exits at the Gulf of Danzig. Most of its tributaries flow to it from the east, rising in the Soviet Union or near the Soviet border. One of them, the Bug, forms about 280 km (174 mi) of the Polish-Soviet border.

The Oder, which with the Nysa (Neisse) River forms most of the border between Poland and the German Democratic Republic (East Germany), is fed by several other rivers and streams, including the Warta, which drains a large section of central and western Poland. The Oder reaches the Baltic Sea through the harbors and bays north of Szczecin.

Much inland area that comprises the lake district and extends across the entire country is poorly drained. Its many lakes add beauty and value to the region, but its swampland has been difficult to reclaim. Most of the lakes are small and shallow, the two largest having areas of little more than 100 sq km (62 mi) each; and nearly a dozen, including some very small ones, have depths of 50 m (164 ft) or more.

PRINCIPAL TOWNS
(estimated population at 31 December 1986)

Town	Population	Town	Population
Warszawa (Warsaw)	1,664,700	Tychy	185,900
Łódź	847,400	Bielsko-Biała	176,900
Kraków (Cracow)	744,000	Ruda Śląska	167,200
Wrocław	640,000	Olsztyn	152,200
Poznań	578,100	Rzeszów	144,900
Gdańsk	468,400	Chorzów	140,500
Szczecin	395,000	Wałbrzych	140,400
Bydoszcz	369,500	Rybnik	139,200
Katowice	367,300	Dąbrowa Górnicza	138,900
Lublin	329,700	Opole	127,500
Sosnowiec	258,100	Elbląg	119,600
Białystok	255,700	Włocławek	117,800
Częstochowa	250,700	Gorzów Wielkopolski	117,600
Gdynia	248,200	Tarnów	117,100
Bytom	239,500	Płock	116,300
Radom	219,100	Zielona Góra	112,200
Gliwice	211,200	Wodzisław Sląski	110,500
Kielce	205,900	Kalisz	105,000
Zabrze	198,900	Koszalin	103,300
Toruń	194,500	Jastrzębie Zdrój	101,400

Area and population

Provinces	Capitals	area sq mi	area sq km	population 1986 estimate
Biała Podlaska	Biała Podlaska	2,065	5,348	297,900
Białystok	Białystok	3882	10,055	671,600
Bielsko	Bielsko Biala	1,430	3,704	873,600
Bydgoszcz	Bydogoszcz	3,996	10,349	1,083,800
Chełm	Chełm	1,493	3,866	240,800
Ciechanów	Ciechanów	2,456	6,362	418,100
Czétochowa	Czéstochowa	2,387	6,103	767,400
Elbląg	Elbląg	2,356	6,103	466,700
Gdansk	Gdansk	2,855	7,394	1,401,500
Gorzów	Gorzów Wielkopolski	3,276	8,484	482,200
Jelenia Góra	Jelenia Góra	1,690	4,378	510,300
Kalis	Kalis	2,514	6,512	696,400
Katowice	Katowice	2,568	6,650	3,916,400
Kielce	Kielce	3,556	9,211	1,107,900
Konin	Konin	1,984	5,139	459,300
Koszalin	Koszalin	3,270	8,470	489,800
Kraków	Kraków	1,256	3,254	1,209,300
Krosno	Krosno	2,202	5,702	475,200
Legnica	Legnica	1,559	4,037	490,600
Leszno	Leszno	1,604	4,154	375,600
Łódź	Łódź	588	1,523	1,149,100
Łomza	Łomza	2,581	6,684	338,700
Lublin	Lublin	2,622	6,792	985,400
Nowy Sącz	Nowy Sącz	2,153	5,576	667,400
Olsztyn	Olsztyn	4,759	12,327	725,700
Opole	Opole	3,295	8,535	1,013,700
Ostrolęka	Ostrołęka	2,509	6,498	384,200
Piła	Piła	3,168	8,205	465,400
Piotrków	Piotrków Trybunalski	2,419	6,266	633,100
Plock	Plock	1,976	5,117	509,300
Poznań	Poznán	3,147	8,151	1,298,000
Przemyśl	Przemyśl	1,713	4,437	395,900
Radom	Radom	2,816	7,294	729,700
Rzeszów	Rzeszów	1,698	4,397	691,300
Siedlce	Siedlce	3,281	8,499	636,500
Sieradz	Sieradz	1,880	4,869	401,200
Skierniewice	Skierniewice	1,529	3,960	409,500
Słupsk	Słupsk	2,878	7,453	396,100
Suwałki	Suwalki	4,050	10,490	449,000
Szczecin	Szczecin	3,854	9,981	942,600
Tarnobrzeg	Tarnobrzeg	2,426	6,283	580,500
Tarnów	Tarnów	1,603	4,151	641,500
Toruń	Toruń	2,065	5,348	640,600
Wałbrzych	Walbrzych	1,609	4,168	735,800
Warszawa	Warszawa	1,463	3,788	2,412,200
Włocławek	Włocławek	1,700	4,402	425,900
Wroctaw	Wroctaw	2,427	6,287	1,113,900
Zamość	Zamość	2,695	6,980	487,900
Zielona Góra	Zielona Góra	3,424	8,868	646,000
Total		120,727	312,683	37,340,500

Base 505077 (547779) 4-82

PORTUGAL

BASIC FACTS

Official Name: Portuguese Republic

Abbreviation: PRT

Area: 92,389 sq km (35,672 sq ft)

Area—World Rank: 106

Population: 10,388,421 (1988) 10,877,000 (2000)

Population—World Rank: 62

Capital: Lisbon

Boundaries: 2,047 km (1,272 mi); Spain 1,215 km (755 mi)

Coastline: 832 km (517 mi)

Longest Distances: 561 km (349 mi) N-S; 218 km (135 mi) E-W

Highest Point: Pico Pt. 2,351 m (7,713 ft)

Lowest Point: Sea Level

Land Use: 32% arable; 6% permanent crops; 6% meadows and pastures; 40% forest and woodland; 16% other

Portugal is one of the smallest countries of Europe as well as one of the most westerly countries of continental Europe, occupying the greater portion of the western littoral of the Iberian Peninsula.

Portugal's boundary with Spain was fixed before that of any other European countries with only a small minor frontier change in 1801. By the Peace of Bedajoz of that year Portugal ceded the nearby township of Olivenca to Spain, where it has remained despite the provision in the Treaty of Vienna of 1815 that the place should be returned to Portugal. The border was further fixed by treaties in 1864 and 1906. Portuguese claims to Olivenca as well as Juromentha, another Alentejo town, still are outstanding, although they never have been pressed in the 20th century.

Geographically, Portugal is not a homogeneous territory, nor is it particularly distinct from Spain. The northeastern part—Tras-os-Montes and Beira Alta—is merely a prolongation of the Castilian Plateau, the Meseta, while farther south, Beira Baixa and the Alentejo region are geographically one with the Spanish region of Estremadura. The central mountain range of the Serra da Estrela divides northern from southern Portugal, just as its prolongation in the Sierras de Gata, Gredos and Guadarrama divides Spain. The two great rivers of Portugal, the Tagus and the Douro, also are the two great rivers of Spain. In the South the Algarve is not geographically different from coastal Andalusia.

For a small country, Portugal has a variety of topographic features and soils. The landforms change from area to area, but the most striking differences are between the North and the South, the Rio Tejo forming a convenient dividing line between the hilly to mountainous regions of the North and the rolling plains of the South. The major geographic regions correspond to the six provinces that existed from the Middle Ages to the 19th century. Three of these regions—Minho, Tras-os-Montes and Beira—lie completely in the North; three-quarters of the Estremadura lies north of the Rio Tejo; while the remainder of Estremadura, Alentejo and Algarve lie in the South.

The Minho is the historic cradle of Portugal. It is divided from Tras-os-Montes by mountain ranges, the Serras do Geres, da Cabreira, de Barroso and do Marão in the East, and bounded by the Minho and Douro rivers in the North and South, respectively. The beautiful river valleys of the Lima, Cavado and Tamega cross it from northeast to southwest, providing some of the finest scenery in Europe. The Geres area is known as the Portuguese Switzerland. Although the soil is not very fertile, the high rainfall allows intensive cultivation, and the region is densely populated despite emigration.

Tras-os-Montes (literally, across the mountains) is the northeasternmost area of Portugal, bounded to the north and east by Spain, to the south by the Rio Douro and to the west by the mountains that separate it from the Minho. The region may be divided into two distinct climatic and geographic parts: the northern terra fria (cold land), which shares the rigors of a continental climate with hot summers and very cold winters; and the terra quente (hot land), which has a more temperate and Mediterranean climate. The former is an extension of the Castilian Plateau, with low rainfall, rocky terrain, infertile soils and poor vegetation. The famous port wine area is on the terraced slopes of the Rio Douro along the region's southern boundary.

The central area south of the Douro and north of the Tagus, bounded by an imaginary line running from just south of the Mondego River inland to a point on the Tagus about 24 km (15 mi) from the Spanish frontier, is the Beiras region. It is a transitional region between the North and the South and is dominated by Coimbra, an ancient town on the Mondego. To the southwest of the town of Guarda lies the Serra da Estrela (the Mountain Range of the Star), which divides the region in two and constitutes the roof of Portugal, rising to 1,981 m (6,500 ft). On the contrary, coastal Beira, known as Beira Litoral, consists of rolling, sandy hills, thickly forested in places.

Beira Litoral and the next region, Estremadura, to the south as far as the Tagus, are unique among the landforms of the Iberian Peninsula in that the area is younger geologically and contains sandstone, limestone and volcanic rock rather than the granite and schist predominant in other areas. The Beira coastal plain, which extends inland up to 48 km (30 mi) from the ocean, contains salt marshes and alluvial deposits and stretches of sand dunes, in some places 3 to 8 km (2 to 5 mi) wide. Around the lagoon known as the Ria de Aveiro is a fertile land reclaimed from the sea and known as Portuguese Holland. Taken as a whole, Beira Litoral, is one of the most fertile regions in the country. The northern part of Beira Baixa is a dry and windswept region similar to Tras-os-Montes.

The region of Estremadura includes the Tejo estuary, the capital city of Lisbon, and the important Tejo valley area known as Ribatejo. The great pilgrimage center of Fatima is here. To the northwest of Lisbon and behind the Costa do Sol, lie the Serra de Sintra, an area of considerable natural beauty celebrated by Shelley and Byron, and in northern Ribatejo, Tomar, considered by Somerset Maugham to be the most beautiful place on earth.

The Alentejo (literally, the land across the Tejo) is a vast area of gently rolling hills, generally rising to about 183 m (600 ft), but occasionally reaching between 274 and 457 m (900 and 1,500 ft). Although Alentejo is monotonously similar in its topographic appearance, usually it is divided into two subregions: Alto Alentejo and Baixo Alentejo. It is a poor and sparsely populated region with low rainfall; virtually no natural vegetation; and infertile soils made more desolate by long, hot summers.

The Algarve, the southernmost region, also is the most Moorish, since it was the last area to be recaptured from the Moors. It is separated from the Alentejo by two mountain ranges: the Serra de Monchique in the West and the Serra de Caldeirao in the East. These mountains also serve to differentiate the interior of the province from its low coastal plain. Its California-like weather permits sea bathing from March through September and accounts for its popularity with tourists.

The Madeiran archipelago is said to have been discovered in 1419 by the Portuguese navigator João Goncalves Zarco, who subsequently undertook its colonization. The archipelago consists of Madeira (Ilha da Madeira; Island of Timber), 55 km (34 mi) long and 23 km (14 mi) wide; the island of Porto Santo, and the uninhabited Desertas and Selvagens. On the main island the land rises sharply from the coast to a height of 1,861 m (6,106 ft) in Pico Ruivo, and settlements usually are at the mouths of ravines or on the gentler slopes. The "Pearl of the Atlantic," as the island is called, is a paradise for naturalists.

The Azorean archipelago is a volcanic mountain chain of nine islands divided into three groups: São Miguel and Santa Maria to the east; Terceira, Pico, Faial, São Jorge and Graciosa in the center; and Flores and Corvo to the northwest. The archipelago is said to have been discovered by Diogo de Silves in 1427. Thermal springs are features on the largest island, San Miguel, on which the capital, Ponta Delgada, is located.

There are several zones of intense seismic activity as well as major geological faults in Portugal. The largest zones are concentrated in the Algarve, the greater Lisbon area and the Rio Tejo estuary as far as Benavente. Smaller zones fan out from Oporto along the rim of the Minho region. Large faults crisscross the Minho region, and larger ones extend into Spain along the northern side of the Serra da Estrela. Faults in Estremadura reach into Lisbon. In southern Portugal there is a triangular fault along the eastern edge of the Serra de Grandola near the coast, west of the Rio Sado, and another extending from Beja in a southwesterly direction. A disastrous earthquake on November 1, 1755, killed an estimated 20,000 people and caused extensive damage in Lisbon, Setubal, Lagos, Portimão and Faro. Seismic activity in the 20th century has centered in northern Minho.

PRINCIPAL TOWNS (population at 1981 census)

Lisboa (Lisbon, the capital)	807,937	Braga	63,033
Porto (Oporto)	327,368	Vila Nova de Gaia	62,469
Amadora	95,518	Barreiro	50,863
Setúbal	77,885	Funchal	44,111
Coímbra	74,616	Almada	42,607

Area and population

Continental Portugal

Districts	**Capitals**	area sq mi	area sq km	population 1986 estimate
Aveiro	Aveiro	1,084	2,808	655,100
Beja	Beja	3,948	10,225	182,000
Braga	Braga	1,032	2,673	756,300
Bragança	Bragança	2,551	6,608	186,400
Castelo Branco	Castelo Branco	2,577	6,675	228,300
Coimbra	Coimbra	1,524	3,947	445,700
Évora	Évora	2,854	7,393	176,900
Faro	Faro	1,915	4,960	337,200
Guarda	Guarda	2,131	5,518	200,400
Leiria	Leiria	1,357	3,515	434,200
Lisboa	Lisbon (Lisboa)	1,066	2,761	2,119,600
Portalegre	Portalegre	2,342	6,065	139,600
Porto	Porto	925	2,395	1,644,400
Santarém	Santarém	2,605	6,747	460,500
Setúbal	Setúbal	1,955	5,064	742,200
Viana do Castelo	Viana do Castelo	871	2,255	264,900
Vila Real	Vila Real	1,671	4,328	265,300
Viseu	Viseu	1,933	5,007	426,500
Azores (Açores) Autonomous Region	Ponta Delgada	868	2,247	252,200
Madeira Autonomous Region	Funchal	306	794	267,400
TOTAL		35,672	92,389	10,185,100

Of the 10 major rivers in Portugal, five have their origins in Spain and form part of the Spanish-Portuguese boundary at one or more points in their courses toward the Atlantic. The remaining five are entirely within Portugal and are short, the longest being the 174-km-long (108-mi-long) Rio Sado. The major river in northern Portugal is the Rio Douro, on whose estuary Oporto is located. This river is 940 km (584 mi) long, of which 740 km (460 mi) are in Spain up to the boundary at Barca d'Alva and 200 km (124 mi) in Portugal. The Douro is navigable by small craft for its full course within Portugal, although only the smallest craft can pass beyond Peso de Regua, 80 km (50 mi) from Oporto. Ocean tides are felt as far as 29 km (18 mi) from the river mouth, which itself is blocked by a sandbar, with the result that an artificial harbor was built at Leixoes to handle oceangoing ships. The river undergoes three sharp drops before reaching Oporto: from 1,463 m (4,800 ft) at its source to 457 m (1,500 ft) at Zamora and to 183 m (600 ft) at Barca d'Alva, always in a deep, winding gorge. The area of the Douro basin in Portugal is 18,653 sq km (7,200 sq mi). The Rio Minho rises in Spanish Galicia. Of its total length of 339 km (211 mi), 74 km (46 mi) form the Spanish-Portuguese boundary. It is navigable by small barges for 45 km (28 mi) upstream.

Portugal
International boundary
District boundary
National capital
District capital
Railroad
Road
0 25 50 Kilometers
0 25 50 Miles
Districts are named after their respective capitals. The Azores districts of Angra do Heroismo, Horta, and Ponta Delgada, and the Madeira Islands district of Funchal are not shown.
NORTH
ATLANTIC
OCEAN
SPAIN
Golfo de Cádiz
Valença
Rio Minho
Rio Lima
Verin
Chaves
Bragança
Benavente
Braga
Guimarães
Fafe
Rio Tâmega
Zamora
Vila Real
Porto
Rio Douro
Barca d'Alva
Salamanca
Aveiro
Viseu
Vilar Formosa
Ciudad-Rodrigo
Rio Mondego
Guarda
Bejar
Coimbra
Covilhã
Figueira da Foz
Rio Zêzere
Plasencia
Castelo Branco
Leiria
Tagus
Tomar
Abrantes
Valencia de Alcántara
Cáceres
Trujillo
Caldas da Rainha
Santarém
Portalegre
Mérida
Mora
Badajoz
Estremoz
Lisbon
Barreiro
Almada
Vendas Novas
Setúbal
Évora
Alcácer do Sal
Rio Sado
Rio Guadiana
Zafra
Fregenal de la Sierra
Moura
Grândola
Sines
Beja
Serpa
Rio Chança
Guadalquivir
Sevilla
Rio
Utrera
Vila Real de Santo Antonio
Huelva
Ayamonte
Portimão
Tavira
Lagos
Sagres
Faro
42
40
38
10
8
6

Tides are felt as far as Moncão, 40 km (25 mi) upstream. There is a dangerous sandbar at the mouth of the Minho. The Minho's basin within Portugal is only 78 sq km (30 sq mi). The Minho has one major tributary in Portugal, the Coura, entering on its left bank. Of the three other rivers in Minho the Lima originates in Spain and the Cavado and the Ave in Portugal. The Lima flows from the Spanish province of Orense to its mouth at Viana do Castelo, forming the boundary for 3.2 km (2 mi). In Portugal its length is 61 km (38 mi) and its basin is 1,101 sq km (425 sq mi). It is navigable for 40 km (25 mi) upstream. The Ave, flowing 85 km (53 mi) from the Serra da Cabreira to its mouth 1.6 km (1 mi) south of Vila do Caonde, has a basin of 1,399 sq km (540 sq mi), but is navigable for only 1.6 km (1 mi) upstream. The Cavado, flowing 117 km (73 mi) from northwestern Tras-os-Montes to its mouth at Esposende, has a basin of 1,606 sq km (620 sq mi) and is navigable for 6.4 km (4 mi).

The principal navigable rivers in Beira are the Rio Mondego and the Rio Vouga, both of Portuguese origin. The Mondego, rising in the Serra da Estrela, makes a meandering course for 219 km (136 mi) before entering the Atlantic at Figueira da Foz. This river is navigable for 85 km (53 mi) upstream, where it meets its principal tributary, the Rio Dão. The Atlantic tide is felt for 19 km (12 mi) upstream. The Mondego basin is 6,813 sq km (2,630 sq mi), including the city of Coimbra. The Vouga flows 135 km (84 mi) to the Ria de Aveiro on the Atlantic coast; 50 km (31 mi) of the Vouga are navigable. The Vouga basin is 3,705 sq km (1,430 sq mi).

The longest river in Portugal as well as in the Iberian Peninsula is the Rio Tejo, with a total length of 999 km (621 mi). In Spanish it is called Tajo and in English Tagus. The Tejo drops from 1,463 m (4,800 ft) at its source to barely 30 m (100 ft) at the boundary. Of its 228-km (142-mi) course in Portugal, only 80 km (50 mi) are navigable by shallow-draft vessels upstream to Santarém, but most river traffic stops at Vila Franca de Xira. The Atlantic tide is felt upriver for 93 km (58 mi). At flood tide, the river rises close to 6 m (20 ft).

The Tejo basin, one of the most fertile regions in Portugal, is 24,922 sq km (9,620 sq mi). Its estuary, the Mar da Palha, encompassing 251 sq km (97 sq mi) in the Lisbon region, is one of the world's great natural harbors. The dry dock at Cacilhas also is one of the world's largest.

In the Alentejo and the Algarve the major rivers are the Sado, the Mira, the Arade and the Guadiana, only the last of which rises in Spain. The Rio Sado flows 174 km (108 mi) northwest, forming the Sado estuary on the Atlantic west of Setubal. The river is navigable for small craft as far as Porto do Rio, 69 km (43 mi) upstream.

The Rio Guadiana, the major river south of the Tejo, rises in Spain and flows for about 547 km (340 mi) to the boundary, then forms the boundary for 64 km (40 mi), flows through the Alentejo for 145 km (90 mi), then forms the border again for 48 km (30 mi) before emptying into the Gulf of Cadiz near the Algarve port of Vila Real de Santo Antonio. The last stretch of the river is navigable for ships displacing up to 1,000 tons, while small craft are able to navigate an additional 24 km (15 mi).

PUERTO RICO

BASIC FACTS

Official Name: Commonwealth of Puerto Rico

Abbreviation: PRO

Area: 9,104 sq km (3,515 sq mi)

Area—World Rank: 148

Population: 3,358,879 (1988) 3,389,000 (2000)

Population—World Rank: 107

Capital: San Juan

Land Boundaries: Nil

Coastline: Undetermined

Longest Distances: 290 km (180 mi) E-W; 82 km (51 Mi) N-S

Highest Point: Punta Mountain 1,338 m (4,390 ft)

Lowest Point: Sea level

Land Use: 8% arable; 7% permanent crops; 38% meadows and pastures; 21% forest and woodland; 26% other

Puerto Rico lies about 80 km (50 mi) east of Hispaniola, from which it is separated by the Mona Passage. It is crossed by mountain ranges, the most prominent being the Cordillera Central, rising to 1,341 m (4,400 ft). The coastal plain is about 24 km (15 mi) wide at its broadest point. About 50 short rivers flow rapidly to the sea. Islands off the coast include Mona and Desecheo to the west and Vieques and Culebra to the east.

Population 1984 estimate

Municipio	population	Municipio	population
Adjuntas	18,900	Juncos	27,000
Aguada	32,400	Lajas	21,300
Aguadilla	55,000	Lares	28,000
Agunas Buenas	23,000	Las Marías	8,600
Aibonito	22,500	Las Piedras	23,100
Añasco	24,400	Loiaz	24,600
Arecibo	87,000	Luquillo	15,400
Arroyo	18,200	Manati	38,000
Barceloneta	19,600	Maricao	6,700
Barranquitas	22,800	Maunabo	11,800
Bayamón	202,500	Mayagüez	101,000
Cabo Rojo	35,000	Moca	29,900
Caguas	121,100	Morovis	21,900
Camuy	26,200	Naguabo	21,300
Canóvanas	32,400	Naranjito	25,100
Carolina	165,700	Orocovis	20,900
Cataño	25,900	Patillas	17,900
Cayey	43,300	Peñuelas	20,200
Ceiba	15,100	Ponce	190,900
Ciales	17,200	Quebradillas	19,700
Cidra	29,600	Rincón	12,400
Coamo	32,200	Rio Grande	37,700
Comerío	18,400	Sabana Grande	21,100
Corozal	29,600	Salinas	26,600
Culebra	1,300	San Germán	34,200
Dorado	26,700	San Juan	428,900

Puerto Rico

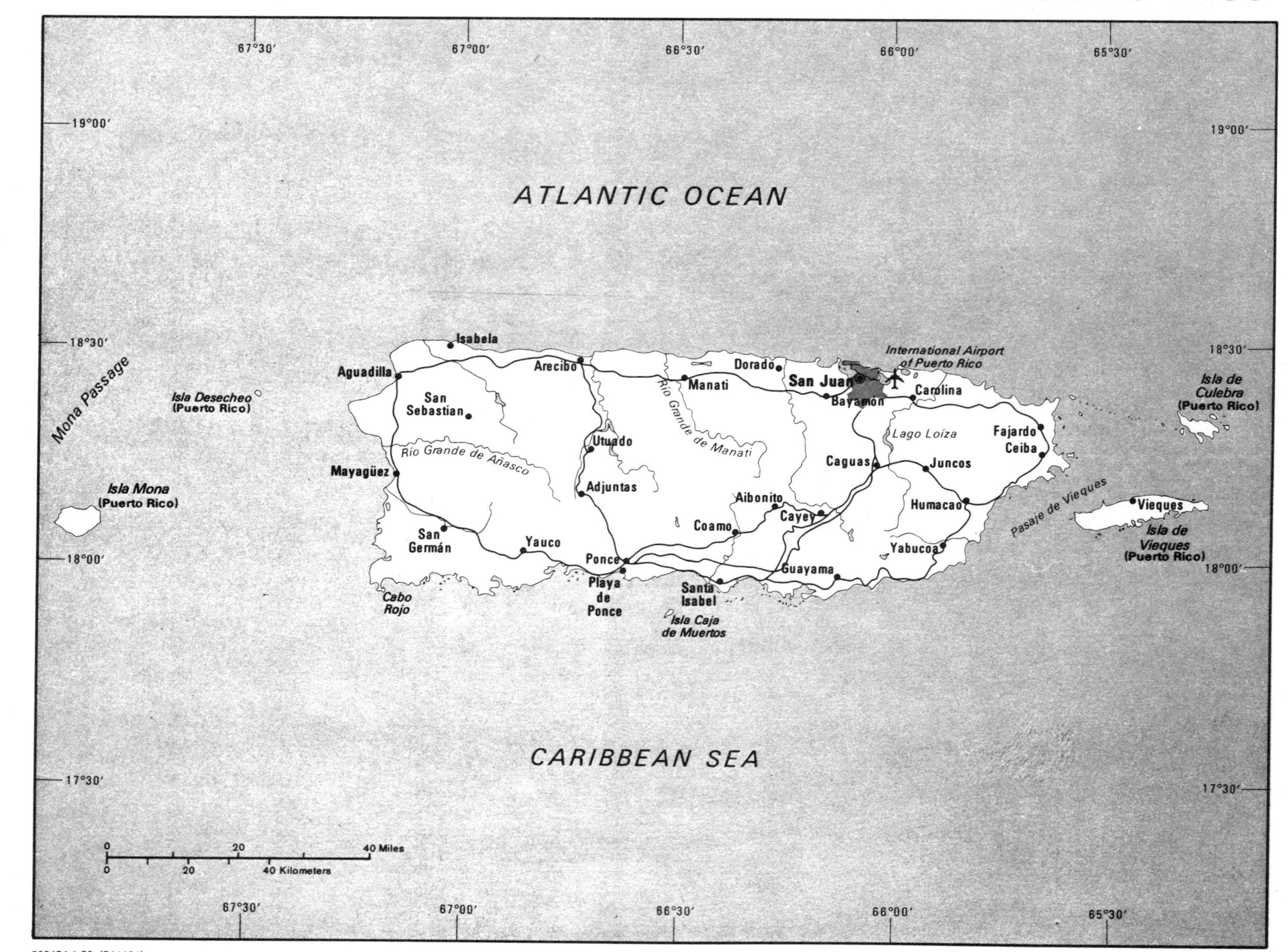

502484 1-76 (541401)
Lambert Conformal Projection
Standard parallels 17°20′ and 22°40′
Scale 1:1,400,000

Population 1984 estimate

Municipio	population	Municipio	population
Fajardo	33,200	San Lorenzo	33,300
Florida	7,600	San Sebastián	36,100
Guánica	18,800	Santa Isabel	19,500
Guayama	40,300	ToaAlta	33,400
Guayanilla	21,000	Toa Baja	77,700
Guaynabo	85,100	Trujillo Alto	50,800
Gurabo	25,000	Utuado	34,600
Hatillo	30,400	Vega Alta	30,000
Hormigueros	15,200	Vega Baja	48,800
Humacao	52,400	Vieques	7,800
Isabela	38,200	Villalba	22,500
Jayuya	15,000	Yabucoa	31,400
Juana Díaz	43,600	Yauco	39,200
		TOTAL	3,270,000

QATAR

BASIC FACTS

Official Name: State of Qatar

Abbreviation: QAT

Area: 11,400 sq km (4,400 sq mi)

Area—World Rank: 143

Population: 328,044 (1988) 560,000 (2000)

Population—World Rank: 147

Capital: Doha

Boundaries: 490 km (350 mi); Saudi Arabia 67 km (42 mi); United Arab Emirates 45 km (28 mi)

Coastline: 378 km (235 mi)

Longest Distances: 161 km (100 mi) N-S; 89 km (55 mi) E-W

Highest Point: Aba Al-Bawl Hill 105 m [344 ft]

Lowest Point: Sea level

Land Use: 5% meadows and pastures; 95% other

Qatar is located on a peninsula extending about 161 km (100 mi) north from the eastern Saudi Arabian mainland and separating the Persian Gulf from the Gulf of Bahrain and its lesser extension, Bahr as Salwa. At various points the northwestern coast is less than 40 km (25 mi) from the main island of Bahrain, and the Hawar Islands immediately off the peninsular coast remains the subject of a territorial dispute between Qatar and Bahrain. The southern shoreline of the Khawr al Udayd forms part of the land boundary that Qatar and Saudi Arabia defined and delimited in 1965.

The land is largely flat desert covered with loose sand and pebbles broken by occasional outcroppings of limestone. The western coast, where most of the oil fields are located, is marked by low cliffs and hills. In the south sand dunes and salt flats predominate. What little natural vegetation and cultivated land exists is confined to the north.

Area and population

Municipalities	area sq mi	area sq km	population 1986 census
Doha	...	...	217,294
al-Guwayrīyah	...	...	1,629
Jeriān al Baṭnah	...	...	2,727
al-Jumaylīyah	...	...	7,217
al-Khawr	...	...	8,993
ar-Rayyān	...	...	91,996
ash-Shamāl	...	...	4,380
Umm aṣ-Ṣilāl	...	...	11,161
al-Wakrah	...	...	23,682
TOTAL	4,400	11,400	369,079

REUNION

BASIC FACTS

Official Name: Department of Reunion

Abbreviation: RUN

Area: 2,544 sq km (982 sq mi)

Area—World Rank: 156

Population: 557,441 (1988) 706,000 (2000)

Populatin—World Rank: 140

Capital: Saint-Denis

Land Boundaries: Nil

Coastline: 201 km (125 mi)

Longest Distances: 55 km (34 mi) N-S; 53 km (33 mi) E-W

Highest Point: Piton des Neiges 3,069 m (10,069 ft)

Lowest Point: Sea level

Land Use: 20% arable; 2% permanent crops; 4% meadows and pastures; 35% forest and woodland; 39% other

Reunion is a small elliptical-shaped island in the South Indian Ocean 1,267 km (750 mi) east of Madagascar. Most of the population and the arable land on the island is concentrated on the narrow plain that encircles the central mountainous massif. The volcanic core of the island is divided into two parts by a high, undulating plateau that extends across the island in a southwest-northeast direction. To the northwest of the plateau is the Piton des Neiges, an eroded peak 3,068 m (10,066 feet) high, surrounded by three large bowl-shaped valleys called "cirques," arranged in a pattern resembling a three-leaf

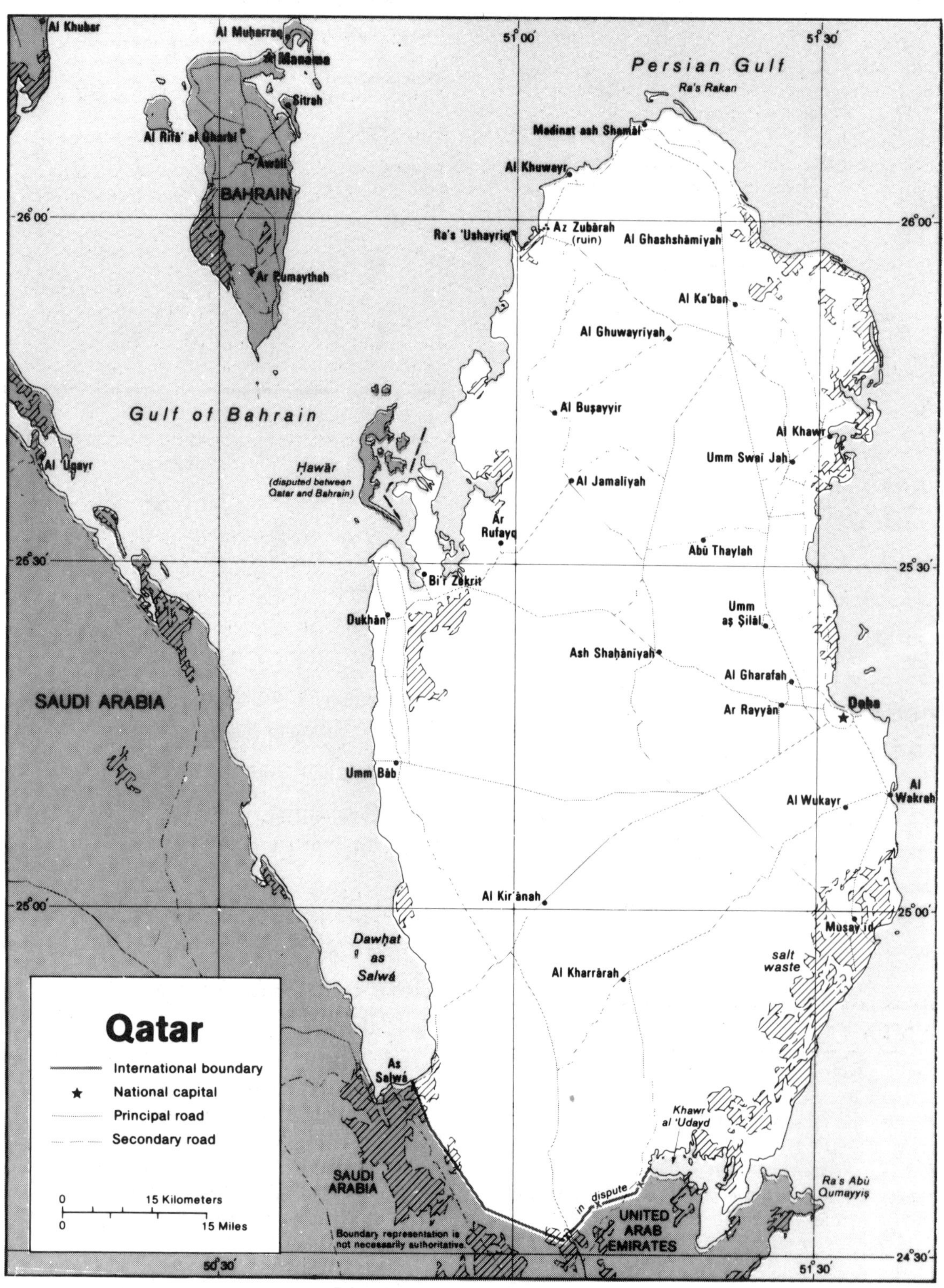
Al Khubar
Al Muḩarraq
Manama
Sitrah
Al Rifā' al Gharbī
'Awālī
BAHRAIN
Ar Rumaythah
51°00'
51°30'
Persian Gulf
Ra's Rakan
Madīnat ash Shamāl
Al Khuwayr
26°00'
Ra's 'Ushayriq
Az Zubārah
(ruin)
Al Ghashshāmīyah
Al Ka'ban
Al Ghuwayrīyah
Gulf of Bahrain
Al Buşayyir
Al Khawr
Umm Swai Jah
Al Jamalīyah
Ḩawār
(disputed between Qatar and Bahrain)
Al 'Uqayr
Ar Rufayq
Abū Thaylah
25°30'
Bi'r Zekrit
Dukhān
Umm aş Şilāl
Ash Shaḩānīyah
Al Gharafah
Ar Rayyān
Doha
SAUDI ARABIA
Umm Bāb
Al Wakrah
Al Wukayr
Al Kir'ānah
25°00'
Muşay'īd
Dawḩat as Salwá
salt waste
Al Kharrārah
Qatar
International boundary
National capital
Principal road
Secondary road
0 15 Kilometers
0 15 Miles
As Salwá
Khawr al 'Udayd
SAUDI ARABIA
in dispute
Ra's Abū Qumayyiş
UNITED ARAB EMIRATES
Boundary representation is not necessarily authoritative
50°30'
51°30'
24°30'

clover. To the southwest of the plateau, amid a barren landscape, is the active volcano, Piton de la Fournaise 2,609 m (8,560 feet), which periodically emits a pyrotechnic display of smoke and glowing lava. The most recent eruption was in 1961. Damage to property and crops, however, is more likely to result from the passage of cyclonic storms across or near the island.

A band of natural vegetation—forest and brush—follows the 1,371 m (4,500 foot) contour line around the island. Above the band the natural vegetation becomes more stunted and sparse. Below that level most of the natural vegetation has been replaced by commercial crops.

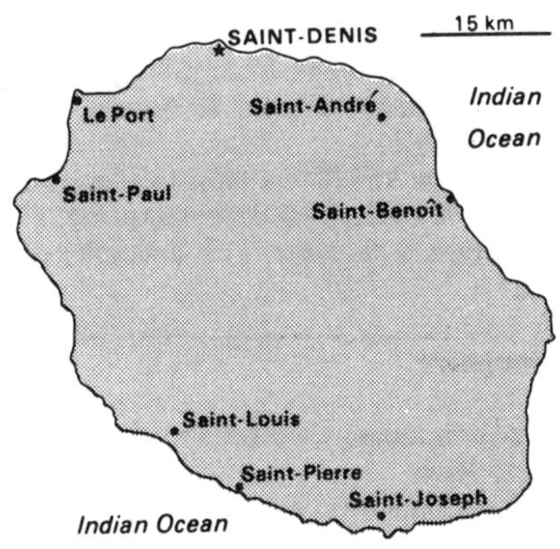

Principal Towns (population at 1982 census): Saint-Denis (capital) 109,068; Saint-Paul 58,410; Saint-Pierre 50,081.

Area and population

Arrondissements	Capitals	area sq mi	area sq km	population 1982 census
Saint-Benoit	Saint-Benoit	284	736	74,312
Saint-Denis	Saint-Denis	164	423	180,647
Saint-Paul	Saint-Paul	180	467	94,378
Saint-Pierre	Saint-Pierre	339	878	166,461
TOTAL		969	2,510	515,798

ROMANIA

BASIC FACTS

Official Name: Socialist Republic of Romania

Abbreviation: ROM

Area: 237,500 sq km (91,699 sq mi)

Area—World Rank: 79

Population: 23,040,883 (1988) 24,098,000 (2000)

Population—World Rank: 37

Capital: Bucharest

Boundaries: 3,153 km (1,959 sq mi); USSR 1,329 km (826 mi); Bulgaria 591 km (367 mi); Yugoslavia 546 km (339 mi); Hungary 442 km (275 mi)

Coastline: 245 km (152 mi)

Longest Distances: 789 km (490 mi) E-W; 475 km (295 mi) N-S

Highest Point: Moldoveanu 2,543 m (8,343 ft)

Lowest Point: Sea level

Land Use: 43% arable; 3% permanent crops; 19% meadows and pastures; 28% forst and woodland; 7% other

Romania, located in southeastern Europe and usually referred to as one of the Balkan states, shares land borders with Bulgaria, Yugoslavia, Hungary, and the Soviet Union and has a shoreline on the Black Sea. The interior of the country is a broad plateau almost surrounded by mountains, which, in turn, are surrounded, except in the north, by plains. The mountains are not unduly rugged, and their gentle slopes plus the rolling interior plateau and the arc of lowlands on the country's periphery provide an unusually large percentage of arable land.

Topographical and Regional Description

All of the mountains and uplands of the country are part of the Carpathian system. The Carpathian Mountains originate in Czechoslovakia, enter Romania in the north from the Soviet Union, and proceed to curl around the country in a semicircle. The ranges in the east are referred to as the Moldavian Carpathians; the slightly higher southern ranges are called the Transylvanian Alps; and the more scattered but generally lower ranges in the west are known as the Bihor Massif. A few peaks in the Moldavian Carpathians rise to nearly 2,286 m (7,500 ft) and several in the Transylvanian Alps reach 2,438 m (8,000 ft) but only a few points in the Bihor Massif approach 1,828 m (6,000 ft).

Lowland areas are generally on the periphery of the country—east, south, and west of the mountains. A plateau, higher than the other lowlands but having elevations averaging only about 365 m (1,200 ft), occupies an area enclosed by the Carpathian ranges.

Moldavia, in the northeast, constitutes about one-fourth of the country's area. It contains the easternmost ranges of the Carpathians and, between the Siretul and Prut rivers, an area of lower hills and plains. The Moldavian Carpathians have maximum elevations of about 2,286 m (7,500 ft) and are the most extensively forested part of the country. The western portion of the mountains contains a range of volcanic origin—the longest of its type in Europe—that is famous for its some 2,000 mineral water springs. Small sections of the hilly country to the northeast also have forests, but most of the lower lands are rolling country, which becomes increasingly flatter in the south. Almost all of the nonforested portions are cultivated.

Walachia, in the south, contains the southern part of the Transylvanian Alps—called the Southern Carpathians by Romanian geographers—and the lowlands that extend between them and the Danube River. West to east it extends from the Iron Gate to Dobruja, which is east of the Danube in the area where the river flows northward for about 160 km (100 mi) before it again turns to the east for its final passage to the sea. Walachia is divided by the Olt River into Oltenia (Lesser Walachia) in the west and Muntenia (Greater Walachia), of which Bucharest is the approximate center, in the east. Nearly all of the Walachian lowlands, except for the marshes along the Danube River, and the seriously eroded foothills of the mountains are cultivated. Grain, sugar beets, and potatoes are grown in all parts of the flatland; the area around Bucharest produces much of the country's garden vegetables; and southern exposures along the mountains are ideally suited for orchards and vineyards.

The Transylvanian Alps have the highest peaks and the steepest slopes in the country; the highest point, with an elevation of about 2,542 m (8,340 ft) above sea level, is 161 km (100 mi) northwest of Bucharest. Among the alpine features of the range are glacial lakes, upland meadows and pastures, and bare rock along the higher ridges. Portions of the mountains are predominantly limestone with characteristic phenomena, such as caves, waterfalls, and underground streams.

Transylvania, the northwestern one-third of the country, includes the historic Transylvanian province and the portions of Maramures, Crisana, and Banat that became part of Romania after World War I. The last three borderland areas are occasionally identified individually.

Nearly all of the lowlands in the west and northwest and the plateau in the central part of the province are cultivated. The western mountain regions are not as rugged as those to the south and east, and average elevations run considerably lower. Many of the intermediate slopes are put to use as pasture or meadowland but, because the climate is colder, there are fewer orchards and vineyards in Transylvania than on the southern sides of the ranges in Walachia. Forests usually have more of the broadleaf deciduous tree varieties than is typical of the higher mountains, but much of the original forest cover has been removed from the gentler Translyvanian slopes.

Dobruja provides Romania's access to the Black Sea. The Danube River forms the region's western border, and its northern side is determined by the northernmost of the three main channels in the Danube delta. The line in the south at which the region has been divided between Romania and Bulgaria is artificial and has been changed several times.

PRINCIPAL TOWNS (estimated population at 1 July 1986)

Bucureşti (Bucharest, the capital)	1,989,823	Oradea	213,846
		Arad	187,744
		Baču	179,877
Braşov	351,493	Sibiu	177,511
Constanşa	327,676	Tîrgu Mureş	158,998
Timişoara	325,272	Piteşti	157,190
Iaşi	313,060	Baia Mare	139,704
Cluj-Napoca	310,017	Buzău	136,080
Galaţi	295,372	Satu Mare	130,082
Craiova	281,044	Piatra-Neamţ	109,393
Brăila	235,620	Botoşani	108,775
Ploieşti	234,886	Reşiţa	105,914

Area and population

Districts	Capitals	area sq mi	area sq km	population 1985 estimate
Alba	Alba Iulia	2,406	6,231	423,600
Arad	Arad	2,954	7,652	502,500
Arges	Piteşti	2,626	6,801	666,300
Bacău	Bacău	2,551	6,606	710,200
Bihor	Oradea	2,909	7,535	653,400
Bistriţa-Năsăud	Bistriţa	2,048	5,305	316,000
Botoşani	Botoşani	1,917	4,965	463,300
Brăila	Brăila	1,824	4,724	398,300
Braşov	Braşov	2,066	5,351	682,400
Buzău	Buzău	2,344	6,072	520,000
Caraş-Severin	Resita	3,283	8,503	404,000
Călăraşi	Călăraşi	1,959	5,074	345,600
Cluj	Cluj-Napoka	2,568	6,650	741,800
Constanţa	Constanţa	2,724	7,055	698,700
Covasna	Stîntu Gheorghe	1,431	3,705	229,500
Dîmboviţa	Tîrgovişte	1,559	4,036	557,900
Dolj	Craiova	2,862	7,413	771,500
Galaţi	Galaţi	1,708	4,425	629,200
Giurgiu	Giurgiu	1,404	3,636	345,500
Gorj	Tîrgu Tiu	2,178	5,641	373,600
Harghita	Miercurea-Ciuc	2,552	6,610	356,600
Hunedoara	Deva	2,709	7,016	554,400
Ialomiţa	Slobozia	1,718	4,449	302,400
Iaşi	Iaşi	2,112	5,469	784,100
Maramureş	Baia Mare	2,400	6,215	538,700
Mehedinţi	Drobeta-Turnu-Severin	1,892	4,900	328,600
Mureş	Tirgu Mureş	2,585	6,696	613,800
Neamţ	Piatra Neamţ	2,274	5,890	566,500
Olt	Slatina	2,126	5,507	531,000
Prahova	Ploieşti	1,812	4,694	861,500
Sălaj	Zalău	1,486	3,850	409,200
Satu Mare	Satu Mare	1,701	4,405	267,200
Sibiu	Sibiu	2,093	5,422	506,300
Suceava	Suceava	3,303	8,555	674,600
Teleorman	Alexandria	2,224	5,760	507,900
Timiş	Timişoara	3,356	8,692	716,400
Tulcea	Tulcea	3,255	8,430	267,100
Vaslui	Vaslui	2,045	5,297	455,300
Vîtcea	Rimnicu Vîlcea	2,203	5,705	424,700
Vrancea	Focşani	1,878	4,863	385,700
Municipality				
Bucharest	Bucharest	654	1,695	2,239,500
TOTAL		91,699	237,500	22,724,800

For nearly 500 years preceding 1878, Dobruja was under Turkish rule. When the Turks were forced to relinquish their control, the largest elements of its population were Romanian and Bulgarian, and it was divided between the two countries. Romania received the larger, but more sparsely populated, northern portion. Between the two world wars Romania held the entire area, but in 1940 Bulgaria regained the southern portion. The 1940 boundaries were reconfirmed after World War II, and since then the Romanian portion has had an area of approximately 15,539 sq km (6,000 sq mi); Bulgaria's has been approximately one-half as large.

Dobruja contains most of the Danube River delta marshland, much of which is not easily exploited for agricultural purposes, although some of the reeds and natural vegetation have limited commercial value. The delta is a natural wildlife preserve, particularly for waterfowl and is large enough so that many species can be protected.

Fishing contributes to the local economy, and 90 percent of the country's catch is taken from the lower Danube

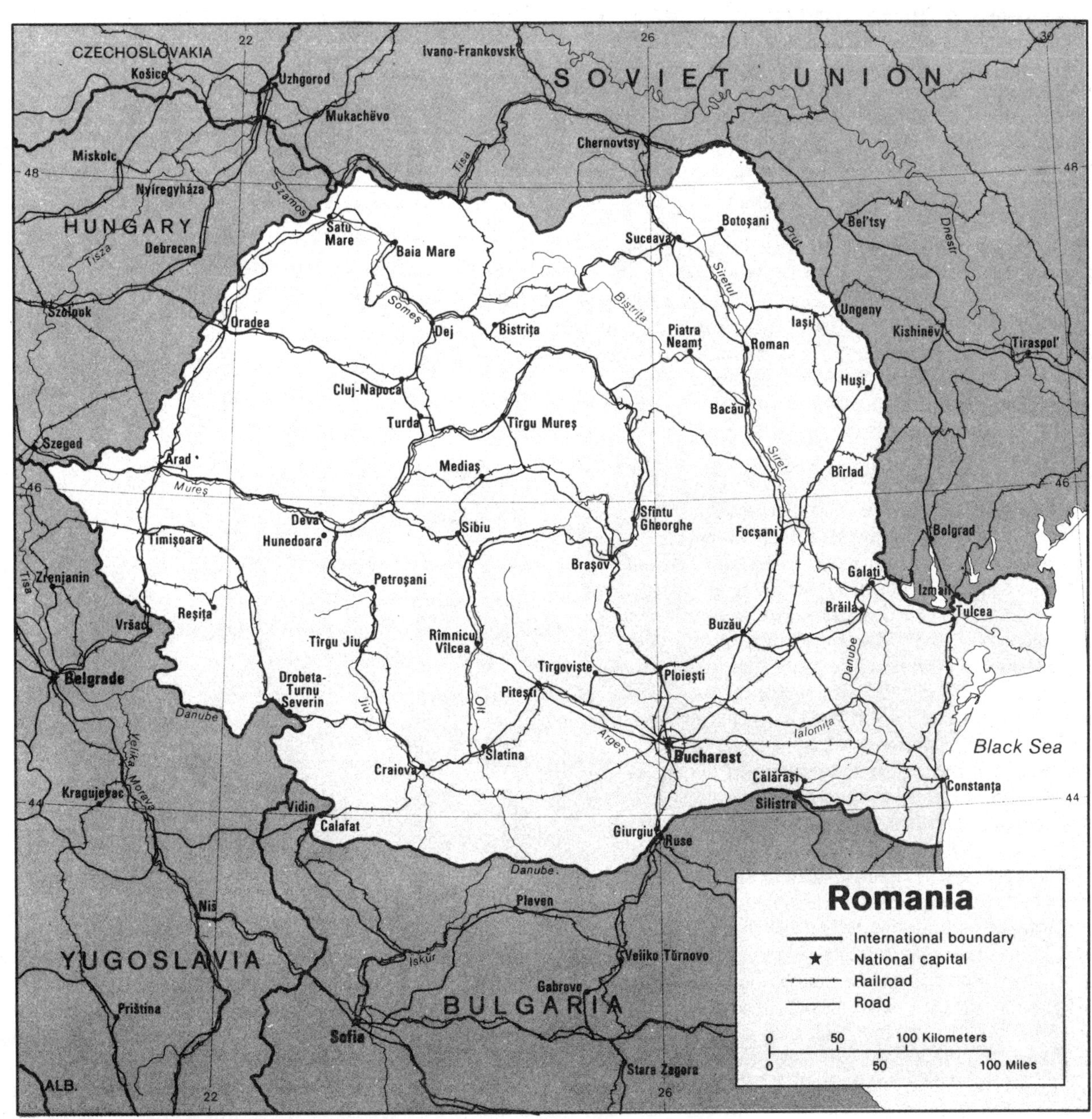

Base 505037 (547584) 8-82

and its delta, from Dobruja's lakes, or off the coast. Willows flourish in parts of the delta, and there are a few deciduous forests in the north-central section. To the west and south, the elevations are higher. The land drains satisfactorily and, although the rainfall average is the lowest in the country, it is adequate for dependable grain crops and vineyards.

Along the southern one-half of the coastline there are pleasant beaches. In summer the dry sunny weather and low humidity make them more attractive tourist resorts.

Bukovina, more isolated than other parts of the country, has a part-Romanian and part-Ukrainian population. Romanian Bukovina is small, totaling only about 8,806 sq km (3,400 sq mi). It was part of Moldavia from the fourteenth century until annexed by Austria in 1775. Romania acquired it from Austria-Hungary in 1918, but after World War II the Soviet Union annexed the 5,438 sq km (2,100 sq mile) northern portion with its largely Ukrainian population.

The approximately 3,366 sq km (1,300 sq mi) of the former province remaining in Romania is picturesque and mountainous. Less than one-third is arable, but domestic animals are kept on hillside pastures and meadows. Steeper slopes are forested.

Drainage

All of Romania's rivers and streams drain to the Black Sea. Except for the minor streams that rise on the eastern slopes of the hills near the sea and flow directly into it, all join the Danube River. Those flowing southward and southeastward from the Transylvanian Alps drain to the Danube directly. Those flowing northward and eastward from Moldavia and Bukovina reach the Danube by way of the Prut River. Most of the Transylvanian streams draining to the north and west flow to the Tisza River, which joins the Danube in Yugoslavia, north of Belgrade.

Romanian tourist literature states that the country has 2,500 lakes, but most are small, and lakes occupy only about 1 percent of the surface area. The largest lakes are along the Danube River and the Black Sea coast. Some of those along the coast are open to the sea and contain salt sea water. These and a few of the fresh water lakes are commercially important for their fish. The many smaller ones scattered throughout the mountains are usually glacial in origin and add much to the beauty of the resort areas.

The Danube drains a basin of 815,850 sq km (315,000 sq mi) that extends eastward from the Black Forest in the Federal Republic of Germany (West Germany) and includes a portion of the southwestern Soviet Union. It is about 2,856 km (1,775 mi) long, including the 1,448 km (900 mi) in or adjacent to Romania, and is fed by more than 300 tributaries, from which it collects an average of about 285,000 cubic ft per minute to discharge into the Black Sea. Much of the Danube delta and a band of up to twenty mi wide along most of the length of the river from the delta to the so-called Iron Gate—where it has cut a deep gorge through the mountains along the Yugoslav border—is marshland.

For descriptive purposes the river is customarily divided into three sections; most of the portion in Romania—from the Iron Gate to the Black Sea—is its lower course. The northern bank of this course, on the Romanian side, is low, flat marshland and, as it approaches its delta, it divides into a number of channels. It also forms several lakes, some of them quite large. At its delta it divides into three major and several minor branches. The delta has an area of about 2,590 sq km (1,000 sq mi) and grows steadily as the river deposits some 2 billion cubic ft of sediment into the sea annually.

RWANDA

BASIC FACTS

Official Name: Republic of Rwanda

Abbreviation: RWN

Area: 26,338 sq km (10,169 sq mi)

Area—World Rank: 131

Population: 7,058,350 (1988) 10,123,000 (2000)

Population—World Rank: 85

Capital: Kigali

Land Boundaries: 893 km (555 mi); Uganda 169 km (105 mi); Tanzania 217 km (135 mi); Burundi 290 km (180 mi); Zaire 217 km (135 mi)

Coastline: Nil

Longest Distances: 248 km (154 mi) NE-SW; 166 km (103 mi) SE-NW

Highest Point: Mount Karisimbi 4,507 m (14,787 ft)

Lowest Point: 950 m (3,100 ft)

Land Use: 29% arable; 11% permanent crops; 18% meadows and pastures; 10% forest and woodland, 32% other

Rwanda is a landlocked country located south of the Equator in east-central Africa. It borders Uganda to the north, Tanzania to the east, Burundi (previously called Urundi) to the south, and Zaire and Lake Kivu to the west.

The geologic base is an irregularly shaped area of the Great East African Plateau. Much of the countryside is covered by grasslands and small farms extending over rolling hills, but there are also areas of swamps and rugged mountains, including volcanic peaks north of Lake Kivu, in the northwest border area. The divide between two of Africa's great watersheds, the Zaire and Nile Basins, extends from north to south through western Rwanda at an average elevation of almost 2,743 m (9,000 ft). On

PRINCIPAL TOWNS (popuulation at 1978 census)

Kigali (capital)	117,749	Ruhengeri	16,025
Butare	21,691	Gisenyi	12,436

Area and population

		area		population
Prefectures	**Capitals**	sq mi	sq km	1983 estimate
Butare	Butare	707	1,830	682,500
Byumba	Byumba	1,925	4,987	623,600
Cyangugu	Cyangugu	859	2,226	343,500
Gikongoro	Gikongoro	846	2,192	401,900
Gisenyi	Gisenyi	925	2,395	566,400
Gitarama	Gitarama	865	2,241	706,200
Kibungo	Kibungo	1,596	4,134	420,200
Kibuye	Kibuye	510	1,320	500,600
Kigali	Kigali	1,255	3,251	835,400
Ruhengeri	Ruhengeri	680	1,762	581,200
TOTAL		10,169	26,338	5,661,400

the western slopes of this Zaire-Nile ridgeline, the land slopes abruptly toward Lake Kivu in the Great Rift Valley on the western border of the country. The eastern slopes are more moderate, with rolling hills extending across the central uplands, at gradually reduced altitudes, to the plains, swamps, and lakes of the eastern border region.

Except for the eastern and western border areas, high altitudes are common. Most of the land is at least 914 m (3,000 ft) above sea level; much of the central plateau has an average altitude of 1,432 m (4,700 ft), and the average for the entire country is about 1,586 m (5,200 ft). The heaviest concentrations of people are located in these central uplands, in the 1,524 to 2,286 m (5,000 to 7,500 ft) altitude levels. Trade winds from the Indian Ocean tend to hold temperatures down, providing the plateaus and rolling hills with a climate that is more healthful for human beings than the higher altitudes of the Zaire-Nile Crest, the lower altitudes in the Rift Valley, or the areas along the eastern border.

There are six regions, from west to east: the narrow Great Rift Valley region along or near Lake Kivu; the volcanic Virunga Mountains and high lava plains of northwestern Rwanda; the steep western slopes of the Zaire-Nile Divide, which extends generally north-south in western Rwanda; the high-altitude area near the ridgeline of this mountain range; the rolling hills and valleys of the central plateaus, which slope eastward from the Zaire-Nile Ridge; and the savannas and marshlands of the eastern and southeastern border areas, which are lower in altitude, warmer, and drier than the central upland plateaus.

Lake Kivu and the Rift Valley

In this westernmost region altitudes range from 792 to 1,609 m (2,600 to 6,000 ft). Volcanic soils, usually richer than the granite-based soil of the central Rwanda plateaus, are found south and north of Lake Kivu. These have been farmed extensively in the north, and some recent agricultural development and resettlement improvement projects were established in the Ruzizi Valley south of the lake. The climate is fairly regular and humid in this southwestern area, and rainfall averages about 127 cm (50 in) per year. The northeastern lakeshore areas near Gisenyi include both volcanic and granite-based soils. The climate is temperate and humid, becoming cooler within a few dozen mi of the lakeshore as elevation increases in the Virunga Mountains to the northwest and the Zaire-Nile Divide to the east.

The Virunga Mountains

Lava plains in the northwest constitute one of the most productive areas. Above these lava plains five volcanic peaks stand within Rwanda or on the Zaire border; three more are within Zaire. Two of these former volcanoes still emit steam and smoke. Trade winds moving against these lofty slopes are forced upward into cooler strata, causing an almost constant mist. This moisture fosters a high-altitude rain forest, which includes a great variety of trees, shrubs, and lichens, varying according to altitude. Upper elevations are too cold for trees, despite the proximity to the Equator, and Mt. Karisimbi, at 4,532 m (14,870 ft) is snowcapped.

The Western Slope

Eat of Lake Kivu a belt averaging 40 km (25 mi) in width makes up the western face of the Zaire-Nile Crest. Within this short horizontal distance the land rises in sharply cut ridges from 1,432 m (4,700 ft) at the lakes shore to an average of 2,743 m (9,000 ft) at the crest of the divide. The rivers are swift torrents that erode deeply during the rainy months and dry up quickly during the two annual dry seasons. Although the percentage of slope on much of this land makes cultivation difficult, food crops and coffee are produced. Where it has not been removed, the natural cover consists of either bushy savanna or forest trees that tend to form a top cover over open aisles (gallery forests).

The Zaire-Nile Crest

Most of the few remaining forests are located on the granite soils of the upper levels of the Zaire-Nile Divide, running north-south in western Rwanda. The average elevation is about 2,743 m (9,000 ft), but some peaks are 244 m (800 ft) higher. The mass of peaks is from 19 to 48 km (12 to 30 mi) wide. Rainfall is heavy, averaging 58 inches annually in much of the chain and is especially heavy in the south near the Burundi border, where more than 178 cm (70 in) is normal. In general, the climate is cooler, more variable, and less comfortable than that of the lower altitudes nearby, and the area has never been densely settled.

The Central Plateaus

East of the mountains lie the central plateaus, covered by rolling hills, becoming progressively lower in altitude as they extend toward the eastern border. Population densities run in the hundreds per sq mile over most of these central uplands. The ancient forests, from which the typical soils of this area were derived, have long since disappeared, and the land has been intensively farmed or grazed. One result has been considerable erosion and soil depletion. Under Belgian guidance thousands of mi of hedges and small dikes were established through erosion control projects, and many rows of trees and shrubs remain

Base 502628 6-75

as visible features of an intensively farmed landscape. Average annual temperatures are near 65° F., and annual rainfall about 127 cm (50 in), both figures varying considerably according to the altitude. By comparison with most tropical locations, the climate of these plateaus is pleasant and is suitable for the cultivation of a wide variety of subsistence crops.

Eastern Swamps and Savannas

Below the farmlands of the central plateaus lie the savannas and swamps of eastern and southeastern Rwanda, at altitudes averaging 1,280 m (4,200 ft). The weather is hotter and drier than that of the upper plateaus. These low hills and broad valleys may receive no precipitation for up to 6 months of the year in some areas, becoming desert-like, although the rainy months may bring as much as 76 cm (30 in) of rain per year.

Kagera National Park (Parc National de la Kagera), along the northeastern border, encloses about 2,590 sq km (1,000 sq mi)—one-tenth of the nation's territory—of protected homeland for large and small African wildlife species. Low, scrubby trees and thorny bushes dot these grasslands. Between its many lakes and marshes, southeastern Rwanda (south of Kigali) also has a semidesert landscape, despite its 76 to 101 cm (30 to 40 in) of rain annually. As everywhere in the country, particularly in the east, precipitation is irregular. Centuries ago this was a pastoral region, which has since been damaged by overgrazing and brush fires. Small numbers of cattle are still kept here, in spite of diseases such as nagana, the cattle disease related to sleeping sickness. This has been a project area for the resettlement programs of recent years. As tsetse fly infestation is reduced, the area offers two possibilities—draining swamplands and irrigating areas previously too dry for most crops.

DRAINAGE

The areas east of the main mountain range, sometimes called the Kagera Basin, drain eventually into the Nile River system. The central uplands and the eastern lakes and swamps are drained by the Nyabarongo River and its main tributaries: the Lukarara, Mwogo, Biruruma, Mukungwa, Base, Nyabugogo, and Akanyaru Rivers. The generalized direction of flow is eastward. Collected runoff moves northward via the Kagera River, which forms much of the eastern boundary, eventually flowing into Lake Victoria.

In these central highlands, east of the Zaire-Nile Crest, erosion is a serious problem, although the average slopes are less steep than those of the west. This is an area of small farms and grazing lands spread over rounded hills between eroded gullies. The various plateaus descend eastward in successive tiers, ending abruptly in central and eastern Rwanda in a series of escarpments. Below, at approximately 914 to 1,524 m (3,000 to 5,000 ft) above sea level, is an irregular basin containing minor elongated valleys, lakes, and swamplands. This basin, less densely populated than the plateaus that border it on three sides, extends from the capital city of Kigali to Lake Rugwero on the southeastern border. Similar swamps and relatively flat lands less desirable for human habitation than the central highlands, extend over hundreds of sq mi in eastern and southeastern border areas.

The Lake Kivu region west of the mountains is part of the large Zaire drainage basin. The western slopes of the Zaire-Nile Divide form a narrow, steep belt of rugged hills and ravines between the mountain crests and the lake, sharply eroded by runoff waters moving swiftly into Lake Kivu or the Ruzizi River Valley. Kivu, with a 1,432 m (4,700-foot) altitude, drains southward into Lake Tanganyika via the Ruzizi, a swift river that descends 701 m (2,300 ft) in less than 120 km (75 mi).

The upper reaches of most rivers in the mountains of the Zaire-Nile Ridge tend to be steep. They become torrents during the rainy season and may be completely dry at other times.

ST. CHRISTOPHER & NEVIS

BASIC FACTS

Official Name: Federation of St. Christopher and Nevis

Abbreviation: SCN

Area: 267 sq km (103 sq mi)

Area—World Rank: 190

Population: 36,738 (1988) 51,000 (2000)

Population—World Rank: 188

Capital: Basseterre

Land Boundaries: Nil

Coastline: 135 km (84 mi)

Longest Distances: N/A

Highest Point: Mount Misery 1,156 m (3,792 ft)

Lowest Point: Sea level

Land Use: 22% arable; 17% permanent crops; 3% meadows and pastures; 17% forest and woodland; 41% other

Shaped like an exclamation mark, St. Kitts and Nevis lie in the northern part of Leeward Islands in the Eastern Caribbean, with Saba and St. Eustatius in the northwest, Barbuda in the northeast, and Antigua to the southwest. They are volcanic islands separated by a channel known as the Narrows some two miles wide. The roughly oval St. Kitts is centered on a mountain range, the highest point of which is Mount Lianuiga (1,156 m, 3,792 ft). The highest peak in Nevis rises to 985 m (3,232 ft).

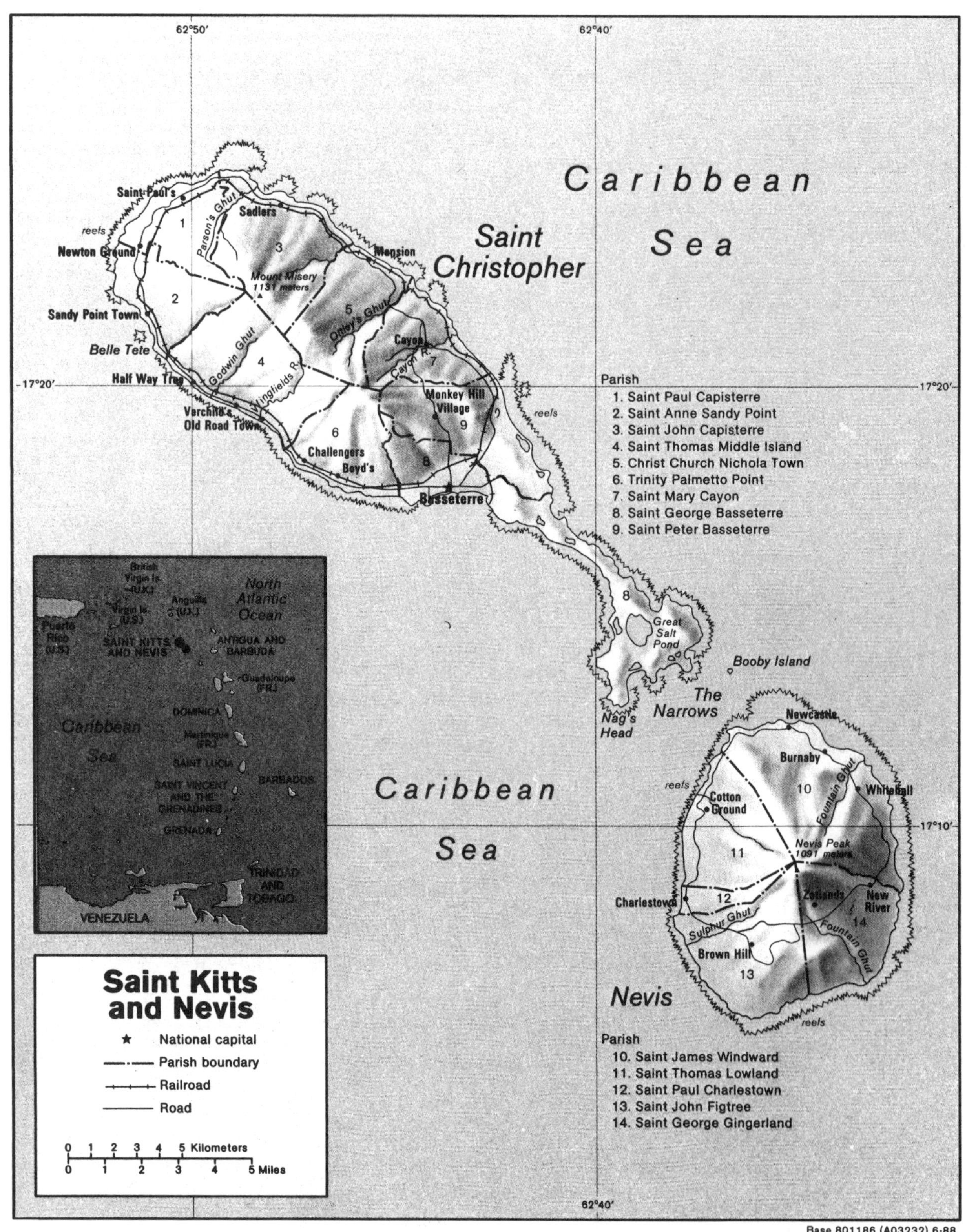

Saint Kitts and Nevis
National capital
Parish boundary
Railroad
Road
0 1 2 3 4 5 Kilometers
0 1 2 3 4 5 Miles
Caribbean Sea
Saint Christopher
Parish
1. Saint Paul Capisterre
2. Saint Anne Sandy Point
3. Saint John Capisterre
4. Saint Thomas Middle Island
5. Christ Church Nichola Town
6. Trinity Palmetto Point
7. Saint Mary Cayon
8. Saint George Basseterre
9. Saint Peter Basseterre
Nevis
Parish
10. Saint James Windward
11. Saint Thomas Lowland
12. Saint Paul Charlestown
13. Saint John Figtree
14. Saint George Gingerland
Saint Paul's
Sadlers
Newton Ground
Mansion
Mount Misery 1131 meters
Sandy Point Town
Belle Tete
Half Way Tree
Verchild's
Old Road Town
Cayon
Monkey Hill Village
Challengers
Boyd's
Basseterre
Parson's Ghut
Godwin Ghut
Ottley's Ghut
Wingfields R.
Cayon R.
reefs
Great Salt Pond
Booby Island
The Narrows
Nag's Head
Newcastle
Burnaby
Whitehall
Cotton Ground
Nevis Peak 1091 meters
Charlestown
Zetlands
New River
Brown Hill
Fountain Ghut
Sulphur Ghut
62°50′
62°40′
17°20′
17°10′
North Atlantic Ocean
British Virgin Is. (U.K.)
Virgin Is. (U.S.)
Anguilla (U.K.)
Puerto Rico (U.S.)
SAINT KITTS AND NEVIS
ANTIGUA AND BARBUDA
Guadeloupe (FR.)
DOMINICA
Martinique (FR.)
SAINT LUCIA
SAINT VINCENT AND THE GRENADINES
BARBADOS
GRENADA
TRINIDAD AND TOBAGO
VENEZUELA
Base 801186 (A03232) 6-88

Area and population		area		population
Islands				1980
Parishes	Capitals	sq mi	sq km	census
Saint Christopher	Basseterre	67.2	174.1	33,881
Christ Church Nichola Town		7.2	18.6	1,989
Saint Anne Sandy Point		4.9	12.8	3,145
Saint George Basseterre		11.1	28.7	14,283
Saint John Capisterre		9.6	24.8	3,163
Saint Mary Cayon		5.8	15.1	3,308
Saint Paul Capisterre		5.3	13.8	2,080
Saint Peter Basseterre		8.0	20.7	2,497
Saint Thomas Middle Island		9.4	24.3	2,255
Trinity Palmetto Point		6.0	15.4	1,161
Nevis	Charlestown	35.7	92.5	9,428
Saint George Gingerland		7.1	18.5	2,295
Saint James Windward		12.0	31.1	1,691
Saint John Figtree		8.2	21.3	2,224
Saint Paul Charlestown		1.4	3.5	1,243
Saint Thomas Lowland		7.0	18.1	1,975
TOTAL		102.9	266.6	43,309

SAINT HELENA

BASIC FACTS

Official Name: St. Helena

Abbreviation: SHN

Area: 412 sq km (159 sq mi)

Area—World Rank: 181

Population: 8,624 (1988) 9,000 (2000)

Population—World Rank: 206

Capital: Jamestown

Land Boundaries: Nil

Coastline: 60 km (37 mi)

Longest Distances: N/A

Highest Point: Queen Mary's Peak 2,060 m (6,760 ft)

Lowest Point: Sea level

Land Use: 7% arable; 7% meadows and pastures; 3% forest and woodland; 83% other

Saint Helena is an island the south Atlantic Ocean about 1,930 km (1,200 mi) from the southwest coast of Africa. Its dependency, Ascension Island, lies 1,131 km (703 mi) northwest of Saint Helena. The islands are volcanic with rugged terrain and small scattered plateaus and cliffs.

The island of Tristan de Cunha, about 2,400 km (1,500 mi) west of Cape Town, also comes under the jurisdiction of Saint Helena. Also in the group are Inaccessible Island (32 km; 20 mi) west of Tristan, the three Nightingale Islands (32 km; 20 mi) to the south, and Gough Island (Diego Alvarez) 350 km (220 mi) to the south.

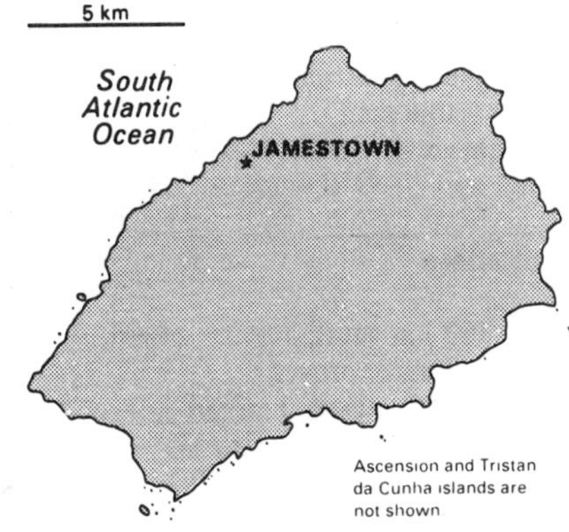

ST. LUCIA

BASIC FACTS

Official Name: St. Lucia

Abbreviation: SLC

Area: 617 sq km (238 sq mi)

Area—World Rank: 171

Population: 136,564 (1988) 185,000 (2000)

Population—World Rank: 167

Capital: Castries

Land Boundaries: Nil

Coastline: 158 km (98 mi)

Longest Distances: N/A

Highest Point: Mount Gimie 950 m (3,117 ft)

Lowest Point: Sea level

Land Use: 8% arable; 20% permanent crops; 5% meadows and pastures; 13% forest and woodland; 54% other

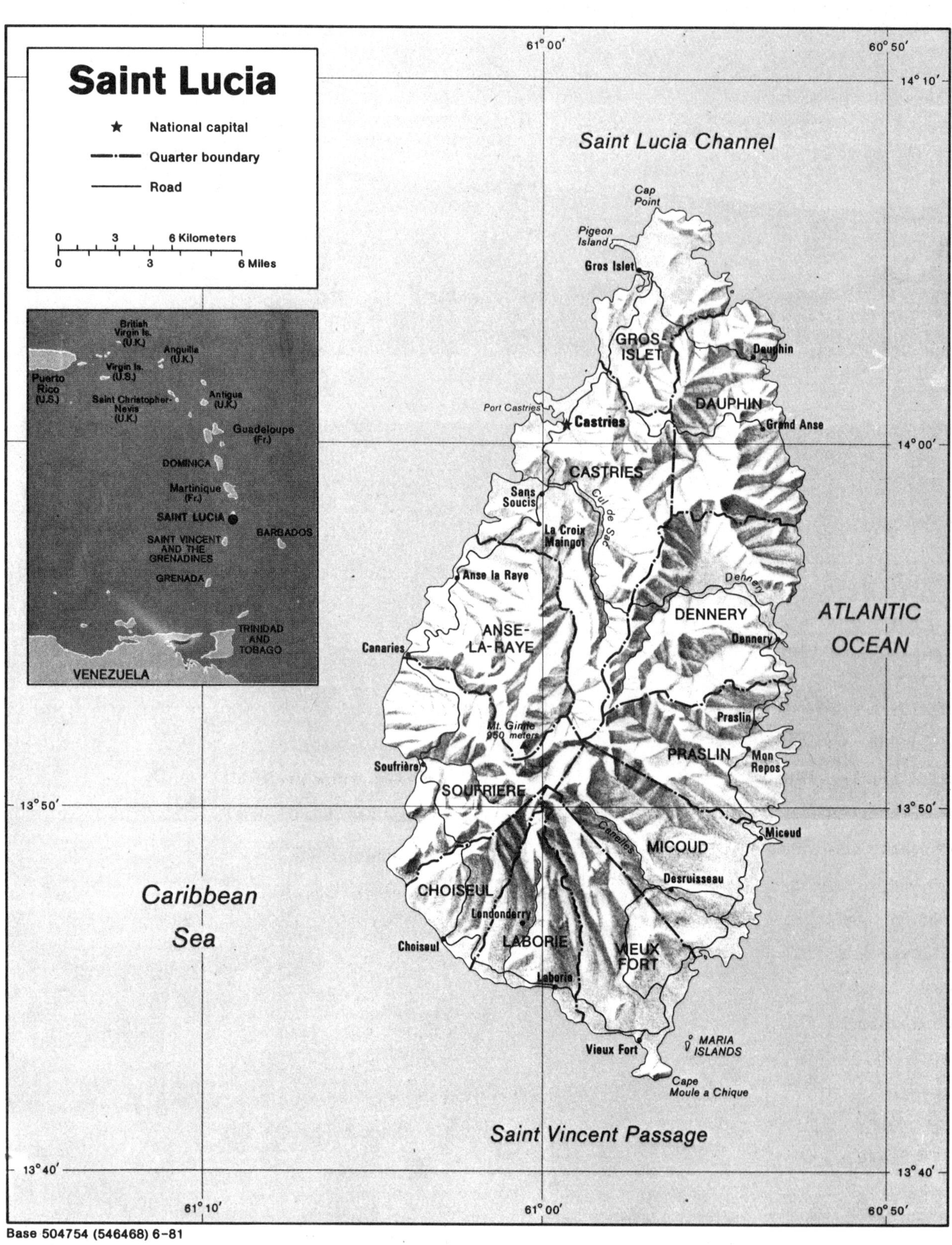

Base 504754 (546468) 6-81

The island of St. Lucia lies in the Windward Islands group in the Caribbean between Martinique and St. Vincent.

The island is of volcanic formation and is relatively hilly, the highest peak, Morne Gimie, being 958 m (3,145 ft).

A number of small rivers flow outward from the central highlands; the principal ones being Dennery, Fond, Piaye, Doree, Canaries, Roseau, and Marquis.

PRINCIPAL TOWNS: Castries (capital), estimated population 52,868 in 1986; other towns Vieux Fort, Soufrière and Gros Islet.

Area and population

		area		population
Quarters	**Capitals**	sq mi	sq km	1986 estimate
Anse-la-Raye	Anse-la-Raye	18.1	46.9	6,111
Canaries	Canaries			2,566
Castries	Castries	30.7	79.5	52,868
Choiseul	Choiseul	12.1	31.3	7,995
Dennery	Dennery	26.9	69.7	11,874
Gros Islet	Gros Islet	39.2	101.5	12,503
Laborie	Laborie	14.6	37.8	8,483
Micoud	Micoud	30.9	80.0	14,678
Soufrière	Soufrière	19.5	50.5	8,972
Vieux Fort	Vieux Fort	16.9	43.8	13,479
TOTAL		238	616	139,529

ST. PIERRE & MIQUELON

BASIC FACTS

Official Name: Department of St. Pierre & Miquelon

Abbreviation: SPM

Area: 242 sq km (93 sq mi)

Area—World Rank: 193

Population: 6,274 (1988) 6,000 (2000)

Population—World Rank: 208

Capital: St. Pierre

Land Boundaries: Nil

Coastline: 120 km (75 mi)

Longest Distances: N/A

Highest Point: Morne de la Grande Montagne 240 m (787 ft)

Lowest Point: Sea level

Land Use: 13% arable; 4% forest and woodland; 83% other

St. Pierre and Miquelon consists of a number of small islands about 25 km (16 mi) from the southern coast of Newfoundland, Canada, in the North Atlantic Ocean. Miquelon and the island of Langlade are linked by a low, sandy isthmus. The terrain is mostly barren rock. Although the archipelago is volcanic in origin, the highest point is under 205 m (673 ft).

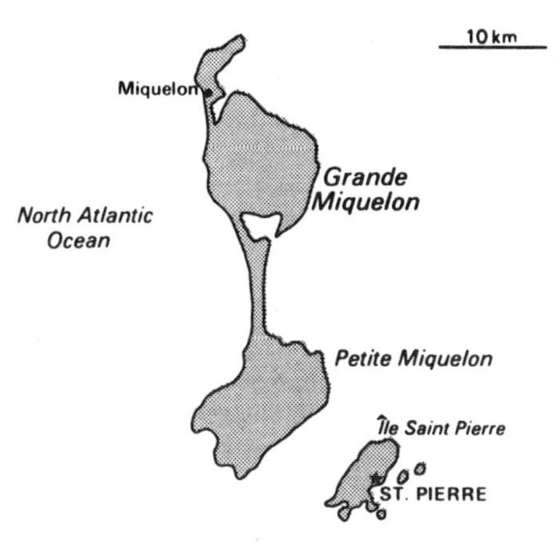

ST. VINCENT & THE GRENADINES

BASIC FACTS

Official Name: St. Vincent & the Grenadines

Abbreviation: SVG

Area: 389 sq km (150 sq mi)

Area—World Rank: 182

Population: 107,425 (1988) 132,000 (2000)

Population—World Rank: 169

Capital: Kingstown

Land Boundaries: Nil

Coastline: 84 km (52 mi)

Longest Distances: N/A

Highest Point: Soufriere 1,234 m (4,048 ft)

Lowest Point: Sea level

Land Use: 38% arable; 12% permanent crops; 6% meadows and pastures; 41% forest and woodland; 3% other

St. Vincent lies at the lower end of the Caribbean chain of Windward Islands some 96 km (60 mi) north of Grenada and 160 km (100 mi) west of Barbados.

Down its whole length the island is dominated by a volcanic range of mountains with four peaks at almost equal distance from each other: Soufriere, Richmond,

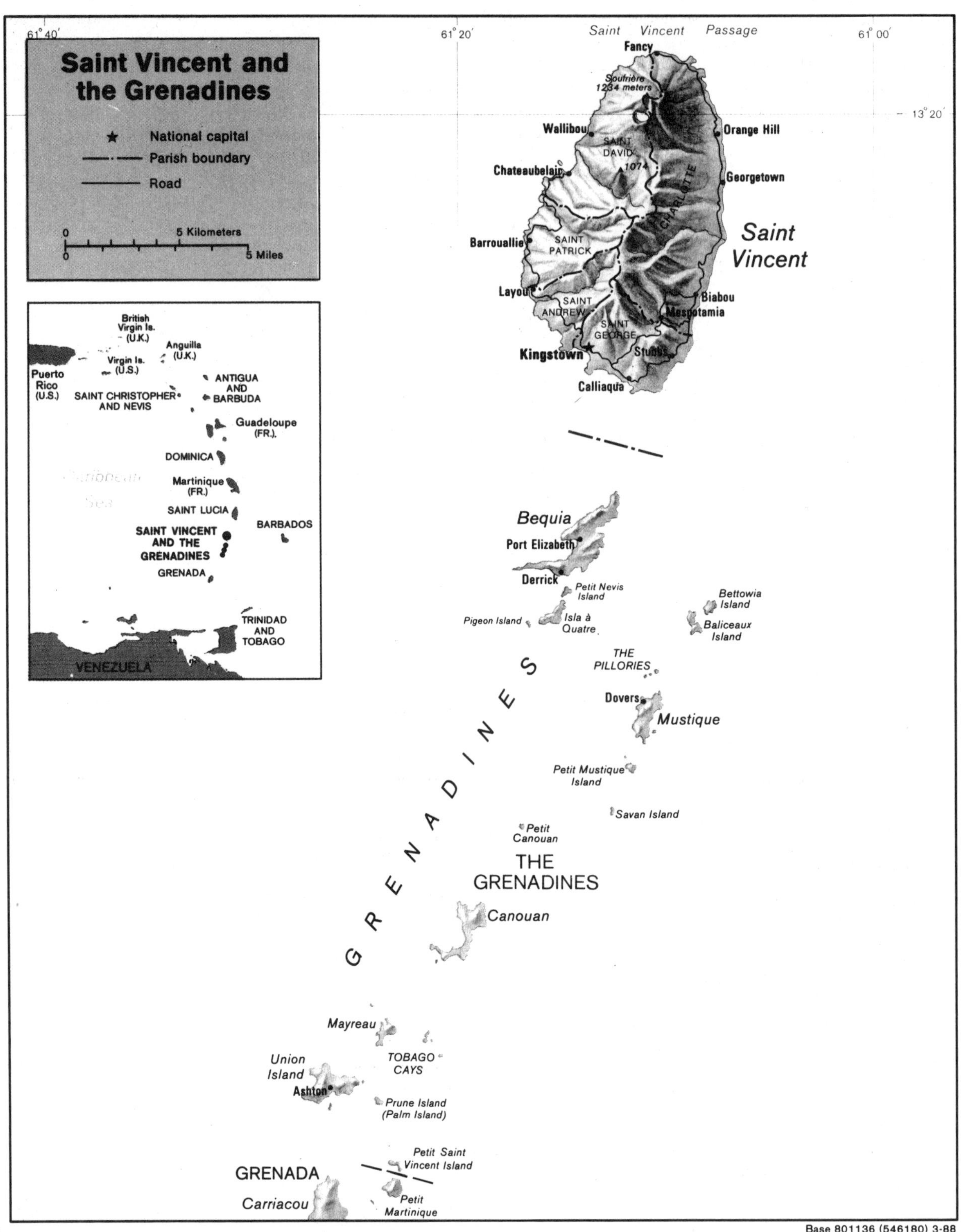

Base 801136 (546180) 3-88

Grand Bonhomme, and St. Andrew. The land slopes gently to the coast on the east in contrast to the rugged terrain on the west. There are many fast flowing rivers.

The Grenadines are a chain of islets between St. Vincent and Grenada. All of them have white beaches noted for their intense beauty and coral reefs with enclosed bays that are ideal for underwater sports. The larger of the Grenadines are Bequia, Canouan, Mustique, and Union while among the smaller are Mayreau, Palm or Prune, Baliceaux, Battawia, and Isle de Quatre. Many of the islets are privately owned and some of them are uninhabited.

Area and population	area		population
Census divisions	sq mi	sq km	1985 estimate
Island of Saint Vincent			
Barrouallie	14.2	36.8	5,187
Bridgetown	7.2	18.6	7,515
Calliaqua	11.8	30.6	19,379
Chateaubelair	30.0	77.7	6,786
Colonaire	13.4	34.7	8,015
Georgetown	22.2	57.5	7,221
Kingstown (city)	1.9	4.9	18,378
Kingstown (suburbs)	6.4	16.6	9,570
Layou	11.1	28.7	6,123
Marriaqua	9.4	24.3	9,341
Sandy Bay	5.3	13.7	3,186
Saint Vincent Grenadines			
Northern Grenadines	9.0	23.3	5,263
Southern Grenadines	7.5	19.4	2,784
TOTAL	150.3	389.3	108,748

SAN MARINO

BASIC FACTS

Official Name: Republic of San Marino

Abbreviation: SNM

Area: 61 sq km (24 sq mi)

Area—World Rank: 205

Population: 22,986 (1988) 23,000 (2000)

Population—World Rank: 196

Capital: San Marino

Land Boundaries: Italy 34 km (21 mi)

Coastline: Nil

Longest Distances: 13.1 km (8.1 mi) NE-SW; 9.1 km (5.7 mi) SE-NW

Highest Point: Monte Titano 739 m (2,425 ft)

Lowest Point: 53 m (174 ft)

Land Use: 17% arable; 83% other

San Marino is the third smallest country in Europe. It is a landlocked state completely surrounded by Italy and located about 24 km (15 mi) southwest of Rimini between the province of Emilia and the region of the Marches.

The capital, San Marino, rests just below the summit of Mount Titano 842 m (2,763 ft). Commercial activity is centered on Borgo Maggiore some 183 m (600 ft) below San Marino and connected to it by a 2.4 km (1.5 mi) winding road. Serravalle is the only other town.

The republic is topographically coextensive with Mount Titano which has three pinnacles, each crowned by old fortifications, one on the north by a castle and the other two by towers. The summit of Mount Titano commands a panoramic view of the Adriatic, only 19 km (12 mi) away. The only agricultural land is at the base of the mountain.

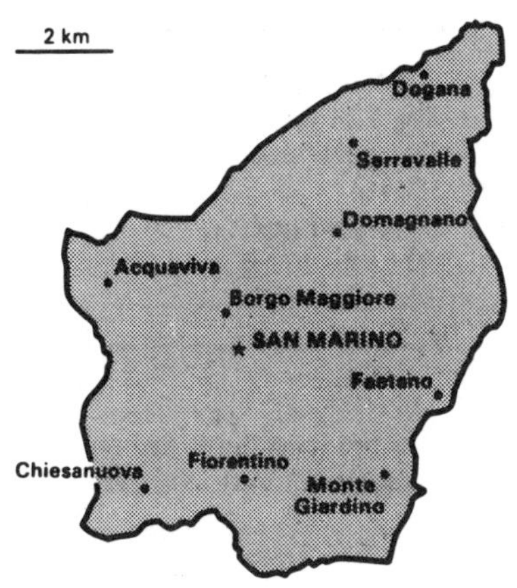

Area and population		area		population
Castles	**Capitals**	sq mi	sq km	1986 estimate
Acquaviva	Acquaviva	1.88	4.86	1,148
Borgo Maggiore	Borgo	3.48	9.01	4,341
Citta	San Marino	2.74	7.09	4,201
Chiesanuova	Chiesanuova	2.11	5.46	712
Domagnano	Domagnano	2.56	6.62	1,840
Faetano	Faetano	2.99	7.75	750
Fiorentino	Fiorentino	2.53	6.56	1,477
Montegiardino	Montegiardino	1.28	3.31	561
Serravalle/ Dogano	Serravalle	4.07	10.53	6,941
TOTAL		23.63*	61.19	21,971

SAO TOME & PRINCIPE

BASIC FACTS

Official Name: Democratic Republic of Sao Tome & Principe

Abbreviation: STP

Area: 1,001 sq km (396 sq mi)

Area—World Rank: 163

Population: 117,430 (1988) 157,000 (2000)

Population—World Rank: 170

Capital: Sao Tome

Land Boundaries: Nil

Coastline: 209 km (130 mi)

Longest Distances: Sao Tome: 49 km (30 mi) NNE-SSW and 29 km (18 mi) ESE-WNW; Principe: 21 km (13 mi) SSE-NNW and 15 km (9 mi) ENE-WSW

Highest Point: Sao Tome Peak 2,024 m (6,640 ft)

Lowest Point: Sea level

Land Use: 1% arable; 36% permanent crops; 1% meadows and pastures; 62% other

Sao Tome & Principe is located in the Gulf of Guinea off the northern coast of Gabon. Sao Tome accounts for 862 sq km (333 sq mi) and Principe for 101 sq km (39 sq mi). Sao Tome is located about 201 km (125 mi) off the northern coast of Gabon. Principe is located about 442 km (275 mi) off the northern coast of Gabon. In addition, there are two small islets, Rolas, crossed by the equator, and Pedras Tinhosas.

Area and population

Islands / Districts	Capitals	area sq mi	area sq km	population 1984 estimate
Principe	São António	55	142	5,671
Paguê	Príncipe	55	142	5,671
São Tomé		332	859	98,693
Aqua Grande	São Tomé	7	17	34,997
Cantagalo	Santana	46	119	11,270
Caué	São João Angolares	103	267	4,972
Lemba	Neves	88	229	8,537
Lobata	Guadalupe	41	105	12,717
Mé-zóchi	Trinidade	47	122	26,200
TOTAL		386	1,001	104,364

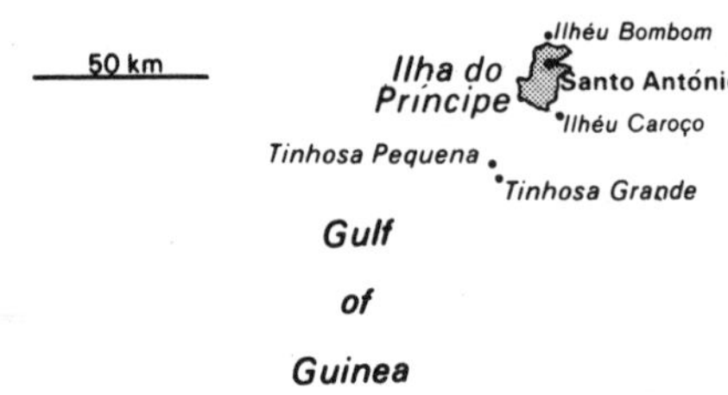

Both islands are active volcanoes and have many craters and lava flows. The highest point in Sao Tome rises 2,024 m (6,640 ft); there are 10 peaks over 1,067 m (3,500 ft). Principe, with a larger plateau area than Sao Tome, rises 948 m (3,110 ft).

SAUDI ARABIA

BASIC FACTS

Official Name: Kingdom of Saudi Arabia

Abbreviation: SAB

Area: 2,240,000 sq km (865,000 sq mi)

Area—World Rank: 12

Population: 15,452,123 (1988) 19,824,000 (2000)

Population—World Rank: 56

Capital: Riyadh

Boundaries: 7,027 km (4,336 sq mi); Jordan 744 km (462 mi); Iraq 895 km (556 mi); Kuwait 163 km (101 mi); Qatar 67 km (42 mi); United Arab Emirates 586 km (364 mi); Oman 676 km (420 mi); South Yemen 830 km (516 mi); Yemen Arab Republic 628 km (390 mi)

Coastline: Persian Gulf 549 km (341 mi); Red Sea and Gulf of Aqaba 1,889 km (1,174 mi)

Longest Distances: 2,295 km (1,426 mi) ESE-WNW; 1,423 km (884 mi) NNE-SSW

Highest Point: 3,133 m (10,279 ft)

Lowest Point: Sea level

Land Use: 1% arable; 39% meadows and pastures; 1% forest and woodland; 59% other

Saudi Arabia occupies approximately 80 percent of the Arabian Peninsula, an area roughly equivalent to the United States east of the Mississippi. Saudi Arabia possesses more undefined than defined boundaries, and as a result the exact size of the country is unknown. Less than 1 percent of the total area is suitable for settled agriculture, and population distribution varies greatly among the towns of the eastern and western coastal areas, the densely populated interior oases, and the vast, almost empty deserts.

The Arabian Peninsula is a unified plateau sloping slightly toward the east, with a major fault line along the

Red Sea coast. In the extreme west a line of rugged mountains parallels the Red Sea coast and forms the watershed for the peninsula. East of this range the land is relatively even; the gradual drop toward the Persian Gulf is broken only by the escarpment of the low Tuwaiq mountains (Jabal Tuwaiq), which extend along a west-facing crescent north and south of Riyadh.

Most of the surface is sand covered, forming the great deserts of the Nafud, the Dahna, and the Rub al Khali (Empty Quarter). Outside these deserts the surface is gravel or, in limited areas in the west-central portion, consists of jumbled beds of lava. The climate of most of the peninsula is debilitatingly hot and dry. In Saudi Arabia only the Asir highlands along the Red Sea receive enough rain to permit a degree of nonirrigated cultivation.

The Arabian Peninsula is an ancient massif composed of stable crystalline rock whose present geologic structure was developing concurrently with the formation of the Alps. Geologic movements caused the entire mass to tilt eastward and the western and southern edges to tilt upward. In the valley created by the fault, called the Great Rift, the Red Sea was formed. The Great Rift runs from the Mediterranean along both sides of the Red Sea and southward through Ethiopia and the lake country of East Africa, gradually disappearing in the area of Mozambique, Zambia, and Zimbabwe. Scientists analyzing color photographs taken by American astronauts on the joint United States-Soviet space mission in July 1975 have detected a vast fan-shaped complex of cracks and fault lines extending north and eastward from the region of the Golan Heights. These are believed to be the final portion of the Great Rift and are presumed to be the result of the slow rotation of the Arabian Peninsula counterclockwise in a way that will in approximately 10 million years close off the Persian Gulf and make it a lake.

On the Peninsula the eastern line of the Great Rift fault is visible in the steep and, in places, very high escarpment, which parallels the Red Sea between the Gulf of Aqaba and the Gulf of Aden. The eastern slope of this escarpment is relatively gentle, dropping to the exposed shield of the extremely ancient landmass that existed before the faulting occurred. A second lower escarpment, the Jabal Tuwaiq, runs roughly north-south through the area of Riyadh. East of this escarpment the shield is covered by layers of sediment increasing in thickness toward the Persian Gulf. Oil is found in the lower strata of these sedimentary layers.

The northern half of the region of the Red Sea escarpment is known as the Hejaz and the more rugged southern half as Asir. In the south a coastal plain, the Tihama Lowlands, rises gradually from the sea to the mountains. Asir extends southward into the borders of mountainous Yemen (Sana). The central plateau, the Najd, extends eastward to the Jabal Tuwaiq and a short distance beyond. A long, narrow strip of sand desert known as the Dahna separates the Najd from eastern Arabia, which slopes gradually eastward to the low-lying sandy coast along the Persian Gulf. North of the Najd a larger sand desert, the Nafud, isolates the heart of the peninsula from the steppes of northern Arabia. South of the Najd lies the largest sand desert in the world, the Rub al Khali.

Natural Regions

Hejaz and Asir

The western coastal escarpment can be considered as two mountain ranges separated by a gap in the vicinity of Mecca. The northern range in the Hejaz seldom exceeds 2,133 m (7,000 ft) and the elevation gradually decreases toward the south to about 609 m (2,000 ft) around Mecca. The mountain wall is rugged, dropping abruptly to the sea and having few and intermittent coastal plains. There is an almost total absence of natural harbors along the Red Sea coast. The western slopes have been stripped of soil by the erosion of infrequent but turbulent rainfalls that have fertilized the plains to the east. The eastern slopes are less steep and are marked by wadis that mark the courses of ancient rivers and still lead the rare rain down into the plains. Scattered oases, drawing water from springs and wells in the vicinity of these wadis, permit some settled agriculture. Of these the largest and most important is Medina.

South of Mecca the mountains are higher, exceeding 2,438 m (8,000 ft) in a number of places; some peaks reach nearly 3,048 m (10,000 ft). The rugged western face of the escarpment drops rather steeply to the coastal plain, the Tihama Lowlands, which averages only about 64 km (40 mi) in width. Along the seacoast is a salty tidal plain of limited agricultural value, backed by potentially rich alluvial plains. The relatively well-watered and fertile upper slopes and the mountains behind are extensively terraced to make possible the maximum use of the land. An average of about twenty inches of rainfall a year permits the cultivation of grain, coffee, qat (a mildly narcotic plant), fruits, and vegetables. The top of the mountain ridge is covered in places by narrow strips of the only natural forest in the country, mainly juniper. Luxuriant undergrowth gives these strips—many only a few dozen ft wide—the character of a tropical rain forest.

The eastern slope of the mountain range in Asir is gentle, melding into a plateau region that drops gradually into the Rub al Khali. Although rainfall is infrequent in this area, a number of fertile wadis, of which the most important are the Bishah and the Tathlith, make oasis agriculture possible on a relatively large scale. A number of extensive lava beds (*harrat*) scar the surfaces of the plateaus east of the mountain ranges in the Hejaz and Asir and give evidence of fairly recent volcanic activity. The largest of these beds is Khaybar, north of Medina.

The Najd

East of the Hejaz and Asir lies the great plateau area of the Najd, the birthplace of the country. This region is mainly rocky plateau interspersed by small, sandy deserts and isolated mountain clumps. The best known of the mountain groups is the Jabal Shammar, northwest of Riyadh and just south of the Nafud desert. This is the home of the pastoral Shammar tribes, which under the leadership of Muhammad ibn Rashid were the most implacable foes of the Sauds in the nineteenth and early twentieth centuries. Their capital was the large oasis of Hail, now a flourishing urban center.

Note: "Jiddah" is the Board of Geographic Names (BGN) approved spelling of Jidda. Also, "Madinah" and "Makkah" are BGN approved spellings for Medina and Mecca, respectively.

PRINCIPAL TOWNS (population at 1974 census)

Riyadh (royal capital)	666,840	Hufuf	101,271
		Tabouk	74,825
Jeddah (administrative capital)	561,104	Buraidah	69,940
		Al-Mobarraz	54,325
Makkah (Mecca)	366,801	Khamis-Mushait	49,581
Ta'if	204,857	Al-Khobar	48,817
Al-Madinah (Medina)	198,186	Najran	47,501
Dammam	127,844	Ha'il (Hayil)	40,502

In conformity with the peninsula as a whole, the plateau slopes toward the east from an elevation of about 1,371 m (4,500 ft) in the west to about 762 m (2,500 ft) at its easternmost limit. A number of wadis cross the region generally in an eastward direction from the Red Sea escarpment toward the Persian Gulf. There is little pattern to these remains of ancient riverbeds; the most important of them are the Rumma, the Surra, and the Dawasir. Rainfall in the region averages less than four inches a year, and several years may elapse between rains. When rain does occur, it may be torrential and cause the wadis to flood, in some cases doing serious damage to settlements and making travel impossible until the water disappears into the gravel and sand base.

The heart of the Najd is the area of the Jabal Tuwaiq, an arc-shaped ridge whose steep west face rises between 121 and 242 m (400 and 800 ft) above the plateau. Many oases exist in this area, which is one of the most densely populated in the country. The most important of these oases are Buraydah, Unayzah, Riyadh, Al Kharj, and Aflaj. Outside this oasis area the Najd is sparsely populated. Sabkah, large salt marshes, are scattered throughout the area.

Northern Arabia

The area north of the Nafud desert is geographically a part of the Syrian desert. It is an upland plateau with a surface of dark-colored rock and gravel and scored by numerous wadis, most trending northeastward toward Iraq. This area, known as Badiet ash Sham, is covered with grass and scrub steppe vegetation and is extensively used for pasture by nomadic and seminomadic herders. The most significant feature of the area is the Wadi Sirhan, a large basin as much as 304 m (1,000 ft) below the surrounding plateau that is the remnant of an ancient inland sea. For thousands of years some of the most heavily traveled caravan routes between the Mediterranean and the central and southern peninsula have passed through the Wadi Sirhan. The most important oases in the area are Jawf and Sakaka, just north of the Nafud.

Eastern Arabia

East of the Dahna lies the rocky Summam plateau, about 75 km (120 mi) wide and dropping in elevation from about 396 m (1,300 ft) in the west to about 243 m (800 ft) in the east. The area is generally barren and has a highly eroded surface of ancient river gorges and isolated buttes.

Farther east the terrain changes abruptly to the flat lowlands of the Persian Gulf coastal plain. This area, about 160 km (100 miles) in width, is generally featureless and covered with gravel or sand. In the north is the gravelly Dibdiba plain and in the south the Jafura sand desert, which reaches the gulf in the vicinity of Abqaiq and Dhahran and merges with the Rub al Khali at its southern end. The coast itself is extremely irregular as sandy plains, marshes, and salt flats merge almost imperceptibly with the sea. As a result, the land surface is unstable; water rises in places almost to the surface, and the sea is shallow and full of shoals and reefs for an extended distance offshore. Only the construction of long moles at Ras Tanura has opened the Saudi coast on the Persian Gulf to seagoing tankers.

Eastern Arabia is sometimes still called Hasa after the great oasis of that name, one of the most potentially fertile areas of the country. Hasa, the largest oasis in the country, actually comprises two neighboring oases, including the town of Hufuf.

Area and population

Regions / Administrative Districts	Capitals	area: sq mi	area: sq km	population: 1985 estimate
al-Gharbīyah (Western)	—	...	...	3,043,189
al-Bāḥah	al-Bāḥah	...	...	...
al-Madīnah	Medina (al-Madīnah)	...	...	...
Makkah	Mecca (Makkah)	...	...	...
al-Janūbiyah (Southern)	—	...	...	625,017
ʿAsīr	Abha	...	...	...
Jīzān	Jīzān	...	...	...
Najrān	Najrān	...	...	...
ash-Shamālīyah (Northern)	—	...	...	679,476
al-Jawf	Sakākah	...	...	...
al-Ḥudūd ash-Shamālīyah (Northern Borders)	ʿArʿar	...	...	...
al-Qurayyāt	an-Nabk	...	...	...
Tabūk	Tabūk	...	...	...
ash-Sharqīyah (Eastern)	—	...	...	3,030,765
ash-Sharqīyah (Eastern)	ad-Dammām	...	...	...
al-Wūsṭā (Central)	—	...	...	3,632,092
Ḥā'il	Ḥā'il	...	...	...
al-Qaṣīm	Buraydah	...	...	...
ar-Riyāḍ	Riyadh (ar-Riyāḍ)	...	...	...
TOTAL		865,000	2,240,000	11,010,539

The Great Deserts

Three great deserts isolate the Najd from north, east, and south as the Red Sea escarpment does from the west. In the north the Nafud—sometimes called the Great Nafud because nafud simply means desert—covers about 64,749 sq km (25,000 square miles) at an elevation of about 914 m (3,000 ft). Longitudinal dunes—scores of miles in length, as much as 91 m (300 ft) high, and separated by valleys as much as ten miles wide—characterize the Nafud. Iron oxide gives the sand a reddish tint, particularly when the sun is low. Within the area are several watering places, and winter rains bring up short-lived but succulent grasses that permit nomadic herding during the winter and spring.

Stretching more than 643 km (400 mi) south from the Nafud in a narrow arc only about 48 km (thirty mi) wide is the Dahna, a narrow band of sand mountains also called the river of sand. Like the Nafud, its sand tends to be reddish in color, particularly in the north where it also shares with the Nafud the longitudinal structure of sand dunes. The Dahna also furnishes the beduin with winter and spring pasture, although water is scarcer than in the Nafud.

The southern portion of the Dahna curves westward following the arc of the Jabal Tuwaiq. At its southern end it merges with the Rub al Khali, one of the most forbidding sand deserts in the world and, until the 1950s, one of the least explored. The topography of this huge area, covering more than 647,490 sq km (250,000 sq mi), is varied. In the west the elevation is about 609 m (2,000 ft) and the sand is fine and soft; in the east the elevation drops to about 183 m (600 ft) and much of the surface is covered by relatively stable sand sheets and salt flats. In places, particularly in the east, longitudinal sand dunes prevail; elsewhere sand mountains as much as 304 m (1,000 ft) in height form complex patterns. Most of the area is totally waterless and uninhabited except for a few wandering beduin tribes.

SENEGAL

BASIC FACTS

Official Name: Republic of Senegal

Abbreviation: SGL

Area: 196,722 sq km (75,955 sq mi)

Area—World Rank: 82

Population: 7,281,022 (1988) 9,765,000 (2000)

Population—World Rank: 82

Capital: Dakar

Boundaries: 3,101 km (1,927 mi); Mauritania 813 km (505 mi); Mali 418 km (260 mi); Guinea 330 km (205 mi); Guinea-Bissau 338 km (210 mi); Gambia 756 km (470 mi)

Coastline: 446 km (277 mi)

Longest Distances: 690 km (429 mi) SE-NW; 406 km (252 mi) NE-SW

Highest Point: 581 m (1,906 ft)

Lowest Point: Sea level

Land Use: 27% arable; 30% meadows and pastures; 31% forest and woodland; 12% other

Senegal extends inland from a 483 km (300-mi) coastline on the Atlantic Ocean, which is marked near its center by the Cap Vert peninsula—the westernmost point in Africa. Most of its land area, is a flat plain—a western segment of the broad savanna that extends across the continent at the southern edge of the Sahara.

Most of the country, including northeastern Senegal—is less than 91 m (300 ft) above sea level. The predominant landscape is a flat expanse of sparse grasses and woody shrubs, remarkable only for the near-total absence of natural landmarks or major changes in elevation. Broken terrain and steep slopes are found only in the extreme southeast, where a few ridges stand above 396 m (1,300 ft). Because gradients are so shallow, all the largest rivers—the Sénégal, Sine, Saloum, and Casamance—are sluggish, marsh-lined streams emptying into ill-defined estuaries along the Atlantic Ocean.

Extensive riverine areas have been converted to farmland, especially in the Sine and Saloum River basins lying east and southeast of the Cap Vert peninsula. Beyond these areas, however, most of the land has little potential except as pasturage.

MAJOR GEOGRAPHIC AREAS

Most of Senegal is an ancient sedimentary basin. Most ridges or other highly visible physiographic features have long since been smoothed into a gently undulating plain by erosion, and ancient valleys or depressions have been filled with alluvium and windblown sand. Volcanic action created the Cap Vert peninsula and the nearby islets. Metamorphic and igneous rock formations appear only in southern Sénégal Oriental Region, in the southeastern part of the country. Except for several minor hills near Thiès, a few mi inland from the Cap Vert peninsula, the southeast is the only area with elevations of more than 91 m (300 ft) above sea level; even there, only a few ridges exceed 396 m (1,300 ft).

Contrasts between geographic regions are primarily functions of climate and the availability of surface and subsurface water supplies. The lower river basins in the west are only a few dozen ft above sea level. Above their floodplains, variation in elevation is so minor that it has little influence on the zonal differences in climate and vegetation. In most inland areas the absence of distinctive terrain features makes it difficult to establish regional or zonal borders except in a most general way. The few minor exceptions include a narrow zone along the coast, which is marked by dunes separating small pools and estuaries.

Coastal Belt

Much of the coastal belt north of the Cap Vert peninsula is covered by small swamps or pools separated by very old dunes that were originally built up by wave and wind action. The belt is still known as the Cayor, a name retained since the sixteenth century, when kings of the Wolof ethnic group dominated this area. This northern

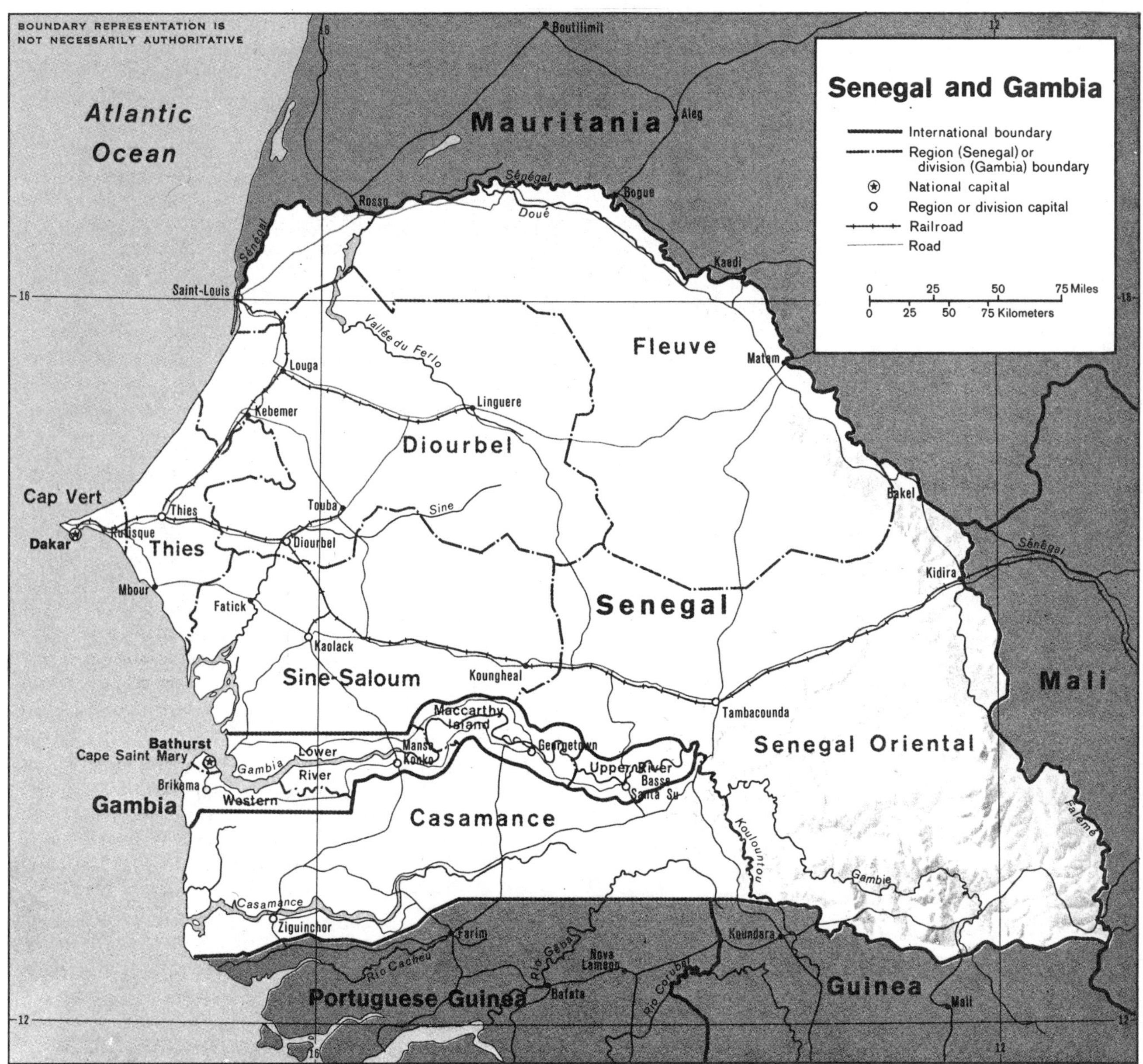

Base 500711 11-72

half of Senegal's shoreline sweeps in a smooth, uninterrupted curve southwestward from the estuary of the Sénégal River to the westernmost point of land on the peninsula, northwest of Dakar. The associated coastal belt of dunes and swampy areas extends inland as much as fifteen mi. The combined influence of the Canary Current, which moves southwestward along the coast, the northeast trade winds, and heavy surf have formed a wide, sandy beach backed by dunes. Some dunes are as much as 30 m (100 ft) high and are interspersed with depressions of clay soil. No streams reach the sea in this belt, but during the short rainy season temporary watercourses empty into these depressions to form a series of freshwater swamps or lakes. During the dry season these damp, fertile bottomlands are green oases of luxuriant growth, surrounded by subdesert conditions. Near the Cap Vert peninsula the beach dunes are not continuous; some of the marshes and lagoons are frequently invaded by the sea and are too salty to be useful for crop production.

South of Dakar the coastal belt narrows considerably. Behind the beaches the ground rises slightly in a series of low, wooded hills. Between them a number of short, seasonal streams find their way to the sea, often through muddy, mangrove-lined marshes and lagoons. North of the estuary of the Saloum River, the coastal belt becomes a wide maze of meandering creeks, channels, and flat, swampy islands, frequently choked by mangrove thickets. Southward, past The Gambia and the Casamance River estuary, the coast appears to have settled during prehistoric times. Creeks and estuaries are clogged by silt and sand. This is an area of salt flats and is unsuitable for agricultural use.

Sénégal River Valley

From the Mali border, where it enters Senegal, the Sénégal River flows in a great arc to the sea, bringing the wealth of its waters to a narrow strip of thirsty land where less than a dozen inches of rain fall annually in a few torrential downpours. The north bank of the river's main channel forms the international boundary with Mauritania. Between low, parched ridges, the river meanders sluggishly through a green, alluvial floodplain, which is 16 to 19 km (ten to twelve mi) wide in its upper reaches and over 64 km (40 mi) wide where the river approaches the sea.

No streams enter the valley, except short washes that flow for a little while after the infrequent rains. The floodplain itself, however, is broken by many marshes and branching channels. Downstream from Dagana, where the river approaches the sea, its various channels form an extensive network in the silt-choked plain, and when the river is high they distribute its waters over a wide area resembling a delta. A feature of the middle reach of the valley above Dagana is the Ile à Morfil—a narrow island several hundred mi long between the river's main channel and the sluggish Doué channel on the opposite side.

Human life in this region is governed by the river's annual flood. Rising water spreads through the whole system of channels, sloughs, and adjacent lowlands until most of the valley is a sheet of water from which the tops of trees appear as green patches; villages stand out as isolated islands. Crops are sown as the flood recedes. Soon after, as the long dry season sets in, the region offers the contrast of a green ribbon winding through a countryside burned brown by the harmattan, a persistent wind from the Sahara.

Western Plains

Except for the dunes in the coastal belt, the only noteworthy elevation in the western plains is a small group of hills that rise to about 61 m (200 ft) above sea level northwest of the town of Thiès. The lowlands extending southeastward from Thiès to Kaolack have developed into an important agricultural area, producing groundnuts (peanuts) and other food crops. Soils are loose and sandy and are easily depleted. The original vegetative cover has been almost completely cleared. During the dry season the land in this area is almost barren except in a few depressions where groundwater is close to the surface. Where the earth is not altogether bare, it is covered with dry grass, dead stubble, stunted bushes, and scattered trees. Shortly after the first rains in June, however, the land comes to life. New grass sprouts, and fields of groundnuts, interspersed with millet and other food crops, form a green landscape.

The Ferlo

An inland continuation of the western plains, the Ferlo is a generally featureless expanse of savanna covering most of the area between the Sénégal River, along the border with Mauritania, and The Gambia. The annual rainfall is so scanty and the thin, sandy soil so porous that water can be obtained during the dry season only from wells located along the few shallow depressions that cross the flat landscape. Nomads move through the area during the short rainy season, when sparse forage is available for their animals. They leave as the rainy season ends. Dried tufts of sun-yellowed grass, scrub, and thorn trees dominate the scene during the rest of the year.

Casamance

Casamance is partially separated from most of Senegal by The Gambia and is slightly different in terms of relief, rainfall, and vegetation. The valleys inland from the coastal belt are flat and subject to flooding each year, but they are separated by areas high enough to escape annual inundations. In the southeastern corner of the region a divide that rises south of the Casamance and Gambie river basins reaches a 61 m (200-foot) elevation. The Casamance River is tidal and salty as far as 120 km (75 mi) inland. As a consequence, low areas along the tidal reaches of the river and along its lower tributaries are often infertile and can support only a limited variety of vegetation unless protected by dikes. Rainfall is heavier, and the rainy season is longer than in the region north of The Gambia. The town of Ziguinchor, for example, usually has three times as much rain as Dakar. Thus, although

Area and population

Regions	Capitals	area sq mi	area sq km	population 1984 estimate
Dakar	Dakar	212	550	1,380,700
Diourbel	Diourbel	1,683	4,359	501,000
Fatick	Fatick	3,064	7,935	506,500
Kaolack	Kaolack	6,181	16,010	741,600
Kolda	Kolda	8,112	21,011	517,600
Louga	Louga	11,270	29,188	493,900
Saint-Louis	Saint-Louis	17,038	44,127	612,100
Tambacounda	Tambacounda	23,012	59,602	355,000
Thiès	Thiès	2,549	6,601	837,900
Ziguinchor	Ziguinchor	2,834	7,339	361,000
TOTAL		75,955	196,722	6,307,300

terrain differences are minimal, vegetation in the coastal area of Casamance is quite different from that in the majority of Senegal's land area and includes mangrove, thick forest, and oil palms. This vegetation shades into wooded or open savanna in the central and eastern parts of the region, where soils are poor and the population is sparse.

The East

A poorly defined plain extends southeastward from the Ferlo to the borders of Mali and Guinea. It straddles a low, poorly defined north-south divide separating the watershed of the Gambie on the west from the narrower Falémé River basin on the east. West of the Falémé confluence with the Sénégal River, ill-defined high ground extends westward to form another minor divide between the Ferlo and the northern half of the Gambie basin. Except for the Falémé River and a few short tributaries, the only perennial watercourses in this area are tributaries of the Gambie. Rainfall and other climatic conditions are about the same as in the Ferlo, except that the annual rainfall increases substantially toward the southern end of the plain, and temperature ranges are wider. Most of the region is poor, seasonal pastureland, dotted with acacia and scrub growth. There are many areas of infertile laterite. The region is thinly populated and, except for the Ferlo, is the least developed part of Senegal.

DRAINAGE

The Sénégal River, which marks the northern border of the country for a considerable distance, carries water from wetter areas farther south into the subarid fringe of the Sahara and eventually to the Atlantic Ocean near Saint-Louis. Over 4,023 km (2,500 mi) long, the Sénégal rises in neighboring Guinea, where sixty to eighty inches of rain fall each year. Its main upper tributary is known as the Bafing River until it is joined in eastern Mali by the Bakoy River, forming the Sénégal River. As it enters Senegal, it is joined from the south by the Falémé River, which also rises in Guinea. Downstream there are no falls or rapids, and the average gradient is only a few inches per mile.

The river's annual floods are fed mainly by the heavy seasonal rains that begin in April in the mountains of Guinea, but flood crests do not arrive in Senegal until mid-September and thereafter take about six weeks to reach the sea. Annual maximums average about forty-five ft above minimums at the Senegal-Mali border, decreasing to well under twelve ft when the crest is within 160 km (100 mi) of the mouth of the river. There, heavy stream deposition and reduced gradient cause the floodwaters to break into a network of distributaries, drawing off so much volume that crests are scarcely perceptible in the river's estuary at Saint-Louis.

During normal low water, ocean tides range nearly 483 km (300 mi) upstream; the river is salty for half that distance, and the system of distributaries becomes a maze of brackish swamps and backwaters. For a month or so during the higher stages of the flood, however, the salty water is forced seaward, and the system is filled with fresh water. In recent years this ebb and flow has been checked at some points by dikes that are opened to admit the fresh water and are later closed to impound it for use during the dry season and to exclude any advancing salt water. The most important arrangement of this kind includes a dam and a gate on the Taoué channel, 160 km (100 mi) inland, not far from Dagana, which control a shallow lake, extending southward about 80 km (50 mi) and averaging about 12 km (8 mi) in width, known as the Lac de Guiers. At highest level the lake waters reach another 64 to 80 km (40 to 50 mi) southeastward into the Ferlo valley, replenishing the water table in an otherwise parched area.

Above Saint-Louis the river forms an estuary that is divided southward by a long sandbar. Known as the Langue de Barbarie, this barrier is not more than half a mile wide and is so low that waves wash across it in some places during rough seas. Throughout the last century its length has varied between 16 and 32 km (10 and 20 mi). At its extremity, where the river meets the open sea, there is a dangerous shifting bar that sometimes prevents the passage of ships for several weeks at a time.

The Saloum River and its major affluent, the Sine River, are sluggish streams feeding into an extensive tidal swamp just north of The Gambia. Only the lower reaches carry water all year, and these are brackish, as the tides penetrate far up the various channels through the swamp.

The middle reaches of the Gambie are within Senegal. Rising in the Fouta Djallon, a highlands area in Guinea, the Gambie enters the southeastern corner of Senegal, swings northwestward, and winds for about 322 m (200 mi) to the border with The Gambia. In addition to a few intermittent tributaries on its northern flank, the river receives the flow of a perennial river, the Koulountou, which runs north from Guinea to join it near the Gambian border.

The Casamance River in southern Senegal drains a narrow basin, less than 32 km (20 mi) wide, between The Gambia and the border with Portuguese Guinea. It is sluggish and swampy for most of its 322 km (200-mi) length. Its main tributary is the Songrougrou River, which joins it from the north about 104 km (sixty-five mi) from the sea. Downstream from this point, the river is a broad estuary, 9 km (six mi) wide at the mouth, and there are several smaller outlets separated by flat islands. Tides penetrate about 160 m (100 mi) inland.

SEYCHELLES

BASIC FACTS

Official Name: Republic of Seychelles

Abbreviation: SEY

Area: 453 sq km (175 sq mi)

Area—World Rank: 178

Population: 68,615 (1988) 70,000 (2000)

Population—World Rank: 181

Capital: Victoria

Land Boundaries: Nil

Coastline: 127 km (79 mi)

Longest Distances: 27 km (17 mi) N-S 11 km (7 mi) E-W

Highest Point: Mount Seychelles 905 m (2,969 ft)

Lowest Point: Sea level

Land Use: 4% arable; 18% permanent crops; 18% forest and woodland; 60% other

Seychelles is located 1770 km (1,100 m) east of Africa and 1126 km (700 m) northeast of Madagascar. Seychelles consist of a core group of high-rising granitic islands and a far-flung group of low coralline atolls. The granitic Seychelles, including Mahé, Silhouette, Praslin and La Digue are situated in a 56 km (35 mi) wide area and are inhabited by most of the colony's population. The coralline Seychelles include the Aldabra Islands and the nearby Cosmoledo Group, the Farquhar Group, and the Amirante Isles and support only a small number of temporary residents.

Physical Environment

The Seychelles consist of 90 to 100 islands scattered across 388,498 sq km (150,000 sq mi) of ocean between 4° and 11°S latitude and roughly between 50° and 60°E longitude. The count on the total number of islands varies depending upon what is considered an island. Some are merely sand cays and shoals barely above the high tide mark. The granitic Seychelles are generally conceded to include thirty-two islands; the remainder are in the coralline group. The total land area of the Seychelles is about 259 sq km (100 sq mi), of which 225 sq km (87 sq mi) are contained in the granitic group.

General Characteristics

Mahé is the largest island in the granitic Seychelles. It is 27 km (seventeen mi) long and five to eight km (three to five mi) wide and constitutes a little more than half—142 sq km (fifty-five square mi)—of the total land area. The other granitic islands, all within 56 km (35 mi) of Mahé, include Praslin, La Digue, Silhouette, Curieuse, Felicité, Frigate, Bird, Cousin and Cousine, and a number of smaller islets and isolated rocks. Their coconut palms, white sand beaches, and great scenic beauty conform to the popular conception of idyllic "South Sea islands."

Area and population

Island Groups	Capital	area sq mi	area sq km	population 1984 estimate
Central (Granitic) group				
La Digue and satellites	—	6	15	2,000
Mahé and satellites	Victoria	61	158	57,400
Praslin and satellites	—	16	42	4,650
Silhouette	—	8	20	200
Other islands	—	2	4	50
Outer (Coralline) islands	—	83	214	400
TOTAL		175	453	64,700

SIERRA LEONE

BASIC FACTS

Official Name: Republic of Sierra Leone

Abbreviation: SLN

Area: 71,740 sq km (27,699 sq mi)

Area—World Rank: 112

Population: 3,963,289 (1988) 4,861,000 (2000)

Population—World Rank: 101

Capital: Freetown

Boundaries: 1,364 km (847 mi) Guinea 652 km (405 mi); Liberia 306 km (190 mi)

Coastline: 406 km (252 mi)

Longest Distances: 338 km (210 mi) N-S 304 km (189 mi) E-W

Highest Point: Bintimani 1,945 m (6,381 ft)

Lowest Point: Sea level

Land Use: 23% arable; 2% permanent crops; 31% meadows and pastures; 29% forest and woodland; 15% other

Sierra Leone, roughly circular in shape,—is a compact country located in the southwestern part of the great bulge of West Africa. Lying between the seventh and tenth parallels north of the equator, it is bounded on the west by the Atlantic Ocean and inland by Guinea and Liberia. Its varied terrain includes the striking, mountainous Sierra Leone Peninsula; a zone of low-lying coastal marshland

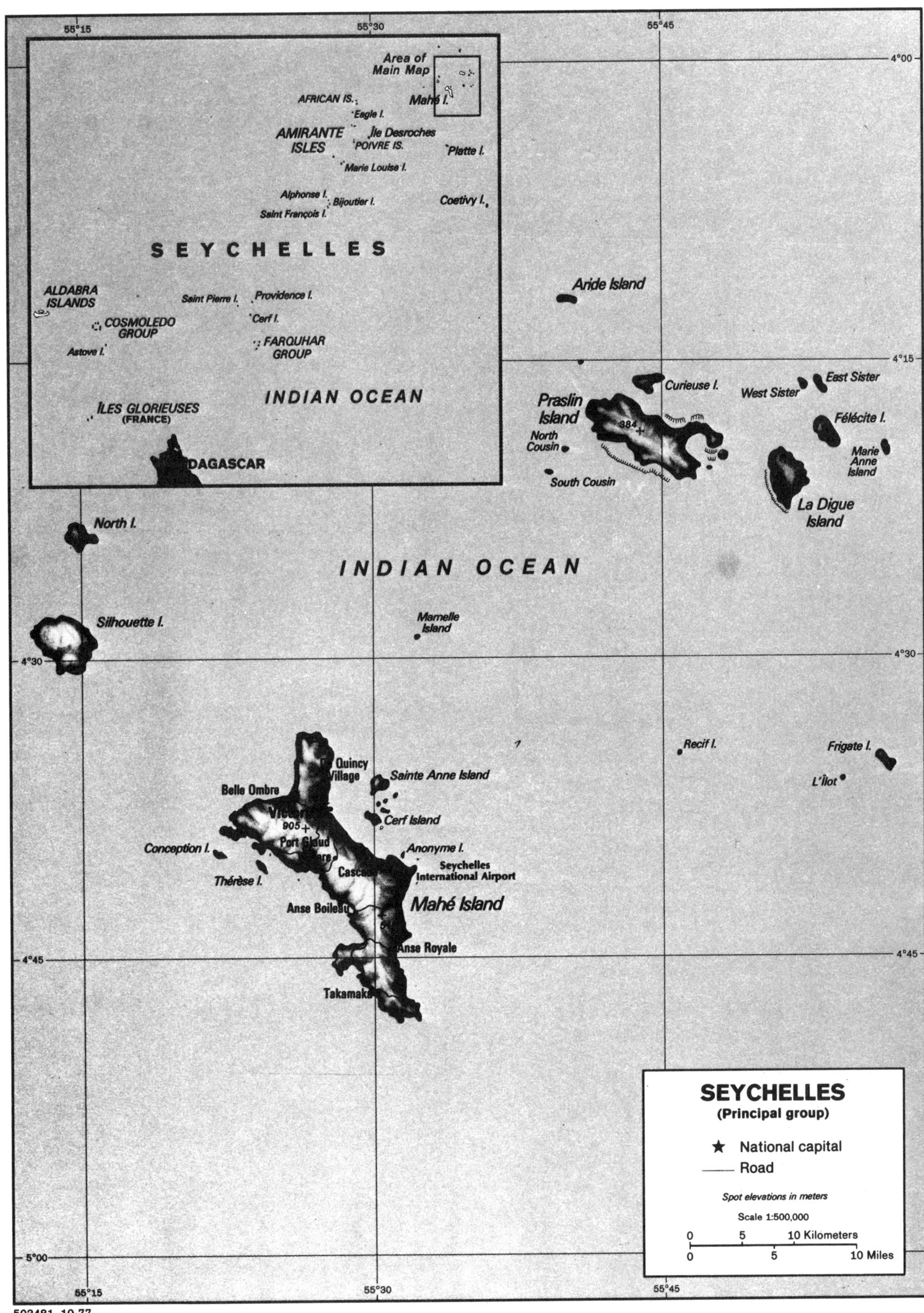

Area of Main Map
AFRICAN IS.
Mahé I.
Eagle I.
AMIRANTE ISLES
Île Desroches
POIVRE IS.
Platte I.
Marie Louise I.
Alphonse I.
Bijoutier I.
Saint François I.
Coetivy I.
SEYCHELLES
ALDABRA ISLANDS
Saint Pierre I.
Providence I.
COSMOLEDO GROUP
Cerf I.
Astove I.
FARQUHAR GROUP
INDIAN OCEAN
ÎLES GLORIEUSES (FRANCE)
MADAGASCAR
Aride Island
Curieuse I.
East Sister
West Sister
Praslin Island
384
Félécite I.
North Cousin
Marie Anne Island
South Cousin
La Digue Island
North I.
INDIAN OCEAN
Silhouette I.
Mamelle Island
Recif I.
Frigate I.
L'Îlot
Quincy Village
Sainte Anne Island
Belle Ombre
905
Cerf Island
Conception I.
Port Glaud
Anonyme I.
Seychelles International Airport
Thérèse I.
Anse Boileau
Mahé Island
Anse Royale
Takamaka
SEYCHELLES
(Principal group)
National capital
Road
Spot elevations in meters
Scale 1:500,000
0 5 10 Kilometers
0 5 10 Miles

503481 10-77

along the Atlantic Ocean; and a wide plains area extending inland to about the middle of the country. East of the plains the land rises to a broad, moderately elevated plateau from which emerge occasional hill masses and mountains that include West Africa's highest point.

Generally fertile soils occur along the coast, but soils of the inland area have deteriorated over large areas as the result of excessive use by man and his destruction of the natural vegetation cover. Mineral deposits of economic significance exist, including iron ore, bauxite, titanium ores, and the more glamorous diamond.

GEOGRAPHIC REGIONS

Natural physical features divide the country into four main geographic regions. In the east a broad area of low plateaus surmounted in places by mountain and hill masses—together with a zone of eroded foothills on the area's western edge—constitutes the Interior Plateaus and Mountains Region. (This region constitutes in fact a western extension of the Guinea Highlands from neighboring Guinea.) To the west of this region lies the almost equally large Interior Low Plains Region, which in turn merges into the narrow Coastal Swampland Region paralleling the Atlantic Ocean. The fourth region, comparatively minute in area but physiographically highly distinctive, includes only the Sierra Leone, or Freetown, Peninsula.

Interior Plateaus and Mountains

This region, which encompasses roughly the eastern half of the country, consists chiefly of a large area of plateaus having elevations of above 304 m (1,000 ft) to about 608 m (2,000 ft). Rising above the relatively flat surface are several mountain masses, including the Loma Mountains, in which is located Loma Mansa (Bintimani), at 1,947 m (6,390 ft) the highest point in West Africa west of the Cameroon Mountains; the Tingi Mountains (sometimes called Tingi Hills), also rising at one point to over 1,829 m (6,000 ft); and the lower Nimini and Sula mountains and Gola, Gori, and Jojina hills. Isolated hills (inselbergs) are also found at many places throughout the plateaus.

In the region's southern section erosion has resulted in a large area of rolling terrain, forty or so miles wide at points and having elevations between 152 and 304 m (500 and 1,000 ft) and scattered low hills. This area now forms the upper basin of the Moa River. The western edge of the plateau exhibits different stages of erosion and in places is characterized by steep-sided river valleys and highly dissected hills. Among the latter the most prominent are the Kangari and Kambui hills. The Kambui Hills, formed of extremely ancient (pre-Cambrian) schists, have important deposits of minerals of economic significance. Such schists are found also in the Sula and other mountains and hills in the region, and they too have associated minerals with them.

The southern part of the region has a somewhat longer rainy season than other areas of Sierra Leone. The rain and the soil conditions make it suitable for growing cocoa and coffee. In general the vegetation has been greatly modified by the practice of bush fallow cultivation, but a considerable amount of secondary forest was still found in this section in the early 1970s. In the region's more northerly part less rainfall is received, and the vegetation is mainly of a derived savanna type.

PRINCIPAL TOWNS

Freetown (capital), population 469,776 (census of December 1985); Koidu 80,000; Bo 26,000; Kenema 13,000; Makeni 12,000.

Area and population		area		population
Provinces Districts	**Capitals**	sq mi	sq km	1985 census
Eastern Province	Kenema	6,005	15,553	960,551
Kailahun	Kailahun	1,490	3,859	233,839
Kenema	Kenema	2,337	6,053	337,055
Kono	Sefadu	2,178	5,641	389,657
Northern Province	Makeni	13,875	35,936	1,262,226
Bombali	Makeni	3,083	7,985	315,914
Kambia	Kambia	1,200	3,108	186,231
Koinaduga	Kabala	4,680	12,121	183,286
Port Loko	Port Loko	2,208	5,719	329,344
Tonkolili	Magburaka	2,704	7,003	247,451
Southern Province	Bo	7,604	19,694	740,510
Bo	Bo	2,015	5,219	268,671
Bonthe (incl. Sherbro)	Bonthe	1,339	3,468	105,007
Moyamba	Moyamba	2,665	6,902	250,514
Pujehun	Pujehun	1,585	4,105	116,318
Western Area	Freetown	215	557	554,243
TOTAL		27,699	71,740	3,517,530

Interior Low Plains

The interior plains range in width from under 48 km (30 mi) near the Liberian border to some 112 or 129 km (70 or 80 mi) in their central and northern parts. Stretching eastward from the coastal swamps and the bordering wet grasslands to the foothills of the plateau region, the plains have elevations mostly from about 30 m to 152 m (100 ft to 500 ft) (where they meet the foothills). Areas of residual hills rise at places to 304 m (1,000 ft) and over, however.

Vegetation over most of the region consists of what is known locally as farm bush, secondary forest, and cultivated crops. A large area in the northeastern part of the region, averaging about 32 km (20 mi) in width and roughly 144 to 160 km (ninety to 100 mi) in length, is covered mostly by swampy grassland; this area is known as the bolilands. Small inland valley swamps, used for rice cultivation, also occur at many other places throughout the region.

Coastal Swampland

This region comprises a zone varying from about 8 to 40 km (5 to 25 mi) in width along the coast. The region is characterized by numerous estuaries whose river channels, as in the case of the Sierra Leone River, continue under the sea and across the continental shelf, indicating submergence of the coast in very recent geological times.

Mangrove swamps line much of the coast, behind which marine and freshwater swamps occupy large areas.

Sierra Leone
International boundary
Province boundary
National capital
Province capital
Railroad
Road
0
50 Kilometers
0
50 Miles
GUINEA
LIBERIA
NORTHERN
EASTERN
SOUTHERN
WESTERN AREA
NORTH ATLANTIC OCEAN
Mamou
Kindia
Coyah
Forécariah
Faranah
Kissidougou
Gueckedou
Falaba
Kabala
Kamakwie
Kamaron
Kambia
Pendembu
Bumbuna
Mange
Makeni
Worodu
Port Loko
Lungi
Lunsar
Magburaka
Sefadu
Pepel
Kupr
Freetown
Waterloo
Yonibana
Yele
Kailahun
Rotifunk
Moyamba
Panguma
Pendembu
Mano
Bo
Kenema
Shenge
Momaligi
Joru
Bonthe
Pujehun
Zimmi
Noway Camp
Sulima
Robertsport
Tubmanburg
Kle
Bong Town
Kakata
Monrovia
Gardnersville
BANANA ISLANDS
TURTLE ISLANDS
Sherbro Island
Lake Mape
Lake Piso
Konkouré
Niger
Kaba
Mongo
Kolente
Rokel
Bagbe
Meli
Pampana
Moa
Jong
Sewa
Kittam
Mano
Gbeya
Lofa
Saint Paul
13
12
11
10
9
8
7

North of Freetown, however, the coast extending to the estuary of the Little Scarcies River is fronted in places by beach ridges and in one section by the cliffs of a low plateau. Soils in this coastal stretch are relatively good and produce cash crops for the Freetown market. Beach ridges also front a long stretch of the coast in the southernmost part of the region but have generally infertile soils. A notable feature of this southern section is the large area of riverain grassland that lies behind the coast.

Sierra Leone Peninsula

The mountainous Sierra Leone Peninsula, on which Freetown is located, is twenty-five mi long and averages about ten mi in width. It is treated as a separate geographic region because its unusual features bear no direct relationship to those of the adjacent coastal region or for that matter to the other two geographic regions. The peninsula consists chiefly of igneous rocks that form—with the nearby Banana Islands—the visible part of a much larger igneous mass submerged beneath the sea. It is believed to have been uplifted in relatively recent geological times judging from its present-day relief and height, which exceeds 609 m (2,000 ft) in several places and reaches a maximum of over 883 m (2,900 ft). Most of the peninsula's hills are included in a forest reserve established to halt erosion and preserve the watershed as a source of water supply for domestic purposes.

Around the base of the mountains is a strip of land about one mile wide consisting of lateritic hardpan. Variations in earlier sea levels have left a number of raised beaches in this strip, the uppermost about 48 m (160 ft) above the present-day sea level. Freetown and other settlements are mainly situated on these raised beaches. The present coast consists of sand ridges and some areas of mangrove swamp.

DRAINAGE

The country drains entirely into the Atlantic Ocean through nine roughly parallel, principal river basins that run generally northeast to southwest and some five small basins of river systems confined to the coastal area. Five of the main basins lie completely within the country, including those of the Rokel (or Seli), Gbangbar, Jong, Sewa (the largest basin, encompassing nearly 14,244 sq km, 5,500 square mi)) and Waanje rivers. To the north of the Rokel basin lie the basins of the Great Scarcies and Little Scarcies rivers, both of which rise in the Fouta Djallon highlands of neighboring Guinea (where they are known respectively as the Kolente and Kaba rivers). In the south are the basins of the Moa River system, which extends into southeastern Guinea and Liberia, and of the Mano River, which with its tributaries drains part of northern Liberia.

The heavy wet season rainfall causes a substantial rise in river levels that in places may be fifty or more ft above the dry season low water mark. The main rivers have cut deeply into the interior plains, resulting in stretches of rapids that, with the seasonal variation in water levels, reduce continuous navigability to 32 or 48 km (20 or 30 mi) of their lower courses; an exception is in the south, where the Kittam and Waanje rivers run for a combined distance of about 80 km (50 mi) behind and parallel to slightly raised coastal ridges that obstruct the rivers' direct entry to the ocean.

Although considerable areas of inland low-lying swampland are periodically flooded, permanent freshwater lakes are few, and all are comparatively small. Only six lakes have an area of one square mile or more, and the country's largest, Lake Mape in Pujehun District, is less than 28 sq km (11 sq mi).

SINGAPORE

BASIC FACTS

Official Name: Republic of Singapore

Abbreviation: SNG

Area: 622 sq km (240 sq mi)

Area—World Rank: 170

Population: 2,645,443 (1988) 3,030,000 (2000)

Population—World Rank: 114

Capital: Singapore

Land Boundaries: Nil

Coastline: 135 km (84 mi)

Longest Distances: 50.7 km (31.5 mi) ENE-WSW; 31.4 km (19.5 mi) SSE-NNW

Highest Point: Timah Hill 166 m (545 ft)

Lowest Point: Sea level

Land Use: 4% arable; 7% permanent crops; 5% forest and woodland; 84% other

Singapore owes its growth and its importance in Southeast Asia to its geographic position rather than to the natural resources of the land. Its area is roughly equivalent to that of Chicago. Almost all of the land area is on the single island of Singapore; about 38 sq km (fifteen square mi) are taken up by fifty-four islets, fewer than half of which are inhabited and the largest of which, Pulau Tekong Besar, is less than 18 sq km (seven square mi). The city of Singapore has an area of 97 sq km (37.6 square mi), a somewhat misleading figure because large areas outside the city limits are heavily urbanized as well. The main island is roughly diamond shaped. Land reclamation has added almost 15 sq km (six square mi) to the total territory since 1966, mostly along the southeast coast.

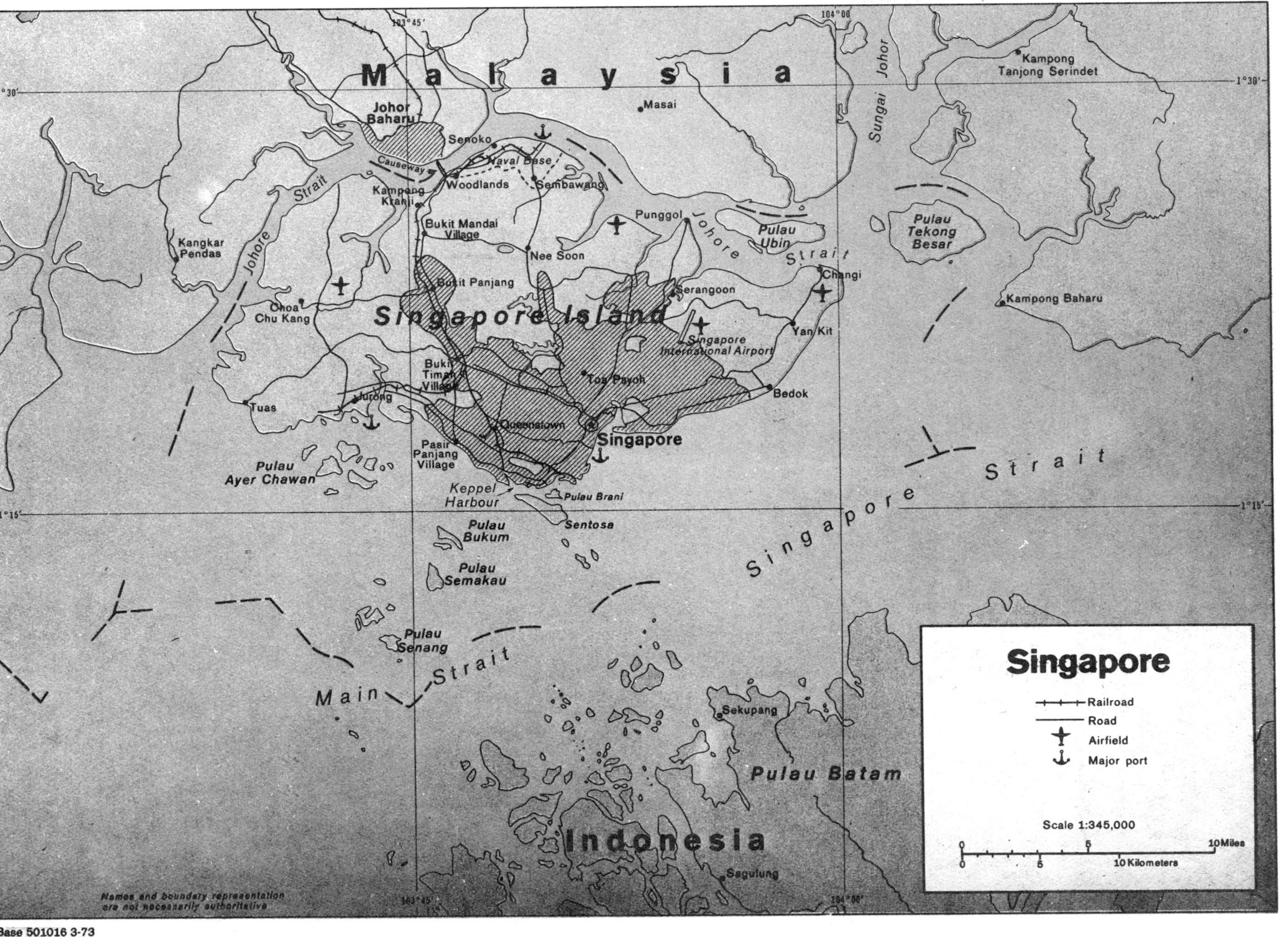

Base 501016 3-73

Singapore is separated by the Johore Strait from the mainland and Malaysia's state of Johor—to which it is connected at the narrowest point of the strait by a causeway seven-tenths of a mile long—and is separated from Indonesia's Pulau Batam and lesser islets by the Singapore Strait where it debouches into the Strait of Malacca—less than five mi wide at its narrowest point. Indonesia and Malaysia hold that the Strait of Malacca and the Singapore Strait fall within the twelve-mile limit of territorial waters and thus are not international waterways. Singapore, whose position as an entrepôt depends on the free passage of international shipping through these waters, maintains the international character of the waterway.

The islands are generally flat and low lying. There is a hilly area in the center of the main island; the highest point, Bukit Timah (Tin Hill), is a little less than 177 m (581 ft) above sea level. The topography of the southwestern part of the island is varied by a series of short cliffs and shallow valleys formed of sedimentary rock. Coastal waters are generally less than 30 m (100 ft) deep. The island is drained by several short rivers, streams, and drainage canals, principally the Singapore, Jurong, Kalang, Kranji, Seletar, and Serangoon rivers. The longest of these, the Seletar River, is only about nine mi long.

Area and population

Census areas	area sq mi	area sq km	population 1984 estimate
Central city area	3	8	157,000
City periphery	17	46	942,800
North	7	19	228,100
Northeast	3	9	301,500
West	7	18	413,200
Suburbs	49	127	754,700
East	7	19	195,000
North	13	34	309,900
West	29	74	249,800
Outlying areas	169	437	674,600
East	46	118	301,100
North	53	137	177,500
West	70	182	196,000
TOTAL	240	622	2,529,100

SOLOMON ISLANDS

BASIC FACTS

Official Name: Solomon Islands

Abbreviation: SMN

Area: 27,556 sq km (10,640 sq mi)

Area—World Rank: 129

Population: 312,196 (1988) 457,000 (2000)

Population—World Rank: 154

Capital: Honiara

Land Boundaries: Nil

Coastline: 4,197 km (2,608 mi)

Longest Distances: 1,688 km (1,049 mi) ESE-WNW; 468 km (291 mi) NNE-SSW

Highest Point: Mount Makarakomburu 2,447 m (8,028 ft)

Lowest Point: Sea level

Land Use: 1% arable; 1% permanent crops; 1% meadows and pastures; 93% forest and woodland; 4% other

The Solomon Islands comprise a double chain of high continental islands formed from the exposed peaks of the submerged mountain chain that extends from Bougainville to northern Vanuatu 1,688 km (1,049 mi) ESE to WNW and 468 km (292 mi) NNE to SSW.

The largest island in the group is Guadalcanal, the fabled site of one of the fiercest battles in World War II, which has an area of 6,475 sq km (2,500 sq mi). Only five other islands are large enough to be named on most maps: Choiseul, New Georgia, Santa Isabel, Malaita, and San Cristobal. Smaller islands are Bellona, Duff, Florida Islands, Gizo, Kolombangara, Ndeni, Ontong Java, Reef Islands, Rennell, Santa Cruz Islands (including Anuta Fetaka, Santa Cruz, Tevai, Tikopia, Utupua, and Vanikoro), Savo, Shortland, Sikaiana, Tulagi, Vella Lavella.

Almost all of the larger islands are volcanic in origin and are covered with steaming jungles and mountain ranges intersected by narrow valleys. The highest peak is the 2,320 meter (7,647 ft) Mt. Popomanishu on Guadalcanal. Guadalcanal also contains the nation's only extensive alluvial plains. Most rivers are short and narrow and impassable except by canoe. Most of the smaller islands are raised, coral or low atolls.

Area and population

Provinces	Capitals	area sq mi	area sq km	population 1986 census
Central Islands	Tulagi	493	1,276	18,522
Guadalcanal	Honiara	2,047	5,302	50,327
Isabel	Buala	1,550	4,014	14,564
Makira	Kira Kira	1,231	3,188	21,646
Malaita	Auki	1,638	4,243	80,183
Temotu	Santa Cruz	358	926	14,683
Western	Gizo	3,310	8,573	55,372
Capital Territory				
Honiara	—	13	34	30,499
TOTAL		10,640	27,556	285,796

SOMALIA

BASIC FACTS

Official Name: Somali Democratic Republic

Abbreviation: SOM

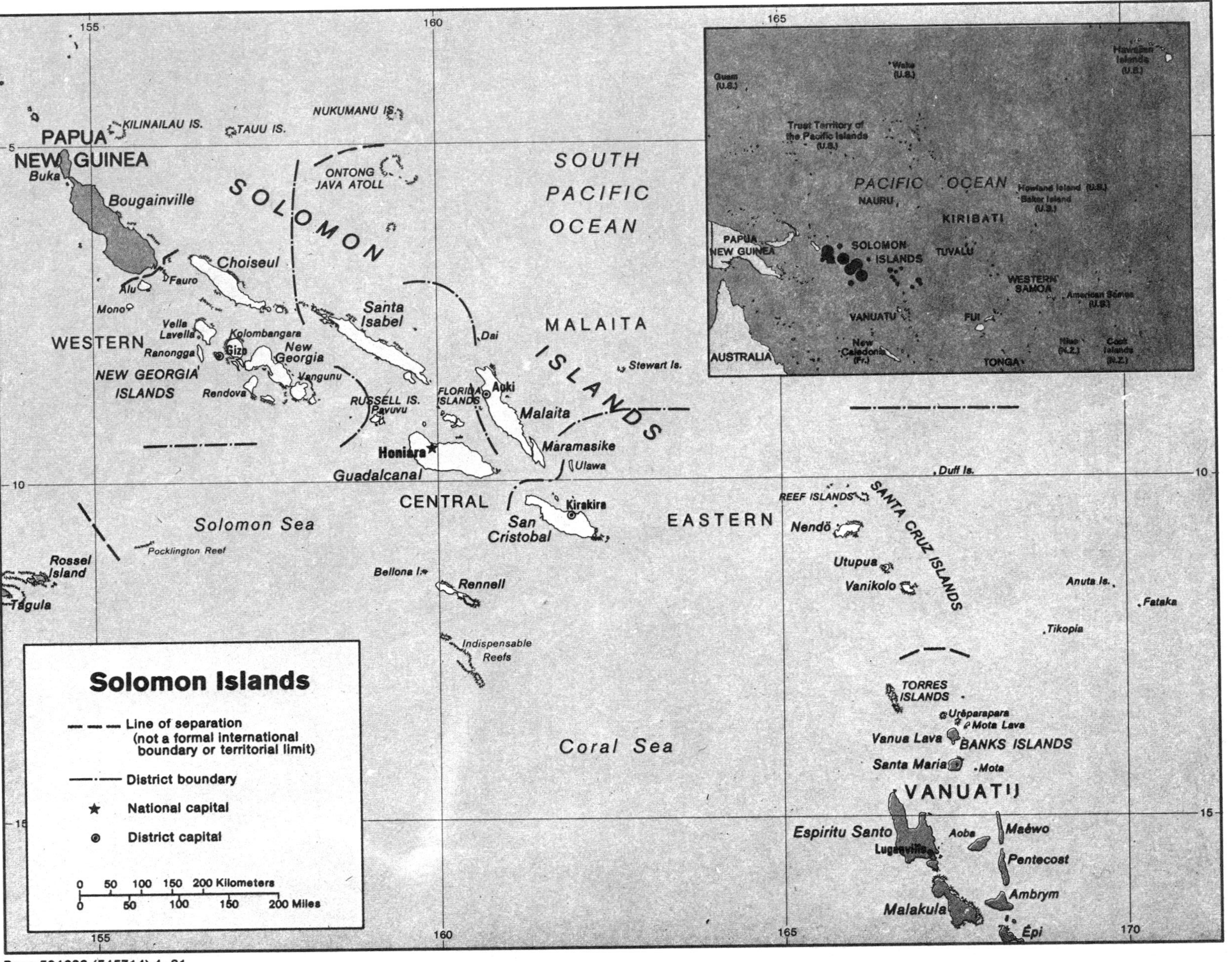

Base 504633 (545714) 4-81

Area: 637,000 sq km (246,000 sq mi)

Area—World Rank: 41

Population: 7,990,085 (1988) 8,827,000 (2000)

Population—World Rank: 87

Capital: Mogadishu

Boundaries: 5,607 km (3,484 mi); Kenya 682 km (424 mi); Ethiopia 1,645 km (1,022 mi); Djibouti 61 km (38 mi)

Coastline: Gulf of Aden 1,046 km (650 mi); Indian Ocean 2,173 km (1,350 mi)

Longest Distances: 1,847 km (1,148 mi) NNE-SSW; 835 km (519 mi) ESE-WNW

Highest Point: Surud Ad 2,407 m (7,897 ft)

Lowest Point: Sea level

Land Use: 2% arable; 46% meadows and pastures; 14% forest and woodland; 38% other

Situated in northeast Africa and the easternmost country of that continent, Somalia has a land area roughly comparable to that of the state of Texas. The country has the shape of the number 7 and is the cap of the geographic region commonly referred to as the Horn of Africa, which includes Ethiopia and Djibouti. Somalia's northern limits, which face the Gulf of Aden, lie about 12° north of the equator. Its eastern and southern bounds face the Indian Ocean; the southernmost point of the country reaches about 1½° below the equator.

PRINCIPAL TOWNS (estimated population in 1981)

Mogadishu (capital)	500,000
Hargeisa	70,000
Kismayu	70,000
Berbera	65,000
Merca	60,000

Area and population

Regions	Capitals	area sq mi	area sq km	population 1980 estimate
Bakool	Xuddur	10,000	27,000	148,700
Banaadir	Mogadishu	400	1,000	520,100
Bari	Boosaaso	27,000	70,000	222,300
Bay	Baydhabo	15,000	39,000	451,000
Galguduud	Dhuusa Mareeb	17,000	43,000	255,900
Gedo	Garbahaarrey	12,000	32,000	235,000
Hiiraan	Beled Weyne	13,000	34,000	219,300
Jubbada Dhexe	Bu'aale	9,000	23,000	147,800
Jubbada Hoose	Kismaayo	24,000	61,000	272,400
Mudug	Gaalkacyo	27,000	70,000	311,200
Nugaal	Garoowe	19,000	50,000	112,200
Sanaag	Ceerigaabo	21,000	54,000	216,500
Shabeellaha Dhexe	Towhar	8,000	22,000	352,000
Shabeellaha Hoose	Marca	10,000	25,000	570,700
Togdheer	Burko	16,000	41,000	383,900
Woqooyi Galbeed	Hargeysa	17,000	45,000	655,000
TOTAL		246,000	637,000	5,074,000

Physical Geography

Physiographically Somalia is a land of limited contrast. In the north, paralleling the Gulf of Aden coast, is a maritime plain that varies in width from about 56 km (35 mi) in the west to as little as one or two mi in the east. Scrub-covered, semiarid, and generally drab in appearance, this plain, known as the Guban (burnt land) because of its heat and dryness during much of the year, is crossed by broad, shallow watercourses that are beds of dry sand except in the rainy seasons. When the rains arrive, however, the vegetation is quickly renewed, and for a time the Guban provides some grazing for nomad livestock.

Away from the gulf the plain rises to the precipitous north-facing cliffs of dissected highlands, which form rugged mountain ranges that extend from the northwestern border with Ethiopia eastward to the tip of the horn, where they end in sheer cliffs at Cape Guardafui. The elevation along the crest of these mountains averages about 1,828 m (6,000 ft) above sea level in the area about 48 km (30 mi) south of Berbera and eastward from that area continues at between 1,828 and 2,133 m (6,000 and 7,000 ft) almost to Cape Guardafui. The country's high point, Surud Ad, which rises to over 2,407 m (7,900 ft) is located near the town of Erigavo.

To the south the mountains descend, often in scarped ledges, through a region of broken mountain terrain, shallow plateau valleys, and usually dry watercourses, known to the Somali as the Ogo. This region merges into an elevated plateau devoid of perennial rivers that, in its especially arid eastern part, which is interspersed with a number of isolated mountain ranges, gradually slopes toward the Indian Ocean; in central Somalia it constitutes the Mudug Plain. A major feature of the eastern section is the long and broad Nugaal Valley, which has an extensive net of intermittent seasonal watercourses. The entire eastern area, whose population consists mainly of pastoral nomads, is in a zone of low and erratic rainfall. It was a major disaster area during the great drought of 1974 and early 1975.

The western part of the plateau is characterized by shallow valleys and dry watercourses. Annual rainfall is greater, however, and in the west are flat areas of arable land that provide a home for dryland farming cultivators. Most important, it is an area of permanent wells to which the predominantly nomadic population returns during the dry seasons. The western plateau slopes gently southward and merges imperceptibly into a zone known as the Haud, a broad, undulating area that constitutes some of the best grazing lands for the Somali nomad despite the lack of appreciable rainfall for over half the year. Enhancing the value of the Haud are the natural depressions that flood during periods of rain to become temporary lakes and ponds.

The Haud, continuing for 160 km (100 mi) or more into the Ogaden region of Ethiopia, is part of the vast Somali Plateau, which lies between the northern Somalian mountains and the highlands of southeast Ethiopia and extends southward and eastward through Ethiopia into central and southwest Somalia. An Anglo-Ethiopian agreement during the colonial era provided for access to

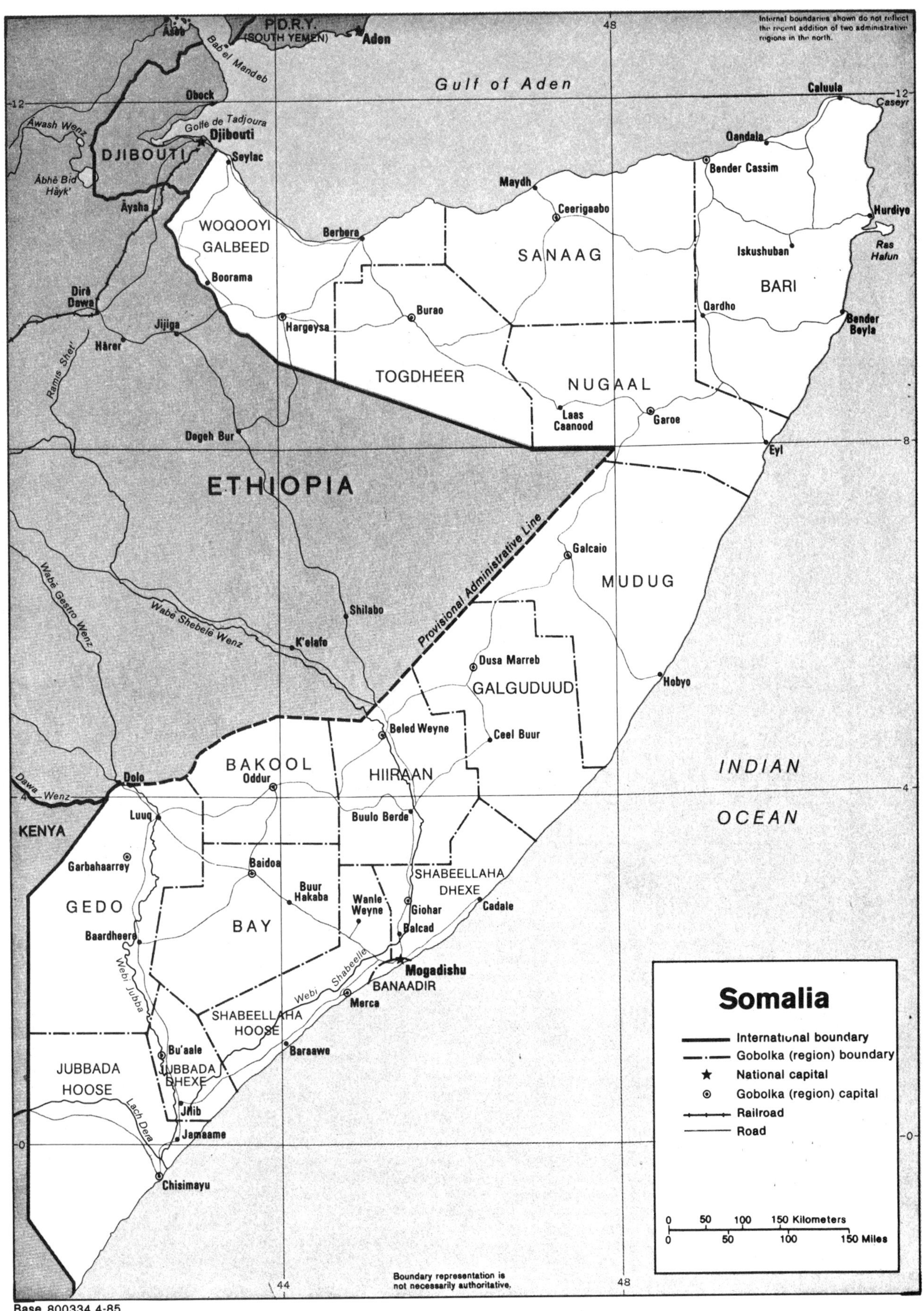

Base 800334 4-85

the Ethiopian part of the plateau by British-protected Somali herdsman from the north. Somalia has laid claim to part of this area, and the newly independent Somali government in 1960 refused to acknowledge the agreement. In the mid-1970s Somali nomads continued to move back and forth with the seasons throughout the Haud in the constant search for forage for their animals.

The physiography of southwestern Somalia is dominated by the country's only two permanent rivers, the Juba and Shabelle. These rivers, which originate in the Ethiopian highland, flow generally southward, cutting wide valleys in the Somali Plateau as it descends toward the sea; the plateau's elevation falls off rapidly in this area. A succeeding large coastal zone, which includes the lower reaches of the rivers and extends from the Mudug Plain to the Kenya border, averages183 m (600 ft) or less above sea level.

The Juba enters the Indian ocean at Kismayo. Although the Shabelle at one time apparently reached the sea near Marka (Merca), its course changed in prehistoric times; in its present-day course it turns south-southwestward near Balad, about 32 km (20 mi) north of Mogadishu, and follows a path parallel to the coast for about 225 km (140 mi). The river is perennial only to a point southwest of Mogadishu; thereafter it consists of swampy areas and dry reaches and is finally lost in the sand of Gelib, not far from the Juba River. During flood seasons in about April to May (the *gu* rainy period) and October to November (*Dayr* rains), the river may be full to near Gelib and occasionally may even break through to the Juba farther south. Favorable rainfall and soil conditions make this entire river region a fertile agricultural area and the center of the country's largest sedentary population.

SOUTH AFRICA

BASIC FACTS

Official Name: Republic of South Africa

Abbreviation: SAF

Area: 1,218,363 sq km (470,412 sq mi)

Area—World Rank: 24

Population: 35,093,971 (1988) 50,494,000 (2000)

Population—World Rank: 27

Capital: Pretoria (administrative); Cape Town (legislative); Bloemfontein (judicial)

Boundaries: 7,637 km (4,746 mi); Botswana 1,778 km (1,105 mi); Zimbabwe 225 km (140 mi); Mozambique 491 km (305 mi); Swaziland 449 km (279 mi); Namibia 1,078 km (670 mi); Lesotho 909 km (565 mi)

Coastline: 2,707 km (1,682 mi)

Longest Distances: 1,821 km (1,132 mi) NE-SW; 1,066 km (662 mi) SE-NW

Highest Point: eNjesuthi 3,446 m (11,306 ft)

Lowest Point: Sea level

Land Use: 10% arable; 1% permanent crops; 65% meadows and pastures; 3% forest and woodland; 21% other

GEOGRAPHICAL FEATURES

South Africa is located at the end of the African continent. Extraterritorial holdings included in the total area consist of Walvis Bay enclave on the coast of Namibia and two small islands—Prince Edward and Marion—southeast of the African mainland. South Africa administers South-West Africa (Namibia), with an area of 824,295 sq km (318,261 sq mi). South Africa considers Transkei, Ciskei, Venda and Bophuthatswana as independent republics and generally excludes them from its geographical area.

Most of the country's borders are defined by natural features. Except for the border with Mozambique and Namibia's southeastern sector, the inland frontier is formed by the course of the Limpopo River in the East and the Orange River in the West. All internal borders are either demarcated or delimited and are undisputed. The western, southern and eastern extremities are marked by 4,300 km (2,672 mi) of coastline formed by the Atlantic and Indian oceans. In the Far South the coastline forms a truncated apex of an inverted triangle, with the Cape Agulhas as its most southerly point.

A number of small landforms off the southwestern coast include Dassen, Robben and Bird islands. All are uninhabited except for Robben Island, which serves as the site of the country's maximum-security prison.

The country's general physiography consists of a broad, centrally depressed plateau edged by a prominent escarpment overlooking marginal slopes that descend to the eastern, southern and western slopes. The mountainous edges of the plateau extend in a sweeping arc from the country's northeastern tip to its southwestern extremity. Collectively these edges are known as the Great Escarpment. The marginal zone tends to be more dissected than the plateau and in places exhibits acute relief features.

Inland from the crest of the Great Escarpment the country consists generally of rolling plains dropping gradually to an altitude of about 900 m (2,952 ft) in the center. Within the plateau there are a number of generally distinctive subregions. The largest of these is the Highveld (*veld* is Afrikaans for grassland), extending from southern Transvaal and encompassing all of Orange Free State and extending southward through northern Cape Province. The undulating land surface lies mostly between 1,200 and 1,800 m (3,937 and 5,906 ft) above sea level. Its northern limit is formed by the Witwatersrand ridge, on which Johannesburg stands at 1,800 m (5,906 ft). Witwatersrand (literally, ridge of white waters), known colloquially as the Rand, is a ridge of gold-bearing rock about 100 km (62 mi) long and 37 km (23 mi) wide in southern Transvaal that is a watershed for a number of northern streams. It was the site of the first gold discovery, in 1886, and subsequently was found to contain the world's largest proven deposits of gold. Of the other subregions, the dissected basalt plateau of Lesotho, surmounting the Highveld in the east, is the highest. To the west lies the Ghaap Plateau, rising imperceptibly westward to the Asbestos and Langeburg ranges overlooking the southern Kalahari. The level Ghaap Plateau lies about 1,500 m (4,921 ft) above sea level and is in fact an outlier of the Highveld, detached from it by the Harts, Vaal and Orange rivers. In northern Transvaal there are the undulating Waterberg (or Palala) Plateau in the Northwest and the Soutpansberg highlands in the North, dipping toward the wide Limpopo trough beyond.

North of the Witwatersrand lies an area known as the Transvaal Middleveld or Bushveld Basin. It is a type of dry savanna, characterized by open grassland with scattered trees and bushes. Its elevation is lower than that of the Highveld and averages between 600 and 900 m (1,959 and 2,952 ft).

The plateau slopes gently westward into the arid Bushmanland and Upper Karoo area, generally known as the Cape Middleveld. It embraces the lower basin of the Orange River west of Kimberley and much of the semiarid tableland of west-central Cape Province. South of the Orange River, the area is characterized by vast depressions and pans. North of the Cape Middleveld, the plateau becomes a sandy plain, the Kalahari Basin, a semiarid southern extension of the great Kalahari Desert, which lies north of South Africa's west-central frontier. The Kalahari is a desert less by virtue of its scanty rainfall than by virtue of its surface sandy deposits, which allow precipitation to sink down too rapidly. It has four fossil rivers—Molopo, Kuruman, Auob and Nossob—which may flow perhaps once in a generation.

Probably the most fundamental topographical feature is the Great Escarpment, a continuous series of mountain ridges that rim the interior plateau, separating it from the marginal areas. The Great Escarpment runs almost unbroken from the Zambezi River in Zimbabwe around the southern edge of the continent and arcs northward, following the western edge of the landmass through Namibia and into Angola. In South Africa the escarpment lies from 55 to 240 km (34 to 149 mi) behind the coastline and has a variety of local names.

In the east and southeast it is known as the Drakensberg Mountains. Traced westward through the Cape Province, it becomes the Stormberg, Suurberg, Sneeuberg, Nuweveld Reeks and Roggeveldberge. The escarpment attains its most majestic form in the great wall of the Natal Drakensberg along the Lesotho border between Mont-aux-Sources in the North and Xalanga Peak in the South. In many places the scrap face is vertical, fully 2,000 m (6,561 ft) from foot to crest. The highest points in the country occur in this stretch: Mont-aux-Sources 3,299 m (10,824 ft), Champagne Castle 3,376 m (11,077 ft) and Giant's Castle 3,313 m. (10,870 ft.). The highest point in the subcontinent, Thabana Ntlengana 3,492 m (11,457 ft) in not on the escarpment itself but on an obscure plateau

PRINCIPAL TOWNS (population at 1985 census)

	City Proper	Metropolitan Area
Cape Town*	776,617	1,911,521
Durban	634,301	982,075
Johannesburg	632,369	1,609,408
Pretoria*	443,059	822,925
Port Elizabeth	272,844	651,993
Umlazi	194,933	n.a.
Roodepoort	141,764	n.a.
Pietermaritzburg	133,809	192,417
Germiston	116,718	n.a.
Boksburg	110,832	n.a.
Bloemfontein*	104,381	232,984
Benoni	94,926	n.a.
East London	84,729	193,819
Kimberley	74,061	149,667
Springs	68,235	n.a.
Vereeniging	60,584	540,142

*Pretoria is the administrative capital, Cape Town the legislative capital and Bloemfontein the judicial capital.

Area and population

Provinces	Capitals	area sq mi	area sq km	population 1983 estimate	population 1985 census
Cape	Cape Town	247,638	641,379	5,374,000	5,041,137
Natal	Pieter-martisburg	22,608	58,555	2,842,000	2,145,018
Orange Free State	Bloem-fontein	49,233	127,513	2,080,000	1,776,903
Transvaal	Pretoria	88,197	228,429	8,950,000	7,532,179
National states					
Gazankulu	Giyani	2,606	6,750	585,000	497,213
KaNgwane	Louieville	1,436	3,720	184,000	392,782
Kwa-Ndebele	Siyabuswa	355	920	200,000	235,855
KwaZulu	Ulundi	12,664	32,800	3,792,000	3,747,015
Lebowa	Lebowakgomo	8,757	22,680	1,869,000	1,835,984
Qwaqwa	Phuthaditjhaba	185	480	306,000	181,559
TOTAL		433,679	1,123,226	26,182,000	23,385,645

block in Lesotho. The Natal Drakensberg is such a formidable wall that in the 250 km (155 mi) between Oliviershoek and Qachas Nex, it is dissected only by a single pass. West of Sneeuberg there is a gap of 80 km (50 mi) in the escarpment, but the Nuweveld rises again after Beaufort West and continues into Namaqualand as the Roggeveld.

Lying between the Great Escarpment and the coast are the Marginal Lands, a zone varying in width from 80 to 240 km (50 to 149 mi) in the East and South to a mere 60 km (37 mi) in the West. From the northern Transvaal area southward through Natal and the Transkei, this zone consists of an inner range of foothills or coastal slopes. In this eastern marginal area, or Lowveld, the fall from the plateau edge to the coastline generally occurs in three descending surfaces, which results in a stepped terrain. Across these steps the chief rivers have cut valleys and canyons to depths of 1,200 m (3,937 ft). In the southern and southwestern Cape Province the Cape Range Mountains dominate the marginal areas. Between them and the Great Escarpment, lies the semiarid Great Karoo, between 450 and 700 m (1,476 and 2,296 ft) in elevation. The Cape folded belt is the core of this subregion. Its dominant feature is a series of parallel mountain ranges towering abruptly over broad, longitudinal valleys. The ranges fall naturally into a western and an eastern group.

Base 504638 10-80 (545702)

In the rugged Hex River mountains to the north of Worcester, they meet in a complex mountain knot where Matroosberg Peak towers to 2,260 m (7,915 ft). In the western folded belt from near Vanrhynsdorp in the North to Cape Hangklip in the South, the ranges stretch north to south and exhibit less intense folding. The western group includes the Sederberge, Bokkeveldberge and Great Winterhoek mountains, with elevations to nearly 2,100 m (6,890 ft). The eastern group includes the Langeberg and the Swartberge. The Langeberg stretches for almost 480 km (298 mi) toward Port Elizabeth and is separated from the Swartberge by the valley of the Little Karoo. The highest point in the Langeberg Mountains is nearly 2,300 m (7,545 ft).

Between the Atlantic coast and the western edge of the Great Escarpment there is a narrow belt of desert, a southward extension of the Namib Desert of Namibia. Except in this area and along the Indian Ocean in northeastern Natal, the coastal plains bordering the seashores are very narrow or entirely absent.

Lack of water is the principal characteristic of the country's hydrology. Not only is it dry country, but all its rivers discharge less than one-third of its annual rainfall. Worldwide 31% of all rain returns to the sea as river discharge, but only 9% in South Africa. The chief reason is the high rate of evaporation to be expected in a sunny country. In certain areas the loose, sandy soils and the nature of surface cover further restrict the runoff. The combined runoff of all South African rivers amounts to 54 billion cu m (1.906 trillion cu ft) only half that of the Zambezi and roughly equal to that of the Nile at Aswan and the Rhine at Rotterdam. The eastern plateau slopes covering 12% of the area, account for some 40% of the total runoff, but it is distributed over a large number of small and strongly dissected river basins, thus limiting its potential use. The Orange River system, on the other hand, drains almost the entire interior plateau, or 47% of the total land area, but accounts for only 22% of the total runoff. Truly perennial rivers occur over one-quarter of the surface, chiefly in the southern and southwestern Cape and on the eastern plateau slopes. Rivers flowing periodically (stationary during the dry season) are found over a further quarter of the surface. Over the entire western interior, rivers are episodic—i.e. they flow only sporadically after infrequent storms, while their beds are dry for the rest of the year. In the absence of lakes or permanent snowfields to stabilize the flow, even the perennial rivers flow irregularly and according to rainfall. Most are seasonal. Further, most South African rivers carry a heavy load of silt. The Orange carries more silt to the sea than any other African river including the Nile, which has a water discharge five times as large.

The Great Escarpment divides the South African river systems into two groups: the plateau rivers and those of the marginal areas. The large interior plateau surface is drained by the northern Limpopo and the central Orange River systems and the latter's tributary, the Vaal. In contrast, the marginal zones below the Great Escarpment are drained by numerous but smaller intermittent streams. The Orange is the giant among the South African rivers. Almost the entire plateau south of the Vaal-Limpopo divide—which runs east to west along the Witwatersrand—is drained by Orange and its tributaries. From the Mont-aux-Sources area it runs 2,250 km (1,397 mi) west to the Atlantic. Its major tributaries are the Caledon and the Vaal. The latter is actually longer and drains a much larger area than the main Orange headstream but contributes only 40% of the combined volume. Some 32 km (20 mi) beyond Kakamas the Orange plunges 147 m (482 ft) over Augrabies Falls to enter the final desolate stretch of its gorge tract. Of the marginal river basins the largest is the Tugela in Natal, which rises only a few hundred meters from the Orange on the Mont-aux-Sources plateau, plunging almost immediately over a spectacular 600 m (1,958 ft) fall to the foot of the escarpment. From here it runs to the Indian Ocean, a mere 250 km (155 mi) to the east. Despite its short course and small drainage basin of 29,000 sq km (11,197 sq mi) the Tugela has a larger volume than the Vaal. In southern Cape Province the mountains have caused the main streams to adopt a trellis drainage pattern. The major rivers in this area are the Gamtoos, Gouritz, Breede and Olifants.

With the exception of Lake Fundudzi in the Soutpansberg, there are no true lakes of any size in the country. The "lakes" at the Wilderness on the southern coast and at St. Lucia, Sibayi and Kosi on the Zululand coast are really freshwater lagoons. Innumerable pans are found in a wide belt from northern Cape Province to southern Transvaal. Pans are shallow stretches of water usually on flat ground and varying in size from a few thousand square meters to hundreds of square kilometers. The largest, Groot Vloer, is 64 km (40 mi) long and 40 km (25 mi) at its widest.

SOVIET UNION

BASIC FACTS

Official Name: Union of Soviet Socialist Republics

Abbreviation: USR

Area: 22,402,200 sq km (8,649,500 sq mi)

Area—World Rank: 1

Population: 286,434,844 (1988) 318,909,000 (2000)

Population—World Rank: 3

Capital: Moscow

Boundaries: 80,302 km (49,899 mi) North Korea 17 km (11 mi); China 7,520 km (4,673 mi); Mongolia 3,441 km (2,138 mi) Afghanistan 2,383 km (1,481 mi); Iran 1,974 km (1,227 mi); Turkey 617 km (383 mi); Romania 1,307 km (812 mi); Hungary 135 km (84 mi); Czechoslovakia 99 km (62 mi); Poland 1,215 km (755 mi); Finland 1,313 km (816 mi); Norway 196 km (122 mi)

Coastline: 60,085 km (37,335 mi) Insular coastline 48,261 km (29,988 mi)

Longest Distances: 10,944 km (6,800 mi) E-W 4,506 km (2,800 mi) N-S

Highest Point: Communism Peak 7,495 m (24,590 ft)

Lowest Point: Vpadina Karagiye 132 m (433 ft) below sea level

Land Use: 10% arable; 17% meadows and pastures; 41% forest and woodland; 32% other

The Union of Soviet Socialist Republics, the world's largest state in area and the third most populous in the world, occupies a dominant position in eastern Europe and northern Asia and covers much of central Asia. It extends over 9,646 km (6,000 mi) from the Baltic Sea in the west to the Bering Strait in the east and somewhat more-than 4,828 km (3,000 mi) from the Pamir Mountains in the south to the Arctic Ocean north of the Arctic Circle. This multinational state, comprising between one-sixth and one-seventh of the landmass of our planet, includes within its borders more than 40 percent of the area of Europe and Asia with over 100 ethnic groups.

The country is outlined in the north by the seas of the Arctic Ocean. To the west there is the Baltic Sea and, as a part of Europe, the Soviet Union has land frontiers with Norway, Finland, Poland, Czechoslovakia, Hungary, and Romania. To the south in Europe there is the Black Sea. Southern boundaries in Asia are with Turkey, Iran, Afghanistan, Communist China, the Mongolian People's Republic, and Korea.

The seas of the Arctic Ocean along the Soviet Union's northern coast include the White Sea, east of Finland, and the Barents Sea, which stretches to the islands of Franz Josef Land in the extreme north. Farther east lie sizable islands, including Novaya Zemlya and Vaygach, beyond which is the Kara Sea reaching the western coast of the Severnaya island group, just north of the Taymyr Peninsula. Beyond are the Laptev Sea and several chains of islands, and farther east are the East Siberian Sea and Wrangel Island. From north to south the Chukchi Sea, Bering Strait, and the Bering Sea separate Siberia from Alaska.

PRINCIPAL TOWNS
(estimated population in '000 at 1 January 1986)

Town	Population	Town	Population
Moskva (Moscow, the capital)	8,703	Zaporozhye	863
Leningrad	4,901	Voronezh	860
Kiyev (Kiev)	2,495	Lvov	753
Tashkent	2,073	Krivoy Rog	691
Baku	1,722	Kishinev	643
Kharkov	1,567	Yaroslavl	630
Minsk	1,510	Karaganda	624
Gorky	1,409	Ustinov (formerly Izhevsk)	620
Novosibirsk	1,405	Frunze	618
Sverdlovsk	1,316	Krasnodar	616
Kuybyshev	1,267	Tolyatti	610
Tbilisi	1,174	Vladivostok	608
Dnepropetrovsk	1,166	Irkutsk	602
Yerevan	1,148	Barnaul	586
Odessa	1,132	Khabarovsk	584
Omsk	1,124	Novokuznetsk	583
Chelyabinsk	1,107	Ulyanovsk	567
Alma-Ata	1,088	Dushanbe	565
Donetsk	1,081	Vilnius	555
Ufa	1,077	Tula	534
Perm	1,066	Penza	532
Kazan	1,057	Orenburg	527
Rostov-on-Don	993	Zhdanov	525
Volgograd	981	Kemerovo	515
Saratov	908	Astrakhan	503
Riga	890	Voroshilovgrad	503
Krasnoyarsk	885		

A major feature of the country's Pacific coastline is the Kamchatka Peninsula with its offshore Karaginsky and Commander islands. To the southwest is the Sea of Okhotsk, separated from the Pacific Ocean proper by the Kuril island chain which runs north to south from Kamchatka to Japan. The large Soviet island of Sakhalin forms the northern limit to the Sea of Japan and lies east of the Soviet mainland.

In Europe the country's coastline south of Finland stretches along the Gulf of Finland and the Baltic Sea, which border the Leningrad region, Estonia (with its two offshore islands of Hiiumaa and Saaremaa), Latvia, Lithuania, and the Kaliningrad (formerly Koenigsberg) region. The Black Sea limits the country on the south along the coasts of the Ukraine, including the Crimea, and the western Caucasus. To the east the Black Sea joins the Sea of Azov, which in turn delimits the eastern Crimea, southern Ukraine and northwestern Caucasus.

In relief, the country consists of a huge plain that extends from the western borders eastward on a broad front up to the moderately elevated Ural Mountains, separating Russia proper from Siberia, and continues across central Siberia. Farther on the land becomes a succession of plateaus and mountains that reach the sea along the country's eastern extremity. Mountain ranges, mostly of great elevation, extend along practically all of the southern border.

The country is rich in rivers and waterways, and its geographers claim that it possesses the greatest hydroelectric power potential in the world. Inland waters include the Caspian Sea, the largest inland body of water in the world, whose shores are shared in the south with Iran; the Aral Sea, just to the east in central Asia; Lake Baikal, in eastern Siberia; Lake Ladoga and Lake Onega, east of Leningrad; and Lake Balkhash, in eastern Kazakhstan.

NATURAL FEATURES

The great plain that comprises so much of the Soviet Union reaches from the western frontiers in Europe to the center of Asia. In the west it is broken by higher ground along the Barents Sea and the border with Finland, extending southward nearly to the Black Sea through the center of European Russia. In the southwest the hilly region of the Ukraine extends eastward to uplands along the Volga River and southward to the mountains of the Caucasus. Farther east, the great plain is interrupted by the chain of the Ural Mountains, which traditionally have been considered a dividing line between Europe and Asia.

South of the Ural Mountains and extending eastward around the Caspian Sea, lowlands extend into Asia as far

Area and population

Soviet Federated Socialist Republic	Capitals	area sq mi	area sq km	population 1986 estimate
Russian S.F.S.R.	Moscow	6,592,800	17,075,400	144,080,000
Soviet Socialist Republics				
Armenian	Yerevan	11,500	29,800	3,362,000
Azerbaijan	Baku	33,400	86,600	6,708,000
Belorussian	Minsk	80,200	207,600	10,008,000
Estonian	Tallinn	17,400	45,100	1,542,000
Georgian	Tbilisi	26,900	69,700	5,234,000
Kazakh	Alma-Ata	1,049,200	2,717,300	16,028,000
Kirgiz	Frunze	76,600	198,500	4,051,000
Latvian	Riga	24,600	63,700	2,622,000
Lithuanian	Vilnius	25,200	65,200	3,603,000
Moldavian	Kishinyov	13,000	33,700	4,147,000
Tadzhik	Dushanbe	55,300	143,100	4,648,000
Turkmen	Ashkhabad	188,500	488,100	3,270,000
Ukrainian	Kiev	233,100	603,700	50,994,000
Uzbek	Tashkent	172,700	447,400	18,487,000
TOTAL LAND AREA		8,600,400	22,274,900	278,784,000
INLAND WATER		49,100	127,300	
TOTAL		8,649,500	22,402,200	

as the foothills of the high ranges of the Pamirs and the Tien Shan, Altai, and Sayan mountains that rise abruptly along the Soviet Union's southern border. Lowlands continue to the north beyond Lake Balkhash and the moderately elevated uplands of Kazakhstan and slope as a wide plain across West Siberia to the shores of the Arctic Ocean. Farther east, beyond the Yenisey River, the land rises to form a hilly plateau region across central Siberia that merges into a complex of mountains and drainage basins extending to the country's eastern seaboard.

Europe's greatest river, the Volga, from a source in the Valdai Hills southeast of Leningrad, flows southeastward past Moscow and then directly southward in the extreme east of Europe into the Caspian Sea. The Volga River, together with its tributaries, is of more importance for transportation purposes than all the other rivers of the country combined. The greatest rivers in point of flow and length, however, are east of the Ural Mountains. These include the Irtysh-Ob, Yenisey, Angara, and the Lena. They all flow northward into the seas of the Arctic Ocean, and their value for transportation purposes is further limited by the fact that in the main their direction crosses, rather than coincides with, the great traffic routes across central Asia.

The country's northern seacoasts are never completely ice free except for the northeast coastal region of the Kola Peninsula with the port of Murmansk, which remains ice free because of the effects of the Gulf Stream. The remainder of the northern coastline is noteworthy only for its insignificant ports and a need for ice-breakers and spotter planes to assist shipping even in the summer when the ice breaks up for a few months. The Pacific seaboard, by contrast, has weather modified by the maritime influence of the Pacific Ocean. Its port city of Vladivostok overlooks Peter the Great Bay which, although icebound in winter, is free of ice in the warm weather months.

The Baltic ports in the far west of the country are closed by ice for several months each winter. The Baltic Sea itself may be viewed as an inland lake rather than an

AUTONOMOUS REPUBLICS

Autonomous Republic	Area (sq km)	Population ('000, 1 January 1986)	Capital	Population of capital ('000, 1 January 1986)
Within RSFSR:				
Bashkir	143,600	3,864	Ufa	1,077
Buryat	351,300	1,013	Ulan-Ude	335*
Chechen-Ingush	19,300	1,222	Grozny	393*
Chuvash	18,300	1,319	Cheboksary	389*
Daghestan	50,300	1,754	Makhachkala	301*
Kabardino-Balkar	12,500	722	Nalchik	227*
Kalmyk	75,900	324	Elista	81*
Karelian	172,400	787	Petrozavodsk	255*
Komi	415,900	1,227	Syktyvkar	213*
Mari	23,200	729	Yoshkar-Ola	231*
Mordovian	26,200	963	Saransk	307*
North Ossetian	8,000	615	Ordzhonikidze	303*
Tatar	68,000	3,531	Kazan	1,057
Tuva	170,500	282	Kyzyl	75*
Udmurt	42,100	1,572	Ustinov (Izhevsk)	620
Yakut	3,103,200	1,001	Yakutsk	180*
Within Azerbaidzhan:				
Nakhichevan	5,500	272	Nakhichevan	37†
Within Georgia:				
Abkhasian	8,600	530	Sukhumi	126*
Adzhar	3,000	383	Batumi	132*
Within Uzbekistan:				
Kara-Kalpak	165,600	1,106	Nukus	139*

*At 1 January 1985.
†At 1 January 1976.

open sea. To reach an open sea, Soviet shipping must go on to the North Sea beyond Denmark and Norway.

To the south the Black Sea, like the Baltic, resembles a lake. In this case it opens into an even greater one, the Mediterranean Sea. Despite its Black Sea location, Odessa, the leading southern Soviet port, is frozen in for about six weeks each winter.

REGIONAL ANALYSIS

Baltic-Belorussia Region

This region is made up of a broad glacial plain that stretches from the shores of the Baltic Sea southward along the Soviet Union's western borders across the lowlands of the Pripet Marshes to the higher elevations of the Ukraine. Its area includes Estonia, Latvia, Lithuania, and the Belorussian union republic and the area on the Baltic Sea around Kaliningrad, once Prussian Koenigsberg. The main urban places include the Baltic port cities of Tallin, Riga, and Kaliningrad. Inland are Vilnius and Minsk, administrative capitals of Lithuanian and the Belorussian union republic, respectively and Brest-Litovsk, an important railway junction for traffic between the Soviet Union and the rest of Europe.

Most of the region is below 152 m (500 ft) in elevation, and a good deal of the land, especially in the south, is either boggy or interspersed with small lakes and swamps. From a low divide that crosses the center of the region, the Western Dvina and Bug rivers and their tributaries drain northwestward to the Baltic Sea. South of the divide, drainage into the Pripet and Dnepr rivers flows to

Soviet Union
International boundary
National capital
Railroad
0 400 800 Kilometers
0 400 800 Miles
Names are not necessarily authoritative.
The United States Government has not recognized the incorporation of Estonia, Latvia, and Lithuania into the Soviet Union. Other boundary representation is not necessarily authoritative.
Kuril Islands (administered by Soviet Union)
Occupied by Soviet Union since 1945, claimed by Japan
North Atlantic Ocean
Norwegian Sea
North Sea
Arctic Ocean
Barents Sea
Kara Sea
Laptev Sea
East Siberian Sea
Chukchi Sea
Bering Sea
Sea of Okhotsk
Sea of Japan
Yellow Sea
East China Sea
Philippine Sea
Baltic Sea
Black Sea
Caspian Sea
Aral Sea
Persian Gulf
Gulf of Bothnia
Svalbard (NOR.)
Franz Josef Land
Novaya Zemlya
Severnaya Zemlya
New Siberian Islands
Wrangel I
Komandorskiye Ostrova
Sakhalin
Arctic Circle
Lake Ladoga
Lake Onega
Lake Baikal
Lake Balkhash
Ozero Issyk-Kul'
NORWAY
SWEDEN
FINLAND
POLAND
ROMANIA
TURKEY
SYRIA
IRAQ
IRAN
AFGHANISTAN
PAK.
CHINA
MONGOLIA
NORTH KOREA
SOUTH KOREA
JAPAN
SAUDI ARABIA
KUWAIT
Estonia
Latvia
Lithuania
Moscow
Leningrad
Arkhangel'sk
Murmansk
Vorkuta
Salekhard
Noril'sk
Yakutsk
Magadan
Petropavlovsk-Kamchatskiy
Khabarovsk
Vladivostok
Irkutsk
Novosibirsk
Omsk
Sverdlovsk
Tashkent
Alma-Ata
Baku
Tbilisi
Yerevan
Kiev
Minsk
Odessa
Ulaanbaatar
Beijing
Tōkyō

the Black sea. A canal between the Pripet and Bug rivers provides barge service across the border with Poland.

Ukraine-Moldavia Region

A land of broad plains and low, rolling hills, with a few wide river valleys, this region extends for roughly 1,287 km (800 mi) across the southern part of Russia in Europe as far as the hilly area north of the Sea of Azov on the western side of the Don river. Politically, the area includes the Ukrainian and the Moldavian union republics, of which the former is by far the larger and more populous. From the edge of the Carpathian Mountains on the west, the land descends to the east. Most of the region is below 457 m (1,500 ft) in elevation and of nearly featureless relief. Major drainage is to the south and east. The Dnepr River flows to the Black Sea through the center of the region, roughly paralleled on the west by the Dniester and Prut rivers and on the east by the Donets River and other tributaries of the Don River.

Transcaucasia

This area is made up of the Transcaucasian union republics of Georgia, Armenia, and Azerbaijan, situated between the Black and Caspian seas and bordering on Turkey and Iran in the south. The region includes most of the high range of the Caucasus Mountain system, some of the summits of which reach 4,572 m (15,000 ft) or more. In Georgia, the Black Sea and the shelter of the mountains assure a moderate, almost subtropical, climate and exceptionally high precipitation for the union. Armenia to the south is very mountainous, and the land under cultivation is arid and requires irrigation. Farther to the east in Azerbaijan, a similar situation exists except for the southern triangle, which is subtropical. In Georgia forest abound, alternating with mountain meadow and steppe soils.

The Russian Soviet Federated Socialist Republic in Europe

Extending eastward from the Baltic-Belorussian-Ukrainian regions and including the land north of Transcaucasia, the Russian Soviet Federated Socialist Republic (RSFSR) in Europe is largely a plain, interrupted by a few ranges of hills and wide river valleys that terminate at the Ural Mountain chain, generally considered as marking the dividing line between Europe and Asia. In the far north a narrow belt of tundra extends inland from the Arctic shoreline. Most of the northern half of the region, however, has natural vegetation, consisting of coniferous forest interspersed with peat bogs and marshes. South of this main belt are deciduous mixed forests that become steppes with trees farther south, the steppe recurring in the region of the northern Caucasus. East of the mouth of the Volga River and northeast of the Caspian Sea begins the Great Central Asian area of semideserts and deserts. The Ural Mountains, which divide Europe and Asia, are

AUTONOMOUS REGIONS

Autonomous Region	Area (sq km)	Population ('000, 1 January 1985)	Capital	Population of capital ('000, 1 January 1985)
Within RSFSR:				
Adygei	7,600	422	Maikop	140
Gorno-Altai	92,600	179	Gorno-Altaisk	40*
Jewish	36,000	207	Birobidzhan	78
Kharachayevo-Cherkess	14,100	390	Cherkessk	102
Khakass	61,900	540	Abakan	147
Within Azerbaidzhan:				
Nagorno-Karabakh	4,400	174	Stepanakert	35*
Within Georgia:				
South Ossetian	3,900	99	Tskhinvali	34*
Within Tadzhikistan:				
Gorno-Badakhshan	63,700	146	Khorog	15*

*At 1 January 1976.

covered by mountain forests as is the Caucasus barrier in the south, which has a belt of warm temperate, broadleaf forest.

The Russian Soviet Federated Socialist Republic in Asia

This broad region stretching eastward from the Ural Mountains north of the Kazakh union republic, Communist China, and the Mongolian People's Republic consists of Siberia proper, which reaches over 3,218 km (2,000 mi) from the Ural Mountains to the Lena River, and the Far East, extending thence to the Pacific seaboard. It covers over half the total area of the Soviet Union and is potentially the richest area.

In relief, the region consists of the west and the north Siberian Lowlands interrupted by the Irtysh, Ob', and Yenisey rivers. The last-named marks the beginning of the Central Siberian Plateau. This plateau is, in turn, interrupted by a large triangular plain, the Central Yakutsk Lowlands, formed by the bend of the Lena River. Beyond the Lena River, the high and confused Verkhoyansk, Cherski, and Kolyma mountain systems range northeast toward the Bering Strait. Along its southern borders the region is delimited by the Altai and Sayan mountains, the Mongolian desert, and the Amur and Ussuri rivers.

Central Asia Region

South of the RSFSR in Asia and extending eastward some 2,414 km (1,500 mi) to the mountainous border of Communist China, this region reaches farther south than any other part of the Soviet Union. It includes as major political units the Kazakh, Turkmen, Uzbek, Tadzhik, and Kirgiz union republics. From the high mountains that rise along the southern and southeastern part of this region, the ground descends to much lower elevations with drainage to the north. Beyond the foothills, the land consists largely of flat to moderately hilly plateau country, crossed by a few major river valleys, and of lowland plains along the Caspian and Aral seas.

SPAIN

BASIC FACTS

Official Name: Spanish State

Abbreviation: SPN

Area: 504,783 sq km (194,898 sq mi)

Area—World Rank: 48

Population: 39,209,765 (1988) 40,747,000 (2000)

Population—World Rank: 25

Capital: Madrid

Boundaries: 5,849 km (3,635 mi) France 647 km (402 mi); Andorra 65 km (40 mi); Gibraltar 1 km (0.6 mi); Portugal 1,232 km (766 mi)

Coastline: Mediterranean Sea 1,670 km (1,038 mi); Atlantic and Bay of Biscay 2,234 km (1,388 mi)

Longest Distances: 1,085 km (764 mi) E-W; 950 km (590 mi) N-S

Highest Point: Teide Peak 3,715 m (12,188 ft)

Lowest Point: Sea Level

Land Use: 31% arable; 10% permanent crops; 21% meadows and pastures; 31% forest and woodland; 7% other

Spain is the third-largest country in Europe, occupying the greater part of the Iberian Peninsula, which it shares with Portugal, and including the Balearic Islands and the Canary Islands.

There are no serious boundary disputes between Spain and its neighbors. The Pyrenees are one of Europe's most effective natural boundaries. The French boundary runs along the highest terrain of the main portion of the range. The frontier with Andorra was delimited in 1863. The border with Portugal follows rivers—the Mino, the Duero, the Tajo (Tagus) and the Guadiana—for about half the length, but neither the rivers nor the land boundaries present significant natural barriers. Gibraltar is British territory, but British sovereignty over the promontory has been contested by Spain for centuries.

Peninsular Spain consists of a central plateau known as the Meseta, which is enclosed by high mountains on the northern, southern, eastern, and part of the western sides. The Meseta also encompasses several mountain systems that are lower than the peripheral mountains. Although Spain thus has physical characteristics that make it to some extent a naturally insular geographic unit, other geographic features and climatic differences tend to compartmentalize the country.

Topographically Spain is divided into four parts: the temperate region to the north and northwest; the marginal mountain ranges; the Meseta and the interior region; and the coastal regions. The categories are far from clear-cut. The North and Northwest, for example, include significant portions of the mountains and coastal regions. Further, the Meseta contains two large low-lying river valleys and is traversed by several major mountain systems.

The North and Northwest comprise Galicia and the provinces adjacent to the Bay of Biscay. Galicia is hilly, having an average elevation of 610 m (2,000 ft). The verdant region is narrower to the east of Galicia, being confined to the oceanside slopes of the mountains. The Río Sil, which flows into the Bay of Biscay just west of Aviles in Asturias, marks the divide between the predominantly granite Galician highlands and the limestones of the Cordillera Cantabrica (Cantabrian Mountains).

Breaks of any size in the ring of highlands that enclose the peninsula occur only in southern Portugal and to the southeast of Portugal along Spain's portion of the Gulf of Cadiz. The highlands of the Northwest near the coast are relatively low, but the hilly countryside blends into the Cordillera Cantabrica. This coastal range extends across the country parallel and adjacent to the Bay of Biscay and descends into it. There are drops exceeding 1,524 m (5,000 ft) within 32 km (20 m) of the shore. Generally, peaks in the Cordillera Cantabrica range from 2,133 to 2,938 m (7,000 to 8,000 ft) with the highest point in the Picos de Europa, about 88 km (55 mi) west-southwest of Santander. The main part of the range extends about 193 km (120 mi) with two major but difficult passes. The lower pass, 984 m (3,230 ft) high, is the route between Santander and Valladolid, and it has 22 rail tunnels in 34 km (21 mi). In the higher pass, 1,283 m (4,210 ft) between Oveido and León, there are 60 tunnels and nine viaducts as the railroad winds through 43 km (27 mi) to cover 11 km (7 mi) as the crow flies.

The Pyrenees extend for about 418 km (260 mi) between the Bay of Biscay and the Mediterranean. Its highest portion is 241 km (150 mi) in length, with an elevation of over 1,524 m (5,000 ft) and its width averages 80 km (50 mi), with 129 km (80 mi) at the widest point. The range descends more abruptly on the French side. The French-Spanish border connects six of the highest peaks, but only one entirely within France reaches 3,048 m (10,000 ft). On the Spanish side three exceed 3,353 m (11,000 ft), the highest, Pico de Aneto, is 3,403 m (11,165 ft). There are few passes across the Pyrenees, but two railroads make the crossing. The route north from Zaragoza includes a long tunnel, while the other, which crosses east of Andorra, is entirely a surface route. Most streams drain to the Spanish side through deep gorges dissecting sparsely populated wastelands. The mountains along the northern two-thirds of the Mediterranean coast are less spectacular.

To the south of Barcelona, the mountains join the coastal end of the Sistema Iberico (Iberian Mountains), which extend southeastward from the Basque provinces and form the southern border of the Río Ebro Valley. The nearly 20,725 sq km (8,000 sq mi) of this region are generally barren and rugged terrain. The Spanish call it the "area of difficulty."

The Sistema Penibetico (Andalusian Mountains) of the lower Mediterranean region and of southernmost Spain

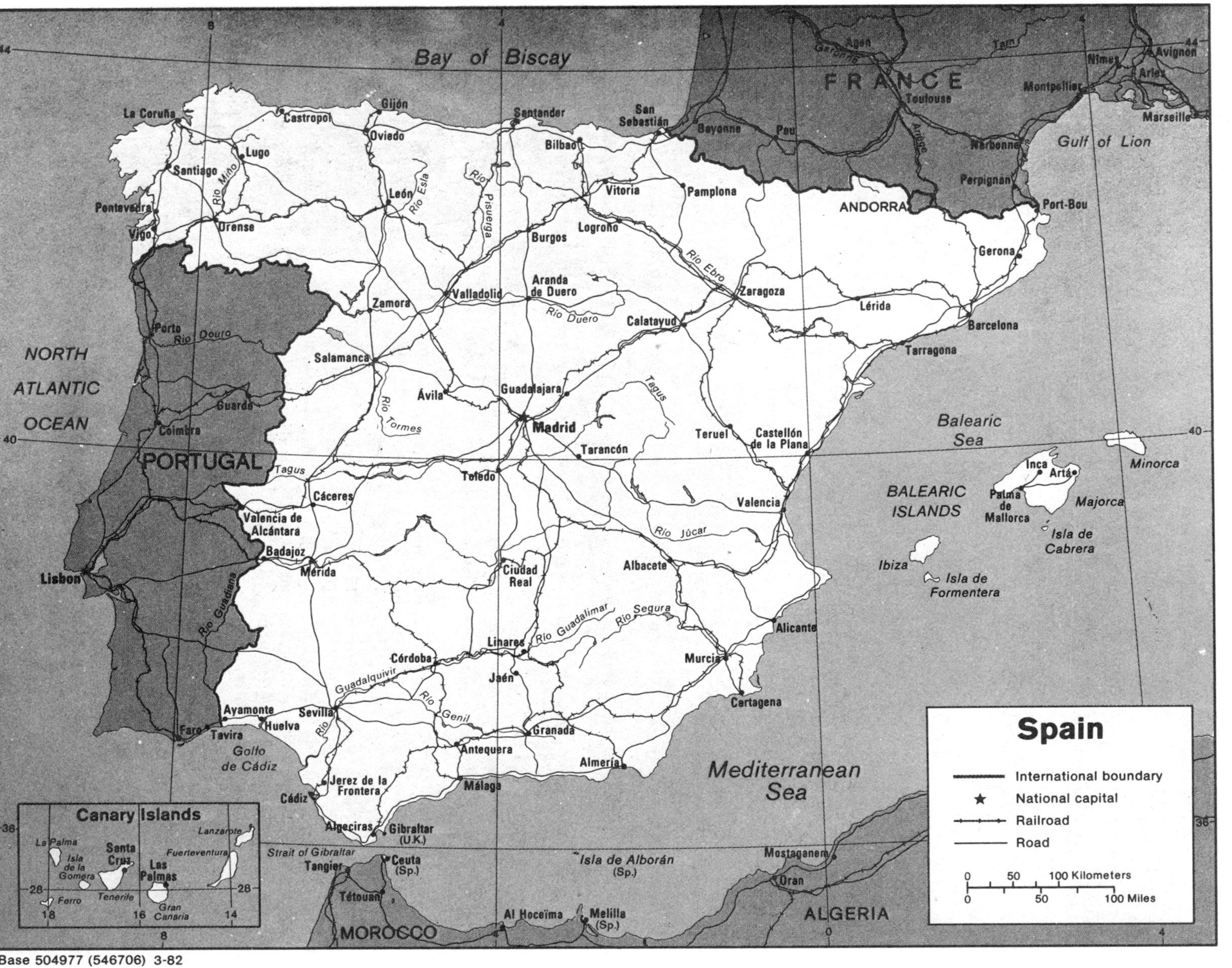

Base 504977 (546706) 3-82

extend southwestward from Cabo de la Nao in Valencia to Gibraltar, a distance of some 579 km (360 mi). The range continues northeastward into the Mediterranean Sea and reappears as the Balearic Islands. The most impressive part of this range is the Sierra Nevada, much of it desolate. It is close to the Mediterranean coast from Gibraltar to Cabo de Gata, and its 3,478 m (11,411 ft) peak, Mulhacen, 32 km (20 mi) southeast of Granada, is the highest point in the peninsula. The coastal Sierra Nevada is separated from a sister range in the North by a geological fault line that runs roughly parallel to the coast. The northern range is equally forbidding, with the exception of a few exotic places, such as Granada.

The Meseta, the vast Spanish tableland, dominates central Spain from the Cordillera Cantabrica in the North to the Sierra Morena in the South and from the Portuguese border in the West to the Sistema Iberico in the east. Generally the Meseta has elevations between 610 and 762 m (2,000 and 2,500 ft), but the Cordillera Carpetovetonica north of the Río Tajo and the lesser ranges south and west of Toledo are smaller in elevation. The northern portion of the Meseta includes most of the Río Duero basin and extends over the flatter lands of the León and Old Castile. The southern Meseta covers the Río Tajo and Río Guadiana basins, extending over most of Extremadura, New Castile, northern Murcia and Andalusia. The Meseta gives way to higher land in the western part of the country between the basins of its three largest rivers. The Cordillera Carpetovetonica has several elevations between 2,133 and 2,591 m (7,000 and 8,500 ft) between the Duero and Tajo basins and northwest of the city of Madrid. The Toledo Mountains between the Tajo and the Guadiana are lower, as is the Sierra Morena in northern Andalusia, with elevations between 152 and 610 m (500 and 2,000 ft). Included in this region are the Río Ebro valley in northeastern Spain and the basin of the Río Guadalquivir in the southwestern Andalusian plain. The Andalusian plain is the only low-lying area that permits easy entry from the sea. Ocean tides in the Guadalquivir reach inland as far as the city of Seville. At Cordova the elevation is only about 152 m (500 ft). The lower Guadalquivir valley is marshy and frequently saline, but around Seville are some of the best agricultural areas of the country. The Ebro basin in the Northeast resembles the Meseta, with a flat, monotonous and barren landscape. Elevations are roughly 305 m (1,000 ft), somewhat higher above Tudela and lower below. Many streams have cut deep ravines in the soft surface rocks.

The northern, or Cantabrian, coast extends about 724 km (450 mi) from France to the northwestern corner of the country. Although the Bay of Biscay is often wild and stormy, with strong westerly or northerly winds, the coastal area is generally pleasant, especially in spring and summer, around San Sebastian and along the Costa Verde of Asturias. The Galician coast from La Coruna to the Portuguese border has a rough terrain, rendering its harbors inaccessible from the interior. The southern Atlantic coast, called Costa de la Luz because of its bright sunshine, is also difficult of access, while the wide mouth of the Guadalquivir is a salt marsh. The Costa del Sol, extending eastward about 274 km (170 mi) from Gibraltar to Cabo de Gata, is relatively narrow, but its exposure to the southern sun has made it a popular resort area. The Levantine coast continues northeast from Cabo de Gata to Cabo de la Nao. This section, known as the Costa Blanca, has white and sunny beaches facing the warm Mediterranean Sea. Above it is the somewhat less famous Costa del Azahar, extending from Cabo de la Nao to the Río Ebro delta, more an agricultural area than a resort. The Catalonian coast extends from the Río Ebro delta to the French border. Its southern portion, the Costa Dorada, extends from the Río Ebro delta to about 98 km (30 mi) north of Barcelona, while the northern section, the Costa Brava, is where the foothills of the Pyrenees meet the sea. Both have a rugged appearance because of their proximity to the mountains.

Spain has some 1,800 rivers and streams, of which only the Río Tajo is more than 965 km (600 mi) long and all but 90 are less than 97 km (60 mi) long. Many of the riverbeds are dry most of the year. The high mountain ranges divert rain-bearing weather systems from the interior of the peninsula. Most of the country slopes to the west and drains toward the Atlantic. The Río Ebro, rising in the Pyrenees, is the only major river flowing into the Mediterranean. There are a few moderately long rivers flowing from the Sistema Iberico, but elsewhere the rivers flowing into the Mediterranean are short coastal ones. The Meseta and the surrounding mountains drain with only minor exceptions to the Atlantic via four major rivers; the Duero, the Tajo, the Guadiana and the Guadalquivir. Lesser rivers are characteristically short, swift and torrential, as well as extremely seasonal and irregular in the volume of water they carry. Thus they are useless for irrigation, and navigation is hampered because their valleys are deep, with steep sides.

Three of the major westward-flowing Meseta rivers—the Duero, the Tajo and the Guadiana—drain nearly half the country. The Duero rises in the Sierra de la Demanda in the vicinity of Soria, forms part of the Spanish-Portuguese border and joins the Atlantic at Porto in Portugal. The Tajo rises in the highlands about 161 km (100 mi) east of Madrid, flows west across central Spain, forms a section of the border with Portugal and empties into the Atlantic at Lisbon (where it is known as the Tagus). The Guadiana rises in La Mancha, its several headstreams uniting in a group of small lakes—Los Ojos de la Guadiana (The Eyes of the Guadiana)—northeast of Ciudad Real. It then flows west to the Portuguese border near Badajoz, turns south along the border, flows through Portugal for about 121 km (75 mi), rejoins the border and follows it to the Gulf of Cadiz. All the Meseta rivers are sluggish most of the year except for a few days each spring and fall, when the raging waters fill the riverbeds. Even the Tajo, the largest of the three, is variable in its volume of water.

South of the Meseta and the Sierra Morena and draining most of the Andalusian plain, the Río Guadalquivir possibly is the country's most valuable and consistent river. Its lower course reaches the Gulf of Cadiz, making Seville the only inland river port. It rises northeast of Granada and is fed by tributaries that drain higher elevations.

PRINCIPAL TOWNS*

(Population at census of 1 March 1981)

Town	Population	Town	Population
Madrid (capital)	3,188,297	Santander	180,328
Barcelona	1,754,900	Jerez de la Frontera	176,238
Valencia	751,734	San Sebastián	175,576
Sevilla (Seville)	653,833	Cartagena	172,751
Zaragoza (Saragossa)	590,750	Salamanca	167,131
Málaga	503,251	Leganés	163,426
Bilbao	433,030	Elche	162,873
Las Palmas de Gran Canaria	366,454	Cádiz	157,766
Valladolid	330,242	Burgos	156,449
Palma de Mallorca	304,422	Terrassa	155,360
L'Hospitalet de Llobregat	294,033	Mostolés	149,649
Murcia	288,631	Alcalá de Henares	142,862
Córdoba	284,737	Almería	140,946
Granada	262,182	Alcorcón	140,657
Vigo	258,724	Santa Coloma de Gramanet	140,588
Gijón	255,969	León	131,132
Alicante	251,387	Huelva	127,806
La Coruña (Corunna)	232,356	Getafe	127,060
Badalona	227,744	Castellón	126,464
Sabadell	194,943	Baracaldo	117,422
Vitoria	192,773	Albacete	117,126
Santa Cruz de Tenerife	190,784	Badajoz	114,361
Oviedo	190,123	La Laguna	112,635
Pamplona	183,126	Tarragona	111,689
		Logroño	110,980
		Lleida (Lérida)	109,573

*Population figures refer to *municipios*, each of which may contain some rural area as well as the urban centre.

The Río Ebro rises in the easternmost part of the Cordillera Cantabrica in Basque country and flows southeast between the Pyrenees and the Sistema Iberico into the Mediterranean in southern Catalonia. Its basin compares with that of the Duero and the Tajo and covers about one-sixth of the country, while in volume of water it compares with the Guadalquivir. Although navigation is treacherous in the gorges through the coastal highlands, the river is navigable by barge in the flat region of its basin during rainy seasons southeastward from Tudela, some 97 km (60 mi) above Zaragoza. The major Galician river, the Río Mino, carries a volume of water equal to or greater than that of the Ebro, although the Mino's course is less than half as long and its basin covers only about a fifth as much area. The Mino rises in north-central Galicia, flows south and southwest until it forms part of the border with Portugal and continues to the Atlantic Ocean. Its flow is more constant than that of the Ebro during the year as well as from year to year.

Rivers in northern Catalonia flow all year, as they are fed from the area south of the Pyrenees that receives abundant precipitation. Along the entire coastal area south of Barcelona, however, the streams are dry most of the year, since they draw water from a lower terrain. The rivers of the North that flow into the Atlantic and Bay of Biscay are short, the Mino being the only exception. Many of them encounter the sea through *rías*—deep estuaries extending from the mountains to the sea, similar to fjords formed as the Atlantic gales scour the sinking land.

In spite of the vast arid areas, Spain has a variety of natural vegetation. However, the Meseta is substantially denuded, and scrub growth has replaced forests except in the North and Northwest. Generally, conifers predominate

Area and population

Autonomous communities	Capitals	area sq mi	area sq km	population 1986 estimate
Andalucía	Seville (Sevilla)	33,694	87,268	6,735,600
Aragón	Zaragoza	18,398	47,650	1,215,600
Asturias	Oviedo	4,079	10,565	1,140,100
Baleares	Palma de Mallorca	1,936	5,014	675,400
Canarias	Santa Cruz de Tenerife	2,796	7,242	1,442,500
Cantabria	Santander	2,042	5,289	527,400
Castilla-La Mancha	Toledo	30,591	79,230	1,670,100
Castilla-León	Valladolid	36,368	94,193	2,602,300
Cataluña	Barcelona	12,328	31,930	6,057,200
Extremadura	Mérida	16,063	41,602	1,084,400
Galicia	Santiago de Compostela	11,365	29,434	2,870,900
La Rioja	Logroño	1,944	5,034	263,100
Madrid	Madrid	3,087	7,995	4,907,100
Murcia	Mrucia	4,370	11,317	1,007,500
Navarra	Pamplona	4,023	10,421	522,500
Pais Vasco	Vitoria	2,803	7,261	2,176,800
Valencia	Valencia	8,998	23,305	3,790,200
TOTAL SPAIN		194,885	504,750	38,688,400
Enclaves in Northern Morocco				
Ceuta	—	7.1	18.5	71,400
Melilla	—	5.4	14	58,600
Chafarinas	—	.24	.61	. . .
Vélez de la Gomera	—	.02	.04	. . .
Alhucemas	—	.004	.01	. . .
TOTAL		194,897.79	504,783.16	38,818,400

in the Pyrenees, at higher elevations in the Meseta mountains and in northeastern Catalonia. Deciduous trees predominate throughout the northern mountains, the middle elevations of the Pyrenees and in a few lower mountain forests of west-central Spain. In the Pyrenees the deciduous zone extends from about 1,295 to 1,600 m (4,250 to 5,250 ft) while hardier pine species survive from 1,600 to 2,133 m (5,250 to 7,000 ft). In the Sierra Nevada, natural vegetation consists largely of steppe grasses and scrub. Lowlands and coastal areas have mostly oak forests, to which pine, walnut, and chestnut have been added by man. Beech replaces oak at higher elevations.

Much of the Cordillera Cantabrica, the higher Pyrenees, the Sierra Nevada and areas west of Cabo de la Nao have silty sand and gravel and extensive areas devoid of soil. Shallow, silty sandy soils cover most of the Meseta and lower elevations of the East and South, and deeper silty sands occur in the Cordillera Carpetovetonica and the Sierra Morena and more extensively in Galicia and in the foothills of the Cordillera Cantabrica. Very deep poorly graded sands and gravels occur to the west of the marshland in the lower Río Guadalquivir valley. Coarse, deep, clayey soils are found in mountain foothills and major river valleys. Very deep clayey sand over silty sand and gravel occurs in the southern foothills of the Cordillera Carpetovetonica. Only about 10% of the country has finer-grained soils. They predominate in the Río Guadalquivir valley and in the low elevations adjacent to Gibraltar. Shallower fine clay and clay and silt mixtures are found in New Castile and Salamanca. Unfortunately, most of the terrain where the finer soils occur is arid or marshy. Alluvial soils make up most of the river valleys. Fine clay and silt are deposited by slower-moving streams. Humus is abundant in mountainous regions but is useless for agriculture. Erosion is particularly severe in the Meseta, where it has left *calveros* (barren places). Saline soils are frequent in the Río Ebro depression and in the Río Guadalquivir marshes.

SRI LANKA

BASIC FACTS

Official Name: Democratic Socialist Republic of Sri Lanka

Abbreviation: SLN

Area: 65,610 sq km (25,332 sq mi)

Area—World Rank: 114

Population: 16,639,695 (1988) 19,752,000 (2000)

Population—World Rank: 47

Capital: Colombo

Land Boundaries: Nil

Coastline: 1,204 km (748 mi)

Longest Distances: 435 km (270 mi) N-S 225 km (140 mi) E-W

Highest Point: Pidurutalagala 2,524 m (8,281 ft)

Lowest Point: Sea Level

Land Use: 16% arable; 17% permanent crops; 7% meadows and pastures; 37% forest and woodland; 23% other

Sri Lanka is an island country in the Indian Ocean, bounded on the west by the Gulf of Mannar and on the east by the Bay of Bengal. It is 434 km (270 mi) long from Point Pedro, its northernmost point, to Dondra Head, its southern tip, and 225 km (140 mi) wide from Colombo on the west coast to Sangamankanda Point on the east coast. Colombo is its capital and chief port.

The island is separated from the southeast tip of the Indian mainland by a narrow strait called Palk Strait. At one time Sri Lanka was geographically a portion of the Indian subcontinent, as evidenced by the continental shelf upon which the island stands. Barely 29 km (18 mi) of shallow sea separates the island from India, a distance easily covered by train ferry. An elevated portion of the continental shelf forms a chain of rocky islands known as Adam's Bridge, located just south of the ferry route. Here Palk Strait widens into the Gulf of Mannar, which in turn merges into the Indian Ocean.

The Geographical Regions

The island is a pear-shaped mass of crystalline rock on which three levels of ground can be distinguished. The first is a coastal belt that rises from sea level to 30 m (100 ft) above and is perhaps 40 km (25 mi) across at its widest point. The second is a belt of rolling plain striated with ridges rising to 152 m (500 ft) in the south. The third is an irregularly shaped mass of hills and mountains having heights over 1,829 m (6,000 ft).

The coastal belt is ringed with palm trees, both coconut and palmyra. Sandy beaches stretch for miles, indented here and there by coastal lagoons, which are useful chiefly for fishing and for collecting salt. Trincomalee, on the northeast, is situated on a natural rock harbor. Not only is it exceptionally beautiful, but also it is large enough and deep enough to accommodate a modern naval fleet. On the other side of the island, on the southern coast, is the harbor of Galle, also of natural rock but much smaller than Trincomalee. Galle has ancient historical associations, as it was the principal landing for trading vessels on the Arabia-Far East spice route.

A coral reef encircles the southern two-thirds of the island and is rich in sea life. Divers obtain pearls from the famous pearl banks in the Gulf of Mannar, off the northwestern coast. The waters surrounding the island are so deep that Sri Lanka is almost unaffected by tidal variations. The daily range is only a few inches. The highest tide recorded was under 1.2 m (4 ft). The ebb and flow of tides usually scoop out accumulated silt deposited in river mounts and carry it out to sea, preventing harbors from closing up; in Sri Lanka, however, because of the near absence of tides, the principal rivers are all blocked by sandbars.

There are many picturesque place names whose derivation can be easily guessed: Foul Point, Elephant Point; Portugal Bay, and Devil's Point, to name a few. Dutch, Portuguese, and British occupation led to such names as Kayts Island, Delft Island, Point Pedro, Back Bay, and Buffalo Island.

Eight to 40 km (5 to 25 mi) inland from the coastal belt the land changes character. Sandy and saline soil gives way to fertile loams, humic soils and, in the north, limestone. It is land lying 30 to 152 m (100 to 500 ft) above sea level and composes the major portion of Sri Lanka. For descriptive purposes, it is best divided into three sections.

The Southwestern Section

The area from the Deduru Oya, north of Chilaw on the western coast, to Hambantota on the southern coast is a series of ridges and valleys. Close to the sea the ridges are low and parallel to the coast, but inland toward the central mountains they are seen against the sky as mountain chains, alternating with long, narrow depressions. Millions of years ago the land was much higher in this section, but rain and rivers wore down all rock but the hardest, leaving the ridges as they are today. This process of erosion transferred the top soil to the valleys, creating some very fertile rice land.

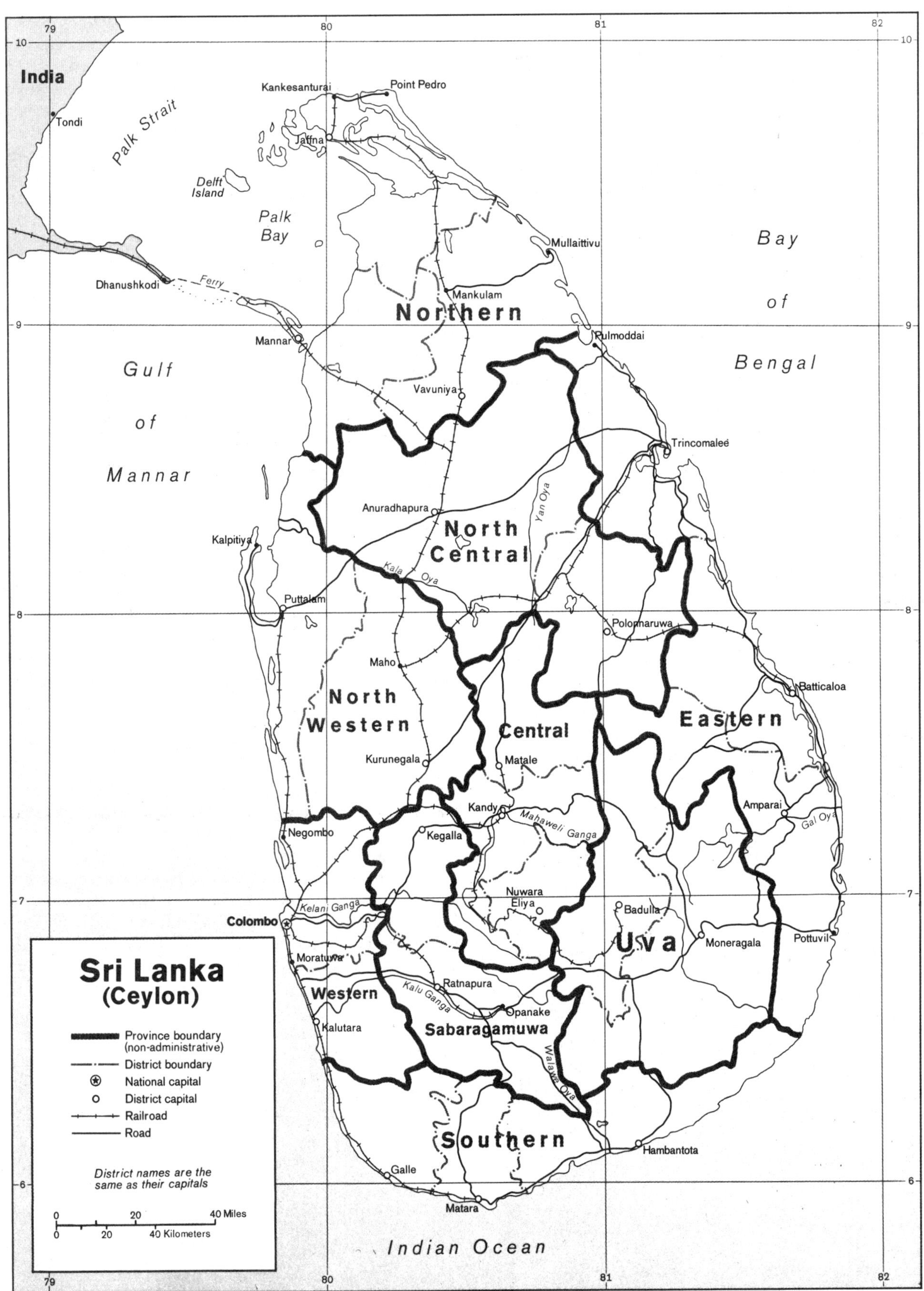

Base 501806 6-74

Although the hills and ridges of the southwestern section approach the central mountain mass, they are distinct from it, set apart by the plain. The land slopes west and south toward the sea, and the numerous rivers that rise in the Bulutota Hills flow westward through the gaps they have made in the ridges. Around Ratnapura the action of rain and running streams loosened gem-bearing gravel from the rocks and deposited it in the valleys, which then were covered with layers of mud and sand to depths of 1.5 to 6 m (5 to 20 ft). The gems are mined by workers with sieves in which they collect the gravel, washing and sorting out the precious stones. Plumbago, or graphite, is found in veins in the rock and is considered economically more important than the gems. The hills and ridges of this section are collectively called the Sabaragamuwa Ridges.

The Southeastern Section

From the plain of the southeastern section, the mountains of the central portion rise up like a wall, giving the appearance of a fortress, for which that area was used for centuries. The plain is studded here and there with rounded hills. The hills are in fact masses of buried rock. Over millions of years a process of erosion and filling has occurred, the tops of the mountains being worn down and the red, lateritic soil building up the plain. Often the hills are almost totally bare and are of unusual and picturesque shape, as evidenced by names such as Elephant Rock and Westminster Abbey.

The soils of the southeastern section, which is bounded roughly by the rivers Walawe Ganga and Mahaweli Ganga, are good for growing cotton, orange trees, and rice. Most of the area is forest and parkland or *talawa* (an open space with scattered trees). The southeastern section is thinly populated, but dried-up tanks and ruined temples survive from ages past. *Talawa* country is now the home of the Veddahs, who practice shifting agriculture.

Toward the coast where the mouths of the Walawe Ganga, the Mahaweli Ganga, and the rivers between them empty into the sea, the plain becomes mostly river basin. As the land is low and flat, flooding can, and does, occur. Because sand and silt have blocked the river mouths, they are navigable only by canoes and small boats.

The Northern Section

Between the Mahaweli Ganga on the east and the Deduru Oya on the west, this section of the plain is very flat, broad at the southern end and tapered toward the Jaffna Peninsula at the northern end. The peninsula is a limestone block, unlike the rest of the island, which is formed from crystalline gneiss and granite. In the wide southern area, near the central mountains, the granite is exposed in long, narrow ridges running north to south. In ancient times the natural caves in these ridges were ancient dwellings for Sinhalese monks who wished to lead an ascetic life. Several rivers cross the plain, flowing northward from their source in the mountains, between the north-south ridges until they leave the 152 m (500 ft) level and turn east and west.

Jaffna Peninsula is the only area in the country having a different composition. The limestone region, of which it is the northernmost part, begins at a line extending from Puttalam on the west coast to Mullaittivu on the east. Only the peninsula and its neighboring islands are of pure limestone, however, as loose stone and the products of erosion cover the larger part of the region to depths of several feet. The surface of the peninsula is dry, since water seeps through into the porous limestone to collect in underground pools or to join the underground streams. The Ceylon Tamils who live in this area manage to cultivate every available acre and successfully grow tobacco and coconuts.

The Hill Country

This region is remarkably different from the rest of the island, consisting not of rolling hills and plain but of high mountain walls, narrow gorges, deep valleys, and lofty plateaus, in which are about sixteen rivers and innumerable waterfalls. The shape of the highest area has been likened to an anchor, the shank pointing northward, the two arms extending southwest and northeast. Elevations of more than 1,524 m (5,000 ft) above sea level are the rule; Adam's Peak at the western tip of the "anchor" rises to 2,243 m (7,360 ft). The remainder of the hill country consists of the Southern and Eastern platforms, the abruptly rising steps by which the area is approached from the south and east; the Uva Basin; the Hatton Plateau; the Dolosbage Group (mountains); the Kandy Plateau; the Knuckles Group (mountains); the Piduru Ridges; and the Matale Valley.

The Southern and Eastern platforms stand like two giant steps in front of the wall of the southern hill country, which in turn rises sharply and suddenly to over 1,524 m (5,000 ft). The wall extends some 80 km (50 mi) between the nine peaks of Namunukula, the highest of which is 2,035 m (6,679 ft), to the summit of Adam's Peak. Two tremendous gaps cut the wall; beyond the point where the wall turns to the north, Haputale Gap and Ella Gap allow access through the mountain barrier onto the slopes of the Uva Basin.

Uva Basin is distinctive for its *patanas,* or grasslands. The basin itself is surrounded by mountain walls on three sides; on the north two great rivers, the Badulla Oya and Uma Oya, drain the valley and work their way to the Mahaweli Ganga lying farther to the north in its own valley. The center of Uva Basin is hilly; the grass-covered slopes resemble the "downs" familiar to the British, and hence the name Uva Downs.

The Hatton Plateau is one of a series of high plains of the hill country and lies west of Uva Basin. Its height is between 914 to 1,219 m (3,000 and 4,000 ft) above sea level. The rivers that flow between its ridges ultimately form the great Mahaweli Ganga. Conditions on the plateau are excellent for growing tea, and nearly all the area is under tea cultivation. Most of the inhabitants are Indian Tamils who live and work on the European-owned and -operated plantations.

Northeast of Hatton Plateau a separate group of mountains, named for its outstanding peak, Dolosbage, is now cleared of its original forest cover and devoted mostly to plantations, especially rubber. The valley of the Mahaweli

Ganga separates this group from the central mass of the hill country, the Piduru Ridges. This formidable, nearly inaccessible mountain fortress is composed of peaks reaching 2,438 m (8,000 ft) and more above sea level, the highest being Pidurutalagala, 2,527 m (8,292 ft). From this and other high peaks great ridges stretch out in all directions, some of the most important being Great Western, Talankanda, a ridge ending in three peaks over 2,346 m (7,700 ft) extending southward, and a ridge extending north toward Kandy. Deep, narrow valleys lie between the ridges and are the principal channels of movement through this rock and forest maze. Four of the main routes lead to Nuwara Eliya, once an English sanatorium, now a thriving tea-producing town, situated in a lofty mountain basin beneath Pidurutalagala.

The Kandy Plateau lies to the north of the Piduru Ridges. It is cut by ridges and valleys and by the Mahaweli Ganga, which here turns in an arc from its northward course to the southeast. In the bend in the river is situated the ancient city of Kandy, above an artificial lake made in 1810 by the last king of Kandy.

The Knuckles Group and Matale Valley are the two northernmost sections of the hill country. The highest point in this group of mountains is Knuckles Peak, so named by the English because of its resemblance to the knuckles of a hand. The peak is 1,863 m (6,112 ft) above sea level and is cut off from the central mountain mass by the Mahaweli Ganga Valley. The Matale Valley is an ancient route to the Kandy Plateau and lies to the west of the Knuckles Group. It is a closed-in valley, being ringed on three sides by high ridges and small hills; on the north it gradually drops away to the northern section of the 152 m (500 ft) plain.

There are many entrances to the hill country along the valleys and through mountain passes, so that communications between the hill country people and the plains people has been possible since ancient times. Cart tracks and footpaths were beaten through dense, wet forests and carved out of rock by thousands of Buddhist, Hindu, and Muslim pilgrims on their trips to Adam's Peak, whose summit bears an indentation resembling a huge human foot. This imprint is the source of many legends regarding its origin. When the British occupied Sri Lanka, they began construction of a network of roads connecting many of the natural entrances. Some of the ancient paths are still in use, but there are also some excellent hardtopped roads and a railroad through this scenic, rugged countryside.

PRINCIPAL TOWNS
(provisional, estimated population at mid-1986)

Colombo (Capital)	683,000	Moratuwa	138,000
Dehiwala-Mount Lavinia	191,000	Kandy	130,000
		Galle	109,000
Jaffna	143,000	Kotte	104,000

Area and population

Districts	Capitals	area sq mi	area sq km	population 1986 estimate
Amparai	Amparai	1,778	4,604	439,000
Anduradhapura	Anuradhapura	2,809	7,275	659,000
Badulla	Badulla	1,090	2,822	668,000
Batticaloa	Batticaloa	1,017	2,633	379,000
Colombo	Colombo	268	695	1,836,000
Galle	Galle	652	1,689	881,000
Hambantota	Hambantota	1,013	2,623	477,000
Jaffna	Jaffna	833	2,158	915,000
Kalutara	Kalutara	624	1,615	892,000
Kandy	Kandy	833	2,158	1,188,000
Kegalle	Kegalle	642	1,663	720,000
Kurunegala	Kurunegala	1,844	4,776	1,333,000
Mannar	Mannar	778	2,014	120,000
Matale	Mateale	768	1,989	392,000
Matara	Matara	481	1,247	717,000
Monaragala	Monaragala	2,188	5,666	320,000
Mullaitivu	Mullaitivu	798	2,066	86,000
Nuwara Eliya	Nuwara Eliya	555	1,437	514,000
Polonnaruwa	Polonnaruwa	1,332	3,449	294,000
Puttalam	Puttalam	1,172	3,036	552,000
Ratnapura	Ratnapura	1,251	3,239	868,000
Trincomalee	Trincomalee	1,048	2,714	292,000
Vavuniya	Vavuniya	1,021	2,645	108,000
TOTAL		25,332	65,610	16,117,000

Rivers

The rivers rise in the high mountains of the hill country and flow over the plateau to the plains in a ring of waterfalls. Some are streams that dry up in the dry season, but others, like the Menik Ganga, have been dammed for irrigation purposes.

The rivers are not large, but they are numerous and are notable for their beauty, both at their sources in rocky gorges of the hill country and on the plains where they spread over the land in shallow lakes. There are sixteen principal rivers; the Mahaweli Ganga is the longest—341 km (206 mi). With the exception of the 167-km-long (104-mi-long) Aruvi Aru, the other chief rivers range from 156 to 100 km (97 to 62 mi) in length. Although they are not useful for navigation, being too wild in the mountains and too shallow on the plains, most supply plenty of fish to the villages along their banks. The rivers effectively drain the uplands but in the lowlands may divide into several streams and, where the land is flat, water may back up and become an obstacle to land development and communications. In such areas serious flooding may occur. Roads and bridges must therefore be massively reinforced, and people build their houses away from the river's edge. The areas around the cities of Colombo, Trincomalee, and Batticaloa are of this type.

Although Sri Lanka has few natural lakes, there are 12,000 bodies of water ranging from tiny ponds to manmade lakes several miles wide that are actually reservoirs. The oldest of these tanks, as they are called locally and in much of southern Asia, is believed to be Basawakkulam, built about 300 B.C. to store water for the ancient city of Anuradhapura on the Malwatu Oya, in what is now North Central Province. Basawakkulam and two companion reservoirs built about 100 B.C. were abandoned and overgrown with dense scrub by the seventeenth century, as was the whole area around and including Anuradhapura. Sources differ as to whether malaria or invasions drove out the inhabitants. The city has been restored, however, and the three ancient reservoirs have been cleaned and made to function again.

The majority of reservoirs are situated in what is known as the dry zone—that is, the north-central region where rainfall is between fifty and seventy-five inches a year. Where the original builders of the complex irrigation system learned their skill is not known, but construction practices appear to have evolved in three cycles. As the people cleared more and more land they needed more extensive irrigation. At first they built simple earthen dams; then they constructed fifty-foot dams that formed bodies of water from 32 to 48 km (20 to 30 mi) in circumference; the third cycle included feeder canals that channeled water from springs in the hills, keeping ponds or reservoirs full and irrigating the land along the way. As the population of the dry zone thinned over the centuries, many of the tanks became choked with underbrush and dried up. They are being restored, and additional modern ones are being constructed as needed.

The waterfalls that ring the high plateau of the hill country are of potential economic value as sources of hydroelectric power. Sixteen or seventeen are locally well known, but countless others are unnamed. Together they are capable of generating more hydroelectric power than the country needs in the foreseeable future.

Inland navigation is made possible by 246 km (153 mi) of canals, most of which were built by the Dutch. They may connect two bodies of water, such as Puttalam Lagoon and Negombo Lake on the west coast, or may channel off floodwaters occurring around Colombo.

SUDAN

BASIC FACTS

Official Name: Republic of the Sudan

Abbreviation: SUD

Area: 2,503,890 sq km (966,757 sq mi)

Area—World Rank: 9

Population: 24,014,495 (1988) 41,905,000 (2000)

Population—World Rank: 32

Capital: Khartoum

Boundaries: 8,536 km (5,303 mi) Egypt 1,275 km (792 mi); Ethiopia 2,266 km (1,408 mi); Kenya 306 km (190 mi); Uganda 435 km (270 mi); Zaire 628 km (390 mi); Central African Republic 1,167 km (725 mi); Chad 1,360 km (845 mi); Libya 383 km (238 mi)

Coastline: Red Sea 716 km (445 mi)

Longest Distances: 2,192 km (1,362 mi) SSE-NNW 1,880 km (1,168 mi) ENE-WSW

Highest Point: Kinyeti 3,187 m (10,456 ft)

Lowest Point: Sea Level

Land Use: 5% arable; 24% meadows and pastures; 20% forest and woodland; 51% other

Sudan—the largest country in Africa—is an immense, sparsely populated plain, with plateaus or mountainous areas near the borders in the west, the southeast, and along the Red Sea coast in the northeast.

The most prevalent landscape is semiarid savanna—a mixture of short grasses, scattered brush, and short trees. Daytime temperatures are high throughout the year, and the dry season ranges from three months in the relatively humid south to nine months in Khartoum, the capital city.

Two contrasts, both associated with the availability of water, are descriptive of Sudan. The southern provinces of Equatoria, Bahr al Ghazal, and Upper Nile receive 76 to 127 cm (30 to 50 in) of rain during the six-to-nine month wet season and produce a rich variety of tall grasses, shrubs, and trees. Permanent swamps in these provinces cover about 129,500 sq km (50,000 sq mi), and there is an excess of water for most of the year. The lush vegetation in the south contrasts sharply with the deserts of Northern Province, where the occasional rains vanish in the parched sand and broad areas are devoid of either vegetation or people.

Narrow belts of irrigated cropland, no more than a few miles wide, bisect the northern savanna and desert along the main Nile River and along the White Nile, the Blue Nile, and the Atbarah rivers. They contrast sharply with the arid savanna or barren desert just beyond the limits of irrigation.

The Land

The Plains

The topography of the country outside the mountains and the Nile valley is generally devoid of contrast, and the flat plain making up most of its huge area distinguishable more by range of vegetation than by peculiarities of terrain. The plain, extending some 804 to 965 km (500 to 600 mi) from east to west and more than 1,609 km (1,000 mi) in its north-south axis, is a part of the broad savanna belt that begins at the southern edge of the Sahara Desert and extends across the African continent. For thousands of square miles, the only features relieving the monotony of the Sudanese plain are low rolling hills—sometimes referred to locally as mountains—or extensive sand dunes created thousands of years ago and partially or entirely fixed by vegetation.

Soils are composed mainly of clay, much of which is impermeable and difficult to cultivate, or sand that contains little clay or humus. Vegetation is a typical savanna mixture of grasses, thorny shrubs (sometimes called scrub), and scattered short trees. The vegetation varies from a lush mixture in the south, where rainfall is relatively heavy for as much as nine months of the year, to sparse grasses and shrubs on sandy soils in the vicinity of 15° to 16° north latitude, near Khartoum. In the east both the plains area and the northern desert are bisected by the Nile River system.

The Northern Desert

A line running east to Atbarah and Port Sudan from the western frontier at 16° north latitude defines the approximate southern limit of desert, which covers the northern

PRINCIPAL TOWNS (population at 1983 census)

Omdurman . . .	526,287	Wadi Medani . .	141,065
Khartoum (capital) .	476,218	Al-Obeid . . .	140,024
Khartoum North . .	341,146	Atbara	73,009
Port Sudan . . .	206,727		

quarter of the Sudan. Northern Province lies almost entirely within the Libyan Desert, which extends into Sudan from the northwest; to the northeast the Nubian Desert covers part of Northern Province and northern Kassala Province.

From the confluence of the White and Blue Nile rivers, the Upper Nile winds northward through this desert area for a distance of 1,287 km (800 mi) inside Sudan and provides the only water for the narrow strips of cultivation along the riverbanks. In the area from Atbarah to Wadi Halfa on the Egyptian frontier almost no rain falls; Wadi Halfa is often completely rainless for years at a time. The settlements along the Nile depend for their livelihood on various types of irrigation or inundating.

The hinterland west of the Nile supports only a few Arab nomads who, with their camels, sheep, or goats, cover great expanses of the parched country in search of grazing, usually south of 18° to 19° north latitude, where a little rain occurs during most years and grass or browse springs to life. Water is available only in scattered oases, such as Al Atrun in the western desert and Well No. 6 on the railway between Wadi Halfa and Abu Hamand. Terrain in this northern desert consists of broad areas of sand and flintrock with occasional hills and outcroppings of basalt, granite, and limestone, often surrounded by banks of sand deposited by the wind.

Mountain Zones

There are four mountain or upland zones. To the northeast lie the Red Sea Hills of Kassala Province; in the west is the Jabal Marrah, a mountain complex sloping to the border with Chad; and in central Sudan in Kordofan Province are the Nuba Mountains, a relatively minor complex rising above the clay plains. The fourth zone includes the Immatong and Dongotona ranges in the extreme south, along the Uganda border.

The Red Sea Hills are eroded outcroppings of base rock rising from a narrow coastal plain, the abruptness of their eastern slope giving rise to gushing torrents during winter rains blown in from the sea. The western slopes of the mountains incline more slowly toward the Nile and receive only light summer rains. North of the Atbarah-Port Sudan railway, the hills extend into the desert—bare of vegetation except in the valleys—but south of the railway increased rainfall permits the growth of a few trees and thorny shrubs.

The area is inhospitable and supports only seminomadic herders, who also cultivate hardy varieties of millet in the wetter valleys. They move their flocks laterally across the mountains or to higher or lower altitudes, depending upon the vagaries of the rainfall at various altitudes. The highest of the Red Sea Hills are above 2,133 m (7,000 ft).

The only major mountain range in western Sudan, the Jabal Marrah, stands in Darfur Province. Rising above 3,048 m (10,000 ft) in elevation, this range forms part of

Area and population

Regions Provinces	Capitals	area sq mi	 sq km	population 1983 census
Aʿālī an-Nīl (Upper Nile)	Malakāl	92,198	792	1,599,605
Aʿālī an-Nīl (Upper Nile)	Nāṣir	45,231	117,148	802,354
Junqulī (Jongley)	Bor	46,781	121,164	797,251
Baḥr al-Ghazāl (Bahr el-Ghazal)	Wāu	77,566	200,894	2,265,510
Baḥr al-Ghazāl al-Gharibīya (Western Bahr el-Ghazal) Baḥr al-Ghazāl ash-Sharqīyah (Eastern Bahr el-Ghazal)	Raga Uwayl	51,960	134,576	1,492,597
al-Buḥayrāh (El Buheyrah)	Rumbek	25,606	66,318	772,913
Dārfūr (Darfur)	al-Fāshir	196,404	508,684	3,093,699
Dārfūr-al-Janūbīyah (Southern Darfur)	Nyala	62,753	162,529	1,765,752
Dāfūr ash-Shamālīyah (Northern Darfur)	al-Fāshir	133,651	346,155	1,327,947
al-Istiwā'īyah (Equatoria)	Jūbā	76,436	197,969	1,406,181
al-Istiwā'īyah al Gharbiyah (Western Equatoria)	Yambio	30,398	78,732	359,056
al-Istiwā'īyah ash-Sharqīyah (Eastern Equatoria)	Jūbā	46,038	119,237	1,047,125
Kurdufān (Kordofan)	al-Ubayyiḍ	146,817	380,255	3,093,294
Kurdufān al-Janūbīyah (Southern Kordofan)	Kāduqlī	61,141	158,355	1,287,525
Kurdufān ash-Shamālīyah (Northern Kordofan)	al-Ubayyiḍ	85,676	221,900	1,805,769
ash-Shamālīyah (Northern)	ad-Dāmir	183,800	476,040	1,083,024
an-Nīl (Nile)	ad-Dāmir	49,167	127,343	649,633
ash-Shamālīyah (Northern)	Dunqulah	134,633	348,697	433,391
ash-Sharqīyah (Eastern)	Kassalā	128,987	334,074	2,208,209
al-Baḥr al-Aḥmar (Red Sea)	Port Sudan	84,912	219,920	695,874
Kassalā (Kassala)	Kassalā	44,075	114,154	1,512,335
al-Wastā (Central)	Wad Madanī	53,675	139,017	4,012,543
an-Nīl al-Abyaḍ (White Nile)	ad-Duwaym	16,149	41,825	933,136
al-Jazīrah (El-Gezira)	Wad Madanī	13,536	35,057	2,023,094
an-Nīl al-Azraq (Blue Nile)	ad-Damazin	23,990	62,135	1,056,313
National Capital				
Kharṭūm (Khartoum)	Khartoum	10,875	28,165	1,802,299
TOTAL		966,757	2,503,890	20,564,364

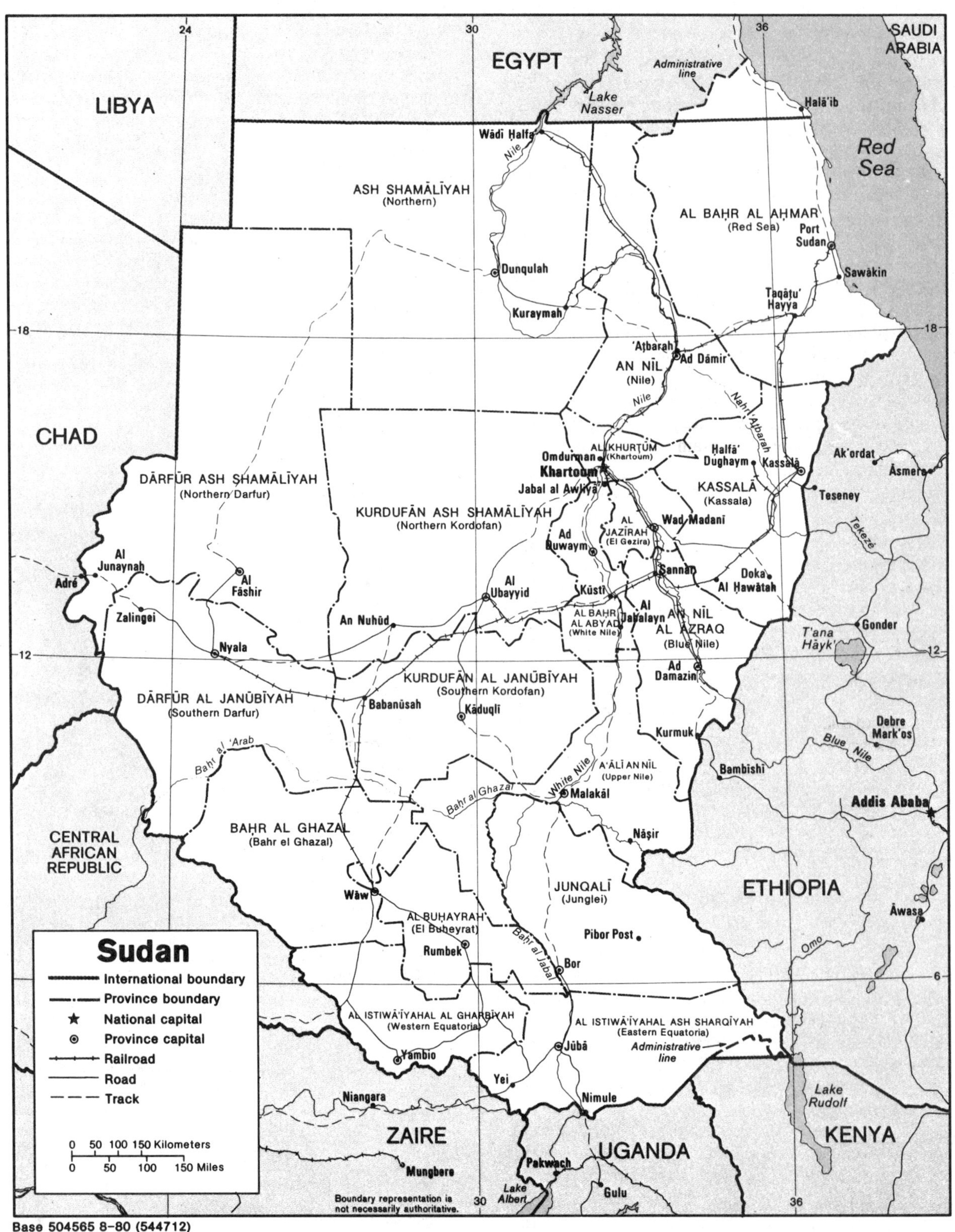

Base 504565 8-80 (544712)

the watershed between the Nile River and Lake Chad drainage basins. The Jabal Marrah is of volcanic origin, and its valleys are relatively fertile. The upper elevations receive a slightly higher rainfall than the surrounding plains, and the relatively richer soil of the valley is more productive.

Some of the rocks and peaks have a sculptured appearance resulting from the action of the rains upon the soft volcanic rock. Much of the eroded rock is deposited by streams on the desert floor below, but on the higher hillsides manmade terraces of ancient origin retain topsoil and water. Although cultivation is generally dependent upon the seasonal rains, some small valleys and terraces are irrigated with water from small perennial mountain streams.

The Nuba Mountains of southern Kordofan Province are scattered granitic masses, rising as much as 914 m (3,000 ft) above a flat clay plain. They are covered in many areas by variations of savanna vegetation. Some slopes were once terraced and then abandoned by subsistence farmers. Water is not as scarce in the mountains as in the surrounding plains. Wells are numerous in the open valleys, and a few short mountain streams continue to flow throughout the year.

The Immatong and Dongotona mountains stand in southern Equatoria Province, the lower Didinga Hills flanking them to the east. The Immatongs are the highest mountains in Sudan, with peaks above 3,048 m (10,000 ft). The Dongotona Mountains, lying east of the Immatongs, reach a maximum height of about 2,529 m (8,300 ft). Both mountain chains have a considerable coverage of rain forest. Below these ranges and the Ironstone Plateau of the southwestern border, foothills and lower plateaus slope generally northward to the Sudd, a vast region of swamps and marshes covering an area of about 7,770 sq km (3,000 sq mi) and extending from eastern Equatoria Province several hundred miles northwestward to the Bahr al Ghazal in Upper Nile Province. The vast swamp and marsh area is as monotonous as the featureless plains farther north, but there is considerable variety of terrain and vegetation in the uplands south of the swamps, particularly near the Uganda and Kenya borders.

Drainage

As the most distant and southernmost source of the Nile River, the White Nile—known in southern Sudan as the Bahr al Jabal—derives much of its water from the lake plateau of east-central Africa. These headwaters include the watersheds around Lake Victoria and Lake Albert, lying on or near the equator, where rainfall exceeds 127 cm (50 in) per year. Much of this water is lost to evaporation before it reaches the Nile tributaries, but a large volume is carried into the swamp areas, including the Sudd, a region of swamps and floating vegetation in south-central Sudan. Losses to evaporation are also heavy in this area. Partially for this reason, the annual input from the White Nile into the upper Nile at Khartoum is only one-fifth of that from the Blue Nile, but it is important because much of the White Nile water arrives during the months when the Blue Nile input is very low.

The Blue Nile rises at Lake Tana in the Ethiopian highlands and makes its way through the mountains for about 804 km (500 mi) before entering Sudan. Torrential summer rains draining into the fast-flowing Blue Nile from these highlands cause the seasonal flood on the lower reaches of the Blue Nile and on the upper Nile—floods upon which half of the people of Sudan are dependent. During floodtime the Blue Nile and its two major tributaries, the Dindar and the Rahad, contribute 70 percent of the water of the upper Nile. During floodtimes the flow of the Blue Nile may be sixty times that of its low water period, and 300 times its low stage during short periods of heavy flooding. During the low water stage on the upper Nile, however, the Blue Nile and the other eastern tributaries may contribute only 20 percent of the total flow.

An important tributary to the upper Nile is the Atbarah River, similar in seasonal behavior to the Blue Nile and also originating in the mountains of Ethiopia. It traverses southern Kassala Province and empties into the Nile at the town of Atbarah.

The gradient of the Nile from Khartoum to Wadi Halfa on the northern border of Sudan is considerably steeper than that of its 1,448 km (900 mi) course south of Khartoum. Along this lower reach are five of the Nile's six cataract areas of swift, rough water.

All perennial streams of significant size in Sudan are part of the Nile system. There are also numerous wadis, or intermittent streams, which flow only part of the year. Some drain into the Nile during the rainy season and stand empty at other times. Others drain into swamps that have no outlet to a river or disappear into the sands of an inland basin during the dry months. For example, the Wadi Howar and the Wadi al Ku, both originating in western Darfur Province, disappear into the desert. Of similar origin, the Wadi Azum eventually reaches the Lake Chad drainage system to the west.

Some of these intermittent streams carry large amounts of water during the rainy season and support local areas of agriculture. The Qash and Baraka rivers flow into Kassala Province from the northern Ethiopian highlands during the months of July, August, and September. The Qash River provides water for important irrigation schemes north of Kassala, and the Baraka feeds the Tawkar delta near the Red Sea coast. The Bahr al Arab in southwestern Sudan is another important seasonal river.

SURINAME

BASIC FACTS

Official Name: Republic of Suriname

Abbreviation: SUR

Area: 163,820 sq km (63,251 sq mi)

Area—World Rank: 87

Population: 394,999 (1988) 579,000 (2000)

Population—World Rank: 146

Capital: Paramaribo

Boundaries: 2,150 km (1,355 mi) French Guiana 467 km (290 mi); Brazil 593 km (368 mi); Guyana 726 km (451 mi)

Coastline: 364 km (226 mi)

Longest Distances: 662 km (411 mi) NE-SW 487 km (303 mi) SE-NW

Highest Point: Juliana Top 1,230 m (3,749 ft)

Lowest Point: Sea Level

Land Use: 97% forest and woodland; 3% other

Suriname is located on the NE coast of South America. It is the smallest independent nation on the continent.

The land is divided into four distinct natural regions: a coastal belt, an intermediate plain, a region of high mountains and a high savanna in the southwest. The coastal plain, beyond which the early settlers seldom penetrated, covers about 16% of the national territory. It is approximately 15 km (10 mi) wide on the eastern border widening to about 80 km (50 mi) in the west. Most of the region is at sea level and diking is necessary to utilize the land. The intermediate plain runs to the edge of the vast rain forest and is about 50 km to 65 km (30 mi to 40 mi) wide. The mountainous rain forest region, rising gradually to an elevation of 1,256 m (4,120 ft) in the Wilhelmina Mountains, makes up about 75% of the national territory but has been only partially explored. The central chain of the Van Asch Van Wijk range runs south to the Tumac-Humac mountains on the Brazilian border with the Wilhelmina and Kayser ranges on the west and the Orange mountains on the east.

Numerous rivers dissect the land, all interconnected by a remarkable system of channels. The principal rivers are the Corantyne, the Nickerie, the Coppename, the Saramacca, the Surinam, the Commewijne and the Marowijne. The largest lake is the Prof. Dr. Ir. W. J. van Blommestein Meer Lake.

Area and population

		area		population
Districts	**Capitals**	sq mi	sq km	1980 census
Brokopondo	Brokopondo	8,278	21,440	20,249
Commewijne	Nieuw Amsterdam	1,587	4,110	14,351
Coronie	Tottness	626	1,620	2,777
Marowijne	Albina	17,753	45,980	23,402
Nickerie	Nieuw Nickerie	24,946	64,610	34,480
Para	Onverwacht	378	980	14,867
Saramacca	Groningen	9,042	23,420	10,335
Suriname		629	1,628	166,494
Town District				
Paramaribo	Paramaribo	12	32	67,905
TOTAL		63,251	163,820	354,860

SWAZILAND

BASIC FACTS

Official Name: Kingdom of Swaziland

Abbreviation: SWZ

Area: 17,364 sq km (6,704 sq mi)

Area—World Rank: 139

Population: 735,302 (1988) 1,098,000 (2000)

Population—World Rank: 139

Capital: Mbabane (administrative); Lobamba (legislative)

Land Boundaries: 544 km (344 mi) Mozambique 108 km (67 mi); South Africa 446 km (277 mi)

Coastline: Nil

Longest Distances: 176 km (109 mi) N-S; 135 km (84 mi) E-W

Highest Point: Emiembe 1,862 m (6,109 ft)

Lowest Point: 21 m (70 ft)

Land Use: 8% arable; 67% meadows and pastures; 6% forest and woodland; 19% other

Landlocked Swaziland is located in Southern Africa. It is topographically part of the South African Plateau and is generally divided into four well-defined regions of nearly equal breadth. From the High Veld in the west, averaging 1,050 to 1,200 m (3,500 ft to 3,900 ft) in elevation, there is a step-like descent eastward through the Middle Veld (450 to 600 m, 1,475 to 1,970 ft) to the Low Veld (150 to 300 m, 490 to 980 ft). To the east of the Low Veld is the Lebombo Range (450 to 825 m, 1,475 to 2,700 ft), which separates the country from the Mozambique coastal plain.

Swaziland is well-watered with four large rivers flowing eastward across it into the Indian Ocean. These are the Komati and the Umbeluzi Rivers in the north, the Great Usutu River in the center and the Ngwavuma River in the south.

Principal Towns (population at August 1982): Mbabane (capital) 38,636; Manzini 13,893.

Area and population

		area		population
Districts	**Capitals**	sq mi	sq km	1986 census
Hhohho	Mbabane	1,378	3,569	179,193
Lubombo	Siteki	2,296	5,947	152,408
Manzini	Manzini	1,571	4,068	190,613
Shiselweni	Nhlangano	1,459	3,780	153,875
TOTAL		6,704	17,364	706,137

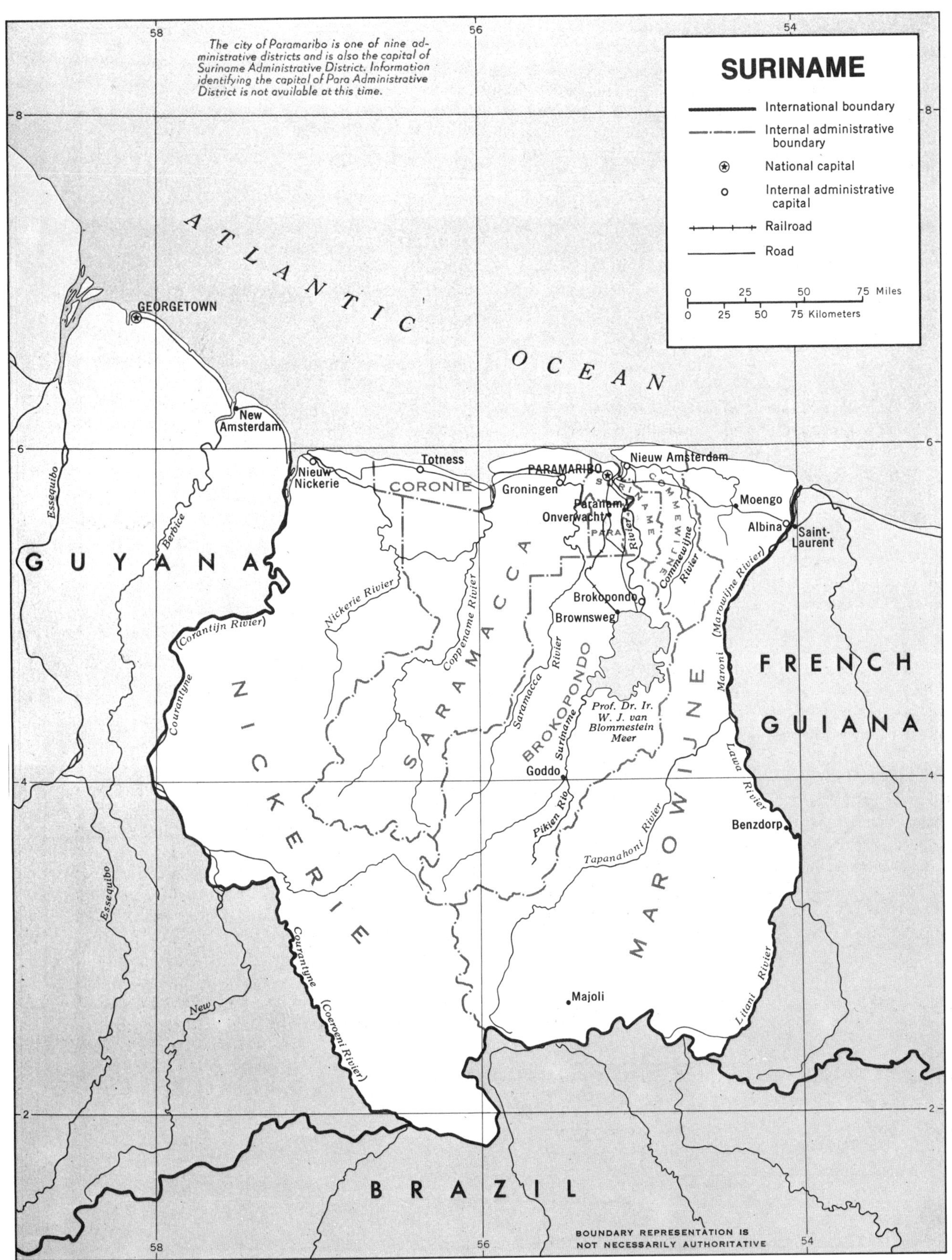

Base 56830 4-68

Swaziland

502744 9-77 (541832)
Lambert Conformal Projection
Standard parallels 6° and 30°
Scale 1:1,150,000

Railroad
Road

SWEDEN

BASIC FACTS

Official Name: Kingdom of Sweden

Abbreviation: SWE

Area: 449,964 sq km (173,732 sq mi)

Area—World Rank: 52

Population: 8,393,071 (1988) 8,539,000 (2000)

Population—World Rank: 71

Capital: Stockholm

Boundaries: 4,595 km (2,855 mi) Finland 586 km (364mi); Norway 1,619 km (1,006 mi)

Coastline: 2,390 km (1,485 mi) Gotland Island 400 km (249 mi) Oland Island 72 km (45 mi)

Longest Distances: 1,574 km (978 mi) N-S 499 km (310 mi) E-W

Highest Point: Mount Kebne 2,111 m (6,926 ft)

Lowest Point: Sea level

Land Use: 7% arable; 2% meadows and pastures; 64% forest and woodland; 27% other

The largest of the Scandinavian countries, Sweden is one of the countries located farthest from the equator. It extends from north to south at roughly the same latitude as Alaska, with about 15% of its total area situated north of the Arctic Circle. Sweden is bounded on the north and northeast by Finland; on the east by the Gulf of Bothnia; on the southeast by the Baltic Sea; on the southwest by the Øresund, the Kattegat and the Skagerrak and on the west by Norway. The two largest Swedish islands are Gotland, with a coastline of 400 km (249 mi), and Öland, with a coastline of 72 km (45 mi).

Sweden is rhombodial in shape, and the most impressive of its geographical features is its length—indeed, the Swedes speak of *vart avlanga land*, our long, drawn-out land. It shares this and many other features with its Siamese twin, Norway, but is a land of lower altitudes and less dissected relief than Norway.

Four topographical divisions can be discerned in the country, although they are of unequal size. The largest is Norrland, which commonly refers to the broad Baltic slopes from the lower reaches of the Dal River northward. This region comprises three-fifths of the country, with its rolling landscape of hills and mountains, forests and large river valleys. In this region are deposits of iron and other ores, which gave rise to Sweden's oldest industrial zone of Bergslagen. Through Värmland and Dalarna, Norrland's landscape merges with those of the faulted lands of Central Sweden, which constitutes the second region. The wooded highland region of Småland in the South is the third region. The fourth region is in southernmost part of the country and is known as Scania or Skåne, a continuation of the fertile plains of Denmark and northern Germany.

PRINCIPAL TOWNS (Estimated Population at 31 December 1987)

Town	Population	Town	Population
Stockholm (capital)	666,810	Borås	100,395
Göteborg (Gothenburg)	431,521	Sundsvall	92,721
Malmö	230,838	Eskilstuna	88,508
Uppsala	159,962	Gävle	87,474
Orebro	119,066	Umeå	86,816
Norrköping	119,001	Lund	84,342
Västerås	118,602	Södertälje	80,263
Linköping	117,563	Halmstad	77,942
Jönköping	108,962	Karlstad	74,892
Helsingborg	106,982	Skellefteå	74,091

Area and population

Counties	Capitals	area sq mi	area sq km	population 1987 estimate
Älvsborg	Vänersborg	4,400	11,395	427,638
Blekinge	Kariskrona	1,136	2,941	150,258
Gävleborg	Gävle	7,024	18,191	287,691
Göteborg och Bohus	Göteborg	1,985	5,141	721,553
Gotland	Visby	1,212	3,140	56,174
Halland	Halmstad	2,106	5,454	242,250
Jämtland	Östersund	19,090	49,443	133,543
Jönköping	Jönköping	3,839	9,944	301,413
Kalmar	Kalmar	4,313	11,170	237,417
Kopparberg	Falun	10,886	28,194	283,191
Kristianstad	Kristianstad	2,350	6,087	280,609
Kronoberg	Växjö	3,266	8,458	173,853
Malmöhus	Malmö	1,907	4,938	753,075
Norrbotten	Luleå	38,191	98,913	261,039
Örebro	Örebro	3,289	8,519	269,620
Östergötland	Linköping	4,078	10,562	394,753
Skaraborg	Mariestad	3,065	7,937	270,111
Södermanland	Nyköping	2,340	6,060	249,479
Stockholm	Stockholm	2,505	6,488	1,593,333
Uppsala	Uppsala	2,698	6,989	254,938
Värmland	Karlstad	6,789	17,584	278,861
Västerbotten	Umeä	21,390	55,401	245,204
Västernorrland	Härnösand	8,370	21,678	261,089
Västmanland	Västerås	2,433	6,302	254,423
TOTAL LAND AREA		158,661	410,929	8,381,515
INLAND WATER		15,071	39,035	
TOTAL		173,732	449,964	

Norrland slopes from the Kjølen Mountains along the Norwegian frontier to the Gulf of Bothnia. Its many rivers, notably the Göta, the Dal, the Angerman, the Ume and the Lule, flowing generally toward the southeast—have incised the plateau surface. Norrland falls readily into three subdivisions. The western highlands follow the Norwegian frontier and are lifted to peaks of over 1,829 m (6,000 ft) of which the highest is Kebnekaise (2,123 m (6,966 ft). Prolonged erosion, however, has reduced much of the surface to a plateau. The depressions of this upland are filled by lakes, most of which lie somewhat more than 305 m (1,000 ft) above the level of the Baltic. The largest of these lakes are Torneträsk 317 sq km (122 sq mi), Luleträsk with Langas 220 sq km (85 sq mi), and the interconnected trio of Hornavan, Uddjaur and Storavan 660 sq km (255 sq mi). There are a number of small icefields above 66°N latitude. Southernmost of the lakes is Kalsjo, to the east of which opens out the lowlands centered in Storsjön, which is the second topographical subdivision of Norrland. The rivers in this region are marked by falls

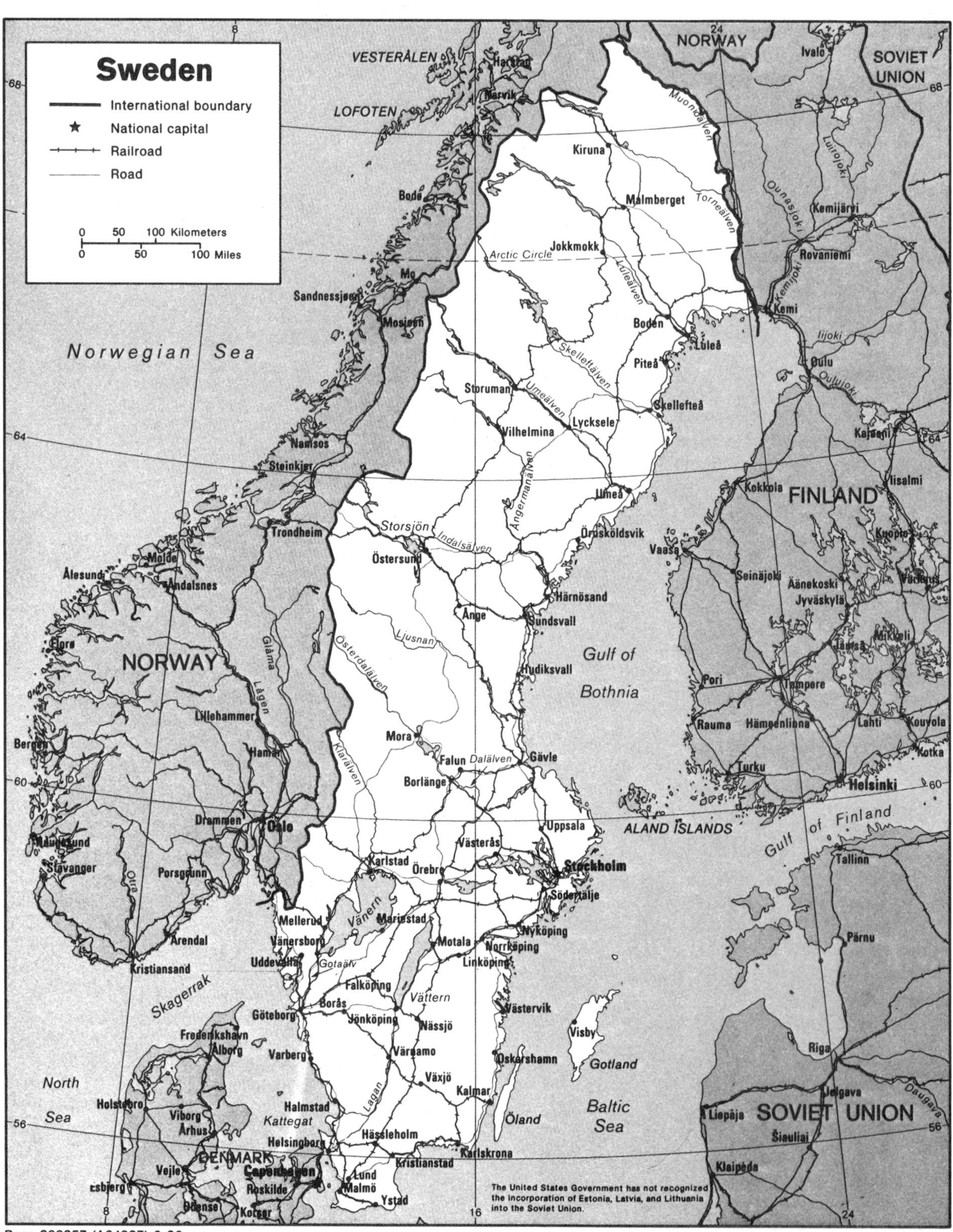

Base 800357 (A04937) 6-86

and rapids. This subregion continues to the Finnish border and is interspersed with spreads of peatland and some of the most extensive forest stands in the country. The piedmont zone yields to the third subdivision, the Bothnian coastal plain, which merges almost imperceptibly into the sea. Both the littoral and estuaries are crowded with islands. The Bothnian coast may be divided into three sections—lower, middle and upper—of which the middle extends from Örnsköldsvik to Skellefteå.

Central Sweden is a shatter zone of lakes and plains. The four principal lakes are Vänern, Vättern, Hjälmaren, and Mälaren, of which Vänern is the second largest in Europe, with an outlet to the west by way of the Göta River. It claims Sweden's largest catchment area. The Trollhättan Falls, on the Göta River, are indicative of the change in level between the lake and the Skagerrak lowlands. Lake Mälaren lies only about 0.6 m (2 ft) above the average level of the Baltic Sea. Throughout the great lakes region are extensive plains such as Uppland (centered on Uppsala), Västmanland, Narke, and East and West Gotland. Archaeological evidence suggests that this lakes and plains region was the core of early Swedish settlements.

To the south of Lake Vättern are the faulted landscapes of Skäne, which, although resembling the Danish landscape, have areas of much more pronounced relief. Skäne differs from much of Sweden not merely in its structure and geological history but also in the fact that its coast is free of islands.

Like other northern countries, Sweden is richly islanded. The archipelago of Stockholm shows the most intense concentration of islands, the outermost of which are separated from their Finnish counterparts by the Åland Sea. In contrast, the western coast archipelago of Bohuslan is a skerry zone where the ice, waves and winds have left the skerries bald in appearance. The Göta River cuts through this rocky back country to the plainlands of Central Sweden. Of Swedish islands, Gotland occupies a special and central place. Although it has a plateau appearance and is skirted with limestone cliffs, it has some of the finest beaches in the Baltic. Its principal town is Visby.

Since the last Ice Age, the land in northern and central Sweden has been rising, in some places up to 0.9 m (3 ft) per century.

SWITZERLAND

BASIC FACTS

Official Name: Swiss Confederation

Abbreviation: SWZ

Area: 41,293 sq km (15,943 sq mi)

Area—World Rank: 121

Population: 6,592,558 (1988) 6,939,000 (2000)

Population—World Rank: 84

Capital: Bern

Land Boundaries: 1,884 km (1,171 mi) West Germany 367 km (228 mi); Liechtenstein 41.2 km (25.6 mi); Austria 164 km (102 mi); Italy 740 km (460 mi); France 573 km (356 mi)

Coastline: Nil

Longest Distances: 364 km (226 mi) E-W 220 km (137 mi) N-S

Highest Point: Monte Rosa (Dufourspitze) 4,634 m (15,203 ft)

Lowest Point: Lago Maggiore 193 m (633 ft)

Land Use: 10% arable; 1% permanent crops; 40% meadows and pastures; 26% forest and woodland; 23% other

Switzerland is a landlocked country in Central Europe forming part of the Alpine arc which stretches almost 1,000 km (621 mi) from Nice in the French Maritime Alps to Vienna in the Austrian Alps. Topographically the country is divided into three regions: the Jura Mountains in the northwest comprising 10% of the territory, the Alps in the south covering 60% of the territory and the Central Plateau or Mittelland covering 30% of the territory and including the fertile plains and rolling hills between the Alps and the Jura.

Switzerland contains the central part of the Alps, roughly a fifth of the total range. The massif is divided by the Rhone and upper Rhine valleys lengthwise and the Reuss and Ticino valleys crosswise creating three groups of ranges: a northeastern group comprising Alpstein-Toggenburg, Glarner, Schwyzer, and East Urner Alps, the northwestern group comprising West Urner, Unterwaldner, Bernese, Fribourg and Vaud Alps, and the southern group comprising Valais, Ticino, and Grison Alps. A young massif in geological terms, the Alps have a complicated stratification system, composed of folds and fold packets which have in many cases been superimposed on each other by thrust pressure and then reduced in height by erosion. The mean altitude is around 1,700 m (5,577 ft) but some 100 summits average 4,000 m (13,123 ft) and the highest peak—Dufourspitze of the Monte Rosa Massif in the Valais alps—reaches 4,634 m (15,204 ft). The Alpine landscape is rich and varied; glacial and fluvial erosion have carved out valleys, terraces and peaks whose continuing evolution makes the Alps a constantly changing region. Measurements taken along the central region show that the Alps continue to grow by an average of one millimeter a year, this growth being balanced, however, by erosion.

On the northwest fringe of the main range, the Pre-Alps, made up mainly of conglomerates, have a less complex structure. They often reach heights of about 2,000 m (6,562 ft).

The Central Plateau was also formed by the Alps. As the Alps and Jura Mountains developed, arms of the sea or lakes persisted in this region. The erosion which set in with the uplifting of the mountains was particularly intensive in the Alps and torrents brought down enormous amounts of sand, gravel and pebbles into the foreland hollows where they were compressed to form new layers of rocks that combined to constitute depth of almost 3 km (1.8 mi) in the Central Plateau. During the various ice ages, the last of which ended around 10,000 B.C., the surface was almost completely covered by the moraines of the Alpine glaciers. Many lateral and terminal moraines occur as elongated hills. Glacial erosion and deposition have created an undulating landscape which favors the formation of lakes. At a mean altitude of 580 m (1,903 ft) this wide plateau stretches from Lake Geneva to Lake Constance and forms a natural setting for communications and cultural activity.

With its delightful landscape of scarps, valleys and plateaus, the Jura Mountains have a much less complex structure than the Alps. Its mean altitude of 700 m (2,296 ft) includes some peaks that rise to around 1,600 m (5,249 ft) and its highest peak, Mont Trendre in the Vaud Jura, reaches 1,679 m (5,508 ft). The Jura can be roughly divided into three types of mountains: the Folded Jura in the south, the Jura Plateau to the north and the Jura Tables to the east. The whole Jura is related to the uplift of the Alps when immense forces crossed the Central Plateau and folded the strata in the Jura region. The Jura folds feature fairly regular, parallel undulations. They exhibit their maximum intensity at the southwestern end of the chain and diminish toward the northwest. In the Jura Tables, the sedimentary layer has been simply lifted and fractured. Over the years valleys have been opened up along the numerous folds through the erosion of running water. From some of these valleys rivers and streams have cut right through the mountain chains creating deep gorges in order to reach the Central Plateau. In other places they form underground courses, or disappear abruptly to reappear in a parallel valley or even at the foot of the range.

Geographically, Switzerland is a land of considerable variety, but this variety is orderly and not chaotic. The mountains are an inseparable and fundamental part of the landscape, bound up with the national psyche. But the thousand valleys that make up the central region form the economic and historical core of the country.

Switzerland forms part of three main river basins: 67.7% of the country is drained by the Rhine into the North Sea; the Rhone (18%) drains into the Mediterranean; the Swiss tributaries of the Po, as the Ticino, and the Adige (9.9%) drain into the Adriatic, and the Inn (4.4%), a tributary of the Danube, flows into the Black Sea.

Lakes are a striking feature of the Swiss landscape, and indeed, no part of the country is further than 15 km (9 mi) from a lake. The major lakes are situated at the foot of the Jura Mountains and the Central Plateau. Lake Geneva (Lake Leman) with an area of 581 sq km (224 sq mi) is the largest while Lake Neuchatel with an area of 215 sq km (83 sq mi) is the largest entirely within Switzerland. Other lakes in this region include Constance, Bienne and Zurich. In the Pre-Alps and the northern and southern edges of the Alps are the Lakes Thun, Brienz, Zug and Lucerne. But Switzerland also boasts of hundreds of smaller lakes, mainly in the Alps. The most beautiful of these are in the Upper Engadine (Grisons)—the Lakes Sils and Silvaplana with their superb tranquil setting.

All the lakes on the Central Plateau between Lakes Constance and Geneva were formed during the Ice Age when glacial ice or moraines created depressions and basins. On the other hand, there are few lakes in the Jura because of the brittle and porous substrata. There is no surface exit from their basins and water oozes away through cracks in the limestone. Lakes Joux and Tailleres are typical of the Jura lakes.

The Swiss Alps have more glaciers than the rest of the range, some 3,000 sq km (1,158 sq mi) in all. Although

PRINCIPAL TOWNS
(estimated population)

	City Proper (1 Jan. 1987)	Conurbation (annual average, 1986)
Bern (Berne, the capital)	137,134	300,316
Zürich	349,549	840,313
Basel (Bâle)	173,160	363,029
Genève (Genf or Geneva)	160,645	384,507
Lausanne	124,206	262,217
Winterthur	84,548	107,812
St Gallen (Saint-Gall)	72,910	125,879
Luzern (Lucerne)	59,904	160,594
Biel (Bienne)	51,341	82,544
Thun	37,074	77,536
La Chaux-de-Fonds	35,726	—
Köniz	35,664	—*
Fribourg	33,935	56,839
Schaffhausen	33,826	53,302
Neuchâtel	32,650	66,142
Chur	30,740	43,163

*Included in the Berne conurbation.

Area and Population

Cantons	Capitals	area sq mi	area sq km	population 1987 estimate
Aargau	Aarau	542	1,405	472,500
Appenzell Ausser-Rhoden	Herisau	94	243	49,500
Appenzell Inner-Rhoden	Appenzell	66	172	13,200
Basel-Landschaft	Liestal	165	428	226,600
Basel-Stadt	Basel	14	37	194,500
Bern	Bern	2,335	6,049	925,300
Fribourg	Fribourg	645	1,670	194,600
Genève	Geneva	109	282	364,100
Glarus	Glarus	264	684	36,500
Graubünden	Chur	2,744	7,106	166,500
Jura	Delémont	323	837	64,800
Luzern	Luzern	576	1,492	305,800
Neuchâtel	Neuchâtel	308	797	155,700
Nidwalder	Stans	107	276	31,100
Obwalden	Sarnen	189	491	27,600
Sankt Gallen	Sankt Gallen	778	2,014	404,200
Schaffhausen	Schaffhausen	115	298	69,800
Schwyz	Schwyz	351	908	103,300
Solothurn	Solothurn	305	791	219,500
Thurgau	Frauenfeld	391	1,013	192,500
Ticino	Bellinzona	1,085	2,811	277,200
Uri	Altdorf	416	1,076	33,400
Valais	Sion	2,018	5,226	232,200
Vaud	Lausanne	1,243	3,219	550,000
Zug	Zug	92	239	81,600
Zürich	Zürich	668	1,729	1,131,100
TOTAL		15,943	41,293	6,523,100

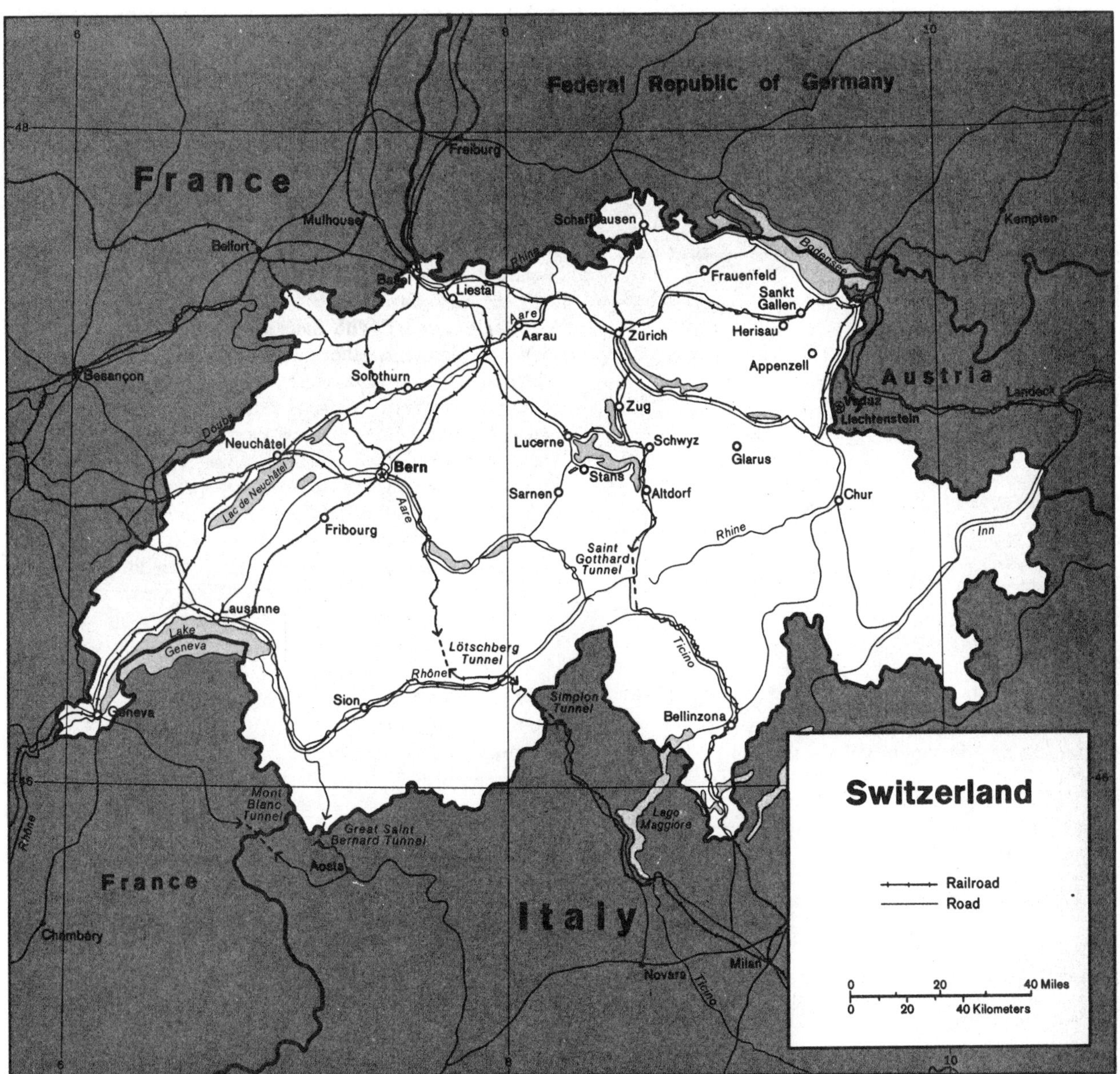

Federal Republic of Germany
France
Austria
Italy
France
Freiburg
Mulhouse
Belfort
Basel
Liestal
Schaffhausen
Frauenfeld
Sankt Gallen
Herisau
Appenzell
Zürich
Aarau
Solothurn
Besançon
Zug
Schwyz
Glarus
Lucerne
Neuchâtel
Bern
Stans
Sarnen
Altdorf
Chur
Fribourg
Vaduz
Liechtenstein
Kempten
Landeck
Lausanne
Lake Geneva
Geneva
Sion
Bellinzona
Lago Maggiore
Aosta
Chambéry
Novara
Milan
Saint Gotthard Tunnel
Lötschberg Tunnel
Simplon Tunnel
Mont Blanc Tunnel
Great Saint Bernard Tunnel
Rhine
Rhône
Aare
Doubs
Ticino
Inn
Bodensee
Lac de Neuchâtel
Switzerland
Railroad
Road
0 20 40 Miles
0 20 40 Kilometers

501438 5-73

some of the glaciers are still advancing, the majority have been retreating during the past century.

One feature peculiar to Switzerland is the *Fohn* which blows through the Alpine valleys bringing a warm wind to the Central Lowlands. The air current is generated when the low temperatures in the north coincides with high temperatures in the south. To the south of the Alps the air becomes cooler as it rises and loses water. This causes monsoon-type rainfall on the southern slopes. When the wind blows down the northern slopes, its temperature rises substantially, sucking up all moisture, for which reason the *Fohn* is described as the "Sahara Air." The temperature can rise by as much as one degree for every one hundred m fall in altitude, so that the local temperatures in the narrow Alpine valleys and foothills reach summer levels as early as March or as late as November.

SYRIA

BASIC FACTS

Official Name: Syrian Arab Republic

Abbreviation: SYR

Area: 185,180 sq km (71,498 sq mi)

Area—World Rank: 84

Population: 11,569,659 (1988) 16,827,000 (2000)

Population—World Rank: 58

Capital: Damascus

Boundaries: 2,419 km (1,503 mi) Turkey 845 km (525 mi); Iraq 596 km (370 mi); Jordan 356 km (221 mi); Israel 80 km (50 mi); Lebanon 359 km (223 mi)

Coastline: 183 km (114 mi)

Longest Distance: 793 km (493 mi) ENE-WSW; 431 km (268 mi) SSE-NNW

Highest Point: Mount Hermon 2,814 m (9,232 ft)

Lowest Point: 200 m (655 ft) below sea level

Land Use: 28% arable land; 3% permanent crops; 46% meadows and pastures; 3% forest and woodland; 20% other

Syria is located in Western Asia and includes a narrow coastal plain stretching south from the Turkish border to Lebanon. Sand dunes cover this littoral, and its flatness is broken only by lateral promontories running down from the mountains to the sea.

The Jabal (mountain) an Nusayriyah, a range paralleling this coastal plain, averages just over 1,212 m (3,976 ft); the highest peak, Nabi Yunis, is about 1,575 m (5,167 ft). The western slopes catch moisture-laden western sea winds and are thus more fertile and more heavily populated than the eastern slopes, which receive only hot, dry winds blowing across the desert. Before reaching the Lebanese border and the Anti-Lebanon Mountains, the Jabal an Nusayriyah range terminates, leaving a corridor—the Homs Gap—through which run the highway and railroad from Homs to the Lebanese port of Tripoli. For centuries the Homs Gap has been a favorite trade and invasion route from the coast to the country's interior and to other parts of Asia. Eastward, the line of the Jabal an Nusayriyah is separated from the Jabal az Zawiyah range and the plateau region by the Al Ghab depression, a fertile, irrigated trench crossed by the meandering Orontes River.

Inland and farther south, the Anti-Lebanon Mountains rise to peaks of over 2,700 m (8,858 ft) on the Syrian-Lebanese frontier and spread in spurs eastward toward the plateau region. The eastern slopes have little rainfall and vegetation and merge eventually with the desert.

In the southwest the lofty Mount Hermon (Jabal ash Shaykh), also on the border between Syria and Lebanon, descends to the Hawran Plateau—frequently referred to as the Hawran—that receives rain-bearing winds from the Mediterranean. All but the lowest slopes of Mount Hermon are uninhabited, however. Volcanic cones, some of which reach over 900 m (2,952 ft) intersperse the open, rolling, once-fertile Hawran Plateau south of Damascus and east of the Anti-Lebanon Mountains. Southeast of the Hawran lies the high volcanic region of the Jabal Druze range, home of the country's Druze population.

The entire eastern plateau region is intersected by a low chain of mountains, the Jabal ar Ruwaq, the Jabal Abu Rujmayn, and the Jabal Bishri, extending northeastward from the Jabal Druze to the Euphrates River. South of these mountains lies a barren desert region known as the Hamad. North of the Jabal ar Ruwaq and east of the city of Homs is another barren area known as the Homs Desert, which has a hard-packed dirt surface.

PRINCIPAL TOWNS (population at 1981 census)

Damascus (capital)	1,112,214	Rakka	87,138
Aleppo	985,413	Hasakeh	73,426
Homs	346,871	Tartous	52,589
Latakia	196,791	Edleb	51,682
Hama	177,208	Dera'a	49,534
Deir ez-Zor	92,091	Suweidiya	43,414

Area and population

Governorates	Capitals	area sq mi	area sq km	population 1987 estimate
Darʿā	Darʾā	1,440	3,730	460,000
Dayr az-Zawr	Dayr az-Zawr	12,765	33,060	487,000
Dimashq	al-larmouk	6,962	18,032	1,127,000
Ḥalab	Aleppo	7,143	18,500	2,269,000
Ḥamāh	Ḥamāh	3,430	8,883	890,000
al-Ḥasakah	al-Ḥasakah	9,009	23,334	810,000
Ḥimṣ	Homs	16,302	42,223	1,007,000
Idlib	Idlib	2,354	6,097	721,000
al-Lādhiqiyah	Latakia	887	2,297	668,000
al-Qunayṭirah	al-Qunayṭirah	719	1,861	34,000
ar-Raqqah	ar-Raqqah	7,574	19,616	422,000
as-Suwaydāʾ	as-Suwaydāʾ	2,143	5,550	240,000
Ṭarṭūṣ	Tartous	730	1,892	542,000
Municipality				
Dimashq	Damascus	41	105	1,292,000
TOTAL		71,498	185,180	10,969,000

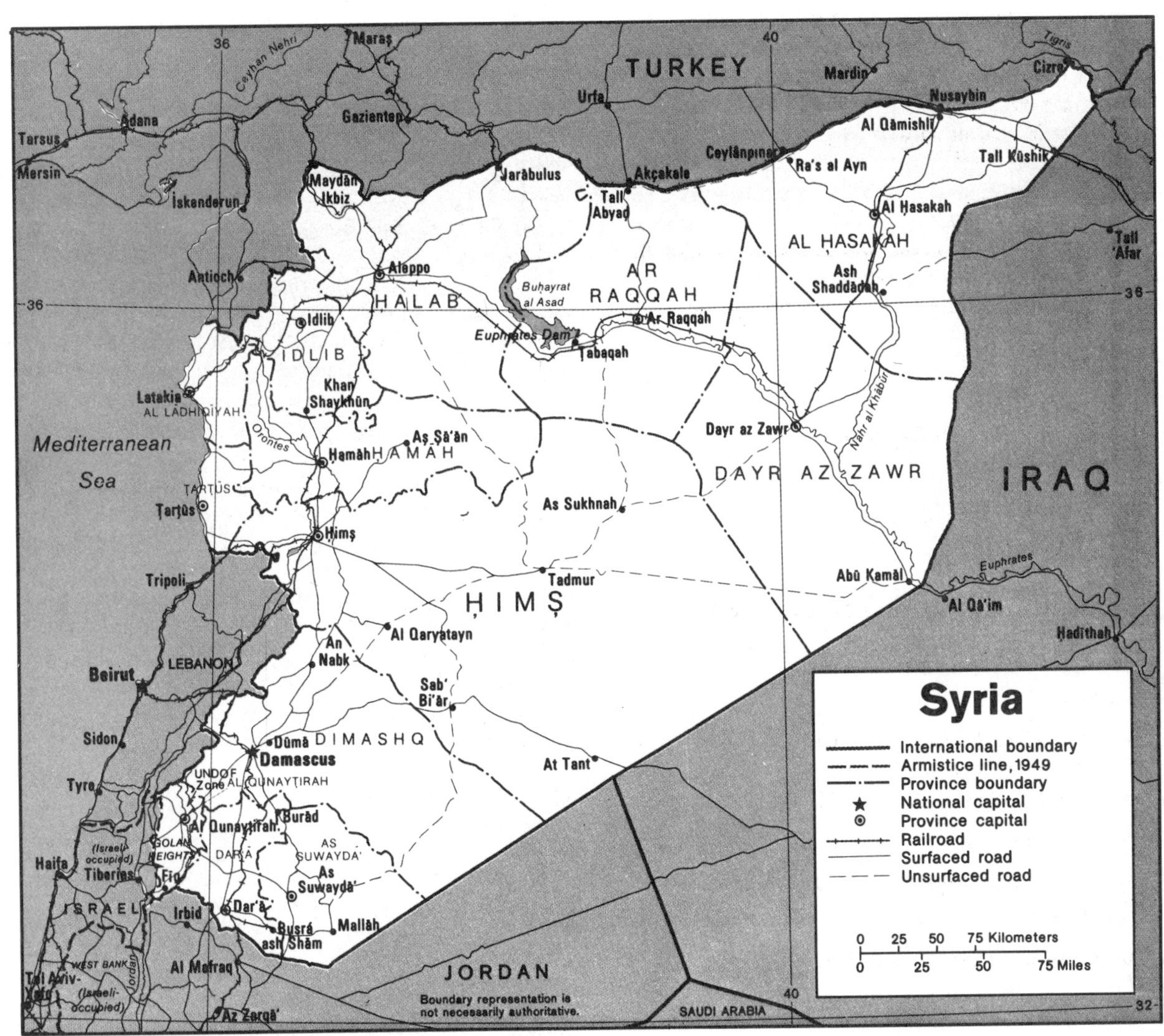

Base 504714 (546149) 3-81

Northeast of the Euphrates River, which originates in the mountains of Turkey and flows diagonally across Syria into Iraq, is the fertile Jazirah region watered by the tributaries of the Euphrates. Oil and natural gas discoveries in the extreme northeastern portion of the Jazirah have significantly enhanced the region's economic potential.

The country's waterways are of vital importance to its agricultural development. The longest and most important river is the Euphrates, which represents more than 80 percent of the country's water resources. Its main left-bank tributaries, the Balikh and the Khabur, are both major rivers and also rise in Turkey. The right-bank tributaries of the Euphrates, however, are small seasonal streams called wadis.

Throughout the arid plateau region east of Damascus, oases, streams, and a few interior rivers that empty into swamps and small lakes provide water for local irrigation. Most important of these is the Barada, a river that rises in the Anti-Lebanon Mountains and disappears into the desert. The Barada creates the Al Ghutah Oasis, site of Damascus. This verdant area, some 30 sq km (11.5 sq mi) has enabled Damascus to prosper since earliest times.

Areas in the Jazirah have been brought under cultivation with the waters of the Khabur River (Nahr al Khabur). The Sinn, a minor river in Al Ladhiqiyah Province, is used to irrigate the area west of the Jabal an Nusayriyah, about 32 km (20 mi) southeast of the port of Latakia. In the south the springs that feed the upper Yarmuk are diverted for irrigation of the Hawran.

Underground water reservoirs that are mainly natural springs are tapped for both irrigation and drinking. The Al Ghab region is richest in underground water resources and contains some nineteen major springs and underground rivers that have a combined yield of thousands of liters per minute.

TAIWAN

BASIC FACTS

Official Name: Republic of China

Abbreviation: TWN

Area: 36,000 sq km (13,900 sq mi)

Area—World Rank: 123

Population: 20,004,391 (1988) 23,569,000 (2000)

Population—World Rank: 41

Capital: Taipei

Land Boundaries: Nil

Coastline: 869 km (540 mi)

Longest Distance: 381 km (237 mi) NNE-SSW; 139 km (86 mi) ESE-WNW

Highest Point: Yu Mountain 3,997 m (13,114 ft)

Lowest Point: Sea level

Land Use: 24% arable; 1% permanent crops; 5% meadows and pastures; 55% forest and woodland; 15% other

Taiwan includes the island of Taiwan, a few small offshore islands, and the larger Pescadores Islands. The Matsu and Quemoy island groups, which still are considered to be part of the mainland province of Fukien, are the remaining components.

The island of Taiwan, 185 km (115 mi) off the mainland, is centered on the Tropic of Cancer. Its geographic position in relation to Asia is roughly comparable to that of Cuba in relation to the North American continent.

The Pescadores, a group of 64 islands totaling 127 sq km (49 square miles) in area and located 40 km (25 miles) west of Taiwan, guard the approaches from the mainland. Quemoy, the largest of a group of 14 small islands totaling 175 sq km (67.7 square miles) in area, lies off Amoy Bay some 136 km (85 miles) west of the Pescadores. Matsu, the largest of a group of 19 small islands totaling 27 sq km (10.5 square miles) in area, is situated 9 km (5.5 miles) off the mainland near Foochow and 183 km (114 miles) north of Quemoy.

Because of its mountainous terrain and heavy rainfall, Taiwan's hydroelectric potential is great. High and rugged mountains occupy more than half of the island and extend

PRINCIPAL TOWNS (estimated population at 31 December 1986)

Town	Population	Town	Population
Taipei (capital)	2,575,180	Fengshan	271,738
Kaohsiung	1,320,552	Chiayi	254,001
Taichung	695,562	Chungli	241,476
Tainan	646,298	Yungho	238,677
Panchiao	491,721	Taoyuan	210,753
Shanchung	358,812	Changhwa	203,541
Keelung	349,616	Pingtun	202,079
Hsinchu	306,088	Hsintien	198,125

Area and population

		area		population
Counties	**Capitals**	sq mi	sq km	1987 estimate
Chang-hua	Chang-hua	415	1,074	1,226,231
Chia-i	Chia-i	734	1,902	564,338
Hsin-chu	Hsin-chu	551	1,428	367,019
Hua-lien	Hua-lien	1,787	4,629	359,340
I-lan	I-lan	825	2,137	448,418
Kao-hsiung	Feng-shan	1,078	2,793	1,080,197
Miao-li	Miao-li	703	1,820	548,187
Nan-t'ou	Nan-t'ou	1,585	4,106	534,920
P'eng-hu	Ma-kung	49	127	100,927
P'ing-tung	P'ing-tung	1,072	2,776	897,714
T'ai-chung	Feng-yuan	792	2,051	1,161,025
T'ai-nan	Hsin-ying	778	2,016	1,003,275
T'ai-pei	Pan-ch'iao	792	2,052	2,727,510
T'ai-tung	T'ai-tung	1,357	3,515	272,477
T'ao-yüan	T'ao-yüan	471	1,221	1,232,209
Yün-lin	Tou-liu	498	1,291	783,526
Municipalities				
Chia-i	—	23	60	254,001
Chi-lung	—	51	133	349,616
Hsin-chu	—	40	104	306,088
Kao-hsiung	—	59	154	1,320,552
T'ai-chung	—	63	163	695,562
T'ai-nan	—	68	176	646,298
Taipei	—	105	272	2,575,180
TOTAL		13,900	36,000	19,454,610

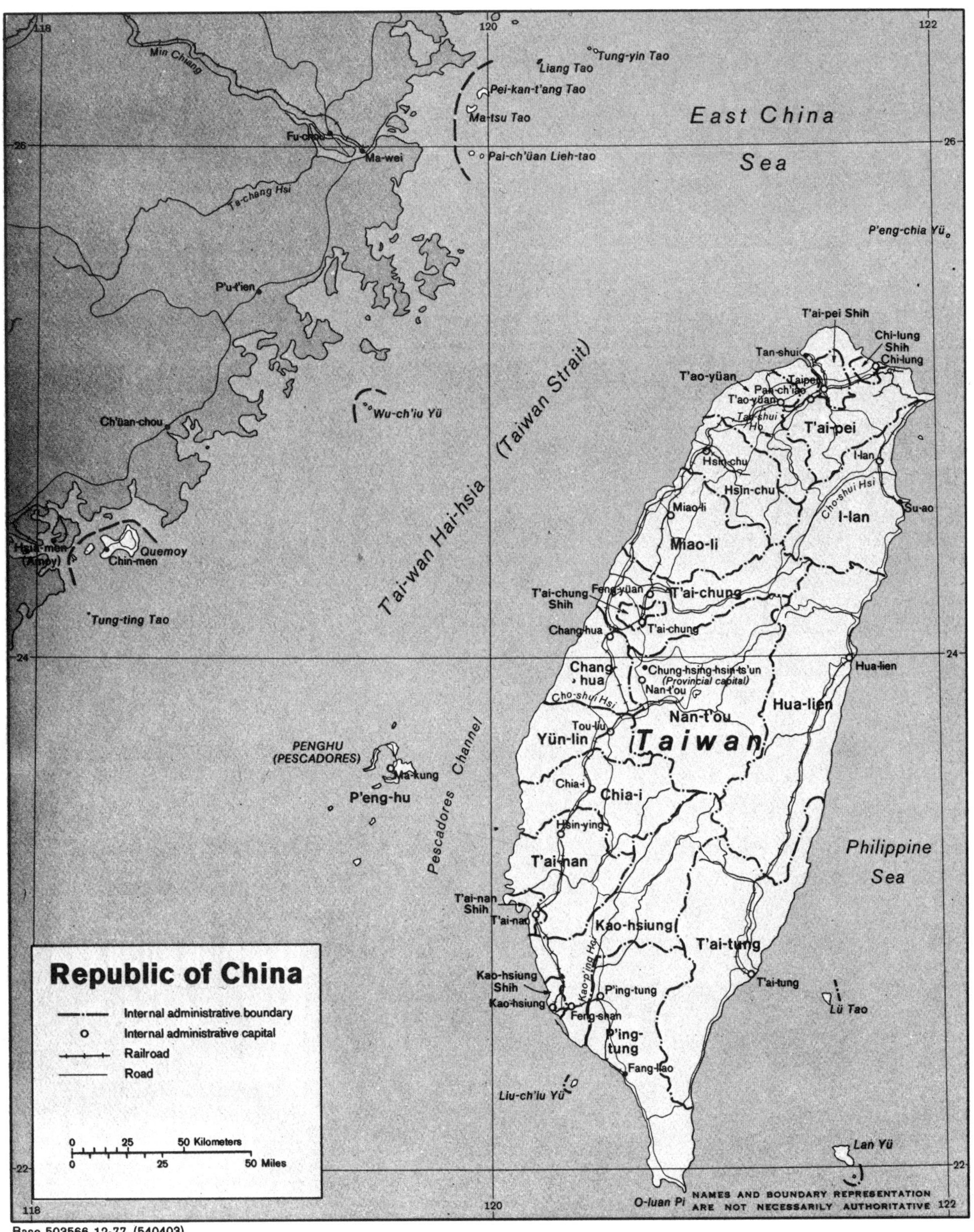

Base 503566 12-77 (540403)

in a north-south direction from its northern tip to its southern extremity. In the far north, detached from the main Central Range, is a short range of volcanic origin called Tatun Shan which rises to over 1,219 m (4,000 feet). In the Central Range there are more than 60 peaks with elevation of over 3,048 m (10,000 feet). The highest of these, Yu Shan, 3,992 m (13,100 feet) in elevation, lies near the absolute center of the island.

On the east coast the Central Range generally drops precipitously to the Pacific Ocean. Near the center of the coast, however, a narrow rift valley about 161 km (100 miles) long intervenes between the Central Range and a lower, but also steep, coastal range. In the west the high mountains are succeeded by foothills which gradually give way to flat alluvial plains where the many short, rapid, mountain streams become meandering rivers. Extensive cultivation is undertaken up the slopes of the foothills, but most of the island's cultivable land lies in the west on the coastal plains.

Taiwan lies astride the earthquake belt, and over 200 shocks are recorded each year. Fortunately, few of these have been severe.

The Pescadores are relatively flat coral reefs which support some agriculture. Quemoy is rocky and boulder-strewn, but it is also partially arable. Matsu, made up of masses of igneous rocks, supports no agriculture and is inhabited only by fishermen in times of peace.

TANZANIA

PRINCIPAL TOWNS (population at 1978 census)

Dar es Salaam	757,346	Mwanza	110,611
Zanzibar	110,669	Tanga	103,409

Area and population

Regions	Capitals	area sq mi	area sq km	population 1987 estimate
Arusha	Arusha	31,698	82,098	1,274,000
Bukoba	Bukoba	10,987	28,456	1,397,000
Dar es Salaam	Dar es Salaam	538	1,393	1,605,000
Dodoma	Dodoma	15,950	41,311	1,239,000
Iringa	Iringa	21,950	56,850	1,167,000
Kigoma	Kigoma	14,301	37,040	828,000
Kilimanjaro	Moshi	5,116	13,250	1,159,000
Lindi	Lindi	25,498	66,040	631,000
Mara	Musoma	8,402	21,760	908,000
Mbeya	Mbeya	23,301	60,350	1,421,000
Morogoro	Morogoro	27,268	70,624	1,202,000
Mtwara	Mtwara	6,452	16,710	916,000
Mwanza	Mwanza	7,600	19,683	1,836,000
Pemba North Pemba South	Wete Chake Chake	380	984	
Pwani	Dar es Salaam	12,566	32,547	600,000
Rukwa	Sumbawanga	26,500	68,635	656,000
Ruvuma	Songea	24,583	63,669	725,000
Shinyanga	Shinyanga	19,598	50,760	1,779,000
Singida	Singida	19,050	49,340	770,000
Tabora	Tabora	29,402	76,150	1,185,000
Tanga	Tanga	10,300	26,677	1,305,000
Zanzibar North Zanzibar South and Central Zanzibar West	Mkokotoni Koani Zanzibar	641	1,660	605,000
TOTAL LAND AREA		342,081	885,987	23,208,000
INLAND WATER		22,800	59,050	
TOTAL		364,881	945,037	

BASIC FACTS

Official Name: United Republic of Tanzania

Abbreviation: TZN

Area: 945,037 sq km (364,881 sq mi)

Area—World Rank: 30

Population: 24,295,250 (1988) 36,008,000 (2000)

Population—World Rank: 34

Capital: Dar es Salaam

Boundaries: 5,114 km (3,178 mi); Uganda 418 km (260 mi); Kenya 769 (478 mi); Mozambique 756 km (470 mi); Malawi 451 km (280 mi); Zambia 322 km (200 mi); Zaire 459 km (285 mi); Burundi 451 km (280 mi); Rwanda 217 km (135 mi)

Coastline: 1,271 km (790 mi)

Longest Distances: 1,223 km (760 mi) N-S; 1,191 km (740 mi) E-W

Highest Point: Kilimanjaro 5,895 m (19,340 ft)

Lowest Point: Sea level

Land Use: 5% arable; 1% permanent crops; 40% meadows and pastures; 47% forest and woodland; 7% other

Tanzania, lying between one and twelve degrees south of the equator, stretches 1,180 km (740 mi) north to south and about 1,200 km (760 mi) east to west. The size of Texas, its total area is 931,082 sq km (363,708 sq mi) including 20,650 sq km (nearly 8,000 sq mi) of inland water.

Most of the country, rising steadily toward the west, consists of extensive rolling plains demarcated by the Rift Valley, a series of immense faults creating both depressions and mountains. Much of it is above 900 m (about 3,000 ft) and some above 1,500 m (nearly 5,000 ft). A small portion, including the islands and the coastal plains, lies below 200 m (600 to 700 ft).

The landscape is extremely varied, changing from coastal mangrove swamps to tropical rain forests and from rolling savannas and high arid plateaus to mountain ranges. It contains both the highest point in Africa, Mount Kilimanjaro, and the lowest, which is the floor of Lake Tanganyika.

The essentially tropical climate is modified by local topography, particularly altitude. The mountain ranges and the area around Lake Victoria (Victoria Nyanza) receive generous amounts of rain, but vast plateau areas in the center of the country are so dry that they cannot support significant cultivation activity, and tsetse fly infestation precludes animal husbandry. Four major ecological regions can be distinguished: high plateaus, mountain lands, lakeshore region, and coastal belt and islands.

High Plateaus

The high plateaus are characterized by monotonous undulating terrain cut slightly by mostly intermittent rivers. There are two major plateaus, the Central Plateau and the Eastern Plateau. The Central Plateau lies between the two branches of the Rift Valley. Its vast expanse forms a huge uplifted basin. Elevation varies from roughly 900 to 1,800 m (3,000 to 5,900 ft) above sea level, the greater part lying at about 1,200 m (4,000 ft). It is a hard dry plain dotted with granitic outcrops.

The Eastern Plateau is in effect a series of lower plateaus descending gradually to the coastal lowlands. In the north it consists basically of the Masai Steppe, an extensive semiarid plain of almost 70,000 sq km (more than 26,000 sq mi). Varying from just under 250 to over 1,000 m (800 to 3,500 ft) above sea level, the steppe is almost a desert with vast areas of dry bush and scanty grass. South of the Uluguru Mountains the plateau broadens to form a rough triangle, the base stretching from Lake Nyasa to the coast. The terrain is broken and toward the coast is characterized by outcrops of isolated hill masses rising sharply from the surrounding land. One of these is the Makonde Plateau in the extreme southeast, a poorly watered tableland of about 3,100 sq km (1,200 sq mi).

Mountain Lands

One of three major mountainous zones extends inland from Tanga to near Lake Manyara. It includes the Usambara and Pare ranges, which together form a wedge-shaped mass reaching a height of almost 2,300 m (7,550 ft) and the Northern Highlands, which contain Mount Kilimanjaro and Mount Meru. Mount Kilimanjaro rises in two peaks united by a saddle. Kibo, the higher peak, is almost 5,900 m (19,300 ft). The so-called glaciers on its top are rapidly decaying remains of a former, more extensive icecap. Meru, the lower peak, rises to 4,560 m (14,960 ft). Both peaks receive considerable amounts of rain on the southern slopes, and tropical rain forest conditions prevail between the altitudes of about 1,700 and 2,900 m (5,600 and 9,500 ft) above sea level on Mount Kilimanjaro and between 1,400 and 1,800 m (4,600 and 6,000 ft) on Mount Meru.

The second zone stretches from the western shore of Lake Natron southward in a series of isolated mountains and mountain chains. They are interspersed with lakes and craters and connected with the northern part of the eastern Rift. Between Lake Natron and Lake Manyara are the Winter Highlands, a volcanic region containing Mount Loolmalassin and the Ngorongoro Crater—roughly 100 to 110 km (sixty to seventy mi) wide—in which is found one of the heaviest concentrations of wildlife in Africa. The shores of Lake Manyara and the nearby Serengeti Plain also teem with wildlife. West of the crater lies Olduvai Gorge, where the paleontological explorations of the late Louis S. B. Leakey, a Kenya-born scientist, and his associates led to the hypothesis that the earliest forms of man may have originated in East Africa.

The third major mountainous region includes the Southern Highlands. They stretch from the Nguru Mountains, about halfway between Dodoma and Dar es Salaam, and the Uluguru Mountains, farther south, to the Livingstone Mountains, which descend sharply toward the eastern shore of Lake Nyasa.

Lakeshore Region

The northern portion of the Central Plateau slopes gently downward to form the large shallow depression containing Lake Victoria, which lies at an elevation of about 1,180 m (3,700 ft). West of the lake are long, narrow rocky hill ranges, which rise above flat lowlands. On the lakeshore are large flooded inlets.

The gradual slope of the land permits agricultural development not possible along the steep embankments of Lakes Tanganyika and Nyasa. The area is densely populated, and the people have a close cultural affinity with those living in the Uganda and Kenya portions of the Lake Victoria basin.

Coastal Belt and Islands

The coastal belt is narrow in the north and south, averaging between sixteen and sixty km (between ten and forty mi) in breadth. It is broader in the center near the lowlands of the Rufiji River valley where it almost reaches the Uluguru mountains.

The 800 km (500 mi) coast is difficult to approach because of numerous coral reefs and shifting sandbars at the mouths of rivers. The inland slopes sufficiently toward the coast to cause most rivers to be unnavigable because of rapids.

The islands are basically coral. Zanzibar, separated from the mainland by a channel thirty-five km (twenty-two mi) wide at its narrowest point, is the largest coralline island on the African coast. It is about eighty km (fifty mi) long and forty km (twenty-five mi) wide with a total area of 1,657 sq km (640 sq mi). Zanzibar rises from a flat eastern plain to a more hilly western area.

Pemba, north of Zanzibar, is smaller; it is sixty-seven km (forty-two mi) long and twenty-two km (fourteen mi) wide with a total area of 984 sq km (380 sq mi). Its topography varies—with small steep hills and valleys. Mafia, forty-three km (twenty-seven mi) long and more than fourteen km (nine mi) wide, is a low island situated about halfway down the coast south of Tanzania.

Drainage

The country's rivers drain into four major basins. Five important rivers and a number of minor ones in the eastern third of Tanzania enter the Indian Ocean directly. Streams around Lake Nyasa empty into the lake and reach the Indian Ocean via the Zambezi River. A number of short rivers (except for the longer Kagera River in northwestern Tanzania) drain into Lake Victoria and ultimately via the Nile River into the Mediterranean Sea. Several rivers in western Tanzania, the longest of which is the Malagarasi, drain into Lake Tanganyika and ultimately via the

TANZANIA
International boundary
National capital
Railroad
Road
International Airport
0 50 100 Miles
0 50 100 Kilometers
UGANDA
KENYA
RWANDA
BURUNDI
ZAIRE
ZAMBIA
MALAWI
MOZAMBIQUE
Lake Edward
Kabale
Lac Kivu
KIGALI
BUJUMBURA
Bukoba
LAKE VICTORIA
Musoma
UKEREWE I.
Mwanza
Biharamulo
Nyeri
Lake Naivasha
NAIROBI
Magadi
Lake Natron
Lake Manyara
Lake Eyasi
Moshi
Arusha
Shinyanga
Nzega
Mombasa
Kigoma
Tabora
Singida
Korogwe
Tanga
Wete
PEMBA I.
Kalemi
LAKE TANGANYIKA
Mkokotoni
Koani
ZANZIBAR I.
Zanzibar
Mpanda
Dodoma
DAR ES SALAAM
Kilosa
Morogoro
Ruvu
INDIAN OCEAN
Mikumi
Kidatu
Iringa
Sumbawanga
Lake Rukwa
MAFIA I.
Ifakara
Lake Mweru
Mbeya
Tunduma
Njombe
Kilwa Kivinje
Lindi
Kasama
LAKE NYASA
Mtwara
Lake Bangweulu
Songea
Masasi
Mzimba
BOUNDARY REPRESENTATION IS NOT NECESSARILY AUTHORITATIVE
Kagera
Mara
Ruvubu
Malagarasi
Wembere
Pangani
Ugalla
Wami
Kisigo
Great Ruaha
Rufiji
Kilombero
Luwegu
Chambeshi
Ruvuma

Zaire River into the Atlantic Ocean. Streams in the northcentral and southwestern sections empty into interior basins.

Many of Tanzania's rivers are shallow or marshy with only seasonal flows. Some, particularly those draining into the Indian Ocean, offer potential for irrigation and hydroelectric power. The use of others is limited by marked variation in flow. The Rufiji River and its tributaries, draining nearly a quarter of the mainland's territory, offer perhaps the greatest potential; some of it has already been realized. For example, the Kilombero River, draining a once marshy valley is now under sufficient control so that some of the area is used for cultivation of sugar, and the Great Ruaha River is the site of a hydroelectric station. The Pangani River, which rises in the northeastern highlands, has three hydroelectric stations.

The lakes provide transportation, are a source of food and livelihood, and offer abundant water supplies for irrigation. The largest lake in Tanzania and in Africa is Lake Victoria. Lake Tanganyika, the world's second deepest lake, has a precipitous coastline and a few poor harbors. Lake Nyasa also has poor harbors. Lake Rukwa to the east of Lake Tanganyika is small and shallow and tends to be brackish. A series of small lakes in the northern part of the country all have salty water—Lake Natron is commercially exploited for salt and soda.

THAILAND

BASIC FACTS

Official Name: Kingdom of Thailand

Abbreviation: THL

Area: 513,115 sq km (198,115 sq mi)

Area—World Rank: 47

Population: 54,588,731 (1988) 64,132,000 (2000)

Population—World Rank: 18

Capital: Bangkok

Boundaries: 7,547 km (4,690 mi); Laos 1,754 km (1,090 mi); Cambodia 803 km (499 mi); Malaysia 576 km (358 mi); Myanmar 1,799 km (1,118 mi)

Coastline: Gulf of Siam 1,875 km (1,165 mi); Andaman Sea 740 km (460 mi)

Longest Distances: 1,556 km (966 mi) N-S; 790 km (491 mi) E-W

Highest Point: Inthanon 2,600 m (8,530 ft)

Lowest Point: Sea level

Land Use: 34% arable; 4% permanent crops; 1% meadows and pastures; 30% forest and woodland; 31% other

PRINCIPAL TOWNS
(population at 1980 census)

Bangkok Metropolis*	4,697,071
Songkhla	172,604
Chon Buri	115,350
Nakhon Si Thammarat	102,123
Chiang Mai	101,594

*Formerly Bangkok and Thonburi. The estimated population of Bangkok Metropolis at 31 December 1986 was 5,468,915 (Source: Ministry of the Interior).

Area and population	area		population
Regions	sq mi	sq km	1986 estimate
Bangkok Metropolis	604	1,565	5,363,378
Central	7,236	18,742	3,552,602
Eastern	14,481	37,507	3,963,061
Northeastern	65,195	168,854	18,060,945
Northern	65,500	169,644	10,391,368
Southern	27,303	70,715	6,441,186
Western	17,795	46,088	4,023,111
TOTAL	198,115	513,115	51,795,651

Thailand lies in the middle of continental Southeast Asia, in an axial position that influenced many aspects of its developing society and culture. The earliest Tai-speaking peoples found their way into the area by following river valleys through the high mountains of southern China and northern Thailand down across the fertile central plain formed by the mighty river, the Mae Nam Chao Phraya. The attraction of the Chao Phraya valley, with its fertile floodplains and tropical monsoon climate so ideally suited to wet-rice cultivation, tended to concentrate settlement in that area rather than in the marginal uplands and mountains of the northern region or in the Khorat plateau to the northeast.

Topography and Drainage

Terrain features include, most conspicuously, high mountains, a central plain, and an upland plateau. The high mountains cover most of the North and extend along the western border with Myanmar and down into the Malay Peninsula. The central plain is effectively the lowland area drained by the Mae Nam Chao Phraya, the country's principal river, and the smaller streams that feed into the delta at the head of the Bight of Bangkok. The upland Khorat plateau is a gently rolling region of low hills and shallow lakes, drained almost entirely by the Mekong River via the Mae Nam Mun. Peninsular Thailand offers a sharply contrasting set of natural features—long coastlines and offshore islands, sandy beaches, and mangrove swamps.

The Chao Phraya and its tributaries drain an estimated one-third of the nation's territory and together, with part of the watershed of the Mekong that empties into the South China Sea, they account for the importance of hydrology in the country's economic development. The network of rivers and man-made canals not only support wet-rice cultivation but provide vitally important waterways for the transport of goods and men.

Regions

Landforms and drainage divide the country into four more or less natural regions, one of them including

Base 800604 (B00150) 2-88

Bangkok and its environs, the country's single major urban area, which in other respects must be considered a distinct region. Each of the four regions differs from the others in population, basic resources, natural features, and level of social and economic development. The diversity of the regions is in fact the most pronounced attribute of Thailand's physical setting.

As described here, the four regions coincide with those used in regional compilations by the Thai government's National Statistical Office but not with those of the Ministry of Agriculture and Cooperatives, which defines a substantially smaller area as the North.

Broadly the North consists of an area of high mountains incised by steep river valleys and upland areas that border the central plain. Its many rivers, including the Ping, Wang, Yom, and Nan, which unite in the lowlands to form the Chao Phraya watershed, provide water resources for numerous small irrigation schemes that support intensive rice cultivation in the alluvial valleys. The mountains are thickly covered by forests yielding valuable timber, but diminishing land reserves, causing exploitative clearing of slopes by a new generation of Thai farmers who have migrated from the central plain and who raise maize and other upland crops, have led to depletion of forest reserves. Intensive farming and population pressure push cultivation efforts up the slopes in vertical ascent patterns harmful to both watersheds and timber resources.

The Center, the historical and contemporary core region of the Central Thai, was long considered the rice bowl of Asia. The highly developed irrigation systems of the region support a concentrated population. Visually the region is dominated by the Chao Phraya and its tributaries—with their waterbound transport and annual floods—and by the cultivated paddy fields that stretch over hundreds of km in an unchanging landscape. Metropolitan Bangkok—the focal point of trade, transport, and industrial activity—is situated on the southern edge of the region at the head of the Gulf of Thailand and encloses part of the delta of the Chao Phraya system. To the southeast lies an area clearly distinguishable from the rest of the Center, consisting mainly of flat lowlands, with marginal soils where rice loses its predominance (partly owing to lack of water), and forest and rainfall are more abundant.

The Northeast is geographically unfavored, beset with many ecological problems, including poor soils. It consists mainly of a dry plateau and a few low hills. The short monsoon season brings heavy flooding in river valleys, but the dry season is long and the prevailing vegetation sparse grass. Mountains ring the plateau on the west and south; the Mekong, whose importance for Thailand lies in the links it provides to the subcontinent, delineates much of the eastern rim. The Mae Nam Mun, the largest river within the Northeast, and its tributary, the Lam Nam Chi, empty into the Mekong.

The South, a narrow peninsular region, is distinctive in climate, terrain, and resources. Its economy is based on rubber cultivation, and rice is grown primarily for subsistence. Rolling to mountainous terrain and the absence of large rivers are conspicuous. North-south mountain barriers and impenetrable tropical forest led to an early initial isolation and separate political development. Rubber plantations, coconut growing, and tin mining by a largely Muslim population distinguish the region economically, ethnically, and politically from the northern half of the country.

TOGO

BASIC FACTS

Official Name: Republic of Togo

Abbreviation: TOG

Area: 56,785 sq km (21,925 sq mi)

Area—World Rank: 115

Population: 3,336,433 (1988) 4,522,000 (2000)

Population—World Rank: 108

Capital: Lome

Boundaries: 1,673 km (1,039 mi); Burkina Faso 126 km (78 mi); Benin 620 km (385 mi); Ghana 877 km (545 mi)

Coastline: Gulf of Guinea 50 km (31 mi)

Longest Distances: 510 km (317 mi) N-S 110 km (68 mi) E-W

Highest Point: Baumann Peak 986 m (3,235 ft)

Lowest Point: Sea level

Land Use: 25% arable; 1% permanent crops; 4% meadows and pastures; 28% forest and woodland; 42% other

Togo is located in West Africa. The country consists primarily of two savanna plains separated by a southwest to northeast range of hills known as the Chaine du Togo. From south to north, the country is composed of six topographical regions: (1) the sandy beaches, estuaries and inland lagoons of the coastal plain; (2) the Ouatchi Plains in the immediate hinterland; (3) the higher Mono tableland; (4) the Chaine du Togo that begins in Benin's Atakora Mountains and ends in Ghana's Akwapim Hills (Togo's highest elevation is found here at Pic Baumann, 986 m, 3,235 ft); (5) the northern sandstone Oti Plateau; and (6) the northwestern granite regions in the vicinity of Dapango.

The Mono Basin and its affluents occupy the southern half of the country. Northern Togo is drained by the Oti, a tributary of the Volta, and the Kara and the Mo Rivers,

PRINCIPAL TOWNS

(estimated population at 1 January 1977)

Lomé (capital)	229,400	Tsevie	15,900
Sokodé	33,500	Anécho	13,300
Palimé	25,500	Mango	10,930*
Atakpamé	21,800	Bafilo	10,100*
Bassari	17,500	Tabligbo	5,120*

*1975 figure.

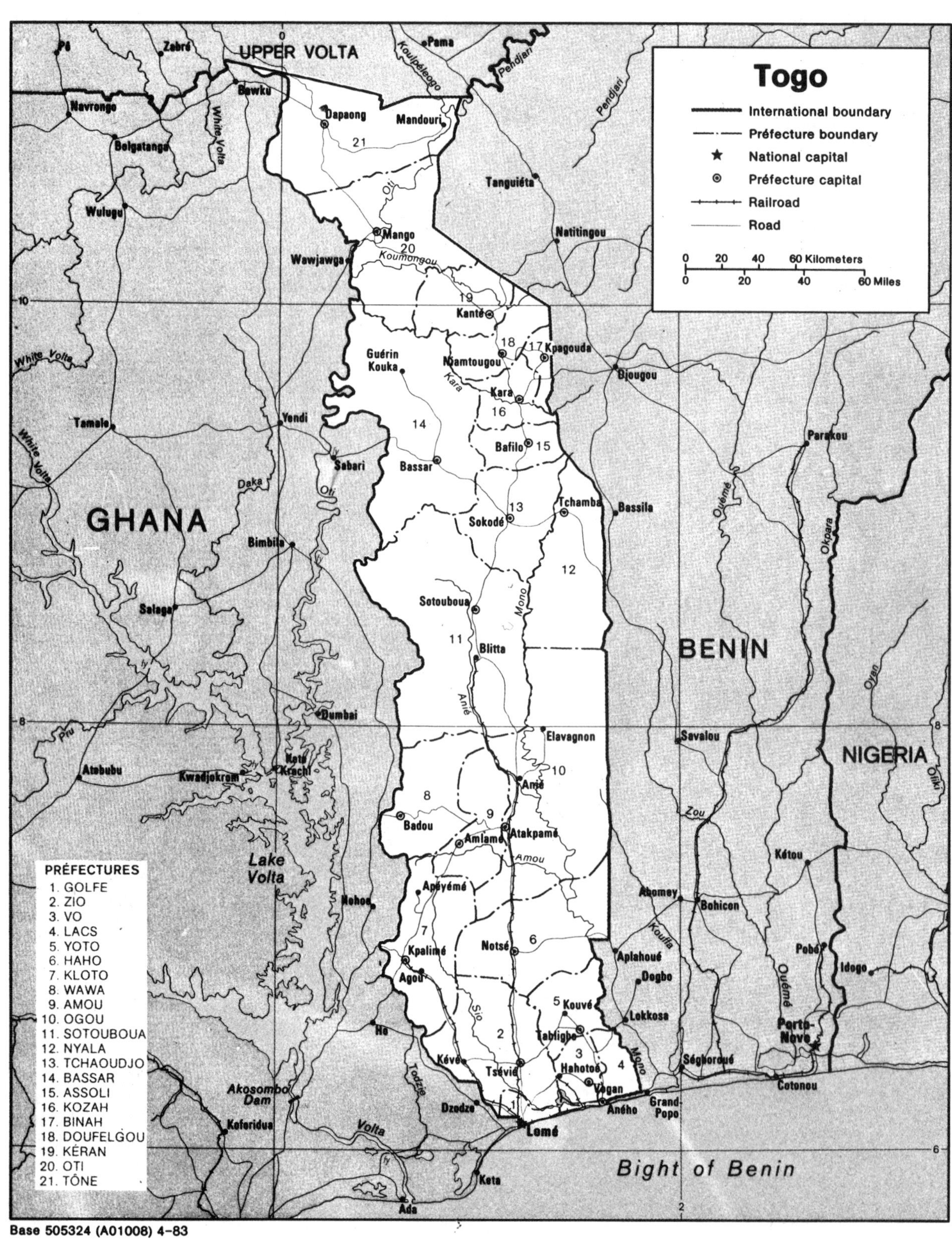

Base 505324 (A01008) 4-83

which drain into the Oti. The only rivers whose mouths are in Togo are the Sio (Chio) and the Haho. Of the many inland lagoons, the largest is Lac Togo.

Area and population

Regions Prefectures	Capitals	area sq mi	area sq km	population 1981 census
Centrale	Sokodé			269,174
Sotouboua	Sotouboua	2,892	7,490	128,617
Tchamba	Tchamba			44,912
Tchaoudjo	Sokodé	2,198	5,692	95,645
De la Kara	Kara			432,626
Assoli	Bafilo	362	938	32,444
Bassar	Bassar	2,444	6,330	118,345
Binah	Pagouda	180	465	50,077
Doufelgou	Niamtougou	432	1,120	66,120
Kéran	Kandé	653	1,692	44,762
Kozah	Kara	419	1,085	120,878
Des Plateaux	Atakpamé			561,656
Amou	Amlamé	1,692	4,382	72,951
Haho	Notsé	1,412	3,658	109,995
Kloto	Kpalimé	1,077	2,790	106,429
Ogou	Atakpamé	2,372	6,145	163,906
Wawa	Badou			108,375
Des Savanes	Dapaong			326,826
Oti	Sansanné-Mango	1,453	3,762	77,747
Tône	Dapaong	1,869	4,840	249,079
Maritime	Lomé			1,039,700
Golfe	Lomé	133	345	438,110
Lacs	Aného	275	712	140,006
Vo	Vogan	290	750	150,313
Yoto	Tabligbo	483	1,250	100,387
Zio	Tsévié	1,289	3,339	210,884
TOTAL		21,925	56,785	2,700,982

TOKELAU

BASIC FACTS

Official Name: Tokelau

Abbreviation: TOK

Area: 12.2 sq km (4.7 sq mi)

Area—World Rank: 212

Population: 1,745 (1988) 1,700 (2000)

Population—World Rank: 213

Capital: None

Land Boundaries: Nil

Coastline: 101 km (63 mi)

Longest Distances: N/A

Highest Point: N/A

Lowest Point: Sea level

Land Use: 100% other

Tokelau consists of three atolls (Atafu, Nukunonu, and Fakaofo) about 480 km (300 mi) north of Western Samoa in the Pacific Ocean. Each atoll has a lagoon encircled by a number of reef-bound islets varying from length from 91 m (100 yards) to 2.5 km (4 mi) and in width from 91 to 366 m (100 to 400 yards) and extending more than 3 m (10 ft) above sea level. All villages are on the leeward side, close to passages through the reefs.

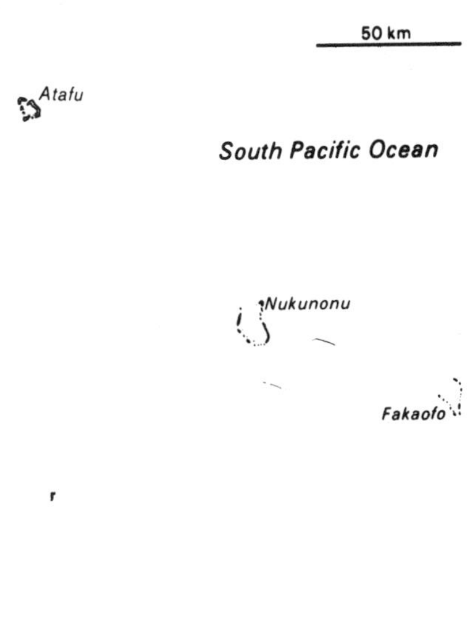

TONGA

BASIC FACTS

Official Name: Kingdom of Tonga

Abbreviation: TNG

Area: 747 sq km (288 sq mi)

Area—World Rank: 167

Population: 99,620 (1988) 101,000 (2000)

Population—World Rank: 174

Capital: Nuku'alofa

Land Boundaries: Nil

Coastline: 560 km (348 mi)

Longest Distances: 631 km (392 mi) NNE-SSW; 209 km (130 mi) ESE-WNW

Highest Point: 1,046 m (3,432 ft)

Lowest Point: Sea level

Land Use: 25% arable; 55% permanent crops; 6% meadows and pastures; 12% forest and woodland; 2% other

Tonga is an archipelago, also known as the Friendl Islands, located in the SW Pacific, about 643 km (400 m

E of Fiji, and about 1,770 km (1,100 mi) from Auckland, New Zealand. It comprises 169 islands, of which only 36 are permanently inhabited. Including the Minerva reefs, which it has recently claimed, the total land area is 749 sq km (289 sq mi), extending 631 km (392 mi) NNE to SSW and 209 km (130 mi) ESE to WNW. The areas of the main islands and island groups are: Tongatapu and Eua (350 sq km, 135 sq mi), Haapai (119 sq km, 46 sq mi), Vavau (143 sq km, 55 sq mi), Niuatoputapu and Tafahi (18 sq km, 7 sq mi) and Niuafoou (52 sq km, 20 sq mi). Tonga's total coastline stretches 560 km (348 mi).

Area and population		area		population
Divisions				1986
Districts	**Capitals**	sq mi	sq km	census
Eua	Ohonua	33.7	87.4	4,393
Eua Foou		...	...	1,995
Eua Proper		...	...	2,398
Haapai	Pangai	42.5	110.0	8,979
Foa		...	...	1,409
Haano		...	...	892
Lulunga		...	...	1,588
Muomua		...	...	897
Pangai		...	...	2,840
Uiha		...	...	1,353
Niuas	Hihifo	27.7	71.7	2,379
Niuafoou		...	...	763
Niuatoputapu		...	...	1,616
Tongatapu	Nukualofa	100.6	260.5	63,614
Kolofoou		...	...	15,782
Kolomotua		...	...	13,117
Kolovai		...	...	4,023
Lapaha		...	...	6,992
Nukunuku		...	...	5,790
Tatakamotonga		...	...	6,778
Vaini		...	...	11,132
Vavau	Neiafu	46.0	119.2	15,170
Hahake		...	...	2,292
Hihifo		...	...	2,095
Leimatua		...	...	2,875
Motu		...	...	1,387
Neiafu		...	...	5,273
Pangaimotu		...	...	1,248
TOTAL LAND AREA		289.5	749.9	94,535
INLAND WATER		11.4	29.6	
TOTAL		300.9	779.5	

The Tonga Islands are dispersed in two parallel chains, volcanic in the west with limestone formations superimposed on them and upifted coral formations in the east. The islands fall into three latitudinal groups: a northern or Vavau group including Hunga, Kapa and Vavau; a central or Haapai group; and a southern or Tongatapu group. The Tongatapu group contains seven major islands; the largest is Tongatapu on which the capital, Nukualofa, is located. Other major islands in this group include Eua, Ata, Atata, Euaki, Kalaau and Kenatea. Of the 36 islands in the Haapi group, only 20 are inhabited and one is an active volcano containing a steaming lake. Of the 34 islands in the Vavau group, 14 are uninhabited. Toward the east the islands are bordered by the Tonga Trench over 11.2 km (7 mi) deep. Most of the uninhabited islands are active or extinct volcanic cones; at least one, Falcon, rises above sea level only during eruptions and disappears at other times.

Except for creeks on Eua and a stream on Niuatoputapu, there is no running water on any of the other islands. Their inhabitants rely on wells and stored rainwater.

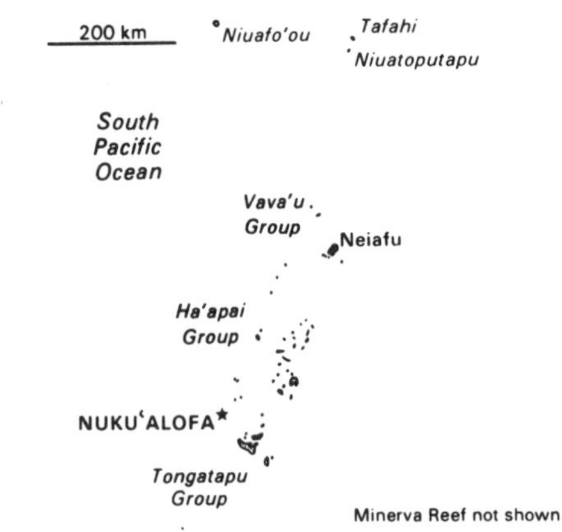

TRINIDAD & TOBAGO

BASIC FACTS

Official Name: Republic of Trinidad & Tobago

Abbreviation: TRT

Area: 5,124 sq km (1,978 sq mi)

Area—World Rank: 151

Population: 1,279,920 (1988) 1,544,000 (2000)

Population—World Rank: 128

Capital: Port-of-Spain

Land Boundaries: Nil

Coastline: 470 km (292 mi)

Longest Distances: Trinidad: 143 km (89 mi) N-S; 61 km (38 mi) E-W Tobago: 42 km (26 mi) NE-SW; 12 km (7.5 mi) NW-SE

Highest Point: El Cerro del Aripo 940 m (3,085 ft)

Lowest Point: Sea level

Land Use: 14% arable; 17% permanent crops; 2% meadows and pastures; 44% forest and woodland; 23% other

The two islands that make up the state of Trinidad and Tobago are situated on the continental shelf of South America and are geographically but not geologically part of the West Indies. Trinidad, much the larger of the two, is at some points within sight of the Venezuelan coast and

Principal towns (population in 1980): Port of Spain (capital) 58,400; San Fernando 34,200; Arima (borough) 24,600.

Area and population

		area		population
Counties	**Capitals**	sq mi	sq km	1986 estimate
Caroni	Chaguanas	213	552	164,800
Nariva/Mayaro	Rio Claro	350	906	32,800
St. Andrew/ St. David	Sangre Grande	364	943	56,900
St. George	. . .	350	907	429,400
St. Patrick	Siparia	255	660	138,000
Tobago	Scarborough	117	303	43,700
Victoria	Princes Town	314	814	215,500
City				
Port-of-Spain	—	4	10	57,400
Boroughs				
Arima	—	3	7	28,200
Point Fortin	—	6	16	
San Fernando	—	2	6	32,600
TOTAL		1,978	5,124	1,199,300

was once a part of the mainland. Tobago, a few miles northeast of Trinidad, is part of a sunken mountain chain related to the continent. The larger island has one moderately high and two low mountain chains but consists for the most part of flatlands and rolling plains; the smaller island is generally mountainous.

Trinidad and Tobago is composed of two major islands and numerous satellite islets located close to South America on the continental shelf. The islands are not geologically part of the Antillean arc, and some writers group Trinidad and Tobago with Curaçao and the other Dutch islands, the offshore islands of Venezuela, and Barbados as the Continental Island group of the West Indies.

Trinidad, second largest of the Commonwealth Caribbean islands, is roughly rectangular in shape with lateral peninsular extensions at the northeast, northwest, and southwest corners. One imaginative observer described it as resembling the boot of a conquistador. The road network is fairly extensive, and on small and compact Trinidad no part of the island is more than a few hours away from any other part. Piarco International Airport, serving coastal Port-of-Spain, is the only major aircraft landing facility on the island currently in regular use; it lies almost one-third of the distance from west to east across the island. Smaller Tobago lies to the northeast of Trinidad and is separated from its sister island by a channel about 32 km (20 mi) in width.

Trinidad is geologically a detached part of the South American continent, separated from it by the Gulf of Paria, an oval-shaped body of water with narrow straits on the north and south bearing respectively the picturesque names of the Dragons Mouth and the Serpents Mouth. The names were reputedly bestowed by Columbus, and an ancient British whimsy is recalled by the fact that they are bordered by Trinidadian counties named for Saint George who slew the dragon and for Saint Patrick who drove the snakes out of Ireland.

Place names reflect the diversity of the country's ethnic background. Scarborough is English in origin, Fyzabad is East Indian, Rio Claro is Spanish, Grande Rivière is French, Arima is Carib, and Rampanalgas is pre-Carib. The Courland River on Tobago recalls the day when the dukes of Courland, a former principality of Latvia, held suzerainty over that island.

On the island of Trinidad the principal mountain system is the Northern Range, a rugged chain that covers the entire northern portion of the island and includes the highest point in the country, the Cerro del Aripo, with an elevation of 940 m (3,085 ft). The Northern Range is geologically an outlier of the Venezuelan portions of the great Andes Mountains. Extending on a slant from northeast to southwest across the middle of the island, the Central Range has average elevations of from 61 to 152 m (200 to 500 ft) and a maximum elevation of 307 m (1,009 ft). Along the southern coast the low and discontinuous Southern Range reaches a maximum elevation of a little less than 304 m (1,000 ft) in the Trinity Hills of the southeast. It was a trinity of adjacent hilltops in the area that allegedly caused Columbus on his third voyage to give the island its name.

The mountains rise steeply in cliffs on the coastal flank of the Northern Range, but on the southern flank they are deeply indented by river valleys as they slope gently to the broad Caroni Plain, the most extensive of the country's lowlands. The Central Range marks the southern limit of the Caroni Plain, and to the south of the Central Range the Naparima Plain on the west and the Nariva Plain on the east are the island's other major lowland areas. Throughout the lowlands the terrain ranges from flat to gently undulating. Because the bulk of the island's population is concentrated in a western lowland belt extending north to south along the Gulf of Paria and the eastern half of the island is sparsely peopled, it is customary to refer to north and south Trinidad but seldom to east or west.

Rivers on the island of Trinidad are numerous but short. The longest are the Ortoire, which extends 50 km (31 mi) eastward to the Atlantic Ocean in the south, and the 40 km (25 mi) Caroni, which runs westward to the Gulf of Paria in the north. Neither is useful for navigation. Although countless rivers and streams provide generally good drainage, heavy seasonal rains often cause flooding.

There are no natural lakes, but extensive swamps occur along the eastern, southern, and western coasts. Some are mangrove swamps separated from the sea by wide sandbars. The most extensive of the swamplands are the Caroni Swamp and the Oropuche Lagoon on the Gulf of Paria and the Nariva Swamp on the Atlantic coast to the east. The waters of most rivers and streams drain ultimately through these swamplands.

The generally fertile soils of the northern and central portions of Trinidad are made up of sedimentary rock often re-sorted as alluvium with a considerable mixture of tropical clays. In the south the soils are sandy, unstable, and considerably less fertile than elsewhere. The island includes numerous small mud volcanoes, gas vents, and natural pitch lakes. Earth tremors are common, and the branch of the University of the West Indies at St. Augustine maintains an important seismic laboratory.

Trinidad has relatively few good harbors. On the north coast the shoreline is heavily indented, but the bays are rockbound, and there is no coastal plain between tidewater

Trinidad and Tobago

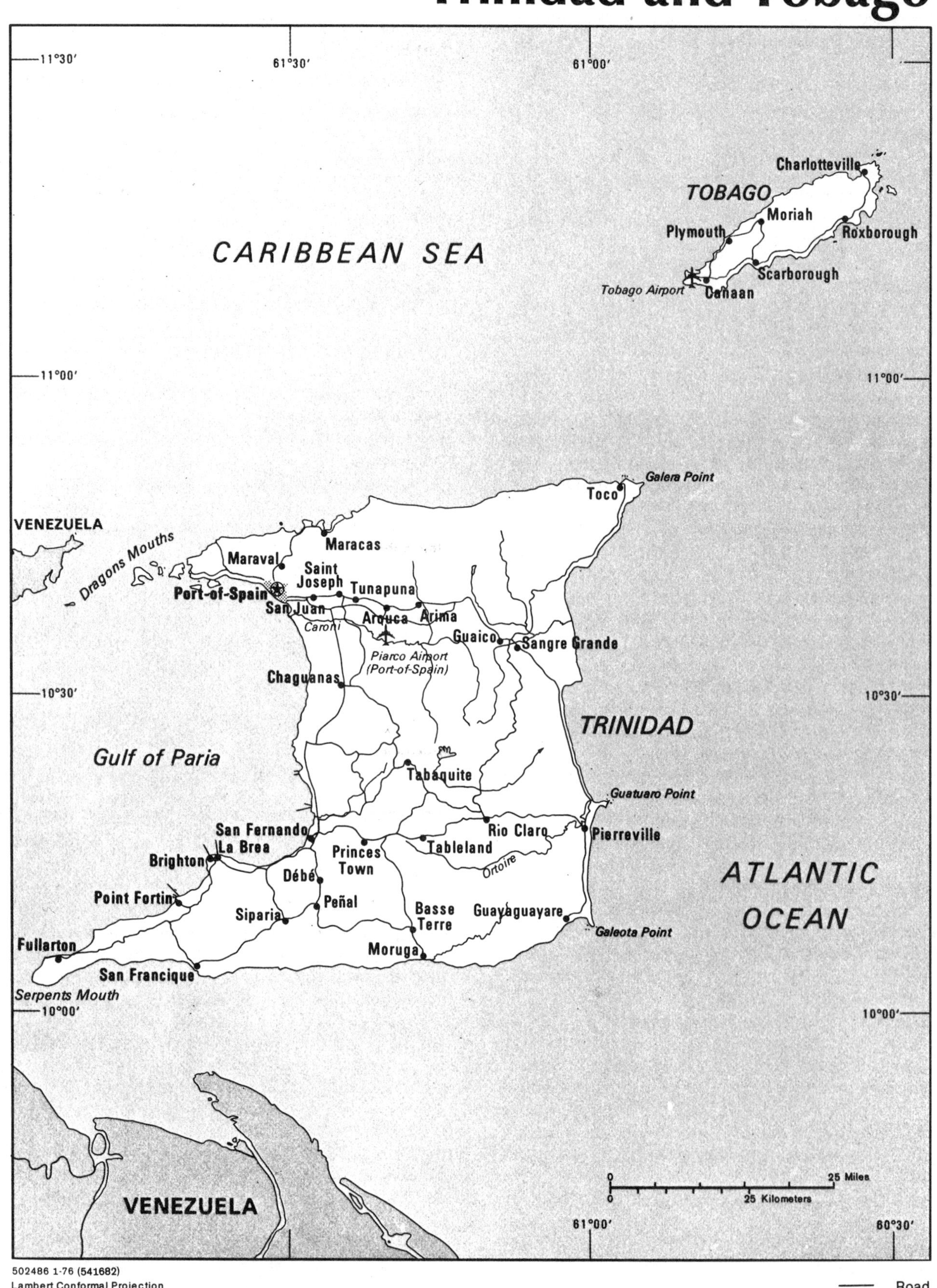

502486 1-76 (541682)
Lambert Conformal Projection
Standard parallels 9°20′ and 14°40′
Scale 1:1,000,000

Road
Airport

and the steep mountain cliffs. On the south the water is shallow, and the bays are too narrow for shipping. The east coast is almost unapproachable by sea because of treacherous Atlantic Ocean currents. On the west, however, the land slopes gently from the Gulf of Paria to an interior of fertile hills and plains. Natural harbors are good, although international traffic exclusive of petroleum has become almost entirely concentrated in the harbor of Port-of-Spain. Chacachacare and Monos islands and most of the remainder of the numerous small islands close to the Trinidad shoreline are located in or near the Dragons Mouth. The country claims a territorial limit of twelve nautical miles except in the Gulf of Paria, where seabed limits have been determined by agreement with Venezuela, but surface and fishing limits are not clearly resolved.

The island of Tobago has an uneven terrain dominated by the Main Ridge, a series of mountains near the northeast coast about 29 km (18 mi) long with elevations reaching a maximum of about 548 m (1,800 ft). South of the Main Ridge are lower hills in which rivers have cut numerous deep and fertile valleys, and the southwestern part of the island consists of an extensive and fairly level coral platform. Most of the limited amount of level land occurs in the southwest, although narrow patches of coastal plain are found elsewhere, most notably around the mouth of the Courland River, which runs westward into the Caribbean Sea between the coral platform and the Main Ridge. Small rivers and streams are numerous, but flooding and erosion are less serious on Tobago than on Trinidad because the upper slopes of the mountains retain much of their original forest cover.

The town of Scarborough is the only important port, but there are several small harbors, and the coastline is indented by numerous inlets and sheltered beaches. In the southwest the coral gardens of Buccoo Reef are major tourist attractions. Tobago has several small satellite islands; the largest are Little Tobago Island, site of a bird sanctuary, and St. Giles, or Melville, Island.

TUNISIA

BASIC FACTS

Official Name: Republic of Tunisia

Abbreviation: TUN

Area: 154,530 sq km (59,664 sq mi)

Area—World Rank: 88

Population: 7,738,026 (1988); 10,751,000 (2000)

Population—World Rank: 75

Capital: Tunis

Boundaries: 2,445 km (1,519 mi) Libya 459 km (285 mi); Algeria 958 km (595 mi)

Coastline: 1,028 km (639 mi)

Longest Distances: 792 km (492 mi) N-S 350 km (217 mi) E-W

Highest Point: Mount Chambi 1,544 m (5,066 mi)

Lowest Point: Chott el Gharsa 17 m (56 ft) below sea level

Land Use: 20% arable; 10% permanent crops; 19% meadows and pastures; 4% forest and woodland; 47% other

Tunisia juts into the Mediterranean Sea midway between Gibraltar and Suez at the point where the narrows between Cape Bon and Sicily divide the Mediterranean into eastern and western basins. As a consequence of this strategic location, the country has since classical times been an important crossroads between Europe and the Middle East.

Together with Algeria, Morocco, and the northwestern portion of Libya known historically as Tripolitania, Tunisia makes up the Maghrib, a region in which fertile coastal lands give way to the great Atlas mountain chain of North Africa and, finally, to the interior expanses of the Sahara Desert. The countries share a common history, ethnic composition, culture, and language.

The Atlas mountain system, which begins in southwestern Morocco, terminates in northeastern Tunisia. Most of northern Tunisia is mountainous, but elevations average less than 300 m (984 ft) and rarely exceed 1,000 m (3,280 ft). The Atlas mountains in Algeria and Morocco, however, reach much higher elevations.

The principal mountain chain, the Dorsale, slants northeastward across the country and plunges into the sea at Cape Bon, an area famed among early Mediterranean navigators. The Dorsale is cut by several transverse depressions, among them the Al Qasrayn (Kasserine) Pass, which figured significantly in the battle for Tunisia during World War II.

Geographic Regions

Tunisia can be divided into three major geographic regions, determined in part by topography and quality of the soils and in particular by the incidence of rainfall, which decreases progressively from north to south. The regions are northern, central, and southern Tunisia.

The Dorsale mountains form a rain-shadow separating northern Tunisia from the remainder of the country. Rainfall tends to be heavier north of this mountain barrier, soils to be richer, and the countryside to be more heavily populated.

Northern Tunisia is a generally mountainous region and is sometimes referred to as the Tell, a term peculiar to

PRINCIPAL COMMUNES (population at 1984 census)

Tunis (capital)	596,654	Gabès	92,258
Sfax (Safaqis)	231,911	Sousse	83,509
Ariana	98,655	Kairouan (Qairawan)	72,254
Bizerta (Bizerte)	94,509	Bardo	65,669
Djerba	92,269	La Goulette	61,609

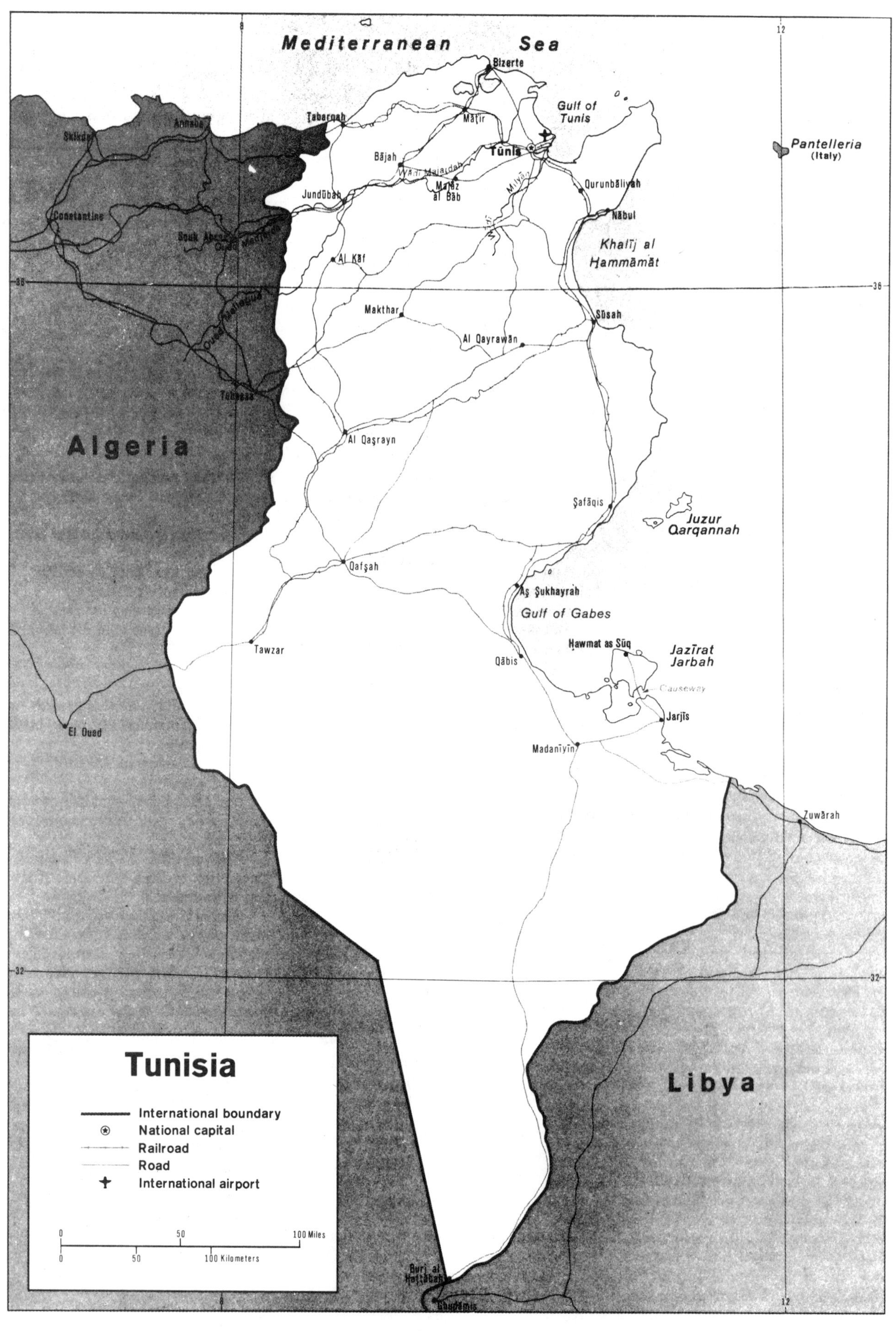
Mediterranean Sea
Bizerte
Gulf of Tunis
Pantelleria
(Italy)
Ţabarqah
Māţir
Bājah
Tūnis
Wādī Majardah
Majāz al Bāb
Qurunbālīyah
Jundūbah
Nābul
Khalīj al Ḩammāmāt
Skikda
Annaba
Constantine
Souk Ahras
Al Kāf
Makthar
Sūsah
Al Qayrawān
Tebessa
Al Qaşrayn
Algeria
Şafāqis
Juzur Qarqannah
Qafşah
Aş Şukhayrah
Gulf of Gabes
Tawzar
Ḩawmat as Sūq
Jazīrat Jarbah
Qābis
Causeway
Jarjīs
El Oued
Madanīyīn
Zuwārah
Libya
Burj al Ḩattabah
Ghadāmis
8
12
36
32
Tunisia
International boundary
National capital
Railroad
Road
International airport
0 50 100 Miles
0 50 100 Kilometers

Area and population

Governorates	Capitals	area sq mi	area sq km	population 1986 estimate
Aryānah	Aryānah	602	1,558	415,800
Bājah	Bājah	1,374	3,558	286,000
Banzart	Banzart	1,423	3,685	412,700
Bin ʿArūs	Bin ʿArūs	294	761	271,600
Jundūbah	Jundūbah	1,198	3,102	379,800
al-Kāf	al-Kāf	1,917	4,965	256,000
Madanīyīn	Madanīyīn	3,316	8,588	324,400
al-Mahdīyah	al-Mahdīyah	1,145	2,966	290,400
al-Munastīr	al-Munastīr	393	1,019	297,700
Nābul	Nābul	1,076	2,788	489,600
Qābis	Qābis	2,770	7,175	264,000
Qafṣah	Qafṣah	3,471	8,990	253,300
al-Qaṣrayn	al-Qaṣrayn	3,114	8,066	322,700
al-Qayrawān	al-Qayrawān	2,591	6,712	451,000
Qibilī	Qibilī	8,527	22,084	104,200
Ṣafāqis	Ṣafāqis	2,913	7,545	627,000
Sīdī Bū Zayd	Sīdī Bū Zayd	2,700	6,994	314,500
Silyānah	Silyānah	1,788	4,631	232,700
Sūsah	Sūsah	1,012	2,621	346,000
Taṭāwin	Taṭāwin	15,015	38,889	109,600
Tawzar	Tawzar	1,822	4,719	73,900
Tūnis	Tunis (Tūnis)	134	346	815,600
Zaghwān	Zaghwān	1,069	2,768	126,400
TOTAL		59,664	154,530	7,464,900

North Africa. It is generally defined as a heavily populated area of high ground located close to the Mediterranean Sea. The region is bisected from east to west by the Majardah River and is divided into subregions made up of the Majardah Valley and the several portions of the Tell. Rising in the mountains of eastern Algeria, the Majardah River flows eastward and discharges into the Gulf of Tunis. Its valley contains Tunisia's best farmland.

Comprising less than one-fourth of the national territory, northern Tunisia is a generally prosperous region. It was heavily settled and exploited by the French, and its physical characteristics resemble those of southern Europe more than those of North Africa.

South of the Dorsale lies central Tunisia, a region of generally poor soils and scanty rainfall. Its interior consists of a predominantly pastoral area made up of the High Steppes and the Low Steppes, the former occurring at greater elevations near the Algerian border. The term *steppes* was used by the French to define the semiarid interior highlands of North Africa, and the area has little other than its name in common with the better known steppes of Central Asia.

The steppe subregions of central Tunisia have scanty natural vegetation and are used for pasturage and the marginal cultivation of cereals. Eastward the Low Steppes give way to a littoral known as the Sahil, an Arabic term meaning coast or shore. Rainfall is not appreciably heavier there, but heavy dew has made possible the intensive cultivation of olives and other tree crops, and there is considerable cereal production. The Sahil is customarily regarded as including the Jarbah and Qarqannah islands in the Gulf of Gabes. The former has been heavily populated since antiquity.

Although natural features and climate make the Sahil part of central Tunisia, it is much more akin socially to the city of Tunis and the fertile and heavily populated coastal zone of northern Tunisia than to the barren steppes of the interior. The Sahil and portions of northern Tunisia combine to make up the economic, political, and cultural heartland of the country.

Southward from the steppes, the arid expanses that make up southern Tunisia commence with an area where elevations are lower and where the landscape is marked by numerous chotts (salt marshes, sometimes seen as *shatts*) that lie below sea level. On higher grounds around these depressions are various oasis settlements and valuable groves of date palms. Farther to the south, the land rises to form the plateaus, tablelands, and occasional eroded hills that make up the Tunisian portion of the Sahara Desert. Fringed by lagoons and salt flats, the narrow gravelly coast of this region formed the traditional access route to the Maghrib from the east. The coast is sparsely settled, and the interior is almost totally empty except for a few nomads and inhabitants of oases that occur along a line of springs at the foot of an interior escarpment.

The Majardah River is the country's only major perennial stream. Other watercourses are seasonal, and the volume of flow even of the Majardah in June and July is less than one-twelfth of that in February, thus minimizing the river's potential as a source of hydroelectric power.

In the central Tunisian steppes, occasional watercourses flow southward out of the Dorsale after heavy rains but evaporate in salt flats without reaching tidewater. The Shatt al Jarid, the largest of southern Tunisia's salt lakes, is dry during half of the year but is flooded to form a shallow salt lake during the winter months. Even intermittent streams are rare in the Sahara Desert, but rich artesian sources make possible numerous well-fed oases. In a large part of the region slightly brackish but usable water from enormous aquifers can be brought to the surface, but at prohibitive cost.

TURKEY

BASIC FACTS

Official Name: Republic of Turkey

Abbreviation: TUR

Area: 779,452 sq km (300,948 sq mi)

Area—World Rank: 36

Population: 54,167,857 (1988); 73,029,000 (2000)

Population—World Rank: 19

Capital: Ankara

Boundaries: 11,068 km (6,878 mi) USSR 610 km (379 mi); Iran 454 km (282 mi) Iraq 331 km (206 mi); Syria 877 km (545 mi); Greece 212 km (132 mi); Bulgaria 269 km (167 mi)

Coastline: 8,315 km (5,167 mi)

Longest Distances: 1,600 km (994 mi) SE-NW 650 km (404 mi) NE-SW

Highest Point: Mount Ararat 5,122 m (16,804 mi)

Lowest Point: Sea level

Land Use: 30% arable; 4% permanent crops; 12% meadows and pastures; 26% forest and woodland; 28% other

Turkey straddles the continents of Asia and Europe and consists of regions in each. About 3 percent is in Europe; this region, known as Thrace (Trakya), is separated from the Asian portion of Turkey by the Bosporus Strait (Istanbul Boğazi or Karadeniz Boğazi), the Sea of Marmara (Marmara Denizi), and the Dardanelles Strait (Çanakkale Boğazi).

The Asian part of the country is known by a variety of names—Asia Minor, Asiatic Turkey, the Anatolian Plateau, and Anatolia (Anadolu). The term *Anatolia* is more frequently used, however, in specific reference to the large, semiarid heartland plateau. The plateau is rimmed by hills and mountains that in many places form a barrier limiting access to the fertile, densely settled coastal regions. Istanbul remains the primary industrial, commercial, and intellectual center, but the Anatolian city of Ankara, which Atatürk and his associates picked as the capital of the new republic, and the whole Anatolian region continues to be viewed as quintessentially Turkish.

Topography and Drainage

Except for a relatively small segment along the Syrian border that is a continuation of the Arabian Platform, Turkey is part of the great Alpine-Himalayan mountain belt. The intensive folding and uplifting of this mountain belt during the Tertiary Period was accompanied by strong volcanic activity and intrusions of igneous rock material, followed by extensive faulting in the Quaternary Period. This faulting is still in progress, and Turkey is one of the more active earthquake regions of the world.

The quakes cause massive damage to buildings and, especially if they occur at night during the winter months, numerous deaths and injuries. For example, a violent earthquake in Erzincan the night of December 28-29, 1939, devastated most of the city and caused an estimated 160,000 deaths. Earthquakes of moderate intensity frequently continue with sporadic shocks over periods of several days or even weeks. The most earthquake-prone region centers on an arc that stretches from the general vicinity of İzmit to the area north of Lake Van on the border with the Soviet Union.

PRINCIPAL TOWNS (population at 1985 census)

Town	Population	Town	Population
Ankara (capital)*	2,235,035	Mersin (İçel)	314,350
Istanbul*	5,475,982	Diyarbakir	305,940
İzmir (Smyrna)*	1,489,772	Antalya	261,114
Adana*	777,554	Erzurum	246,053
Bursa	612,510	Malatya	243,138
Gaziantep	478,635	Samsun	240,674
Konya	439,181	İzmit (Kocaeli)	233,338
Kayseri	373,937	Kahramanmaraş	210,371
Eskişehir	366,765	Sivas	198,553

*Within municipal boundaries.

The terrain is structurally complex. A central massif composed of uplifted blocks and downfolded troughs, covered by recent deposits and giving the appearance of a plateau with rough terrain, is wedged between two folded mountain ranges that converge in the east—the Pontic Mountains along the Black Sea and the Taurus Mountains bounding the Mediterranean Sea and the Arabian Platform. True lowland is confined to the Ergene Plain in Thrace, extending along rivers that discharge into the Aegean Sea or the Sea of Marmara, and to a few narrow coastal strips along the Black Sea and Mediterranean coasts.

Nearly 85 percent of the land lies above 450 m (1,476 ft) and the median altitude of the country is 1,128 m (3,701 ft). Even more important is the fact that in Asian Turkey flat or gently sloping land is rare and largely confined to the deltas of the Kilil Irmak, the coastal plains of Antalya and Adana, and the valley floors of the Gediz and the Menderes rivers, as well as to some interior high plains in Anatolia, mainly around Tuz Gölü (Salt Lake) and the basin of Konya (Konya Ovasi). Moderately sloping land surface is limited almost entirely to Thrace and to the hill-land of the Arabian Platform along the border with Syria.

Over 80 percent of the land surface is rough, broken, and mountainous and is therefore of limited agricultural value. These features are more accentuated in the eastern part of the country where the Taurus and Pontic ranges converge into a lofty mountain region with a median altitude of over 1,500 m (4,921 ft) reaching its highest altitude along the borders with the Soviet Union and Iran. Turkey's highest mountain peak, Mount Ararat (Ağri Daği)—about 5,166 m (16,949 ft)—is situated near the tripoint where the boundaries of the three countries meet.

Natural Regions

The Aegean Coastlands

The western portion of the region consists mainly of rolling plateau country that is well suited for agriculture, receiving about fifty-two cm of rainfall annually. The region contains the cities of Istanbul and Edirne and is densely populated. The Bosporus is about twenty-five km (15 mi) long and averages 1.5 km (1 mi) in width but narrows in places to less than 500 m (1,640 ft). Both banks, Asiatic and European, rise steeply from the water

Area and population

Geographical regions	area sq mi	area sq km	population 1985 census
Akdeniz kiyisi (Mediterranean Coast)	22,933	59,395	4,653,426
Bati Anadolu (West Anatolia)	29,742	77,031	3,538,253
Doğu Anadolu (East Anatolia)	68,074	176,311	6,290,086
Güneydoğu Anadolu (Southeast Anatolia)	15,347	39,749	2,413,593
İç Anadolu (Central Anatolia)	91,254	236,347	12,193,155
Karadeniz kiyisi (Black Sea Coast)	31,388	81,295	6,652,172
Marmara ve Ege kiyilari (Marmara and Aegean coasts)	33,035	85,560	9,834,576
Trakya (Thrace)	9,175	23,764	5,089,197
TOTAL	300,948	779,452	50,664,458

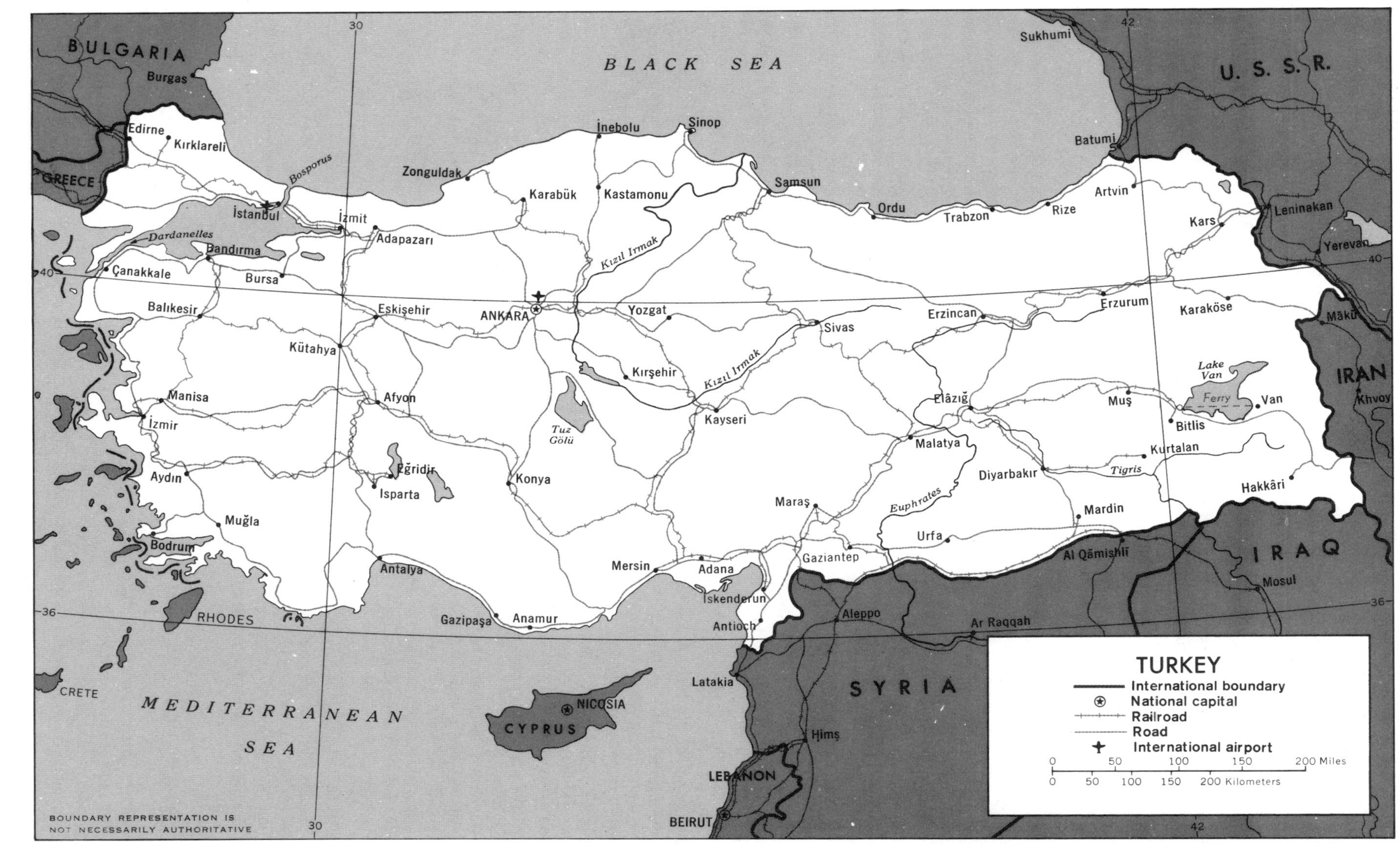
TURKEY
International boundary
National capital
Railroad
Road
International airport
0 50 100 150 200 Miles
0 50 100 150 200 Kilometers
BOUNDARY REPRESENTATION IS NOT NECESSARILY AUTHORITATIVE
BLACK SEA
MEDITERRANEAN SEA
BULGARIA
GREECE
U. S. S. R.
IRAN
IRAQ
SYRIA
LEBANON
CYPRUS
NICOSIA
RHODES
CRETE
Burgas
Edirne
Kırklareli
Bosporus
İstanbul
Dardanelles
Bandırma
Çanakkale
Bursa
İzmit
Adapazarı
Zonguldak
Karabük
İnebolu
Kastamonu
Sinop
Samsun
Ordu
Trabzon
Rize
Artvin
Batumi
Sukhumi
Leninakan
Yerevan
Kars
Kızıl Irmak
ANKARA
Eskişehir
Balıkesir
Kütahya
Manisa
İzmir
Afyon
Aydın
Muğla
Bodrum
Eğridir
Isparta
Antalya
Konya
Tuz Gölü
Kırşehir
Yozgat
Kayseri
Sivas
Erzincan
Erzurum
Karaköse
Mākū
Khvoy
Lake Van
Ferry
Van
Bitlis
Muş
Kurtalan
Tigris
Hakkâri
Elâzığ
Malatya
Diyarbakır
Euphrates
Mardin
Urfa
Maraş
Gaziantep
Adana
Mersin
İskenderun
Antioch
Gazipaşa
Anamur
Aleppo
Ar Raqqah
Al Qāmishlī
Mosul
Latakia
Hims
BEIRUT
30
36
40
42

and form a succession of cliffs, coves, and nearly landlocked bays. Most of the shores are densely wooded and are marked by numerous small towns and villages. The Dardanelles Strait is approximately forty km (25 mi) long and increases in width toward the south. Unlike along the Bosporus, there are few settlements of any kind along the shores of the Dardanelles.

The Aegean region in Asia has fertile soils and a typically Mediterranean climate with mild, wet winters and hot, dry summers. The lowlands contain about half of the country's agricultural wealth in the broad, cultivated valleys, the most important of which are the İzmit Valley, the Bursa Plains, and the Plains of Troy. The area is densely populated, particularly around Bursa and İzmir.

The Black Sea Region

The region has a steep and rocky coast, and rivers cascade through gorges of the coastal ranges. A few larger rivers that have cut back through the Pontic Mountains have tributaries that flow in broad elevated basins. Access inland from the coast is limited to a few narrow valleys, and the coast therefore has always been isolated from the interior.

The narrow coastal ribbon running between Zonguldak and Rize, widening here and there into a fertile delta, is an area of concentrated cultivation. All available areas, including mountain slopes wherever they are not too steep, are put to use. The mild, damp climate favors commercial farming, and the western part of this region is the center of Turkey's heavy industry.

The Mediterranean Coastlands

The plains of the Mediterranean coast are rich in agricultural resources. The fertile, humid soils and the warm climate make these areas ideal for growing citrus fruits and grapes, cereals and, in irrigated areas, rice and cotton. Summers are hot, and droughts are not uncommon.

The plains around Adana are largely reclaimed floodlands. In the western part of the Mediterranean coastal region, rivers have not cut valleys to the sea; movement inland therefore is restricted. The backland is mainly karst and rises sharply from the coast to elevations of 2,800 m (9,186 ft). There are few major cities along this coast, but the triangular plain of Antalya is extensive enough to support the rapidly growing city and port of the same name, which is an important trading center.

The Central Plateau

The plateaulike, arid highlands of Anatolia are considered the heartland of the country. Akin to the steppes of the Soviet Union, the region varies in altitude from 600 to 1,200 m (1,968 to 3,937 ft) west to east. Wooded areas are confined to the northwest and northeast, and cultivation is restricted to the neighboring rivers where the valleys are sufficiently wide. Irrigation is practiced wherever water is available; the deeply entrenched river courses-make it difficult to raise water to the surrounding agricultural land, however. For the most part, the region is bare and monotonous and is used for grazing.

Rainfall is limited and in Ankara amounts to less than twenty-five cm annually. Wheat and barley are the most important crops, but the yields are irregular, and crops fail entirely in years of drought. Stockraising also is important, but overgrazing has caused soil erosion in the plateau, and during the frequent dust storms of summer a fine yellow powder blows across the plains. In bad years there are severe losses of stock. Locusts occasionally ravage the eastern area in April and May. An area of extreme heat and virtually no rainfall in summer, the Central Plateau is cold in winter, with heavy, lasting snows, and villages may be isolated by severe snowstorms.

The Eastern Highlands

Eastern Turkey is rugged country with higher elevations, a more severe climate, and greater precipitation than the central plateau. In the extreme east at Kars, winter temperatures have been known to fall as low as minus 39.6°C.

From the highlands in the north, sometimes called Turkey's Siberia, to the mountains of Kurdistan (land of the Kurds) in the south that descend toward the Mesopotamian plain in Iraq, vast stretches of this eastern region consist only of wild or barren wasteland. Many of the peaks are extinct volcanoes that are 3,000 to 5,000 m (9,842 to 16,494 ft) in height. Fertile basins, such as the Mus Valley west of Lake Van and various river corridors, lie at the foot of the lofty ranges.

TURKS & CAICOS ISLANDS

BASIC FACTS

Official Name: Turks & Caicos Islands

Abbreviation: TCI

Area: 500 km (193 mi)

Area—World Rank: 174

Population: 9,295 (1988) 13,000 (2000)

Population—World Rank: 203

Capital: Grand Turk (Cockburn Town)

Land Boundaries: Nil

Coastline: about 300 km (186 mi)

Longest Distances: N/A

Highest Point: 48 m (157 ft)

Lowest Point: Sea level

Land Use: 2% arable; 98% other

Turks and Caicos Islands consist of more than 30 islands forming the southeastern end of the Bahamas chain of islands about 145 km (90 mi) north of Haiti. Eight of the islands are inhabited: Grand Turk, Salt Cay, South Caicos, Middle Caicos, North Caicos, Providenciales, Pine Cay, and Parrot Cay.

Turks and Caicos Islands

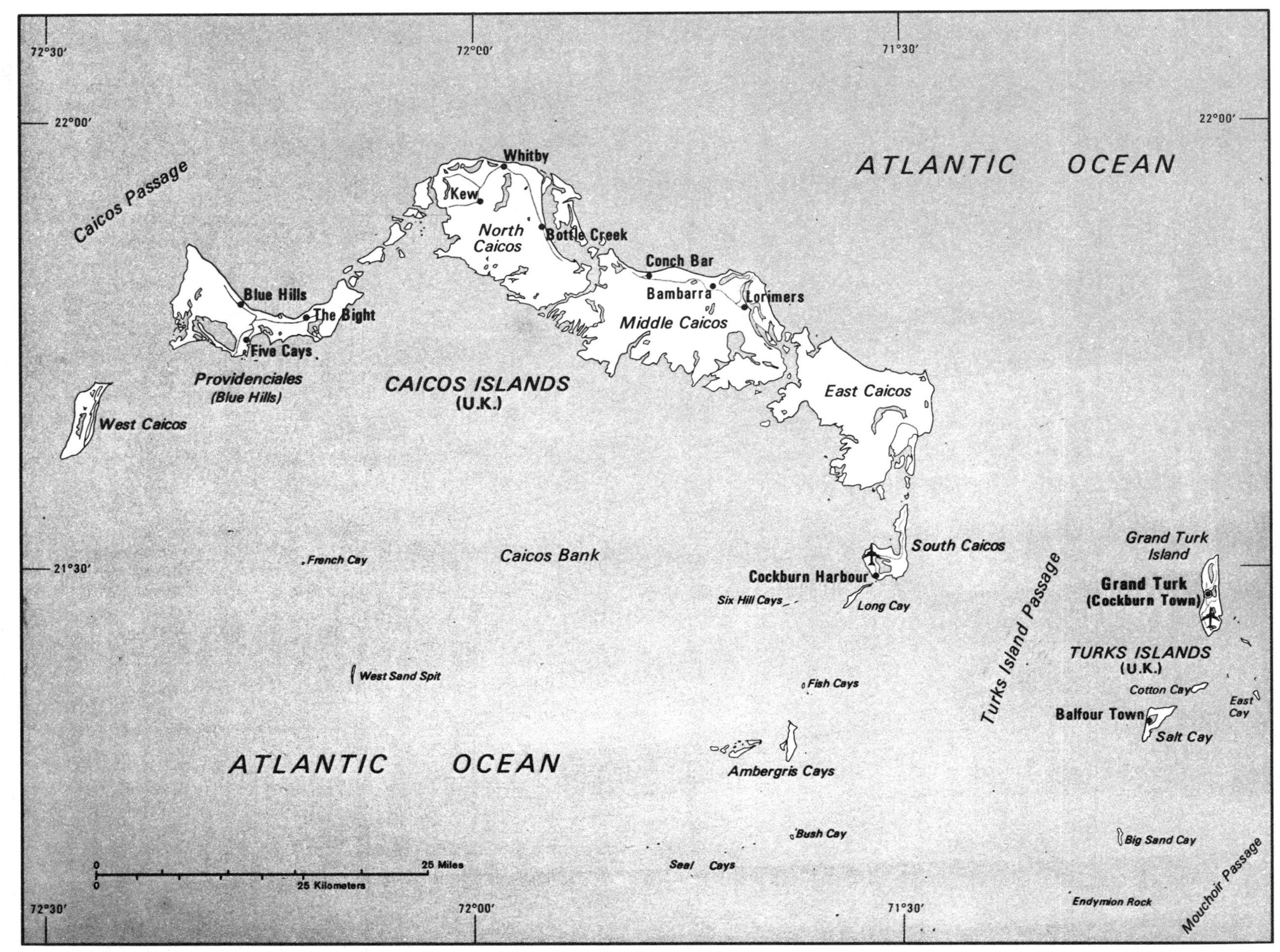

502495 1-76 (541657)
Transverse Mercator Projection
Scale 1:700,000

Road
Airport

TUVALU

BASIC FACTS

Official Name: Tuvalu

Abbreviation: TVL

Area: 24 sq km (9.3 sq mi)

Area—World Rank: 208

Population: 8,475 (1988) 8,000 (2000)

Population—World Rank: 204

Capital: Funafuti

Land Boundaries: Nil

Coastline: 24 km (15 mi)

Longest Distances: 677 km (421 mi) SSE-NNW; 146 km (91 mi) ENE-WSW

Highest Point: 5 m (15 ft)

Lowest Point: Sea level

Land Use: 100% other

Tuvalu is a widely scattered group of nine islands in the western Pacific bounded by Fiji to the south, Kiribati to the north, and Solomon Islands to the west. The islands are Funafuti, Nanomea, Nanumanga, Niulakita, Niutao, Nui, Nukufetau, Nukulailai, and Vaitupu. The islands extend 560 km (350 mi) north to south and cover an area of 26 sq km (10 sq mi).

Area and population

Islands	area sq mi	area sq km	population 1985 census
Funafuti	0.91	2.36	2,810
Nanumaga	1.00	2.59	672
Nanumea	1.38	3.57	879
Niulakita	0.16	0.41	74
Niutao	0.82	2.12	904
Nui	1.27	3.29	604
Nukufetau	1.18	3.06	694
Nukulaelae	0.64	1.66	315
Vaitupu	1.89	4.90	1,231
TOTAL	9.25	23.96	8,229

UGANDA

BASIC FACTS

Official Name: Republic of Uganda

Abbreviation: UGN

Area: 241,040 sq km (93,070 sq mi)

Area—World Rank: 77

Population: 16,446,906 (1988) 22,400,000 (2000)

Population—World Rank: 50

Capital: Kampala

Land Boundaries: 2,558 km (1,590 mi); Sudan 435 km (270 mi); Kenya 772 km (480 mi); Tanzania 418 km (260 mi); Rwanda 169 km (105 mi); Zaire 764 km (475 mi)

Coastline: Nil

Longest Distances: 787 km (489 mi) NNE-SSW; 486 km (302 mi) ESE-WNW

Highest Point: Margherita Peak 5,109 m (16,763 ft)

Lowest Point: 650 m (2,000 ft)

Land Use: 23% arable; 9% permanent crops; 25% meadows and pastures; 30% forest and woodland; 13% other

Situated astride the equator, Uganda is a landlocked country bordered by Sudan to the north, Kenya to the east, Tanzania and Rwanda to the south, and Zaire to the west. Soils and climatic conditions are excellent; about 78 percent of the land is potentially agriculturally productive, although only about 25 percent is used at any one time for cultivation and grazing.

The country is predominantly a plateau 914 to 1,524 m (3,000 to 5,000 ft) above sea level. The main mountain masses and other relief features are located on the borders. On the west are the western rift valley, the Ruwenzori Mountains, or Mountains of the Moon as they are often called, and the Mufumbiro volcanoes.

Lake Victoria includes much of the southern boundary, and the eastern boundary is mostly demarcated by highlands, which are dominated by volcanic Mount Elgon. The lowest altitude is 609 km (2,000 ft) in the valley of the Albert Nile near Nimule on the Sudan border; the highest is the 5,109 m (16,763 ft) of Margherita Peak, the summit of the Ruwenzori Range. Despite this large range of altitude, most of the country is a fairly regular plateau.

The country lies on the great plateau of east-central Africa, at no point less than 805 km (500 mi) in a direct line from the Indian Ocean. There are no natural geographic regions; the prominent relief feature is that of the plateau, dissected by numerous rivers, swamps, and lakes. Over most of the country the landscape has a sameness, but the row of volcanoes along the east, and the western rift valley system, flanked by highlands, offer sharp contrasts with the open vistas and level horizons of the central plateaus. The main mountain masses of the Ruwenzori, the Mufumbiro, and Mount Elgon are situated on the western and eastern borders.

Highlands

In the extreme southeast are the Mufumbiro volcanoes, of which only the northern slope is in Uganda. From these volcanic highlands an area of more than 1,524 m (5,000 ft) above sea level extends northeastward through Kigezi District into western Ankole District. The Mufum-

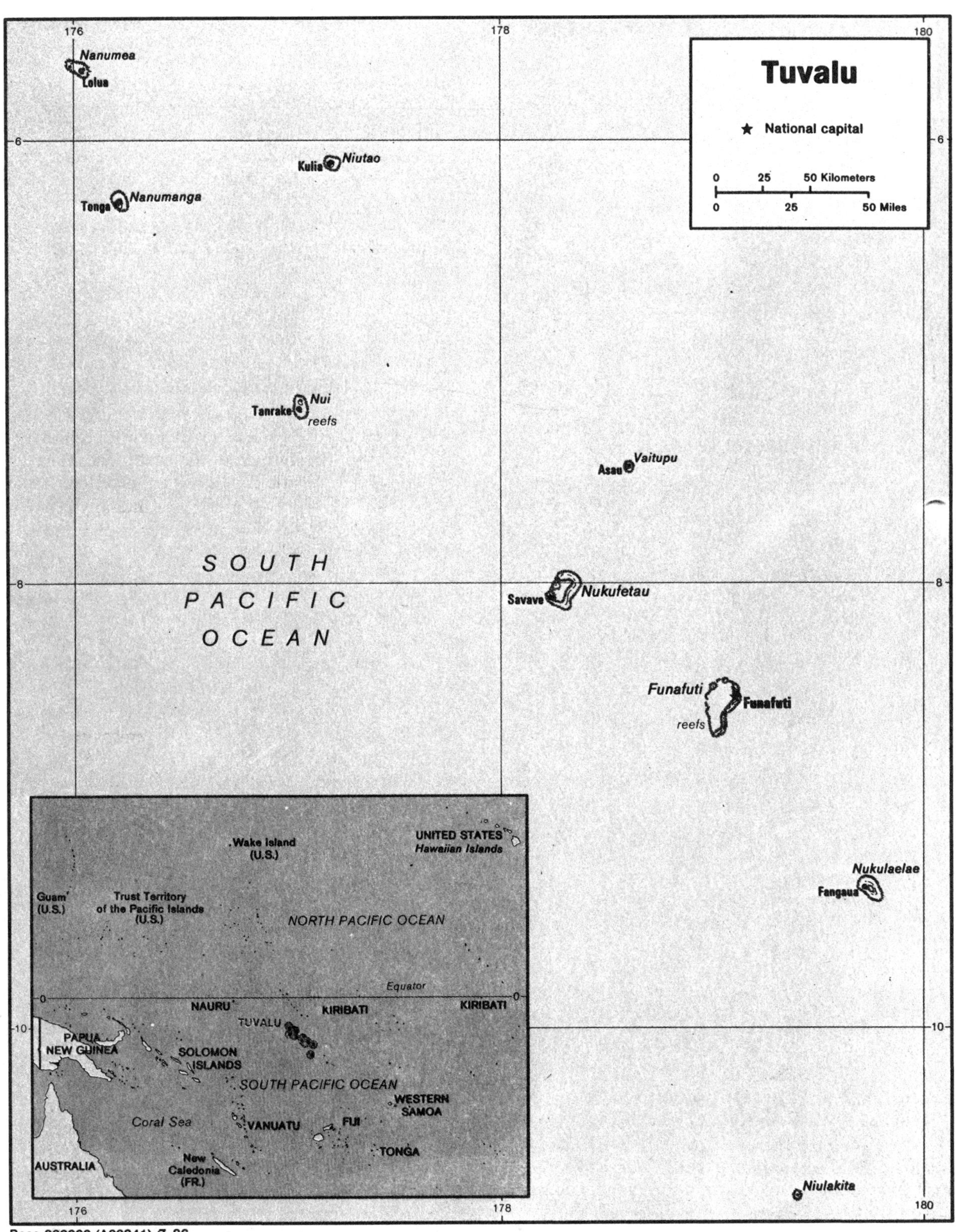

Base 800202 (A00341) 7-86

PRINCIPAL TOWNS (population at 1969 census)

Kampala (capital)	330,700	Mbale	23,544
Jinja and Njeru	52,509	Entebbe	21,096
Bugembe planning area	46,884	Gulu	18,170

1980 (preliminary census results): Kampala 458,423; Jinja 45,060; Masaka 29,123; Mbale 28,039; Mbarara 23,155; Gulu 14,958.

Area and population

Provinces / Districts	Capitals	area sq mi	area sq km	population 1985 estimate
Busoga	Jinja	7,030	18,200	1,408,600
Iganga	Bulamogi	5,060	13,110	755,100
Jinja	Jinja	280	730	253,400
Kamuli	Namwendwa	1,680	4,350	400,100
Central	Kampala	2,470	6,400	1,298,800
Kampala	Kampala	70	180	560,800
Mpigi	Mpigi	2,400	6,220	738,000
Eastern	Mbale	8,400	21,760	2,316,000
Kapchorwa	Kaptanya	670	1,740	83,100
Kumi	Kumi	1,100	2,860	273,100
Mbale	Bunkoko	980	2,550	647,400
Soroti	Soroti	3,880	10,060	545,300
Tororo	Sukulu	1,760	4,550	767,100
Karamoja	Moroto	10,550	27,320	405,600
Kotido	Kotido	5,100	13,210	194,700
Moroto	Katikekile	5,450	14,110	210,900
Nile	Arua	6,070	15,730	921,500
Arua	Olaki	3,020	7,830	543,300
Moyo	Moyo	1,930	5,010	119,600
Nebbi	Nebbi	1,120	2,890	258,600
North Buganda	Bombo	13,030	33,750	1,802,800
Luwero	Luwero	3,550	9,200	477,800
Mubende	Bageza	3,980	10,310	616,500
Mukono	Kawuga Mukono	5,500	14,240	708,500
Northern	Gulu	16,070	41,610	1,459,200
Apac	Apac	2,510	6,490	369,000
Gulu	Bungatira	4,530	11,740	305,500
Kitgum	Labongo	6,230	16,140	354,100
Lira	Lira	2,810	7,250	430,600
South Buganda	Masaka	8,220	21,300	1,071,200
Masaka	Kaswa Bukoto	6,300	16,330	741,600
Rakai	Byakabanda	1,920	4,970	329,600
Southern	Mbarara	8,290	21,480	2,270,500
Bushenyi	Bumbaire	2,080	5,400	600,300
Kabale	Rubale	960	2,490	503,700
Mbarara	Kakika	4,180	10,840	829,100
Rukungiri	Kagunga	1,060	2,750	337,400
Western	Butebe	12,910	33,440	1,725,600
Bundibugyo	Busaru	900	2,340	134,500
Hoima	Hoima	3,820	9,900	358,400
Kabarole	Karambe	3,230	8,360	630,500
Kasese	Rukoki	1,240	3,200	342,400
Masindi	Nyangeya	3,720	9,640	259,800
TOTAL LAND AREA		76,080	197,040	14,679,800
INLAND WATER		16,990	44,000	
TOTAL		93,070	241,040	

biro Range includes 3,645 m (11,960-ft) Mount Sabinio, the meeting point of Uganda, Rwanda, and Zaire. Its highest mountain is Mahavura at 4,127 m (13,540 ft).

These highlands are separated from the Ruwenzori Mountains, also on the western border, by a low valley containing Lake George and its outlet into Lake Edward, the Kazinga Channel. The Ruwenzori Range is about 80 km (50 mi) long, and rises into a number of peaks of more than 4,267 m (14,000 ft) of which the highest is the Margherita, 5,109 m (16,763 ft). Above the 4,267 m (14,000-ft) level the mountains are capped with snow and large glaciers.

To the east, the approach to the Kenya borderlands is marked by volcanic centers and hills. Mount Elgon, between Sebei District and Kenya, is 4,321 m (14,178 ft) at its highest point. Mount Debasien, in Karamoja District, is 3,068 m (10,067 ft) while Mount Moroto, still further north, is 3,083 m (10,116 ft). Mount Morungole near the northeast border is 2,750 m (9,022 ft) and Mount Zulia in the extreme northeast is 2,148 m (7,048 ft) high. West of the border mountains are a number of other smaller mountain masses, including the Labwor Hills, ranging from 1,798 to 2,530 m (5,900 to 8,300 ft), more or less isolated from each other and rising abruptly out of the plains. On the northern border are the southern outlines of the Imatong Mountains of the Sudan, all about 1,828 m (6,000 ft) high.

Lakes

Lakes Albert, Edward, and George are troughs in the western rift valley system, while Lakes Victoria and Kyoga are shallow basins on the plateau. Lake Albert and Lake Edward are shared by Uganda and Zaire, but Lake George, connected with Lake Edward by the Kazinga Channel, is wholly within Uganda. Of Lake Victoria's 69,484 sq km (26,828 sq mi), 20,430 (11,749) are in Uganda, the remainder being divided between Kenya and Tanzania.

Lakes Kyoga and Kwania, in the center of the country, with an area of 2,046 sq km (790 sq mi), are surrounded by an even larger area of swamp. Lake Salisbury, to the northeast of Lake Kyoga, provides an outlet for the waters north of Mount Elgon to the Nile system. West of Lake Victoria, in the south, is a group of some six lakes connected by swamp.

All of the Lakes are relatively shallow, the maximum depth recorded in Lake Victoria is 82 m (270 ft) that in Lake Albert 51 m (168 ft), in Lake Edward 117 m (384 ft) in Lakes Kyoga and Kwania 7.3 m (24 ft), and in Lake George 3 m (10 ft). Though the lakes are less productive than the land, they do support a flourishing fishing industry. The fish are not thoroughly exploited because inshore fishing is the general practice. Lake Victoria contributes the greatest share of the catch and has many species of economic value. Yet it ranks low among the lakes in its yield per sq mi because much of it lies more than a day's trip away from the land. Fishing takes place at about 200 landings, the largest of which is at Jinja, but there are others of importance concentrated around the Entebbe peninsula, at Bukakata on the western shore, and at Mjanji near the Kenya border.

Lake Kyoga was the last of the major lakes to be exploited for fish before 1950. Its irregular shoreline now encourages extensive fishing, and all parts of the lake lie within easy reach of the shore. Fishing activity is limited mainly by inadequate roads and the inaccessibility of the large market towns.

The waters of Lake Albert are more extensively fished than those of Lake Victoria but less so than those of Lake Kyoga. Most fishing is done from the northern shore, and the middle of the lake is barely touched. There are many fish to be found in Lake George, and commercial fishing under

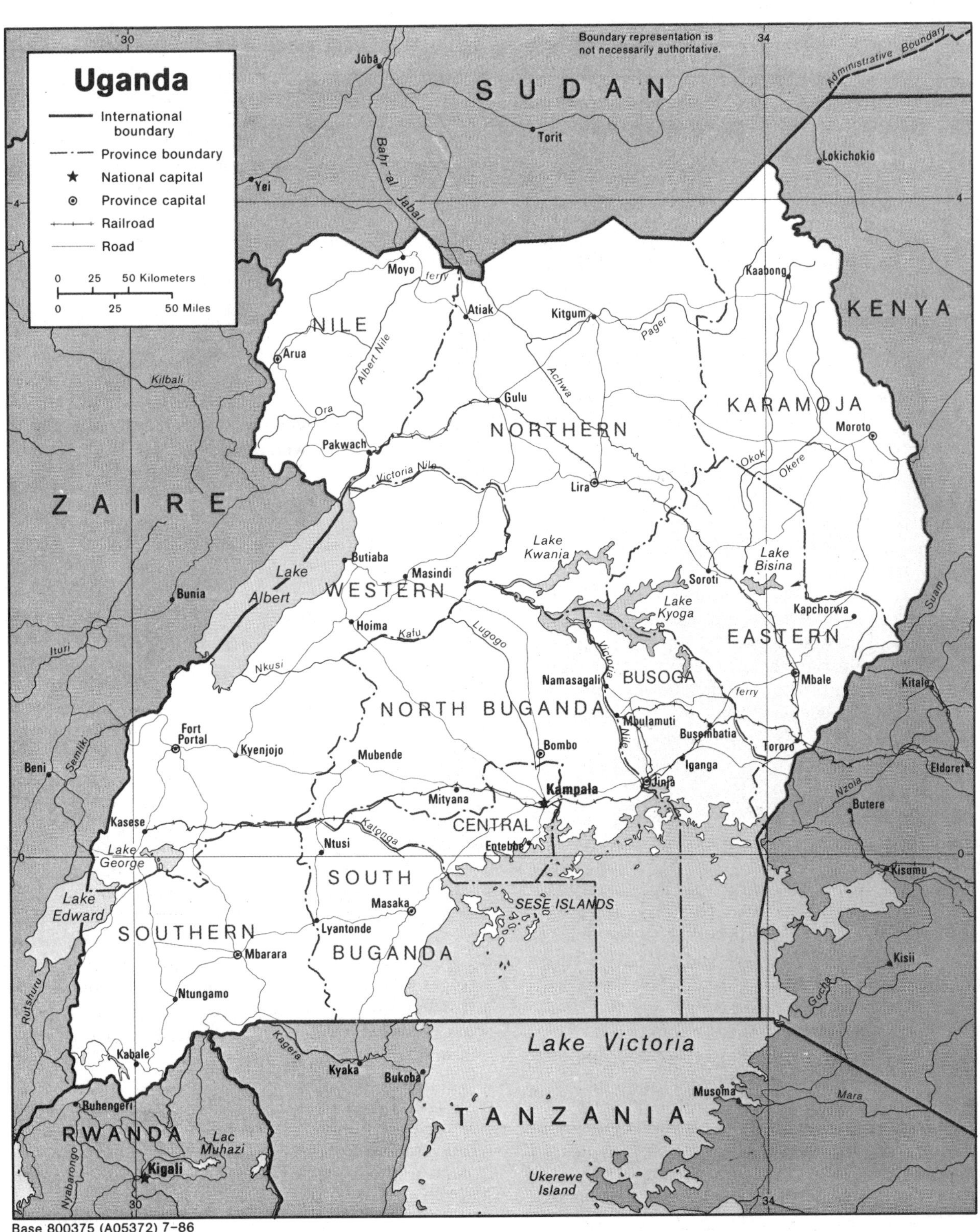

Base 800375 (A05372) 7-86

close government control began there in 1950. The catch from the more extensive waters of Lake Edward is similar in size to that from Lake George. There are still areas of these lakes which are untouched, however, because of lack of transport facilities and accessible markets.

Rivers

The country is within the upper basin of the White Nile. From the Owen Falls, at Jinja, to the point at which the Albert Nile crosses the northern border with the Sudan, there is a descent of over 518 m (1,700 ft), accomplished for the most part by a series of falls and rapids.

The 459 km (285 mi) Victoria Nile runs from Lake Victoria at Jinja over Owen Falls north into Lake Kyoga. It leaves Kyoga in a westward direction, and changes to a northerly course at its junction with the Kafu River, which runs into it from the west. After Karuma Falls the river follows a westward course over Murchison Falls (a drop of 40 m (130 ft) through a narrow cleft) into Lake Albert. From Lake Albert it flows, as the Albert Nile, in a northerly direction, leaving Uganda at Nimule on the Sudan border.

The innumerable other rivers are, for the most part, sluggish. Some are not much more than vegetation-covered swamps. The Katonga and Kafu Rivers of the west are swampy but have a greater flow than many. The Katonga runs into a swamp at the northeast corner of Lake Victoria. The Kafu flows into the western end of Lake Kwania, but its headwaters connect with those of the Nkusi, flowing westward into the southern end of Lake Albert. Other major rivers are the Aswa, Pager, and Dopeth-Okok of the northeast and the Mpongo, a tributary of the Kafu. Only in the hill regions and on the slopes of the western rift valley are clear, running streams commonly found. Permanent streams and rivers usually carry water only in the wet seasons, creating a need for catchment projects.

Drainage

In recent geologic times the unwarping and faulting that created the western rift valley led to a piling up of waters in the relatively downwarped zone to the east that now forms the basin of Lake Victoria. The Lake overflows at a low point near Jinja to form the Victoria Nile, flowing into Lake Kyoga. The waters of the Lake Kyoga basin are carried by the Victoria Nile to Lake Albert and then, by the Albert Nile, to the north.

From a watershed about 16 km (10 mi) from the north shore of Lake Victoria, the rivers empty into Lake Kyoga; south of it they drain into Lake Victoria.

Somewhere in the swamps between the western boundary and the Victoria Nile is a divide, to the east of which the waters flow through Lakes Victoria and Kyoga, and to the west of which steeply graded streams drain into rivers that flow toward the western rift lakes. With its apparent wealth of waterways and lakes, and a good rainfall pattern over most of the land, Uganda's water resources fall short of requirements and water storage systems are needed in some areas. Most of the 5,180 sq km (2,000 sq mi) of swamp lie in the lowland area bordering the Nile; reclamation remains to be carried out in this area, but in the upland areas of the southwest a number of swampy areas have been reclaimed.

UNITED ARAB EMIRATES

BASIC FACTS

Official Name: United Arab Emirates

Abbreviation: UAE

Area: 77,700 sq km (30,000 sq mi)

Area—World Rank: 110

Population: 1,980,354 (1988) 5,849,000 (2000)

Population—World Rank: 123

Capital: Abu Dhabi

Boundaries: 1,940 km (1,206 mi) Oman 513 km (319 mi); Saudi Arabia 586 km (364 mi) Qatar 64 km (40 mi)

Coastline: 777 km (483 mi)

Longest Distances: 544 km (338 mi) NE-SW 361 km (224 mi) SE-NW

Highest Point: N/A

Lowest Point: Sea level

Land Use: 2% meadows and pastures; 98% other

United Arab Emirates consist of seven Persian Gulf emirates—Abu Dhabi, Ajman, Dubai, Al Fujayrah, Ra's al Khaymah, Ash Shariqah, and Umm al Qaywayn as well as numerous islands including three occupied by Iran: Jazireh-ye Abu Musa, Jazireh-ye Tonb-e Bozorg (Greater Tonb) and Jazireh-ye Tonb-e Kuchek (Lesser Tonb).

Because of disputed claims to some of the islands, the lack of precise information as to the the size of many of the islands, and many undefined land boundaries, the exact size of the country is unknown. Although by 1976 all territorial disputes between and among the seven emirates had theoretically been resolved, the several enclaves, some of which were under so-called shared or joint administration, retained considerable potential for interemirate disputes.

Despite the nearly 644 km (400 mi) of coastline, the country had only one large natural harbor, Dubai. There are numerous small islands, reefs, and shifting sandbars that menace navigation, and strong tides and occasional windstorms further complicate ship movements near the shore.

South and west of Abu Dhabi the vast rolling sand dunes merge into the Rub al Khali (Empty Quarter) of Saudi Arabia. Large sections of the Abu Dhabi coast are salt marshes *(sabkhat)* that extend for mi inland. Inland

United Arab Emirates

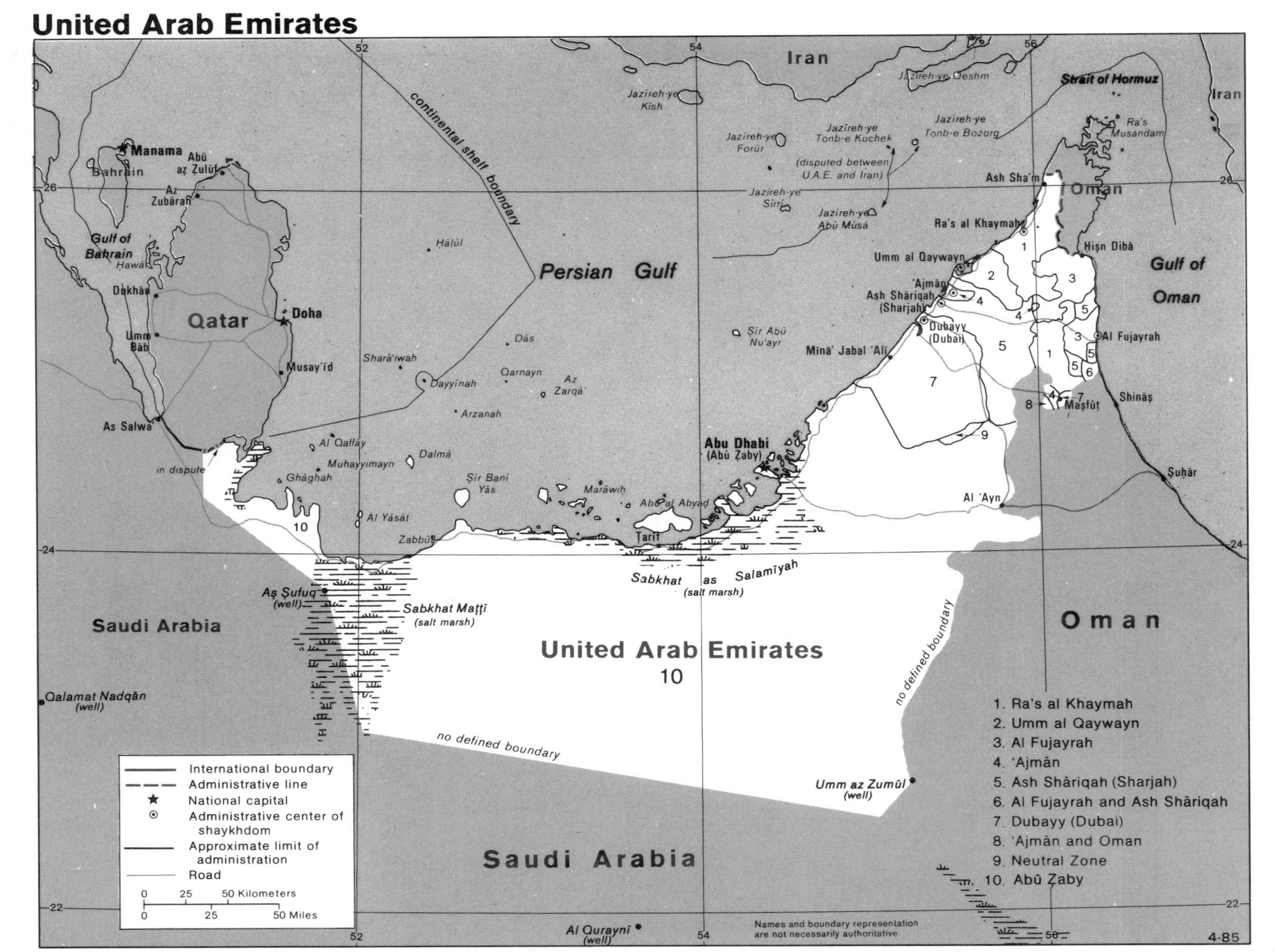

from the Gulf of Oman coast the terrain is sharply different. The western Hajar Mountains—rising in places to 2,133 or 2,438 m (7,000 or 8,000 ft)—run down close to the shore in many places. Ras al Khaymah, Fujayrah, and the eastern part of the Sharjah are hilly or mountainous regions topographically distinct from Abu Dhabi and Dubai, which together account for more than 87 percent of the territory. Only in a few scattered areas is agriculture possible.

Area and population

		area		population
Emirates	**Capitals**	sq mi	sq km	1985 census
Abu Dhabi (Abū Ẓaby)	Abu Dhabi	26,000	67,350	670,125
Ajman (ʿAjmān)	Ajman	100	250	64,318
Dubai (Dubayy)	Dubai	1,510	3,900	419,104
Fujairah (Al-Fujayrah)	Fujairah	440	1,150	54,425
Ras al-Khaimah (Raʾs al-Khaymah)	Ras al-Khaimah	660	1,700	116,470
Sharjah (Ash-Shāriqah)	Sharjah	1,000	2,600	268,722
Umm al-Qaiwain (Umm al-Qaywayn)	Umm al-Qaiwain	290	750	29,229
TOTAL		30,000	77,700	1,622,393

UNITED KINGDOM

BASIC FACTS

Official Name: United Kingdom of Great Britain & Northern Ireland

Abbreviation: UK

Area: 244,110 sq km (94,251 sq mi)

Area—World Rank: 76

Population: 56,935,845 (1988); 58,392,000 (2000)

Population—World Rank: 16

Capital: London

Boundaries: 8,352 km (5,190 mi) Ireland: 434 km (270 mi)

Coastline: 7,918 km (4,920 mi)

Longest Distances: 965 km (600 mi) N-S; 485 km (300 mi) E-W

Highest Point: Ben Nevis 1,343 m (4,406 ft)

Lowest Point: Holme Fen 3 m (9 ft) below sea level

Land Use: 29% arable; 48% meadows and pastures; 9% forest and woodland; 14% other

GEOGRAPHICAL FEATURES

The United Kingdom is located off the northwestern coast of Europe between the Atlantic Ocean on the north and northwest and the North Sea on the east. It is separated from the continent by the Strait of Dover and the English Channel, 34 km (21 mi) wide at its narrowest point, and from the Irish Republic by the Irish Sea and St. George's Channel. Its total area—the same as that of West Germany, New Zealand or Uganda, and half that of France—is shared by four constituent units: England (130,373 sq km; 50,337 sq mi); Wales (20,767 sq km; 8,018 sq mi); and Scotland (78,775 sq km; 30,415 sq mi), forming Great Britain, and Northern Ireland (14,120 sq km; 5,452 sq mi), on the island of Ireland separated from Great Britain by the North Channel. There are also several island groups and hundreds of small single islands, of which the best known are the Orkney Islands, the Shetland Islands, the Outer Hebrides, Skye, Mull, Islay, Arran, the Isle of Man, the Isles of Scilly and the Channel Islands. The distance from the southern coast of the extreme North of Scotland is just under 1,000 km (622 mi), and the widest part of Great Britain is under 500 km (311 mi). No place in Britain is more than 120 km (75 mi) from tidal water. The prime meridian of 0° passes through the Old Observatory at Greenwich, near London.

The seas surrounding the British Isles are shallow, usually less than 90 m (295 ft or 49 fath.) because the islands lie on the continental shelf. However, to the northwest, along the edge of the shelf, the sea floor plunges abruptly 180 m (591 ft) to 900 m (2,953 ft).

Most of England consists of low plains and rolling downs, particularly in the South and the Southeast, where the land does not rise higher than 305 m (1,000 ft) at any point. The major hill regions are in the North, the West and the Southwest. Running from east to west on the Scottish border are the Cheviot Hills, and from north to south from the Scottish border to Derbyshire in central England is the Pennine Range. South the Pennines lie the Central Midlands, a plains region with low, rolling hills and fertile valleys. The eastern coast is low-lying, much of it less than 5 m (15 ft) above sea level and protected by embankments against inundations from gales and unusually high tides. The highest point in England is Scafell Pike 978 m (3,210 ft) in the celebrated Lake District in the Northwest. The longest of the rivers, flowing from the Central Highlands to the sea, are the Severn 338 km (210 mi) in the West and the Thames 322 km (200 mi) in the Southeast. Other rivers include the Humber, the Tees, the Tyne and the Tweed in the East, the Avon and the Exe in the South and the Mersey in the West.

Scotland has three distinct topographical regions: the Northern Highlands, occupying almost the entire northern half of the country and containing the highest point in the British Isles, Ben Nevis 1,343 m (4,406 ft); the Central Lowlands, with an average elevation of 152 m (500 ft) and containing the valleys of the Tay, Forth and Clyde rivers; and the Southern Uplands, rising to their peak at Merrick 842 m (2,764 ft) with moorland cut by many valleys and rivers.

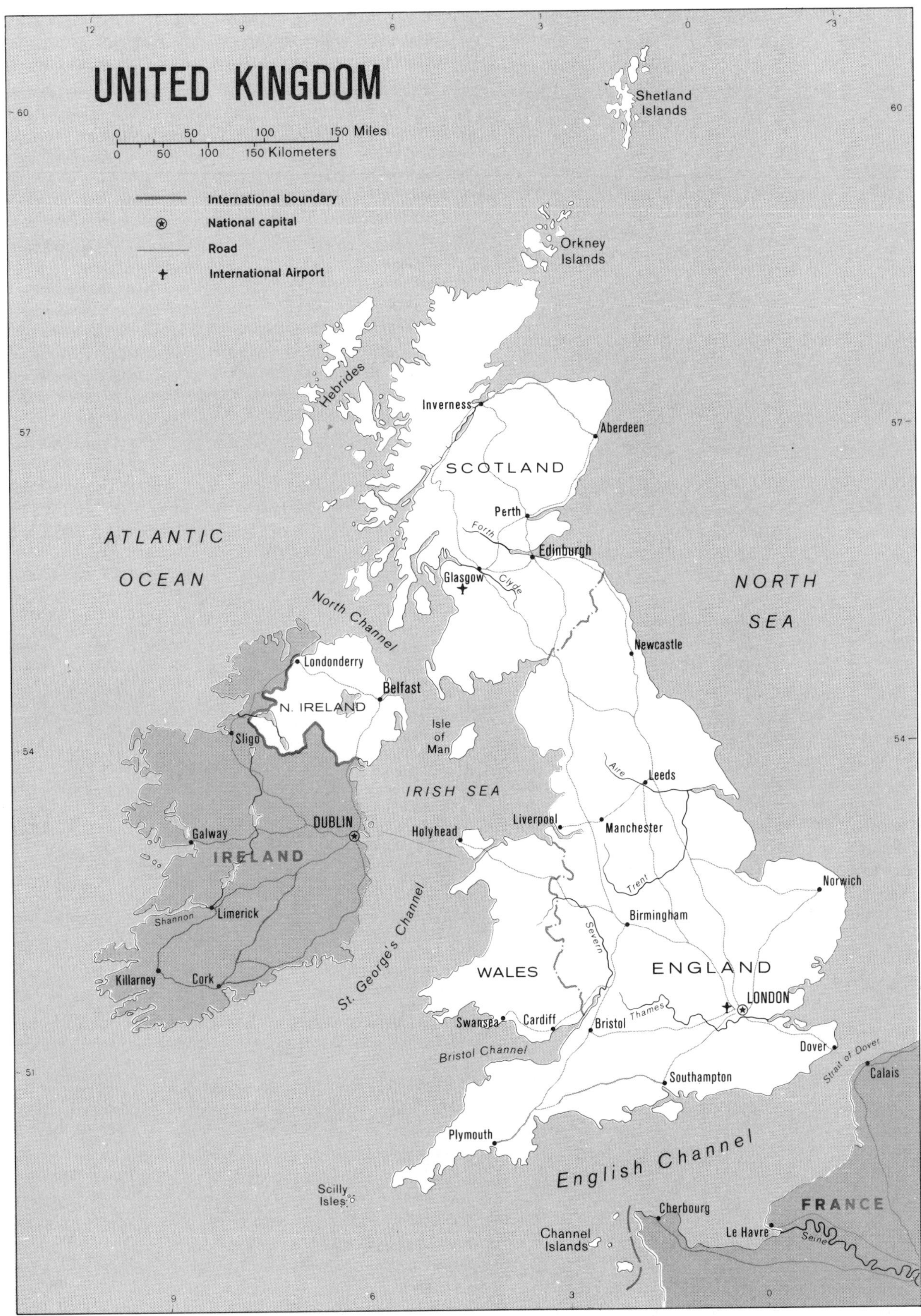
UNITED KINGDOM
0 50 100 150 Miles
0 50 100 150 Kilometers
International boundary
National capital
Road
International Airport
Shetland Islands
Orkney Islands
Hebrides
Inverness
Aberdeen
SCOTLAND
Perth
Forth
Edinburgh
Glasgow
Clyde
ATLANTIC
OCEAN
NORTH
SEA
North Channel
Newcastle
Londonderry
Belfast
N. IRELAND
Sligo
Isle of Man
IRISH SEA
Leeds
Aire
DUBLIN
Holyhead
Liverpool
Manchester
Galway
IRELAND
Trent
Norwich
Limerick
Shannon
St. George's Channel
Birmingham
Severn
Killarney
Cork
WALES
ENGLAND
LONDON
Swansea
Cardiff
Bristol
Thames
Bristol Channel
Dover
Strait of Dover
Calais
Southampton
Plymouth
English Channel
Scilly Isles
Cherbourg
FRANCE
Le Havre
Seine
Channel Islands

518634 5-77

PRINCIPAL CITIES* (population estimates at mid-1986, '000)

City	Population	City	Population
Greater London (capital)	6,775.2	Leicester	281.1
Birmingham	1,004.1	Cardiff	279.5
Glasgow	725.1	Nottingham	277.8
Leeds	710.9	Walsall	261.8
Sheffield	534.3	Bolton	261.6
Liverpool	483.0	Kingston upon Hull	258.0
Bradford	463.1	Plymouth	256.0
Manchester	451.4	Rotherham	252.1
Edinburgh	438.2	Wolverhampton	251.9
Bristol	391.5	Stoke-on-Trent	246.9
Kirklees	376.6	Salford	239.3
Wirral	334.8	Barnsley	222.2
Coventry	310.4	Oldham	220.0
Wakefield	309.3	Aberdeen	216.3
Wigan	306.6	Trafford	216.3
Belfast	303.6	Derby	216.2
Sandwell	301.1	Tameside	215.2
Dudley	300.9	Gateshead	207.3
Sefton	298.0	Rochdale	206.6
Sunderland	297.7	Renfrew	203.2
Stockport	289.9	Solihull	202.2
Doncaster	289.3	Southampton	200.5
Newcastle upon Tyne	281.4		

*Local authority areas after 1974 local government reorganization with populations greater than 200,000.

Source: Office of Population Censuses and Surveys; General Register Offices for Scotland and Northern Ireland.

Wales is a country of hills and mountains, with extensive tracts of high plateau and shorter stretches of mountain ranges deeply dissected by river valleys. The Cambrian Mountains occupy almost the entire area and include Wales' highest point, Mount Snowdon 1,085 m (3,560 ft). There are narrow coastal plains in the South and West and small lowland areas in the North, including the valley of the Dee.

Northern Ireland consists mostly of low-lying plateaus and hills, generally from 152 to 183 m (500 to 600 ft) high. The Mourne Mountains in the Southeast include the highest point in Northern Ireland, Slieve Donard 852 m (2,796 ft). In a central depression lies Lough Neagh, the largest lake in the kingdom.

The coastline is long and rugged and heavily indented with towering cliffs and headlands and numerous bays and inlets, among them the deep and narrow lochs and wide firths of Scotland. Many river estuaries serve as fine harbors. There is a marked movement of tides.

ISLE OF MAN

The Isle of Man lies in the Irish Sea between the Cumbrian coast of England and Northern Ireland. The Capital is Douglas. South Man is bisected by a central valley. There are hills in the north.

THE CHANNEL ISLANDS

The Channel Islands consist of Jersey, Guernsey, Alderney, Sark, Herm, Jethou, Brechou and Lihou. Jersey is the largest of the Channel Islands and it lies to the southeast of Guernsey, the second largest. Jersey is a gently rolling plain with low rugged hills along the north coast. Guernsey is mostly level with low hills in southwest.

Area and population

Countries	Capitals	area sq mi	area sq km	population 1986 estimate
England	London	50,363	130,439	47,254,500
Counties				
Avon		520	1,346	946,600
Bedfordshire		477	1,235	521,000
Berkshire		486	1,259	734,100
Buckinghamshire		727	1,883	612,900
Cambridgeshire		1,316	3,409	635,200
Cheshire		899	2,329	946,500
Cleveland		225	583	557,600
Cornwall		1,376	3,564	448,200
Cumbria		2,629	6,810	486,600
Derbyshire		1,016	2,631	916,800
Devon		2,591	6,711	999,000
Dorset		1,025	2,654	638,200
Durham		941	2,436	599,600
East Sussex		693	1,795	689,700
Essex		1,418	3,672	1,512,100
Gloucestershire		1,020	2,643	517,100
Greater London		610	1,579	6,775,200
Greater Manchester		497	1,287	2,579,500
Hampshire		1,458	3,777	1,527,700
Hereford & Worcester		1,516	3,927	654,500
Hertfordshire		631	1,634	985,700
Humberside		1,356	3,512	848,500
Isle of Wight		147	381	124,600
Kent		1,441	3,731	1,500,900
Lancashire		1,183	3,064	1,380,700
Leicestershire		986	2,553	875,000
Lincolnshire		2,284	5,915	567,300
Merseyside		252	652	1,467,600
Norfolk		2,073	5,368	727,800
Northamptonshire		914	2,367	554,400
Northumberland		1,943	5,032	301,000
North Yorkshire		3,208	8,309	699,800
Nottinghamshire		836	2,164	1,006,400
Oxfordshire		1,007	2,608	574,700
Shropshire		1,347	3,490	392,700
Somerset		1,332	3,451	448,900
South Yorkshire		602	1,560	1,297,900
Staffordshire		1,049	2,716	1,021,000
Suffolk		1,466	3,797	628,600
Surrey		648	1,679	1,011,400
Tyne and Wear		208	540	1,135,500
Warwickshire		765	1,981	480,700
West Midlands		347	899	2,632,300
West Sussex		768	1,989	694,700
West Yorkshire		787	2,039	2,053,100
Wiltshire		1,344	3,480	545,200
Northern Ireland	Belfast	5,452	14,120	1,567,000
Scotland	Edinburgh	30,418	78,783	5,121,000
Regions				
Borders		1,814	4,698	101,800
Central		1,042	2,700	271,800
Dumfries and Galloway		2,481	6,425	146,800
Fife		509	1,319	343,800
Grampian		3,379	8,752	502,800
Highland		10,092	26,137	200,800
Lothian		683	1,770	741,900
Strathclyde		5,318	13,773	2,344,600
Tayside		2,951	7,643	392,400
Island areas (TOTAL)		2,149	5,566	74,300
Wales	Cardiff	8,019	20,768	2,821,000
Counties				
Clwyd		937	2,427	399,600
Dyfed		2,227	5,768	339,000
Gwent		531	1,376	441,800

Area and population

Countries	Capitals	area sq mi	area sq km	population 1986 estimate
Gwynedd		1,494	3,869	234,600
Mid Glamorgan		393	1,018	534,500
Powys		1,960	5,077	112,400
South Glamorgan		161	416	395,700
West Glamorgan		316	817	363,400
TOTAL		94,251	244,110	56,763,500

UNITED STATES

BASIC FACTS

Official Name: United States of America

Abbreviation: USA

Area: 9,529,063 sq km (3,679,192 sq mi)

Area—World Rank: 4

Population: 246,042,565 (1988) 267,995,000 (2000)

Population—World Rank: 4

Capital: Washington D.C.

Boundaries: 29,695 km (18,441 mi); Canada 6,416 km (3,984 mi); Mexico 3,326 km (2,065 mi)

Coastline: 19,924 km (12,373 mi)

Longest Distances: Conterminous US: 4,662 km (2,897 mi) ENE-WSW; 4,583 km (2,848 mi) SSE-NNW; Alaska 3,639 km (2,261 mi) E-W; 2,185 km (1,358 mi) N-S; Hawaii 2,536 km (1,576 mi) N-S; 2,293 km (1,425 mi) E-W

Highest Point: Mount McKinley 6,194 m (20,320 ft)

Lowest Point: Death Valley 86 m (282 ft) below sea level

Land Use: 20% arable; 26% meadows and pastures; 29% forest and woodland; 25% other

GEOGRAPHICAL FEATURES

Located on the continent of North America, the United States is the fourth-largest country in the world. The United States comprises three physically separated territories: the conterminous United States, comprising 48 states; Alaska; and Hawaii. The conterminous United States extends 4,662 km (2,897 mi) east-northeast to west-southwest and 4,582 km (2,848 mi) south-southeast to north-northwest. The geographical center of the conterminous states is in Smith County, Kansas. Alaska extends 3,639 km (2,261 mi) east–west and 2,285 km (1,420 mi) north–south. Hawaii extends 2,356 km (1,576 mi) north–south and 2,293 km (1,425 mi) east–west. The geographic center of the United States including Alaska and Hawaii is in Butte County, South Dakota.

The various topographic divisions of the United States are determined by the mountain ranges that traverse the country from north to south. On the basis of this structure, the country is divided into 34 natural regions or physiographic provinces grouped into 11 major divisions. Almost half the country consists of plains. The vast central plain extends northward across Canada to the Arctic Ocean and southward to the coastal plain that borders the Gulf of Mexico. Along the northern part of the Atlantic coast the coastal plain narrows and becomes fragmented where it slopes northeastward under the Atlantic Ocean. Other extensive plains occur in the structural basins of the western mountains, such as the broad valley of the Sacramento and San Joaquin rivers in California and the Wyoming Basin.

About a quarter of the country is made up of plateaus, of which the six major ones are the Piedmont Province; the interior low plateaus; and the Ozark, Appalachian, Columbia and Colorado plateaus. The four eastern ones are older than those in the West, which are semiarid and therefore have been subject to greater erosional and weathering processes.

Mountains make another quarter of the country. The central plains are flanked on the east by the Appalachian Highlands and on the west by the Rocky Mountain system The shorelines in the West and in the East also reveal structural differences. The Atlantic coast south of New England is a plain with extensive sandy beaches, while the Pacific coast is mountainous with relatively fewer beaches. The coast of southeastern Alaska is mountainous, with glaciers and fjords. There are similar differences

Major Divisions and Physiographic Provinces of the United States.

Major Divisions	Provinces	Characteristics
Laurentian Upland	Superior Upland	An upland with altitudes up to 2,000 ft, but without much local relief; drainage irregular; many lakes; extends far north in Canada around both sides of Hudson Bay.
	Continental Shelf	Submarine plain sloping seaward to depth of about 600 ft; the submerged part of the Coastal Plain.
Atlantic Plain	Coastal Plain	Broad plain rising inland; shores mostly sandy beaches backed by estuaries and marshes; mud flats at mouth of Mississippi River; some limestone bluffs on west coast of Florida; inland ridges parallel the coast; altitudes less than 500 ft; along Atlantic coast the surface slopes northeast and northern valleys are tidal inlets.

Major Divisions and Physiographic Provinces of the United States.

Major Divisions	Provinces	Characteristics
	Piedmont Province	Rolling upland; altitudes 500 to 2,000 ft in the south; surface slopes northeast (like the Coastal Plain) and altitudes at the north are below 500 ft
	Blue Ridge Province	Easternmost ridge of the Appalachian Highlands; altitudes above 5,000 ft
	Valley and Ridge Province	Parallel valleys and mountainous ridges; altitudes mostly between 1,000 and 3,000 ft; lower to the north like the Coastal Plain and Piedmont provinces.
Appalachian Highlands	Appalachian Plateaus	Plateau; surface mostly 2,000 to 3,000 ft; slopes west; deeply incised by winding stream valleys; considerable local relief; hillsides steep.
	New England Province	Mostly hilly upland with altitudes below 1,500 ft; locally mountainous with altitudes above 5,000 ft; coast irregular and rocky.
	Adirondack Province	Mountains rising to more than 5,000 ft
	St. Lawrence Valley	Rolling lowland with altitudes below 500 ft
	Central Lowland	Vast plain, 500 to 2,000 ft; the agricultural heart of the country.
Interior Plains	Great Plains Province	Western extension of the Central Lowland rising westward from 2,000 to 5,000 ft; semiarid.
	Interior Low Plateaus	Plateaus; less than 1,000 ft; rolling uplands with moderate relief.
	Ozark Plateaus	Rolling upland; mostly above 1,000 ft
Interior Highlands	Ouachita Province	Like the Valley and Ridge Province; altitudes 500 to 2,000 ft
	Southern Rocky Mountains	A series of mountain ranges and intermontane basins, mostly trending north; high part of the continental divide; altitudes 5,000 to 14,000 ft
	Wyoming Basin	Elevated semiarid basins; isloated low mountains; altitudes mostly between 5,000 and 7,000 ft
Rocky Mountain System	Middle Rocky Mountains	An assortment of different kinds of mountains with differing trends and semiarid intermontane basins; features here resemble those of the neighboring provinces; altitudes mostly 5,000 to about 12,000 ft
	Northern Rocky Mountains	Linear blocky mountains and basins in the east; highly irregular granitic mountains without linear trends in the southwest; altitudes mostly between 4,000 and 7,000 ft
	Colorado Plateau	Highest plateaus in the country; surface mostly above 5,000 ft and up to 11,000 ft; canyons; semiarid.
Intermontane Plateaus	Basin and Range Province	Mostly elongate, blocky mountains separated by desert basins and trending north; pattern more irregular in the south; altitudes from below sea level (Death Valley, Salton Sea) to more than 12,000 ft, but relief between mountains and adjoining basins generally no more than about 5,000 ft; most basins in the north are without exterior drainage.
	Columbia Plateau	Mostly a plateau of lava flows; altitudes mostly below 5,000 ft; semiarid but crossed by two major rivers, the Columbia and Snake.
	Cascade-Sierra Nevada	Northerly trending mountains; Cascades a series of volcanos; Sierra Nevada a blocky mass of granite with steep eastern slope and long gentle western slope; altitudes up to 14,000 ft; western slopes humid, eastern slopes semiarid.
Pacific Mountain System	Pacific Border Province	Coastal Ranges with altitudes mostly below 2,000 ft and separated from the high Cascade-Sierra Nevada Province by troughs less than 500 ft in altitude.
	Lower California Province	Northern end of the granitic ridge forming the Lower California Peninsula.
	Southeastern Coast Mountains	Rugged coastal mountains up to 9,000 ft in altitude; glaciers, fjords.
	Glaciered Coast	Coastal Mountains up to 20,000 ft in altitude; 5,000 sq mi of glaciers; perpetual snow above 2,500 ft
	South-central Alaska	Mountain ranges and troughs in arcs curving around the Gulf of Alaska; altitudes up to 20,000 ft
Alaska	Alaska Peninsula and Aleutian Islands	Chiefly a chain of volcanoes; altitudes mostly less than 6,000 ft except at the north; bordered on the south by the Aleutian trough, an oceanic trench 20,000 ft deep.
	Interior Alaska	Mostly the Yukon River valley between Alaska Range on the south and Brooks Range on the north; deltaic flat at the west; mostly dissected upland in the east with uplands 1,000 to 2,000 ft higher than the rivers.
	Seward Peninsula and Bering Coast Uplands	Rugged plateau, mostly (1,000 to 2,500 ft) in altitude; most ground is permanently frozen.
	Arctic Slope	Long slope north from the Brooks Range to broad coastal plain bordering the Arctic Ocean; permanently frozen ground.
Hawaii	Hawaii	Oceanic volcanic islands in part bordered by coral reef; altitudes up to (13,000 ft).
Puerto Rico	Puerto Rico	Oceanic island bordered on north by a trench more than 30,000 ft deep; an east–west-trending ridge extends the length of the island; long slope to north; shorter, more precipitous slope to south; shores partly bordered by coral reef.

Source: Charles B. Hunt, *Physiography of the United States.*

in the offshore areas. Off the Atlantic coast the Continental Shelf is more than 161 km (100 mi) wide but is quite narrow along the Pacific coast. The southern shore of the Alaska Peninsula and the Aleutian Islands are bordered by oceanic trenches.

The Coastal Plain with a width of 161 to 322 km (100 to 200 mi), extends along the Atlantic seaboard and the Gulf coast and makes up about 10% of the land area. This plain continues seaward as the Continental Shelf for another 161 to 322 km, where it plunges downward to depths of more than 3.2 km (2 mi). The Coastal Plain is noted for its low topographic relief and extensive marshy tracts. More than half the plain is below 152 m (500 ft). The surface of the plain slopes northward, and the valleys in the north are drowned to form the Chesapeake, Delaware and New York bays and Long Island Sound. The shore, more than 4,827 km (3,000 mi) long, has numerous sandy beaches. There also are swamps, such as the Dismal Swamp in North Carolina and Virginia; mud flats, such as at the delta of the Mississippi River; and coral banks, as in Florida. Where the Coastal Plain borders the Piedmont Province is a line of falls called the Fall Line, where most of the seaboard cities are located.

The Coastal Plain is divided into five sections:

1. The Embayed section, stretching from Cape Cod south to the Neuse River in North Carolina.
2. Cape Fear Arch, extending from the Newuse River south to the Santee River.
3. The Sea Islands Downwarp, extending between Cape Fear Arch and the Peninsular Arch in Florida. Here the coastline is marked by many islands and salt marshes.
4. The Peninsular Arch, of which Florida is the highest part and that also includes the submerged Bahama Shelf to the southeast, the Continental Shelf to the east and the Florida Shelf to the west. Florida is a land of lakes, of which Lake Okeechobee covers 1,943 sq km (750 sq mi). To the south of Lake Okeechobee is the Everglades.
5. The Gulf Coast sections, including the Mississippi River Alluvial Plain.

South of the Fall Line hills of the East Gulf Coastal Plain is the Black Belt, named for its characteristic soil, one of the most fertile areas in the region. The Mississippi River Alluvial Plain is about 805 km (500 mi) long and 80 to 161 km (50 to 100 mi) wide, with altitudes of between 30 and 84 m (100 and 275 ft). The plain is divided into five basins: the Atchafalaya Basin, the Yazoo Basin, the Tensas Basin, the St. Francis River Basin and the Black River Basin. The Coastal Plain ends inland at the escarpment formed by the Balcones fault zone, which marks the edge of the Edwards Plateau.

The Appalachian Highlands is the mountainous part of the eastern United States including the Piedmont, Blue Ridge, Valley and Ridge, Appalachian Plateaus, Adirondack and New England provinces and the St. Lawrence Lowland. The highlands have an area equal to that of the Coastal Plain. Compared to the Rocky Mountains the Appalachians are not high and show less angular and more rounded landforms. The three highest mountain peaks are Mount Mitchell, in North Carolina 2,037 m (6,684 ft); Clingman's Dome, in Tennessee 2,024 m (6,642 ft); and Mount Mansfield, in Vermont 1,916 m (6,288 ft). Twelve other peaks are over 731 m (2,400 ft). The major basins are Connecticut River, Newark, Gettysburg, Culpepper, Richmond, Danville and Deep River. Rivers are numerous, and lakes abound in the glaciated areas. The Piedmont stretches from southernmost New York to Alabama, with a maximum width of about 201 km (125 mi) and altitudes between 30 and 305 m (100 and 1,000 ft). Blue Ridge Province extends from Georgia to Pennsylvania, ranging in width from 8 to 80 km (5 to 50 mi). In places it is a single ridge; elsewhere, a complex of closely spaced ridges. Valley and Ridge Province extends the entire length of the Appalachian Highlands and is divided into three sections. The Appalachian Plateaus form the western border of the highlands, reaching 914 m (3,000 ft) at the Allegheny Front and along the Hudson River Valley. The Adirondack Mountains in northern New York State have a domelike structure more than 161 km (100 mi) in diameter. New England Province is considerably more elevated, particularly the Green Mountains in Vermont, the Hoosac Mountains in Massachusetts and the White Mountains of New Hampshire and Maine. Isolated mountains that rise above the general level of the surrounding terrain are called monadnocks. Principal lowlands and basins include Narragansett, Boston, Bangor Bowl and Aroostook Plain. One characteristic feature of the province is the rocky coast.

The great central plain between the Appalachian Plateaus and the Rocky Mountains includes the core Central Lowland, with the Superior Upland to the north; the Interior Low Plateaus, the Ozark Plateaus and Ouachita Province to the south; and the Great Plains to the west. The Central Lowland covers about 15% of the United States in 16 states. At the north of the province is the Great Lakes section, south of it the Till Plain section and farther to the northwest the Small Lakes section. Much of the province was glaciated, except for the Driftless section and the Osage Plain. The Superior Upland is part of the Precambrian Shield of Canada. The effect of glaciation in this province is seen in the ''10,000 lakes'' of Minnesota. The Interior Low Plateaus include Lexington Plain and Nashville Basin and parallel the Appalachian Highlands, extending 965 km (600 mi) from Alabama to Ohio.

The Ouachita Mountains and the Ozark Plateaus have lower average altitudes than the Appalachians and rarely are higher than 792 m (2,600 ft). Between these two regions is the Arkansas River, at an altitude of 152 m (500 ft). Great Plains Province covers about 12% of the country and covers the semiarid part of the Interior Plains. The plains slope eastward from about 1,676 m (5,500 ft) at the ft of the Rocky Mountains to about 610 m (2,000 ft) at the eastern boundary. The valleys of the streams that flow eastward from the Rocky Mountains to the Missouri and Mississippi rivers are broad, steep-sided and shallow, and between them are smooth uplands and nearly flat plains. At the northern part of the Great Plains is the Missouri Plateau, which has a more prominent relief than the plains. Here there are numerous dome mountains rising up to 610 m (2,000 ft), of which the highest are the Black Hills. Northwest of the Black Hills is a volcanic neck about 305 m (1,000 ft) high known as Devil's Tower, a national monument. South of the Missouri Plateau is the High Plains section. On the west the High Plains are separated from the Rocky Mountains by a basin known as the Colorado Piedmont section. To the south is the Raton section, with high mesas or plateaus, some as high as 2,133 m (7,000 ft). Parts of the High Plains are sandy, with extensive dunes. In southeastern New Mexico and the Panhandle of Texas are the Staked Plains, the Llano Estacado, one of the most nearly level parts of the United States. The Staked Plains end westward at the Pecos Valley.

The Rocky Mountains are the backbone of the Western Hemisphere, stretching from Alaska to Patagonia. Within the United States they constitute the Continental Divide, separating the Interior Plains from the Intermontane Plateaus and the Pacific Mountain systems. The Rocky Mountains contain three provinces: Northern, Middle and Southern. Together with the Wyoming Basin they cover about 10% of the country.

The Rocky Mountains are characterized by high peaks, many over 4,267 m (14,000 ft); rugged relief; extensive forests; and spectacular scenery. The Southern Rockies can be crossed only through high passes, all above 2,743 m (9,000 ft). They consist of a series of ranges: on the east, Laramie Range, Front Range, Wet Mountains and Sangre de Cristo Range; and on the west, Park Range, Gore Range, Sawatch Range, Elk Mountains, San Juan Mountains and Jemez Mountains. The two groups of ranges are separated in the north by the basin of the North Platte River, North Park, the valley of the Arkansas River and South Park and in the south by the broad San Luis Valley and the trough of the Rio Grande. The Middle Rockies include ranges such as Bighorn, Owl Creek, Wind River, Uinta and Wasatch. Most of Yellowstone National Park is an elevated lava plateau. The Middle Rockies include some of the most impressive mountains in the western United States, but only some parts are higher than 3,353 m (11,000 ft). The Wyoming Basin is an elevated depression separating the Middle and Southern Rockies and consists of a series of basins, such as the Washakie Basin, and the basins along the Green, North Platte and Wind rivers. Yellowstone Volcanic Plateau contains thousands of hot springs and geysers, many of which erupt intermittently, although a few, like Old Faithful, erupt periodically. Some discharge clear water; others, highly colored water; and one group, known as the paint pots, a pasty mud.

The Northern Rockies include a great many mountain ranges, such as the Okanogan Highlands, the Clearwater Mountains, the Garnet Range, the Bitteroot Mountains and the Little Belt Mountains. The principal valleys—or grabens—are the Rocky Mountain Trench, which extends along the Kalispell Valley, and the Purcell Trench, along the lower course of the Kootenai River on the western side of the Purcell Mountains.

The Colorado Plateau, between the Rocky Mountains and Basin and Range Province, is the most colorful part of the United States, with spectacular geological features. It is characterized by arid areas of bare rock and sparse vegetation and brightly colored desert scenery. The general plateau surface is higher than 1,524 m (5,000 ft), and some peaks reach 3,353 m (11,000 ft). Striking topographic features include volcanoes, cinder cones, volcanic necks, mesas and dome mountains. The high southwestern part of the plateau is referred to as the Grand Canyon section. The western part of the section is divided into a series of blocks by the Grand Wash, Hurricane and Sevier faults. The southern rim of the plateau in New Mexico and western Arizona is known as the Datil section. Its principal structural features are the Zuni Mountains and Mount Taylor Volcano. To the north of the Grand Canyon and the Datil section is the depression known as the Navajo section. The deepest part of this depression, San Juan Basin, is separated from a shallower basin that lies to the west under Black Mesa by the Defiance Upwarp. North of the Navajo section is the Canyon Lands section where, as the name implies, canyons are dominant. This section has four large upwarps: the Uncompaghre Upwarp, the Monument Upwarp, the Circle Cliffs Upwarp and the San Rafael Swell. North of the Canyon Lands section is Uinta Basin, which forms an embayment between the Middle and the Southern Rocky Mountains. At the western edge of the Colorado Plateau is the High Plateaus section. The Colorado Plateau illustrates the benefits of aridity, because without it, the plateau would have lost all or most of its spectacular features.

Basin and Range Province covers desert basins and ranges between the Colorado Plateau and the Columbia Plateau. The basins may range in altitude from below sea level (as at Death Valley and Salton Sea) to about 1,524 m (5,000 ft). The mountain ranges number over 200, running in parallel ridges between 3,048 and 3,962 m (10,000 and 13,000 ft) in height. The province is characterized by colorful vegetation, a warm desert climate and large numbers of lakes. The Great Basin is centered in Nevada and has five subdivisions: Central Area, Bonneville Basin, Lahontan Basin, Lave and Lake Area, and Southern Area. Bonneville Basin includes Great Salt Lake, Utah Lake and Sevier Lake; Lahontan Basin, Pyramid Lake and Lake Winnemucca; Lava and Lake Area, Honey Lake, Eagle Lake, Goose Lake and Klamath Lake; and Southern Area, Lake Mead, Walker Lake, Mono Lake and Owens Lake. Death Valley, which covers over 518 sq km (200 sq mi) and is more than 76 m (250 ft) below sea level, is a dry, salt-encrusted playa. The Sonoran Desert section, including the Mojave Desert in southeastern California and the deserts in southwestern Arizona, extends into Mexico along the eastern shore of the Gulf of California. The boundary between the Mojave Desert and the Great Basin is along the Garlock fault, while the southern boundary of the Sonoran Desert section is along the San Andreas fault at or near the foot of the San Gabriel and San Bernardino mountains. The Salton Trough is the emerged part of the 1,609 km long (1,000 mi long) trough occupied by the Gulf of California. Three major faults extend northwestward into California from this depression: San Andreas, San Jacinto and Elsinore. The northeastern side of the trough, along the San Andreas fault, is marked by the Gila and Chocolate mountains and other features. The Mexican Highlands section, which includes the Rio Grande Valley, consists of mountain ranges separated by wide valleys, the latter occupying two-thirds of the area. The Sacramento Mountains section forms the eastern border of Basin and Range Province and adjoins the Great Plains. At the north of the mountains is Estancia Basin, with the Sandia and Manzano mountains to the west and the Pedernal Hills to the east. South of Estancia Valley is Chupadera Mesa, and to its south lies the Sacramento Range, which extends more than 129 km (80 mi) to the Texas border, whence it becomes the Guadalupe and Delaware mountains. To the west of the range lies Tularosa Basin.

PHYSIOGRAPHIC DIAGRAM
Adapted from a map compiled by
Erwin Raisz, 1954
The United States

The Columbia Plateau includes the Walla Walla section, the Harney and Peyette section, the Snake River Plain and the Blue Mountains section, all in the Northwest. The plateau is characterized by a semiarid climate and dry canyons and is dominated by the Columbia and Snake rivers and their plains. The lowest part of the Walla Walla section is Pasco Basin, in which the Columbia River is joined by the Snake. The Snake River crosses the uplift in Hells Canyon, which is deeper than Grand Canyon. On the eastern wall of the canyon, Seven Devils Mountain reaches a peak of 2,868 m (9,410 ft) and He Devil Mountain 2,438 m (8,000 ft). The Harney Lake section is a volcanic plain at the southwestern corner of the plateau and also is known as the Great Sand Desert or the Great Sage Plain. The Payette section occupies two basins: One is the basin of the Snake River between the Idaho batholith and the Owyhee Mountains; the other is south of the Owyhee Mountains. The Snake River Plain is a continuation of the trough represented by the Payette section. The plain is interrupted by many volcanic cones, among them the Craters of the Moon. East of the Snake River Plain is the Yellowstone Plateau, where volcanic activity still is strong.

The Pacific Mountain system comprises some 518,135 sq km (200,000 sq mi) along the coast, extending east to the deserts in Basin and Range Province and the Columbia Plateau. The mountains, which are among the highest in the country, account for the deserts, which are in their shadow. The mountains include granitic mountains, such as the Sierra Nevada and Klamath mountains; volcanic mountains such as the Cascade Range; complexly folded and faulted mountains such as the Transverse Ranges and Olympic Mountains; moderately folded and much faulted ranges as the Oregon Coast Range; and dome mountains such as the Marysville Buttes of the California Central Valley. The mountains are arranged in the form of three links, with holes between. These mountains are among the highest, roughest and the most scenic in the United States. The mountainous coast is part of the circum-Pacific volcanic belt and also part of a seismically active belt subject to frequent and severe earthquakes. The major divisions of Pacific Border and Lower California provinces are (1) the Cascade-Sierra Mountains, (2) the Pacific Border, and (3) the Lower California Peninsular Range. The Sierra Nevada is a huge block mountain about 563 km (350 mi) long and roughly 97 km (60 mi) wide. The mountain reaches its maximum height along the eastern scarp and from the summit slopes westward. It is crossed by six passes: Donner Pass, 2,161 m (7,089 ft); Emerald Pass, 2,164 m (7,100 ft); Carson Pass, 2,613 m (8,573 ft); Ebbets Pass, 2,661 m (8,730 ft); Sonora Pass, 2,934 m (9,626 ft); and Tioga Pass, 3,030 m (9,941 ft). In its entire length the Sierra Nevada is not crossed by any river. The Cascade Mountains are divided into southern, middle, high and western sections. In the Cascades is Crater Lake, once a volcano known as Mount Mazama. Pacific Border Province comprises the California Trough, the California Coast Ranges, the Transverse Ranges, the Klamath Mountains, the Oregon Coast Range, the Olympic Mountains and the Puget Trough. The highest ranges in the Tranverse Ranges are the San Gabriel and the San Bernardino Mountains, both located along the San Andreas fault. They also include Los Angeles Basin, which is the only coastal plain in the province. The Klamath Mountains section is a mountainous coastal area between the coastal ranges of California and Oregon and through which the Klamath River and the Rogue River have cut meandering gorges. The Oregon Coast Range consists of irregular hills and low mountains along the coast of Oregon and southwestern Washington. The highest summits are less than 1,219 m (4,000 ft), and the general relief is less than 610 m (2,000 ft). The hillsides are rounded and the valleys are open. Puget Sound is the southern part of a depression that extends from Willamette Valley in Oregon northward, barely 40 km (25 mi) wide. The Lower California Province, between the Salton Trough and the coast, is the northern end of Baja California.

Unlike the Atlantic coast, there are only half a dozen plains along the Pacific coast. The coast is mountainous and characterized by steep bluffs and elevated marine terraces. The shoreline is straight and fully exposed to the surf, without barrier beaches or lagoons. Up to Point Conception the Continental Shelf is barely 80 km (50 mi) wide, but south of this point the width increases to 241 km (150 mi). Two major mountains ridges extend about 2,414 km (1,500 mi) westward from the coast into the Pacific Ocean.

Alaska is the largest state, with a land area about one-fifth that of the other 49 states combined. There are seven physiographic provinces: (1) the southeastern coastal mountains, including the southeastern panhandle, with numerous islands and fjords; (2) the glaciered coast, another mountainous coastal area; (3) south-central Alaska; (4) the Alaska Peninsula and the Aleutian Islands; (5) interior Alaska, including the Yukon Valley, the Kilbuck and Kuskokwim Mountains and the Brooks Range; (6) the Seward Peninsula and Bering Coast Uplands; and (7) the Arctic Slope. Alaska is characterized by great relief, from

PRINCIPAL CITIES (estimated population at mid-1984)

City	Population	City	Population
New York	7,262,700	Nashville-Davidson	473,670
Los Angeles	3,259,340	Austin	466,550
Chicago	3,009,530	Oklahoma City	446,120
Houston	1,728,910	Kansas City, Mo.	441,170
Philadelphia	1,642,900	Fort Worth	429,550
Detroit	1,086,220	St Louis	426,300
San Diego	1,015,190	Atlanta	421,910
Dallas	1,003,520	Long Beach	396,280
San Antonio	914,350	Portland, Ore	387,870
Phoenix	894,070	Pittsburgh	387,490
Baltimore	752,800	Miami	373,940
San Francisco	749,000	Tulsa	373,750
Indianapolis	719,820	Honolulu	372,330
San Jose	712,080	Cincinnati	369,750
Memphis	652,640	Albuquerque	366,750
Washington, DC		Tucson	358,850
(capital)	626,000	Oakland	356,960
Jacksonville	609,860	Minneapolis	356,840
Milwaukee	605,090	Charlotte	352,070
Boston	573,600	Omaha	349,270
Columbus	566,030	Toledo	340,680
New Orleans	554,500	Virginia Beach	333,400
Cleveland	535,830	Buffalo	324,820
Denver	505,000	Sacramento	323,550
El Paso	491,800	Newark	316,240
Seattle	486,200		

Mount McKinley, at 6,193 m (20,320 ft) the highest point in North America, to the Aleutian Trench, at 7,620 m (25,000 ft) below sea level the lowest point around North America. The state has hundreds of volcanoes and thousands of glaciers as well as a vast area permanently frozen. The major mountain ranges are found in South-central Alaska: Chugach, Wrangell, Talkeetna, the Alaska Range and the Aleutian Range.

The Alaska Peninsula and the Aleutian Islands form an arc consisting of more than 75 volcanoes extending 2,414 km (1,500 mi) from Mount Spurr opposite Cook Inlet, to Buldir Volcano between the Rat Islands and the Near Islands. Interior Alaska lies between the Alaska Range on the south and the Arctic slope on the north, roughly half the state. Interior Alaska is mostly the drainage basin of the Yukon River and its Koyukuk, Tanana and Porcupine tributaries. The lower Yukon and Lower Kuskokwim rivers cross an extensive low deltaic flat containing numerous large swamps. The Arctic Slope is a nearly featureless coastal plain containing Point Barrow, the northernmost point in the United States.

The Hawaiian Islands are basaltic volcanoes near the middle of the Pacific Ocean along a northwest-trending ridge that divides two oceanic deeps, each of which descends more than 5,486 m (18,000 ft) below sea level. There are five principal islands and four smaller ones. Hawaii, the most easterly and largest of the islands, has peaks more than 3,962 m (13,000 ft) in altitude. The islands are hilly, especially toward the east. The coastlines are mostly rocky and rough. At the exposed northeastern shores erosion has produced seacliffs 914 m (3,000 ft) high. Only Oahu and Niihau have large coastal plains. There is only one harbor, Pearl Harbor, west of Honolulu on Oahu, and there is a bay at Hilo on Hawaii. The volcanoes at Mauna, Loa, and Kilauea on Hawaii still are active and erupt every few years.

Rivers in the Appalachian Highlands are short, but they are numerous and large in terms of discharge. The central United States drains in three directions—to the Mississippi River, to the St. Lawrence River and to Hudson Bay. Most of the region drains into the Mississippi, which is one of the world's great rivers both in length and discharge. Its major tributaries are the Tennessee, Ohio, Missouri and Arkansas rivers. The Arkansas flows east from Colorado to central Kansas. The Tennessee River, with its headwaters in Valley and Ridge Province, flows through the Appalachian Plateaus around the Cincinnati Arch to join the Ohio and Mississippi rivers. In the Rocky Mountains and Wyoming Province the major rivers are the Laramie; the North Platte and its tributary the Sweetwater; and the Bighorn, which drains most of the Wyoming Basin. The principal streams of the Colorado Plateau are the Colorado, the Green, the Gunnison, the Dolores, the San Juan, the Price and the San Rafael. The Great Basin and the Mojave Desert are curiously lands of lakes, both dry and perennial, including Sevier Lake, Utah Lake and Great Salt lake, fed by the Sevier, Provo and Jordan rivers, respectively. Other lakes, such as Pyramid, Walker, Summer and Albert, contain high percentages of dissolved solids. One of the most important rivers in the Great Basin is the Humboldt of Nevada. The Columbia Plateau is dominated by the Columbia, Snake and Owyhee rivers. In Pacific Mountain Province the Sacramento, Klamath and Willamette are wide rivers, although they are short and drain small areas. Lakes are numerous in the Cascades and in the Sierra Nevada. The largest are Lake Chelan in Washington, Crater Lake, Lake Tahoe and Yosemite Lake. In Alaska the Yukon, one of the longest rivers in the world, is navigable by river steamer most of the way to its headwaters in Canada. Its principal tributaries are the Koyukuk, Tanana and Porcupine rivers. The Kuskokwim, the other major river, rises in the Alaska Range and flows southwest. Two rivers, the Stikine and the Taku, cross the Coast Mountains and discharge into the Alexander Archipelago.

Longest Rivers in the United States

Rivers	Length Km (Mi.)	Discharge Rate 1,000 cms (1,000 cfs)
Mississippi-Missouri-Jefferson-Beaverhead-Red Rock	6,018 (3,740)	12.83 (453)
Yukon-Lewes-Teslin-Misutlin	3,186 (1,980)	6.8 (240)
Rio Grande	3,025 (1,880)	0.08 (2.7)
Arkansas	2,333 (1,450)	1.19 (42)
Colorado	2,333 (1,450)	0.10 (3.7)
Atchafalaya-Red	2,253 (1,400)	5.13 (181)
Columbia-Snake	2,124 (1,320)	5.35 (189)
Brazos	2,108 (1,310)	0.15 (5.3)
Ohio-Allegheny	2,108 (1,310)	6.71 (237)
Mississippi (Upper)	1,883 (1,170)	2.78 (98)

The largest lake entirely within the United States is Lake Michigan, with an area of 58,031 sq km (22,400 sq mi) and a maximum depth of 281 m (923 ft). The second-largest is Great Salt Lake in Utah, with an area of 4,352 sq km (1,680 sq mi) and a maximum depth of 15 m (48 ft). The Great Lakes other than Lake Michigan (Superior, Huron, Erie and Ontario) are shared with Canada.

The highest mountains are all in Alaska, including McKinley south peak 6,193 m (20,320 ft) and McKinley north peak 5,934 m (19,470 ft); Foraker 5,303 m (17,400 ft); Bona 5,029 m (16,500 ft); Sanford 4,950 m (16,240 ft); Churchill 4,767 m (15,640 ft); University 4,581 m (15,030 ft); Bear 4,526 m (14,850 ft) and Hunter 4,441 m (14,570 ft). Mount Whitney, in California, is the highest peak in the lower 48 at 4,416 m (14,490 ft), followed by 21 others over 3,048 m (10,000 ft), all in California, Colorado and the western coast. In the Rockies are Ebert 4,398 m (14,430 ft), Harvard 4,395 m (14,420 ft), Massive 4,395 m (14,420 ft), La Plata 4,380 m (14,370 ft), Blanca 4,365 m (14,320 ft), Uncompahgre 4,361 m (14,310 ft), Crestone 4,355 m (14,290 ft), Lincoln 4,355 m (14,290 ft), Antero 4,349 m (14,270 ft), Grays 4,349 m

(14,270 ft), Torreys 4,349 m (14,270 ft), Castle 4,346 m (14,260 ft), Evans 4,346 m (14,260 ft), Longs 4,346 m (14,260 ft) and Qwandary 4,346 m (14,260 ft). In the Cascade are Rainier 4,392 m (14,410 ft) and Shasta 4,316 m (14,160 ft), and in Sierra Nevada, Williamson 4,380 m (14,370 ft).

Certain regions of the United States are subject to severe seismic activity and others to moderate seismic activity. One seismic belt extends along the western edge of the Middle Rocky Mountains and across the southeastern part of the Northern Rocky Mountains to the Great Plains. This belt has been the site of two major earthquakes, one at Helena in 1925 and the other at Hebgen Lake in 1959. In the Colorado Plateau epicenters of major earthquakes form lines along the western edges of the plateau and along its southeastern edge along the Rio Grande Valley. Epicenters of minor earthquakes form a line along the southwestern edge of the plateau and along its northern and eastern sides where it joins the Rocky Mountains. In Basin and Range Province, epicenters are concentrated along the western and eastern parts of the Great Basin and also in the Salton Trough and along the western part of the Sonoran Desert, in the Rio Grande section of the Mexican Highlands and in western Texas. Earthquakes are more frequent on the Pacific Coast, which has experienced over 200 in the past 100 years, including destructive ones in 1857, 1872 and 1906. Earthquakes are frequent in two belts in Alaska. One forms an arc along the Aleutians, the Alaska Peninsula and the Alaska Range to the vicinity of Fairbanks. The other begins in the St. Elias Mountains and extends southward to the Puget Trough. A household may feel half a dozen earthquakes annually in Anchorage, which was virtually destroyed in the 1964 earthquake. Near the coast, earthquakes are accompanied by huge waves, sometimes called tsunamis. The most active volcano in the United States is Mount St. Helens, which erupted violently in 1980, spewing volcanic ash over much of the Northwest.

Except for Puerto Rico, U.S. territories and dependencies consist of small islands. They include American Samoa, Guam, Virgin Islands, Baker Island, Johnson Atoll, Kingman Reef, Midway Islands, Navassa Island, Palmyra Atoll and Wake Island. Since July 18, 1947, the United States has administered the Trust Territory of the Pacific Islands, but has recently entered into a new relationship with three of the four political units: the Northern Mariana Islands, the Federated States of Micronesia and the Republic of the Marshall Islands. Palau continues to be administered as a Trust Territory.

American Samoa comprises the seven islands of Tutuila, Tau, Olosega, Ofu, Aunuu, Rose and Swain's with a total land area of 199 sq km (77 sq mi) and a coastline of 116 km (72 mi). The islands lie in the south central Pacific Ocean about 3,700 km (2,300 mi) southwest of Hawaii. All the islands are volcanic with rugged peaks and limited coastal plains and two are coral atolls. The capital is Pago Pago on Tutuila. The 1988 population was 39,254.

Guam is the southernmost and largest of the Mariana Islands about 2,170 km (1,347 mi) west of Honolulu. It has a total land area of 541 sq km (209 sq mi). The island is of volcanic origin and is surrounded by coral reefs. A relatively flat coraline limestone plateau—the source of the island's fresh water—is flanked by steep coastal cliffs and narrow coastal plains in the north, low rising hills in the center and mountains in the south. The capital is Agana.

Virgin Islands comprises 68 islands about 64 km (40 mi) east of Puerto Rico along the Anegada Passage—a key shipping lane for the Panama Canal—with a total land area of 352 sq km (136 sq mi). Only three of the islands are important in size: St. Croix, St. Thomas and St. John. Although rarely affected by hurricanes, the islands are subject to frequent severe droughts, floods, earthquakes and lack of natural fresh water resources. The terrain is mostly hilly to rugged and mountainous with little level land. About three-fourths of St. John is a national park. The capital is Charlotte Amalie. The 1988 population of the Virgin Islands was 112,636.

Marshall Islands consist of two groups of islands, the Ratak and Ralik chains comprising 31 atolls, including Bikini, Eniwetak and Kwajalein, with a total land area of 181.3 sq km (70 sq mi). The islands lie in the area of the Pacific Ocean known as Micronesia about 3,200 km (1,988 mi) southwest of Hawaii and 2,100 km (1,304 mi) southeast of Guam and include two archipelagic chains of 30 atolls and 1,152 islands. The capital is the Dalap-Uliga-Darrit municipality on Majuro Island. The population in 1988 was 40,609.

The Federated States of Micronesia forms the archipelago of the Caroline Islands about 800 km (497 mi) east of the Philippines. The territory includes (from west to east) Yap, Truk, Pohnpei (formerly, Ponape), the Kosrae. The terrain varies from high mountains to low, coral atolls. There are volcanic outcroppings on Pohnpei, Kosrae and Truk. The total land area is 702 sq km (271 sq mi). The capital is Kolonia but a new capital is being built. The 1988 population was 86,094.

Northern Mariana Islands comprise 16 islands in the western Pacific Ocean, about 5,635 km (3,499 mi) west southwest of Honolulu with a total land area of 477 sq km (184 sq mi). The southern islands are limestone with level terraces and fringing coral reefs while the northern islands are volcanic. Mount Pagan on Pagan Island is an active volcano which last erupted in May 1981. The capital is Saipan. The 1988 population was 20,591.

Palau (also known as Belau) consists of more than 200 islands in a chain about 650 km (403 mi) long, 7,150 km (4,440 mi) southwest of Hawaii and 1,160 km (720 mi) south of Guam The total land area is 458 sq km (177 sq mi). The terrain varies from the high mountainous main island of Babelthuap to low, coral islands, usually fringed by large barrier reefs. The capital is Koror but a new capital is being built in eastern Babelthuap. The 1988 population was 14,106.

Baker Island is a small island with a land area 1.5 sq km (0.6 sq mi) located 2,575 km (1,599 mi) southwest of Honolulu in the North Pacific Ocean, just north of the equator, about halfway between Hawaii and Australia. It is a low, nearly level coral island. The island is an unincorporated territory administered by the Fish and Wildlife Service of the US Department of the Interior as a National Wildlife Refuge. Public use is restricted to scientists on a permit only basis. The island is uninhabited.

Howland Island is a small island with a total land area of 1.7 sq km (0.6 sq mi) about 2,575 km (1,599 mi) southwest of Honolulu in the North Pacific Ocean, just north of the equator, about halfway between Honolulu and Australia. It is nearly level sandy coral island with a fringing reef and depressed central area. There is no fresh water. The island, which is uninhabited, is administered by the Fish and Wildlife Service of the U.S. Department of Interior as a National Wildlife Refuge. Public use is limited to scientists with permits.

Jarvis Island is a small island, 4.6 sq km (2.8 sq mi) in area 2,090 km (1,298 mi) south of Honolulu in the South Pacific Ocean, just south of the equator. It is a sandy, coral island with narrow fringing reef. The island, which is uninhabited, is administered by the Fish and Wildlife Service of the U.S. Department of the Interior as a National Wildlife Refuge.

Johnston Atoll is a small island with a land area of 2.8 sq km (1.0 sq mi) about 1,328 km (825 mi) west southwest of Honolulu in the North Pacific Ocean. It comprises the natural islands of Johnston Island and Sand Island and the man-made islands of North Island (Akau) and East Island (Hikina). The islands are flat with a maximum elevation of 4 meters. The islands are closed to the public. As a former nuclear weapons test site, they are administered by the U.S. Defense Nuclear Agency cooperatively with the Fish and Wildlife Service.

Kingman Reef is a small island about 1 sq km (0.38 sq mi) in area about 1,600 km (993 mi) south southwest of Honolulu in the North Pacific Ocean. It a flat barren coral atoll with a maximum elevation of one meter. The island, which is uninhabited, is administered by the U.S. Navy.

Midway Islands is a group of islands, the largest of which are the Eastern Island and the Sand Island, with a total land area of 5.2 sq km (2 sq mi) about 2,350 km (1,459 mi) west northwest of Honolulu at the western end of the Hawaiian Islands group. It is a low, flat coral island. It is administered by the U.S. Navy. The 1987 population was 1,500.

Navassa Island is a small island, about 5.2 sq km (2 sq mi) in area, in the Caribbean Sea between Cuba, Haiti and Jamaica. It is mostly exposed rock with enough grassland to support a large goat population. It has a central limestone plateau ringed by vertical white cliffs. The island is uninhabited except for transient Haitian fishermen. The climate is tropical marine. The island is administered by the U.S. Coast Guard.

Palmyra Atoll is a small island group with a land area of 11.9 sq km (4.6 sq mi) about 1,600 km (993 mi) south southwest of Honolulu in the North Pacific Ocean. It comprises about 50 small islets covered with dense vegetation. The maximum elevation is about two meters. The island, which is uninhabited, is privately owned, but administered by the Office of Territorial and International Affairs, U.S. Department of the Interior.

Wake Island is a small island, about 6.5 sq km (2.5 sq mi) in area, 3,700 km (2,298 mi) west of Honolulu in the North Pacific Ocean. It is an atoll consisting of three coral islands built upon an underwater volcano. The central lagoon is a former crater and the islands are part of the rim. Under an agreement with the U.S. Department of the Interior, the island is administered by the U.S. Air Force. The 1988 population was 302.

Area and population

States	Capitals	area sq mi	area sq km	population 1986 estimate
Alabama	Montgomery	51,705	133,915	4,053,000
Alaska	Juneau	591,004	1,530,693	534,000
Arizona	Phoenix	114,000	295,259	3,317,000
Arkansas	Little Rock	53,187	137,754	2,372,000
California	Sacramento	158,706	411,047	26,981,000
Colorado	Denver	104,091	269,594	3,267,000
Connecticut	Hartford	5,018	12,997	3,189,000
Delaware	Dover	2,044	5,294	633,000
Florida	Tallahassee	58,664	151,939	11,675,000
Georgia	Atlanta	58,910	152,576	6,104,000
Hawaii	Honolulu	6,471	16,760	1,062,000
Idaho	Boise	83,564	216,430	1,003,000
Illinois	Springfield	57,871	149,885	11,553,000
Indiana	Indianapolis	36,413	94,309	5,504,000
Iowa	Des Moines	56,275	145,752	2,851,000
Kansas	Topeka	82,277	213,096	2,461,000
Kentucky	Frankfort	40,409	104,659	3,728,000
Louisiana	Baton Rouge	47,752	123,677	4,501,000
Maine	Augusta	33,265	86,156	1,174,000
Maryland	Annapolis	10,460	27,091	4,463,000
Massachusetts	Boston	8,284	21,455	5,832,000
Michigan	Lansing	97,102	251,493	9,145,000
Minnesota	St. Paul	86,614	224,329	4,214,000
Mississippi	Jackson	47,689	123,514	2,625,000
Missouri	Jefferson City	69,697	180,514	5,066,000
Montana	Helena	147,046	380,847	819,000
Nebraska	Lincoln	77,355	200,349	1,598,000
Nevada	Carson City	110,561	286,352	963,000
New Hampshire	Concord	9,279	24,032	1,027,000
New Jersey	Trenton	7,787	20,168	7,620,000
New Mexico	Santa Fe	121,593	314,924	1,479,000
New York	Albany	52,735	136,583	17,772,000
North Carolina	Raleigh	52,669	136,412	6,331,000
North Dakota	Bismarck	70,702	183,117	679,000
Ohio	Columbus	44,787	115,998	10,752,000
Oklahoma	Oklahoma City	69,956	181,185	3,305,000
Oregon	Salem	97,073	251,418	2,698,000
Pennsylvania	Harrisburg	46,043	119,251	11,889,000
Rhode Island	Providence	1,212	3,139	975,000
South Carolina	Columbia	31,113	80,582	3,378,000
South Dakota	Pierre	77,116	199,730	708,000
Tennessee	Nashville	42,144	109,152	4,803,000
Texas	Austin	266,807	691,027	16,682,000
Utah	Salt Lake City	84,899	219,887	1,665,000
Vermont	Montpelier	9,614	24,900	541,000
Virginia	Richmond	40,767	105,586	5,787,000
Washington	Olympia	68,139	176,479	4,463,000
West Virginia	Charleston	24,231	62,758	1,919,000
Wisconsin	Madison	66,215	171,496	4,785,000
Wyoming	Cheyenne	97,809	253,324	507,000
District				
Dist. of Columbia	—	69	179	626,000
TOTAL		3,679,192	9,529,063	241,077,000

URUGUAY

BASIC FACTS

Official Name: Oriental Republic of Uruguay

Abbreviation: URG

Area: 176,215 sq km (68,037 sq mi)

Area—World Rank: 86

Population: 2,976,138 (1988) 3,364,000 (2000)

Population—World Rank: 110

Capital: Montevideo

Boundaries: 2,063 km (1,282 mi); Brazil 1,003 km (623 mi); Argentina 495 km (308 mi)

Coastline: 565 km (351 mi)

Longest Distances: 555 km (345 mi) NNW-SSE; 504 km (313 mi) ENE-WSW

Highest Point: Mount Catedral 514 m (1,686 ft)

Lowest Point: Sea level

Land Use: 8% arable; 78% meadows and pastures; 4% forest and woodland; 10% other

Uruguay is the smallest country in South America. With the exception of Canada, it is the only country in the Western Hemisphere located entirely outside the tropics. All of its land surface lies to the east of the North American continent excluding Newfoundland and the capital city of Montevideo is about 965 km (600 mi) east of Bermuda. Its time zone is two hours earlier than Eastern Standard Time.

The perimeter of the wedge-shaped country, often described as compact and homogeneous, is devoid of pronounced extrusions or intrusions. The low plateau of the interior is marked by gentle up-and-down features except in the northeast, where the eroded hill ranges have sharp edges. The coastal lowlands have gently uneven characteristics featured by rivers and streams, sand dunes, small hill systems, and isolated knolls.

Compactness of the country and the lack of high relief features have combined to make easier the construction of a good transportation network, in which all major roads and railroads originate along the frontiers and converge on Montevideo. This convergence is at once a cause and a consequence of the development of a settlement pattern that has resulted in the accumulation of almost half the population in the capital city, a concentration unrivaled in the Western Hemisphere. In addition, population density decreases in proportion to the distance from Montevideo, and the principal cities of the interior center are located along access routes leading to Montevideo.

A transitional geological buffer between the landmasses of Argentina and Brazil, abundantly watered Uruguay is in a technical sense almost an island. Nearly all of its eastern perimeter is marked by a large tidal lagoon on the north and by the Atlantic Ocean. The southern coast is bordered by the great Río de la Plata estuary, and on the west the country is separated from the northern part of Argentina by the Río Uruguay. About two-thirds of the northern border corresponds to the course of major and minor rivers. Some border questions between Uruguay and its two contiguous neighbors, Argentina and Brazil, remain unsettled. None, however, represents serious problems, and irredentism is not an issue.

About 70 percent of the country lies in the La Plata drainage basin, an enormous area drained by the Uruguay and Paraná river systems.

Most of the countryside is made up of wide expanses of undulating prairie, an extension of the humid pampa of Argentina, broken in places by low ranges of hills pointing southwestward from the highlands of southern Brazil. Narrow lowlands of the Río de la Plata littoral and the Río Uruguay flood plain are devoted primarily to farming. The remainder of the country is a low, broken plateau divided into cattle and sheep ranches.

The country represents a transition from the pampas of Argentina to the hilly uplands and valley intersections of southern Brazil's Paraná plateau. In general, it is a country of gentle hills and hollows.

This rolling characteristic is general but less evident close to the eastern, southern, and western borders. In the northeast corner, adjacent to Laguna Merín, is a low plain area where rice is grown under irrigation. Directly southward, a narrow alluvial Atlantic coastal plain is broken by sand dunes, marshes, and coastal lagoons. Infertile stretches of sandy soil extend inland for distances of up to 8 km (5 mi). At Punta del Este, the coastline leaves the Atlantic to veer westward sharply for more than 322 km (200 mi) along the Río de la Plata estuary to the mouth of the Río Uruguay, the estuary's westernmost extremity. The littoral is somewhat broader here and merges almost imperceptibly with the grasslands and hills of the interior.

Soils consist of sands, clays, loess, and alluvium deposited by the numerous streams. The black soil is rich in potassium and enriched by decay of the lush cover of vegetation. Soils of similar composition are found on the flood plain of the Río Uruguay, which forms the country's western frontier. Along this flood plain, however, in the portion to the north of the mouth of the Río Negro there are extensions of the interior hill ranges that encroach on the flood plain.

The remaining three-fourths or more of the country consists of a rolling plateau, featured by ranges of low hills that become more prominent in the north as they merge into the highlands of southern Brazil. The geological foundation of most of the region is made up of gneiss, red sandstone, and granite, and an extension of the basaltic plateau of Brazil reaches southward in a broad band west of the Río Negro. Corridors between the hill ranges are floored with clay and sedimentary deposits.

Legend holds that when the ship of the discoverer Magellan first made landfall in the Río de la Plata, on sight-ing the conical hill west of the site of what was to become Montevideo, his lookout called out in Portuguese "Monte vide eu" (I see a mountain). This quite possibly is how the capital city acquired its name, but the country as a whole is so lacking in lofty relief features that its highest peak is Cerro Mirador in Maldonado Department near the southern coast. Its height is 501 m (1,644 ft).

The most important of the *cuchillas* (hill ranges) are the Cuchilla Grande and the Cuchilla de Haedo. Only in these and in the Cuchilla Santa Ana along the Brazilian frontier do altitudes with any frequency exceed 183 m (600 ft). Both of the two major ranges extend southwestward from Brazil, one on the eastern flank and one on the western, in directions roughly defining the course of the country's principal river, the Río Negro. To the east, the Cuchilla Grande and its several spur ranges form the country's most extensive hill system and its most important drainage divide. Here, ridges are frequently 305 m (1,000 ft) or more in elevation. West of the river in the northern portion of the Cuchilla de Haedo, near the Brazilian border, is the site of the Uruguayan countryside's most nearly rugged terrain. Elevations are low, but eroded layers of sandstone and laval deposits sometimes overlaid by basalt give sharp outlines to the topography.

The rolling character of the land surface derives from the presence of dozens of these *cuchilla* ranges. Insignificant as relief features, they have been important factors in the development of the country. In many instances the ridges serve as administrative boundaries and as routes followed by the roads and railroads, which frequently tend to avoid the underbrush-clogged lowlands along the rivers and streams. In addition, they contribute to a drainage system much more effective than that of the Argentine pampas, where the uncompromising flatness of the land prevents much runoff after rains and causes surface water to accumulate.

Hydrography

The name of Uruguay, first applied to the Río Uruguay, may have been taken from the Guaraní Indian word *uruguä,* meaning a kind of mussel. Another explanation is that it derives from Guaraní word-components meaning birds that come from the water. Both are indicative of a well-watered land. Most of the country's frontiers are riverine; rivers, streams, and occasional watercourses form an elaborate tracery over the entire country; lakes and lagoons are numerous; and a generally high water table makes well-digging easy.

The largest of the rivers is the Río Uruguay, but the country can claim only partial title to it. The river marks the entire western boundary with Argentina and extends farther to the north as a portion of the Argentina-Brazil frontier. It is flanked by low banks, and disastrous floods sometimes inundate large areas. The Río de la Plata, marking the entire southern boundary, is actually an estuary of the Atlantic Ocean and is saline except at its western extremity, where the Paraná and the Uruguay spew enormous quantities of fresh water into it.

There are three systems of internal rivers and streams: those flowing westward into the Río Uruguay, those flowing eastward either into the Atlantic Ocean or into tidal lagoons, and those flowing southward into the Río de la Plata.

Most of the rivers draining westward originate in the inland hills and descend through narrow valleys to join the Río Uruguay. The longest and most important, however, is the Río Negro, which rises in southern Brazil. It bisects the country as it flows southwestward over a distance of some 804 km (500 mi) to join the Río Uruguay south of the town of Fray Bentos. Its principal tributary, the country's second most important river, is the Río Yi, which flows a distance of 225 km (140 mi) from its source in the Cuchilla Grande. Other major western rivers originate in the Cuchilla de Haedo and join the Río Uruguay north of Paysandú.

The rivers flowing eastward originate in the Cuchilla Grande and are shallower and more variable in flow. They empty into lagoons, located behind the coastal dunes, or directly into the Atlantic through small estuaries. The largest of these, the Río Cebollatí, flows into Laguna Merín. The irregularity of the country's drainage pattern is illustrated by the fact that the headwaters of this eastward-draining river lie some 80 km (50 mi) to the west of that of the westward-draining Río Yi.

In 1937 work began on a Río Negro dam at a site a short distance upstream from the town of Paso de los Toros in the central part of the country. The resulting reservoir, called the Embalse del Río Negro, with a length of 140 km (87 mi) and a surface area of over 10,359 sq km (4,000 sq mi), is the largest manmade lake in South America. At a later date a second dam was constructed on

PRINCIPAL TOWNS (population at 1985 census)

Town	Population	Town	Population
Montevideo (capital)	1,246,500	Rivera	55,400
Salto	77,400	Melo	39,600
Paysandú	75,200	Tacuarembó	38,600
Las Piedras	61,300	Minas	33,700
		Mercedes	33,300

Area and population

		area		population
Departments	**Capitals**	sq mi	sq km	1985 census
Artigas	Artigas	4,605	11,928	68,400
Canelones	Canelones	1,751	4,536	359,700
Cerro Largo	Melo	5,270	13,648	78,000
Colonia	Colonia del Sacramento	2,358	6,106	112,100
Durazno	Durazno	4,495	11,643	54,700
Flores	Trinidad	1,986	5,144	24,400
Florida	Florida	4,022	10,417	65,400
Lavalleja	Minas	3,867	10,016	61,700
Maldonado	Maldonado	1,851	4,793	93,000
Montevideo	Montevideo	205	530	1,309,100
Paysandú	Paysandú	5,375	13,922	104,500
Río Negro	Fray Bentos	3,584	9,282	47,500
Rivera	Rivera	3,618	9,370	88,400
Rocha	Rocha	4,074	10,551	68,500
Salto	Salto	5,468	14,163	107,300
San José	San José de Mayo	1,927	4,992	91,900
Soriano	Mercedes	3,478	9,008	77,500
Tacuarembó	Tacuarembó	5,961	15,438	82,600
Treinta y Tres	Trienta y Tres	3,679	9,529	45,500
TOTAL LAND AREA		67,574	175,016	2,940,200
INLAND WATER		463	1,199	
TOTAL AREA		68,037	176,215	

Base 501488 7-73

the river a few mi downstream from the first. The two dams are sites of major hydroelectric power installations.

The largest of the eastern coastal lagoons is the Laguna Merín. Serving as a part of the northeastern frontier with Brazil and with its northern extremity located entirely in Brazil, it discharges its waters through that country's São Gonçalo Canal. Strung along the coast between the Brazilian border of Punta del Este are a half-dozen other lagoons with surface areas of 39 to 181 sq km (15 to 70 sq mi) as well as numerous smaller ones. Some are fresh water, while others have direct tidal connection with the Atlantic and as a consequence are brackish.

Some 1,246 km (775 mi) of the country's rivers are reported to be navigable. This figure, however, apparently includes 378 km (235 mi) along the Río de la Plata estuary. On the Río Uruguay oceangoing vessels of up to fourteen ft in draft can ascend 225 km (140 mi) to Paysandú, and vessels with drafts of up to nine ft can continue 96 km (60 mi) farther to Salto, where passage is blocked by falls. Much of the Río Negro is obstructed by sandbanks and shallows, but it is navigable by coastal vessels for about 72 km (45 mi) upstream from its mouth. Rivers navigable for some distance by shallow-draft small craft include the Yi and Quequay Grande in the west, the Cebollatí in the east, and the Santa Lucia in the south.

Geographical Regions

The land is homogeneous, and changes in topography, climate, and vegetation are limited in character and moderate in degree. The two principal regions of the country are the Agricultural Lowland and the Pastoral Plateau Interior.

The Agricultural Lowland starts in Maldonado Department and follows the littoral of the Río de la Plata westward for the full extent of the estuary in a band averaging about 80 km (50 mi) in depth. It is sometimes regarded as continuing northward along the Río Uruguay in a narrowing band as far as the river town of Salto. Rarely, the sand dunes and swamps of the Atlantic plain are also included. Soil, relief, and vegetation vary considerably within the Agricultural Lowland, which merges gradually with the rolling grasslands of the interior, but it is an area of varied and intensive agriculture in which most of the population lives. The Pastoral Plateau Interior makes up the remaining three-fourths or more of the country. In general, sheep ranches occupy the territory to the north of the Río Negro, and cattle ranches prevail to the river's south.

VANUATU

BASIC FACTS

Official Name: Republic of Vanuatu

Abbreviation: VAN

Area: 12,190 sq km (4,707 sq mi)

Area—World Rank: 141

Population: 154,691 (1988) 215,000 (2000)

Population—World Rank: 166

Capital: Port-Vila

Land Boundaries: Nil

Coastline: 2,528 km (1,570 mi)

Longest Distances: N/A

Highest Point: Mount Tabwemasana 1,879 m (6,165 ft)

Lowest Point: Sea Level

Land Use: 1% arable; 5% permanent crops; 2% meadows and pastures; 1% forest and woodland; 91% other

Vanuatu is a chain of 13 large and about 70 small islands stretching from south of the Solomon Islands to the east of New Caledonia with a combined land area of 14,763 sq km (5,700 sq mi). The Y-shaped archipelago whose open end is in the north has a total coastline of 2,528 km (1,570 mi). The 13 large islands are Vanua Lava, Banks, Espiritu Santo, Maew, Pentecost, Aoba, Malekula, Ambrim, Epi, Efate, Eromanga, Tana, and Aneityum. These are all high islands, quite mountainous and containing extensive rain forests. The other islands are of the coral atoll or almost atoll types. There are at least three active volcanoes: Tanna, Ambrym, and Lopevi.

Area and population

Local Government Regions	Capitals	area sq mi	area sq km	population 1987 estimate
Ambrym	Eas	257	666	8,100
Ambae/Maéwo	Longana	270	699	11,780
Banks/Torres	Sola	341	882	6,400
Éfaté	Vila	356	923	28,590
Épi	Ringdove	172	446	3,090
Malekula	Lakatoro	793	2,053	18,850
Paama	Liro	23	60	2,420
Pentecost	Loltong	193	499	11,780
Santo/Malo	Luganville	1,640	4,248	26,310
Shepherd	Morua	33	86	5,160
Taféa	Isangel	629	1,628	22,400
TOTAL		4,707	12,190	144,880

VATICAN CITY

BASIC FACTS

Official Name: State of the Vatican

Abbreviation: VAN

Area: 0.438 sq km (0.27 sq mi)

Area—World Rank: 216

Population: 752

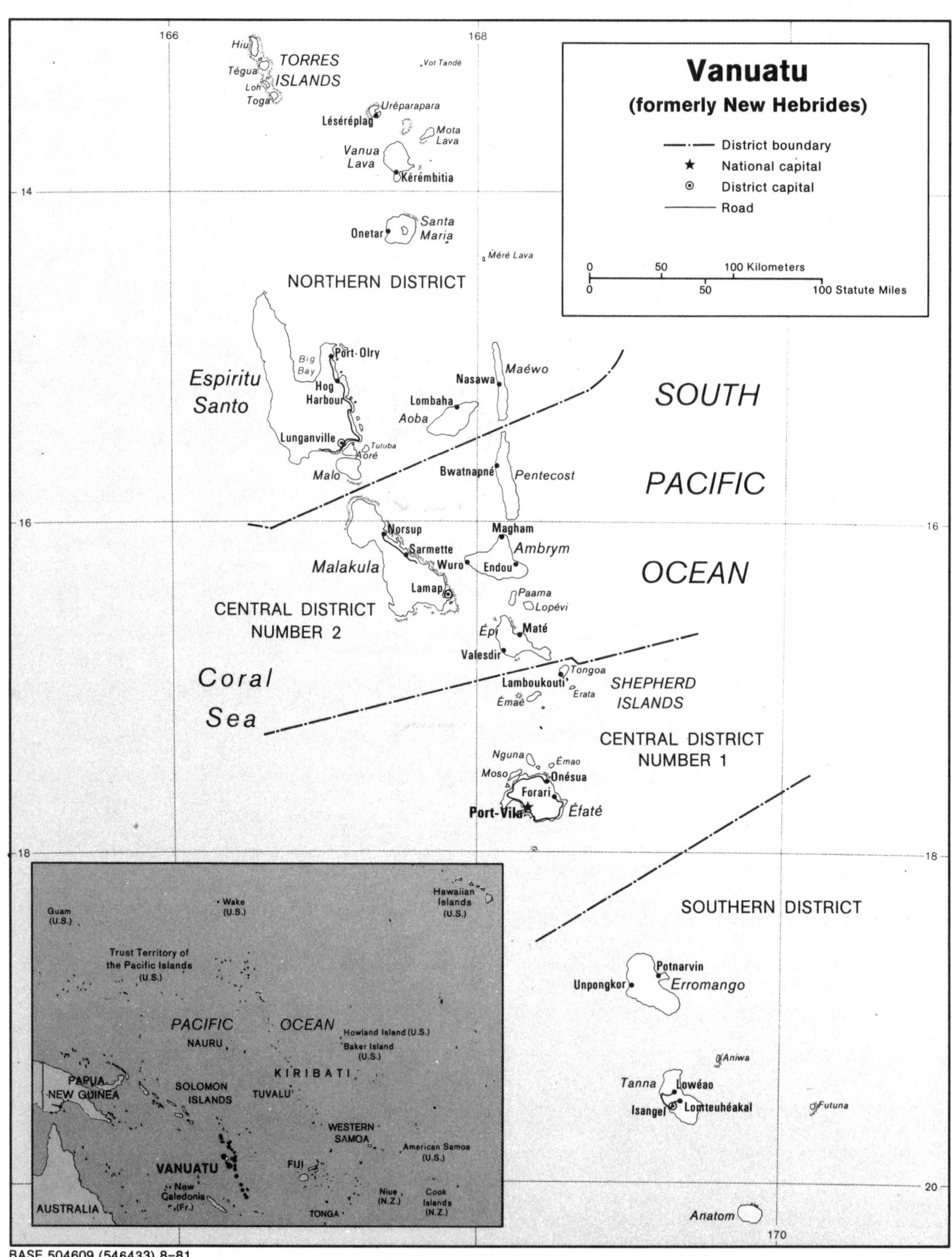

BASE 504609 (546433) 8-81

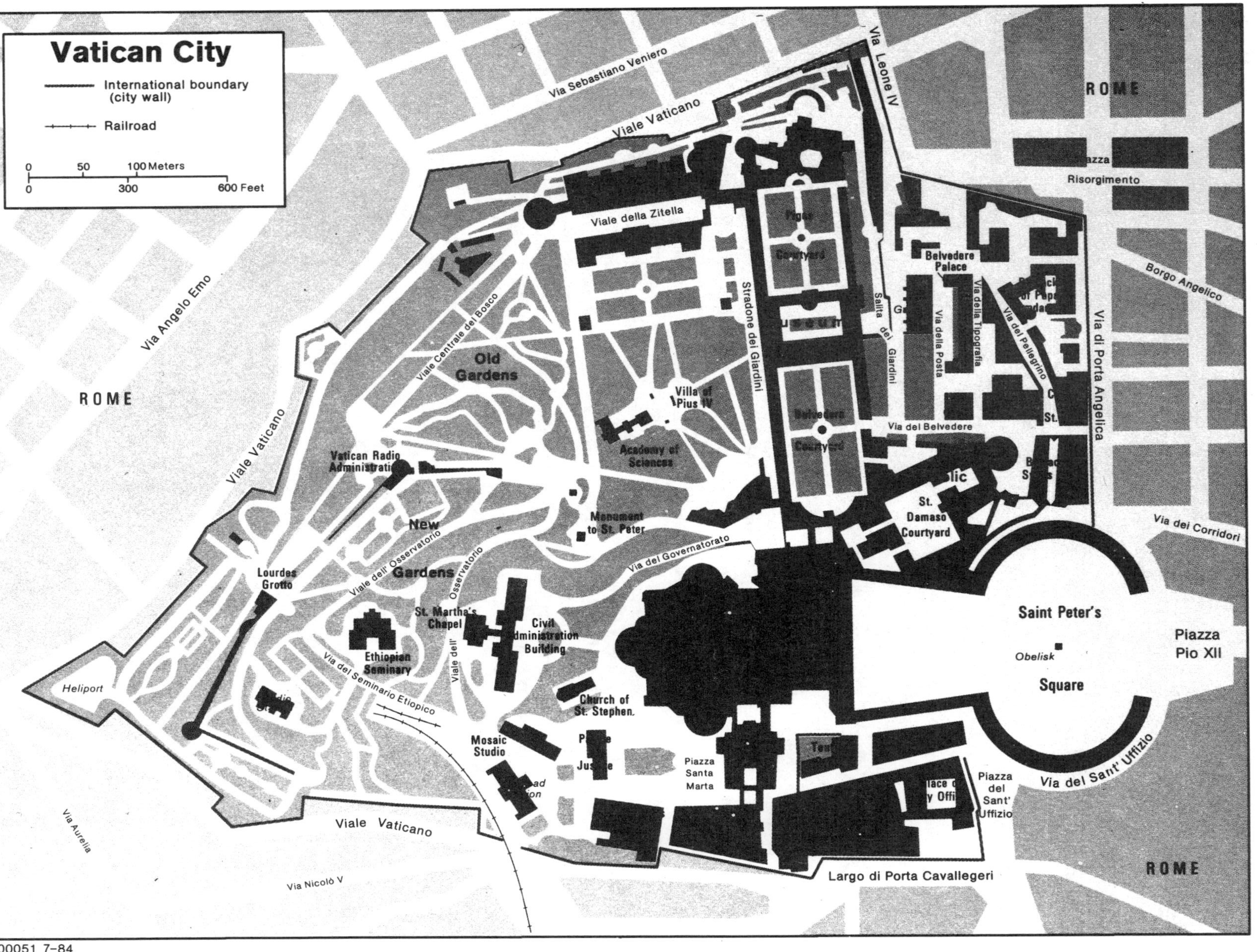

Vatican City
International boundary (city wall)
Railroad
0 50 100 Meters
0 300 600 Feet
ROME
Via Sebastiano Veniero
Viale Vaticano
Via Leone IV
Risorgimento
Viale della Zitella
Belvedere Palace
Borgo Angelico
Via della Tipografia
Via della Posta
Via del Pellegrino
Via di Porta Angelica
Stradone dei Giardini
Salita dei Giardini
Via Angelo Emo
Viale Centrale del Bosco
Old Gardens
Villa of Pius IV
Via del Belvedere
Academy of Sciences
Vatican Radio
Monument to St. Peter
St. Damaso Courtyard
Via dei Corridori
New Gardens
Viale dell' Osservatorio
Via del Governatorato
Lourdes Grotto
St. Martha's Chapel
Civil Administration Building
Saint Peter's Square
Obelisk
Piazza Pio XII
Ethiopian Seminary
Viale dell' Osservatorio
Heliport
Via del Seminario Etiopico
Church of St. Stephen.
Mosaic Studio
Piazza Santa Marta
Piazza del Sant' Uffizio
Via del Sant' Uffizio
Viale Vaticano
Via Aurelia
Via Nicolò V
Largo di Porta Cavallegeri

800051 7-84

Population—World Rank: 216

Capital: Vatican City

Land Boundaries: Italy 3.2 km (1.98 mi)

Coastline: Nil

Longest Distances: N/A

Highest Point: 75 m (245 ft)

Lowest Point: 19 m (62 ft)

Land Use: 100% other

Located in Rome, the Vatican City is the smallest state in the world lying on the west bank of the Tiber River and to the west of the Castel Sant'Angelo. On the west and south it is bounded by the Leonine Wall. Vatican City comprises the following buildings and landmarks: St. Peter's Square; St. Peter's Basilica, the largest Christian church in the world, a quadrangular area north of the square containing administrative buildings and the Belvedere Park; the pontifical palaces to the west of Belvedere Park; and the Vatican Gardens which occupy about half the area. Outside Vatican proper, the pontiff exercises extraterritoriality over a number of churches and palaces in Rome, notably the Lateran Basilica and Palace in the Piazza San Giovanni, the Palace of San Callisto at the foot of the Janiculum hill, and the basilicas of Santa Maria Maggiore and San Paolo fuori le Mura. Extraterritoriality outside the city of Rome extends to the papal villa and its environs (almost 40 ha; 100 ac) at Castel Gandolfo and to the area (about 420 ha; 1,037 ac) at Santa Maria de Galeria, some 19.3 km (12 mi) from Rome, the site of the Vatican radio station established in 1957. The principality has no rivers or hills.

VENEZUELA

BASIC FACTS

Official Name: Republic of Venezuela

Abbreviation: VZL

Area: 912,050 sq km (352,144 sq mi)

Area—World Rank: 32

Population: 18,775,780 (1988) 24,715,000 (2000)

Population—World Rank: 42

Capital: Caracas

Boundaries: 8,293 km (5,154 mi); Guyana 743 km (462 mi); Brazil 2,000 km (1,243 mi); Colombia 2,050 km (1,274 mi)

Coastline: 3,500 km (2,175 mi)

Longest Distances: 1,487 km (924 mi) WNW-ESE; 1,175 km (730 mi) NNE-SSW

Highest Point: Bolivar Peak 5,007 m (16,427 ft)

Lowest Point: Sea Level

Land Use: 3% arable; 1% permanent crops; 20% meadows and pastures; 39% forest and woodland; 37% other

Geographers customarily divide Venezuela into four continental regions, plus the numerous small islands near the Caribbean coast. There is no consensus with respect to the extent and boundaries of these regions, and various names have been assigned to them. According to one representative system of classification, however, they may be identified as the Maracaibo Lowlands, the Northern Mountains, the Orinoco Lowlands, and the Guayana Highlands.

The Maracaibo Lowlands comprise almost 51,800 sq km (20,000 sq mi) of level plains around Lake Maracaibo and the Gulf of Venezuela, extend south and west to the Venezuelan Andes, and include the Paraguaná Peninsula. An open oval in shape, the region is delimited on the west by the curve of the Sierra de Perija (also called Serrania de las Motilones) and on the south and east by the Cordillera de Mérida (also called Sierra de Mérida) and the Segovia Highlands. Some geographers regard the Maracaibo Lowlands as a part of a larger region that also includes the Orinoco Delta and a discontinuous narrow coastal plain fronting on the Caribbean. To this tropical region they give the name Coastal Lowlands. A majority, however, prefer to regard the Orinoco Delta as part of the Orinoco Lowlands and the northern coastal strip as part of the Northern Mountains.

Broken by several gaps, the Northern Mountains and their spur ranges extend close to the northern coastline from the Colombian border on the west to the Paria Peninsula on the east. At the border the Eastern Cordillera of the Andes divides into the Sierra de Perija, which extends northwestward along the border toward the Gulf of Venezuela, and the Cordillera de Mérida, which extends east of Lake Maracaibo. The Cordillera de Mérida chain broadens northward to form the Segovia Highlands, consisting of heavily dissected plateaus decreasing in altitude from 1,828 m (6,000 ft) at the latitude of the city of Barquisimeto at their southern extremity to 183 m (600 ft) in the north before descending to the coastline.

Before rising as the Cordillera de Mérida, the Colombian Andes fall off abruptly into the Táchira Gap in the extreme west of Venezuela. The Andean spine crosses this gap as a relatively low saddle, dividing rivers flowing northward to Lake Maracaibo from those flowing southward across the Orinoco Lowlands. This saddle, across which the Pan American Highway passes between San Cristóbal in Venezuela and Cúcuta in Colombia, has served historically as an important communication route between the two countries. There is some question as to the region to which the southern slopes and piedmonts of the Northern Mountains properly belong. Topographically they are parts of the mountains, and their economic ties are with mountain urban centers. They are parts of the Orinoco drainage basin, however, and administratively they belong to the Orinoco Lowlands states. They may, accordingly, be considered a transitional zone between the two geographical regions.

The fertile valley of Táchira, almost self-sufficient agriculturally, connects readily with Colombia and with the plains to the south but does not have easy access to the coast or to the cities of the Central Highlands of the Northern Mountains. This geographic isolation contributed to making it a traditional seat of strong regionalism. The Cordillera de Mérida, which varies in width between 13 and 64 km (8 and 40 mi), contains intermontane basins and valleys from which mountain slopes rise to extensive areas of high-plateau grassland and, finally, to the only permanently snowcapped peaks in the country. The highest, Pico Bolívar at over 24,998 m (16,400 ft), is located in Mérida State, which is called "The Roof of Venezuela." The piedmont band on the north side of this chain, like those of the other Northern Mountains chains, is narrow and more abrupt but much more accessible than the band flanking the Orinoco Lowlands.

Geographically this branch of the Andes terminates at the Yaracuy Gap at the eastern end of the Cordillera de Mérida. A coastal range then runs 483 km (300 mi) eastward to the Paria Peninsula. Topographically, however, some geographers refer to the entire mountain complex of northern Venezuela as the Venezuelan Andes.

The coastal range is actually composed of two parallel ranges separated by an intermediate lateral depression that reaches a maximum width of about 48 km (30 mi). This depression is the core area of the country; it has the densest population, the most intensive agriculture, and the best developed transportation network. The coastal range, where altitudes often exceed 1,524 m (5,000 ft) and peaks reach to from 2,133 to 2,743 m (7,000 to 9,000 ft) is flanked on the north by narrow coastal plains except at points where the mountain slopes descend directly to the Caribbean.

There are few easy passages through the two ranges. The principal pass through the coastal mountains has easy gradients and permits a highway to connect the city of Valencia with the Caribbean coast. A zigzagging highway from Caracas to its port of La Guaira was replaced in 1953 by a superhighway tunneling in part through the coastal range. To the south the historic pass of Carabobo gives Valencia direct access to the Orinoco Lowlands. A second pass cuts through the same range to the east of Lake Valencia.

One part of the coastal range terminates at Cape Codera on the Caribbean, but remnants of the parallel range continue an additional 80 km (50 mi) eastward to end near the Unare River. Farther eastward beyond an extensive lowland gap the Eastern Highlands (also called the Cumaná Highlands) rise in a broad block commencing between the old port cities of Barcelona and Cumaná and extending eastward to terminate in coastal marches near the Gulf of Paria. To the north, narrow ridges associated with this system extend laterally along the spines of the twin peninsulas of Araya and Paria. To the south, rounded hills with altitudes of less than 304 m (1,000 ft) extend into the Orinoco Lowlands. At the core of the Eastern Highlands peaks reach 2,438 m (8,000 ft), but most of the system is made up of relatively low, dissected uplands with a scattering of shallow erosion and rift valleys.

The Orinoco Lowlands are the great plains that extend from the Colombian border south of the Andes to the Atlantic and lie between the Northern Mountains and the Orinoco River. The region, covering 259,000 sq km (100,000 sq mi) is nearly 1,609 km (1,000 mi) long and between 130 and 647 km (50 and 250 mi) wide. It is commonly referred to as the llanos (plains), although at the Orinoco Delta there is a large jungle swamp area that is hot and humid.

The region tilts gradually southwestward from the Northern Mountains to the Orinoco and northeastward from the Colombian border to the Atlantic, but altitudes never exceed 965 m (600 ft), gradients are almost imperceptible, and rivers are sluggish and meandering. North of the Apure River, a major tributary of the Orinoco, rivers flowing out of the Northern Mountains have cut shallow valleys in the western llanos, leaving eroded mesalike ridges that give the land a gently rolling appearance. South of the Apure the terrain is flatter and altitudes lower. There are a few small forests, but most of the plains are natural pastureland. Open range is unbroken by fences, and cattle ranches reach up to 404,686 ha (1 million ac) in size.

The Guayana Highlands, rising almost immediately south of the Orinoco River are considered to be the oldest land areas of the country, and erosion over the centuries has caused unusual formations. Comprising about 57 percent of the national territory, the 517,988 sq km (200,000 sq mi) highlands consist principally of plateau areas scored by swiftly running tributaries of the Orinoco. The most conspicuous topographical feature of the region is the Gran Sabana, a deeply eroded high plateau some 36,260 sq km (14,000 sq mi) in extent that rises deep in the interior from lower highland areas in abrupt cliffs up to 762 m (2,500 ft) in height. From its rolling surface emerge massive perpendicular, flat-topped bluffs reaching considerable altitudes, called *tepuis* by the Venezuelans; the loftiest of these, Mount Roraima at the Brazil-Guyana tripoint, is over 2,743 m (9,000 ft) above sea level. The most famous *tepui* is the one that contains Angel Falls, the world's highest waterfall. Elsewhere the region consists of plateaus varying from 304 to 914 m (1,000 to 3,000 ft) in elevation; occasional peaks rise above 2,133 m (7,000 ft), except in the south of Amazonas Territory where swampy flatlands drain northward into the Orinoco

PRINCIPAL TOWNS

(metropolitan areas, estimated population at 30 June 1987)

Town	Population	Town	Population
Caracas (capital)	3,247,698	Mérida	250,442
Maracaibo	1,295,421	Ciudad Bolívar	250,069
Valencia	1,134,623	Cumaná	247,051
Maracay	857,982	Guarena-Guatire	219,312
Barquisimeto	718,197	Cabimas	213,733
Ciudad Guayana	466,418	Acarigua-Araure	205,065
Barcelona/Puerto La Cruz	417,501	La Victoria	183,271
San Cristóbal	338,188	Valera	182,238
Departamento Vargas	320,576	Barinas	166,069
Maturín	252,645	Punto Fijo	162,843
		Los Teques	159,866
		Lagunillas	130,245

near its headwaters. Relatively deep sinkholes are also found in the Guayana Highlands.

In addition to the mainland there are seventy-two islands of varied size and description. The most important by far is Margarita, the principal island in the small group that constitutes the state of Nueva Esparta (New Sparta), named for the bravery of its soldiers during the wars of independence (1810–21). Rocky, mountainous, and with little rainfall, it is nevertheless heavily populated (about 150,000 inhabitants), and its valleys are intensively farmed. Two other inhabited islands complete Nueva Esparta. The better known is Cubagua, once famous for its pearl fisheries and the site of the ruins of Nueva Cádiz, the first Spanish settlement in South America. The other islands vary from rocks, to coral reefs, to sandbars. The most distant, the minute island of Aves, situated 483 km (300 mi) miles north of Margarita, is near the Leeward Islands and was transferred from Spain in 1865.

Hydrography

Although there are more than 1,000 rivers in the country, the river systems are dominated by the Orinoco. Flowing more than 2,574 km (1,600 mi) to the Atlantic from its source in the Guayana Highlands at the Brazilian border, it is the world's eighth largest river and the largest in South America after the Amazon. Its flow varies substantially by season, the high water level in August exceeding by as much as forty feet the low levels of March and April. When the river is low, Atlantic tidal effects can reach Ciudad Bolívar, 418 km (260 mi) upstream.

Through most of the Orinoco's course the gradient is almost imperceptible. As it passes through the central part of Amazonas Territory, it divides its waters. Through the Casiquiare Channel it sends one-third of its volume through the Negro River to the Amazon along navigable waterways. Accordingly an inland waterway from the Amazon to the mouth of the Orinoco is navigable by light craft except where two falls separate the upper and lower reaches of the Orinoco near the town of Puerto Ayacucho.

The Orinoco, together with its 436 tributaries, provides drainage for about four-fifths of the country. It is five miles wide in some parts, and its delta has fifty mouths. It gathers the interior runoff from the Northern Mountains, most of the water from the Guayana Highlands, and the seasonal waters of the llanos. In addition it provides drainage for the eastern slopes of the Colombian Andes through the Meta and other Colombian rivers. The runoff from the Northern Mountains is carried to the Orinoco by sluggish, meandering rivers that are subject to extensive seasonal flooding over the alluvial flats of the northern llanos. As a result semiaquatic conditions prevail during the height of the rainy season. With the conclusion of the rains the waters recede, leaving stagnant pools that evaporate under the hot sun, and the land becomes parched.

Most of the rivers rising in the Northern Mountains flow southeastward to the Apure River, a tributary of the Orinoco. From its headwaters in the Cordillera de Mérida, the Apure crosses the llanos in a generally eastward direction. Few rivers flow into it from the poorly drained region to the south, and much of the southeast extremity is swamp forest.

Area and population

States	Capitals	area sq mi	area sq km	population 1987 estimate
Anzoátegui	Barcelona	16,700	43,300	820,274
Apure	San Fernando de Apure	29,500	76,500	239,296
Aragua	Maracay	2,700	7,014	1,196,817
Barinas	Barinas	13,600	35,200	428,453
Bolívar	Cuidad Bolívar	91,900	238,000	895,607
Carabobo	Valencia	1,795	4,650	1,443,464
Cojedes	San Carlos	5,700	14,800	178,255
Falcón	Coro	9,600	24,800	599,017
Guárico	San Jan de Los Morros	25,091	64,986	457,133
Lara	Barquisimeto	7,600	19,800	1,155,411
Mérida	Mérida	4,400	11,300	580,277
Miranda	Los Teques	3,070	7,950	1,837,762
Monagas	Maturín	11,200	28,900	475,487
Nueva Esparta	La Asunción	440	1,150	254,643
Portuguesa	Guanare	5,900	15,200	553,898
Sucre	Cumaná	4,600	11,800	705,515
Táchira	San Cristóbal	4,300	11,100	800,879
Trujillo	Trujillo	2,900	7,400	526,183
Yaracuy	San Felipe	2,700	7,100	356,355
Zulia	Maracaibo	24,400	63,100	2,071,058
Other federal entities				
Amazonas	Puerto Ayacucho	67,900	175,750	76,889
Delta Amacuro	Tucupita	15,500	40,200	89,458
Dependencias Federales	—	50	120	. . .
Distrito Federal	Caracas	745	1,930	2,530,026
TOTAL		352,144	912,050	18,272,157

The other major Venezuelan river is the fast-flowing Caroní, which originates in the Gran Sabana and flows northward to join the Orinoco at Cuidad Guayana after gathering the waters of the other major Gran Sabana streams. It is believed to have one of the highest hydroelectric power potentials of any river in Latin America, and it provides half of Venezuela's hydroelectric power. Electricity generated by the Caroní River is transforming the Guayana Highlands region into an industrial area.

Most of the rivers flowing from the Northern Mountains into Lake Maracaibo and the Caribbean are short and rapid, with deeply etched valleys. An important exception is the Tuy River, which passes slowly through the intermontane depression of the Central Highlands to drain the country's most prosperous agricultural lands. A minor drainage system is provided by the Unare River, which flows into the Caribbean through the gap between the Central and Eastern Highlands. It drains the lowlands of the Unare Gap and a portion of the Central Highlands.

Because of the country's near-equatorial location and the absence of mountains with elevations sufficient to permit much snow accumulation or retention, the rivers are almost wholly dependent on rainfall for their flow. The seasonal character of the rainfall in the Northern Mountains and the consequent irregularity of the flow of rivers southward out of them have discouraged agricultural development of the piedmont and the northern fringes of the llanos. The taming of these rivers by construction of dams to permit irrigated farming during the dry season—a program started during the 1950s—has begun to show an important effect on the country's economy.

Venezuela

502489 1-76 (541398)
Lambert Conformal Projection
Standard parallels 10°00 and 2°40
Scale 1:8,000,000
Boundary representation is not necessarily authoritative

Lake Maracaibo, covering about 16,316 sq km (6,300 sq mi) is the largest body of water in Latin America and provides convenient transportation of coffee and other agricultural products from the surrounding plains and the Andean mountain slopes. In the north it is directly connected with the Gulf of Venezuela by an island-dotted narrows some 40 km (25 mi) miles in length. Although the lake has an average depth of thirty feet and is navigable to its southern end, the exit from the narrows into the Gulf of Venezuela was originally blocked by a sandbar only seven feet beneath the surface at low tide. Because of the importance of Maracaibo petroleum production, the government has dredged a channel to permit passage of tankers through the narrows to the loading port of Maracaibo. The connection of the lake with the sea, however, makes its waters brackish in parts and unfit for either drinking or irrigation.

Second in importance among the hundreds of lakes is Lake Valencia, a body of water 29 km (18 mi) in length and 16 km (10 mi) in width located about 96 km (60 mi) southwest of Caracas in the heart of the country's best agricultural lands. Originally the lake drained southward toward the Orinoco, but excessive forest clearing on surrounding mountain slopes and overplanting of adjacent level ground during the colonial period caused rainwater to run off in spates, whereas the waters of the lake had previously been replenished by gradual underground seepage. As a consequence by 1800 its waters had subsided to a point where it was left without a surface outlet. The wars of independence caused forestry and crop planting to be neglected and permitted second-growth forest to appear. A quarter of a century later the lake had risen and once more had found an outlet. In the twentieth century, however, forest clearing and intensive agriculture have been resumed, and the drilling of numerous artesian wells in the vicinity has reduced the lake's natural water supply. During the late 1960s it was again without a surface outlet.

Two artificial bodies of water are the Guárico Reservoir, formed in 1956 by containment of the Guárico River, and the reservoir being formed behind the lower Orinoco dam complex. In addition there are numerous tarns in the Cordillera de Mérida, lagoons scattered among coastal lowlands, and lagoons and a sizable lake in the swampy land east of Lake Maracaibo.

VIETNAM

BASIC FACTS

Official Name: Socialist Republic of Vietnam

Abbreviation: SRV

Area: 331,653 sq km (128,052 sq mi)

Area—World Rank: 60

Population: 65,185,278 (1988) 82,310,000 (2000)

Population—World Rank: 12

Capital: Hanoi

Boundaries: 6,127 km (3,807 mi); China 1,281 km (796 mi); Cambodia 982 km (610 mi); Laos 1,555 km (966 mi)

Coastline: 2,309 km (1,435 mi)

Longest Distances: 1,650 km (1,025 mi) N-S; 600 km (373 mi) E-W

Highest Point: Fan Si Pan 3,143 m (10,312 ft)

Lowest Point: Sea Level

Land Use: 22% arable; 2% permanent crops; 1% meadows and pastures; 40% forest and woodland; 35% other

Vietnam has four major topographical divisions: the Red River Coastal Plain, the mountain system known as the Chaine Annamitique, the Mekong Delta, and the Central Lowlands. The dominant feature of the coastal plain is the Red River (or Tonkin) Delta. It includes, however, the strip of coastal lowlands extending south from the Delta. The Delta itself, drained by a vast network of canals and waterways, is a rich rice-growing area. Far more extensive but less productive are the forested highlands, projecting southward from the Yunnan plateau.

The Mekong Delta occupies the southern two-fifths of the country, and its fertile alluvial plains, favored by heavy rainfall, make it one of the great ricegrowing areas of the world. The Chaîne Annamitique, with several high plateaus, dominates the area northward from the Mekong Delta. The Central Lowlands consist of a fertile, narrow, coastal strip along the eastern slopes of the Chaîne Annamitique.

The Coastal Plain

The Red River Delta is a flat, triangular region, smaller but more intensively developed than is the Mekong River Delta. It stretches some 241 km (150 mi) inland and about 120 km (75 mi) along the coast, south of Haiphong. Almost entirely built up of alluvium, the Delta was formerly an extension of the Gulf of Tonkin, which has since been filled by the deposits of the rivers which run into the basin. The coast is indented by the mouths of the Red River and lesser streams to the south.

The Red River, rising in Yunnan Province in China, has a total length of about 1,167 km (725 mi). Its two major tributaries, the Song Lo (sometimes called the Lo River, sometimes Rivière Claire, and sometimes Clear River) and the Black River (Song Da in Vietnamese), give it a large flow of water—during the rainy season as much as 22,653 cu m (800,000 cubic feet) per second, or twice as much as the maximum flow of the Nile River in Egypt.

The entire delta region, backed by the steep rises of the forested highlands, has only minor variations in relief, chiefly in the west. Most of it is no more than 3 m (10 ft) above sea level, and much of it is 1 m (3 ft) or less. The area is subject to frequent flooding.

The Highlands

The highlands and plateaus enclosing the Red River Delta contrast sharply with the plains region both in topography and in level of development. The sparsely populated forested highlands are inhabited mainly by Mongoloid tribal minority groups. Typically, these peoples live in small villages set in forested clearings. Settlements are found both in the valley bottoms and at higher levels, in areas where the production of rice and other crops supports relative large villages in some locations.

One branch of the mountains and highlands, projecting southward from the plateau in Yunnan Province, extends along the country's entire border with Laos and, except at the northeastern tip of Laos, separates the Red River basin from that of the Mekong River. Elevations along this branch range from 914 to 3,048 m (3,000 to 10,000 ft), with Vietnam's highest peak, Fan Si Pan, rising to 3,149 m (10,308 ft), in the extreme northwest and about 32 km (20 mi) southwest of Lao Cai. The southern portion of the branch, known as the Chaîne Annamitique, continues southward along Vietnam's boundary with Laos and Cambodia. Another branch, unnamed but sometimes referred to as the Northern Highlands, extends along the border with China, terminating in a series of islands northeast of Haiphong in the Gulf of Tonkin.

Mekong Delta

The 4,506 km (2,800 mi) long Mekong is one of the 12 great rivers of the world and, together with the Amur, the Hwang Ho (Yellow), and the Chang Chiang (Yangtze), provides the Pacific drainage of the Asian Continent. From its source in the high plateau of Tibet, not far from the headwaters of the Yangtze of China and the Salween of Burma, it flows through Tibet and China to the northern border of Laos. There it separates Burma from Laos and, farther downstream, Laos from Thailand. Flowing through Cambodia, it bifurcates at the capital, Phnom Penh, at the apex of the delta, the broad base of which is in Vietnam to the south and east on the South China Sea. The southern branch, the Song Hau Giang flows directly to the sea; the larger northern branch splits into four parts about 80 km (50 mi) before reaching the sea.

The heavily silted Mekong is navigable by seagoing craft of shallow draft only. These can proceed as far as the rapids at Kompong Cham, about 112 km (70 mi) above Phnom Penh. Thereafter, frequent portages are required, and the river journey to Luang Prabang in central Laos takes about 37 days in the dry season and 27 days in the wet season.

A tributary, which enters the Mekong at Phnom Penh, drains the Tonle Sap. This large fresh-water lake, which was once an arm of the sea, covers approximately 2,589 sq km (1,000 sq mi) in central Cambodia and serves as a regulating reservoir to stabilize the flow of water through the lower Mekong. When the river is in flood, its silted delta outlets are unable to carry off the floodwaters, and they back up into the Tonle Sap, expanding the lake to cover as much as 4 times its low-water area. As the river flood subsides, the water reverses its course and flows from the lake to the sea. The effect is to reduce significantly the danger of serious floods in the Mekong Delta. The first major flood in 30 years occurred in October 1961, when the volume of water was too great for the Tonle Sap to hold. The disaster was worsened by being unexpected.

The delta proper, approximately 67,340 sq km (26,000 sq mi) in area, was built up by the five branches of the Mekong, which total about 483 km (300 mi) in length, and the system of three smaller rivers—the Song Vam Co, the Song Sai Gon and the Song Dong Nai. The low, level plain, nowhere more than 3 m (10 ft) above sea level, is very fertile, and cultivated land extends to the immediate shoreline in the vicinity of the river mouths. So much sediment is brought to the sea that the coastline to the south is advancing as much as 76 m (250 ft) per year. More than 23,309 sq km (9,000 sq mi) of delta land are under rice cultivation. Drainage is effected chiefly by tidal action which differs greatly from place to place. The difference between high and low tides is about 1 m (3 ft) on the Song Hau Giang and double that along the northern outlets of the Mekong and the Song Sai Gon. The southernmost tip of the delta, known as the Mui Bai Bung (Ca Mau Peninsula), is covered with dense jungles and the shoreline by mangrove swamps.

The people of the delta region live in villages made up largely of bamboo houses with thatched roofs which are built directly on the ground. Village streets usually follow a regular pattern, and the entire community ordinarily is enclosed by a bamboo fence. Levees and dykes built for flood control are used extensively as village sites, and are often strung out along riverbanks and roads. During the flood period, the only dry land is that forming the banks of the canals and rivers. On such pieces of land, often isolated by the floodwaters from other communities and neighbors, stand the farmhouses with their small garden plots for growing vegetables and a few fruit trees.

Chaîne Annamitique

The Chaîne Annamitique is the southernmost spur, over 1,207 km (750 mi) in length, of the rugged mountains which originate in Tibet and China. It extends southeastward, forming the border between Vietnam and Laos and, later, between Vietnam and Cambodia, until it reaches the Mekong Delta where it terminates about 80 km (50 mi) north of Ho Chi Minh City. The Chaîne is irregular in height and form and gives off numerous spurs which divide the coastal strip into a series of compartments and render north-south communication difficult. The southern reaches of these mountains are quite extensive and form an effective natural barrier for the containment of the people who live in the Mekong basin.

The northern portion of the Chaîne is narrow and very rugged; within the southern portion is formed a plateau area, known as the Central Highlands, which is about 161 km (100 mi) wide and 322 km (200 mi) in length. The peaks of the Chaîne Annamitique range in height from about 1,524 m (5,000 ft) to the 2,597 (8,521 ft) height of Ngoc Ang, which is about 120 km (75 mi) inland from Mui Batangan.

Base 800470 (546744) 12-85

In the mountain and plateau areas the settlement patterns are varied. Most typical among the *Montagnards* are the simple bamboo structures, built on pilings and having thatched roofs. Some tribes often construct extended family long houses, with a long communal house located centrally in the village. Along the hillsides, villages are usually made up of clusters of dwellings, fairly close together.

The Central Highlands, an area covering approximately 51,800 sq km (20,000 sq mi), consists of two distinct parts. The northern part, called Cao Nguyen Dac Lac, extends some 281 km (175 mi) north from the vicinity of Ban Me Thuot to the Ngoc Ang peak. Irregular in shape, it varies in elevation from about 182 to 487 m (600 to 1,600 ft), with a few peaks rising much higher. This area of approximately 13,985 sq km (5,400 sq mi) is covered mainly with bamboo and tropical broadleaf forests interspersed with farms and rubber plantations. The southern portion of the Central Highlands, much of it over 914 m (3,000 ft) above sea level, includes about 10,359 sq km (4,000 sq mi) of usable land. Da Lat, a modern hill city in the center of the area, is overlooked by Monts Lang Bian, with an elevation of 2,249 m (7,380 ft). The forest growth is predominantly of broadleaf evergreens at higher elevations and bamboo on the lower slopes. Coffee, tea, tobacco and temperate-climate vegetables flourish in the fertile soil.

The sparsely settled plateaus of the Central Highlands, with their extensive forests and rich soil, are particularly important to Vietnam for potential expansion room from the densely populated lowlands.

Central Lowlands

The Central Lowlands extend along the sea from the Mekong Delta northward to the Demarcation Line. On the landward side, the Chaîne Annamitique rises precipitously above the Lowlands and, in some areas, is nearly 64 km (40 mi) inland; elsewhere it veers shoreward and at several points crowds into the sea. In general, the land is fertile and is extensively cultivated. The chief crop is rice, and considerable sugarcane is also grown. Fishing is good along the entire coast and is important both as an industry and for local subsistence in the southern section near the Mekong Delta.

From the Mekong Delta an infertile coastal strip, generally narrow and covered with shifting sand dunes, extends northeastward some 160 km (100 mi) to Mui Dinh. This region has less rainfall than any other part of Vietnam.

From Mui Dinh northward the coastal plain remains narrow for about 160 km (100 mi) to Mui Dieu where a mountain spur presses against the shore. In this section there are occasional stretches of quite fertile land where rice is grown.

From Mui Dieu to Vung Da Nang, about 402 km (250 mi) north, lie the most extensive and fertile plains of the Central Lowlands coast where two rice crops a year are grown. From Vung Da Nang to Hue, about 80 km (50 mi) farther north, mountain spurs jut into the sea at several places. From Hue to the 17th parallel—80 km (50 mi) beyond—much of the shore is fringed by a narrow line of sand dunes backed by an intensively cultivated flat fertile area.

In the Central Lowlands two distinct types of village predominate. The fishing village, strung out along the coastal plain, usually consists of a close-knit group of dwellings located in a sheltered cove or bay. In the second type of village, fishing is not the major economic activity

PRINCIPAL TOWNS (estimated population in 1973)

Hanoi (capital)	2,674,400*	Qui Nhon	213,757
Ho Chi Minh City (formerly Saigon)	3,419,978†	Hué	209,043
		Can Tho	182,424
Haiphong	1,279,067†	Mytho	119,892
Da Nang	492,194	Cam Ranh	118,111
Nha Trang	216,227	Vungtau	108,436
		Dalat	105,072

*At 31 December 1983

†population census, 1 October 1979.

Area and population

Provinces	Capitals	area sq mi	area sq km	population 1979 census
An Giang	Long Xuyen	1,349	3,493	1,532,362
Bac Thai	Thai Nguyen	2,521	6,530	815,105
Ben Tre	Ben Tre	859	2,225	1,041,838
Binh Tri Thien	Hue	7,081	18,340	1,901,713
Cao Bang	Cao Bang	3,261	8,445	479,823
Cuu Long	Vihn Long	1,488	3,854	1,504,215
Dac Lac	Buon Me Thoat	7,645	19,800	490,198
Dong Nai	Bien Hoa	2,926	7,578	1,304,799
Dong Thap	Cao Lamh	1,309	3,391	1,182,787
Gia Lai-Cong Tum	Cong Tum	9,860	25,536	595,906
Ha Bac	Bac Giang	1,780	4,609	1,662,671
Ha Nam Ninh	Nam Dinh	1,453	3,763	2,781,409
Ha Son Binh	Hanoi	2,308	5,978	1,537,190
Ha Tuyen	Ha Giang	5,219	13,518	782,453
Hai Hung	Hai Duong	986	2,555	2,145,662
Hau Giang	Can Tho	2,365	6,126	2,232,891
Hoang Lien Son	Lao Cai	5,734	14,852	778,217
Kien Giang	Rach Gia	2,455	6,358	994,673
Lai Chau	Lai Chau	6,586	17,068	322,077
Lam Dong	Da lat	3,835	9,933	396,657
Lang Son	Lang Son	3,161	8,187	484,657
Long An	Tan An	1,681	4,355	957,264
Minh Hai	Bac Lieu	2,972	7,697	1,219,595
Nghe Tinh	Vinh	8,688	22,502	3,111,989
Nghia Binh	Qui Nhon	4,595	11,900	2,095,354
Phu Khanh	Nha Trang	3,785	9,804	1,188,637
Quang Nam-Da Nang	Da Nang	4,629	11,989	1,529,520
Quang Ninh	Hai Duong	2,293	5,938	750,055
Son La	Son La	5,586	14,468	487,793
Song Be	Thu Dau Mo	3,807	9,859	659,093
Tay Ninh	Ho Chi Minh City	1,556	4,030	684,006
Thai Binh	Thai Binh	577	1,495	1,506,235
Thanh Hoa	Thanh Hoa	4,300	11,138	2,532,261
Thuan Hai	Phan Thiet	4,392	11,374	938,255
Tien Giang	My Tho	918	2,377	1,264,498
Vinh Phu	Viet Tri	1,786	4,626	1,488,348
Municipalities				
Haiphong	—	585	1,515	1,279,067
Hanoi	—	826	2,139	2,570,905
Ho Chi Minh City	—	787	2,029	3,419,978
Special Zone				
Vung Tau-Con Dao	—	108	279	91,610
TOTAL		128,052	331,653	52,741,766

and its formation follows a pattern similar to that of the delta with houses more dispersed over a broader area.

VIRGIN ISLANDS (BRITISH)

BASIC FACTS

Official Name: Virgin Islands (British)

Abbreviation: BVI

Area: 150 sq km (58 sq mi)

Area—World Rank: 199

Population: 12,075 (1988) 15,000 (2000)

Population—World Rank: 201

Capital: Road Town

Land Boundaries: Nil

Coastline: 80 km (50 mi)

Longest Distances: N/A

Highest Point: Mount Sage 527 m (1,730 ft)

Lowest Point: Sea Level

Land Use: 20% arable; 7% permanent crops; 33% meadows and pastures; 7% forest and woodland; 33% other

The British Virgin Islands consist of more than 40 mountainous islands, of which only 15 are inhabited, lying at the northern end of the Leeward Islands, about 100 km (60 mi) to the east of Puerto Rico. The coral islands are relatively flat. The volcanic islands are steep and hilly.

WALLIS AND FUTUNA

BASIC FACTS

Official Name: Territory of the Wallis and Futuna Islands

Abbreviation: WFI

Area: 274 sq km (106 sq mi)

Area—World Rank: 189

Population: 14,254 (1988) 26,000 (2000)

Population—World Rank: 199

Capital: Mata-Utu

Land Boundaries: Nil

Coastline: 129 km (80 mi)

Longest Distance: N/A

Highest Point: Mount Singavi 762 m (2,500 ft)

Lowest Point: Sea level

Land Use: 5% arable; 20% permanent crops; 75% other

Wallis and Futuna Islands comprise two groups of islands: the Wallis (or Uvea) Island and 22 islands on the surrounding reef and, to the southeast, Futuna (or Hooru) comprising the two islands of Futuna and Alofi. They are located northeast of Fiji and west of Western Samoa. The islands are mostly volcanic.

WESTERN SAMOA

BASIC FACTS

Official Name: Independent State of Western Samoa

Abbreviation: WSM

Area: 2,831 sq km (1,093 sq mi)

Area—World Rank: 154

Population: 178,045 (1988) 173,000 (2000)

Population—World Rank: 164

Capital: Apia

Land Boundaries: Nil

Coastline: 371 km (231 mi)

Longest Distances: 153 km (93 mi) ESE-WNW; 39 km (24 mi) NNE-SSW

Highest Point: Mauga Silisili 1,858 m (6,096 ft)

Virgin Islands

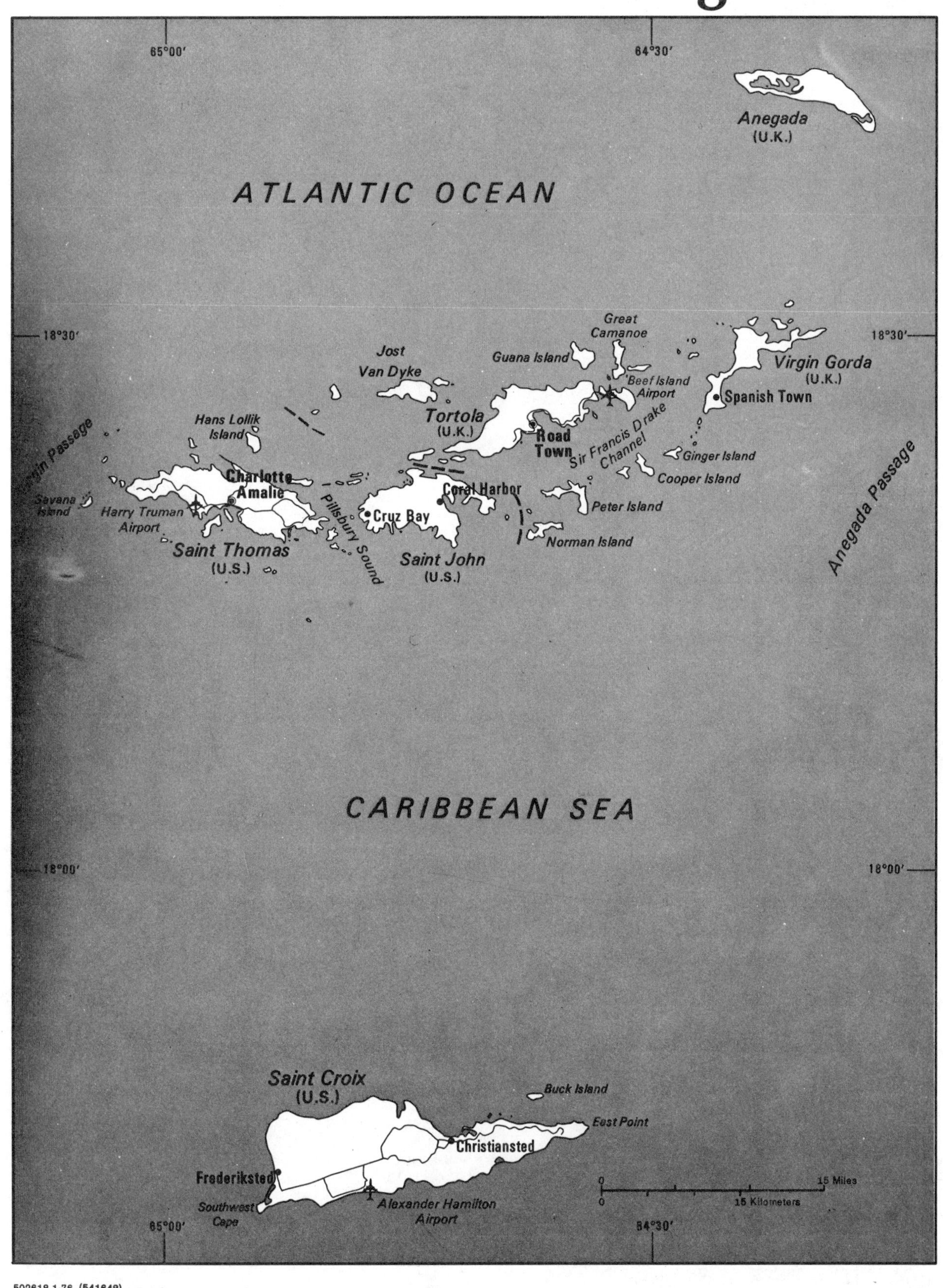

502618 1-76 (541648)
Lambert Conformal Projection
Standard parallels 17°20′ and 22°40′
Scale 1:600,000

Road

Airport

Line of Separation
(not a formal territorial boundary or territorial limit)

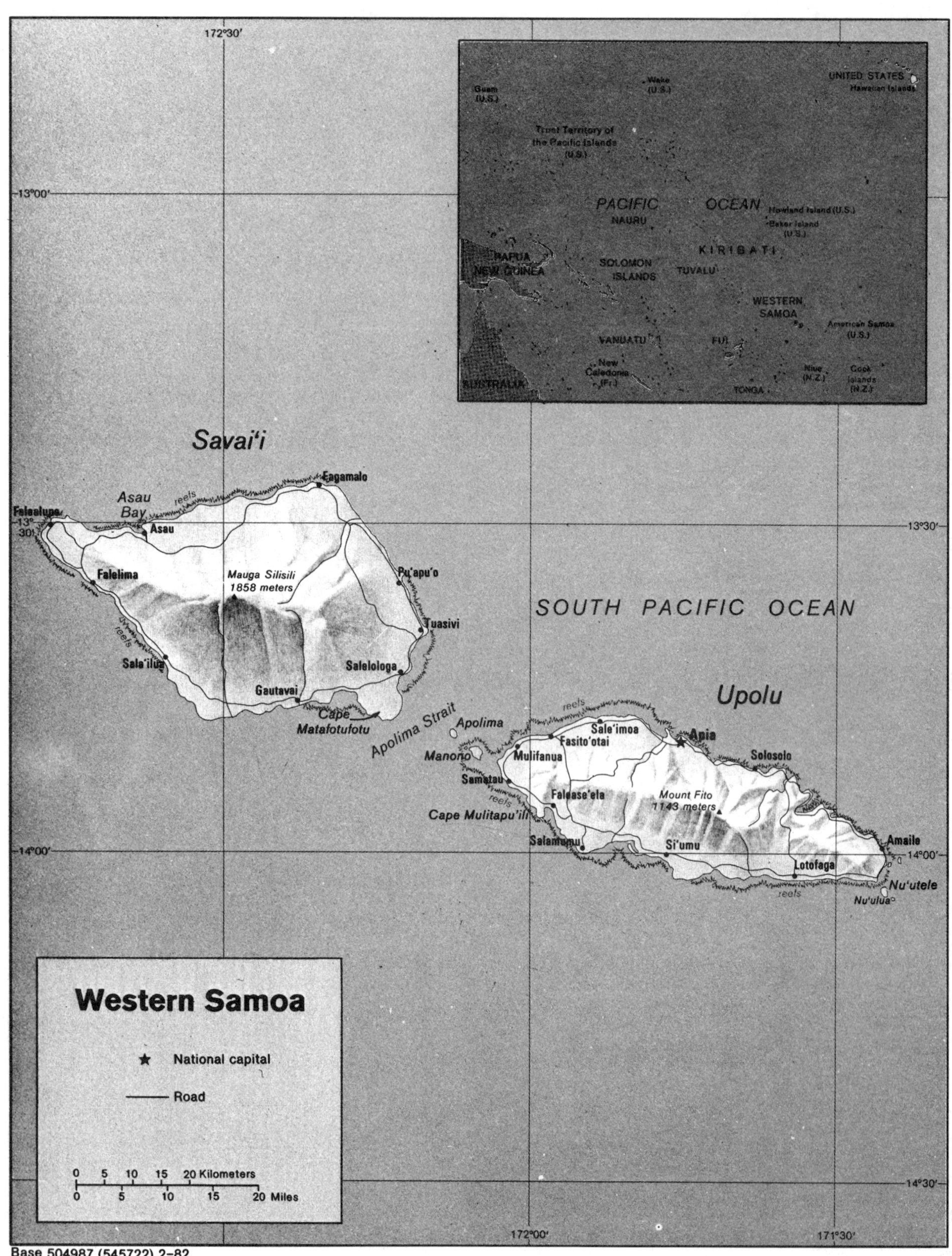

Base 504987 (545722) 2-82

Lowest Point: Sea level

Land Use: 19% arable; 24% permanent crops; 47% forest and woodland; 10% other

Western Samoa is located in the Pacific Ocean, 2,574 km (1,600 mi) NE of Auckland, New Zealand. It consists of two large islands, Upolu and Savaii, and the smaller islands of Apolima, Fanuatapu, Manono, Namua, Nuusafee and Nuutele. Of these, only Upolu and Savaii, separated by 18 km (11 mi), are inhabited.

The islands are of volcanic origin, and the coasts are surrounded by coral reefs. There are many dormant volcanoes; the most recent period of volcanic activity was between 1905 and 1911. Rugged mountain ranges form the core of both Savaii (1,709 sq km, 660 sq mi) and Upolu (1,113 sq km, 430 sq mi). The highest elevations are 1,857 meters (6,094 ft) on Savaii and 1,099 meters (3,608 ft) on Upolu. Because of large areas laid waste by lava flows on Savaii, the island supports less population than the smaller Upolu.

Area and Population	area		population
Islands Political Districts	sq mi	sq km	1981 census
Savaii	659	1,707	43,150
Fa'aseleleaga			11,876
Gaga'emauga			3,893
Gaga'ifomauga			5,304
Lealataua			1,934
Palauli			9,234
Satupa'itea			5,391
Vaisigano			5,518
Upolu	432	1,119	113,199
A'ana			13,149
A'ana-i-Sisifo			3,363
Aiga-i-le-Tai			3,960
Aleipata			4,236
Anoama'a			7,816
Fagaloa			1,519
Falealili			4,727
Faleata			16,821
Gaga'emauga			2,750
Lefaga			3,776
Lepa and Lotofaga			3,058
Safata			6,711
Sagaga			12,253
Vaimauga			29,060
TOTAL	1,093	2,831	156,349

YEMEN (ADEN)

BASIC FACTS

Official Name: People's Democratic Republic of Yemen

Abbreviation: PDRY

Area: 336,869 sq km (130,066 sq mi)

Area—World Rank: 59

Population: 2,452,620 (1988) 3,191,000 (2000)

Population—World Rank: 118

Capital: Aden

Boundaries: 2,909 km (1,808 mi); Saudi Arabia 830 km (516 mi); Oman 288 km (179 mi); Yemen Arab Republic 581 km (361 mi)

Coastline: 1,210 km (752 mi)

Longest Distances: 1,127 km (700 mi) NE-SW; 412 km (256 mi) SE-NW

Highest Point: 2,516 m (8,255 ft)

Lowest Point: Sea level

Land Use: 1% arable; 27% meadows and pastures; 7% forest and woodland; 65% other

Yemen (Aden) is located at the southwestern edge of the Arabian Peninsula. In addition it possesses the islands of Perim and Kamaran in the Bab Al Mandab and the 3,626-sq-km (1,400-square-mile) island of Socotra in the Gulf of Aden. The political status of Perim was decided by a plebiscite. Socotra had been ruled by the sultan of Qishn, and at independence possession of Socotra passed to the new national government.

Physically Yemen (Aden) encompasses the irregular southern end of the Arabian plateau, which is formed by ancient granites and partly covered by sedimentary limestones and sands. The coastal area is flat and sandy and varies from five to ten miles in width. Farther inland the country is mountainous; the hills reach some 2,438 m (8,000 feet) in the extreme west and gradually taper off to the east. The tableland is interspersed with deep valleys and wadis, or riverbeds, which are usually quite dry without any vegetation. The topography remains largely the same until it reaches the sands of the Rub al Khali, or Empty Quarter, in Saudi Arabia.

A distinct topographic feature in the eastern part of Yemen (Aden) is the Wadi Hadramaut (which means "death is present"), a narrow valley located in the central part of the country and extending eastward to the Gulf of Aden from the area north of the desert of Ramlat as Sabatayn. The wadi runs parallel to the coast some 160 to 193 km (100 to 120 mi) inland. Surrounded by desolate hills and desert, the broad upper and middle parts of the Hadramaut, with their alluvial soil and floodwaters, are relatively fertile and are inhabited by a farming population; the lower eastern part of the valley, which turns southward to the sea, is barren and largely uninhabited.

Area and population		area		population
Governorates	Capitals	sq mi	sq km	1984 estimate
Abyān	Zinjibār	8,297	21,489	412,574
ʿAdan	Aden	2,695	6,980	386,364
Hadramawt	al-Mukallā	59,991	155,376	651,469
Lahij	Lahij	4,928	12,766	362,809
al-Mahrah	al-Ghaydah	25,618	66,350	80,722
Shabwah	ʿAtāq	28,536	73,908	214,767
TOTAL		130,066	336,869	2,108,705

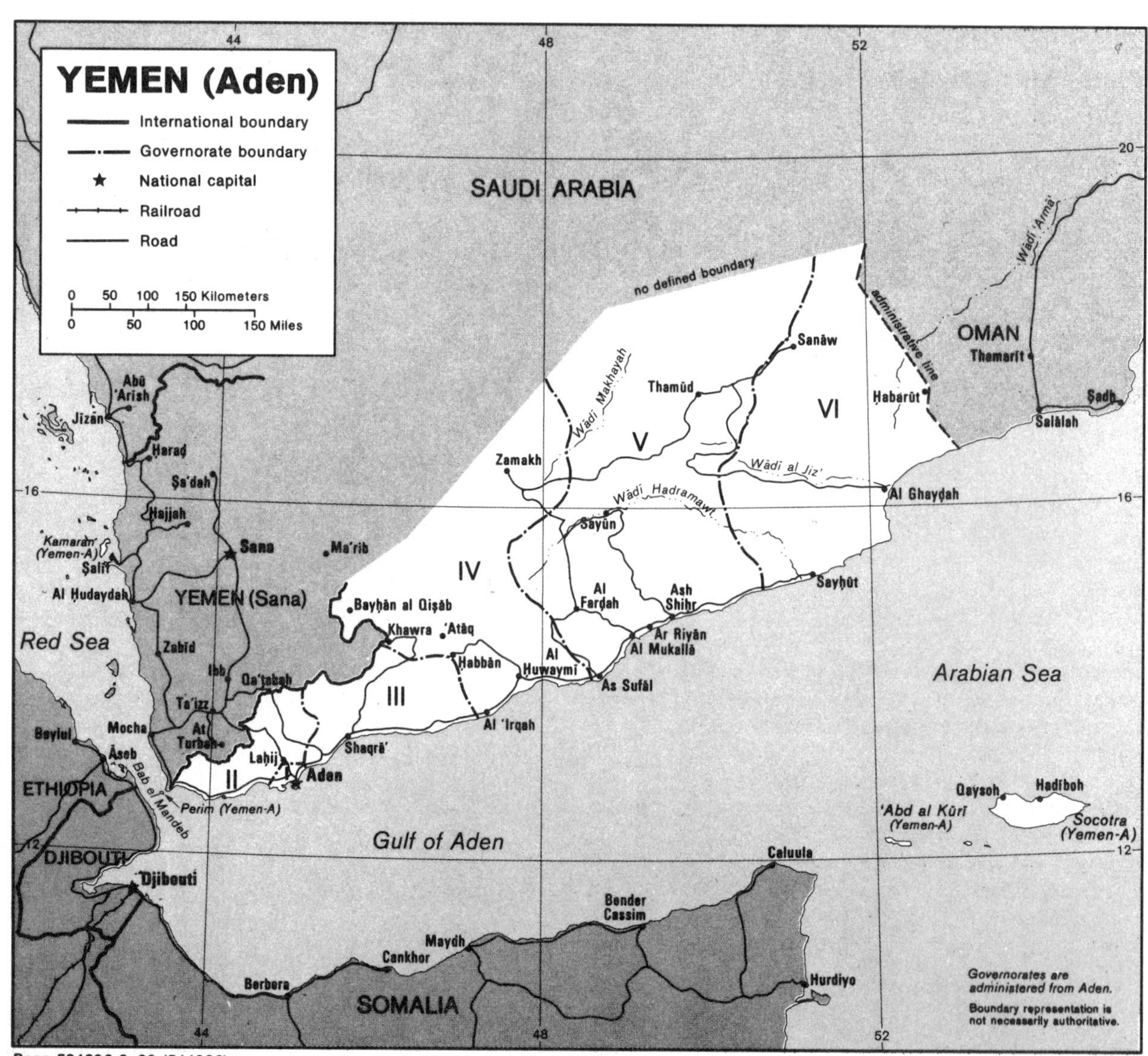

Base 504396 3-80 (544986)

YEMEN (SANA)

BASIC FACTS

Official Name: Yemen Arab Republic

Abbreviation: YAR

Area: 195,000 sq km (75,300 sq mi)

Area—World Rank: 83

Population: 6,732,420 (1988) 7,550,000 (2000)

Population—World Rank: 72

Capital: Sana

Boundaries: 1,661 km (1,032 mi); Saudi Arabia 628 km (390 mi); PDRY 581 km (361 mi)

Coastline: 452 km (281 mi)

Longest Distances: 540 km (336 mi) N-S; 418 km (260 MI0 E-W

Highest Point: Mount Nabi Shuayb 3,760 m (12,336 ft)

Lowest Point: Sea level

Land Use: 14% arable; 36% meadows and pastures; 8% forest and woodland; 42% other

Yemen (Sana) is located in the southwest corner of the Arabian Peninsula. The northern and eastern parts of the country border Saudi Arabia. The eastern boundaries, however, have never been demarcated. In the south and southeast Yemen adjoins the People's Republic of Southern Yemen, but here, too, much of the border remains undefined.

Yemen has two principal geographic regions: the Tihama, a 418-km (260-mi) coastal plain, and the mountainous interior. The Tihama is a narrow, hot, humid, semidesert, and almost waterless strip that extends the entire seacoast from Maydi on the northern frontier with Saudi Arabia to the Bab al Mandab at the country's southern limits and occupies approximately 10 percent of the country. Seven major wadis flow eastward from their sources in the central-western slopes of the interior highlands and permit limited agricultural activity. Hodeida (Al Hudaydah), the country's major port and most important commercial center, is located in the western area of the Tihama.

The mountainous interior was formed as a result of block faulting along a north-south axis parallel to the Red Sea and along an east-west axis parallel to the Gulf of Aden. The central mountain range created by this uplift begins in the vicinity of Taizz. Average elevation in the interior ranges between 2,133 and 3,048 m (7,000 and 10,000 ft), and the highest peak, Nabi Shuayb, located near Sana, reaches 3,760 m (12,337 ft). Seven major wadis—deeply eroded riverbeds—run through the region, creating deep gorges as they proceed to the Tihama. The eastern watershed is largely the product of three major wadis. The area of this dip slope is called the Mashriq, which at its eastern limits melds into the forbidding great desert, the Rub al Khali, or Empty Quarter, most of which is in Saudi Arabia. South and east of Dhamar and Ibb deep and narrow wadis—the Bana and the Tuban—cause very steep gradients. The rocky spars, deep ravines, and sharp, steep ridges render access to the interior difficult, a fact that has contributed to the physical and political isolation of the country.

PRINCIPAL TOWNS (population at 1981 census) Sanʿa (capital) 277,818; Hodeida, 126,386; Taiz 119,573.

Area and population

Governorates	Capitals	area sq mi	area sq km	population 1986 census
al-Bayḍā'	al-Bayḍā'	4,310	11,170	381,249
Dhamār	Dhamār	3,430	8,870	812,981
Ḥajjah	Ḥajjah	3,700	9,590	897,814
al-Ḥudaydah	al-Ḥudaydah	5,240	13,580	1,294,359
Ibb	Ibb	2,480	6,430	1,511,879
al-Jawf	al-Jawf	. . .	. . .	87,299
al-Maḥwit	al-Maḥwit	830	2,160	322,226
Maʾrib	Maʾrib	15,400	39,890	121,437
Ṣaʿdah	Ṣaʿdah	4,950	12,810	344,152
Ṣanʿāʾ	Ṣanʿāʾ	7,840	20,310	1,856,876
Taʿizz	Taʿizz	4,020	10,420	1,643,901
TOTAL		52,210	135,230	9,274,173

YUGOSLAVIA

BASIC FACTS

Official Name: Socialist Federal Republic of Yugoslavia

Abbreviation: SFRY

Area: 255,804 sq km (98,766 sq mi)

Area—World Rank: 72

Population: 23,580,148 (1988) 25,641,000 (2000)

Population—World Rank: 33

Capital: Belgrade

Boundaries: 5,061 km (3,145 mi); Austria 324 km (201 mi); Hungary 623 km (387 mi); Romania 557 km (346 mi); Bulgaria 536 km (333 mi); Greece 262 km (163 mi); Albania 465 km (289 mi); Italy 202 km (126 mi)

Coastline: 2,092 km (1,300 mi)

Longest Distances: 978 km (608 mi) ESE-WNW; 501 km (311 mi) NNE-SSW

Highest Point: Triglav 2,864 m (9,396 ft)

Lowest Point: Sea level

Land Use: 28% arable; 3% permanent crops; 25% meadows and pastures; 36% forest and woodland; 8% other

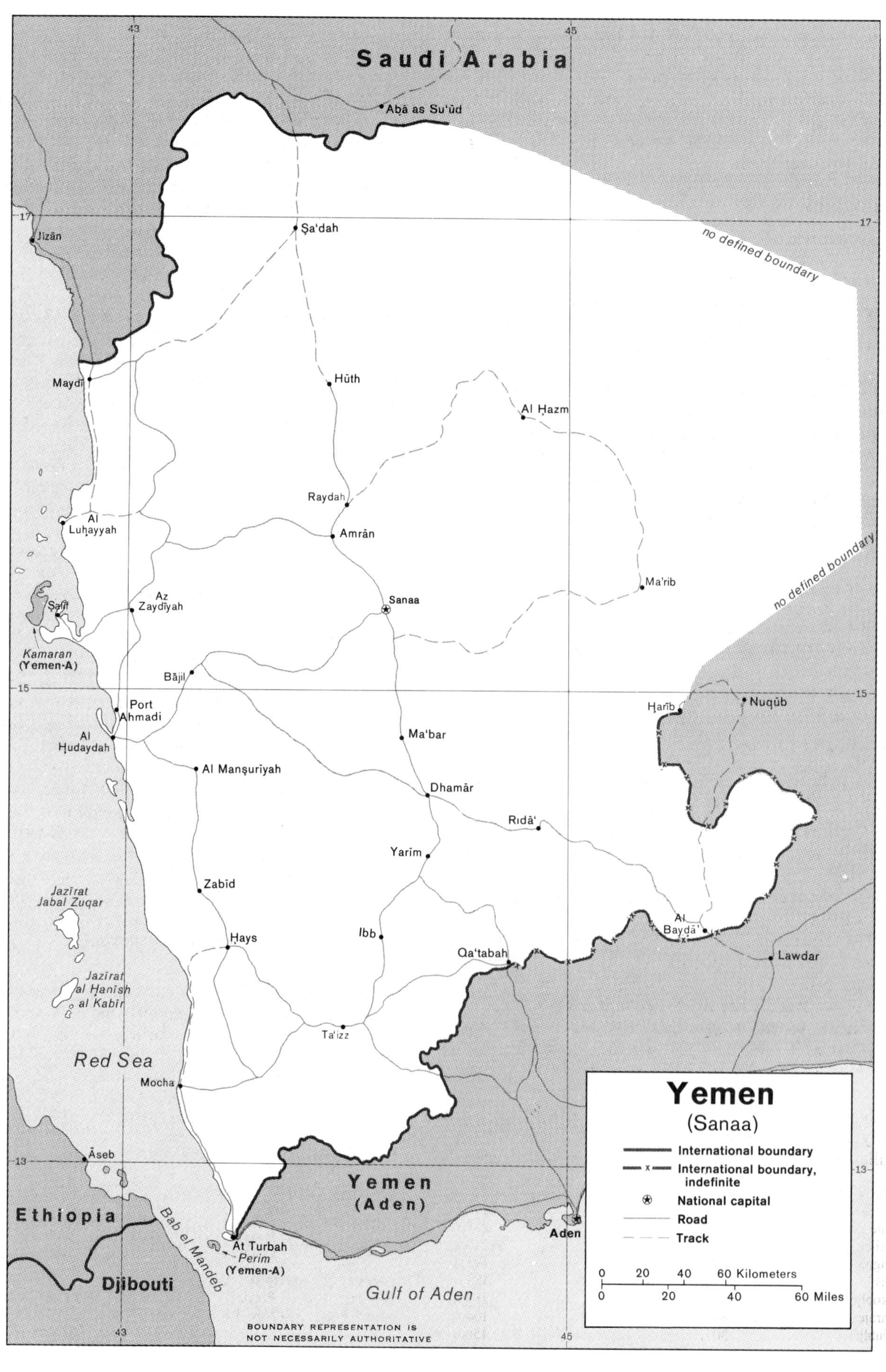

Saudi Arabia
Abâ as Su'ûd
Jīzān
Şa'dah
no defined boundary
Maydī
Hūth
Al Ḩazm
Raydah
Al Luḩayyah
Amrān
Ma'rib
no defined boundary
Şalīf
Az Zaydīyah
Sanaa
Kamaran (Yemen-A)
Bājil
Port Ahmadi
Ḩarīb
Nuqūb
Al Ḩudaydah
Ma'bar
Al Manşurīyah
Dhamār
Rıdā'
Yarīm
Jazīrat Jabal Zuqar
Zabīd
Al Bayḑā'
Ḩays
Ibb
Qa'tabah
Lawdar
Jazīrat al Ḩanīsh al Kabīr
Ta'izz
Red Sea
Mocha
Yemen
(Sanaa)
International boundary
International boundary, indefinite
National capital
Road
Track
Āseb
Yemen (Aden)
Ethiopia
Bab el Mandeb
At Turbah
Perim (Yemen-A)
Aden
Djibouti
Gulf of Aden
0 20 40 60 Kilometers
0 20 40 60 Miles
BOUNDARY REPRESENTATION IS NOT NECESSARILY AUTHORITATIVE
43
45
17
15
13

In geography and population, Yugoslavia is the largest Balkan nation and the seventh largest in Europe (excluding the Soviet Union). Geographically it consists of the coastal and interior highlands and mountains and the Pannonian Plains in the north and northeast. Demographically it contains numerous ethnic groups and the adherents of three major religions.

Topography

The greater part of the country is hilly or mountainous; about 60 percent of the total land area consists of hills and ridges from 200 to 1,000 m (656 to 3,280 ft) in elevation, and another 20 percent consists of high mountains and ranges over 1,000 m (3,280 ft) high. Tall mountains are a dominant feature of the landscape in the south and southeast as well as in the northeast near the Austrian border.

Geological fault lines are widespread in the mountains south of the Sava and Danube rivers. These structural seams in the earth's crust periodically shift, causing earth tremors and occasional earthquakes. The most vulnerable region lies in the general area between Banja Luka and Skopje. An earthquake in Skopje in July 1963 killed over 1,000 people and demolished almost all of the city's buildings in three quick shocks that lasted only thirty seconds. As of mid-1989 the disaster remained the worst in the nation's history. Although the epicenter of the quake was at or near the city's center, the government rebuilt Skopje on the same spot, thus continuing over 2,000 years of urban life. In October 1969 an earthquake destroyed over 70 percent of the buildings in Banja Luka, but only nine people out of a population of over 70,000 were killed.

Coastal and Interior Highlands

The highland region is the most extensive and rugged area of the country. It extends in a northwest-southeast direction for about 960 km (590 mi) from the Austrian to the Greek border. The west and northwest sections of this extensive region contain mountains resembling the higher Austrian Alps to the north, having sharp peaks and ridges. Interspersed among them are deep gorges and narrow valleys.

The Julian Alps, which occupy the westernmost corner of the country, are among the most rugged in Europe and contain many summits that exceed 1,800 m (5,905 ft). One peak, Triglav, has an elevation of 2,846 m (9,337 ft), the highest in the country. Eastward these mountains possess less well-defined ridges, and their crests decrease in height to about 1,000 m (3,280 ft) in the vicinity of Maribor.

South of these mountains the rough terrain changes to hilly areas interspersed with flat valleys. Many of these valleys enlarge into basins whose elevations are generally less than 450 m (1,476 ft). One of the largest of these extended basin areas is located near Ljubljana. Farther to the south the rugged terrain extends into the limestone ranges of the Dinaric Alps. This region, frequently referred to as karst or karstland, is distinctive because of the underground drainage channels that have been formed by the long-term seepage of water down through the soluble limestone. This action leaves the surface dry and over the years has formed many large depressions in the high coastal plateaus.

The coastal area is studded with more than 600 offshore islands, most of which are rocky and hilly. The Dinaric chain consists of ridges that parallel the coast and range from 700 m (2,296 ft) to over 2,200 m (7,217 ft) in height. The flat depressions formed in the limestone hills vary considerably in size. Most are quite narrow, but some have been elongated to as much as 60 km (37 mi). The general surface configuration of the karst area is rocky, featuring many desolate cliffs that support little vegetation.

The east central part of the country is an extensive, dissected region with crests and ridges between 500 m (1,640 ft) and 1,800 m (5,905 ft) in elevation. In the north the foothills of the mountains merge with the upland terraces of the Pannonian Plains; in the south there are karst-like uplands, and the ridges have greater ranges of elevation. The highest peaks are located in the interstream areas of the mountains near the Albanian border.

The highlands in the eastern section include several systems of rugged mountains and hill groups. Most of these are separated by broad valleys and fairly flat though narrow basins. Summits in most of the region are sharp-crested and rise between 600 and 1,200 m (1,958 and 3,937 ft). The highest and most rugged terrain in this section of the country is in the southern portion, west of Skopje and along the Bulgarian border.

Large areas of rolling hills are located in the vicinity of Skopje and Pristina. These hills, as well as those further to the north above the Nisava River, are marked by numerous corridors formed by low valleys and small narrow basins. In some places these valleys are connected by narrow defiles and rise rather abruptly into the steep slopes of the foothills of the nearby mountains.

Pannonian Plains

The Pannonian Plains are a southward extension of the Great Hungarian Plains (Pannonia) and consist of the valleys of the middle and lower Drava, the middle and lower

PRINCIPAL TOWNS (population at 1981 census)

Beograd (Belgrade, the capital)	1,470,073	Novi Sad	257,685
Osijek	867,646	Split	235,922
Zagreb	768,700	Priština	210,040
Niš	643,470	Rijeka	193,044
Skoplje (Skopje)	506,547	Maribor	185,699
Sarajevo	448,519	Banja Luka	183,618
Ljubljana	305,211	Kragujevac	164,823
		Subotica	154,611

Area and population		area		population
Socialist republics	**Capitals**	sq mi	sq km	1986 estimate
Bosnia and Hercegovina	Sarajevo	19,741	51,129	4,356,000
Croatia	Zagreb	21,829	56,538	4,665,000
Macedonia	Skopje	9,928	25,713	2,041,000
Montenegro	Titogräd	5,333	13,812	619,000
Serbia	Belgrade	21,609	55,968	5,803,000
Slovenia	Ljubljana	7,819	20,251	1,934,000
Autonomous provinces				
Kosovo	Priština	4,203	10,887	1,804,000
Vojvodina	Novi Sad	8,304	21,506	2,049,000
TOTAL		98,776	255,804	23,271,000

Base 504660 (545723) 1-81

Sava, and lower Tisa, and the middle Danube rivers. Occupying approximately 20 percent of the country's land area, this region extends about 480 km (298 mi) northwest-southeast and has a maximum north-south width of about 200 km (124 mi). This region, the most fertile in the nation, is the area occupied by the ancient Pannonian Sea, which disappeared as the runoff from the surrounding mountains gradually filled it with rich alluvial deposits. It is a sedimentary region containing wide valley basins, alluvial plains, sandy dunes, and low, rolling hills covered with fertile loam.

In general the area is low and flat. Few elevations reach more than 100 m (328 ft) above sea level. The major rivers are located on broad flood plains. The western portion of the plains rise to rolling hills that reach elevations of 150 m (492 ft) or more. Between the Drava and Sava rivers a narrow ridge, averaging about 240 m (787 ft) in height, extends to the southeast and bisects that portion of the plain lying to the west of the Danube River.

Drainage Systems

In the Pannonian Plains drainage is to the Black Sea through the Danube and its four large tributaries: The Drava, Tisa, Sava, and Morava. The Danube in some places has a width of over three km (1.8 mi) and depths up to fifteen m (49 ft) or more at high water. Large fluctuations between high and low water are typical. High water generally lasts from March or April until May or June; the low water period usually lasts from mid-July to late October.

The largest portion of the broad interior highlands area is drained by the Danube and the upper reaches of its right bank tributaries. In the southeast the Vardar River and its tributaries drain that area into the Aegean Sea; the Neretva River is the principal source of drainage from this area into the Adriatic Sea.

Drainage on the western slopes of the karst area is provided by short rapid surface streams and underground drainage into the Adriatic. The bulk of the water collects in depressions among the limestone hills and seeps into underground channels. Occasionally the water emerges in springs, but most of it moves into the Adriatic below sea level, then rises to the surface of the sea.

ZAÏRE

BASIC FACTS

Official Name: Republic of Zaïre

Abbreviation: ZAR

Area: 2,344,885 sq km (905,365 sq mi)

Area—World Rank: 11

Population: 33,293,946 (1988) 42,980,000 (2000)

Population—World Rank: 28

Capital Kinshasa

Boundaries: 10,160 km (6,314 mi); Central African Republic 1,577 km (980 mi); Sudan 628 km (390 mi); Uganda 764 km (475 mi); Rwanda 217 km (135 mi); Burundi 233 km (145 mi); Tanzania 459 km (285 mi); Zambia 2,107 km (1,309 mi); Angola 2,258 km (1,420 mi); Cabinda 225 km (140 mi); Congo 1,625 km (1,010 mi)

Coastline: 40 km (25 mi)

Longest Distances: 2,276 km (1,414 mi) SSE-NNW; 2,236 km (1,389 mi) ENE-WSW

Highest Point: Margherita Peak 5,109 m (16,763 ft)

Lowest Point: Sea level

Land Use: 3% arable; 4% meadows and pastures; 78% forest and woodland; 15% other

Zaïre lies almost entirely within the Zaïre River Basin and constitutes about two-thirds of it, giving the country a certain geographic unity. The Zaïre River itself and the many streams, large and small, draining into it (and ultimately into the Atlantic Ocean) provide the most significant system of inland waterways in Africa. In places where navigation is not possible, the rapids and falls that prevent it furnish hydroelectric potential surpassing that of any country in the world.

Geographic Regions

Zaïre's hydrographic unity notwithstanding, it is possible to distinguish four major geographic regions, defined in terms of terrain and patterns of natural vegetation. The core region is the Central Zaïre Basin, a large depression—its average elevation is about 400 m (1,312 ft)—often referred to as the *cuvette* (saucer or shallow bowl). Its roughly 800,000 sq km (308,881 sq mi) constitute about a third of Zaïre's territory, and it encompasses most of the country's more than 1 million sq km (385,102 sq mi) of equatorial forest. A substantial proportion of the forest within the *cuvette* is swamp, and still more of it consists of a mixture of marshy and firm land.

North and south of the *cuvette* lie higher plains and, occasionally, hills covered with varying mixtures of savanna grasses and woodlands. The Southern Uplands region, like the *cuvette,* constitutes about a third of Zaïre's territory. The northern part of the region slopes from south to north, starting at about 1,000 m (3,280 ft) and falling to about 500 m (1,640 ft) as it approaches the *cuvette*. Its vegetation cover is more varied than that of the Northern Uplands: in some areas woodland is dominant; in others, savanna grasses. South of the *cuvette* along the streams flowing into the Kasaï River are extensive gallery forests. In the far southeast most of Shaba is characterized by somewhat higher plateaus and low mountains. The westernmost section of Zaïre, a partly forested panhandle reaching the Atlantic Ocean, is an extension of the Southern Uplands that drops sharply to a very narrow shore.

PRINCIPAL TOWNS (estimated population at 1 July 1976)

Kinshasa (capital)	2,443,876
Kananga (formerly Luluabourg)	704,211
Lumbumbashi (Elisabethville)	451,332
Mbuji-Mayi	382,632
Kisangani (Stanleyville)	339,210
Bukavu (Costermanville)	209,051
Kikwit.	172,450
Matadi	162,396
Mbandaka (Coquilhatville)	149,118
Likasi (Jadotville)	146,394*

*1970 estimate.

Area and population

		area		population
Regions	**Capitals**	sq mi	sq km	1984 census
Bandundu	Bandundu	114,154	295,658	3,682,845
Bas-Zaire	Matadi	20,819	53,920	1,971,520
Equateur	Mbandaka	155,712	403,293	3,405,512
Haut-Zaire	Kisangani	194,302	503,239	4,206,069
Kasai Occidental	Kananga	60,605	156,967	2,287,416
Kasai Oriental	Mbuji-Mayi	64,949	168,216	2,402,603
Kivu	Bukavu	99,098	256,662	5,187,865
Shaba (Katanga)	Lubumbashi	191,879	496,965	3,874,019
Neutral City				
Kinshasa	3,848	9,965	2,653,558	
TOTAL		905,365	2,344,885	29,671,407

In the much narrower Northern Uplands the cover is largely savanna, and woodlands are rarer. The average altitude of this region is about 600 m (1,968 ft), but it rises as high as 900 m (2,952 ft) where it meets the western edge of the Eastern Highlands.

The Eastern Highlands region is the highest and most rugged portion of the country. It extends for more than 1,500 km (932 mi) from above Lake Albert to the southern tip of Shaba below Lubumbashi and varies in width from eighty to 560 km (50 to 348 mi). Its hills and mountains range in altitude from about 1,000 m (3,280 ft) to more than 5,000 m (16,494 ft). The western arm of the Great Rift Valley forms a natural eastern boundary to this region. The eastern border of Zaïre extends through the valley and its system of lakes, which are separated from each other by plains situated between high mountain ranges.

In this region changes in elevation bring marked changes in vegetation, which ranges from montane savanna to heavy montane forest. The Ruwenzori Mountains (mountains of the moon) between lakes Edward and Albert constitute the highest range in Africa. That and their location on the equator make for a varied and spectacular flora. Not so high, and spectacular in a different fashion, are the Virunga Mountains north of Lake Kivu, site of a number of active volcanoes.

In the northeast a narrow belt of the highlands running from just north of Lake Kivu to the northwestern edge of Lake Edward lies in the Nile River Basin. Lakes Kivu and Tanganyika thus drain via tributaries of the Zaïre to the Atlantic Ocean, and lakes Edward and Albert drain via the Nile into the Mediterranean. A small part of southeasternmost Zaïre lies in the Zambezi River Basin, which drains into the Indian Ocean.

ZAMBIA

BASIC FACTS

Official Name: Republic of Zambia

Abbreviation: ZAM

Area: 752,614 sq km (290,586 sq mi)

Area—World Rank: 38

Population: 7,546,177 (1988) 11,237,000 (2000)

Population—World Rank: 79

Capital: Lusaka

Land Boundaries: 5,627 km (3,496 mi); Tanzania 322 km (200 mi); Malawi 746 km (464 mi); Mozambique 424 km (263 mi); Zimbabwe 739 km (459 mi); Botswana 100 m (109.4 yards); Namibia 203 km (126 mi); Angola 1,086 km (675 mi); Zaïre 2,107 km (1,309 mi)

Coastline: Nil

Longest Distances: 1,206 km (749 mi) E-W; 815 km (506 mi) N-S

Highest Point: 2,164 m (7,100 ft)

Lowest Point: 329 m (1,081 ft)

Land Use: 7% arable; 47% meadows and pastures; 27% forest and woodland; 19% other

Most of Zambia lies on a portion of the great plateau that dominates central and southern Africa's landmass. Some of that plateau is undulating, some relatively flat, but most of it ranges between 900 and 1,500 m (2,952 and 4,921 ft), the higher sections above 1,200 m (3,937 ft) occurring, for the most part, in the north. The few areas of still higher land (some of it above 2,000 m (6,551 ft) are found in the northeast, chiefly among the borders with Tanzania (the Mbala Highlands) and Malawi (the Mafingi Mountains). The significant areas of lower land are rift valleys in the east (the Luangwa River valley) and in the south (the middle Zambezi River valley), both bounded by escarpments.

Most of Zambia's streams—except for those in part of the eastern lobe—ultimately drain into the Indian Ocean via the Zambezi River and its main tributaries. In addition to those streams that enter the Zambezi directly, there are

PRINCIPAL TOWNS (mid–1987 estimates)

Lusaka (capital)	818,994	Chingola	187,310
Kitwe	449,442	Luanshya	160,667
Ndola	418,142	Livingstone	94,637
Mufulira	192,323	Kalulushi	89,065
Kabwe (Broken Hill)	190,752	Chililabombwe	79,010

Base 505275 (A00820) 1-83

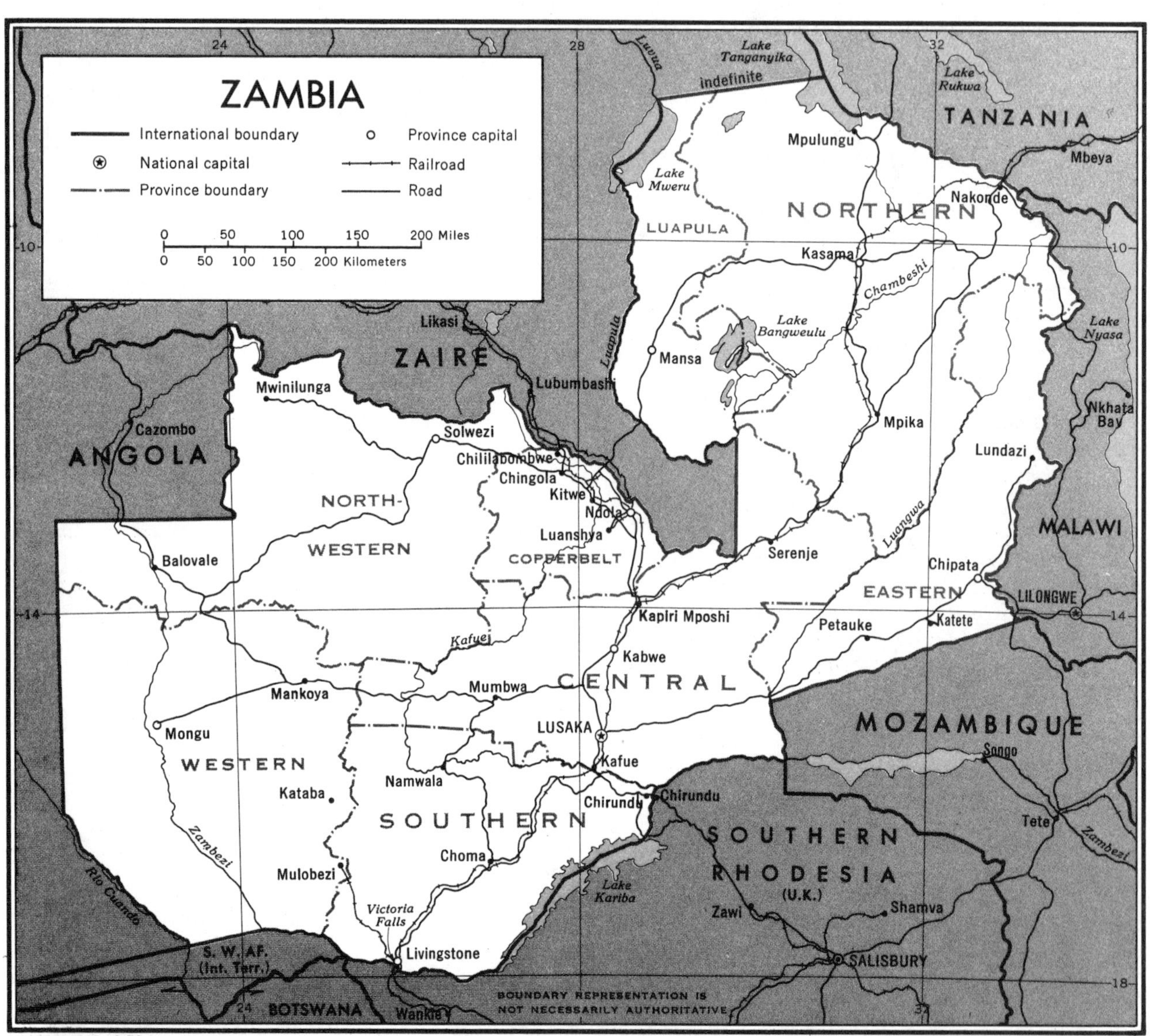

Base 503205 12-76

three main tributary systems—those of the Kafue, Luangwa, and Lusemfwa rivers.

The upper Zambezi, running roughly from north to south, has a low gradient, and the area through which it passes is marked by floodplains and swamps. After turning eastward, the Zambezi flows over Victoria Falls and through the middle Zambezi valley, much of it occupied by the great man-made Lake Kariba. Much of the upper Kafue River is also characterized by a low gradient, and extensive swamps are common. The Kafue Flats, through which the river passes after turning east, are too poorly drained to provide cultivable land but furnish good pasturage. The Lusemfwa drains that small portion of Zambia between the line of rail and the Luangwa valley and joins the Luangwa River more steeply graded than most of Zambia's streams.

The network of rivers and streams of northeastern Zambia drains via the Zaïre River to the Atlantic Ocean. The Chambeshi River collects most of the water in the region, its many channels then discharging into Lake Bangweulu or directly into the Luapula River. In either case the water finds its way via the Luapula to Lake Mweru and thence to a tributary of the Zaïre. Other streams enter Lake Mweru or Lake Tanganyika (whose waters also enter the Zaïre system) directly. A very small portion of the far northeast drains inland to Lake Rukwa in Tanzania.

Lake Bangweulu and several smaller bodies of water are part of the Bangweulu swamp complex. The main swamp is permanently flooded; the periphery (a belt about forty km (24 mi) wide is flooded during and immediately after the rainy season.

The flow of all watercourses in Zambia is affected by the clear demarcation between rainy and dry seasons. Most small streams dry up sometime between May and October, and even the larger rivers show a substantial difference between maximum (occurring variously between February and May) and minimum discharges.

Area and population

		area		population
Provinces	**Capitals**	sq mi	sq km	1980 census
Central	Kabwe	36,446	94,395	513,835
Copperbelt	Ndola	12,096	31,328	1,248,888
Eastern	Chipata	26,682	69,106	656,381
Luapula	Mansa	19,524	50,567	412,798
Lusaka	Lusaka	8,454	21,896	693,878
Northern	Kasama	57,076	147,826	677,894
North-Western	Solwezi	48,582	125,827	301,677
Southern	Livingstone	32,928	85,283	666,469
Western	Mongu	48,798	126,386	487,988
TOTAL		290,586	752,614	5,679,808

ZIMBABWE

BASIC FACTS

Official Name: Republic of Zimbabwe

Abbreviation: ZIM

Area: 390,759 sq km (150,873 sq mi)

Area—World Rank: 55

Population: 9,728,547 (1988) 11,943,000 (2000)

Population—World Rank: 69

Capital: Harare

Land Boundaries: 2,988 km (1,857 mi); Mozambique 1,223 km (760 mi); South Africa 225 km (140 mi); Botswana 813 km (505 mi); Zambia 727 km (452 mi)

Coastline: Nil

Longest Distances: 852 km (529 mi) WNW-ESE; 710 km (441 mi) NNE-SSW

Highest Point: Inyangani 2,592 m (8,504 ft)

Lowest Point: 162 m (530 ft)

Land Use: 7% arable; 12% meadows and pastures; 62% forest and woodland; 19% other

The granite plateau that forms most of Zimbabwe is partially delineated in several border areas by massive ancient faults (rifts) associated with the larger faults of the East African Rift Valley system. Within Zimbabwe much of the plateau surface is an extensive veld, a rolling plain covered with a mixture of grasses and open woodlands. The veld slopes gently downward from a central upland region to considerably lower plains areas near the borders. The highest area of grass and woodlands, the highveld, is between 80 to 160 km (50 and 100 mi) wide and extends from northeast to southwest for 643 km (400 mi) across the center of the landmass.

Through most of its length the highveld is between 1,219 and 1,524 m (4,000 and 5,000 ft) above sea level, forming a broad watershed, or divide, that separates the territory's major river basins. Its surface features vary from relatively smooth or rolling to rough, almost mountainous terrain. Here—and in lower altitude veld regions as well—rocky hills and buttes, known locally as kopjes, interrupt the otherwise flat or rolling landscapes of grassland, woods, and cultivated fields.

This central upland is also marked by the Great Dyke, a massive extrusion of ancient lava that extends generally from northeast to southwest for 482 km (300 mi). Only a few miles wide in most segments, it rises as much as 457 m (1,500 ft) above the surrounding highveld in a series of eroded ridges. Various segments of this long series of ridges are known as the Doro Range, the Selondi Range, the Mashona Hills, and the Umvukwe Range. The Umvukwe is the northernmost and highest segment of the Great Dyke, reaching to more than 1,706 m (5,600 ft) above sea level.

In north-central Zimbabwe the broad expanse of the highveld breaks up into several arms, one of which joins a lesser watershed ridgeline extending southeastward from Harare to the mountains around Umtali along the eastern border. This eastern mountain complex is the highest area

Base 505385 (544636) 10-82

PRINCIPAL TOWNS (population at census of August 1982)

Harare (Salisbury)	656,000	Hwange (Wankie)	39,200
Bulawayo	413,800	Masvingo (Fort Victoria)	30,600
Chitungwiza	172,600	Zvishavane (Shabani)	26,800
Gweru (Gwelo)	78,900	Chinhoyi (Sinoia)	24,300
Mutare (Umtali)	69,600	Redcliff	22,000
Kwekwe (Que Que)	47,600	Marondera (Marandellas)	20,300
Kadoma (Gatooma)	44,600		

in the territory. Most peaks are between 1,828 and 2,368 m (6,000 and 8,000 ft) high; the loftiest, Mount Inyangi, reaches 2,595 m (8,514 ft).

On its northwestern and southeastern slopes, the highveld merges imperceptibly into the medium altitude wooded grasslands (middleveld), generally defined as the area between 914 and 1,219 m (3,000 and 4,000 ft) in elevation. On the southeastern flank of the highveld, the middleveld is no more than 120 km (75 mi) wide in most areas, narrowing to less than 80 km (50 mi) northeast of Fort Victoria. On the opposite slope in western and northwestern Zimbabwe, it covers a much larger area, having a breadth of 210 km (150 mi) or more in the Gwaai and Shangani river valleys northwest of Bulawayo.

In the southeast the lowveld, which is generally considered to include the land below 914 m (3,000 ft) extends from the nominal edge of the middleveld to the southern and southeastern borders, covering nearly one-fifth of the territory. In the northwest and the north the lowveld is divided into three major sections, partially separated by escarpments and local ranges of hills. These sections slope directly to the Zambezi River or to the 281-km (175-mi) shoreline of Lake Kariba.

DRAINAGE

Runoff from the entire territory, excluding only a small area in the southwest, is carried by three rivers that reach the Indian Ocean via Mozambique. Two of them originate outside Zimbabwe, the Zambezi along the Zambia-Angola border and the Limpopo in South Africa. The headwaters of the third major stream, the Sabi, are south of Harare on the eastern slopes of the highveld.

Tributaries of the Zambezi River, which marks much of the northern border, collect runoff waters from nearly all of the land in the western, northern, and northeastern areas of the country. As the longest of all African rivers flowing to the Indian Ocean, the Zambezi is already carrying a heavy volume of water as it approaches the westernmost part of Zimbabwe. In this area the river drops over Victoria Falls, a cataract 106 m (350 ft) high at its maximum and nearly a mile wide. The average annual flow over Victoria Falls is 37 million acre-feet, varying widely from year to year.

Area and population

Provinces	Capitals	area: sq mi	area: sq km	population: 1982 census
Manicaland	Mutare	13,463	34,870	1,099,202
Mashonaland Central	Bindura	10,534	27,284	563,407
Mashonaland East	Harare	9,627	24,934	1,495,984
Mashonaland West	Chinhoyi	23,346	60,467	858,962
Masvingo (Victoria)	Masvingo	17,108	44,310	1,031,697
Matabeleland North	Bulawayo	28,393	73,537	885,339
Matabeleland South	Gwanda	25,633	66,390	519,636
Midlands	Gweru	22,767	58,967	1,091,844
TOTAL		150,873	390,759	7,546,071

At Kariba, 483 km (300 mi) below Victoria Falls, a major dam on the Zambezi (completed in 1959) created Lake Kariba. This is one of largest man-made lakes in the world, reaching 281 km (175 mi) upstream and flooding 5,180 sq km (2,000 sq mi) of territory in Zimbabwe and Zambia.

The Zambezi system draws water from all of Zimbabwe lying northwest of the central ridgeline of the highveld except for an area of low rainfall near the southwestern border, where the Nata River and a few other small intermittent streams carry the annual runoff to an internal drainage basin in Botswana.

Runoff from most of the land on the eastern and southeastern slopes of the central ridgeline flows generally southeastward through numerous tributaries of the Limpopo and Sabi rivers. Although it is shorter and carries a lesser volume of water than the Zambezi, the Limpopo is also one of Africa's major streams. Its gradient is relatively even except in the Beitbridge area, where it flows through an extensive gorge marked by rapids and several waterfalls. Numerous tributaries carry a heavy volume of water from upper levels of the veld, where annual rainfall is heavier than in the lowveld areas near the Limpopo itself.

Southeastern Zimbabwe from the towns of Shabani and Triangle northward to the ridgeline between Marandellas and Umtali (marked by the Umtali-Harare rail line) is drained by the Sabi River and its tributaries. From the upper levels of its watershed west of Umtali, the Sabi flows southward into the lowveld. There it is joined by the Lundi, which collects water from the Shabani and Fort Victoria areas; it turns eastward at the border, eventually reaching the Indian Ocean in east-central Mozambique.

PART II

THE EARTH AND ITS CONTINENTS

Earth

The earth is the fifth largest planet of the solar system and the third in order from the sun. It is located at an average distance of 149,637 million km (93 million mi) from the sun which it orbits in one year or 365.2425 days. The earth rotates from west to east, that is perpendicular to the plane of the equator. The period of one complete rotation is 23 hours, 56 minutes and 4 seconds. The equatorial diameter of the earth is 12,753 km (7,926 mi) and its polar diameter is 12,711 km (7,900 mi). The circumference of the earth around the equator is 40,064 km (24,900 mi) while its circumference around the poles is 40,000 km (24,860 mi). The earth is not a true sphere but rather a geoid—an ellipsoid flattened at the poles. Small bulges produce four corners, at Ireland, off Peru, south of Africa and near New Guinea. Earth's total surface area is estimated at 510,934,000 sq km (197,272,000 sq mi) of which the land area accounts for 148,148,000 sq km (57,200,000 sq mi) or about 29%. The average land elevation is 823 m (2,700 ft) and the average ocean depth 3,810 m (12,500 ft). The age of the earth is estimated at 3.7 billion years.

Land boundaries: 442,000 km (274,482 mi)
Coastline: 359,000 km (222,939 mi)
Maritime claims:
Contiguous zone: generally 24 nm, but varies from 4 nm to 24 nm
Continental shelf: generally 200 nm, but some are 200 meters (656 ft) in depth
Exclusive fishing zone: most are 200 nm, but varies from 12 nm to 200 nm
Extended economic zone: 200 nm, only Madagascar claims 150 nm
Territorial sea: generally 12 nm, but varies from 3 nm to 200 nm

Disputes: 13 international land boundary disputes—Argentina-Uruguay, Bangladesh-India, Brazil-Paraguay, Brazil-Uruguay, Cambodia-Vietnam, China-India, China-USSR, Ecuador-Peru, El Salvador-Honduras, French Guiana-Suriname, Guyana-Suriname, Guyana-Venezuela, Qatar-UAE
Climate: two large areas of polar climates separated by two rather narrow temperate zones from a wide equatorial band of tropical to subtropical climates
Terrain: highest elevation is Mt. Everest at 8,848 m (29,030 ft) and lowest elevation is the Dead Sea at 392 m (1,286 ft) below sea level; greatest ocean depth is the Marianas Trench at 10,924 m (33,775 ft)
Land use: 10% arable land; 1% permanent crops; 24% meadows and pastures; 31% forest and woodland; 34% other; includes 1.6% irrigated

CONTINENTS

AFRICA is the second largest continent on the globe accounting for 20% of the land surface. It straddles the Northern and Southern Hemispheres. Length 7,996 km (4,970 mi); Breadth 7,401 km (4,600 mi); Land Area 30,244,049 sq km (11,677,239 sq mi); Coastline 25,905 km (16,100 mi); Most southerly point Cape Agulhas; Most easterly point Cape Hafun; Most westerly point Cape Almadies; Most northerly point Ras el Sekka; Highest point Mt. Kibo 5,895 m (19,340 ft); Lowest point Lake Assal, Djibouti, 155 m (510 ft) below sea level.

ASIA is the largest continent on the globe accounting for 29.7% of the land surface. Land Area 44,391,188 sq km (17,139,455 sq mi); Most northerly point Cape Chelyuskin; Most easterly point Cape Dezhneva; Most southerly point Cape Piai in Malaysia; Most Westerly point Cape Baba in Turkey; Highest point Mt. Everest, 8,848 m (29,030 ft); Lowest point Dead Sea, 395 m (1,296 ft) below sea level.

AUSTRALIA is the smallest continent on the globe accounting for 5.1% of the land surface. Land Area 7,686,884 sq km (2,967,909 sq mi); Highest point Mt. Kosciusko, 2,230 m (7,316 ft); Lowest point Lake Eyre 16 m (52 ft) below sea level.

EUROPE is the sixth largest continent accounting for 7% of the land surface. Land surface 10,354,636 sq km (3,997,929 sq mi); Most northerly point North Cape, Norway; Most southerly point Cape Tarifa, Spain; Most westerly point Dunmore Head, Eire; Most Easterly point Ural River; Highest point El'brus, USSR 5,642 m (18,510 ft); Lowest point Caspian Sea 28 m (92 ft) below sea level.

NORTH AMERICA is the third largest continent accounting for 16.3% of the land surface. Land Area 24,247,039 sq km (9,361,791 sq mi); Most northerly point Cape Morris Jessup, Greenland; Most easterly point SE coast of Labrador; Most southerly point SE Panama; Most westerly point Attu Island in Aleutian Islands; Highest point Mt McKinley, 6,193 m (20,320 ft); Lowest point Death Valley 85 m (282 ft) below sea level.

SOUTH AMERICA is the fourth largest continent accounting for 12% of the land surface. Land Area 17,821,028 sq km (6,880,706 sq mi); Length 7,240 km (4,500 mi); Breadth 5,149 km (3,200 mi); Most northerly point Point Gallinas, Colombia; Most easterly point Point Coqueiros, Brazil; Most southerly point Cape Horn; Most westerly point Point Parinas, Peru; Highest point Aconcagua, Argentina 6,959 m (22,831 ft); Lowest point Salinas Grandes, Argentina, 40 m (131 ft) below sea level.

ANTARCTICA is the fifth largest continent with a land area of 14 million sq km (8.694 million sq mi) accounting for 9.6% of the earth's land surface. Located around the South Pole, it is covered with 98% thick continental ice sheet with average elevations between 2,000 and 4,000 m (6,562 and 13,124 ft). Ice-free coastal areas include parts of southern Victoria Land, Wilkes Land, and the scientific research areas of Graham Land and Ross Island on McMurdo Sound. Glaciers form ice shelves along half of coastline. Highest point: Vincent Massif, 5,139 m (16,866 ft).

OCEANS

Oceans cover 361,132,000 sq km (139,433,065 sq mi), or 70%, of the earth's total surface to an average depth of 3,810 m (12,500 ft). The deepest part of the oceans is found in the Marianas Trench in the Western Pacific, 11,430 m (37,500 ft). The total amount of water on the earth's surface is estimated at 340 million cubic miles.

Arctic Ocean

Total area: 14,056,000 km (5,427,022 sq mi); includes Baffin Bay, Barents Sea, Beaufort Sea, Chukchi Sea, East Siberian Sea, Greenland Sea, Hudson Bay, Hudson Strait, Kara Sea, Laptev Sea, and other tributary water bodies.

Comparative area: slightly more than 1.5 times the size of the US; smallest of the world's four oceans (after Pacific Ocean, Atlantic Ocean, and Indian Ocean)

Coastline: 45,389 km (28,186 mi)

Climate: persistent cold and relatively modest annual temperature ranges; winters characterized by continuous darkness, cold and stable weather conditions, and clear skies; summers characterized by continuous daylight, damp and foggy weather, and weak cyclones with rain or snow

Terrain: central surface covered by a perennial drifting polar icepack which averages about 3 m (9.8 ft) in thickness, although pressure ridges may be three times that size; clockwise drift pattern in the Beaufort Gyral Stream, but nearly straight line movement from the New Siberian Islands (USSR) to Denmark Strait (between Greenland and Iceland); the ice pack is surrounded by open seas during the summer, but more than doubles in size during the winter and extends to the encircling land masses; the ocean floor is about 50% continental shelf (highest percentage of any ocean) with the remainder a central basin interrupted by three submarine ridges (Alpha Cordillera, Nansen Cordillera, and Lomonsov Ridge); maximum depth is 4,665 m (15,305 ft) in the Fram Basin

Natural resources: sand and gravel aggregates, placer deposits, polymetallic nodules, oil and gas fields, fish, marine mammals (seals, whales)

Environment: endangered marine species include walruses and whales; ice islands occasionally break away from northern Ellesmere Island; icebergs calved from western Greenland and extreme northeastern Canada; maximum snow cover in March or April about 20 to 50 cm (7.8 in to 20 in) over the frozen ocean and lasts about 10 months; permafrost in islands; virtually icelocked from October to June; fragile ecosystem slow to change and slow to recover from disruptions or damage

Note: major chokepoint is the southern Chukchi Sea (northern access to the Pacific Ocean via the Bering Strait); ships subject to superstructure icing from October to May; strategic location between North America and the USSR; shortest marine link between the extremes of eastern and western USSR; floating research stations operated by the US and USSR

Communications

Ports: Churchill (Canada), Murmansk (USSR), Prudhoe Bay (US)

Telecommunications: no submarine cables

Note: sparse network of air, ocean, river, and land routes; the Northwest Passage (North America) and Northern Sea Route (Asia) are important waterways

Atlantic Ocean

Total area: 82,217,000 sq km (31,743,983 sq mi); includes Baltic Sea, Black Sea, Caribbean Sea, Davis Strait, Denmark Strait, Drake Passage, Gulf of Mexico, Mediterranean Sea, North Sea, Norwegian Sea, Weddell Sea, and other tributary water bodies

Comparative area: slightly less than nine times the size of the US; second-largest of the world's four oceans (after the Pacific Ocean, but larger than Indian Ocean or Arctic Ocean)

Coastline: 111,866 km (69,469 mi)

Climate: tropical cyclones (hurricanes) develop off the coast of Africa near Cape Verde and move westward into the Caribbean Sea and southeastern US; hurricanes can occur from May to December, but are most frequent from August to November

Terrain: surface usually covered with sea ice in Labrador Sea, Denmark Strait, and Baltic Sea from October to June; clockwise warm water gyre (broad, circular system of currents) in the north Atlantic, counterclockwise warm water gyre in the south Atlantic; the ocean floor is dominated by the Mid-Atlantic Ridge, a rugged north-south centerline for the entire Atlantic basin; maximum depth is 8,605 m (28,233 ft) in the Puerto Rico Trench

Natural resources: oil and gas fields, fish, marine mammals (seals and whales), sand and gravel aggregates, placer deposits, polymetallic nodules, precious stones

Environment: endangered marine species include the manatee, seals, sea lions, turtles, and whales; municipal sludge pollution off eastern US, southern Brazil, and eastern Argentina; oil pollution in Caribbean Sea, Gulf of Mexico, Lake Maracaibo, Mediterranean Sea, and North Sea; industrial waste and municipal sewage pollution in Baltic Sea, North Sea, and Mediterranean Sea; icebergs common in Davis Strait, Denmark Strait, and the northwestern Atlantic from February to August and have been spotted as far south as Bermuda and the Madeira Islands; icebergs from Antarctica occur in the extreme southern Atlantic

Note: ships subject to superstructure icing in extreme north Atlantic from October to May and extreme south Atlantic from May to October; persistent fog can be a hazard to shipping from May to September; major choke points include the Dardanelles, Strait of Gibraltar, access to the Panama and Suez Canals; strategic straits include the Dover Strait, Straits of Florida, Mona Passage, The Sound (Øresund), and Windward Passage; north Atlantic shipping lanes subject to icebergs from February to August; the Equator divides the Atlantic Ocean into the North Atlantic Ocean and South Atlantic Ocean

Communications

Ports: Alexandria (Egypt), Algiers (Algeria), Antwerp (Belgium), Barcelona (Spain), Buenos Aires (Argentina), Casablanca (Morocco), Colón (Panama), Copenhagen (Denmark), Dakar (Senegal), Gdańsk (Poland), Hamburg (FRG), Helsinki (Finland), Las Palmas (Canary Islands, Spain), Le Havre (France), Leningrad (USSR), Lisbon (Portugal), London (UK), Marseille (France), Montevideo (Uruguay), Montreal (Canada), Naples (Italy), New Orleans (US), New York (US), Oran (Algeria), Oslo (Norway), Piraiéves (Greece), Rio de Janeiro (Brazil), Rotterdam (Netherlands), Stockholm (Sweden)

Telecommunications: numerous submarine cables with most between continental Europe and the UK, North America and the UK, and in the Mediterranean

Note: Kiel Canal and St. Lawrence Seaway are two important waterways

Indian Ocean

Total area: 73,600,000 sq km (28,460,960 sq mi); Arabian Sea, Bass Strait, Bay of Bengal, Java Sea, Persian Gulf, Red Sea, Strait of Malacca, Timor Sea, and other tributary water bodies

Comparative area: slightly less than eight times the size of the US; third largest ocean (after the Pacific Ocean and Atlantic Ocean, but larger than the Arctic Ocean)

Coastline: 66,526 km (41,312 mi)

Climate: northeast monsoon (December to April), southwest monsoon (June to October); tropical cyclones occur during May/June and October/November in the north Indian Ocean and January/February in the south Indian Ocean

Terrain: surface dominated by counterclockwise gyre (broad, circular system of currents) in the south Indian Ocean; unique reversal of surface currents in the north Indian Ocean—low pressure over southwest Asia from hot, rising, summer air results in the southwest monsoon and southwest-to-northeast winds and currents, while high pressure over northern Asia from cold, falling, winter air results in the northeast monsoon and northeast-to-southwest winds and currents; ocean floor is dominated by the Mid-Indian Ocean Ridge and subdivided by the Southeast Indian Ocean Ridge, Southwest Indian Ocean Ridge, and Ninety East Ridge; maximum depth is 7,258 m (23,813 ft) in the Java Trench

Natural resources: oil and gas fields, fish, shrimp, sand and gravel aggregates, placer deposits, polymetallic nodules

Environment: endangered marine species include the dugong, seals, turtles, and whales; oil pollution in the Arabian Sea, Persian Gulf, and Red Sea

Note: major choke points include Bab el Mandeb, Strait of Hormuz, Strait of Malacca, southern access to the Suez Canal, and the Lombok Strait; ships subject to superstructure icing in extreme south near Antarctica from May to October

Communications

Ports: Bombay (India), Calcutta (India), Colombo (Sri Lanka), Durban (South Africa), Fremantle (Australia), Jakarta (Indonesia), Melbourne (Australia), Richard's Bay (South Africa)

Telecommunications: no submarine cables

Pacific Ocean

Total area: 165,384,000 km (63,854,762 sq mi); includes Arafura Sea, Banda Sea, Bellingshausen Sea, Bering Sea, Bering Strait, Coral Sea, East China Sea, Gulf of Alaska, Makassar Strait, Philippine Sea, Ross Sea, Sea of Japan, Sea of Okhotsk, South China Sea, Tasman Sea, and other tributary water bodies

Comparative area: slightly less than 18 times the size of the US; the largest ocean (followed by the Atlantic Ocean, Indian Ocean, and Arctic Ocean); covers about one-third of the global surface; larger than the total land area of the world

Coastline: 135,663 km (84,246 mi)

Climate: the western Pacific is monsoonal—a rainy season occurs during the summer months when moisture-laden winds blow from the ocean over the land and a dry season during the winter months when dry winds blow from the Asian land mass back to the ocean

Terrain: surface in the northern Pacific dominated by a clockwise, warm water gyre (broad, circular system of currents) and in the southern Pacific by a counterclockwise, cool water gyre; sea ice occurs in the Bering Sea and Sea of Okhotsk during winter and reaches maximum northern extent from Antarctica in October; the ocean floor in eastern Pacific is dominated by the East Pacific Rise, while the western Pacific is dissected by deep trenches; the world's greatest depth is 10,924 m (33,774 ft) in the Marianas Trench

Natural resources: oil and gas fields, polymetallic nodules, sand and gravel aggregates, placer deposits, fish

Environment: endangered marine species include the dugong, seal lion, sea otter, seals, turtles, and whales; oil pollution in Philippine Sea and South China Sea; dotted with low coral islands and rugged volcanic islands in the southwestern Pacific Ocean; subject to tropical cyclones (typhoons) in southeast and east Asia from May to December (most frequent from July to October); tropical cyclones (hurricanes) may form south of Mexico and strike Central America and Mexico from June to October (most common in August and September); southern shipping lanes subject to icebergs from Antarctica; occasional El Niño phenomenon occurs off the coast of Peru when the trade winds slacken and the warm Equatorial Countercurrent moves south killing the plankton—primary food source for anchovies which move to better feeding grounds, but resident marine birds starve by the thousands

Note: the major choke points are the Bering Strait, Panama Canal, Luzon Strait, and the Singapore Strait; the Equator divides the Pacific Ocean into the North Pacific Ocean and the South Pacific Ocean; ships subject to superstructure icing in extreme north from October to May and in extreme south from May to October; persistent fog in the northern Pacific from June to December is a hazard to shipping; surrounded by a zone of violent volcanic and earthquake activity sometimes referred to as the Pacific Ring of Fire

Communications

Ports: Bangkok (Thailand), Hong Kong, Los Angeles (US), Manila (Philippines), Pusan (South Korea), San Francisco (US), Seattle (US), Shanghai (China), Singapore, Sydney (Australia), Vladivostok (USSR), Wellington (NZ), Yokohama (Japan)

Telecommunications: several submarine cables with network focused on Guam and Hawaii.

SEAS

ADRIATIC SEA or **GULF OF VENICE** Arm of the Mediterranean Sea between Italy and Balkan Peninsula; 805 km (500 mi) long, 177 km (110 mi) wide.

AEGEAN SEA Arm of the Mediterranean Sea between Greece and Turkey; 666 km (400 mi) long, 322 km (200 mi) wide.

ALAND SEA Sea between Ahvenanmaa archipelago and Swedish mainland.

AMUNDSEN SEA Arm of South Pacific Ocean off Marie Byrd Land, Antarctica.

ANDAMAN SEA Arm of the Bay of Bengal E of Andoman and Nicobar Islands; 564,620 sq km (218,000 sq mi).

ARABIAN SEA Part of Indian Ocean between India and Arabia.

ARAFURA SEA Sea between N Australia and Indonesia; 1,287 km (800 mi) long, 563 km (350 mi) wide.

ARAL SEA Inland Sea between Kazakh SSR and Uzbek SSR in SW USSR; 66,457 sq km (25,659 sq mi), 450 km (280 mi) long; 68 m (223 ft) maximum depth.

AZOV SEA Sea between Ukrainian SSR and Rostov Oblast in USSR; 37,599 sq km (14,517 sq mi), 322 km (200 mi) long.

BALI SEA Sea between Kangean Islands and Bali, Indonesia.

BALTIC SEA Arm of the Atlantic Ocean connecting with the North Sea; 1,699 km (1,056 mi) long, 42,229 sq km (163,050 sq mi); Maximum depth 469 m (1,539 ft).

BANDA SEA Sea in E Malay archipelago NE of Timor and SE of Celebes; 738,150 sq km (285,000 sq mi); Maximum depth 6,400 m (21,000 ft).

BARENTS SEA Part of Arctic Ocean N of Norway and USSR; 1,370,359 sq km (529,096 sq mi).

BEAUFORT SEA Part of Arctic Ocean NE of Alaska; Maximum depth 4,572 m (15,000 ft).

BELLINGSHAUSEN SEA Inlet of South Pacific Ocean N of Antarctica.

BERING SEA Part of N Pacific Ocean connecting with the Arctic Ocean by Bering Strait; 2,292,150 sq km (885,000 sq mi); Maximum depth 4,773 m (15,659 ft).

BINGO SEA Expansion of the Inland Sea, Japan.

BISMARCK SEA Part of the West Pacific Ocean enclosed by the Bismarck Archipelago; 805 km (500 mi) long.

BLACK SEA or **EUXINE SEA** Sea between Europe and Asia connecting with Aegean Sea through the Bosporus; 466,200 sq km (180,000 sq mi); Maximum depth 2,210 m (7,250 ft).

CAMOTES SEA Body of water in Visayan Islands, Philippines.

CANDIA SEA See Sea of Crete.

CARIBBEAN SEA Arm of the Atlantic Ocean bounded by the West Indies and South America; 2,176,910 sq km (1,049,000 sq mi); Maximum depth 7,534 m (24,720 ft).

CASPIAN SEA Inland salt lake between Europe and Asia, the largest in the world; 371,994 sq km (143,550 sq mi), 1,200 km (746 mi) long and 434 km (270 mi) wide.

CELEBES SEA Part of the Pacific Ocean connected with the Java Sea by the Makasar Strait; 427,350 sq km (165,000 sq mi), 676 km (420 mi) long, 837 km (520 mi) wide.

CERAM SEA Section of W Pacific Ocean in Indonesia; 420 km (250 mi) by 129 km (80 mi).

CHINA SEA Part of Pacific Ocean reaching from Japan to S end of Malay Peninsula, often divided into East China Sea (1,249,157 sq km; 482,300 sq mi; Maximum depth 2,781 m; 9,126 ft) and South China Sea (2,319,096 sq km; 895,400 sq mi; Maximum depth 4,572 m; 15,000 ft).

CHINNERETH, SEA OF See Sea of Galilee.

CHUKCHI SEA Part of Arctic Ocean N of Bering Strait.

CORAL SEA Part of Pacific Ocean between Queensland, Australia and the New Hebrides and New Caledonia.

CRETE, SEA OF Part of the E Mediterranean Sea N of Crete.

DEAD SEA Salt lake on the boundary between Israel and Jordan; 82 km (51 mi) by 18 km (11 mi), 1,020 sq km (394 sq mi); Lowest point on earth's surface 397 m (1,302 ft) below the level of the Mediterranean Sea.

EAST SIBERIAN SEA Part of the Arctic Ocean from New Siberian Islands to Wrangel Island.

FLORES SEA Sea between E end of Java Sea and W end of Banda Sea; 241 km (150 mi) wide.

GALILEE, SEA OF, also **SEA OF TIBERIAS** Freshwater lake in N Israel; 21 km (13 mi) by 11.2 km (7 mi); 212 m (696 ft) below sea level.

IZMIT Inlet of E Sea of Marmara NW coast of Turkey in Asia.

IZMIR Inlet of the Aegean Sea W Turkey opposite Chios Island; 64 km (40 mi) long.

JOSEPH BONAPARTE GULF Inlet of the Timor Sea NE Western Australia; 362 km (225 mi) by 161 km (100 mi).

KALAMATA See Gulf of Messina.

KALLONI Inlet of the Aegean Sea in S Lesbos Island.

KANDALAKSHA Inlet of NW White Sea on NW coast of RSFSR, USSR.

KARA BOGAZ GOL Inlet of E Caspian Sea on coast of Turkmen SSR, USSR; 161 km (100 mi) by 137 km (85 mi).

KASSANDRA See Toronaic Gulf.

KULA GULF Between NW New Georgia Island and Kolombangara Island in Solomon Islands; 27 km (17 mi) by 16 km (10 mi).

KUSADASI GULF Inlet of E Aegean Sea W coast of Turkey NE of W coast of Samos; 72 km (45 mi) by 32 km (20 mi).

KUTCH or **GULF OF CUTCH** Inlet of the Arabian Sea, Gujerat, W India.

LACONIA Inlet of the Mediterranean Sea S Greece 48 km (30 mi) long.

LADRILLERO Inlet of Pacific Ocean W of Wellington Island of SW coast of Chile.

LAGONOY Inlet of the Pacific Ocean in SE Luzon, Philippines.

LEYTE Inlet of the Pacific Ocean E of Leyte and S of Samar, E Philippines.

LIAO-TUNG N part of the Gulf of Chihli, NE China.

LINGAYEN GULF Inlet of the S China Sea on NW coast of Philippines 56 km (35 mi) by 37 km (23 mi).

LIONS Inlet of the Mediterranean Sea S coast of France.

MALIAKOS Inlet of the Aegean Sea E coast of Greece, a W extension of the Atalante Channel.

MALI KVARNER Gulf NE Adriatic Sea between Cres Island and Losinj Island on W and Pag Island and Rag Island on E.

MANDAR Strait on SW coast of Celebes Island, Indonesia, N of Makasar.

MANNAR Part of Indian Ocean W of Sri Lanka SE of the S tip of India.

MARACAIBO See Gulf of Venezuela.

MARTABAN Inlet of the Bay of Bengal on coast of Lower Myanmar.

MESSINIA Inlet of the Mediterranean Sea of SW coast of Peloponnesus, Greece.

MEXICO SE coast of N America; 1,592 sq km (615 sq mi); 1,609 km (1,000 mi) by 1,247 km (775 mi); Maximum depth 3,732 m (12,245 ft).

MILAZZO Inlet of the Mediterranean Sea N coast of Sicily 25 km (16 mi).

MONTIJO Inlet of Pacific Ocean SW coast of Panama.

MORROSQUILLO Inlet of Caribbean Sea NW coast of Colombia.

MOSQUITO Inlet of the Caribbean Sea N coast of Panama.

MURCIELAGOS Inlet of the Pacific Ocean NW coast of Costa Rica.

NASSAU Body of water S Tierra del Fuego, S Chile.

NICOYA Inlet of Pacific Ocean NW central coast of Costa Rica.

OB Inlet of the Arctic Ocean E of Yamal Peninsula, USSR; 885 km (550 mi) by 80 km (50 mi).

OMAN Arm of the Arabian Sea extending between N Oman, SE Arabian Peninsula and SE Iran; 547 km (340 mi) by 370 km (230 mi).

ORISTANO Inlet of the Mediterranean Sea on W central coast of Sardinia; 16 km (10 mi) long.

OROSEI Inlet of the Tyrrhenian Sea E coast of Sardinia, Italy.

PALMAS SW coast of Sardinia, Italy.

PANAMA Inlet of the Pacific Ocean S coast of Panama.

PANAY Inlet of NE Sulu Sea, Philippines; 72 km (45 mi) long.

PAPUA Inlet of the Coral Sea, SE coast of Papua New Guinea.

PARIA Inlet of the Atlantic Ocean between the W coast of the island of Trinidad and the Venezuelan mainland.

PARITA Inlet of the Gulf of Panama on the W.

PATRAS, also **GULF OF CALYDON** Inlet of the Ionian Sea W of Greece.

PENAS Inlet of S. Pacific Ocean on SW coast of Chile.

PERSIAN GULF, also **ARABIAN GULF** Arm of the Arabian Sea; 885 km (550 mi) long and 322 km (200 mi) wide; 229,992 sq km (88,800 sq mi); Average depth 100 m (328 ft).

QUEEN MAUD GULF Gulf between SE Victoria Island and the mainland Northwest Territories, Canada.

RAGAY GULF N arm of the Sibuyan Sea in SE Luzon, Philippines; 96 km (60 mi) by 51 km (32 mi).

RIGA Inlet of the NE Baltic Sea extending S into Latvia, USSR; 160 km (100 mi) by 96 km (60 mi).

ROSAS Inlet of the Mediterranean Sea NE coast of Spain.

ST. FLORENT Inlet, NW coast of Corsica, France.

ST. LAWRENCE Gulf of the Atlantic Ocean off E coast of Canada, between New Foundland and the Canadian mainland of Quebec.

ST. MALO Inlet of the English Channel NW coast of France.

ST. VINCENT South Australia, E of Yorke Peninsula; 160 km (100 mi) long.

SAKHALIN Inlet of Sea of Okhotsk between N end of Sakhalin Island and mainland USSR.

SALERNO Inlet of the Tyrrhenian Sea SW coast of Italy.

SALONIKA Arm of the NW Aegean Sea extending into NE coast of Greece.

SALUM Inlet of the Mediterranean Sea; 442 km (275 mi) W of Alexandria, Egypt.

SAN BLAS Inlet of the Caribbean Sea on the N coast of Panama.

SAN JORGE (1) Inlet of the Atlantic Ocean S Argentina. (2) Inlet of the Mediterranean Sea on E coast of Spain.

SAN JOSE Inlet of Gulf of San Mateas, S Argentina.

SAN MATEAS Inlet of the Atlantic Ocean in SE Rio Negro Province S central Argentina.

SAN MIGUEL Inlet of the Gulf of Panama extending E into SE Panama.

SANT EUFEMIA Inlet of the Tyrrhenian Sea, W coast of Calabria, S Italy.

SARONIC GULF or **GULF OF AEGINA** Inlet of the Aegean Sea on SE coast of Greece.

SAROS GULF Inlet of the NE Aegean Sea extending E into SW Turkey.

SHELIKHOV GULF Gulf in Khabarovsk Krai, USSR; 322 km (200 mi) wide.

SIAM also **GULF OF THAILAND** Inlet of the South China Sea; 619 km (385 mi) by 619 km (385 mi).

SIDRA Inlet of the Mediterranean Sea N central coast of Libya.

SINGITIC GULF Inlet of the Aegean Sea, NE coast of Greece.

SORSOGON GULF Landlocked body of water in Luzon, Philippines; 30 km (19 mi) by 5 to 13 km (3 to 8 mi.)

SPENCER GULF also **SPENCER'S GULF** Inlet, S Australia, between Yorke and Eyre Peninsulas; 322 km (200 mi) by 145 km (90 mi).

SPEZIA Inlet of the Gulf of Genoa, NW Italy.

SQUILLACE Inlet of the Ionian Sea, S Italy.

STRYMONIC GULF also, **GULF OF RENDINA** or **GULF OF ORFANI** Inlet of the Aegean Sea NE coast of Greece.

SUEZ NW arm of the Red Sea joined to the Mediterranean Sea by the Suez Canal.

TADJOURA also **GULF OF TAJURA** Inlet of the Gulf of Aden, E of Djibouti.

TAGANROG NE arm of the Sea of Azov, S USSR.

TARANTO also **GULF OF ATRENTUM** Inlet of the Ionian Sea, SE coast of Italy; 112 km (70 mi) long.

TEHUANTEPEC Inlet of the Pacific Ocean, SE Mexico.

TOLO Inlet of the Banda Sea E central coast of Celebes Island, Indonesia.

TOMINI also **GULF OF GORONTALO** Inlet of the Molucca Sea extending into the coast of N Celebes, Indonesia; 386 km (240 mi) long.

TONKIN Arm of the South China Sea E of N Vietnam and W of Hainan Island, S China; 483 km (300 mi) long.

TORONAIC, also **GULF OF CASSANDRA** Inlet of the N Aegean Sea between the Sithonia and Cassandra peninsulas, NE Greece.

TRINIDAD Inlet of the Pacific Ocean SW of S Wellington Island, off SW coast of Chile.

TUNIS Inlet of the Mediterranean Sea NE coast of Tunisia.

URABA Bay NW coast of Colombia at the end of the Gulf of Darien.

VALENCIA Inlet of the Mediterranean Sea E coast of Spain.

VAN DIEMEN Inlet of the Arafura Sea N Northern Territory, Australia.

VELLA GULF Open water area in Solomon Islands in W Pacific Ocean.

VENEZUELA, also **GULF OF MARACAIBO** Inlet of the Caribbean Sea NW Venezuela.

VENICE North section of the Adriatic Sea.

VOLOS, also **PAGASAEAN GULF** Inlet of the Aegean Sea, E Thessaly, E coast of Greece.

WRIGLEY GULF Inlet of South Pacific Ocean E of Hobbs Coast in Marie Byrd Land, Antarctica.

RANKINGS

Section 1: Oceans and Seas
Section 2: Ocean Depths
Section 3: Islands
Section 4: Rivers (1) by length
(2) by area of drainage
(3) by discharge rates
Section 5: Highest Peaks
Section 6: Lakes (1) by area
(2) by depth
(3) by volume of water
Section 7: Largest Countries by Area
Section 8: Highest Cities
Section 9: Coldest Places
Section 10: Warmest Places
Section 11: Volcanoes
Section 12: Highest Waterfalls

1 OCEANS & SEAS

Rank/Name	Area (sq mi, 000)
1 Pacific Ocean	70,017
2 Atlantic Ocean	36,415
3 Indian Ocean	28,617
4 Arctic Ocean	44,732
5 Coral Sea	1,850
6 Arabian Sea	1,492
7 South China Sea	1,423
8 Caribbean Sea	1,063
9 Mediterranean Sea	971
10 Bering Sea	890
11 Bay of Bengal	839
12 Sea of Okhotsk	614
13 Gulf of Mexico	596
14 Gulf of Guinea	592
15 Barents Sea	542
16 Norwegian Sea	534
17 Gulf of Alaska	512
18 Hudson Bay	476
19 Greenland Sea	465
20 Arafura Sea	400
21 Philippine Sea	400
22 Sea of Japan	378
23 East Siberian Sea	348
24 Kara Sea	341
25 China Sea	290
26 Solomon Sea	278
27 Banda Sea	268
28 Baffin Bay	266
29 Laptev Sea	251
30 Timor Sea	237
31 Andaman Sea	232
32 North Sea	232
33 Chukchi Sea	225
34 Great Australian Bight	187
35 Beaufort Sea	184
36 Celebes Sea	182
37 Black Sea	178
38 Red Sea	174
39 Java Sea	167
40 Sulu Sea	162
41 Yellow Sea	161
42 Baltic Sea	149
43 Gulf of Carpentaria	120
44 Molucca Sea	119
45 Persian Gulf	93
46 Gulf of Siam	92
47 Gulf of St. Lawrence	92
48 Gulf of Aden	85
49 Makassar Strait	75
50 Ceram Sea	72

2 OCEAN DEPTHS

Rank/Name	Ocean	Depth (ft)
1 Mariana Trench	Pacific	38,635
2 Philippine Trench	Pacific	37,720
3 Tonga Trench	Pacific	37,166
4 Izu Trench	Pacific	36,850
5 Kermadec Trench	Pacific	34,728
6 Kuril Trench	Pacific	34,678
7 New Britain Trench	Pacific	31,657
8 Puerto Rico Trench	Atlantic	31,037
9 Bonin Trench	Pacific	29,816
10 Japan Trench	Pacific	29,157
11 South Sandwich Trench	Atlantic	28,406
12 Palau Trench	Pacific	27,972
13 Peru-Chile Trench	Pacific	27,687
14 Yap Trench	Pacific	27,552
15 Aleutian Trench	Pacific	26,775
16 Roanche Gap	Atlantic	26,542
17 Cayman Trench	Atlantic	26,519
18 New Hebrides Trench	Pacific	25,971
19 Ryukyu Trench	Pacific	25,597
20 Java Trench	Indian	24,744
21 Diamantina Trench	Indian	24,249
22 Mid America Trench	Pacific	22,297
23 Brazil Basin	Atlantic	22,274
24 Ob Trench	Indian	21,785
25 Vema Trench	Indian	19,482
26 Agulhas Basin	Indian	19,380
27 Ionian Basin	Mediterranean Sea	17,306
28 Eurasia Basin	Arctic	16,122

3 ISLANDS

Rank/Name	Body of Water	Area (sq mi)
AFRICA		
1 Madagascar	Indian O	227,000
2 Kerguelen	Indian O	2,320
3 Reunion	Indian O	969
4 Bioko, Equatorial Guinea	Bight of Bonny	779
5 Tenerife, Canary Is	Atlantic O	745
6 Mauritius	Indian O	720
7 Fuerte Ventura, Canary Is	Atlantic O	642
8 Zanzibar	Indian O	640
9 Gran Canaria, Canary Is	Atlantic O	592
10 Grande Comore, Comoros	Indian	443
11 Sao Tiago, Cape Verde Is	Atlantic O	384
12 Pemba, Zanzibar	Indian	380
13 Dahlac, Ethiopia	Red Sea	347
14 Sao Tome	Gulf of Guinea	330
15 Lanzarote, Canary Is	Atlantic O	302
16 Santo Antao, Cape Verde	Atlantic O	302
17 Madeira	Atlantic O	286
18 La Palma, Canary Is	Atlantic O	256
19 Boa Vista, Cape Verde	Atlantic O	239
20 Djerba, Tunisia	Gulf of Gabes	197

ANTARCTICA

1 Alexander	Bellingshausen	16,700
2 Berkner	Weddell Sea	1,500

ASIA

1 Borneo	South China Sea	285,000
2 Sumatra	Andaman Sea	164,000
3 Honshu, Japan	Pacific O	88,000
4 Celebes, Indonesia	Celebes Sea	67,400
5 Java	Indian O	50,000
6 Luzon, Philippines	Pacific O	40,400
7 Mindanao, Philippines	Pacific O	36,500
8 Hokkaido, Japan	Pacific O	30,100
9 Sakhalin, USSR	Sea of Okhotsk	29,500
10 Sri Lanka	Indian O	25,200
11 Taiwan	Pacific O	13,800
12 Kyushu, Japan	Pacific O	13,800
13 Hainan, China	South China Sea	13,100
14 Timor, Indonesia	Timor Sea	10,200
15 Halmahera, Indonesia	Molucca Sea	6,950
16 Shikoku, Japan	Pacific O	6,860
17 Ceram, Indonesia	Banda Sea	6,620
18 Flores, Indonesia	Flores Sea	5,520
19 October Revolution, USSR	Arctic O	5,470
20 Sumbawa, Indonesia	Indian O	5,160
21 Samar, Philippines	Pacific O	5,050
22 Negros, Philippines	Sulu Sea	4,900
23 Palawan, Philippines	South China Sea	4,550
24 Kotelnyy, USSR	Arctic O	4,500
25 Panay, Philippines	Sulu Sea	4,450
26 Bangka, Indonesia	Java Sea	4,370
27 Bolshevik, USSR	Arctic O	4,350
28 Sumba, Indonesia	Indian O	4,310
29 Mindoro, Philippines	South China Sea	3,760
30 Cyprus	Mediterranean	3,570

EUROPE

1 Great Britain	North Sea	84,400
2 Iceland	Atlantic O	39,700
3 Ireland	Atlantic O	32,500
4 Novaya Zemlya	North Kara Sea	18,900
5 West Spitsbergen, Svalbard	Arctic O	15,300
6 Novaya Zemlya,	Barents Sea	12,800
7 Sicily	Mediterranean	9,810
8 Sardinia, Italy	Mediterranean	9,190
9 North East Land, Svalbard	Barents Sea	5,790
10 Corsica, France	Mediterranean	3,370
11 Crete, Greece	Mediterranean	3,190
12 Zealand, Denmark	Baltic Sea	2,710
13 Edge, Svalbard	Barents Sea	1,940
14 Vendsyssel-Thy, Denmark	Skagerrak Strait	1,810
15 Euboea, Greece	Aegean Sea	1,410
16 Majorca, Spain	Mediterranean	1,400
17 Vaygach, USSR	Kara Sea	1,310
18 Kolguyev, USSR	Barents Sea	1,240
19 Gotland, Sweden	Baltic Sea	1,160
20 Fyn, Denmark	Baltic Sea	1,150

NORTH AMERICA

1 Greenland	Atlantic O	823,000
2 Baffin	Baffin Bay	196,000
3 Victoria, Canada	Viscount Melville Sound	83,900
4 Ellesmere, Canada	Arctic O	75,800
5 Newfoundland	Atlantic O	42,000
6 Cuba	Caribbean Sea	40,500
7 Hispaniola	Atlantic O	29,200
8 Banks, Canada	Arctic O	27,000
9 Devon, Canada	Baffin Bay	21,300
10 Axel Heiberg, Canada	Arctic O	16,700
11 Melville, Canada	Viscount Melville Sound	16,300
12 Southampton, Canada	Husdon Bay	15,900
13 Prince of Wales, Canada	Viscount Melville Sound	12,900
14 Vancouver, Canada	Pacific O	12,100
15 Somerset, Canada	Lancaster Sound	9,570
16 Bathurst, Canada	Viscount Melville Sound	6,190
17 Prince Patrick, Canada	Arctic O	6,120
18 King William, Canada	Queen Maud Gulf	5,060
19 Ellef Ringnes, Canada	Arctic O	4,360
20 Bylot, Canada	Baffin Bay	4,270
21 Jamaica	Caribbean Sea	4,240
22 Cape Breton, Canada	Atlantic O	3,980
23 Prince Charles, Canada	Foxe Basin	3,680
24 Kodiak, Alaska	Pacific O	3,670
25 Puerto Rico	Atlantic O	3,350
26 Disco, Greenland	Davis Strait	3,310
27 Anticosti, Quebec	Gulf of St. Lawrence	3,070
28 Cornwallis, Canada	Barrow Strait	2,700
29 Prince of Wales, Alaska	Pacific O	2,590
30 Graham, Canada	Pacific O	2,460

OCEANIA

1 New Guinea	Pacific O	305,000
2 South Island, New Zealand	Pacific O	58,200
3 North Island, New Zealand	Pacific O	44,200
4 Tasmania, Australia	Indian O	24,900
5 New Britain, Papua New Guinea	Bismarck Sea	14,600
6 New Caledonia	Coral Sea	6,470
7 Dolak, Irian	Barat Arafura Sea	4,160
8 Hawaii	Pacific O	4,040
9 Viti Levu, Fiji	Pacific O	4,010
10 Bougainville, Papua New Guinea	Pacific O	3,880

11 New Ireland, Papua New Guinea	Pacific O	3,340
12 Melville, Australia	Timor Sea	2,400
13 Guadalcanal, Solomon Is	Solomon Sea	2,170
14 Vanua Levu, Fiji	Pacific O	2,140
15 Espiritu Santo	Coral Sea	1,930

SOUTH AMERICA

1 Grande de Tierra del Fuego	Atlantic O	18,700
2 Marajo, Brazil	Atlantic O	18,500
3 Bananal, Brazil	Araguaia R	7,720
4 Chiloe, Chile	Pacific O	3,240
5 Wellington, Chile	Trinidad Gulf	2,610
6 East Falkland	Atlantic O	2,440
7 Santa Ines, Chile	Pacific O	2,120
8 Caviana, Brazil	Atlantic O	1,920
9 Grande de Gurupa, Brazil	Amazon R	1,880
10 West Falkland	Atlantic O	1,680
11 Hoste, Chile	Pacific O	1,590
12 Isabela, Ecuador	Pacific O	1,450
13 South Georgia, Falkland Is	Atlantic O	1,450
14 Riesco, Chile	Strait of Magellan	1,200
15 Magdalena, Chile	Moraleda Channel	998
16 Navarino, Chile	Nassau Bay	955
17 Clarence, Chile	Strait of Magellan	863
18 Mexiana, Brazil	Atlantic O	592
19 Sao Luis, Brazil	Atlantic O	465
20 Santa Cruz, Ecuador	Pacific O	358

4 RIVERS

I BY LENGTH

Rank/Name	Length in mi
1 Nile	4,160
2 Amazon	4,080
3 Mississippi	3,740
4 Yangtze	3,720
5 Yenisei	3,650
6 Amur	3,590
7 Ob-Irtysh	3,360
8 Plata	3,030
9 Yellow	3,010
10 Zaire	2,880
11 Lena	2,730
12 Mackenzie	2,630
13 Mekong	2,600
14 Niger	2,550
15 Murray	2,330
16 Volga	2,290
17 Madeira	1,990
18 Yukon	1,980
19 Ob	1,970
20 St. Lawrence	1,900
21 Rio Grande	1,880
22 Syr	1,880
23 Jurua	1,860
24 Purus	1,860
25 Lower Tunguska	1,860
26 Indus	1,790
27 Danube	1,770
28 Brahmaputra	1,770
29 Salween	1,750
30 Sao Francisco	1,730
31 Para	1,710
32 Tarim	1,710
33 Vilyuy	1,650
34 Zambezi	1,650
35 Amu Darya	1,630
36 Paraguay	1,610
37 Nelson	1,600
38 Ural	1,570
39 Ganges	1,560
40 Shebeli	1,550
41 Ubangi	1,530
42 Ishim	1,520
43 Shatt al-Arab	1,510
44 Arkansas	1,450
45 Colorado	1,450
46 Olenek	1,420
47 Caqueta	1,420
48 Dnieper	1,420
49 Aldan	1,410
50 Atchafalaya	1,400

II BY AREA OF DRAINAGE BASINS

Rank/Name	Drainage Basin (sq mi 000)
1 Amazon	2,375
2 Zaire	1,476
3 Mississippi	1,247
4 Plata	1,197
5 Ob-Irtysh	1,154
6 Nile	1,082
7 Yenisei	1,011
8 Lena	961
9 Niger	808
10 Amur	792
11 Yangtze	705
12 Mackenzie	697
13 Volga	525
14 Zambezi	514
15 St. Lawrence	503
16 Madeira	463
17 Indus	450
18 Shatt al-Arab	427
19 Paraguay	425
20 Nelson	414
21 Murray	408
22 Negro	386
23 Ganges	368
24 Kasai	349
25 Orinoco	340

Rank/Name	
26 Yukon	328
27 Para	323
28 Mekong	313
29 Salado	309
30 Okovanggo	303
31 Danube	298
32 Ubangi	298
33 Yellow	297
34 Ob	295
35 Aldan	281
36 Shari	270
37 Orange	261
38 Ohio	258
39 Columbia	253
40 Kolima	248
41 Colorado	246
42 Sao Francisco	236
43 Guapore	232
44 Brahmaputra	224
45 Panjnad	206
46 Sungari	202
47 Kama	196
48 Dnieper	195
49 Lower Tunguska	183
50 Amu	180

III BY DISCHARGE RATES

Rank/Name	Discharge rate (1,000 cubic ft/second)
1 Amazon	6,180
2 Zaire	1,377
3 Negro	1,236
4 Yangtze	1,137
5 Orinoco	890
6 Plata	809
7 Madeira	770
8 Yenisey	636
9 Brahmaputra	575
10 Lena	569
11 Zambezi	565
12 Mekong	501
13 St. Lawrence	460
14 Mississippi	453
15 Irrawaddy	447
16 Ganges	441
17 Pearl	409
18 Ob-Irtysh	360
19 Salween	353
20 Kasai	351
21 Amur	346
22 Mackenzie	343
23 Fly	318
24 Para	305
25 Magdalena	283
26 Volga	271
27 Ubangi	265
28 Ohio	257
29 Caqueta	247
30 Yukon	240
31 Indus	235
32 Danube	227
33 Tapajos-Juruena	212
34 Niger	201
35 Uruguay	194
36 Columbia	189
37 Atchafalaya	181
38 Aldan	177
39 Caroni	177
40 Putumayo	177
41 Ob	174
42 Ogooue	165
43 Paraguay	155
44 Jurua	141
45 Pechora	141
46 Fraser	137
47 Kama	128
48 Columbia	120
49 Northern Dvina	120
50 Lower Tunguska	118

5 HIGHEST PEAKS

Rank/Name	Country	Height(ft)
AFRICA		
1 Kibo, Kilimanjaro	Tanzania	19,340
2 Mawensi, Kilimanjaro	Tanzania	17,100
3 Batian	Kenya	17,050
4 Nelion	Kenya	17,020
5 Margherita	Uganda-Zaire	16,760
6 Alexandra	Uganda-Zaire	16,700
7 Albert	Zaire	16,690
8 Savoia	Uganda	16,330
9 Elena	Uganda	16,300
10 Elizabeth	Uganda	16,170
11 Philip	Uganda	16,140
12 Moebius	Uganda	16,130
13 Vittorio Emanuele	Uganda	16,040
14 Ensonga	Uganda	15,960
15 Edward	Uganda	15,890
ANTARCTICA		
1 Vinson	Antarctica	16,860
2 Tyree	Antarctica	16,290
3 Shinn	Antarctica	15,750
4 Gardner	Antarctica	15,370
5 Epperly	Antarctica	15,100
ASIA		
1 Everest	Nepal-China	29,030
2 K2	Jammu-Kashmir	28,250
3 Kanchenjunga	Nepal-India	28,170
4 Lhotse	China-Nepal	27,890
5 Kanchenjunga S peak	India-Nepal	27,800
6 Makalu	China-Nepal	27,790
7 Kanchenjunga W peak	India-Nepal	27,620

8 Lhotse Shar	China-Nepal	27,500
9 Dhaulagiri	Nepal	26,810
10 Cho Oyu	China-Nepal	26,750
11 Manaslu	Nepal	26,660
12 Nanga Parbat	Jammu-Kashmir	26,660
13 Annapurna I	Nepal	26,500
14 Gasherbrum	Jammu-Kashmir	26,400
15 Broad	Jammu-Kashmir	26,400
16 Gasherbrum II	Jammu-Kashmir	26,360
17 Gosainthan	Tibet, China	26,290
18 Broad, Middle peak	Jammu-Kashmir	26,250
19 Gasherbrum III	Jammu-Kashmir	26,090
20 Annapurna II	Nepal	26,040
21 Gasherbrum IV	Jammu-Kashmir	26,000
22 Gyachung Kang	China-Nepal	25,990
23 Nanga Parbat II	Jammu-Kashmir	25,950
24 Kangbachen	India-Nepal	25,930
25 Manaslu E	Nepal	25,900
26 Distaghil Sar	Jammu-Kashmir	25,870
27 Nuptse	Nepal	25,850
28 Himachuli	Nepal	25,800
29 Khiangyang Kish, Karakorum	Jammu-Kashmir	25,760
30 Ngojumba Ri	China-Nepal	25,720
31 Dakura	Nepal	25,710
32 Masherbrum E	Jammu-Kashmir	25,660
33 Nanda Devi W peak	India	25,650
34 Nanga Parbat, N peak	Jammu-Kashmir	25,650
35 Chomo Lonzo	China-Nepal	25,640
36 Masherbrum W peak	Jammu-Kashmir	25,610
37 Rakaposhi	Jammu-Kashmir	25,550
38 Batura Mustagh	Jammu-Kashmir	25,540
39 Gasherbrum	Jammu-Kashmir	25,500
40 Kamet	China-India	25,500

EUROPE

1 Elbrus W peak	USSR	18,480
2 Elbrus E peak	USSR	18,360
3 Shkhara	USSR	17,060
4 Dykh	USSR	17,050
5 Koshtan	USSR	16,880
6 Shkhara	USSR	16,880
7 Pushkina	USSR	16,730
8 Dzangi	USSR	16,570
9 Kazbek	USSR	16,560
10 Katyn	USSR	16,310
11 Shota Rustaveli	USSR	16,270
12 Mizhirgi W	USSR	16,170
13 Mizhirgi E	USSR	16,140
14 Kundyum	USSR	16,010
15 Gestola	USSR	15,930
16 Tetnuld	USSR	15,920
17 Mont Blanc	France-Italy	15,770
18 Dzhimariy	USSR	15,680
19 Adish	USSR	15,570
20 Ushba	USSR	15,450

NORTH AMERICA

1 Mt. McKinley S	USA	20,320
2 Logan Central peak	Canada	19,520
3 Logan West peak	Canada	19,470
4 McKinley N	USA	19,470
5 Logan East peak	Canada	19,420
6 Citlatepetl	Mexico	18,700
7 Logan N peak	Canada	18,270
8 Saint Elias	USA-Canada	18,010
9 Popocatepetl	Mexico	17,890
10 Foraker	USA	17,400
11 Ixtacihuatl	Mexico	17,340
12 Queen	Canada	17,300
13 Lucania	Canada	17,150
14 King	Canada	16,970
15 Steele	Canada	16,640
16 Bona	USA	16,500
17 Blackburn	USA	16,390
18 Blackburn SE	USA	16,290
19 Sanford	USA	16,240
20 Wood	Canada	15,880

OCEANIA

1 Ngga Pulu	Indonesia	16,500
2 Daam	Indonesia	16,150
3 Pilimsit	Indonesia	15,750
4 Trikora	Indonesia	15,580
5 Mandala	Indonesia	15,420
6 Wilhelm	Papua New Guinea	15,400
7 Wisnumurti	Indonesia	15,080
8 Yamin	Indonesia	14,860
9 Kubor	Papua New Guinea	14,300
10 Herbert	Papua New Guinea	14,000

SOUTH AMERICA

1 Aconcagua	Argentina	22,840
2 Ojos del Salado	Argentina-Chile	22,560
3 Bonete	Argentina	22,550
4 Pissis	Argentina	22,240
5 Huascaran S	Peru	22,210
6 Mercedario	Argentina	22,210
7 Llullaillaco	Argentina-Chile	22,100
8 Libertador	Argentina	22,050
9 Ojos del Salado	Argentina-Chile	22,050

Rank/Name	Country	Height (ft)
10 Tupungato	Argentina-Chile	21,900
11 Gonzalez	Argentina-Chile	21,850
12 Huascaran N	Peru	21,840
13 Muerto	Argentina-Chile	21,820
14 Yerupaja N	Peru	21,760
15 Incahuasi	Argentina-Chile	21,700
16 Galan	Argentina	21,650
17 Tres Cruces	Argentina-Chile	21,540
18 Gonzalez N	Argentina-Chile	21,490
19 Sajama	Bolivia	21,460
20 Yerupaja	Peru	21,380

6 LAKES

I BY AREA

Rank/Name	Country	Area (sq mi)
1 Caspian Sea	Iran-USSR	143,240
2 Superior	Canada-USA	31,760
3 Aral Sea	USSR	24,900
4 Victoria	Uganda-Tanzania-Kenya	24,300
5 Huron	Canada-USA	23,000
6 Michigan	USA	22,400
7 Tanganyika	Burundi-Tanzania-Zaire-Zambia	12,350
8 Baikal	USSR	12,160
9 Great Bear	Canada	12,100
10 Great Slave	Canada	11,030
11 Erie	Canada-USA	9,920
12 Winnipeg	Canada	9,420
13 Malawi	Malawi-Tanzania-Mozambique	8,680
14 Ontario	Canada-USA	7,440
15 Ludoga	USSR	7,000
16 Balkhash	USSR	6,560
17 Maracaibo	Venezuela	5,020
18 Chad	Chad-Cameroon-Niger-Nigeria	4,000
19 Patos	Brazil	3,920
20 Onega	USSR	3,750
21 Nicaragua	Nicaragua	3,150
22 Titicaca	Bolivia-Peru	3,100
23 Athabasca	Canada	3,060
24 Eyre	Australia	2,970
25 Reindeer	Canada	2,570
26 Rudolf	Ethiopia-Kenya	2,470
27 Issyk Kul	USSR	2,410
28 Torrens	Australia	2,230
29 Albert	Zaire-Uganda	2,160
30 Vanern	Sweden	2,160
31 Netilling	Canada	2,140
32 Winnipegosis	Canada	2,070
33 Bangweulu	Zambia	1,930
34 Nipigon	Canada	1,870
35 Gairdner	Australia	1,840
36 Manitoba	Canada	1,800
37 Koko	China	1,720
38 Kyoga	Uganda	1,710
39 Mweru	Zaire	1,680
40 Great Salt	USA	1,680
41 Peipus	USSR	1,660
42 Woods	Canada	1,580
43 Taymyr	USSR	1,540
44 Khanka	China	1,540
45 Orumiyeh	Iran	1,500
46 Dubawnt	Canada	1,480
47 Van	Turkey	1,440
48 Tana	Ethiopia	1,390
49 Poyang	China	1,290
50 Uvs	Mongolia	1,290

II BY DEPTH (ft)

Rank/Name	Country	Depth (ft)
1 Baikal	USSR	5,315
2 Tanganyika	Burundi-Tanzania-Zaire-Zambia	4,825
3 Caspian Sea	Iran-USSR	3,363
4 Malawi	Malawi-Tanzania-Mozambique	2,316
5 Issyk Kul	USSR	2,303
6 Great Slave	Canada	2,015
7 Matana	Indonesia	1,936
8 Crater	USA	1,932
9 Toba	Indonesia	1,736
10 Sarez	USSR	1,657
11 Tahoe	USA	1,645
12 Chelan	USA	1,605
13 Kivu	Rwanda-Zaire	1,575
14 Quesnel	Canada	1,560
15 Adams	Canada	1,500
16 Mjosa	Norway	1,473
17 Manapuri	New Zealand	1,453
18 Nahuel Huapi	Argentina	1,437
19 Dead Sea	Israel-Jordan	1,421
20 Tazawa	Japan	1,394
21 Great Bear	Canada	1,356
22 Como	Italy	1,352
23 Superior	Canada-USA	1,333
24 Hawea	New Zealand	1,286
25 Maggiore	Italy-Switzerland	1,221
26 Chilko	Canada	1,200
27 Pend Oreille	USA	1,200
28 Shikotsu	Japan	1,191
29 Powell	Canada	1,174
30 Llanquihue	Chile	1,148
31 Garda	Italy	1,135

32 Towada	Japan	1,096
33 Telestskoya	USSR	1,066
34 Eutsuk	Canada	1,060
35 Geneva	France-Switzerland	1,017
36 Morar	Scotland	1,017
37 Kurile	USSR	1,004
38 Walker	USA	1,000
39 Titicaca	Bolivia-Peru	997
40 Argentino	Argentina	984
41 Iliamna	USA	980
42 Tyrifjorden	Norway	968
43 Lugano	Italy-Switzerland	945
44 Takla	Canada	941
45 Ohrid	Albania-Yugoslavia	938
46 Atlin	Canada	930
47 Nuyakuk	USA	930
48 Michigan	USA	923
49 Harrison	Canada	916
50 Te Anau	New Zealand	906

III BY VOLUME OF WATER (cubic miles)

1 Caspian Sea	Iran-USSR	16,021
2 Baikal	USSR	5,517
3 Tanganyika	Burundi-Tanzania-Zaire	4,277
4 Superior	Canada-USA	2,941
5 Malawi	Malawi-Tanzania-Mozambique	1,473
6 Michigan	USA	1,185
7 Huron	Canada-USA	849
8 Victoria	Kenya-Tanzania-Uganda	604
9 Great Bear	Canada	550
10 Great Slave	Canada	501
11 Issyk Kul	USSR	417
12 Ontario	Canada-USA	398
13 Aral Sea	USSR	348
14 Ladoga	USSR	218
15 Titicaca	Bolivia-Peru	198
16 Erie	Canada-USA	116
17 Winnipeg	Canada	89
18 Kivu	Rwanda-Zaire	80
19 Onega	USSR	70
20 Maracaibo	Venezuela	67

7 LARGEST COUNTRIES

AREA

Rank/Name	**Area (1,000 sq mi)**	**% of World-Land Area**
1 USSR	8,649	14.9
2 Canada	3,852	6.6
3 China	3,769	6.5
4 USA	3,683	6.3
5 Brazil	3,286	5.7
6 Australia	2,966	5.1
7 India	1,184	2.0
8 Argentina	1,073	1.9
9 Sudan	967	1.7
10 Zaire	906	1.6
11 Algeria	885	1.5
12 Greenland	840	1.4
13 Saudi Arabia	830	1.4
14 Mexico	781	1.3
15 Indonesia	741	1.3
16 Libya	731	1.3
17 Iran	636	1.1
18 Mongolia	604	1.0
19 Peru	496	0.86
20 Niger	482	0.83
21 Angola	481	0.83
22 Mali	479	0.83
23 Ethiopia	472	0.81
24 South Africa	471	0.81
25 Chad	459	0.79
26 Colombia	440	0.76
27 Bolivia	424	0.73
28 Mauritania	398	0.69
29 Egypt	387	0.67
30 Tanzania	365	0.63
31 Nigeria	357	0.62
32 Venezuela	352	0.61
33 Namibia	318	0.55
34 Turkey	315	0.54
35 Pakistan	310	0.54
36 Mozambique	309	0.53
37 Chile	292	0.50
38 Zambia	291	0.50
39 Myanmar	262	0.45
40 Afghanistan	250	0.43
41 Somalia	246	0.42
42 Central African Republic	241	0.41
43 Botswana	232	0.40
44 Madagascar	227	0.39
45 Kenya	225	0.39
46 France	213	0.37
47 Thailand	198	0.34
48 Spain	195	0.34
49 Cameroon	184	0.32
50 Papua New Guinea	178	0.31

8 HIGHEST CITIES

Rank/City	Elevation (ft)
1 Potosi, Bolivia	13,045
2 Lhasa, Tibet	12,002
3 La Paz, Bolivia	11,736
4 Cuzco, Peru	11,152
5 Quito, Ecuador	9,249
6 Sucre, Bolivia	9,154
7 Toluca de Lerdo, Mexico	8,793
8 Bogota, Columbia	8,675
9 Cochabamba, Bolivia	8,393
10 Pachuca de Soto, Mexico	7,900
11 Addis Ababa, Ethiopia	7,900
12 Asmara, Ethiopia	7,789
13 Arequipa, Peru	7,559
14 Mexico City, Mexico	7,546
15 Netzahualcoyotl, Mexico	7,474
16 Sining, China	7,363
17 Sana, North Yemen	7,260
18 Simla, India	7,225
19 Puebla, Mexico	7,094
20 Manizales, Colombia	7,021
21 Santa Fe, USA	6,996
22 Guanajuato	6,726
23 Morelia, Mexico	6,368
24 Kunming, China	6,211
25 Durango, Mexico	6,198
26 Aguascalientes, Mexico	6,195
27 Leon, Mexico	6,185
28 San Luis Potosi, Mexico	6,158
29 Harar, Ethiopia	6,089
30 Cheyenne, USA	6,062
31 Colorado Springs, USA	6,012
32 Celaya, Mexico	5,932
33 Kabul, Afghanistan	5,903
34 Taxco, Mexico	5,840
35 Hamadan, Iran	5,824
36 Butte, USA	5,755
37 Johannesburg, South Africa	5,750
38 Kokiu, China	5,709
39 Irapuato, Mexico	5,656
40 Mixco, Guatemala	5,551
41 Queretaro, Mexico	5,528
42 Nairobi, Kenya	5,453
43 Germiston, South Africa	5,450
44 Boulder, USA	5,430
45 Windhoek, South Africa	5,428
46 Uruapan de Progreso, Mexico	5,361
47 Lakewood, USA	5,355
48 Taif, Saudi Arabia	5,348
49 Aurora, USA	5,342
50 Denver, USA	5,280

9 COLDEST PLACES

Rank/Place/Country	Lowest Recorded Temperature F
1 Eismitte, Greenland	−85
2 Yakutsk, USSR	−84
3 Fairbanks, USA	−66
4 Aklavik, Canada	−62
5 Ulaanbaatar, Mongolia	−48
6 Harbin, China	−43
7 Kuusamo, Finland	−40
8 Haparanda, Sweden	−34
9 Krakow, Poland	−28
10 Cluj, Rumania	−26
11 Prerov, Czechoslovakia	−23
12 Debrecen, Hungary	−22
13 Trondheim, Norway	−22
14 Erzurum, Turkey	−22
15 Kushiro, Japan	−19
16 Pyongyang, North Korea	−19
17 Nuremburg, West Germany	−18
18 Sofia, Bulgaria	−17
19 Innsbruck, Austria	−16
20 Belgrade, Yugoslavia	−14
21 Lyon, France	−13
22 Kermanshah, Iran	−13
23 Aarhus, Denmark	−12
24 Seoul, South Korea	−12
25 Zurich, Switzerland	−12

10 WARMEST PLACES

Rank/Place/Country	Highest Recorded Temperature in F
1 Arouane, Mali	130
2 Cloncurry, Australia	127
3 Abadan, Iran	127
4 Wadi Halfa, Sudan	127
5 Fort Flatters, Algeria	124
6 Aswan, Egypt	124
7 Mosul, Iraq	124
8 Cufra, Libya	122
9 Multan, Pakistan	122
10 Gabes, Tunisia	122
11 Faya, Chad	121
12 Nema, Mauritania	120
13 Marrakech, Morocco	120
14 Dhahran, Saudi Arabia	120
15 Lucknow, India	119
16 Kuwait	119
17 Sharjah, United Arab Emirates	118
18 Phoenix, Arizona	118
19 Ouagadougou, Burkina	118
20 Djibouti	117
21 Guaymas, Mexico	117
22 Chicao, Mozambique	117
23 Berbera, Somalia	117
24 Seville, Spain	117
25 Aleppo, Syria	117

11 VOLCANOES

Rank/Volcano/Country	Height (ft)
AFRICA	
1 Kilimanjaro, Tanzania	19,340
2 Cameroon, Cameroons	13,354
3 Teide, Canary Is	12,198
4 Nyirangongo, Zaire	11,400
5 Nyamuragira, Zaire	10,028
6 Fogo, Cape Verde Is	9,281
7 Karthala, Comoro Is	8,000
8 Piton de la Faournaise, Reunion	5,981
9 Erta-Ale, Ethiopia	1,650
ANTARCTICA	
1 Erebus, Ross Island	12,450
2 Big Ben, Heart Island	9,007
3 Melbourne, Victoria Land	8,500
4 Deception Island, South Shetland I	1,890
ASIA-OCEANIA	
1 Klyuchevskaya, USSR	15,584
2 Kerintji, Sumatra	12,467
3 Fuji, Japan	12,388
4 Rindjani, Indonesia	12,224
5 Tolbachik, USSR	12,080
6 Semeru, Java	12,060
7 Ichinskaya, USSR	11,880
8 Kronotskaya, USSR	11,575
9 Kroyakskaya, USSR	11,339
10 Slamet, Java	11,247
11 Raung, Java	10,932
12 Shiveluch, USSR	10,771
13 Dempo, Sumatra	10,364
14 Ardjuno-Welirang, Java	10,354
15 Agung, Bali	10,308
16 Sundoro, Java	10,285
17 Tjiremai, Java	10,098
18 One-Take, Japan	10,049
19 Mayon, Philippines	9,991
20 Papandajan, Java	9,802
21 Gede, Java	9,705
22 Zhupanovsky, USSR	9,705
23 Apo, Philippines	9,690
24 Merapi, Java	9,551
25 Berzymianny, USSR	9,514
26 Marapi, Sumatra	9,485
27 Tambora, Indonesia	9,353
28 Raupehu, New Zealand	9,175
29 Peuetsagoe, Sumatra	9,121
30 Avachinskaya, USSR	9,026
31 Balbi, Solomon Is	9,000
32 Geureudong, Sumatra	8,497
33 Asama, Japan	8,300
34 Sumbing, Sumatra	8,225
35 Tandikat, Sumatra	8,166
36 Niigata Yakeyama, Japan	8,111
37 Yake Dake, Japan	8,064
38 Canalaon, Philippines	8,015
39 Sinabung, Sumatra	7,913
40 Bromo, Java	7,848
41 Idjen, Java	7,828
42 Alaid, Kuril Is	7,662
43 Ulawun, New Britain	7,532
44 Ngauruhoe, New Zealand	7,515
45 Guntur, Java	7,379
46 Bamus, New Britain	7,338
47 Chokai, Japan	7,300
48 Galunggung, Java	7,113
49 Amburombu, Indonesia	7,051
50 Sorikmerapi, Sumatra	7,037
51 Butak Petarangan, Java	6,890
52 Sibajak, Sumatra	6,870
53 Tokachi, Japan	6,813
54 Azuma, Japan	6,700
55 Tangkuban Prahu, Java	6,637
56 Tongariro, New Zealand	6,458
57 Zheltovskaya, USSR	6,401
58 Catarman, Philippines	6,371
59 Kaba, Sumatra	6,358
60 Sangeang Api, Indonesia	6,351
61 Nasu, Japan	6,210
62 Tiatia, Kuril Is	6,013
63 Manam, Papua New Guinea	6,000
64 Soputan, Celebes	5,994
65 Siau, Indonesia	5,853
66 Kelud, Java	5,679
67 Batur, Bali	5,636
68 Temate, Indonesia	5,627
69 Lewotobi, Indonesia	5,591
70 Kirisima, Japan	5,577
71 Lamongan, Java	5,482
72 Keli Mutu, Indonesia	5,460
73 Akita Komaga Take, Japan	5,449
74 Lli Boleng, Indonesia	5,443
75 Gamkunoro, Indonesia	5,364
76 Aso, Japan	5,223
77 Lewotobi Laki Laki, Indonesia	5,217
78 Lokon Empung, Celebes	5,187
79 Bulusan, Philippines	5,115
80 Sarycheva, Kuril Is	4,960
81 Me-akan, Japan	4,931
82 Ibu, Indonesia	4,921
83 Karkar, Papua New Guinea	4,920
84 Karymskaya, USSR	4,869
85 Popevi, New Hebrides	4,755
86 Ambrym, New Hebrides	4,376
87 Mahawu, Celebes	4,367
88 Awu, Indonesia	4,350
89 Lli Lewotolo, Indonesia	4,348
90 Tongkoko, Celebes	3,770
91 Lli Werung, Indonesia	3,678
92 Komaga take, Japan	3,669
93 Sakurazima, Japan	3,668
94 Langila, New Britain	3,586
95 Dukono, Indonesia	3,566
96 Lamington, Papua New Guinea	3,500
97 Lolobau, New Britain	3,058
98 Suwanosezima, Japan	2,640
99 O-Sima, Japan	2,550

100 Usu, Japan	2,400
101 White Island, New Zealand	1,075

CENTRAL AMERICA-CARIBBEAN

1 Tajumulco, Guatemala	13,845
2 Tacana, Guatemala	13,428
3 Acatenango, Guatemala	12,992
4 Fuego, Guatemala	12,582
5 Santiaguito, Guatemala	12,362
6 Atitlan, Guatemala	11,565
7 Irazu, Costa Rica	11,260
8 San Pedro, Guatemala	9,921
9 Poas, Coasta Rica	8,930
10 Pacaya, Guatemala	8,346
11 Izalco, El Salvador	7,749
12 San Miguel, El Salvador	6,994
13 Rincon de la Vieja, Costa Rica	6,234
14 El Viejo (San Cristobal) Nicaragua	5,840
15 Ometepe, Nicaragua	5,106
16 Arenal, Costa Rica	5,092
17 La Soufriere, Guadeloupe	4,813
18 Pelee, Martinique	4,583
19 Momotombo, Nicaragua	4,199
20 Conchagua, El Salvador	4,100
21 Soufriere, St. Vincent	4,048
22 Telica, Nicaragua	3,409

SOUTH AMERICA

1 Guallatiri, Chile	19,882
2 Lascar, Chile	19,652
3 Cotopaxi, Ecuador	19,347
4 El Misti, Peru	19,098
5 Tupungatito, Chile	18,504
6 Tolima, Colombia	18,002
7 Sangay, Ecuador	17,159
8 Tungurahua, Ecuador	16,512
9 Cotacachi, Ecuador	16,204
10 Pichincha, Ecuador	15,696
11 Purace, Colombia	15,604
12 Lautaro, Chile	11,098
13 Llaima, Chile	10,239
14 Villarrica, Chile	9,318
15 Hudson, Chile	8,580
16 Shoshuenco, Chile	7,743
17 Puyehue, Chile	7,349
18 Calbuco, Chile	6,611
19 Alcedo, Galapagos Is	3,599

MID-PACIFIC

1 Mauna Kea, Hawaii	13,796
2 Mauna Loa, Hawaii	13,680
3 Kilauea, Hawaii	4,077

MID-ATLANTIC RANGE

1 Beerenberg, Jan Mayen Is	7,470
2 Tristan da Cunha	6,760
3 Askja, Iceland	4,954
4 Hekla, Iceland	4,892
5 Katla, Iceland	3,182
6 Leirhnukur, Iceland	2,145
7 Krafla, Iceland	2,145
8 Surtsey, Iceland	568

EUROPE

1 Etna, Sicily	11,053
2 Vesuvius, Italy	4,190
3 Stromboli, Italy	3,038
4 Thera, Greece	1,824
5 Vulcano, Italy	1,637

NORTH AMERICA

1 Citlaltepec, Mexico	18,700
2 Popocatepetl, Mexico	17,887
3 Rainier, Washington, USA	14,410
4 Wrangell, Alaska, USA	14,163
5 Colima, Mexico	14,003
6 Torbert, Alaska, USA	11,413
7 Spurr, Alaska, USA	11,069
8 Baker, Washington, USA	10,779
9 Lassen, California, USA	10,457
10 Redoubt, Alaska, USA	10,197
11 Lliamna, Alaska, USA	10,092
12 Mt. St. Helens, Washington, USA	9,677
13 Shishaldin, Aleutian Is., USA	9,387
14 Veniaminof, Alaska, USA	8,225
15 Pavlof, Aleutian Is., USA	8,215
16 Griggs, Alaska, USA	7,600
17 Paricutin, Mexico	7,451
18 Mageik, Alaska, USA	7,244
19 Douglas, Alaska, USA	7,064
20 Chiginagak, Alaska, USA	7,031
21 Katmai, Alaska, USA	6,715
22 Kukak, Alaska, USA	6,700
23 Makushin, Aleutina Is., USA	6,680
24 Pogromni, Alaska, USA	6,588
25 Martin, Alaska, USA	6,050
26 Trident, Alaska, USA	6,010
27 Tanaga, Aleutian Is., USA	5,925
28 Great Sitkin, Aleutian Is., USA	5,710
29 Cleveland, Aleutian Is., USA	5,675
30 Gareloi, Aleutian Is., USA	5,334
31 Korovin, Aleutian Is., USA	4,852
32 Kanaga, Aleutian Is., USA	4,416
33 Ariakchak, Alaska, USA	4,400
34 Akutan, Aleutian Is., USA	4,275
35 Kiska, Aleutian Is., USA	4,275
36 Augustine, Alaska, USA	3,927
37 Little Sitkin, Aleutian Is., USA	3,897
38 Okmok, Aleutian Is., USA	3,519
39 Seguam, Alaska, USA	3,458

12 HIGHEST WATERFALLS

Rank/Waterfall	Country	Height (ft)
AFRICA		
1 Tugela	South Africa	1,350
2 Shire	Malawi	1,200
3 Kaloba	Zaire	1,115
4 Mtarazi	Mozambique-Zimbabwe	1,000
5 Kalambo	Tanzania-Zambia	704
6 Maletsunyane	Lesotho	630
7 Fincha	Ethiopia	508
8 Aughrabies	South Africa	482
9 Baratieri	Ethiopia	459
10 Magwa	South Africa	450
ASIA		
1 Mawsmai	India	1,148
2 Kalupis	Malaysia	1,100
3 Thylliejlongwa	India	997
4 Jog	India	829
5 Kurundu Oya	Sri Lanka	620
6 Diyaluma	Sri Lanka	560
7 Tondano	Celebes, Indonesia	492
8 Ilya Muromets	USSR	463
9 Bambarakanda	Sri Lanka	461
10 Nachi	Japan	430
EUROPE		
1 Ormeli	Norway	1,847
2 Tysse	Norway	1,749
3 Vestre Mardola	Norway	1,535
4 Gavarnie	France	1,385
5 Verma	Norway	1,250
6 Austerbo	Norway	1,247
7 Serio	Italy	1,034
8 Rembesdals	Norway	984
9 Tyssestrengene	Norway	984
10 Staubbach	Switzerland	980
11 Ostre Mardola, Upper	Norway	974
12 Vettis	Norway	902
13 Valur	Morway	892
14 Mollius	Norway	883
15 Austerkrok	Norway	843
16 Seculejo	Spain	820
17 Skykkje	Norway	820
18 Ostre Mardola, Lower	Norway	722
19 Feigum	Norway	715
20 Rogaland	Norway	689
21 Teverone	Italy	680
22 Maradals	Norway	656
23 Aurstaupet	Norway	633
24 Voring	Norway	597
25 Stauber	Switzerland	590
NORTH AMERICA		
1 Ribbon	USA	1,612
2 Della	Canada	1,443
3 Yosemite, Upper	USA	1,430
4 Takakkaw	Canada	1,200
5 Silver Strand	USA	1,170
6 Basaseachic	Mexico	1,020
7 Twin	Canada	900
8 Hunlen	Canada	830
9 Fairy	USA	700
10 Feather	USA	640
11 Bridalveil	USA	620
12 Panther	Canada	600
13 Nevada	USA	594
14 Multnomah	USA	542
15 Sentinel	USA	500
16 Helmcken	Canada	450
17 Bridal Veil	Canada	400
18 Illilouette	USA	370
19 Comet	USA	320
20 Yosemite, Lower	USA	320
OCEANIA		
1 Kahiwa	Hawaii, USA	1,750
2 Cleve-Garth	New Zealand	1,476
3 Ahui	French Polynesia	1,148
4 Wallomombi	Australia	1,100
5 Wallaman	Australia	970
6 Elizabeth Grant	Australia	900
7 Helena	New Zealand	830
8 Sutherland, Upper	New Zealand	815
9 Barron	Australia	770
10 Sutherland, Middle	New Zealand	751
11 Tiavi	Samoa	600
12 Tully	Australia	550
13 Bowen	New Zealand	520
14 Stirling	New Zealand	480
15 Akaka	Hawaii, USA	418
SOUTH AMERICA		
1 Angel, Upper	Venezuela	2,648
2 Itatinga	Brazil	2,060
3 Cuquenan	Guyana-Venezuela	2,000
4 Pilao	Brazil	1,719
5 Roraima	Guyana	1,500
6 King George VI	Guyana	1,200
7 Candelas	Colombia	984
8 Sewerd	Peru	877
9 King Edward VIII	Guyana	840
10 Kaieteur	Guyana	741
11 Casca d'Anta	Brazil	666
12 Sakaika	Guyana	629
13 Angel, Lower	Venezuela	564
14 Tequendama	Bogota	515
15 Wakowaieng	Guyana	440
16 Fagundes	Brazil	413
17 Kumarow	Guyana	400
18 Itiquira	Brazil	394

CLIMATES OF THE WORLD

Temperature Distribution

The distribution of temperature over the world and its variations through the year depend primarily on the amount of distribution of the radiant energy received from the sun in different regions. This in turn depends mainly on latitude but is greatly modified by the distribution of continents and oceans, prevailing winds, oceanic circulation, topography, and other factors.

Maps showing average temperatures over the surface of the earth for January and for July are given in figures 1 and 2.

In the winter of the Northern Hemisphere, it will be noted, the poleward temperature gradient (that is, the rate of fall in temperature) north of latitude 15° is very steep over the interior of North America. This is shown by the fact that the lines indicating changes in temperature come very close together. The temperature gradient is also steep toward the cold pole over Asia—the area marked −50°. In western Europe, to the east of the Atlantic Ocean and the North Atlantic Drift, and in the region of prevailing westerly winds, the temperature gradient is much more gradual, as indicated by the fact that the isotherms, or lines of equal temperature, are far apart. In the winter of the Southern Hemisphere, as shown on the map for July (a winter month south of the Equator), the temperature gradient toward the South Pole is very gradual, and the isothermal deflections from the east-west direction (that is, the dipping of the isothermal lines) are of minor importance because continental effects are largely absent.

In the summers of the two hemispheres—July in the north and January in the south—the temperature gradients poleward are very much diminished as compared with those during the winter. This is especially marked over the middle and higher northern latitudes because of the greater warming of the extensive interiors of North America and Eurasia than of the smaller land areas in middle and higher southern latitudes.

Distribution of Precipitation

Whether precipitation (see the map, fig. 3) occurs as rain or snow or in the rarer forms of hail or sleet depends largely on the temperature climate, which may be influenced more by elevation than by latitude, as in the case of the perpetually snowcapped mountain peaks and glaciers on the Equator in both South America and Africa.

The quantity of precipitation is governed by the amount of water vapor in the air and the nature of the process that leads to its condensation into liquid or solid form through cooling. Air may ascend to great elevations through local convection, as in thunderstorms and in tropical regions generally; it may be forced up over topographical elevations across the prevailing wind direction, as on the southern or windward slopes of the Himalayas in the path of the southwest monsoon in India; or it may ascend more or less gradually in migratory low-pressure formations such as those that govern the main features of weather in the United States.

The areas of heaviest precipitation on the map (fig. 3) are generally located, as would be expected, in tropical regions, where because of high temperature the greatest amount of water vapor may be present in the atmosphere and the greatest evaporation takes place—although only where conditions favor condensation can rainfall occur. Outstanding exceptions are certain regions in high latitudes, such as southern Alaska, western Norway, and southern Chile, where relatively warm, moist winds from the sea undergo forced ascent over considerable elevations.

In marked contrast to the rainy regions just named are the dry polar regions, where the water-vapor content of the air is always very low because of the low temperature and very limited evaporation. The dry areas in the subtropical belts of high atmospheric pressure (in the vicinity of latitude 30° on all continents, and especially from the extreme western Sahara over a broad, somewhat broken belt to the Desert of Gobi) and the arid strips on the lee sides of mountains on whose windward slopes precipitation is heavy to excessive, are caused by conditions which, even though the temperature may be high, are unfavorable to the condensation of whatever water vapor may be present in the atmosphere.

In the tables following are data on mean maximum and minimum temperatures for January, April, July, and October, with extremes recorded in the period of record, and monthly and annual precipitation for about 800 selected stations well distributed over the earth.

North America

North America is nearly all within middle and northern latitudes. Consequently it has a large central area in which the continental type of climate with marked seasonal temperature is to be found.

Along the coasts of northern Alaska, western Canada, and the northwestern part of the United States, moderate midsummer temperatures are in marked contrast to those prevailing in the interior east of the mountains. (Note, for example, the great southward dip of the 60° isotherm along the west coast in fig. 2.) Again, the mild midwinter temperatures in the coastal areas stand out against the severe conditions to be found from the Great Lakes region northward and northwestward (fig. 1).

In the West Indian region, temperature conditions are subtropical; and in Mexico and Central America, climatic zones depend on elevation, ranging from subtropical to temperate in the higher levels.

The prevailing westerly wind movement carries the continental type of climate eastward over the United States, so that the region of maritime climate along the Atlantic Ocean is very narrow.

From the Aleutian Peninsula to northern California west of the crests of the mountains, there is a narrow strip where annual precipitation is over 40 inches; it exceeds 100 inches locally on the coast of British Columbia (see fig. 3). East of this belt there is an abrupt fall-off in precipitation to less than 20 inches annually over the western half of the continent from Lower California northward, and to even less than 5 inches in parts of what used to be called the "Great American Desert," in the southwestern part of the United States.

In the eastern part of the continent—that is, from the southeastern part of the United States northeastward to Newfoundland—the average annual precipitation is more than 40 inches. Rainfall in the West Indies, southern Mexico, and Central America is generally abundant. It is very spotty, however, varying widely even within short distances, especially from the windward to the leeward sides of the mountains.

The northern areas are, of course, very cold; but the midwinter low temperatures fall far short of the records set in the cold-pole area of northeastern Siberia, where the vast extent of land becomes much colder than the partly ice-covered area of northern Canada.

South America

A large part of South America lies within the Tropics and has a characteristically tropical climate. The remaining rather narrow southern portion is not subject to the extremes of heat and cold that are found where wide land areas give full sway to the continental type of climate with its hot summers and cold winters, as in North America and Asia. Temperature anomalies unusual for a given latitude are to be found mainly at the elevated levels of the Andean region stretching from the Isthmus of Panama to Cape Horn.

The Antarctic Current and its cool Humboldt branch skirting the western shores northward to the Equator, together with the prevailing on-shore winds, exert a strong cooling influence over the coastal regions of all the western countries of South America except Colombia. On the east the southerly moving Brazilian current from tropical waters has the opposite, or warming, effect except along southern Argentina.

In the northern countries of South America the sharply contrasted dry and wet seasons are related to the regime of the trade winds. In the dry season (corresponding to winter in the Northern Hemisphere) these winds sweep the entire region, while the wet season (corresponding to summer in the Northern Hemisphere) calms and variable winds prevail. In the basin of the Amazon River the rainfall is related to the equatorial belt of low pressure and to the trade winds, which give the maximum amounts of rainfall in the extreme west, where they ascend the Andean slopes.

The desert areas on the west coast of South America, extending from the Equator to the latitude of Santiago, are due primarily to the cold Humboldt or Peruvian Current and upwelling coastal water. The moist, cool ocean ais is warmed in passing over the land, with a consequent decrease in relative humidity, so that the dew point is not reached and condensation of vapor does not occur until the incoming air has reached high elevations in the Andes, where temperatures are very much lower than along the coast.

In southern Chile the summer has moderate rainfall, and winters are excessively wet. The conditions that prevail farther north are not present here, and condensation of moisture from the ocean progresses from the shores up to the crests of the Andes. By the time the air passes these elevations, however, the moisture has been so depleted that the winds in the leeward slopes are dry, becoming more and more so as they are warmed on reaching lower levels. The mountains can be looked upon as casting a great "rain shadow"—an area of little rain—over southern Argentina.

Europe

In Europe there is no extensive north-south mountain system such as is found in both the Americas, and the general east-west direction of the ranges in the south allows the conditions in the maritime west to change rather gradually toward Asia. Generally rainfall is heaviest on the western coast, where locally it exceeds 60 inches annually, and diminishes toward the east—except in the elevated Alpine and Caucasus regions—to less than 20 inches in eastern Russia. There is a well-defined rain shadow in Scandinavia, with over 60 inches of rain in western Norway and less than 20 inches in eastern Sweden.

Over much of Europe rainfall is both abundant and rather evenly distributed throughout the year. The chief feature of seasonal distribution of precipitation is the marked winter maximum and the extremely dry, even droughty, summers in most of the Mediterranean lands.

Isothermal lines have the general direction of the parallels of latitude except in winter, when the waters of the western ocean, warmed by the Gulf Stream, give them a north-south trend. Generally there are no marked dips in isotherms due to elevation and continental type of climate such as are found in North America. In Scandinavia, however, the winter map shows an abrupt fall in temperature from the western coast of Norway to the eastern coast of Sweden and thence a continued fall eastward, under a type of exposure more and more continental in contrast to the oceanic exposure on the west.

Asia

The vast extent of Asia gives full opportunity for continental conditions to develop a cold area of high barometric pressure in winter and a low-pressure, hot area in summer, the former northeast of the Himalayas and the latter stretching widely from west to east in the latitude of northern India. (See the area marked 90° on the map, fig. 2.) These distributions of pressure give to India the well-known monsoon seasons, during which the wind comes from one direction for several months, and also affect the yearly distribution of rainfall over eastern Asia.

In winter, the air circulation is outward over the land from the cold pole, and precipitation is very light over the entire continent. In summer, on the contrary, there is an inflow of air from the oceans; even the southeast trade winds flow across the Equator and merge into the southwest monsoon which crosses India. This usually produces abundant rain over most of that country, with excessively heavy amounts when the air is forced to rise, even to moderate elevations, in its passage the land. At Cherrapunji (4,455 feet), on the southern side of the Khasi Hills in Assam, the average rainfall in a winter month is about 1 inch, while in both June and July it is approximately 100 inches. However, this heavy summer rainfall meets an impassable barrier in the Himalaya Mountains, while the much lighter summer monsoon rainfall over Japan and eastern Asia does not extend far into China because of lesser elevations. Consequently, while the southeast quadrant of Asia, including the East Indies, also with monsoon winds, has heavy to excessive annual rainfall, the remainder of the continent is dry, with vast areas receiving less than 10 inches annually.

North of the Himalayas the low plains are excessively cold in winter and temperatures rise high in summer. At Verkhoyansk in the cold-pole area, and north of the Arctic Circle, the mean temperature in January is about –59° F, and in July approximately 64°; the extreme records are a maximum of 98°, from readings at 1 p.m., and a minimum of –90°.

In southwestern Asia the winter temperature control is still the interior high-pressure area, and temperatures are generally low, especially at high elevations; in summer at low elevations excessively high maxima are recorded, as, for example, in the Tigris-Euphrates Valley.

Africa

Africa, like South America, lies very largely within the Tropics. There too, temperature distribution is determined mainly by altitude. Moreover, along the southern portion of the western coast the cool Benguela Current moves northward, and on the eastern coast are the warm tropical currents of the Indian Ocean, which create conditions closely paralleling those found around the South American Continent. In the strictly tropical areas of Africa conditions are characterized by prevailing low barometric pressure, with conventional rainfall and alternate northward and southward movement of the heat equator, while in both the north and the south the ruling influences are the belts of high barometric pressure.

Except in the Atlas Mountains in the northwest where the considerable elevations set up a barrier in the path of trade winds and produce moderate rainfall, the desert conditions typified by the Sahara extend from the Atlantic to the Red Sea and from the Mediterranean southward well beyond the northern Tropic to about the latitudes of southern Arabia.

South of the Sahara, rainfall increases rapidly, becoming abundant to heavy from the west coast to the central lakes, with annual maxima of over 80 inches in the regions bordering the eastern and western extremes of the Guinea coast. This marked increase in precipitation does not extend to the eastern portion of the middle region of the continent, where the annual amounts received are below 40 inches and decrease to less than 10 inches on the coasts of Somalia. Also to the south of the central rainy area there is a rapid fall in precipitation toward the arid regions of Southwest Africa, where conditions are similar to those in Somalia.

The heavy rainfall over sections of Ethiopia from June to October, when more than 40 inches fall and bring the overflowing of the otherwise arid Nile Valley, is one of the earth's outstanding features of seasonal distribution of rainfall.

Moist equatorial climate is typified by conditions in the Democratic Republic of the Congo; arid torrid climate by those of the United Arab Republic and the Sahara; and moderate plateau climate by those found in parts of Ethiopia, Kenya, and Tanzania.

Australia

In the southern winter the high-pressure belt crosses the interior of Australia, and all except the southernmost parts of the continent are dry. In summer, on the other hand, this pressure belt has moved south of the continent, still giving dry conditions over the southern and western areas. Thus the total annual precipitation is less than 20 inches except in the extreme southwest and in a strip circling from southeast to northwest. The average annual precipitation is even less than 10 inches in a large south-central area.

In the south the winter precipitation is the cyclonic type; the heavy summer rains of the north are of monsoon origin; and those of the eastern borders are in large part orographic, owing to the presence of the highlands in the immediate vicinity of the coasts. In the outer border of the rainfall strip along the coastal region, the mean annual rainfall is over 40 inches and in many localities over 60 inches. This is true for the monsoon rains in the north.

Because of the location of Australia, on both sides of the southern Tropic, temperatures far below freezing are to be found only in a small part of the continent, in the south at high elevations. In the arid interior extreme maximum temperatures are very high, ranking with those of the hottest regions on the earth.

TEMPERATURE AND PRECIPITATION DATA FOR REPRESENTATIVE WORLD-WIDE STATIONS

COUNTRY AND STATION	LATITUDE	LONGITUDE	ELEVATION	TEMPERATURE: LENGTH OF RECORD	AVERAGE DAILY: JANUARY MAXIMUM	JANUARY MINIMUM	APRIL MAXIMUM	APRIL MINIMUM	JULY MAXIMUM	JULY MINIMUM	OCTOBER MAXIMUM	OCTOBER MINIMUM	EXTREME: MAXIMUM	EXTREME: MINIMUM	AVERAGE PRECIPITATION: LENGTH OF RECORD	JANUARY	FEBRUARY	MARCH	APRIL	MAY	JUNE	JULY	AUGUST	SEPTEMBER	OCTOBER	NOVEMBER	DECEMBER	YEAR
	° '	° '	FEET	YEAR	°F	°F	°F	°F	°F	°F	°F	°F	°F	°F	YEAR	IN.	IN.	IN.	IN.	IN.	IN.	IN.	IN.	IN.	IN.	IN.	IN.	IN.
					NORTH AMERICA																							
United States (Conterminous):																												
Albuquerque, N. Mex.	35 03N	106 37W	5,311	30	46	24	69	42	91	66	71	45	104	-16	30	0.4	0.4	0.5	0.5	0.8	0.6	1.2	1.3	1.0	0.8	0.4	0.5	8.4
Asheville, N.C.	35 26N	82 32W	2,140	30	48	28	67	42	84	61	68	45	99	- 7	30	4.2	4.0	4.8	4.0	3.7	3.5	5.9	4.9	3.6	3.1	2.8	3.6	48.1
Atlanta, Ga.	33 39N	84 26W	1,010	30	52	37	70	50	87	71	72	52	103	- 9	30	4.4	4.5	5.4	4.5	3.2	3.8	4.7	3.6	3.3	2.4	3.0	4.4	47.2
Austin, Tex.	30 18N	97 42W	597	30	60	41	78	57	95	74	82	60	109	- 2	30	2.4	2.6	2.1	3.6	3.7	3.2	2.2	1.9	3.4	2.8	2.1	2.5	32.5
Birmingham, Ala.	33 34N	86 45W	620	30	57	36	76	50	93	71	79	52	107	-10	30	5.0	5.3	6.0	4.5	3.4	4.0	5.2	4.9	3.3	3.0	3.5	5.0	53.1
Bismark, N. Dak.	46 46N	100 45W	1,647	30	20	0	55	32	86	58	59	34	114	-45	30	0.4	0.4	0.8	1.2	2.0	3.4	2.2	1.7	1.2	0.9	0.6	0.4	15.2
Boise, Idaho	43 34N	116 13W	2,838	30	36	22	63	37	91	59	65	38	112	-28	30	1.3	1.3	1.3	1.2	1.3	0.9	0.2	0.2	0.4	0.8	1.2	1.3	11.4
Brownsville, Tex.	25 54N	97 26W	16	30	71	52	82	66	93	76	85	67	104	12	30	1.4	1.5	1.0	1.6	2.4	3.0	1.7	2.8	5.0	3.5	1.3	1.7	26.9
Buffalo, N.Y.	42 56N	78 44W	705	30	31	18	53	34	80	59	60	41	99	-21	30	2.8	2.7	3.2	3.0	3.0	2.5	2.6	3.1	3.1	3.0	3.6	3.0	35.6
Cheyenne, Wyo.	41 09N	104 49W	6,126	30	37	14	56	30	85	55	63	32	100	-38	30	0.5	0.6	1.2	1.9	2.5	2.1	1.8	1.4	1.1	0.8	0.6	0.5	15.0
Chicago, Ill.	41 47N	87 45W	607	30	33	19	57	41	84	67	63	47	105	-23	30	1.9	1.6	2.7	3.0	3.7	4.1	3.4	3.2	2.7	2.8	2.2	1.9	33.2
Des Moines, Iowa	41 32N	93 39W	938	30	29	11	59	38	87	65	66	43	110	-30	30	1.3	1.1	2.1	2.5	4.1	4.7	3.1	3.7	2.9	2.1	1.8	1.1	30.5
Dodge City, Kans.	37 46N	99 58W	2,582	30	42	20	66	41	93	68	71	46	109	-26	30	0.6	0.7	1.2	1.8	3.2	3.0	2.3	2.4	1.5	1.4	0.6	0.5	19.2
El Paso, Tex.	31 48N	106 24W	3,918	30	56	30	78	49	95	69	79	50	109	- 8	30	0.5	0.4	0.4	0.3	0.4	0.7	1.3	1.2	1.1	0.9	0.3	0.5	8.0
Indianapolis, Ind.	39 44N	86 17W	792	30	37	21	61	40	86	64	67	44	107	-25	30	3.1	2.3	3.4	3.7	4.0	4.6	3.5	3.0	3.2	2.6	3.1	2.7	39.2
Jacksonville, Fla.	30 25N	81 39W	20	30	67	45	80	58	92	73	80	62	105	10	30	2.5	2.9	3.5	3.6	3.5	6.3	7.7	6.9	7.6	5.2	1.7	2.2	53.6
Kansas City, Mo.	39 07N	94 36W	742	30	40	23	66	46	92	71	72	49	113	-22	30	1.4	1.2	2.5	3.6	4.4	4.6	3.2	3.8	3.3	2.9	1.8	1.5	34.2
Las Vegas, Nev.	36 05N	115 10W	2,162	30	54	32	78	51	104	76	80	53	117	8	30	0.5	0.4	0.4	0.2	0.1	*	0.5	0.5	0.3	0.2	0.3	0.4	3.8
Los Angeles, Calif.	33 56N	118 23W	97	30	64	45	67	52	76	62	73	57	110	23	30	2.7	2.9	1.8	1.1	0.1	0.1	*	*	0.2	0.4	1.1	2.4	12.8
Louisville, Ky.	38 11N	85 44W	477	30	44	27	66	43	89	67	70	46	107	-20	30	4.1	3.3	4.6	3.8	3.9	4.0	3.4	3.0	2.6	2.3	3.2	3.2	41.4
Miami, Fla.	25 48N	80 16W	7	30	76	58	83	66	89	75	85	71	100	28	30	2.0	1.9	2.3	3.9	6.4	7.4	6.8	7.0	9.5	8.2	2.8	1.7	59.9
Minneapolis, Minn.	44 53N	93 13W	834	30	22	2	56	33	84	61	61	37	108	-34	30	0.7	0.8	1.5	1.9	3.2	4.0	3.3	3.2	2.4	1.6	1.4	0.9	24.9
Missoula, Mont.	46 55N	114 05W	3,190	30	28	10	57	31	85	49	58	30	105	-33	30	0.9	0.9	0.7	1.0	1.9	1.9	0.9	0.7	1.0	1.0	0.9	1.1	12.9
Nasheville, Tenn.	36 07N	86 41W	590	30	49	31	71	48	91	70	74	49	107	-15	30	5.5	4.5	5.2	3.7	3.7	3.3	3.7	2.9	2.9	2.3	3.3	4.2	45.2
New Orleans, La.	29 59N	90 15W	3	30	64	45	78	58	91	73	80	61	102	7	30	3.8	4.0	5.3	4.6	4.4	4.4	6.7	5.3	5.0	2.8	3.3	4.1	53.7
New York, N.Y.	40 47N	73 58W	132	30	40	27	60	43	85	68	66	50	106	-15	30	3.3	2.8	4.0	3.4	3.7	3.3	3.7	4.4	3.9	3.1	3.4	3.3	42.3
Oklahoma City, Okla.	35 24N	97 36W	1,285	30	46	28	71	49	93	72	74	52	113	-17	30	1.3	1.4	2.0	3.1	5.2	4.5	2.4	2.5	3.0	2.5	1.6	1.4	30.9
Phoenix, Ariz.	33 26N	112 01W	1,117	30	64	35	84	50	105	75	87	55	118	16	30	0.7	0.9	0.7	0.3	0.1	0.1	0.8	1.1	0.7	0.5	0.5	0.9	7.[illegible]
Pittsburgh, Pa.	40 27N	80 00W	747	30	40	25	63	42	85	65	65	45	103	-20	30	2.8	2.3	3.5	3.4	3.8	4.0	3.6	3.5	2.7	2.5	2.3	2.5	36.9
Portland, Maine	43 39N	70 19W	47	30	32	12	53	32	80	57	60	37	103	-39	30	4.4	3.8	4.3	3.7	3.4	3.2	2.9	2.4	3.5	3.2	4.2	3.9	42.9
Portland, Oreg.	45 36N	122 36W	21	30	44	33	62	42	79	56	63	45	107	- 3	30	5.4	4.2	3.8	2.1	2.0	1.7	0.4	0.7	1.6	3.6	5.3	6.4	37.2
Reno, Nev.	39 30N	119 47W	4,404	30	45	16	65	31	89	46	69	29	106	-19	30	1.2	1.0	0.7	0.5	0.5	0.4	0.3	0.2	0.2	0.5	0.6	1.1	7.2
Salt Lake City, Utah	40 46N	111 58W	4,220	30	37	18	63	36	94	60	65	38	107	-30	30	1.4	1.2	1.6	1.8	1.4	1.0	0.6	0.9	0.5	1.2	1.3	1.2	14.1
San Francisco, Calif.	37 37N	122 23W	8	30	55	42	64	47	72	54	71	51	106	20	30	4.0	3.5	2.7	1.3	0.5	0.1	*	*	0.2	0.7	1.6	4.1	18.7
Sault Ste. Marie, Mich.	46 28N	84 22W	721	30	23	8	46	30	76	54	55	38	98	-37	30	2.1	1.5	1.8	2.2	2.8	3.3	2.5	2.9	3.8	2.8	3.3	2.3	31.3
Seattle, Wash.	47 27N	122 18W	400	30	44	33	58	40	76	54	60	44	100	0	30	5.7	4.2	3.8	2.4	1.7	1.6	0.8	1.0	2.1	4.0	5.4	6.3	39.0
Sheridan, Wyo.	44 46N	106 58W	3,964	30	34	9	56	31	87	56	62	33	106	-41	30	0.6	0.7	1.4	2.2	2.6	2.6	1.2	0.9	1.2	1.1	0.8	0.6	15.9
Spokane, Wash.	47 38N	117 32W	2,356	30	31	19	59	36	86	55	60	38	108	-30	30	2.4	1.9	1.5	0.9	1.2	1.5	0.4	0.4	0.8	1.6	2.2	2.4	17.2
Washington, D.C.	38 51N	77 03W	14	30	44	30	66	46	87	69	68	50	106	-15	30	3.0	2.5	3.2	3.2	4.1	3.2	4.2	4.9	3.8	3.1	2.8	2.8	40.8
Wilmington, N.C.	34 16N	77 55W	28	30	58	37	74	51	89	71	76	55	104	5	30	2.9	3.4	4.0	2.9	3.5	4.3	7.7	6.9	6.3	3.0	3.1	3.4	51.4

See footnotes at end of table.

TEMPERATURE AND PRECIPITATION DATA FOR REPRESENTATIVE WORLD-WIDE STATIONS

COUNTRY AND STATION	LATITUDE	LONGITUDE	ELEVATION	TEMPERATURE: LENGTH OF RECORD	AVERAGE DAILY: JANUARY MAXIMUM	JANUARY MINIMUM	APRIL MAXIMUM	APRIL MINIMUM	JULY MAXIMUM	JULY MINIMUM	OCTOBER MAXIMUM	OCTOBER MINIMUM	EXTREME MAXIMUM	EXTREME MINIMUM	AVERAGE PRECIPITATION: LENGTH OF RECORD	JANUARY	FEBRUARY	MARCH	APRIL	MAY	JUNE	JULY	AUGUST	SEPTEMBER	OCTOBER	NOVEMBER	DECEMBER	YEAR
	° ′	° ′	FEET	YEAR	°F	°F	°F	°F	°F	°F	°F	°F	°F	°F	YEAR	IN.	IN.	IN.	IN.	IN.	IN.	IN.	IN.	IN.	IN.	IN.	IN.	IN.
United States, Alaska:																												
Anchorage	61 13N	149 52W	85	30	21	4	44	28	65	50	42	28	86	-38	30	0.8	0.7	0.5	0.4	0.5	1.0	1.9	2.6	2.5	1.9	1.0	0.9	14.7
Annette	55 02N	131 34W	110	30	38	30	50	37	63	51	51	42	90	- 4	30	11.4	8.5	9.6	9.1	7.1	5.7	6.0	7.5	9.9	16.9	14.7	12.1	118.5
Barrow	71 18N	156 47W	31	30	- 9	-23	7	- 7	45	33	21	12	78	-56	30	0.2	0.2	0.1	0.1	0.1	0.4	0.8	0.9	0.6	0.5	0.2	0.2	4.3
Bethel	60 47N	161 48W	125	30	11	- 4	34	18	62	48	38	25	90	-52	30	1.1	1.1	1.0	0.6	1.0	1.2	2.0	4.2	2.6	1.5	1.1	1.0	18.4
Cold Bay	55 12N	162 43W	96	30	33	23	38	28	54	45	45	36	78	- 9	30	2.3	3.2	1.8	1.5	2.3	2.0	1.8	4.3	4.3	4.6	3.8	2.6	34.5
Fairbanks	64 49N	147 52W	436	30	- 1	-21	42	17	72	48	35	17	99	-66	30	0.9	0.5	0.4	0.3	0.7	1.4	1.8	2.2	1.1	0.9	0.6	0.5	11.3
Juneau	58 22N	134 35W	12	30	30	20	45	31	63	48	47	37	89	-21	30	4.0	3.1	3.3	2.9	3.2	3.4	4.5	5.0	6.7	8.3	6.1	4.2	54.7
King Salmon	58 41N	156 39W	49	30	21	6	41	25	63	47	43	29	88	-40	30	1.1	1.0	1.0	0.6	1.0	1.4	2.1	3.4	3.1	2.2	1.5	1.0	19.4
Nome	64 30N	165 26W	13	30	12	- 3	28	14	55	44	35	24	84	-47	30	1.0	0.9	0.9	0.8	0.7	0.9	2.3	3.8	2.7	1.7	1.2	1.0	17.9
St. Paul Island	57 09N	170 13W	22	30	30	21	33	24	49	42	41	33	64	-26	30	1.8	1.2	1.1	1.0	1.3	1.2	2.3	3.3	3.1	3.2	2.5	1.8	23.8
Shemya	52 43N	174 06E	122	30	34	29	38	33	49	44	42	38	63	16	30	2.5	2.3	2.6	2.1	2.4	1.3	2.2	2.1	2.3	2.8	2.7	2.1	27.4
Yakutat	59 31N	139 40W	28	30	34	20	45	29	61	48	49	35	86	-24	30	10.9	8.2	8.7	7.2	8.0	5.1	8.4	10.9	16.6	19.6	16.1	12.3	132.0
Canada:																												
Aklavik, N.W.T.	68 14N	135 00W	30	22	-10	-26	19	- 2	66	47	25	15	93	-62	22	0.5	0.5	0.4	0.5	0.5	0.8	1.4	1.4	0.9	0.9	0.8	0.4	9.0
Alert, N.W.T.	82 31N	62 20W	95	9	-19	-29	- 8	-18	44	36	2	- 7	67	-53	10	0.2	0.3	0.3	0.3	0.5	0.6	0.5	1.1	1.0	0.9	0.2	0.4	6.3
Calgary, Alta.	51 06N	114 01W	3,540	55	24	2	53	27	76	47	54	29	97	-49	55	0.5	0.5	0.8	1.0	2.3	3.1	2.5	2.3	1.5	0.7	0.7	0.6	16.7
Charlottetown, P.E.I.	46 17N	63 08W	181	65	26	10	43	30	73	58	54	41	98	-27	65	3.8	3.0	3.2	2.8	2.7	2.6	3.0	3.4	3.4	4.1	3.8	4.0	39.8
Chatham, N.B.	47 00N	65 27W	109	50	23	2	47	28	77	56	55	37	102	-43	50	3.4	2.7	3.3	3.0	3.2	3.6	3.9	4.0	3.1	4.0	3.4	3.2	40.8
Churchhill, Man.	58 45N	94 04W	94	30	-11	-27	24	4	64	43	34	20	96	-57	30	0.5	0.6	0.9	0.9	0.9	1.9	2.2	2.7	2.3	1.4	1.0	0.7	16.0
Edmonton, Alta.	53 34N	113 31W	2,219	71	16	- 3	52	28	74	50	51	30	99	-57	71	0.9	0.7	0.7	1.0	1.9	3.2	3.3	2.4	1.3	0.8	0.9	0.9	18.0
Fort Nelson, B.C.	58 50N	122 35W	1,253	12	1	-15	47	25	74	51	43	25	98	-61	13	0.9	1.2	0.7	0.8	1.4	2.5	2.4	1.5	1.3	1.0	1.4	1.2	16.3
Fort Simpson, N.W.T.	61 45N	121 14W	554	42	-10	-27	38	14	74	50	36	21	97	-70	42	0.7	0.7	0.5	0.7	1.4	1.5	2.0	1.5	1.3	1.1	0.9	0.8	13.1
Frobisher Bay, N.W.T.	63 45N	68 33W	110	18	- 9	-23	16	- 1	53	39	29	18	76	-49	10	0.7	0.9	0.8	0.8	0.7	0.9	1.5	2.0	1.8	1.1	1.1	1.0	13.3
Gander, Nfld.	48 57N	54 34W	496	14	27	13	40	27	71	52	51	37	96	-17	14	2.6	3.3	2.8	2.6	2.6	2.8	3.6	3.6	3.7	4.1	4.2	3.7	39.6
Halifax, N.S.	44 39N	63 34W	83	75	32	15	47	31	74	55	57	41	99	-21	71	5.4	4.4	4.9	4.5	4.1	4.0	3.8	4.4	4.1	5.4	5.3	5.4	55.7
Kapuskasing, Ont.	49 25N	82 28W	743	19	10	-14	43	19	75	50	47	31	101	-53	19	2.0	1.1	1.6	1.8	2.1	2.3	3.4	2.9	3.5	2.5	2.4	1.9	27.5
Knob Lake, Que.	54 48N	66 49W	1,712	30	- 3	-21	30	12	64	46	37	25	88	-59	10	1.9	1.9	1.4	1.6	1.7	3.3	3.3	4.4	3.4	2.9	2.4	1.5	29.7
Montreal, Que.	45 30N	73 34W	187	67	21	6	50	33	78	61	54	40	97	-35	77	3.8	3.0	3.5	2.6	3.1	3.4	3.7	3.5	3.7	3.4	3.5	3.6	40.8
North Bay, Ont.	46 21N	79 25W	1,216	17	22	2	48	28	78	56	49	36	99	-46	23	2.0	1.5	1.8	2.2	2.5	3.2	3.2	2.7	3.7	3.2	2.7	2.1	30.8
Ottowa, Ont.	45 19N	75 40W	374	65	21	3	51	31	81	58	54	37	102	-38	65	2.9	2.2	2.8	2.7	2.5	3.5	3.4	2.6	3.2	2.9	3.0	2.6	34.3
Penticton, B.C.	49 28N	119 36W	1,129	32	32	21	61	35	84	53	59	38	105	-16	32	1.0	0.7	0.7	0.7	1.1	1.2	0.8	0.8	1.0	0.8	0.9	1.1	10.8
Port Arthur, Ont.	48 22N	89 19W	644	62	17	- 4	44	26	74	52	50	34	104	-42	59	0.9	0.8	1.0	1.5	2.1	2.8	3.6	2.8	3.4	2.5	1.5	0.9	23.8
Prince George, B.C.	53 53N	122 41W	2,218	27	23	3	54	27	75	44	52	30	102	-58	27	1.8	1.2	1.4	0.8	1.3	2.1	1.6	1.9	2.0	2.0	1.9	1.9	19.9
Prince Rupert, B.C.	54 17N	130 23W	170	26	39	30	50	37	62	49	53	42	90	- 3	26	9.8	7.6	8.4	6.7	5.3	4.1	4.8	5.1	7.7	12.2	12.3	11.3	95.3
Quebec, Que.	46 48N	71 23W	239	72	18	2	44	29	76	57	51	37	97	-34	72	3.5	2.7	3.0	2.4	3.1	3.7	4.0	4.0	3.6	3.4	3.2	3.2	39.8
Regina, Sask.	50 26N	104 40W	1,884	55	10	-11	50	26	79	51	52	27	110	-56	49	0.5	0.3	0.7	0.7	1.8	3.3	2.4	1.8	1.3	0.9	0.6	0.4	14.7
Resolute, N.W.T.	74 43N	94 59W	220	13	-20	-33	- 1	-16	45	35	11	0	61	-61	7	0.1	0.1	0.2	0.2	0.5	0.8	0.9	1.1	0.8	0.5	0.2	0.1	5.5
St. John, N.B.	45 17N	66 04W	119	61	28	11	43	32	69	54	54	41	93	-24	61	4.1	3.1	3.7	3.2	3.1	3.2	3.1	3.6	3.7	4.1	3.9	3.8	42.6
St. Johns, Nfld.	47 32N	52 44W	211	68	30	18	41	29	69	51	53	40	93	-21	58	5.3	5.1	4.6	3.8	3.9	3.1	3.1	4.0	3.7	4.8	5.7	6.0	53.1
Saskatoon, Sask.	52 08N	106 38W	1,690	38	9	-11	49	26	77	52	51	27	104	-55	38	0.9	0.5	0.7	0.7	1.4	2.6	2.4	1.9	1.5	0.9	0.5	0.6	14.6
The Pas, Man.	53 49N	101 15W	890	27	1	-18	45	21	75	54	45	26	100	-54	27	0.6	0.5	0.7	0.8	1.4	2.2	2.2	2.1	2.0	1.2	1.0	0.8	15.5
Toronto, Ont.	43 40N	79 24W	379	105	30	16	50	34	79	59	56	40	105	-26	105	2.7	2.4	2.6	2.5	2.9	2.7	3.0	2.7	2.9	2.4	2.8	2.6	32.2
Vancouver, B.C.	49 17N	123 05W	127	43	41	32	58	40	74	54	57	44	92	2	41	8.6	5.8	5.0	3.3	2.8	2.5	1.2	1.7	3.6	5.8	8.3	8.8	57.4

Whitehorse, Y.T.	60 43N	135 04W	2,303	10	13	- 3	41	22	67	45	41	28	91	-62	10	0.6	0.5	0.6	0.4	0.6	1.0	1.6	1.5	1.3	0.7	1.0	0.8	10.6
Winnipeg, Man.	49 54N	97 14W	783	66	7	-13	48	27	79	55	51	31	108	-54	66	0.9	0.9	1.2	1.4	2.3	3.1	3.1	2.5	2.3	1.5	1.1	0.9	21.2
Yellow Knife, N.W.T.	62 28N	114 27W	674	13	- 8	-23	29	9	69	52	36	26	90	-60	13	0.8	0.6	0.7	0.4	0.7	0.6	1.5	1.4	1.0	1.3	1.0	0.8	10.8
Greenland:																												
Angmagssalik	65 36N	37 33W	95	30	23	10	35	16	54	37	35	25	77	-26	38	2.9	2.4	2.6	2.1	2.0	1.8	1.5	2.1	3.3	4.7	3.0	2.7	31.1
Danmarkshaven	76 46N	19 00W	7	2	- 1	-15	6	-13	47	34	13	2	63	-42	2	1.2	0.7	0.7	0.1	0.2	0.2	0.5	0.6	0.3	0.3	1.0	0.7	6.0
Eismitte	70 53N	40 42W	9,843	1	-33	-53	-14	-37	19	1	-23	-42	27	-85	1	0.6	0.2	0.3	0.2	0.1	0.1	0.1	0.4	0.3	0.5	0.5	1.0	4.3
Godthaab	64 10N	51 43W	66	40	19	10	31	20	52	38	35	26	76	-20	45	1.4	1.7	1.6	1.2	1.7	1.4	2.2	3.1	3.3	2.5	1.9	1.5	23.5
Ivigtut	61 12N	48 10W	98	48	24	12	38	24	57	42	40	29	86	-20	50	3.3	2.6	3.4	2.5	3.5	3.2	3.1	3.7	5.9	5.7	4.6	3.1	44.6
Jacobshavn	69 13N	51 02W	104	32	8	- 7	24	6	51	40	31	20	71	-46	52	0.4	0.4	0.5	0.5	0.6	0.8	1.2	1.4	1.3	0.9	0.7	0.5	9.2
Nord	81 36N	16 40W	118	8	-15	-28	- 5	-18	44	35	3	- 6	61	-60	8	0.8	0.8	0.5	0.3	0.1	0.3	1.0	1.4	1.2	0.6	1.4	0.5	8.9
Scoresbysund	70 29N	21 58W	56	12	12	- 3	22	6	49	36	25	15	63	-42	12	1.8	1.4	0.9	1.4	0.4	0.8	1.5	0.7	1.7	1.4	1.1	1.9	15.0
Thule	76 31N	68 44W	251	12	- 4	-17	10	- 7	46	38	19	8	63	-44	12	0.4	0.3	0.2	0.2	0.3	0.2	0.7	0.6	0.6	0.7	0.5	0.2	4.9
Upernivik	72 47N	56 07W	59	40	- 1	-13	15	- 1	48	35	29	21	69	-44	50	0.4	0.5	0.7	0.6	0.6	0.5	0.9	1.1	1.1	1.1	1.1	0.6	9.2
Mexico:																												
Acapulco	16 50N	99 56W	10	8	85	70	87	71	89	75	88	74	97	60	40	0.3	*	0.0	*	1.4	12.8	9.1	9.3	13.9	6.7	1.2	0.4	55.1
Chihuahua	28 42N	105 57W	4,429	9	65	36	81	51	89	66	79	51	102	12	22	0.2	0.4	0.3	0.2	0.2	1.7	3.6	3.7	3.3	0.9	0.5	0.4	15.4
Guadalajara	20 41N	103 20W	5,194	26	73	45	85	53	79	60	78	56	101	26	33	0.4	0.2	0.2	0.2	1.1	8.8	9.4	8.5	7.2	2.2	0.8	0.7	39.7
Guaymas	27 57N	110 55W	58	9	74	57	84	65	96	82	91	75	117	41	41	0.5	0.2	0.2	0.1	*	0.1	1.7	2.7	2.1	0.7	0.3	0.8	9.4
La Paz	24 07N	110 17W	85	9	74	54	86	58	96	73	90	68	108	31	12	0.2	0.1	0.0	0.0	0.0	0.2	0.4	1.2	1.4	0.6	0.5	1.1	5.7
Lerdo	25 30N	103 32W	3,740	10	72	45	86	57	90	68	82	58	105	23	14	0.4	0.1	0.2	0.3	0.8	1.5	1.5	1.3	2.0	0.8	0.8	0.5	10.2
Manzanillo	19 04N	104 20W	26	17	86	68	87	67	93	76	91	76	103	54	17	0.1	0.2	*	0.0	0.1	4.7	5.7	6.4	14.5	5.1	0.9	1.8	39.5
Mazatlan	23 11N	106 25W	256	10	71	61	76	65	86	77	85	76	93	52	46	0.8	0.5	0.2	0.1	0.1	1.5	5.9	8.3	8.0	2.6	0.9	1.3	30.2
Merida	20 58N	89 38W	72	22	83	62	92	69	92	73	87	71	106	51	40	1.2	0.9	0.7	0.8	3.2	5.6	5.2	5.6	6.8	3.8	1.3	1.3	36.5
Mexico City	19 26N	99 04W	7,340	42	66	42	78	52	74	54	70	50	92	24	48	0.2	0.3	0.5	0.7	1.9	4.1	4.5	4.3	4.1	1.6	0.5	0.3	23.0
Monterrey	25 40N	100 18W	1,732	11	68	48	84	62	90	71	80	64	107	25	33	0.6	0.7	0.8	1.3	1.3	3.0	2.3	2.4	5.2	3.0	1.5	0.8	22.9
Salina Cruz	16 12N	95 12W	184	10	85	72	88	76	89	76	87	75	98	62	22	*	0.4	0.6	0.5	3.3	11.6	4.5	5.5	7.1	4.0	0.9	0.1	38.5
Tampico	22 16N	97 51W	78	12	75	59	83	69	89	75	85	71	104	34	12	1.5	1.2	1.0	1.5	1.9	8.7	4.9	4.8	10.8	5.0	2.0	1.6	44.9
Vera Cruz	19 12N	96 08W	52	10	77	66	83	72	87	74	85	73	98	53	40	0.9	0.6	0.6	0.8	2.6	10.4	4.1	11.1	13.9	6.9	3.0	1.0	65.7
CENTRAL AMERICA																												
British Honduras:																												
Belize	17 31N	88 11W	17	27	81	67	86	74	87	75	86	72	97	49	33	5.4	2.4	1.5	2.2	4.3	7.7	6.4	6.7	9.6	12.0	8.9	7.3	74.4
Canal Zone:																												
Balboa Heights	08 57N	79 33W	118	34	88	71	90	74	87	74	85	73	97	63	46	1.0	0.4	0.7	2.9	8.0	8.4	7.1	7.9	8.2	10.1	10.2	4.8	69.7
Cristobal	09 21N	79 54W	35	36	84	76	86	77	85	76	86	75	97	66	73	3.4	1.5	1.5	4.1	12.5	13.9	15.6	15.3	12.7	15.8	22.3	11.7	130.3
Costa Rica:																												
San Jose	09 56N	84 08W	3,760	8	75	58	79	62	77	62	77	60	92	49	34	0.6	0.2	0.8	1.8	9.0	9.5	8.3	9.5	12.0	11.8	5.7	1.6	70.8
El Salvador:																												
San Salvador	13 42N	89 13W	2,238	39	90	60	93	65	89	65	87	65	105	45	39	0.3	0.2	0.4	1.7	7.7	12.9	11.5	11.7	12.1	9.5	1.6	0.4	70.0
Guatemala:																												
Guatemala City	14 37N	90 31W	4,855	6	73	53	82	58	78	60	76	60	90	41	29	0.3	0.1	0.5	1.2	6.0	10.8	8.0	7.8	9.1	6.8	0.9	0.3	51.8
Honduras:																												
Tela	15 46N	87 27W	41	4	82	67	87	72	88	73	86	71	96	58	20	8.9	5.1	2.6	3.3	4.3	5.0	6.4	9.4	7.7	13.5	15.9	14.0	96.1
WEST INDIES																												
Bridgetown, Barbados	13 08N	59 36W	181	35	83	70	86	72	86	74	86	73	95	61	22	2.6	1.1	1.3	1.4	2.3	4.4	5.8	5.8	6.7	7.0	8.1	3.8	50.3
Camp Jacob, Guadaloupe	16 01N	61 42W	1,750	19	77	64	79	65	81	68	81	68	92	54	21	9.2	6.1	8.1	7.3	11.5	14.1	17.6	15.3	16.4	12.4	12.3	10.1	140.4
Ciudad Trujillo, Dominican Rep.	18 29N	69 54W	57	26	84	66	85	69	88	72	87	72	98	59	25	2.4	1.4	1.9	3.9	6.8	6.2	6.4	6.3	7.3	6.0	4.8	2.4	55.8
Fort-de-France, Martinique	14 37N	61 05W	13	22	83	69	86	71	86	74	87	73	96	56	31	4.7	4.3	2.9	3.9	4.7	7.4	9.4	10.3	9.3	9.7	7.9	5.9	80.4
Hamilton, Bermuda	32 17N	64 46W	151	59	68	58	71	59	85	73	79	69	99	40	62	4.4	4.7	4.8	4.1	4.6	4.4	4.5	5.4	5.2	5.8	5.0	4.7	57.6
Havana, Cuba	23 08N	82 21W	80	25	79	65	84	69	89	75	85	73	104	43	72	2.8	1.8	1.8	2.3	4.7	6.5	4.9	5.3	5.9	6.8	3.1	2.3	48.2
Kingston, Jamaica	17 58N	76 48W	110	33	86	67	87	70	90	73	88	73	97	56	59	0.9	0.6	0.9	1.2	4.0	3.5	1.5	3.6	3.9	7.1	2.9	1.4	31.5
La Guerite, St. Christopher																												
(St. Kitts)	17 20N	62 45W	157	19	80	71	83	73	86	76	85	75	91	61	21	4.1	2.0	2.3	2.3	3.8	3.6	4.4	5.2	6.0	5.4	7.3	4.5	50.9
Nassau, Bahamas	25 05N	77 21W	12	35	77	65	81	69	88	75	85	73	94	41	57	1.4	1.5	1.4	2.5	4.6	6.4	5.8	5.3	6.9	6.5	2.8	1.3	46.4
Port-au-Prince, Haiti	18 33N	72 20W	121	42	87	68	89	71	94	74	90	72	101	58	70	1.3	2.3	3.4	6.3	9.1	4.0	2.9	5.7	6.9	6.7	3.4	1.3	53.3
Saint Clair, Trinidad	10 40N	61 31W	67	49	87	69	90	69	88	71	89	71	101	52	97	2.7	1.6	1.8	2.1	3.7	7.6	8.6	9.7	7.6	6.7	7.2	4.9	64.2
Saint Thomas, Virgin Is.	18 20N	64 58W	11	9	82	71	85	74	88	77	87	76	92	63	9	2.5	1.9	1.7	2.2	4.6	3.2	3.2	4.1	6.9	5.6	3.9	3.9	43.7
San Juan, Puerto Rico	18 26N	66 00W	13	30	81	67	84	69	87	74	87	73	94	60	30	4.7	2.9	2.2	3.7	7.1	5.7	6.3	7.1	6.8	5.8	6.5	5.4	64.2

TEMPERATURE AND PRECIPITATION DATA FOR REPRESENTATIVE WORLD-WIDE STATIONS

COUNTRY AND STATION	LATITUDE	LONGITUDE	ELEVATION	TEMPERATURE: LENGTH OF RECORD	AVERAGE DAILY: JANUARY MAXIMUM	JANUARY MINIMUM	APRIL MAXIMUM	APRIL MINIMUM	JULY MAXIMUM	JULY MINIMUM	OCTOBER MAXIMUM	OCTOBER MINIMUM	EXTREME MAXIMUM	EXTREME MINIMUM	AVERAGE PRECIPITATION: LENGTH OF RECORD	JANUARY	FEBRUARY	MARCH	APRIL	MAY	JUNE	JULY	AUGUST	SEPTEMBER	OCTOBER	NOVEMBER	DECEMBER	YEAR
	° ′	° ′	FEET	YEAR	°F	°F	°F	°F	°F	°F	°F	°F	°F	°F	YEAR	IN.	IN.	IN.	IN.	IN.	IN.	IN.	IN.	IN.	IN.	IN.	IN.	IN.
SOUTH AMERICA																												
Argentina:																												
Bahia Blanca	38 43S	62 16W	95	33	88	62	71	51	57	39	71	48	109	18	46	1.7	2.2	2.5	2.3	1.2	0.9	1.0	1.0	1.6	2.2	2.1	1.9	20.6
Buenos Aires	34 35S	58 29W	89	23	85	63	72	53	57	42	69	50	104	22	70	3.1	2.8	4.3	3.5	3.0	2.4	2.2	2.4	3.1	3.4	3.3	3.9	37.4
Cipolletti	38 57S	67 59W	889	9	89	56	72	40	55	29	72	43	107	9	24	0.4	0.4	0.7	0.4	0.6	0.6	0.5	0.3	0.6	0.9	0.5	0.5	6.4
Corrientes	27 28S	58 50W	177	39	93	71	81	63	71	53	82	60	112	30	40	4.7	4.5	5.3	5.6	3.3	1.9	1.7	1.5	2.8	4.7	5.2	5.2	46.4
La Quiaca	22 06S	65 36W	11,345	23	70	41	69	32	60	16	71	32	95	0	25	3.5	2.6	1.8	0.3	*	0.0	*	*	0.1	0.3	1.0	2.7	12.3
Mendoza	32 53S	68 49W	2,625	23	90	60	73	47	59	35	76	50	109	15	46	0.9	1.2	1.1	0.5	0.4	0.3	0.2	0.3	0.5	0.7	0.7	0.7	7.5
Parana	31 44S	60 31W	210	12	91	67	77	58	62	45	75	54	113	21	23	3.1	3.1	3.9	4.9	2.6	1.2	1.2	1.6	2.4	2.8	3.7	4.5	35.0
Puerto Madryn	42 47S	65 01W	26	50	81	57	70	46	55	36	68	45	104	10	50	0.4	0.6	0.7	0.5	0.9	0.6	0.6	0.4	0.6	0.7	0.4	0.6	7.0
Santa Cruz	50 01S	68 32W	39	12	70	48	57	39	41	28	58	39	94	1	20	0.6	0.3	0.3	0.6	0.4	0.5	0.4	0.5	0.3	0.3	0.4	0.7	5.3
Santiago del Estero	27 46S	64 18W	653	28	97	69	82	59	70	44	87	59	116	19	20	3.4	3.0	3.0	1.3	0.6	0.3	0.2	0.2	0.5	1.4	2.5	4.1	20.4
Ushuaia	54 50S	68 20W	26	16	57	41	48	33	39	25	52	35	85	- 6	21	2.0	2.6	1.9	2.1	1.5	1.2	1.2	1.1	1.3	1.6	1.5	1.9	19.9
Bolivia:																												
Concepcion	16 15S	62 03W	1,607	5	85	66	86	62	81	54	88	62	101	32	16	7.2	4.7	4.4	1.8	2.0	1.5	1.1	0.9	1.2	2.9	5.0	5.9	38.6
La Paz	16 30S	68 08W	12,001	31	63	43	65	40	62	33	66	40	80	26	50	4.5	4.2	2.6	1.3	0.5	0.3	0.4	0.5	1.1	1.6	1.9	3.7	22.6
Sucre	19 03S	65 17W	9,344	5	63	48	63	45	61	37	65	46	88	25	52	7.3	4.9	3.7	1.6	0.2	0.1	0.2	0.3	1.0	1.6	2.6	4.3	27.8
Brazil:																												
Barra do Corda	05 35S	45 28W	266	9	89	71	89	71	92	64	94	72	103	45	9	6.7	8.7	8.0	6.1	2.3	1.0	0.7	0.7	1.0	2.5	3.9	5.7	47.2
Bela Vista	22 06S	56 22W	525	13	91	67	85	61	77	49	87	61	108	20	20	6.6	4.9	4.4	4.3	5.0	2.8	1.3	1.8	2.9	5.4	5.8	7.0	52.2
Belem	01 27S	48 29W	42	16	87	72	87	73	88	71	89	71	98	61	20	12.5	14.1	14.1	12.6	10.2	6.7	5.9	4.4	3.5	3.3	2.6	6.1	96.0
Brasilia	15 51S	47 56W	3,481	3	80	65	82	62	78	51	82	64	93	46	3	9.0	7.8	4.8	3.4	1.4	*	0.0	*	1.3	4.9	9.7	11.7	54.0
Conceicao do Araguaia	08 15S	49 12W	53	5	88	70	91	68	95	63	93	68	102	55	5	14.9	12.1	10.8	4.1	1.9	0.4	*	0.5	1.5	6.6	4.9	8.6	66.2
Corumba	19 00S	57 39W	381	8	94	73	92	73	84	64	93	70	106	33	11	7.3	5.9	5.1	4.6	2.9	1.9	0.3	1.2	2.6	4.0	5.6	7.1	48.5
Florianopolis	27 35S	48 33W	96	17	83	72	74	64	68	57	73	63	102	32	25	7.6	5.6	6.3	4.1	3.6	3.5	2.2	3.7	4.3	5.1	3.5	4.3	53.1
Goias	15 58S	50 04W	1,706	11	86	63	91	63	89	56	94	63	104	41	11	12.5	9.9	10.2	4.6	0.4	0.3	0.0	0.3	2.3	5.3	9.4	9.5	64.8
Guarapuava	25 16S	51 30W	3,592	10	79	61	73	55	66	47	74	53	94	23	5	8.7	5.8	5.4	4.5	4.6	6.5	2.7	3.6	4.6	6.9	6.6	6.1	65.8
Manaus	03 08S	60 01W	144	11	88	75	87	75	89	75	92	76	101	63	25	9.8	9.1	10.3	8.7	6.7	3.3	2.3	1.5	1.8	4.2	5.6	8.0	71.3
Natal	05 46S	35 12W	52	18	87	76	86	73	82	69	85	75	100	61	18	1.9	4.8	7.0	9.2	7.1	8.7	7.7	3.8	1.4	0.8	0.7	1.1	54.2
Parana	12 26S	48 06W	853	19	90	58	90	58	91	48	94	58	105	37	19	11.3	9.3	9.4	4.0	0.5	*	0.1	0.2	1.1	5.0	9.1	12.2	62.3
Porto Alegre	30 02S	51 13W	33	22	87	67	78	60	66	49	74	57	105	25	22	3.5	3.2	3.9	4.1	4.5	5.1	4.5	5.0	5.2	3.4	3.1	3.5	49.1
Quixeramobim	05 12S	39 18W	653	9	92	79	86	76	88	74	93	77	100	63	13	0.7	5.0	6.6	5.0	7.0	1.7	0.7	0.6	0.4	0.6	0.7	0.6	29.6
Recife	08 04S	34 53W	97	27	86	77	85	75	80	71	84	75	94	50	56	2.1	3.3	6.3	8.7	10.5	10.9	10.0	6.0	2.5	1.0	1.0	1.1	63.4
Rio de Janeiro	22 55S	43 12W	201	38	84	73	80	69	75	63	77	66	102	46	84	4.9	4.8	5.1	4.2	3.1	2.1	1.6	1.7	2.6	3.1	4.1	5.4	42.6
Salvador (Bahia)	13 00S	38 30W	154	25	86	74	84	74	79	69	83	71	100	50	20	2.6	5.3	6.1	11.2	10.8	9.4	7.2	4.8	3.3	4.0	4.5	5.6	74.8
Santarem	02 30S	54 42W	66	22	86	73	85	73	87	71	91	73	99	65	22	6.8	10.9	13.2	12.9	11.3	6.9	4.1	1.7	1.5	1.9	2.3	4.1	77.9
Sao Paulo	23 37S	46 39W	2,628	44	77	63	73	59	66	53	68	57	100	32	24	8.8	7.8	6.0	2.2	3.0	2.4	1.5	2.1	3.5	4.6	6.0	9.4	57.3
Sena Madureira	09 04S	68 39W	443	12	92	69	91	68	91	63	93	69	100	41	17	11.2	11.3	10.2	9.4	4.1	2.2	1.1	1.5	4.0	7.0	7.5	11.7	81.2
Uaupes	00 08S	67 05W	272	15	88	72	88	72	85	70	89	71	100	52	10	10.3	7.7	10.0	10.6	12.0	9.2	8.8	7.2	5.1	6.9	7.2	10.4	105.4
Uruguaiana	29 46S	57 07W	246	15	91	69	78	59	66	48	77	55	108	27	12	3.6	3.6	5.6	5.1	3.7	4.2	3.2	2.8	3.6	4.1	2.9	4.1	46.6
Chile:																												
Ancud	41 47S	73 52W	184	30	62	51	57	47	50	42	55	45	82	30	46	3.1	3.7	5.3	7.4	9.9	11.0	10.3	9.4	6.5	4.2	4.7	4.6	80.1
Antofagasta	23 42S	70 24W	308	22	76	63	70	58	63	51	66	55	86	37	32	0.0	0.0	0.0	*	*	0.1	0.2	0.1	*	0.1	*	0.0	0.5
Arica	18 28S	70 20W	95	15	78	64	74	60	66	54	69	58	93	39	25	*	0.0	0.0	0.0	0.0	0.0	0.0	*	0.0	0.0	0.0	*	*
Cabo Raper	46 50S	75 38W	131	8	58	46	54	44	47	38	51	40	72	28	10	7.8	5.8	7.1	7.7	7.5	7.9	9.5	7.5	5.6	7.0	6.7	7.0	87.1

Los Evangelistas	52 23S	75 07W	190	16	50	44	48	41	43	36	45	39	66	19	27	11.7	10.0	11.3	11.4	9.6	9.4	9.4	8.6	9.2	8.8	9.9	10.1	119.4
Potrerillos	26 30S	69 27W	9,350	7	65	49	63	47	57	40	61	44	75	20	7	*	*	0.3	*	0.7	*	0.5	0.3	0.2	0.2	0.0	*	2.2
Puerto Aisen	42 24S	72 42W	33	8	63	50	55	43	45	37	55	42	93	18	11	7.8	7.8	8.3	7.5	14.7	10.4	11.1	11.1	6.5	7.8	7.0	7.9	107.9
Punta Arenas	53 10S	70 54W	26	15	58	45	50	39	40	31	51	38	86	11	15	1.5	0.9	1.3	1.4	1.3	1.6	1.1	1.2	0.9	1.1	0.7	1.4	14.4
Santiago	33 27S	70 42W	1,706	14	85	53	74	45	59	37	72	45	99	24	58	0.1	0.1	0.2	0.5	2.5	3.3	3.0	2.2	1.2	0.6	0.3	0.2	14.2
Valdivia	39 48S	73 14W	16	29	73	52	62	46	52	41	63	44	97	19	60	2.6	2.9	5.2	9.2	14.2	17.7	15.5	12.9	8.2	5.0	4.9	4.1	102.4
Valparaiso	33 01S	71 38W	135	30	72	56	67	52	60	47	65	50	94	32	41	0.1	*	0.3	0.6	4.1	5.9	3.9	2.9	1.3	0.4	0.2	0.2	19.9
Colombia:																												
Andagoya	05 06N	76 40W	197	8	90	75	90	75	89	74	90	74	97	62	15	25.0	21.4	19.5	26.1	25.5	25.8	23.3	25.3	24.6	22.7	22.4	19.5	281.1
Bogota	04 42N	74 08W	8,355	10	67	48	67	51	64	50	66	50	75	30	49	2.3	2.6	4.0	5.8	4.5	2.4	2.0	2.2	2.4	6.3	4.7	2.6	41.8
Cartagena	10 28N	75 30W	39	6	84	73	87	76	88	78	87	77	98	61	10	0.4	0.0	0.4	0.9	3.4	3.4	3.0	0.6	0.5	10.8	8.9	4.5	36.8
Ipiales	00 50N	77 42W	9,680	9	61	50	60	49	57	42	62	49	77	32	13	3.1	2.3	3.5	3.5	2.8	1.9	1.3	1.1	1.4	3.1	3.3	2.6	29.9
Tumaco	01 49N	78 45W	7	10	82	75	84	76	82	75	82	75	90	64	10	16.9	11.7	9.6	14.6	17.4	12.0	7.7	7.3	7.3	5.9	4.9	7.0	122.3
Ecuador:																												
Cuenca	02 53S	78 39W	8,301	7	69	50	69	50	65	47	70	49	81	29	10	2.0	1.8	3.2	4.3	4.3	1.7	0.9	1.1	1.6	3.1	1.8	2.5	28.3
Guayaquil	02 10S	79 53W	20	5	87	72	88	72	84	67	86	68	98	52	10	8.3	11.4	11.5	8.1	2.1	0.4	0.2	*	*	*	0.1	1.1	43.2
Quito	00 08S	78 29W	9,222	54	67	46	69	47	71	44	71	46	86	25	33	3.9	4.4	5.6	6.9	5.4	1.7	0.8	1.2	2.7	4.4	3.8	3.1	43.9
French Guiana:																												
Cayenne	04 56N	52 27W	20	38	84	74	86	75	88	73	91	74	97	65	51	14.4	12.3	15.8	18.9	21.7	15.5	6.9	2.8	1.2	1.3	4.6	10.7	126.1
Guyana:																												
Georgetown	06 50N	58 12W	6	54	84	74	85	76	85	75	87	76	93	68	35	8.0	4.5	6.9	5.5	11.4	11.9	10.0	6.9	3.2	3.0	6.1	11.3	88.7
Lethem	03 24N	59 38W	270	3	91	73	91	74	87	73	92	76	97	63	9	1.2	1.4	1.3	5.7	11.5	11.9	14.8	9.4	3.4	2.3	4.3	1.3	68.5
Paraguay:																												
Asuncion	25 17S	57 30W	456	15	95	71	84	65	74	53	86	62	110	29	30	5.5	5.1	4.3	5.2	4.6	2.7	2.2	1.5	3.1	5.5	5.9	6.2	51.8
Bahia Negra	20 14S	58 10W	318	20	92	74	87	68	79	61	90	69	106	35	20	5.4	5.3	4.9	2.9	2.3	1.6	1.5	0.6	2.3	4.2	5.3	4.3	40.6
Peru:																												
Arequipa	16 21S	71 34W	8,460	13	67	49	67	48	67	47	68	47	82	25	37	1.3	1.8	0.7	0.2	*	*	*	*	0.0	*	*	0.4	4.4
Cajamarca	07 09S	78 30W	8,662	9	71	48	70	47	70	41	71	47	79	25	9	3.6	4.2	4.6	3.4	1.7	9.5	9.2	9.3	2.3	2.3	1.9	3.2	28.2
Cusco	13 33S	71 59W	10,866	13	68	45	71	40	70	31	72	43	86	16	12	6.4	5.9	4.3	2.0	0.6	0.2	0.2	0.4	1.0	2.6	3.0	5.4	32.0
Iquitos	03 45S	73 13W	384	5	90	71	87	71	88	68	90	70	100	54	5	9.1	10.4	9.4	13.6	10.7	5.7	6.4	5.2	10.5	7.3	9:1	10.3	107.7
Lima	12 05S	77 03W	394	15	82	66	80	63	67	57	71	58	93	49	15	0.1	*	*	*	0.2	0.2	0.3	0.3	0.3	0.1	0.1	*	1.6
Mollendo	17 00S	72 07W	80	10	79	66	76	63	67	57	70	59	90	50	10	*	0.1	*	*	0.1	0.1	*	0.2	0.2	0.1	0.1	*	0.9
Surinam:																												
Paramaribo	05 49N	55 09W	12	35	85	72	86	73	87	73	91	73	99	62	75	8.4	6.5	7.9	9.0	12.2	11.9	9.1	6.2	3.1	3.0	4.9	8.8	91.0
Uruguay:																												
Artigas	30 24S	56 23W	384	13	91	65	77	55	65	45	75	54	107	24	50	4.3	3.9	4.7	5.1	4.1	4.1	2.8	3.0	4.0	4.7	3.8	4.1	48.6
Montevideo	34 52S	56 12W	72	56	83	62	71	53	58	43	68	49	109	25	56	2.9	2.6	3.9	3.9	3.3	3.2	2.9	3.1	3.0	2.6	2.9	3.1	37.4
Venezuela:																												
Caracas	10 30N	66 56W	3,418	30	75	56	81	60	78	61	79	61	91	45	46	0.9	0.4	0.6	1.3	3.1	4.0	4.3	4.3	4.2	4.3	3.7	1.8	32.9
Ciudad Bolivar	08 07N	63 32W	197	10	90	72	93	75	90	75	93	75	100	64	10	1.4	0.8	0.7	1.0	3.8	5.5	6.3	7.1	3.6	4.0	2.8	1.3	38.3
Maracaibo	10 39N	71 36W	20	12	90	73	92	76	94	76	92	76	102	66	36	0.1	*	0.3	0.8	2.7	2.2	1.8	2.2	2.8	5.9	3.3	0.6	22.7
Merida	08 36N	71 10W	5,293	14	73	56	75	60	76	59	75	60	90	48	14	2.5	1.5	3.6	6.7	9.8	7.3	4.7	5.7	6.7	9.5	8.2	3.4	69.7
Santa Elena	04 36N	61 07W	2,976	10	82	61	82	63	81	61	84	61	95	48	10	3.2	3.2	3.2	5.7	9.6	9.5	9.1	7.6	5.3	4.9	4.9	4.5	70.7
PACIFIC ISLANDS																												
Easter Is. (Isla de Pascua)	27 10S	109 26W	98	4	77	64	78	63	70	58	73	58	88	46	10	4.8	3.7	4.6	4.2	4.6	4.3	3.5	3.0	2.7	3.7	4.6	4.9	48.6
Mas a Tierra (Juan Fernandez)	33 37S	78 52W	20	25	72	60	68	57	60	50	61	51	86	39	29	0.8	1.2	1.6	3.4	5.9	6.4	5.8	4.4	2.9	1.9	1.6	1.0	36.9
Seymour Is. (Galapagos Is.)	00 28S	90 18W	36	3	86	72	87	75	81	69	81	67	93	58	3	0.8	1.4	1.1	0.7	*	*	*	*	*	*	*	*	4.0
ATLANTIC ISLANDS																												
Fernando de Noronha	03 50S	32 25W	148	32	84	75	82	75	81	73	82	75	93	63	32	1.7	4.7	7.4	10.5	10.5	7.3	5.4	1.9	0.7	0.3	0.4	0.5	51.3
Cumberland Bay, South Georgia	54 16S	36 30W	8	23	48	35	42	29	34	23	41	28	84	- 3	24	3.3	4.3	5.3	5.4	5.2	4.9	5.5	5.3	3.5	2.6	3.4	3.0	51.7
Laurie Is., South Orkneys	60 44S	44 44W	13	48	35	29	31	21	20	4	30	19	54	-40	46	1.4	1.5	1.9	1.6	1.2	1.0	1.3	1.3	1.1	1.1	1.3	1.0	15.7
Stanley, Falkland Isles	51 42S	57 51W	6	25	56	42	49	37	40	31	48	35	76	12	41	2.8	2.3	2.5	2.6	2.6	2.1	2.0	2.0	1.5	1.6	2.0	2.8	26.8

See footnotes at end of table.

TEMPERATURE AND PRECIPITATION DATA FOR REPRESENTATIVE WORLD-WIDE STATIONS

COUNTRY AND STATION	LATITUDE	LONGITUDE	ELE-VATION	TEMPERATURE: LENGTH OF RECORD	AVERAGE DAILY: JANUARY MAXIMUM	JANUARY MINIMUM	APRIL MAXIMUM	APRIL MINIMUM	JULY MAXIMUM	JULY MINIMUM	OCTOBER MAXIMUM	OCTOBER MINIMUM	EXTREME: MAXIMUM	EXTREME: MINIMUM	AVERAGE PRECIPITATION: LENGTH OF RECORD	JANUARY	FEBRUARY	MARCH	APRIL	MAY	JUNE	JULY	AUGUST	SEPTEMBER	OCTOBER	NOVEMBER	DECEMBER	YEAR
	° '	° '	FEET	YEAR	°F	°F	°F	°F	°F	°F	°F	°F	°F	°F	YEAR	IN.	IN.	IN.	IN.	IN.	IN.	IN.	IN.	IN.	IN.	IN.	IN.	IN.
EUROPE																												
Albania:																												
Durres	41 19N	19 28E	23	10	51	42	63	55	83	74	68	58	95	21	10	3.0	3.3	3.9	2.2	1.6	1.9	0.5	1.9	1.7	7.1	8.5	7.3	42.9
Andorra:																												
Les Escaldes	42 30N	01 31E	3,543	5	43	29	59	39	78	55	61	42	91	0	9	1.5	1.7	2.9	2.4	4.7	3.1	2.2	3.4	3.1	3.5	3.3	2.5	34.3
Austria:																												
Innsbruck	47 16N	11 24E	1,909	34	34	20	60	39	78	55	58	40	97	-16	35	2.1	1.8	1.5	2.2	2.9	4.1	5.1	4.5	3.1	2.4	2.2	1.9	33.8
Vienna (Wien)	48 15N	16 22E	664	50	34	26	57	41	75	59	55	44	98	-14	100	1.5	1.4	1.8	2.0	2.8	2.7	3.0	2.7	2.0	2.0	1.9	1.8	25.6
Bulgaria:																												
Sofiya (Sofia)	42 42N	23 20E	1,805	30	34	22	62	41	82	57	63	42	99	-17	27	1.3	1.1	1.7	2.3	3.3	3.2	2.4	2.0	2.3	2.1	1.9	1.4	25.0
Varna	43 12N	27 55E	115	30	40	30	59	43	84	63	67	50	107	-12	20	1.5	0.9	1.2	1.2	1.8	2.6	1.9	1.2	1.5	1.9	1.9	2.0	19.6
Cyprus:																												
Nicosia	35 09N	33 17E	716	40	58	42	74	50	97	69	81	58	116	23	64	2.9	2.0	1.3	0.8	1.1	0.4	*	*	0.2	0.9	1.7	3.0	14.6
Czechoslovakia:																												
Praha (Prague)	50 05N	14 25E	662	40	34	25	55	40	74	58	54	44	98	-16	70	0.9	0.8	1.1	1.5	2.4	2.8	2.6	2.2	1.7	1.2	1.2	0.9	19.3
Prerov	49 27N	17 27E	702	20	34	25	57	38	77	55	56	40	100	-23	21	1.3	1.1	1.1	2.0	2.4	2.9	3.5	3.2	2.0	2.4	1.5	1.4	24.8
Denmark:																												
Copenhagen (Kobenhavn)	55 41N	12 33E	43	30	36	29	50	37	72	55	53	42	91	- 3	30	1.6	1.3	1.2	1.7	1.7	2.1	2.2	3.2	1.9	2.1	2.2	2.1	23.3
Aarhus	56 08N	10 12E	161	21	35	27	51	37	70	54	53	42	87	-12	21	2.3	1.5	1.4	1.8	1.2	2.2	2.5	3.3	3.2	2.6	2.5	2.1	26.6
Finland:																												
Helsinki	60 10N	24 57E	30	20	27	17	43	31	71	57	45	37	89	-23	50	2.2	1.7	1.7	1.7	1.9	2.0	2.3	3.3	2.8	2.9	2.7	2.4	27.6
Kuusamo	65 57N	29 12E	843	20	17	2	35	18	68	50	36	27	90	-40	20	1.1	1.1	1.1	1.1	1.4	2.3	2.8	3.0	2.1	2.1	1.6	1.1	20.8
Vaasa	63 05N	21 36E	13	18	26	16	41	28	69	55	44	36	89	-29	19	1.1	0.8	0.8	1.0	1.4	1.8	2.4	2.5	2.7	2.3	1.7	1.1	19.6
France:																												
Ajaccio (Corsica)	41 52N	08 35E	243	46	56	40	66	48	85	64	72	55	103	23	86	3.0	2.3	2.6	2.2	1.6	0.9	2.8	0.7	1.7	3.8	4.4	3.1	29.1
Bordeaux	44 50N	00 43W	157	51	48	35	63	44	80	58	66	47	102	9	47	2.7	2.8	2.9	2.6	2.5	2.3	2.0	1.9	2.2	3.0	3.9	3.9	32.7
Brest	48 19N	04 47W	56	56	49	40	57	44	70	56	61	49	95	7	56	3.5	3.0	2.5	2.5	1.9	2.0	2.0	2.2	2.3	3.6	4.2	4.4	34.1
Cherbourg	49 39N	01 38W	30	47	47	40	54	43	67	57	59	50	91	14	47	3.3	2.9	2.7	2.0	1.9	1.8	1.9	3.0	2.9	4.6	5.1	5.2	37.3
Lille	50 35N	03 05W	141	40	42	33	58	40	75	55	59	45	96	0	40	2.5	1.9	2.5	2.0	2.4	2.2	2.8	2.3	2.6	3.0	3.0	3.2	30.3
Lyon	45 42N	04 47E	938	70	41	30	61	42	80	58	61	45	105	-13	70	1.4	1.4	1.8	2.1	2.8	2.9	2.8	2.9	3.1	3.1	2.6	1.9	28.8
Marseille	43 18N	05 23E	246	72	53	38	59	41	78	58	76	57	101	9	102	1.9	1.5	1.8	2.0	1.9	1.0	0.6	0.9	2.6	3.7	3.1	2.2	23.2
Paris	48 49N	02 29E	164	66	42	32	60	41	76	55	59	44	105	1	118	1.5	1.3	1.5	1.7	2.0	2.1	2.1	2.0	2.0	2.2	2.0	1.9	22.3
Strasbourg	48 35N	07 46E	465	20	40	31	59	41	78	57	58	43	101	- 8	20	1.6	1.4	1.7	2.6	2.6	3.1	3.4	3.4	3.1	2.7	2.0	1.9	29.5
Toulouse	43 33N	01 23E	538	47	47	35	62	43	82	59	66	48	111	1	47	1.9	1.7	2.3	2.7	2.9	2.4	1.5	2.1	2.3	2.2	2.4	2.3	26.7
Germany:																												
Berlin	52 27N	13 18E	187	50	35	26	55	38	74	55	55	41	96	-15	40	1.9	1.3	1.5	1.7	1.9	2.3	3.1	2.2	1.9	1.7	1.7	1.9	23.1
Bremen	53 05N	08 47E	52	50	37	30	53	38	71	55	54	43	94	- 7	80	1.9	1.6	1.8	1.5	2.1	2.6	3.2	2.8	2.1	2.2	2.0	2.2	26.0
Frankfurt A/M	50 07N	08 40E	338	50	37	29	58	41	75	56	56	43	100	- 7	80	1.7	1.3	1.6	1.5	2.0	2.5	2.8	2.6	1.9	2.2	2.0	2.0	24.1
Hamburg	53 33N	09 58E	66	50	35	28	51	39	69	56	53	44	92	- 4	80	2.1	1.9	2.0	1.8	2.1	2.7	3.4	3.2	2.5	2.6	2.1	2.5	28.9
Munchen (Munich)	48 09N	11 34E	1,739	50	33	23	54	37	72	54	53	40	92	-14	80	1.7	1.4	1.9	2.7	3.7	4.6	4.7	4.2	3.2	2.2	1.9	1.9	34.1
Munster	51 58N	07 38E	207	50	39	29	56	38	73	54	56	42	96	-17	40	2.6	1.9	2.2	2.0	2.2	2.7	3.3	3.1	2.5	2.7	2.4	2.9	30.5
Nurnberg	49 27N	11 03E	1,050	50	35	26	56	38	74	55	55	41	99	-18	80	1.5	1.2	1.3	1.7	2.2	2.5	3.1	3.1	2.1	2.1	1.9	1.7	24.4
Gibraltar:																												
Windmill Hill	36 06N	05 21W	400	12	58	50	64	55	77	66	70	61	97	35	12	4.6	3.4	3.7	2.5	1.4	0.2	*	0.1	0.8	3.5	4.1	5.4	29.7

Greece:																												
Athinai (Athens)	37 58N	23 43E	351	72	54	42	67	52	90	72	74	60	109	20	80	2.2	1.6	1.4	0.8	0.8	0.6	0.2	0.4	0.6	1.7	2.8	2.8	15.8
Iraklion (Crete)	35 20N	25 08E	98	21	60	48	70	54	85	72	77	62	114	32	22	3.7	3.0	1.6	0.9	0.7	0.1	*	0.1	0.7	1.7	2.7	4.0	19.2
Rodhos (Rhodes)	36 26N	28 15E	289	10	59	51	67	59	83	74	76	68	104	30	6	5.7	3.9	2.6	1.7	0.5	0.3	0.0	*	0.4	1.7	5.2	6.7	28.5
Thessaloniki (Salonika)	40 37N	22 57E	78	9	49	37	66	49	90	70	73	56	107	15	26	1.5	1.5	1.6	1.9	2.0	1.2	1.0	0.7	1.2	2.4	2.1	1.9	19.0
Hungary:																												
Budapest	47 31N	19 02E	394	50	35	26	62	44	82	61	61	45	103	-10	50	1.5	1.5	1.7	2.0	2.7	2.6	2.0	1.9	1.8	2.1	2.4	2.0	24.2
Debrecen	47 36N	21 39E	430	50	33	21	61	39	81	57	60	41	102	-22	80	1.2	1.1	1.4	1.8	2.4	2.8	2.5	2.3	1.8	2.2	2.0	1.6	23.1
Iceland:																												
Akureyri	65 41N	18 05W	16	23	34	26	40	30	57	47	43	34	83	- 8	26	1.7	1.5	1.7	1.3	0.6	0.9	1.3	1.6	1.9	2.3	1.9	1.9	18.6
Reykjavik	64 09N	21 56W	92	25	36	28	43	33	58	48	44	36	74	4	30	4.0	3.1	3.0	2.1	1.6	1.7	2.0	2.6	3.1	3.4	3.6	3.7	33.9
Ireland:																												
Cork	51 54N	08 29W	56	27	48	38	55	41	68	53	58	44	85	15	35	4.9	3.6	3.3	2.6	2.9	2.0	2.9	3.1	2.9	3.9	4.5	4.7	41.3
Dublin	53 22N	06 21W	155	30	47	35	54	38	67	51	57	43	86	8	35	2.7	2.2	2.0	1.9	2.3	2.0	2.8	3.0	2.8	2.7	2.7	2.6	29.7
Shannon Airport	52 41N	08 55W	8	9	46	36	55	41	66	53	58	45	87	12	12	3.8	3.0	2.0	2.2	2.4	2.1	3.1	3.0	3.0	3.4	4.2	4.3	36.5
Italy:																												
Ancona	43 37N	13 32E	52	30	46	36	62	50	83	68	67	55	102	18	30	2.6	1.7	1.6	2.3	2.1	1.9	1.5	1.5	3.5	3.7	2.5	3.0	28.0
Cagliari (Sardinia)	39 15N	09 03E	3	30	56	43	66	50	86	67	72	58	102	25	25	2.2	1.5	1.5	1.2	1.5	0.5	0.1	0.4	1.0	3.0	1.8	2.3	17.0
Genova (Genoa)	44 24N	08 55E	318	10	50	41	65	53	82	70	73	58	100	18	10	3.9	4.0	3.3	3.4	4.6	1.4	1.6	2.3	4.7	6.1	7.2	4.1	46.6
Napoli (Naples)	40 51N	14 15E	82	30	54	40	65	52	84	70	71	60	101	24	30	3.7	3.2	3.0	2.6	1.8	1.8	0.6	0.7	2.8	5.1	4.5	5.4	35.2
Palermo (Sicily)	38 07N	13 19E	354	10	58	47	67	53	86	71	75	62	113	31	30	3.8	3.4	2.4	1.9	1.1	0.6	0.2	0.6	2.0	3.7	4.1	4.5	28.3
Rome	41 48N	12 36E	377	10	54	39	68	46	88	64	73	53	104	20	30	3.3	2.9	2.0	2.0	1.9	0.7	0.4	0.7	2.8	4.3	4.4	4.1	29.5
Taranto	40 28N	17 17E	56	10	55	43	59	50	89	70	73	58	108	26	10	1.6	0.9	1.3	0.8	1.0	0.6	0.4	0.7	1.0	2.2	1.8	1.9	14.2
Venezia (Venice)	45 26N	12 23E	82	10	43	33	63	49	82	67	65	52	97	14	30	2.0	2.1	2.4	2.8	3.2	3.3	2.6	2.6	2.6	3.7	3.5	2.6	33.4
Luxembourg:																												
Luxembourg	49 37N	06 03E	1,096	7	36	29	58	40	74	55	56	43	99	-10	100	2.3	2.0	1.9	2.1	2.4	2.5	2.8	2.6	2.4	2.7	2.7	2.8	29.2
Malta:																												
Valletta	35 54N	14 31E	233	90	59	51	66	56	84	72	76	66	105	34	90	3.3	2.3	1.5	0.8	0.4	0.1	*	0.2	1.3	2.7	3.6	3.9	20.3
Monaco:																												
Monaco	43 44N	07 25E	180	60	54	46	61	53	77	70	67	60	93	27	60	2.4	2.3	3.1	2.2	2.1	1.4	0.7	1.1	2.3	4.7	4.3	3.5	30.1
Netherlands:																												
Amsterdam	52 23N	04 55E	5	29	40	34	52	43	69	59	56	48	95	3	29	2.0	1.4	1.3	1.6	1.8	1.8	2.6	2.7	2.8	2.8	2.6	2.2	25.6
Norway:																												
Bergen	60 24N	05 19E	141	49	43	27	55	34	72	51	57	38	89	3	75	7.9	6.0	5.4	4.4	3.9	4.2	5.2	7.3	9.2	9.2	8.0	8.1	78.8
Kristiansand	58 10N	07 59E	175	11	32	25	50	35	71	53	53	39	90	-14	56	5.0	3.6	3.6	2.7	2.5	2.8	3.5	5.3	4.7	6.2	5.7	6.4	52.0
Oslo	59 56N	10 44E	308	44	30	20	50	34	73	56	49	37	93	-21	56	1.7	1.3	1.4	1.6	1.8	2.4	2.9	3.8	2.5	2.9	2.3	2.3	26.9
Tromso	69 39N	18 57E	335	47	30	22	37	27	59	48	40	33	83	- 1	75	4.1	3.8	3.3	2.4	2.1	2.1	2.3	2.9	4.7	4.5	4.0	3.9	40.1
Trondheim	63 25N	10 27E	417	44	31	22	45	32	66	51	46	36	95	-22	65	3.1	2.7	2.6	2.0	1.7	1.9	2.4	3.0	3.4	3.7	2.8	2.8	32.1
Vardo	70 22N	31 06E	43	40	27	19	34	26	53	44	38	32	80	-11	56	2.5	2.5	2.3	1.5	1.3	1.3	1.5	1.7	1.9	2.5	2.1	2.4	23.5
Poland:																												
Gdansk (Danzig)	54 24N	18 40E	36	36	33	25	49	37	70	56	53	42	94	-16	35	1.2	1.0	1.3	1.5	1.8	2.3	2.8	2.6	2.1	1.8	1.8	1.5	21.7
Krakow	50 04N	19 57E	723	35	32	22	55	38	76	57	56	41	97	-28	35	1.1	1.3	1.4	1.8	2.8	4.0	4.5	3.8	2.7	2.2	1.7	1.3	28.6
Warsaw	52 13N	21 02E	294	25	30	21	54	38	75	56	54	41	98	-22	113	1.2	1.1	1.3	1.5	1.9	2.6	3.0	3.0	1.9	1.7	1.4	1.4	22.0
Wroclaw (Breslau)	51 07N	17 05E	482	50	35	25	55	39	74	57	55	42	98	-26	40	1.5	1.1	1.5	1.7	2.4	2.4	3.4	2.7	1.8	1.7	1.5	1.5	23.2
Portugal:																												
Braganca	41 49N	06 47W	2,395	11	46	31	59	39	80	54	62	42	103	10	11	11.9	6.9	7.7	3.7	3.0	1.6	0.5	0.6	1.5	3.0	6.3	7.1	53.8
Lagos	37 06N	08 38W	46	21	61	47	67	52	83	64	73	58	107	28	17	3.2	2.6	2.8	1.4	0.8	0.2	*	*	0.4	1.5	2.6	2.8	18.3
Lisbon	38 43N	09 08W	313	75	56	46	64	52	79	63	69	57	103	29	75	3.3	3.2	3.1	2.4	1.7	0.7	0.2	0.2	1.4	3.1	4.2	3.6	27.0
Romània:																												
Bucuresti (Bucharest)	44 25N	26 06E	269	41	33	20	63	41	86	61	65	44	105	-18	41	1.5	1.1	1.7	1.6	2.5	3.8	2.3	1.8	1.5	1.6	1.9	1.5	22.8
Cluj	46 47N	23 40E	1,286	15	31	18	58	38	79	56	60	41	100	-26	16	1.3	1.2	1.0	2.1	3.3	3.3	2.6	3.3	2.0	1.7	1.0	1.2	24.0
Constanta	44 11N	28 39E	13	20	37	25	55	42	79	63	62	49	101	-13	39	1.2	1.2	1.1	1.1	1.3	1.7	1.3	1.1	1.1	1.4	1.2	1.4	15.1
Spain:																												
Almeria	36 51N	02 28W	213	20	61	47	69	54	85	69	76	62	108	34	20	0.9	1.0	0.7	0.9	0.7	0.2	*	0.1	0.6	0.9	1.5	1.1	8.6
Barcelona	41 24N	02 09E	312	20	56	42	64	51	81	69	71	58	98	24	30	1.2	2.1	1.9	1.8	1.8	1.3	1.2	1.7	2.6	3.4	2.7	1.8	23.5
Burgos	42 20N	03 42W	2,825	29	42	30	57	38	77	53	61	43	99	0	29	1.5	1.5	2.1	1.9	2.4	1.7	0.8	0.7	1.4	2.0	2.2	2.0	20.2
Madrid	40 25N	03 41W	2,188	30	47	33	64	44	87	62	66	48	102	14	30	1.1	1.7	1.7	1.7	1.5	1.2	0.4	0.3	1.2	1.9	2.2	1.6	16.5
Sevilla	37 29N	05 59W	98	26	59	41	73	51	96	67	78	57	117	27	26	2.2	2.9	3.3	2.3	1.3	0.9	0.1	0.1	1.1	2.6	3.7	2.8	23.3
Valencia	39 28N	00 23W	79	26	58	41	67	51	83	68	73	57	107	20	29	0.9	1.5	0.9	1.2	1.1	1.3	0.4	0.5	2.2	1.6	2.5	1.3	15.4
Sweden:																												
Abisko	68 21N	18 49E	1,273	11	20	6	33	19	61	45	35	24	82	-30	11	0.7	0.6	0.5	0.5	0.7	1.8	1.6	1.8	1.2	1.0	0.6	0.6	11.7

TEMPERATURE AND PRECIPITATION DATA FOR REPRESENTATIVE WORLD-WIDE STATIONS

COUNTRY AND STATION	LATITUDE	LONGITUDE	ELE-VATION	TEMPERATURE: LENGTH OF RECORD	AVERAGE DAILY JANUARY MAXIMUM	JANUARY MINIMUM	APRIL MAXIMUM	APRIL MINIMUM	JULY MAXIMUM	JULY MINIMUM	OCTOBER MAXIMUM	OCTOBER MINIMUM	EXTREME MAXIMUM	EXTREME MINIMUM	AVERAGE PRECIPITATION: LENGTH OF RECORD	JANUARY	FEBRUARY	MARCH	APRIL	MAY	JUNE	JULY	AUGUST	SEPTEMBER	OCTOBER	NOVEMBER	DECEMBER	YEAR
	° '	° '	FEET	YEAR	°F	°F	°F	°F	°F	°F	°F	°F	°F	°F	YEAR	IN.	IN.	IN.	IN.	IN.	IN.	IN.	IN.	IN.	IN.	IN.	IN.	IN.
Sweden cont'd:																												
Goteberg	57 42N	11 58E	55	39	35	27	48	36	69	56	51	42	88	-13	61	2.5	2.0	2.0	1.7	1.9	2.2	2.8	3.7	3.1	3.1	2.7	2.8	30.5
Haparanda	65 50N	24 09E	30	20	22	10	38	23	71	53	39	30	89	-34	20	2.2	1.6	1.2	1.5	1.4	1.7	2.1	2.8	2.6	2.8	2.5	2.0	24.4
Karlstad	59 23N	13 30E	164	30	30	20	49	32	73	56	49	38	93	-21	30	1.9	1.2	1.2	1.4	1.9	1.9	2.6	3.1	2.9	2.4	2.4	1.9	24.8
Sarna	61 41N	13 07E	1,504	20	19	4	42	23	69	46	42	28	91	-51	20	1.6	0.8	0.9	1.2	1.6	2.8	3.6	3.3	2.6	2.3	1.8	1.8	24.3
Stockholm	59 21N	18 04E	146	30	31	23	45	32	70	55	48	39	97	-26	30	1.5	1.1	1.1	1.5	1.6	1.9	2.8	3.1	2.1	2.1	1.9	1.9	22.4
Visby (Gotland)	57 39N	18 18E	36	30	35	28	44	33	67	55	50	41	88	1	30	1.7	1.1	1.2	1.4	1.1	1.4	2.0	2.7	1.7	1.9	2.1	2.0	20.3
Switzerland:																												
Berne	46 57N	07 26E	1,877	30	35	26	56	39	74	56	55	42	96	- 9	77	1.9	2.0	2.6	3.0	3.7	4.4	4.4	4.3	3.5	3.5	2.7	2.5	38.5
Geneve (Geneva)	46 12N	06 09E	1,329	30	39	29	58	41	77	58	58	44	101	- 1	125	1.9	1.8	2.2	2.5	3.0	3.1	2.9	3.6	3.6	3.8	3.1	2.4	33.9
Zurich	47 23N	08 33E	1,617	23	38	28	57	39	76	55	57	42	98	-12	23	2.3	1.9	2.9	3.4	4.0	4.9	5.0	4.6	3.3	3.2	2.5	2.9	40.9
Turkey:																												
Edirne (Adrianople)	41 39N	26 34E	154	18	41	28	66	44	88	63	70	49	107	- 8	18	2.2	1.9	1.7	1.9	1.7	2.1	1.5	1.1	1.1	2.1	2.9	3.0	23.2
Istanbul (Constantinople)	40 58N	28 50E	59	18	45	36	61	45	81	65	67	54	100	17	18	3.7	2.3	2.6	1.9	1.4	1.3	1.7	1.5	2.3	3.8	4.1	4.9	31.5
United Kingdom:																												
Belfast	54 35N	05 56W	57	7	42	34	53	38	65	52	55	44	82	14	30	4.2	2.8	2.3	2.4	2.3	2.5	3.5	3.5	3.4	3.8	3.6	3.9	38.2
Birmingham	52 29N	01 56W	535	30	42	35	53	40	69	54	55	45	92	11	30	2.9	2.1	1.7	2.2	2.5	1.8	2.8	2.7	2.3	2.9	3.2	2.6	29.7
Cardiff	51 28N	03 10W	203	30	45	36	55	41	69	54	57	45	91	2	30	4.6	3.0	2.3	2.5	3.0	2.2	3.4	3.9	3.6	4.5	4.6	4.3	41.9
Dublin	53 22N	06 21W	155	30	47	35	54	38	67	51	57	43	86	8	35	2.7	2.2	2.0	1.9	2.3	2.0	2.8	3.0	2.8	2.7	2.7	2.6	29.7
Edinburgh	55 55N	03 11W	441	30	43	35	50	39	65	52	53	44	83	15	30	2.5	1.6	1.6	1.6	2.2	1.9	3.1	3.1	2.6	2.9	2.4	2.1	27.6
London	51 29N	00 00	149	30	44	35	56	40	73	55	58	44	99	9	30	2.0	1.5	1.4	1.8	1.8	1.6	2.0	2.2	1.8	2.3	2.5	2.0	22.9
Liverpool	53 24N	03 04W	198	30	44	36	52	41	66	55	55	46	87	15	30	2.7	1.9	1.5	1.6	2.2	2.0	2.8	3.1	2.6	3.0	3.0	2.5	28.9
Perth	56 24N	03 27W	77	30	43	32	53	38	68	51	55	41	89	0	30	3.1	2.2	1.9	1.7	2.3	2.0	3.1	2.9	2.8	3.3	2.7	2.7	30.7
Plymouth	50 21N	04 07W	87	30	47	40	54	43	66	55	58	49	88	16	30	4.3	3.0	2.6	2.3	2.5	2.0	2.6	2.9	2.8	3.8	4.6	4.4	37.8
Wick	58 26N	03 05W	119	30	42	35	48	38	59	50	52	43	80	8	30	2.9	2.1	1.8	2.1	1.8	2.0	2.6	2.6	2.9	3.2	3.1	2.9	30.0
U.S.S.R.:																												
Arkhangelsk	64 33N	40 32E	22	23	9	2	36	23	64	51	36	30	91	-49	25	1.2	1.1	1.1	0.7	1.3	1.9	2.6	2.7	2.2	1.9	1.6	1.3	19.8
Astrakhan	46 21N	48 02E	45	10	23	14	57	40	85	69	56	40	99	-22	25	0.5	0.5	0.4	0.6	0.6	0.7	0.5	0.4	0.6	0.4	0.6	0.6	6.4
Dnepropetrovsk	48 27N	35 04E	259	18	25	16	53	39	80	62	56	40	101	-25	17	1.4	1.1	1.2	1.4	1.8	3.0	1.9	1.6	1.0	1.8	1.6	1.6	19.4
Kaunas	54 54N	23 53E	118	19	26	18	49	34	72	53	50	38	96	-23	19	1.6	1.3	1.3	1.8	2.0	3.2	3.3	3.5	1.9	1.9	1.6	1.6	25.0
Kirov	58 36N	49 41E	594	20	6	- 2	41	27	72	55	37	29	92	-43	29	1.2	1.0	0.9	0.9	1.9	2.5	2.1	2.9	2.3	2.0	1.6	1.3	20.6
Kursk	51 45N	36 12E	773	15	19	11	47	35	74	58	48	36	91	-23	20	1.5	1.3	1.2	1.5	2.2	2.5	3.2	2.3	1.6	1.8	1.5	1.7	22.3
Leningrad	59 56N	30 16E	16	26	23	12	45	31	71	57	45	37	91	-36	95	1.0	0.9	0.9	1.0	1.6	2.0	2.5	2.8	2.1	1.8	1.4	1.2	19.2
Lvov	49 50N	24 01E	978	9	31	22	53	38	77	59	55	43	97	-29	35	1.3	1.5	1.8	2.0	2.8	3.7	4.1	3.1	2.4	2.1	0.8	1.6	28.2
Minsk	53 54N	27 33E	738	12	22	13	47	33	70	54	47	36	92	-27	20	1.4	1.5	1.3	1.5	2.0	2.8	3.0	3.1	1.6	1.5	1.5	1.7	22.9
Moskva (Moscow)	55 46N	37 40E	505	15	21	9	47	31	76	55	46	34	96	-27	11	1.5	1.4	1.1	1.9	2.2	2.9	3.0	2.9	1.9	2.7	1.7	1.6	24.8
Odessa	46 29N	30 44E	214	20	28	22	52	41	79	65	57	47	99	-13	15	1.0	0.7	0.7	1.1	1.1	1.9	1.6	1.4	1.1	1.4	1.1	1.1	14.3
Riga	56 57N	24 06E	67	30	29	20	48	35	72	56	49	39	93	-20	57	1.3	1.0	1.1	1.2	1.7	2.4	3.0	3.0	2.1	2.0	1.9	1.5	22.2
Saratov	51 32N	46 03E	197	14	15	7	50	35	82	64	48	36	102	-27	15	1.0	1.0	0.8	1.0	1.3	1.8	1.2	1.3	1.1	1.4	1.4	1.2	14.5
Sevastopol	44 37N	33 31E	75	20	39	30	55	42	79	65	63	50	97	- 4	30	1.1	1.1	1.1	0.9	0.6	1.1	0.8	0.6	1.1	1.5	1.2	1.1	12.2
Stalingrad	48 42N	44 31E	136	8	15	4	52	36	84	65	53	37	106	-30	12	0.9	1.0	0.6	0.6	1.0	1.9	0.9	0.8	0.7	1.0	1.5	1.3	12.2
Stavropol	45 02N	41 58E	1,886	18	26	17	50	37	76	60	55	42	95	-22	41	1.4	1.1	1.5	2.4	3.0	4.1	3.0	2.0	2.5	2.3	1.8	1.8	26.9
Tallin	59 26N	24 48E	146	15	27	18	42	31	70	55	47	38	89	-19	63	1.1	1.0	0.9	1.1	1.7	1.9	2.1	2.7	2.3	2.1	1.9	1.5	20.2
Tbilisi	41 43N	44 48E	1,325	10	39	26	61	44	83	65	64	48	95	6	10	0.7	0.8	1.3	1.6	3.6	3.1	2.2	1.7	1.9	1.3	2.0	1.2	21.4
Ust'Shchugor	64 16N	57 34E	279	15	4	-14	35	17	65	49	33	23	90	-67	15	1.1	0.8	0.8	0.7	1.4	2.2	3.0	3.2	2.4	2.2	1.5	1.3	20.6
Ufy	54 43N	55 56E	571	20	6	- 3	44	30	75	58	41	31	99	-42	23	1.6	1.3	1.2	0.9	1.6	2.4	2.6	2.2	1.8	2.3	2.2	2.3	22.5

Yugoslavia:																												
Beograd (Belgrade)	44 48N	20 28E	453	16	37	27	64	45	84	61	65	47	107	-14	16	1.6	1.3	1.6	2.2	2.6	2.8	1.9	2.5	1.7	2.7	1.8	1.9	24.6
Skopje	41 59N	21 28E	787	10	40	26	67	42	88	60	65	43	105	-11	10	1.5	1.2	1.3	1.5	1.9	1.9	1.3	1.1	1.1	2.6	2.3	1.8	19.5
Split	43 31N	16 26E	420	14	51	29	65	50	87	68	69	55	100	17	51	3.1	2.5	3.2	3.0	2.5	2.1	1.2	1.6	2.9	4.4	4.2	4.4	35.1
OCEAN ISLANDS																												
Bjornoya, Bear Island	74 31N	19 01E	49	10	26	17	27	16	44	36	36	29	71	-25	25	1.6	1.3	1.3	0.9	0.8	0.7	0.8	1.2	1.8	1.7	1.4	1.6	15.1
Gronfjorden, Spitzbergen	78 02N	14 15E	23	19	10	- 4	15	- 3	46	38	25	17	60	-57	15	1.4	1.3	1.1	0.9	0.5	0.4	0.6	0.9	1.0	1.2	0.9	1.5	11.7
Horta, Azores	38 32N	28 38W	200	30	62	54	64	55	76	65	71	62	88	38	30	4.5	4.1	4.2	3.0	2.9	2.0	1.5	1.9	3.2	4.4	4.1	4.5	40.3
Jan Mayen	71 01N	08 28W	131	5	31	21	31	22	46	38	39	29	60	-18	29	2.1	1.7	1.6	1.4	0.9	0.9	1.4	1.8	2.5	2.5	2.2	2.2	21.2
Lerwick, Shetland Island	60 08N	01 11W	269	30	42	35	46	37	58	49	50	42	71	17	30	4.5	3.4	2.9	2.7	2.2	2.2	2.7	2.9	3.7	4.3	4.5	4.5	40.5
Matochkin Shar, Novaya Zemlya	73 16N	56 24E	61	9	8	- 6	13	- 1	47	36	30	21	68	-41	9	0.6	0.6	0.6	0.4	0.3	0.4	1.4	1.5	1.5	0.6	0.6	0.4	8.9
Ponta Delgada, Azores	37 45N	25 40W	118	30	62	54	64	55	76	64	71	61	85	37	30	4.0	3.5	3.5	2.5	2.3	1.4	1.0	1.2	2.9	3.6	3.7	3.0	32.6
Stornoway, Hebrides	58 11N	06 21W	34	30	44	37	49	39	61	51	53	44	78	11	15	6.4	3.2	3.2	3.1	2.5	2.4	3.0	4.3	4.7	6.2	4.6	5.5	49.1
Thorshavn, Faeroes	62 02N	06 45W	82	50	42	33	45	36	56	47	58	40	70	8	50	6.6	5.2	4.8	3.6	3.4	2.5	3.1	3.5	4.7	5.9	6.3	6.6	56.2
AFRICA																												
Algeria:																												
Adrar	27 52N	00 17W	938	15	69	39	92	60	115	82	92	63	124	25	15	*	*	0.1	*	*	*	*	*	*	0.2	0.2	*	0.6
Alger (Algiers)	36 46N	03 03E	194	25	59	49	68	55	83	70	74	63	107	32	25	4.4	3.3	2.9	1.6	1.8	0.6	*	0.2	1.6	3.1	5.1	5.4	30.0
Bone	36 54N	07 46E	66	26	59	46	67	52	85	69	75	61	115	32	26	5.6	4.1	2.9	2.2	1.5	0.6	0.1	0.3	1.2	3.0	4.3	5.2	31.0
El Golea	30 35N	02 53E	1,247	15	63	37	84	56	107	79	87	60	120	23	15	0.1	0.3	0.5	*	*	*	*	*	*	0.3	0.4	0.3	1.9
Fort Flatters	28 06N	06 42E	1,224	15	67	38	90	59	110	78	92	63	124	19	15	0.3	0.1	0.1	0.2	*	*	0.0	*	*	*	0.2	0.2	1.1
Tamanrasset	22 42N	05 31E	4,593	15	67	39	86	56	95	71	85	59	102	20	15	0.2	*	*	0.2	0.4	0.1	0.1	0.4	0.1	*	*	*	1.5
Touggourt	33 07N	06 04E	226	26	62	38	83	55	107	77	84	59	122	26	26	0.2	0.4	0.5	0.2	0.2	0.2	*	*	0.1	0.3	0.5	0.3	2.9
Angola:																												
Cangamba	13 41S	19 52E	4,331	6	84	62	89	58	82	46	87	59	109	20	7	8.9	7.4	6.8	1.8	0.1	0.0	0.0	0.2	0.2	1.6	5.1	8.5	40.6
Luanda	08 49S	13 13E	194	27	83	74	85	75	74	65	79	71	98	58	59	1.0	1.4	3.0	4.6	0.5	*	*	*	0.1	0.2	1.1	0.8	12.7
Mocamedes	15 12S	12 09E	10	15	79	65	82	66	68	56	74	61	102	44	21	0.3	0.4	0.7	0.5	*	*	*	*	*	*	0.1	0.1	2.1
Nova Lisboa	12 48S	15 45E	5,577	14	78	58	78	57	77	47	81	58	90	36	14	8.7	7.8	9.8	5.7	0.4	0.0	*	*	0.6	5.5	9.6	8.9	57.0
Botswana:																												
Francistown	21 13S	27 30E	3,294	20	88	65	83	56	75	41	90	61	107	24	28	4.2	3.1	2.8	0.7	0.2	0.1	*	*	*	0.9	2.3	3.4	17.7
Maun	19 59S	23 25E	3,091	20	90	66	87	58	77	42	95	64	110	24	20	4.3	3.8	3.5	1.1	0.2	*	0.0	0.0	*	0.5	1.9	2.8	18.2
Tsabong	26 03S	22 27E	3,156	10	94	65	83	51	71	34	88	54	107	15	14	2.0	1.9	1.9	1.3	0.4	0.4	0.1	*	0.2	0.7	1.1	1.5	11.5
Cameroon:																												
Ngaoundere	07 17N	13 19E	3,601	9	87	55	87	64	82	63	82	61	102	46	10	*	*	1.1	5.5	7.0	8.4	10.6	9.6	9.2	5.3	0.5	*	57.2
Yaounde	03 53N	11 32E	2,526	11	85	67	85	66	80	66	81	65	96	57	11	0.9	2.6	5.8	6.7	7.7	6.0	2.9	3.1	8.4	11.6	4.6	0.9	61.2
Central African Republic:																												
Bangui	04 22N	18 34E	1,270	5	90	68	91	71	85	69	87	69	101	57	5	1.0	1.7	5.0	5.3	7.4	4.5	8.9	8.1	5.9	7.9	4.9	0.2	60.8
Ndele	08 24N	20 39E	1,939	3	99	67	98	73	86	69	90	68	109	58	3	0.2	1.3	0.6	1.7	8.4	6.1	8.3	10.1	10.7	7.8	0.6	0.0	55.8
Chad:																												
Am Timan	11 02N	20 17E	1,430	3	98	56	105	68	89	70	96	67	113	43	3	0.0	0.0	0.1	1.2	4.3	5.0	7.3	12.3	5.8	1.2	0.0	0.0	37.2
Fort Lamy	12 07N	15 02E	968	5	93	57	107	74	92	72	97	70	114	47	5	0.0	0.0	0.0	0.1	1.2	2.6	6.7	12.6	4.7	1.4	0.0	0.0	29.3
Largeau (Faya)	18 00N	19 10E	837	5	84	54	104	69	109	76	103	72	121	37	5	0.0	0.0	0.0	0.0	*	0.0	*	0.7	*	0.0	0.0	0.0	0.7
Congo, Democratic Republic of the:																												
Albertville	05 54S	29 12E	2,493	5	85	66	83	67	82	58	87	67	92	50	20	4.2	4.7	6.3	8.4	3.3	0.3	0.1	0.3	0.8	2.8	7.9	6.3	45.4
Kinspasa (Leopoldville)	04 20S	15 18E	1,066	8	87	70	89	71	81	64	88	70	97	58	12	5.3	5.7	7.7	7.7	6.2	0.3	0.1	0.1	1.2	4.7	8.7	5.6	53.3
Luluabourg	05 54S	22 25E	2,198	3	85	68	86	68	85	63	85	68	94	57	14	5.4	5.6	7.7	7.6	3.3	0.8	0.5	2.3	4.6	6.5	9.1	8.9	62.3
Stanleyville	00 26N	25 14E	1,370	8	88	69	88	70	84	67	86	68	97	61	14	2.1	3.3	7.0	6.2	5.4	4.5	5.2	6.5	7.2	8.6	7.8	3.3	67.1
Congo, Republic of:																												
Brazzaville	04 15S	15 15E	1,043	15	88	69	91	71	82	63	89	70	98	54	18	6.3	4.9	7.4	7.0	4.3	0.6	*	*	2.2	5.4	11.5	8.4	58.0
Ouesso	01 37N	16 04E	1,132	4	88	69	91	71	85	69	87	69	106	60	4	2.4	3.6	6.4	3.2	5.8	4.6	2.9	3.7	7.9	10.0	5.7	2.4	58.6
Pointe Noire (Loango)	04 39S	11 48E	164	7	85	73	87	74	78	66	83	72	93	59	7	5.4	6.7	6.4	8.0	3.9	0.0	0.0	0.0	0.4	4.1	6.6	6.6	48.1
Dahomey:																												
Cotonou	06 21N	02 26E	23	5	80	74	83	78	78	74	80	75	95	65	10	1.3	1.3	4.6	4.9	10.0	14.4	3.5	1.5	2.6	5.3	2.3	0.5	52.4
Ethiopia:																												
Addis Ababa	09 20N	38 45E	8,038	15	75	43	77	50	69	50	75	45	94	32	37	0.5	1.5	2.6	3.4	3.4	5.4	11.0	11.8	7.5	0.8	0.6	0.2	48.7
Asmara	15 17N	38 55E	7,628	9	74	44	78	51	71	53	72	53	88	31	17	*	*	0.4	1.5	1.5	1.3	6.7	5.0	1.3	0.3	0.4	*	18.4

TEMPERATURE AND PRECIPITATION DATA FOR REPRESENTATIVE WORLD-WIDE STATIONS

COUNTRY AND STATION	LATITUDE	LONGITUDE	ELEVATION	TEMPERATURE: LENGTH OF RECORD	AVERAGE DAILY: JANUARY MAXIMUM	JANUARY MINIMUM	APRIL MAXIMUM	APRIL MINIMUM	JULY MAXIMUM	JULY MINIMUM	OCTOBER MAXIMUM	OCTOBER MINIMUM	EXTREME MAXIMUM	EXTREME MINIMUM	AVERAGE PRECIPITATION: LENGTH OF RECORD	JANUARY	FEBRUARY	MARCH	APRIL	MAY	JUNE	JULY	AUGUST	SEPTEMBER	OCTOBER	NOVEMBER	DECEMBER	YEAR
	° ′	° ′	FEET	YEAR	°F	°F	°F	°F	°F	°F	°F	°F	°F	°F	YEAR	IN.	IN.	IN.	IN.	IN.	IN.	IN.	IN.	IN.	IN.	IN.	IN.	IN.
Ethiopia cont'd:																												
Diredawa	09 02N	41 45E	3,937	8	81	58	91	69	90	68	89	67	100	49	8	0.8	0.8	3.3	3.0	2.8	1.5	4.3	3.8	2.2	0.5	0.3	0.8	24.1
Gambela	08 15N	34 35E	1,345	26	98	64	98	71	87	69	92	67	111	48	30	0.2	0.4	1.4	3.2	5.9	6.7	8.5	9.5	7.3	3.5	1.8	0.4	48.8
French Territory of Afars and Issas (F.T.A.I.):																												
Djibouti	11 36N	43 09E	23	16	84	73	90	79	106	87	92	80	117	63	46	0.4	0.5	1.0	0.5	0.2	*	0.1	0.3	0.3	0.4	0.9	0.5	5.1
Gabon:																												
Libreville	00 23N	09 26E	115	11	87	73	89	73	83	68	86	71	99	62	21	9.8	9.3	13.2	13.4	9.6	0.5	0.1	0.7	4.1	13.6	14.7	9.8	98.8
Mayoumba	03 25S	10 38E	200	8	84	73	86	73	78	68	82	72	91	60	8	6.5	9.3	6.2	10.2	2.3	0.1	0.0	0.2	2.6	9.3	10.7	4.6	62.0
Gambia:																												
Bathurst	13 21N	16 40W	90	9	88	59	91	65	86	74	89	72	106	45	9	0.1	0.1	*	*	0.4	2.3	11.1	19.7	12.2	4.3	0.7	0.1	51.0
Ghana:																												
Accra	05 33N	00 12W	88	17	87	73	88	76	81	73	85	74	100	59	65	0.6	1.3	2.2	3.2	5.6	7.0	1.8	0.6	1.4	2.5	1.4	0.9	28.5
Kumasi	06 40N	01 37W	942	10	88	66	89	71	82	70	86	70	100	51	10	0.8	2.3	5.7	5.1	7.5	7.9	4.3	3.1	6.8	7.1	3.7	0.8	55.2
Guinea:																												
Conakry	09 31N	13 43W	23	7	88	72	90	73	83	72	87	73	96	63	10	0.1	0.1	0.4	0.9	6.2	22.0	51.1	41.5	26.9	14.6	4.8	0.4	169.0
Kouroussa	10 39N	09 53W	1,217	9	93	60	99	73	87	69	90	69	109	39	10	0.4	0.3	0.9	2.8	5.3	9.7	11.7	13.6	13.4	6.6	1.3	0.4	66.4
Ifni (now in Morocco):																												
Sidi Ifni	29 27N	10 11W	148	14	66	52	71	59	75	64	75	62	124	40	14	1.0	0.6	0.5	0.6	0.1	0.1	*	*	0.4	0.1	0.9	1.8	6.1
Ivory Coast:																												
Abidjan	05 19N	04 01W	65	13	88	73	90	75	83	73	85	74	96	59	10	1.6	2.1	3.9	4.9	14.2	19.5	8.4	2.1	2.8	6.6	7.9	3.1	77.1
Bouake	07 42N	05 00W	1,194	12	91	68	92	70	85	68	89	68	104	57	10	0.4	1.5	4.1	5.8	5.3	6.0	3.1	4.6	8.2	5.2	1.5	1.0	46.7
Kenya:																												
Mombasa	04 03S	39 39E	52	45	87	75	86	76	81	71	84	74	96	61	54	1.0	0.7	2.5	7.7	12.6	4.7	3.5	2.5	2.5	3.4	3.8	2.4	47.3
Nairobi	01 16S	36 48E	5,971	15	77	54	75	58	69	51	76	55	87	41	17	1.5	2.5	4.9	8.3	6.2	1.8	0.6	0.9	1.2	2.1	4.3	3.4	37.7
Liberia:																												
Monrovia	06 18N	10 48W	75	6	89	71	90	72	80	72	86	72	97	62	4	0.2	0.1	4.4	11.7	13.4	36.1	24.2	18.6	29.9	25.2	8.2	2.9	174.9
Libya:																												
Banghazi (Benghazi)	32 06N	20 04E	82	46	63	50	74	58	84	71	80	66	109	37	46	2.6	1.6	0.8	0.2	0.1	*	*	*	0.1	0.7	1.8	2.6	10.5
Cufra	24 12N	23 21E	1,276	7	69	43	90	62	101	75	90	64	122	26	7	*	0.0	0.0	0.0	*	0.0	0.0	0.0	0.0	0.0	0.0	*	*
Sabhah	27 01N	14 26E	1,457	3	64	41	89	60	102	74	91	64	120	24	10	*	*	*	*	0.1	0.1	0.0	0.0	0.0	*	*	*	0.3
Tarabulus (Triopoli)	32 54N	13 11E	72	47	61	47	72	57	85	71	80	65	114	33	56	3.2	1.8	1.1	0.4	0.2	0.1	*	*	0.4	1.6	2.6	3.7	15.1
Malagasy Republic:																												
Diego Suarez	12 17S	49 17E	100	11	88	75	88	75	84	69	86	72	98	63	31	10.6	9.5	7.6	2.2	0.3	0.2	0.2	0.3	0.3	0.7	1.1	5.8	38.7
Tananarive	18 55S	47 33E	4,500	44	79	61	76	58	68	48	80	54	95	34	62	11.8	11.0	7.0	2.1	0.7	0.3	0.3	0.4	0.7	2.4	5.3	11.3	53.4
Tulear	23 20S	43 41E	20	27	92	72	89	64	81	58	86	65	108	43	15	3.1	3.2	1.4	0.3	0.7	0.4	0.1	0.2	0.3	0.7	1.4	1.7	13.5
Malawi:																												
Karonga	09 57S	33 56E	1,596	8	86	71	85	70	81	59	91	66	99	51	8	7.1	7.0	10.8	6.2	1.7	0.1	*	*	0.0	0.3	0.3	4.7	38.3
Zomba	15 23S	35 19E	3,141	27	80	65	78	62	72	53	85	64	95	41	29	12.1	9.9	10.1	2.7	0.7	0.4	0.3	0.3	0.2	1.0	4.3	10.9	52.9
Mali:																												
Araouane	18 54N	03 33W	935	8	81	48	110	67	111	79	103	70	130	37	10	*	*	0.0	0.0	0.0	0.2	0.2	0.5	0.6	0.1	0.1	*	1.7
Bamako	12 39N	07 58W	1,116	11	91	61	103	76	89	71	93	71	117	47	10	*	*	0.1	0.6	2.9	5.4	11.0	13.7	8.1	1.7	0.6	*	44.1
Gao	16 16N	00 03W	902	15	83	58	105	77	97	80	100	78	116	44	19	*	0.0	*	0.1	0.4	1.0	2.9	5.4	1.5	0.2	*	0.0	11.5

Mauritania:																												
Atar	20 31N	13 04W	761	7	84	54	97	67	106	81	98	72	117	39	10	*	0.0	*	*	*	0.1	0.3	1.2	1.1	0.1	*	*	2.8
Nema	16 36N	07 16W	883	9	86	62	105	79	99	78	101	79	120	47	10	0.1	*	*	*	0.7	1.1	2.3	4.7	2.1	0.7	*	0.1	11.6
Nouakchott	18 07N	15 36W	69	5	85	57	90	64	89	74	91	71	115	44	10	*	0.1	*	*	*	0.1	0.5	4.1	0.9	0.4	0.1	*	6.2
Morocco:																												
Casablanca	33 35N	07 39W	164	48	63	45	69	52	79	65	76	58	110	31	40	2.1	1.9	2.2	1.4	0.9	0.2	0.0	*	0.3	1.5	2.6	2.8	15.9
Marrakech	31 36N	08 01W	1,509	35	65	40	79	52	101	67	83	57	120	27	31	1.0	1.1	1.3	1.2	0.6	0.3	0.1	0.1	0.4	0.9	1.2	1.2	9.4
Rabat	34 00N	06 50W	213	35	63	46	71	52	82	63	77	58	118	32	29	2.6	2.5	2.6	1.7	1.1	0.3	*	*	0.4	1.9	3.3	3.4	19.8
Tangier	35 48N	05 49W	239	35	60	47	65	51	80	64	72	59	106	28	35	4.5	4.2	4.8	3.5	1.7	0.6	*	*	0.9	3.9	5.8	5.4	35.3
Mozambique:																												
Beira	19 50S	34 51E	28	37	89	75	86	71	77	61	87	71	109	48	39	10.9	8.4	10.1	4.2	2.2	1.3	1.2	1.1	0.8	5.2	5.3	9.2	59.9
Chicoa	15 36S	32 21E	899	8	96	65	93	63	86	55	101	68	117	32	8	7.8	5.7	4.4	0.6	*	*	*	*	*	1.1	2.6	5.2	27.4
Lourenco Marques	25 58S	32 36E	194	42	86	71	83	66	76	55	82	64	114	45	42	5.1	4.9	4.9	2.1	1.1	0.8	0.5	0.5	1.1	1.9	3.2	3.8	29.9
Niger:																												
Agades	16 59N	07 59E	1,706	8	86	50	105	70	104	75	101	68	115	40	10	0.0	0.0	*	*	0.2	0.3	1.9	3.7	0.7	0.0	0.0	0.0	6.8
Bilma	18 41N	12 55E	1,171	9	81	45	101	63	108	75	101	62	116	29	10	0.0	0.0	0.0	*	*	0.0	0.1	0.5	0.3	0.0	0.0	0.0	0.9
Niamey	13 31N	02 06E	709	10	93	58	108	77	94	74	101	74	114	47	10	*	*	0.2	0.3	1.3	3.2	5.2	7.4	3.7	0.5	*	0.0	21.6
Nigeria:																												
Enugu	06 27N	07 29E	763	11	90	72	91	74	83	71	87	71	99	55	33	0.7	1.1	2.6	5.9	10.4	11.4	7.6	6.7	12.8	9.8	2.1	0.5	71.5
Kaduna	10 35N	06 26E	2,113	18	89	59	95	72	83	68	89	66	105	46	34	*	0.1	0.5	2.5	5.9	7.1	8.5	11.9	10.6	2.9	0.1	*	50.1
Lagos	06 27N	03 24E	10	32	88	74	89	77	83	74	85	74	104	60	47	1.1	1.8	4.0	5.9	10.6	18.1	11.0	2.5	5.5	8.1	2.7	1.0	72.3
Maiduguri	11 51N	13 05E	1,162	15	90	54	104	72	90	73	96	68	112	43	40	*	*	*	0.3	1.6	2.7	7.1	8.7	4.2	0.7	*	0.0	25.3
Portuguese Guinea:																												
Bolama	11 34N	15 26W	62	31	88	67	91	73	84	74	87	74	106	59	37	*	*	*	*	0.8	7.8	23.1	27.6	16.9	8.0	1.6	0.1	85.9
Rhodesia:																												
Bulawayo	20 09S	28 37E	4,405	15	81	61	79	56	70	45	85	59	99	28	50	5.6	4.3	3.3	0.7	0.4	0.1	*	*	0.2	0.8	3.2	4.8	23.4
Salisbury	17 50S	31 08E	4,831	15	78	60	78	55	70	44	83	58	95	32	50	7.7	7.0	4.6	1.1	0.5	0.1	*	0.1	0.2	1.1	3.8	6.4	32.6
Senegal:																												
Dakar	14 42N	17 29W	131	25	79	64	81	65	88	76	89	76	109	53	26	*	*	*	*	*	0.7	3.5	10.0	5.2	1.5	0.1	0.3	21.3
Kaolack	14 08N	16 04W	20	9	93	60	103	68	91	75	93	74	114	48	10	*	0.0	*	*	0.3	2.6	6.9	10.7	7.0	2.7	0.1	*	30.3
Sierra Leone:																												
Freetown/Lungi	08 37N	13 12W	92	8	87	73	88	76	82	73	85	72	98	62	8	0.4	0.2	1.2	3.1	9.5	14.3	29.2	36.5	22.3	14.2	5.5	1.2	137.6
Somalia:																												
Berbera	10 26N	45 02E	45	30	84	68	89	77	107	88	92	76	117	58	30	0.3	0.1	0.2	0.5	0.3	*	*	0.1	*	0.1	0.2	0.2	2.0
Mogadishu (Mogadiscio)	02 02N	45 21E	39	13	86	73	90	78	83	73	86	76	97	59	21	*	*	*	2.3	2.3	3.8	2.5	1.9	1.0	0.9	1.6	0.5	16.9
South Africa, Republic of:																												
Capetown	33 54S	18 32E	56	19	78	60	72	53	63	45	70	52	103	28	18	0.6	0.3	0.7	1.9	3.1	3.3	3.5	2.6	1.7	1.2	0.7	0.4	20.0
Durban	29 50S	31 02E	16	15	81	69	78	64	72	52	75	62	107	39	78	4.3	4.8	5.1	3.0	2.0	1.3	1.1	1.5	2.8	4.3	4.8	4.7	39.7
Kimberley	28 48S	24 46E	3,927	19	91	64	77	52	65	36	83	54	103	20	57	2.4	2.5	3.1	1.5	0.7	0.2	0.2	0.3	0.6	1.0	1.6	2.0	16.1
Port Elizabeth	33 59S	25 36E	190	14	78	61	73	55	67	45	70	54	104	31	84	1.2	1.3	1.9	1.8	2.4	1.8	1.9	2.0	2.3	2.2	2.2	1.7	22.7
Port Nolloth	29 14S	16 52E	23	20	67	53	66	50	62	45	64	49	107	31	64	0.1	0.1	0.2	0.2	0.3	0.3	0.3	0.3	0.2	0.1	0.1	0.1	2.3
Pretoria	25 45S	28 14E	4,491	13	81	60	75	50	66	37	80	55	96	24	12	5.0	4.3	4.5	1.7	0.9	0.6	0.3	0.2	0.8	2.2	5.2	5.2	30.9
Walvis Bay	22 56S	14 30E	24	20	73	59	75	55	70	47	67	51	104	25	20	*	0.2	0.3	0.1	0.1	*	*	0.1	*	*	*	*	0.9
Southwest Africa:																												
Keetmanshoop	26 35S	18 08E	3,295	17	95	65	85	57	70	42	87	55	108	26	45	0.8	1.1	1.4	0.6	0.2	*	*	*	0.1	0.2	0.3	0.4	5.2
Windhoek	22 34S	17 06E	5,669	30	85	63	77	55	68	43	84	59	97	25	60	3.0	2.9	3.1	1.6	0.3	*	*	*	0.1	0.4	0.9	1.9	14.3
Spanish Sahara:																												
Semara	26 46N	11 31W	1,509	6	73	47	88	58	99	66	88	61	121	37	6	0.1	*	0.0	*	*	0.0	0.0	*	1.0	*	0.4	0.0	1.5
Villa Cisneros	23 42N	15 52W	35	12	71	56	74	60	78	65	80	65	107	48	14	*	*	*	*	0.1	0.0	*	0.2	1.4	0.1	0.2	1.0	3.0
Sudan:																												
El Fasher	13 38N	25 21E	2,395	17	88	50	102	64	96	70	99	64	113	33	17	*	0.0	*	*	0.3	0.7	4.5	5.3	1.2	0.2	0.0	0.0	12.2
Khartoum	15 37N	32 33E	1,279	46	90	59	105	72	101	77	104	75	118	41	46	*	*	*	*	0.1	0.3	2.1	2.8	0.7	0.2	*	0.0	6.2
Port Sudan	19 37N	37 13E	18	30	81	68	89	71	106	83	93	76	117	50	40	0.2	0.1	*	*	*	*	0.3	0.1	*	0.4	1.7	0.9	3.7
Wadi Halfa	21 55N	31 20E	410	39	75	46	98	62	106	74	98	67	127	28	39	*	*	*	*	*	0.0	*	*	*	*	*	0.0	*
Wau	07 42N	28 03E	1,443	38	96	64	99	72	89	69	93	69	115	50	38	*	0.2	0.9	2.6	5.3	6.5	7.5	8.2	6.6	4.9	0.6	*	43.3
Tanzania:																												
Dares Salaam	06 50S	39 18E	47	44	83	77	86	73	83	66	85	69	96	59	49	2.6	2.6	5.1	11.4	7.4	1.3	1.2	1.0	1.2	1.6	2.9	3.6	41.9
Iringa	07 47S	35 42E	5,330	14	76	59	75	59	72	52	80	57	90	42	24	6.8	5.1	7.1	3.5	0.5	*	*	*	0.1	0.2	1.5	4.5	29.3
Kigoma	04 53S	29 38E	2,903	26	80	67	81	67	83	63	84	69	100	53	18	4.8	5.0	5.9	5.1	1.7	0.2	0.1	0.2	0.7	1.9	5.6	5.3	36.5
Togo:																												
Lome	06 10N	01 15E	72	5	85	72	86	74	80	71	83	72	94	58	15	0.6	0.9	1.9	4.6	5.7	8.8	2.8	0.4	1.4	2.4	1.1	0.4	31.0

TEMPERATURE AND PRECIPITATION DATA FOR REPRESENTATIVE WORLD-WIDE STATIONS

COUNTRY AND STATION	LATITUDE ° '	LONGITUDE ° '	ELEVATION FEET	Temperature: LENGTH OF RECORD YEAR	January Maximum °F	January Minimum °F	April Maximum °F	April Minimum °F	July Maximum °F	July Minimum °F	October Maximum °F	October Minimum °F	Extreme Maximum °F	Extreme Minimum °F	Precipitation: LENGTH OF RECORD YEAR	January IN.	February IN.	March IN.	April IN.	May IN.	June IN.	July IN.	August IN.	September IN.	October IN.	November IN.	December IN.	Year IN.
Tunisia:																												
Gabes	33 53N	10 07E	7	50	61	43	74	54	89	71	81	62	122	27	50	0.9	0.7	0.8	0.4	0.3	*	*	0.1	0.5	1.2	1.2	0.6	6.7
Tunis	36 47N	10 12E	217	50	58	43	70	51	90	68	77	59	118	30	50	2.5	2.0	1.6	1.4	0.7	0.3	0.1	0.3	1.3	2.0	1.9	2.4	16.5
Uganda:																												
Kampala	00 20N	32 36E	4,304	15	83	65	79	64	77	62	81	63	97	53	15	1.8	2.4	5.1	6.9	5.8	2.9	1.8	3.4	3.6	3.8	4.8	3.9	46.2
Lira	02 15N	32 54E	3,560	14	91	61	86	64	81	61	86	61	100	50	14	0.7	1.0	3.5	6.9	7.9	4.9	6.4	10.0	8.3	6.1	3.2	1.8	60.7
United Arab Republic:																												
Alexandria	31 12N	29 53E	105	45	65	51	74	59	85	73	83	68	111	37	61	1.9	0.9	0.4	0.1	*	*	*	*	*	0.2	1.3	2.2	7.0
Aswan	24 02N	32 53E	366	46	74	50	96	66	106	79	98	71	124	35	11	*	*	*	*	*	*	0.0	0.0	0.0	*	*	*	*
Cairo	29 52N	31 20E	381	42	65	47	83	57	96	70	86	65	117	34	42	0.2	0.2	0.2	0.1	0.1	*	0.0	0.0	*	*	0.1	0.2	1.1
Upper Volta:																												
Bobo Dioulasso	11 10N	04 15W	1,411	11	92	58	99	71	87	69	90	70	115	46	10	0.1	0.2	1.1	2.1	4.6	4.8	9.8	12.0	8.5	2.5	0.7	0.0	46.4
Ouagadougou	12 22N	01 31W	991	10	92	60	103	79	91	74	95	74	118	48	15	*	0.1	0.5	0.6	3.3	4.8	8.0	10.9	5.7	1.3	*	0.0	35.2
Zambia:																												
Balovale	13 34S	23 06E	3,577	8	82	65	84	61	81	47	91	64	108	38	9	8.5	6.9	5.8	1.2	*	0.0	0.0	*	0.3	2.3	4.4	8.9	38.3
Kasama	10 12S	31 11E	4,544	10	79	61	79	60	76	50	87	62	95	39	10	10.7	9.9	10.9	2.8	0.5	*	*	*	*	0.8	6.4	9.5	51.5
Lusaka	15 25S	28 19E	4,191	10	78	63	79	59	73	49	88	64	100	39	10	9.1	7.5	5.6	0.7	0.1	*	*	0.0	*	0.4	3.6	5.9	32.9
ATLANTIC ISLANDS:																												
Funchal, Madeira Island	32 38N	16 55W	82	30	66	56	67	58	75	66	74	65	103	40	30	2.5	2.9	3.1	1.3	0.7	0.2	*	*	1.0	3.0	3.5	3.3	21.5
Georgetown, Ascension Island	07 56S	14 25W	55	29	85	73	88	75	84	72	83	71	95	65	45	0.2	0.4	0.7	1.1	0.5	0.5	0.5	0.4	0.3	0.3	0.2	0.1	5.2
Hutts Gate, St. Helena	15 57S	05 40W	2,062	30	68	60	69	61	62	55	61	54	82	50	30	2.1	3.1	4.2	3.1	2.8	3.2	4.3	2.6	2.2	1.7	1.2	1.6	32.1
Las Palmas, Canary Islands	28 11N	15 28W	20	45	70	58	71	61	77	67	79	67	99	46	48	1.4	0.9	0.9	0.5	0.2	*	*	*	0.2	1.1	2.1	1.6	8.6
Porto da Praia, Cape Verde Is.	14 54N	23 31W	112	25	77	68	79	69	83	75	85	76	94	56	25	0.1	*	*	*	0.0	*	0.2	3.8	4.5	1.2	0.3	0.1	10.2
Santa Isabel, Fernando Po	03 46N	08 46E	-----	2	87	67	89	70	84	69	86	70	102	61	16	1.3	2.5	4.2	7.2	9.4	11.1	7.4	6.6	9.6	10.4	3.5	1.7	74.9
Sao Tome, Sao Tome	00 20N	06 43E	16	10	86	73	86	73	82	69	84	71	91	56	10	3.2	4.2	5.9	5.0	5.3	1.1	*	*	0.9	4.3	4.6	3.5	38.0
Tristan da Cunha	37 03S	12 19W	75	5	66	59	64	57	57	50	59	51	75	38	5	3.5	3.5	6.4	4.7	7.1	5.9	6.1	6.9	7.9	5.8	4.3	4.0	66.1
INDIAN OCEAN ISLANDS:																												
Agalega Island	10 26S	56 40E	10	3	86	77	87	77	83	75	84	75	91	69	2	5.9	10.1	4.9	6.9	13.2	8.9	8.7	3.2	1.8	4.2	7.0	10.0	84.7
Cocos (Keeling) Island	12 05S	96 53E	15	36	86	77	85	78	82	76	84	76	94	68	38	5.4	7.7	8.5	10.4	7.9	9.0	8.7	4.8	3.7	3.3	4.2	4.6	78.2
Heard Island	53 01S	73 23E	16	5	41	35	39	33	34	27	35	28	58	13	5	5.8	5.8	5.7	6.1	5.8	3.9	3.6	2.2	2.5	3.7	4.0	5.1	54.3
Hellburg, Reunion Island	21 04S	55 22E	3,070	5	74	59	73	56	65	48	69	51	84	40	11	22.4	8.0	16.4	7.2	5.3	4.4	3.1	3.0	2.0	2.3	3.5	12.9	90.5
Port Victoria, Seychelles	04 37S	55 27E	15	60	83	76	86	77	81	75	83	75	92	67	64	15.2	10.5	9.2	7.2	6.7	4.0	3.3	2.7	5.1	6.1	9.1	13.4	92.5
Royal Alfred Observatory, Mauritius	20 06S	57 32E	181	40	86	73	82	70	75	62	80	64	95	50	43	8.5	7.8	8.7	5.0	3.8	2.6	2.3	2.5	1.4	1.6	1.8	4.6	50.6
ASIA – FAR EAST																												
China																												
Canton	23 10N	113 20E	59	26	65	49	77	65	91	77	85	67	101	31	36	0.9	1.9	4.2	6.8	10.6	10.6	8.1	8.5	6.5	3.4	1.2	0.9	63.6
Chanasha	28 15	112 58E	161	14	45	35	70	56	94	78	75	59	109	16	26	1.9	3.7	5.3	5.7	8.2	8.7	4.4	4.3	2.7	3.0	2.7	1.5	52.1
Chungking	29 30N	106 33E	855	27	51	42	73	59	93	76	71	61	111	28	60	0.7	0.8	1.5	3.8	5.7	7.1	5.6	4.7	5.8	4.3	1.9	0.8	42.9

China cont'd:																												
Hankow	30 35N	114 17E	75	29	46	34	69	55	93	78	74	60	108	9	55	1.8	1.9	3.6	5.8	7.0	9.0	7.0	4.1	3.0	3.1	1.9	1.2	49.4
Harbin (Ha-erh-pin)	45 45N	126 38E	476	35	7	-14	54	31	84	65	54	31	102	-43	38	0.2	0.2	0.4	0.9	1.7	3.7	6.6	4.7	2.3	1.2	0.5	0.2	22.6
Kashgar	39 24N	76 07E	4,296	27	33	12	71	48	92	68	71	43	106	-15	18	0.6	0.1	0.5	0.2	0.3	0.2	0.4	0.3	0.1	0.1	0.2	0.3	3.2
Kunming	25 02N	102 43E	6,211	32	61	37	76	51	77	62	70	53	91	22	31	0.4	0.5	0.7	0.8	4.3	6.3	8.8	8.6	5.0	3.0	1.7	0.4	40.5
Lanchow	36 06N	103 55E	5,105	8	33	7	65	40	84	61	62	39	100	- 3	4	0.2	0.2	0.2	0.5	0.8	0.7	3.3	5.1	2.2	0.6	0.0	0.3	14.1
Mukden (Shen-yang)	41 47N	123 24E	138	40	20	- 2	60	36	87	69	62	39	103	-28	42	0.2	0.2	0.7	1.2	2.6	3.8	7.0	6.3	2.9	1.7	0.9	0.4	28.2
Shanghai	31 12N	121 26E	16	56	47	32	67	49	91	75	75	56	104	10	81	1.9	2.4	3.3	3.6	3.8	7.0	5.8	5.5	5.2	2.9	2.1	1.5	45.0
Tientsin	39 10N	117 10E	13	24	33	16	68	45	90	73	68	48	109	- 3	25	0.2	0.1	0.4	0.5	1.1	2.4	7.6	6.0	1.7	0.6	0.4	0.2	21.0
Urumchi	43 45N	87 40E	2,972	6	13	-7	60	36	82	58	50	31	112	-30	6	0.6	0.3	0.5	1.5	1.1	1.5	0.7	1.0	0.6	1.7	1.6	0.4	11.5
Hong Kong:	22 18N	114 10E	109	50	64	56	75	67	87	78	81	73	97	32	50	1.3	1.8	2.9	5.4	11.5	15.5	15.0	14.2	10.1	4.5	1.7	1.2	85.1
Japan:																												
Kushiro	43 02N	144 12E	315	41	30	8	44	31	66	55	58	40	87	-19	41	1.8	1.4	2.8	3.6	3.8	4.1	4.4	4.9	6.6	4.0	3.1	2.0	42.9
Miyako	39 38N	141 59E	98	30	43	23	58	37	77	62	66	46	99	1	30	2.9	3.0	3.2	3.5	4.5	5.0	5.0	7.2	9.5	6.8	3.0	2.6	56.2
Nagasaki	32 44N	129 53E	436	59	49	36	66	50	85	73	72	58	98	22	59	2.8	3.3	4.9	7.3	6.7	12.3	10.1	6.9	9.8	4.5	3.7	3.2	75.5
Osaka	34 47N	135 26E	49	60	47	32	65	47	87	73	72	55	102	19	60	1.7	2.3	3.8	5.2	4.9	7.4	5.9	4.4	7.0	5.1	3.0	1.9	52.6
Tokyo	35 41N	139 46E	19	60	47	29	63	46	83	70	69	55	101	17	60	1.9	2.9	4.2	5.3	5.8	6.5	5.6	6.0	9.2	8.2	3.8	2.2	61.6
Korea:																												
Pusan	35 10N	129 07E	6	36	43	29	62	47	81	71	70	54	97	7	36	1.7	1.4	2.7	5.5	5.2	7.9	11.6	5.1	6.8	2.9	1.6	1.2	53.6
P'yongyang	39 01N	125 49E	94	43	27	8	61	38	84	69	65	43	100	-19	43	0.6	0.4	1.0	1.8	2.6	3.0	9.3	9.0	4.4	1.8	1.6	0.8	36.4
Seoul	37 31N	126 55E	34	22	32	15	62	41	84	70	67	45	99	-12	22	1.2	0.8	1.5	3.0	3.2	5.1	14.8	10.5	4.7	1.6	1.8	1.0	49.2
Mongolia:																												
Ulan Bator	47 54N	106 56E	4,287	13	- 2	-27	45	18	71	50	44	17	97	-48	15	*	*	0.1	0.2	0.3	1.0	2.9	1.9	0.8	0.2	0.2	0.1	7.7
Taiwan:																												
Tainan	22 57N	120 12E	53	13	72	55	82	67	89	77	86	70	95	39	13	0.7	0.7	1.1	3.2	6.3	15.6	16.0	15.8	8.4	1.2	0.9	0.6	70.5
Taipei	25 04N	121 32E	21	12	66	53	77	64	92	76	80	68	101	32	12	3.8	5.3	4.3	5.3	6.9	8.8	8.8	8.7	8.2	5.5	4.2	2.9	72.7
Union of Soviet Socialist Republics:																												
Alma-Ata	43 16N	76 53E	2,543	19	23	7	56	38	81	60	55	35	100	-30	27	1.3	0.9	2.2	4.0	3.7	2.6	1.4	1.2	1.0	2.0	1.9	1.3	23.5
Chita (Tchita)	52 02N	113 30E	2,218	10	-10	-27	42	19	75	51	38	18	99	-52	24	0.1	0.1	0.1	0.4	1.1	1.8	3.3	3.3	1.2	0.5	0.2	0.2	12.3
Dubinka	69 07N	87 00E	141	5	-23	-31	6	-10	59	47	19	11	84	-62	5	0.3	0.4	0.2	0.3	0.6	1.9	1.5	2.1	1.8	0.9	0.4	0.3	10.7
Irkutsk	52 16N	104 19E	1,532	10	3	-15	42	20	70	50	41	21	98	-58	38	0.5	0.4	0.3	0.6	1.3	2.2	3.1	2.8	1.7	0.7	0.6	0.6	14.9
Kazalinsk	45 46N	62 06E	207	10	16	5	58	27	90	65	57	35	108	-27	19	0.4	0.4	0.5	0.5	0.6	0.2	0.2	0.3	0.3	0.4	0.5	0.6	4.9
Khabarovsk	48 28N	135 03E	165	7	- 2	-13	41	28	75	63	48	34	91	-46	8	0.3	0.2	0.3	0.7	2.0	3.5	4.1	3.3	3.0	0.7	0.6	0.5	19.2
Kirensk	57 47N	108 07E	938	18	-14	-28	38	15	74	51	10	- 4	95	-71	19	0.8	0.5	0.5	0.5	1.0	1.8	2.1	2.1	1.7	1.0	1.0	1.0	14.0
Krasnoyarsk	56 01N	92 52E	498	10	3	-10	34	23	67	55	34	26	103	-47	8	0.1	0.2	0.1	0.2	1.0	1.4	1.2	2.1	1.7	0.9	0.5	0.4	9.8
Markovo	64 45N	170 50E	85	15	-19	-29	5	- 8	59	47	16	9	84	-72	16	0.2	0.2	0.3	0.1	0.3	0.8	1.0	1.9	1.1	0.4	0.4	0.3	7.0
Narym	58 50N	81 39E	197	13	- 7	-18	35	19	71	56	35	25	94	-61	14	0.8	0.5	0.8	0.5	1.3	2.6	2.4	2.7	1.7	1.4	1.1	0.9	16.8
Okhotsk	59 21N	143 17E	18	19	- 6	-17	29	10	57	48	33	21	78	-50	25	0.1	0.1	0.2	0.4	0.9	1.6	2.2	2.6	2.4	1.0	0.2	0.1	11.8
Omsk	54 58N	73 20E	279	19	- 1	-14	39	21	74	56	40	27	102	-56	22	0.6	0.3	0.3	0.5	1.2	2.0	2.0	2.0	1.1	1.0	0.7	0.8	12.5
Petropavlovsk	52 53N	158 42E	286	7	23	11	35	25	56	47	46	34	84	-29	35	3.0	2.2	3.4	2.5	2.2	2.0	3.1	3.2	3.8	3.9	3.6	3.0	35.9
Salehkard	66 31N	66 35E	60	18	-13	-21	18	4	61	49	26	20	85	-65	27	0.3	0.3	0.3	0.3	0.7	1.3	1.9	2.0	1.5	0.7	0.5	0.4	10.2
Semipalatinsk	50 24N	80 13E	709	10	8	- 7	45	26	81	57	46	30	101	-47	10	0.9	0.5	0.5	0.6	1.2	1.5	1.1	1.3	0.7	1.2	1.1	1.0	11.6
Sverdlovsk	56 49N	60 38E	894	21	6	- 5	42	26	70	54	37	28	94	-45	29	0.5	0.4	0.5	0.7	1.9	2.7	2.6	2.7	1.6	1.2	1.1	0.8	16.7
Tashkent	41 20N	69 18E	1,569	19	37	21	65	47	92	64	65	41	106	-19	19	2.1	1.1	2.6	2.3	1.4	0.5	0.2	0.1	0.1	1.2	1.5	1.6	14.7
Verkhoyansk	67 34N	133 51E	328	24	-54	-63	19	-10	66	47	12	- 3	98	-90	44	0.2	0.2	0.1	0.2	0.3	0.9	1.1	1.0	0.5	0.3	0.3	0.2	5.3
Vladivostok	43 07N	131 55E	94	14	13	0	46	34	71	60	55	41	92	-22	53	0.3	0.4	0.7	1.2	2.1	2.9	3.3	4.7	4.3	1.9	1.2	0.6	23.6
Yakutsk	62 01N	129 43E	535	19	-45	-53	27	6	73	54	23	11	97	-84	22	0.3	0.2	0.1	0.3	0.4	1.1	1.6	1.3	1.1	0.5	0.4	0.3	7.4
ASIA – SOUTHEAST																												
Brunei:																												
Brunei	04 55N	114 55E	10	5	85	76	87	77	87	76	86	77	99	70	12	14.6	7.6	7.8	9.8	10.9	9.5	9.0	7.3	11.8	14.5	15.2	13.0	131.0
Burma:																												
Mandalay	21 59N	96 06E	252	20	82	55	101	77	93	78	89	73	111	44	20	0.1	0.1	0.2	1.2	5.8	6.3	2.7	4.1	5.4	4.3	2.0	0.4	32.6
Moulmein	16 26N	97 39E	150	43	89	65	95	77	83	74	88	75	103	52	60	0.2	0.2	0.5	3.0	19.9	37.1	47.5	44.2	27.1	8.5	1.7	0.3	190.2
Cambodia:																												
Phnom Penh	11 33N	104 51E	39	37	88	71	95	76	90	76	87	76	105	55	49	0.3	0.4	1.4	3.1	5.7	5.8	6.0	6.1	8.9	9.9	5.5	1.7	54.8
Indonesia:																												
Batavia (Jakarta)	06 11S	106 50E	26	80	84	74	87	75	87	73	87	74	98	66	78	11.8	11.8	8.3	5.8	4.5	3.8	2.5	1.7	2.6	4.4	5.6	8.0	70.8
Manokwari	00 53S	134 03E	10	5	86	73	86	74	86	74	87	74	93	68	40	12.0	9.4	13.2	11.1	7.8	7.2	5.4	5.6	4.9	4.7	6.5	10.3	98.1

TEMPERATURE AND PRECIPITATION DATA FOR REPRESENTATIVE WORLD-WIDE STATIONS

COUNTRY AND STATION	LATITUDE	LONGITUDE	ELE-VATION	TEMPERATURE											AVERAGE PRECIPITATION													
				LENGTH OF RECORD	AVERAGE DAILY JANUARY		APRIL		JULY		OCTOBER		EXTREME		LENGTH OF RECORD	JANUARY	FEBRUARY	MARCH	APRIL	MAY	JUNE	JULY	AUGUST	SEPTEMBER	OCTOBER	NOVEMBER	DECEMBER	YEAR
					MAXIMUM	MINIMUM	MAXIMUM	MINIMUM	MAXIMUM	MINIMUM	MAXIMUM	MINIMUM	MAXIMUM	MINIMUM														
	° '	° '	FEET	YEAR	°F	°F	°F	°F	°F	°F	°F	°F	°F	°F	YEAR	IN.	IN.	IN.	IN.	IN.	IN.	IN.	IN.	IN.	IN.	IN.	IN.	IN.
Indonesia cont'd:																												
Mapanget	01 32N	124 55E	264	21	85	73	86	73	87	73	89	72	97	65	63	18.6	13.8	12.2	8.0	6.4	6.5	4.8	4.0	3.3	4.9	8.9	14.7	106.1
Penfui	10 10S	123 39E	335	21	87	75	89	72	88	70	92	72	101	58	63	15.2	13.7	9.2	2.6	1.2	0.4	0.2	0.0	0.0	0.7	3.3	9.1	55.7
Pontianak	00 00N	109 20E	13	20	87	74	89	75	89	74	89	75	96	68	63	10.8	8.2	9.5	10.9	11.1	8.7	6.5	8.0	9.0	14.4	15.3	12.7	125.1
Tabing	00 52S	100 21E	19	21	87	74	87	75	87	74	86	74	94	68	63	13.9	10.1	12.2	14.5	12.8	11.7	10.5	13.7	16.2	20.1	20.5	19.2	175.4
Tarakan	03 19N	117 33E	20	19	85	73	86	75	87	74	87	74	94	67	31	10.9	10.2	14.0	13.9	13.5	12.6	10.3	12.4	11.6	14.3	15.2	13.4	152.3
Laos:																												
Vientiane	17 58N	102 34E	559	13	83	58	95	73	89	75	88	71	108	32	27	0.2	0.6	1.5	3.9	10.5	11.9	10.5	11.5	11.9	4.3	0.6	0.1	67.5
Malaya, Fed.:																												
Kuala Lumpur	03 06N	101 42E	111	19	90	72	91	74	90	72	89	73	99	64	19	6.2	7.9	10.2	11.5	8.8	5.1	3.9	6.4	8.6	9.8	10.2	7.5	96.1
Singapore	01 18N	103 50E	33	39	86	73	88	75	88	75	87	74	97	66	64	9.9	6.8	7.6	7.4	6.8	6.8	6.7	7.7	7.0	8.2	10.0	10.1	95.0
North Borneo:																												
Sanda Kan	05 54N	118 03E	38	45	85	74	89	76	89	75	88	75	99	70	46	19.0	10.9	8.6	4.5	6.2	7.4	6.7	7.9	9.3	10.2	14.5	18.5	123.7
Philippine Islands:																												
Davao	07 07N	125 38E	88	15	87	72	91	73	88	73	89	73	97	65	34	4.8	4.5	5.2	5.8	9.2	9.1	6.5	6.5	6.7	7.9	5.3	6.1	77.6
Manila	14 31N	121 00E	49	61	86	69	93	73	88	75	88	74	101	58	75	0.9	0.5	0.7	1.3	5.1	10.0	17.0	16.6	14.0	7.6	5.7	2.6	82.0
Sarawak:																												
Kuching	01 29N	110 20E	85	5	85	72	90	73	90	72	89	73	98	64	19	24.0	20.1	12.9	11.0	10.3	7.1	7.7	9.2	8.6	10.5	14.1	18.2	153.7
Thailand:																												
Bangkok	13 44N	100 30E	53	10	89	67	95	78	90	76	88	76	104	50	10	0.2	1.1	1.1	2.3	5.2	6.0	6.9	9.2	14.0	9.9	1.8	0.1	57.8
Viet Nam																												
Hanoi	21 03N	105 52E	20	12	68	58	80	70	92	79	84	72	108	41	12	0.8	1.2	2.5	3.6	4.1	11.2	11.9	15.2	10.0	3.5	2.6	2.8	69.4
Saigon	10 49N	106 39E	33	31	89	70	95	76	88	75	88	74	104	57	33	0.6	0.1	0.5	1.7	8.7	13.0	12.4	10.6	13.2	10.6	4.5	2.2	78.1
ASIA – MIDDLE EAST																												
Aden:																												
Riyan	14 39N	49 19E	83	13	82	67	88	74	92	77	88	72	111	57	13	0.3	0.1	0.6	0.2	*	0.1	0.1	0.1	*	*	0.7	0.3	2.5
Afghanistan:																												
Kabul	34 30N	69 13E	5,955	9	36	18	66	43	92	61	73	42	104	- 6	45	1.3	1.5	3.6	3.3	0.9	0.2	0.1	0.1	*	0.4	0.6	0.6	12.6
Kandhar	31 36N	65 40E	3,462	7	56	31	83	50	102	66	85	44	111	14	7	3.1	1.7	0.8	0.3	0.2	*	0.1	*	0.0	*	*	0.8	7.0
Ceylon:																												
Colombo	06 54N	79 52E	22	25	86	72	88	76	85	77	85	75	99	59	40	3.5	2.7	5.8	9.1	14.6	8.8	5.3	4.3	6.3	13.7	12.4	5.8	92.3
East Pakistan:																												
Dacca	23 46N	90 23E	24	60	77	56	92	74	88	79	88	75	108	43	61	0.3	1.2	2.4	5.4	9.6	12.4	13.0	13.3	9.8	5.3	1.0	0.2	73.9
India:																												
Ahmadabad	23 03N	72 37E	180	45	85	58	104	75	93	79	97	73	118	36	45	*	0.1	0.1	*	0.4	3.7	12.2	8.1	4.2	0.4	0.1	*	29.3
Bangalore	12 57N	77 40E	2,937	60	80	57	93	69	81	66	82	65	102	46	60	0.2	0.3	0.4	1.6	4.2	2.9	3.9	5.0	6.7	5.9	2.7	0.4	34.2
Bombay	19 06N	72 51E	27	60	88	62	93	74	88	75	93	73	110	46	60	0.1	0.1	0.1	*	0.7	19.1	24.3	13.4	10.4	2.5	0.5	0.1	71.2
Calcutta	22 32N	88 20E	21	60	80	55	97	76	90	79	89	74	111	44	60	0.4	1.2	1.4	1.7	5.5	11.7	12.8	12.9	9.9	4.5	0.8	0.2	63.0
Cherrapunji	25 15N	91 44E	4,309	35	60	46	71	59	72	65	72	61	87	33	35	0.7	2.1	7.3	26.2	50.4	106.1	96.3	70.1	43.3	19.4	2.7	0.5	425.1
Hyderabad	17 27N	78 28E	1,741	50	85	59	101	75	87	73	88	68	112	43	45	0.3	0.4	0.5	1.2	1.1	4.4	6.0	5.3	6.5	2.5	1.1	0.3	29.6
Jalpaiguri	26 32N	88 43E	272	50	74	50	90	68	89	77	87	70	104	36	55	0.3	0.7	1.3	3.7	11.8	25.9	32.2	25.3	21.2	5.6	0.5	0.2	128.7
Lucknow	26 45N	80 52E	400	60	74	47	101	71	92	80	91	67	119	34	60	0.8	0.7	0.3	0.3	0.8	4.5	12.0	11.5	7.4	1.3	0.2	0.3	40.1

Madras	13 04N	80 15E	51	60	85	67	95	78	96	79	90	75	113	57	60	1.4	0.4	0.3	0.6	1.0	1.9	3.6	4.6	4.7	12.0	14.0	5.5	50.0
Mormugao	15 22N	73 49E	157	10	86	70	88	79	83	75	86	75	98	59	30	*	*	*	0.7	2.6	29.6	31.2	15.9	9.5	3.8	1.3	0.2	94.8
New Delhi	28 35N	77 12E	695	10	71	43	97	68	95	80	93	64	115	31	75	0.9	0.7	0.5	0.3	0.5	2.9	7.1	6.8	4.6	0.4	0.1	0.4	25.2
Silchar	24 49N	92 48E	95	60	78	52	88	69	90	77	88	72	103	41	53	0.8	2.1	7.9	14.3	15.6	21.7	19.7	19.7	14.4	6.5	1.4	0.4	124.5
Indian Ocean Islands:																												
Port Blair, Andaman Is.	11 40N	92 43E	261	60	84	72	89	75	84	75	84	74	97	62	60	1.8	1.1	1.1	2.4	15.1	21.7	15.4	16.3	17.4	12.5	10.5	7.9	123.2
Amini Divi, Laccadive Is.	11 07N	72 44E	13	29	86	74	92	80	86	77	86	77	99	65	30	0.7	*	*	1.5	3.7	14.3	12.0	7.7	6.3	5.8	2.6	1.3	56.0
Minicoy, Maldive Is.	08 18N	73 00E	9	20	85	73	87	80	85	76	85	76	98	63	50	1.8	0.7	0.9	2.3	7.0	11.6	8.9	7.8	6.3	7.3	5.5	3.4	63.5
Car Nicobar, Nicobar Is.	09 09N	92 49E	47	13	86	77	90	77	86	77	85	75	95	66	30	3.9	1.2	2.1	3.5	12.5	12.4	9.3	10.2	12.9	11.6	11.4	7.8	98.8
Iran:																												
Abadan	30 21N	48 13E	10	12	64	44	90	62	112	81	98	63	127	24	10	1.5	1.7	0.6	0.8	0.1	0.0	0.0	0.0	0.0	0.1	1.0	1.8	7.6
Esfahan (Isfahan)	32 37N	51 41E	5,238	45	47	25	72	46	98	67	78	47	108	- 4	45	0.7	0.6	0.8	0.6	0.3	*	0.1	*	*	0.1	0.4	0.7	4.4
Kermanshah	34 19N	47 07E	4,331	15	45	23	68	38	99	56	79	38	108	-13	15	2.6	2.3	2.8	2.2	1.6	*	*	*	*	0.4	2.0	2.4	16.4
Rezaiyeh	37 32N	45 05E	4,364	3	32	17	67	45	91	64	67	47	99	-11	3	1.9	2.3	2.0	1.7	1.2	0.5	*	0.1	0.2	1.5	0.8	1.6	13.8
Tehran	35 41N	51 19E	3,937	24	45	27	71	49	99	72	76	53	109	- 5	33	1.8	1.5	1.8	1.4	0.5	0.1	0.1	0.1	0.1	0.3	0.8	1.2	9.7
Iraq:																												
Baghdad	33 20N	44 24E	111	15	60	39	85	57	110	76	92	61	121	18	15	0.9	1.0	1.1	0.5	0.1	*	*	*	*	0.1	0.8	1.0	5.5
Basra	30 34N	47 47E	8	10	64	45	85	63	104	81	94	64	123	24	10	1.4	1.1	1.2	1.2	0.2	0.0	*	*	*	*	1.4	0.8	7.3
Mosul	36 19N	43 09E	730	26	54	35	77	49	109	72	88	51	124	12	29	2.8	3.1	2.1	1.9	0.7	*	*	*	*	0.2	1.9	2.4	15.2
Israel:																												
Haifa	32 48N	35 02E	23	16	65	49	77	58	88	75	85	68	112	27	30	6.9	4.3	1.6	1.0	0.2	*	*	*	0.1	1.0	3.7	7.3	26.2
Jerusalem	31 47N	35 13E	2,654	19	55	41	73	50	87	63	81	59	107	26	50	5.1	4.7	2.9	0.9	0.1	*	0.0	0.0	*	0.3	2.2	3.5	19.7
Tel Aviv	32 06N	34 46E	33	10	64	50	70	57	82	72	79	65	102	34	10	4.9	2.7	2.0	0.7	0.1	0.0	0.0	0.0	0.1	0.4	4.1	6.1	21.1
Jammu/Kashmir:																												
Srinagar	33 58N	74 46E	5,458	50	41	24	67	45	88	64	74	41	106	- 4	50	2.9	2.8	3.6	3.7	2.4	1.4	2.3	2.4	1.5	1.2	0.4	1.3	25.9
Jordan:																												
Amman	31 58N	35 59E	2,547	25	54	39	73	49	89	65	81	57	109	21	25	2.7	2.9	1.2	0.6	0.2	0.0	0.0	0.0	*	0.2	1.3	1.8	10.9
Kuwait:																												
Kuwait	29 21N	48 00E	16	14	61	49	83	68	103	86	91	73	119	33	10	0.9	0.9	1.1	0.2	*	0.0	0.0	0.0	0.0	0.1	0.6	1.1	5.1
Lebanon:																												
Beirut	33 54N	35 28E	111	62	62	51	72	58	87	73	81	69	107	30	71	7.5	6.2	3.7	2.2	0.7	0.1	*	*	0.2	2.0	5.2	7.3	35.1
Nepal:																												
Katmandu	27 42N	85 22E	4,423	27	65	36	84	53	84	69	80	56	99	27	9	0.6	1.6	0.9	2.3	4.8	9.7	14.7	13.6	6.1	1.5	0.3	0.1	56.2
Oman and Muscat:																												
Muscat	23 37N	58 35E	15	23	77	66	90	78	97	87	93	80	116	51	38	1.1	0.7	0.4	0.4	*	0.1	*	*	0.0	0.1	0.4	0.7	3.9
Pakistan (West):																												
Karachi	24 48N	66 59E	13	43	77	55	90	73	91	81	91	72	118	39	59	0.5	0.4	0.3	0.1	0.1	0.7	3.2	1.6	0.5	0.1	0.1	0.2	7.8
Multan	30 11N	71 25E	400	60	68	42	95	68	102	86	94	64	122	29	60	0.4	0.4	0.4	0.3	0.3	0.6	2.0	1.8	0.5	0.1	0.1	0.2	7.1
Rawalpindi	33 35N	73 03E	1,676	60	62	38	86	59	98	77	89	57	118	25	60	2.5	2.5	2.7	1.9	1.3	2.3	8.1	9.2	3.9	0.6	0.3	1.2	36.5
Saudi Arabia:																												
Dhahran	26 16N	50 10E	78	10	69	54	90	69	107	86	95	73	120	40	10	1.1	0.6	0.4	0.2	0.1	0.0	0.0	0.0	0.0	0.0	0.2	0.9	3.5
Jidda	21 28N	39 10E	20	5	84	66	91	70	99	79	95	73	117	49	5	0.2	*	*	*	*	0.0	*	*	*	*	1.0	1.2	2.5
Riyadh	24 39N	46 42E	1,938	3	70	46	89	64	107	78	94	61	113	19	3	0.1	0.8	0.9	1.0	0.4	*	0.0	*	0.0	0.0	*	*	3.2
Syria:																												
Deir Ez Zor	35 21N	40 09E	699	5	53	35	80	52	105	78	86	56	114	16	8	1.6	0.8	0.3	0.8	0.1	*	0.0	0.0	0.0	0.2	1.5	0.9	6.2
Dimashq (Damascus)	33 30N	36 20E	2,362	13	53	36	75	49	96	64	81	54	113	21	7	1.7	1.7	0.3	0.5	0.1	*	*	0.0	0.7	0.4	1.6	1.6	8.6
Halab (Aleppo)	36 14N	37 08E	1,280	8	50	34	75	48	97	69	81	54	117	9	10	3.5	2.5	1.5	1.1	0.3	0.1	0.0	*	*	1.0	2.2	3.3	15.5
Trucial Kingdoms:																												
Sharjah	25 20N	55 24E	18	11	74	54	86	65	100	82	92	71	118	37	12	0.9	0.9	0.4	0.2	0.0	0.0	0.0	0.0	0.0	0.0	0.4	1.4	4.2
Turkey:																												
Adana	36 59N	35 18E	82	21	57	39	74	51	93	71	84	58	109	19	31	4.3	4.0	2.5	1.6	2.0	0.7	0.2	0.2	0.7	1.9	2.4	3.8	24.3
Ankara	39 57N	32 53E	2,825	26	39	24	63	40	86	59	69	44	104	-13	24	1.3	1.2	1.3	1.3	1.9	1.0	0.5	0.4	0.7	0.9	1.2	1.9	13.6
Erzurum	39 54N	41 16E	6,402	16	24	8	50	32	78	53	59	37	93	-22	16	1.4	1.6	2.0	2.5	3.1	2.1	1.3	0.9	1.1	2.3	1.8	1.1	21.2
Izmir (Smyrna)	38 27N	27 15E	92	39	55	39	70	49	92	69	76	55	108	12	58	4.4	3.3	3.0	1.7	1.3	0.6	0.2	0.2	0.8	2.1	3.3	4.8	25.5
Samsun	41 17N	36 19E	131	24	50	38	59	45	79	65	69	56	103	20	27	2.9	2.6	2.7	2.3	1.8	1.5	1.5	1.3	2.4	3.2	3.5	2.4	29.1
Yemen:																												
Kamaran I.	15 20N	42 37E	20	26	82	74	89	79	98	85	93	82	105	66	21	0.2	0.2	0.1	0.1	0.1	*	0.5	0.7	0.1	0.1	0.4	0.9	3.4

See footnotes at end of table.

TEMPERATURE AND PRECIPITATION DATA FOR REPRESENTATIVE WORLD-WIDE STATIONS

COUNTRY AND STATION	LATITUDE	LONGITUDE	ELEVATION	TEMPERATURE: LENGTH OF RECORD	AVERAGE DAILY: JANUARY MAXIMUM	JANUARY MINIMUM	APRIL MAXIMUM	APRIL MINIMUM	JULY MAXIMUM	JULY MINIMUM	OCTOBER MAXIMUM	OCTOBER MINIMUM	EXTREME MAXIMUM	EXTREME MINIMUM	AVERAGE PRECIPITATION: LENGTH OF RECORD	JANUARY	FEBRUARY	MARCH	APRIL	MAY	JUNE	JULY	AUGUST	SEPTEMBER	OCTOBER	NOVEMBER	DECEMBER	YEAR
	° ′	° ′	FEET	YEAR	°F	°F	°F	°F	°F	°F	°F	°F	°F	°F	YEAR	IN.	IN.	IN.	IN.	IN.	IN.	IN.	IN.	IN.	IN.	IN.	IN.	IN.
								AUSTRALIA & PACIFIC ISLANDS																				
Australia:																												
Adelaide	34 57S	138 32E	20	86	86	61	73	55	59	45	73	51	118	32	104	0.8	0.7	1.0	1.8	2.7	3.0	2.6	2.6	2.1	1.7	1.1	1.0	21.1
Alice Springs	23 48S	133 53E	1,791	62	97	70	81	54	67	39	88	58	111	19	30	1.7	1.3	1.1	0.4	0.6	0.5	0.3	0.3	0.3	0.7	1.2	1.5	9.9
Bourke	30 05S	145 58E	361	63	99	70	82	55	65	40	85	56	125	25	72	1.4	1.5	1.1	1.1	1.0	1.1	0.9	0.8	0.8	0.9	1.2	1.4	13.2
Brisbane	27 25S	153 05E	17	53	85	69	79	61	68	49	80	60	110	35	91	6.4	6.3	5.7	3.7	2.8	2.6	2.2	1.9	1.9	2.5	3.7	5.0	44.7
Broome	17 57S	122 13E	56	41	92	79	93	72	82	58	91	72	113	40	50	6.3	5.8	3.9	1.2	0.6	0.9	0.2	0.1	*	*	0.6	3.3	22.9
Burketown	17 45S	139 33E	30	31	93	77	91	69	82	55	93	70	110	40	53	8.2	6.3	5.2	1.0	0.2	0.3	*	*	*	0.4	1.5	4.4	27.5
Canberra	35 18S	149 11E	1,886	23	82	55	67	44	52	33	68	43	109	14	25	1.9	1.7	2.2	1.6	1.8	2.1	1.8	2.2	1.6	2.2	1.9	2.0	23.0
Carnarvon	24 53S	113 40E	13	43	88	72	84	66	71	51	78	61	118	37	57	0.4	0.7	0.7	0.6	1.5	2.4	1.6	0.7	0.2	0.1	*	0.2	9.1
Cloncurry	20 40S	140 30E	622	32	99	77	90	67	77	51	95	68	127	35	59	4.4	4.2	2.4	0.7	0.5	0.6	0.3	0.1	0.3	0.5	1.3	2.7	18.0
Esperance	33 50S	121 55E	14	44	77	60	72	54	62	45	68	50	117	31	60	0.7	0.7	1.2	1.8	3.3	4.1	4.0	3.8	2.7	2.2	1.0	0.9	26.4
Laverton	28 40S	122 23E	1,510	30	96	69	81	57	64	41	82	55	115	25	30	0.8	0.8	1.6	0.8	0.9	0.7	0.6	0.5	0.2	0.3	0.8	0.8	8.8
Melbourne	37 49S	144 58E	115	88	78	57	68	51	56	42	67	48	114	27	88	1.9	1.8	2.2	2.3	2.1	2.1	1.9	1.9	2.3	2.6	2.3	2.3	25.7
Mundiwindi	23 52S	120 10E	1,840	15	101	64	87	61	70	41	89	58	112	22	15	1.0	1.9	2.0	0.8	0.6	0.9	0.1	0.3	0.3	0.5	0.5	1.2	10.1
Perth	31 56S	115 58E	64	44	85	63	76	57	63	48	70	53	112	31	63	0.3	0.4	0.8	1.7	5.1	7.1	6.7	5.7	3.4	2.2	0.8	0.5	34.7
Port Darwin	12 25S	130 52E	104	58	90	77	92	76	87	67	93	77	105	55	70	15.2	12.3	10.0	3.8	0.6	0.1	*	0.1	0.5	2.0	4.7	9.4	58.7
Sydney	33 52S	151 02E	62	87	78	65	71	58	60	46	71	56	114	35	87	3.5	4.0	5.0	5.3	5.0	4.6	4.6	3.0	2.9	2.8	2.9	2.9	46.5
Thursday Island	10 35S	142 13E	200	31	87	77	86	77	82	73	86	76	98	64	49	18.2	15.8	13.9	8.0	1.6	0.5	0.4	0.2	0.1	0.3	1.5	7.0	67.5
Townsville	19 15S	146 46E	18	31	87	76	84	70	75	59	83	71	110	39	67	10.9	11.2	7.2	3.3	1.3	1.4	0.6	0.5	0.7	1.3	1.9	5.4	45.7
William Creek	28 55S	136 21E	247	39	96	69	80	55	65	41	84	56	119	25	30	0.5	0.6	0.3	0.3	0.3	0.5	0.2	0.3	0.3	0.5	0.5	0.7	5.0
Windorah	25 26S	142 36E	390	29	101	74	86	59	70	43	91	61	116	26	50	1.4	1.6	1.6	0.9	0.8	0.8	0.5	0.4	0.5	0.6	0.9	1.4	11.4
Tasmania:																												
Hobart	42 53S	147 20E	177	70	71	53	63	48	52	40	63	46	105	28	100	1.9	1.5	1.8	1.9	1.8	2.2	2.1	1.9	2.1	2.3	2.4	2.1	24.0
New Zealand:																												
Auckland	37 00S	174 47E	23	36	73	60	67	56	56	46	63	52	90	33	92	3.1	3.7	3.2	3.8	5.0	5.4	5.7	4.6	4.0	4.0	3.5	3.1	49.1
Christchurch	43 29S	172 32E	118	52	70	53	62	45	50	35	62	44	96	21	64	2.2	1.7	1.9	1.9	2.6	2.6	2.7	1.9	1.8	1.7	1.9	2.2	25.1
Dunedin	45 55S	170 12E	4	77	66	50	59	45	48	37	59	42	94	23	77	3.4	2.8	3.0	2.8	3.2	3.2	3.1	3.0	2.7	3.0	3.2	3.5	36.9
Wellington	41 17S	174 46E	415	66	69	56	63	51	53	42	60	48	88	29	79	3.2	3.2	3.2	3.8	4.6	4.6	5.4	4.6	3.8	4.0	3.5	3.5	47.4
PACIFIC ISLANDS:																												
Canton, Phoenix Is.	02 46S	171 43W	9	12	88	78	89	78	89	78	90	78	98	70	30	2.6	2.2	2.5	3.6	4.3	2.6	2.6	2.5	1.2	1.1	1.6	2.6	29.4
Guam, Marianas Is.	13 33N	144 50E	361	30	84	72	86	73	87	72	86	73	95	54	30	4.6	3.5	2.6	3.0	4.2	5.9	9.0	12.8	13.4	13.1	10.3	6.1	88.5
Honolulu, Hawaii	21 20N	157 55W	7	30	79	66	80	68	85	73	84	72	93	56	30	3.8	3.3	2.9	1.3	1.0	0.3	0.4	0.9	1.0	1.8	2.2	3.0	21.9
Iwo Jima, Bonin Is.	24 47N	141 19E	353	15	71	64	77	69	86	78	84	76	95	46	17	3.2	2.5	2.1	3.7	4.9	4.0	6.4	6.5	4.6	5.9	4.8	4.3	52.8
Madang, New Guinea	05 12S	145 47E	19	12	87	75	88	74	88	74	88	75	98	62	20	12.1	11.9	14.9	16.9	15.1	10.8	7.6	4.8	5.3	10.0	13.3	14.5	137.2
Midway Is.	28 13N	177 23W	29	21	69	62	71	64	81	74	79	72	92	46	20	4.6	3.7	3.1	2.5	1.9	1.3	2.9	3.9	3.7	3.7	3.6	4.2	40.7
Naha, Okinawa	26 12N	127 39E	96	30	67	56	76	64	89	77	81	69	96	41	30	5.3	5.4	6.1	6.1	8.9	10.0	7.1	10.0	7.1	6.6	5.9	4.3	82.8
Noumea, New Caledonia	22 16S	166 27E	246	24	86	72	83	70	76	62	80	65	99	52	52	3.7	5.1	5.7	5.2	4.4	3.7	3.6	2.6	2.5	2.0	2.4	2.6	43.5
Pago Pago, Samoa	14 19S	170 43W	29	2	87	75	87	76	83	74	85	75	98	67	41	24.5	20.5	19.2	16.5	15.4	12.3	10.0	8.2	13.1	14.9	19.2	19.8	193.6
Ponape, Caroline Is.	06 58N	158 13E	123	30	86	75	86	75	87	73	87	72	96	67	30	11.1	9.7	14.6	20.0	20.3	16.7	16.2	16.3	15.8	16.0	16.9	18.3	191.9
Port Moresby, New Guinea	09 29S	147 09E	126	20	89	76	87	75	83	73	86	75	98	64	38	7.0	7.6	6.7	4.2	2.5	1.3	1.1	0.7	1.0	1.4	1.9	4.4	39.8
Rabaul, New Guinea	04 13S	152 11E	28	19	90	73	90	73	89	73	92	73	100	65	24	14.8	10.4	10.2	10.0	5.2	3.3	5.4	3.7	3.5	5.1	7.1	10.1	88.8
Suva, Fiji Is.	18 08S	178 26E	20	43	86	74	84	73	79	68	81	70	98	55	43	11.4	10.7	14.5	12.2	10.1	6.7	4.9	8.3	7.7	8.3	9.8	12.5	117.1

Tahiti, Society Is.	17 33S	149 36W	7	23	89	72	89	72	86	68	87	70	93	61	27	13.2	11.5	6.5	6.8	4.9	3.2	2.6	1.9	2.3	3.4	6.5	11.9	74.7
Tulagi, Solomon Is.	09 05S	160 10E	8	20	88	76	88	76	86	76	87	76	96	68	37	14.3	15.8	15.0	10.0	8.1	6.8	7.6	8.7	8.0	8.7	10.0	10.4	123.4
Wake Is.	19 17N	166 39E	11	30	82	73	83	74	87	77	86	77	92	64	30	1.1	1.4	1.5	1.9	2.0	1.9	4.6	7.1	5.2	5.3	3.1	1.8	36.9
Yap, Caroline Is.	9 31N	138 08E	62	30	85	76	87	77	88	75	88	75	97	69	30	7.9	4.6	5.4	6.4	9.5	10.7	13.8	14.7	14.0	13.2	11.2	10.2	121.6
ANTARCTICA																												
Byrd Station	80 01S	119 32W	5,095	6	10	- 2	-11	-30	-25	-45	-15	-33	31	-82	6	0.4	0.4	0.2	0.3	0.4	0.5	0.7	0.7	0.3	0.7	0.0	0.3	4.9
Ellsworth	77 44S	41 07W	139	6	22	12	-10	-25	-21	-35	- 2	-15	36	-70	6	0.3	0.2	0.3	0.6	0.2	0.2	0.2	0.2	0.3	0.4	0.5	0.2	3.6
McMurdo Station	77 53S	166 48W	8	10	30	21	- 1	-13	- 9	-24	2	-12	42	-59	10	0.5	0.7	0.4	0.4	0.4	0.3	0.2	0.3	0.4	0.2	0.2	0.3	4.3
South Pole Station	89 59S	000 00W	9,186	5	-16	-23	-66	-79	-67	-81	-55	-64	6	-107	5	*	0.1	0.0	0.0	0.0	0.0	0.0	0.0	0.0	*	0.0	*	0.1
Wilkes	66 16S	110 31E	31	7	34	28	17	9	8	- 3	16	6	46	-35	7	0.5	0.4	1.7	1.1	1.4	1.2	1.3	0.8	1.5	1.2	0.8	0.3	12.2

NOTES

1. "Length of Record" refers to average daily maximum and minimum temperatures and precipitation. A standard period of the 30 years from 1931-1960 had been used for locations in the United States and some other countries. The length of record of extreme maximum and minimum temperatures includes all available years of data for a given location and is usually for a longer period.

2. * - Less than 0.05"

3. Except for Antarctica, amounts of solid precipitation such as snow or hail have been converted to their water equivalent. Because of the frequent occurrence of blowing snow, it has not been possible to determine the precise amount of precipitation actually falling in Antarctica. The values shown are the average amounts of solid snow accumulating in a given period as determined by snow markers. The liquid content of the accumulation is undetermined.

WORLDWIDE EXTREMES OF TEMPERATURE AND PRECIPITATION RECORDED BY CONTINENTAL AREA

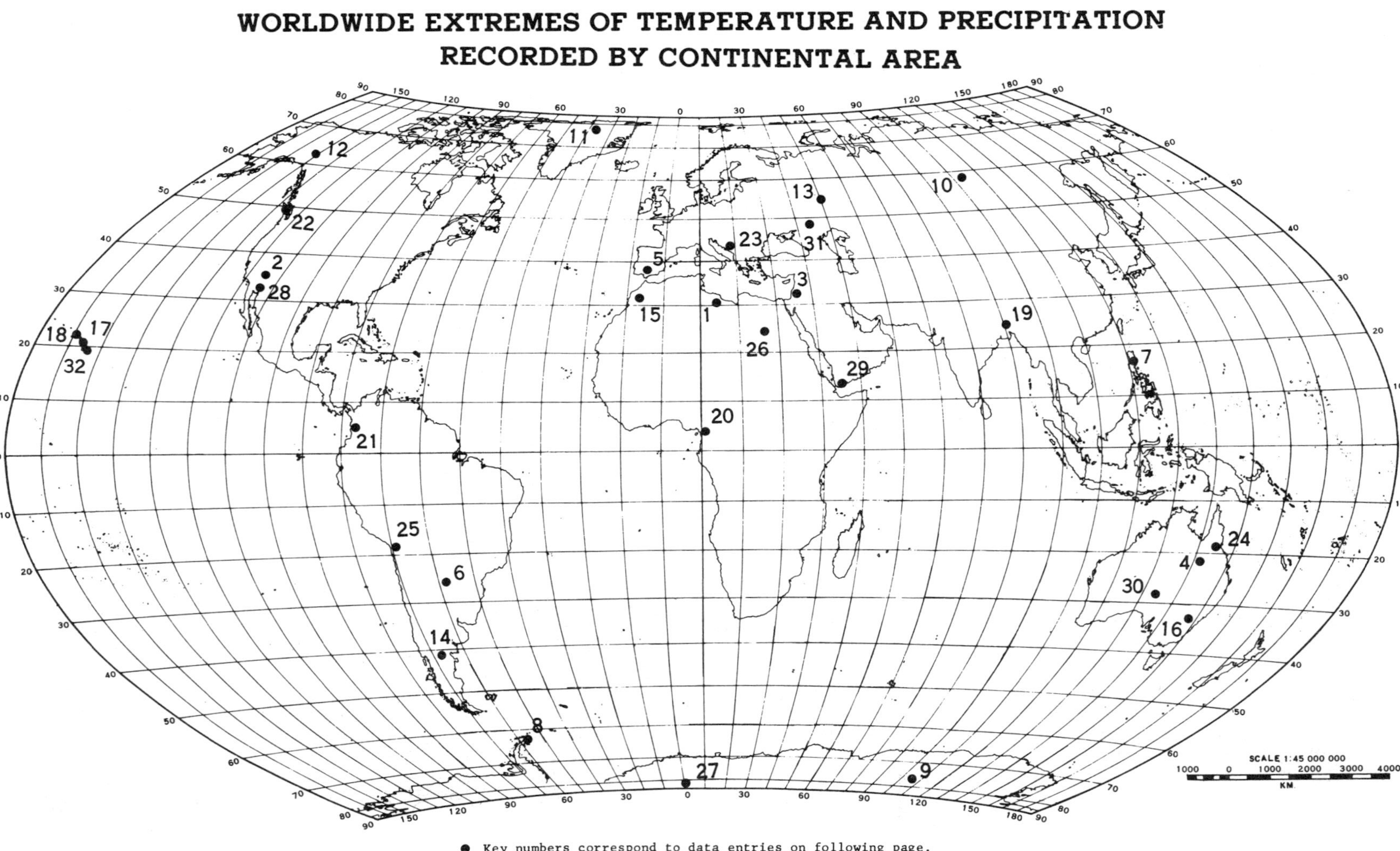

● Key numbers correspond to data entries on following page.

AVERAGE JANUARY TEMPERATURE (F°)

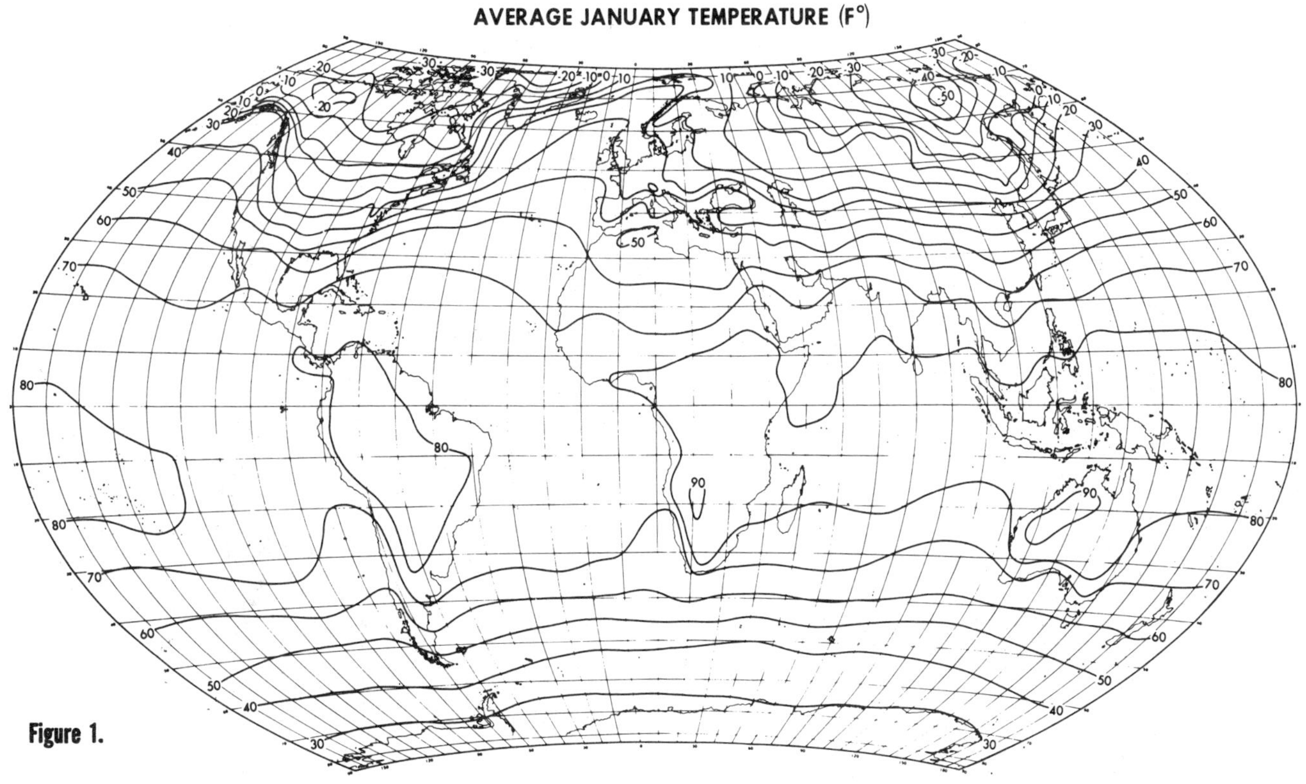

Figure 1.

AVERAGE JULY TEMPERATURE (F°)

Figure 2.

GENERAL PATTERN OF ANNUAL WORLD PRECIPITATION (INCHES)

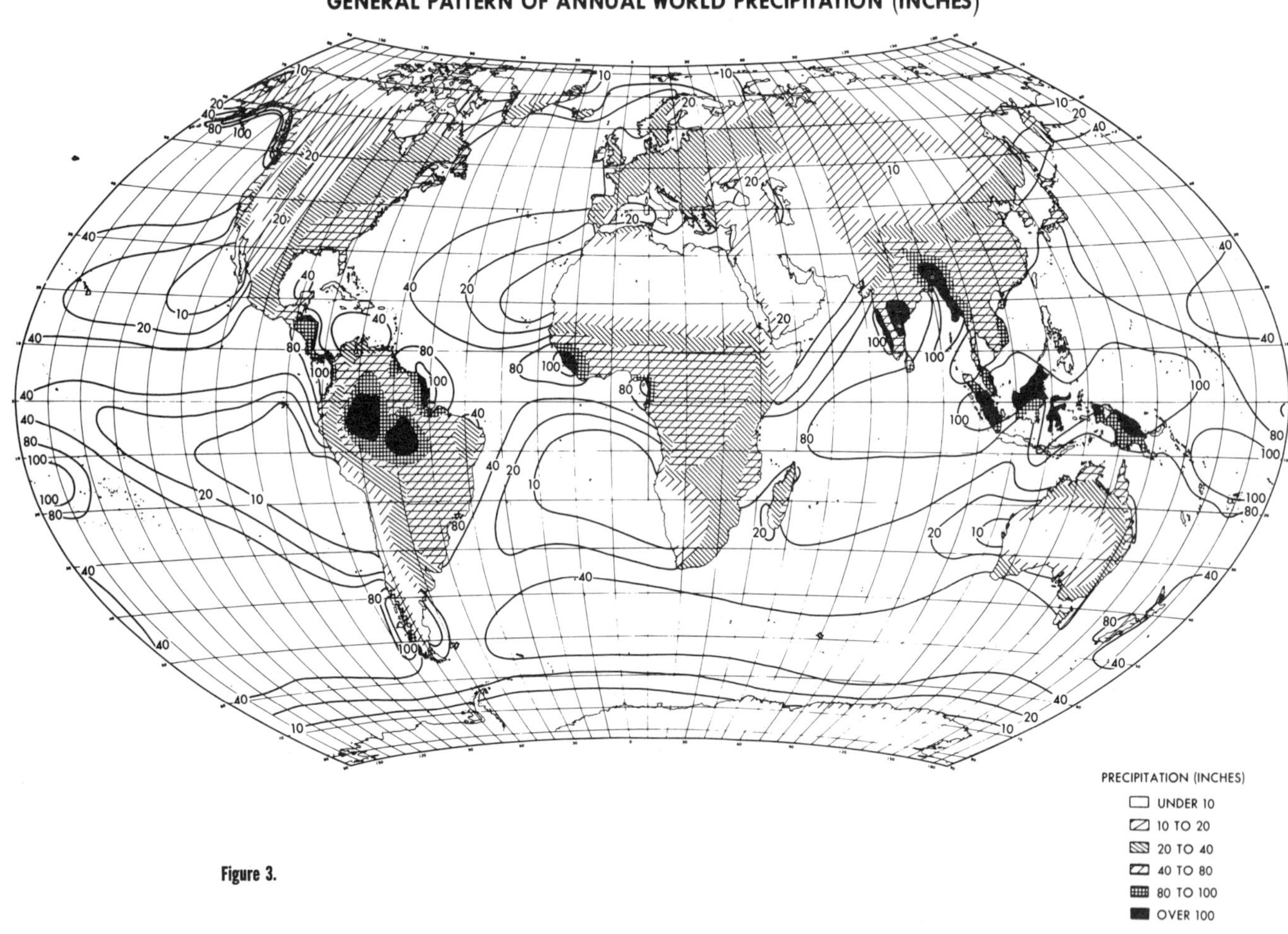

Figure 3.

SOURCES

Area Handbooks 110 vols (GPO)

Background Notes (State Department) Bound together as *Countries of the World* (2 vols, Gale, 1988)

Chambers World Gazetteer edited by David Munro (Chambers/Cambridge, 1988)

Encyclopedia of the Third World edited by George Thomas Kurian (3 vols, Facts On File, 1986)

Encyclopedia of the First World edited by George Thomas Kurian (3 vols, Facts On File, 1989)

Encyclopedia of the Second World edited by George Thomas Kurian (Facts On File, 1989)

Europa Yearbook: A World Survey (2 vols, Europa Publications, 1988)

New Cyclopedia of Names edited by Clarence Barnhart. (3 vols, Appleton-Century-Crofts, 1954)

Post Reports (State Department) Bound together as *Cities of the World* (4 vols, Gale, 1987)

The Statesman's Yearbook: World Gazatteer edited by John Paxton. (St. Martin's, 1979)

Webster's New Geographical Dictionary (Merriam-Webster, 1980)

World Factbook (Annual, GPO)

Worldmark Encyclopedia of the Nations (5 vols, Wiley, 1986)

World Facts and Figures edited by Victor Showers (Wiley, 1979)

Index